ESSENTIALS OF
Biology

Fifth Edition

Sylvia S. Mader

Michael Windelspecht

Mc
Graw
Hill
Education

ESSENTIALS OF BIOLOGY, FIFTH EDITION

Published by McGraw-Hill Education, 2 Penn Plaza, New York, NY 10121. Copyright © 2018 by McGraw-Hill Education. All rights reserved. Printed in the United States of America. Previous editions © 2015, 2012, and 2010. No part of this publication may be reproduced or distributed in any form or by any means, or stored in a database or retrieval system, without the prior written consent of McGraw-Hill Education, including, but not limited to, in any network or other electronic storage or transmission, or broadcast for distance learning.

Some ancillaries, including electronic and print components, may not be available to customers outside the United States.

This book is printed on acid-free paper.

1 2 3 4 5 6 7 8 9 0 LMN 21 20 19 18 17

ISBN 978–1–259–66026–9
MHID 1–259–66026–5

Senior Vice President, Products & Markets: *G. Scott Virkler*
Vice President, General Manager, Products & Markets: *Marty Lange*
Vice President, Content Design & Delivery: *Betsy Whalen*
Managing Director: *Lynn Breithaupt*
Executive Brand Manager: *Michelle Vogler*
Director of Development: *Rose M. Koos*
Product Developer: *Anne Winch*
Director of Digital Content: *Michael Koot, PhD*
Market Development Manager: *Jenna Paleski*
Marketing Manager: *Britney Ross*
Program Manager: *Angie FitzPatrick*
Content Project Manager (Core): *Jayne Klein*
Content Project Manager (Assessment): *Brent dela Cruz*
Senior Buyer: *Sandy Ludovissy*
Designer: *David W. Hash*
Cover Image: *© Antonio Cali66/EyeEm/Getty Images*
Content Licensing Specialists: *Lori Hancock/Lorraine Buczek*
Compositor: *Aptara*
Typeface: *10/13 STIX MathJax Main*
Printer: *LSC Communications*

All credits appearing on page are considered to be an extension of the copyright page.

Library of Congress Cataloging-in-Publication Data

Names: Mader, Sylvia S., author. | Windelspecht, Michael, 1963- , author.
Title: Essentials of biology / Sylvia S. Mader, Michael Windelspecht ; with
    contributions by Dave Cox, Lincoln Land Community College, Gretel Guest,
    Durham Technical Community College.
Description: Fifth edition | New York, NY : McGraw-Hill Education, 2016.
Identifiers: LCCN 2016042242| ISBN 9781259660269 (alk. paper) | ISBN
    1259660265 (alk. paper)
Subjects: LCSH: Biology--Textbooks.
Classification: LCC QH308.2 .M24 2016 | DDC 570--dc23 LC record available athttps://lccn.loc.gov/2016042242

The Internet addresses listed in the text were accurate at the time of publication. The inclusion of a website does not indicate an endorsement by the authors or McGraw-Hill Education, and McGraw-Hill Education does not guarantee the accuracy of the information presented at these sites.

www.mhhe.com

Brief Contents

1 Biology: The Science of Life 1

PART I The Cell

2 The Chemical Basis of Life 21
3 The Organic Molecules of Life 38
4 Inside the Cell 56
5 The Dynamic Cell 79
6 Energy for Life 95
7 Energy for Cells 110

PART II Genetics

8 Cellular Reproduction 125
9 Meiosis and the Genetic Basis of Sexual Reproduction 145
10 Patterns of Inheritance 160
11 DNA Biology 184
12 Biotechnology and Genomics 207
13 Genetic Counseling 221

PART III Evolution

14 Darwin and Evolution 235
15 Evolution on a Small Scale 251
16 Evolution on a Large Scale 265

PART IV Diversity of Life

17 The Microorganisms: Viruses, Bacteria, and Protists 286
18 The Plants and Fungi 311
19 The Animals 336

PART V Plant Structure and Function

20 Plant Anatomy and Growth 372
21 Plant Responses and Reproduction 392

PART VI Animal Structure and Function

22 Being Organized and Steady 414
23 The Transport Systems 431
24 The Maintenance Systems 450
25 Digestion and Human Nutrition 465
26 Defenses Against Disease 496
27 The Control Systems 513
28 Sensory Input and Motor Output 536
29 Reproduction and Embryonic Development 556

PART VII Ecology

30 Ecology and Populations 580
31 Communities and Ecosystems 600
32 Human Impact on the Biosphere 626

About the Authors

© Jacqueline Baer Photography

Dr. Sylvia S. Mader Sylvia Mader has authored several nationally recognized biology texts published by McGraw-Hill. Educated at Bryn Mawr College, Harvard University, Tufts University, and Nova Southeastern University, she holds degrees in both Biology and Education. Over the years she has taught at University of Massachusetts, Lowell; Massachusetts Bay Community College; Suffolk University; and Nathan Mayhew Seminars. Her ability to reach out to science-shy students led to the writing of her first text, *Inquiry into Life*, which is now in its fifteenth edition. Highly acclaimed for her crisp and entertaining writing style, her books have become models for others who write in the field of biology.

Dr. Mader enjoys taking time to visit and explore the various ecosystems of the biosphere. Her several trips to the Florida Everglades and Caribbean coral reefs resulted in talks she has given to various groups around the country. She has visited the tundra in Alaska, the taiga in the Canadian Rockies, the Sonoran Desert in Arizona, and tropical rain forests in South America and Australia. A photo safari to the Serengeti in Kenya resulted in a number of photographs for her texts. She was thrilled to think of walking in Darwin's footsteps when she journeyed to the Galápagos Islands with a group of biology educators. Dr. Mader was also a member of a group of biology educators who traveled to China to meet with their Chinese counterparts and exchange ideas about the teaching of modern-day biology.

© Ricochet Creative Productions LLC

Dr. Michael Windelspecht As an educator, Dr. Windelspecht has taught introductory biology, genetics, and human genetics in the online, traditional, and hybrid environments at community colleges, comprehensive universities, and military institutions. For over a decade he served as the Introductory Biology Coordinator at Appalachian State University, where he directed a program that enrolled over 4,500 students annually.

He received degrees from Michigan State University (BS, zoology–genetics) and the University of South Florida (PhD, evolutionary genetics) and has published papers in areas as diverse as science education, water quality, and the evolution of insecticide resistance. His current interests are in the analysis of data from digital learning platforms for the development of personalized microlearning assets and next generation publication platforms. He is currently a member of the National Association of Science Writers and several science education associations. He has served as the keynote speaker on the development of multimedia resources for online and hybrid science classrooms. In 2015 he won the DevLearn HyperDrive competition for a strategy to integrate student data into the textbook revision process.

As an author and editor, Dr. Windelspecht has over 20 reference textbooks and multiple print and online lab manuals. He has founded several science communication companies, including Ricochet Creative Productions, which actively develops and assesses new technologies for the science classroom. You can learn more about Dr. Windelspecht by visiting his website at www.michaelwindelspecht.com.

Preface

This Fifth Edition of *Essentials of Biology* provides nonscience majors with a fundamental understanding of the science of biology. The overall focus of this edition addresses the learning styles of modern students, and in the process, increases their understanding of the importance of science in their lives.

Students in today's world are being exposed, almost on a daily basis, to exciting new discoveries and insights that, in many cases, were beyond our predictions even a few short years ago. It is our task, as instructors, not only to make these findings available to our students, but to enlighten students as to why these discoveries are important to their lives and society. At the same time, we must provide students with a firm foundation in those core principles on which biology is founded, and in doing so, provide them with the background to keep up with the many discoveries still to come.

In addition to the evolution of the introductory biology curriculum, students and instructors are increasingly requesting digital resources to utilize as learning resources. McGraw-Hill Education has long been an innovator in the development of digital resources, and this text, and its authors, are at the forefront of the integration of these technologies into the science classroom.

The authors identified several goals that guided the preparation of this new edition:

1. Updating of chapter openers and Connections content to focus on issues and topics important in a nonscience majors classroom
2. Utilization of the data from the LearnSmart adaptive learning platforms to identify content areas within the text that students demonstrated difficulty in mastering
3. Refinement of digital assets to provide a more effective assessment of learning outcomes to enable instructors in the flipped, online, and hybrid teaching environments
4. Development of a new series of videos and websites to introduce relevancy and engage students in the content

Relevancy

The use of real world examples to demonstrate the importance of biology in the lives of students is widely recognized as an effective teaching strategy for the introductory biology classroom. Students want to learn about the topics they are interested in. The development of relevancy-based resources is a major focus for the authors of the Mader series of texts. Some examples of how we have increased the relevancy content of this edition include:

- A series of new chapter openers to introduce relevancy to the chapter. The authors chose topics that would be of interest to a nonscience major, and represent what would typically be found on a major news source.
- The development of new relevancy-based videos, BioNow, that offer relevant, applied classroom resources to allow students to feel that they can actually do and learn biology themselves.

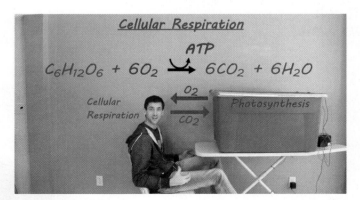

- A website, RicochetScience.com, managed by Dr. Windelspecht, that provides updates on news and stories that are interesting to nonscience majors. The Biology101 project links these resources to the major topics of the text. The site also features videos and tutorial animations to assist the students in recognizing the relevancy of what they are learning in the classroom.

- In addition, the author's website, michaelwindelspecht.com, contains videos and articles on how the *Essentials of Biology* text may be easily adapted for use in a topics-based course, or in the hybrid, online, and flipped classroom environments.

Engaging Students

Today's science classroom relies heavily on the use of digital assets, including animations and videos, to engage students and reinforce difficult concepts. *Essentials of Biology* includes two resources specifically designed for the introductory science class to help you achieve these goals.

BioNow Videos

The BioNow series of videos, narrated and produced by educator Jason Carlson, provide a relevant, applied approach that allows your students to feel they can actually do and learn biology themselves. While tying directly to the content of your course, the videos help students relate their daily lives to the biology you teach and then connect what they learn back to their lives.

Each video provides an engaging and entertaining story about applying the science of biology to a real situation or problem. Attention is taken to use tools and techniques that any regular person could perform, so your students see the science as something they could do and understand.

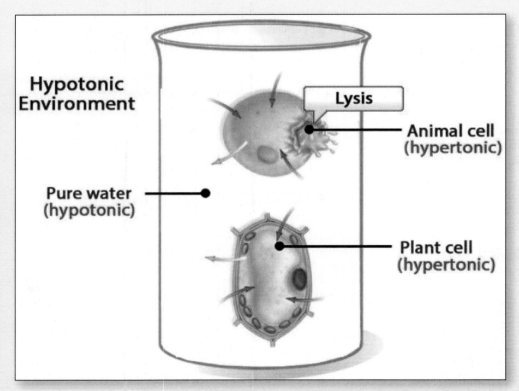

Tutorial Videos

The author, Michael Windelspecht, has prepared a series of tutorial videos to help students understand some of the more difficult topics in each chapter. Each video explores a specific figure in the text. During the video, important terms and processes are called out, allowing you to focus on the key aspects of the figure.

For students, these act as informal office hours, where they can review the most difficult concepts in the chapter at a pace which helps them learn. Instructors of hybrid and flipped courses will find these useful as online supplements.

Overview of Content Changes to *Essentials of Biology,* Fifth Edition

A number of the chapters in this edition now include references and links to new BioNow relevancy videos that have been designed to show students how the science of biology applies to their everyday lives. All of these are available in the instructor and student resources section within Connect. In addition, within the end of chapter material, the Connecting the Concepts content has been included in the Summarize section to better help the students understand the connections within the chapter.

Chapter 1: Biology: The Science of Life contains an updated chapter opener on species that have been recently discovered. The levels of biological organization now includes a description of species. The content on challenges facing science (Section 1.4) now includes more content on biodiversity loss, emerging diseases, and climate change.

Part I *The Cell*

The chapter opener for **Chapter 2: The Chemical Basis of Life** has been updated to include recent discoveries associated with the search for the precursors of life on Titan and comets. **Chapter 4: Inside the Cell** starts with a discussion of the importance of stem cells. **Chapter 6: Energy for Life** contains a new figure (Fig. 6.5) on the absorption spectrum of the major photosynthetic pigments.

Part II *Genetics*

Chapter 8: Cellular Reproduction starts with a new chapter opener on the *p53* gene and cancer. The material on mitosis in a plant cell (Fig. 8.6) has been expanded to make it more similar to the coverage of the animal cells. The content on the treatments of cancer (Section 8.5) has been expanded to include immunotherapy. The first section of **Chapter 10: Patterns of Inheritance** now explains why earlobes and dimples should not be used as examples of Mendelian traits in humans. The material on non-Mendelian genetics (Section 10.3) includes eye color in humans as an example of genetic interactions. **Chapter 11: DNA Biology** begins with a new chapter opener on the possibilities of synthetic DNA. The chapter has a new figure on semi-conservative replication (Fig. 11.6). **Chapter 12: Biotechnology and Genomics** begins with a new chapter opener on CRISPR and genome editing. The section on biotechnology (Section 12.1) now includes a discussion on genetic sequencing and genome editing (CRISPR, Fig. 12.4). The material on biotechnology products (Section 12.3) includes new examples of both plant and animal products. **Chapter 13: Genetic Counseling** was renamed to indicate a focus on how DNA changes and the processes of genetic testing and gene therapy. The section on genetic testing (Section 13.3) includes content on genetic sequencing for individuals and the reliability of OTC genetic tests.

Part III *Evolution*

Chapter 14: Darwin and Evolution now begins with a chapter opener on the evolution of antibiotic resistance, including both MRSA and *Shigella.* Figure 14.11 has been updated to better demonstrate Wallace's contribution to the study of biogeography. In **Chapter 15: Evolution on a Small Scale,** a new chapter opener now describes how changes in a single gene have allowed humans to live at high elevations. A new figure on the types of selection (Fig. 15.1) has been added. The examples of directional selection now focus on studies of coloration in guppies. **Chapter 16: Evolution on a Large Scale** starts with a new chapter opener on the evolution of the birds. The geological timescale (Table 16.1) has been updated.

Part IV *Diversity of Life*

In **Chapter 17: The Microorganisms: Viruses, Bacteria, and Protists,** a new opener on the Ebola outbreak in Africa has been included. A new connection piece on the world's largest virus has been added. The content on eukaryotic supergroups (Table 17.1) has been updated to reflect recent classification changes and a new figure (Fig. 17.20) added. The entire chapter has been reorganized according to eukaryotic supergroups. **Chapter 18: The Plants and Fungi** contains a new illustration of fungal evolution (Fig. 18.19). **Chapter 19: The Animals** starts with a new opener on canine evolution. New figures illustrate the general characteristics of animals (Fig. 19.1) and the general evolution of animals (Fig. 19.4). For the insects (Section 19.4), a new connection piece explores why mosquitoes are disease vectors. In the section on human evolution (Section 19.5), the diagram of human evolution (Fig. 19.38) has been updated, and a new illustration added (Fig. 19.41) on the migration of *Homo erectus.* Additional content has been added on both Neandertals and Denisovans.

Part VI *Animal Structure and Function*

Chapter 22: Being Organized and Steady contains a new chapter opener on the homeostatic requirements of pop icon Taylor Swift during performances. In **Chapter 23: The Transport Systems,** a new chapter opener on synthetic blood is included. The content on nutrition and the digestive system (previously in Chapter 24) has been combined in **Chapter 25: Digestion and Human Nutrition.** The chapter opener now explores the relationship between gluten and celiac disease. A new section (Section 25.4) is included that outlines how nutritional information is updated and how to interpret nutrition labels on food. **Chapter 26: Defenses Against Disease** begins with a look at the development of a vaccine against the Zika virus. **Chapter 29: Reproduction and Embryonic Development** has a new chapter opener on in-vitro fertilization (IVF) using genetic material from three parents. The introductory content on the differences between sexual and asexual reproduction have been separated into distinct headings. A new reading has been added on how Zika virus contributes to birth defects.

Part VII *Ecology*

Chapter 30: Ecology and Populations contains a new chapter opener on population growth in the asian carp. The levels of biological organization have been updated (Fig. 30.1) to reflect changes introduced in Chapter 1. The human population statistics have been updated throughout to reflect 2015 data. The information on predator-prey dynamics has been updated to include more current research on hare-lynx populations. **Chapter 31: Communities and Ecosystems** contains a new opener on the consequences of global climate change. New figures (Fig. 31.27) illustrate projections of global temperature increases and the influence of climate change in the United States (Fig. 31.28). A new map of terrestrial biomes (Fig. 31.29) has been added. The chapter opener for **Chapter 32: Human Impact on the Biosphere** now examines the Flint water crisis.

McGraw-Hill Connect®
Learn Without Limits

Connect is a teaching and learning platform that is proven to deliver better results for students and instructors.

Connect empowers students by continually adapting to deliver precisely what they need, when they need it, and how they need it, so your class time is more engaging and effective.

73% of instructors who use Connect require it; instructor satisfaction increases by 28% when Connect is required.

Connect's Impact on Retention Rates, Pass Rates, and Average Exam Scores

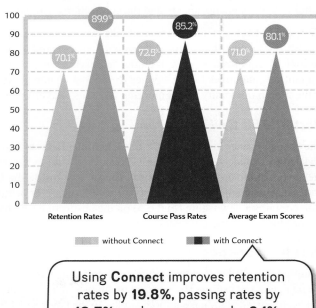

Retention Rates — 70.1%, 89.9%
Course Pass Rates — 72.5%, 85.2%
Average Exam Scores — 71.0%, 80.1%

without Connect | with Connect

Using **Connect** improves retention rates by **19.8%**, passing rates by **12.7%**, and exam scores by **9.1%**.

Analytics

Connect Insight®

Connect Insight is Connect's new one-of-a-kind visual analytics dashboard that provides at-a-glance information regarding student performance, which is immediately actionable. By presenting assignment, assessment, and topical performance results together with a time metric that is easily visible for aggregate or individual results, Connect Insight gives the user the ability to take a just-in-time approach to teaching and learning, which was never before available. Connect Insight presents data that helps instructors improve class performance in a way that is efficient and effective.

Impact on Final Course Grade Distribution

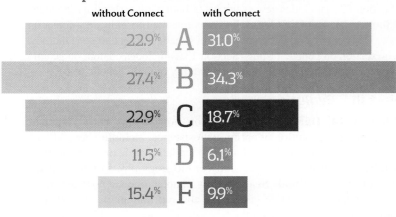

without Connect		with Connect
22.9%	A	31.0%
27.4%	B	34.3%
22.9%	C	18.7%
11.5%	D	6.1%
15.4%	F	9.9%

Adaptive

THE **ADAPTIVE** READING EXPERIENCE DESIGNED TO TRANSFORM THE WAY STUDENTS READ

More students earn **A's** and **B's** when they use McGraw-Hill Education **Adaptive** products.

SmartBook®

Proven to help students improve grades and study more efficiently, SmartBook contains the same content within the print book, but actively tailors that content to the needs of the individual. SmartBook's adaptive technology provides precise, personalized instruction on what the student should do next, guiding the student to master and remember key concepts, targeting gaps in knowledge and offering customized feedback, and driving the student toward comprehension and retention of the subject matter. Available on tablets, SmartBook puts learning at the student's fingertips—anywhere, anytime.

Over **8 billion questions** have been answered, making McGraw-Hill Education products more intelligent, reliable, and precise.

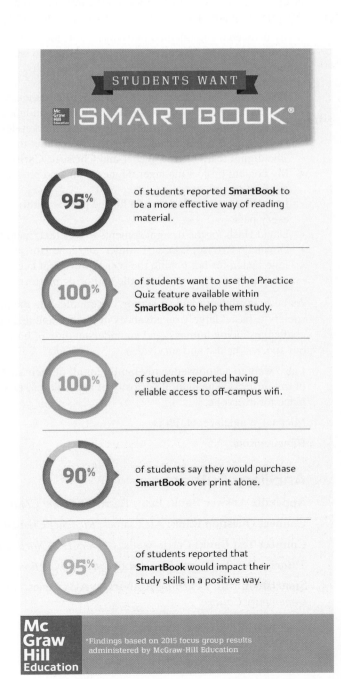

STUDENTS WANT

SMARTBOOK®

95% of students reported **SmartBook** to be a more effective way of reading material.

100% of students want to use the Practice Quiz feature available within **SmartBook** to help them study.

100% of students reported having reliable access to off-campus wifi.

90% of students say they would purchase **SmartBook** over print alone.

95% of students reported that **SmartBook** would impact their study skills in a positive way.

Mc Graw Hill Education

*Findings based on 2015 focus group results administered by McGraw-Hill Education

www.mheducation.com

Acknowledgments

Dr. Sylvia Mader is one of the icons of science education. Her dedication to her students, coupled to her clear, concise writing style, has benefited the education of thousands of students over the past four decades. As an educator, it is an honor to continue her legacy and to bring her message to the next generation of students.

As always, I had the privilege to work with a phenomenal group of people on this edition. I would especially like to thank you, the numerous instructors who have shared emails with me or have invited me into your classrooms, both physically and virtually, to discuss your needs as instructors and the needs of your students. You are all dedicated and talented teachers, and your energy and devotion to quality teaching is what drives a textbook revision.

Many dedicated and talented individuals assisted in the development of *Essentials of Biology,* Fifth Edition. I am very grateful for the help of so many professionals at McGraw-Hill who were involved in bringing this book to fruition. Therefore, I would like to thank the following:

- The product developer, Anne Winch, for her patience and impeccable ability to keep me focused.
- My brand manager, Michelle Vogler, for her guidance and reminding me why what we do is important.
- My marketing manager, Britney Ross, and market development manager, Jenna Paleski, for placing me in contact with great instructors, on campus and virtually, throughout this process.
- The digital team of Eric Weber and Christine Carlson for helping me envision the possibilities in our new digital world.
- My content project manager, Jayne Klein, and program manager, Angie Fitzpatrick, for calmly steering this project throughout the publication process.
- Lori Hancock and Jo Johnson for the photos within this text. Biology is a visual science, and your contributions are evident on every page.
- David Hash for the design elements in this text, including one of the most beautiful textbook covers in the business.
- Dawnelle Krouse, Lauren Timmer, and Jane Hoover who acted as my proofreaders and copyeditor for this edition.
- Jane Peden for her behind the scenes work that keeps us all functioning.
- Inkling for providing a dynamic authoring platform, and Aptara for all of their technical assistance.

As both an educator and an author, communicating the importance of science represents one of my greatest passions. Our modern society is based largely on advances in science and technology over the past few decades. As I present in this text, there are many challenges facing humans, and an understanding of how science can help analyze, and offer solutions to, these problems is critical to our species' health and survival.

I also want to acknowledge my family for all of their support. My wife and partner Sandy has never wavered in her energy and support of my projects. The natural curiosity of my children, Devin and Kayla, has provided me with the motivation to make this world a better place for everyone.

Michael Windelspecht, Ph.D.

Blowing Rock, NC

Ancillary Authors

Appendix Answer Bank: Betsy Harris, *Ricochet Creative Productions*

Connect Question Bank: Alex James, *Ricochet Creative Productions*

Connect Test Bank: Carrie Wells, *University of North Carolina, Charlotte;* Jennifer Wiatrowski, *Pasco-Hernando Community College*

Tutorial Development: *Ricochet Creative Productions*

SmartBook with Learning Resources: Alex James, *Ricochet Creative Productions;* Patrick Galliart, *North Iowa Area Community College*

Contents

CHAPTER 1

Biology: The Science of Life   1

1.1 The Characteristics of Life   2
Life Requires Materials and Energy   2
Living Organisms Maintain an Internal Environment   4
Living Organisms Respond   5
Living Organisms Reproduce and Develop   5
Living Organisms Have Adaptations   6

1.2 Evolution: The Core Concept of Biology   6
Natural Selection and Evolutionary Processes   7
Organizing the Diversity of Life   9

1.3 Science: A Way of Knowing   11
Start with an Observation   11
Develop a Hypothesis   12
Make a Prediction and Perform Experiments   12
Develop a Conclusion   13
Scientific Theory   14
An Example of a Controlled Study   14
Publishing the Results   15

1.4 Challenges Facing Science   16
Biodiversity and Habitat Loss   16
Emerging and Reemerging Diseases   17
Climate Change   18

PART I   The Cell

CHAPTER 2

The Chemical Basis of Life   21

2.1 Atoms and Atomic Bonds   22
Atomic Structure   23
The Periodic Table 23
Isotopes   24
Arrangement of Electrons in an Atom   25

Types of Chemical Bonds   26
Chemical Formulas and Reactions   28

2.2 Water's Importance to Life   29
The Structure of Water   29
Properties of Water   29

2.3 Acids and Bases   33
Acidic Solutions (High H+ Concentration)   33
Basic Solutions (Low H+ Concentration)   34
pH and the pH Scale   34
Buffers and pH   35

CHAPTER 3

The Organic Molecules of Life   38

3.1 Organic Molecules   39
The Carbon Atom   39
The Carbon Skeleton and Functional Groups   40

3.2 The Biological Molecules of Cells   41
Carbohydrates   42
Lipids   45
Proteins   48
Nucleic Acids   51

CHAPTER 4

Inside the Cell   56

4.1 Cells Under the Microscope   57

4.2 The Plasma Membrane   59
Functions of Membrane Proteins   61

4.3 The Two Main Types of Cells   62
Prokaryotic Cells   62

4.4 Eukaryotic Cells   64
Nucleus and Ribosomes   66
Endomembrane System   68
Vesicles and Vacuoles   69
Energy-Related Organelles   69
The Cytoskeleton and Motor Proteins   72

Centrioles   72

Cilia and Flagella   73

4.5 Outside the Eukaryotic Cell   74

Cell Walls   74

Extracellular Matrix 74

Junctions Between Cells   75

CHAPTER **5**

The Dynamic Cell   79

5.1 What Is Energy? 80

Measuring Energy   80

Energy Laws   80

5.2 ATP: Energy for Cells 82

Structure of ATP   82

Use and Production of ATP   82

The Flow of Energy   84

5.3 Metabolic Pathways and Enzymes   85

An Enzyme's Active Site   86

Energy of Activation   87

5.4 Cell Transport   88

Passive Transport: No Energy Required   88

Active Transport: Energy Required   91

Bulk Transport   92

CHAPTER **6**

Energy for Life   95

6.1 Overview of Photosynthesis   96

Plants as Photosynthesizers   97

The Photosynthetic Process   98

6.2 The Light Reactions—Harvesting Energy   99

Photosynthetic Pigments   100

The Light Reactions: Capturing Solar Energy   100

6.3 The Calvin Cycle Reactions—Making Sugars   103

Overview of the Calvin Cycle 103

Reduction of Carbon Dioxide   104

The Fate of G3P   104

6.4 Variations in Photosynthesis   105

C_3 Photosynthesis   105

C_4 Photosynthesis   105

CAM Photosynthesis   106

Evolutionary Trends   106

CHAPTER **7**

Energy for Cells   110

7.1 Cellular Respiration   111

Phases of Complete Glucose Breakdown   112

7.2 Outside the Mitochondria: Glycolysis   113

Energy-Investment Step   113

Energy-Harvesting Steps   114

7.3 Outside the Mitochondria: Fermentation   115

Lactic Acid Fermentation   115

Alcohol Fermentation   116

7.4 Inside the Mitochondria   117

Preparatory Reaction   117

The Citric Acid Cycle   118

The Electron Transport Chain   119

7.5 Metabolic Fate of Food   121

Energy Yield from Glucose Metabolism   121

Alternative Metabolic Pathways   121

PART II   Genetics

CHAPTER **8**

Cellular Reproduction   125

8.1 The Basics of Cellular Reproduction   126

Chromosomes   126

Chromatin to Chromosomes   127

8.2 The Cell Cycle: Interphase, Mitosis, and Cytokinesis   128

Interphase   128

M (Mitotic) Phase   129

8.3 The Cell Cycle Control System   134

Cell Cycle Checkpoints   134

Internal and External Signals   134

Apoptosis   135

© Pascal Goetgheluck/
SPL/Science Source

8.4   The Cell Cycle and Cancer   136

Proto-Oncogenes and Tumor Suppressor Genes   136

Other Genetic Changes and Cancer   138

8.5   Characteristics of Cancer   139

Characteristics of Cancer Cells   139

Cancer Treatment   140

Prevention of Cancer   141

CHAPTER **9**

Meiosis and the Genetic Basis of Sexual Reproduction   145

9.1   An Overview of Meiosis   146

Homologous Chromosomes   146

The Human Life Cycle   147

Overview of Meiosis   147

9.2   The Phases of Meiosis   148

The First Division—Meiosis I   151

The Second Division—Meiosis II   151

9.3   Meiosis Compared with Mitosis   152

Meiosis I Compared with Mitosis   152

Meiosis II Compared with Mitosis   153

Mitosis and Meiosis Occur at Different Times   154

9.4   Changes in Chromosome Number   154

Down Syndrome   155

Abnormal Sex Chromosome Number   156

CHAPTER **10**

Patterns of Inheritance   160

10.1   Mendel's Laws   161

Mendel's Experimental Procedure   161

One-Trait Inheritance   162

Two-Trait Inheritance   166

Mendel's Laws and Probability   167

Mendel's Laws and Meiosis   168

10.2   Mendel's Laws Apply to Humans   169

Family Pedigrees   169

Genetic Disorders of Interest   170

10.3   Beyond Mendel's Laws   173

Incomplete Dominance   173

Multiple-Allele Traits   174

Polygenic Inheritance   174

Gene Interactions   176

Pleiotropy   177

Linkage   177

10.4   Sex-Linked Inheritance   178

Sex-Linked Alleles   179

Pedigrees for Sex-Linked Disorders   179

X-Linked Recessive Disorders   180

© Eye of Science/Science Source

CHAPTER **11**

DNA Biology   184

11.1   DNA and RNA Structure and Function   185

Structure of DNA   86

Replication of DNA   189

RNA Structure and Function   190

11.2   Gene Expression   191

From DNA to RNA to Protein   192

Review of Gene Expression   196

11.3   Gene Regulation   197

Levels of Gene Expression Control   197

CHAPTER **12**

Biotechnology and Genomics   207

12.1   Biotechnology   208

Recombinant DNA Technology   208

DNA Sequencing   209

Polymerase Chain Reaction   209

DNA Analysis   210

Genome Editing   211

12.2   Stem Cells and Cloning   212

Reproductive and Therapeutic Cloning   212

12.3   Biotechnology Products   214

Genetically Modified Bacteria   214

Genetically Modified Plants   214

Genetically Modified Animals   215

12.4   Genomics and Proteomics   216

Sequencing the Bases of the Human Genome   216

Proteomics and Bioinformatics   218

CHAPTER 13

Genetic Counseling   221

13.1 Gene Mutations   222
Causes of Gene Mutations   222
Types and Effects of Mutations   223

13.2 Chromosomal Mutations   224
Deletions and Duplications   224
Translocation   225
Inversion   226

13.3 Genetic Testing   226
Analyzing the Chromosomes   227
Testing for a Protein   228
Testing the DNA   228
Testing the Fetus   229
Testing the Embryo and Egg   230

13.4 Gene Therapy   232
Ex Vivo Gene Therapy   232
In Vivo Gene Therapy   232

PART III   Evolution

CHAPTER 14

Darwin and Evolution   235

14.1 Darwin's Theory of Evolution   236
Before Darwin   237
Darwin's Conclusions   238
Natural Selection and Adaptation   240
Darwin and Wallace   243

14.2 Evidence of Evolutionary Change   244
Fossil Evidence   244
Biogeographical Evidence   246
Anatomical Evidence   246
Molecular Evidence   248

CHAPTER 15

Evolution on a Small Scale   251

15.1 Natural Selection   252
Types of Selection   253
Sexual Selection   254

Adaptations Are Not Perfect   255
Maintenance of Variations   255

15.2 Microevolution   257
Evolution in a Genetic Context   257
Causes of Microevolution   260

CHAPTER 16

Evolution on a Large Scale   265

16.1 Speciation and Macroevolution   266
Defining Species   266
Models of Speciation   269

16.2 The Fossil Record   272
The Geological Timescale   272
The Pace of Speciation   274
Causes of Mass Extinctions   275

16.3 Systematics   275
Linnaean Classification   277
Phylogenetic Trees   278
Cladistics and Cladograms   280
The Three-Domain System   281

PART IV   Diversity of Life

CHAPTER 17

The Microorganisms: Viruses, Bacteria, and Protists   286

17.1 The Viruses   287
Structure of a Virus   287
Viral Reproduction   288
Plant Viruses   289
Animal Viruses   289

17.2 Viroids and Prions   292

17.3 The Prokaryotes   293
The Origin of the First Cells   293
Bacteria   294
Archaea   299

17.4 The Protists   301
Evolution of Protists   301
Classification of Protists   301

CHAPTER 18

The Plants and Fungi   311

18.1 Overview of the Plants   312
An Overview of Plant Evolution   312
Alternation of Generations   314

18.2 Diversity of Plants   315
Nonvascular Plants   315
Vascular Plants   316
Gymnosperms   320
Angiosperms   321
Economic Benefits of Plants   324
Ecological Benefits of Plants   324

18.3 The Fungi   325
General Biology of a Fungus 325
Fungal Diversity   326
Ecological Benefits of Fungi   329
Economic Benefits of Fungi   330
Fungi as Disease-Causing Organisms   331

CHAPTER 19

Both Water and Land: Animals   337

19.1 Evolution of Animals   337
Ancestry of Animals   338
The Evolutionary Tree of Animals   338
Evolutionary Trends   339

19.2 Sponges and Cnidarians: The Early Animals   341
Sponges: Multicellularity   341
Cnidarians: True Tissues   342

19.3 Flatworms, Molluscs, and Annelids: The Lophotrochozoans   343
Flatworms: Bilateral Symmetry   343
Molluscs   344
Annelids: Segmented Worms   345

19.4 Roundworms and Arthropods: The Ecdysozoans   347
Roundworms: Pseudocoelomates   347
Arthropods: Jointed Appendages   348

19.5 Echinoderms and Chordates: The Deuterostomes   353
Echinoderms   353
Chordates   354
Fishes: First Jaws and Lungs   356
Amphibians: Jointed Vertebrate Limbs   358
Reptiles: Amniotic Egg   358
Mammals: Hair and Mammary Glands   360

19.6 Human Evolution   363
Evolution of Humanlike Hominins   365
Evolution of Modern Humans   367

PART V   Plant Structure and Function

CHAPTER 20

Plant Anatomy and Growth   372

20.1 Plant Cells and Tissues   373
Epidermal Tissue   373
Ground Tissue   374
Vascular Tissue   374

20.2 Plant Organs   375
Monocots Versus Eudicots   376

20.3 Organization of Leaves, Stems, and Roots   377
Leaves   377
Stems   378
Roots   382

20.4 Plant Nutrition   385
Adaptations of Roots for Mineral Uptake   386

20.5 Transport of Nutrients   387
Water Transport in Xylem   387
Sugar Transport in Phloem   388

CHAPTER 21

Plant Responses and Reproduction   392

21.1 Plant Hormones   393
Auxins   393
Gibberellins   394
Cytokinins   395

Abscisic Acid   395

Ethylene   396

21.2 Plant Responses   396

Tropisms   397

Photoperiodism   398

21.3 Sexual Reproduction in Flowering Plants   399

Overview of the Plant Life Cycle   399

Flowers   400

From Spores to Fertilization   401

Development of the Seed in a Eudicot   403

Monocots Versus Eudicots   404

Fruit Types and Seed Dispersal   404

Germination of Seeds   405

21.4 Asexual Reproduction and Genetic Engineering in Plants   407

Propagation of Plants in a Garden   407

Propagation of Plants in Tissue Culture   407

Genetic Engineering of Plants   408

PART VI   Animal Structure and Function

CHAPTER **22**

Being Organized and Steady   414

22.1 The Body's Organization   415

Epithelial Tissue Protects   417

Connective Tissue Connects and Supports   419

Muscular Tissue Moves the Body   421

Nervous Tissue Communicates   422

22.2 Organs and Organ Systems   423

Transport and Protection   424

Maintenance of the Body   424

Control   424

Sensory Input and Motor Output   425

Reproduction   425

22.3 Homeostasis   426

Organ Systems and Homeostasis   426

Negative Feedback   427

CHAPTER **23**

The Transport Systems   431

23.1 Open and Closed Circulatory Systems   432

Open Circulatory Systems   433

Closed Circulatory Systems   433

Comparison of Vertebrate Circulatory Pathways   433

23.2 Transport in Humans   435

The Human Heart   435

Blood Vessels   437

Lymphatic System   440

Capillary Exchange in the Tissues   441

23.3 Blood: A Transport Medium   442

Plasma   442

Formed Elements   442

Cardiovascular Disorders   445

© Anthony Mercieca/Science Source

CHAPTER **24**

The Maintenance Systems   450

24.1 Respiratory System   451

The Human Respiratory Tract   451

Breathing   453

Lungs and External Exchange of Gases   454

Transport and Internal Exchange of Gases   455

24.2 Urinary System   457

Human Kidney   457

Problems with Kidney Function   461

CHAPTER **25**

Digestion and Human Nutrition   465

25.1 Digestive System   466

Complete and Incomplete Digestive Systems   466

The Digestive Tract   466

Accessory Organs   467

Digestive Enzymes   473

25.2 Nutrition   475

Introducing the Nutrients   475

25.3   The Classes of Nutrients   476
Carbohydrates   476
Lipids   478
Proteins   479
Minerals   480
Vitamins   482
Water   483

25.4   Understanding Nutrition Guidelines   484
Updating Dietary Guidelines 484
Visualizing Dietary Guidelines   484
Dietary Supplements 485
The Bottom Line   487

25.5   Nutrition and Health   487
Body Mass Index   488
Disorders Associated with Obesity   490
Eating Disorders   492

CHAPTER **26**

Defenses Against Disease   496

26.1   Overview of the Immune System   497
Lymphatic Organs   497
Cells of the Immune System   499

26.2   Nonspecific Defenses and Innate Immunity   499
Barriers to Entry   499
The Inflammatory Response   500
The Complement System   501
Natural Killer Cells   501

26.3   Specific Defenses and Adaptive Immunity   502
B Cells and the Antibody Response   502
T Cells and the Cellular Response   503

26.4   Immunizations   506

26.5   Disorders of the Immune System   508
Allergies   508
Autoimmune Diseases   509
AIDS   509

CHAPTER **27**

The Control Systems   513

27.1   Nervous System   514
Examples of Nervous Systems   515
The Human Nervous System   515
Neurons   516
The Nerve Impulse   516
The Synapse   518
Drug Abuse   518
The Central Nervous System   520
The Peripheral Nervous System   523

27.2   Endocrine System   526
The Action of Hormones   526
Hypothalamus and Pituitary Gland   527
Thyroid and Parathyroid Glands   529
Adrenal Glands   530
Pancreas   531

CHAPTER **28**

Sensory Input and Motor Output   536

28.1   The Senses   537
Chemical Senses   537
Hearing and Balance   538
Vision   542
Cutaneous Receptors and Proprioceptors   544

28.2   The Motor Systems   545
Types of Skeletons   546
The Human Skeleton   547
Skeletal Muscle Structure and Physiology   548

CHAPTER **29**

Reproduction and Embryonic Development   556

29.1   How Animals Reproduce   557
Asexual Versus Sexual Reproduction   557
Sexual Reproduction   557

29.2 Human Reproduction   559
Male Reproductive System   559
Female Reproductive System   562
Control of Reproduction   564
Infertility   566
Sexually Transmitted Diseases   568

29.3 Human Embryonic Development   570
Fertilization   571
Early Embryonic Development   572
Later Embryonic Development   573
Placenta   575
Fetal Development and Birth   575

PART VII ECOLOGY

CHAPTER **30**

Ecology and Populations   580

30.1 The Scope of Ecology   581
Ecology: A Biological Science   582

30.2 The Human Population   583
Present Population Growth   583
Future Population Growth   584
More-Developed Versus Less-Developed
    Countries   585
Comparing Age Structures   586
Population Growth and Environmental Impact   587

30.3 Characteristics of Populations   588
Distribution and Density   588
Population Growth   588
Patterns of Population Growth   590
Factors That Regulate Population Growth   592

30.4 Life History Patterns and Extinction   595
Extinction   595

CHAPTER **31**

Communities and Ecosystems   600

31.1 Ecology of Communities   601
Community Composition and Diversity   602
Ecological Succession   603
Interactions in Communities   604
Community Stability   608

31.2 Ecology of Ecosystems   610
Autotrophs   610
Heterotrophs   610
Energy Flow and Chemical Cycling   611
Chemical Cycling   614

31.3 Ecology of Major Ecosystems   619
Primary Productivity   620

CHAPTER **32**

Human Impact on the Biosphere   626

32.1 Conservation Biology   627

32.2 Biodiversity   628
Direct Values of Biodiversity   629
Indirect Values of Biodiversity   630

**32.3 Resources and Environmental
Impact   632**
Land   633
Water   635
Food   637
Energy   639
Minerals   641
Other Sources of Pollution   641

32.4 Sustainable Societies   643
Today's Society   644
Characteristics of a Sustainable Society   644

Appendix A Periodic Table of the Elements
    & The Metric System   A-1

Appendix B Answer Key   B-1

Glossary   G-1

Index   I-1

(anglerfish): © Theodore Pietsch/University of Washington; (spider): © MSc. Rafael Fonseca-Ferreira; (*Dendrogramma*): © Jean Just, Reinhardt Mobjerg Kristensen and Jorgen Olesen; (elephant shrew): © Dr. Galen Rathbun and Dr. Jack Dumbacher

1

Biology: The Science of Life

The Diversity of Life

Life on Earth takes on a staggering variety of forms, often with appearances and behaviors that may be strange to humans. As we will see in this chapter, one of the ways that biologists classify life is by species. So how many species are there on the planet? The truth is, we really don't know. Recent estimates suggest that there may be around 8.7 million species on the planet, but many scientists believe that number is probably much higher, especially when the bacteria are factored in. So far, less than 2 million species have been identified, and most of those are insects.

However, new species, such as those shown here, are being discovered all the time. While investigating the impacts of the 2010 oil spills in the Gulf of Mexico, researchers discovered the anglerfish *Lasiognathus dinema* (*top, left*). Recently, two new species of *Dendrogramma* were discovered off the coast of Australia (*bottom, left*). This genus is so unique that it does not fit into any current classification. A new eyeless cave spider, *Iandumoema smeagol*, named after the Lord of the Rings character, is so specialized that it is believed to be found in a limited number of caves (*top, right*). New mammals have also recently been discovered, such as the world's smallest elephant shrew, *Macroscelides micus* (*bottom, right*).

As we will learn in this chapter, although life is diverse, it also shares a number of important characteristics.

As you read through this chapter, think about the following questions:

1. What are the general characteristics that separate living organisms from nonliving things?
2. How do species fit into the biological levels of organization?
3. What are some of the challenges facing science today?

OUTLINE

1.1 The Characteristics of Life   2

1.2 Evolution: The Core Concept of Biology   6

1.3 Science: A Way of Knowing   11

1.4 Challenges Facing Science   16

bacteria   human

plant   fungi

Figure 1.1 Diversity of life.

Biology is the study of life in all of its diverse forms.

(bacteria): © Science Photo Library RF/Getty Images; (human): © Purestock/
Superstock RF; (plant): © McGraw-Hill Education; (fungi): © Jorgen Bausager/
Getty RF

1.1 The Characteristics of Life

Learning Outcomes

Upon completion of this section, you should be able to

1. Explain the basic characteristics that are common to all living organisms.
2. Distinguish between the levels of biological organization.
3. Summarize how the terms *homeostasis, metabolism,* and *adaptation* relate to all living organisms.
4. Contrast chemical cycling and energy flow within an ecosystem.

As we observed in the chapter opener, life is diverse (**Fig. 1.1**). Life may be found everywhere on the planet, from thermal vents at the bottom of the ocean to the coldest reaches of Antarctica. **Biology** is the scientific study of life. Biologists study not only life's diversity but also the characteristics that are shared by all living organisms. These characteristics include levels of organization, the ability to acquire materials and energy, the ability to maintain an internal environment, the ability to respond to stimuli, the ability to reproduce and develop, and the ability to adapt and evolve to changing conditions. By studying these characteristics, we gain insight into the complex nature of life, which helps us distinguish living organisms from nonliving things. In the next sections, we will explore these characteristics in more detail.

The complex organization of life begins with atoms, the basic units of matter. Atoms combine to form small molecules, which join to form larger molecules within a **cell,** the smallest, most basic unit of life. Although a cell is alive, it is made from nonliving molecules (**Fig. 1.2**).

The majority of the organisms on the planet, such as the bacteria and most protists, are single-celled. Plants, fungi, and animals are **multicellular** organisms and are therefore composed of many types of cells, which often combine to form **tissues.** Tissues make up **organs,** as when various tissues combine to form a heart or a leaf. Organs work together in **organ systems;** for example, the heart and blood vessels form the cardiovascular system. Various organ systems work together within complex organisms.

The organization of life extends beyond the individual organism. A **species** is a group of similar organisms that are capable of interbreeding. All of the members of a species within a particular area belong to a **population.** When populations interact, such as the humans, zebras, and trees in Figure 1.2, they form a **community.** At the **ecosystem** level, communities interact with the physical environment (soil, atmosphere, etc.). Collectively, the ecosystems on the planet are called the **biosphere,** the zone of air, land, and water at the surface of the Earth where living organisms are found.

Life Requires Materials and Energy

Life from single cells to complex organisms cannot maintain organization or carry on necessary activities without an outside source of materials and energy. Food provides nutrient molecules, which are used as building blocks or energy sources. **Energy** is the capacity to do work, and it takes work to maintain the organization of the cell and the organism. When cells use nutrient molecules to make their parts and products, they carry out a sequence of chemical reactions. The term **metabolism** encompasses all the chemical reactions that occur in a cell.

Figure 1.2 **Levels of biological organization.**

All life is connected by levels of biological organization that extend from atoms to the biosphere.

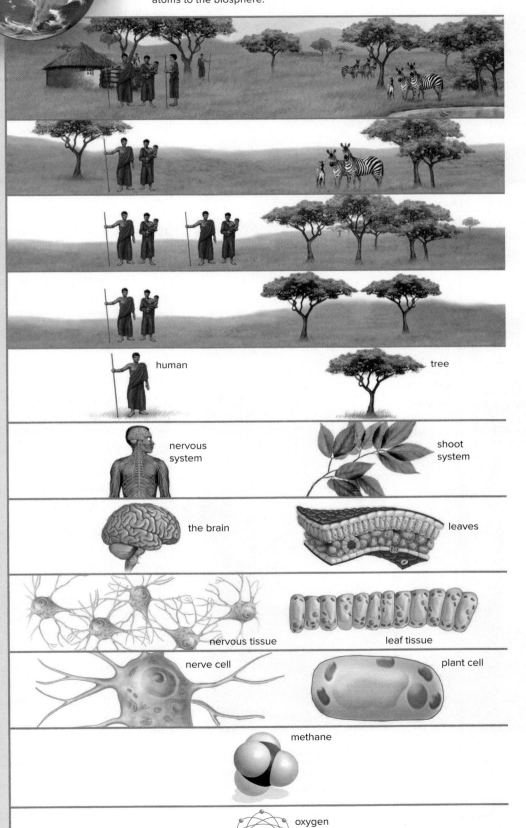

Biosphere
Regions of the Earth's crust, waters, and atmosphere inhabited by living organisms

Ecosystem
A community plus the physical environment

Community
Interacting populations in a particular area

Population
Organisms of the same species in a particular area

Species
A group of similar, interbreeding organisms

Organism
An individual; complex individuals contain organ systems

human

tree

Organ System
Composed of several organs working together

nervous system

shoot system

Organ
Composed of tissues functioning together for a specific task

the brain

leaves

Tissue
A group of cells with a common structure and function

nervous tissue

leaf tissue

Cell
The structural and functional unit of all living organisms

nerve cell

plant cell

Molecule
Union of two or more atoms of the same or different elements

methane

Atom
Smallest unit of an element; composed of electrons, protons, and neutrons

oxygen

Figure 1.3 **Acquiring nutrient materials and energy.**

All organisms, including this mongoose eating a snake, require nutrients and energy.

© Gallo Images–Dave Hamman/Getty RF

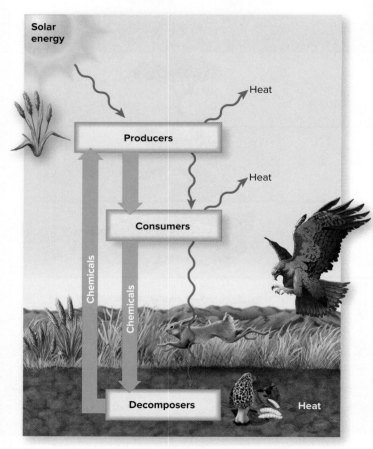

Figure 1.4 **Chemical cycling and energy flow in an ecosystem.**

In an ecosystem, chemical cycling (aqua arrows) and energy flow (red arrows) begin when plants use solar energy and inorganic nutrients to produce their own food. Chemicals and energy are passed from one population to another in a food chain. Eventually, energy dissipates as heat. With the death and decomposition of organisms, chemicals are returned to living plants once more.

The ultimate source of energy for nearly all life on Earth is the sun. Plants and certain other organisms are able to capture solar energy and carry on **photosynthesis,** a process that transforms solar energy into the chemical energy of nutrient molecules. For this reason, these organisms are commonly called producers. Animals and plants get energy by metabolizing (**Fig. 1.3**), or breaking down, the nutrient molecules made by the producers.

The energy and chemical flow between organisms also defines how an ecosystem functions (**Fig. 1.4**). Within an ecosystem, chemical cycling and energy flow begin when producers, such as grasses, take in solar energy and inorganic nutrients to produce food (organic nutrients) by photosynthesis. Chemical cycling (aqua arrows) occurs as chemicals move from one population to another in a food chain, until death and decomposition allow inorganic nutrients to be returned to the producers once again. Energy (red arrows), on the other hand, flows from the sun through plants and the other members of the food chain as they feed on one another. The energy gradually dissipates and returns to the atmosphere as heat. Because energy does not cycle, ecosystems could not stay in existence without solar energy and the ability of photosynthetic organisms to absorb it.

Energy flow and nutrient cycling in an ecosystem largely determine where different ecosystems are found in the biosphere. The two most biologically diverse ecosystems—tropical rain forests and coral reefs—occur where solar energy is very abundant and nutrient cycling is continuous.

The availability of energy and nutrients also determines the type of biological communities that occur within an ecosystem. One example of an ecosystem in North America is the grasslands, which are inhabited by populations of rabbits, hawks, and various types of grasses, among many others. The energy input and nutrient cycling of a grassland are less than those of a rain forest, so the community structure and food chains of these ecosystems differ.

Living Organisms Maintain an Internal Environment

For metabolic processes to continue, living organisms need to keep themselves stable with regard to temperature, moisture level, acidity, and other factors critical to maintaining life. This is called **homeostasis,** or the maintenance of internal conditions within certain physiological boundaries.

Many organisms depend on behavior to regulate their internal environment. A chilly lizard may raise its internal temperature by basking in the sun on a hot rock. When it starts to overheat, it scurries for cool shade. Other organisms have control mechanisms that do not require any conscious activity. When you are studying and forget to eat lunch, your liver releases stored sugar to keep your blood sugar level within normal limits. Many of the organ systems of our bodies are involved in maintaining homeostasis.

Living Organisms Respond

Living organisms find energy and/or nutrients by interacting with their surroundings. Even single-celled organisms can respond to their environment. The beating of microscopic hairs or the snapping of whiplike tails moves them toward or away from light or chemicals. Multicellular organisms can manage more complex responses. A monarch butterfly can sense the approach of fall and begin its flight south, where resources are still abundant. A vulture can smell meat a mile away and soar toward dinner.

The ability to respond often results in movement: The leaves of a plant turn toward the sun, and animals dart toward safety. Appropriate responses help ensure survival of the organism and allow it to carry on its daily activities. Altogether, we call these activities the *behavior* of the organism.

Living Organisms Reproduce and Develop

Life comes only from life. Every living organism has the ability to **reproduce,** or make another organism like itself. Bacteria and other types of single-celled organisms simply split in two. In multicellular organisms, the reproductive process usually begins with the pairing of a sperm from one partner and an egg from the other partner. The union of sperm and egg, followed by many cell divisions, results in an immature individual, which grows and develops through various stages to become an adult.

An embryo develops into a whale or a yellow daffodil or a human because of the specific set of **genes,** or genetic instructions, inherited from its parents (**Fig. 1.5**). In all organisms, the genes are located on long molecules of **DNA (deoxyribonucleic acid),** the genetic blueprint of life. Variations in genes account for the differences between species and individuals. These differences are the result of **mutations,** or inheritable changes in the genetic information. Mutation provides an important source of variation in the genetic information. However, not all mutations are bad—the observable differences in eye and hair color are examples of mutations.

By studying DNA, scientists are able to understand not only the basis for specific traits, like susceptibility for certain types of cancer, but also the evolutionary history of the species. Reproduction involves the passing of genetic information from a parent to its offspring. Therefore, the information found within the DNA represents a record of our molecular heritage. This includes not only a record of the individual's lineage, but also how the species is related to other species.

DNA provides the blueprint or instructions for the organization and metabolism of the particular organism. All cells in a multicellular organism contain the same set of genes, but only certain ones are turned on in each type of specialized cell. Through the process of **development,** cells express specific genes to distinguish themselves from other cells, thus forming tissues and organs.

DNA

Figure 1.5   Reproduction is a characteristic of life.

Whether they are single-celled or multicellular, all organisms reproduce. Offspring receive a copy of their parents' DNA and therefore a copy of the parents' genes.

(photo): © Purestock/Superstock RF; (DNA): © David Mack/SPL/Science Source

Living Organisms Have Adaptations

Adaptations are modifications that make organisms suited to their way of life. Some hawks have the ability to catch fish; others are best at catching rabbits. Hawks can fly, in part, because they have hollow bones to reduce their weight and flight muscles to depress and elevate their wings. When a hawk dives, its strong feet take the first shock of the landing, and its long, sharp claws reach out and hold onto the prey. Hawks have exceptionally keen vision, which enables them not only to spot prey from great heights but also to estimate distance and speed.

Humans also have adaptations that allow them to live in specific environments. Humans who live at extreme elevations in the Himalayas (over 13,000 feet, or 4,000 meters) have an adaptation that reduces the amount of hemoglobin produced in the blood. Hemoglobin is important for the transport of oxygen. Normally, as elevation increases, the amount of hemoglobin increases, but too much hemoglobin makes the blood thick, which can cause health problems. In some high-elevation populations, a mutation in a single gene reduces the risk.

Evolution, or the manner in which species become adapted to their environment, is discussed in the next section of this chapter.

CONNECTING THE CONCEPTS

1.1

All living organisms, from bacteria to humans, share the same basic characteristics of life.

Check Your Progress 1.1

1. List the basic characteristics common to all life.
2. List, in order starting with the least organized, the levels of biological organization.
3. Explain how chemical cycling and energy flow occur at both the organism and the ecosystem levels of organization.

1.2 Evolution: The Core Concept of Biology

Learning Outcomes

Upon completion of this section, you should be able to

1. Define the term *evolution*.
2. Explain the process of natural selection and its relationship to evolutionary processes.
3. Summarize the general characteristics of the domains and major kingdoms of life.

Despite diversity in form, function, and lifestyle, organisms share the same basic characteristics. As mentioned, they are all composed of cells organized in a similar manner. Their genes are composed of DNA, and they carry out the same metabolic reactions to acquire energy and maintain their organization. The unity of living organisms suggests that they are descended from a common ancestor—the first cell or cells.

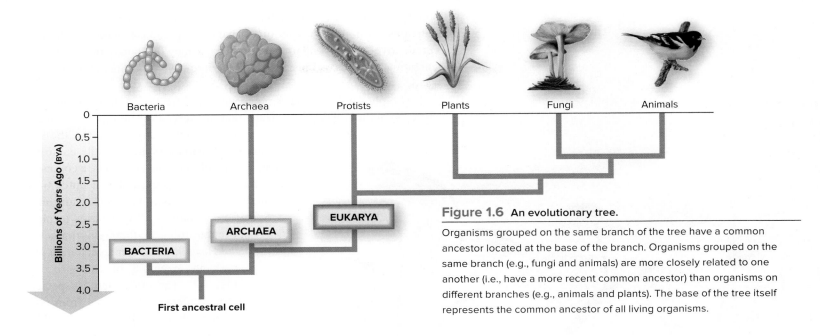

Figure 1.6 An evolutionary tree.

Organisms grouped on the same branch of the tree have a common ancestor located at the base of the branch. Organisms grouped on the same branch (e.g., fungi and animals) are more closely related to one another (i.e., have a more recent common ancestor) than organisms on different branches (e.g., animals and plants). The base of the tree itself represents the common ancestor of all living organisms.

An evolutionary tree is like a family tree (**Fig. 1.6**). Just as a family tree shows how a group of people have descended from one couple, an evolutionary tree traces the ancestry of life on Earth to a common ancestor. One couple can have diverse children, and likewise a population can be a common ancestor to several other groups, each adapted to a particular set of environmental conditions. **Evolution** is the process in which populations change over time to adapt to their environment, and pass on these changes to the next generation. Evolution is considered the unifying concept of biology because it explains so many aspects of biology, including how living organisms arose from a single ancestor and the tremendous diversity of life on the planet.

Natural Selection and Evolutionary Processes

In the nineteenth century, two naturalists—Charles Darwin and Alfred Russel Wallace—came independently to the conclusion that evolution occurs by means of a process called natural selection. Charles Darwin is the more famous of the two because he wrote a book called *On the Origin of Species,* which presented much data to substantiate the occurrence of evolution by natural selection. Since that time, evolution has become the core concept of biology because the theory explains so many different types of observations in every field of biology.

The process of **natural selection** is the mechanism of evolutionary change and is based on how a population changes in response to its environment. Environments may change due to the influence of living factors (such as a new predator) or nonliving factors (such as temperature). As the environment changes over time, some individuals of a species may possess certain adaptations that make them better suited to the new environment. Individuals of a species that are better adapted to their environment tend to live longer and produce more offspring than other individuals. This differential

Connections: Health

How does evolution affect me personally?

In the presence of an antibiotic, resistant bacteria are selected to reproduce over and over again, until the entire population of bacteria becomes resistant to the antibiotic. In 1959, a new antibiotic called methicillin became available to treat bacterial (staph) infections that were already resistant to penicillin. In 1974, 2% of the staph infections were classified as MRSA (methicillin-resistant *Staphylococcus aureus*), but by 2004 the number had risen to 63%. In response, the Centers for Disease Control and Prevention conducted an aggressive campaign to educate health-care workers about preventing MRSA infections. The program was very successful, and between 2005 and 2008 the number of MRSA infections in hospitals declined by 28%. However, MRSA remains an important concern of the medical community.

reproductive success, called natural selection, results in changes in the characteristics of a population over time. That is, adaptations that result in higher reproductive success tend to increase in frequency in a population from one generation to the next. This change in the frequency of traits in populations is called evolution.

The phrase "common descent with modification" sums up the process of evolution because it means that, as descent occurs from common ancestors, modifications occur that cause the organisms to be adapted (suited) to the environment. As a result, one species can be a common ancestor to several species, each adapted to a particular set of environmental conditions. Specific adaptations allow species to play particular roles in their environment. The Hawaiian honeycreepers are a remarkable example of this process (**Fig. 1.7**). The more than 50 species of honeycreepers all evolved from one species of finch, which likely originated in North America and arrived in the Hawaiian islands between 3 and 5 million years ago. Modern honeycreepers have an assortment of bill shapes adapted to different types of food. Some honeycreeper species have curved, elongated bills used for drinking flower nectar. Others have strong, hooked bills suited to digging in tree bark and seizing wood-boring insects or short, straight, finchlike bills for feeding on small seeds and fruits. Even with such dramatic differences in feeding habits and bill shapes, honeycreepers still share certain characteristics, which stem from their common finch ancestor. The various honeycreeper species are similar in body shape and size, as well as mating and nesting behavior.

The study of evolution encompasses all levels of biological organization. Indeed, much of today's evolution research is carried out at the molecular level, comparing the DNA of different groups of organisms to determine how they are related. Looking at how life has changed over time, from its origin to the current day, helps us understand why there are so many different kinds of organisms and why they have the characteristics they do. An understanding of evolution by natural selection also has practical applications, including the prevention and treatment of disease.

Today, we know that, because of selection, resistance to antibiotic drugs has become increasingly common in a number of bacterial species, including those that cause tuberculosis, gonorrhea, and staph infections. Antibiotic drugs, such as penicillin, kill susceptible bacteria. However, some bacteria in the body of a patient undergoing antibiotic treatment may be unharmed by the drug. Bacteria can survive antibiotic drugs in many different ways. For example, certain bacteria can endure treatment with penicillin because they break down the drug, rendering it harmless. If even one bacterial cell lives because it is antibiotic-resistant, then its descendants will inherit this drug-defeating ability. The more antibiotic drugs are used, the more natural selection favors resistant bacteria, and the more often antibiotic-resistant infections will occur.

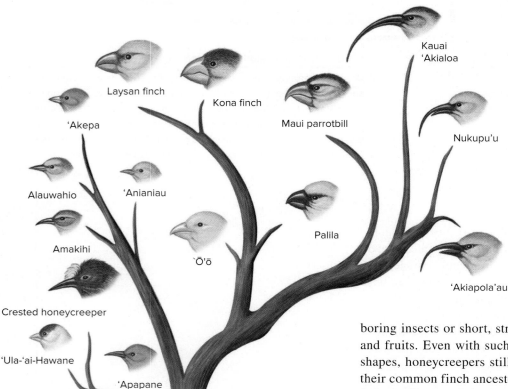

Figure 1.7 Evolution of Hawaiian honeycreepers.

Hawaiian honeycreepers, descendants of a single ancestral species, display an amazing diversity of bill shapes and sizes.

Organizing the Diversity of Life

Think of an enormous department store, offering thousands of different items for sale. The various items are grouped in departments—electronics, apparel, furniture, and so on—to make them easy for customers to find. Because life is so diverse, it is helpful to have a system that groups organisms into categories. Two areas of biology help us group organisms into categories: **Taxonomy** is the discipline of identifying and naming organisms according to certain rules, and **systematics** makes sense out of the bewildering variety of life on Earth by classifying organisms according to their presumed evolutionary relationships. As systematists learn more about evolutionary relationships between species, the taxonomy of a given organism may change. Systematists are even now making observations and performing experiments that will one day bring about changes in the classification system adopted by this text.

Categories of Classification

The classification categories, from least inclusive to most inclusive, are species, **genus, family, order, class, phylum, kingdom,** and **domain** (**Table 1.1**). Each successive category above species contains more types of organisms than the preceding one. Species placed within one genus share many specific characteristics and are the most closely related, while species placed in the same domain share only general characteristics. For example, all species in the genus *Pisum* look pretty much the same—that is, like pea plants—but species in the plant kingdom can be quite varied, as is evident when we compare grasses with trees. By the same token, only modern humans are in the genus *Homo,* but many types of species, from tiny hydras to huge whales, are members of the animal kingdom. Species placed in different domains are the most distantly related. For now, we will focus on the general characteristics of the domains and kingdoms of life.

Domains

The most inclusive and general levels of classification are the domains (**Table 1.2**). Biochemical evidence (obtained from the study of DNA and proteins) suggests that there are only three domains of life: **domain Bacteria, domain Archaea, and domain Eukarya.** Both domain Archaea and domain Bacteria contain prokaryotes. Prokaryotes are single-celled, and they lack the membrane-bound nucleus found in the eukaryotes of domain Eukarya.

Prokaryotes are structurally simple but metabolically complex. Archaea live in aquatic environments that lack oxygen or are too salty, too hot, or too acidic for most other organisms. Perhaps these environments are similar to those of the primitive Earth and archaea are representative of the first cells that evolved. Bacteria are found almost everywhere—in the water, soil, and atmosphere, as well as on our skin and in our mouths and large intestines. Although some bacteria cause diseases, others perform useful services, both environmental and commercial. For example, they are used to conduct genetic research in our laboratories (the *E. coli* in Table 1.2 is one example), to produce innumerable products in our factories, and to purify water in our sewage treatment plants.

Kingdoms

Systematicists are just beginning to understand how to categorize domain Archaea and domain Bacteria into kingdoms. Currently, there are four kingdoms

Table 1.1 Levels of Biological Organization

Category	Human	Corn
Domain	Eukarya	Eukarya
Kingdom	Animalia	Plantae
Phylum	Chordata	Anthophyta
Class	Mammalia	Liliopsida
Order	Primates	Commelinales
Family	Hominidae	Poaceae
Genus	Homo	Zea
*Species**	H. sapiens	Z. mays

* To specify an organism, you must use the full binomial name, such as *Homo sapiens.*

Table 1.2 Domains and Kingdoms of Life

Domain	Kingdom	Example
Archaea		8,330× — Capable of living in extreme environments. *Methanosarcina mazei,* a methane-generating prokaryote.
Bacteria		6,600× — Structurally simple but metabolically diverse. *Escherichia coli,* a prokaryote found in our intestinal tracts.
Eukarya	Protists*	250× — Diverse group of eukaryotes, many single-celled. *Euglena,* an organism with both plant and animal-like characteristics.
Eukarya	Plants	Multicellular photosynthesizers. The bristlecone pine, *Pinus longaeva,* one of the oldest organisms on the planet.
Eukarya	Animals	Multicellular organisms that ingest food. *Homo sapiens*—humans.
Eukarya	Fungi	Multicellular decomposers. *Amanita*—a mushroom.

* Many systematists are suggesting that kingdom Protista be subdivided to better reflect the evolutionary relationships of these organisms.

(Archaea): © Eye of Science/Science Source; (Bacteria): © A. B. Dowsett/SPL/ Science Source; (protists): © blickwinkel/Fox/Alamy; (plants): © Brenda Tharp/ Science Source; (animals): © Radius Images/Getty RF; (fungi): © Corbis RF

within domain Eukarya. **Protists** (kingdom Protista) are a very diverse group of eukaryotic organisms, some of which are single-celled and others multicellular. Some protists are photosynthetic, some are decomposers, and some ingest their food. As we will see in Section 17.4, systematicists recognize that there are considerable differences among the members of this kingdom, and efforts are underway to introduce a new classification scheme for the protists. Among the **fungi** (kingdom Fungi) are the familiar molds and mushrooms that, along with many types of bacteria, help decompose dead organisms. **Plants** (kingdom Plantae) are well known as multicellular photosynthesizers. **Animals** (kingdom Animalia) are multicellular organisms that ingest their food.

The three domains and the four kingdoms within the domain Eukarya are depicted in the evolutionary tree in Figure 1.6. This tree, which is largely based on the DNA of organisms, shows that the domain Bacteria was the first to arise in the history of life, followed by the domain Archaea and finally the domain Eukarya. The domain Archaea is more closely related to the domain Eukarya than either is to the domain Bacteria. Among the Eukarya, the protists gave rise to the kingdoms of plants, fungi, and animals. Later in this text, we will return to this evolutionary tree, and the evolution of each kingdom will be discussed separately.

Scientific Naming

Biologists give each living organism a two-part scientific name called a **binomial name.** For example, the scientific name for the garden pea is *Pisum sativum;* our own species is *Homo sapiens.* The first word is the genus, and the second word is the specific epithet of a species within a genus. The genus may be abbreviated, such as *P. sativum* or *H. sapiens.* Scientific names are universally used by biologists to avoid confusion. Common names tend to overlap, and often they are from the languages of the people who use the names. But scientific names are based on Latin, a universal language that not too long ago was well known by most scholars. Table 1.2 provides some examples of binomial names. Now that we know the general groups to which organisms are classified, and how scientists assign scientific names, let's return to our discussion of evolution and the process by which diversity arises.

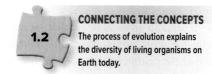

CONNECTING THE CONCEPTS

1.2 The process of evolution explains the diversity of living organisms on Earth today.

Check Your Progress 1.2

1. List the eight classification categories, from least to most inclusive.

2. Describe the process of natural selection, and explain its relationship to evolutionary change.

3. Explain why the concept of descent with modification is important in understanding the evolutionary process.

1.3 Science: A Way of Knowing

Learning Outcomes

Upon completion of this section, you should be able to

1. Identify the steps of the scientific method.
2. Describe the basic requirements for a controlled experiment.
3. Distinguish between a theory and a hypothesis.

Biology is the scientific study of life. Religion, aesthetics, ethics, and science are all ways that humans have of finding order in the natural world. Science differs from the other disciplines by its process, which often involves the use of the scientific method (**Fig. 1.8**). The scientific method acts as a guideline for scientific studies. Scientists often modify or adapt the process to suit their particular area of study.

Start with an Observation

Scientists believe that nature is orderly and measurable—that natural laws, such as the law of gravity, do not change with time—and that a natural event, or *phenomenon,* can be understood more fully through **observation**—a formal

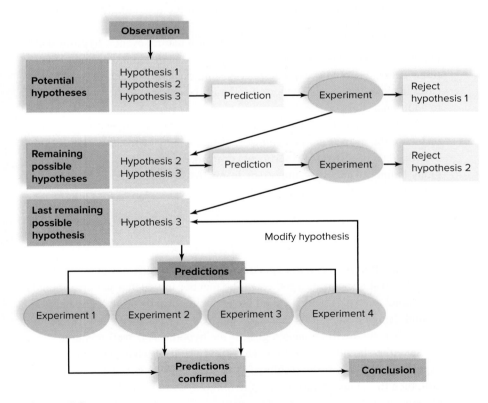

Figure 1.8 **Flow diagram for the scientific method.**

On the basis of new and/or previous observations, a scientist formulates a hypothesis. The hypothesis is used to develop predictions to be tested by further experiments and/or observations, and new data either support or do not support the hypothesis. Following an experiment, a scientist often chooses to retest the same hypothesis or to test a related hypothesis. Conclusions from many different but related experiments may lead to the development of a scientific theory. For example, studies pertaining to development, anatomy, and fossil remains all support the theory of evolution.

way of watching the natural world. Scientists rely on their senses (sight, hearing, touch) to make observations, but also extend the ability of their senses by using instruments; for example, a microscope enables them to see objects that could never be seen by the naked eye. Finally, scientists may expand their understanding even further by taking advantage of the knowledge and experiences of other scientists. For instance, they may look up past studies on the Internet or at the library, or they may write or speak to others who are researching similar topics.

Develop a Hypothesis

After making observations and gathering knowledge about a phenomenon, a scientist uses inductive reasoning. **Inductive reasoning** occurs whenever a person uses creative thinking to combine isolated facts into a cohesive whole. Chance alone can help a scientist arrive at an idea. The most famous case pertains to the antibiotic penicillin, which was discovered in 1928. While examining a petri dish of bacteria that had accidentally become contaminated with the mold *Penicillium,* Alexander Fleming observed an area around the mold that was free of bacteria. Fleming had long been interested in finding cures for human diseases caused by bacteria, and he was very knowledgeable about antibacterial substances. So when Fleming saw the dramatic effect of *Penicillium* mold on bacteria, he reasoned that the mold might be producing an antibacterial substance.

We call such a possible explanation for a natural event a **hypothesis.** A hypothesis is based on existing knowledge, so it is much more informed than a mere guess. Fleming's hypothesis was supported by further study. In most cases, a hypothesis is not supported and must be either modified and subjected to additional study or rejected.

All of a scientist's past experiences, no matter what they might be, may influence the formation of a hypothesis. But a scientist considers only hypotheses that can be tested by experiments or further observations. Moral and religious beliefs, while very important to our lives, differ among cultures and through time and are not always testable.

Make a Prediction and Perform Experiments

Scientists often perform an **experiment,** which is a series of procedures designed to test a specific hypothesis. The manner in which a scientist intends to conduct an experiment is called the **experimental design.** A good experimental design ensures that scientists are testing what they want to test and that their results will be meaningful. If the hypothesis is well prepared, then the scientist should be able to make a **prediction** of what the results of the experiment will be. If the results of the experiment do not match the prediction, then the scientist must revisit the initial hypothesis and design a new set of experiments.

Experiments can take many forms, depending on the area of biology that the scientist is examining. For example, experiments in the laboratory may be confined to tubes and beakers, whereas ecological studies may require large tracts of land. However, in all experimental designs, the researcher attempts to keep all of the conditions constant except for an **experimental variable,** which is the factor in the experiment that is being deliberately changed. The result is called the **responding variable** (or dependent variable) since its value is based on the experimental variable.

To ensure that the results will be meaningful, an experiment contains both test groups and a **control group.** A test group is exposed to the experimental variable, but the control group is not. If the control group and test groups show the same results, the experimenter knows that the hypothesis predicting a difference between them is not supported.

Scientists often use **model systems** and model organisms to test a hypothesis. Some common model organisms are shown in **Figure 1.9.** Model organisms are chosen because they allow the researcher to control aspects of the experiment, such as age and genetic background. Cell biologists may use mice for modeling the effects of a new drug. Like model organisms, model systems allow the scientist to control specific variables and environmental conditions in a way that may not be possible in the natural environment. For example, ecologists may use computer programs to model how human activities will affect the climate of a specific ecosystem. While models provide useful information, they do not always answer the original question completely. For example, medicine that is effective in mice should ideally be tested in humans, and ecological experiments that are conducted using computer simulations need to be verified by actual field experiments. Biologists, and all other scientists, continuously revise their experiments to better understand how different factors may influence their original observation.

Drosophila melanogaster

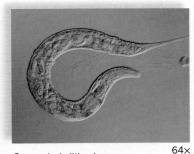
Caenorhabditis elegans 64×

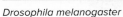
Arabidopsis thaliana

Mus musculus

Figure 1.9 **Model organisms used in scientific studies.**

Drosophila melanogaster is used as a model organism in the study of genetics. *Mus musculus* is used in the study of medicine. *Caenorhabditis elegans* is used by developmental biologists, and *Arabidopsis thaliana* is used by botanists to understand plant genetics.

(*D. melanogaster*): © Graphic Science/Alamy; (*C. elegans*): © Sinclair Stammers/Science Source; (*A. thaliana*): Wildlife GmbH/Alamy; (*M. musculus*): © Steve Gorton/Getty Images

Collect and Analyze the Data

The results of an experiment are referred to as the **data.** Mathematical data are often displayed in the form of a graph or table. Sometimes studies rely on statistical data. Statistical analysis allows a scientist to detect relationships in the data that may not be obvious on the surface. Let's say an investigator wants to know if eating onions can prevent women from getting osteoporosis (weak bones). The scientist conducts a survey asking women about their onion-eating habits and then correlates these data with the condition of their bones. Other scientists critiquing this study would want to know the following: How many women were surveyed? How old were the women? What were their exercise habits? What criteria were used to determine the condition of their bones? And what is the probability that the data are in error? The greater the variance in the data, the greater the probability of error. In any case, even if the data do suggest a correlation, scientists would want to know if there is a specific ingredient in onions that has a direct biochemical or physiological effect on bones. After all, correlation does not necessarily mean causation. It could be that women who eat onions eat lots of vegetables and have healthier diets overall than women who do not eat onions. In this way, scientists are skeptics who always pressure one another to keep investigating.

Develop a Conclusion

Scientists must analyze the data in order to reach a **conclusion** about whether a hypothesis is supported or not. Because science progresses, the conclusion of one experiment can lead to the hypothesis for another experiment (see Fig. 1.8).

State Hypothesis:
Antibiotic B is a better treatment for
ulcers than antibiotic A.

Perform Experiment:
Groups were treated the same
except as noted.

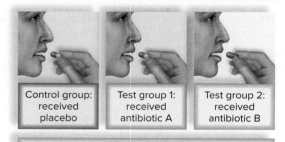

| Control group: received placebo | Test group 1: received antibiotic A | Test group 2: received antibiotic B |

Collect Data:
Each subject was examined
for the presence of ulcers.

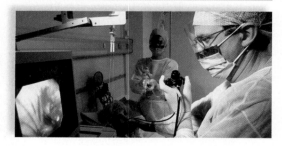

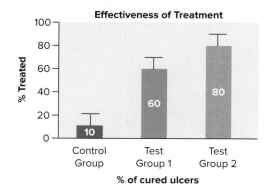

Effectiveness of Treatment

% Treated

Control Group: 10
Test Group 1: 60
Test Group 2: 80

% of cured ulcers

In other words, results that do not support one hypothesis can often help a scientist formulate another hypothesis to be tested. Scientists report their findings in scientific journals, so that their methodology and data are available to other scientists. Experiments and observations must be *repeatable*—that is, the reporting scientist and any scientist who repeats the experiment must get the same results, or else the data are suspect.

Scientific Theory

The ultimate goal of science is to understand the natural world in terms of **scientific theories,** which are accepted explanations for how the world works. Some of the basic theories of biology are the cell theory, which says that all organisms are composed of cells; the gene theory, which says that inherited information dictates the form, function, and behavior of organisms; and the theory of evolution, which says that all organisms have a common ancestor and that each organism is adapted to a particular way of life.

The theory of evolution is considered the unifying concept of biology because it pertains to many different aspects of organisms. For example, the theory of evolution enables scientists to understand the history of life, the variety of organisms, and the anatomy, physiology, and development of organisms. The theory of evolution has been a very fruitful scientific theory, meaning that it has helped scientists generate new testable hypotheses. Because the theory of evolution has been supported by so many observations and experiments for over 150 years, some biologists refer to this theory as a **principle,** a term sometimes used for a theory that is generally accepted by an overwhelming number of scientists. Other scientists prefer the term **law** instead of *principle.*

An Example of a Controlled Study

We now know that most stomach and intestinal ulcers (open sores) are caused by the bacterium *Helicobacter pylori.* Let's say investigators want to determine which of two antibiotics is best for the treatment of an ulcer. When clinicians do an experiment, they try to vary just the experimental variables—in this case, the medications being tested. Each antibiotic is administered to an independent test group, but the control group is not given an antibiotic. If by chance the control group shows the same results as one of the test groups, the investigators may conclude that that the antibiotic in that test group is ineffective, because it does not show a significant difference in treatment to the control group. The study depicted in **Figure 1.10** shows how investigators may study this hypothesis:

Hypothesis: Newly discovered antibiotic B is a better treatment for ulcers than antibiotic A, which is in current use.

Figure 1.10 **A controlled laboratory experiment to test the effectiveness of two medications in humans.**

In this study, a large number of people were divided into three groups. The control group received a placebo and no medication. One of the test groups received medication A, and the other test group received medication B. The results are depicted in a graph, and it shows that medication B was a more effective treatment than medication A for the treatment of ulcers.

In any experiment, it is important to reduce the number of possible variables (differences). In this experiment, those variables may include factors such differences in sex, weight, or previous illnesses among the individuals. Therefore, the investigators *randomly* divide a large group of volunteers equally into experimental groups. The hope is that any differences will be distributed evenly among the three groups. The larger the number of volunteers (the sample size), the greater the chance of reducing the influence of external variables. This is why many medical studies often involve thousands of individuals.

The three groups are to be treated like this:

Control group: Subjects with ulcers are not treated with either antibiotic.
Test group 1: Subjects with ulcers are treated with antibiotic A.
Test group 2: Subjects with ulcers are treated with antibiotic B.

After the investigators have determined that all volunteers do have ulcers, they will want the subjects to think they are all receiving the *same* treatment. This is an additional way to protect the results from any influence other than the medication. To achieve this end, the subjects in the control group can receive a **placebo,** a treatment that appears to be the same as that administered to the other two groups but actually contains no medication. Overall, the goal of a placebo is to analyze whether other undetermined factors may be influencing the study.

The Results

After 2 weeks of administering the same amount of medication (or placebo) in the same way, the stomach and intestinal linings of each subject are examined to determine if ulcers are still present. Endoscopy, a procedure depicted in the lower photograph in Figure 1.10, is one way to examine a patient for the presence of ulcers. This procedure, which is performed under sedation, involves inserting an endoscope—a small, flexible tube with a tiny camera on the end—down the throat and into the stomach and the upper part of the intestine. Then, the doctor can see the lining of these organs and can check for ulcers. Tests performed during an endoscopy can also determine if *H. pylori* is present.

Because endoscopy is somewhat subjective, it is probably best if the examiner is not aware of which group the subject is in; otherwise, the examiner's prejudice may influence the examination. When neither the patient nor the technician is aware of the specific treatment, it is called a *double-blind* study.

In this study, the investigators may decide to determine the effectiveness of the medication by the percentage of people who no longer have ulcers. So, if 20 people out of 100 still have ulcers, the medication is 80% effective. The difference in effectiveness is easily read in the graph portion of Figure 1.10.

Conclusion: On the basis of their data, the investigators conclude that their hypothesis has been supported.

Publishing the Results

Scientific studies are customarily published in a scientific journal (**Fig. 1.11**), so that all aspects of a study are available to the scientific community. Before information is published in scientific journals, it is typically reviewed by experts, who ensure that the research is credible, accurate, unbiased, and well executed. Another scientist should be able to read about an experiment in a scientific journal, repeat the experiment in a different location, and get the same (or very similar) results. Some articles are rejected for publication by reviewers when they believe there is something questionable about the design of an experiment or the manner in which it was conducted. This process

Figure 1.11 **Scientific publications.**

Scientific journals, such as *Science,* are scholarly journals in which researchers share their findings with other scientists. Scientific magazines, such as *New Scientist* and *National Geographic,* contain articles that are usually written by reporters for a broader audience.

of rejection is important in science, since it causes researchers to critically review their hypotheses, predictions, and experimental designs, so that their next attempt will more adequately address their hypotheses. Often, it takes several rounds of revision before research is accepted for publication in a scientific journal.

Scientific magazines (Fig. 1.11), such as *New Scientist* or *National Geographic,* differ from scientific journals in that they report scientific findings to the general public. The information in these articles is usually obtained from articles first published in scientific journals.

As mentioned previously, the conclusion of one experiment often leads to another experiment. The need for scientists to expand on findings explains why science changes and the findings of yesterday may be improved upon tomorrow.

CONNECTING THE CONCEPTS

1.3 The scientific method is the process by which scientists study the natural world and develop explanations for their observations.

Check Your Progress 1.3

1. Summarize the purpose of each step in the scientific method.
2. Explain why a controlled study is important in research.
3. Explain why publishing scientific studies is important.

1.4 Challenges Facing Science

Learning Outcomes

Upon completion of this section, you should be able to

1. Distinguish between science and technology.
2. Summarize some of the major challenges currently facing science.

As we have learned in this chapter, science is a systematic way of acquiring knowledge about the natural world. Science is a slightly different endeavor than technology. **Technology** is the application of scientific knowledge to the interests of humans. Scientific investigations are the basis for the majority of our technological advances. It is often the case that a new technology, such as your cell phone or a new drug, is based on years of scientific investigations. In this section, we are going to explore some of the challenges facing science, technology, and society.

Biodiversity and Habitat Loss

Biodiversity is the total number and relative abundance of species, the variability of their genes, and the different ecosystems in which they live. The biodiversity of our planet has been estimated to be around 8.7 million species (not counting bacteria), and so far, approximately 2.3 million have been identified and named. **Extinction** is the disappearance of a species or larger classification category. It is estimated that presently we are losing hundreds of species every year due to human activities and that as much as 38% of all species, including most primates, birds, and amphibians, may be in danger of extinction before the end of the century. Many biologists are alarmed about the present rate of extinction and hypothesize that it may eventually rival the rates of the five mass extinctions that occurred during our planet's history. The last mass extinction, about 65 million years ago, caused many plant and animal species, including the dinosaurs, to become extinct.

The two most biologically diverse ecosystems—tropical rain forests and coral reefs—are home to many organisms. These ecosystems are also threatened by human activities. The canopy of the tropical rain forest alone supports a variety of organisms, including orchids, insects, and monkeys. Coral reefs, which are found just offshore of the continents and islands near the equator, are built up from calcium carbonate skeletons of sea animals called corals. Reefs provide a habitat for many animals, including jellyfish, sponges, snails, crabs, lobsters, sea turtles, moray eels, and some of the world's most colorful fishes. Like tropical rain forests, coral reefs are severely threatened as the human population increases in size. Some reefs are 50 million years old, yet in just a few decades, human activities have destroyed an estimated 25% of all coral reefs and seriously degraded another 30%. At this rate, nearly three-quarters could be destroyed within 40 years. Similar statistics are available for tropical rain forests.

The destruction of healthy ecosystems has many unintended effects. For example, we depend on ecosystems for food, medicines (**Fig. 1.12**), and various raw materials. Draining of the natural wetlands of the Mississippi and Ohio Rivers and the construction of levees have worsened flooding problems, making once fertile farmland undesirable. The destruction of South American rain forests has killed many species that may have yielded the next miracle drug and has decreased the availability of many types of lumber. We are only now beginning to realize that we depend on ecosystems even more for the services they provide. Just as chemical cycling occurs within a single ecosystem, so all ecosystems keep chemicals cycling throughout the biosphere. The workings of ecosystems ensure that the environmental conditions of the biosphere are suitable for the continued existence of humans. And several studies show that ecosystems cannot function properly unless they remain biologically diverse. We will explore the concept of biodiversity in greater detail in Chapters 30 through 32 of the text.

Emerging and Reemerging Diseases

Over the past decade, avian influenza (H5N1 and H7N9), swine flu (H1N1), severe acute respiratory syndrome (SARS), and Middle East respiratory syndrome (MERS) have been in the news. These are called **emerging diseases** because they are relatively new to humans. Where do emerging diseases come from? Some of them may result from new and/or increased exposure to animals or insect populations that act as vectors for disease. Changes in human behavior and use of technology can also result in new diseases. SARS is thought to have arisen in Guandong, China, due to the consumption of civets, a type of exotic cat considered a delicacy. The civets were possibly infected by exposure to horseshoe bats sold in open markets. Legionnaires' disease emerged in 1976 due to bacterial contamination of a large air-conditioning system in a hotel. The bacteria thrived in the cooling tower used as the water source for the air-conditioning system. In addition, globalization results in the transport all over the world of diseases that were previously restricted to isolated communities. The first SARS cases were reported in southern China in

Figure 1.12 **The importance of biodiversity.**

Snails of the genus *Conus* are known to produce powerful painkillers. Unfortunately, their habitat on coral reefs is threatened by human activity.

© Franco Banfi/Waterframe/Age fotostock

November of 2002. By the end of February 2003, SARS had reached nine countries/provinces, mostly through airline travel.

Some pathogens mutate and change hosts, jumping from birds to humans, for example. Before 1997, avian flu was thought to affect only birds. A mutated strain jumped to humans in the 1997 outbreak. To control that epidemic, officials killed 1.5 million chickens to remove the source of the virus. New forms of avian influenza (bird flu) are being discovered every few years.

Reemerging diseases are also a concern. Unlike an emerging disease, a reemerging disease has been known to cause disease in humans for some time, but generally has not been considered a health risk due to a relatively low level of incidence in human populations. However, reemerging diseases can cause problems. An excellent example is the Ebola outbreak in West Africa of 2014–2015. Ebola outbreaks have been known since 1976, but generally have only affected small groups of humans. The 2014–2015 outbreak was a much larger event. Although the exact numbers may never be known, it is estimated that over 28,000 people were infected, with over 11,000 fatalities. The outbreak has disrupted the societies of several West African nations.

Both emerging and reemerging diseases have the potential to cause health problems for humans across the globe. Scientists investigate not only the causes of these diseases (for example, the viruses) but also their effects on our bodies and the mechanisms by which they are transmitted. We will take a closer look at viruses in Section 17.1 of the text.

Climate Change

The term **climate change** refers to changes in the normal cycles of the Earth's climate that may be attributed to human activity. Climate change is primarily due to an imbalance in the chemical cycling of the element carbon. Normally, carbon is cycled within an ecosystem. However, due to human activities, more carbon dioxide is being released into the atmosphere than is being removed. In 1850, atmospheric CO_2 was at about 280 parts per million (ppm); today, it is over 400 ppm. This increase is largely due to the burning of fossil fuels and the destruction of forests to make way for farmland and pasture. Today, the amount of carbon dioxide released into the atmosphere is about twice the amount that remains in the atmosphere. It's believed that most of this dissolves in the oceans, which is increasing their acidity. The increased amount of carbon dioxide (and other gases) in the atmosphere is causing a rise in temperature called **global warming.** These gases allow the sun's rays to pass through, but they absorb and radiate heat back to Earth, a phenomenon called the *greenhouse effect*.

There is a consensus among scientists from around the globe that climate change and global warming are causing significant changes in many of the Earth's ecosystems and represent one of the greatest challenges of our time. Throughout this text, we will return to how climate change is affecting ecosystems, reducing biodiversity, and contributing to human disease.

CONNECTING THE CONCEPTS

1.4 Science benefits society by providing us with information to make informed decisions.

Check Your Progress 1.4

1. Explain the relationship between science and technology.
2. Summarize why biodiversity loss, emerging diseases, and climate change are current challenges of science.

SUMMARIZE

An understanding of the diversity of life on Earth is essential for the well-being of humans. The process of science helps us increase our knowledge of the natural world.

1.1 All living organisms, from bacteria to humans, share the same basic characteristics of life.

1.2 The process of evolution explains the diversity of living organisms on Earth today.

1.3 The scientific method is the process by which scientists study the natural world and develop explanations for their observations.

1.4 Science benefits society by providing us with information to make informed decisions.

1.1 The Characteristics of Life

All organisms share the following characteristics of life:

- Organization: The levels of biological organization extend as follows: atoms and molecules → cells → tissues → **organs** → **organ systems** → organisms → **species** →**populations** → **communities** → **ecosystems** → **biosphere.** In an ecosystem, populations interact with one another and with the physical environment.

- Acquire materials and energy from the environment: Organisms need an outside source of nutrients and energy for **metabolism.** While chemicals cycle within an ecosystem, **photosynthesis** is needed to capture energy for use within the ecosystem.

- Maintain an internal environment: **Homeostasis** means staying just about the same despite changes in the external environment.

- Respond to stimuli: Organisms react to internal and external events.

- **Reproduce** and develop: The genetic information of life is carried in the molecules of **deoxyribonucleic acid (DNA)** found in every cell. Reproduction passes copies of an organism's **genes** to the next generation. This information directs the **development** of the organism over time.

- Have **adaptations** that make them suitable for their environment.

1.2 Evolution: The Core Concept of Biology

Evolution, or the change in a species over time, explains the unity and diversity of life. **Natural selection** is the process that results in evolution. Descent from a common ancestor explains why organisms share some characteristics, and adaptation to various ways of life explains the diversity of life-forms. An evolutionary tree is a diagram that may be used to describe how groups of organisms are related to one another.

Life may be classified into large groups called domains and kingdoms. The three domains are

- **Domain Archaea: prokaryotes** that live in extreme environments
- **Domain Bacteria:** the majority of prokaryotes
- **Domain Eukarya: eukaryotes** (plants, animals, fungi, protists)

Within domain Eukarya are four kingdoms: Protista (**protists**); **Fungi;** Plantae (**plants**); and Animalia (**animals**).

Systematics is the classification of organisms based on evolutionary relationships. The categories include species, **genus, family, order, class, phylum, kingdom,** and **domain.** Taxonomy is involved in naming organisms. A **binomial name** (such as *Homo sapiens*) consists of the genus (*Homo*) and the specific epithet (*sapiens*).

1.3 Science: A Way of Knowing

The scientific process uses **inductive reasoning** and includes a series of systematic steps known as the scientific method:

- **Observations,** which use the senses and may also include studies done by others

- A **hypothesis** that leads to a **prediction**

- Experiments that support or refute the hypothesis

- A **conclusion** reached by analyzing **data** to determine whether the results support or do not support the hypothesis

A hypothesis confirmed by many different studies becomes known as a **scientific theory.** Scientific theories are also referred to as **laws** or **principles.**

Experimental design is important in the scientific method. In an experiment, a single **experimental variable** is varied to measure the influence on the **responding variable.** Experiments should utilize **control groups.** Control groups may be given a **placebo** to ensure that the experiment is valid. Often, scientists use model organisms and **model systems** in their experimental designs.

1.4 Challenges Facing Science

Scientific findings often lead to the development of a **technology** that can be of service to humans.

There are a number of important issues facing science in today's world. These include **emerging diseases;** human influence on ecosystems, which is resulting in a loss of **biodiversity** and **extinction;** and **global warming,** which is contributing to **climate change.**

ASSESS

Testing Yourself

Choose the best answer for each question.

1.1 The Characteristics of Life

1. A modification that helps equip organisms for their way of life is a(n)
   a. homeostasis.   c. adaptation.
   b. natural selection.   d. extinction.

2. All of the chemical reactions that occur in a cell are called
   a. homeostasis.   c. heterostasis.
   b. metabolism.   d. cytoplasm.

3. Which of the following represents the lowest level of biological organization that still may be considered alive?
   a. tissue   c. cell
   b. molecule   d. population

4. The region of the Earth's surface where all organisms are found is called the
   a. ecosystem.   c. community.
   b. population.   d. biosphere.

1.2 Evolution: The Core Concept of Biology

5. The mechanism by which species undergo evolutionary change is called
   a. behavior   c. natural selection.
   b. homeostasis.   d. systematics.

6. Which of the following is not a domain?
   a. Archaea   c. Plantae
   b. Eukarya   d. Bacteria

7. A binomial name indicates
   a. the domain of the organism.
   b. the genus and species (specific epithet).
   c. the kingdom.
   d. the age of the organism.

1.3 Science: A Way of Knowing

8. A hypothesis cannot be formed without which of the following?
   a. experimentation   c. data
   b. observation   d. theory

9. Information collected from a scientific experiment is known as
   a. a scientific theory.   c. data.
   b. a hypothesis.   d. a conclusion.

10. Placebos are often used in which of the following?
    a. data analysis   c. test groups
    b. control groups   d. model organisms

1.4 Challenges Facing Science

11. _____ is the application of scientific investigations for the benefit of humans.
    a. Bioethics   c. Adaptation
    b. Evolution   d. Technology

12. Human influence can be associated with which of the following challenges facing science?
    a. loss of biodiversity
    b. climate change
    c. emerging diseases
    d. All of these are correct.

ENGAGE

BioNOW

Want to know how this science is relevant to your life? Check out the BioNow video below.

   • Characteristics of Life

How do you exhibit the general characteristics of life in your daily activities?

Thinking Critically

1. Based on the accompanying evolutionary tree, which prokaryotic domain gave rise to the domain Eukarya? Which kingdom in domain Eukarya gave rise to plants, animals, and fungi?

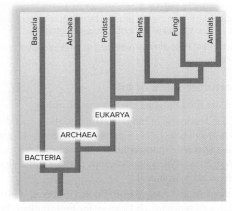

2. You are a scientist working at a pharmaceutical company and have developed a new cancer medication that has the potential for use in humans. Outline a series of experiments, including the use of a model, to test whether the cancer medication works.

3. Scientists are currently exploring the possibility that life may exist on some of the planets and moons of our solar system. Suppose that you were a scientist on one of these research teams and were tasked with determining whether a new potential life-form exhibited the characteristics of behavior or adaptation. What would be your hypothesis? What types of experiments would you design?

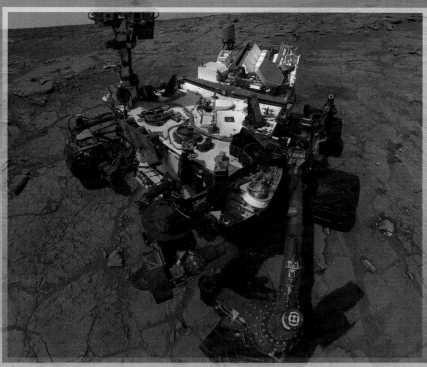

© NASA/JPL-Caltech/MSSS

The Chemical Basis of Life

The Building Blocks of Life

You may never have heard of Enceladus or Europa, but they are both at the frontline of our species' effort to understand the nature of life. Enceladus is one of Saturn's moons, and Europa orbits Jupiter. Why are these moons so special? Because scientists believe that both of these moons contain water, and plenty of it. While both Enceladus and Europa are far from the sun, the gravitational pull of their parent planets means that beneath the frozen surface of both of these moons are oceans of liquid water. And as we will see, water has an important relationship with life.

At other locations in our solar system, scientists are looking for evidence of the chemicals that act as the building blocks of life. For example, on Titan, a moon of Saturn, NASA's *Cassini-Huygens* space probe has detected the presence of these building blocks, including lakes of methane and ammonia and vast deposits of hydrogen and carbon compounds called hydrocarbons.

Even more recently, the *Rosetta* space probe, launched by the European Space Agency (ESA), completed its 10-year mission to land a probe on the surface of a comet. Some of the early data from this mission support the hypothesis that comets may contain the organic building blocks of life. NASA has recently announced missions to Europa and Mars that will continue the search for signs of life in our solar system. Many of these searches focus on the presence of water. The information obtained from these missions will help us better understand how life originated on our planet.

In this chapter, we will explore the building blocks of all matter—the atoms—and the importance of water to life as we know it.

As you read through this chapter, think about the following questions:

1. What are the basic characteristics that define life?
2. What evidence would you look for on Enceladus, Europa or Mars that would tell you that life may have existed there in the past?
3. Why is water considered to be so important to life?

OUTLINE

2.1 Atoms and Atomic Bonds   22

2.2 Water's Importance to Life   29

2.3 Acids and Bases   33

BEFORE YOU BEGIN

Before beginning this chapter, take a few moments to review the following discussions.

Section 1.1 What are the basic characteristics of all living organisms?

Figure 1.2 How do molecules relate to cells in the levels of biological organization?

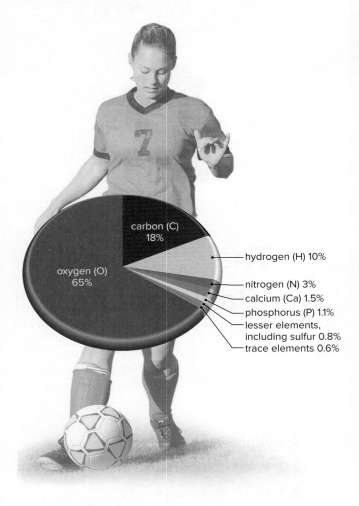

Figure 2.1 **Elements in living organisms.**

If analyzed at the level of atoms, all living organisms, including humans, are mostly composed of oxygen, carbon, hydrogen, nitrogen, calcium, and phosphorus.

(photo): © Brand X Pictures/PunchStock RF

2.1 Atoms and Atomic Bonds

Learning Outcomes

Upon completion of this section, you should be able to

1. Distinguish among the types, locations, and charges of subatomic particles.
2. Relate how the arrangement of electrons determines an element's reactivity.
3. Explain how isotopes are useful in the study of biology.
4. Contrast ionic and covalent bonds.
5. Identify the reactants and products in a chemical equation.

As you are studying right now, everything around you, including your desk and computer, is made of matter. **Matter** may be defined as anything that takes up space and has mass. Matter can exist as a solid, liquid, gas, or plasma. Living organisms, such as ourselves, and nonliving things, such as the air we breathe, are all made of matter.

All matter is composed of elements. An **element** is a substance that cannot be broken down into other substances by ordinary chemical means. There are only 92 naturally occurring elements, and each of these differs from the others in its chemical or physical properties, such as density, solubility, melting point, and reactivity.

While all of the elements are present on Earth, the proportion of each element differs between living organisms and nonliving things. Four elements—carbon, hydrogen, nitrogen, and oxygen—make up about 96% of the body weight of most organisms (**Fig. 2.1**), from simple, one-celled life-forms to complex, multicellular plants and animals. Other elements, such as phosphorus, calcium, and sulfur, may also be found in abundance in living organisms. A number of elements, including minerals such as zinc and chromium, are found at very low, or *trace,* levels. Regardless of their abundance and function in living organisms, the basic building blocks of each element share some common characteristics.

Connections: Scientific Inquiry

Where do elements come from?

We are all familiar with elements. Iron, sodium, oxygen, and carbon are all common in our lives, but where do they originate?

© NASA/ JPL-Caltech/ STSci

Normal chemical reactions do not produce elements. The majority of the heavier elements, such as iron, are produced only by the intense chemical and physical reactions within stars. When these stars reach the end of their lives, they explode, producing a supernova. Supernovas scatter the heavier elements into space, where they eventually are involved in the formation of planets.

The late astronomer and philosopher Carl Sagan (1934–1996) frequently referred to humans as "star stuff." In many ways, we, and all other living organisms, are formed from elements that originated within the stars.

Atomic Structure

The *atomic theory* states that elements consist of tiny particles called **atoms.** Because each element consists of only one kind of atom, the same name is given to an element and its atoms. This name is represented by one or two letters, called the **atomic symbol.** For example, the symbol H stands for an atom of hydrogen, and the symbol Na (for *natrium* in Latin) stands for an atom of sodium.

If we could look inside a single atom, we would see that it is made mostly of three types of subatomic particles: **neutrons,** which have no electrical charge; **protons,** which have a positive charge; and **electrons,** which have a negative charge. Protons and neutrons are located within the center of an atom, which is called the *nucleus,* while electrons move about the nucleus.

Figure 2.2 shows the arrangement of the subatomic particles in a helium atom, which has only two electrons. In Figure 2.2*b*, the circle represents the approximate location of the electrons based on their energy state. However, electrons are in a constant state of motion, so their estimated location is often shown as a cloud (Fig. 2.2*a*). Overall, most of an atom is empty space. In fact, if we could draw an atom the size of a baseball stadium, the nucleus would be like a gumball in the center of the stadium, and the electrons would be tiny specks whirling about in the upper stands. Usually, we can only indicate where the electrons are expected to be. In our analogy, the electrons might very well stray outside the stadium at times.

Since atoms are a form of matter, you might expect each atom to have a certain mass. In effect, the **mass number** of an atom is just about equal to the sum of its protons and neutrons. Protons and neutrons are assigned one mass unit each. Electrons, being matter, have mass, but they are so small that their mass is assumed to be zero in most calculations. The term *mass* is used, rather than *weight,* because mass is constant but weight is associated with gravity and thus varies depending on an object's location in the universe.

All atoms of an element have the same number of protons. This is called the atom's **atomic number.** The number of protons makes an atom unique and may be used to identify which element the atom belongs to. As we will see, the number of neutrons may vary between atoms of an element. The average of the mass numbers for these atoms is called the **atomic mass.**

The atomic number tells you the number of positively charged protons. If the atom is electrically neutral, then the atomic number also indicates the number of negatively charged electrons. To determine the usual number of neutrons, subtract the number of protons from the atomic mass and take the closest whole number.

The Periodic Table

Once chemists discovered a number of the elements, they began to realize that the elements' chemical and physical characteristics recur in a predictable manner. The periodic table (**Fig. 2.3**) was developed as a way to display the elements, and therefore the atoms, according to these characteristics.

In a periodic table, the atomic number is written above the atomic symbol. The atomic mass is written below the atomic symbol. For example, the carbon atom is shown in this way:

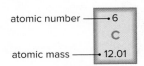

atomic number —— 6
C
atomic mass —— 12.01

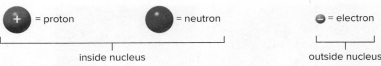

= proton = neutron = electron

inside nucleus | outside nucleus

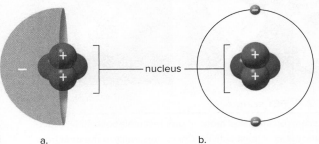

nucleus

a. | b.

Figure 2.2 **Two models of helium (He).**

Atoms contain subatomic particles, which are located as shown in these two simplified models of helium. Protons are positively charged, neutrons have no charge, and electrons are negatively charged. Protons and neutrons are within the nucleus, and electrons are outside the nucleus. **a.** This model shows electrons as a negatively charged cloud around the nucleus. **b.** In this model, the average location of electrons is represented by a circle.

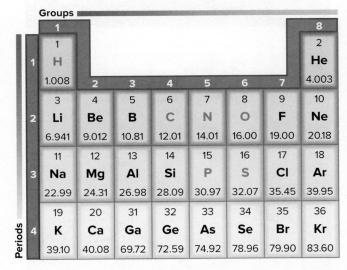

Figure 2.3 **A portion of the periodic table.**

In the periodic table, the elements, and therefore the atoms that compose them, are in the order of their atomic numbers but arranged in periods (horizontal rows) and groups (vertical columns). All the atoms in a particular group have certain chemical characteristics in common. The elements highlighted in red make up the the majority of matter in organic molecules (Chapter 3). A full periodic table is provided in Appendix A.

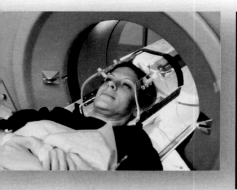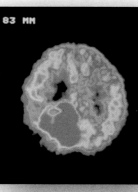

Figure 2.4 PET scan.

In a PET scan, red indicates areas of greatest metabolic activity, and blue means areas of least activity. Computers analyze the data from different sections of an organ—in this case, the human brain.

© National Institutes of Health

a.

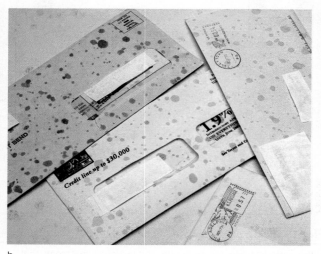

b.

Figure 2.5 High levels of radiation.

a. The Fukushima nuclear facility. Following a tsunami, an accident at this facility released radioactive isotopes into the environment. **b.** Radiation can also be used to sterilize items, such as mail and packages, to protect us from biological agents, such as anthrax.

(a): © DigitalGlobe/Getty Images; (b): © Getty Images

Every atom is in a particular period (the horizontal rows) and in a particular group (the vertical columns). The atoms in group 8 are called the *noble gases* because they are gases that rarely react with another atom, for reasons we will discuss later in this section. In Figure 2.3, notice that helium (He) and neon (Ne) are noble gases.

Isotopes

Isotopes are atoms of the same element that differ in the number of neutrons. In other words, isotopes have the same number of protons, but they have different mass numbers. In some cases, a nucleus with excess neutrons is unstable and may decay and emit radiation. Such an isotope is said to be radioactive. However, not all isotopes are radioactive. The radiation given off by radioactive isotopes can be detected in various ways. Most people are familiar with the use of a Geiger counter to detect radiation. However, other methods to detect radiation exist that are useful in medicine and science.

Uses of Radioactive Isotopes

The importance of chemistry to biology and medicine is nowhere more evident than in the many uses of radioactive isotopes. For example, radioactive isotopes can be used as tracers to detect molecular changes or to destroy abnormal or infectious cells. Since both radioactive isotopes and stable isotopes contain the same number of electrons and protons, they essentially behave the same in chemical reactions. Therefore, a researcher can use a small amount of radioactive isotope as a tracer to detect how a group of cells or an organ is processing a certain element or molecule. For example, by giving a person a small amount of radioactive iodine (iodine-131), it is possible to determine whether the thyroid gland is functioning properly. Another example is a procedure called positron-emission tomography (PET), which utilizes tracers to determine the comparative activity of tissues. A radioactively labeled glucose tracer that emits a positron (a subatomic particle that is the opposite of an electron) is injected into the body. Positrons emit small amounts of radiation, which may be detected by sensors and analyzed by a computer. The result is a color image that shows which tissues took up glucose and are thus metabolically active (**Fig. 2.4**). A number of conditions, such as tumors, Alzheimer disease, epilepsy, or a stroke, may be detected using PET scans.

Radioactive substances in the environment can cause harmful chemical changes in cells, damage DNA, and cause cancer. The release of radioactive particles following a nuclear power plant accident can have far-reaching and long-lasting effects on human health. For example, a 2011 Pacific tsunami caused a release of radioactive cesium-137 from the Fukushima nuclear facility (**Fig. 2.5***a*). But the effects of radiation can also be put to good use. Packets of radioactive isotopes can be placed in the body, so that the subatomic particles emitted destroy only cancer cells, with little risk to the rest of the body.

Radiation from radioactive isotopes has been used for many years to sterilize medical and dental equipment. Since the terrorist attacks in 2001, mail that is destined for the White House and congressional offices in Washington, DC, is irradiated to protect against dangerous biological agents, such as anthrax (Fig. 2.5*b*).

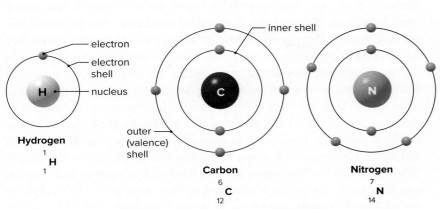

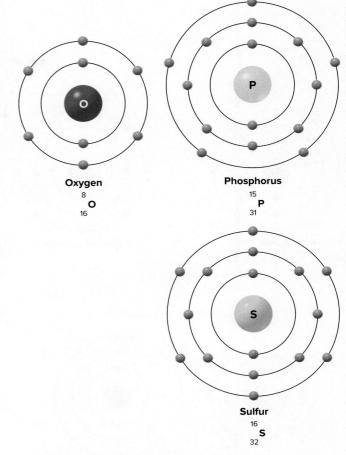

Figure 2.6 **Atoms of six important elements.**

Electrons orbit the nucleus at particular energy levels called electron shells. The first shell contains up to two electrons, and each shell thereafter has an increasing number of electrons.

Connections: Health

Does irradiation add radioactive particles to food?

No, the process of food irradiation exposes certain types of foods to a form of radiation called ionizing radiation. While ionizing radiation is very useful in killing bacteria on foods, it does not accumulate in or on the food. If you shine a light on a wall, the wall will not accumulate the light, nor will the wall emit light when you turn your light out. The same is true of ionizing radiation and the food irradiation process. Rather, food irradiation helps protect our food supply against disease-causing bacteria, such as *Salmonella* and *Escherichia coli* O157:H7.

Arrangement of Electrons in an Atom

Electrons in atoms are much like the blades of a ceiling fan. When the fan is moving, it is difficult to see the individual blades, and all you see is a whirling blur. When the fan is stopped, each blade has a specific location and can be seen. Likewise, the electrons of an atom are constantly moving. Although it is not possible to determine the precise location of an individual electron at any given moment, it is useful to construct models of atoms that show electrons at discrete energy levels about the nucleus (**Fig. 2.6**). These energy levels are commonly called **electron shells.** Since the nucleus of an atom is positively charged, negatively charged electrons require an increasing amount of energy to push them farther away from the nucleus. Electrons in outer electron shells, therefore, contain more energy than those in inner electron shells.

Each electron shell contains a certain number of electrons. In the models shown in Figure 2.6, the electron shells are drawn as concentric rings about the nucleus. These shells are used to represent the energy of the electrons, and not necessarily their physical location. The first shell closest to the nucleus can contain two electrons; thereafter, shells increase in the number of electrons that they may contain (second shell = 8, third = 16, etc.). In atoms with more than one electron shell, the lower level is generally filled with electrons first, before electrons are added to higher levels.

The sulfur atom, with an atomic number of 16, has two electrons in the first shell, eight electrons in the second shell, and six electrons in the third, or outer, shell. Notice in the periodic table (see Fig. 2.3) that sulfur is in the third period. In other words, the period tells you how many shells an atom has. Also note that sulfur is in group 6. The group tells you how many electrons an atom has in its outer shell.

If an atom has only one shell, the outer shell is complete when it has two electrons. If an atom has two or more shells, the outer shell is most stable when it has eight electrons; this is called the **octet rule.** We mentioned that atoms in group 8 of the periodic table are called the noble gases because they do not ordinarily undergo reactions. Atoms with fewer than eight electrons in the outer shell react with other atoms in such a way that each has a completed outer shell after the reaction. Atoms can give up, accept, or share electrons in order to have eight electrons in the outer shell. In other words, the number of electrons in an atom's outer shell, called the **valence shell,** determines its chemical reactivity. The size of an atom is also important. Both carbon (C) and silicon (Si) atoms are in group 4, and therefore they have four electrons in their valence shells. This means they can bond with as many as four other atoms in order to achieve eight electrons in their outer shells. But carbon in period 2 has two shells, and silicon in period 3 has three shells. The smaller atom, carbon, can bond to other carbon atoms and form long-chained molecules, while the larger silicon atom is unable to bond to other silicon atoms. This partially explains why carbon, and not silicon, plays an important role in building the molecules of life. Overall, the chemical properties of atoms—that is, the ways they react—are largely determined by the arrangement of their electrons.

Types of Chemical Bonds

A group of atoms bonded together is called a **molecule.** When a molecule contains atoms of more than one element, it can be called a **compound.** Compounds and molecules contain two primary types of chemical bonds: ionic and covalent. The type of bond that forms depends on whether two bonded atoms share electrons or whether one has given electrons to the other. For example, in hydrogen gas (H_2), the two hydrogen atoms are sharing electrons in order to fill the valence shells of both atoms. When sodium chloride (NaCl) forms, however, the sodium atom (Na) gives an electron to the chlorine (Cl) atom, and in that way each atom has eight electrons in the outer shell.

Ionic Bonding

An **ionic bond** forms when two atoms are held together by the attraction between opposite charges. The reaction between sodium and chlorine atoms is an example of how an ionic bond is formed. Consider that sodium (Na), with only one electron in its third shell, usually gives up an electron (**Fig. 2.7***a*). Once it does so, the second shell, with eight electrons, becomes its outer shell. Chlorine (Cl), on the other hand, tends to take on an electron, because its outer shell has seven electrons. If chlorine gets one more electron, it has a completed outer shell. So, when a sodium atom and a chlorine atom react, an electron is transferred from sodium to chlorine. Now both atoms have eight electrons in their outer shells.

This electron transfer causes these atoms to become **ions,** or charged atoms. The sodium ion has one more proton than it has electrons; therefore, it has a net charge of +1 (symbolized by Na^+). The chloride ion has one more electron than it has protons; therefore, it has a net charge of −1 (symbolized by Cl^-). Negatively charged ions often have names that end in "ide," and thus Cl^- is called a chloride ion. In the periodic table (see Fig. 2.3), atoms in groups 1 and 2 and groups 6 and 7 become ions when they react with other atoms. Atoms in groups 2 and 6 always transfer two electrons. For example, calcium becomes Ca^{2+}, while oxygen becomes O^{2-}.

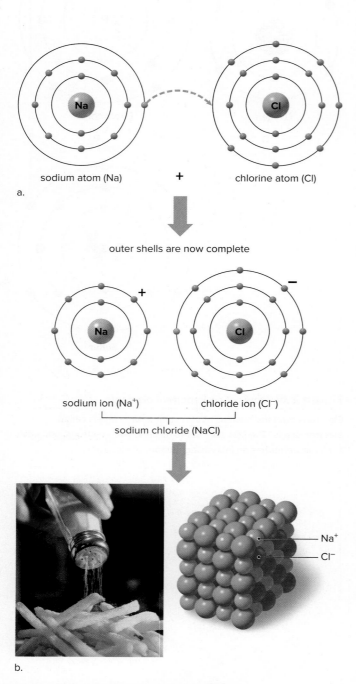

sodium atom (Na) + chlorine atom (Cl)

a.

outer shells are now complete

sodium ion (Na^+) chloride ion (Cl^-)

sodium chloride (NaCl)

Na^+
Cl^-

b.

Figure 2.7 **Formation of sodium chloride.**

a. During the formation of sodium chloride, an electron is transferred from the sodium atom to the chlorine atom. At the completion of the reaction, each atom has eight electrons in the outer shell, but each also carries a charge as shown. **b.** In a sodium chloride crystal (commonly called salt), attraction between the Na^+ and Cl^- ions causes the atoms to assume a three-dimensional lattice shape.

Ionic compounds are often found as **salts,** solid substances that usually separate and exist as individual ions in water. A common example is sodium chloride (NaCl), or table salt. A sodium chloride crystal illustrates the solid form of a salt (Fig. 2.7*b*). When sodium chloride is placed in water, the ionic bonds break, causing the Na^+ and Cl^- ions to dissociate. Ionic compounds are most commonly found in this dissociated (ionized) form in biological systems because these systems are 70–90% water.

Covalent Bonding

A **covalent bond** results when two atoms share electrons in order to have a completed outer shell. In a hydrogen atom, the outer shell is complete when it contains two electrons. If hydrogen is in the presence of a strong electron acceptor, it gives up its electron to become a hydrogen ion (H^+). But if this is not possible, hydrogen can share with another atom, and thereby have a completed outer shell. For example, one hydrogen atom can share with another hydrogen atom. In this case, the two orbitals overlap and the electrons are shared between them—that is, you count the electrons as belonging to both atoms:

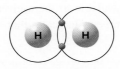

Hydrogen gas (H₂)

Rather than drawing an orbital model like the one above, scientists often use simpler ways to indicate molecules. A *structural formula* uses straight lines, as in H—H. The straight line is used to indicate a pair of shared electrons. A *molecular formula* omits the lines that indicate bonds and simply shows the number of atoms involved, as in H_2.

Sometimes, atoms share more than two electrons to complete their octets. A double covalent bond occurs when two atoms share two pairs of electrons, as in this molecule of oxygen gas:

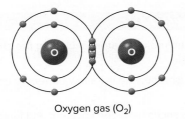

Oxygen gas (O₂)

In order to show that oxygen gas (O_2) contains a double bond, the structural formula is written as O=O to indicate that two pairs of electrons are shared between the oxygen atoms.

It is also possible for atoms to form triple covalent bonds, as in nitrogen gas (N_2), which can be written as N≡N. Single covalent bonds between atoms are quite strong, but double and triple bonds are even stronger.

A single atom may form bonds with more than one other atom. For example, the molecule methane results when carbon binds to four hydrogen atoms (**Fig. 2.8***a*). In methane, each bond actually points to one corner of a four-sided structure called a tetrahedron (Fig. 2.8*b*). The best model to show this arrangement is a ball-and-stick model (Fig. 2.8*c*). Space-filling models (Fig. 2.8*d*) come closest to showing the actual shape of a molecule. The shapes of molecules help dictate the functional roles they play in organisms.

Methane (CH₄)

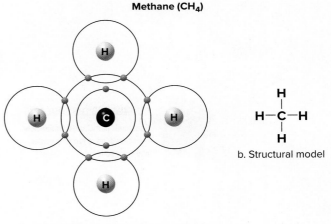

a. Electron model showing covalent bonds

b. Structural model

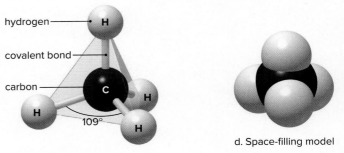

c. Ball-and-stick model

d. Space-filling model

Figure 2.8 Shapes of covalently bonded molecules.

An electron model (**a**) and a structural model (**b**) show that methane (CH₄) contains one carbon atom bonded to four hydrogen atoms. **c.** The ball-and-stick model shows that, when carbon bonds to four other atoms, as in methane, each bond actually points to one corner of a tetrahedron. **d.** The space-filling model is a three-dimensional representation of the molecule.

Chemical Formulas and Reactions

Chemical reactions are very important to organisms. In a chemical reaction, the molecules are often represented by a chemical formula, such as the one below for the energy molecule glucose.

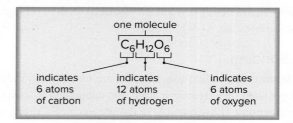

Notice how the chemical formula for glucose indicates the type and quantity of each element that is found in the molecule. Chemical formulas do not indicate the arrangement of these elements. As we will see later, they are sometimes several different structures that can be formed based on a chemical formula.

We have already noted that the process of photosynthesis enables plants to make molecular energy available to themselves and other organisms. An overall equation for photosynthesis indicates that some bonds are broken and others are formed:

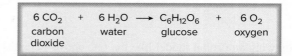

This equation says that six molecules of carbon dioxide react with six molecules of water to form one glucose molecule and six molecules of oxygen. The **reactants** (molecules that participate in the reaction) are shown on the left of the arrow, and the **products** (molecules formed by the reaction) are shown on the right. Notice that the equation is "balanced"—that is, the same number of each type of atom occurs on both sides of the arrow.

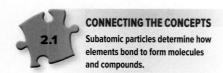

CONNECTING THE CONCEPTS

2.1 Subatomic particles determine how elements bond to form molecules and compounds.

Check Your Progress 2.1

1. Describe the structure of an atom, including the charge of each subatomic particle.

2. Define the term *isotope*, and list a few beneficial uses of radioactive isotopes.

3. Explain the differences between covalent and ionic bonds.

4. Summarize the octet rule, and explain how it relates to an element—s reactivity.

5. Distinguish between reactants and products in a chemical equation.

2.2 Water's Importance to Life

> **Learning Outcomes**
>
> Upon completion of this section, you should be able to
>
> 1. Describe the general structure of a water molecule.
> 2. List the properties of water that are important to life.
> 3. Understand the importance of hydrogen bonds to the properties of water.

Life began in water, and water is the single most important molecule on Earth. All organisms are 70–90% water; their cells consist of membranous compartments enclosing aqueous solutions. The structure of a water molecule gives it unique properties. These properties play an important role in how living organisms function.

The Structure of Water

The electrons shared between two atoms in a covalent bond are not always shared equally. Atoms differ in their **electronegativity**—that is, their affinity for electrons in a covalent bond. Atoms that are more electronegative tend to hold shared electrons more tightly than do those that are less electronegative. This unequal sharing of electrons causes the bond to become **polar,** meaning that the atoms on both sides of the bond are partially charged, even though the overall molecule itself bears no net charge.

For example, in a water molecule, oxygen shares electrons with two hydrogen atoms. Oxygen is more electronegative than hydrogen, so the two bonds are polar. The shared electrons spend more time orbiting the oxygen nucleus than the hydrogen nuclei, and this unequal sharing of electrons makes water a polar molecule. The covalent bonds are angled, and the molecule is bent roughly into a ∧ shape. The point of the ∧ (oxygen) is the negative (–) end, and the two hydrogens are the positive (+) end (**Fig. 2.9***a*).

The polarity of water molecules causes them to be attracted to one another. The positive hydrogen atoms in one molecule are attracted to the negative oxygen atoms in other water molecules. This attraction is called a **hydrogen bond,** and each water molecule can engage in as many as four hydrogen bonds (Fig. 2.9*b*). The covalent bond is much stronger than a hydrogen bond, but the large number of hydrogen bonds in water make for a strong attractive force. The properties of water are due to its polarity and its ability to form hydrogen bonds.

Properties of Water

We often take water for granted, but without water, life as we know it would not exist. The properties of water that support life are solvency, cohesion and adhesion, high surface tension, high heat capacity, and varying density.

Water Is a Solvent

Because of its polarity and hydrogen-bonding ability, water dissolves a great number of substances. Molecules that are attracted to water are said to be **hydrophilic** (*hydro,* water; *phil,* love). Polar and ionized molecules are usually hydrophilic. Nonionized and nonpolar molecules that are not attracted to water are said to be **hydrophobic** (*hydro,* water; *phob,* fear).

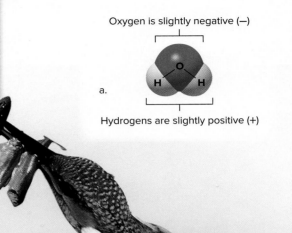

Oxygen is slightly negative (—)

a.

Hydrogens are slightly positive (+)

b. ------ hydrogen bond

Figure 2.9 **The structure of water.**

The properties of water play an important role in all life, including the kingfisher shown here. **a.** The space-filling model shows the ∧ shape of a water molecule. Oxygen attracts the shared electrons more than hydrogen atoms do, and this causes the molecule to be polar: The oxygen carries a slightly negative charge and the hydrogens carry a slightly positive charge. **b.** The positive hydrogens form hydrogen bonds with the negative oxygen in nearby molecules.

(photo): © Arco Images/GmbH/Alamy

When a salt such as sodium chloride (NaCl) is put into water, the negative ends of the water molecules are attracted to the sodium ions, and the positive ends of the water molecules are attracted to the chloride ions. This attraction causes the sodium ions and the chloride ions to break up, or dissociate, in water:

The salt NaCl dissociates in water.

Water may also dissolve polar nonionic substances, such as long chains of glucose, by forming hydrogen bonds with them. When ions and molecules disperse in water, they move about and collide, allowing reactions to occur. The interior of our cells is composed primarily of water, and the ability of water to act as a solvent allows the atoms and molecules within each cell to readily interact and participate in chemical reactions.

Water Molecules Are Cohesive and Adhesive

Cohesion refers to the ability of water molecules to cling to each other due to hydrogen bonding. Because of cohesion, water exists as a liquid under ordinary conditions of temperature and pressure. The strong cohesion of water molecules is apparent because water flows freely, yet the molecules do not separate from each other. **Adhesion** refers to the ability of water molecules to cling to other polar surfaces. This ability is due to water's polarity. The positive and negative poles of water molecules cause them to adhere to other polar surfaces.

Due to cohesion and adhesion, liquid water is an excellent transport system (**Fig. 2.10**). Both within and outside the cell, water assists in the transport

Figure 2.10 Cohesion and adhesion of water molecules.

Water's properties of cohesion and adhesion make it an excellent transport medium in both trees and humans.

(tree): © Paul Davies/Alamy; (man): © Asiaselects/Getty RF

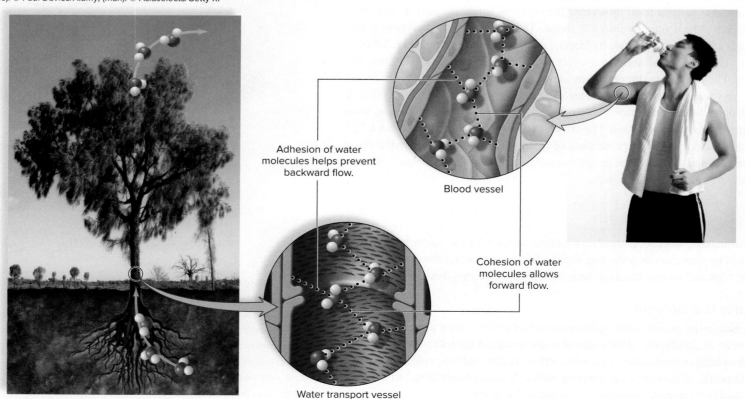

Adhesion of water molecules helps prevent backward flow.

Blood vessel

Cohesion of water molecules allows forward flow.

Water transport vessel

of nutrients and waste materials. Many multicellular animals contain internal vessels in which water assists the transport of materials. The liquid portion of blood, which transports dissolved and suspended substances within the body, is 90% water. The cohesion and adhesion of water molecules allow blood to fill the tubular vessels of the cardiovascular system, making transport possible. Cohesion and adhesion also contribute to the transport of water in plants. Plants have their roots anchored in the soil, where they absorb water, but the leaves are uplifted and exposed to solar energy. Water evaporating from the leaves is immediately replaced with water molecules from transport vessels that extend from the roots to the leaves. Because water molecules are cohesive, a tension is created that pulls a water column up from the roots. Adhesion of water to the walls of the vessels also helps prevent the water column from breaking apart.

Water Has a High Surface Tension

Because the water molecules at the surface are more strongly attracted to each other than to the air above, water molecules at the surface cling tightly to each other. Thus, we say that water exhibits surface tension. The stronger the force between molecules in a liquid, the greater the surface tension. Hydrogen bonding is the main force that causes water to have a high surface tension. If you slowly fill a glass with water, you may notice that the level of the water forms a small dome above the top of the glass. This is due to the surface tension of water.

Connections: Scientific Inquiry

How do some insects walk on water?

Anyone who has visited a pond or stream has witnessed insects walking on the surface of the water. These insects, commonly called water striders, have evolved this ability by adapting to two properties of water—its surface tension and the fact that water is a polar molecule. By trapping small air bubbles in the hairs on their legs, water striders are able to remain buoyant and not break the surface tension of the water molecules. Many water striders also secrete a nonpolar wax, which further repels the water molecules and keeps the insect afloat.

© Martin Shields/Alamy

Water Has a High Heat Capacity

The many hydrogen bonds that link water molecules allow water to absorb heat without greatly changing in temperature. Water's high heat capacity is important not only for aquatic organisms but for all organisms. Because the temperature of water rises and falls slowly, terrestrial organisms are better able to maintain their normal internal temperatures and are protected from rapid temperature changes.

Water also has a high heat of vaporization: It takes a great deal of heat to break the hydrogen bonds in water so that it becomes gaseous and evaporates into the environment. If the heat given off by our metabolic activities were to go directly into raising our body temperature, death would follow. Instead, the heat is dispelled as sweat evaporates (**Fig. 2.11**).

Figure 2.11 **Heat capacity and heat of vaporization.**

At room temperature, water is a liquid. **a.** Water takes a large amount of heat to vaporize at 100°C. **b.** It takes much body heat to vaporize sweat, which is mostly liquid water, and the vaporization helps keep our bodies cool when the temperature rises.

(a): © Jill Bratten/McGraw-Hill Education; (b): © Clerkenwell/Getty RF

a.

b.

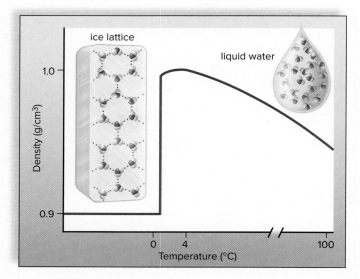

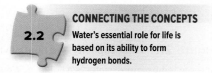

Figure 2.12 Properties of ice.

The geometric requirements of hydrogen bonding of water molecules cause ice to be less dense than liquid water.

CONNECTING THE CONCEPTS

2.2 Water's essential role for life is based on its ability to form hydrogen bonds.

Because of water's high heat capacity and high heat of vaporization, temperatures along the majority of Earth's coasts are moderate. During the summer, the ocean absorbs and stores solar heat; during the winter, the ocean releases it slowly. In contrast, the interior regions of the continents experience abrupt changes in temperature.

Ice Is Less Dense Than Water

Unlike other substances, water expands as it freezes, which explains why cans of soda burst when placed in a freezer and how roads in northern climates become bumpy because of "frost heaves" in the winter. Since water expands as it freezes, ice is less dense than liquid water, and therefore ice floats on liquid water (**Fig. 2.12**).

Connections: Ecology

Why is it important that ice is less dense than water?

If ice were more dense than water, it would sink, and ponds, lakes, and perhaps even the ocean would freeze solid, making life impossible in the water as well as on land. Instead, bodies of water always freeze from the top down. When a body of water freezes on the surface, the ice acts as an insulator to prevent the water below it from freezing. This protects aquatic organisms, so that they can survive the winter. As ice melts in the spring, it draws heat from the environment, helping prevent a sudden change in temperature that might be harmful to life.

© Corbis RF

Check Your Progress 2.2

1. Explain how the structure of a water molecule gives it unique properties.

2. Describe how the properties of water make it an important molecule for life.

3. Explain how hydrogen bonds relate to the properties of water.

2.3 Acids and Bases

Learning Outcomes

Upon completion of this section, you should be able to

1. Distinguish between an acid and a base.
2. Interpret the pH scale.
3. Explain the purpose of a buffer.

As shown in the equation below, when water dissociates (breaks apart), it releases an equal number of hydrogen ions (H^+) and hydroxide ions (OH^-).

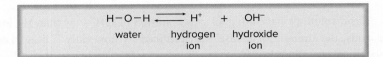

$$H-O-H \rightleftharpoons H^+ + OH^-$$

water   hydrogen   hydroxide
                  ion   ion

We can determine whether a solution is acidic or basic by examining the proportion of hydrogen and hydroxide ions in the solution.

Acidic Solutions (High H^+ Concentration)

Lemon juice, vinegar, tomato juice, and coffee are all acidic solutions. Acidic solutions have a sharp or sour taste, and therefore we sometimes associate them with indigestion. What do they have in common? To a chemist, **acids** are substances that dissociate in water, releasing hydrogen ions (H^+). For example, an important acid is hydrochloric acid (HCl), which dissociates in this manner:

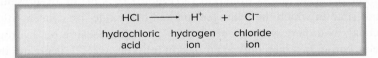

$$HCl \longrightarrow H^+ + Cl^-$$

hydrochloric   hydrogen   chloride
     acid   ion   ion

Acidic solutions have a higher concentration of H^+ ions than OH^- ions. The acidity of a substance depends on how fully it dissociates in water. HCl dissociates almost completely; therefore, it is called a strong acid. If hydrochloric acid is added to a beaker of water, the number of hydrogen ions (H^+) increases greatly.

Connections: Scientific Inquiry

How strong is the acid in your stomach?

Within the gastric juice of the stomach is hydrochloric acid (HCl), which has a pH value between 1.0 and 2.0. This acid makes the contents of your stomach around 1 million times more acidic than water and 100 times more acidic than vinegar. While theoretically the gastric juice in your stomach is able to dissolve metals, such as steel, in reality the contents of your stomach are exposed to these extreme pH values for only a short period of time before moving into the remainder of the intestinal tract, where the acid levels are quickly neutralized.

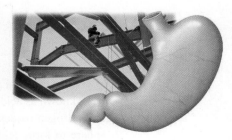

(photo): © Corbis RF

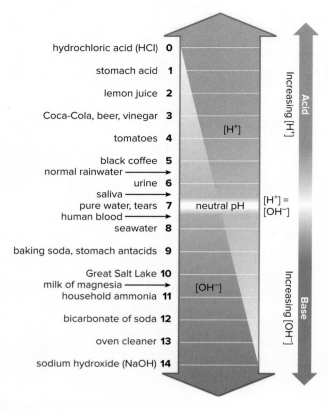

Figure 2.13 The pH scale.

The proportionate amount of hydrogen ions to hydroxide ions is indicated by the diagonal line. Any solution with a pH above 7 is basic, while any solution with a pH below 7 is acidic.

Basic Solutions (Low H⁺ Concentration)

Milk of magnesia and ammonia are common basic (alkaline) solutions that most people are familiar with. Basic solutions have a bitter taste and feel slippery when in water. To a chemist, **bases** are substances that either take up hydrogen ions (H^+) or release hydroxide ions (OH^-). For example, an important base is sodium hydroxide (NaOH), which dissociates in this manner:

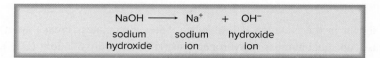

Basic solutions have a higher concentration of OH^- ions than H^+ ions. Like acids, the strength of a base is determined by how fully it dissociates. Dissociation of sodium hydroxide is almost complete; therefore, it is called a strong base. If sodium hydroxide is added to a beaker of water, the number of hydroxide ions increases.

Many strong bases, such as ammonia, are useful household cleansers. Ammonia has a poison symbol and carries a strong warning not to ingest the product. Neither acids nor bases should be tasted, because they are quite destructive to cells.

pH and the pH Scale

pH[1] is a mathematical way of indicating the number of hydrogen ions in a solution. The **pH scale** is used to indicate the acidity or basicity of a solution. The pH scale ranges from 0 to 14 (**Fig. 2.13**). A pH of 7 represents a neutral state in which the hydrogen ion and hydroxide ion concentrations are equal, as in pure water. A pH below 7 is acidic because the hydrogen ion concentration, commonly expressed in brackets as $[H^+]$, is greater than the hydroxide concentration, $[OH^-]$. A pH above 7 is basic because $[OH^-]$ is greater than $[H^+]$. Further, as we move down the pH scale from pH 7 to pH 0, each unit has 10 times the acidity $[H^+]$ of the previous unit. As we move up the scale from 7 to 14, each unit has 10 times the basicity $[OH^-]$ of the previous unit.

The pH scale was devised to eliminate the use of cumbersome numbers. For example, the hydrogen ion concentrations of several solutions are given on the left, and the pH is on the right:

$[H^+]$ (moles per liter)		pH
0.000001	= 1×10^{-6}	6 (acid)
0.0000001	= 1×10^{-7}	7 (neutral)
0.00000001	= 1×10^{-8}	8 (base)

The effect of pH on organisms is dramatically illustrated by the phenomenon known as acid precipitation. When fossil fuels are burned, sulfur dioxide and nitrogen oxides are produced, and they combine with water in the atmosphere to form acids. These acids then come in contact with organisms and objects, leading to damage or even death.

---

[1] pH is defined as the negative log of the hydrogen ion concentration, or $-\log[H^+]$. A log is the power to which 10 must be raised to produce a given number.

Figure 2.14 Acidosis.

Acidosis occurs when the body in unable to buffer high H^+ ion concentrations in the blood. Some of the symptoms of acidosis are shown here.

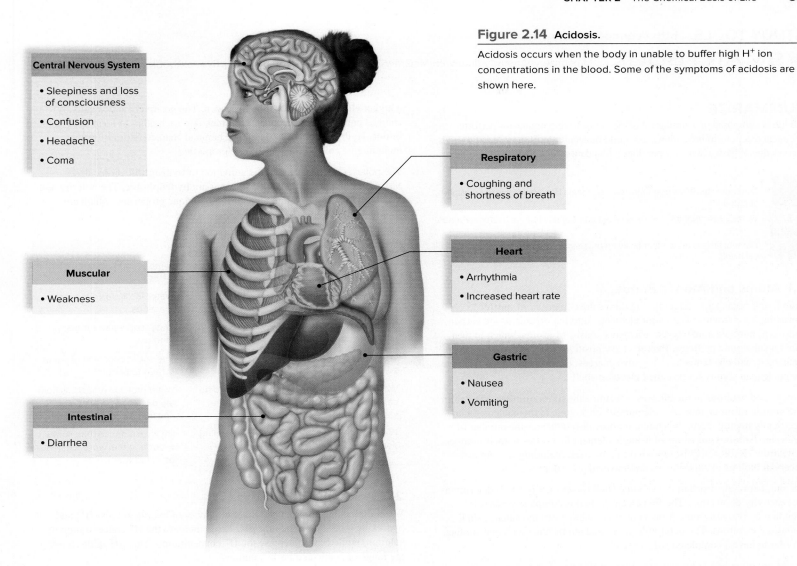

Central Nervous System

- Sleepiness and loss of consciousness
- Confusion
- Headache
- Coma

Respiratory

- Coughing and shortness of breath

Muscular

- Weakness

Heart

- Arrhythmia
- Increased heart rate

Gastric

- Nausea
- Vomiting

Intestinal

- Diarrhea

Buffers and pH

A **buffer** is a chemical or a combination of chemicals that keeps pH within established limits. Buffers resist pH changes because they can take up excess hydrogen ions (H^+) or hydroxide ions (OH^-). In the human body, pH needs to be kept within a narrow range in order to maintain homeostasis. Diseases such as diabetes and congestive heart failure may bring on a condition called acidosis, in which the body is unable to buffer the excessive production of H^+ ions. If left untreated, acidosis can cause a number of health problems (**Fig. 2.14**) and may result in coma or death.

CONNECTING THE CONCEPTS

2.3 Homeostasis is dependent on living organisms' ability to maintain specific pH levels.

Check Your Progress 2.3

1. Distinguish between an acid and a base.
2. Generalize what information is given by the pH of a solution.
3. Summarize how buffers are used to regulate the pH of the human body.

STUDY TOOLS   http://connect.mheducation.com

 ■SMARTBOOK®   Maximize your study time with McGraw-Hill SmartBook®, the first adaptive textbook.

SUMMARIZE

All life is composed of a similar set of elements. Living organisms perform chemical reactions to break down and build molecules to fit their specific needs. Knowledge of basic chemistry provides a foundation for understanding biology.

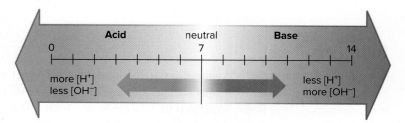

2.1   Subatomic particles determine how elements bond to form molecules and compounds.

2.2   Water's essential role for life is based on its ability to form hydrogen bonds.

2.3   Homeostasis is dependent on living organisms' ability to maintain specific pH levels.

2.1 Atoms and Atomic Bonds

Both living organisms and nonliving things are composed of **matter** consisting of **elements.** The major elements in living organisms are carbon, hydrogen, nitrogen, and oxygen. Elements consist of **atoms,** which in turn contain subatomic particles. **Protons** have positive charges, **neutrons** are uncharged, and **electrons** have negative charges. Electrons are found outside the nucleus in energy levels called **electron shells.**

Protons and neutrons in the nucleus determine the **mass number** of an atom. The **atomic number** indicates the number of protons in the nucleus. In an electrically neutral atom, the atomic number also indicates the number of electrons. **Isotopes** are atoms of a single element that differ in their number of neutrons. Radioactive isotopes have many uses, including serving as tracers in biological experiments and medical procedures.

The number of electrons in the **valence shell** (outer energy level) determines the reactivity of an atom. The first electron shell is complete when it is occupied by two electrons. Atoms are most stable when the valence shell contains 8 electrons. The **octet rule** states that atoms react with one another in order to have a completed valence shell.

Atoms are often bonded together to form **molecules.** If the elements in a molecule are different, it is called a **compound.** Molecules and compounds may be held together by ionic or covalent bonds. **Ionic bonds** are formed by an attraction between oppositely charged ions. **Ions** form when atoms lose or gain one or more electrons to achieve a completed outer shell. **Covalent bonds** occur when electrons are shared between two atoms. There are single covalent bonds (sharing one pair of electrons), double (sharing two pairs of electrons), and triple (sharing three pairs of electrons).

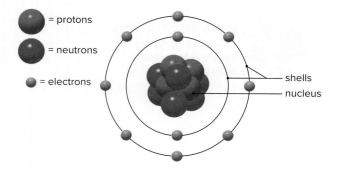

= protons

= neutrons

= electrons

shells
nucleus

2.2 Water's Importance to Life

The two covalent O—H bonds of water are **polar** because the electrons are not shared equally between oxygen and hydrogen. This is because oxygen has a higher **electronegativity** than hydrogen. The positively and negatively charged ends of the molecules are attracted to each other to form **hydrogen bonds.** Hydrogen bonds explain the chemical characteristics of water, including its **cohesive** and **adhesive** properties.

The polarity of water causes the attraction of **hydrophilic** molecules. Molecules that are not attracted to water are **hydrophobic.** The polarity and hydrogen bonding in water account for its unique properties, which are summarized in Table 2.1:

Table 2.1 Properties of Water

Properties	Chemical Reason(s)	Effect
Water is a solvent.	Polarity	Water facilitates chemical reactions.
Water is cohesive and adhesive.	Hydrogen bonding; polarity	Water serves as a transport medium.
Water has a high surface tension.	Hydrogen bonding	The surface tension of water is hard to break.
Water has a high heat capacity and heat of vaporization.	Hydrogen bonding	Water protects organisms from rapid changes in temperature and from overheating.
Water varies in density.	Hydrogen bonding	Ice is less dense than liquid water and therefore ice floats on liquid water.

2.3 Acids and Bases

Water dissociates to produce an equal number of hydrogen ions (H^+) and hydroxide ions. This is neutral **pH. Acids** increase the H^+ concentration of a solution, while **bases** decrease the H^+ concentration. This **pH scale** shows the range of acidic and basic solutions:

Acid   neutral   Base

0   7   14

more [H^+]   less [H^+]
less [OH^-]   more [OH^-]

Cells are sensitive to pH changes. Biological systems contain **buffers** that help keep the pH within a normal range.

ASSESS

Testing Yourself

Choose the best answer for each question.

2.1 Atoms and Atomic Bonds

1. The mass number of an atom depends primarily on the number of
   a. protons and neutrons.   c. neutrons and electrons.
   b. positrons.   d. protons and electrons.

2. The most abundant element by weight in the human body is
   a. carbon.
   c. oxygen.
   b. hydrogen.
   d. nitrogen.

3. In the following equation, name the reactants: $NaCl \rightarrow Na + Cl$.
   a. Na
   c. NaCl
   b. Cl
   d. Both a and b are correct.

4. A covalent bond in which electrons are not shared equally is called
   a. polar.
   c. nonpolar.
   b. normal.
   d. neutral.

5. Refer to Figure 2.3. Which element has the same number of valence electrons as nitrogen (N)?
   a. carbon (C)
   c. neon (Ne)
   b. phosphorus (P)
   d. oxygen (O)

2.2 Water's Importance to Life

6. Water flows freely but does not separate into individual molecules because water is
   a. cohesive.
   c. hydrophobic.
   b. hydrophilic.
   d. adhesive.

7. Compounds having an affinity for water are said to be
   a. cohesive.
   c. hydrophobic.
   b. hydrophilic.
   d. adhesive.

8. Water freezes from the top down because
   a. water has a high surface tension.
   b. ice has a high heat capacity.
   c. ice is less dense than water.
   d. ice is less cohesive than water.

9. Water can absorb a large amount of heat without much change in temperature because it has a high
   a. surface tension.
   b. heat capacity.
   c. hydrogen ion (H^+) concentration.
   d. hydroxide ion (OH^-) concentration.

2.3 Acids and Bases

10. A pH of 3 is _____.
    a. basic
    c. acidic
    b. neutral
    d. a buffer

11. _____ contribute hydrogen ions (H^+) to a solution
    a. Bases
    c. Acids
    b. Isotopes
    d. Compounds

12. To maintain a constant pH, many organisms use _____ to regulate the hydrogen ion concentration.
    a. buffers
    c. acids
    b. bases
    d. isotopes

ENGAGE

BioNOW

Want to know how this science is relevant to your life? Check out the BioNow video below.

- Properties of Water

What characteristic of water do you think is most important to a living organism, such as yourself?

Thinking Critically

1. On a hot summer day, you decide to dive into a swimming pool. Before you begin your dive, you notice that the surface of the water is smooth and continuous. After the dive, you discover that some water droplets are clinging to your skin and that your skin temperature feels cooler. Explain these observations based on the properties of water.

2. Like carbon, silicon has four electrons in its outer shell, yet life evolved to be carbon-based. What is there about silicon's structure that might prevent it from sharing with four other elements and prevent it from forming the many varied shapes of carbon-containing molecules?

3. Antacids are a common over-the-counter remedy for heartburn, a condition caused by an overabundance of H^+ ions in the stomach. Based on what you know regarding pH, how do the chemicals in antacids work?

4. Acid precipitation is produced when atmospheric water is polluted by sulfur dioxide and nitrous oxide emissions. These emissions are mostly produced by the burning of fossil fuels, particularly coal. In the atmosphere, these compounds are converted to sulfuric and nitric acids, and are absorbed into water droplets in the atmosphere. Eventually they fall back to Earth as acid precipitation. How do you think acid precipitation can negatively influence an ecosystem?

3

The Organic Molecules of Life

© Kim Scott/Ricochet Creative Productions LLC

OUTLINE

3.1 Organic Molecules   39

3.2 The Biological Molecules of Cells   41

BEFORE YOU BEGIN

Before beginning this chapter, take a few moments to review the following discussions.

Section 2.1 What is a covalent bond?

Section 2.2 What is the difference between hydrophobic and hydrophilic molecules?

Figure 2.8 How are molecules represented in a diagram?

Not All Cholesterol Is Bad

Without cholesterol, your body would not function well. What is it about cholesterol that can cause problems in your body? Although blood tests now distinguish between "good" cholesterol and "bad" cholesterol, in reality there is only one molecule called cholesterol. Cholesterol is essential to the normal functioning of cells. It is a part of a cell's outer membrane, and it serves as a precursor to hormones, such as the sex hormones testosterone and estrogen. Without cholesterol, your cells would be in major trouble.

So when does cholesterol become "bad"? Your body packages cholesterol in high-density lipoproteins (HDLs) or low-density lipoproteins (LDLs), depending on such factors as genetics, your diet, your activity level, and whether you smoke. Cholesterol packaged as HDL is primarily on its way from the tissues to the liver for recycling. This reduces the likelihood of the formation of deposits called plaques in the arteries; thus, HDL cholesterol is considered "good." LDL is carrying cholesterol to the tissues; high levels of LDL cholesterol contribute to the development of plaques, which can result in heart disease, making LDL the "bad" form. When cholesterol levels are measured, the ratio of LDL to HDL is determined. Exercising, improving your diet, and not smoking can all lower LDL levels. In addition, a variety of prescription medications can help lower LDL levels and raise HDL levels. These lipoproteins are just one example of the organic molecules that our bodies need to function correctly.

In this chapter, you will learn about the structure and function of the major classes of organic molecules, including carbohydrates, lipids, proteins, and nucleic acids.

As you read through this chapter, think about the following questions:

1. What class of biological molecules does cholesterol belong to?
2. Is cholesterol a hydrophobic or hydrophilic molecule?
3. What distinguishes cholesterol from other biological molecules?

3.1 Organic Molecules

Learning Outcomes

Upon completion of this section, you should be able to

1. Distinguish between organic and inorganic molecules.
2. Recognize the importance of functional groups in determining the chemical properties of an organic molecule.

The study of chemistry can be divided into two major categories: organic chemistry and inorganic chemistry. The difference between the two is relatively simple. Organic chemistry is the study of organic molecules. An **organic** molecule contains atoms of carbon and hydrogen. Organic molecules make up portions of cells, tissues, and organs (**Fig. 3.1**). An inorganic molecule does not contain a combination of carbon and hydrogen. Water (H_2O) and table salt (NaCl) are examples of inorganic molecules. This chapter focuses on the diversity and functions of organic molecules in cells, also commonly referred to as **biological molecules,** or simply biomolecules.

The Carbon Atom

A microscopic bacterial cell can contain thousands of different organic molecules. In fact, there appears to be an almost unlimited variation in the structure of organic molecules. What is there about carbon that makes organic molecules so diverse and so complex? Recall that carbon, with a total of six electrons, has four electrons in the outer shell (see Fig. 2.8). In order to acquire four electrons

Connections: Health

Is there a connection between organic molecules and organic produce?

© Purestock/Superstock RF

As you just learned, organic molecules are those that contain carbon and hydrogen. Vegetables and fruits are all organic in that they contain these elements. The use of the term *organic* when related to farming means that there are certain production standards in place when growing the food. Normally, it means no pesticides or herbicides with harsh chemicals were used and the crop was grown as naturally as possible. So, as you can see, there are two different ways that the term *organic* may be used.

Figure 3.1 Organic molecules have a variety of functions.

a. Carbohydrates form fiber that provides support to plants. **b.** Proteins help form the cell walls of bacteria. **c.** Lipids, such as oils, are used for energy storage for plants. **d.** Nucleic acids form DNA which acts to store genetic information.

(a): © SuperStock/Alamy; (b): © Science Photo Library/Alamy Stock Photo RF; (c): © Zeljko Radojko/Getty Images; (d): Design Pics/Bilderbuch RF

a.

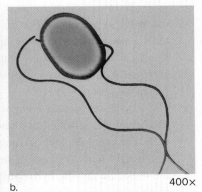
b. 400×

c.

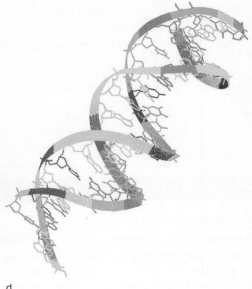
d.

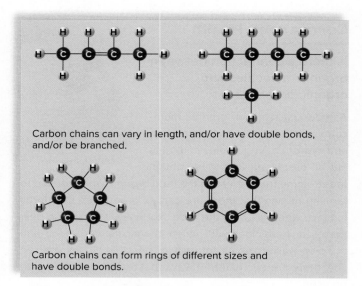

Carbon chains can vary in length, and/or have double bonds, and/or be branched.

Carbon chains can form rings of different sizes and have double bonds.

Figure 3.2 Hydrocarbons are highly versatile.

Hydrocarbons contain only hydrogen and carbon. Even so, they can be quite varied, according to the number of carbons, the placement of any double bonds, possible branching, and possible ring formation.

Functional Groups		
Group	**Structure**	**Found In**
Hydroxyl	R—O—H	Alcohols, sugars
Carboxyl	R—C (=O)(O—H)	Amino acids, fatty acids
Amino	R—N (H)(H)	Amino acids, proteins
Sulfhydryl	R—S—H	Amino acid cysteine, proteins
Phosphate	R—O—P (H)(O)(O)—O—H	ATP, nucleic acids
R = remainder of molecule		

Figure 3.3 Common functional groups.

Molecules with the same carbon skeleton can still differ according to the type of functional group attached. These functional groups help determine the chemical reactivity of the molecule. In this illustration, the remainder of the molecule, the hydrocarbon chain, is represented by an R.

to complete its outer shell, a carbon atom almost always shares electrons, and typically with the elements hydrogen, nitrogen, and oxygen—the elements that make up most of the weight of living organisms (see Section 2.1).

Because carbon is small and needs to acquire four electrons, carbon can bond with as many as four other elements. Carbon atoms most often share electrons with other carbon atoms. The C—C bond is stable, and the result is that carbon chains can be quite long. Hydrocarbons are chains of carbon atoms that are also bonded only to hydrogen atoms. Any carbon atom of a hydrocarbon molecule can start a branch chain, and a hydrocarbon can turn back on itself to form a ring compound (**Fig. 3.2**). Carbon can also form double bonds with other atoms, including another carbon atom.

The versatile nature of carbon means that it can form a variety of molecules with the same chemical formula (types of atoms) but different structures. Molecules with different structures, but the same combinations of atoms, are called **isomers.** The chemistry of carbon leads to a huge structural diversity of organic molecules. Since structure dictates function, this structural diversity means that these molecules have a wide range of diverse functions.

The Carbon Skeleton and Functional Groups

The carbon chain of an organic molecule is called its skeleton, or backbone. This terminology is appropriate because, just as your skeleton accounts for your shape, so does the carbon skeleton of an organic molecule account for its shape. The reactivity of an organic molecule is largely dependent on the attached functional groups (**Fig. 3.3**). A **functional group** is a specific combination of bonded atoms that always has the same chemical properties and therefore always reacts in the same way, regardless of the particular carbon skeleton to which it is attached. Notice in Figure 3.3 the letter R attached to each functional group. This stands for the remainder of the molecule, and it indicates where the functional group attaches to the hydrocarbon chain.

The functional groups of an organic molecule therefore help determine its chemical properties. For example, organic molecules, such as fats and proteins, containing carboxyl groups (—COOH) are both polar (hydrophilic) and weakly acidic. Phosphate groups contribute to the structure of nucleic acids, such as DNA. Proteins and amino acids possess the nitrogen-containing amino functional group (—NH$_2$).

Because cells are composed mainly of water, the ability to interact with and be soluble in water profoundly affects the activity of organic molecules in cells. For example, hydrocarbons are largely **hydrophobic** (not soluble in water), but if a number of —OH functional groups are added (such as in glucose), the molecule may be **hydrophilic** (water-soluble). The functional groups also identify the types of reactions that the molecule will undergo. For example, fats are formed by the interaction of molecules containing alcohols and carboxyl groups, and proteins are formed when the amino and carboxyl functional groups of nearby amino acids are linked.

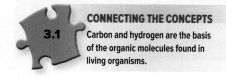

CONNECTING THE CONCEPTS

3.1 Carbon and hydrogen are the basis of the organic molecules found in living organisms.

3.2 The Biological Molecules of Cells

Learning Outcomes

Upon completion of this section, you should be able to

1. Summarize the structure and function of each category of carbohydrates.

2. Summarize the structure and function of each category of lipids.

3. Summarize the structure and function of proteins.

4. Summarize the two categories of nucleic acids, and describe their biological functions.

Despite their great diversity, biological molecules are grouped into only four categories: carbohydrates, lipids, proteins, and nucleic acids. You are very familiar with these molecules because certain foods are known to be rich in carbohydrates (**Fig. 3.4**), lipids (**Fig. 3.5**), or proteins (**Fig. 3.6**). When you digest these foods, they break down into smaller molecules, or subunits. Your body then takes these subunits and builds from them the large macromolecules that make up your cells. Foods also contain nucleic acids, the type of biological molecule that forms the genetic material of all living organisms.

Figure 3.4 Carbohydrates.

Carbohydrates are used in animals as a short-term energy source. A diet loaded with simple carbohydrates may make you prone to type 2 diabetes and other illnesses. In contrast, moderate amounts of complex carbohydrates (fiber) have a variety of health benefits.

© John Thoeming/McGraw-Hill Education

Figure 3.5 Lipid foods.

Lipids are primarily associated with the long-term storage of energy. However, some are involved in forming hormones and components of our cells. Saturated fats are associated with an increased risk of cardiovascular disease. Choose vegetable oils over the animal fat in lard, butter, and other dairy products.

© John Thoeming/McGraw-Hill Education

Figure 3.6 Protein foods.

Proteins are involved in almost all of the functions in your body, but you do not require a large protein source in every meal. Small servings can provide you with all the amino acids you need. Some forms of protein, such as beef, contain high amounts of fat. Fish, however, contains beneficial oils that lower the incidence of cardiovascular disease.

© John Thoeming/McGraw-Hill Education

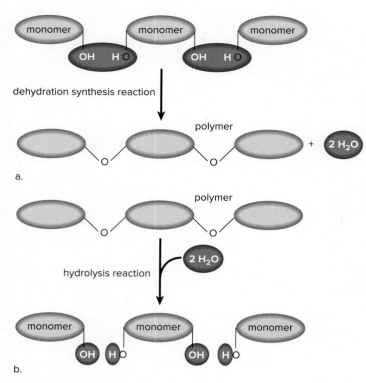

Figure 3.7 **Synthesis and breakdown of polymers.**

a. In cells, synthesis often occurs when monomers bond during a dehydration synthesis reaction (removal of H_2O). **b.** Breakdown occurs when the monomers in a polymer separate because of a hydrolysis reaction (the addition of H_2O).

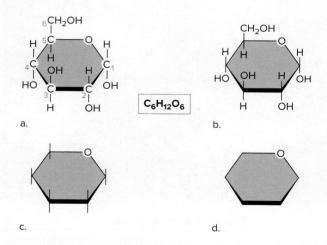

Figure 3.8 Glucose.

Each of these structural formulas represents glucose. **a.** The carbon skeleton (with carbon atoms numbered) and all attached groups are shown. **b.** The carbons in the skeleton are omitted and only the functional groups are shown **c.** The carbons in the skeleton and functional groups are omitted. **d.** Only the ring shape of the molecule remains.

Many of the biological molecules are composed of a large number of similar building blocks, called **monomers.** When multiple monomers are joined together they form a **polymer.** A protein can contain hundreds of amino acid monomers, and a nucleic acid can contain hundreds of nucleotide monomers. How can polymers get so large? Just as a train increases in length when boxcars are hitched together one by one, so a polymer gets longer as monomers bond to one another.

The most common type of chemical reaction that is used to build a polymer from a group of monomers is called a **dehydration synthesis reaction.** It is called this because the equivalent of a water (H_2O) molecule, meaning both an —OH (hydroxyl group) and an H (hydrogen atom), is removed as the reaction occurs (**Fig. 3.7***a*).

To break down a biological molecule, a cell uses an opposite type of reaction. During a **hydrolysis** (*hydro,* water; *lysis,* break) **reaction,** an —OH group from water attaches to one monomer, and an H from water attaches to the other monomer (Fig. 3.7*b*). In other words, water is used to break the bond holding monomers together.

Carbohydrates

In living organisms, **carbohydrates** are almost universally used as an immediate energy source. However, for many organisms, such as plants and fungi, they also have structural functions. Carbohydrates may exist either as saccharide (sugar) monomers or as polymers of saccharides. Typically, the sugar glucose is a common monomer of carbohydrate polymers. The term *carbohydrate* may refer to a single sugar molecule (monosaccharide), two bonded sugar molecules (disaccharide), or many sugar molecules bonded together (polysaccharide).

Monosaccharides: Energy Molecules

Because **monosaccharides** have only a single sugar molecule, they are also known as simple sugars. A simple sugar can have a carbon backbone consisting of three to seven carbons. Monosaccharides, and carbohydrates in general, often possess many polar —OH functional groups, which make them soluble in water. In a water environment, such as that within our cells, carbohydrates often form a ringlike structure, as you can see by examining the structural formula for glucose (**Fig. 3.8**).

Glucose, with six carbon atoms, has a molecular formula of $C_6H_{12}O_6$. Glucose has two important isomers, called *fructose* and *galactose,* but even so, we usually think of glucose when we see the formula $C_6H_{12}O_6$. That's because glucose has a special place in the chemistry of organisms. Photosynthetic organisms, such as plants and bacteria, manufacture glucose using energy from the sun. This glucose is used as the preferred immediate source of energy for nearly all types of organisms. In other words, glucose has a central role in the energy reactions of cells.

Ribose and **deoxyribose,** with five carbon atoms, are significant because they are found in the nucleic acids RNA and DNA, respectively. RNA and DNA are discussed later in this section.

Disaccharides: Varied Uses

A **disaccharide** contains two monosaccharides linked together by a dehydration synthesis reaction. Some common disaccharides are *maltose, sucrose,* and *lactose.*

Maltose is a disaccharide that contains two glucose subunits. The brewing of beer relies on maltose, usually obtained from barley. During the production of beer, yeast breaks down the maltose and then uses the glucose as an energy source in a process called *fermentation.* A waste product of this reaction is ethyl alcohol (**Fig. 3.9**).

Sucrose, a disaccharide acquired from sugar beets and sugarcane, is of special interest because we use it as a sweetener. Our bodies digest sucrose into its two monomers, glucose and fructose. Later, the fructose is changed to glucose, our usual energy source. If the body doesn't need more energy at the moment, the glucose can be metabolized to fat. While glucose is the energy source of choice for animal cells, fat is the body's primary energy storage form. That's why eating lots of sugary desserts can make you gain weight.

Lactose is a disaccharide commonly found in milk. Lactose contains a glucose molecule combined with a galactose molecule. Individuals who are *lactose intolerant* are not able to break down the disaccharide lactose. The disaccharide then moves through the intestinal tract undigested, where the normal intestinal bacteria use it as an energy source. Symptoms of lactose intolerance include abdominal pain, gas, bloating, and diarrhea.

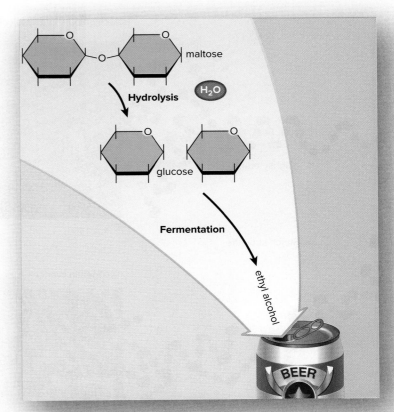

Figure 3.9 **Breakdown of maltose, a disaccharide.**

Maltose is the energy source for yeast during the production of beer. Yeasts differ as to the amount of maltose they convert to alcohol, so selection of the type of yeast is important for the correct result.

Connections: Health

What is high-fructose corn syrup?

Many beverages made commercially, including many cola drinks, contain high-fructose corn syrup (HFCS). In the 1980s, a commercial method was developed for converting the glucose in corn syrup to the much sweeter-tasting fructose. Nutritionists are not in favor of eating highly *processed* foods that are rich in sucrose, HFCS, and white starches. They say these foods provide "empty" calories, meaning that, although they supply energy, they don't supply any of the vitamins, minerals, and fiber needed in the diet. In contrast, minimally processed foods provide glucose, starch, and many other types of nutritious molecules.

© Richard Hutchings/
McGraw-Hill Education

Polysaccharides as Energy Storage Molecules

Polysaccharides are polymers of monosaccharides, usually glucose. Some types of polysaccharides function as short-term energy storage molecules because they are much larger than a monosaccharide and are relatively insoluble. Polysaccharides cannot easily pass through the plasma membrane and are kept (stored) within the cell.

Plants store glucose as **starch.** For example, the cells of a potato contain starch granules, which act as an energy storage location during winter for

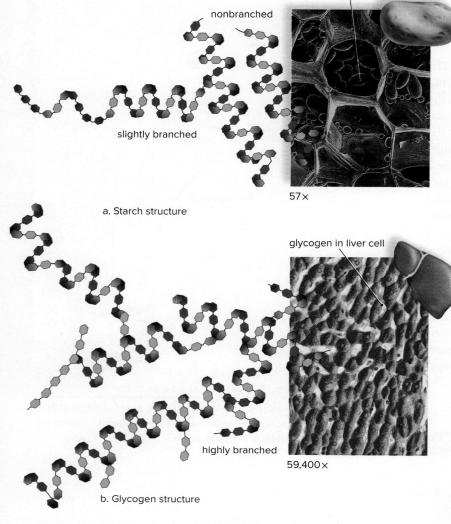

starch granule in potato cell

nonbranched

slightly branched

57×

a. Starch structure

glycogen in liver cell

highly branched

59,400×

b. Glycogen structure

Figure 3.10 **Starch and glycogen structure and function.**

a. Glucose is stored in plants as starch. Starch is a chain of glucose molecules that can be nonbranched or branched. The electron micrograph shows the location of starch in potato cells. **b.** Glucose is stored in animals as glycogen. Glycogen is a highly branched polymer of glucose molecules. The electron micrograph shows glycogen deposits in a portion of a liver cell.

(photos): (a): © Dr. Jeremy Burgess/Science Source; (b): © Don W. Fawcett/Science Source

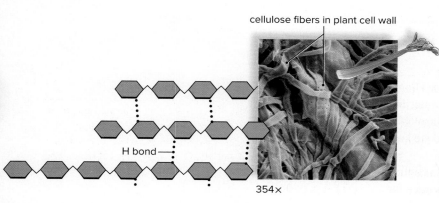

cellulose fibers in plant cell wall

H bond

354×

growth in the spring. Notice in **Figure 3.10***a* that starch exists in two forms—one is nonbranched and the other is slightly branched.

Animals store glucose as **glycogen,** which is more highly branched than starch (Fig. 3.10*b*). Branching subjects a polysaccharide to more attacks by hydrolytic enzymes; therefore, branching makes a polysaccharide easier to break down.

The storage and release of glucose from liver cells are controlled by hormones, such as insulin. We will take a much closer look at these hormones, and how the body regulates glucose levels, in Section 27.2.

Polysaccharides as Structural Molecules

Some types of polysaccharides function as structural components of cells. One example is **cellulose,** which is the most abundant of all the carbohydrates. Plant and algal cell walls all contain cellulose, and therefore it may be found in all of the tissues of a plant. Many commercial products, from wood to paper, are made from cellulose.

The bonds joining the glucose subunits in cellulose (**Fig 3.11**) are different from those found in starch and glycogen (see Fig. 3.10). As a result, the molecule does not spiral or have branches. The long glucose chains are held parallel to each other by hydrogen bonding to form strong microfibrils, which are grouped into fibers. The fibers crisscross within plant cell walls for even more strength.

The different bond structure means that the digestive systems of animals can't hydrolyze cellulose, but some microorganisms have this ability. Cows and other ruminants (cud-chewing animals) have an internal pouch, where microorganisms break down cellulose to glucose. In humans, cellulose has the benefit of serving as dietary fiber, which maintains regular elimination and digestive system health.

Chitin is a polymer of glucose molecules. However, in chitin, each glucose subunit has an amino group (—NH$_2$) attached to it. Since the functional groups attached to organic molecules determine their properties, chitin is chemically different from other glucose polymers, such as cellulose, even though the linkage between the glucose molecules is very similar. Chitin is found in a variety of organisms, including animals and fungi. In animals such as insects, crabs, and lobsters, chitin is found in the external skeleton, or exoskeleton. Even though chitin, like cellulose, is not digestible by humans, it still has many good uses. Seeds are coated with chitin, and this protects them from attack by soil fungi. Because chitin also has antibacterial and antiviral properties, it is processed and used in medicine as a wound dressing and suture material. Chitin is even useful during the production of cosmetics and various foods.

Figure 3.11 **Cellulose structure and function.**

In plant cell walls, each cellulose fiber contains several microfibrils. Each microfibril contains many polymers of glucose hydrogen-bonded together. The micrograph shows cellulose fibers in a plant cell wall.

(photo): © Cheryl Power/Science Source

Connections: Health

What is the difference between soluble and insoluble fiber?

Fiber, also called roughage, is composed mainly of the undigested carbohydrates that pass through the digestive system. Most fiber is derived from the structural carbohydrates of plants. These include such materials as cellulose, pectins, and lignin. Fiber is not truly a nutrient, since we do not use it directly for energy or cell building, but it is an extremely important component of our diet. Fiber not only adds bulk to material in the intestines, keeping the colon functioning normally, but also binds many types of harmful chemicals in the diet, including cholesterol, and prevents them from being absorbed. There are two basic types of fiber—insoluble and soluble. Soluble fiber dissolves in water and acts in the binding of cholesterol. Soluble fiber is found in many fruits, as well as oat grains. Insoluble fiber provides bulk to the fecal material and is found in bran, nuts, seeds, and whole-wheat foods.

© McGraw-Hill Education

Lipids

Although molecules classified as **lipids** are quite varied, they have one characteristic in common: They are all hydrophobic and insoluble in water. You may have noticed that oil and water do not mix. For example, salad dressings are rich in vegetable oils. Even after shaking, the vegetable oil will separate out from the water. This is due to the fact that lipids possess long, nonpolar hydrocarbon chains and a relative lack of hydrophilic functional groups.

Lipids are very diverse, and they have varied structures and functions. **Fats** (such as bacon fat, lard, and butter) and **oils** (such as corn oil, olive oil, and coconut oil) are some well-known lipids. You may wonder about the differences between these terms. In general, fats are solid at room temperature, while oils are liquid at room temperature. In animals, fats are used for both insulation and long-term energy storage. They are used to insulate marine mammals from cold arctic waters and to protect our internal organs from damage. Instead of fats, plants use oils for long-term energy storage. In animals, the secretions of oil glands help waterproof skin, hair, and feathers.

Fats and Oils: Long-Term Energy Storage

Fats and oils contain two types of subunit molecules: glycerol and fatty acids (**Fig. 3.12**). **Glycerol** contains three —OH groups. The —OH groups are polar; therefore, glycerol is soluble in water. A **fatty acid** has a long chain of carbon atoms bonded only to hydrogen, with a carboxyl group at one end. A fat or an oil forms when the carboxyl portions of three fatty acids react with the —OH groups of glycerol. This is a dehydration synthesis reaction because, in addition to a fat molecule, three molecules of water result. Fats and oils are degraded during a hydrolysis reaction, in which water is added to the molecule. Because three long fatty acids are attached to the glycerol molecule, fats and oils are also called **triglycerides.** This structure can pack a lot of energy into one molecule. Thus, it is logical that fats and oils are the body's primary long-term energy storage molecules.

Fatty acids are the primary components of fats and oils. Most of the fatty acids in cells contain 16 or 18 carbon atoms per molecule, although smaller or

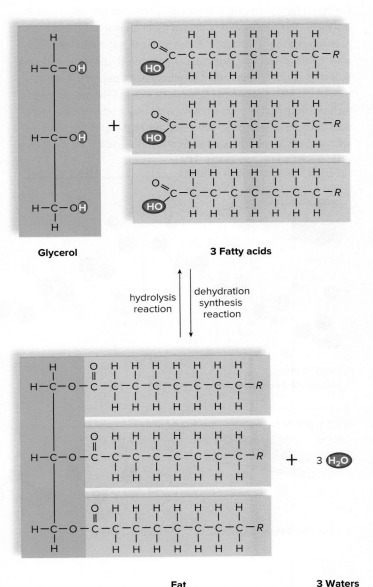

Figure 3.12 Synthesis and breakdown of fat.

Following a dehydration synthesis reaction, glycerol is bonded to three fatty acid molecules, and water is given off. Following a hydrolysis reaction, the bonds are broken due to the addition of water. R represents the remainder of the molecule, which in this case is a continuation of the hydrocarbon chain, composed of 16 or 18 carbons.

Connections: Health

What are omega-3 fatty acids?

Not all fats are bad. In fact, some of them are essential to our health. A special class of unsaturated fatty acids, the omega-3 fatty acids (also called n-3 fatty acids), are considered both essential and developmentally important nutrients. The name *omega-3* is derived from the location of the double bond in the carbon chain. The three important omega-3 fatty acids are linolenic acid (ALA), docosahexaenoic acid (DHA), and eicosapentaenoic acid (EPA). Omega-3 fatty acids are a major component of the fatty acids in the brain, and adequate amounts of them appear to be important in children and young adults. A diet that is rich in these fatty acids also offers protection against cardiovascular disease, and research is ongoing with regard to other health benefits. DHA may reduce the risk of Alzheimer disease. DHA and EPA may be manufactured from ALA in small amounts within our bodies. Some of the best sources of omega-3 fatty acids are cold-water fish, such as salmon and sardines. Flax oil, also called linseed oil, is an excellent plant-based source of omega-3 fatty acids.

larger ones are also found. Fatty acids are either saturated or unsaturated (**Fig. 3.13**). **Unsaturated fatty acids** have double bonds in the carbon chain wherever the number of hydrogens is less than two per carbon atom (Fig. 3.13*a*). **Saturated fatty acids** have no double bonds between the carbon atoms (Fig. 3.12*b*). The carbon chain is saturated, so to speak, with all the hydrogens it can hold. Saturation or unsaturation of a fatty acid determines its chemical and physical properties.

In general, oils are liquids at room temperature because they contain unsaturated fatty acids. Notice in Figure 3.13*a* that the double bond creates a bend in the fatty acid chain. Such kinks prevent close packing between the hydrocarbon chains and account for the fluidity of oils. On the other hand, butter, which contains mostly saturated fatty acids, is a solid at room temperature. The saturated fatty acid chains can pack together more tightly because they have no kinks (Fig. 3.13*b*).

A unsaturated fat may be called a **trans fats** if it contains a C=C bond with the hydrogen atoms located on opposite sides of the bond (Fig 3.13*c*). Trans fats are often formed during the processing of foods, such as margarine, baked goods, and fried foods, to make the product more solid. The label "partially-hydrogenated oils" typically indicates the presence of trans fats in a food product.

In our bodies, saturated fats and trans fats tend to stick together in the blood, forming plaque. The accumulation of plaque in the blood vessels causes a disease called atherosclerosis. Atherosclerosis contributes to high blood pressure and heart attacks. Unsaturated oils, particularly *monounsaturated* (one double bond) oils but also *polyunsaturated* (many double bonds) oils,

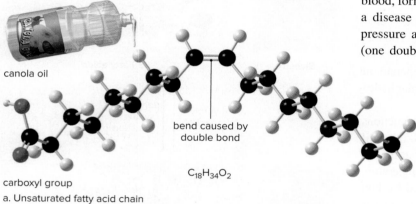

canola oil

bend caused by
double bond

$C_{18}H_{34}O_2$

carboxyl group

a. Unsaturated fatty acid chain

Figure 3.13 Fatty acids.

A fatty acid has a carboxyl group attached to a long hydrocarbon chain. **a.** If there is one double bond between adjacent carbon atoms in the chain, the fatty acid is monounsaturated. **b.** If there are no double bonds, the fatty acid is saturated. A diet high in saturated fats appears to contribute to diseases of the heart and blood vessels. **c.** In processed foods, the attached hydrogens in the double bonds may be in the trans configuration. If so, the fatty acid is a trans fatty acid. Trans fats are also linked to cardiovascular disease.

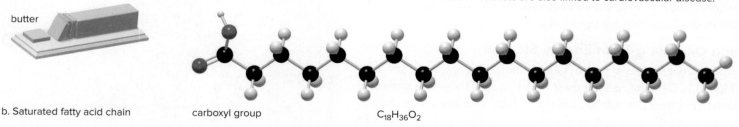

butter

carboxyl group

$C_{18}H_{36}O_2$

b. Saturated fatty acid chain

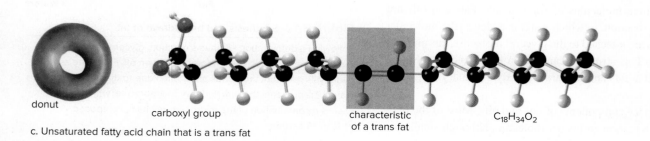

donut

carboxyl group

characteristic
of a trans fat

$C_{18}H_{34}O_2$

c. Unsaturated fatty acid chain that is a trans fat

have been found to be protective against atherosclerosis because they do not stick together as much in the blood. These healthy oils are found in abundance in olive oil, canola oil, and certain fish.

Phospholipids: Membrane Components

Phospholipids, as implied by their name, contain a phosphate functional group. Essentially, a phospholipid is constructed like a triglyceride, except that, in place of the third fatty acid attached to glycerol, there is a charged phosphate group. The phosphate group is usually bonded to another polar functional group, indicated by *R* in **Figure 3.14a**. Thus, one end of the molecule is hydrophilic and water soluble. This portion of the molecule is the polar head. The hydrocarbon chains of the fatty acids, or the nonpolar tails, are hydrophobic and not water soluble.

Because phospholipids have both hydrophilic (polar) heads and hydrophobic (nonpolar) tails, they tend to arrange themselves so that only the polar heads interact with the watery environment outside and the nonpolar tails crowd inward away from the water. Between two compartments of water, such as the outside and inside of a cell, phospholipids become a bilayer in which the polar heads project outward and the nonpolar tails project inward (Fig 3.14*b*). The bulk of the plasma membrane that surrounds cells consists of a fairly fluid phospholipid bilayer, as do all the other membranes in the cell. A plasma membrane is essential to the structure and function of a cell, and thus phospholipids are vital to humans and other organisms.

Steroids: Four Fused Rings

Steroids are lipids that possess a unique carbon skeleton made of four fused rings, shown in orange in **Figure 3.15**. Unlike other lipids, steroids do not contain fatty acids, but they are similar to other lipids because they are insoluble in water. Steroids are also very diverse. The types of steroids differ primarily in the types of functional groups attached to their carbon skeleton.

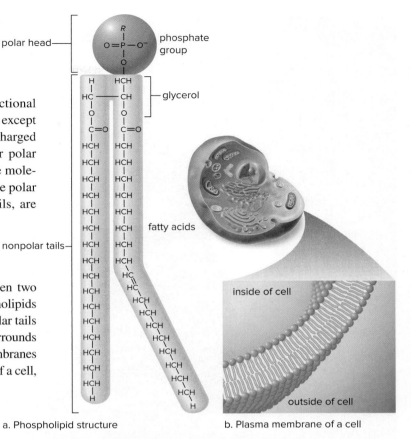

a. Phospholipid structure   b. Plasma membrane of a cell

Figure 3.14 **Phospholipids form membranes.**

a. Phospholipids are constructed like fats, except that, in place of the third fatty acid, they have a charged phosphate group. The hydrophilic (polar) head group is soluble in water, whereas the two hydrophobic (nonpolar) tail groups are not. **b.** This causes the molecules to arrange themselves as a bilayer in the plasma membrane that surrounds a cell.

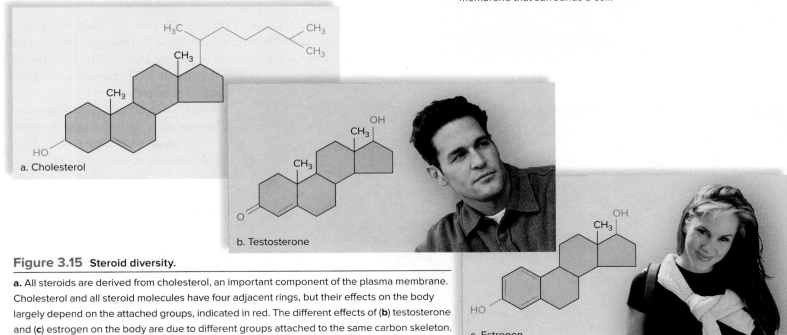

a. Cholesterol

b. Testosterone

c. Estrogen

Figure 3.15 **Steroid diversity.**

a. All steroids are derived from cholesterol, an important component of the plasma membrane. Cholesterol and all steroid molecules have four adjacent rings, but their effects on the body largely depend on the attached groups, indicated in red. The different effects of (**b**) testosterone and (**c**) estrogen on the body are due to different groups attached to the same carbon skeleton.

(photos): (b, c): © Corbis RF

Cholesterol (Fig. 3.15*a*) is a component of an animal cell's plasma membrane, and it is the precursor of other steroids, such as the sex hormones testosterone and estrogen. The male sex hormone, testosterone, is formed primarily in the testes, and the female sex hormone, estrogen, is formed primarily in the ovaries (Fig. 3.15*b,c*). Testosterone and estrogen differ only in the functional groups attached to the same carbon skeleton, yet they have a profound effect on the bodies and the sexuality of humans and other animals.

Anabolic steroids, such as synthetic testosterone, can be used to increase muscle mass. The result is usually unfortunate, however. The presence of the steroid in the body upsets the normal hormonal balance: The testes atrophy (shrink and weaken), and males may develop breasts; females tend to grow facial hair and lose hair on their head. Because steroid use gives athletes an unfair advantage and destroys their health—heart, kidney, liver, and psychological disorders are common—anabolic steroids are banned by professional athletic associations.

Proteins

Proteins are of primary importance in the structure and function of cells. Here are some of their many functions:

Support Some proteins are structural proteins. Examples include the protein in spiderwebs; keratin, the protein that contributes to hair and fingernails; and collagen, the protein that lends support to skin, ligaments, and tendons (**Fig. 3.16***a*).

Metabolism Many proteins are **enzymes.** They bring reactants together and thereby act as catalysts, speeding up chemical reactions in cells. Enzymes are specific for particular types of reactions and can function at body temperature.

Transport Channel and carrier proteins in the plasma membrane allow substances to enter and exit cells. Other proteins transport molecules in the blood of animals—for example, hemoglobin, found in red blood cells, is a complex protein that transports oxygen (Fig. 3.16*b*).

Defense Some proteins, called antibodies, combine with disease-causing agents to prevent those agents from destroying cells and causing diseases and disorders.

Regulation Hormones are regulatory proteins. They serve as intercellular messengers that influence the metabolism of cells. For example, the hormone insulin regulates the concentration of glucose in the blood, while human growth hormone (hGH) contributes to determining the height of an individual.

Motion The contractile proteins actin and myosin allow parts of cells to move and cause muscles to contract (Fig. 3.16*c*). Muscle contraction enables animals to move from place to place and substances to move through the body. It also regulates body temperature.

a. Structural proteins

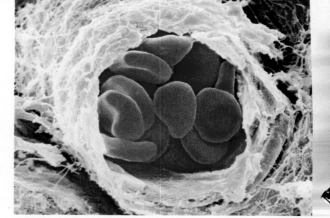

b. Transport proteins

1,740× c. Contractile proteins

Figure 3.16 Types of protein.

a. The protein in hair, fingernails, and spiderwebs (keratin) is a structural protein, as is collagen. **b.** Hemoglobin, a major protein in red blood cells, is involved in transporting oxygen. **c.** The contractile proteins actin and myosin cause muscles to move.

(a): (woman): © Reuters/Corbis; (web): © Chris Cheadle/Getty Images; (b): © P. Motta & S. Correr/Science Source; (c): © Duomo/Corbis

The structures and functions of cells differ according to the types of protein they contain. Muscle cells contain actin and myosin; red blood cells contain hemoglobin; support cells produce the collagen they secrete. While proteins are often viewed as energy molecules, we will see in Section 7.5 that they are not a preferred energy source for the cell.

Amino Acids: Monomers of Proteins

Proteins are polymers, and their monomers are called **amino acids.** Amino acids have a unique carbon skeleton, in which a central carbon atom bonds to a hydrogen atom, two functional groups, and a variable side chain, or *R* group (**Fig. 3.17a**). The name *amino acid* is appropriate because one of the two functional groups is an —NH_2 (amino group) and the other is a —COOH (an acid, or carboxyl, group, see Fig. 3.3).

There are 20 different amino acids, which differ by their particular *R* group. The *R* groups range in complexity from a single hydrogen atom to a complicated ring structure. The unique chemical properties of an amino acid depend on the chemical properties of the *R* group. For example, some *R* groups are polar, some are charged, and some are hydrophobic. The amino acid cysteine, for instance, has an *R* group that ends with a sulfhydryl (—SH) group, which can connect one chain of amino acids to another by a disulfide bond, —S—S—. Four amino acids commonly found in cells are shown in Figure 3.17b.

Peptides

A **peptide** is formed when two amino acids are joined by a dehydration synthesis reaction between the carboxyl group of one and the amino group of another (**Fig. 3.18**).

The resulting covalent bond between two amino acids is called a **peptide bond.** The atoms associated with the peptide bond share the electrons unevenly because oxygen is more electronegative than nitrogen.

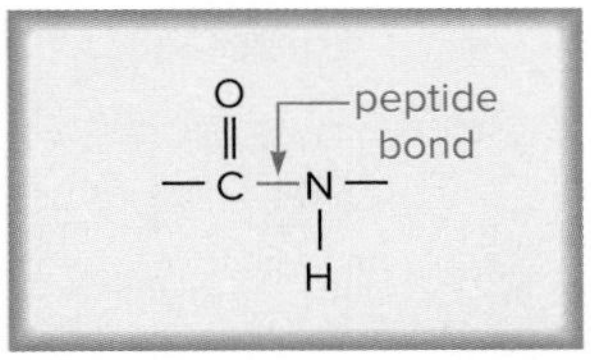

Therefore, the peptide bond is polar, and hydrogen bonding is possible between the C=O of one amino acid and the N—H of another amino acid in a polypeptide.

A **polypeptide** is a chain of many amino acids joined by peptide bonds. As we will see next, proteins are polypeptide chains that have folded into complex shapes. A protein may contain one or more polypeptide chains. While some proteins are small, others are composed of a very large number of amino acids. Ribonuclease, a small protein that breaks down RNA, contains barely more than 100 amino acids. But some proteins are very large, such as titin, which contains over 33,000 amino acids! Titin is an integral part of the structure of your muscles; without it, your muscles would not function properly.

The amino acid sequence determines a protein's final three-dimensional shape, and thus its function. Each polypeptide has its own normal sequence. Proteins that have an abnormal sequence of amino acids often have the wrong shape and cannot function properly.

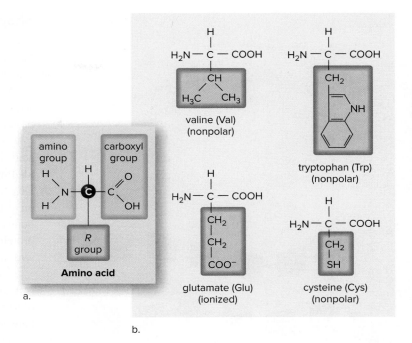

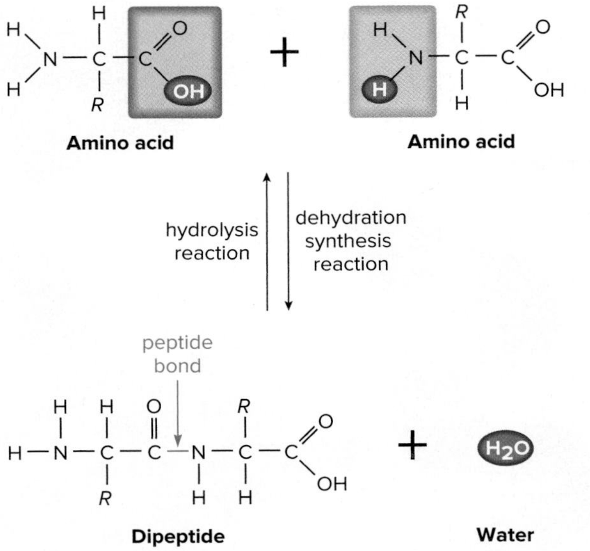

Figure 3.17 **Amino acids.**

a. Structure of an amino acid. **b.** Proteins contain arrangements of 20 different kinds of amino acids, four of which are shown here. Amino acids differ by the particular *R* group (blue) attached to the central carbon. Some *R* groups are nonpolar and hydrophobic, some are polar and hydrophilic, and some are ionized and hydrophilic.

Figure 3.18 **Synthesis and degradation of peptide.**

Following a dehydration synthesis reaction, a peptide bond joins two amino acids, and water is given off. Following a hydrolysis reaction, the bond is broken due to the addition of water.

Shape of Proteins

All proteins have multiple levels of structure. These levels are called the primary, secondary, tertiary, and quaternary structures (**Fig. 3.19**). A protein's sequence of amino acids is called its *primary structure*. Consider that an almost infinite number of words can be constructed by varying the number and sequence of the 26 letters in our alphabet. In the same way, many different proteins can result by varying the number and sequence of just 20 amino acids.

The *secondary structure* of a protein results when portions of the amino acid chain take on a certain orientation in space, depending on the number and identity of the amino acids present in the chain. Portions of the polypeptide can have the spiral shape called an alpha helix, or a polypeptide chain can turn back on itself, like an accordion, a shape called a beta pleated sheet. Hydrogen bonds between nearby peptide bonds maintain the secondary structure of a protein. Any polypeptide may contain one or more secondary structure regions within the same chain.

Proteins also have a *tertiary structure*. The tertiary structure of a protein is its overall three-dimensional shape that results from the folding and twisting of its secondary structure. The tertiary structure is held in place by interactions between the *R* groups of amino acids making up the helices and beta pleated sheets within the polypeptide. Several types of interactions are possible. For

Figure 3.19 Levels of protein organization.

All proteins have a primary structure. Examples of secondary structure include helices (e.g., keratin, collagen) or pleated sheets (e.g., silk). Globular proteins always have a tertiary structure, and most have a quaternary structure (e.g., hemoglobin, enzymes).

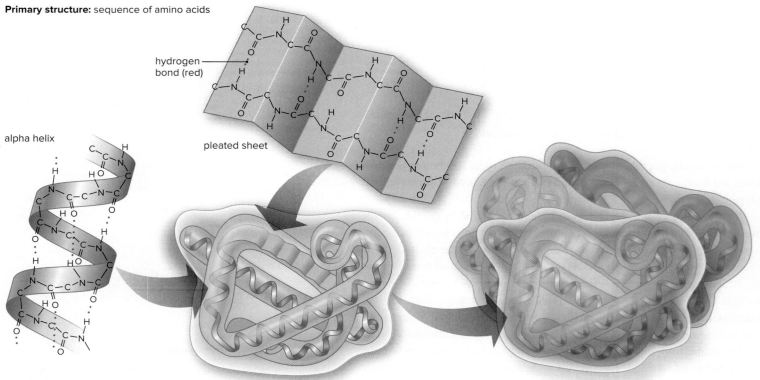

Primary structure: sequence of amino acids

Secondary structure: alpha helix and pleated sheet

Tertiary structure: globular shape

Quaternary structure: more than one polypeptide

example, ionic bonds, hydrogen bonds, and even covalent bonds can occur between R groups. Also, a tight packing of side chains can occur when hydrophobic R groups come together to avoid contact with water.

The tertiary structure of a protein determines its function, and this structure can be affected by the environmental conditions, such as pH and temperature. Acids and bases may interrupt interactions between R groups and affect a protein's structure. Likewise, a rise in temperature can disrupt the shape of an enzyme. For example, frying an egg disrupts the tertiary structure of the proteins in it, causing the protein albumin to become solid and change color. The protein has been **denatured** (broken down and inactivated) and has lost its function.

Connections: Scientific Inquiry

How do perms and relaxers work on hair?

Hair, composed of proteins, can be altered chemically to be curly or straight. For instance, a perm contains chemicals that break and re-form the bonds between the functional groups in the amino acids to form disulfide bonds, resulting in spirals and making curls. A relaxer will break the bonds between disulfide bond regions to straighten the proteins. Neither are permanent, because they do not alter the genes controlling the original shape of the proteins that make up the hair.

© Amos Morgan/Getty RF

Some proteins, such as hemoglobin and insulin, have a *quaternary structure* because they contain more than one polypeptide chain. Each polypeptide chain in such a protein has its own primary, secondary, and tertiary structures. The quaternary structure is determined by how the individual polypeptide chains interact. The quaternary structure can also affect a protein's function. If hemoglobin's quaternary structure is disrupted, it can no longer carry oxygen in the blood.

The overall shapes of many proteins may be classified as fibrous or globular. Fibrous proteins adopt a rodlike structure. *Keratin,* the fibrous protein in hair, fingernails, horns, reptilian scales, and feathers, has many helical regions. *Collagen,* the protein that gives shape to the skin, tendons, ligaments, cartilage, and bones of animals, also contains many helical secondary structures and adopts a rodlike shape. Thus, most fibrous proteins have structural roles. Globular proteins have a rounded or irregular, three-dimensional tertiary structure. Enzymes are globular proteins, as is hemoglobin.

Nucleic Acids

DNA (*deoxyribonucleic acid*) and **RNA (*ribonucleic acid*)** are the **nucleic acids** found in cells. Early investigators called them nucleic acids because they were first detected in the nucleus. DNA acts as the location within the cell where the genetic information is stored. Each DNA molecule contains many genes, which specify the sequence of the amino acids in proteins. RNA is the molecule that aids in transcribing and translating DNA into proteins.

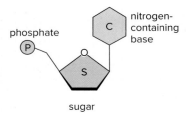

a. Nucleotide

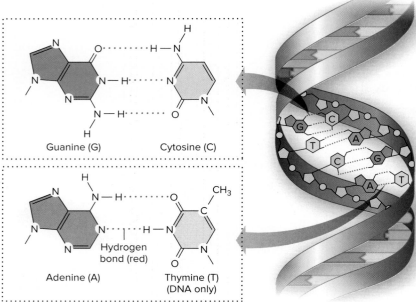

b. DNA structure with base pairs: G with C and A with T

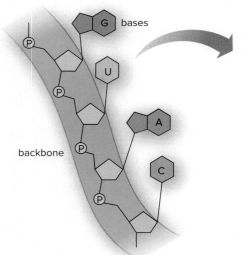

c. RNA structure with bases G, U, A, C

Figure 3.20 DNA and RNA structure.

a. Structure of a nucleotide. **b.** Structure of DNA and its bases.
c. Structure of RNA, in which uracil replaces thymine.

Nucleic acids are polymers in which the monomer is called a **nucleotide.** All nucleotides, whether they are in DNA or RNA, have three parts: a phosphate (a —PO_4 functional group) , a 5-carbon sugar, and a nitrogen-containing base (**Fig. 3.20***a*). The sugar is deoxyribose in DNA and ribose in RNA, which accounts for their names. Deoxyribose has one less oxygen than does ribose. Each nucleotide of DNA contains one of four nitrogen-containing bases: adenine (A), guanine (G), cytosine (C), or thymine (T).

As you can see in Figure 3.20*b*, DNA is a double helix, meaning that two strands spiral around one another. In each strand, the backbone of the molecule is composed of phosphates bonded to sugars, and the bases project to the inside. Interestingly, the base guanine (G) is always paired with cytosine (C), and the base adenine (A) is always paired with thymine (T). This is called *complementary base pairing.* Complementary base pairing holds the two strands together and is very important when DNA makes a copy of itself, a process called replication.

RNA differs from DNA not only by its sugar but also because it uses the base uracil (U) instead of thymine. Whereas DNA is double-stranded, RNA is single-stranded (Fig. 3.20*c*). Complementary base pairing also allows DNA to pass genetic information to RNA. The information is stored in the sequence of bases. DNA has a triplet code called codons, and every three bases stands for one of the 20 amino acids in cells. Once you know the sequence of bases in a gene, you know the sequence of amino acids in a polypeptide. As a result of the Human Genome Project, we now have a complete sequence of each of the 3.2 billion bases, and roughly 23,000 genes, in humans. As we will see throughout this text, this information has led to some important insights not only into our evolutionary past but also into the development of treatments for many diseases.

ATP: An Energy Molecule

In addition to being one of the subunits of nucleic acids, the nucleotide adenine has a derivative with a metabolic function that is very important to most cells. When adenosine (adenine plus ribose) is modified by the addition of three phosphate groups, it becomes adenosine triphosphate (ATP), which acts as an energy carrier in cells. ATP will be discussed in more detail in Section 5.2.

Relationship Between Proteins and Nucleic Acids

We have learned that the functional group of an amino acid determines its behavior and that the order of the amino acids within a polypeptide determines its shape. The shape of a protein determines its function. The structure and function of cells are determined by the types of proteins they contain. The same is true for organisms—that is, they differ with regard to their proteins. We have also learned that DNA, the genetic material, bears instructions for the sequence of amino acids in polypeptides. The proteins of organisms differ because their genes differ.

Sometimes very small changes in a gene cause large changes in the protein encoded by the gene, resulting in illness. For example, in sickle-cell disease, the affected individual's red blood cells are sickle shaped because, in one location on the protein, the amino acid valine (Val) appears where the amino acid glutamate (Glu) should be (**Fig. 3.21**). This substitution makes red blood cells lose their normally round, flexible shape and become hard and jagged. When these abnormal red blood cells go through the small blood vessels, they may clog the flow of blood. This condition can cause pain, organ damage, and a low red blood cell count (called anemia), and it can be fatal if untreated. What's the root of the problem? The affected individual inherited a faulty DNA code for an amino acid in just one of hemoglobin's polypeptides.

CONNECTING THE CONCEPTS

3.2   Organic molecules, such as carbohydrates, lipids, proteins, and nucleic acids, have a wide variety of biological functions.

Check Your Progress 3.2

1. Describe the different classes of carbohydrates, and give an example of a structural and energy carbohydrate.
2. List the classes of lipids and provide a function for each.
3. Summarize the roles of proteins in the body.
4. Describe the levels of protein structure.
5. Explain how nucleic acids differ in structure from other biological molecules.
6. Describe the relationship between nucleic acids and proteins.

Figure 3.21 Sickle-cell disease.

One of the amino acid chains in hemoglobin is 146 amino acids long. Sickle-cell disease, characterized by sickled red blood cells, results when valine occurs instead of glutamate at the sixth amino acid position.

(photos): © Eye of Science/Science Source

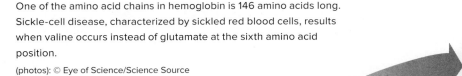

H_2N— Val — His — Leu — Thr — Pro — Glu — Glu —

Normal hemoglobin

H_2N— Val — His — Leu — Thr — Pro — Val — Glu —

Sickle-cell hemoglobin

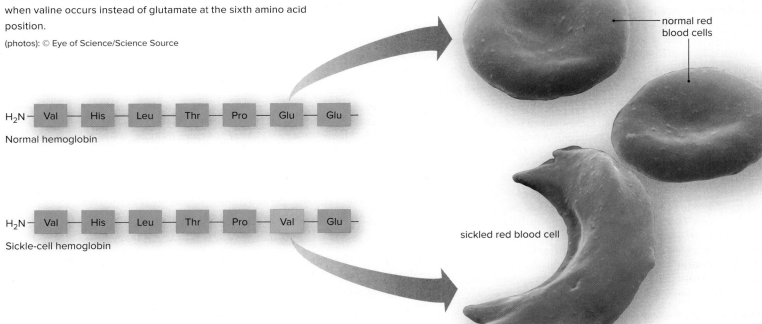

normal red blood cells

sickled red blood cell

ISMARTBOOK® Maximize your study time with McGraw-Hill SmartBook®, the first adaptive textbook.

SUMMARIZE

The major classes of organic molecules, including carbohydrates, lipids, proteins, and nucleic acids, are essential for life.

3.1 Carbon and hydrogen are the basis of the organic molecules found in living organisms.

3.2 Organic molecules, such as carbohydrates, lipids, proteins, and nucleic acids, have a wide variety of biological functions.

3.1 Organic Molecules

In order to be **organic,** a molecule must contain carbon and hydrogen. The chemistry of carbon contributes to the diversity of organic molecules. Functional groups determine the chemical reactivity of organic molecules. Some functional groups are nonpolar (**hydrophobic**), and others are polar (**hydrophilic**). The organic molecules in cells are called **biological molecules.**

Isomers are molecules with a similar composition of atoms, but different structures.

3.2 The Biological Molecules of Cells

Biological molecules are synthesized by **dehydration synthesis reactions** and degraded by **hydrolysis reactions.** Living organisms are composed of four types of biological molecules: carbohydrates, lipids, proteins, and nucleic acids. Polysaccharides, proteins, and nucleic acids are **polymers** constructed when their particular type of **monomer** forms long chains. Lipids are varied and do not have a particular monomer. The characteristics of these molecules are summarized in **Table 3.1.**

Carbohydrates

Carbohydrates are primarily used as energy molecules, although some have structural characteristics. **Glucose** is a **monosaccharide** that serves as blood sugar and as a monomer of starch, glycogen, and cellulose. Its isomers are fructose and galactose. **Ribose** and **deoxyribose** are monosaccharides found in nucleic acids.

Sucrose is a **disaccharide** (glucose and fructose), which we know as table sugar. Other disaccharides are maltose and lactose.

Polysaccharides (all polymers of glucose) include **starch,** which stores energy in plants; **glycogen,** which stores energy in animals; and cellulose. **Cellulose** makes up the structure of plant cell walls. **Chitin** is a polysaccharide that contains amino groups.

Lipids

Lipids are hydrophobic molecules that often serve as long-term energy storage molecules. **Fats** and **oils,** which are composed of **glycerol** and **fatty acids,** are called **triglycerides.** Triglycerides made of **saturated fatty acids** (having no double bonds) are solids and are called fats. Triglycerides composed of **unsaturated fatty acids** (having double bonds) are liquids and are called oils. **Trans fats** are unsaturated fats with a unique form of chemical bond in the fatty acid chain.

Phospholipids have the same structure as triglycerides, except that a group containing phosphate takes the place of one fatty acid. Phospholipids make up the plasma membrane, as well as other cellular membranes.

Steroids are lipids with a unique structure composed of four fused hydrocarbon rings. **Cholesterol,** a steroid, is a component of the plasma membrane. The sex hormones testosterone and estrogen are steroids.

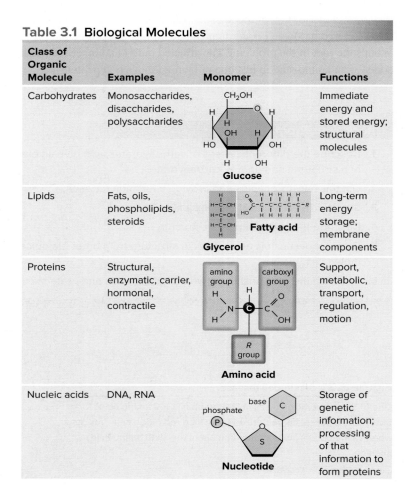

Table 3.1 Biological Molecules

Class of Organic Molecule	Examples	Monomer	Functions
Carbohydrates	Monosaccharides, disaccharides, polysaccharides	Glucose	Immediate energy and stored energy; structural molecules
Lipids	Fats, oils, phospholipids, steroids	Glycerol / Fatty acid	Long-term energy storage; membrane components
Proteins	Structural, enzymatic, carrier, hormonal, contractile	Amino acid	Support, metabolic, transport, regulation, motion
Nucleic acids	DNA, RNA	Nucleotide	Storage of genetic information; processing of that information to form proteins

Proteins

Proteins are polymers of **amino acids.** A **peptide** is composed of two amino acids linked by a **peptide bond. Polypeptides** contain many amino acids. Proteins have a wide variety of functions, including acting as **enzymes** to accelerate chemical reactions.

A protein has several levels of structure. When a protein loses this structure, it is said to be **denatured.** The levels of structure include

- Primary structure is the primary sequence of amino acids.
- Secondary structure is a helix or pleated sheet.
- Tertiary structure forms due to folding and twisting of the secondary structure.
- Quaternary structure occurs when a protein has more than one polypeptide.

Nucleic Acids

Cells and organisms differ because of their proteins, which are coded for by genes composed of nucleic acids. **Nucleic acids** are polymers of **nucleotides.**

DNA (deoxyribonucleic acid) contains the sugar deoxyribose and is the chemical that makes up our genes. DNA is double-stranded and shaped like a double helix. The two strands of DNA are joined by complementary base pairing: Adenine (A) pairs with thymine (T), and guanine (G) pairs with cytosine (C).

RNA (ribonucleic acid) serves as a helper to DNA during protein synthesis. Its sugar is ribose, and it contains the base uracil in place of thymine. RNA is single-stranded and thus does not form a double helix. **Table 3.2** compares the structures of DNA and RNA. A modified nucleic acid, ATP, is used as a main energy carrier for most cells.

Table 3.2 DNA Structure Compared with RNA Structure

	DNA	RNA
Sugar	Deoxyribose	Ribose
Bases	Adenine, guanine, thymine, cytosine	Adenine, guanine, uracil, cytosine
Strands	Double-stranded with base pairing	Single-stranded
Helix	Yes	No

ASSESS

Testing Yourself

Choose the best answer for each question.

3.1 Organic Molecules

1. Which of the following is an organic molecule?
   - **a.** CO_2
   - **b.** H_2O
   - **c.** $C_6H_{12}O_6$
   - **d.** O_2
   - **e.** More than one of these are correct.

2. Carbon chains can vary in
   - **a.** length.
   - **b.** number of double bonds.
   - **c.** branching pattern.
   - **d.** All of these are correct.

3. Organic molecules containing carboxyl groups are
   - **a.** nonpolar.
   - **b.** acidic.
   - **c.** basic.
   - **d.** More than one of these are correct.

4. Which of the following determines the chemical reactivity of biological molecules?
   - **a.** isomers
   - **b.** functional groups
   - **c.** number of carbon atoms
   - **d.** length of the hydrocarbon chain

3.2 The Biological Molecules of Cells

5. An amino acid is to a protein as a _____ is to a nucleic acid.
   - **a.** nucleotide
   - **c.** monosaccharide
   - **c.** fatty acid
   - **d.** triglyceride

6. Biomolecules are polymers that are formed when _____ are joined by a _____ reaction.
   - **a.** monomers, dehydration synthesis
   - **b.** multimers, dehydration synthesis
   - **c.** subunits, reduction
   - **d.** monomers, hydrolysis

7. Which of the following is a monosaccharide?
   - **a.** glucose
   - **b.** cellulose
   - **c.** lactose
   - **d.** sucrose

For questions 8–15, match the items to those in the key. Answers can be used more than once.

Key:

- **a.** carbohydrate
- **b.** lipid
- **c.** protein
- **d.** nucleic acid

8. Cellulose, the major component of plant cell walls
9. Keratin, found in hair, fingernails, horns, and feathers
10. Steroids such as cholesterol and sex hormones
11. Composed of nucleotides
12. Insoluble in water due to hydrocarbon chains
13. Sometimes undergoes complementary base pairing
14. May contain pleated sheets and helices
15. May be a ring of six carbon atoms attached to hydroxyl groups
16. A triglyceride contains
    - **a.** glycerol and three fatty acids.
    - **b.** glycerol and two fatty acids.
    - **c.** protein and three fatty acids.
    - **d.** a fatty acid and three sugars.
17. Variations in three-dimensional shapes among proteins are due to bonding between the
    - **a.** amino groups.
    - **b.** ion groups.
    - **c.** R groups.
    - **d.** H atoms.
18. The polysaccharide found in plant cell walls is
    - **a.** glucose.
    - **b.** starch.
    - **c.** maltose.
    - **d.** cellulose.

ENGAGE

Thinking Critically

1. The chapter opener discusses why not all cholesterol is bad—yet many physicians refer to lipoproteins as being "good" and "bad." Why do you think this is done? Why is it not accurate to describe any organic molecule as "good" or "bad"?

2. In order to understand the relationship between enzyme structure and function, researchers often study mutations that swap one amino acid for another. In one enzyme, function is retained if a particular amino acid is replaced by one that has a nonpolar R group, but function is lost if the amino acid is replaced by one with a polar R group. Why might that be?

3. Scientists have observed that, while two species might have the same protein, the more distantly related they are, the more likely the sequence of the amino acids has changed. What might account for this observation? Why would this be an important thing to study?

4

Inside the Cell

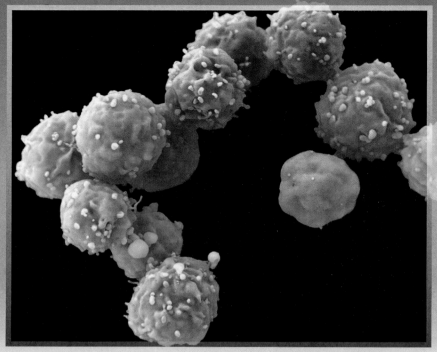

© Steve Gschmeissner/Science Source

OUTLINE

4.1 Cells Under the Microscope   57

4.2 The Plasma Membrane   59

4.3 The Two Main Types of Cells   62

4.4 Eukaryotic Cells   64

4.5 Outside the Eukaryotic Cell   74

BEFORE YOU BEGIN

Before beginning this chapter, take a few moments to review the following discussions.

Section 1.1 What are the basic characteristics of life?

Section 2.2 What properties of water enable it to support life?

Section 3.2 What are the basic roles of carbohydrates, fats, proteins, and nucleic acids in living organisms?

The Diversity and Unity of Cells

One of the new frontiers in medicine is the use of human stem cells to regenerate tissue that has been damaged by a disease. Recently, researchers have shown that some forms of a vision disorder called macular degeneration, type 2 diabetes, and even Alzheimer disease may be treated by using stem cells.

But what is a stem cell? Unlike most cells of our bodies, which are specialized and can divide only a limited number of times, stem cells are unspecialized and have the potential to divide almost indefinitely. As a stem cell divides, it can develop into a wide variety of cell types. This is possible because a stem cell contains all of the genetic information and internal structures to become almost any other type of cell in the body; it just hasn't begun the process of specialization. In other words, a stem cell is a cell that doesn't know what it wants to be when it grows up, but once it is placed in a tissue, such as the eye or pancreas, it starts to assume the functions of the neighboring cells. Often, it can compensate for nearby dysfunctional cells and help correct a disorder or disease.

Cells are the fundamental units of life. Although the diversity of organisms is incredible, the cells of all organisms share many similarities. In this chapter, we will explore the two main evolutionary classes of cells and see how they are similar as well as different. This overview of cell structure will help you understand how biological organisms, such as yourself, function.

As you read through this chapter, think about the following questions:

1. What are some of the differences between a eukaryotic stem cell and a prokaryotic bacterial cell?
2. What is the purpose of compartmentalization in a eukaryotic cell?

4.1 Cells Under the Microscope

Learning Outcomes

Upon completion of this section, you should be able to

1. Explain why microscopes are needed to see most cells.
2. Summarize the relationship between a cell's surface-area-to-volume ratio and its size.

A **cell** is the fundamental unit of life. In fact, all life is made of cells. However, cells are extremely diverse in their shape and function. Our own bodies are composed of several hundred cell types, and each type is present in billions of copies. For example, there are nerve cells to conduct information, muscle cells that allow movement, gland cells that secrete hormones, and bone cells to provide shape. As we will see, the structure of each of these is specialized to perform its particular function.

While cells are complex, they are tiny—most require a microscope to be seen (**Fig. 4.1**). The light microscope, invented in the seventeenth century, allows us to see cells, but not much of their complexity. That's because the properties of light limit the amount of detail a light microscope can reveal. Electron microscopes, invented in the 1930s, overcome this limit by using beams of electrons instead of beams of light as their source of illumination. An electron

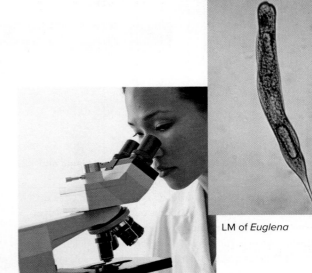

LM of *Euglena*   470×

TEM of a cell showing numerous organelles

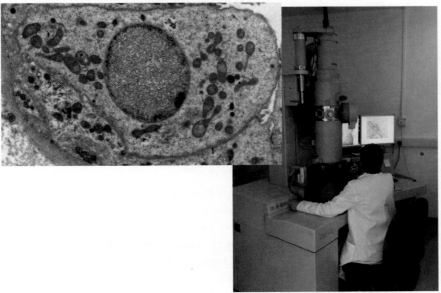

Scientist using a light microscope

Scientist using an electron microscope

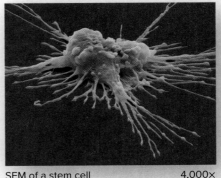

SEM of a stem cell   4,000×

Figure 4.1 Using microscopes to see cells.

Scientists use many types of microscopes to view cells, including the light microscope, the transmission electron microscope, and the scanning electron microscope. The light microscope and the transmission electron microscope reveal the insides of cells, while the scanning electron microscope shows three-dimensional surface features. Pictures resulting from the use of the light microscope are called light micrographs (LM), and those that result from the use of an electron microscope are called either transmission electron micrographs (TEM) or scanning electron micrographs (SEM).

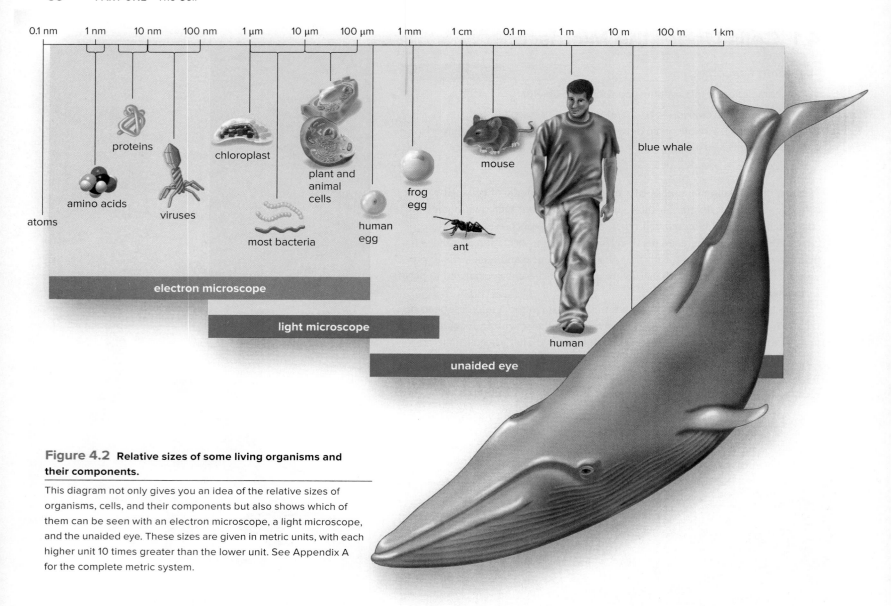

Figure 4.2 Relative sizes of some living organisms and their components.

This diagram not only gives you an idea of the relative sizes of organisms, cells, and their components but also shows which of them can be seen with an electron microscope, a light microscope, and the unaided eye. These sizes are given in metric units, with each higher unit 10 times greater than the lower unit. See Appendix A for the complete metric system.

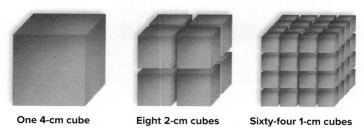

One 4-cm cube   Eight 2-cm cubes   Sixty-four 1-cm cubes

Figure 4.3 Surface-area-to-volume relationships.

While all three groups of cubes have the same volume, the cubes on the right have four times the surface area.

microscope enables us to see the surface features and fine details of cells, and even some of the larger molecules within them. A newer type of microscope, called the scanning probe microscope, physically scans the surface of the specimen and is able to distinguish objects around a nanometer in size!

Figure 4.2 compares the visual ranges of the electron microscope, the light microscope, and the unaided eye.

Why are cells so small? Cells need to be able to rapidly exchange materials with the external environment. Therefore, a cell needs a surface area large enough to allow efficient movement of nutrients into the cell and waste materials out of the cell. Let's use a simple cube as an example (**Fig. 4.3**). Cutting a larger cube into smaller cubes changes the **surface-area-to-volume ratio.** The higher the ratio of surface area to internal volume, the faster the exchange of materials with the environment. Therefore, small cells, not large cells, are most likely to have an adequate surface area for exchanging wastes and nutrients.

Many cells also possess adaptations that increase the surface-area-to-volume ratio. For example, cells in your small intestine have tiny, fingerlike projections (microvilli), which increase the surface area available for absorbing nutrients. Without this adaptation, your small intestine would have to be hundreds of meters long!

Connections: Scientific Inquiry

What is the largest known cell?

While the ostrich egg is frequently used as an example of the largest single cell, in fact, the record is currently held by the nerve cells of two of the most impressive animals on the planet, the giant squid (*Architeuthis dux*) and the colossal squid (*Mesonychoteuthis hamiltoni*). These deep-sea-living relatives of the octopus can grow to over 14 meters (46 feet) in length and weigh as much as a ton. The nerve cells of these squids span the length of their bodies, making them the longest and largest cells currently identified. Scientists often use these nerve cells as models for studying how the nervous system transmits information over long distances.

© Mark Mitchell/REX/
Newscom

CONNECTING THE CONCEPTS
4.1 The microscopic size of a cell maximizes its surface-area-to-volume ratio.

Check Your Progress 4.1

1. Explain why microscopes are needed to view most cells.
2. Identify the types of microscope needed to view the following: frog egg, animal cell, amino acid, chloroplast, plant cell.
3. Explain why a large surface-area-to-volume ratio is needed for the proper functioning of cells.

4.2 The Plasma Membrane

Learning Outcomes

Upon completion of this section, you should be able to

1. Recognize the key components of the cell plasma membrane.
2. Explain how the plasma membrane regulates the passage of molecules into and out of the cell.
3. Describe the diverse functions of the proteins embedded in the plasma membrane.

All cells have an outer membrane called the **plasma membrane,** which acts as the boundary between the outside and inside of a cell. The integrity and function of the plasma membrane are vital to a cell because this membrane acts much like a gatekeeper, regulating the passage of molecules and ions into and out of the cell.

The structure of the plasma membrane plays an important role in its function. In all cells, the plasma membrane consists of a phospholipid bilayer

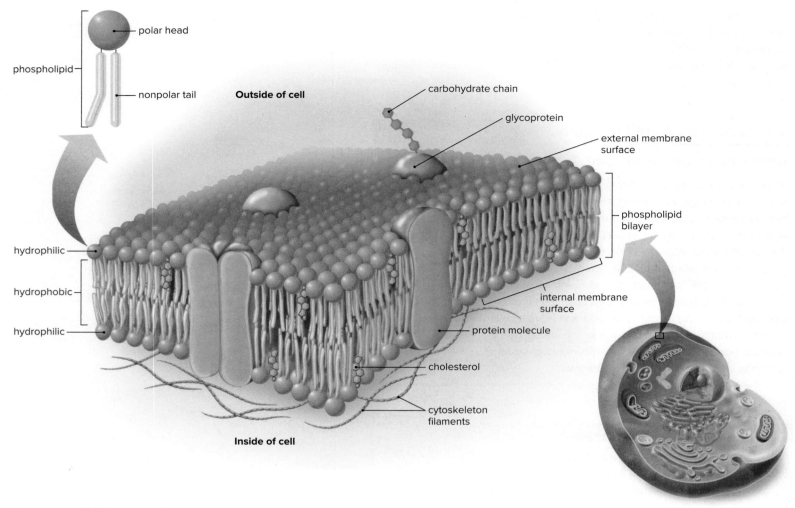

polar head

phospholipid

nonpolar tail

Outside of cell

carbohydrate chain

glycoprotein

external membrane surface

phospholipid bilayer

hydrophilic

hydrophobic

hydrophilic

internal membrane surface

protein molecule

cholesterol

cytoskeleton filaments

Inside of cell

Figure 4.4 A model of the plasma membrane.

The plasma membrane is composed of a phospholipid bilayer. The polar heads of the phospholipids are at the surfaces of the membrane; the nonpolar tails make up the interior of the membrane. Proteins embedded in the membrane have various functions (see Fig. 4.5).

with numerous proteins embedded in it (**Fig. 4.4**). In the bilayer, the polar (hydrophilic) heads of the phospholipids are oriented in two directions. In the outer layer of the membrane, the phospholipid heads face toward the external environment. On the interior layer of the membrane, the heads of phospholipid molecules are directed toward the interior cytoplasm of the cell. The nonpolar (hydrophobic) tails of the phospholipids point toward each other, in the space between the layers, where there is no water. Cholesterol molecules, present in the plasma membrane of some cells, lend support to the membrane, giving it the general consistency of olive oil.

The structure of the plasma membrane (Fig. 4.4) is often referred to as the *fluid-mosaic model,* since the protein molecules embedded in the membrane have a pattern (a mosaic) within the fluid phospholipid bilayer. The actual pattern of proteins varies according to the type of cell, but it may also vary within the membrane of an individual cell over time. For example, the plasma membrane of a red blood cell contains over 50 different types of proteins, and they can vary in their location on the surface, forming a mosaic pattern.

Short chains of sugars are attached to the outer surface of some of these proteins, forming *glycoproteins.* The sugar chain helps a protein perform its particular function. For example, some glycoproteins are involved in establishing the identity of the cell. They often play an important role in the immune response against disease-causing agents entering the body.

Functions of Membrane Proteins

The proteins embedded in the plasma membrane have a variety of functions.

Channel Proteins

Channel proteins form a tunnel across the entire membrane, allowing only one or a few types of specific molecules to move readily through the membrane (**Fig. 4.5***a*). For example, *aquaporins* are channel proteins that allow water to enter or exit a cell. Without aquaporins in the kidneys, your body would soon dehydrate.

Transport Proteins

Transport proteins are also involved in the passage of molecules and ions through the membrane. They often combine with a substance and help it move across the membrane, with an input of energy (Fig. 4.5*b*). For example, a transport protein conveys sodium and potassium ions across a nerve cell membrane. Without this transport protein, nerve conduction would be impossible.

Cell Recognition Proteins

Cell recognition proteins are glycoproteins (Fig. 4.5*c*). Among other functions, these proteins enable our bodies to distinguish between our own cells and the cells of other organisms. Without this distinction, pathogens would be able to freely invade the body.

Receptor Proteins

A receptor protein has a shape that allows a specific molecule, called a signal molecule, to bind to it (Fig. 4.5*d*). The binding of a signal molecule causes the receptor protein to change its shape and thereby bring about a cellular response. For example, the hormone insulin binds to a receptor protein in liver cells, and thereafter these cells store glucose.

Enzymatic Proteins

Some plasma membrane proteins are enzymatic proteins that directly participate in metabolic reactions (Fig. 4.5*e*). Without enzymes, some of which are attached to the various membranes of a cell, the cell would never be able to perform the degradative and synthetic reactions that are important to its function.

Junction Proteins

Proteins are also involved in forming various types of junctions (Fig. 4.5*f*) between cells. The junctions assist cell-to-cell adhesion and communication. The adhesion junctions in your bladder keep the cells bound together as the bladder swells with urine.

CONNECTING THE CONCEPTS

4.2 All cells have a plasma membrane that consists of phospholipids and embedded proteins.

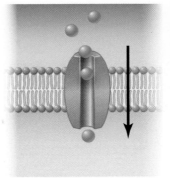

a. Channel protein

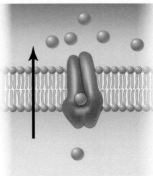

b. Transport protein

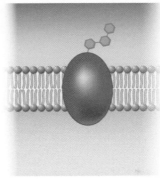

c. Cell recognition protein

d. Receptor protein

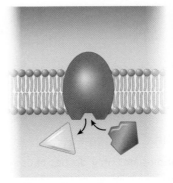

e. Enzymatic protein

f. Junction proteins

Figure 4.5 **Membrane protein diversity.**

Each of these types of proteins provides a function for the cell.

Check Your Progress 4.2

1. Identify the basic components that make up the structure of the plasma membrane.
2. Explain why the plasma membrane is described as a fluid-mosaic model.
3. Distinguish among the types of membrane proteins by function.

4.3 The Two Main Types of Cells

Learning Outcomes

Upon completion of this section, you should be able to

1. Identify the characteristics common to all cells.
2. Distinguish between prokaryotic and eukaryotic cells.
3. Identify the structures of a prokaryotic cell.

The idea that all organisms are composed of cells and that cells come only from preexisting cells are the two central tenets of the **cell theory.** We can identify some characteristics that are common to all cells:

- A plasma membrane made of phospholipids that regulates the movement of materials into and out of the cell
- A semifluid interior, called the cytoplasm, where chemical reactions occur
- Genetic material (DNA) that provides the information needed for cellular activities, including growth and reproduction

Cells are divided into two main types, according to the way their genetic material is organized. The **prokaryotic cells** (Greek; *pro,* before, and *karyon,* kernel or nucleus) lack a membrane-bound nucleus. Their DNA is located in a region of the cytoplasm called the nucleoid. The other type of cells, **eukaryotic cells** (Greek; *eu,* true), have a nucleus that houses their DNA. We will explore eukaryotic cell structure in Section 4.4.

Prokaryotic Cells

Around 3.5 billion years ago, the first cells to appear on Earth were prokaryotes. Today, prokaryotes are classified as being part of either domain Archaea or domain Bacteria (see Table 1.2). Prokaryotic cells are generally much smaller in size and simpler in structure than eukaryotic cells. Their small size and simple structure allow them to reproduce very quickly and effectively; they exist in great numbers in the air, in bodies of water, in the soil, and even on you. Prokaryotes are an extremely successful group of organisms.

Bacteria are well known because they cause some serious diseases, including tuberculosis, throat infections, and gonorrhea. Even so, the biosphere would not long continue without bacteria. Many bacteria decompose dead remains and contribute to ecological cycles. Bacteria also assist humans in another way—we use them to manufacture all sorts of products, from industrial chemicals to food products and drugs. The active cultures found in a container of yogurt are bacteria that are beneficial to us. There are also billions of bacteria teeming within your intestines! We are dependent on bacteria for many things, including the synthesis of some vitamins we can't make ourselves.

Our knowledge about how DNA specifies the sequence of amino acids in proteins was greatly advanced by experiments utilizing *Escherichia coli* (*E. coli*), a bacterium that lives in the human large intestine. The fact that this information applies to all prokaryotes and eukaryotes, even ourselves, reveals the remarkable unity of life and gives evidence that all organisms share a common descent.

Bacterial Structure

In bacteria, the cytoplasm is surrounded by a plasma membrane. Most bacteria possess a cell wall, and sometimes also a capsule (**Fig. 4.6**). The cytoplasm

Connections: Health

Doesn't *E. coli* cause intestinal problems?

The majority of *E. coli* in our bodies not only do not cause disease but actually benefit us by providing small amounts of vitamin K and some B vitamins. These *E. coli* even defend your intestinal tract against infection by other bacteria. However, one strain of *E. coli* (called O157:H7) produces a toxin (called Shiga toxin) that causes the intestines to become inflamed and secrete fluids, causing diarrhea. It typically is obtained from eating undercooked meat products. Still, even friendly strains of *E. coli* can cause problems in humans if they enter the urinary or genital tract.

Figure 4.6 A prokaryotic cell.

The structures in this diagram are characteristic of most prokaryotic cells.

(photo): © Sercomi/Science Source

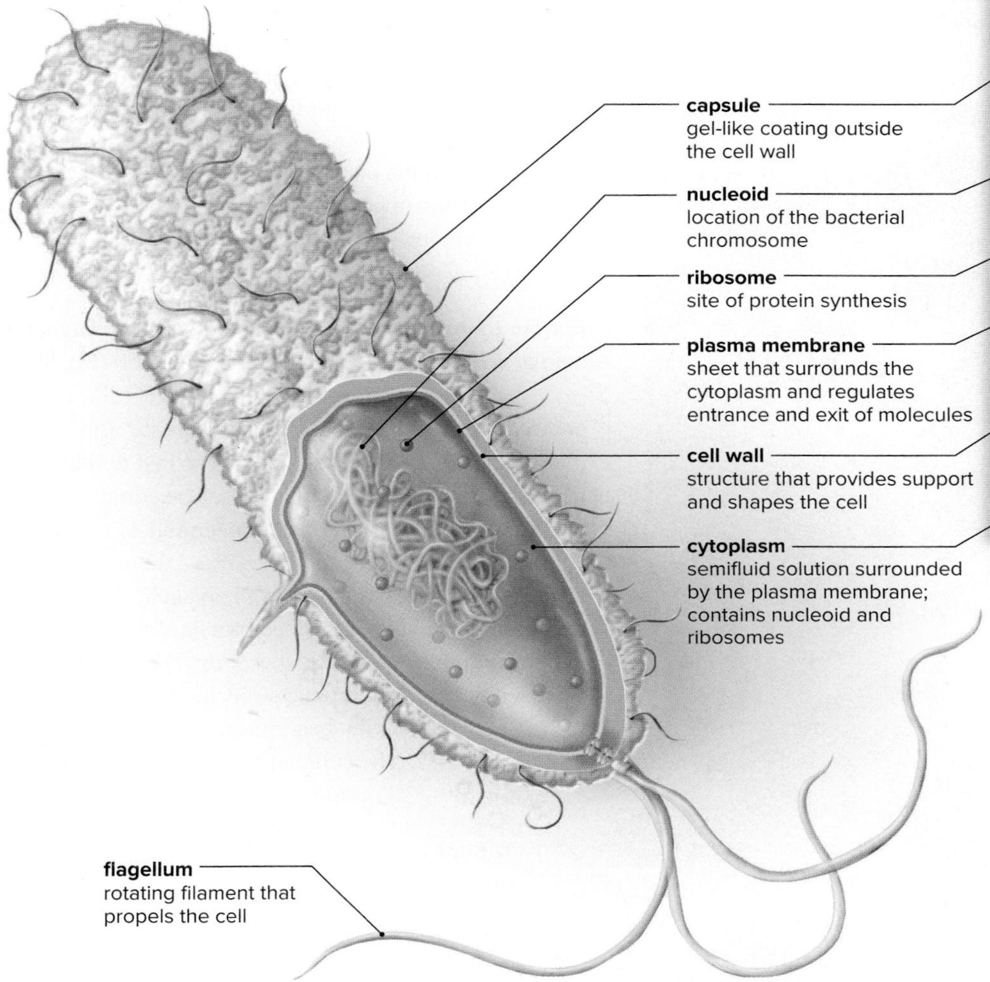

capsule
gel-like coating outside
the cell wall

nucleoid
location of the bacterial
chromosome

ribosome
site of protein synthesis

plasma membrane
sheet that surrounds the
cytoplasm and regulates
entrance and exit of molecules

cell wall
structure that provides support
and shapes the cell

cytoplasm
semifluid solution surrounded
by the plasma membrane;
contains nucleoid and
ribosomes

flagellum
rotating filament that
propels the cell

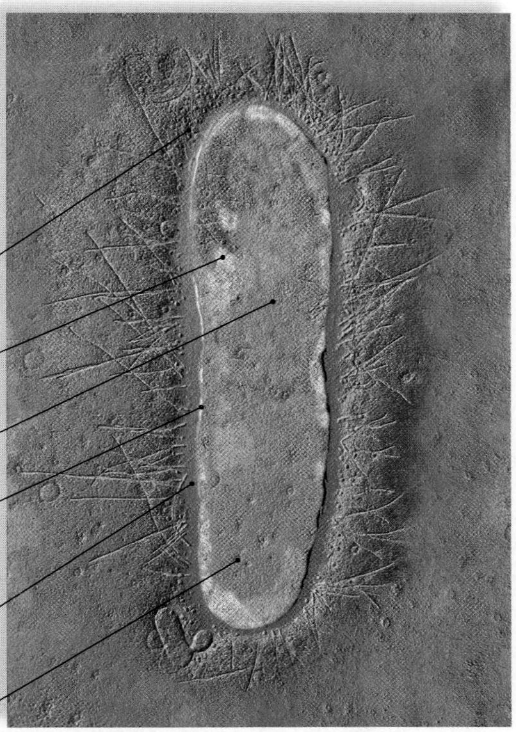

Escherichia coli   32,000×

contains a variety of enzymes. Enzymes are organic catalysts that speed up the many types of chemical reactions that are required to maintain an organism. As we have discussed, the plasma membranes of prokaryotes and eukaryotes have a similar structure (see Section 4.2). The **cell wall** maintains the shape of the cell, even if the cytoplasm should happen to take up an abundance of water. The **capsule** is a protective layer of polysaccharides lying outside the cell wall.

In bacteria, the DNA is located in a single circular, coiled chromosome that resides in a region of the cell called the **nucleoid.** The many proteins specified by bacterial DNA are synthesized on tiny structures called **ribosomes.** A bacterial cell contains thousands of ribosomes.

Some bacteria have **flagella** (sing., flagellum), which are tail-like appendages that allow the bacteria to propel themselves. A bacterial flagellum does not move back and forth like a whip. Instead, it moves the cell in a rotary motion. Sometimes flagella occur only at the ends of a cell, and other times they are dispersed randomly over the surface. We will take a closer look at the structure of prokaryotic cells, and their methods of reproduction, in Section 17.3

CONNECTING THE CONCEPTS

4.3 Prokaryotic cells are simple cells whose DNA is not enclosed by a membrane.

Check Your Progress 4.3

1. List the basic characteristics of all cells.

2. Identify the major distinctions between a prokaryotic cell and a eukaryotic cell.

3. Describe the role of the nucleoid, cell wall, ribosomes, and flagella in a prokaryotic cell.

Figure 4.7 Structure of a typical animal cell.

a. False-colored TEM of an animal cell. **b.** Generalized drawing of the same cell.

(a): © Alfred Pasieka/Science Source

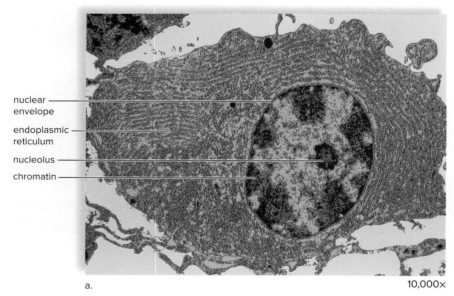

nuclear envelope

endoplasmic reticulum

nucleolus

chromatin

a.   10,000×

vesicle formation

vesicle

centrioles (in centrosome)

rough ER

mitochondrion

ribosome (attached to rough ER)

smooth ER

lysosome

cytoplasm

plasma membrane

Golgi apparatus

nucleus:
— nuclear envelope
— nuclear pore
— nucleolus
— chromatin

cytoskeleton:
— filaments
— microtubules

— polyribosome (in cytoplasm)

— ribosome (in cytoplasm)

b.

4.4 Eukaryotic Cells

Learning Outcomes

Upon completion of this section, you should be able to

1. Identify the general function of the organelles in a eukaryotic cell.
2. State the components of the endomembrane system, and list their functions.
3. Identify the energy roles that chloroplasts and mitochondria play in a cell.
4. Relate the specific components of the cytoskeleton to their diverse roles within the cell.

For our tour of a eukaryotic cell, we will be using a typical animal cell (**Fig. 4.7**) and a general plant cell (**Fig. 4.8**).

You should notice that eukaryotic cells are highly compartmentalized. These compartments are formed by membranes that create internal spaces that divide the labor necessary to conduct life functions. The compartments of a eukaryotic cell, typically called **organelles,** carry out specialized functions that together allow the cell to be more efficient and successful. Nearly all organelles are surrounded by a membrane with embedded proteins, many of which are **enzymes** (molecules that speed up chemical reactions). These enzymes make products specific to that organelle, but their action benefits the whole cell system.

The cell can be seen as a system of interconnected organelles that work together to conduct and regulate life processes. For example, the **nucleus** is a compartment that houses the genetic material within eukaryotic chromosomes, which contain hereditary information. The nucleus communicates with ribosomes in the cytoplasm, and the organelles of the endomembrane system—notably, the endoplasmic reticulum (ER) and the Golgi apparatus—communicate with one another. Production of specific molecules takes place inside or on the surface of organelles. These products are then transported around the cell by transport **vesicles,** membranous sacs that enclose the molecules and keep them separate from the cytoplasm. For example, the endoplasmic reticulum communicates with the Golgi apparatus by means of transport vesicles.

Vesicles move around by means of an extensive network of protein fibers called the cytoskeleton, which also maintains cell shape and assists with cell movement. These protein fibers allow the vesicles to move from one organelle to another. Organelles are also moved from place to place using this transport system. Think of the cytoskeleton as a three-dimensional road system inside cells used to transport important cargo from place to place.

The energy-related organelles—chloroplasts in plants and mitochondria in both plant and animal cells—are responsible for generating the majority of the energy needed to perform cellular processes.

As we review the specific functions of each organelle, you should refer back to Figures 4.7 and 4.8 so that you better understand how these organelles act as components of a living cell.

Figure 4.8 Structure of a typical plant cell.

a. False-colored TEM of a plant cell. **b.** Generalized drawing of the same cell.

(a): © Biophoto Associates/Science Source

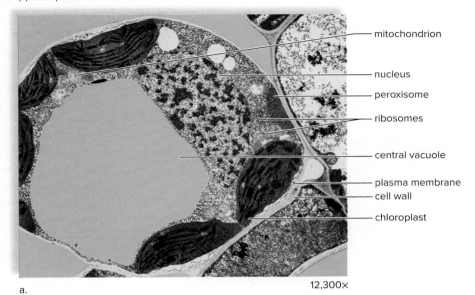

mitochondrion
nucleus
peroxisome
ribosomes
central vacuole
plasma membrane
cell wall
chloroplast

a.
12,300×

energy organelles:
chloroplast
mitochondrion

central vacuole

cell wall of adjacent cell
cell wall
plasma membrane
cytoplasm
plasmodesmata

centrosome

nucleus:
nuclear envelope
nuclear pore
nucleolus
chromatin

ribosome (attached to rough ER)

endomembrane system:
rough ER
smooth ER
lysosome
Golgi apparatus
vesicle

cytoskeleton:
microtubule
filaments

ribosome (in cytoplasm)

b.

Nucleus and Ribosomes

The nucleus stores genetic information, and the ribosomes in the cytoplasm use this information to carry out the manufacture of proteins.

The Nucleus

Because of its large size, the nucleus is one of the most noticeable structures in the eukaryotic cell (**Fig. 4.9**). The nucleus contains **chromatin** within a semifluid matrix called the nucleoplasm. Chromatin is a network of DNA, protein, and a small amount of RNA. Just before the cell divides, the chromatin condenses and coils into rodlike structures called **chromosomes.** All the cells of an organism contain the same number of chromosomes, except for the egg and sperm, which usually have half this number.

The DNA within a chromosome is organized into genes, each of which has a specific sequence of nucleotides. These nucleotides may code for a polypeptide, or sometimes regulatory RNA molecules. For now, it is important to recognize that the information in the DNA is processed to produce messenger RNA (mRNA). As its name suggests, mRNA acts as a messenger between the DNA and the ribosome, where polypeptide chains are formed. We will take a closer look at how this information is processed in Section 11.2. Because proteins are important in determining the structure and function of a cell, the nucleus may be thought of as the command center of the cell.

Within the nucleus is a dark structure called a **nucleolus.** In the nucleolus, a type of RNA called ribosomal RNA (rRNA) is produced. Proteins join with rRNA to form the subunits of ribosomes. The assembled ribosomal subunits are then sent out of the nucleus into the cytoplasm, where they join and assume their role in protein synthesis.

Figure 4.9 Structure of the nucleus.

The nuclear envelope contains pores that allow substances to pass from the nucleus to the cytoplasm.

(photo): © Biophoto Associates/Science Source

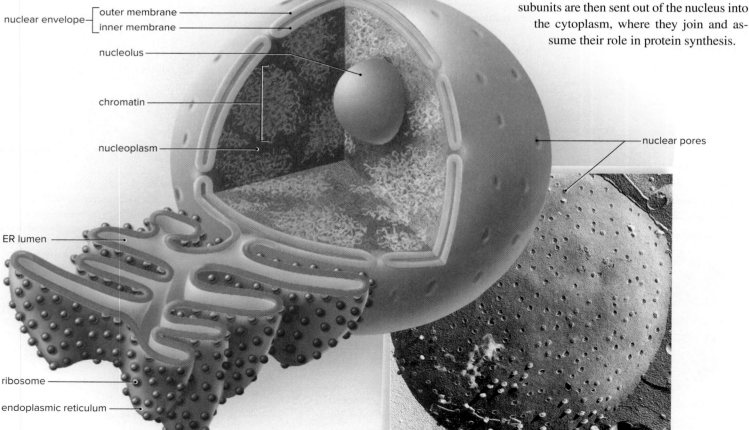

nuclear envelope
outer membrane
inner membrane
nucleolus
chromatin
nucleoplasm
ER lumen
ribosome
endoplasmic reticulum
nuclear pores

30,000×

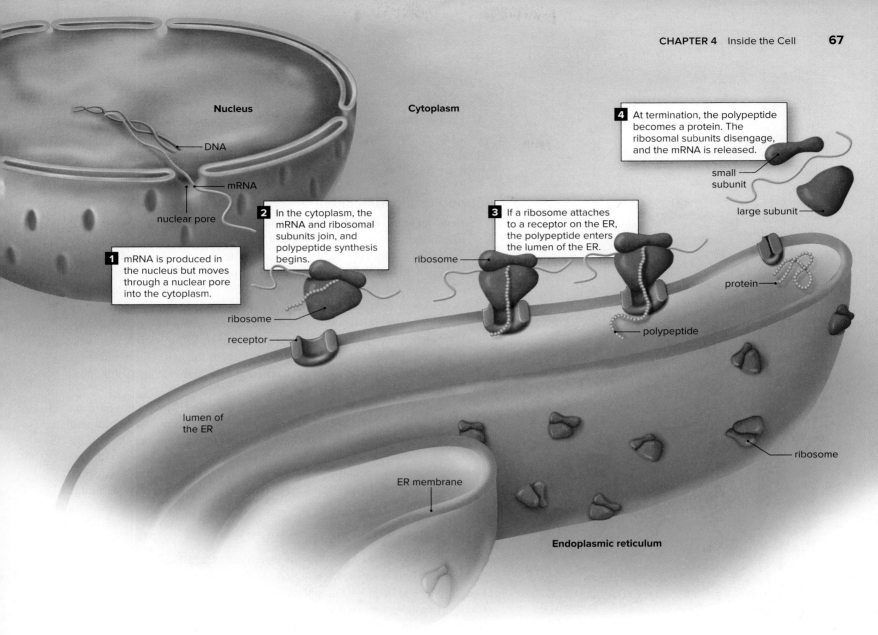

4 At termination, the polypeptide becomes a protein. The ribosomal subunits disengage, and the mRNA is released.

Nucleus

Cytoplasm

DNA

mRNA

nuclear pore

2 In the cytoplasm, the mRNA and ribosomal subunits join, and polypeptide synthesis begins.

1 mRNA is produced in the nucleus but moves through a nuclear pore into the cytoplasm.

ribosome

receptor

lumen of the ER

3 If a ribosome attaches to a receptor on the ER, the polypeptide enters the lumen of the ER.

ribosome

polypeptide

small subunit

large subunit

protein

ribosome

ER membrane

Endoplasmic reticulum

The nucleus is separated from the cytoplasm by a double membrane of phospholipids known as the **nuclear envelope.** Located throughout the nuclear envelope are *nuclear pores* that allow the nucleus to communicate with the cytoplasm. The nuclear pores are of sufficient size (100 nm) to permit the passage of ribosomal subunits and RNA molecules out of the nucleus into the cytoplasm, as well as the passage of proteins from the cytoplasm into the nucleus.

Ribosomes

Ribosomes are found in both prokaryotes and eukaryotes. In both types of cells, ribosomes are composed of two subunits, one large and one small. Each subunit has its own mix of proteins and rRNA. The ribosome acts as a workbench, and it is here that the information contained within the mRNA from the nucleus is used to synthesize a polypeptide chain (see Section 11.2). Proteins may contain one or more polypeptide chains.

In eukaryotic cells, some ribosomes occur freely within the cytoplasm. Other ribosomes are attached to the endoplasmic reticulum (ER), an organelle of the endomembrane system. After the ribosome binds to a receptor at the ER, the polypeptide being synthesized enters the lumen (interior) of the ER, where it may be further modified (**Fig. 4.10**) and then assume its final shape.

Figure 4.10 The nucleus, ribosomes, and endoplasmic reticulum (ER).

After mRNA leaves the nucleus, it attaches itself to a ribosome, and polypeptide synthesis begins. When a ribosome combines with a receptor at the ER, the polypeptide enters the lumen of the ER through a channel in the receptor. Exterior to the ER, the ribosome splits, releasing the mRNA while a protein takes shape inside the ER lumen.

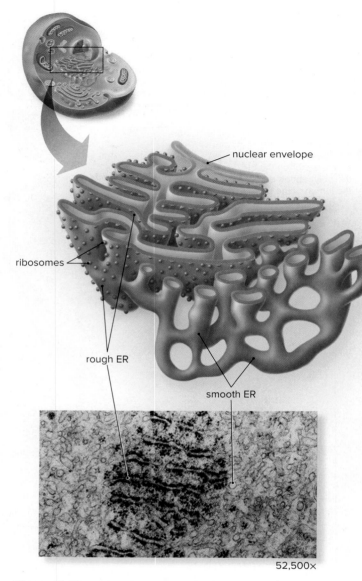

Figure 4.11 Endoplasmic reticulum (ER).

The rough ER consists of flattened saccules and has ribosomes present on its surface. The ribosomes synthesize polypeptides, which then enter the rough ER for modification. The smooth ER lacks ribosomes and is more tubular in structure. Lipids are synthesized by the smooth ER, which can have other functions as well.

(photo): © Martin M. Rotker/Science Source

Endomembrane System

The **endomembrane system** consists of the nuclear envelope, the membranes of the endoplasmic reticulum (ER), the Golgi apparatus, and numerous vesicles. This system helps compartmentalize the cell, so that particular enzymatic reactions are restricted to specific regions. Transport vesicles carry molecules from one part of the system to another.

Endoplasmic Reticulum

The **endoplasmic reticulum (ER)** consists of an interconnected system of membranous channels and saccules (flattened vesicles). It is physically continuous with the outer membrane of the nuclear envelope (**Fig. 4.11**).

Rough ER is studded with ribosomes on the side of the membrane that faces the cytoplasm; therefore, rough ER is able to synthesize polypeptides. It also modifies the polypeptides after they have entered the central enclosed region of the ER, called the lumen, where proteins take shape. The rough ER forms transport vesicles, which take proteins to other parts of the cell. Often, transport vesicles are on their way to the plasma membrane or the Golgi apparatus (described below).

Smooth ER, which is continuous with rough ER, does not have attached ribosomes. Smooth ER synthesizes lipids, such as phospholipids and steroids. The functions of smooth ER are dependent on the particular cell. In the testes, it produces testosterone and, in the liver, it helps detoxify drugs. Regardless of any specialized function, smooth ER also forms transport vesicles that carry molecules to other parts of the cell, notably the Golgi apparatus (**Fig. 4.12**).

Golgi Apparatus

The **Golgi apparatus,** named for its discoverer, Camillo Golgi, consists of a stack of slightly curved, flattened saccules resembling pancakes. The Golgi apparatus may be thought of as a transfer station. First, it receives transport vesicles sent to it by rough and smooth ER. The molecules within the vesicles are modified as they move between saccules. For example, sugars may be added to or removed from proteins within the saccules of the Golgi. Finally, the Golgi apparatus sorts the modified molecules and packages them into new transport vesicles according to their particular destinations. Outgoing transport vesicles may return to the ER or proceed to the plasma membrane, where they discharge their contents during secretion. In animal cells, some of the vesicles that leave the Golgi are lysosomes, which are discussed next.

Lysosomes

Lysosomes are vesicles, produced by the Golgi apparatus, that digest molecules and even portions of the cell itself. Sometimes, after engulfing molecules outside the cell, a vesicle formed at the plasma membrane fuses with a lysosome. Lysosomal enzymes then digest the contents of the vesicle. In Tay-Sachs disease, a genetic disorder, lysosomes in nerve cells are missing an enzyme for a particular lipid molecule. The cells become so full of storage lipids that they lose their ability to function. In all cases, the individual dies, usually in childhood. Research is currently underway on using advances in medicine, such as gene therapy (see Section 13.4), to provide the missing enzyme for these children.

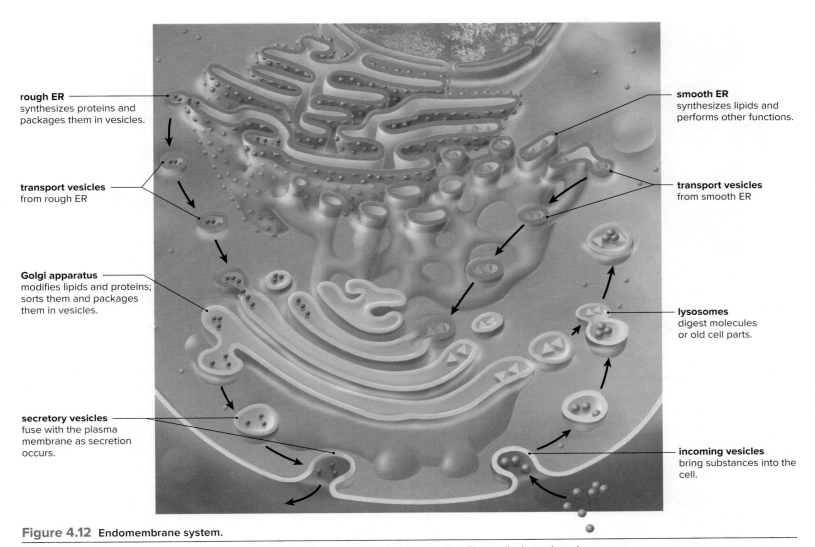

rough ER
synthesizes proteins and
packages them in vesicles.

transport vesicles
from rough ER

Golgi apparatus
modifies lipids and proteins;
sorts them and packages
them in vesicles.

secretory vesicles
fuse with the plasma
membrane as secretion
occurs.

smooth ER
synthesizes lipids and
performs other functions.

transport vesicles
from smooth ER

lysosomes
digest molecules
or old cell parts.

incoming vesicles
bring substances into the
cell.

Figure 4.12 Endomembrane system.

The organelles in the endomembrane system work together to carry out the functions noted. Plant cells do not have lysosomes.

Vesicles and Vacuoles

Vacuoles, like vesicles, are membranous sacs, but vacuoles are larger than vesicles and are more specialized. For example, the contractile vacuoles of aquatic protists are involved in removing excess water from the cell. Some protists have large digestive vacuoles for breaking down nutrients (**Fig. 4.13***a*). Plant vacuoles usually store substances, such as nutrients or ions. These vacuoles contain not only water, sugars, and salts but also pigments and toxic molecules (Fig. 4.13*b*). The pigments are responsible for many of the red, blue, and purple colors of flowers and some leaves. The toxic substances help protect a plant from herbivorous animals.

Energy-Related Organelles

Chloroplasts and mitochondria are the two eukaryotic organelles that specialize in energy conversion. **Chloroplasts** use solar energy to synthesize carbohydrates. **Mitochondria** (sing., mitochondrion) break down carbohydrates to produce **adenosine triphosphate (ATP)** molecules. The production of ATP is of great importance because ATP serves as a carrier of energy in cells. The energy of ATP is used whenever a cell synthesizes molecules, transports molecules, or carries out a special function, such as muscle contraction or nerve conduction. Without a constant supply of ATP, no cell could exist for long.

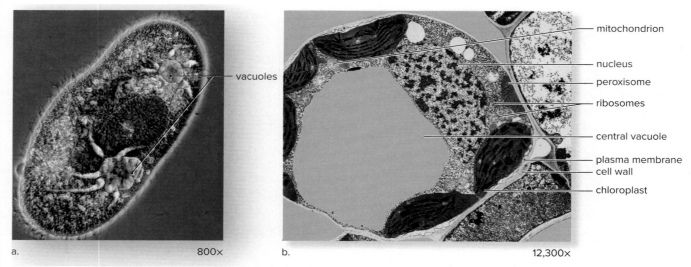

a.   800×

— vacuoles

— mitochondrion

— nucleus

— peroxisome

— ribosomes

— central vacuole

— plasma membrane
— cell wall

— chloroplast

b.   12,300×

Figure 4.13   Vacuoles.

a. Contractile vacuoles of a protist. **b.** A large central vacuole of a plant cell.

(a): © Roland Birke/Getty Images; (b): © Biophoto Associates/Science Source

Figure 4.14   Chloroplast structure.

a. Generalized drawing of a chloroplast, the organelle that carries out photosynthesis, showing some of the internal structures. **b.** Electron micrograph of a chloroplast.

(b): © Science Source

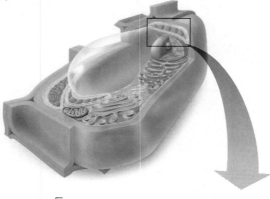

double
membrane ⎰ outer membrane
     ⎱ inner membrane

a.

thylakoid   thylakoid
space   membrane

thylakoid

Chloroplasts

The chloroplast, an organelle found in plants and algae, is the location where carbon dioxide gas, water, and energy from the sun are used to produce carbohydrates by the process of **photosynthesis.** The chloroplast is quite large, having twice the width and as much as five times the length of a mitochondrion. Chloroplasts are bound by a double membrane, which includes an outer membrane and an inner membrane. The large inner space, called the *stroma,* contains a concentrated mixture of enzymes and disclike sacs called *thylakoids.* A stack of thylakoids is called a *granum.* The lumens of thylakoid sacs form a large internal compartment called the *thylakoid space* (**Fig. 4.14**). The pigments that capture solar energy are located in the membrane of the thylakoids, and the enzymes that synthesize carbohydrates are in the stroma. The carbohydrates produced by chloroplasts serve as organic nutrient molecules for plants and, ultimately, for all living organisms on the planet. We will take a closer look at photosynthesis in Chapter 6.

The discovery that chloroplasts have their own DNA and ribosomes supports an accepted theory that chloroplasts are derived from photosynthetic bacteria that entered a eukaryotic cell in the distant past. This process is called *endosymbiosis.* We will take a closer look at this process when we explore the evolution of the protists (see Section 17.4).

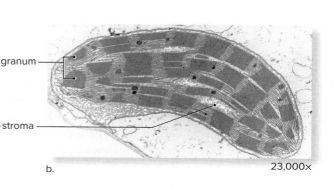

— granum

— stroma

b.   23,000×

Mitochondria

Mitochondria are much smaller than chloroplasts, and they are usually visible only under an electron microscope. We think of mitochondria as having a shape like that shown in **Figure 4.15,** but actually they often change shape, becoming longer and thinner or shorter and broader. Mitochondria can form long, moving chains (like locomotives on a train), or they can remain fixed in one location (often where energy is most needed). For example, they are packed between the contractile elements of cardiac cells (in the heart) and wrapped around the interior of a sperm's flagellum.

Like chloroplasts, mitochondria are bound by a double membrane. The inner membrane is highly convoluted into folds, called *cristae,* that project into the interior space, called the *matrix.* Cristae increase the surface area of the inner membrane so much that, in a liver cell, they account for about one-third of the total membrane in the cell.

Mitochondria are often called the powerhouses of the cell because they produce most of the ATP the cell utilizes. The matrix contains a highly concentrated mixture of enzymes that assists the breakdown of carbohydrates and other nutrient molecules. These reactions supply the chemical energy that permits ATP synthesis to take place on the cristae. The complete breakdown of carbohydrates, which also involves the cytoplasm, is called **cellular respiration** because oxygen is needed and carbon dioxide is given off. We will take a closer look at cellular respiration in Chapter 7.

The matrix also contains mitochondrial DNA and ribosomes. The presence of mitochondrial DNA and ribosomes is evidence that mitochondria and chloroplasts have similar origins and are derived from bacteria that took up residence in an early eukaryotic cell. Like the origin of chloroplasts, the origin of mitochondria is an example of endosymbiosis. All eukaryotic cells (with a few rare exceptions) have mitochondria, but only photosynthetic organisms (plants and algae) have chloroplasts.

Connections: Scientific Inquiry

How do we know that mitochondria and chloroplasts were once bacteria?

The DNA found in these organelles is structured differently than that found in the nucleus. Both mitochondria and chloroplasts contain a single, circular chromosome—similar to those found in prokaryotes. The genes located on this chromosome are very closely related to prokaryotic genes in both structure and function. In addition, both mitochondria and chloroplasts reproduce in a manner very similar to the process of binary fission in bacteria. These observations, coupled with detailed analyses of mitochondrial and chloroplast DNA, have strongly suggested that both of these organelles arose from an ancient symbiotic event (also called endosymbiosis) that played an important role in the evolution of the eukaryotic cell.

Figure 4.15 **Mitochondrion structure.**

a. Generalized drawing that reveals the internal structure of a mitochondrion, the organelle that is involved in cellular respiration. **b.** Electron micrograph of a mitochondrion.

(b): © Dr. Keith R. Porter

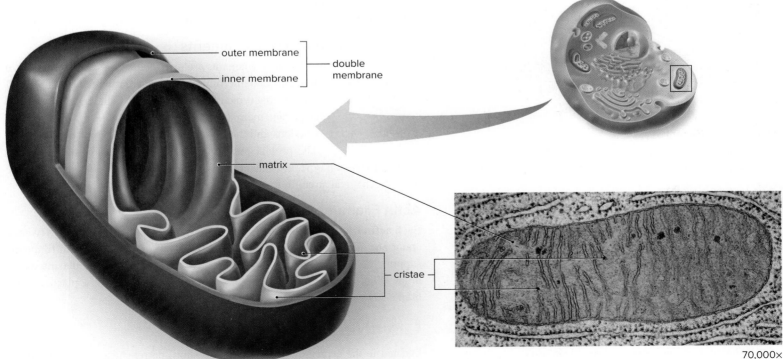

outer membrane · inner membrane · double membrane · matrix · cristae

70,000×

a.   b.

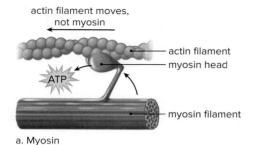

a. Myosin

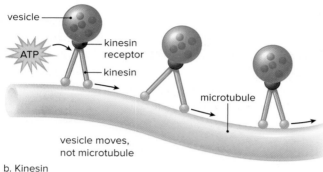

b. Kinesin

Figure 4.16 Motor proteins.

a. Myosin creates movement by detaching and reattaching to actin to pull the actin filament along the myosin filament. **b.** The motor protein kinesin carries organelles along microtubule tracks. One end binds to an organelle, and the other end attaches, detaches, and reattaches to the microtubule. ATP supplies the energy for both myosin and kinesin.

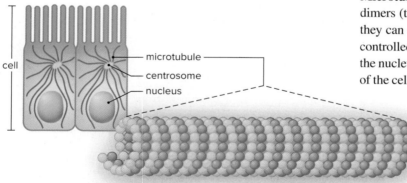

Figure 4.17 Microtubules.

Microtubules are located throughout the cell and radiate outward from the centrosome.

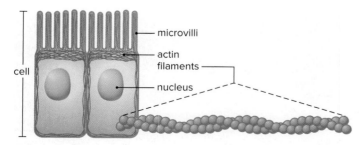

Figure 4.18 Actin filaments.

Actin filaments are organized into bundles or networks just under the plasma membrane, where they lend support to the shape of a cell.

The Cytoskeleton and Motor Proteins

The **cytoskeleton** is a network of interconnected protein filaments and tubules that extends from the nucleus to the plasma membrane in eukaryotic cells. Much as bones and muscles give an animal structure and produce movement, the elements of the cytoskeleton maintain cell shape and, along with *motor proteins,* allow the cell and its organelles to move. But unlike an animal's skeleton, the cytoskeleton is highly dynamic—its elements can be quickly assembled and disassembled as appropriate. The cytoskeleton includes microtubules, intermediate filaments, and actin filaments.

Motor Proteins

Motor proteins associated with the cytoskeleton are instrumental in allowing cellular movements. The major motor proteins are myosin, kinesin, and dynein.

Myosin often interacts with actin filaments when movement occurs. For example, myosin is interacting with actin filaments when cells move in an amoeboid fashion and/or engulf large particles. During animal cell division, actin, in conjunction with myosin, pinches the original cell into two new cells. When a muscle cell contracts, myosin pulls actin filaments toward the middle of the cell (**Fig. 4.16***a*).

Kinesin and dynein move along microtubules much as a car travels along a highway. First, an organelle, perhaps a vesicle, combines with the motor protein, and then the protein attaches, detaches, and reattaches farther along the microtubule. In this way, the organelle moves from one place to another in the cell (Fig. 4.16*b*). Kinesin and dynein are acting similarly when transport vesicles take materials from the Golgi apparatus to their final destinations.

Microtubules

Microtubules are small, hollow cylinders composed of 13 long chains of tubulin dimers (two tubulin molecules at a time; **Fig. 4.17**). Microtubules are dynamic; they can easily change their length by removing tubulin dimers. This process is controlled by the **centrosome,** a microtubule organizing center, which lies near the nucleus. Microtubules radiating from the centrosome help maintain the shape of the cell and act as tracks along which organelles and other materials can move.

Intermediate Filaments

Intermediate filaments are intermediate in size between actin filaments and microtubules. They are ropelike assemblies of proteins that typically run between the nuclear envelope and the plasma membrane. The network they form supports both the nucleus and the plasma membrane.

The protein making up intermediate filaments differs according to the cell type. Intermediate filaments made of the protein keratin give great mechanical strength to skin cells.

Actin Filaments

Each **actin filament** consists of two chains of globular actin monomers twisted about one another in a helical manner to form a long filament. Actin filaments support the cell, forming a dense, complex web just under the plasma membrane (**Fig. 4.18**). Actin filaments also support projections of the plasma membrane, such as microvilli.

Centrioles

Located in the centrosome, **centrioles** are short, barrel-shaped organelles composed of microtubules (**Fig. 4.19**). It's possible that centrioles give rise to basal

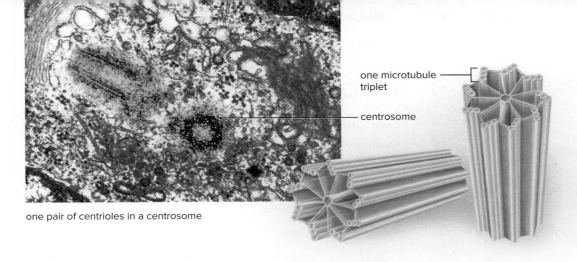

one microtubule triplet

centrosome

Figure 4.19 Centrioles.

A pair of centrioles lies to one side of the nucleus in an animal cell.

(photo): © Don W. Fawcett/Science Source

one pair of centrioles in a centrosome

bodies, which are located at the base of cilia and flagella and are believed to or-ganize the microtubules in these structures. The centrioles are also involved in organizing microtubules during cell division. However, some eukaryotes, such as plants and fungi, lack centrioles (although they have centrosomes), suggesting that centrioles are not necessary for the assembly of cytoplasmic microtubules.

Cilia and Flagella

Cilia and flagella (sing., cilium, flagellum) are whiplike projections of cells (**Fig. 4.20**). **Cilia** move stiffly, like an oar, and **flagella** move in an undulating, snakelike fashion. Cilia are short (2–10 µm), and flagella are longer (usually no more than 200 µm).

Some single-celled pro-tists utilize cilia or flagella to move about. In our bodies, cili-ated cells are critical to respira-tory health and our ability to reproduce. The ciliated cells that line our respiratory tract sweep debris trapped within mucus back up into the throat, which helps keep the lungs clean. Simi-larly, ciliated cells move an egg along the uterine tube, where it can be fertil-ized by a flagellated sperm cell.

CONNECTING THE CONCEPTS

4.4 Eukaryotic cells all possess a nucleus and internal organelles with specialized functions.

Check Your Progress 4.4

1. List the components of the nucleus and ribosomes, and give a function for each.

2. List the components of the endomembrane system, and list the function of each component.

3. Summarize the special functions of vacuoles, and hypothesize what might happen if they are not present in a cell.

4. Compare and contrast the structure and function of the two energy-related organelles of a eukaryotic cell.

5. Describe the functions of the cytoskeleton proteins and the motor proteins in a eukaryotic cell.

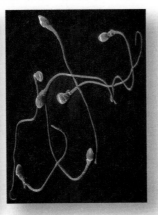

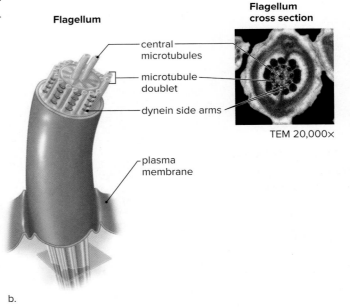

Flagellum

Flagellum cross section

central microtubules

microtubule doublet

dynein side arms

TEM 20,000×

plasma membrane

cilia in bronchial wall of lungs

flagella of sperm

a.

b.

Figure 4.20 Cilia and flagella.

a. Cilia in the bronchial wall and the flagella of sperm are organelles capable of movement. The cilia in the bronchi of our lungs sweep mucus and debris back up into the throat, where it can be swallowed or ejected. The flagella of sperm allow them to swim to the egg. **b.** Cilia and flagella have a distinct pattern of microtubules bounded by a plasma membrane.

(a): (cilia): © Kallista Images/Getty Images; (sperm flagella): © David M. Phillips/Science Source; (b): (flagellum cross section): © Steve Gschmeissner/Science Source

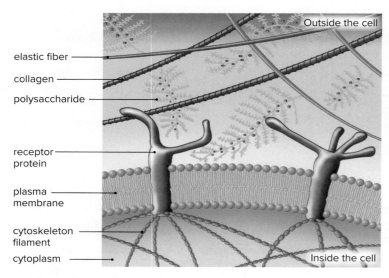

Outside the cell

elastic fiber

collagen

polysaccharide

receptor protein

plasma membrane

cytoskeleton filament

cytoplasm

Inside the cell

Figure 4.21 **Animal cell extracellular matrix.**

The extracellular matrix supports an animal cell and affects its behavior.

4.5 Outside the Eukaryotic Cell

Learning Outcomes

Upon completion of this section, you should be able to

1. Describe the structure of a plant cell wall.
2. State the purpose of the extracellular matrix.
3. Distinguish among the types of junctions present between eukaryotic cells.

A cell does not consist only of its plasma membrane and internal contents. Most cells also have extracellular structures formed from materials the cell produces and transports across its plasma membrane. These structures may either provide support, or allow for interaction with other cells.

Cell Walls

A **cell wall** provides support to the cell. Cells walls are found in many eukaryotic cells, including those of plants, fungi, and most protists but not those of animals. The composition of the cell wall differs between plants and fungi, but in this section we will focus on the plant cell wall.

A *primary cell wall* contains cellulose fibrils and noncellulose substances, and these allow the wall to stretch when the cell is growing. Adhesive substances are abundant outside the cell wall in the middle lamella, a layer that holds two plant cells together. For added strength, some plant cells have a *secondary cell wall* that forms inside the primary cell wall. The secondary wall has a greater quantity of cellulose fibrils, which are laid down at right angles to one another. Lignin, a substance that adds strength, is a common ingredient of secondary cell walls.

Extracellular Matrix

Animal cells do not have a cell wall, but they do have an extracellular matrix outside the cell. The **extracellular matrix (ECM)** is a meshwork of fibrous proteins and polysaccharides in close association with the cell that produced them (**Fig. 4.21**). Collagen and elastin are two well-known proteins in the extracellular matrix. Collagen resists stretching, and elastin provides resilience. The polysaccharides play a dynamic role by directing the migration of cells along collagen fibers during development. Other ECM proteins bind to receptors in a cell's plasma membrane, permitting communication between the extracellular matrix and the cytoskeleton within the cytoplasm of the cell.

The extracellular matrices of tissues vary greatly. They may be quite flexible, as in cartilage, or rock solid, as in bone. The rigidity of the extracellular matrix is influenced mainly by the number and types of protein fibers present and how they are arranged. The extracellular matrix of bone is very hard because, in addition to the components already mentioned, mineral salts—notably, calcium salts—are deposited outside the cell.

Junctions Between Cells

Three types of junctions are found between certain cells: adhesion junctions, tight junctions, and gap junctions. The type of junction between two cells depends on whether or not the cells need to be able to exchange materials and whether or not they need to be joined together very tightly.

In **adhesion junctions,** internal cytoplasmic plaques, firmly attached to the cytoskeleton within each cell, are joined by intercellular filaments (**Fig. 4.22***a*). The result is a sturdy but flexible sheet of cells. In some organs—such as the heart, stomach, and bladder, where tissues must stretch—adhesion junctions hold the cells together.

Adjacent cells are even more closely joined by **tight junctions,** in which plasma membrane proteins actually attach to each other, producing a zipperlike fastening (Fig. 4.22*b*). The cells of tissues that serve as barriers are held together by tight junctions; for example, urine stays within kidney tubules because the cells of the tubules are joined by tight junctions.

A **gap junction** allows cells to communicate. A gap junction is formed when two identical plasma membrane channels join (Fig. 4.22*c*). The channel of each cell is lined by six plasma membrane proteins that allow the junction to open and close. A gap junction lends strength to the cells, but it also allows small molecules and ions to pass between them. Gap junctions are important in heart muscle and smooth muscle because they permit the flow of ions that is required for the cells in these tissues to contract as a unit.

In a plant, living cells are connected by **plasmodesmata** (sing., plasmodesma), numerous narrow, membrane-lined channels that pass through the cell wall (**Fig. 4.23**). Cytoplasmic strands within these channels allow the direct exchange of some materials between adjacent plant cells and eventually among all the cells of a plant. The plasmodesmata allow only water and other small molecules to pass freely from cell to cell.

CONNECTING THE CONCEPTS

4.5 Cells are held together by specialized junctions.

Check Your Progress 4.5

1. Explain why some eukaryotic cells have cell walls.

2. Describe the composition of the extracellular matrix of an animal cell.

3. Compare the structure and function of adhesion, tight, and gap junctions.

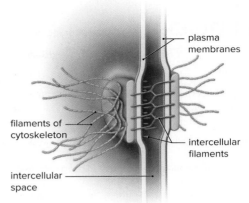

a. Adhesion junction

plasma membranes

filaments of cytoskeleton

intercellular filaments

intercellular space

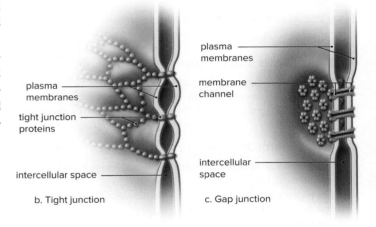

plasma membranes

tight junction proteins

intercellular space

b. Tight junction

plasma membranes

membrane channel

intercellular space

c. Gap junction

Figure 4.22 **Junctions between cells of the intestinal wall.**

a. In adhesion junctions, intercellular filaments run between two cells. **b.** Tight junctions between cells form an impermeable barrier because their adjacent plasma membranes are joined. **c.** Gap junctions allow communication between two cells because adjacent plasma membrane channels are joined.

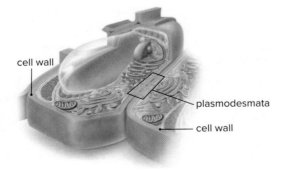

cell wall

plasmodesmata

cell wall

Figure 4.23 **Plasmodesmata.**

Plant cells are joined by membrane-lined channels, called plasmodesmata, that contain cytoplasm. Through these channels, water and other small molecules can pass from cell to cell.

(photo): © Biophoto Associates/Science Source

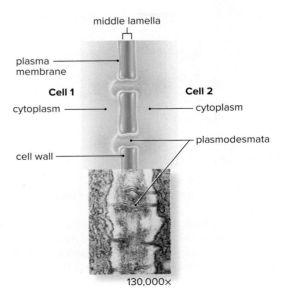

middle lamella

plasma membrane

Cell 1

cytoplasm

Cell 2

cytoplasm

plasmodesmata

cell wall

130,000×

STUDY TOOLS

 SMARTBOOK® Maximize your study time with McGraw-Hill SmartBook®, the first adaptive textbook.

SUMMARIZE

Cells are the fundamental units of all living organisms. There are two types of cells: prokaryotic cells and eukaryotic cells.

4.1 The microscopic size of a cell maximizes its surface-area-to-volume ratio.

4.2 All cells have a plasma membrane that consists of phospholipids and embedded proteins.

4.3 Prokaryotic cells are simple cells whose DNA is not enclosed by a membrane.

4.4 Eukaryotic cells all possess a nucleus and internal organelles with specialized functions.

4.5 Cells are held together by specialized junctions.

4.1 Cells Under the Microscope

- **Cells** are microscopic in size. Although a light microscope allows you to see cells, it cannot reveal the detail that an electron microscope can.
- The overall size of a cell is regulated by the **surface-area-to-volume ratio.**

4.2 The Plasma Membrane

- The **plasma membrane** of both prokaryotes and eukaryotes is a phospholipid bilayer.
- The phospholipid bilayer regulates the passage of molecules and ions into and out of the cell.
- The fluid-mosaic model of membrane structure shows that the embedded proteins form a mosaic (varying) pattern.
- The types of embedded proteins are channel, transport, cell recognition, receptor, enzymatic, and junction proteins.

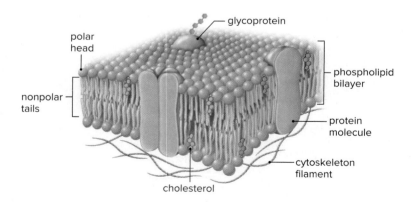

4.3 The Two Main Types of Cells

- The **cell theory** states that all life is made of cells.
- Some of the differences between **prokaryotic cells** and **eukaryotic cells** are presented in **Table 4.1**.
- All cells have a plasma membrane, cytoplasm, and genetic material.
- Prokaryotic cells lack a nucleus but possess a **nucleoid** region where the genetic material is located. Prokaryotic cells are surrounded by a **cell wall** and a **capsule,** and they often move with the use of flagella. The cytoplasm contains **ribosomes** for protein synthesis.

Table 4.1 Differences Between Prokaryotic and Eukaryotic Cells

Characteristic	Prokaryotic Cells	Eukaryotic Cells
Size	1–5 μm	Typically > 50 μm
DNA Location	Nucleoid region	Nucleus
Organelles	No	Yes
Chromosomes	One, circular	Multiple, linear

4.4 Eukaryotic Cells

Eukaryotic cells, which are much larger than prokaryotic cells, contain **organelles,** compartments that are specialized for specific cellular functions. A summary of these organelles and other structures is provided in **Table 4.2**.

Table 4.2 Summary of Eukaryotic Organelles

Organelle	Description
Plasma membrane	Encloses the cytoplasm; regulates interactions with the external environment
Nucleus	Contains the genetic material (DNA); nucleolus is the site of ribosome formation
Ribosomes	Location where polypeptides and proteins are formed
Vesicles	Small sacs that move materials between organelles in the endomembrane system
Rough ER	Component of the endomembrane system that has ribosomes attached; synthesizes proteins
Smooth ER	Endomembrane system organelle where lipids and some carbohydrates are synthesized; detoxifies some chemicals
Golgi apparatus	Processing and packaging center
Lysosome	Vesicle that contains enzymes that break down incoming molecules and cellular components
Chloroplast	Site of photosynthesis and carbohydrate formation (not found in animals)
Mitochondrion	Site of cellular respiration and ATP synthesis
Cell wall	Layer of cellulose that supports cells (not found in animals)
Cytoskeleton	Internal framework of protein fibers; moves organelles and maintains cell shape
Flagella and cilia	Involved in moving the cell or moving materials along the surface of the cell

Nucleus and Ribosomes

- The nucleus houses **chromatin,** which contains DNA, the genetic material. During division, chromatin becomes condensed into **chromosomes.**
- The **nucleolus** is an area within the nucleus where ribosomal RNA is produced. Proteins are combined with the rRNA to form the subunits of ribosomes, which exit the **nuclear envelope** through nuclear pores.
- Ribosomes in the cytoplasm synthesize polypeptides using information transferred to the DNA by mRNA.

Endomembrane System

- The **endoplasmic reticulum** (**ER**) is part of the endomembrane system. **Rough ER** has ribosomes, which produce polypeptides, on its surface. These polypeptides enter the ER, are modified, and become proteins, which are then packaged in transport vesicles.
- **Smooth ER** synthesizes lipids, but it also has various metabolic functions, depending on the cell type. It can also form transport vesicles.
- The **Golgi apparatus** is a transfer station that receives transport vesicles and modifies, sorts, and repackages proteins into transport vesicles that fuse with the plasma membrane as secretion occurs.
- **Lysosomes** are produced by the Golgi apparatus. They contain enzymes that carry out intracellular digestion.

Vacuoles and Vesicles

- Vacuoles are large, membranous sacs specialized for storage, contraction, digestion, and other functions.
- Vesicles are small, membranous sacs.

Energy-Related Organelles

- **Chloroplasts** capture the energy of the sun and carry on **photosynthesis,** which produces carbohydrates.
- **Mitochondria** are the site of **cellular respiration.** They break down carbohydrates (and other organic molecules) to produce **adenosine triphosphate (ATP).**

The Cytoskeleton and Motor Proteins

- The **cytoskeleton** maintains cell shape and allows the cell and the organelles to move.
- **Microtubules** radiate from the **centrosome** and are present in cytoplasm. They also occur in centrioles, cilia, and flagella.
- Intermediate filaments support the nuclear envelope and the plasma membrane. They give mechanical strength to skin cells, for example.
- **Actin filaments** are long, thin, helical filaments that support the plasma membrane and projections of the cell.
- Motor proteins allow cellular movements to occur and move vesicles and organelles within the cell.

Centrioles

- **Centrioles** are present in animal cells, but not plant cells. They appear to be involved in microtubule formation.

Cilia and Flagella

- **Cilia** and **flagella** are hairlike projections that allow some cells to move.

4.5 Outside the Eukaryotic Cell

A **cell wall** provides support for plant, fungi, and some protist cells. In plants, cellulose is the main component of the cell wall.

Animal cells have an **extracellular matrix** (**ECM**) that contains proteins and polysaccharides produced by the cells; it helps support cells and aids in communication between them.

Cells are connected by a variety of junctions:
- **Adhesion junctions** and **tight junctions,** if present, help hold cells together.
- **Gap junctions** allow the passage of small molecules between cells.
- Small, membrane-lined channels called **plasmodesmata** span the cell wall and contain strands of cytoplasm, which allow materials to pass from one cell to another.

ASSESS

Testing Yourself

Choose the best answer for each question.

4.1 Cells Under the Microscope

1. Which of the following can be viewed only with an electron microscope?
   a. virus
   b. chloroplast
   c. bacterium
   d. human egg
2. As a cell increases in size, its surface-area-to-volume ratio
   a. increases.
   b. decreases.
   c. stays the same.

4.2 The Plasma Membrane

3. The plasma membrane is said to be a fluid-mosaic model because it contains
   a. waxes suspended within a mosaic of phospholipids.
   b. a mosaic of proteins suspended within a phospholipid bilayer.
   c. a polysaccharide mosaic suspended within a protein bilayer.
   d. a mosaic of phospholipids suspended within a protein bilayer.
4. Which of the following types of proteins allow materials to move into, or out of, the cell?
   a. receptor proteins
   b. junction proteins
   c. enzymatic proteins
   d. channel proteins

4.3 The Two Main Types of Cells

5. Which of the following is not found in a prokaryotic cell?
   a. cytoplasm
   b. ribosome
   c. plasma membrane
   d. mitochondrion
   e. nucleoid
6. The ____ is located outside the cell wall in a prokaryotic cell.
   a. nucleoid
   b. nucleus
   c. capsule
   d. ribosome

4.4 Eukaryotic Cells

7. Which of these structures is involved in protein synthesis?
   a. ribosome
   b. plasma membrane
   c. mitochondrion
   d. microtubule

8. Label the indicated structures in this diagram:

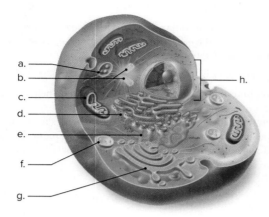

a. ____
b. ____
c. ____
d. ____
e. ____
f. ____
g. ____
h. ____

9. Which of the following is found in a plant cell, but not in an animal cell?
   a. chloroplast
   b. Golgi apparatus
   c. mitochondrion
   d. ribosome
   e. plasma membrane

10. The endomembrane system consists of all of the following, except
    a. lysosomes.
    b. Golgi apparatus.
    c. mitochondria.
    d. endoplasmic reticulum.

11. The majority of adenosine triphosphate (ATP) needed by the cell is produced by the
    a. nucleus.
    b. chloroplasts.
    c. mitochondria.
    d. ribosomes.

12. _____ move materials between the organelles of the endomembrane system.
    a. Vesicles
    b. RNA
    c. Ribosomes
    d. Nuclear pores

13. Centrioles are made of
    a. intermediate fibers
    b. actin filaments
    c. microtubules
    d. All of these are correct.

4.5 Outside the Eukaryotic Cell

14. Which of the following is not part of the cytoskeleton?
    a. intermediate filaments
    b. actin filaments
    c. microtubules
    d. centrioles

15. _____ are involved in the movement of the cell.
    a. Cilia
    b. Flagella
    c. Centrioles
    d. Intermediate filaments
    e. Both a and b are correct.

ENGAGE

BioNOW

Want to know how this science is relevant to your life? Check out the BioNow video below.

- Cell Size

Why would a larger surface-area-to-volume ratio increase metabolic efficiency?

Thinking Critically

1. Eggs come in different sizes, from the small egg of a hummingbird to the large egg of an ostrich. Based on what you know about the surface-area-to-volume ratio of cells, which of these eggs would have the higher metabolic rate, and why?

2. *Giardia lamblia* is a protist that is commonly known for contaminating water supplies and causing diarrhea. While *Giardia* is a eukaryote, its cells lack mitochondria. What would be the overall effect on a eukaryotic cell of a lack of mitochondria? How might have *Giardia* adapted for this potential difficulty?

3. One aspect of the new science of synthetic biology involves the laboratory design of cells to perform specific functions. Suppose that you wanted to make a protein for use in a drug trial. Design a cell that would build the protein and export it from the cell.

© Axel Fassio/Getty RF

5

The Dynamic Cell

Red Hot Chili Peppers

Have you ever bitten into a hot pepper and had the sensation that your mouth is on fire? This is because the chili pepper plant produces a chemical, called capsaicin, that binds to a protein found in the plasma membrane of pain receptors in your mouth. One of the important functions of a plasma membrane is to control what molecules move into and out of the cell. In the membrane are channel proteins that allow the movement of calcium ions across the membrane. When these channels are open, movement of the calcium ions into the cell causes the pain receptor to send a signal to the brain. The brain then interprets this signal as pain or discomfort, such as a burning sensation. Lots of things can trigger these channels, including temperature, acidic pH, heat, and chemicals such as capsaicin.

As long as the capsaicin is present, the pathway will remain active and signals will be sent to the brain. So the quickest way to alleviate the pain is to remove the capsaicin and close the channel protein. Unfortunately, since capsaicin is lipid-soluble, drinking cool water does very little to alleviate the pain. However, drinking milk, or eating bread or rice, often helps remove the capsaicin. Often the first bite is the worst, since the capsaicin causes an initial opening of all of the channels simultaneously. The receptors can become desensitized to capsaicin, which is why later bites of the same pepper don't produce the same results. Interestingly, as we will see at the start of Chapter 18, plants produce capsaicin as a weapon in a long-standing evolutionary battle with the fungi.

In this chapter, we will explore not only how cells move materials in and out, but also the basic properties of energy and how cells use metabolic pathways and enzymes to conduct the complex reactions needed to sustain life.

As you read through this chapter, think about the following questions:

1. How do transport and channel proteins function in a plasma membrane?
2. What type of transport are the calcium channels performing?

OUTLINE

5.1 What Is Energy? 80
5.2 ATP: Energy for Cells 82
5.3 Metabolic Pathways and Enzymes 85
5.4 Cell Transport 88

BEFORE YOU BEGIN

Before beginning this chapter, take a few moments to review the following discussions.

Section 1.1 Why is metabolism important to all life?
Section 3.2 What are the four levels of structure associated with the three-dimensional shape of a protein?
Section 4.2 What are the roles of proteins in the plasma membrane of a cell?

79

Potential energy is stored energy of position or configuration.

Kinetic energy is energy of motion.

Potential energy is being converted to kinetic energy.

Diving Board

Figure 5.1 Potential energy versus kinetic energy.

Food contains potential energy, which a diver can convert to kinetic energy in order to climb a ladder. The diver converts the potential energy associated with height to kinetic energy when she jumps. With every conversion to kinetic energy, some potential energy is lost as heat.

(left and center): © Patrik Giardino/Corbis; (right): © Joe McBride/ Corbis

Connections: Health

How much energy do you need per day to sustain life?

The measure of the minimum energy requirement needed to sustain life is called the basal metabolic rate, or BMR. The BMR is responsible for activities such as maintaining body temperature, heartbeat, and basic nervous functions. The BMR varies widely, depending on the age and sex of the individual, as well as body mass, genetics, and activity level. BMR values may be as low as 1,200 and as high as 2,000 kilocalories per day. There are numerous online calculators that can help you estimate your personal BMR value.

5.1 What Is Energy?

Learning Outcomes

Upon completion of this section, you should be able to

1. Distinguish between potential and kinetic forms of energy.
2. Describe the two laws of thermodynamics.
3. Summarize how the laws of thermodynamics and the concept of entropy relate to living organisms.

In Section 1.1, we discussed that living organisms must acquire and use energy. So what is energy? **Energy** is defined as the capacity to do work—to make things happen. Without a source of energy, life, including humans, would not exist on our planet. Our biosphere gets its energy from the sun, and thereafter one form of energy is changed to another form as life processes take place. The two basic forms of energy are potential energy and kinetic energy. **Potential energy** is stored energy, and **kinetic energy** is the energy of motion. Potential energy is constantly being converted to kinetic energy, and vice versa. An example is shown in **Figure 5.1.** The food a diver has for breakfast contains chemical energy, which is a form of potential energy. As the diver climbs the ladder to the diving platform, the potential energy of food is converted to the kinetic energy of motion, a type of mechanical energy. By the time she reaches the top of the platform, kinetic energy has been converted to the potential energy of location. As she begins her dive, this potential energy is converted to the kinetic energy of motion again. But with each conversion, some energy is lost as heat and other unusable forms.

Measuring Energy

Chemists use a unit of measurement called the *joule* to measure energy, but it is common to measure food energy in terms of calories. A **calorie** is the amount of heat required to raise the temperature of 1 gram of water by 1 degree Celsius. This isn't much energy, so the caloric value of food is listed in nutrition labels and diet charts in terms of **kilocalories** (kcals, or 1,000 calories). On food labels, and in scientific studies, an uppercase C (Calorie) indicates 1,000 calories.

Energy Laws

Two **energy laws** govern energy flow and help us understand the principles of energy conversion. Collectively, these are called the laws of thermodynamics.

Conservation of Energy

The first law of thermodynamics, also called the law of conservation of energy, tells us *energy cannot be created or destroyed, but it can be changed from one form to another.* Relating the law to the example of the diver we previously discussed, we know that she had to acquire energy by eating food before she could climb the ladder and that energy conversions occurred before she completed the dive. At the cellular level, energy is often changed between forms. For example, your muscles often store energy as the complex carbohydrate glycogen, which then may be converted to kinetic energy in muscle contraction.

Entropy

The second law of thermodynamics tells us *energy cannot be changed from one form to another without a loss of usable energy.* Many forms of energy are usable, such as the energy of the sun, food, and ATP. **Heat** is diffuse energy and the least usable form. Every energy conversion results in a loss of usable energy in the form of heat. In our example, much of the potential energy stored in the food is lost as heat as the diver converts it into the kinetic energy of motion. The generation of this heat allows us to maintain a constant body temperature, but eventually the heat is lost to the environment.

Every energy transformation leads to an increase in the amount of disorganization or disorder. The term **entropy** refers to the relative amount of disorganization. The only way to bring about or maintain order is to add more energy to a system. To take an example from your own experience, a tidy room is more organized and less stable than a messy room, which is disorganized and more stable (**Fig. 5.2***a*). In other words, your room is much more likely to stay messy than it is to stay tidy. Why? Unless you continually add energy to keep your room organized and neat, it will inevitably become less organized and messy.

All energy transformations, including those in cells, lead to an increase in entropy. Figure 5.2*b* shows a process that occurs in cells because it proceeds from a more ordered state to a more disordered state. Just as a tidy room tends to become messy, hydrogen ions (H^+) that have accumulated on one side of a membrane tend to move to the other side unless they are prevented from doing so by the addition of energy. Why? Because when hydrogen ions are distributed equally on both sides of the membrane, no additional energy is needed to keep them that way, and the entropy, or disorder, of their arrangement has increased. The result is a more stable arrangement of H^+ ions in the cell.

What about reactions in cells that apparently proceed from disorder to order? For example, plant cells can make glucose out of carbon dioxide and water. How do they do it? In order to overcome the natural tendency toward disorder, energy input is required, just as energy is required to organize a messy room. Likewise, energy provided by the sun allows plants to make glucose, a highly organized molecule, from the more disorganized water and carbon dioxide. Even this process, however, involves a loss of some potential energy. When light energy is converted to chemical energy in plant cells, some of the sun's energy is always lost as heat. In other words, the organization of a cell has a constant energy cost that also results in an increase in the entropy of the universe.

Entropy: Second Law of Thermodynamics

a.

• more organized	• less organized
• more potential energy	• less potential energy
• less stable	• more stable
• less entropy	• more entropy

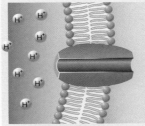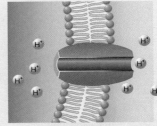

Unequal distribution of hydrogen ions	Equal distribution of hydrogen ions

b.

Figure 5.2 Cells and entropy.

The second law of thermodynamics states that entropy (disorder) always increases. Therefore, **(a)** a tidy room tends to become messy and disorganized, and **(b)** hydrogen ions (H^+) on one side of a membrane tend to move to the other side, so that the ions are equally distributed. Both processes result in a loss of potential energy and an increase in entropy.

(photos: both): © Keith Eng, 2008

Check Your Progress 5.1

1. Contrast potential energy with kinetic energy; give an example of potential energy being changed to kinetic energy.

2. Describe the two laws of thermodynamics.

3. Explain how cells avoid entropy (disorder) and maintain their organization.

CONNECTING THE CONCEPTS

5.1 The laws of thermodynamics determine how living organisms use energy.

5.2 ATP: Energy for Cells

Learning Outcomes

Upon completion of this section, you should be able to

1. Summarize the role of adenosine triphosphate (ATP) in a cell.
2. Describe the phases of the ATP cycle.
3. Describe the flow of energy between photosynthesis and cellular respiration.

Adenosine triphosphate (ATP) is the energy currency of cells. Just as you use coins to purchase all sorts of products, a cell uses ATP to carry out nearly all of its activities, including synthesizing proteins, transporting ions across the plasma membranes, and causing organelles and flagella to move. Cells use the readily accessible energy supplied by ATP to provide energy wherever it is needed.

Structure of ATP

ATP is a nucleotide (see Section 3.2), the type of molecule that serves as a monomer for the construction of DNA and RNA. ATP's name, adenosine triphosphate, means that it contains the sugar ribose, the nitrogen-containing base adenine, and three phosphate groups (**Fig. 5.3**). The three phosphate groups (shown as (P) in diagrams and formulas) are negatively charged and repel one another. Like trying to push together two negative ends of a battery, it takes energy to overcome the repulsion of the phosphate groups and link them by chemical bonds. This is why the bonds between the phosphate groups are high-energy bonds.

However, these linked phosphate groups also make the molecule unstable. ATP easily loses the phosphate group at the end of the chain because the breakdown products, ADP (adenosine diphosphate) and the separate phosphate group, are more stable than ATP. This reaction is written as ATP → ADP + (P). Energy is released as ATP breaks down. ADP can also lose another phosphate group to become AMP (adenosine monophosphate).

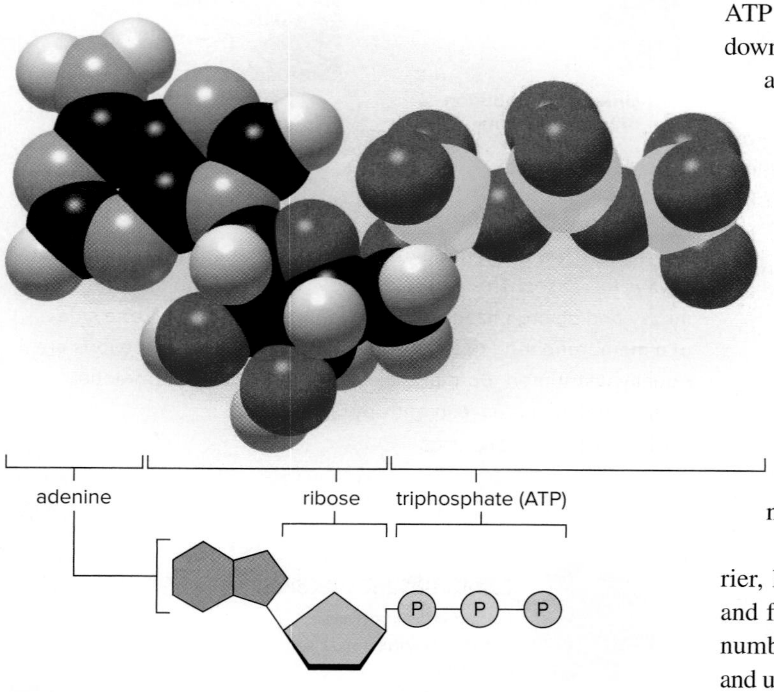

Use and Production of ATP

The continual breakdown and regeneration of ATP is known as the ATP cycle (**Fig. 5.4**). ATP holds energy for only a short period of time before it is used in a reaction that requires energy. Then ATP is rebuilt from ADP + (P). Each ATP molecule undergoes about 10,000 cycles of synthesis and breakdown every day. Our bodies use some 45 kg (about 99 lb) of ATP daily (assuming minimal activity!), and the amount on hand at any one moment is sufficiently high to meet current metabolic needs for only about 1 minute.

ATP's instability, the very feature that makes it an effective energy carrier, keeps it from being an energy storage molecule. Instead, carbohydrates and fats are the preferred energy storage molecules of cells due to their large number of H—C bonds. Their energy is extracted during cellular respiration and used to rebuild ATP, mostly within mitochondria. You will learn in Chapter 7 that the breakdown of one molecule of glucose permits the building of around 38 molecules of ATP. During cellular respiration, only 39% of the potential energy of glucose is converted to the potential energy of ATP; the rest is lost as heat.

Figure 5.3 ATP.

ATP, the universal energy currency of cells, is composed of the nucleotide adenine, the sugar ribose, and three linked phosphate groups (called a triphosphate).

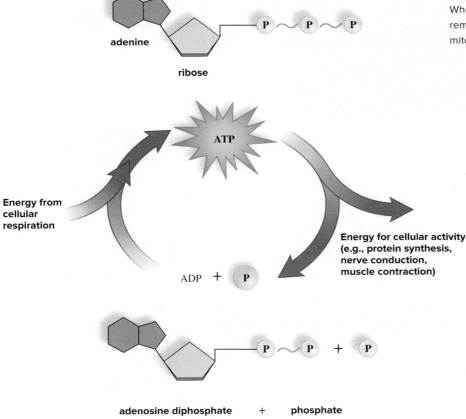

Figure 5.4 **The ATP cycle.**

When ATP is used as an energy source, a phosphate group is removed by hydrolysis. ATP is primarily regenerated in the mitochondria by cellular respiration.

The production of ATP is still worthwhile for the cell for the following reasons:

1. ATP releases energy quickly, which facilitates the speed of enzymatic reactions.
2. When ATP becomes ADP + ⓟ, the amount of energy released is usually just enough for a biological purpose. Breaking down an entire carbohydrate or fat molecule would be wasteful, since it would release much more energy than is needed.
3. The structure of ATP allows its breakdown to be easily *coupled* to an energy-requiring reaction, as described next.

Coupled Reactions

Many metabolic reactions require energy. Energy can be supplied when a reaction that requires energy (e.g., building a protein) occurs in the vicinity of a reaction that gives up energy (e.g., ATP breakdown). These are called coupled reactions, and they allow the energy-releasing reaction to provide the energy needed to start the energy-requiring reaction. Usually, the energy-releasing reaction is the breakdown of ATP, which generally releases more energy (and heat) than the amount consumed by the energy-requiring reaction. This increases entropy, but both reactions will proceed. The simplest way to represent a coupled reaction is like this:

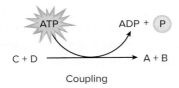

Coupling

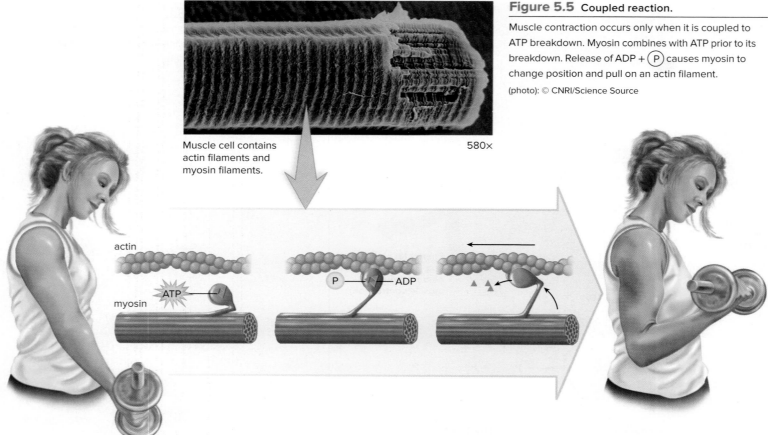

Muscle cell contains actin filaments and myosin filaments.

580×

Figure 5.5 Coupled reaction.

Muscle contraction occurs only when it is coupled to ATP breakdown. Myosin combines with ATP prior to its breakdown. Release of ADP + (P) causes myosin to change position and pull on an actin filament.

(photo): © CNRI/Science Source

actin

ATP

P — ADP

myosin

Connections: Health

What is creatine, and is it safe?

In humans, creatine is found in muscle cells as creatine phosphate (sometimes called phosphocreatine). Its function is to provide a brief, quick recharge of ATP molecules in muscle cells. Over the past several years, creatine supplements have gained popularity with athletes for muscle-building as a performance enhancement. While there is some evidence that creatine supplements may increase performance for some individuals, there have been no long-term FDA studies on the potential detrimental effects of creatine use on body organs—such as the kidneys, which are responsible for the secretion of excess creatine in the urine. In addition, creatine supplements can produce dangerous interactions when used with over-the-counter drugs, such as acetaminophen and sometimes even caffeine. Additional studies are needed to determine if creatine supplements are safe.

© McGraw-Hill Education

This reaction tells you that coupling occurs, but it does not show how coupling is achieved. Typically, the reaction transfers a phosphate group from ATP to one of the molecules in the reaction. This may either cause the molecule to change shape (and start a new function) or energize the molecule. In either case, the molecule has the ability to perform a function within the cell as a result of the coupled reaction. For example, when polypeptide synthesis occurs at a ribosome, an enzyme transfers a phosphate group from ATP to each amino acid in turn, and this transfer activates the amino acid, causing it to bond with another amino acid.

Figure 5.5 shows how ATP breakdown provides the energy necessary for muscle contraction. During muscle contraction, myosin filaments pull actin filaments to the center of the cell, and the muscle shortens. First, myosin combines with ATP, and only then does ATP break down to ADP + (P). The release of ADP + (P) from the molecule causes myosin to change shape and pull on the actin filament.

The Flow of Energy

In the biosphere, the activities of chloroplasts and mitochondria enable energy to flow from the sun to the majority of life on the planet (the exception may be some organisms that live near deep-sea vents). During photosynthesis, the chloroplasts in plants capture solar energy and use it to convert water and carbon dioxide to carbohydrates, which serve as food for themselves and for other organisms. During cellular respiration, mitochondria complete the breakdown of carbohydrates and use the released energy to build ATP molecules.

Notice in **Figure 5.6** that cellular respiration requires oxygen and produces carbon dioxide and water, the very molecules taken up by chloroplasts. It is actually the cycling of molecules between chloroplasts and mitochondria that allows a flow of energy from the sun through all living organisms. This flow of energy maintains the levels of biological organization from molecules to organisms to ecosystems. In keeping with the energy laws, useful energy is lost with each chemical transformation, and eventually the solar energy captured by plants is lost in the form of heat. In this way, living organisms are dependent upon an input of solar energy.

Like all life, humans are also involved in this cycle. We inhale oxygen and eat plants and their stored carbohydrates or other animals that have eaten plants. Oxygen and nutrient molecules enter our mitochondria, which produce ATP and release carbon dioxide and water.

CONNECTING THE CONCEPTS

5.2 The energy currency of the cell is ATP, which is used by cells to power their cellular functions.

Without a supply of energy-rich foods, we could not produce the ATP molecules needed to maintain our bodies and carry on activities.

Figure 5.6 Flow of energy.

Chloroplasts convert solar energy to the chemical energy stored in nutrient molecules. Mitochondria convert this chemical energy to ATP molecules, which cells use to perform chemical, transport, and mechanical work.

(leaves): © Comstock/PunchStock RF; (woman): © Karl Weatherly/Getty RF

Check Your Progress 5.2

1. Describe how ATP is produced, and explain why ATP cannot be used as an energy storage molecule.

2. Illustrate a coupled reaction, and explain the role of ATP in a coupled reaction.

3. Describe how cellular respiration and photosynthesis are connected.

5.3 Metabolic Pathways and Enzymes

Learning Outcomes

Upon completion of this section, you should be able to

1. Illustrate how metabolic reactions are catalyzed by specific enzymes.
2. Identify the role that enzymes play in metabolic pathways.
3. Explain the induced fit model of enzymatic action.
4. Detail the processes that inhibit enzyme activity.
5. Relate the role of enzymes in lowering the energy of activation needed for a reaction.

Life is a series of controlled chemical reactions. Just as a cell phone is built in a series of steps in a factory, the chemical reactions in a cell are linked to occur in a particular order. In the pathway, one reaction leads to the next reaction, which leads to the next reaction, and so forth, in an organized, highly structured manner. This is called a metabolic, or biochemical, pathway. Metabolic pathways begin with a particular **reactant** and terminate with an end **product.** For example, in cells, glucose is broken down by a metabolic pathway, called cellular respiration (see Chapter 7), which consists of a series of reactions that produces the end products of energy (ATP), CO_2, and H_2O.

Another example of a metabolic pathway is shown below. In this diagram, the letters A–F are reactants, and the letters B–G are products. Notice how the product from the previous reaction becomes the reactant of the next reaction. In the first reaction, A is the reactant and B is the product. Then B becomes the reactant in the next reaction of the pathway, and C is the product. This process continues until the final product (G) forms.

$$A \xrightarrow{e_1} B \xrightarrow{e_2} C \xrightarrow{e_3} D \xrightarrow{e_4} E \xrightarrow{e_5} F \xrightarrow{e_6} G$$

In this diagram, the letters e_1–e_6 represent **enzymes,** which are usually protein molecules that function as organic catalysts to speed chemical reactions. Enzymes can only speed reactions that are possible to begin with. In the cell, an enzyme is similar to a mutual friend who causes two people to meet and interact, because an enzyme brings together particular molecules and causes them to react with one another.

The reactant molecules that the enzyme acts on are called its **substrates.** An enzyme converts substrates into products. The substrates and products of an enzymatic reaction vary greatly. Many enzymes facilitate the breakdown of a substrate into multiple products. Or an enzyme may convert a single substrate into a single product. Still others may combine two or more substrates into a single product.

An Enzyme's Active Site

In most instances, only one small part of the enzyme, called the **active site,** accommodates the substrate(s) (**Fig. 5.7**). At the active site, the substrate fits into the enzyme seemingly as a key fits a lock; thus, most enzymes can fit only one substrate. However, the active site undergoes a slight change in shape in order to accommodate the substrate(s). This mechanism of enzyme action is called the **induced fit model** because the enzyme is induced (caused) to undergo a slight alteration to achieve optimal fit .

The change in the shape of the active site facilitates the reaction that occurs next. After the reaction has been completed, the products are released, and the active site returns to its original state, ready to bind to another substrate molecule. Only a very small amount of each enzyme is needed in a cell because enzymes are not used up by the reactions.

Figure 5.7 Enzymatic action.

An enzyme has an active site where the substrates and enzyme fit together in such a way that the substrates are oriented to react. Following the reaction, the products are released, and the enzyme is free to act again.

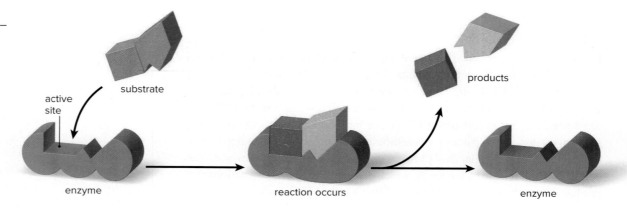

active site

substrate

products

enzyme

reaction occurs

enzyme

Enzyme Inhibition

Enzyme inhibition occurs when an active enzyme is prevented from combining with its substrate. Enzyme inhibitors are often poisonous to certain organisms. Cyanide, for example, is an inhibitor of the enzyme cytochrome *c* oxidase, which performs a vital function in cells because it is involved in making ATP. Cyanide is a poison because it binds the enzyme, blocking its activity. But some enzyme inhibitors are useful drugs. In another example, penicillin is a poison for bacteria, but not humans, because it blocks the active site of an enzyme unique to bacteria. Many other antibiotic drugs also act as enzyme inhibitors.

The activity of almost every enzyme in a cell is regulated by **feedback inhibition.** In the simplest case, when a product is in abundance, it competes with the substrate for the enzyme's active site. As the product is used up, inhibition is reduced, and then more product can be produced. In this way, the concentration of the product always stays within a certain range.

Most metabolic pathways are regulated by more complex types of feedback inhibition (**Fig. 5.8**). In these instances, when the end product is plentiful, it binds to a site other than the active site of the first enzyme in the pathway. This binding changes the shape of the active site, preventing the enzyme from binding to its substrate. Without the activity of the first enzyme, the entire pathway shuts down.

Energy of Activation

Molecules frequently do not react with one another unless they are activated in some way. In the lab, activation is very often achieved by heating a mixture to increase the number of effective collisions between molecules.

The energy needed to cause molecules to react with one another is called the **energy of activation (E_a).** The energy of activation acts as a metabolic speed bump; it limits how fast a reaction can proceed from reactants to products. Enzymes lower the amount of activation energy needed in a reaction (**Fig. 5.9**). In this way they act as catalysts that speed up the overall rate of the reaction. Enzymes do not change the amount of energy in the products or reactants, they simply alter the rate of the reaction.

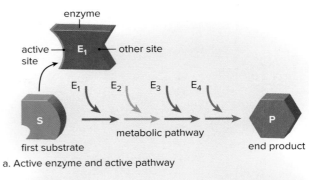

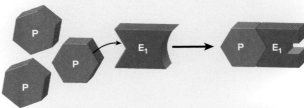

a. Active enzyme and active pathway

b. Feedback inhibition

c. Inactive enzyme and inactive pathway

Figure 5.8 Feedback inhibition.

a. This type of feedback inhibition occurs when the end product (P) of an active enzyme pathway is plentiful and **(b)** binds to the first enzyme (E_1) of the pathway at a site other than the active site. This changes the shape of the active site, so that **(c)** the substrate (S) can no longer bind to the enzyme. Now the entire pathway becomes inactive.

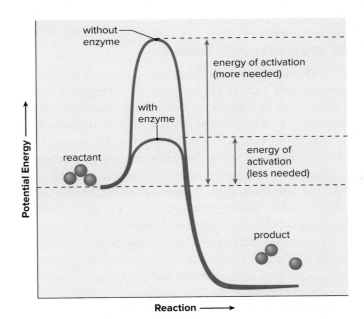

Figure 5.9 Energy of activation (E_a).

Enzymes speed the rate of reactions because they lower the amount of energy required for the reactants to react. Even reactions like this one, in which the energy of the product is less than the energy of the reactant, speed up when an enzyme is present.

CONNECTING THE CONCEPTS

5.3 Metabolic pathways are organized sets of chemical reactions in a cell that are regulated by enzymes.

Check Your Progress 5.3

1. Explain the benefit of metabolic pathways in cells.
2. Describe how the induced fit model explains the binding of a substrate to an enzyme's active site.
3. Summarize the benefit of using feedback inhibition to control metabolic pathways.

5.4 Cell Transport

Learning Outcomes

Upon completion of this section, you should be able to

1. Categorize the various ways in which materials can move across plasma membranes.
2. List the types of passive transport that can be used by cells.
3. Explain osmosis and the effect it has on cells in environments of different tonicities.
4. Describe how active transport is accomplished.
5. Define the various forms of bulk transport that can move materials into or out of a cell.

The plasma membrane regulates the passage of molecules into and out of the cell. This function is crucial because the life of the cell depends on the maintenance of its normal composition. The plasma membrane can carry out this function because it is selectively permeable, meaning that certain substances can freely pass through the membrane, some are transported across, and others are prohibited from entering or leaving.

Basically, substances enter a cell in one of three ways: passive transport, active transport, or bulk transport. Although there are different types of passive transport, in all of them substances move from an area of higher concentration to an area of lower concentration, and no energy is required. Active transport moves substances against a concentration gradient (from low to high concentration) and requires both a transport protein and lots of ATP. Bulk transport requires energy, but movement of the large substances involved is independent of concentration gradients.

Connections: Health

What causes cystic fibrosis?

In 1989, scientists determined that defects in a gene on chromosome 7 cause cystic fibrosis (CF). This gene, called *CFTR* (cystic fibrosis conductance transmembrane regulator), codes for a protein that is responsible for the movement of chloride ions across the membranes of cells that produce mucus, sweat, and saliva. Defects in this gene cause an improper water-salt balance in the excretions of these cells, which in turn leads to the symptoms of CF. Currently, there are over 1,400 known mutations in the CF gene. This tremendous amount of variation accounts for the differences in the severity of the symptoms in CF patients. By knowing the precise gene that causes the disease, scientists have been able to develop new treatment options for people with CF. At one time, people with CF rarely saw their 20th birthday; now it is routine for them to live into their 30s and 40s. New treatments, such as gene therapy, are being explored for patients with CF.

Passive Transport: No Energy Required

Diffusion

One form of passive transport is diffusion. During **diffusion,** molecules move down their concentration gradient until equilibrium is achieved and they are distributed equally. Diffusion does not need to occur across a membrane; it occurs because molecules are in motion, but it is a *passive* form of transport because energy is not expended. In cells, diffusion may occur across a plasma membrane. Some small, noncharged molecules, such as oxygen and carbon dioxide, are able to slip between the phospholipid molecules making up the plasma membrane.

Diffusion is a physical process that can be observed with any type of molecule. For example, when a crystal of dye is placed in water (**Fig. 5.10**), the dye and water molecules move in various directions, but their net movement, which is the sum of their motions, is toward the region of lower concentration. Eventually, the dye is dispersed, with the dye particles being equally distributed on either side of the membrane, and there is no net movement of dye in either direction.

A **solution** contains both a solute and a solvent. In this case, the dye is called the **solute,** and the water is called the **solvent.** Solutes are usually solids or gases, and solvents are usually liquids.

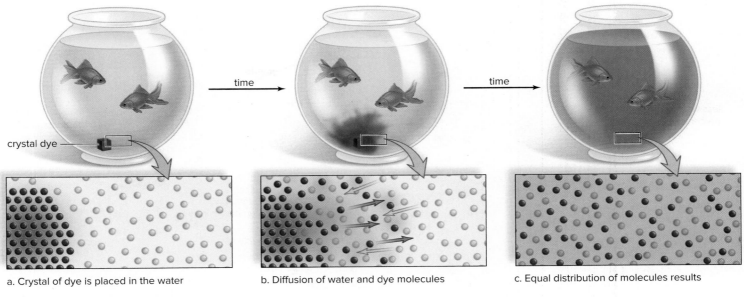

a. Crystal of dye is placed in the water   b. Diffusion of water and dye molecules   c. Equal distribution of molecules results

Figure 5.10 **Simple diffusion demonstration.**

Diffusion is spontaneous, and no chemical energy is required to bring it about. **a.** When a dye crystal is placed in water, it is concentrated in one area. **b.** The dye dissolves in the water, and there is a net movement of dye molecules from a higher to a lower concentration. There is also a net movement of water molecules from a higher to a lower concentration. **c.** Eventually, the water and the dye molecules are equally distributed throughout the container.

Facilitated Diffusion

Facilitated diffusion (**Fig. 5.11**) occurs when an ion or a molecule diffuses across a membrane with assistance of a channel protein or carrier protein. While water may diffuse across a membrane because of its size, it is a polar molecule, so to move water across the plasma membrane cells use channel proteins called *aquaporins*. This is why water can cross the membrane much more quickly than expected. Glucose and amino acids are assisted across the plasma membrane by carrier proteins that change shape as they pass through. Channel proteins and carrier proteins are very specific to the molecule they assist across the membrane.

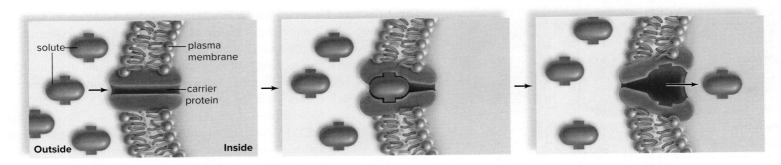

Figure 5.11 **Facilitated diffusion.**

In facilitated diffusion, a carrier protein in the plasma membrane allows molecules to move from areas of high concentration to areas of low concentration.

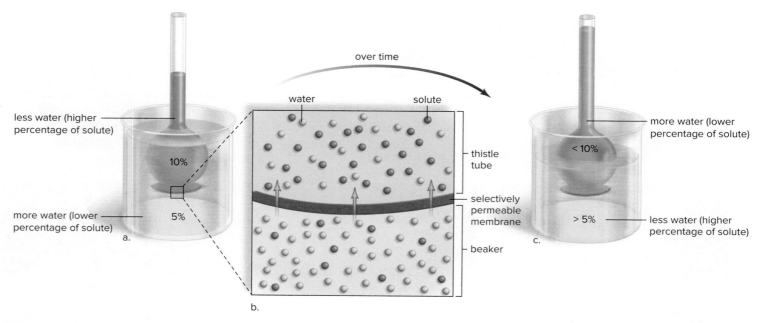

Figure 5.12 Osmosis demonstration.

a. A thistle tube, covered at the broad end by a semipermeable membrane, contains a 10% salt solution. The beaker contains only a 5% solution. **b.** The solute (green circles) is unable to pass through the membrane, but the water molecules (blue circles) pass through in both directions. Therefore, there is a net movement of water toward the inside of the tube, where the percentage of water molecules is lower than on the outside. **c.** The level of the solution rises in the tube because of the incoming water.

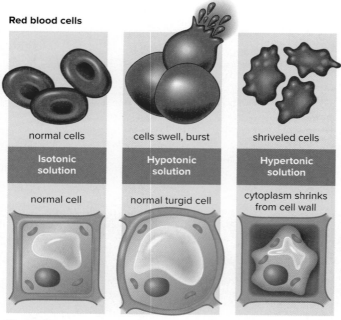

Figure 5.13 Osmosis in animal and plant cells.

In an isotonic solution, cells neither gain nor lose water. In a hypotonic solution, cells gain water. Red blood cells swell to bursting, and plant cells become turgid. In a hypertonic solution, cells lose water. Red blood cells shrivel, and plant cell cytoplasm shrinks away from the cell wall.

Osmosis

Osmosis is the diffusion of water across a semipermeable membrane, from an area of higher concentration to an area of lower concentration. Osmosis often occurs through the channel proteins called *aquaporins*.

An example of osmosis is shown in **Figure 5.12.** A tube containing a 10% salt solution and covered at one end by a membrane is placed in a beaker that contains a 5% salt solution (Fig. 5.12*a*). Water can cross the membrane, but the solute (dissolved salt) cannot. Therefore, there will be a net movement of water molecules across the membrane from the beaker to the inside of the tube. Theoretically, the solution inside the tube will rise until there is an equal concentration of water on both sides of the membrane (Fig. 5.12*b,c*).

The Effect of Osmosis on Cells When a solute is added to a cell's surroundings, the semipermeable plasma membrane often limits movement of the solute back and forth. Instead, water moves back and forth across the membrane by osmosis. This movement of water can affect the size and shape of cells. In the laboratory, cells are normally placed in **isotonic solutions** (*iso*, same as) in which the cell neither gains nor loses water—that is, the concentration of water is the same on both sides of the membrane. A 0.9% solution of the salt sodium chloride (NaCl) is known to be isotonic to red blood cells; therefore, intravenous solutions used in medical settings usually have this concentration.

Cells placed in a **hypotonic solution** (*hypo*, less than) gain water (**Fig. 5.13**). Outside the cell, the concentration of solute is less, and the concentration of water is greater, than inside the cell. Animal cells placed in a hypotonic solution expand and sometimes burst. The term *lysis* refers to disrupted cells; *hemolysis,* then, is disrupted red blood cells.

When a plant cell is placed in a hypotonic solution, the large central vacuole gains water, and the plasma membrane pushes against the rigid cell wall as the plant cell becomes *turgid.* The plant cell does not burst because the cell wall does not give way. Turgor pressure in plant cells is extremely important in maintaining a plant's erect position.

Cells placed in a **hypertonic solution** (*hyper,* more than) lose water. Outside the cell, the concentration of solute is higher, and the concentration of water is lower, than inside the cell. Animal cells placed in a hypertonic solution shrink. For example, meats are sometimes preserved by being salted. Bacteria are killed not by the salt but by the lack of water in the meat.

When a plant cell is placed in a hypertonic solution, the plasma membrane pulls away from the cell wall as the large central vacuole loses water. This is an example of *plasmolysis,* shrinking of the cytoplasm due to osmosis. Cut flowers placed into salty water will wilt due to plasmolysis. The dead plants you may see along a roadside could have died due to exposure to a hypertonic solution during the winter, when salt was used on the road.

Active Transport: Energy Required

During **active transport,** molecules or ions move through the plasma membrane against their concentration gradient—meaning that they move from areas of low concentration to high concentration. Cells use active transport to accumulate specific molecules or ions on one side of the plasma membrane (**Fig. 5.14**).

For example, iodine collects in the cells of the thyroid gland; glucose is completely absorbed from the digestive tract by the cells lining the digestive tract; and sodium can be almost completely withdrawn from urine by cells lining the kidney tubules. In these instances, molecules move against their concentration gradients, a situation that requires both a transport protein and much ATP. Therefore, cells involved in active transport, such as kidney cells, have many mitochondria near their plasma membranes to generate the ATP that is needed.

Proteins involved in active transport are often called *pumps* because, just as a water pump uses energy to move water against the force of gravity, proteins use energy to move a substance against its concentration gradient. One type of pump that is active in all animal cells, but is especially associated with nerve and muscle cells, moves sodium ions (Na^+) to the outside of the cell and potassium ions (K^+) to the inside of the cell. These two events are linked, and the carrier protein is called a **sodium-potassium pump.**

Connections: Health

Why can't you drink seawater?

Seawater is hypertonic to our cells. It contains approximately 3.5% salt, whereas our cells contain 0.9%. Once the salt entered your blood, your cells would shrivel up and die as they lost water trying to dilute the excess salt. Your kidneys can only produce urine that is slightly less salty than seawater, so you would dehydrate providing the amount of water necessary to rid your body of the salt.

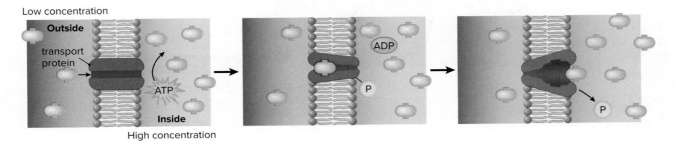

Figure 5.14 Active transport.

During active transport, a transport protein uses energy to move a solute across the plasma membrane toward a higher concentration. Note that the transport protein changes shape during the process.

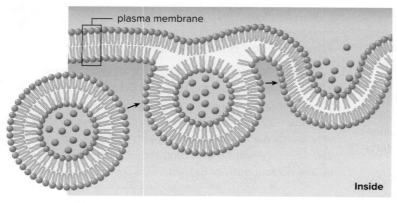

plasma membrane

Inside

a. Exocytosis

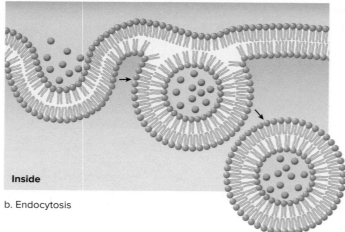

Inside

b. Endocytosis

Figure 5.15 **Bulk transport.**

During **(a)** exocytosis and **(b)** endocytosis, vesicle formation transports substances out of or into a cell, respectively.

Bulk Transport

Macromolecules, such as polypeptides, polysaccharides, and polynucleotides, are often too large to be moved by transport proteins. Instead, vesicle formation takes them into or out of a cell. For example, digestive enzymes and hormones use molecules transported out of the cell by **exocytosis** (**Fig. 5.15**a). In cells that synthesize these products, secretory vesicles accumulate near the plasma membrane. These vesicles release their contents when the cell is stimulated to do so.

When cells take in substances by vesicle formation, the process is known as **endocytosis** (Fig. 5.15b). If the material taken in is large, such as a food particle or another cell, the process is called *phagocytosis*. Phagocytosis is common in single-celled organisms, such as amoebas. It also occurs in humans. Certain types of human white blood cells are amoeboid—that is, they are mobile, like an amoeba, and are able to engulf debris, such as worn-out red blood cells and bacteria. When an endocytotic vesicle fuses with a lysosome, digestion occurs. In Section 26.3, we will see that this process is a necessary and preliminary step toward the development of immunity to bacterial diseases.

Pinocytosis occurs when vesicles form around a liquid or around very small particles. White blood cells, cells that line the kidney tubules and the intestinal wall, and plant root cells all use pinocytosis to ingest substances.

During **receptor-mediated endocytosis,** receptors for particular substances are found at one location in the plasma membrane. This location is called a *coated pit* because there is a layer of proteins on its intracellular side (**Fig. 5.16**).

Receptor-mediated endocytosis is selective and much more efficient than ordinary pinocytosis. It is involved when moving cholesterol from the blood into the cells of the body. Problems with this process can cause a disease called familial hypercholesterolemia, which results in elevated blood cholesterol levels.

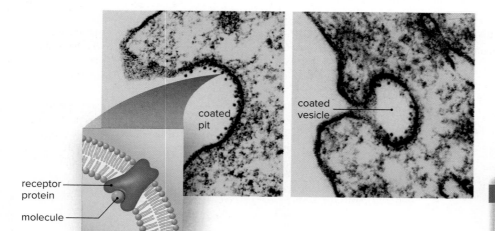

coated pit

coated vesicle

receptor protein

molecule

Figure 5.16 **Receptor-mediated endocytosis.**

During receptor-mediated endocytosis, molecules first bind to specific receptor proteins that are in a coated pit. The vesicle that forms contains the molecules and their receptors.

(photos: both): © Mark Bretscher

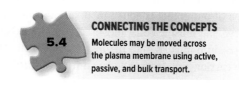

CONNECTING THE CONCEPTS

5.4

Molecules may be moved across the plasma membrane using active, passive, and bulk transport.

Check Your Progress 5.4

1. Explain the similarities and differences between diffusion, facilitated diffusion, and osmosis.

2. Describe the effects of osmosis on cells.

3. Explain why a cell would use active transport.

4. Differentiate between the types of bulk transport.

SMARTBOOK® Maximize your study time with McGraw-Hill SmartBook®, the first adaptive textbook.

SUMMARIZE

All living organisms require energy in order to live. Materials must move in and out of cells to produce the energy for life.

5.1 The laws of thermodynamics determine how living organisms use energy.

5.2 The energy currency of the cell is ATP, which is used by cells to power their cellular functions.

5.3 Metabolic pathways are organized sets of chemical reactions in a cell that are regulated by enzymes.

5.4 Molecules may be moved across the plasma membrane using active, passive, and bulk transport.

5.1 What Is Energy?

- **Energy** is the capacity to do work. **Potential energy** can be converted to **kinetic energy,** and vice versa. Solar and chemical energy are forms of potential energy; mechanical energy is a form of kinetic energy. Two energy laws hold true universally:
  - Energy cannot be created or destroyed, but it can be transferred or transformed. We can measure energy in many ways; food energy is measured in **calories.** 1,000 calories equals a **kilocalorie.**
  - When energy is converted from one form to another, some is lost as **heat.** Therefore, the **entropy,** or disorder, of the universe is increasing, and only a constant input of energy maintains the organization of living organisms.

5.2 ATP: Energy for Cells

Energy flows from the sun through chloroplasts and mitochondria, which produce ATP. Because ATP has three linked phosphate groups, it is a high-energy molecule that tends to break down to ADP + Ⓟ, releasing energy. ATP breakdown is coupled to various energy-requiring cellular reactions, including protein synthesis, active transport, and muscle contraction. Cellular respiration provides the energy for the production of ATP. The following diagram summarizes the ATP cycle:

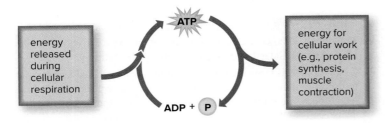

5.3 Metabolic Pathways and Enzymes

A metabolic pathway is a series of reactions that proceeds in an orderly, step-by-step manner. **Reactants** enter a metabolic pathway and then are modified to produce **products.**

Each reaction requires an **enzyme** that is specific to its substrate. An enzyme brings **substrates** together at the enzyme's **active site** according to the **induced fit model.** By doing so, they speed reactions by lowering the **energy of activation.** The activity of most enzymes and metabolic pathways is regulated by **feedback inhibition.**

5.4 Cell Transport

- The plasma membrane is semipermeable; some substances can freely cross the membrane, and some must be assisted across if they are to enter the cell.

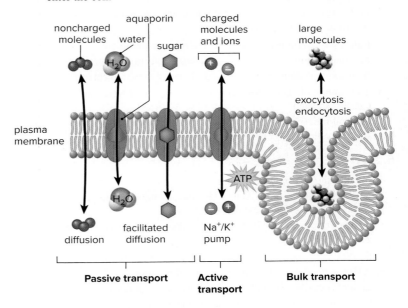

Passive transport requires no metabolic energy and moves substances from an area of higher concentration to an area of lower concentration.

In **diffusion,** molecules move from the area of higher concentration to the area of lower concentration until the concentration of molecules is the same at both sites. Some molecules, such as dissolved gases, cross plasma membranes by diffusion. In a **solution,** the **solute** is dissolved in the **solvent.**

- In **facilitated diffusion,** molecules diffuse across a plasma membrane through a channel protein (aquaporins are channel proteins for water) or with the assistance of carrier proteins.
- **Osmosis** is the diffusion of water (often using aquaporins) across a membrane toward the area of lower water concentration (higher solute concentration). Water moves from areas of low solute concentration (**hypotonic**) toward areas of high solute concentration (**hypertonic**). **Isotonic** solutions have equal solute concentrations.

Active transport requires metabolic energy (ATP) and moves substances across a membrane from an area of lower concentration to an area of higher concentration. In many cases, a transport protein acts as a pump that causes a substance to move against its concentration gradient. For example, the **sodium-potassium pump** carries Na^+ to the outside of the cell and K^+ to the inside of the cell.

Bulk transport requires vesicle formation and metabolic energy. It occurs independently of concentration gradients.

- **Exocytosis** transports macromolecules out of a cell via vesicle formation and often results in secretion.
- **Endocytosis** transports macromolecules into a cell via vesicle formation.
- **Receptor-mediated endocytosis** makes use of receptor proteins in the plasma membrane.

ASSESS

Testing Yourself

Choose the best answer for each question.

5.1 What Is Energy?

1. Which of the following is correct regarding the laws of energy?
   a. Energy cannot be created or destroyed.
   b. Energy can be changed from one form to another.
   c. Energy is lost as heat when it is converted from one form to another.
   d. All of these are correct.

2. Entropy is a term used to indicate the relative amount of
   a. organization.
   b. disorganization.
   c. enzyme action.
   d. None of these are correct.

3. The amount of energy needed to raise one gram of water one degree Celsius is called
   a. a calorie.
   b. a kilocalorie.
   c. ATP.
   d. heat.

5.2 ATP: Energy for Cells

4. Identify the incorrect statement:
   a. When ATP becomes ADP + Ⓟ, the amount of energy released is enough for a biological purpose.
   b. The structure of ATP allows its breakdown to be easily coupled.
   c. ATP can be easily recycled.
   d. ATP consists of an adenine, a sugar, and two phosphates.

5. ATP is a good source of energy for a cell because
   a. it is versatile—it can be used in many types of reactions.
   b. its breakdown is easily coupled with energy-requiring reactions.
   c. it provides just the right amount of energy for cellular reactions.
   d. All of these are correct.

5.3 Metabolic Pathways and Enzymes

6. In the induced fit model of enzyme action,
   a. the substrate exactly fits multiple enzymes.
   b. many substrates fit every enzyme.
   c. the enzyme changes shape slightly to accommodate the substrate.
   d. the enzyme doesn't fit the substrate until it is induced by another substrate.

7. Enzymes catalyze reactions by
   a. bringing the reactants together, so that they can bind.
   b. lowering the activation energy.
   c. Both a and b are correct.
   d. Neither a nor b is correct.

8. The active site of an enzyme
   a. is identical in structure for all enzymes.
   b. is the part of the enzyme where its substrate can fit.
   c. can be used only once.
   d. is not affected by environmental factors, such as pH and temperature.

5.4 Cell Transport

For questions 9–12, match the items to the correct answers in the key. Each answer may include more than one item.

Key:

a. diffusion
b. facilitated diffusion
c. osmosis
d. active transport

9. Movement of molecules from an area of high concentration to an area of low concentration.
10. Requires energy input.
11. Movement of water across a membrane.
12. Uses a membrane protein.
13. The movement of water across the plasma membrane of a cell is facilitated by
    a. aquaporins.
    b. substrate-mediated endocytosis.
    c. receptor-mediated diffusion.
    d. channel proteins.
    e. Both a and d are correct.
14. Seawater is _____ to blood because it contains a(n) _____ concentration of solutes.
    a. isotonic, higher
    b. hypotonic, higher
    c. isotonic, equal
    d. hypertonic, higher
15. A coated pit is associated with
    a. simple diffusion.
    b. osmosis.
    c. receptor-mediated endocytosis.
    d. pinocytosis.

ENGAGE

BioNOW

Want to know how this science is relevant to your life? Check out the BioNow videos below:
- Saltwater Filter
- Energy Part I: Energy Transfers

1. **Saltwater Filter:** Explain how the potato uses the principles of diffusion to measure the salt concentration in the branch samples.

2. **Energy Transfers:** Explain how both laws of thermodynamics apply to the experiments in this video.

Thinking Critically

1. How might a knowledge of the structure of an active site of an enzyme allow you to build a drug to regulate a metabolic pathway?

2. Cystic fibrosis is a genetic disorder that results in a defective membrane transport protein. The defective protein closes chloride ion channels in membranes, preventing chloride ions from being exported out of cells. This results in the development of a thick mucus on the outer surfaces of cells. This mucus clogs the ducts that carry digestive enzymes from the pancreas to the small intestine, clogs the airways in the lungs, and promotes lung infections. Why do you think the defective protein results in a thick, sticky mucus outside the cells, instead of a loose, fluid covering?

3. Some prokaryotes, especially archaea, are capable of living in extreme environments, such as deep-sea vents, where temperatures can reach 80°C (176°F). Few organisms can survive at this temperature. What adaptations might archaea possess that allow them to survive in such extreme heat?

© Corbis RF

Colors of Fall

Taking a walk through the woods in the fall, as the leaves begin to change color, can be an enjoyable and relaxing form of exercise. But why are the leaves changing color? Leaves contain several different types of pigment, including the green chlorophylls and the yellow to red carotenoids. These pigments play an important role in the absorption of solar energy for photosynthesis. In the fall, when the lower temperatures signal a change in the seasons, the supply of water and nutrients to the leaves declines, and the chlorophyll begins to degrade. Now the carotenoids, which were always present but masked by the chlorophylls, become visible, in the process causing the change in leaf color.

Interestingly, the same carotenoids act as antioxidants in our diets. Antioxidants protect us against free radicals, dangerous molecules that otherwise degrade the DNA and proteins within a cell. Antioxidants, such as beta-carotene, are common in red and yellow vegetables.

In this chapter, we will see how these pigments are involved in the process of photosynthesis and how plants capture solar energy to manufacture glucose, their food supply.

As you read through this chapter, think about the following questions:

1. What is the role of the photosynthetic pigments in photosynthesis?
2. How do plants convert solar energy to chemical energy?
3. What is the role of water in photosynthesis?

6

Energy for Life

OUTLINE

6.1 Overview of Photosynthesis   96

6.2 The Light Reactions—Harvesting Energy   99

6.3 The Calvin Cycle Reactions—Making Sugars   103

6.4 Variations in Photosynthesis   105

BEFORE YOU BEGIN

Before beginning this chapter, take a few moments to review the following discussions.

Section 1.1 What is the role of the producers in an ecosystem?

Section 4.4 What is the structure of a chloroplast?

Section 5.2 How is ADP energized to form ATP?

6.1 Overview of Photosynthesis

Learning Outcomes

Upon completion of this section, you should be able to

1. Describe the function of the chloroplast in photosynthesis.
2. State the overall chemical equation for photosynthesis.
3. Recognize what is meant by the terms *reduction* and *oxidation*.

Photosynthesis transforms solar energy into the chemical energy of a carbohydrate. Photosynthetic organisms, including plants, algae, and cyanobacteria, produce an enormous amount of carbohydrates (**Fig. 6.1**). If the carbohydrates were instantly converted to coal, and the coal loaded into standard railroad cars (each car holding about 50 tons), the photosynthesizers of the biosphere would fill more than 100 cars *per second* with coal.

It is no wonder, then, that photosynthetic organisms are able to sustain themselves and, with the exception of a few types of bacteria (see Section 17.3), all of the other living organisms on Earth.[1] To appreciate this, consider that most

---

[1] A few types of bacteria are chemosynthetic organisms, which obtain the necessary energy to produce their own organic nutrients by oxidizing inorganic compounds.

Figure 6.1 Photosynthetic organisms.

Photosynthetic organisms include plants, such as flowers and mosses, that typically live on land; photosynthetic protists, such as the *Euglena*, diatoms, and kelp shown here, which typically live in water; and cyanobacteria, a type of bacterium that lives everywhere.

(kelp): © Chuck Davis/Stone/Getty Images; (diatoms): © Ed Reschke; (*Euglena*): © M.I. (Spike) Walker/Alamy; (sunflower): © Corbis RF; (moss): © Steven P. Lynch; (cyanobacteria): © John Hardy, University of Washington, Stevens Point Department of Biology

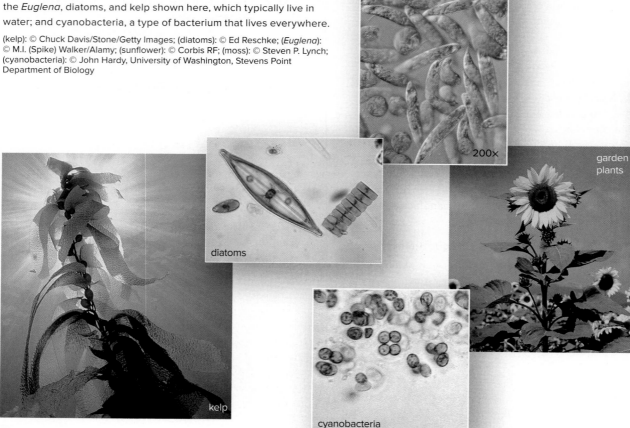

food chains lead back to plants. In other words, producers, which have the ability to synthesize carbohydrates, feed not only themselves but also consumers, which must take in preformed organic molecules.

The energy stored in the bodies of plants is important not only for the plant, but for humans. For example, coal is a fuel source that is formed from the partially fossilized remains of plants that lived millions of years ago in shallow seas around the world. It is just one example of a fossil fuel. The wood of trees is also commonly used as fuel. In addition, the fermentation of plant materials produces alcohol, which can be used to fuel automobiles directly or as a gasoline additive.

Connections: Scientific Inquiry

Where did the oxygen in the atmosphere come from?

The atmosphere of the early Earth was lacking in oxygen but rich in carbon dioxide and other gases. The first photosynthetic organisms are believed to have appeared around 3.5 billion years ago (BYA). Most scientists think that these were the cyanobacteria (Fig. 6.1). For almost a billion years, these organisms churned out oxygen as a by-product of photosynthesis. Then, beginning around 1.5 BYA, single-celled photosynthetic eukaryotes (such as algae and diatoms) began to make a contribution. It was not until between 400 and 450 MYA that land plants participated in the production of oxygen.

Plants as Photosynthesizers

The green portions of plants, particularly the leaves, carry on photosynthesis. Carbon dioxide (CO_2), the raw material for photosynthesis, enters the leaf through small openings called *stomata* (sing., *stoma*). Spaces within the leaf temporarily store CO_2 until it is needed for photosynthesis. The roots of a plant absorb water, which then moves in vascular tissue up the stem to the leaves, where it enters leaf veins. Special cells, called **mesophyll cells,** conduct photosynthesis. CO_2 and water diffuse into these cells, and then into the **chloroplasts,** the organelles that carry out photosynthesis (**Fig. 6.2**).

In a chloroplast, a double membrane surrounds a fluid-filled area called the **stroma.** A separate membrane system within the stroma forms flattened sacs called **thylakoids,** which in some places are stacked to form **grana** (sing., granum), so named because early microscopists thought they looked like piles of seeds. The space within each thylakoid is connected to the space within every other thylakoid, thereby forming an inner compartment within chloroplasts called the *thylakoid space.*

Chlorophyll and other pigments reside within the membranes of the thylakoids. These pigments are capable of absorbing solar energy, the energy that drives photosynthesis. A thylakoid membrane also contains protein complexes that convert solar energy into a chemical form. This energy is used in the stroma, where enzymes reduce CO_2 to carbohydrate.

Humans, and indeed nearly all organisms, release CO_2 into the air as a waste product from metabolic processes such as cellular respiration. This is some of the same CO_2 that enters a leaf through the stomata and is converted to carbohydrate. Carbohydrate, in the form of glucose, is the chief energy source for most organisms. Thus, the relationship between producers (e.g., plants) and consumers (e.g., animals) is a fundamental part of the intricate web of life.

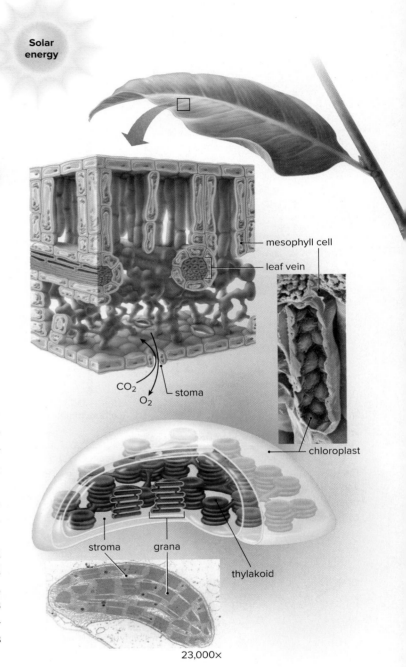

Solar energy

mesophyll cell

leaf vein

chloroplast

CO_2

O_2

stoma

stroma

grana

thylakoid

23,000×

Figure 6.2 Leaves and photosynthesis.

The raw materials for photosynthesis are carbon dioxide and water. Water enters a leaf by way of leaf veins. Carbon dioxide enters a leaf by way of stomata. Chloroplasts have two major parts. The grana are made up of thylakoids, membranous disks that contain photosynthetic pigments, such as chlorophyll. These pigments absorb solar energy. The stroma is a fluid-filled space where carbon dioxide is enzymatically processed to form carbohydrates, such as glucose.

(mesophyll cell):© Dr. David Furness, Keele University/Science Source; (chloroplast):© Science Source

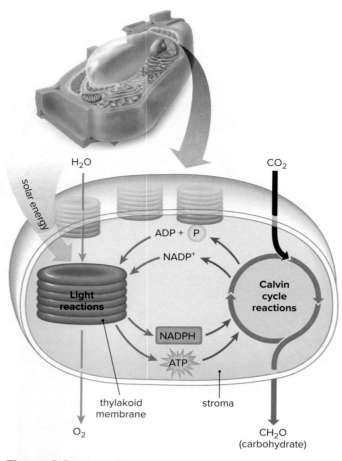

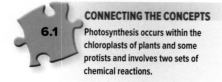

Figure 6.3 The photosynthetic process.

The process of photosynthesis consists of the light reactions and the Calvin cycle reactions. The light reactions, which produce ATP and NADPH, occur in a thylakoid membrane. These molecules are used by the Calvin cycle in the stroma to reduce carbon dioxide to a carbohydrate.

CONNECTING THE CONCEPTS

6.1

Photosynthesis occurs within the chloroplasts of plants and some protists and involves two sets of chemical reactions.

Check Your Progress 6.1

1. Describe the location of the stroma, grana, and thylakoid in a chloroplast.

2. Summarize the reactions of photosynthesis in a chemical equation.

3. Distinguish between a reduction and an oxidation reaction.

The Photosynthetic Process

Carbon dioxide (CO_2) and water (H_2O), the reactants for photosynthesis, contain much less energy than the end product, carbohydrate (CH_2O). Sunlight provides the needed energy to make photosynthesis occur, but how does it happen? During photosynthesis, electrons are removed from H_2O and energized by solar energy. An enzyme helper called $NADP^+$ then transfers these high-energy electrons to CO_2, thus forming carbohydrate. These electrons are accompanied by hydrogen ions (H^+), and that's why it is possible to track the movement of electrons during photosynthesis by following the movement of the hydrogen ions.

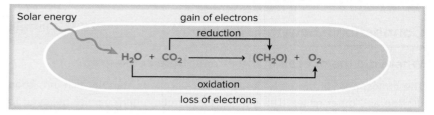

Reduction occurs when a molecule gains electrons (e^-) and hydrogen ions (H^+), and the molecule is said to be reduced. In the equation above, CO_2 is reduced when it becomes CH_2O. **Oxidation** occurs when a molecule gives up (e^-) and hydrogen ions, and the molecule is said to be oxidized. In our equation, water is oxidized to form O_2. In a cell, reduction reactions are often coupled with oxidation reactions. Collectively, these coupled reactions are commonly called **redox reactions.**

   Notice that if you multiply CH_2O by 6, you get $C_6H_{12}O_6$, which is the formula for glucose. While glucose is often described as the final product of photosynthesis, as we will see, the reduction of CO_2 produces a molecule called G3P (glyceraldehyde 3-phosphate). G3P molecules are often combined to form glucose shortly after the photosynthetic reactions.

Two Sets of Reactions

An overall equation for photosynthesis tells us the beginning reactant and the end products of the pathway. But much goes on in between. The word *photosynthesis* suggests that the process requires two sets of reactions. *Photo,* which means light, refers to the reactions that capture solar energy, and *synthesis* refers to the reactions that produce a carbohydrate. The two sets of reactions are called the **light reactions** and the **Calvin cycle** reactions (**Fig. 6.3**).

Light Reactions   The following events occur in the thylakoid membrane during the light reactions:

- Chlorophyll within the thylakoid membranes absorbs solar energy and energizes electrons.
- Water is oxidized, releasing electrons, hydrogen ions (H^+), and oxygen.
- ATP is produced from ADP + Ⓟ with the help of an electron transport chain.
- $NADP^+$, an enzyme helper, accepts electrons (is reduced) and becomes NADPH.

Calvin Cycle Reactions   The following events occur in the stroma during the Calvin cycle reactions:

- CO_2 is taken up by one of the molecules in the cycle.
- ATP and NADPH from the light reactions reduce CO_2 to a carbohydrate (G3P).

In the following sections, we discuss the details of these two reactions.

6.2 The Light Reactions—Harvesting Energy

Learning Outcomes

Upon completion of this section, you should be able to

1. Describe the function of photosynthetic pigments.
2. Explain the flow of electrons in the light reactions of photosynthesis.
3. Explain how ATP and NADPH are generated in the light reactions.

During the light reactions of photosynthesis, the pigments within the thylakoid membranes absorb solar energy. Solar energy (also called radiant energy) can be described in terms of its wavelength and its energy content in an electromagnetic spectrum. **Figure 6.4** lists the types of solar energy, from the type with the shortest wavelength, gamma rays, to that with the longest, radio waves. Notice how *visible light* occupies only a small portion of this spectrum.

Visible light contains various wavelengths of light, as can be proven by passing it through a prism; the different wavelengths appear to us as the colors of the rainbow, ranging from violet (the shortest wavelength) to blue, green, yellow, orange, and red (the longest wavelength). The energy content is highest for violet light and lowest for red light.

Less than half of the solar radiation that hits the Earth's atmosphere ever reaches its surface, and most of this radiation is within the visible-light range. Higher-energy wavelengths are screened out by the ozone layer in the atmosphere, and lower-energy wavelengths are screened out by water vapor and CO_2 before they reach the Earth's surface. Both the organic molecules within organisms and certain life processes, such as vision and photosynthesis, are adapted to utilize the radiation that is most prevalent in the environment.

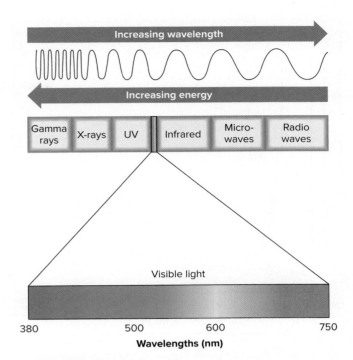

Figure 6.4 The electromagnetic spectrum.

Solar energy exists in a range of wavelengths extending from the very short wavelengths of gamma rays through the very long wavelengths of radio waves. Visible light, which drives photosynthesis, is expanded to show its component colors. The colors differ according to wavelength and energy content.

Connections: Scientific Inquiry

How do the grow-lights in a greenhouse work?

The grow-lights in commercial greenhouses do not emit all of the wavelengths of visible light. Instead, these lamps have been designed to emit more strongly in the blue and red wavelengths of light. Since these are the preferred wavelengths of the chlorophyll molecules, the growers can optimize plant growth. You will also notice that the colors of your clothes may appear slightly different under a grow-light. This is because the light reaching the pigments in the cloth does not contain all of the wavelengths, so not all the colors are reflected back to your eyes.

© Natig Aghayev/Shutterstock RF

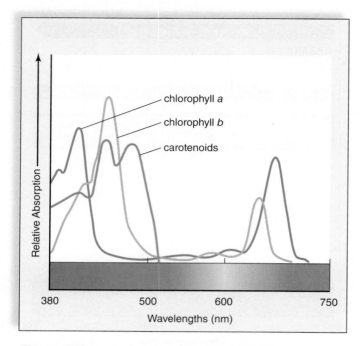

Figure 6.5 Photosynthetic pigments and photosynthesis.

The photosynthetic pigments in chlorophylls *a* and *b* and the carotenoids absorb certain wavelengths within visible light. This is their absorption spectrum.

Photosynthetic Pigments

The pigments found within most types of photosynthesizing cells are primarily **chlorophylls** and **carotenoids.** These pigments are capable of absorbing portions of visible light. Both chlorophyll *a* and chlorophyll *b* absorb violet, blue, and red wavelengths better than those of other colors (**Fig 6.5**).

Because green light is reflected and only minimally absorbed by chlorophylls, leaves appear green to us. Accessory pigments, such as the carotenoids, appear yellow or orange because they are able to absorb light in the violet-blue-green range, but not the yellow-orange range. As mentioned in the chapter opener, these pigments and others become noticeable in the fall when chlorophyll breaks down and the other pigments are uncovered (**Fig. 6.6**).

The Light Reactions: Capturing Solar Energy

Much as a solar energy panel captures the sun's energy and stores it in a battery, the light reactions (**Fig 6.7**) capture the sun's energy and store it in the form of a hydrogen ion (H^+) gradient. This hydrogen ion gradient is used to produce ATP molecules. The light reactions also produce an important molecule called NADPH.

The light reactions use two **photosystems,** called photosystem I (PS I) and photosystem II (PS II). The photosystems are named for the order in which they were discovered, not for the order in which they participate in the photosynthetic process. Here is a summary of how the pathway works:

- *Both photosystems receive photons.* PS II and PS I each consist of a pigment complex (containing chlorophyll and carotenoid molecules) and an electron acceptor. The pigment complex serves as an "antenna" for gathering solar energy. The energy is passed from one pigment to the other until it is concentrated in a particular pair of chlorophyll *a* molecules, called the reaction center.

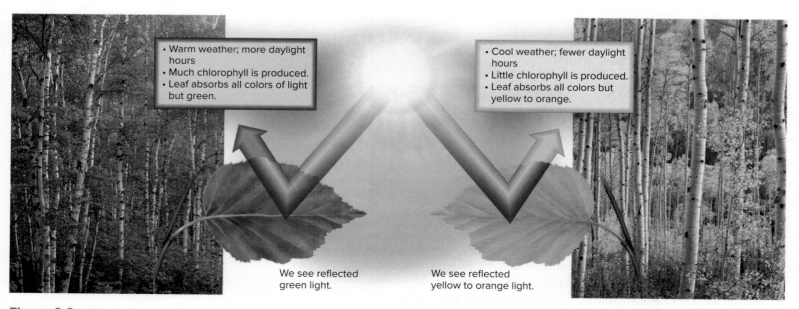

Figure 6.6 Leaf colors.

During the summer, leaves appear green because the chlorophylls absorb other portions of the visible spectrum better than they absorb green. During the fall, chlorophyll breaks down, and the remaining carotenoids cause leaves to appear yellow to orange because they do not absorb these colors.

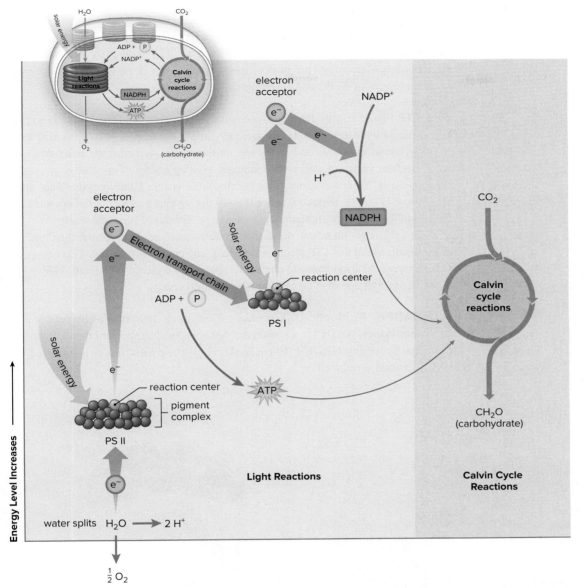

Figure 6.7 The electron pathway in the light reactions.

Energized electrons leave PS II and pass down an electron transport chain, leading to the formation of ATP. Energized electrons (replaced from PS II) leave PS I and pass to NADP$^+$, which then combines with H$^+$ to become NADPH. The electrons from PS II are replaced with electrons from water as it splits to release oxygen.

- *PS II splits water.* Due to the absorption of solar energy, electrons (e$^-$) in the reaction center of PS II become so energized that they escape and move to a nearby electron-acceptor molecule. Replacement electrons are removed from water, which splits by a process called *photolysis,* releasing oxygen and two hydrogen ions (H$^+$). The electron acceptor sends energized electrons, received from the reaction center, down an electron transport chain.

- *The electron transport chain establishes an energy gradient.* In the **electron transport chain,** a series of carriers pass electrons from one to the other, releasing energy stored in the form of a hydrogen ion (H$^+$) gradient. Later, ATP is produced (see below).

- *PS I produces NADPH.* When the PS I pigment complex absorbs solar energy, energized electrons leave its reaction center and are captured by a different electron acceptor. Low-energy electrons from the electron transport chain adjacent to PS II replace those lost by PS I. The electron acceptor in PS I passes its electrons to an NADP$^+$ molecule. NADP$^+$ accepts electrons and a hydrogen ion and is reduced, becoming NADPH.

The Electron Transport Chain

PS I, PS II, and the electron transport chain are molecular complexes located within the thylakoid membrane (**Fig. 6.8**). The molecular complexes are densely clustered to promote efficient electron transfer between carriers. Also present is an ATP synthase complex.

ATP Production During photosynthesis, the thylakoid space acts as a reservoir for hydrogen ions (H⁺). First, each time water is split, two H⁺ remain in the thylakoid space. Second, as the electrons move from carrier to carrier down the electron transport chain, the electrons give up energy. This energy is used to pump H⁺ from the stroma into the thylakoid space. Therefore, there are many more H⁺ in the thylakoid space than in the stroma, and a H⁺ gradient has been established. This gradient contains a large amount of potential energy.

The H⁺ then flow down their concentration gradient, across the thylakoid membrane at the ATP synthase complex, and energy is released. This causes the enzyme **ATP synthase** to change its shape and produce ATP from ADP + Ⓟ. The production of ATP captures the released energy.

NADPH Production Enzymes often have nonprotein helpers called **coenzymes.** NADP⁺ is a coenzyme that accepts electrons and then a hydrogen ion, becoming NADPH. When NADPH gives up electrons to a substrate, the substrate is reduced.

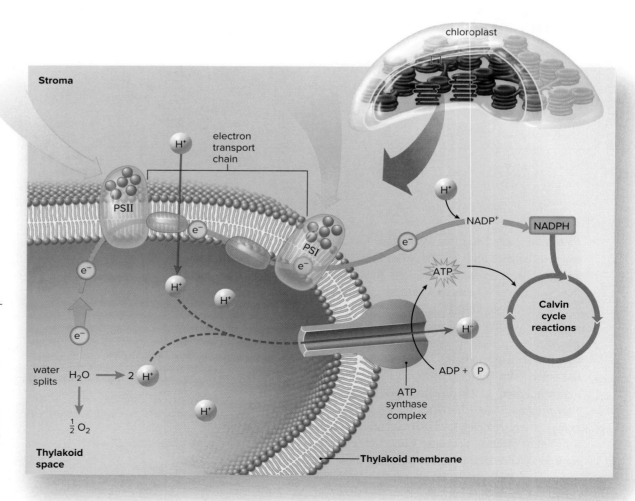

Figure 6.8 The electron transport chain.

Molecular complexes of the electron transport chain within the thylakoid membrane pump hydrogen ions from the stroma into the thylakoid space. When hydrogen ions flow back out of the space into the stroma through the ATP synthase complex, ATP is produced from ADP + Ⓟ. NADP⁺ accepts two electrons and joins with H⁺ to become NADPH.

We have seen that, during the light reactions, NADP$^+$ receives electrons at the end of the electron pathway in the thylakoid membrane, and then it picks up a hydrogen ion to become NADPH (see Fig.6.8).

Check Your Progress 6.2

1. Explain why plants have more than one photosynthetic pigment.
2. Summarize how the light reactions capture solar energy, and identify the outputs of these reactions.
3. Describe the path of electrons from water to NADPH.
4. Explain why water is needed in the light reactions.

CONNECTING THE CONCEPTS

6.2 The light reactions of photosynthesis capture solar energy to produce the **ATP** and **NADPH** needed to synthesize carbohydrates.

6.3 The Calvin Cycle Reactions— Making Sugars

Learning Outcomes

Upon completion of this section, you should be able to

1. Summarize the three stages of the Calvin cycle, and describe the major event that occurs during each stage.
2. Describe how ATP and NADPH are utilized in the manufacture of carbohydrates by the Calvin cycle.
3. Summarize how the output of the Calvin cycle is used to make other carbohydrates.

The ATP and NADPH generated by the light reactions power the Calvin cycle reactions. The Calvin cycle is often referred to as the light-independent reactions, since these reactions do not require an input of solar energy but, rather, rely on the outputs of the light reactions. The cycle is named for Melvin Calvin, who, with colleagues, used the radioactive isotope ^{14}C as a tracer to discover the reactions that make up the cycle.

Overview of the Calvin Cycle

The Calvin cycle reactions occur in the stroma of chloroplasts (see Fig. 6.2). This series of reactions (**Fig. 6.9**) utilizes carbon dioxide (CO$_2$) from the atmosphere and consists of these three steps: (1) CO$_2$ fixation, (2) CO$_2$ reduction, and (3) regeneration of the first substrate, RuBP (ribulose-1,5-bisphosphate). The reactions produce molecules of G3P (glyceraldehyde 3-phosphate), which plants use to produce glucose and other types of organic molecules. The Calvin cycle requires a lot of energy, which is supplied by ATP and NADPH from the light reactions.

Fixation of Carbon Dioxide

Carbon dioxide (CO$_2$) fixation is the first step of the Calvin cycle. During this reaction, CO$_2$ from the atmosphere is attached to RuBP, a 5-carbon molecule.

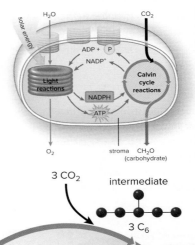

Figure 6.9 The Calvin cycle reactions.

The Calvin cycle is divided into three portions: CO$_2$ fixation, CO$_2$ reduction, and regeneration of RuBP. Because five G3P are needed to re-form three RuBP, it takes three turns of the cycle to achieve a net gain of one G3P. Two G3P molecules are needed to form glucose. Each ● is a carbon atom (C).

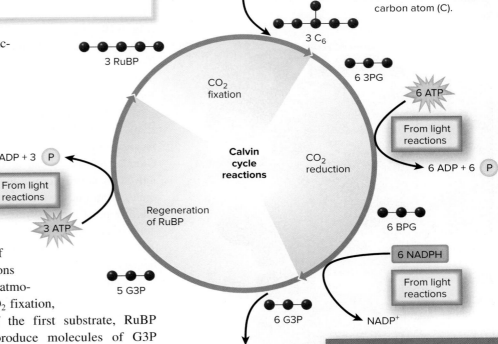

Key Molecules of the Calvin Cycle	
RuBP	ribulose-1,5-bisphosphate
3PG	3-phosphoglycerate
BPG	1,3-bisphosphoglycerate
G3P	glyceraldehyde 3-phosphate

Connections: Ecology

Can forests absorb enough carbon dioxide to offset climate change?

© PhotoLink/ Getty RF

When speaking of carbon absorption by forests, you are actually talking about the process of carbon dioxide fixation in the Calvin cycle. The U.S. Department of Agriculture estimates that, worldwide, forests absorb over 750 million metric tons of carbon dioxide per year. Unfortunately, this is only about 30% of the carbon dioxide emitted by human activity. While other factors are involved in the cycling of carbon (see Section 31.2), it is unlikely that forests alone can absorb enough carbon to balance the equation and reduce global climate change.

Figure 6.10   The fate of G3P.

G3P is the first reactant in a number of plant cell metabolic pathways, such as those that lead to products other than carbohydrates, such as oils (blue box), and those that are carbohydrates (green boxes). Two G3Ps are needed to produce glucose, the molecule that is often considered the end product of photosynthesis. However, other carbohydrates, such as sucrose (a transport sugar), starch (the storage form of glucose), and cellulose (a structural carbohydrate), may be produced.

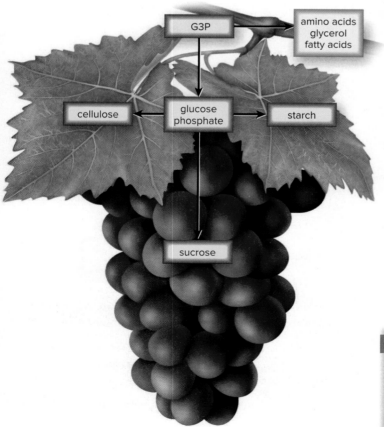

The enzyme for this reaction is called **RuBP carboxylase (rubisco)**, and the result is a 6-carbon molecule that soon thereafter splits into two 3-carbon molecules.

Reduction of Carbon Dioxide

Reduction of CO_2 is the sequence of reactions that uses NADPH and some of the ATP from the light reactions. NADPH and ATP supply the needed electrons and energy for CO_2 reduction, respectively. Electrons are added from NADPH, and through a series of reactions carbon dioxide is reduced to form G3P, a carbohydrate.

Regeneration of RuBP

The reactions of the Calvin cycle are cyclic, which means that some of the product returns to the start of the reactions. The final stage of the Calvin cycle (Figure 6.9) uses ATP from the light reactions to regenerate RuBP so that it can participate again in carbon fixation.

Notice that the Calvin cycle reactions in Figure 6.9 are multiplied by three because it takes three turns of the Calvin cycle to allow one G3P to exit for making glucose and other carbohydrates. Why? For every three turns of the Calvin cycle, five molecules of G3P are used to re-form three molecules of RuBP, which are needed to begin the next cycle of the reactions. In the process, some of the ATP produced by the light reactions is used.

The Fate of G3P

Compared with animal cells, algae and plants have enormous biochemical capabilities. From a G3P molecule, they can make all the molecules they need (**Fig. 6.10**). A plant can utilize the hydrocarbon skeleton of G3P to form fatty acids and glycerol, which are combined in plant oils. We are all familiar with the corn oil, sunflower oil, and olive oil used in cooking. Also, when nitrogen is added to the hydrocarbon skeleton derived from G3P, amino acids are formed.

Notice also that glucose phosphate is among the organic molecules that result from G3P metabolism. Glucose is the molecule that plants and animals most often metabolize to produce ATP molecules to meet their energy needs. Glucose phosphate can be combined with fructose (and the phosphate removed) to form sucrose, the molecule that plants use to transport carbohydrates from one part of the plant to another. Glucose phosphate is also the starting point for the synthesis of starch and cellulose. Starch is the storage form of glucose. Some starch is stored in chloroplasts, but most starch is stored in amyloplasts in plant roots. Cellulose is a structural component of plant cell walls; it serves as fiber in our diets because we are unable to digest it.

CONNECTING THE CONCEPTS

6.3   The Calvin cycle uses the outputs of the light reactions to reduce CO_2 and produce carbohydrates.

Check Your Progress 6.3

1. Briefly summarize the role of the Calvin cycle in photosynthesis.
2. Identify the source of the NADPH and ATP used in the Calvin cycle.
3. Explain the importance of G3P to plants.

6.4 Variations in Photosynthesis

Learning Outcomes

Upon completion of this section, you should be able to

1. Define C_4 photosynthesis, and explain why some plants must use this type of photosynthesis.
2. Describe both the advantages and the disadvantages of C_4 photosynthesis over C_3 photosynthesis.
3. Compare and contrast the leaf structure of a C_3 plant with that of a C_4 plant.
4. Explain CAM photosynthesis, and describe the conditions under which plants can use it.

Plants are physically adapted to their environments. In the cold, windy climates of the north, evergreen trees have small, narrow leaves that look like needles. In the warm, wet climates of the south, some evergreen trees have large, flat leaves to catch the rays of the sun.

C_3 Photosynthesis

Plants are metabolically adapted to their environments. Where temperature and rainfall tend to be moderate, plants carry on **C_3 photosynthesis** and are therefore called C_3 plants. Most plants are C_3 plants, and the reactions that we discussed in Section 6.3 are characteristic of this type of photosynthesis.

In a C_3 plant (**Fig. 6.11**), the first detectable molecule after CO_2 fixation (see Fig. 6.9) is a molecule composed of three carbon atoms, which accounts for the name C_3. In a C_3 leaf, mesophyll cells, arranged in parallel rows, contain well-formed chloroplasts. The Calvin cycle, including CO_2 fixation, occurs in the chloroplasts of mesophyll cells. The majority of plant species carry out C_3 photosynthesis.

If the weather is hot and dry, stomata close, preventing the loss of water, which might cause the plant to wilt and die. However, closing the stomata also prevents CO_2 from entering the leaf and traps O_2, a by-product of photosynthesis, within the leaf spaces, where it can diffuse back into mesophyll cells. In C_3 plants, this O_2 competes with CO_2 for the active site of *rubisco,* the first enzyme of the Calvin cycle, and thus less C_3 is produced. This competition is called *photorespiration* and it results in a reduced efficiency of photosynthesis in C_3 plants in hot, dry conditions.

C_4 Photosynthesis

Some plants have evolved an adaptation that allows them to be successful in hot, dry conditions. These plants carry out **C_4 photosynthesis** instead of C_3 photosynthesis. In a C_4 plant, the first detectable molecule following CO_2 fixation is a molecule composed of four carbon atoms (thus, C_4). C_4 plants are able to avoid the uptake of O_2 by rubisco. Let's explore how C_4 plants do this. The anatomy of a C_4 plant is different from that of a C_3 plant (**Fig. 6.12**). In a C_4 leaf, chloroplasts are located in the mesophyll cells, but they are also located in bundle sheath cells, which surround the leaf vein. Further, the mesophyll cells are arranged concentrically around the bundle sheath cells, shielding the bundle sheath cells from O_2 in the leaf spaces.

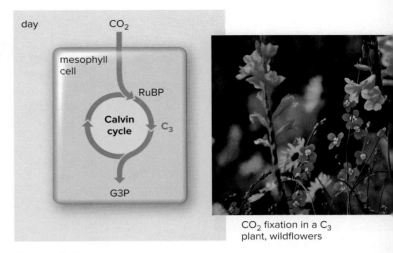

CO$_2$ fixation in a C$_3$ plant, wildflowers

Figure 6.11 **Carbon dioxide fixation in C$_3$ plants.**

In C$_3$ plants, such as these wildflowers, CO$_2$ is taken up by the Calvin cycle directly in mesophyll cells, and the first detectable molecule is a C$_3$ molecule (red).

© Brand X Pictures/PunchStock RF

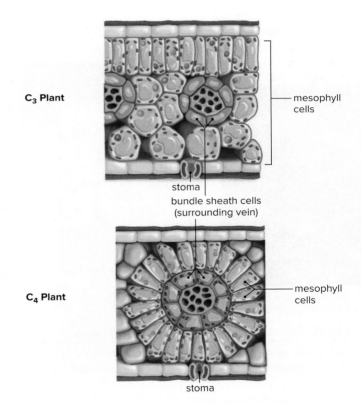

Figure 6.12 **Comparison of C$_3$ and C$_4$ plant anatomy.**

In C$_3$ plants, mesophyll cells are arranged in rows and contain chloroplasts. The Calvin cycle, including CO$_2$ fixation, occurs in mesophyll cells. In C$_4$ plants, mesophyll cells are arranged in rings around bundle sheath cells, and both contain chloroplasts. CO$_2$ fixation occurs within mesophyll cells, and the Calvin cycle occurs in the bundle sheath cells.

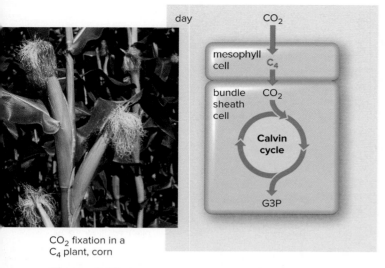

CO$_2$ fixation in a
C$_4$ plant, corn

Figure 6.13 **Carbon dioxide fixation in C$_4$ plants.**

In C$_4$ plants, such as corn, CO$_2$ fixation results in C$_4$ molecules (red)
in mesophyll cells. The C$_4$ molecule releases CO$_2$ to the Calvin cycle
in bundle sheath cells.

(photo): © USDA/Doug Wilson, photographer

CO$_2$ fixation in a
CAM plant, pineapple

Figure 6.14 **Carbon dioxide fixation in a CAM plant.**

CAM plants, such as pineapple, fix CO$_2$ at night, forming a C$_4$
molecule (red). The C$_4$ molecule releases CO$_2$ to the Calvin cycle
during the day.

(photo): © S. Alden/PhotoLink/Getty RF

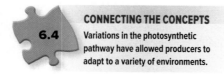

CONNECTING THE CONCEPTS

6.4

Variations in the photosynthetic
pathway have allowed producers to
adapt to a variety of environments.

In C$_4$ plants, the Calvin cycle occurs only in the bundle sheath cells and
not in the mesophyll cells. Because the bundle sheath cells are not accessible
to leaf spaces, the CO$_2$ needed for the Calvin cycle is not directly taken from
the air. Instead, CO$_2$ is fixed in mesophyll cells by a C$_3$ molecule, and a C$_4$
molecule forms. The C$_4$ molecule is modified and then pumped into bundle
sheath cells (**Fig. 6.13**). Then CO$_2$ enters the Calvin cycle. This represents
partitioning of CO$_2$ fixation and the Calvin cycle in separate spaces.

It takes energy to pump molecules, and you would think that the C$_4$ pathway would be disadvantageous. But in hot, dry climates, the net photosynthetic
rate of C$_4$ plants, such as sugarcane, corn, and Bermuda grass, is two to three
times that of C$_3$ plants, such as wheat, rice, and oats. Why do C$_4$ plants enjoy
such an advantage? When the weather is hot and dry and stomata close, rubisco
is not exposed to O$_2$, and yield is maintained.

When the weather is moderate, C$_3$ plants ordinarily have the advantage,
but when the weather becomes hot and dry, C$_4$ plants have their chance to take
over, and we can expect them to predominate. In the early summer, C$_3$ plants,
such as Kentucky bluegrass and creeping bent grass, predominate in lawns in
the cooler parts of the United States, but in midsummer, crabgrass, a C$_4$ plant,
begins to take over.

CAM Photosynthesis

Another type of photosynthesis is called **CAM photosynthesis,** which stands
for crassulacean-acid metabolism. It gets its name from the Crassulaceae, a
family of flowering succulent (water-containing) plants that live in warm, arid
regions. CAM photosynthesis was first discovered in these plants, but now it is
known to be prevalent among most succulent plants that grow in desert environments, including cacti.

Whereas a C$_4$ plant represents partitioning in space (that is, CO$_2$ fixation
occurs in mesophyll cells, and the Calvin cycle occurs in bundle sheath cells),
CAM photosynthesis uses partitioning in time. During the night, CAM plants
use C$_3$ molecules to fix CO$_2$, forming C$_4$ molecules. These molecules are stored
in large vacuoles in mesophyll cells. During the day, the C$_4$ molecules release
CO$_2$ to the Calvin cycle when NADPH and ATP are available from the light
reactions (**Fig. 6.14**).

The primary advantage of this partitioning again relates to the conservation of water. CAM plants open their stomata only at night; therefore, only at
that time is atmospheric CO$_2$ available. During the day, the stomata close. This
conserves water, but CO$_2$ cannot enter the plant.

Photosynthesis in a CAM plant is minimal because of the limited amount of
CO$_2$ fixed at night, but it does allow CAM plants to live under stressful conditions.

Evolutionary Trends

C$_4$ plants most likely evolved in, and are adapted to, areas of high light intensi-
ties, high temperatures, and limited rainfall. However, C$_4$ plants are more
sensitive to cold, and C$_3$ plants probably do better than C$_4$ plants below 25°C
(77°F).

CAM plants, on the other hand, compete well with either of the other
types of plant when the environment is extremely arid. Surprisingly, CAM is
quite widespread and has evolved in 30 families of flowering plants, including
cacti, stonecrops, orchids, and bromeliads. It is also found among nonflower-
ing plants, such as some ferns and cone-bearing trees.

Connections: Scientific Inquiry

What is known about C₄ plant evolution?

Scientists are trying to understand the evolutionary triggers that gave rise to the C_4 plants. They do know that the first evidence of C_4 plants dates back to around 25 MYA, which makes them a relatively young group of organisms. The exact location in which they first evolved is still a matter of debate, but most scientists agree it was probably a hot, dry environment. Despite being newcomers on the evolutionary scene, C_4 plants are very successful. Although only 4% of plant species are C_4 plants, over 20% of total annual plant growth is conducted by these plants. Of the 18 most problematic weed plants on the planet, 14 are C_4 plants.

STUDY TOOLS   http://connect.mheducation.com

SMARTBOOK® Maximize your study time with McGraw-Hill SmartBook®, the first adaptive textbook.

SUMMARIZE

Photosynthesis captures the energy from the sun to produce carbohydrates, which serve as the primary energy source for the majority of life on the planet.

6.1 Photosynthesis occurs within the chloroplasts of plants and some protists and involves two sets of chemical reactions.

6.2 The light reactions of photosynthesis capture solar energy to produce the ATP and NADPH needed to synthesize carbohydrates.

6.3 The Calvin cycle uses the outputs of the light reactions to reduce CO_2 and produce carbohydrates.

6.4 Variations in the photosynthetic pathway have allowed producers to adapt to a variety of environments.

6.1 Overview of Photosynthesis

Cyanobacteria, algae, and plants carry on **photosynthesis,** a process in which water is oxidized and carbon dioxide is reduced using solar energy. The combination of a **reduction** and an **oxidation** reaction is called a **redox reaction.** The end products of photosynthesis include carbohydrate and oxygen:

$$\text{solar energy} + CO_2 + H_2O \longrightarrow (CH_2O) + O_2$$

In plants, photosynthesis takes place in the **chloroplasts** of **mesophyll cells.** A chloroplast is bound by a double membrane and contains two main components: (1) the liquid **stroma** and (2) the membranous **grana** made up of **thylakoids.** During photosynthesis, the **light reactions** take place using **chlorophyll** molecules in the thylakoid membrane, and the **Calvin cycle** reactions take place in the stroma:

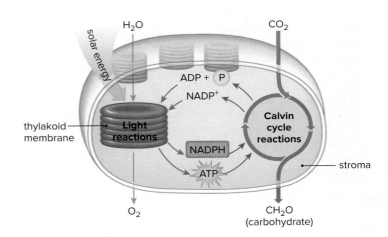

6.2 The Light Reactions—Harvesting Energy

- The light reactions use solar energy in the visible-light range. Pigment complexes, such as **chlorophyll** and **carotenoids,** in two **photosystems** (PS II and PS I) absorb various wavelengths of light. The collected solar energy results in high-energy electrons at the reaction centers of the photosystems.

- Solar energy enters PS II, and energized electrons are picked up by an electron acceptor. The oxidation (splitting) of water replaces these electrons in the reaction center. Oxygen (O_2) is released to the atmosphere, and hydrogen ions (H^+) remain in the thylakoid space.

- As electrons pass from one acceptor to another in an **electron transport chain,** the release of energy allows the carriers to pump H^+ into the thylakoid space. The buildup of H^+ establishes an electrochemical gradient.

- When solar energy is absorbed by PS I, energized electrons leave and are ultimately received by NADP⁺ (a **coenzyme**), which also combines with H⁺ from the stroma to become NADPH. Electrons from PS II replace those lost by PS I.
- When H⁺ flows down its concentration gradient through the channel present in **ATP synthase** complexes, ATP is synthesized from ADP and Ⓟ by ATP synthase.

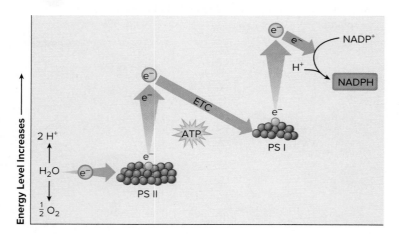

6.3 The Calvin Cycle Reactions—Making Sugars

The Calvin cycle consists of the following stages:

- *CO_2 fixation:* The enzyme **RuBP carboxylas**e fixes CO_2 to RuBP, producing a 6-carbon molecule that immediately breaks down to two C_3 molecules.
- *CO_2 reduction:* CO_2 (incorporated into an organic molecule) is reduced to carbohydrate (CH_2O). This step requires the NADPH and some of the ATP from the light reactions.
- *Regeneration of RuBP:* For every three turns of the Calvin cycle, the net gain is one G3P molecule; the other five G3P molecules are used to re-form three molecules of RuBP. This step also requires the energy of ATP. G3P is then converted to all the organic molecules a plant needs. It takes two G3P molecules to make one glucose molecule.

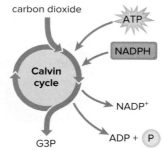

6.4 Variations in Photosynthesis

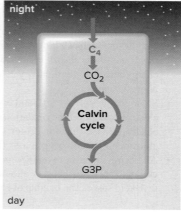

C₄ Photosynthesis

When the weather is hot and dry, plants using **C_3 photosynthesis** are at a disadvantage because stomata close and O_2 from photosynthesis competes with CO_2 for the active site of RuBP carboxylase. C_4 plants avoid this drawback. In **C_4 photosynthesis,** CO_2 fixation occurs in mesophyll cells, and the Calvin cycle reactions occur in bundle sheath cells. In mesophyll cells, a C_3 molecule fixes CO_2, and the result is a C_4 molecule, for which this type of photosynthesis is named. A modified form of this molecule is pumped into bundle sheath cells, where CO_2 is released to the Calvin cycle. This represents partitioning of CO_2 fixation and the Calvin cycle in space.

CAM Photosynthesis

Plants that used **CAM photosynthesis** live in hot, dry environments. At night when stomata remain open, a C_3 molecule fixes CO_2 to produce a C_4 molecule. The next day, CO_2 is released and enters the Calvin cycle within the same cells. This represents a partitioning of CO_2 fixation and the Calvin cycle in time: Carbon dioxide fixation occurs at night, and the Calvin cycle occurs during the day.

Photosynthesis produces the carbohydrates that the majority of living organisms on Earth use as a source of energy.

ASSESS

Testing Yourself

Choose the best answer for each question.

6.1 Overview of Photosynthesis

1. The raw materials for photosynthesis are
   a. oxygen and water.
   b. oxygen and carbon dioxide.
   c. carbon dioxide and water.
   d. carbohydrates and water.
   e. carbohydrates and carbon dioxide.
2. During photosynthesis, carbon dioxide is _____ and water is _____.
   a. reduced, oxidized
   b. oxidized, reduced
   c. reduced, reduced
   d. oxidized, oxidized

6.2 The Light Reactions—Harvesting Energy

3. When electrons in the reaction center of PS II are passed to an energy-acceptor molecule, they are replaced by electrons that have been given up by
   a. oxygen.
   b. glucose.
   c. carbon dioxide.
   d. water.
4. During the light reactions of photosynthesis, ATP is produced when hydrogen ions move
   a. down a concentration gradient from the thylakoid space to the stroma.
   b. against a concentration gradient from the thylakoid space to the stroma.
   c. down a concentration gradient from the stroma to the thylakoid space.
   d. against a concentration gradient from the stroma to the thylakoid space.
5. Oxygen is generated by the
   a. light reactions.
   b. Calvin cycle.
   c. light reactions and the Calvin cycle reactions.
   d. closing of stomata.

6. The enzyme that produces ATP from ADP + $\textcircled{P}$ in the thylakoid is
   a. RuBP carboxylase.
   d. ATP synthase.
   b. G3P.
   e. coenzyme A.
   c. ATPase.

6.3 The Calvin Cycle Reactions—Making Sugars

7. Energy for the Calvin cycle is supplied by _____ from the light reactions.
   a. carbon dioxide and water
   d. ATP and water
   b. ATP and NADPH
   e. electrons and hydrogen ions
   c. carbon dioxide and ATP

8. Carbon dioxide fixation occurs when CO_2 combines with
   a. ATP.
   c. G3P.
   b. NADPH.
   d. RuBP.

9. The output of the Calvin cycle that is used to generate carbohydrates is
   a. NADPH.
   c. G3P.
   b. CO_2.
   d. RuBP.

6.4 Variations in Photosynthesis

10. In C_4 plants, CO_2 fixation occurs
    a. by adding electrons and a hydrogen ion when splitting water in bundle sheath cells.
    b. by joining CO_2 to a C_3 molecule and pumping it into bundle sheath cells.
    c. by splitting water only in bundle sheath cells.
    d. by oxidizing a C_4 molecule with O_2 derived from bundle sheath cells.

11. A leaf of a C_4 plant differs from that of a C_3 plant because
    a. stomata cannot close.
    b. the mesophyll has no leaf spaces.
    c. bundle sheath cells are protected from leaf spaces by a ring of mesophyll cells.
    d. C_4 plant leaves are narrower to allow more O_2 to diffuse away.

ENGAGE

BioNOW

Want to know how this science is relevant to your life? Check out the BioNow video below:

- Energy Part II: Photosynthesis

Why does this experiment only need to produce light in the wavelengths at the two ends of the visible light spectrum?

Thinking Critically

1. In 1882, T. W. Engelmann carried out an ingenious experiment to demonstrate that chlorophyll absorbs light in the blue and red portions of the spectrum. He placed a single filament of a green alga in a drop of water on a microscope slide. Then he passed light through a prism and onto the string of algal cells. The slide also contained aerobic bacterial cells. After some time, he peered into the microscope and saw the bacteria clustered around the regions of the algal filament that were receiving blue light and red light, as shown in the following illustration. Why do you suppose the bacterial cells were clustered in this manner?

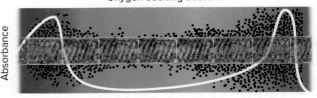

2. Artificial leaves are beginning to be developed by biotechnology companies. Outline the basic characteristics that an artificial leaf would need to have to manufacture a sugar.

3. Why do broad, thin leaves provide an advantage for photosynthesis?

4. How might climate change influence the distribution of C3 and C4 plants in North America?

5. The products of photosynthesis are increasingly being used to generate carbohydrates, which in turn are used to produce ethanol as a biofuel. The conversion of food crops, such as corn, into ethanol has created concerns about the reserves of food in the world. Based on your understanding of photosynthesis, what types of other plants might make a good choice for ethanol production in an increasingly warmer world?

7

Energy for Cells

© Pool/Getty Images Sports

OUTLINE

7.1 Cellular Respiration   111

7.2 Outside the Mitochondria:
Glycolysis   113

7.3 Outside the Mitochondria:
Fermentation   115

7.4 Inside the Mitochondria   117

7.5 Metabolic Fate of Food   121

BEFORE YOU BEGIN

Before beginning this chapter, take a few moments to review the following discussions.

Section 3.2 What are the roles of carbohydrates in living organisms?

Section 4.4 What is the function of the mitochondria in a cell?

Section 5.2 Why does the ATP cycle resemble a rechargeable battery?

Metabolic Demands of Athletes

During a typical basketball game, such as the 2016 NCAA championship game between Villanova and North Carolina, the starting players run an average of 4 to 5 kilometers (about 3 miles). However, unlike the steady pace maintained by marathoners, basketball players experience periods of intense activity (sprinting), followed by brief periods of rest. This start-and-stop nature of the game means that the athletes' muscles are constantly switching between aerobic and anaerobic metabolism.

During aerobic metabolism, the muscle cells use oxygen in order to completely break down glucose and produce ATP, the energy "battery" of the cell. This process, which uses carbohydrates and produces carbon dioxide as a waste product, occurs in the cytoplasm and mitochondria of the cell. However, running short, fast sprints quickly depletes oxygen levels in the blood and drives the muscles into anaerobic metabolism. Without oxygen, glucose cannot be completely broken down and is changed to lactate, which is responsible for the muscle "burn" we feel after intense exercise. Once oxygen is restored, usually after a brief rest, the body is able to return to aerobic metabolism. Professional athletes are conditioned to reduce the amount of time they are in anaerobic metabolism and to maximize their aerobic metabolism.

In this chapter, we will not only examine these pathways but also explore how nutrients such as fats and proteins can be used to generate ATP for our cellular activities.

As you read through this chapter, think about the following questions:

1. What are the differences between the aerobic and anaerobic pathways?
2. What is the role of oxygen in these pathways?
3. How are other organic nutrients, such as fats and proteins, used instead of glucose for energy?

7.1 Cellular Respiration

Learning Outcomes

Upon completion of this section, you should be able to

1. Explain the role of cellular respiration in a cell.
2. State the overall reaction for glucose breakdown.
3. List the four phases of cellular respiration, and identify the location of each within the cell.

No matter what you are doing, from playing basketball to studying, your cells are using energy. **Adenosine triphosphate (ATP)** molecules power activities at the cellular level. ATP molecules are produced during **cellular respiration** by the breakdown of organic molecules, primarily the carbohydrate glucose, with the participation of the mitochondria within eukaryotic cells. Cellular respiration is aptly named because, just as you take in oxygen (O_2) and give off carbon dioxide (CO_2) during breathing, so do the mitochondria in your cells (**Fig. 7.1**) during cellular respiration. In fact, cellular respiration, which occurs in all the cells of the body, is the reason you breathe.

Oxidation of substrates is a fundamental part of cellular respiration. Oxidation represents the removal of electrons and hydrogen ions from a molecule, which are then used to reduce a second molecule, completing a redox reaction. Cellular respiration is an excellent example of a redox reaction. During cellular respiration, glucose is oxidized to form carbon dioxide (CO_2). In the process, oxygen is reduced to form water (H_2O). In this case, the redox reaction releases energy, which is then used by the cell.

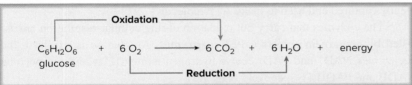

$$\text{Oxidation}$$
$$\underset{\text{glucose}}{C_6H_{12}O_6} + 6\,O_2 \longrightarrow 6\,CO_2 + 6\,H_2O + \text{energy}$$
$$\text{Reduction}$$

This reaction should look familiar. Notice that if you reverse the arrow in this equation, you obtain the overall reaction for photosynthesis (see Section 6.1). In cellular respiration, the oxidation of carbohydrates produces energy and carbon dioxide. In photosynthesis, the reduction of carbon dioxide, using energy from the sun, produces carbohydrate.

Figure 7.1 Cellular respiration.

Glucose from our food and the oxygen we breathe are requirements for cellular respiration. Carbon dioxide and water are released as by-products. The process begins in the cytoplasm but is completed in the mitochondria.

(photo): © Peter Cade/Getty Images

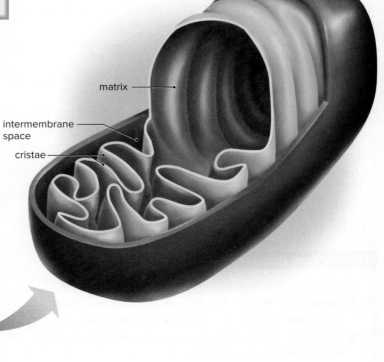

double membrane — outer membrane / inner membrane

matrix

intermembrane space

cristae

oxygen

H_2O

CO_2

glucose

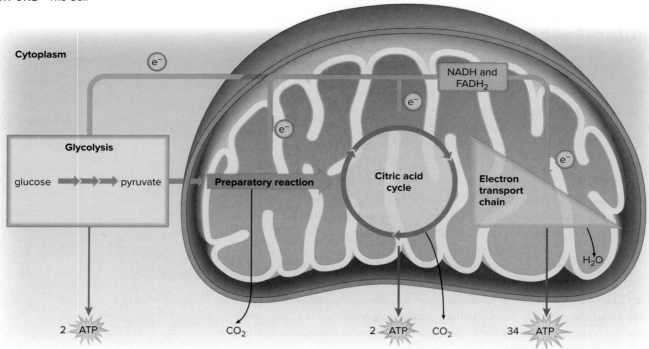

Figure 7.2 The four phases of complete glucose breakdown.

The enzymatic reactions of glycolysis take place in the cytoplasm. The preparatory reaction, the citric acid cycle, and the electron transport chain occur in mitochondria.

Phases of Complete Glucose Breakdown

The breakdown of glucose releases a lot of energy. However, unlike the burning of sugar on a stove, which releases most of the energy as heat, the cell controls the reactions so that the glucose molecules are broken down slowly. This allows the energy in the bonds of the glucose molecule to be more efficiently captured and used to make ATP molecules.

The enzymes that carry out oxidation during cellular respiration are assisted by nonprotein helpers called **coenzymes.** As glucose is oxidized, the coenzymes NAD^+ and FAD receive hydrogen atoms ($H^+ + e^-$) and become NADH and $FADH_2$, respectively (**Fig. 7.2**).[1]

There are four phases to cellular respiration. Notice the inputs and outputs of each phase (Fig. 7.2).

- **Glycolysis,** which occurs in the cytoplasm of the cell, is the breakdown of glucose to two molecules of pyruvate. Energy is invested to activate glucose, two ATP are gained, and oxidation results in NADH, which will be used later for additional ATP production.
- The **preparatory (prep) reaction** takes place in the matrix of mitochondria. Pyruvate is broken down to a 2-carbon acetyl group carried by coenzyme A (CoA). Oxidation of pyruvate yields not only NADH but also CO_2.
- The **citric acid cycle** also takes place in the matrix of mitochondria. As oxidation occurs, NADH and $FADH_2$ result and more CO_2 is released. The citric acid cycle is able to produce two ATP per glucose molecule.
- The **electron transport chain** is a series of electron carriers in the cristae of mitochondria. NADH and $FADH_2$ give up electrons to the chain. Energy is released and captured as the electrons move from a higher energy to a lower energy state. Later, this energy will be used for the production of ATP. Oxygen (O_2) is the final electron acceptor. It then combines with hydrogen ions (H^+) to produce water (H_2O).

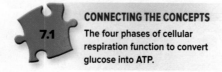

CONNECTING THE CONCEPTS

7.1 The four phases of cellular respiration function to convert glucose into ATP.

Check Your Progress 7.1

1. Explain why cellular respiration is needed by the cell.

2. Explain the similarities between the overall reaction for cellular respiration and that for photosynthesis.

3. Summarize the four stages of cellular respiration, and identify their locations in the cell.

---

[1]NAD = nicotinamide adenine dinucleotide; FAD = flavin adenine dinucleotide.

7.2 Outside the Mitochondria: Glycolysis

Learning Outcomes

Upon completion of this section, you should be able to

1. Explain the role of glycolysis in a cell.
2. Distinguish between the energy-investment and energy-harvesting steps of glycolysis.
3. Summarize how the metabolic pathway of glycolysis partially breaks down glucose.

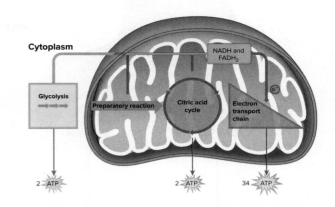

In both eukaryotes and prokaryotes, glycolysis takes place within the cytoplasm of the cell. During glycolysis, glucose, a 6-carbon molecule, is broken down to two molecules of pyruvate, a 3-carbon molecule. Glycolysis is divided into (1) the energy-investment step, when some ATP is used to begin the reactions, and (2) the energy-harvesting steps, when both NADH and ATP are produced. **Figure 7.3** outlines the basic steps of glycolysis, including some of the more important intermediate molecules. As you examine this figure, notice how it includes an initial investment of ATP, followed by an ATP-harvesting phase.

Energy-Investment Step

During the energy-investment step, two ATP transfer phosphate groups to substrates, and two ADP + P result. Why does this happen? Much the way your car battery starts the engine, the transfer of phosphates activates the substrates, so that they can undergo the energy-harvesting reactions.

3PG	3-phosphoglycerate
BPG	1,3-bisphosphoglycerate
G3P	glyceraldehyde 3-phosphate

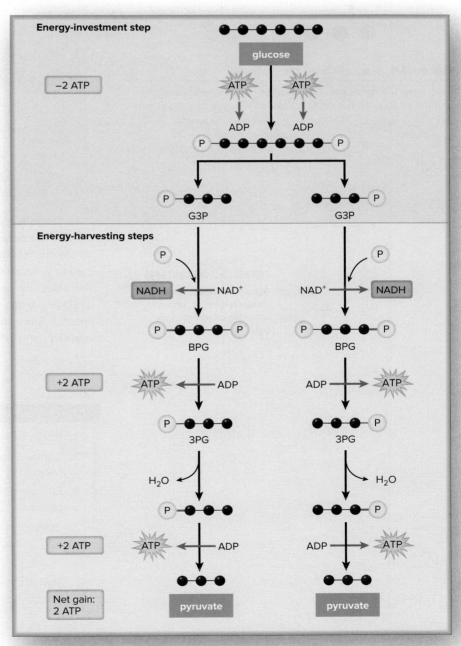

Figure 7.3 Glycolysis.

This metabolic pathway begins with a 6-carbon glucose and ends with two 3-carbon pyruvate molecules. A net gain of two ATP can be calculated by subtracting the ATP molecules expended during the energy-investment step from those produced during the energy-harvesting steps. Each ● is a carbon atom (C).

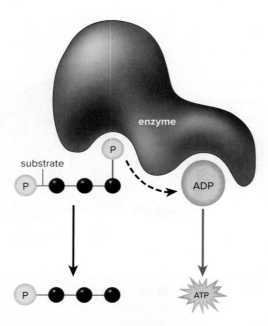

Figure 7.4 Substrate-level ATP synthesis.

At an enzyme's active site, ADP acquires an energized phosphate group from an activated substrate, and ATP results. The net gain of two ATP from glycolysis is the result of substrate-level ATP synthesis.

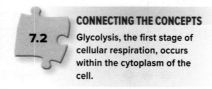

CONNECTING THE CONCEPTS
7.2
Glycolysis, the first stage of cellular respiration, occurs within the cytoplasm of the cell.

Energy-Harvesting Steps

During the energy-harvesting steps, electrons and hydrogen ions are removed from the substrates and captured by NAD^+, producing two NADH.

As the reactions proceed, the energized phosphate groups on the intermediates are used to synthesize four ATP. This is accomplished by a process called substrate-level ATP synthesis (**Fig. 7.4**), which basically involves a transfer of a phosphate group from the intermediate molecule to ADP, thus forming ATP.

What is the net gain of ATP from glycolysis? Notice in Figure 7.3 that 2 ATP are used to get started, and 4 ATP are produced. Therefore, there is a net gain of 2 ATP from glycolysis. The overall inputs and outputs of glycolysis are provided below.

Glycolysis	
Inputs	**Outputs**
glucose	2 pyruvate
2 NAD^+	2 NADH
2 ATP	2 ADP
4 ADP + 4 P	4 ATP
	2 ATP net

How the reactions proceed from here is determined by whether oxygen is present. In a eukaryotic cell, if oxygen is available, the end product of glycolysis (pyruvate), enters the aerobic reactions within mitochondria, where it is used to generate more ATP (see Section 7.4). If oxygen is not available, pyruvate enters the anaerobic fermentation pathways (see Section 7.3). In humans, if oxygen is not available, pyruvate is reduced to lactate (see the chapter opener). Since the anaerobic pathways also occur in the cytoplasm, we will consider them next.

Check Your Progress 7.2

1. Briefly explain the role of glycolysis in a cell.

2. Contrast the energy-investment step of glycolysis with the energy-harvesting steps.

3. Summarize the net yield of ATP per glucose molecule that results from glycolysis.

7.3 Outside the Mitochondria: Fermentation

The complete breakdown of glucose requires an input of oxygen to keep the electron transport chain working. If oxygen is limited, pyruvate molecules accumulate in the cell, and intermediates, such as NAD^+ and FAD, cannot be recycled. To correct for this, cells may enter anaerobic pathways, such as **fermentation,** following glycolysis. There are two basic forms of fermentation—lactic acid and alcohol (**Fig. 7.5**).

The overall inputs and outputs of fermentation are shown below:

Fermentation	
Inputs	**Outputs**
glucose	2 lactate or 2 alcohol and 2 CO_2
2 ATP	2 ADP
4 ADP + 4 P	4 ATP
	2 ATP net

Now let us take a closer look at the two different forms of fermentation.

Lactic Acid Fermentation

In animals and some bacteria, the pyruvate formed by glycolysis accepts two hydrogen atoms and is reduced to lactate. In the water environment of the cell, this forms lactic acid. Notice in Figure 7.5 that two NADH pass hydrogen atoms to pyruvate, reducing it. Why is it beneficial for pyruvate to be reduced to lactate when oxygen is not available? The answer is that this reaction regenerates NAD^+, which can then pick up more electrons during the earlier reactions of glycolysis. This regeneration of NAD^+ keeps glycolysis going, during which ATP is produced by substrate-level ATP synthesis.

The two ATP produced by the anaerobic processes of glycolysis and fermentation represent only a small fraction of the potential energy stored in a glucose molecule. Following fermentation, most of this potential energy is still waiting to be released. Despite the low yield of only two ATP, the anaerobic pathways are essential. They can provide a rapid burst of ATP, and muscle cells are more apt than other cells to carry on fermentation. When our muscles are working vigorously over a short period of time, as when we run, fermentation is a way to produce ATP even though oxygen is temporarily in limited supply.

Connections: Health

Does lactic acid cause soreness in your muscles after exercise?

While the accumulation of lactic acid in muscles does cause the "burn" you may feel after prolonged exercise, it does not usually cause long-term muscle soreness. Instead, the tenderness that occurs a day or two after intense exercise is probably being caused by the response of anti-inflammatory components of the immune system to damage to the muscle tissues.

© Bill Robbins/Getty Images

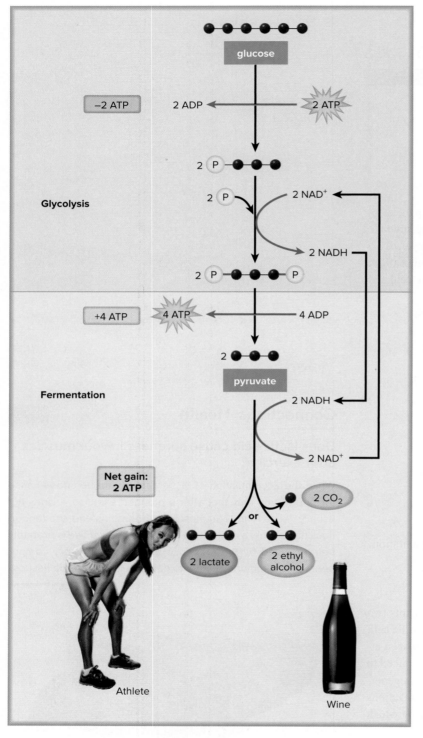

When we stop running, our bodies are in *oxygen deficit,* as indicated by the fact that we continue to breathe very heavily for a time. Recovery is complete when enough oxygen is present to completely break down glucose. Blood carries away the lactate formed in muscles and transports it to the liver, where it is reconverted to pyruvate. Some of the pyruvate is oxidized completely, and the rest is converted back to glucose.

Alcohol Fermentation

In other organisms (bacteria and fungi), pyruvate is reduced to produce alcohol. As was the case with lactic acid fermentation, the electrons needed to reduce the pyruvate are supplied by NADH molecules. In the process, NAD^+ molecules are regenerated for use in glycolysis (Fig. 7.5). However, unlike lactic acid fermentation, alcohol fermentation releases small amounts of CO_2.

Yeasts (a type of fungi) are good examples of microorganisms that generate ethyl alcohol and CO_2 when they carry out fermentation. When yeasts are used to leaven bread, the CO_2 makes the bread rise. When yeasts are used to ferment grapes for wine production or to ferment wort—derived from barley—for beer production, ethyl alcohol is the desired product. However, the yeasts are killed by the very product they produce.

Figure 7.5 The anaerobic pathways.

Fermentation consists of glycolysis followed by a reduction of pyruvate by NADH. This regenerates NAD^+, which returns to the glycolytic pathway to pick up more hydrogen atoms.

(runner): © Michael Svoboda/Getty RF.; (wine): © C Squared Studios/Getty RF

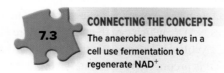

CONNECTING THE CONCEPTS

7.3

The anaerobic pathways in a cell use fermentation to regenerate NAD^+.

Check Your Progress 7.3

1. Explain the difference between aerobic respiration and anaerobic fermentation.

2. Explain why fermentation is needed when oxygen is not present.

3. Describe the difference between lactic acid and alcohol fermentation.

7.4 Inside the Mitochondria

Learning Outcomes

Upon completion of this section, you should be able to

1. Identify the roles of the preparatory reaction and the citric acid cycle in the breakdown of glucose.
2. Specify how the electron transport chain produces most of the ATP during cellular respiration.
3. Identify the inputs and outputs of each pathway of aerobic cellular respiration.
4. Explain the role of oxygen in cellular respiration.

If oxygen is present following glycolysis, the cell will enter the aerobic phases of cellular respiration. These reactions all occur inside the mitochondria (**Fig. 7.6**).

Preparatory Reaction

Occurring in the matrix, the preparatory (prep) reaction is so named because it prepares the outputs of glycolysis (pyruvate molecules) for use in the citric acid cycle. Recall that glycolysis splits each glucose into two pyruvate molecules (see Fig. 7.3); thus, the preparatory reaction occurs twice per glucose molecule. During the preparatory reaction (**Fig. 7.7, green box**),

- Pyruvate is oxidized, and a CO_2 molecule is given off. This is part of the CO_2 we breathe out.
- NAD^+ accepts electrons and hydrogen ions, forming NADH.
- The product, a 2-carbon acetyl group, is attached to coenzyme A (CoA), forming **acetyl-CoA.**

Therefore,

- Per glucose molecule, the outputs are two CO_2, two NADH, and two acetyl-CoA.

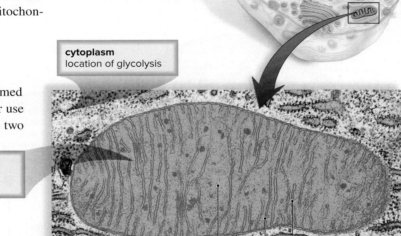

cytoplasm
location of glycolysis

mitochondrion
location of aerobic respiration

outer membrane   matrix   cristae   85,000×

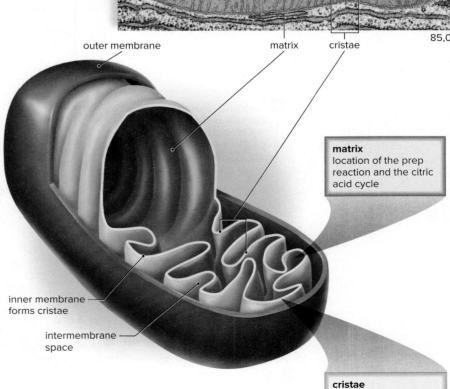

matrix
location of the prep reaction and the citric acid cycle

inner membrane
forms cristae

intermembrane
space

cristae
location of the electron transport chain

Figure 7.6 Mitochondrion structure and function.

A mitochondrion is bound by a double membrane. The inner membrane folds inward to form the shelflike cristae. Glycolysis takes place in the cytoplasm outside the mitochondrion. The preparatory (prep) reaction and the citric acid cycle occur within the mitochondrial matrix. The electron transport chain is located on the cristae of each mitochondrion.

(photo): © Keith R. Porter/Science Source

The Citric Acid Cycle

The citric acid cycle is a cyclical metabolic pathway located in the matrix of mitochondria (Fig. 7.7). It was originally called the *Krebs cycle* to honor the scientist who first studied it. At the start of the citric acid cycle, the 2-carbon acetyl group carried by CoA joins with a 4-carbon molecule, producing a 6-carbon citrate molecule. The CoA is released and is used again in the preparatory reaction.

During the citric acid cycle,

- The acetyl group is oxidized, in the process forming CO_2.
- Both NAD^+ and FAD accept electrons and hydrogen ions, resulting in NADH and $FADH_2$.
- Substrate-level ATP synthesis occurs (see Fig. 7.4), and an ATP results.

Because the citric acid cycle turns twice for each original glucose, the inputs and outputs of the citric acid cycle per glucose are as follows:

Citric acid cycle	
Inputs	**Outputs**
2 acetyl-CoA	4 CO_2
6 NAD^+	6 NADH
2 FAD	2 $FADH_2$
2 ADP + 2 Ⓟ	2 ATP

Figure 7.7 **The preparatory reaction and the citric acid cycle.**

Each acetyl-CoA from the preparatory reaction enters the citric acid cycle. The net result of one turn of this cycle of reactions is the oxidation of the acetyl group to two CO_2 and the formation of three NADH and one $FADH_2$. Substrate-level ATP synthesis occurs, and the result is one ATP. For each glucose, the citric acid cycle turns twice, doubling each of these amounts.

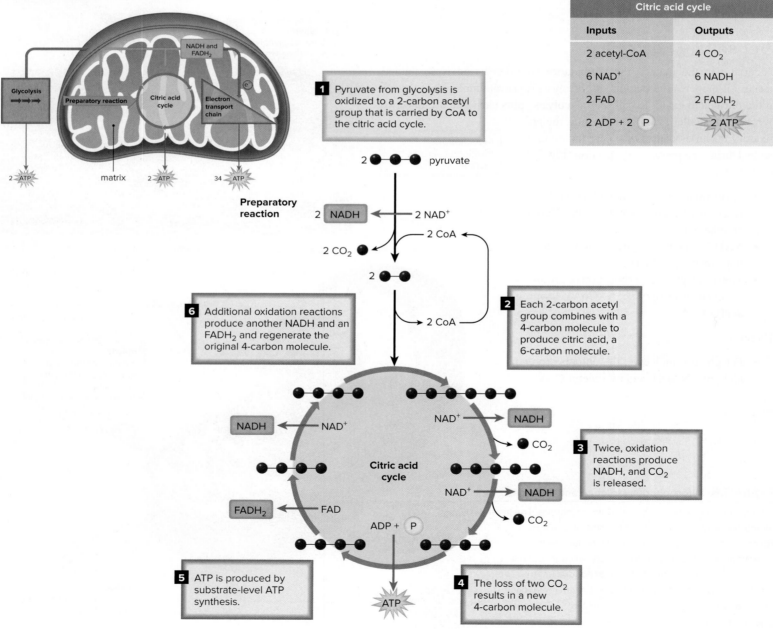

The Electron Transport Chain

The electron transport chain located in the cristae of mitochondria is a series of carriers that pass electrons from one to another. NADH and $FADH_2$ deliver electrons to the chain. Consider that the hydrogen atoms attached to NADH and $FADH_2$ consist of an electron (e^-) and a hydrogen ion (H^+). The carriers of the electron transport chain accept only e^- and not H^+.

In **Figure 7.8**, notice that high-energy electrons are being transferred by NADH and $FADH_2$ from glycolysis, the preparatory reaction, and the citric acid cycle. The electron transport chain is a series of redox reactions that remove the high-energy electrons from NADH and $FADH_2$. As these reactions occur, energy is released and captured for ATP production.

The final acceptor of electrons is oxygen (O_2). Your breathing provides the O_2 for cellular respiration. Overall, the role of oxygen in cellular respiration is to act as the final acceptor of electrons in the electron transport chain. Why can oxygen play this role? Because oxygen attracts electrons to a greater degree than do the carriers of the chain. Once oxygen accepts electrons, it combines with H^+ to form a water molecule, which is one of the end products of cellular respiration (see the equation in Section 7.1).

Once NADH has delivered electrons to the electron transport chain, NAD^+ is regenerated and can be used again. In the same manner, FAD is regenerated and can be used again. The recycling of coenzymes, and for that matter ADP, increases cellular efficiency, since it does away with the need to synthesize new NAD^+, FAD, and ADP each time the reactions occur.

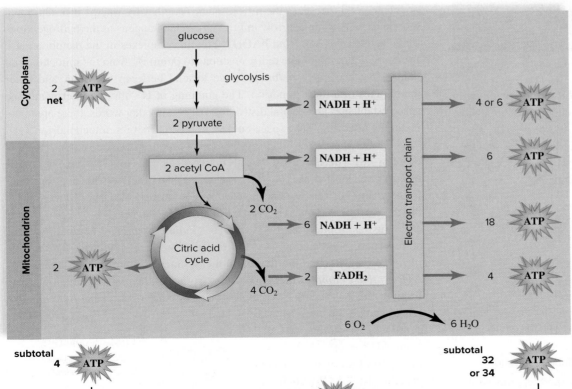

Figure 7.8 The electron transport chain.

An electron transport chain receives NADH and $FADH_2$, each carrying high-energy electrons, from glycolysis, the preparatory reaction, and the citric acid cycle. The majority of ATP production occurs in the electron transport chain.

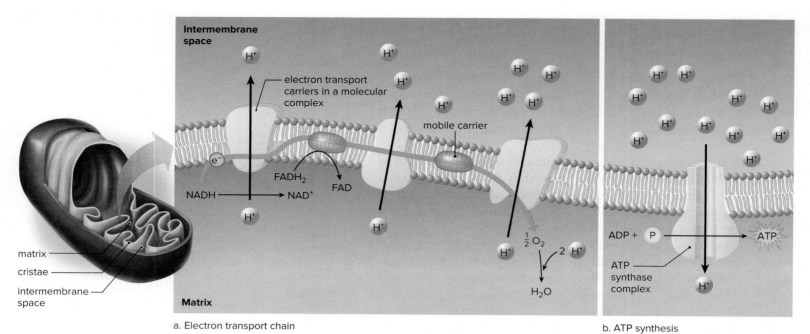

a. Electron transport chain

b. ATP synthesis

Figure 7.9 The organization of cristae.

Molecular complexes that contain the electron transport carriers are located in the cristae, as are ATP synthase complexes. **a.** As electrons move from one carrier to the other, hydrogen ions (H^+) are pumped from the matrix into the intermembrane space. **b.** As hydrogen ions flow back down a concentration gradient through an ATP synthase complex, ATP is synthesized by the enzyme ATP synthase.

Connections: Health

Why is carbon monoxide so dangerous?

Carbon monoxide (CO) is a colorless, odorless gas that is produced by the incomplete burning of organic material, such as fossil fuels and wood. Carbon monoxide binds more strongly than oxygen to the hemoglobin in your red blood cells. When this happens, the red blood cells cannot transport adequate oxygen to the mitochondria of your cells, and ATP production shuts down. Over 400 people die annually in the United States due to carbon monoxide poisoning. For a list of ways to prevent CO poisoning, visit the CDC website at http://www.cdc.gov/co/default.htm.

© Hisham F. Ibrahim/ Getty RF

Check Your Progress 7.4

1. Summarize the roles of the preparatory reaction, citric acid cycle, and electron transport chain in cellular respiration.

2. Explain how oxygen is used in cellular respiration.

3. List the inputs and outputs of the preparatory reaction, citric acid cycle, and electron transport chain.

Generating ATP

The electron transport chain is a series of electron carriers (mostly proteins) located within the inner mitochondrial membrane. ATP synthesis is carried out by ATP synthase complexes, also located in this membrane (**Fig. 7.9**).

The carriers of the electron transport chain accept electrons from NADH or $FADH_2$ and then pass them from one to another by way of two additional mobile carriers (orange arrow in Fig. 7.9). What happens to the hydrogen ions (H^+) carried by NADH and $FADH_2$? Protein complexes in the membrane use the energy released by the redox reactions to pump H^+ from the mitochondrial matrix into the intermembrane space located between the outer and inner membranes of a mitochondrion. The pumping of H^+ into the intermembrane space establishes an unequal distribution of H^+; in other words, there are many H^+ in the intermembrane space but few in the matrix of a mitochondrion.

Notice that NADH and $FADH_2$ do not produce the same amount of ATP (see Fig 7.8). This is because NADH delivers electrons to the first carrier of the electron transport chain (Fig 7.9), whereas $FADH_2$ delivers its electrons later in the chain. For this reason, the electrons from each NADH produce 3 ATP, but the electrons from each $FADH_2$ produce only 2 ATP.

Just as in photosynthesis, the H^+ gradient contains a large amount of stored energy that can be used to drive forward ATP synthesis. The cristae of mitochondria (like the thylakoid membranes of chloroplasts) contain an *ATP synthase complex* that allows H^+ to return to the matrix. The flow of H^+ through the ATP synthase complex brings about a conformational change, which causes the enzyme ATP synthase to synthesize ATP from ADP + ⓟ. ATP leaves the matrix by way of a channel protein. This ATP remains in the cell and is used for cellular work.

CONNECTING THE CONCEPTS

7.4

If oxygen is present, cells will use the aerobic pathways of the citric acid cycle and electron transport chain within the mitochondria to generate ATP.

7.5 Metabolic Fate of Food

Learning Outcomes

Upon completion of this section, you should be able to

1. Calculate the amount of ATP produced by each glucose molecule entering cellular respiration.
2. Recognize how alternate metabolic pathways allow proteins and fats to be used for ATP production.

Energy Yield from Glucose Metabolism

How much ATP is actually produced per molecule of glucose? **Figure 7.10** calculates the potential ATP yield for the complete breakdown of glucose to CO_2 and H_2O. For each glucose, there is a net gain of two ATP from glycolysis, which takes place in the cytoplasm. The citric acid cycle, which occurs in the matrix of mitochondria, accounts for two ATP per glucose. This means that a total of four ATP form due to substrate-level ATP synthesis outside the electron transport chain.

Most of the ATP produced comes from the electron transport chain and the ATP synthase complex. For each glucose, 10 NADH and 2 $FADH_2$ take electrons from glycolysis and the citric acid cycle to the electron transport chain. The maximum number of ATP produced by the chain is therefore 34 ATP, and the maximum number produced by both the chain and substrate-level ATP synthesis is 38. However, not all cells produce the maximum yield. Metabolic differences cause some cells to produce 36 or fewer ATP. Still, a yield of 36–38 ATP represents about 40% of the energy that was initially available in the glucose molecule. The rest of the energy is lost in the form of heat.

Alternative Metabolic Pathways

You probably know that you can obtain energy from foods other than carbohydrates. For example, your body has the means of obtaining energy from a meal of pepperoni pizza (**Fig. 7.11**), not all of which is carbohydrates. Let's start with the fats in the cheese and meat. When a fat is used as an energy source, it breaks down to glycerol and three fatty acid chains (see Fig. 3.12). Because glycerol is a carbohydrate, it enters the process of cellular respiration during glycolysis. Fatty acids can be metabolized to acetyl groups, which enter the citric acid cycle. A fatty acid with a chain of 18 carbons can make three times the number of acetyl groups that a glucose molecule does. For this reason, fats are an efficient form of stored energy. The complete breakdown of glycerol and fatty acids results in many more ATP per fat molecule than does the breakdown of a glucose molecule.

So how about the proteins in the pepperoni? Only the hydrocarbon backbone of amino acids can be used by the cellular respiration pathways. The amino group becomes ammonia (NH_3), which becomes part of urea, the primary excretory product of humans. Just where the hydrocarbon backbone from an amino acid begins degradation to produce ATP depends on its length. Figure 7.11 shows that the hydrocarbon backbone from an amino acid can enter cellular respiration pathways at pyruvate, at acetyl-CoA, or during the citric acid cycle.

The smaller molecules in Figure 7.11 can also be used to synthesize larger molecules. In such instances, ATP is used instead of generated. For example, some substrates of the citric acid cycle can become amino acids through the addition of an amino group, and these amino acids can be employed

Phase	NADH	FADH$_2$	ATP Yield
Glycolysis	2	–	2
Prep reaction	2	–	–
Citric acid cycle	6	2	2
Electron transport chain	10 ——————→ 2 ——→		30 4
Total ATP			38

Figure 7.10 **Calculating ATP energy yield per glucose molecule.**

Substrate-level ATP synthesis during glycolysis and the citric acid cycle account for 4 ATP. The electron transport chain produces a maximum of 34 ATP, and the maximum total is thus 38 ATP. Some cells, however, produce only 36 ATP per glucose, or even less.

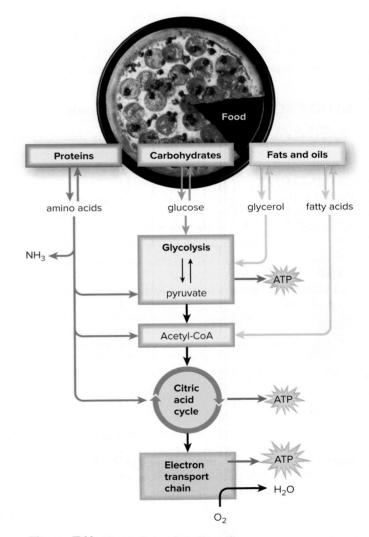

Figure 7.11 **Alternative metabolic pathways.**

The carbohydrates, proteins, and fats in pizza can be used to generate ATP.

(photo): © C Squared Studios/Getty RF

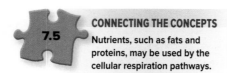

CONNECTING THE CONCEPTS

7.5

Nutrients, such as fats and proteins, may be used by the cellular respiration pathways.

to synthesize proteins. Similarly, substrates from glycolysis can become glycerol, and acetyl groups can be used to produce fatty acids. When glycerol and three fatty acids join, the result is a molecule of fat (triglyceride). This explains why you can gain weight from eating carbohydrate-rich foods.

Connections: Health

What is chromium?

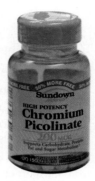

Chromium is a trace mineral found naturally in many vegetables. Most commonly, people are familiar with chromium as a component of over-the-counter "fat-burning" medications. Chromium is known to have an influence on the activity of insulin, allowing the body to use carbohydrates and fats more efficiently as energy sources. However, most people are not deficient in chromium, and scientific studies have not shown a conclusive link between increased levels of chromium and weight loss. While chromium supplements are sometimes prescribed to treat some medical conditions, such as type 2 diabetes, it is important to note that the FDA does not regulate dietary supplements.

© McGraw-Hill Education

Check Your Progress 7.5

1. Summarize the amount of ATP produced during cellular respiration in the presence of oxygen.

2. Explain why a fat molecule produces more ATP than a glucose molecule.

3. Explain how cells use proteins for energy.

STUDY TOOLS   **http://connect.mheducation.com**

SMARTBOOK®   Maximize your study time with McGraw-Hill SmartBook®, the first adaptive textbook.

SUMMARIZE

Cellular respiration converts the organic molecules in food into ATP, the universal energy required by all living organisms. Pathways exist for the production of ATP in both aerobic and anaerobic environments.

7.1 The four phases of cellular respiration function to convert glucose into ATP.

7.2 Glycolysis, the first stage of cellular respiration, occurs within the cytoplasm of the cell.

7.3 The anaerobic pathways in a cell use fermentation to regenerate NAD⁺.

7.4 If oxygen is present, cells will use the aerobic pathways of the citric acid cycle and electron transport chain within the mitochondria to generate ATP.

7.5 Nutrients, such as fats and proteins, may be used by the cellular respiration pathways.

7.1 Cellular Respiration

Cellular respiration produces **adenosine triphosphate (ATP)** from organic molecules, particularly glucose. During cellular respiration, glucose is oxidized to CO_2, which we exhale. Oxygen (O_2), which we breathe in, is reduced to H_2O. When glucose is oxidized, energy is released. The following equation gives an overview of these events:

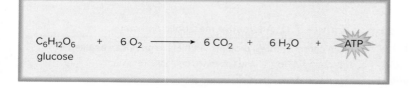

Coenzymes, such as NAD^+ and FAD, assist in the process. Cellular respiration consists of a series of reactions (glycolysis, preparatory reaction, citric acid cycle, and electron transport chain) that capture the energy of oxidation and use it to produce ATP molecules.

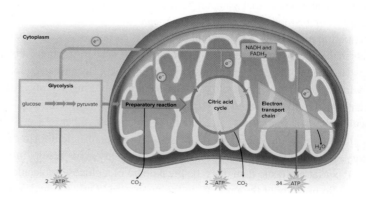

7.2 Outside the Mitochondria: Glycolysis

Glycolysis, the breakdown of glucose to two molecules of pyruvate, is a series of enzymatic reactions that occur in the cytoplasm. During glycolysis,

- Glucose is oxidized by the removal of hydrogen ions (H^+) and electrons (e^-).
- When NAD^+ accepts electrons, NADH results.

Breakdown releases enough energy to immediately give a net gain of two ATP by substrate-level ATP synthesis. Following glycolysis, if oxygen is available, pyruvate from glycolysis enters mitochondria. If oxygen is absent, pyruvate enters the fermentation pathways.

7.3 Outside the Mitochondria: Fermentation

Fermentation involves glycolysis followed by the reduction of pyruvate by NADH, either to lactate or to alcohol and CO_2. The reduction of pyruvate regenerates NAD^+, which can accept more hydrogen atoms during glycolysis.

- Although fermentation results in only two ATP, it still provides a quick burst of energy for short-term, strenuous muscular activity.
- The accumulation of lactate puts an individual in oxygen deficit, which is the amount of oxygen needed when lactate is completely metabolized to CO_2 and H_2O.

7.4 Inside the Mitochondria

In the presence of oxygen, the aerobic reactions of cellular respiration occur.

Preparatory Reaction

During the preparatory reaction in the matrix, pyruvate is oxidized, releasing CO_2.

- NAD^+ accepts hydrogen ions and electrons, forming NADH.
- An acetyl group, the end product, combines with CoA, forming **acetyl-CoA.**
- This reaction takes place twice for each molecule of glucose.

The Citric Acid Cycle

Acetyl groups enter the citric acid cycle, a series of reactions occurring in the mitochondrial matrix. During one turn of the cycle, oxidation results in two CO_2, three NADH, and one $FADH_2$. One turn also produces one ATP. There are two turns of the cycle per glucose molecule.

The Electron Transport Chain

The final stage of cellular respiration involves the electron transport chain located in the cristae of the mitochondria. The chain is a series of electron carriers that accept high-energy electrons (e^-) from NADH and $FADH_2$ and pass electrons along until they are finally low-energy electrons received by oxygen, which combines with H^+ to produce water.

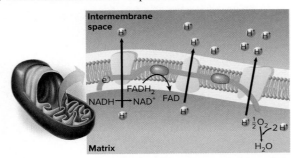

The carriers of the electron transport chain are located in molecular complexes on the cristae of mitochondria. These carriers capture energy from the passage of electrons and use it to pump H^+ into the intermembrane space of the mitochondrion. When H^+ flows down its gradient into the matrix through ATP synthase complexes, energy is released and used to form ATP molecules from ADP and (P).

7.5 Metabolic Fate of Food

Energy Yield from Glucose Metabolism

Of the maximum 38 ATP formed by complete glucose breakdown, 4 are the result of substrate-level ATP synthesis, and the rest are produced as a result of the electron transport chain and ATP synthase. Most cells produce fewer than 36 ATP per glucose molecule.

Alternative Metabolic Pathways

Besides carbohydrates, glycerol and fatty acids from fats and amino acids from proteins can undergo cellular respiration by entering glycolysis and/or the citric acid cycle. These metabolic pathways also provide substrates for the synthesis of fats and proteins.

ASSESS

Testing Yourself

Choose the best answer for each question.

7.1 Cellular Respiration

1. During cellular respiration, _____ is oxidized and _____ is reduced.
   a. glucose, oxygen
   b. glucose, water
   c. oxygen, water
   d. water, oxygen
   e. oxygen, carbon dioxide

2. Cellular respiration requires _____, and _____ are produced.
   a. water and carbon dioxide, oxygen and glucose
   b. oxygen and carbon dioxide, glucose and water
   c. oxygen and glucose, carbon dioxide and water
   d. water and glucose, oxygen and carbon dioxide
   e. ATP and glucose, oxygen and water

7.2 Outside the Mitochondria: Glycolysis

3. During the energy-investment step of glycolysis, which of the following is consumed?
   a. ATP
   b. NADH
   c. ADP + (P)
   d. NAD^+

4. Which of the following is not true?
   a. Glycolysis can occur anaerobically.
   b. Glycolysis results in the production of pyruvate.
   c. Glycolysis can occur in the cytoplasm of the cell or the mitochondria.
   d. Glycolysis yields a net product of two ATP.

7.3 Outside the Mitochondria: Fermentation

5. When animals carry out fermentation, they produce _____, while yeasts produce _____.
   a. lactate, NADH
   b. lactate, ethyl alcohol and CO_2
   c. NADH, ethyl alcohol and CO_2
   d. $FADH_2$, lactate
   e. ethyl alcohol and CO_2, lactate

6. Fermentation does not yield as much ATP as cellular respiration does because fermentation
   a. generates mostly heat.
   b. makes use of only a small amount of the potential energy in glucose.
   c. creates by-products that require large amounts of ATP to break down.
   d. creates ATP molecules that leak into the cytoplasm and are broken down.

7.4 Inside the Mitochondria

7. The citric acid cycle results in the release of
   a. carbon dioxide.
   b. pyruvate.
   c. oxygen.
   d. water.

8. Which of the following reactions occur in the matrix of the mitochondria?
   a. glycolysis and the preparatory reaction
   b. the preparatory reaction and the citric acid cycle
   c. the citric acid cycle and the electron transport chain
   d. the electron transport chain and glycolysis

9. The final electron acceptor in the electron transport chain is
   a. NADH.
   b. FADH$_2$.
   c. oxygen.
   d. water.
10. The citric acid cycle produces _____ ATP molecule(s) per glucose molecule:
    a. 1
    b. 2
    c. 4
    d. 16
11. NAD$^+$ and FAD act as _____ carriers in cellular respiration.
    a. oxygen
    b. carbon dioxide
    c. electron
    d. carbon
12. The preparatory reaction converts _____ to _____ for use in the citric acid cycle.
    a. glucose; ATP
    b. ATP; carbon dioxide
    c. pyruvate; acetyl-CoA
    d. NADH; NAD$^+$

7.5 Metabolic Fate of Food

13. Which of the following do not enter the cellular respiration pathways?
    a. fats
    b. amino acids
    c. nucleic acids
    d. carbohydrates
14. The metabolic process that produces the most ATP molecules is
    a. glycolysis.
    b. the citric acid cycle.
    c. the electron transport chain.
    d. fermentation.
15. Under optimal conditions, how many ATP may be produced per molecule of glucose?
    a. 2
    b. 32
    c. 20
    d. 38

ENGAGE

BioNOW

Want to know how this science is relevant to your life? Check out the BioNow video below:

- Energy Part III: Cellular Respiration

Refer to the equation for cellular respiration in Section 7.1. In these experiments, what are the sources of CHO and oxygen for these reactions, and what process is producing the CO_2 and energy needed by the plants?

Thinking Critically

1. The compound malonate, a substrate of the citric acid cycle, can be poisonous at high concentrations because it can block cellular respiration. How might a substrate of the citric acid cycle block respiration when it is present in great excess?
2. Insulin resistance can lead to diabetes. Insulin normally causes cells to take in glucose from the blood, but insulin resistance causes cells to fail to respond properly to this signal. The pancreas then compensates by secreting even more insulin into the body. Researchers have found a connection among a high-fat diet, decreased mitochondrial function, and insulin resistance in some elderly patients. What might be the connection among these three conditions?
3. You observe that prokaryotic cells sometimes have indentations in the plasma membrane where metabolic pathways take place. How could this observation be used to support your hypothesis that mitochondria evolved when a prokaryote cell was engulfed by a nucleated cell?
4. Rotenone is a broad-spectrum insecticide that inhibits the electron transport chain. Why might it be toxic to humans?

8

© Yuri_Arcurs/Getty RF

Cellular Reproduction

Why Don't Elephants Get Cancer?

Elephants, like humans, are long-lived organisms. Yet, despite the fact that they are long-lived, recent studies have indicated that only about 5% of elephants will develop cancer in their lifetime, compared to up to 25% of humans.

Cancer results from a failure to control the cell cycle, a series of steps that all cells go through prior to initiating cell division. A gene called *p53* is an important component of that control mechanism. This gene belongs to a class of genes called tumor suppressor genes. At specific points in the cell cycle, the proteins encoded by tumor suppressor genes check the DNA for damage. If the cells are damaged, they are not allowed to divide.

Humans only have one pair of *p53* genes (one from each parent). Luckily, you only need a single copy to regulate the cell cycle. But if both genes are not functioning properly, because of inheritance of a defective gene or mutations that inactivate a gene, then they can't act as gatekeepers and, instead, will allow cells to divide continuously. Unrestricted cell growth is a characteristic of cancer. In humans, mutations in the *p53* gene are associated with increased risk of a number of cancers. Interestingly, elephants have 20 pairs of *p53* genes in their genome, which means that it is very unlikely that their cells will lose all of this gene's functionality.

In this chapter, we will explore how cell division is regulated, as well as how genes such as *p53* function in the regulation of the cycle.

As you read through this chapter, think about the following questions:

1. What are the roles of the checkpoints in the cell cycle?
2. At what checkpoint do you think *p53* is normally active?
3. Why would a failure at cell cycle checkpoints result in cancer?

OUTLINE

8.1 The Basics of Cellular Reproduction 126

8.2 The Cell Cycle: Interphase, Mitosis, and Cytokinesis 128

8.3 The Cell Cycle Control System 134

8.4 The Cell Cycle and Cancer 136

8.5 The Characteristics of Cancer 139

BEFORE YOU BEGIN

Before beginning this chapter, take a few moments to review the following discussions.

Section 3.2 What is the structure of DNA?

Section 4.4 What are the roles of the microtubules and nucleus in a eukaryotic cell?

Section 4.5 What is a cell wall, and what types of organisms is it found in?

a. Children grow.

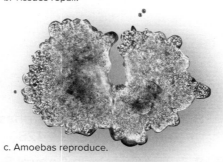

b. Tissues repair.

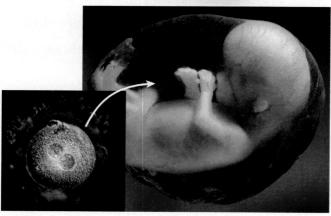

c. Amoebas reproduce.

d. Zygotes develop.

Figure 8.1 Cellular reproduction.

Cellular reproduction occurs **(a)** when small children grow to be adults and **(b)** when our tissues repair themselves. Cellular reproduction also occurs **(c)** when single-celled organisms reproduce and **(d)** when zygotes, the product of sperm and egg fusion, develop into multicellular organisms capable of independent existence.

8.1 The Basics of Cellular Reproduction

Learning Outcomes

Upon completion of this section, you should be able to

1. Summarize the purpose of cellular reproduction.
2. Understand the relationship of sister chromatids to chromosomes.
3. Explain the roles of histones and the nucleosome in the compaction of the chromatin.

We humans, like other multicellular organisms, begin life as a single cell. However, because of cellular reproduction, we become an organism consisting of trillions of cells in less than 10 months. Even after we are born, cellular reproduction doesn't stop—it continues as we grow (**Fig. 8.1***a*), and when we are adults, it replaces worn-out or damaged tissues (Fig. 8.1*b*). Right now, your body is producing thousands of new red blood cells, skin cells, and cells that line your respiratory and digestive tracts. If you suffer a cut, cellular reproduction helps repair the injury.

Cellular reproduction is also necessary for the reproduction of certain organisms. In bacteria, and some protists such as amoebas (Fig. 8.1*c*), cellular reproduction involves a simple splitting of the cell into two new cells. This process, called **binary fission,** is a form of *asexual reproduction* because it produces new cells that are identical to the original parent cell. Sexual reproduction (see Chapter 9), which involves a sperm and an egg, produces offspring that are different from the parents.

The cell theory (see Section 4.3) emphasizes the importance of cellular reproduction when it states that all cells come from preexisting cells. Therefore, you can't have a new cell without a preexisting cell, and you can't have a new organism without a preexisting organism (Fig. 8.1*d*). Cellular reproduction is necessary for the production of both new cells and new organisms.

Cellular reproduction involves two important processes: growth and cell division. During growth, a cell duplicates its contents, including the organelles and its DNA. Then, during cell division, the DNA and other cellular contents of the *parent cell* are distributed to the *daughter cells*. These terms have nothing to do with gender; they are simply a way to designate the beginning cell and the resulting cells. Both processes are heavily regulated to prevent runaway cellular reproduction, which, as we will see, can have serious consequences.

Chromosomes

An important event that occurs in preparation for cell division is **DNA replication,** the process by which a cell copies its DNA (see Section 11.1). Once this is complete, a full copy of all the DNA may be passed on to both daughter cells by the process of cell division. This DNA contains all of the instructions that each cell needs to live and perform not only its internal functions, but its role as part of a tissue or organ in a multicellular organism.

However, a human cell contains an astonishing 2 meters (about 6.6 feet) of DNA, and moving this large quantity of DNA is no easy task. To make cell division easier, DNA and associated proteins are packaged into a set of **chromosomes,** which allow the DNA to be distributed to the daughter cells. The packaging of the DNA into chromosomes and the process of DNA replication are the work of proteins and enzymes in the nucleus of the cell.

Chromatin to Chromosomes

When a eukaryotic cell is not undergoing cell division, the DNA and associated proteins have the appearance of thin threads called **chromatin.** Closer examination reveals that chromatin is periodically wound around a core of eight protein molecules, so that it looks like beads on a string. The protein molecules are **histones,** and each bead is called a **nucleosome (Fig. 8.2)**. Chromatin normally adopts a zigzag structure, and then it is folded into loops for further compaction. This looped chromatin more easily fits within the nucleus.

Just before cell division occurs, the chromatin condenses multiple times into large loops, which produces highly compacted chromosomes. The chromosomes are often 10,000 times more compact that the chromatin, which also allows them to be viewed with a light microscope.

Prior to cell division, the chromosomes are duplicated. A duplicated chromosome is composed of two identical halves, called **sister chromatids,** held together at a constricted region called a **centromere.** Each sister chromatid contains an identical DNA double helix.

It is also important to note that every species has a characteristic number of chromosomes; a human cell has 23 pairs of chromosomes, for a total chromosome number of 46. Sexual reproduction provides each of us with one copy of each chromosome from each parent.

CONNECTING THE CONCEPTS

8.1 The DNA of living organisms is packaged into chromosomes, which are distributed to new cells by the process of cell division.

Check Your Progress 8.1

1. Explain why cellular reproduction is necessary for life.
2. Distinguish between a sister chromatid and a chromosome.
3. Summarize why histones and nucleosomes are needed by a cell.

Connections: Scientific Inquiry

Is the number of chromosomes related to the complexity of the organism?

© 2008 Robbin Moran

In eukaryotes, such as humans, the number of chromosomes varies considerably. A fruit fly has 8 chromosomes, and humans have 46. So at first glance, you may think that a relationship exists between chromosome number and complexity. However, yeasts (one-celled eukaryotes) have 32 chromosomes, and horses have 64. Some ferns, as shown here, have 1,252 chromosomes. Thus, there is no definite relationship between the number of chromosomes and the complexity of the organism.

Figure 8.2 Chromosome compaction.

a. Histones are responsible for packaging chromatin, so that it fits into the nucleus. Prior to cell division, chromatin is further condensed into chromosomes, so that the genetic material can be easily divided between the daughter cells. **b.** A micrograph of nucleosomes on a DNA strand.

(nucleosomes): © Don W. Fawcett/Science Source

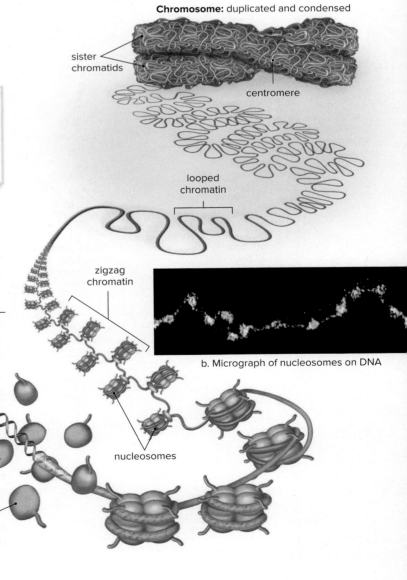

Chromosome: duplicated and condensed

sister chromatids

centromere

looped chromatin

zigzag chromatin

b. Micrograph of nucleosomes on DNA

nucleosomes

histones

a. Levels of chromatin compaction

8.2 The Cell Cycle: Interphase, Mitosis, and Cytokinesis

We have already seen that cellular reproduction involves duplication of cell contents followed by cell division. For cellular reproduction to be orderly, you would expect the first event to occur before the second event, and that's just what happens during the cell cycle. The **cell cycle** is an orderly sequence of stages that take place between the time a new cell has arisen from the division of the parent cell to the point when it has given rise to two daughter cells. It consists of **interphase,** the time when the cell performs its usual functions; a period of nuclear division called **mitosis;** and division of the cytoplasm, or **cytokinesis.**

Interphase

As **Figure 8.3** shows, most of the cell cycle is spent in interphase. The amount of time the cell takes for interphase varies widely. Embryonic cells complete the entire cell cycle in just a few hours. A rapidly dividing mammalian cell, such as an adult stem cell, typically takes about 24 hours to complete the cell cycle and spends 22 hours in interphase.

DNA replication occurs in the middle of interphase and serves as a way to divide interphase into three phases: G_1, S, and G_2. G_1 is the phase before DNA replication, and G_2 is the phase following DNA synthesis. Originally, G stood for "gap," but now that we know how metabolically active the cell is, it is better to think of G as standing for "growth." Protein synthesis is very much a part of these growth stages.

During G_1, a cell doubles its organelles (such as mitochondria and ribosomes) and accumulates materials that will be used for DNA replication. At this point, the cell integrates internal and external signals and "decides" whether to continue with the cell cycle (see Section 8.3). Some cells, such as muscle cells, typically remain in interphase, and cell division is permanently arrested. These cells are said to have entered a G_0 phase. If DNA damage occurs, many cells in the G_0 phase can reenter the cell cycle and divide again to repair the damage. But a few cell types, such as nerve cells, almost never divide again once they have entered G_0.

Following G_1, the cell enters the S phase. The S stands for "synthesis," and certainly DNA synthesis is required for DNA replication. At the beginning of the S phase, each chromosome has one **chromatid** consisting of a single DNA double helix. At the end of this stage, each chromosome is composed of two *sister chromatids,* each having one double helix. The two chromatids of each chromosome remain attached at the centromere. DNA replication produces the duplicated chromosomes.

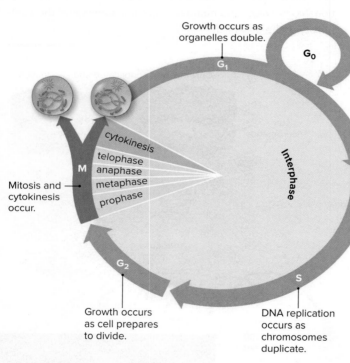

Growth occurs as organelles double.

G_1

G_0

cytokinesis
telophase
anaphase
metaphase
prophase

M

Interphase

Mitosis and cytokinesis occur.

G_2

S

Growth occurs as cell prepares to divide.

DNA replication occurs as chromosomes duplicate.

Figure 8.3 The cell cycle.

Cells go through a cycle consisting of four phases: G_1, S, G_2, and M. Interphase includes the G_1, S, and G_2 phases. Some cells can exit G_1 and enter a G_0 phase.

The G_2 phase extends from the completion of DNA replication to the onset of mitosis. During this stage, the cell synthesizes the proteins that will be needed for cell division, such as the protein found in microtubules. The microtubules, part of the cytoskeleton, play an important role in cell division.

M (Mitotic) Phase

Cell division occurs during the M phase, which encompasses both division of the nucleus and division of the cytoplasm. The type of nuclear division associated with the cell cycle is called *mitosis,* which accounts for why this stage is called the M phase.

During mitosis, the duplicated nuclear contents of the parent cell are distributed equally to the daughter cells. As a result of mitosis, the daughter nuclei are identical to the parent cell and to each other—they all have the same number and kinds of chromosomes.

Recall that, during the S phase, the DNA of each chromosome is replicated to produce a duplicated chromosome that contains two identical sister chromatids, which remain attached at the centromere (**Fig. 8.4**). Each chromatid is a single DNA double helix containing the same sequence of base pairs as the original chromosome. During mitosis, the sister chromatids of each chromosome separate and are now called *daughter chromosomes.* Because each original chromosome goes through the same process of DNA replication followed by separation of the sister chromatids, the daughter nuclei produced by mitosis are genetically identical to each other and to the parent nucleus. Thus, if the parent nucleus has four chromosomes, each daughter nucleus also has four chromosomes of exactly the same type. One way to keep track of the number of chromosomes in drawings is to count the number of centromeres, because every chromosome has a centromere.

Most eukaryotic cells have an even number of chromosomes because each parent has contributed half of the chromosomes to the new individual.

The Spindle

While it may seem easy to separate the chromatids of only 4 duplicated chromosomes, imagine the task when there are 46 chromosomes, as in humans, or 78, as in dogs. Certainly, it is helpful if chromosomes are highly condensed before the task begins, but some mechanism is needed to complete separation in an organized manner. Most eukaryotic cells rely on a **spindle,** a structure of the cytoskeleton, to pull the chromatids apart. A spindle has spindle fibers made of microtubules that are able to assemble and disassemble. First, the microtubules assemble to form the spindle that takes over the center of the cell and separates the chromatids. Later, the microtubules disassemble.

A **centrosome** is the primary microtubule organizing center of a cell. In an animal cell, each centrosome has two barrel-like structures, called **centrioles,** and an array of microtubules called an *aster.* Plant cells have centrosomes, but they are not clearly visible because they lack centrioles. Centrosome duplication occurs at the start of the S phase of the cell cycle and has been completed by G_2. During the first part of the M phase, the centrosomes separate and move to opposite sides of the nucleus, where they form the poles of the spindle. As the nuclear envelope breaks down, spindle fibers take over the center of the cell. Some overlap at the *spindle equator,* which is midway between the poles. Others attach to duplicated chromosomes in a way that ensures the separation

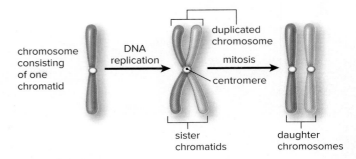

Figure 8.4 **Overview of mitosis.**

DNA replication forms the sister chromatids, which are separated to form daughter chromosomes by mitosis.

of the sister chromatids and their proper distribution to the daughter cells. Whereas the chromosomes will be inside the newly formed daughter nuclei, a centrosome will be just outside it.

Phases of Mitosis in Animal and Plant Cells

Traditionally, mitosis is divided into a sequence of events, even though it is a continuous process. We will describe mitosis as having four phases: **prophase, metaphase, anaphase,** and **telophase.** With the exception of the movement of the centrioles, the descriptions of the phases of mitosis in this section apply to both animal (**Fig. 8.5**) and plant cells (**Fig. 8.6**).

Before studying the descriptions of these phases in Figure 8.5, recall that, before mitosis begins, DNA has been replicated. Each chromatid contains a double helix of DNA, and each chromosome consists of two sister chromatids attached at a centromere. Notice that, in Figure 8.5, some chromosomes are colored red and some are colored blue. The red chromosomes were inherited from one parent, the blue chromosomes from the other parent.

Figure 8.5 **Phases of mitosis in animal cells.**

The images on the top row are micrographs of an animal cell during mitosis. In the lower images, the red chromosomes were inherited from one parent and the blue chromosomes from the other parent.

(all): © Ed Reschke

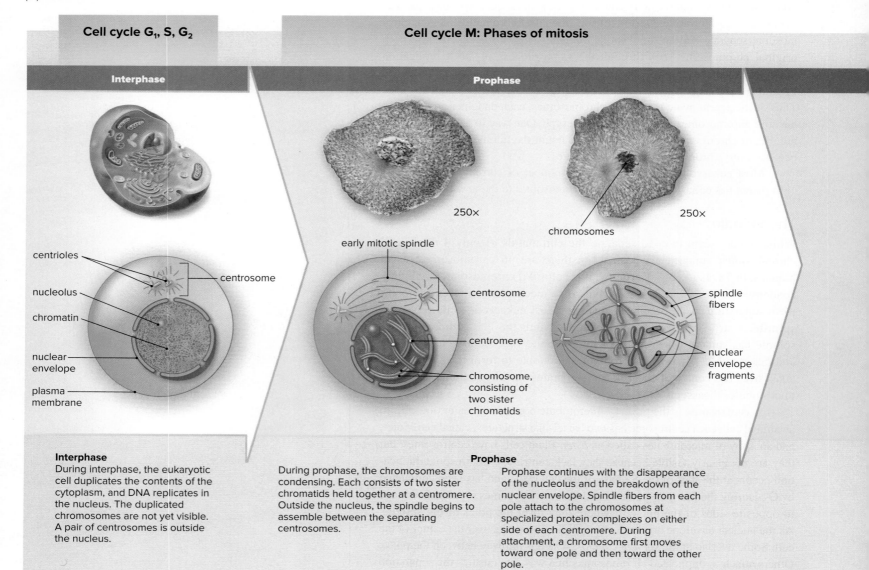

Cell cycle G₁, S, G₂ — Interphase
Interphase
During interphase, the eukaryotic cell duplicates the contents of the cytoplasm, and DNA replicates in the nucleus. The duplicated chromosomes are not yet visible. A pair of centrosomes is outside the nucleus.

Cell cycle M: Phases of mitosis — Prophase
Prophase
During prophase, the chromosomes are condensing. Each consists of two sister chromatids held together at a centromere. Outside the nucleus, the spindle begins to assemble between the separating centrosomes.

Prophase
Prophase continues with the disappearance of the nucleolus and the breakdown of the nuclear envelope. Spindle fibers from each pole attach to the chromosomes at specialized protein complexes on either side of each centromere. During attachment, a chromosome first moves toward one pole and then toward the other pole.

During mitosis, a spindle arises that will separate the sister chromatids of each duplicated chromosome. Once separated, the sister chromatids become daughter chromosomes. In this way, the resulting daughter nuclei are identical to each other and to the parent nucleus. In Figure 8.5, the daughter nuclei not only have four chromosomes but they each have one red short, one blue short, one red long, and one blue long, the same as the parent cell had. In other words, each daughter nucleus is genetically the same as the original parent nucleus. Mitosis is usually followed by division of the cytoplasm, or cytokinesis. Cytokinesis begins during telophase and continues after the nuclei have formed in the daughter cells. The cell cycle is now complete, and the daughter cells enter interphase, during which the cell will grow and DNA will replicate once again.

Cytokinesis in Animal and Plant Cells

In most cells, cytokinesis follows mitosis. When mitosis occurs, but cytokinesis doesn't occur, the result is a multinucleated cell. Some organisms, such as fungi and slime molds, and certain structures in plants (such as the embryo sac) are multinucleated.

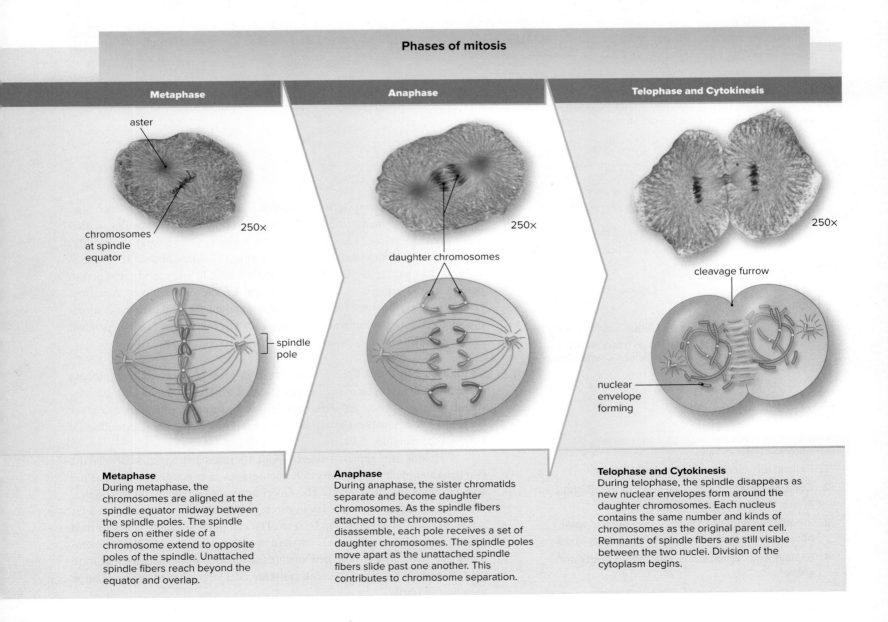

Phases of mitosis

| Metaphase | Anaphase | Telophase and Cytokinesis |

aster

chromosomes at spindle equator

250×

spindle pole

daughter chromosomes

250×

cleavage furrow

250×

nuclear envelope forming

Metaphase
During metaphase, the chromosomes are aligned at the spindle equator midway between the spindle poles. The spindle fibers on either side of a chromosome extend to opposite poles of the spindle. Unattached spindle fibers reach beyond the equator and overlap.

Anaphase
During anaphase, the sister chromatids separate and become daughter chromosomes. As the spindle fibers attached to the chromosomes disassemble, each pole receives a set of daughter chromosomes. The spindle poles move apart as the unattached spindle fibers slide past one another. This contributes to chromosome separation.

Telophase and Cytokinesis
During telophase, the spindle disappears as new nuclear envelopes form around the daughter chromosomes. Each nucleus contains the same number and kinds of chromosomes as the original parent cell. Remnants of spindle fibers are still visible between the two nuclei. Division of the cytoplasm begins.

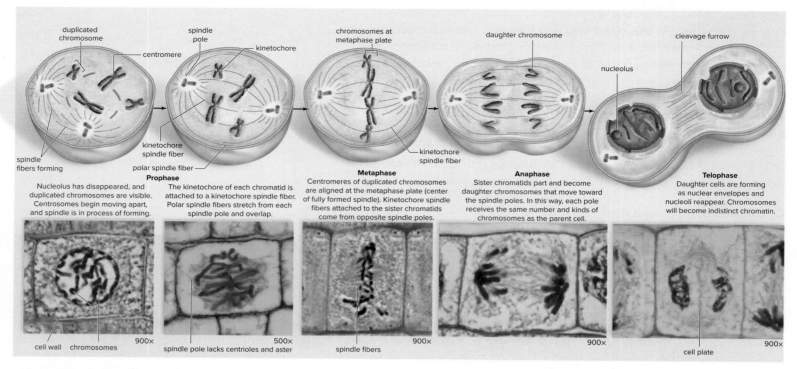

Figure 8.6 Mitosis in a plant cell.

The chromosomes are stained blue, and microtubules of the spindle fibers are stained pink in these photos of a dividing plant from the genus Allium.

(prophase, metaphase, anaphase, telophase): © Kent Wood/Science Source; (prometaphase): © Ed Reschke

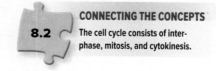

CONNECTING THE CONCEPTS

8.2   The cell cycle consists of interphase, mitosis, and cytokinesis.

Check Your Progress 8.2

1. Explain the importance of the cell cycle.

2. Compare and contrast what is occurring during each phase of interphase: G₁, S, and G₂.

3. Identify the phases of mitosis, and explain what happens to the chromosomes during each phase.

4. Explain why cytokinesis is different in plants and animals.

Cytokinesis in Animal Cells   In animal cells, a cleavage furrow, which is an indentation of the membrane between the two daughter nuclei, begins as anaphase draws to a close. The cleavage furrow deepens when a band of actin filaments, called the contractile ring, slowly forms a circular constriction between the two daughter cells. The action of the contractile ring can be likened to pulling a drawstring ever tighter around the middle of a balloon. A narrow bridge between the two cells is visible during telophase, and then the contractile ring continues to separate the cytoplasm until there are two independent daughter cells (**Fig. 8.7**).

Cytokinesis in Plant Cells   Cytokinesis in plant cells occurs by a process different from that seen in animal cells (**Fig. 8.8**). The rigid cell wall that surrounds plant cells does not permit cytokinesis by furrowing. Instead, cytokinesis in plant cells involves the building of new plasma membranes and cell walls between the daughter cells.

Cytokinesis is apparent when a small, flattened disk appears between the two daughter plant cells. In electron micrographs, it is possible to see that the disk is composed of vesicles. The Golgi apparatus produces these vesicles, which move along microtubules to the region of the disk. As more vesicles arrive and fuse, a cell plate can be seen. The **cell plate** is simply newly formed plasma membrane that expands outward until it reaches the old plasma membrane and fuses with it. The new membrane releases the molecules that form the new plant cell walls. These cell walls are later strengthened by the addition of cellulose fibrils.

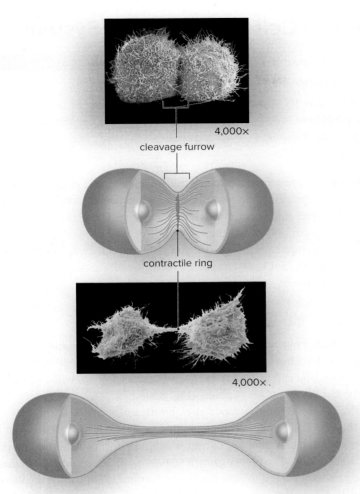

cleavage furrow

contractile ring

4,000×

4,000×

Figure 8.7 Cytokinesis in animal cells.

A single cell becomes two cells by a furrowing process. A cleavage furrow appears as early as anaphase, and a contractile ring, composed of actin filaments, gradually gets smaller until there are two cells.

(top): © National Institutes of Health (NIH)/USHHS; (bottom): © Steve Gschmeissner/SPL/Getty RF

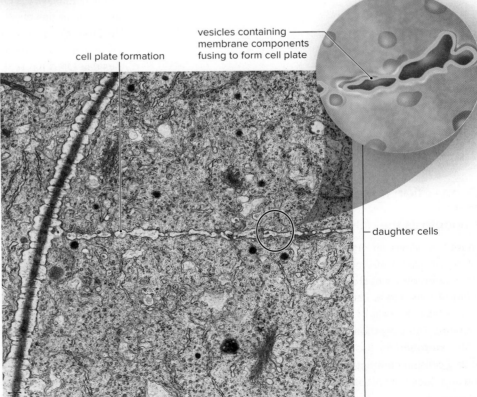

vesicles containing membrane components fusing to form cell plate

cell plate formation

daughter cells

Figure 8.8 Cytokinesis in plant cells.

During cytokinesis in a plant cell, a cell plate forms midway between two daughter nuclei and extends to the original plasma membrane.

© Biophoto Associates/Science Source

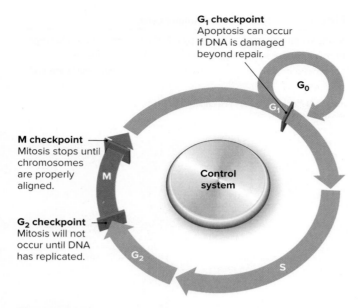

G₁ checkpoint
Apoptosis can occur if DNA is damaged beyond repair.

M checkpoint
Mitosis stops until chromosomes are properly aligned.

G₂ checkpoint
Mitosis will not occur until DNA has replicated.

G₀

G₁

Control system

M

G₂

S

Figure 8.9 Cell cycle checkpoints.

Three important checkpoints are designated. Internal and external signals determine whether the cell is ready to proceed past cell cycle checkpoints.

Connections: Health

Is apoptosis the only cause of cell death?

Apoptosis allows an organism to control cell death. It is an important aspect of embryological development. The opposite of apoptosis is necrosis, or unprogrammed cell death, the death of cells due to disease or injury. For example, during a heart attack, the cells of the heart are deprived of oxygen and nutrients. The cells then burst and die, causing inflammation of the surrounding tissues, which can cause the death of nearby cells. In contrast, apoptosis does not cause inflammation and does not spread to surrounding cells.

8.3 The Cell Cycle Control System

Learning Outcomes

Upon completion of this section, you should be able to

1. Summarize the role of checkpoints in the cell cycle.
2. Explain how checkpoints are regulated by internal and external signals.
3. Describe the process of apoptosis.

In order for a cell to reproduce successfully, the cell cycle must be controlled. The importance of cell cycle control can be appreciated by comparing the cell cycle to the events that occur in a washing machine. The washer's control system starts to wash only when the tub is full of water, doesn't spin until the water has been emptied, and so forth. Similarly, the cell cycle's control system ensures that the G_1, S, G_2, and M phases occur in order and start only when the previous phase has been successfully completed.

Cell Cycle Checkpoints

Just as a washing machine will not begin to agitate the load until the tub is full, the cell cycle has **checkpoints** that can delay the cell cycle until certain conditions are met. The cell cycle has many checkpoints, but we will consider only three: the G_1, G_2, and M checkpoints (**Fig. 8.9**).

The *G_1 checkpoint* is especially significant because, if the cell cycle passes this checkpoint, the cell is committed to divide. If the cell does not pass this checkpoint, it can enter G_0, during which it performs its normal functions but does not divide. The proper growth signals, such as certain growth factors, must be present for a cell to pass the *G_1* checkpoint. Additionally, the integrity of the cell's DNA is checked. If the DNA is damaged, the protein p53 can stop the cycle at this checkpoint and initiate DNA repair. If repair is not possible, the protein can cause the cell to undergo programmed cell death, or **apoptosis.**

The cell cycle halts momentarily at the *G_2 checkpoint* until the cell verifies that DNA has replicated. This prevents the initiation of the M phase unless the chromosomes are duplicated. Also, if DNA is damaged, as from exposure to solar radiation or X-rays, arresting the cell cycle at this checkpoint allows time for the damage to be repaired, so that it is not passed on to daughter cells.

Another cell cycle checkpoint occurs during the mitotic stage (*M checkpoint*). The cycle hesitates between metaphase and anaphase to make sure the chromosomes are properly attached to the spindle and will be distributed accurately to the daughter cells. The cell cycle does not continue until every chromosome is ready for the nuclear division process.

Internal and External Signals

The checkpoints of the cell cycle are controlled by internal and external signals. These signals are typically molecules that stimulate or inhibit cellular functions. Researchers have identified a series of internal signals called **cyclins.** The levels of these proteins increase and decrease as the cell cycle progresses; therefore they act as cellular timekeepers. The appropriate cyclin has to be present at the correct levels for the cell to proceed from the G_1 phase to the S phase, and from the G_2 phase into the M phase. In addition, enzymes called **kinases** remove phosphate from ATP and add it to another molecule. The addition of the energized phosphate from ATP often acts as an off/on switch for cellular activities. Kinases are active in the removal of the nuclear membrane and the condensation of the chromosomes early in prophase.

Some external signals, such as growth factors and hormones, stimulate cells to go through the cell cycle. Growth factors also stimulate tissue repair. Even cells that are arrested in G_0 will finish the cell cycle if growth factors stimulate them to do so. For example, epidermal growth factor (EGF) stimulates skin in the vicinity of an injury to finish the cell cycle, thereby repairing the damage. Hormones act on tissues at a distance, and some signal cells to divide. For example, at a certain time in the menstrual cycle of females, the hormone estrogen stimulates cells lining the uterus to divide and prepares the lining for implantation of a fertilized egg.

The cell cycle can be inhibited by cells coming into close contact with other cells. In the laboratory, eukaryotic cells will divide until they line a container in a one-cell-thick sheet. Then they stop dividing, due to a phenomenon termed *contact inhibition*. Contact inhibition prevents cells from overgrowing within the body. When tissues are damaged, cells divide to repair the damage. But once the tissue has been repaired, contact inhibition prevents cell overgrowth by halting the cell cycle.

Some years ago, it was noted that mammalian cells in cell cultures divide about 70 times, and then they die. Cells seem to "remember" the number of times they have already divided, and they stop when they have reached the usual number of cell divisions. It's as if *senescence,* the aging of cells, is dependent on an internal battery-operated clock that runs down and then stops. We now know that senescence is due to the shortening of telomeres. A **telomere** is a repeating DNA base sequence (TTAGGG) at the ends of chromosomes that can be as long as 15,000 base pairs. Telomeres have been likened to the protective caps on the ends of shoelaces because they ensure chromosomal stability. Each time a cell divides, a portion of a telomere is lost. When telomeres become too short, the cell is "old" and dies by the process of apoptosis.

Apoptosis

Apoptosis is often defined as programmed cell death because the cell progresses through a typical series of events that bring about its destruction (**Fig. 8.10**). The cell rounds up and loses contact with its neighbors. The nucleus fragments, and the plasma membrane develops blisters, called blebs. Finally, the cell breaks into fragments, which are engulfed by white blood cells and/or neighboring cells. Oddly enough, healthy cells routinely harbor the enzymes, called caspases, that bring about apoptosis. These enzymes are ordinarily held in check by inhibitors but are unleashed by either internal or external signals.

Cell division and apoptosis are two opposing processes that keep the number of cells in the body at an appropriate level. In other words, cell division increases and apoptosis decreases the number of **somatic cells** (body cells). Both the cell cycle and apoptosis are normal parts of growth and development. An organism begins as a single cell that repeatedly undergoes the cell cycle to produce many cells, but eventually some cells must die for the organism to take shape. For example, when a tadpole becomes a frog, the tail disappears as apoptosis occurs. In humans, the fingers and toes of an embryo are at first webbed, but later they are freed from one another as a result of apoptosis.

Apoptosis occurs all the time, particularly if an abnormal cell that could become cancerous appears. Death through apoptosis prevents a tumor from developing.

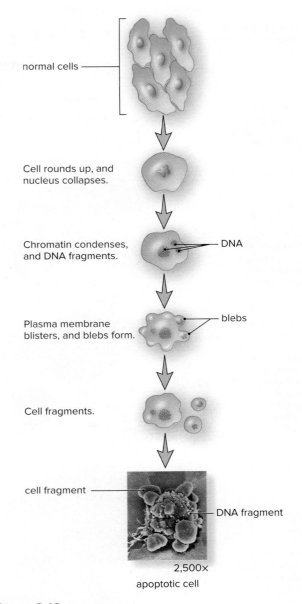

normal cells

Cell rounds up, and nucleus collapses.

Chromatin condenses, and DNA fragments. — DNA

Plasma membrane blisters, and blebs form. — blebs

Cell fragments.

cell fragment —

— DNA fragment

2,500×
apoptotic cell

Figure 8.10 **Apoptosis.**

Apoptosis is a sequence of events that results in a fragmented cell. The fragments are phagocytized by white blood cells and neighboring tissue cells.

© Steve Gschmeissner/Science Source

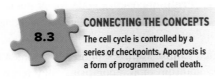

CONNECTING THE CONCEPTS

8.3 The cell cycle is controlled by a series of checkpoints. Apoptosis is a form of programmed cell death.

Check Your Progress 8.3

1. Explain the significance of checkpoints in the cell cycle.

2. Explain how signals are used for cell cycle control, and describe the two types of signals used by cells.

3. Predict what might happen to an organism if apoptosis were not regulated.

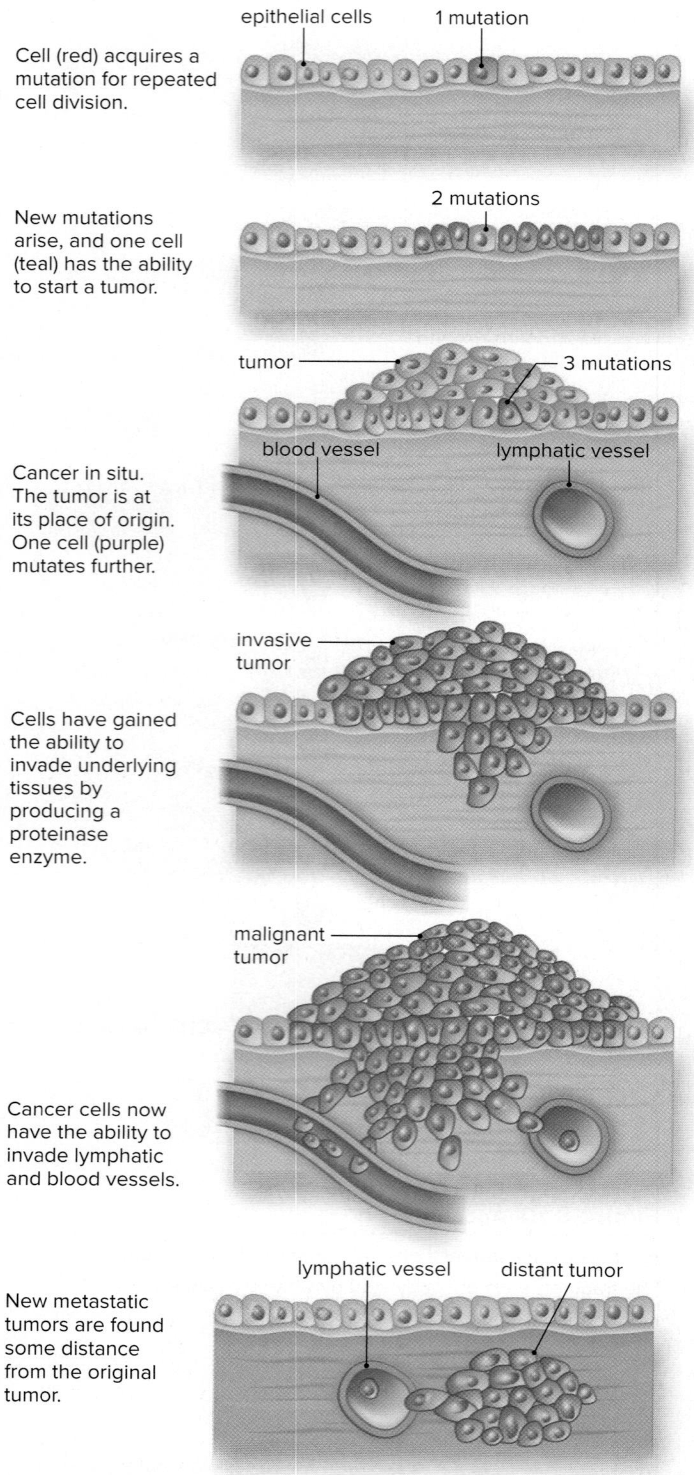

Cell (red) acquires a mutation for repeated cell division.

epithelial cells   1 mutation

New mutations arise, and one cell (teal) has the ability to start a tumor.

2 mutations

tumor   3 mutations

blood vessel   lymphatic vessel

Cancer in situ. The tumor is at its place of origin. One cell (purple) mutates further.

invasive tumor

Cells have gained the ability to invade underlying tissues by producing a proteinase enzyme.

malignant tumor

Cancer cells now have the ability to invade lymphatic and blood vessels.

lymphatic vessel   distant tumor

New metastatic tumors are found some distance from the original tumor.

Figure 8.11 **Development of cancer.**

The development of cancer requires several mutations, each contributing to the development of a tumor. Normally, cells exhibit contact inhibition and controlled growth. They stop dividing when they are bordered by other cells. Tumor cells lack contact inhibition and pile up, forming a tumor. Cancer cells do not fulfill the functions of a specialized tissue, and they have abnormal chromosomes. As more mutations accumulate, the cancer cells may spread and form tumors in other parts of the body.

8.4 The Cell Cycle and Cancer

Learning Outcomes

Upon completion of this section, you should be able to

1. Distinguish between proto-oncogenes and tumor suppressor genes in regard to the development of cancer.
2. Explain the role of telomerase in stem cells and cancer cells.
3. Summarize how chromosomal rearrangements may cause some forms of cancer.
4. Identify the relationship between certain genes and cancer.

Cancer is a genetic disease caused by a lack of control in the cell cycle (see Fig. 8.9). The development of cancer requires several mutations, each propelling cells toward the development of a tumor. These mutations disrupt the many redundant regulatory pathways that prevent normal cells from taking on the characteristics of cancer cells (see Section 8.5). Because of these cumulative events, it often takes several years for cancer to develop, but the likelihood of cancer increases as we age.

Carcinogenesis is the development of cancer. **Figure 8.11** illustrates the development of colon cancer, a form of cancer that has been studied in detail. This figure shows that a single abnormal cell begins the process toward the development of a tumor. Along the way, the most aggressive cell becomes the dominant cell of the tumor. As additional mutations occur, the tumor cells release a growth factor, which causes neighboring blood vessels to branch into the cancerous tissue, a process called *angiogenesis*. Additional mutations allow cancer cells to produce enzymes that degrade the basement membrane and invade underlying tissues. Cancer cells are motile—able to travel through the blood or lymphatic vessels to other parts of the body—where they start distant tumors, a process called *metastasis*.

Cells that are already highly specialized, such as nerve cells and cardiac muscle cells, seldom become cancer cells because they rarely divide. Carcinogenesis is more likely to begin in cells that have the capacity to enter the cell cycle. Fibroblasts and cells lining the cavities of the lungs, liver, uterus, and kidneys are able to divide when stimulated to do so. Adult stem cells continue to divide throughout life. These include blood-forming cells in the bone marrow and basal cells of the skin and digestive tract. Continuous division of these cells is required because blood cells live only a short while, and the cells that line the intestines and the cells that form the outer layer of the skin are continually sloughed off.

Proto-Oncogenes and Tumor Suppressor Genes

Recall that the cell cycle (see Fig. 8.9) consists of interphase followed by mitosis. Special proteins help regulate the cell cycle at checkpoints. When cancer develops, the cell cycle occurs repeatedly, due in large part to mutations in two types of genes. As shown in **Figure 8.12,**

1. **Proto-oncogenes** code for proteins that promote the cell cycle and inhibit apoptosis. They are often likened to the gas pedal of a car because they accelerate the cell cycle.
2. **Tumor suppressor genes** code for proteins that inhibit the cell cycle and promote apoptosis. They are often likened to the brakes of a car

because they slow down the cell cycle and stop cells from dividing inappropriately. They are called tumor suppressors because tumors may occur when mutations cause these genes to become nonfunctional.

Proto-Oncogenes Become Oncogenes

When proto-oncogenes mutate, they become cancer-causing genes called **oncogenes.** Proto-oncogenes promote the cell cycle, and oncogenes accelerate the cell cycle. Therefore, a mutation that causes a proto-oncogene to become an oncogene is a "gain of function" mutation.

A **growth factor** is a chemical signal that activates a cell-signaling pathway by bringing about phosphorylation of a signaling protein (Fig. 8.12a). The cell-signaling pathway then activates numerous proteins, many of which promote the cell cycle (Fig. 8.12b). Some proto-oncogenes code for a growth factor or for a receptor protein that receives a growth factor. When these proto-oncogenes become oncogenes, receptor proteins are easy to activate and may even be stimulated by a growth factor produced by the receiving cell. For example, the *RAS* proto-oncogenes promote mitosis when a growth factor binds to a receptor. When *RAS* proto-oncogenes become oncogenes, they promote mitosis even when growth factors are not present. *RAS* oncogenes are involved in 20–30% of human cancers.

Tumor Suppressor Genes Become Inactive

When tumor suppressor genes mutate, their products no longer inhibit the cell cycle or promote apoptosis. Therefore, these mutations can be called "loss of function" mutations (Fig. 8.12).

The *p53* tumor suppressor gene produces a protein that checks the DNA for damage before it proceeds through the G_1 checkpoint. If there are breaks in the DNA, the cell is instructed to enter into G_0 phase, and if these repairs cannot be repaired, the cell undergoes apoptosis. Failure of the *p53* gene to perform this function allows cells with DNA damage to rapidly divide, potentially leading to cancer (see chapter opener).

Cancer develops gradually; multiple mutations usually occur before a cell becomes cancerous. As you can imagine, the conversion of proto-oncogenes into oncogenes, coupled with the inactivation of tumor suppressor genes, causes the cell cycle to continue unabated, just as a car with a sticking gas pedal and faulty brakes soon careens out of control.

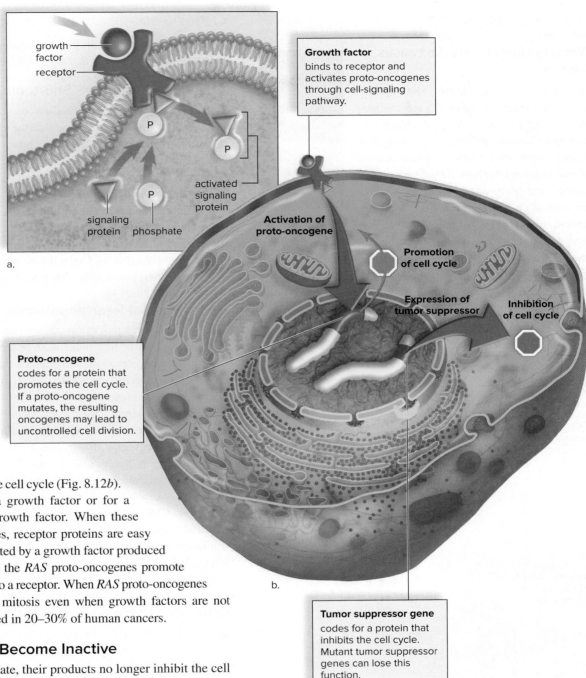

Growth factor
binds to receptor and activates proto-oncogenes through cell-signaling pathway.

growth factor
receptor

P

signaling protein   phosphate

P

activated signaling protein

P

a.

Activation of proto-oncogene

Promotion of cell cycle

Expression of tumor suppressor

Inhibition of cell cycle

Proto-oncogene
codes for a protein that promotes the cell cycle. If a proto-oncogene mutates, the resulting oncogenes may lead to uncontrolled cell division.

b.

Tumor suppressor gene
codes for a protein that inhibits the cell cycle. Mutant tumor suppressor genes can lose this function.

Figure 8.12 Role of proto-oncogenes and tumor suppressor genes in the cell cycle.

a. A growth factor binds to a receptor, stimulating a signal transduction pathway (green arrow). **b.** The signal transduction pathway activates the expression of proto-oncogenes that promote the cell cycle. Tumor suppressor genes limit cell division by inhibiting the cell cycle (red arrow).

Connections: Health

What in cigarette smoke causes cancer?

There are over 7,000 known chemicals in tobacco smoke. Of these, 70 are carcinogens, or cancer-causing agents. Many carcinogens are also mutagens, meaning that they promote mutations in DNA, including those in genes such as tumor suppressor genes and proto-oncogenes. Included in this list of chemicals are arsenic (a heavy metal), benzene (an additive in gasoline), and cadmium (a metal found in batteries). In addition to the carcinogens, cigarette smoke contains high levels of carbon monoxide and hydrogen cyanide, both of which reduce the body's ability to process oxygen. Exposure to these chemicals is estimated to reduce life expectancy by 13.2 years for males, 14.5 years for females.

© Glow Images RF

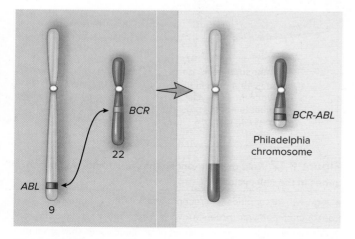

Figure 8.13 **A translocation can create the Philadelphia chromosome.**

The Philadelphia chromosome results when a portion of chromosome 9 is swapped for a portion of chromosome 22. This creates the *BCR-ABL* oncogene on chromosome 22. The Philadelphia chromosome is responsible for 95% of cases of chronic myelogenous leukemia.

Other Genetic Changes and Cancer

In addition to mutations in proto-oncogenes and tumor supressor genes, there are other genetic changes that may also contribute to cancer development.

Absence of Telomere Shortening

The telomeres at the end of the chromosomes play an important role in regulating cell division. Just as the caps at the ends of shoelaces protect them from unraveling, so telomeres promote chromosomal stability, so that replication can occur. Each time a cell divides, some portion of a telomere is lost; when telomeres become too short, the chromosomes cannot be replicated properly and the cell cycle is stopped.

Embryonic cells and certain adult cells, such as stem cells and germ cells, have an enzyme, called *telomerase,* that can rebuild telomeres. The gene that codes for telomerase is turned on in cancer cells. If this happens, the telomeres do not shorten, and cells divide over and over again. Telomerase is believed to become active only after a cell has already started proliferating wildly.

Chromosomal Rearrangements

When the chromosomes of cancer cells become unstable, portions of the DNA double helix may be lost, duplicated, or scrambled. For example, a portion of a chromosome may break off and reattach to another chromosome. These events, called *translocations* (see Section 13.2), may lead to cancer, especially if they disrupt genes that regulate the cell cycle.

A phenomenon called a *Philadelphia chromosome* is the result of a translocation between chromosomes 9 and 22 (**Fig. 8.13**). The *BCR* (*b*reakpoint *c*luster *r*egion) and *ABL* genes are fused, creating an oncogene, called *BCR-ABL,* that promotes cell division. This translocation causes nearly 95% of cases of chronic myelogenous leukemia (CML), a cancer of the bone marrow. Recently, researchers have used a drug called imatinib (Gleevec) to successfully treat CML. This drug inhibits the activity of the protein coded for by *BCR-ABL.* Encouraged by this success, scientists are now seeking to develop similar drugs that can inhibit the products of other oncogenes. Although cancer is usually a somatic disease, meaning that it develops only in body cells, some individuals may inherit a predisposition for developing some forms of cancer.

Other Cell Cycle Genes Associated with Cancer

BRCA1* and *BRCA2 In 1990, DNA studies of large families in which females tended to develop breast cancer identified the first allele associated with that disease. Scientists named that gene *b*reast *ca*ncer 1, or *BRCA1*. Other studies found that some forms of breast cancer were due to a faulty allele of another gene, called *BRCA2*. Both alleles are part of mutant tumor suppressor genes that are inherited in an autosomal recessive manner. If one mutated allele is inherited from either parent, a mutation in the other allele is required before the predisposition to cancer is increased. Because the first mutated gene is inherited, it is present in all cells of the body, and then cancer is more likely wherever the second mutation occurs, for example, in the breast or ovaries.

***RB* Gene** The *RB* gene is also a tumor suppressor gene. It takes its name from its association with an eye tumor called a *retinoblastoma, which first appears as a white mass in the retina. A tumor in one eye is most common because it takes mutations in both alleles before cancer can develop. Children who inherit a mutated allele are more likely to have tumors in both eyes.

***RET* Gene**   An abnormal allele of the *RET* gene, which predisposes a person to thyroid cancer, can be passed from parent to child. *RET* is a proto-oncogene known to be inherited in an autosomal dominant manner—only one mutated allele is needed to increase a predisposition to cancer. The remaining mutations necessary for thyroid cancer to develop are acquired (not inherited).

Testing for Genes Associated with Cancer

Genetic tests can detect the presence of specific alleles in many of the genes mentioned in this section. The advances in personal genomics are also allowing people to be able to be screened for specific genes associated with a family history of cancer (see Section 13.3) Persons who have inherited certain alleles of these genes may decide to have additional testing done, or sometimes elective surgery, to detect the presence of cancer.

Check Your Progress 8.4

1. Contrast the activity of proto-oncogenes with that of tumor suppressor genes.
2. Summarize how chromosomal influences, such as telomeres and translocations, may cause cancer.
3. Explain the difference in how the *RB* and *RET* genes contribute to cancer.

8.5 Characteristics of Cancer

Learning Outcomes

Upon completion of this section, you should be able to

1. Describe the characteristics of cancer cells.
2. Summarize the types of treatment for cancer.
3. Describe the factors that reduce the risk of cancer.

Cancers are classified according to their location. *Carcinomas* are cancers of the epithelial tissue that lines organs; *sarcomas* are cancers arising in muscle or connective tissue (especially bone or cartilage); and *leukemias* are cancers of the blood. In this section, we'll consider some general characteristics of cancer cells and the treatment and prevention of cancer.

Characteristics of Cancer Cells

As we have seen, carcinogenesis is a gradual process that requires the accumulation of multiple mutations over time. Therefore, it may be decades before cancer develops and is diagnosed. Cancer cells have the following characteristics:

   Cancer cells lack differentiation. Cancer cells lose their specialization and do not contribute to the functioning of a body part. A cancer cell does not look like a differentiated epithelial, muscle, nervous, or connective tissue cell; instead, it looks distinctly abnormal (**Fig. 8.14**). As mentioned, normal cells can enter the cell cycle about 70 times, and then they die. Cancer cells can enter the cell cycle repeatedly; in this way, they are immortal.

Connections: Health

Can transposons cause cancer?

Transposons are small, mobile sequences of DNA that have the ability to move throughout the genome. Also known as "jumping genes," transposons are known to cause mutations as they move (see Section 13.1). While the chances of a transposon disrupting the activity of a proto-oncogene or a tumor supressor gene in a specific cell are very small (our genome has over 3.4 billion nucleotides), there have been cases when a transposon has caused a loss of cell cycle control and been a factor in the development of cancer. Transposon activity may be associated with the development of other diseases, such as some forms of hemophilia and muscular dystrophy.

CONNECTING THE CONCEPTS
8.4 Failure of the cell cycle controls may lead to cancer.

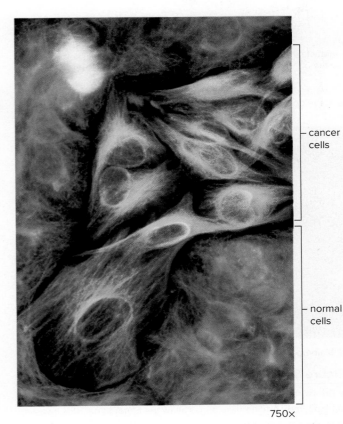

750×

Figure 8.14 Cancer cells.

In this false-colored image, cancer cells (yellow and red) have an abnormal shape compared to normal cells (blue and green).

© Nancy Kedersha/Immunogen/SPL/Science Source

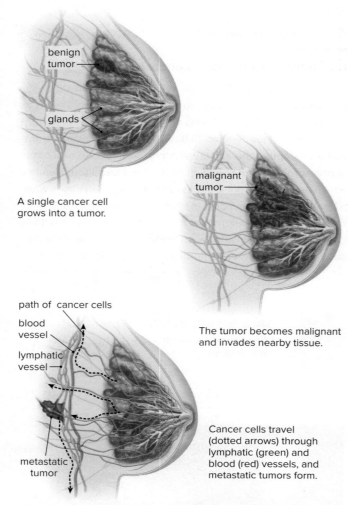

A single cancer cell grows into a tumor.

The tumor becomes malignant and invades nearby tissue.

Cancer cells travel (dotted arrows) through lymphatic (green) and blood (red) vessels, and metastatic tumors form.

a. Development of breast cancer and metastatic tumors

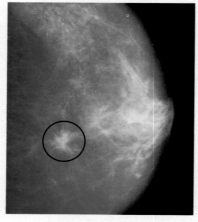

b. Mammogram showing tumor

Figure 8.15 Development of breast cancer.

a. Breast cancer begins when a single abnormal cell grows into a benign tumor. Additional changes allow the tumor to become invasive, or malignant, and invade nearby tissue. Metastasis occurs when the tumor invades blood and lymphatic vessels and new tumors develop some distance from the original tumor. **b.** A mammogram (X-ray image of the breast) can find a tumor (circled) too small to be felt.

(b): © Breast Scanning Unit, Kings College Hospital, London/SPL/Science Source

Cancer cells have abnormal nuclei. The nuclei of cancer cells are enlarged and may contain an abnormal number of chromosomes. The chromosomes are also abnormal; some parts may be duplicated, or some may be deleted. In addition, gene amplification (extra copies of specific genes) is seen much more frequently than in normal cells.

Cancer cells do not undergo apoptosis. Ordinarily, cells with damaged DNA undergo apoptosis, preventing tumors from developing. Cancer cells, however, do not respond normally to the signals to initiate apoptosis, and thus they continue to divide.

Cancer cells form tumors. Normal cells anchor themselves to a substratum and/or adhere to their neighbors. Then, they exhibit contact inhibition and stop dividing. Cancer cells, on the other hand, have lost all restraint; they pile on top of one another and grow in multiple layers, forming a **tumor.** They have a reduced need for growth factors, and they no longer respond to inhibitory signals. As cancer develops, the most aggressive cell becomes the dominant cell of the tumor.

Cancer cells undergo metastasis and promote angiogenesis. A **benign tumor** is usually contained within a capsule and therefore cannot invade adjacent tissue. Additional changes may cause cancer cells to produce enzymes that allow tumor cells to escape the capsule and invade nearby tissues. The tumor is then called a **malignant tumor,** meaning that it is invasive and may spread. Cells from a malignant tumor may travel through the blood or lymph to start new tumors elsewhere within the body. This process is known as **metastasis (Fig. 8.15).**

An actively growing tumor can grow only so large before it becomes unable to obtain sufficient nutrients to support further growth. Additional mutations in cancer cells allow them to secrete factors that promote **angiogenesis,** the formation of new blood vessels. Additional nutrients and oxygen reach the tumor, allowing it to grow larger. Some modes of cancer treatment are aimed at preventing angiogenesis. The patient's prognosis (probable outcome) depends on (1) whether the tumor has invaded surrounding tissues and (2) whether there are metastatic tumors in distant parts of the body.

Cancer Treatment

Cancer treatments either remove the tumor or interfere with the cancer cells' ability to reproduce. For many solid tumors, removal by surgery is often the first line of treatment. When the cancer is detected at an early stage, surgery may be sufficient to cure the patient by removing all cancerous cells.

Because cancer cells are rapidly dividing, they are susceptible to radiation therapy and chemotherapy. The goal of radiation is to kill cancer cells within a specific tumor by directing high-energy beams at the tumor. DNA is damaged to the point that replication can no longer occur, and the cancer cells undergo apoptosis. Chemotherapy is a way to kill cancer cells that have spread throughout the body. Like radiation, most chemotherapeutic drugs lead to the death of cells by damaging their DNA or interfering with DNA synthesis. Others interfere with the functioning of the mitotic spindle. The drug vinblastine, first obtained from a flowering plant called a periwinkle, prevents the spindle from forming. Taxol, extracted from the bark of the Pacific yew tree, prevents the spindle from functioning as it should. Unfortunately, radiation and chemotherapy often damage cells other than cancer cells, leading to side effects, such as nausea and hair loss.

Immunotherapy treatment involves using the patient's own immune system cells (see Section 26.3) to target cancer cells for destruction. These treatments involve identifying distinct differences between cancer cells and the normal cells of the body. In many cases, these treatments target slight differences in the composition of proteins in the plasma membranes of cancer cells and normal cells. Research in these areas is focusing on the development of monoclonal antibodies that act specifically on cancer cells and vaccines that prime the immune system to identify and destroy specific cancer cells.

Prevention of Cancer

Evidence is clear that the risk of certain types of cancer can be reduced by adopting protective behaviors and following recommended dietary guidelines.

Protective Behaviors

To lower the risk of developing certain cancers, people are advised to avoid smoking, sunbathing, and excessive alcohol consumption.

Cigarette smoking accounts for about 30% of all cancer deaths. Smoking is responsible for 90% of lung cancer cases among men and 80% among women. People who smoke two or more packs of cigarettes a day have lung cancer mortality rates 15 to 25 times greater than those of nonsmokers. Smokeless tobacco (chewing tobacco or snuff) increases the risk of cancers of the mouth, larynx, throat, and esophagus.

Many skin cancers are sun-related. Sun exposure is a major factor in the development of the most dangerous type of skin cancer, melanoma, and the incidence of this cancer increases in people living near the equator. Similarly, excessive exposure to radon gas[1] in homes increases the risk of lung cancer, especially in cigarette smokers. It is best to test your home and, if necessary, take the proper remedial actions.

Cancers of the mouth, throat, esophagus, larynx, and liver occur more frequently among heavy drinkers, especially when accompanied by tobacco use (cigarettes, cigars, or chewing tobacco).

Protective Diet

The risk of some forms of cancer is up to 40% higher among obese men and women compared with people of normal weight. Thus, weight loss in these groups can reduce cancer risk. In addition, the following dietary guidelines are recommended:

- Increase your consumption of foods that are rich in vitamins A and C (**Fig. 8.16**). Beta-carotene, a precursor of vitamin A, is found in dark green, leafy vegetables, carrots, and various fruits. Vitamin C is present in citrus fruits. These vitamins are called antioxidants, because in cells they prevent the formation of free radicals (organic ions having an unpaired electron),

Connections: Health

What are some of the other options for treating cancer?

Research into the prevention and treatment of cancer is evolving rapidly. Each year, researchers present new technologies and procedures to detect and eliminate cancer cells in the body. Several organizations, including the American Cancer Society (http://www.cancer.org/treatment/index) maintain current lists of resources for the treatment of cancer.

Figure 8.16   The right diet helps prevent cancer.

Foods that help protect against cancer include fruits and dark green, leafy vegetables, which are rich in vitamins A and C, and vegetables from the cabbage family, including broccoli.

(leafy greens): © Ingram Publishing/Superstock RF; (blueberries): © Purestock/Superstock RF; (oranges): © Foodcollection RF; (broccoli): © Creative Studio Heinemann/Getty RF

8.5   CONNECTING THE CONCEPTS
Cancer cells have distinct characteristics, including a lack of differentiation and the ability to move to other areas of the body.

---

[1]Radon gas results from the natural radioactive breakdown of uranium in soil, rocks and water

which can damage DNA. Vitamin C also prevents the conversion of nitrates and nitrites into carcinogenic nitrosamines in the digestive tract.

- Avoid salt-cured or pickled foods because they may increase the risk of stomach and esophageal cancers. Smoked foods, such as ham and sausage, contain chemical carcinogens similar to those in tobacco smoke. Nitrites are sometimes added to processed meats (e.g., hot dogs and cold cuts) and other foods to protect them from spoilage; as mentioned, nitrites can be converted to cancer-causing nitrosamines in the digestive tract.
- Include in the diet vegetables from the cabbage family, which includes cabbage, broccoli, brussels sprouts, kohlrabi, and cauliflower. Eating these vegetables may reduce the risk of gastrointestinal and respiratory tract cancers.

Check Your Progress 8.5

1. List the general characteristics of a cancer cell.
2. Summarize the ways that cancer may be treated.
3. Explain how lifestyle choices may reduce the risk of cancer.

SUMMARIZE

Cells in our bodies undergo cellular reproduction in order to grow, repair, and maintain ourselves. When this cell cycle is not regulated, diseases like cancer can result.

8.1 The DNA of living organisms is packaged into chromosomes, which are distributed to new cells by the process of cell division.

8.2 The cell cycle consists of interphase, mitosis, and cytokinesis.

8.3 The cell cycle is controlled by a series of checkpoints. Apoptosis is a form of programmed cell death.

8.4 Failure of the cell cycle controls may lead to cancer.

8.5 Cancer cells have distinct characteristics, including a lack of differentiation and the ability to move to other areas of the body.

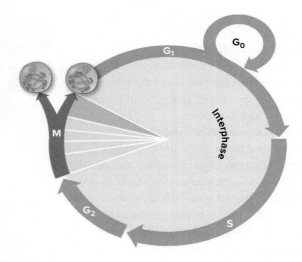

8.1 The Basics of Cellular Reproduction

Cellular reproduction occurs not only in the growth of organisms from a single-celled zygote to a multicellular organism but also in the process of **binary fission** in single-celled organisms. It is responsible for the growth and repair of tissues. Cellular reproduction involves two processes: growth, during which a cell grows and doubles its contents; and cell division, during which the parent cell's contents are split between two daughter cells.

Prior to cell division, the DNA of a cell undergoes **DNA replication.** During cell division, one-half of the cytoplasm and a copy of the cell's DNA (packaged as **chromosomes**) are passed on to each of two daughter cells.

When a cell is not dividing, its DNA is wound around **histones** to form **nucleosomes,** then arranged in a zigzag configuration to form **chromatin.** Just before cell division, duplicated chromatin condenses into large loops to form duplicated chromosomes consisting of two chromatids, called **sister chromatids,** held together at a **centromere.**

8.2 The Cell Cycle: Interphase, Mitosis, and Cytokinesis

The **cell cycle** consists of interphase and the M (mitotic) phase, which consists of mitosis and cytokinesis.

Interphase has the following phases:

- G_1: The cell doubles its organelles and accumulates materials that will be used for DNA replication.
- S: The cell replicates its DNA.
- G_2: The cell synthesizes the proteins that will be needed for cell division.

Cell division consists of mitosis and cytokinesis.

Mitosis has four phases:

- During **prophase,** the chromosomes are condensing. Outside the nucleus, the **spindle** begins to assemble between the separating **centrosomes** (which contain **centrioles** in animals). Prophase continues with the disappearance of the nucleolus and the breakdown of the nuclear envelope. Spindle fibers from each pole attach to the chromosomes in the region of a centromere.
- During **metaphase,** the chromosomes are aligned at the spindle equator midway between the spindle poles.
- During **anaphase,** the sister chromatids separate and become daughter chromosomes. As the microtubules attached to the chromosomes disassemble, each pole receives a set of daughter chromosomes.

- During **telophase,** the spindle disappears as new nuclear envelopes form around the daughter chromosomes. Each nucleus contains the same number and kinds of chromosomes as the original parent nucleus. Division of the cytoplasm begins.

Cytokinesis differs for animal and plant cells:

- In animal cells, a furrowing process involving actin filaments divides the cytoplasm.
- In plant cells, a **cell plate** forms, from which the plasma membrane and cell walls develop.

8.3 The Cell Cycle Control System

Checkpoints are points at which chemical signals promote or inhibit the continuance of the cell cycle. Important checkpoints are these three:

- At G_1, the cell can enter G_0 or undergo apoptosis if DNA is damaged beyond repair. If the cell cycle passes this checkpoint, the cell is committed to complete the cycle.
- At G_2, the cell checks to make sure DNA has been replicated properly.
- At the mitotic (M) checkpoint, the cell makes sure the chromosomes are properly aligned and ready to be partitioned to the daughter cells.

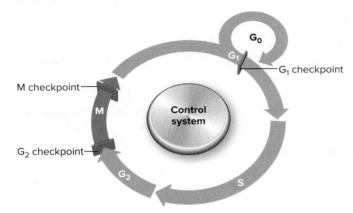

Both internal and external signals act at checkpoints to control the cell cycle. **Cyclins** and **kinases** are internal signals that promote either DNA replication or mitosis. Growth factors and hormones are external signals that promote the cell cycle in connection with growth. Inhibitory external signals are responsible for contact inhibition. Internally, when **telomeres** are too short, the cell cycle stops because the chromosomes become unstable.

When DNA cannot be repaired, **apoptosis** occurs as enzymes bring about destruction of the nucleus and the rest of the cell. Apoptosis reduces the number of **somatic cells** in the body.

8.4 The Cell Cycle and Cancer

Cancer is a genetic disease caused by lack of control in the cell cycle. Several genetic mechanisms determine how fast the cell proceeds through the cell cycle.

Proto-Oncogenes and Tumor Suppressor Genes

- **Proto-oncogenes** ordinarily promote the cell cycle and inhibit apoptosis. Many are associated with **growth factors.** Mutation results in a gain of function for these activities.
- **Tumor suppressor genes** ordinarily suppress the cell cycle and promote apoptosis. Mutation results in loss of function for these activities.

The development of cancer is a gradual process. As control of gene expression is further compromised, cancer cells develop additional characteristics that lead to angiogenesis and metastasis.

Other Genetic Changes and Cancer

Other genetic influences contribute to cancer development:

- When telomerase is present, **telomeres** don't shorten and cells can continue to divide.
- Chromosomal rearrangements, such as translocations, may disrupt genes that regulate the cell cycle.
- Certain inherited genes are known to cause cancer: *BRCA1* and *BRCA2* are mutant genes associated with breast cancer; a mutant *RB* gene is associated with the development of eye tumors; a mutant *RET* gene predisposes a person to thyroid cancer.

Genetic tests are available to help physicians diagnose cancer. These genes are turned off in normal cells but are active in cancer cells. For example, if a test for the presence of telomerase is positive, the cell is cancerous.

8.5 Characteristics of Cancer

Carcinogenesis is the development of cancer. Cancer results when control of the cell cycle is lost and cells divide uncontrollably. Cancer cells are undifferentiated, have abnormal nuclei, do not undergo apoptosis, and form both **benign** and **malignant tumors.** A tumor may stimulate **angiogenesis,** allowing the tumor to obtain more nutrients and grow larger. Tumors may also undergo **metastasis,** which causes cancer to spread throughout the body. Certain behaviors, such as avoiding smoking and sunbathing and adopting a diet rich in fruits and vegetables, are protective against cancer.

ASSESS

Testing Yourself

Choose the best answer for each question.

8.1 The Basics of Cellular Reproduction

1. A duplicated chromosome is composed of identical
    a. chromosome arms.
    b. sister chromatids.
    c. nucleosomes.
    d. homologues.
2. The type of cell division that is responsible for cell division in amoebas, or other single-celled organisms, is called
    a. binary fission.
    b. DNA replication.
    c. mitosis.
    d. apoptosis.

8.2 The Cell Cycle: Interphase, Mitosis, and Cytokinesis

3. _____ is nuclear division, whereas _____ is division of the cytoplasm.
    a. Cytokinesis, mitosis
    b. Apoptosis, mitosis
    c. Mitosis, apoptosis
    d. Mitosis, cytokinesis

For questions 4–6, match the items to those in the key. Answers can be used more than once or not at all.

Key:

    a. G_1 phase of interphase
    b. G_2 phase of interphase
    c. S phase of interphase

4. This phase follows mitosis.

5. DNA is replicated during this phase.

6. During this phase, the cell produces the proteins that will be needed for cell division.

7. Label the phases of the cell cycle on the following diagram. Include anaphase, cytokinesis, G_1 phase, G_2 phase, metaphase, prophase, S phase, and telophase.

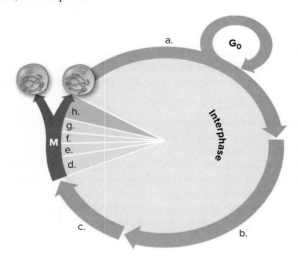

For questions 8–11, match the items to those in the key. Answers can be used more than once or not at all.

Key:

  a. prophase

  b. metaphase

  c. anaphase

  d. telophase

8. The nucleolus disappears, and the nuclear membrane breaks down.

9. The spindle disappears, and the nuclear envelope forms.

10. Sister chromatids separate.

11. Chromosomes are aligned on the spindle equator.

8.3 The Cell Cycle Control System

12. Which cell cycle checkpoint allows damaged DNA to be repaired before it is passed on to daughter cells?

  a. G_1 **c.** S

  b. G_2 **d.** mitotic

13. Kinases and cyclins are

  a. forms of microtubules.

  b. internal signals that control cell division.

  c. external signals that control cell division.

  d. drugs used to prevent cancer.

8.4 The Cell Cycle and Cancer

14. Which of the following statements is true?

  a. The gene product of a proto-oncogene inhibits the cell cycle.

  b. Proto-oncogenes are mutant versions of oncogenes that cause cancer.

  c. When mutated tumor suppressor genes lose their function, cancer may result.

  d. When tumor suppressor genes mutate, they become oncogenes.

  e. All of these are correct.

15. Which of the following is not a tumor suppressor gene?

  a. *RET*   **c.** *BRCA1*

  b. *RB*   **d.** *BRCA2*

8.5 Characteristics of Cancer

16. Which of the following is not a feature of cancer cells?

  a. exhibiting contact inhibition

  b. having enlarged nuclei

  c. stimulating the formation of new blood vessels

  d. being capable of traveling through blood and lymph

17. Which of these is a behavior that could help prevent cancer?

  a. maintaining a healthy weight

  b. eating more dark green, leafy vegetables, carrots, and various fruits

  c. not smoking

  d. maintaining estrogen levels through hormone replacement therapy

  e. All of these are correct.

ENGAGE

BioNOW

Want to know how this science is relevant to your life? Check out the BioNow video below:

  • Cell Division

What was the purpose of the rooting hormone in this experiment, and what part of the cell cycle do you think it was targeting?

Thinking Critically

1. The survivors of the atomic bombs that were dropped on Hiroshima and Nagasaki have been the subjects of long-term studies of the effects of ionizing radiation on cancer incidence. The frequencies of different types of cancer in these individuals varied across the decades. In the 1950s, high levels of leukemia and cancers of the lung and thyroid gland were observed. The 1960s and 1970s brought high levels of breast and salivary gland cancers. In the 1980s, rates of colon cancer were especially high.

  a. Why do you suppose the rates of different types of cancer varied across time?

  b. Propose a study of survivors of the 2011 Fukushima nuclear disaster in Japan that could be used to test your answer to part a.

2. In the chapter opener, it was mentioned that the genome of elephants includes multiple copies of tumor suppressor genes. Humans are also long-lived, but have only a single pair of these genes. Propose an evolutionary scenario by which the changes in the elephant genome occurred over time.

3. BPA is a chemical compound that has historically been used in the manufacture of plastic products. However, cells often mistake BPA molecules for hormones that accelerate the cell cycle. Because of this, BPA is associated with an increased risk of certain cancers.

  a. How might BPA interact with the cell cycle and its checkpoints?

  b. Why do you think that very small concentrations of BPA might have a large effect on a cell?

© Image Source/Age fotostock RF

Meiosis and the Genetic Basis of Sexual Reproduction

Meiosis and the Genetic Variation of the Species

What are the chances that there will ever be another human like you on this planet? Because of a process called meiosis, two individuals can create offspring that are genetically different from themselves and from each other. This process creates an enormous amount of diversity; in humans, more than 70 trillion different genetic combinations are possible from the mating of just two individuals! In other words, meiosis ensures that you are unique; statistically it is unlikely that anyone will ever be the same genetically as you. In animals, meiosis begins the process that produces cells called gametes, which play an important role in sexual reproduction. In humans, sperm are the male gametes and eggs are the female gametes. The processes by which they are formed are called spermatogenesis and oogenesis, respectively.

Males and females differ in the ways that they form gametes. While meiosis in the two sexes is very similar, there are some important differences in how spermatogenesis and oogenesis occur. One major difference pertains to the age at which the process begins and ends. In males, sperm production does not begin until puberty but then continues throughout a male's lifetime. In females, the process of producing eggs has started before the female is even born and ends around the age of 50, a time called menopause.

Another difference concerns the number of gametes that can be produced. In males, sperm production is unlimited, whereas females typically produce only one egg a month. Many more eggs start the process than ever finish it. In this chapter, you will see how meiosis is involved in providing the variation so important in the production of gametes. We will focus on the role of meiosis in animals but will return to how plants use meiosis in Section 21.3.

As you read through this chapter, think about the following questions:

1. Why is meiosis sometimes called reduction division?
2. What is the role of meiosis in introducing new variation?
3. Why is genetic variation necessary for the success of a species?

OUTLINE

9.1 An Overview of Meiosis   146

9.2 The Phases of Meiosis   151

9.3 Meiosis Compared with Mitosis   152

9.4 Changes in Chromosome Number   154

BEFORE YOU BEGIN

Before beginning this chapter, take a few moments to review the following discussions.

Section 4.4 What is the role of the cytoskeleton in a eukaryotic cell?

Section 8.2 At what point of the cell cycle is the DNA replicated?

Section 8.2 What events occur during prophase, metaphase, telophase, and anaphase of mitosis?

9.1 An Overview of Meiosis

Learning Outcomes

Upon completion of this section, you should be able to

1. Explain the purpose of meiosis.
2. Describe the human life cycle.
3. Define the terms *diploid, haploid, sister chromatid,* and *homologous chromosomes.*
4. Describe the processes of synapsis and crossing-over, and explain why these processes occur during meiosis.

Humans, like most other animals, practice sexual reproduction, in which two parents pass chromosomes to their offspring. But why are the genetic and physical characteristics of siblings different from those of each other and from those of either parent? The answer is **meiosis,** a type of nuclear division that is important in sexual reproduction.

Meiosis serves two major functions: (1) reducing the chromosome number and (2) shuffling the chromosomes and genes to produce genetically different gametes, called sperm (males) and eggs (females). Since we can't predict which sperm will fertilize a given egg, random fertilization introduces another level of additional variation. In the end, each offspring is unique and has a different combination of chromosomes and genes than either parent. But how does meiosis bring about the distribution of chromosomes to offspring in a way that ensures not only the correct number of chromosomes but also a unique combination of chromosomes and genes? For the answer, let's start by examining the chromosomes.

Homologous Chromosomes

Geneticists and genetic counselors can visualize chromosomes by looking at a picture of the chromosomes called a **karyotype (Fig. 9.1).** Notice how the chromosomes occur in pairs. This is because we inherit one of each chromosome from each parent. Therefore, both males and females normally have 23 pairs of chromosomes. These twenty-three pairs of chromosomes, or 46 altogether, constitute the **diploid (2n) number** of chromosomes in humans. Half this number is the **haploid (n) number** of chromosomes. Notice that the haploid number (n) identifies the number of different chromosomes in a cell, while the diploid number is due to the fact that we inherit one set of 23 from each parent.

Of these 23 pairs, 22 of these, called **autosomes,** are the same in both males and females. The remaining pair are called the **sex chromosomes,** because they contain the genes that determine gender. The larger sex chromosome is the X chromosome, and the smaller is the Y chromosome. Females have two X chromosomes; males have a single X and Y.

The members of a chromosome pair are called **homologous chromosomes,** or **homologues,** because they have the same size,

Figure 9.1 Homologous chromosomes.

In body cells, the chromosomes occur in pairs called homologous chromosomes. In this karyotype of chromosomes from a male, human cell the pairs have been numbered. These chromosomes are duplicated, and each one is composed of two sister chromatids.

© CNRI/SPL/ Science Source

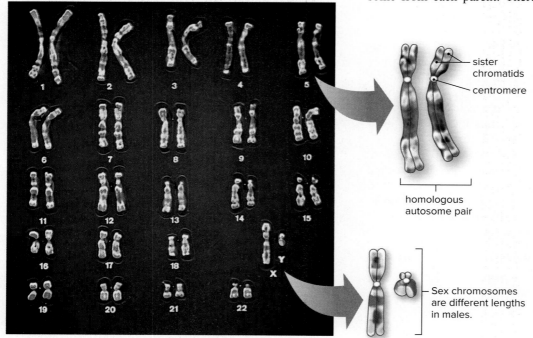

sister chromatids

centromere

homologous autosome pair

Sex chromosomes are different lengths in males.

shape, and location of the centromere (Fig. 9.1). When chromosomes are stained and viewed under a microscope, homologous chromosomes have the same characteristic banding pattern. One homologue of each pair was contributed by each parent.

Homologous chromosomes contain the same types of genes arranged in the same order. A gene is a set of instructions responsible for a specific trait, for example, whether you have freckles or not. Just as your mother may have freckles and your father may not, each homologue may have different versions of a gene. Alternate versions of a gene for a particular trait are called **alleles.** The location of the alleles of a gene on homolgous chromosomes is the same, but the information within the alleles may be slightly different. Your mother's allele for freckles is on one homologue, and your father's allele for no freckles is on the other homologue. We will take a closer look at alleles and patterns of inheritance in Chapter 10.

The Human Life Cycle

The term **life cycle** in sexually reproducing organisms refers to all the reproductive events that occur from one generation to the next. The human life cycle involves two types of nuclear division: mitosis and meiosis (**Fig. 9.2**).

Connections: Scientific Inquiry

Where do new alleles come from?

While meiosis is responsible for shuffling the genetic information in a parent cell so that the daughter cells are genetically different, it does not generate new alleles. Instead, genetic variation, in the form of new alleles of a gene, occurs primarily because of mutations in the DNA. These mutations change the sequence of nucleotides in the DNA. These changes may alter the structure of the protein encoded by the DNA, producing a slightly different variation of the trait (for example, attached or unattached earlobes). Other factors play a role, as we will discuss in Chapter 11.

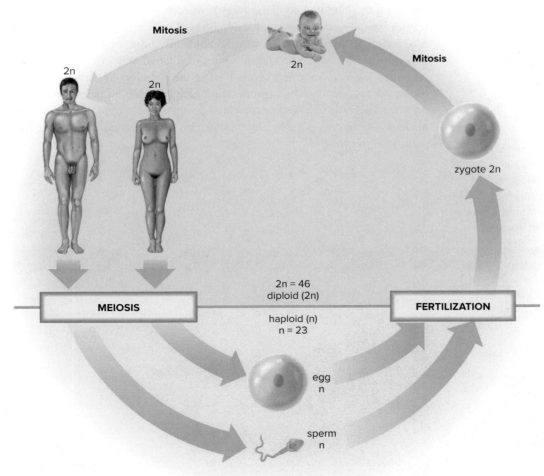

Figure 9.2 Life cycle of humans.

Meiosis is part of gamete production in both males (sperm) and females (eggs). When a haploid sperm fertilizes a haploid egg, the zygote is diploid. The zygote undergoes mitosis as it eventually develops into a newborn child. Mitosis continues throughout life during growth and repair.

Connections: Scientific Inquiry

Does meiosis always produce gametes?

While meiosis is the first step in the production of gametes in animals, the life cycle of other organisms, such as plants, is slightly different. In plants, meiosis still reduces the number of chromosomes in half, but instead of producing gametes, it produces another reproductive structure called a spore. Spores are an important part of the plant life cycle (see Fig. 21.11). Although plants do use gametes, such as pollen, for sexual reproduction, they are produced by mitosis, not meiosis.

© Getty Images/National Geographic RF

During development and after birth, mitosis is involved in the continued growth of the child and the repair of tissues at any time. As a result of mitosis, each somatic (body) cell has the diploid number of chromosomes.

During sexual reproduction, meiosis reduces the chromosome number from the diploid to the haploid number in such a way that the gametes (sperm and egg) have one chromosome derived from each homologous pair of chromosomes. In males, meiosis is a part of **spermatogenesis,** which occurs in the testes and produces sperm. In females, meiosis is a part of **oogenesis,** which occurs in the ovaries and produces eggs. After the sperm and egg join during fertilization, the resulting cell, called the **zygote,** again has a diploid number of homologous chromosomes. The zygote then undergoes mitosis and differentiation of cells to become a fetus, and eventually a new human.

Meiosis is important because, if it did not halve the chromosome number, the gametes would contain the same number of chromosomes as the body cells and the number of chromosomes would double with each new generation. Within a few generations, the cells of sexually reproducing organisms would be nothing but chromosomes! But meiosis, followed by fertilization, keeps the chromosome number constant in each generation.

Overview of Meiosis

Meiosis results in four daughter cells because it consists of two divisions, called **meiosis I** and **meiosis II** (**Fig. 9.3**). Before meiosis I begins, each chromosome has duplicated and is composed of two sister chromatids. During meiosis I, the homologous chromosomes of each pair come together and line up side by side in an event called **synapsis.** Synapsis results in a **tetrad,** an association of four chromatids (two homologous chromosomes consisting of two chromatids each). The chromosomes of a tetrad stay in close proximity until they separate. Later in meiosis I, when the homologous chromosomes of each pair separate, one chromosome from each homologous pair goes to each daughter nucleus. No rules restrict which chromosome goes to which daughter nucleus. Therefore, all possible combinations of chromosomes may occur within the gametes.

Figure 9.3 **Overview of meiosis.**

Following duplication of chromosomes, the diploid (2n) parent cell undergoes two divisions, meiosis I and meiosis II. During meiosis I, homologous chromosomes separate, and during meiosis II, sister chromatids separate. The final daughter cells are haploid (n). (The blue chromosomes were originally inherited from one parent, and the red chromosomes were originally inherited from the other parent.)

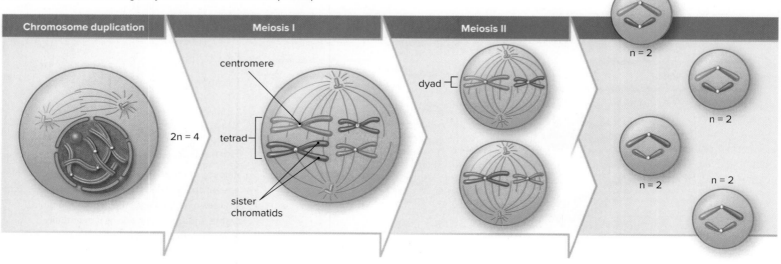

The cell does not reenter interphase between meiosis I and meiosis II, so there is no duplication of the chromosomes. Why? One of the roles of meiosis is to reduce the chromosome number; therefore, following meiosis I, the daughter nuclei each have half the number of chromosomes, but the chromosomes are still duplicated. The chromosomes are called **dyads** because each one is composed of two sister chromatids. During meiosis II, the sister chromatids of each dyad separate and become daughter chromosomes. The resulting four new daughter cells have the haploid number of chromosomes. If the parent cell has four chromosomes, then following meiosis each daughter cell has two chromosomes. (Remember that counting the centromeres tells you the number of chromosomes in a nucleus.)

Recall that another purpose of meiosis is to produce genetically different gametes. Notice in Figure 9.3 that the daughter cells (on the right) do not have the same combinations of chromosomes as the original parent cell. Why not? The homologous chromosomes of each pair separated during meiosis I. Other chromosome combinations are possible in addition to those depicted. It's possible for the daughter cells to have chromosomes from only one parent (in our example, all red or all blue). The daughter cells from meiosis will complete either spermatogenesis (and become sperm) or oogenesis (and become eggs).

Crossing-Over

Meiosis not only reduces the chromosome number but also shuffles the genetic information between the homologous chromosomes. When a tetrad forms during synapsis, the **nonsister chromatids** may exchange genetic material, an event called **crossing-over** (**Fig. 9.4**a). Crossing-over occurs between nonsister chromatids of the two chromosomes in a tetrad. Notice in Figure 9.4b how one of the blue sister chromatids is exchanging information with the red nonsister chromatid of its tetrad. Crossing-over between the nonsister chromatids shuffles the alleles and serves as the way that meiosis brings about genetic recombination in the gametes.

Recall that the homologous chromosomes carry the same genes for traits, such as the presence of freckles or a susceptibility to a disease, but the alleles may be different. Therefore, when the nonsister chromatids exchange genetic material, the sister chromatids then have a different combination of alleles, and the resulting gametes will be genetically different. Thus, in Figure 9.4b, even though two of the gametes will have the same chromosomes, these chromosomes may not have the

Figure 9.4 Synapsis and Crossing-Over.

a. During prophase I, homologous chromosomes line up side by side and form a tetrad, consisting of two chromosomes each, containing two chromatids, for a total of four chromatids. **b.** When homologous chromosomes are in synapsis, the nonsister chromatids exchange genetic material. This illustration shows only one crossover per chromosome pair, but the average is slightly more than two per homologous pair in humans. Following crossing-over, the sister chromatids of a dyad may no longer be identical and instead may have different combinations of alleles.

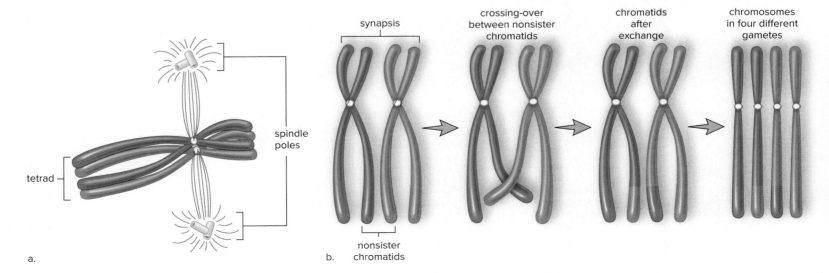

synapsis

crossing-over between nonsister chromatids

chromatids after exchange

chromosomes in four different gametes

spindle poles

tetrad

a.

b. nonsister chromatids

same combination of alleles as before because of crossing-over. Crossing-over increases the diversity of the gametes and, therefore, of the offspring.

At fertilization, new combinations of chromosomes can occur, and a zygote can have any one of a vast number of combinations of chromosomes. In humans, $(2^{23})^2$, or 70,368,744,000,000, chromosomally different zygotes are possible, even assuming no crossing-over.

The Importance of Meiosis

Meiosis is important for the following reasons:

- It helps keep the chromosome number constant by producing haploid daughter cells that become the gametes. When a haploid sperm fertilizes a haploid egg, the new individual has the diploid number of chromosomes.
- It introduces genetic variations because (1) crossing-over can result in different types of alleles on the sister chromatids of a homologue, and (2) every possible combination of chromosomes can occur in the daughter cells.

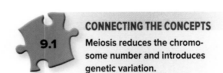

CONNECTING THE CONCEPTS

9.1   Meiosis reduces the chromosome number and introduces genetic variation.

Check Your Progress 9.1

1. Explain why organisms use meiosis rather than mitosis.
2. Summarize the role of mitosis and meiosis in the human life cycle.
3. Explain the relationship between sister chromatids and homologous chromosomes.
4. Describe the two means by which meiosis results in genetically different gametes, and explain why this is important.

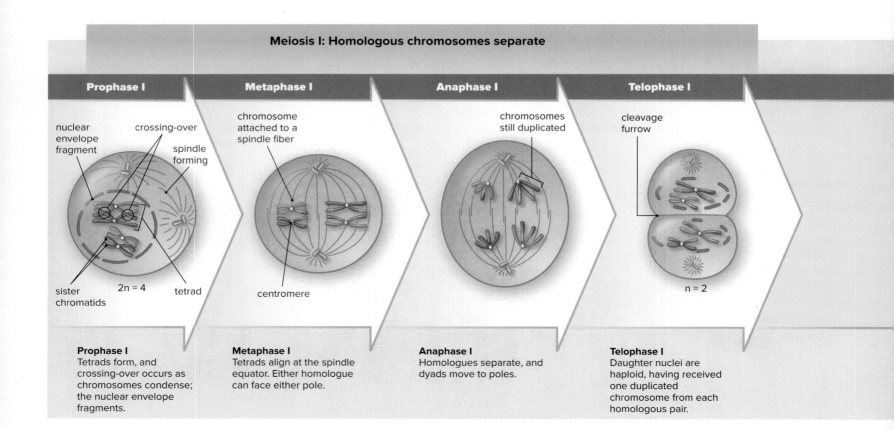

Meiosis I: Homologous chromosomes separate

Prophase I	Metaphase I	Anaphase I	Telophase I

nuclear envelope fragment — crossing-over — spindle forming

sister chromatids — 2n = 4 — tetrad

chromosome attached to a spindle fiber — centromere

chromosomes still duplicated

cleavage furrow — n = 2

Prophase I
Tetrads form, and crossing-over occurs as chromosomes condense; the nuclear envelope fragments.

Metaphase I
Tetrads align at the spindle equator. Either homologue can face either pole.

Anaphase I
Homologues separate, and dyads move to poles.

Telophase I
Daughter nuclei are haploid, having received one duplicated chromosome from each homologous pair.

9.2 The Phases of Meiosis

Learning Outcomes

Upon completion of this section, you should be able to

1. List the phases of meiosis, and briefly describe the events that occur during each phase.
2. Contrast the alignment of chromosomes during metaphase I and metaphase II of meiosis.

The same four stages of mitosis—prophase, metaphase, anaphase, and telophase—occur during both meiosis I and meiosis II. However, meiosis I and meiosis II differ from each other and from mitosis in the way chromosomes align during metaphase (**Fig. 9.5**).

The First Division—Meiosis I

To help you recall the events of meiosis, keep in mind what meiosis accomplishes in humans—namely, the production of gametes that have a reduced chromosome number and are genetically different from each other and from the parent cell. During prophase I, the nuclear envelope fragments and the nucleolus disappears as the spindle appears. As the chromosomes condense, homologues undergo synapsis to produce tetrads. Crossing-over between nonsister chromatids occurs during synapsis, and this "shuffles" the alleles on chromosomes.

During metaphase I, the tetrads attach to the spindle and align at the spindle equator, with each homologue facing an opposite spindle pole. It does

Figure 9.5 Meiosis

Meiosis I and meiosis II both consist of prophase, metaphase, anaphase, and telophase. Meiosis I shuffles the genetic combinations using crossing-over (prophase I) and alignment of the homologous chromosomes (metaphase I). The end result of meiosis II is four haploid (n) cells.

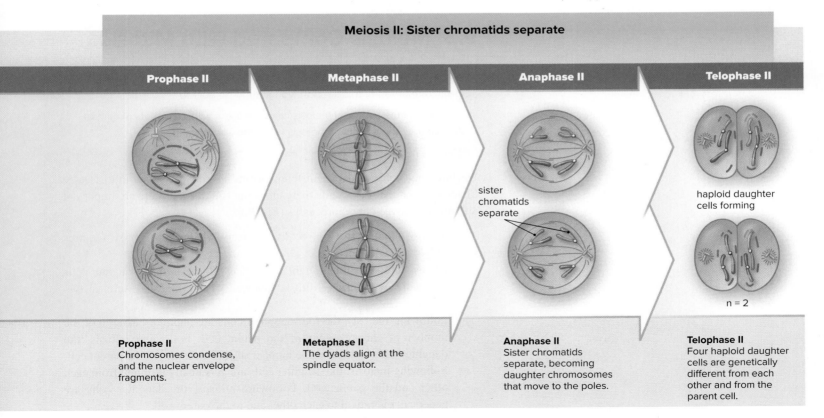

Meiosis II: Sister chromatids separate

Prophase II
Chromosomes condense, and the nuclear envelope fragments.

Metaphase II
The dyads align at the spindle equator.

Anaphase II
Sister chromatids separate, becoming daughter chromosomes that move to the poles.

Telophase II
Four haploid daughter cells are genetically different from each other and from the parent cell.

not matter which homologous chromosome faces which pole; therefore, all possible combinations of chromosomes will occur in the gametes. In effect, metaphase of meiosis I shuffles the chromosomes into new combinations.

The homologous chromosomes then separate during anaphase I. Following re-formation of the nuclear envelopes during telophase and cytokinesis, the daughter nuclei are haploid: Each daughter cell contains only one chromosome from each homologous pair. The chromosomes are now dyads, and each still has two sister chromatids. No replication of DNA occurs between meiosis I and II, a period called **interkinesis.**

The Second Division—Meiosis II

Essentially, the events of meiosis II are the same as those of mitosis, except that the cells are haploid. At the beginning of prophase II, a spindle appears while the nuclear envelope fragments and the nucleolus disappears. Dyads are present, and each attaches to the spindle. During metaphase II, the dyads are lined up at the spindle equator, with sister chromatids facing opposite spindle poles. During anaphase II, the sister chromatids of each dyad separate and move toward the poles. Both poles receive the same number and kinds of chromosomes. In telophase II, the spindle disappears as nuclear envelopes form.

During cytokinesis, the plasma membrane pinches off to form two complete cells, each of which has the haploid number (n) of chromosomes. The gametes are genetically dissimilar because they can contain different combinations of chromosomes and because crossing-over changes which alleles are together on a chromosome. Because both cells from meiosis I undergo meiosis II, four daughter cells are produced from the original diploid parent cell. Each daughter cell contains a unique combination of genes.

CONNECTING THE CONCEPTS

9.2 Meiosis consists of two divisions, which introduce variation and reduce the chromosome number in the daughter cells.

Check Your Progress 9.2

1. Explain how prophase I and metaphase I introduce variation.

2. Compare the number and types of chromosomes that are present at metaphase I and metaphase II of meiosis.

3. Summarize the differences between meiosis I and meiosis II.

9.3 Meiosis Compared with Mitosis

Learning Outcomes

Upon completion of this section, you should be able to

1. Contrast the events of meiosis I and meiosis II with the events of mitosis.
2. Contrast the events of meiosis I with the events of meiosis II.

We have now observed two different forms of cell division—mitosis (see Chapter 8) and meiosis. While there are some similarities, there are important differences between these processes (**Fig. 9.6**):

- Meiosis requires two consecutive nuclear divisions, but mitosis requires only one nuclear division.
- Meiosis produces four daughter nuclei, and there are four daughter cells following cytokinesis. Mitosis followed by cytokinesis results in two daughter cells.
- Following meiosis, the four daughter cells are haploid, having half the number of chromosomes of the parent cell. Following mitosis, the daughter cells have the same number of chromosomes as the parent cell.
- Following meiosis, the daughter cells are genetically different from each other and the parent cell. Following mitosis, the daughter cells are genetically identical to each other and to the parent cell.

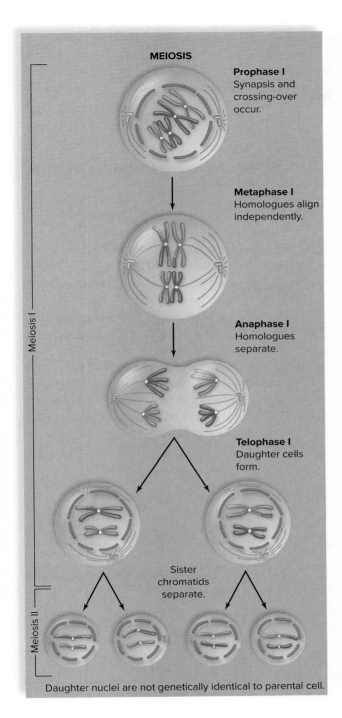

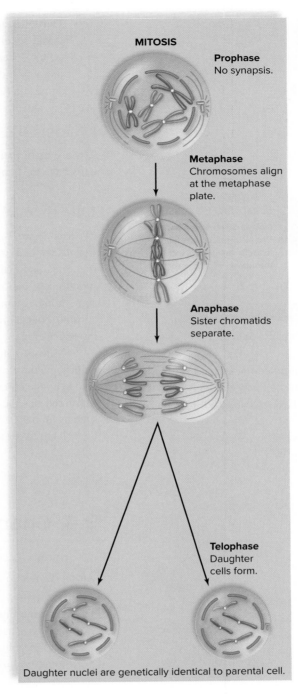

Figure 9.6 **Meiosis compared with mitosis.**

Notice that meiosis produces four haploid cells, whereas mitosis produces two diploid cells.

Meiosis I Compared with Mitosis

The following events distinguish meiosis I from mitosis:

- During prophase I of meiosis, synapsis occurs. During synapsis, tetrads form and crossing-over occurs. These events do not occur during mitosis.
- During metaphase I of meiosis, tetrads align at the spindle equator, with homologous chromosomes facing opposite spindle poles. The paired chromosomes have a total of four chromatids. During metaphase in mitosis, dyads align separately at the spindle equator.

- During anaphase I of meiosis, the homologous chromosomes of each tetrad separate, and dyads (with centromeres intact) move to opposite poles. Sister chromatids do not separate during anaphase I. During anaphase of mitosis, sister chromatids separate, becoming daughter chromosomes that move to opposite poles.

Meiosis II Compared with Mitosis

The events of meiosis II are like those of mitosis, except that in meiosis II the cells have the haploid number of chromosomes.

Mitosis and Meiosis Occur at Different Times

Meiosis occurs only at certain times in the life cycle of sexually reproducing organisms, and only in specialized tissues. In humans, meiosis occurs only in the testes and ovaries, where it is involved in the production of gametes. The functions of meiosis are to provide gamete variation and to keep the chromosome number constant generation after generation. With fertilization, the full chromosome number is restored. Because unlike gametes fuse, fertilization introduces great genetic diversity into the offspring.

Mitosis is much more common because it occurs in all tissues during embryonic development and during growth and repair. The function of mitosis is to keep the chromosome number constant in all the cells of the body, so that every cell has the same genetic material. These differences allow organisms, such as humans, to produce both the cells needed for the body to function and the cells used for reproduction (see Fig. 9.2).

CONNECTING THE CONCEPTS

9.3 Meiosis differs from mitosis in that meiosis produces four genetically unique cells with half the number of chromosomes.

Check Your Progress 9.3

1. Compare the number of cells produced by mitosis and meiosis.
2. Identify the number of chromosomes found in the daughter cells produced by mitosis and meiosis.
3. Summarize the major differences among the stages of meiosis I, meiosis II, and mitosis.

9.4 Changes in Chromosome Number

Learning Outcomes

Upon completion of this section, you should be able to

1. Define *nondisjunction,* and briefly explain how it may bring about an abnormal chromosome number.
2. List the causes and symptoms of Down syndrome.
3. List the syndromes that may result from an abnormal sex chromosome number, and briefly explain the cause of each.

The normal number of chromosomes in human cells is 46 (2n = 46), but occasionally humans are born with an abnormal number of chromosomes because the chromosomes fail to separate correctly, called **nondisjunction,** during meiosis. If nondisjunction occurs during meiosis I, both members of a homologous pair go into the same daughter cell. If it occurs during meiosis II, the sister chromatids will fail to separate and both daughter chromosomes will go into the same gamete (**Fig. 9.7**). If an egg that ends up with 24 chromosomes instead of 23 is fertilized with a normal sperm, the result is a *trisomy,* so called because one type of chromosome is present in three copies. If an egg that has 22 chromosomes instead of 23 is fertilized by a normal sperm, the result is a *monosomy,* so called because one type of chromosome is present in a single copy.

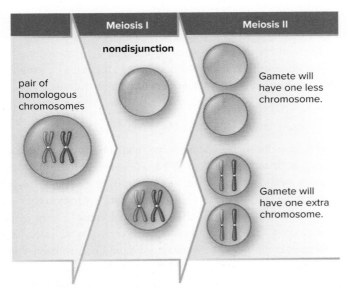

a. Nondisjunction during meiosis I

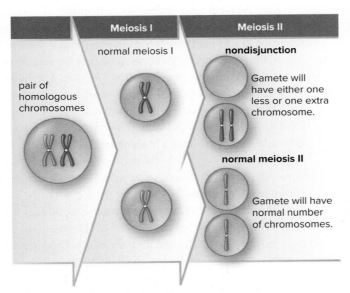

b. Nondisjunction during meiosis II

Figure 9.7 **Nondisjunction during meiosis.**

Because of nondisjunction, gametes either lack a chromosome or have an extra chromosome. Nondisjunction can occur **(a)** during meiosis I if homologous chromosomes fail to separate or **(b)** during meiosis II if the sister chromatids fail to separate completely.

Down Syndrome

Down syndrome, also called *trisomy 21,* is a condition in which an individual has three copies of chromosome 21 (**Fig. 9.8**). In most instances, the egg contained two copies of this chromosome instead of one. However, in around 20% of cases, the sperm contributed the extra chromosome 21.

Down syndrome is easily recognized by the following characteristics: short stature; an eyelid fold; stubby fingers; a wide gap between the first and second toes; a large, fissured tongue; a round head; a palm crease; and, at times, mental disabilities, which can sometimes be severe.

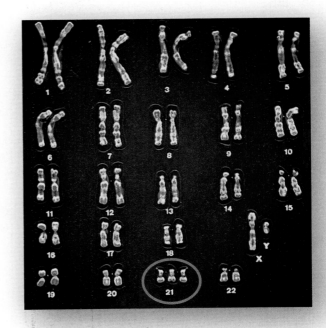

Figure 9.8 **Down syndrome.**

Down syndrome occurs when the egg or the sperm has an extra chromosome 21 due to nondisjunction in either meiosis I or meiosis II. When fertilized, the resulting cell has three copies of chromosome 21 (a trisomy; see red circle).

© CNRI/SPL/Science Source

Connections: Health

Why is the age of a female a factor in Down syndrome?

There is a difference in the timing of meiosis between males and females. Following puberty, males produce sperm continuously throughout their lives. In contrast, meiosis for females begins about five months after being conceived. However, the process is paused at prophase I of meiosis. Only after puberty are a few of these cells allowed to continue meiosis as part of the female menstrual cycle. Since long periods of time may occur between the start and completion of meiosis, there is a greater chance that nondisjunction will occur, and thus there is a greater chance of producing a child with Down syndrome as a female ages.

The chance of a woman having a Down syndrome child increases rapidly with age, starting at about age 40. The frequency of Down syndrome is 1 in 800 births for mothers under 40 years of age and 1 in 80 for mothers over 40. However, since women under 40 have more children as a group, most Down syndrome babies are born to them than to women over 40.

Abnormal Sex Chromosome Number

Nondisjunction during oogenesis or spermatogenesis can result in gametes that have too few or too many X or Y chromosomes. Figure 9.7 can be used to illustrate nondisjunction of the sex chromosomes during oogenesis if we assume that the chromosomes shown represent X chromosomes.

Just as an extra copy of chromosome 21 causes Down syndrome, additional or missing X or Y chromosomes cause certain syndromes. Newborns with an abnormal sex chromosome number are more likely to survive than are those with an abnormal autosome number. This is because normally females, like males, have only one functioning X chromosome. The other X chromosome is inactivated by a process called **X-inactivation,** producing an inactive chromosome called a *Barr body.* If a person has more than two X chromosomes (for example, an XXX female), multiple X chromosomes may be inactivated in her cells. X inactivation is a form of epigenetic inheritance (see Section 11.2).

In humans, the presence of a Y chromosome, not the number of X chromosomes, almost always determines maleness. The *SRY* (*sex-determining region Y*) gene located on the short arm of the Y chromosome produces a hormone called testis-determining factor, which plays a critical role in the development of male genitals. No matter how many X chromosomes are involved, an individual with a Y chromosome is a male, assuming that a functional *SRY* is on the Y chromosome. Individuals lacking a functional *SRY* on their Y chromosome have Swyer syndrome, and are also known as "XY females."

A person with **Turner syndrome** (45, XO) is a female. The number 45 indicates the total number of chromosomes the individual has, and the O signifies the absence of a second sex chromosome. Turner syndrome females are short, with a broad chest and webbed neck. The ovaries, uterine tubes, and uterus are very small and underdeveloped. Turner females do not undergo puberty or menstruate, and their breasts do not develop (**Fig. 9.9***a*). However, some have given birth following in vitro fertilization using donor eggs. These women usually have normal intelligence and can lead fairly normal lives if they receive hormone supplements.

A person with **Klinefelter syndrome** (47, XXY) is a male. A Klinefelter male has two or more X chromosomes in addition to a Y chromosome. The extra X chromosomes become Barr bodies. In males with Klinefelter syndrome, the testes and prostate gland are underdeveloped. There is no facial hair, but some breast development may occur (Fig. 9.9*b*). Affected individuals generally have large hands and feet and very long arms and legs. They are usually slow to learn but are not mentally handicapped unless they inherit more than two X chromosomes. As with Turner syndrome, it is best for parents to know as soon as possible that their child has Klinefelter syndrome because much can be done to help the child lead a normal life.

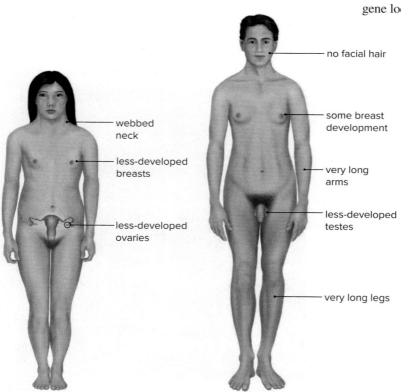

a. A female with Turner (XO) syndrome   b. A male with Klinefelter (XXY) syndrome

Figure 9.9 **Abnormal sex chromosome number.**

a. A female with Turner syndrome (XO) has a short, thick neck; short stature; and a lack of breast development. **b.** A male with Klinefelter syndrome (XXY) has immature sex organs and some development of the breasts.

Connections: Scientific Inquiry

Are individuals who are XYY more aggressive?

While the presence of a Y chromosome is the determining factor in the development of male sexual characteristics, the presence of two Y chromosomes, called *Jacobs syndrome* (XYY), does not appear to make an individual have a more aggressive personality. Early studies of Jacobs syndrome in the 1960s suggested that XYY males may be more prone to violent crimes; however, the study was conducted using a very small sample size, and subsequent studies quickly dispelled the link. While XYY males are known to have great height, and sometimes learning problems, there is very little evidence suggesting an increase in aggressive behavior. It is estimated that 1 in 1,000 males is an XYY individual.

X   Y   Y

(XYY): © Biophoto Associates/Science Source

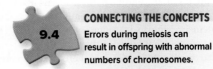

9.4

CONNECTING THE CONCEPTS

Errors during meiosis can result in offspring with abnormal numbers of chromosomes.

Check Your Progress 9.4

1. Describe how a zygote can receive an abnormal chromosome number.
2. Contrast monosomy with trisomy.
3. Distinguish between a chromosomal abnormality in the sex chromosomes and one in the autosomes.

STUDY TOOLS   http://connect.mheducation.com

SMARTBOOK® Maximize your study time with McGraw-Hill SmartBook®, the first adaptive textbook.

SUMMARIZE

Meiosis occurs during sexual reproduction. It is often the basis for genetic diversity within a species.

9.1 Meiosis reduces the chromosome number and introduces genetic variation.

9.2 Meiosis consists of two divisions, which introduce variation and reduce the chromosome number in the daughter cells.

9.3 Meiosis differs from mitosis in that meiosis produces four genetically unique cells with half the number of chromosomes.

9.4 Errors during meiosis can result in offspring with abnormal numbers of chromosomes.

9.1 An Overview of Meiosis

Meiosis is a type of nuclear division that reduces the chromosome number and shuffles the genes between chromosomes. It produces haploid daughter cells with a unique combination of chromosomes and alleles.

Homologous Chromosomes

Homologous chromosomes (**homologues**) are the same size with the same centromere position, and they contain the same genes. The **alleles** of a gene may differ between homologues. Diploid cells in humans have 22 homologous pairs of **autosomes** and 1 pair of **sex chromosomes,** for a total of 46 chromosomes. Males are XY and females are XX. In humans, the **haploid number (n)** of chromosomes is 23, and the **diploid number (2n)** is 46.

The Human Life Cycle

The human **life cycle** has two types of nuclear division:

- Mitosis ensures that every body cell has 23 pairs of chromosomes. It occurs during growth and repair.

- **Meiosis** is involved in the formation of gametes by either spermatogenesis (males) or oogenesis (females). It ensures that the gametes are haploid and have 23 chromosomes, 1 from each of the pairs of chromosomes.

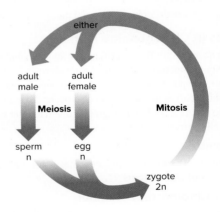

either

adult male   adult female

Meiosis   **Mitosis**

sperm n   egg n

zygote 2n

Overview of Meiosis

Meiosis has two nuclear divisions called **meiosis I** and **meiosis II**. Meiosis results in four haploid (n) daughter cells.

- **Meiosis I: Synapsis** and the formation of **tetrads** lead to separation of homologous chromosomes. Because no restrictions govern which member of a pair goes to which spindle pole, the gametes will contain all possible combinations of chromosomes.
- **Meiosis II:** The chromosomes are still duplicated at the beginning of meiosis II and are referred to as **dyads.** Sister chromatids later separate, forming daughter chromosomes. The daughter cells are haploid. If the

parent cell has four chromosomes, each of the daughter cells has two chromosomes. All possible combinations of chromosomes occur among the daughter cells.

- **Crossing-over: Crossing-over** between **nonsister chromatids** during meiosis I shuffles the alleles on the chromosomes, so that each sister chromatid of a homologue has a different mix of alleles.
- **The importance of meiosis:** Meiosis is important because the chromosome number stays constant between the generations of individuals and because the daughter cells, and therefore the gametes, are genetically different. When fertilization occurs, the offspring are genetically unique.

9.2 The Phases of Meiosis

The First Division—Meiosis I

- **Prophase I:** Tetrads form, and crossing-over occurs as chromosomes condense; the nuclear envelope fragments.
- **Metaphase I:** Tetrads align at the spindle equator. Either homologue can face either pole.
- **Anaphase I:** Homologues of each tetrad separate, and dyads move to the poles.
- **Telophase I:** Daughter nuclei are haploid, having received one duplicated chromosome from each homologous pair.
- Following telophase I, the daughter cells enter **interkinesis.** There is no duplication of the DNA during interkinesis.

The Second Division—Meiosis II

- **Prophase II:** Chromosomes condense, and the nuclear envelope fragments.
- **Metaphase II:** The haploid number of dyads align at the spindle equator.
- **Anaphase II:** Sister chromatids separate, becoming daughter chromosomes that move to the poles.
- **Telophase II:** Four haploid daughter cells are genetically different from each other and from the parent cell.

9.3 Meiosis Compared with Mitosis

Table 9.1 summarizes the differences between meiosis and mitosis.

9.4 Changes in Chromosome Number

Nondisjunction accounts for the inheritance of an abnormal number of chromosomes. Nondisjunction can occur during meiosis I if homologous chromosomes fail to separate so that both go into one daughter cell and the other daughter cell receives neither. It can also occur during meiosis II if chromatids fail to separate so that both go into one daughter cell and the other daughter cell receives neither.

Down Syndrome

In **Down syndrome,** an example of an autosomal syndrome, the individual inherits three copies of chromosome 21.

Abnormal Sex Chromosome Number

Examples of syndromes caused by inheritance of an abnormal number of sex chromosomes are **Turner syndrome** (XO) and **Klinefelter syndrome** (XXY).

- **Turner syndrome:** An XO individual inherits only one X chromosome. This individual survives because, even in an XX individual, **X-inactivation** produces Barr bodies for all but one X chromosome in the cell.
- **Klinefelter syndrome:** An XXY individual inherits two (or more) X chromosomes and a Y chromosome. This individual survives because one or more X chromosomes become a Barr body.

Table 9.1 Comparison of Important Events in Meiosis and Mitosis

Meiosis I	Mitosis
Prophase I	*Prophase*
Pairing of homologous chromosomes; crossing-over	No pairing of chromosomes; no crossing-over
Metaphase I	*Metaphase*
Tetrads at spindle equator	Dyads at spindle equator
Anaphase I	*Anaphase*
Homologues of each tetrad separate, and dyads move to poles.	Sister chromatids separate, becoming daughter chromosomes that move to the poles.
Telophase I	*Telophase*
Two haploid daughter cells, not identical to the parent cell	Two diploid daughter cells, identical to the parent cell
Meiosis II	**Mitosis**
Prophase II	*Prophase*
No pairing of chromosomes	No pairing of chromosomes
Metaphase II	*Metaphase*
Haploid number of dyads at spindle equator	Diploid number of dyads at spindle equator
Anaphase II	*Anaphase*
Sister chromatids separate, becoming daughter chromosomes that move to the poles.	Sister chromatids separate, becoming daughter chromosomes that move to the poles.
Telophase II	*Telophase*
Four haploid daughter cells, different from each other genetically and from the parent cell	Two diploid daughter cells, genetically identical to the parent cell

ASSESS

Testing Yourself

Choose the best answer for each question.

9.1 The Basics of Meiosis

1. Mitosis _____ chromosome number, whereas meiosis _____ the chromosome number of the daughter cells.
   - **a.** maintains, increases
   - **b.** increases, maintains
   - **c.** increases, decreases
   - **d.** maintains, decreases
2. A human cell contains _____ pair(s) of sex chromosomes.
   - **a.** 1
   - **b.** 2
   - **c.** 22
   - **d.** 23

9.2 The Phases of Meiosis

For questions 3–10, match the items to those in the key.

Key:

a. prophase I
b. metaphase I
c. anaphase I
d. telophase I
e. prophase II
f. metaphase II
g. anaphase II
h. telophase II

3. A cleavage furrow forms, resulting in haploid nuclei. Each chromosome contains two chromatids.

4. Tetrads form, and crossing-over occurs.

5. Dyads align at the spindle equator.

6. Four haploid daughter cells are created.

7. Homologous chromosomes separate and move to opposite poles.

8. Sister chromatids separate.

9. Tetrads align on the spindle equator.

10. Chromosomes in haploid nuclei condense.

11. Crossing-over
    a. results in genetic recombination between nonhomologous chromosomes.
    b. results in genetic recombination between sister chromatids.
    c. occurs only between members of a tetrad.
    d. occurs between adjacent tetrads.

9.3 Meiosis Compared with Mitosis

For questions 12–14, match the items to those in the key.

Key:

a. mitosis
b. meiosis
c. both mitosis and meiosis

12. produce(s) daughter cells with different genetic combinations

13. produce(s) haploid daughter cells

14. consist(s) of prophase, metaphase, anaphase, and telophase

9.4 Changes in Chromosome Number

15. Nondisjunction can result in
    a. monosomy.
    b. unequal cytoplasm.
    c. trisomy.
    d. Both a and c are correct.

16. An individual with Turner syndrome is conceived when a normal gamete unites with
    a. an egg that experienced nondisjunction during oogenesis.
    b. a sperm that experienced nondisjunction during spermatogenesis.
    c. Either a or b is correct.
    d. Neither a nor b is correct.

ENGAGE

Thinking Critically

1. Although most men with Klinefelter syndrome are infertile, some are able to father children. It was found that most fertile individuals with Klinefelter syndrome exhibit mosaicism, in which some cells are normal (46, XY) but others contain the extra chromosome (47, XXY). How might mosaicism come about? What effects might result?

2. In the nineteenth century, physicians noticed that people with Down syndrome were often the youngest children in large families. Some physicians suggested that the disorder was due to "maternal reproductive exhaustion." What might be the reasons for the relationship between advanced maternal age and incidence of Down syndrome?

3. As you can now understand, meiosis creates a very large amount of genetic variability within a population. Why might this be advantageous?

10

Patterns of Inheritance

© Jill Braaten

OUTLINE

10.1 Mendel's Laws   161

10.2 Mendel's Laws Apply to Humans   169

10.3 Beyond Mendel's Laws   173

10.4 Sex-Linked Inheritance   178

BEFORE YOU BEGIN

Before beginning this chapter, take a few moments to review the following discussions.

Section 3.2 What are the roles of proteins and nucleic acids in a cell?

Section 8.2 What is a sister chromatid?

Section 9.2 How does meiosis introduce new genetic combinations?

Warning for Phenylketonurics

Have you ever looked at the back of a can of diet soda and wondered what the warning "Phenylketonurics: Contains Phenylalanine" means? Phenylalanine is an amino acid that is found in almost all foods containing protein. It is also found in the artificial sweetener aspartame. For most people, phenylalanine is an important amino acid, and one that does not present any problems since any excess is broken down by enzymes in the body. However, some people, called phenylketonurics, lack a functional copy of this enzyme, and thus are unable to break down the phenylalanine. The excess may accumulate in the body, causing a variety of nervous system disorders. An estimated 1 in 10,000 people in the United States are phenylketonurics.

Like the rest of us, you are the product of your family tree. The DNA you inherit from your parents directly affects the proteins that enable your body to function properly. Rare genetic disorders like phenylketonuria pique our curiosity about how traits are inherited from one generation to the next. In this chapter, you will learn that the process of meiosis can be used to predict the inheritance of a trait, and that the genetic diversity produced through meiosis can sometimes lead to phenylketonuria.

You will also learn about patterns of inheritance first described by Mendel, and will see that certain traits, such as phenylketonuria, are recessive, meaning that it takes two nonfunctional copies of that gene to cause the disorder. This chapter will introduce you to observable patterns of inheritance, including human genetic disorders that are linked to specific genes on our chromosomes.

As you read through the chapter, think about the following questions:

1. How does the collection of chromosomes we inherit from our parents affect our bodies' appearance and function?
2. What patterns of inheritance apply to phenylketonuria?
3. How does meiosis help predict the probability of producing gametes and inheriting a trait?

10.1 Mendel's Laws

Learning Outcomes

Upon completion of this section, you should be able to

1. Explain Mendel's laws of inheritance.
2. Distinguish between genotype and phenotype.
3. Distinguish between dominant and recessive traits.
4. Apply Mendel's laws to solve and interpret monohybrid and dihybrid genetic crosses.
5. Recognize and explain the relationship between Mendel's laws and meiosis.

The science of genetics explains patterns of inheritance (why you share similar characteristics with your relatives) and variations between offspring from one generation to the next (why you have a different combination of traits than your parents). Virtually every culture in history has attempted to explain observed inheritance patterns. An understanding of these patterns has always been important to agriculture, animal husbandry (the science of breeding animals), and medicine.

Mendel's Experimental Procedure

Gregor Mendel was an Austrian monk who, after performing a series of ingenious experiments in the 1860s, developed several important laws on patterns of inheritance (**Fig. 10.1***a*). Mendel studied science and mathematics at the University of Vienna, and at the time of his research in genetics, he was a substitute natural science teacher at a local high school. His background in mathematics prompted him to use a statistical basis for his breeding experiments. He prepared for his experiments carefully and conducted preliminary studies with various animals and plants. When Mendel began his work, most plant and animal breeders acknowledged that both sexes contribute equally to a new individual. However, they were unable to account for the presence of variations (or differences) among the members of a family, generation after generation. Mendel's models of heredity account for such variations. In addition, Mendel's models are compatible with the theory of evolution, which states that various combinations of traits are tested by the environment, and those combinations that lead to reproductive success are the ones that are passed on.

Mendel's experimental organism was the garden pea, *Pisum sativum*. The garden pea was a good choice, since it was easy to cultivate and had a short generation time. And although peas normally self-pollinate (pollen goes only to the same flower), they could be cross-pollinated by hand. Many varieties of peas were available, and Mendel initially grew 22 of them. For his experiments, he chose 7 varieties that produced easily identifiable differences (Fig. 10.1*b*). When these varieties self-pollinated, they were *true-breeding*— meaning that the offspring were like the parent plants and like each other. In contrast to his predecessors, Mendel studied the inheritance of relatively simple and easily detected traits, such as seed shape, seed color, and flower color, and he observed no intermediate characteristics among the offspring. **Figure 10.2** shows Mendel's procedure.

a.

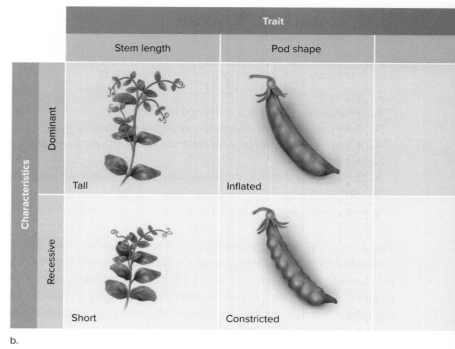

b.

Figure 10.1 Mendel working in his garden.

a. Mendel grew and tended the garden peas, *Pisum sativum,* he used for his experiments. **b.** Mendel selected these seven traits for study. Before he did any crosses, he made sure the parent plants bred true. All the offspring had either the dominant trait or the recessive trait.

(a): © Ned M. Seidler/National Geographic Creative

As Mendel followed the inheritance of individual traits, he kept careful records. He used his understanding of the mathematical laws of probability to interpret his results and to arrive at a theory that has since been supported by innumerable experiments. This theory is called the *particulate theory of inheritance,* because it is based on the existence of minute particles we now call genes. Inheritance involves the reshuffling of the same genes from generation to generation.

One-Trait Inheritance

After ensuring that his pea plants were true-breeding—for example, that his plants with green pods always produced green-pod offspring and his yellow-pod plants always produced yellow-pod offspring—Mendel was ready to perform a cross between these two strains. Mendel called the original parents the *P generation,* the first-generation offspring the *F₁* (for filial) *generation,* and the second-generation offspring the *F₂ generation.*

1 Cut away anthers.

2 Brush on pollen from another plant.

stigma
anther
ovary

Pollen grains containing sperm are produced in the anther. When pollen grains are brushed onto the stigma, sperm fertilizes eggs in the ovary. Fertilized eggs are located in ovules, which develop into seeds.

a. Flower structure

3 The results of cross from a parent that produces round, yellow seeds × parent that produces wrinkled yellow seeds.

b. Cross-pollination

Figure 10.2 Garden pea anatomy and traits.

a. In the garden pea, pollen grains produced in the anther contain sperm, and ovules in the ovary contain eggs. **b.** When Mendel performed crosses, he brushed pollen from one type of pea plant onto the stigma of another plant. After fertilization, the ovules developed into seeds (peas).

Trait				
Seed shape	Seed color	Flower position	Flower color	Pod color
Round	Yellow	Axial	Purple	Green
Wrinkled	Green	Terminal	White	Yellow

The diagram in **Figure 10.3** representing Mendel's F$_1$ cross is called a Punnett square (named after Reginald Punnett, an early-twentieth-century geneticist). In a **Punnett square,** all possible types of sperm are lined up vertically, and all possible types of eggs are lined up horizontally, or vice versa, so that every possible combination of gametes the offspring may inherit occurs within the square.

As Figure 10.3 shows, when Mendel crossed green-pod plants with yellow-pod plants, all the F$_1$ offspring resembled the green-pod parent. This did not mean that the other characteristic, yellow pods, had disappeared permanently. Notice that when Mendel allowed the F$_1$ plants to self-pollinate, three-fourths of the F$_2$ generation produced green pods and one-fourth produced yellow pods, a 3:1 ratio. Therefore, the F$_1$ plants had been able to pass on a factor for the yellow color—it didn't just disappear. Mendel recognized that the F$_1$ plants produced green pods because the green color was dominant over the yellow color. This type of cross is commonly called a **monohybrid cross,** since it involves two individuals who have two different alleles (they are heterozygous, Gg) for a single trait.

Mendel's mathematical approach led him to interpret his results differently than previous breeders had done. He reasoned that a 3:1 ratio among the F$_2$ offspring was possible only if (1) the F$_1$ parents contained two separate copies of each hereditary factor, one dominant and the other recessive; (2) the factors separated when the gametes were formed, and each gamete carried only one copy of each factor; and (3) random fusion of all possible gametes occurred upon fertilization. Only in this way would the yellow color reoccur in the F$_2$ generation.

Figure 10.3 One-trait cross.

The P generation plants differ for one trait—the color of the pods. All the F$_1$ generation plants produced green pods, but the factor for yellow pods has not disappeared because one-fourth of the F$_2$ generation produced yellow pods. The 3:1 ratio allowed Mendel to deduce that individuals have two discrete genetic factors for each trait.

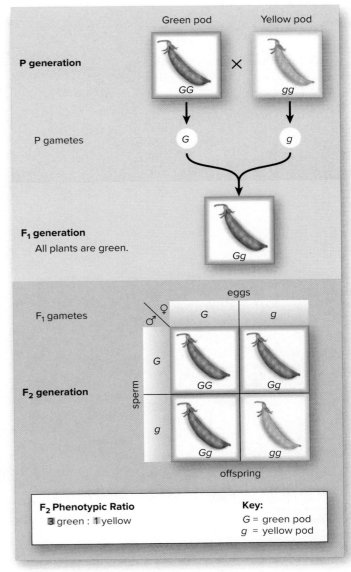

Green pod   Yellow pod

P generation GG × gg

P gametes G g

F$_1$ generation
All plants are green. Gg

F$_1$ gametes

eggs

	G	g
G	GG	Gg
g	Gg	gg

sperm

F$_2$ generation

offspring

F$_2$ Phenotypic Ratio
3 green : **1** yellow

Key:
G = green pod
g = yellow pod

Connections: Scientific Inquiry

Why so many peas?

Mendel did not have an obsession with pea plants. Instead, as a mathematician, he recognized that, in order for his results to be statistically significant, he needed to examine a large sample. Peas are ideal for this type of analysis, since each pea is an individual zygote. In his ini-

© Nigel Cattlin/ Science Source

tial experiments, Mendel examined the shapes and colors of over 15,300 seeds and 1,700 pea pods! Only by doing so could he be assured that the observed 3:1 ratios were not simply the result of randomness.

One-Trait Testcross

Mendel's experimental use of simple dominant and recessive characteristics allowed him to test his interpretation of his crosses. To see if the F_1 generation carries a recessive factor, Mendel crossed his F_1 plants with green pods with true-breeding plants with yellow pods. He reasoned that half the offspring would produce green pods and half would produce yellow pods (**Fig. 10.4** *a*). And, indeed, those were the results he obtained; therefore, his hypothesis that factors segregate when gametes are formed was supported.

This is an example of a **testcross.** While the procedure has been largely replaced by more detailed genetic analyses (see Section 12.1), it is still used as an example of how to determine whether an individual with the dominant trait has two dominant factors for a particular trait. This is not possible to tell by observation because an individual can exhibit the dominant appearance while having only one dominant factor. Figure 10.4*b* shows that if the green-pod parent plant has two dominant factors, all the offspring will produce green pods.

After doing one-trait crosses, Mendel arrived at his first law of inheritance, called the **law of segregation,** which is a cornerstone of the particulate theory of inheritance. The law of segregation states the following:

- Each individual has two factors for each trait.
- The factors segregate (separate) during the formation of the gametes.
- Each gamete contains only one factor from each pair of factors.
- Fertilization gives each new individual two factors for each trait.

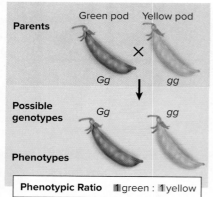

a.

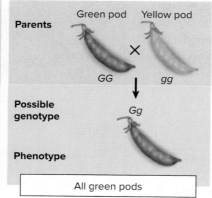

b.

Figure 10.4 One-trait testcross.

Crossing an individual with a dominant appearance (phenotype) with a recessive individual indicates the genotype. **a.** If a parent with the dominant phenotype has only one dominant factor, the results among the offspring are 1:1. **b.** If a parent with the dominant phenotype has two dominant factors, all offspring have the dominant phenotype.

The Modern Interpretation of Mendel's Work

In the early twentieth century, scientists noted the parallel behavior of Mendel's particulate factors and chromosomes and proposed the *chromosomal theory of inheritance,* which states that chromosomes are carriers of genetic information. Today, we recognize that each trait is controlled by alleles (alternate forms of a gene) that occur on the chromosomes at a particular location, called a **locus.** The **dominant allele** is so named because of its ability to mask the expression of the other allele, called the **recessive allele.** In many cases, the dominant allele is identified by an uppercase (capital) letter, and the recessive allele by the same letter but lowercase (small). While the alleles on one homologue can be the alternates of the alleles on the other homologue (**Fig. 10.5**), the sister chromatids have the same types of alleles. These alleles act as the blueprints for traits at the physiological and cellular level. We now know that these instructions may be modified slightly by environmental influences (see Section 10.3) and by regulation of gene expression (see Section 11.2).

Connections: Scientific Inquiry

Why are some alleles dominant and others recessive?

Genes contain information for the production of a specific protein. In the pea plants Mendel used, there is a gene for flower color that has two variations, or alleles. The dominant allele produces a purple flower color. However, the recessive allele codes for a variation of the protein that does not produce the correct pigmentation. In other words, it has lost its function. This results in a white color. In a flower with one dominant and one recessive allele (a heterozygote), the dominant purple color masks the recessive white color. As you will see later in this chapter, not all dominant alleles completely mask the recessive allele.

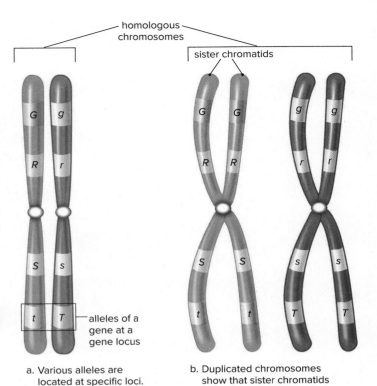

a. Various alleles are located at specific loci.

b. Duplicated chromosomes show that sister chromatids have identical alleles.

Figure 10.5 **Homologous chromosomes.**

a. The letters represent alleles—that is, alternate forms of a gene. Each allelic pair, such as *Gg* or *Tt,* is located on homologous chromosomes at a particular gene locus. **b.** Sister chromatids carry the same alleles in the same order.

As you learned in Chapter 9, meiosis is the type of cell division that reduces the chromosome number. During meiosis I, the homologues separate. Therefore, the process of meiosis explains Mendel's law of segregation and why a gamete has only one allele for each trait (see Fig. 10.8). When fertilization occurs, the resulting offspring again have two alleles for each trait, one from each parent.

Genotype versus Phenotype

Different combinations of alleles for a trait can give an organism the same outward appearance. For instance, a recessive allele in humans (*a*) causes albinism. However, if you are *AA* or *Aa* and not *aa,* you have normal pigmentation. For this reason, it is necessary to distinguish between the alleles present in an organism and the appearance of that organism.

The word **genotype** refers to the combination of alleles in a cell or organism. Genotype may be indicated by letters or by short, descriptive phrases. When an organism has two identical alleles for a trait, it is termed **homozygous.** A person who is *homozygous dominant* for normal pigmentation possesses two dominant alleles (*AA*). All gametes from this individual will contain an allele for normal pigmentation (*A*). Likewise, all gametes produced by a *homozygous recessive* parent (*aa*) contain an allele for albinism (*a*). Individuals who have two different alleles (*Aa*), are said to be **heterozygous.**

Table 10.1 Genotype versus Phenotype

Allele Combination	Genotype	Phenotype
AA	Homozygous dominant	Normal pigmentation
Aa	Heterozygous	Normal pigmentation
aa	Homozygous recessive	Albinism

The word **phenotype** refers to the physical appearance of the individual. An organism's phenotype is mostly determined by its genotype. The homozygous dominant (*AA*) individual and the heterozygous (*Aa*) individual both show the dominant phenotype and have normal pigmentation, while the homozygous recessive individual (*aa*) shows the recessive phenotype of albinism (**Table 10.1**).

Two-Trait Inheritance

Mendel performed a second series of crosses in which he crossed true-breeding plants that differed in two traits (**Fig. 10.6**). For example, he crossed tall plants having green pods (*TTGG*) with short plants having yellow pods (*ttgg*). The F$_1$ plants showed both dominant characteristics (tall with green pods).

The question Mendel had was how these alleles would segregate in the F$_2$ generation. There were two possible results that could occur in the F$_2$ generation:

1. If the dominant factors (*TG*) always go together into the F$_1$ gametes, and the recessive factors (*tg*) always stay together, then two phenotypes result among the F$_2$ plants—tall plants with green pods and short plants with yellow pods.
2. If the four factors segregate into the F$_1$ gametes independently, then four phenotypes result among the F$_2$ plants—tall plants with green pods, tall plants with yellow pods, short plants with green pods, and short plants with yellow pods.

As was the case with the one-trait, or monohybrid, cross (see Fig. 10.3), Mendel allowed the F$_1$ generation to interbreed to produce an F$_2$ generation.

Figure 10.6 shows that Mendel observed four phenotypes among the F$_2$ plants, supporting the second hypothesis. Therefore, Mendel formulated his second law of heredity—the **law of independent assortment**—which states the following:

- Each pair of factors segregates (assorts) independently of the other pairs.
- All possible combinations of factors can occur in the gametes.

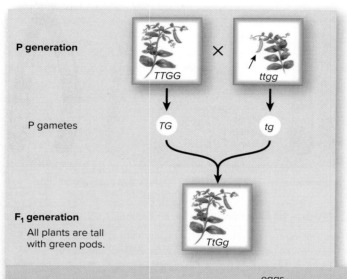

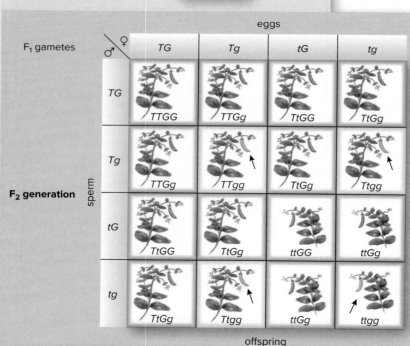

F$_2$ Phenotypic Ratio

9 ▪ tall plant, green pod
3 ▫ tall plant, yellow pod
3 ▪ short plant, green pod
1 ▪ short plant, yellow pod

Key:

T = tall plant
t = short plant
G = green pod
g = yellow pod

Figure 10.6 Two-trait cross by Mendel.

P generation plants differ in two regards—stem length and pod color. The F$_1$ generation shows only the dominant phenotypes, but all possible phenotypes appear among the F$_2$ generation. The 9:3:3:1 ratio allowed Mendel to deduce that factors segregate into gametes independently of other factors. (Arrow indicates yellow pods.)

The type of cross illustrated in Figure 10.6 is also referred to as a **dihybrid cross,** since it is a cross for two traits, with each individual involved in the cross having two different alleles for each trait. Figure 10.6 illustrates that *the expected phenotypic results of a dihybrid cross are always 9:3:3:1 when both parents are heterozygous for the two traits.* However, it is important to note that ratios other than 9:3:3:1 are possible if the parents are not heterozygous for both traits.

Two-Trait Testcross

The fruit fly *Drosophila melanogaster,* less than one-fifth the size of a housefly, is a favorite subject for genetic research because it has several mutant characteristics that are easily determined. The *wild-type* fly—the type you are most likely to find in nature—has long wings and a gray body, while some mutant flies have short (vestigial) wings and black (ebony) bodies. The key for a cross involving these traits is L = long wings, l = short wings, G = gray body, and g = black body.

A two-trait testcross can be used to determine whether an individual is homozygous dominant or heterozygous for either of the two traits. Because it is not possible to determine the genotype of a long-winged, gray-bodied fly by inspection, this genotype may be represented as $L\_G\_$.

In a two-trait testcross, an individual with the dominant phenotype for both traits is crossed with an individual with the recessive phenotype for both traits because this individual has a *known* genotype. For example, a long-winged, gray-bodied fly is crossed with a short-winged, black-bodied fly. The heterozygous parent fly (*LlGg*) forms four different types of gametes. The homozygous parent fly (*llgg*) forms only one type of gamete:

P:	*LlGg*	×	*llgg*
Gametes:	*LG, Lg, lG, lg*		*lg*

As **Figure 10.7** shows, all possible combinations of phenotypes occur among the offspring. This 1:1:1:1 phenotypic ratio shows that the $L\_G\_$ fly is heterozygous for both traits and has the genotype *LlGg*. A Punnett square can also be used to predict the chances of an offspring having a particular phenotype (Fig. 10.7). What are the chances of an offspring with long wings and a gray body? The chances are one in four, or 25%. What are the chances of an offspring with short wings and a gray body? The chances are also one in four, or 25%.

Mendel's Laws and Probability

The importance of Mendel's work is that it made a connection between observed patterns of inheritance and probability. When we use a Punnett square to calculate the results of genetic crosses, we assume that each gamete contains one allele for each trait (law of segregation) and that collectively the gametes have all possible combinations of alleles (law of independent assortment). Further, we assume that the male and female gametes combine at random—that is, all possible sperm have an equal chance to fertilize all possible eggs. Under these circumstances, we can use the rules of probability to calculate the expected phenotypic ratio. The *rule of multiplication* says that the chance of two (or more) independent events occurring together is the product of their chances of occurring separately. For example, the chance of getting tails when you toss a coin is ½. The chance of getting two tails when you toss two coins at once is ½ × ½ = ¼.

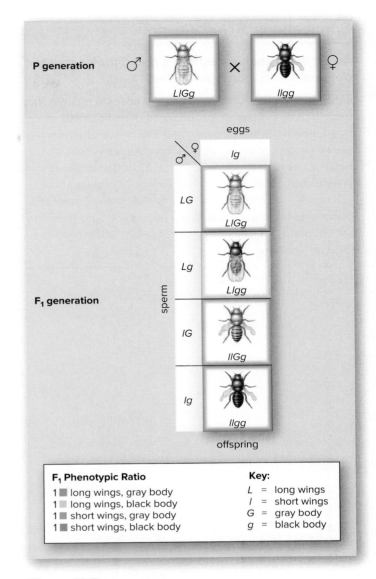

Figure 10.7 Two-trait testcross.

If a fly heterozygous for both traits is crossed with a fly that is recessive for both traits, the expected ratio of phenotypes is 1:1:1:1.

Connections: Scientific Inquiry

Aren't traits such as earlobes and dimples due to dominant and recessive alleles?

For a considerable amount of time, geneticists thought that traits such as earlobe shape (attached or unattached), shape of the hairline (widow's peak), and the presence of dimples were examples of simple dominant and recessive inheritance. However, as we learn more about the 23,000 or so genes that make us human, we have discovered that these traits are actually under the control of multiple genes. These interactions complicate the expression of the phenotype, and thus they are not good examples of simple dominant and recessive allele combinations.

Let's use the rule of multiplication to calculate the expected results in Figure 10.7. Because each allele pair separates independently, we can treat the cross as two separate one-trait crosses:

$$Ll \times ll: \quad \text{Probability of } ll \text{ offspring} \quad = \quad \tfrac{1}{2}$$
$$Gg \times gg: \quad \text{Probability of } gg \text{ offspring} \quad = \quad \tfrac{1}{2}$$

The probability of the offspring's genotype being *llgg* is:

$$\tfrac{1}{2}\,ll \quad \times \quad \tfrac{1}{2}\,gg \quad = \quad \tfrac{1}{4}\,llgg$$

And the same results are obtained for the other possible genotypes among the offspring in Figure 10.7.

Mendel's Laws and Meiosis

Today, we are aware that Mendel's laws relate to the process of meiosis. In **Figure 10.8,** a human cell has two pairs of homologous chromosomes, recognized by length—one pair of homologues is short and the other is long. In these diagrams, the coloring is used to signify which parent the chromosome was inherited from; one homologue of each pair is the "paternal" chromosome, and the other is the "maternal" chromosome. When the homologues separate (segregate), each gamete receives one member from each pair. The homologues

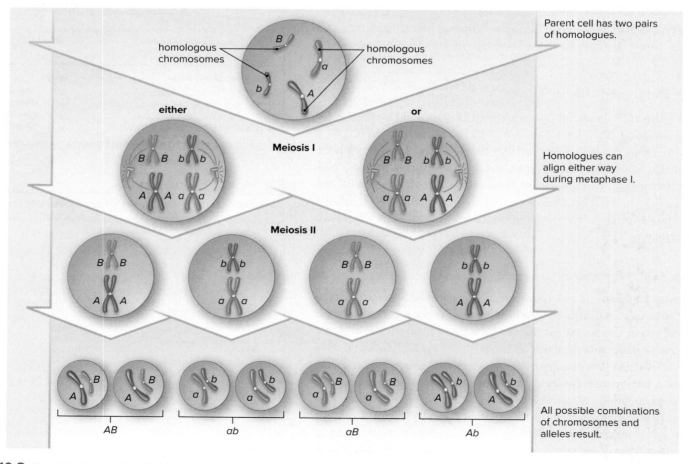

Figure 10.8 Mendel's laws and meiosis.

A human cell has 23 pairs of homologous chromosomes (homologues), of which 2 pairs are represented here. The homologues, and the alleles they carry, segregate independently during gamete formation. Therefore, all possible combinations of chromosomes and alleles occur in the gametes.

separate (assort) independently; it does not matter which member of each pair goes into which gamete. In the simplest of terms, a gamete in Figure 10.8 can receive one short and one long chromosome of either color. Therefore, all possible combinations of chromosomes are in the gametes.

In Figure 10.8, the alleles *A* and *a* are on one pair of homologues, and the alleles *B* and *b* are on the other pair of homologues. Because there are no restrictions as to which homologue goes into which gamete, a gamete can receive either an *A* or an *a* and either a *B* or a *b* in any combination because the chromosomes the alleles are located on assort independently. In the end, collectively, the gametes will have all possible combinations of alleles.

CONNECTING THE CONCEPTS

10.1 Mendel established two laws that may be used to explain patterns of inheritance.

Check Your Progress 10.1

1. Summarize what Mendel's experiments explained regarding patterns of inheritance.
2. Compare the phenotypic and genotypic ratios of a cross between two heterozygotes.
3. Solve the following (using Figure 10.6): What is the probability that a cross between two individuals with a *TtGg* genotype will produce a short plant with green pods?
4. Explain how meiosis relates to Mendel's experiments with dihybrid crosses.

10.2 Mendel's Laws Apply to Humans

Learning Outcomes

Upon completion of this section, you should be able to

1. Interpret a pedigree to determine if the pattern of inheritance is autosomal dominant or autosomal recessive.
2. List the characteristics of autosomal dominant and autosomal recessive pedigrees.
3. List some common genetic disorders, state the symptoms of each, and describe the inheritance pattern that each one exhibits.

Many traits and disorders in humans, and other organisms, are genetic in origin and follow Mendel's laws. These traits are often controlled by a single pair of alleles on the autosomal chromosomes. An **autosome** is any chromosome other than a sex (X or Y) chromosome. The patterns of inheritance associated with the sex chromosomes are discussed in Section 10.4.

Family Pedigrees

A **pedigree** is a chart of a family's history with regard to a particular genetic trait. In a pedigree, males are designated by squares and females by circles. Shaded circles and squares represent individuals expressing the genetic trait (often a disorder). A line between a square and a circle represents a union. A vertical line going downward leads directly to a single child; if there are more children, they are placed off a horizontal line. To identify specific individuals in a pedigree, generations are often numbered down the left side, and individuals are then identified from left to right. For example, in **Figure 10.9,** the male at the top is identified as individual I-2.

The analysis of a pedigree is useful for genetic counselors. Once a pattern of inheritance has been established, the counselor can then determine the chances that any child born to the couple will have the abnormal phenotype.

Pedigrees for Autosomal Disorders

A family pedigree for an *autosomal recessive disorder* is shown in Figure 10.9. In this pattern, a child may have the recessive phenotype of the disorder even though neither parent is affected. These heterozygous parents are sometimes referred to as *carriers* because, although they are unaffected, they are capable

Connections: Scientific Inquiry

Are all dominant alleles "normal"?

In most examples, the dominant allele represents the "normal" allele and the recessive allele represents the altered version. However, this is not always the case. For example, one form of polydactyly—the presence of extra fingers or

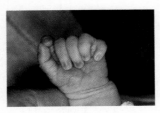

© Science Photo Library/Science Source

toes—is due to the inheritance of a dominant allele. The "normal" allele is recessive, meaning that an individual only needs to inherit one copy of the polydactyly allele in order to display the phenotype. Polydactyly is also an example of a trait that has variable expressivity, meaning that for one person the presence of the allele may cause an extra finger but in another person it may be an extra toe.

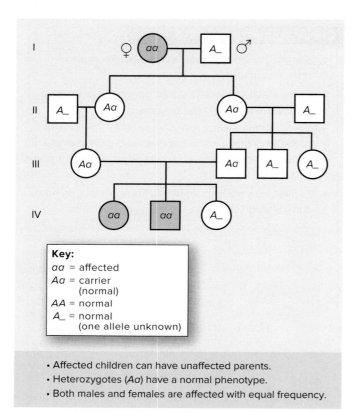

Key:
aa = affected
Aa = carrier
    (normal)
AA = normal
*A_* = normal
    (one allele unknown)

• Affected children can have unaffected parents.
• Heterozygotes (*Aa*) have a normal phenotype.
• Both males and females are affected with equal frequency.

Figure 10.9   Autosomal recessive pedigree.

The list gives ways to recognize an autosomal recessive disorder. How would you know that individual III-1 is heterozygous?

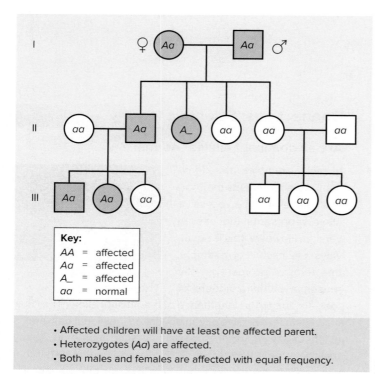

Key:
AA = affected
Aa = affected
*A_* = affected
aa = normal

• Affected children will have at least one affected parent.
• Heterozygotes (*Aa*) are affected.
• Both males and females are affected with equal frequency.

Figure 10.10   Autosomal dominant pedigree.

The list gives ways to recognize an autosomal dominant disorder. Why can't the genotype of individual II-3 be determined from this pedigree?

of having a child with the genetic disorder. If the family pedigree suggests that the parents are carriers for an autosomal recessive disorder, the counselor might suggest confirming this using the appropriate genetic test. Then, if the parents so desire, it would be possible to do prenatal testing of the fetus for the genetic disorder.

Figure 10.9 lists other ways that a counselor may recognize an autosomal recessive pattern of inheritance. Notice that, in this pedigree, cousins III-1 and III-2 are the parents of three children, two of whom have the disorder. Generally, reproduction between closely related individuals is more likely to bring out recessive traits since there is a greater chance that the related individuals carry the same alleles.

This pedigree also shows that "chance has no memory"; therefore, each child born to heterozygous parents has a 25% chance of having the disorder. In other words, if a heterozygous couple has four children, it is possible that each child might have the condition.

Figure 10.10 shows an *autosomal dominant* pattern of inheritance. In this pattern, a child can be unaffected even when the parents are heterozygous and therefore affected. Figure 10.10 lists other ways to recognize an autosomal dominant pattern of inheritance. This pedigree illustrates that, when both parents are unaffected, all their children are unaffected. Why? Because neither parent has a dominant gene that causes the condition to be passed on.

Genetic Disorders of Interest

Although many conditions are caused by interactions of genes and the environment (see Section 10.3), there are many human disorders caused by single gene mutations. A few examples of autosomal disorders are discussed in this section. Some of these disorders are recessive, and therefore an individual must inherit two affected alleles before having the disorder. Others are dominant, meaning that it takes only one affected allele to cause the disorder.

Methemoglobinemia

Methemoglobinemia is a relatively harmless autosomal recessive disorder that results from an accumulation of methemoglobin, an alternative form of hemoglobin, in the blood. Since methemoglobin is blue instead of red, the skin of people with the disorder appears bluish-purple in color (**Fig. 10.11**).

A persistent and determined physician finally solved the age-old mystery of what causes methemoglobinemia through blood tests and pedigree analysis of a family with the disorder, the Fugates of Troublesome Creek in Kentucky. On a hunch, the physician tested the Fugates for the enzyme diaphorase, which normally converts methemoglobin back to hemoglobin, and found that they indeed lacked the enzyme. Next, he treated a patient with the disorder in a simple but rather unconventional manner—by injecting a dye called methylene blue! The dye is a strong reducing agent capable of donating electrons to methemoglobin, converting it back into hemoglobin. The results were striking and immediate—the patient's skin quickly turned pink again.

---

[1] An autosomal recessive pedigree (Fig. 10.9) shows that affected children can have unaffected parents, that heterozygotes have a normal phenotype, and that both males and females are affected equally. The individual III-1 is heterozygous because she had affected children. An autosomal dominant pedigree

[2] (Fig. 10.10) shows that affected children have at least one affected parent, that heterozygotes are affected, and that males and females are equally affected. The individual II-2 is heterozygous because he had one unaffected child. The genotype of individual II-3 cannot be determined because there are no unaffected children present.

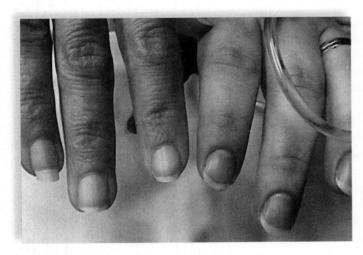

Figure 10.11 Methemoglobinemia.

The hand of the woman on the right appears blue due to methemoglobinemia. Individuals with the disorder lack the enzyme diaphorase and are unable to clear methemoglobin from the blood.

© Division of Medical Toxicology, University of Virginia

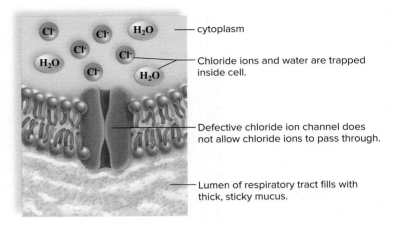

cytoplasm

Chloride ions and water are trapped inside cell.

Defective chloride ion channel does not allow chloride ions to pass through.

Lumen of respiratory tract fills with thick, sticky mucus.

Figure 10.12 Cystic fibrosis.

Cystic fibrosis is due to a faulty channel protein that is supposed to regulate the flow of chloride ions through the plasma membrane.

Cystic Fibrosis

Cystic fibrosis is an autosomal recessive disorder that occurs among all ethnic groups, but it is the most common lethal genetic disorder among Caucasians in the United States. Research has demonstrated that chloride ions (Cl^-) fail to pass through a plasma membrane channel protein in the cells of these patients (**Fig. 10.12***a*). Ordinarily, after chloride ions have passed through the membrane, sodium ions (Na^+) and water follow. It is believed that the lack of water passing out of the cells then causes abnormally thick mucus in the bronchial tubes and pancreatic ducts, thus interfering with the function of the lungs and pancreas. To ease breathing in affected children, the thick mucus in the lungs must be loosened periodically, but still the lungs become infected frequently. Clogged pancreatic ducts prevent digestive enzymes from reaching the small intestine, and to improve digestion, patients take digestive enzymes mixed with applesauce before every meal.

Alkaptonuria

Black urine disease, or *alkaptonuria,* is a rare genetic disorder that follows an autosomal recessive inheritance pattern. People with alkaptonuria lack a functional copy of the *homogentisate oxygenase* (*HGD*) gene found on chromosome 3. The HGD enzyme normally breaks down a compound called homogentisic acid. When the enzyme is missing, homogentisic acid accumulates in the blood and is passed into the urine. The compound turns black on exposure to air, giving the urine a characteristic color and odor (**Fig. 10.13**). Homogentisic acid also accumulates in joint spaces and connective tissues, leading to darkening of the tissues and eventually arthritis by adulthood.

Sickle-Cell Disease

Sickle-cell disease is an autosomal recessive disorder in which the red blood cells are not biconcave disks, like normal red blood cells, but are irregular in shape (**Fig. 10.14**). In fact, many are sickle-shaped. A single base change in the globin gene causes the hemoglobin in affected individuals to differ from normal hemoglobin by a single amino acid. The abnormal hemoglobin molecules stack up and form insoluble rods, and the red blood cells become sickle-shaped.

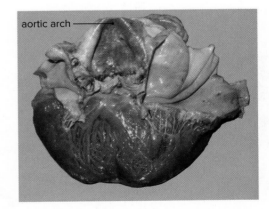

aortic arch

Figure 10.13 Alkaptonuria.

Alkaptonuria is caused by the inability to metabolize a compound called homogentisic acid. Excess homogentisic acid accumulates in the urine and connective tissues (note the dark staining), and can result in connective tissue disorders.

© Biophoto Associates/Science Source

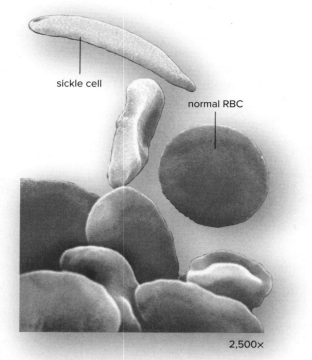

Figure 10.14 Sickle-cell disease.

Persons with sickle-cell disease have sickle-shaped red blood cells because of an abnormal hemoglobin molecule.

© Dr. Gopal Murti/SPL/Science Source

Because sickle-shaped cells can't pass along narrow capillaries as well as disk-shaped cells can, they clog the vessels and break down. This is why persons with sickle-cell disease suffer from poor circulation, anemia, and low resistance to infection. Internal hemorrhaging leads to further complications, such as jaundice, episodic pain in the abdomen and joints, and damage to internal organs.

Sickle-cell heterozygotes have normal blood cells but carry the sickle-cell trait. Most experts believe that persons that are heterozygous for the sickle-cell trait are generally healthy and do not need to restrict their physical activity. However, there occasionally may be problems if they experience dehydration or mild oxygen deprivation.

Osteogenesis imperfecta

Osteogenesis imperfecta is an autosomal dominant genetic disorder that results in weakened, brittle bones. Although at least nine types of the disorder are known, most are linked to mutations in two genes necessary to the synthesis of a type I collagen—one of the most abundant proteins in the human body. Collagen has many roles, including providing strength and rigidity to bone and forming the framework for most of the body's tissues. Osteogenesis imperfecta leads to a defective type I collagen that causes the bones to be brittle and weak. Because the mutant collagen can cause structural defects even when combined with normal type I collagen, osteogenesis imperfecta is generally considered to be dominant. Osteogenesis imperfecta, which has an incidence of approximately 1 in 5,000 live births, affects all racial groups similarly and has been documented as long as 300 years ago. Some historians think that the Viking chieftain Ivar Ragnarsson, who was known as Ivar the Boneless and was often carried into battle on a shield, had this condition. In most cases, the diagnosis is made in young children who visit the emergency room frequently due to broken bones. Some children with the disorder have an unusual blue tint in the sclera, the white portion of the eye; reduced skin elasticity; weakened teeth; and occasionally heart valve abnormalities. Currently, the disorder is treatable with a number of drugs that help increase bone mass, but these drugs must be taken long-term.

Huntington Disease

Huntington disease is a dominant neurological disorder that leads to progressive degeneration of neurons in the brain (**Fig. 10.15**). The disease is caused by a single mutated copy of the gene for a protein called huntingtin. Most patients appear normal until they are middle-aged and have already had children, who may also have the inherited disorder. There is no effective treatment, and death usually occurs 10 to 15 years after the onset of symptoms.

Several years ago, researchers found that the gene for Huntington disease is located on chromosome 4. A test was developed for the presence of the gene, but few people want to know if they have inherited the

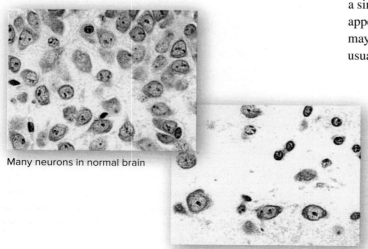

Many neurons in normal brain

Loss of neurons in Huntington brain

Figure 10.15 Huntington disease.

Huntington disease is characterized by increasingly serious psychomotor and mental disturbances because of a loss of neurons in the brain.

(both): © P. Hemachandra Reddy, Ph.D.

gene because there is no cure. But now we know that the disease stems from an unusual mutation. Extra codons cause the huntingtin protein to have a series of extra glutamines. Whereas the normal version of huntingtin has stretches of between 10 and 25 glutamines, mutant huntingtin may contain 36 or more. Because of the extra glutamines, the huntingtin protein changes shape and forms large clumps inside neurons. Even worse, it attracts and causes other proteins to clump with it. One of these proteins, called CBP, helps nerve cells survive. Researchers hope to combat the disease by boosting CBP levels.

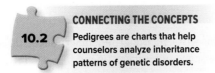

CONNECTING THE CONCEPTS

10.2 Pedigrees are charts that help counselors analyze inheritance patterns of genetic disorders.

10.3 Beyond Mendel's Laws

Learning Outcomes

Upon completion of this section, you should be able to

1. Understand how incomplete dominance and codominance deviate from traditional Mendelian expectations.
2. Explain the characteristics of polygenic and multifactorial traits.
3. Describe how epistatic interactions and pleiotropy influence phenotypes.
4. Explain what is meant by the term *linkage*.

Since Mendel's time, variations in the dominant/recessive relationship have been discovered. Some alleles are neither dominant nor recessive, and some genes have more than two alleles. Furthermore, some traits are affected by more than one pair of genes and by the environment. Together, these variations make it clear that the concept of the genotype alone cannot account for all the observable traits of an organism.

Incomplete Dominance

Incomplete dominance occurs when the heterozygote is intermediate between the two homozygotes. For example, when a curly-haired individual has children with a straight-haired individual, their children have wavy hair. When two wavy-haired persons have children, the expected phenotypic ratio among the offspring is 1:2:1—one curly-haired child to two with wavy hair to one with straight hair. We can explain incomplete dominance by assuming that the dominant allele encodes for a gene product that is not completely capable of masking the recessive allele.

Another example of incomplete dominance in humans is familial hypercholesterolemia. This disease is due to a variation in the number of LDL-cholesterol receptor proteins in the plasma membrane. The number of receptors is related to the ability of the body to remove cholesterol from the blood. The fewer the number of receptors, the greater the concentration of cholesterol in the blood (**Fig. 10.16**). A person with two mutated alleles lacks LDL-cholesterol receptors. A person with only one mutated allele has half the normal number of receptors, and a person with two normal alleles has the usual number of receptors. People with the full number of receptors do not have familial hypercholesterolemia. When receptors are completely absent, excessive cholesterol is deposited in various places in the body, including under the skin.

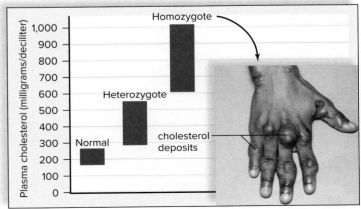

Figure 10.16 Incomplete dominance.

In incomplete dominance, the heterozygote is intermediate between the homozygotes. Familial hypercholesterolemia is an example in humans. Persons with one mutated allele have an abnormally high level of cholesterol in the blood, and those with two mutated alleles have a higher level still.

© Medical-On-Line/Alamy

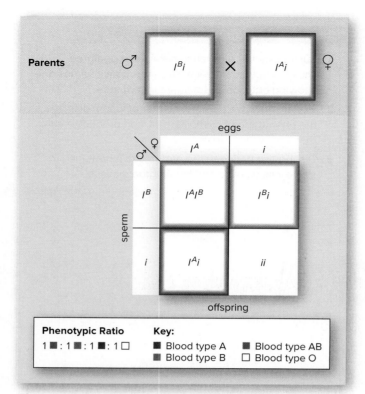

Figure 10.17 Inheritance of ABO blood type.

A mating between individuals with type A blood and type B blood can result in any one of the four blood types. Why? Because the parents are *I*^B*i* and *I*^A*i*. The *i* allele is recessive; both *I*^A and *I*^B are dominant.

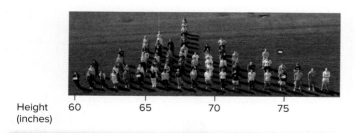

Height 60   65   70   75
(inches)

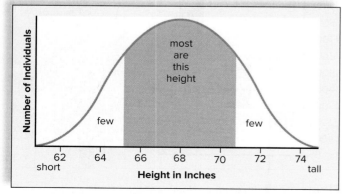

Figure 10.18 Height in humans, a polygenic trait.

When you record the heights of a large group of people chosen at random, the values follow a bell-shaped curve because multiple gene pairs control the trait. Environmental influences, such as nutrition, also affect the phenotypes.

© David Hyde and Wayne Falda/McGraw-Hill Education

The presence of excessive cholesterol in the blood contributes to the development cardiovascular disease (see Section 23.3). Therefore, those with no LDL-cholesterol receptors die of cardiovascular disease as children. Individuals with half the number of receptors may die when young or after they have reached middle age.

Multiple-Allele Traits

In ABO blood group inheritance, three alleles determine the presence or absence of antigens on red blood cells and therefore blood type:

I^A = A antigen on red blood cells
I^B = B antigen on red blood cells
i = Neither A nor B antigen on red blood cells

Each person has only two of the three possible alleles, and both I^A and I^B are dominant over i. Therefore, there are two possible genotypes for type A blood ($I^A I^A$ and $I^A i$) and two possible genotypes for type B blood ($I^B I^B$ and $I^B i$). Type O blood can only result from one genotype (ii) because the i allele is recessive. But I^A and I^B are fully expressed in the presence of each other. Therefore, if a person inherits one of each of these alleles, that person will have a fourth blood type, AB. **Figure 10.17** shows that matings between individuals with certain genotypes can produce individuals with phenotypes different than those of the parents.

The possible genotypes and phenotypes for blood type are as follows:

Phenotype	Genotype
A	$I^A I^A$ or $I^A i$
B	$I^B I^B$ or $I^B i$
AB	$I^A I^B$
O	ii

Notice that human blood type inheritance is also an example of **codominance,** another type of inheritance that differs from Mendel's findings because more than one allele is fully expressed. When an individual has blood type AB, both A and B antigens appear on the red blood cells. The two different capital letters signify that both alleles are coding for an antigen.

Polygenic Inheritance

Polygenic traits are those that are governed by multiple genes, each with several sets of alleles. Each of these alleles has a quantitive effect on the phenotype. Generally speaking, each dominant allele produces a gene product that has an additive effect, while recessive alleles produce little or no effect. The result is a continuous variation of phenotypes, resulting in a distribution of these phenotypes that resembles a bell-shaped curve. The more genes involved, the more continuous the variations and distribution of the phenotypes. One example is human height, where genomic studies have indicated that there may be as many as 700 different alleles that influence phenotypes. The combination of these genes produces minor additive effects and a characteristic bell-shaped distribution of phenotypes (**Fig. 10.18**).

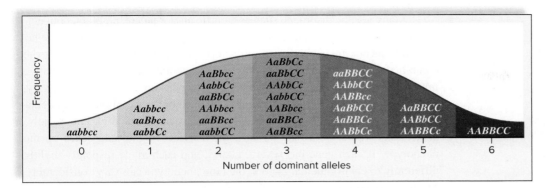

Frequency

Number of dominant alleles

Figure 10.19 **Skin color is a polygenic trait.**

Skin color is controlled by many pairs of alleles, which results in a range of phenotypes. The vast majority of people have skin colors in the middle range, whereas fewer people have skin colors in the extreme range.

Environmental Influences

In many cases, environmental effects also influence the range of phenotypes associated with polygenic traits. This interaction produces a **multifactorial trait.** For example, in the case of height (**Fig. 10.18**), differences in nutrition are associated with variations in the height phenotype.

Another excellent example of a multifactorial trait in humans is skin color. Our skin is responsible for assisting in the production of vitamin D, which acts as a hormone to increase bone density. In response to UV radiation from the sun, cells in the skin called melanocytes produce a protective pigmentation called melanin. In northern geographic areas (such as northern Europe), less melanin is produced so that vitamin D production is increased. In geographic areas closer to the equator, more melanin is produced to protect the skin from UV damage.

In addition to environmental influences, skin color is a polygenic trait that is influenced by a collection of over 100 different genes. For our discussion here, we will use a simple model of only three pairs of alleles (*Aa* and *Bb* and *Cc*). In this model, each dominant allele represents an allele that makes a contribution to the pigment to the skin. When a very dark person reproduces with a very light person, the children have medium-brown skin. When two people with the genotype *AaBbCc* reproduce with one another, individuals may range in skin color from very dark to very light. The distribution of these phenotypes typically follows a bell-shaped curve, meaning that few people have the extreme phenotypes and most people have the phenotype that lies in the middle. A bell-shaped curve is a common identifying characteristic of a polygenic trait (**Fig. 10.19**).

The interaction of the environment (sunlight exposure) with these genotypes produces additional variations in the phenotype. For example, individuals who are *AaBbCc* may vary in their skin color, even though they possess the same genotype, and several possible phenotypes fall between the two extremes.

Temperature is another example of an environmental factor that can influence the phenotypes of plants and animals. Primroses have white flowers when grown above 32°C but red flowers when grown at 24°C. The coats of Siamese cats and Himalayan rabbits are darker in color at the ears, nose, paws, and tail. Himalayan rabbits are known to be homozygous for the allele *ch,* which is involved in the production of melanin. Experimental evidence indicates that the enzyme encoded by this gene is active only at a low temperature, and therefore black fur occurs only at the extremities where body heat is lost to the environment. When the animal is placed in a warmer environment, new fur on these body parts is light in color.

Connections: Scientific Inquiry

Is skin color a good indication of a person's race?

The simple answer to this question is no, an individual's skin color is not a true indicator of his or her genetic heritage. People of the same skin color are not necessarily genetically related. Because a wide variety of genes control skin color, and

© Maria Taglienti-Molinari/Brand X Pictures/ Jupiterimages RF

the environment plays a strong role in the phenotype, a person's skin color does not necessarily indicate his or her ancestors' origins. Scientists are actively investigating the genetic basis of race and have found some interesting relationships between a person's genetic heritage and his or her response to certain medications. This will be explored in greater detail in Chapter 22.

Many genetic disorders, such as cleft lip and/or palate, clubfoot, congenital dislocations of the hip, hypertension, diabetes, schizophrenia, and even allergies and cancers, are most likely multifactorial because they are likely due to the combined action of many genes plus environmental influences. The relative importance of genetic and environmental influences on a phenotype can vary, and often it is a challenge to determine how much of the variation in the phenotype may be attributed to each factor. This is especially true for complex polygenic traits where there may be an additive effect of multiple genes on the phenotype. If each gene has several alleles, and each allele responds slightly differently to environmental factors, then the phenotype can vary considerably. Multifactorial traits are a real challenge for drug manufacturers, since they must determine the response to a new drug based on genetic factors (for example, the ethnic background of the patient) and environmental factors (such as diet).

In humans, the study of multifactorial traits contributes to our understanding of the nature versus nurture debate. Researchers are trying to determine what percentage of various traits is due to nature (inheritance) and what percentage is due to nurture (the environment). Some studies use twins separated from birth, because if identical twins in different environments share a trait, that trait is most likely inherited. Identical twins are more similar in their intellectual talents, personality traits, and levels of lifelong happiness than are fraternal twins separated from birth. NASA conducted an experiment involving the astronaut twins Scott and Mark Kelly on the International Space Station to study how the space environment influences genetics. Biologists conclude that all behavioral traits are partly determined by inheritance and that the genes for these traits act together in complex combinations to produce a phenotype that is modified by environmental influences.

Gene Interactions

As we have already noted, genes rarely act alone to produce a phenotype. In many cases, multiple genes may be part of a metabolic pathway consisting of the interaction of multiple enzymes and proteins.

An example of a trait determined by gene interaction is human eye color (**Fig. 10.20**). Multiple pigments are involved in producing eye color, including melanin. While a number of factors (including structure of the eye) control minor variations in eye color (such as green or hazel eyes), a gene called *OCA2* is involved in producing the melanin that establishes the brown/blue basis of eye color. Individuals who lack *OCA2* completely (as occurs in albinism) have

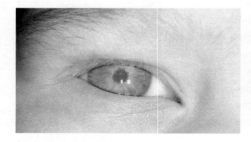  

Figure 10.20 Gene interactions and eye color.

Individuals with red eyes (*left*) lack the ability to produce eye pigments. A presence of a dominant OCA2 allele produces the common brown eye color (*center*). Blue eyes (*right*) may occur from either being homozygous recessive for an *OCA2* allele, or due to an interaction with a second gene, *HERC2*.

Figure 10.21 **Marfan syndrome, multiple effects of a single human gene.**

Individuals with Marfan syndrome exhibit defects in the connective tissue throughout the body. Important changes occur in the skeleton, heart, blood vessels, eyes, lungs, and skin. All these conditions are due to mutation of the gene *FBN1*, which codes for a constituent of connective tissue.

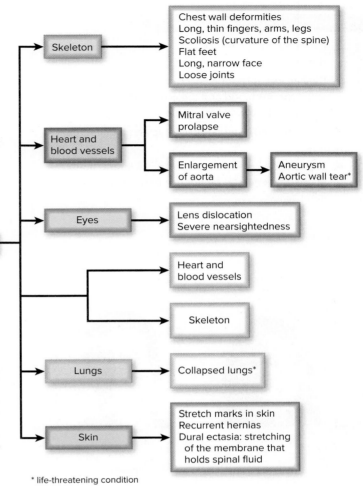

* life-threatening condition

red eyes. A recessive form of *OCA2* results in blue eyes in homozygous recessive individuals, whereas a single dominant *OCA2* allele produces brown eyes (the most common eye color).

However, there is another gene, *HERC2,* that can override the instructions of the *OCA2* gene. If individuals are homozygous for a recessive allele of *HERC2,* they will have blue eyes regardless of the genotype associated with *OCA2*. Geneticists call this type of interaction, where one gene can override another, an **epistatic interaction.**

Pleiotropy

Pleiotropy occurs when a single gene has more than one effect. Often, this leads to a *syndrome,* a group of symptoms that appear together and indicate the presence of a particular genetic mutation. For example, persons with Marfan syndrome have disproportionately long arms, legs, hands, and feet; a weakened aorta; and poor eyesight (**Fig. 10.21**). All of these characteristics are due to the production of abnormal connective tissue. Marfan syndrome has been linked to a mutated gene (*FBN1*) on chromosome 15 that ordinarily specifies a functional protein called fibrillin. This protein is essential for the formation of elastic fibers in connective tissue. Without the structural support of normal connective tissue, the aorta can burst, particularly if the person is engaged in a strenuous sport, such as volleyball or basketball. Flo Hyman may have been the best American woman volleyball player ever, but she fell to the floor and died when only 31 years old because her aorta gave way during a game. Now that coaches are aware of Marfan syndrome, they are on the lookout for it among very tall basketball players.

Many other disorders, including sickle-cell disease and porphyria, are also examples of pleiotropic traits. Porphyria is caused by a chemical insufficiency in the production of hemoglobin, the pigment that makes red blood cells red. The symptoms of porphyria are photosensitivity, strong abdominal pain, port-wine-colored urine, and paralysis in the arms and legs. Many members of the British royal family in the late 1700s and early 1800s suffered from this disorder, which can lead to epileptic convulsions, bizarre behavior, and coma. The vampire legends are most likely also based on individuals with a specific form of porphyria.

Linkage

Not long after Mendel's study of inheritance, investigators began to realize that many different types of alleles are on a single chromosome. A chromosome doesn't contain just one or two alleles; it contains a long series of alleles in a definite sequence. The sequence is fixed because each allele has its own locus on a chromosome. All the alleles on one chromosome form a **linkage group** because they tend to be inherited together. For example, **Figure 10.22** shows some of the traits in the chromosome 19 linkage group of humans.

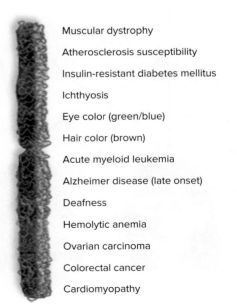

Figure 10.22 **A selection of traits located on human chromosome 19 (not to scale).**

These traits form a single linkage group. In human genetics, genes are sometimes named after the disease they are associated with, not their function in the body.

When we do two-trait crosses (see Fig. 10.6), we are assuming that the alleles are on nonhomologous chromosomes and therefore are not linked. Alleles that are linked do not show independent assortment and therefore do not follow the typical Mendelian genotypic and phenotypic ratios. Historically, researchers used the differences between the expected Mendelian ratios and observed ratios of linked genes to construct genetic maps of the chromosomes. These maps indicated the relative distance between the genes on a chromosome. Today, genetic maps of chromosomes are obtained by direct sequencing of the chromosomes and other biochemical procedures.

CONNECTING THE CONCEPTS

10.3

Variations from expected Mendelian inheritance patterns may occur due to gene interactions, environmental influences, and linkage of genes on chromosomes.

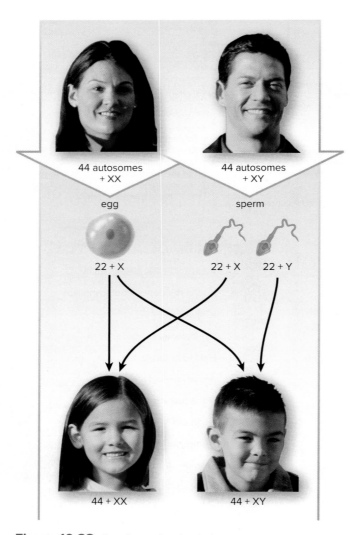

Figure 10.23 **Sex determination in humans.**

Sperm determine the sex of an offspring because sperm can carry an X or a Y chromosome. Males inherit a Y chromosome. Genes on the Y chromosome also affect the gender of an individual.

(all): © Ryan McVay/Getty RF

10.4 Sex-Linked Inheritance

Learning Outcomes

Upon completion of this section, you should be able to

1. Explain differences in the inheritance of sex-linked traits between male and female offspring.
2. Solve and interpret genetic crosses that exhibit sex-linked inheritance.
3. Identify disorders that are associated with sex-linked inheritance.

Normally, both males and females have 23 pairs of chromosomes; 22 pairs are called autosomes, and 1 pair is the **sex chromosomes.** The sex chromosomes are so named because they differ between the sexes. In humans, males have the sex chromosomes X and Y, and females have two X chromosomes (**Fig. 10.23**).

The much shorter Y chromosome contains fewer than 200 genes, and most of these genes are concerned with sex differences between men and women. One of the genes on the Y chromosome, *SRY,* does not have a copy on the X chromosome. If the functional *SRY* gene is present, the individual becomes a male, and if it is absent, the individual becomes a female. Thus, female sex development is the "default setting."

In contrast, the X chromosome is quite large and contains nearly 2,000 genes, most of which have nothing to do with the gender of the individual. By tradition, the term **X-linked** refers to such genes carried on the X chromosome. Examples of X-linked traits in humans include hemophilia and red-green color blindness. The Y chromosome does not carry these genes, which makes for an interesting inheritance pattern.

It would be logical to suppose that a sex-linked trait is passed from father to son or from mother to daughter, but this is not always the case. For X-linked traits, a male always receives an X-linked allele from his mother, from whom he inherited an X chromosome. Because the X and Y chromosomes are not homologous, the Y chromosome from the father does not carry an allele for the trait. Usually, a sex-linked genetic disorder is recessive. Therefore, a female must receive two alleles, one from each parent, in order to develop the condition.

Sex-Linked Alleles

For X-linked traits, the allele on the X chromosome is shown as a letter attached to the X chromosome. For example, following is the key for red-green color blindness, a well-known X-linked recessive disorder:

$$X^B = \text{normal vision} \qquad X^b = \text{color blindness}$$

The possible genotypes and phenotypes in both males and females are as follows:

Genotype	Phenotype
$X^B X^B$	Female who has normal color vision
$X^B X^b$	Carrier female with normal color vision
$X^b X^b$	Female who is color-blind
$X^B Y$	Male who has normal vision
$X^b Y$	Male who is color-blind

The second genotype is a *carrier* female because, although a female with this genotype has normal color vision, she is capable of passing on an allele for color blindness. Color-blind females are rare because they must receive the allele from both parents. Color-blind males are more common because they need only one recessive allele to be color-blind. The allele for color blindness has to be inherited from their mother because it is on the X chromosome. Males inherit their Y chromosome from their father.

Now let us consider a mating between a man with normal vision and a heterozygous woman (**Fig. 10.24**). What is the chance that this couple will have a color-blind daughter? A color-blind son? All the daughters will have normal color vision because they all receive an X^B from their father. The sons, however, have a 50% chance of being color-blind, depending on whether they receive an X^B or an X^b from their mother. The inheritance of a Y chromosome from their father cannot offset the inheritance of an X^b from their mother. Because the Y chromosome doesn't have an allele for the trait, it can't possibly prevent color blindness in a son. Note in Figure 10.24 that the phenotypic results for sex-linked traits are given separately for males and females.

Pedigrees for Sex-Linked Disorders

Figure 10.25 shows a pedigree for an *X-linked recessive disorder*. Recall that sons inherit X-linked recessive traits from their mother because their only X chromosome comes from their mother. More males than females have the disorder because recessive alleles on the X chromosome are always expressed in males—the Y chromosome lacks the allele. Females who have the condition inherited the allele from both their mother and their father, and all the sons of such a female will have the condition.

If a male has an X-linked recessive condition, his daughters are all carriers even if his partner is normal. Therefore, X-linked recessive conditions often appear to pass from grandfather to grandson. Figure 10.25 lists other ways to recognize an X-linked recessive disorder.

Still fewer traits are known to be *X-linked dominant*. If a condition is X-linked dominant, daughters of affected males have a 100% chance of having the condition. Females can pass an X-linked dominant allele to both sons and daughters. If a female is heterozygous and her partner is normal, each child has

Connections: Scientific Inquiry

What is an example of a sex-linked trait in humans?

Perhaps the best-known example of an X-linked trait in humans is hemophilia A. This disease is due to a recessive mutation in one of the genes associated with blood clotting. Since the gene is located on the X chromosome, males who inherit the defective allele from their mothers have hemophilia. Females must inherit a defective gene from both their mother and their father. Hemophilia A has been a problem in the royal families of England, Spain, Russia, and Germany for several centuries.

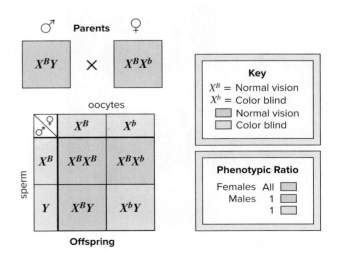

Figure 10.24 X-linked inheritance.

Some forms of color blindness are associated with a gene on the X chromosome, and therefore follow a sex-linked pattern of inheritance.

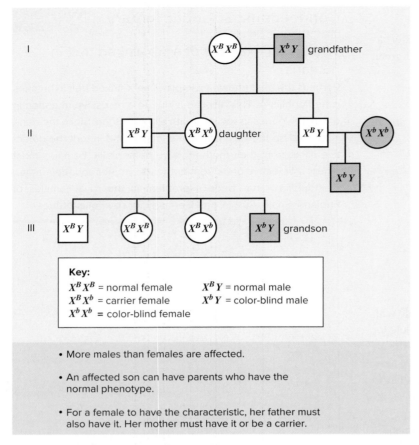

Key:
$X^B X^B$ = normal female $X^B Y$ = normal male
$X^B X^b$ = carrier female $X^b Y$ = color-blind male
$X^b X^b$ = color-blind female

- More males than females are affected.

- An affected son can have parents who have the normal phenotype.

- For a female to have the characteristic, her father must also have it. Her mother must have it or be a carrier.

Figure 10.25   X-linked recessive pedigree.

The list gives ways of recognizing an X-linked recessive disorder such as color blindness.

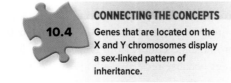

a 50% chance of escaping an X-linked dominant disorder, depending on which of the mother's X chromosomes is inherited.

Still fewer genetic disorders involve genes carried on the Y chromosome. A counselor would recognize a *Y-linked pattern of inheritance* because Y-linked disorders are present only in males and are passed directly from a father to *all* sons but not to daughters.

X-Linked Recessive Disorders

Color Blindness

Red-green color blindness is a common X-linked recessive disorder (see Fig. 10.24). About 8% of Caucasian men see brighter greens as tans, olive greens as browns, and reds as reddish-browns. A few cannot tell reds from greens at all; they see only yellows, blues, blacks, whites, and grays.

Duchenne Muscular Dystrophy

Duchenne muscular dystrophy is an X-linked recessive disorder characterized by wasting away of the muscles. The absence of a protein, now called dystrophin, is the cause of the disorder. Much investigative work determined that dystrophin is involved in the release of calcium from the sarcoplasmic reticulum in muscle fibers. The lack of dystrophin causes calcium to leak into the cell, which promotes the action of an enzyme that dissolves muscle fibers. When the body attempts to repair the tissue, fibrous tissue forms (**Fig. 10.26**), and this cuts off the blood supply, so that more and more cells die.

Symptoms such as waddling gait, toe walking, frequent falls, and difficulty in rising may appear as soon as the child starts to walk. Muscle weakness intensifies until the individual is confined to a wheelchair. Death usually occurs by age 20; therefore, affected males are rarely fathers. The recessive allele remains in the population through passage from carrier mother to carrier daughter.

As therapy, immature muscle cells can be injected into muscles, and for every 100,000 cells injected, dystrophin production occurs in 30–40% of muscle fibers.

fibrous tissue   normal tissue

Figure 10.26   Muscular dystrophy.

In muscular dystrophy, the calves enlarge because fibrous tissue develops as muscles waste away due to a lack of the protein dystrophin.

(left, right): © Dr. Rabi Tawil; (center): © Muscular Dystrophy Association

CONNECTING THE CONCEPTS

10.4   Genes that are located on the X and Y chromosomes display a sex-linked pattern of inheritance.

Check Your Progress 10.4

1. Explain why males produce two different gametes with respect to sex chromosomes.

2. Explain the differences between autosomal recessive and X-linked recessive patterns of inheritance.

3. List two examples of sex-linked disorders.

SUMMARIZE

Our traits are determined not only by the genes we inherit from our parents, but also by environmental factors and interactions between genes.

10.1 Mendel established two laws that may be used to explain patterns of inheritance.

10.2 Pedigrees are charts that help counselors analyze inheritance patterns of genetic disorders.

10.3 Variations from expected Mendelian inheritance patterns may occur due to gene interactions, environmental influences, and linkage of genes on chromosomes.

10.4 Genes that are located on the X and Y chromosomes display a sex-linked pattern of inheritance.

10.1 Mendel's Laws

In 1860, Gregor Mendel, an Austrian monk, developed two laws of heredity based on crosses utilizing the garden pea. Mendelian crosses are often visualized using a **Punnett square.**

Law of Segregation

From his **monohybrid crosses** (involving a single trait), Mendel established the **law of segregation,** which states the following:

- Each individual has two factors for each trait.
- The factors segregate (separate) during the formation of the gametes.
- Each gamete contains only one factor from each pair of factors.
- Fertilization gives each new individual two factors for each trait.

In genetics today, an allele is a variation of a gene. Each allele exists at a specific locus on a chromosome. Homologous chromosomes have their alleles at the same loci. **Genotype** refers to the alleles of the individual, and **phenotype** refers to the physical characteristics associated with these alleles. **Dominant alleles** mask the expression of **recessive alleles. Homozygous** dominant individuals have the dominant phenotype (i.e., FF = freckles). Homozygous recessive individuals have the recessive phenotype (i.e., ff = no freckles). **Heterozygous** individuals have the dominant phenotype (i.e., Ff = freckles). A **testcross** may be used to determine if an individual with a homozygous phenotype is homozygous dominant or heterozygous.

Law of Independent Assortment

Using **dihybrid crosses** (involving two traits), Mendel was able to develop the **law of independent assortment,** which states the following:

- Each pair of factors segregates (assorts) independently of the other pairs.
- All possible combinations of factors can occur in the gametes.

In genetics today,

- Each pair of homologues separates independently of the other pairs.
- All possible combinations of chromosomes and their alleles occur in the gametes.
- Mendel's laws are consistent with the manner in which homologues and their alleles separate during meiosis.

Common Autosomal Genetic Crosses

A Punnett square assumes that all types of sperm can fertilize all types of eggs and gives these results:

Aa	×	Aa	3:1 phenotypic ratio
Aa	×	aa	1:1 phenotypic ratio
$AaBb$	×	$AaBb$	9:3:3:1 phenotypic ratio
$AaBb$	×	$aabb$	1:1:1:1 phenotypic ratio

10.2 Mendel's Laws Apply to Humans

Many traits and disorders in humans follow Mendel's laws. These traits are often controlled by a single pair of alleles on the **autosomes.**

Family Pedigrees

A **pedigree** is a visual representation of the history of a genetic disorder in a family. Constructing pedigrees helps genetic counselors decide whether a genetic disorder that runs in a family is autosomal recessive or autosomal dominant.

Human Autosomal Disorders

Some common human autosomal disorders include:

- Methemoglobinemia (inability to clear methemoglobin from the blood)
- Cystic fibrosis (faulty regulator of chloride ion passage through plasma membrane)
- Alkaptonuria (inability to metabolize homogentisic acid)
- Sickle-cell disease (sickle-shaped red blood cells)
- Osteogenesis imperfecta (defective collagen tissue)
- Huntington disease (abnormal huntingtin protein)

10.3 Beyond Mendel's Laws

In some patterns of inheritance, the alleles are not just dominant or recessive.

Incomplete Dominance

In **incomplete dominance,** the heterozygote is intermediate between the two homozygotes. For example, in hypercholesterolemia, heterozygous individuals have blood cholesterol levels between the extremes of homozygous dominant and homozygous recessive individuals.

Multiple-Allele Traits

The multiple-allele inheritance pattern is exemplified in humans by blood type inheritance. Every individual has two out of three possible alleles: I^A, I^B, and i. Both I^A and I^B are expressed; therefore, this is also a case of **codominance.**

Polygenic Inheritance

A **polygenic** trait is controlled by more than one set of alleles. The dominant alleles have an additive effect on the phenotype.

Allele A
Allele B   Additive effect of dominant
Allele C   alleles on phenotype

Many traits, especially those governed by multiple genes, are modified by the environment. These traits are called **multifactorial traits.**

Gene Interactions

Epistatic interactions between genes occur when one gene has the ability to override the instructions of another gene. These interactions often occur in biochemical pathways.

Pleiotropy

In **pleiotropy,** one gene (consisting of two alleles) has multiple effects on the body. For example, all the disorders common to Marfan syndrome are due to a mutation that leads to a defect in the composition of connective tissue.

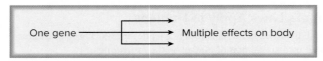

One gene ──────→ Multiple effects on body

Linkage

Alleles on the same chromosome are linked, and they are generally inherited together as a **linkage group.**

10.4 Sex-Linked Inheritance

Males produce two different types of gametes—those that contain an X chromosome and those that contain a Y. The chromosome from the male gamete determines the gender of the new individual: XX = female, XY = male.

Alleles that are carried on the X chromosome are said to be **X-linked.** Y-linked alleles are found on the Y chromosome. Since males are XY, they need to inherit only one recessive allele on their X chromosome to have a recessive genetic disorder.

- $X^B X^b \times X^B Y$: All daughters will be normal, even though they have a 50% chance of being carriers, but sons have a 50% chance of being color-blind.
- $X^B X^B \times X^b Y$: All children are normal (daughters will be carriers).

Human X-Linked Disorders

- Color blindness (inability to detect colors normally)
- Duchenne muscular dystrophy (absence of dystrophin leading to muscle weakness)

ASSESS

Testing Yourself

Choose the best answer for each question.

10.1 Mendel's Laws

1. In a testcross of $H\_\_$ and hh individuals, you obtain 20 offspring; 10 of these have the dominant phenotype, and 10 have the recessive phenotype. Based on this, the genotype of the $H\_\_$ individual is
   a. heterozygous.
   b. homozygous dominant.
   c. homozygous recessive
   d. None of these are correct.

2. Which of the following is not a component of the law of segregation?
   a. Each gamete contains one factor from each pair of factors in the parent.
   b. Factors segregate during gamete formation.
   c. Following fertilization, the new individual carries two factors for each trait.
   d. Each individual has one factor for each trait.

3. When using a Punnett square to predict offspring ratios, we assume that
   a. each gamete contains one allele of each gene.
   b. the gametes have all possible combinations of alleles.
   c. male and female gametes combine at random.
   d. All of these are correct.

4. Fill in the blank spaces in the following Punnett square.

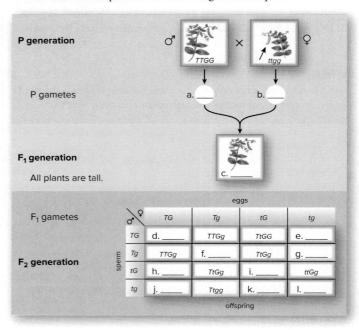

10.2 Mendel's Laws Apply to Humans

For questions 5–7, identify whether the term applies to an (a) autosomal dominant or (b) autosomal recessive trait.

5. Affected children always have at least one affected parent.
6. Affected children can have unaffected parents.
7. Two affected parents may have unaffected children.

For questions 8–10, classify the condition as (a) autosomal dominant or (b) autosomal recessive.

8. methemoglobinemia
9. Huntington disease
10. osteogenesis imperfecta
11. Cystic fibrosis (CF) is an autosomal recessive disorder in humans. If two unaffected individuals have a child with CF, what is the chance that their second child will have CF?
    a. 25%   d. 100%
    b. 50%   e. None of these are correct.
    c. 75%

10.3 Beyond Mendel's Laws

12. When one physical trait is affected by two or more pairs of alleles, the pattern is called
    a. incomplete dominance.   d. multiple allele.
    b. codominance.   e. polygenic inheritance.
    c. homozygous dominant.

13. When two monohybrid round squashes are crossed, the offspring ratio is ¼ flat to ½ oblong to ¼ round. Squash shape, therefore, is determined by incomplete dominance. What offspring ratio would you expect from a cross between a squash plant with oblong fruit and one with round fruit?
    a. all oblong   d. ¾ round to ¼ oblong
    b. all round   e. ½ oblong to ½ round
    c. ¾ oblong to ¼ round

14. If a man with blood type AB marries a woman with blood type B whose father was type O, what phenotypes could their children have?
    a. A only
    b. A, AB, B, and O
    c. AB only
    d. A, AB, and B
    e. O only

10.4 Sex-Linked Inheritance

15. A mother who is heterozygous for an X-linked recessive trait has children with a man who has the trait. What percentage of their daughters will exhibit the trait?
    a. 0%
    b. 25%
    c. 50%
    d. 75%
    e. 100%

ENGAGE

BioNOW

Want to know how this science is relevant to your life? Check out the BioNow video below:

• Glowfish Genetics

How might an experiment be designed to test whether the pattern of inheritance in the glowfish is dominant or codominant?

Thinking Critically

1. Multiple gene pairs may interact to produce a single phenotype. In garden peas, genes *C* and *P* are required for pigment production in flowers. Gene *C* codes for an enzyme that converts a compound into a colorless intermediate product. Gene *P* codes for an enzyme that converts the colorless intermediate product into anthocyanin, a purple pigment. A flower, therefore, will be purple only if it contains at least one dominant allele for each of the two genes (*C__P__*). What phenotypic ratio would you expect in the F_2 generation following a cross between two double heterozygotes (*CcPp*)?

2. Assume for a moment that you are working in a genetics lab with the fruit fly *Drosophila melanogaster,* a model organism for genetic research. You want to determine whether a newly found *Drosophila* characteristic is dominant or recessive.
    a. Explain how you would construct such a cross, as well as the expected outcomes that would indicate dominance.
    b. Why would working with *Drosophila* be easier than working with humans?

3. Technology that can separate X-bearing and Y-bearing sperm offers prospective parents the opportunity to choose the sex of their child. First, the sperm are dosed with a DNA-staining chemical. Because the X chromosome has slightly more DNA than the Y chromosome, it takes up more dye. When a laser beam shines on the sperm, the X-bearing sperm shine more brightly than the Y-bearing sperm.
    a. How might this technology be beneficial for couples that have a history of X-linked genetic disorders?
    b. How might this technology be misused?

11

DNA Biology

(child): © IT Stock/PunchStock RF; (orchids): © Medioimages/PunchStock RF; (DNA helix): © Radius Images/Alamy RF; (mushrooms): © IT Stock/age fotostock RF

OUTLINE

11.1 DNA and RNA Structure and Function 185

11.2 Gene Expression 191

11.3 Gene Regulation 197

BEFORE YOU BEGIN

Before beginning this chapter, take a few moments to review the following discussions.

Section 3.2 What are the general structures of DNA and RNA molecules?

Sections 4.3 and **4.4** What are the differences between prokaryotic and eukaryotic cells?

Section 8.1 What is the end result of DNA replication?

Synthetic DNA

Deoxyribonucleic acid, or DNA, is the genetic material of all life on Earth. Each of the organisms in the photo above, and all of the estimated 2.3 million species on the planet, use a combination of four nucleotides (A, C, G, and T) to code for 20 different amino acids. The combination of these amino acids produces the tremendous variation that gives us the diversity of life.

However, a biotechnology company named Synthorx has recently expanded this genetic alphabet by creating two new forms of nucleotides, called X and Y. When these nucleotides are incorporated into the DNA of *E.coli* bacteria, the number of potential amino acids that can be coded for increases from 20 to 172.

What does this mean for us? Scientists believe that with this innovation they will be able to develop new drugs (including antibiotics) and vaccines that can help protect us from some disease. It may also be possible to develop genetically modified organisms that resist pathogens, grow faster, or produce foods with higher nutritional content.

In this chapter, we will explore the structure and function of the genetic material and examine how the instructions contained within DNA are expressed to form the proteins that create the characteristics of life.

As you read through this chapter, think about the following questions:

1. How is information contained within a molecule of DNA?
2. What are the differences between DNA and RNA?
3. How does the flow of genetic information from DNA to protein to trait work?

11.1 DNA and RNA Structure and Function

Learning Outcomes

Upon completion of this section, you should be able to

1. Understand how scientists determined that DNA was the genetic material.
2. Describe the structure of a DNA molecule.
3. List the steps involved in the replication of DNA.
4. Compare and contrast the structure of RNA with that of DNA.
5. List the three major types of RNA, and describe their functions.

Mendel knew nothing about DNA. It took many years for investigators to come to the conclusion that Mendel's factors, now called genes, are on the chromosomes. Then, researchers wanted to show that DNA, not protein, is responsible for heredity. One experiment, by Alfred Hershey and Martha Chase, involved the use of a virus that attacks bacteria, such as *E. coli* (**Fig. 11.1**). A virus is composed of an outer capsid made of protein and an inner core of DNA. The use of radioactive tracers showed that the viral DNA, but not protein, enters bacteria and directs the formation of new viruses. By the early 1950s, investigators had learned that genes are composed of DNA and that mutated genes result in errors of metabolism. Therefore, DNA in some way must control the cell.

While scientists did not yet know the actual structure of a DNA molecule, they did know its chemical components. The name *deoxyribonucleic acid* is derived from the chemical components of DNA's building blocks, the nucleotides. They also knew that DNA had the following characteristics:

- Variability to account for differences in the wide variety of life on the planet.
- The ability to replicate so that every cell gets a copy during cell division.
- Storage of the information needed to control the cell.
- The ability to change or mutate, to allow for the evolution of new species.

Figure 11.1 **DNA as the genetic material.**

An early experiment by Hershey and Chase helped determine that DNA is the genetic material. Their experiment involved a virus that infects bacteria, such as *E. coli*. They wanted to know which part of the virus—the capsid made of protein or the DNA inside the capsid—enters the bacterium. Radioactive tracers showed that DNA, not protein, enters the bacterium and guides the formation of new viruses. Therefore, DNA must be the genetic material.

© Lee Simon/Science Source

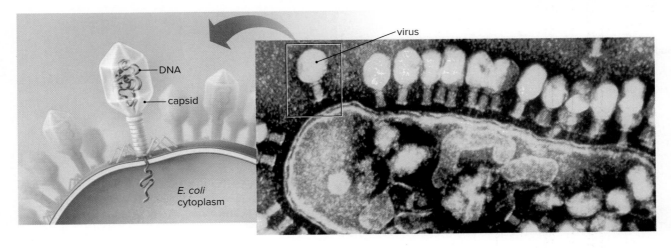

Structure of DNA

Once researchers knew that DNA was the genetic material, they were racing against time and each other to determine the structure of DNA. They believed that whoever discovered it first would get a Nobel Prize. How James Watson and Francis Crick determined the structure of DNA (and eventually received a Nobel Prize) resembles the solving of a mystery, in which each clue was added to the total picture until the breathtaking design of DNA—a double helix—was finally revealed. To achieve this success, Watson and Crick particularly relied on studies done by Erwin Chargaff and Rosalind Franklin.

Chargaff's Rules

Before Erwin Chargaff began his work, it was known that DNA contains four different types of nucleotides, which contain different nitrogen-containing bases (**Fig. 11.2**). The bases **adenine (A)** and **guanine (G)** are purines with a double ring, and the bases **thymine (T)** and **cytosine (C)** are pyrimidines with a single ring. With the development of new chemical techniques in the 1940s, Chargaff decided to analyze in detail the base content of DNA in various species.

In contrast to accepted beliefs, Chargaff found that each species has its own percentages of each type of nucleotide. For example, in a human cell, 31% of bases are adenine; 31% are thymine; 19% are guanine; and 19% are cytosine. In all the species Chargaff studied, the amount of A always equaled the amount of T, and the amount of G always equaled the amount of C. These relationships are called *Chargaff's rules:*

- The amounts of A, T, G, and C in DNA varies from species to species.
- In each species, the amounts of A and T are equal (A = T), as are the amounts of G and C (G = C).

Figure 11.2 Nucleotide composition of DNA and RNA.

a. All nucleotides contain phosphate, a 5-carbon sugar, and a nitrogen-containing base, such as cytosine (C). **b.** Structure of phosphate. **c.** In DNA, the sugar is deoxyribose; in RNA, the sugar is ribose. **d.** In DNA, the nitrogen-containing bases are adenine, guanine, cytosine, and thymine; in RNA, the bases are adenine, guanine, cytosine, and uracil.

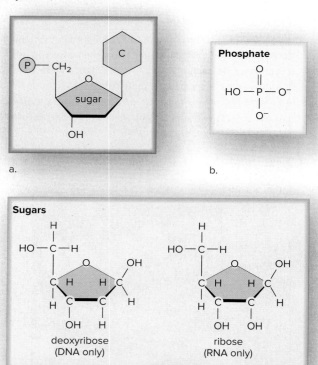

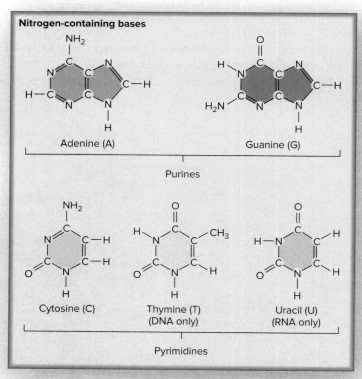

Figure 11.3 X-ray diffraction pattern of DNA.

a. Rosalind Franklin X-rayed DNA using crystallography techniques. **b.** The resulting pattern indicated that DNA is a double helix (see X pattern in the center) and that part of the molecule is repeated over and over again (see the dark portions at top and bottom). Watson and Crick said that this repeating feature is the paired bases.

(both): © Science Source

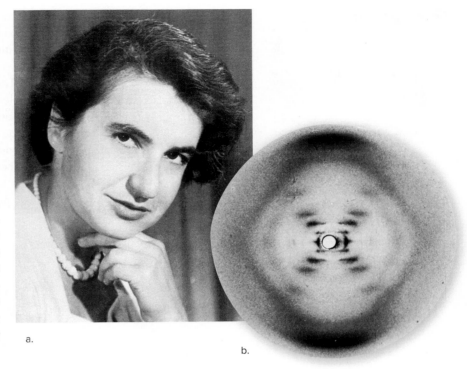

a.

b.

Chargaff's data suggest that DNA has a means to be stable, in that A can pair only with T and G can pair only with C. His data also show that DNA can be variable as required for the genetic material. Today, we know that the paired bases may occur in any order and the amount of variability in their sequences is overwhelming. For example, suppose a chromosome contains 140 million base pairs. Since any of the four possible nucleotide pairs can be present at each pair location, the total number of possible nucleotide pair sequences is $4^{140 \times 10^6}$, or $4^{140,000,000}$.

Franklin's X-Ray Diffraction Data

Rosalind Franklin was a researcher at King's College in London in the early 1950s (**Fig. 11.3a**). She was studying the structure of DNA using X-ray crystallography. When a crystal (a solid substance whose atoms are arranged in a definite manner) is X-rayed, the X-ray beam is diffracted (deflected), and the pattern that results shows how the atoms are arranged in the crystal.

First, Franklin made a concentrated, viscous solution of DNA and then saw that it could be separated into fibers. Under the right conditions, the fibers were enough like a crystal that, when they were X-rayed, a diffraction pattern resulted. The X-ray diffraction pattern of DNA shows that DNA is a double helix. The helical shape is indicated by the crossed (X) pattern in the center of the photograph in Figure 11.3b. The dark areas at the top and bottom of the photograph indicate that some portion of the helix is repeated many times.

The Watson and Crick Model

In 1951, James Watson, having just earned a Ph.D., began an internship at the University of Cambridge, England. There, he met Francis Crick, a British physicist who was interested in molecular structures. Together, they set out to determine the structure of DNA and to build a model that would explain how DNA varies from species to species, replicates, stores information, and undergoes mutation.

Based on the available data, they knew the following:

1. DNA is a polymer of four types of nucleotides with the bases adenine (A), guanine (G), cytosine (C), and thymine (T).
2. Based on Chargaff's rules, A = T and G = C.
3. Based on Franklin's X-ray diffraction photograph, DNA is a double helix with a repeating pattern.

Using these data, Watson and Crick built a model of DNA out of wire and tin (**Fig. 11.4**). The model showed that the deoxyribose sugar–phosphate molecules are bonded to one another to make up the sides of a twisted ladder. The

Figure 11.4 Watson and Crick model of DNA.

James Watson (left) and Francis Crick with their model of DNA.

© A. Barrington Brown/Science Source

nitrogenous bases make up the rungs of the ladder—they project into the middle and hydrogen bond with bases on the other strand. Indeed, the pairing of A with T and G with C—called **complementary base pairing**—results in rungs of a consistent width, as elucidated by the X-ray diffraction data.

Figure 11.5 shows that DNA is a double-stranded molecule that wraps around itself to form a double helix. The two strands are antiparallel and run in opposite directions, as best seen in Figure 11.5*a*. Notice that in the double helix, each strand has a 5′ end where a free ⓟ appears and a 3′ end where a free —OH group appears. In Figure 11.5*b*, the carbon atoms in deoxyribose are numbered. The 5′ carbon has an attached ⓟ group, and the 3′ carbon has an attached —OH group, which is circled and colored pink for easy recognition.

The double-helix model of DNA permits the base pairs to be in any order, a necessity for genetic variability between species. Also, the model suggests that complementary base pairing may play a role in the replication of DNA. As Watson and Crick pointed out in their original paper, published in *Nature* in 1953, "It has not escaped our notice that the specific pairing we have postulated immediately suggests a possible copying mechanism for the genetic material."

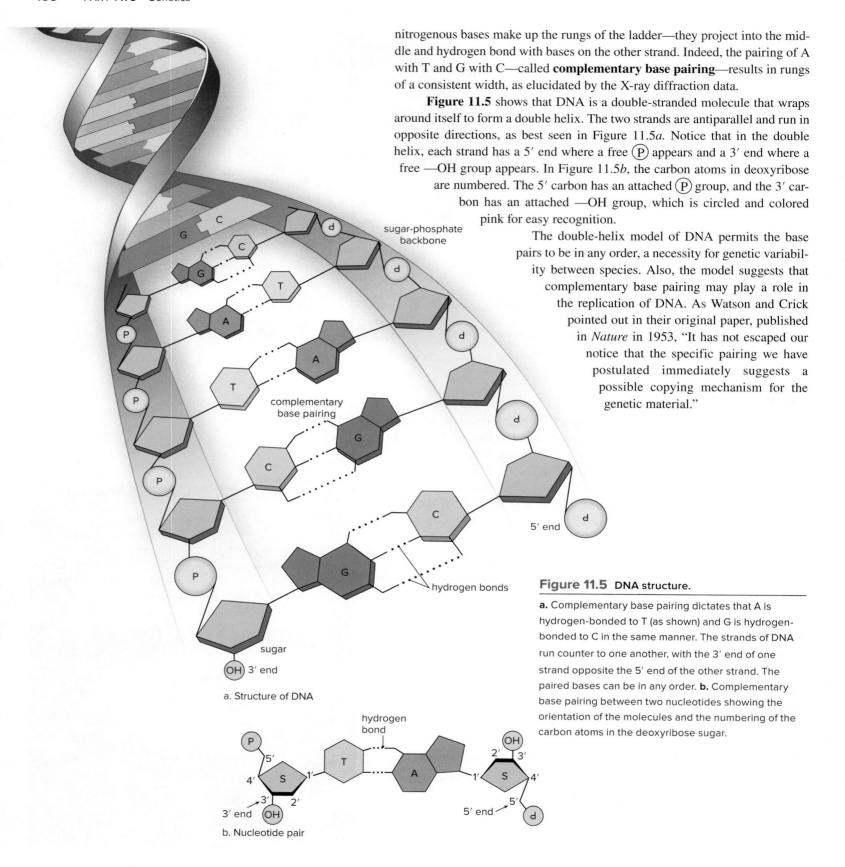

Figure 11.5 DNA structure.

a. Complementary base pairing dictates that A is hydrogen-bonded to T (as shown) and G is hydrogen-bonded to C in the same manner. The strands of DNA run counter to one another, with the 3′ end of one strand opposite the 5′ end of the other strand. The paired bases can be in any order. **b.** Complementary base pairing between two nucleotides showing the orientation of the molecules and the numbering of the carbon atoms in the deoxyribose sugar.

a. Structure of DNA

b. Nucleotide pair

Replication of DNA

Cells need to make identical copies of themselves for the growth and repair of tissues. Before division occurs, each new cell requires an exact copy of the parent cell's DNA, so that it can pass on its genetic information to the next generation of cells. **DNA replication** refers to the process of making an identical copy of a DNA molecule. DNA replication occurs during the S phase of the cell cycle (see Fig. 8.3).

During DNA replication, the two DNA strands, which are held together by hydrogen bonds, are separated and each old strand of the parent molecule serves as a **template** for a new strand in a daughter molecule (**Fig. 11.6**). This process is referred to as **semiconservative,** since one of the two old strands is conserved, or present, in each daughter molecule.

To begin replication, the DNA double helix must separate and unwind. This is accomplished by breaking the hydrogen bonds between the nucleotides, then unwinding the helix structure using an enzyme called **helicase.** At this point, new nucleotides are added to the parental template strand. Nucleotides, ever present in the nucleus, will complementary base-pair onto the now single-stranded parental strand. The addition of the new strand is completed using an enzyme complex called **DNA polymerase.** The daughter strand is synthesized by DNA polymerase in a 5′–3′ direction, as shown in Figure 11.6. Any breaks in the sugar-phosphate backbone are sealed by the enzyme **DNA ligase.**

In Figure 11.6, the backbones of the parent molecule (original double strand) are blue. Following replication, the daughter molecules each have a green backbone (new strand) and a blue backbone (old strand). A daughter DNA double helix has the same sequence of base pairs as the parent DNA double helix had. Complementary base pairing has allowed this sequence to be maintained.

In eukaryotes, DNA replication begins at numerous sites, called origins of replication, along the length of the chromosome. At each origin of replication, a replication fork forms, allowing replication to proceed in both directions. Around each replication fork, a "replication bubble" forms. It is within

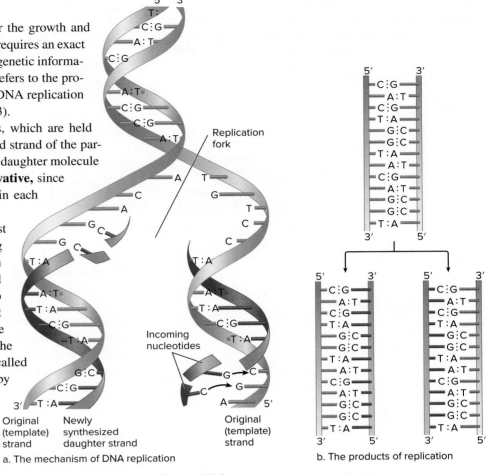

a. The mechanism of DNA replication

b. The products of replication

Figure 11.6 **Semiconservative replication.**

a. After the DNA double helix unwinds, each parental strand serves as a template for the formation of a new daughter strand. Complementary free nucleotides hydrogen bond to a matching base (e.g., A with T; G with C) in each parental strand and are joined to form a complete daughter strand. **b.** Two helices, each with a daughter and a parental strand (semiconservative), are produced during replication.

Connections: Scientific Inquiry

How long does it take to copy the DNA in one human cell?

The enzyme DNA polymerase in humans can copy approximately 50 base pairs per second. If only one DNA polymerase were used to copy human DNA, it would take almost 3 weeks. However, multiple DNA polymerases copy the human genome by starting at many different places. All 3 billion base pairs can be copied in about 8 hours in a rapidly dividing cell.

© Comstock Images/
Jupiter Images RF

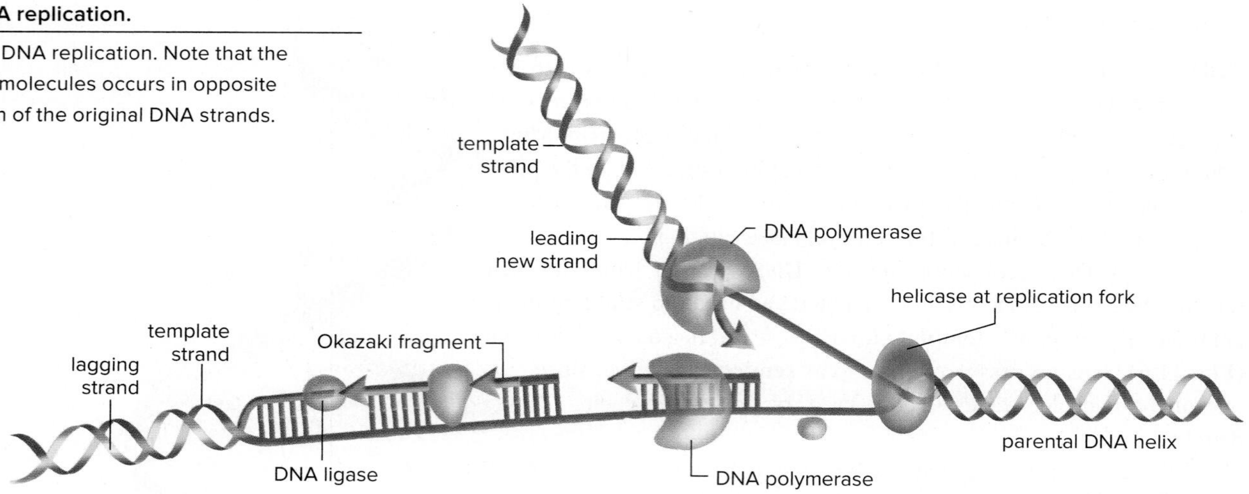

Figure 11.7 Eukaryotic DNA replication.

The major enzymes involved in DNA replication. Note that the synthesis of the daughter DNA molecules occurs in opposite directions due to the orientation of the original DNA strands.

template strand

leading new strand — DNA polymerase

helicase at replication fork

lagging strand

template strand

Okazaki fragment

DNA ligase

DNA polymerase

parental DNA helix

the replication bubble that the process of DNA replication occurs. Replication proceeds along each strand in opposite directions until the entire double helix is copied. (**Fig. 11.7**). Although eukaryotes replicate their DNA at a fairly slow rate—500 to 5,000 base pairs per minute—there are many individual origins of replication throughout the DNA molecule. Therefore, eukaryotic cells complete the replication of the diploid amount of DNA (in humans, over 3 billion base pairs) in a matter of hours!

RNA Structure and Function

Ribonucleic acid (RNA) is made up of nucleotides containing the sugar ribose, thus accounting for its name. The four nucleotides that make up an RNA molecule have the following bases: adenine (A), **uracil (U),** cytosine (C), and guanine (G). Notice that, in RNA, uracil replaces the thymine found in DNA (**Fig. 11.8**).

RNA, unlike DNA, is single-stranded, but the single RNA strand sometimes doubles back on itself, allowing complementary base pairing to occur. The similarities and differences between these two types of nucleic acid molecules are listed in **Table 11.1.**

In general, RNA is a helper to DNA, allowing protein synthesis to occur according to the genetic information that DNA provides. There are three major types of RNA, each with a specific function in protein synthesis.

Messenger RNA

Messenger RNA (mRNA) is produced in the nucleus of a eukaryotic cell, as well as in the nucleoid region of a prokaryotic cell. DNA serves as a template for the formation of mRNA during a process called *transcription*. Which DNA genes are transcribed into mRNA is highly regulated in each type of cell and accounts for the specific functions of all cell types. Once formed, mRNA carries genetic information from DNA in the nucleus to the ribosomes in the cytoplasm, where protein synthesis occurs through a process called translation.

Transfer RNA

Transfer RNA (tRNA) is also produced in the cell nucleus of eukaryotes. Appropriate to its name, tRNA transfers amino acids present in the cytoplasm to the ribosomes. There are 20 different amino acids, and each has its own tRNA molecule. At the ribosome, a process called *translation* joins the amino acids to form a polypeptide chain.

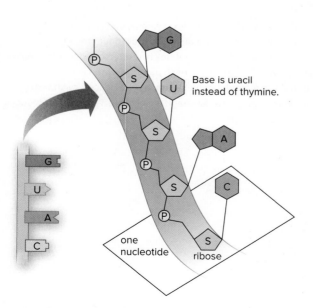

Base is uracil instead of thymine.

one nucleotide

ribose

Figure 11.8 Structure of RNA.

Like DNA, RNA is a polymer of nucleotides. In an RNA nucleotide, the sugar ribose is attached to a phosphate molecule and to a base, either G, U, A, or C. Notice that, in RNA, the base uracil (U) replaces thymine as one of the bases. RNA is usually single-stranded, whereas DNA is double-stranded.

Ribosomal RNA

In eukaryotic cells, **ribosomal RNA (rRNA)** is produced in the nucleolus of the nucleus, where a portion of DNA serves as a template for its formation. Ribosomal RNA joins with proteins made in the cytoplasm to form the subunits of **ribosomes,** one large and one small. Each subunit has its own mix of proteins and rRNA. The subunits leave the nucleus and come together in the cytoplasm when protein synthesis is about to begin.

Proteins are synthesized at the ribosomes, which look like small granules in low-power electron micrographs. Ribosomes in the cytoplasm may be free floating or in clusters called polyribosomes. Often, they are found attached to the edge of the endoplasmic reticulum (ER). Proteins synthesized by ribosomes attached to the ER normally are used by the ER. Proteins synthesized by free ribosomes or polyribosomes are used in the cytoplasm; a protein is carried in a transport vesicle to the Golgi apparatus for modification and then transported to the plasma membrane, where it can leave the cell.

Table 11.1 Comparison of DNA and RNA

SIMILARITIES OF DNA AND RNA	
Both are nucleic acids.	
Both are composed of nucleotides.	
Both have a sugar-phosphate backbone.	
Both have four different types of bases.	

DIFFERENCES BETWEEN DNA AND RNA	
DNA	**RNA**
Found in nucleus	Found in nucleus and cytoplasm
Genetic material	Helper to DNA
Sugar is deoxyribose.	Sugar is ribose.
Bases are A, T, C, G.	Bases are A, U, C, G.
Double-stranded	Single-stranded
DNA is transcribed (to give a variety of RNA molecules).	mRNA is translated (to make proteins).

CONNECTING THE CONCEPTS

11.1 DNA is the genetic material, it contains the instructions for carrying out the processes of life. RNA is involved in the processing of that information.

Check Your Progress 11.1

1. Summarize the structure of DNA, focusing on the complementary base pairing and antiparallel configuration.
2. Describe the process of DNA replication, including the functions of the major enzymes involved.
3. Compare and contrast the structures of DNA and RNA.
4. Summarize the functions of mRNA, rRNA, and tRNA.

11.2 Gene Expression

Learning Outcomes

Upon completion of this section, you should be able to

1. Summarize the steps involved in gene expression.
2. Explain how the genetic code is used in the expression of a gene.
3. Describe the process of transcription and its role in gene expression.
4. Summarize the modifications that occur during mRNA processing.
5. Describe the process of translation and its role in gene expression.

In the early twentieth century, the English physician Sir Archibald Garrod observed that family members often have the same metabolic disorder, and he said it was likely they all lacked the same functioning enzyme in a metabolic pathway. He introduced the phrase "inborn error of metabolism" to describe this relationship. Garrod's findings were generally overlooked until 1940, when George Beadle and Edward Tatum devised a way to confirm his hypothesis. These investigators performed a series of experiments utilizing red bread mold and found that each of their mutant molds was indeed unable to produce a particular enzyme. This led them to propose that one gene directs the synthesis of one enzyme (or protein). Today we know that genes are also responsible for specifying any type of protein in a cell, not just enzymes, so Beadle and Tatum's finding has been modified to "one gene–one polypeptide."

From DNA to RNA to Protein

It's one thing to know that genes specify proteins and another to explain how they do it. Molecular genetics, which largely began when Watson and Crick discovered the structure of DNA in the 1950s, is able to explain exactly how genes control the building of specific types of proteins. Consider that, in eukaryotes, DNA resides in the nucleus but RNA is found both in the nucleus and in the cytoplasm, where protein synthesis occurs. This means that DNA must pass its genetic information to mRNA, which then actively participates in protein synthesis. The *central dogma of molecular biology* states that genetic information flows from DNA to RNA to protein (**Fig. 11.9**).

Gene expression has occurred when a gene's product—the protein it specifies—is functioning in a cell. Specifically, gene expression requires two processes, called transcription and translation. In eukaryotes, transcription takes place in the nucleus, and translation takes place in the cytoplasm. During **transcription,** a portion of DNA serves as a template for mRNA formation. During **translation,** the sequence of mRNA bases (which are complementary to those in the template DNA) determines the sequence of amino acids in a polypeptide. So, in effect, molecular genetics tells us that genetic information lies in the sequence of the bases in DNA, which through mRNA determines the sequence of amino acids in a protein. tRNA assists mRNA during protein synthesis by bringing amino acids to the ribosomes. Proteins differ from one another by the sequence of their amino acids, and proteins determine the structure and function of cells and indeed the phenotype of the organism.

It is important to note that the central dogma addresses the processing of genetic information for the purpose of producing a protein. However, scientists now recognize that there are segments of DNA that are expressed to form RNA molecules that have more of a regulatory role (such as micro-RNAs), and do not directly carry information that codes for proteins. We will explore this again in Section 11.3.

The Genetic Code

The information contained in DNA and RNA is written in a chemical language different from that found in the protein specified by the DNA and RNA. The cell needs a way to translate one language into the other, and it uses the genetic code.

But how can the four bases of RNA (A, C, G, U) provide enough combinations to code for the 20 amino acids found in proteins? If the code were a singlet code (one base standing for an amino acid), only four amino acids could be encoded. If the code were a doublet (any two bases standing for one amino acid), it still would not be possible to code for 20 amino acids. However, by using a triplet code, the four bases can supply 64 different combinations, far more than needed to code for 20 different amino acids.

Each three-letter (three-base) unit of an mRNA molecule is called a **codon,** and it codes for a single amino acid (**Fig. 11.10**). Sixty-one codons correspond to a particular amino acid; the remaining three are stop codons that signal the end of a polypeptide. The codon that stands for the amino acid methionine is also used as a start codon that signals the initiation of translation. Most amino acids have more than one codon, which offers some protection against possibly harmful mutations that might change the sequence of the amino acids in a protein.

Cracking the genetic code was no simple matter. Researchers performed a series of experiments in which they added artificial mRNA to a medium containing bacterial ribosomes and a mixture of amino acids. By comparing the bases in

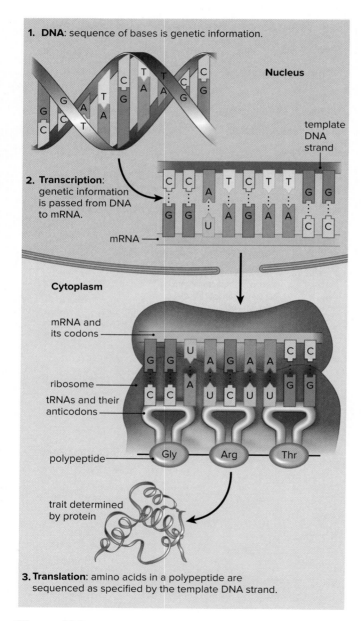

Figure 11.9 Flow of genetic information.

1. Genetic information (in the genes) consists of a particular sequence of bases. **2.** The process of transcription in the nucleus passes this genetic information to an mRNA molecule. The mRNA moves to a ribosome in the cytoplasm. **3.** During the process of translation at a ribosome in the cytoplasm, the genetic information results in a particular sequence of amino acids in a protein. The activity or inactivity of this protein contributes to a trait and often determines whether the phenotype is normal or not normal.

Second base

	U	C	A	G	
U	UUU ⎤ phenylalanine (Phe) UUC ⎦ UUA ⎤ leucine (Leu) UUG ⎦	UCU ⎤ UCC ⎥ serine (Ser) UCA ⎥ UCG ⎦	UAU ⎤ tyrosine (Tyr) UAC ⎦ **UAA** stop **UAG** stop	UGU ⎤ cysteine (Cys) UGC ⎦ **UGA** stop UGG tryptophan (Trp)	U C A G
C	CUU ⎤ CUC ⎥ leucine (Leu) CUA ⎥ CUG ⎦	CCU ⎤ CCC ⎥ proline (Pro) CCA ⎥ CCG ⎦	CAU ⎤ histidine (His) CAC ⎦ CAA ⎤ glutamine (Gln) CAG ⎦	CGU ⎤ CGC ⎥ arginine (Arg) CGA ⎥ CGG ⎦	U C A G
A	AUU ⎤ AUC ⎥ isoleucine (Ile) AUA ⎦ **AUG** methionine (Met) **(start)**	ACU ⎤ ACC ⎥ threonine (Thr) ACA ⎥ ACG ⎦	AAU ⎤ asparagine (Asn) AAC ⎦ AAA ⎤ lysine (Lys) AAG ⎦	AGU ⎤ serine (Ser) AGC ⎦ AGA ⎤ arginine (Arg) AGG ⎦	U C A G
G	GUU ⎤ GUC ⎥ valine (Val) GUA ⎥ GUG ⎦	GCU ⎤ GCC ⎥ alanine (Ala) GCA ⎥ GCG ⎦	GAU ⎤ aspartic acid (Asp) GAC ⎦ GAA ⎤ glutamic acid (Glu) GAG ⎦	GGU ⎤ GGC ⎥ glycine (Gly) GGA ⎥ GGG ⎦	U C A G

First base (left axis) · **Third base** (right axis)

Figure 11.10 **Messenger RNA codons.**

Notice that, in this chart, each of the codons is composed of three letters that stand for three bases. As an example, find the rectangle where C is the first base and A is the second base. U, C, A, or G can be the third base. CAU and CAC are codons for histidine; CAA and CAG are codons for glutamine.

the mRNA with the resulting polypeptide, they were able to learn the code. For example, an mRNA with a sequence of repeating guanines (GGG′GGG′…) encoded a string of glycine amino acids, so the researchers concluded that the mRNA codon GGG specifies the amino acid glycine in a protein.

The genetic code is considered to be almost universal for all life on Earth. This suggests an evolutionary aspect to the genetic code: that the code dates back to the very first organisms on Earth and that all life is related.

Connections: Scientific Inquiry

Which types of organisms use a different genetic code?

While almost all organisms use the genetic code presented in Figure 11.10, there are a few exceptions. Interestingly, in the mitochondria of cells, the stop codon UGA codes for the amino acid tryptophan, and AUA is methionine, not leucine. There are similar differences in the chloroplasts of plants. Still, the number of exceptions is very small, and thus the genetic code may be considered "universal." However, scientists are actively looking for signs of life on other planets in our solar system, which may cause us to revise our use of the term "universal."

Transcription

During transcription of DNA, a strand of RNA forms that is complementary to a portion of DNA. While all three classes of RNA are formed by transcription, we will focus on transcription to create mRNA.

mRNA Is Formed   Transcription begins when the enzyme **RNA polymerase** binds tightly to a **promoter,** a region of DNA with a special nucleotide sequence that marks the beginning of a gene. RNA polymerase opens up the DNA helix just in front of it, so that complementary base pairing can occur. Then the enzyme adds new RNA nucleotides that are complementary to those in the template DNA strand, and an mRNA molecule results (**Fig. 11.11**).

The resulting **mRNA transcript** is a complementary copy of the sequence of bases in the template DNA strand. Once transcription is completed, the mRNA is ready to be processed before it leaves the nucleus for the cytoplasm.

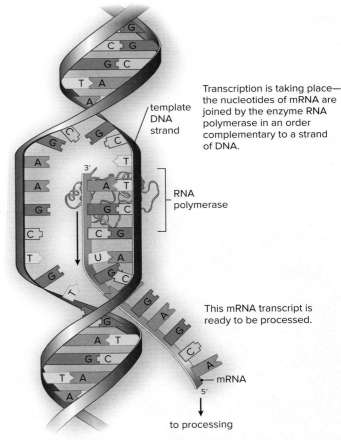

Transcription is taking place—the nucleotides of mRNA are joined by the enzyme RNA polymerase in an order complementary to a strand of DNA.

template DNA strand

RNA polymerase

This mRNA transcript is ready to be processed.

mRNA

to processing

Figure 11.11 Transcription to form mRNA.

During transcription, complementary RNA is made from a DNA template. A portion of DNA unwinds and unzips at the point of attachment of RNA polymerase. A strand of RNA, such as mRNA, is produced when complementary bases join in the order dictated by the sequence of bases in the template DNA strand.

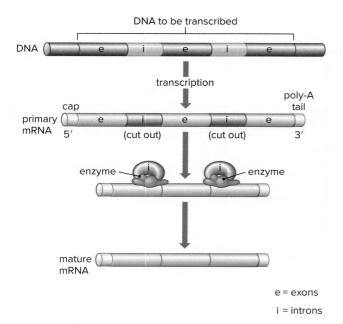

e = exons

i = introns

Figure 11.12 mRNA processing.

Primary mRNA results when both exons and introns are transcribed from DNA. A cap and a poly-A tail are attached to the ends of the mRNA transcript, and the introns are removed, so that only the exons remain. This mature mRNA molecule moves into the cytoplasm of the cell, where translation occurs.

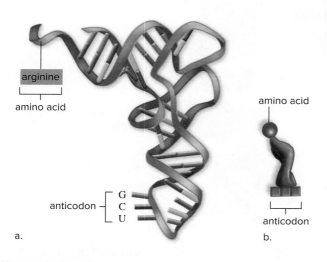

Figure 11.13 tRNA structure and function.

a. A tRNA, which is single-stranded but folded, has an amino acid attached to one end and an anticodon at the other end. The anticodon is complementary to a codon. **b.** The tRNA will be represented like this in the following illustrations.

mRNA Is Processed   The newly synthesized *primary mRNA* must be processed in order for it to be used properly. Processing occurs in the nucleus of eukaryotic cells. Three steps are required: capping, the addition of a poly-A tail, and splicing (**Fig. 11.12**). After processing, the mRNA is called a *mature mRNA* molecule.

The first nucleotide of the primary mRNA is modified by the addition of a cap that is composed of an altered guanine nucleotide. On the 3′ end, enzymes add a poly-A tail, a series of adenosine nucleotides. These modifications provide stability to the mRNA; only mRNA molecules that have a cap and tail remain active in the cell.

Most genes in humans are interrupted by segments of DNA that do not code for protein. These portions are called *introns* because they are intervening segments. The other portions of the gene, called *exons,* contain the protein-coding regions of the gene. In *mRNA splicing,* the introns are removed and the exons joined together. The result is a mature mRNA molecule consisting of continuous exons. Scientists now know that the genetic sequences in the introns are not "junk" and that many play a regulatory function in gene expression.

Ordinarily, processing brings together all the exons of a gene. In some instances, however, cells use only certain exons rather than all of them to form the mature RNA transcript. The result is a different protein product in each cell. In other words, this *alternative mRNA splicing* increases the number of protein products that can be made from a single gene.

After the mRNA strand is processed, it passes from the cell nucleus into the cytoplasm for translation.

Translation: An Overview

Translation is the second step by which gene expression leads to protein (polypeptide) synthesis. Translation requires several enzymes, mRNA, and the other two types of RNA: transfer RNA and ribosomal RNA.

Transfer RNA Takes Amino Acids to the Ribosomes   Each tRNA is a single-stranded nucleic acid that doubles back on itself such that complementary base pairing results in the cloverleaf-like shape shown in **Figure 11.13.** There is at least one tRNA molecule for each of the 20 amino acids found in proteins. The amino acid binds to one end of the molecule. The opposite end of the molecule contains an **anticodon,** a group of three bases that is complementary to a specific codon of mRNA.

During translation, the order of codons in mRNA determines the order in which tRNAs bind at the ribosomes. When a tRNA–amino acid complex comes to a ribosome, its anticodon pairs with an mRNA codon. For example, if the codon is ACC, what is the anticodon, and what amino acid will be attached to the tRNA molecule? From Figure 11.10, we can determine this:

Codon (mRNA)	Anticodon (tRNA)	Amino Acid (protein)
ACC	UGG	Threonine

After translation is complete, a protein contains the sequence of amino acids originally specified by DNA. This is the genetic information that DNA stores and passes on to each cell during the cell cycle, then to the next generation of individuals. DNA's sequence of bases determines the proteins in a cell, and the proteins determine the function of each cell.

Ribosomes Are the Site of Protein Synthesis   Ribosomes are the small structural bodies where translation occurs. Ribosomes are composed of many proteins and several ribosomal RNAs (rRNAs). In eukaryotic cells, rRNA is produced in a nucleolus within the nucleus. There, it joins with proteins manufactured in the cytoplasm to form two ribosomal subunits, one large and one small. The subunits leave the nucleus and join together in the cytoplasm to form a ribosome just as protein synthesis begins.

A ribosome has a binding site for mRNA as well as binding sites for two tRNA molecules at a time (Fig. 11.13). These binding sites facilitate complementary base pairing between tRNA anticodons and mRNA codons. The P binding site is for a tRNA attached to a *peptide*, and the A binding site is for a newly arrived tRNA attached to an *amino acid*. The E site is for tRNA molecules *exiting* the ribosome.

Translation Has Three Phases

Polypeptide synthesis has three phases: initiation, an elongation cycle, and termination. Enzymes are required for each of the steps to occur, and energy is needed for the first two steps.

Initiation Initiation is the step that brings all of the translation components together (**Fig. 11.14**):

- The small ribosomal subunit attaches to the mRNA in the vicinity of the start codon (AUG).
- The anticodon of the initiator tRNA–methionine complex pairs with this codon.
- The large ribosomal subunit joins to the small subunit.

Elongation Cycle During the elongation cycle, the polypeptide chain increases in length one amino acid at a time (**Fig. 11.15**):

- The tRNA at the P site contains the growing peptide chain.
- This tRNA passes its peptide to tRNA–amino acid at the A site. The tRNA at the P site enters the E site.
- During *translocation*, the tRNA-peptide moves to the P site, the empty tRNA in the E site exits the ribosome, and the codon at the A site is ready for the next tRNA–amino acid.

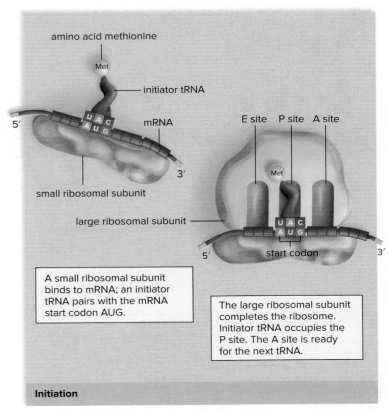

A small ribosomal subunit binds to mRNA; an initiator tRNA pairs with the mRNA start codon AUG.

The large ribosomal subunit completes the ribosome. Initiator tRNA occupies the P site. The A site is ready for the next tRNA.

Initiation

Figure 11.14 Initiation.

During initiation, the components needed for translation are assembled as shown.

Figure 11.15 Elongation.

Note that the polypeptide is already at the P site. During elongation, polypeptide synthesis occurs as amino acids are added one at a time to the growing chain.

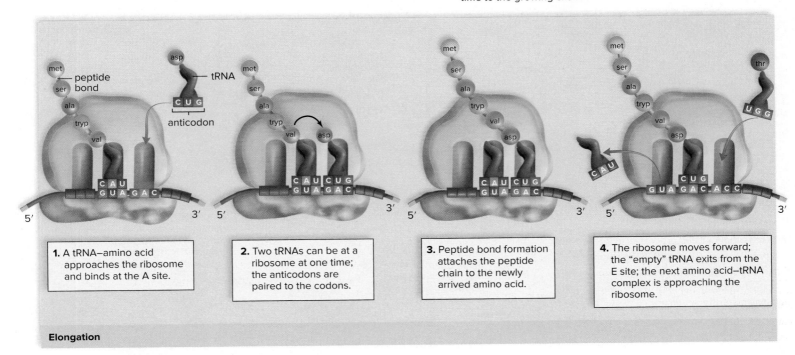

1. A tRNA–amino acid approaches the ribosome and binds at the A site.

2. Two tRNAs can be at a ribosome at one time; the anticodons are paired to the codons.

3. Peptide bond formation attaches the peptide chain to the newly arrived amino acid.

4. The ribosome moves forward; the "empty" tRNA exits from the E site; the next amino acid–tRNA complex is approaching the ribosome.

Elongation

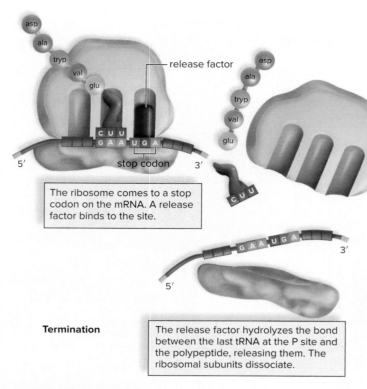

The ribosome comes to a stop codon on the mRNA. A release factor binds to the site.

Termination

The release factor hydrolyzes the bond between the last tRNA at the P site and the polypeptide, releasing them. The ribosomal subunits dissociate.

Figure 11.16 **Termination.**

During termination, the finished polypeptide chain is released, and the translation components disassemble.

The complete cycle—complementary base pairing of new tRNA, transfer of the growing peptide chain, and translocation—is repeated at a rapid rate. The outgoing tRNA is recycled and can pick up another amino acid in the cytoplasm to take to the ribosome.

Termination   Termination occurs when a stop codon (see Fig. 11.10) appears at the A site. Then, the polypeptide and the assembled components that carried out protein synthesis are separated from one another (**Fig. 11.16**).

- A protein complex called a **release factor** binds to the stop codon and cleaves the polypeptide from the last tRNA.
- The mRNA, ribosomes, and tRNA molecules can then be used for another round of translation.

Review of Gene Expression

Genes are segments of DNA that code for proteins. A gene is expressed when its protein product has been made. **Figure 11.17** reviews transcription and mRNA processing in the nucleus, as well as translation during protein synthesis in the cytoplasm of a eukaryotic cell. Note that it is possible for the cell to produce many copies of the same protein at the same time. As soon as the initial portion of mRNA has been translated by one ribosome and the ribosome has begun to move down the mRNA, another ribosome attaches to the same mRNA, forming a complex called a **polyribosome** (see step 7 in Fig. 11.17).

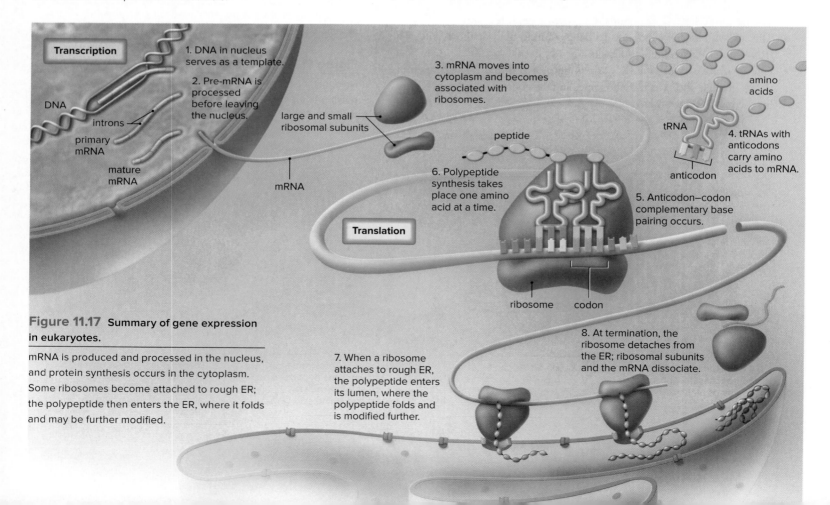

Figure 11.17 **Summary of gene expression in eukaryotes.**

mRNA is produced and processed in the nucleus, and protein synthesis occurs in the cytoplasm. Some ribosomes become attached to rough ER; the polypeptide then enters the ER, where it folds and may be further modified.

Transcription

1. DNA in nucleus serves as a template.

2. Pre-mRNA is processed before leaving the nucleus.

DNA

introns

primary mRNA

mature mRNA

mRNA

3. mRNA moves into cytoplasm and becomes associated with ribosomes.

large and small ribosomal subunits

peptide

amino acids

tRNA

4. tRNAs with anticodons carry amino acids to mRNA.

anticodon

6. Polypeptide synthesis takes place one amino acid at a time.

Translation

5. Anticodon–codon complementary base pairing occurs.

ribosome   codon

7. When a ribosome attaches to rough ER, the polypeptide enters its lumen, where the polypeptide folds and is modified further.

8. At termination, the ribosome detaches from the ER; ribosomal subunits and the mRNA dissociate.

Some ribosomes remain free in the cytoplasm, and others become attached to rough ER. In the latter case, the polypeptide enters the lumen of the ER by way of a channel, where it can be further processed by the addition of sugars. Transport vesicles carry the protein to other locations in the cell, including the Golgi apparatus, which may modify it further and package it in a vesicle for transport out of the cell or cause it to become embedded in the plasma membrane. Proteins have innumerable functions in cells, from enzymatic to structural. Proteins also have functions outside the cell. Together, they account for the structure and function of cells, tissues, organs, and the organism.

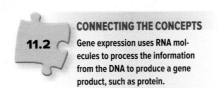

CONNECTING THE CONCEPTS

11.2 Gene expression uses RNA molecules to process the information from the DNA to produce a gene product, such as protein.

Check Your Progress 11.2

1. Describe the events that are involved in the expression of the information contained in the DNA to a functional protein.
2. Explain the purpose of the genetic code and how it fits into the central dogma of biology.
3. Summarize why mRNA is modified following transcription.
4. Explain how ribosomes and tRNA are involved in translation.

11.3 Gene Regulation

Learning Outcomes

Upon completion of this section, you should be able to

1. Summarize the operation of the *lac* operon in prokaryotes.
2. List the levels of control of gene expression in eukaryotes.
3. Distinguish between euchromatin and heterochromatin.
4. Explain the role of transcription factors in eukaryotic gene regulation.

Levels of Gene Expression Control

The human body consists of many types of cells, which differ in structure and function. Each cell type must contain its own mix of proteins that makes it different from all other cell types. Therefore, only certain genes are active in cells that perform specialized functions, such as nerve, muscle, gland, and blood cells.

Some of these active genes are called housekeeping genes because they govern functions that are common to many types of cells, such as glucose metabolism. But the activity of some genes accounts for the specialization of cells. In other words, gene expression is controlled in a cell, and this control accounts for its specialization (**Fig. 11.18**). Let's begin by examining a simple system of controlling gene expression in prokaryotes.

Gene Expression in Prokaryotes

The bacterium *Escherichia coli* lives in the human large intestine and can quickly adjust its production of enzymes according to what a person eats. If you drink a glass of milk, *E. coli* immediately begins to make three enzymes needed to metabolize lactose. The transcription of all three enzymes is under the control of one promoter, a short DNA sequence where RNA polymerase attaches. French microbiologists François Jacob and Jacques Monod called such a cluster of bacterial genes, along with the DNA sequences that control their transcription, an **operon.** They received a Nobel Prize in 1961 for their investigations because they were the first to show how gene expression is controlled—specifically, how the *lac* operon is controlled in lactose metabolism (**Fig. 11.19**).

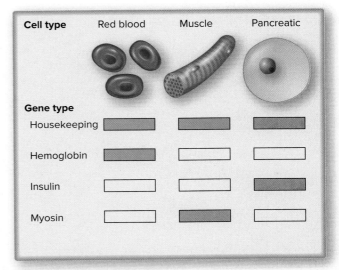

Figure 11.18 **Gene expression in specialized cells.**

Each cell in your body contains a complete set of genes. While housekeeping genes are expressed in all cells, the expression of other genes is limited by the function of the cell.

Figure 11.19 The *lac* operon.

a. The regulatory gene codes for a repressor, which bonds to the operator and prevents RNA polymerase from attaching to the promoter. Therefore, transcription of the lactose-metabolizing genes does not occur.
b. When lactose is present, it binds to the repressor, changing its shape. The repressor becomes inactive and can no longer bind to the operator. Now RNA polymerase binds to the promoter, and the lactose-metabolizing genes are expressed.

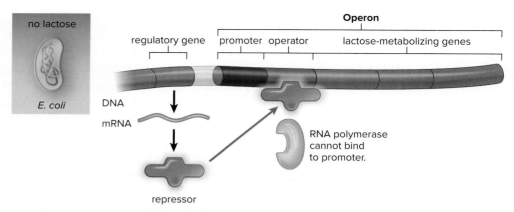

a. **Lactose is absent—operon is turned off.**
   Enzymes needed to metabolize lactose are not produced.

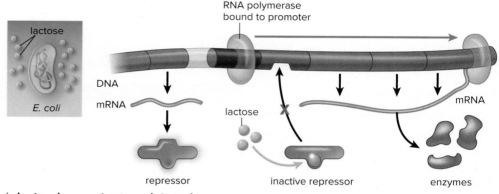

b. **Lactose is present—operon is turned on.**
   Enzymes needed to metabolize lactose are produced.

Jacob and Monod proposed that a **regulatory gene** located outside the operon codes for a **repressor**—a protein that, in the *lac* operon, normally binds to the *operator,* which lies next to the promoter. When the repressor is attached to the operator, transcription of the lactose-metabolizing genes does not take place because RNA polymerase is unable to bind to the promoter. The *lac* operon is normally turned off in this way because lactose is usually absent (Fig. 11.19*a*).

What turns the operon on when lactose is present? Lactose binds to the repressor and it changes shape. Now the repressor is unable to bind to the operator and RNA polymerase is able to bind to the promoter (Fig. 11.19*b*). Transcription of the genes needed for lactose metabolism occurs. When lactose is present and glucose, the preferred sugar, is absent, a protein called CAP (not shown in Fig. 11.19) assists in the binding of RNA polymerase to the promoter. This further ensures that the lactose-metabolizing enzymes are transcribed when they are needed.

Other bacterial operons, such as those that control amino acid synthesis, are usually turned on. For example, in the *trp* operon, the regulatory gene codes for a repressor that ordinarily is unable to attach to the operator. Therefore, the genes needed to make the amino acid tryptophan are ordinarily expressed. When tryptophan is present, it binds to the repressor. A change in shape activates the repressor and allows it to bind to the operator. Now the operon is turned off.

Gene Expression in Eukaryotes

In bacteria, a single promoter serves several genes that make up a transcription unit, while in eukaryotes, each gene has its own promoter where RNA polymerase binds. Bacteria rely mostly on transcriptional control, but eukaryotes employ a variety of mechanisms to regulate gene expression. These mechanisms affect whether a gene is expressed, the speed with which it is expressed, and how long it is expressed.

Some mechanisms of gene expression occur in the nucleus; others occur in the cytoplasm (**Fig. 11.20**). In the nucleus, chromatin condensation, DNA transcription, and mRNA processing all play a role in determining which genes are expressed in a particular cell type.

In the cytoplasm, mRNA translation into a polypeptide at the ribosomes can occur right away or be delayed. The mRNA can last a long time or be destroyed immediately, and the same holds true for a protein. These mechanisms control the quantity of gene product and/or how long it is active.

Chromatin Condensation   Eukaryotes utilize chromatin condensation as a way to keep genes turned on or off. The more tightly chromatin is compacted, the less often genes within it are expressed. Darkly staining portions of chromatin, called **heterochromatin,** represent tightly compacted, inactive chromatin. A dramatic example of this is the *Barr body* in mammalian females. Females have a small, darkly staining mass of condensed chromatin adhering to the inner edge of the nuclear envelope. This structure is an inactive X chromosome.

How do we know that Barr bodies are inactive X chromosomes that are not producing gene product? Suppose 50% of the cells in a female have one X chromosome active, and 50% have the other X chromosome active. Wouldn't the body of a heterozygous female be a mosaic, with "patches" of genetically different cells? This is exactly what happens. For example, human females who are heterozygous for an X-linked recessive form of ocular albinism have patches of pigmented and nonpigmented cells at the back of the eye. And women who are heterozygous for the hereditary absence of sweat glands have patches of skin lacking sweat glands. The female calico cat also provides dramatic support for a difference in X-inactivation in its cells (**Fig. 11.21**).

Connections: Scientific Inquiry

Why does X-inactivation occur?

While the X chromosome contains over 1,600 genes, the majority of these are not involved in anything related to sex. Therefore, in humans and other mammals, a female with two X chromosomes normally produces twice as much gene product (protein) as a male. Too much protein can negatively influence many metabolic pathways, so in order to regulate the amount of protein produced, females inactivate all but one of their X chromsomes. Most females possess a single Barr body, but the cells of XXX females have two Barr bodies, and males with Klinefelter syndrome (XXY) have a single Barr body in their cells, even though they are males.

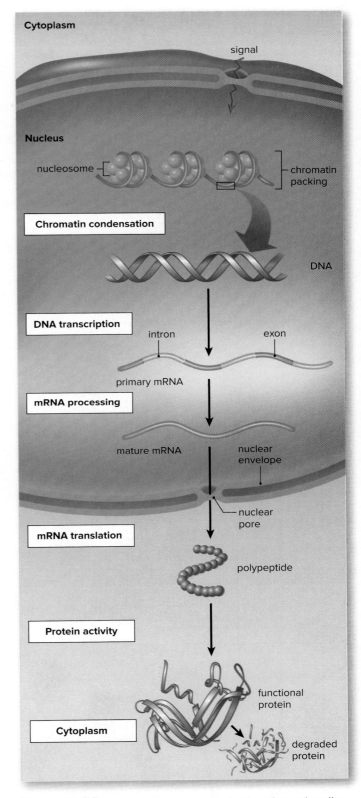

Figure 11.20 Control of gene expression in eukaryotic cells.

Gene expression is controlled at various levels in eukaryotic cells. Also, external signals (red arrow), such as growth factors, may alter gene expression.

Figure 11.21 X-inactivation in mammalian females.

In cats, the alleles for black or orange coat color are carried on the X chromosome. Random X-inactivation occurs in females. Therefore, in heterozygous females, 50% of the cells have an allele for black coat color, and 50% of cells have an allele for orange coat color. The result is calico or tortoiseshell cats that have coats with patches of both black and orange. A calico cat, but not a tortoiseshell cat, has patches of white due to the activity of another gene.

© Photodisc/Getty RF

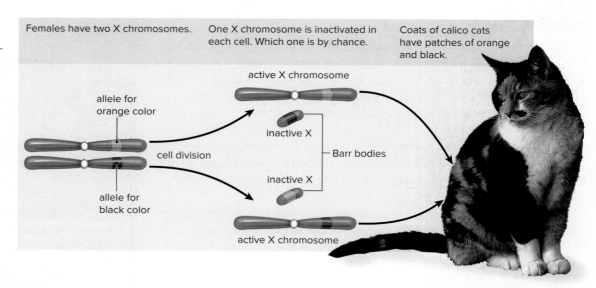

Females have two X chromosomes.

One X chromosome is inactivated in each cell. Which one is by chance.

Coats of calico cats have patches of orange and black.

active X chromosome

allele for orange color

inactive X

Barr bodies

cell division

inactive X

allele for black color

active X chromosome

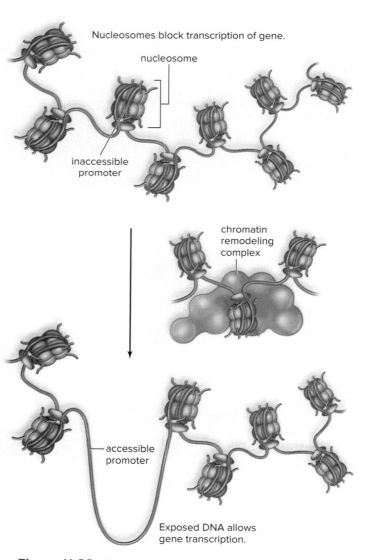

Nucleosomes block transcription of gene.

nucleosome

inaccessible promoter

chromatin remodeling complex

accessible promoter

Exposed DNA allows gene transcription.

Figure 11.22 Histone displacement.

In euchromatin, a chromatin remodeling complex pushes aside the histone portions of nucleosomes, so that transcription can begin.

When heterochromatin undergoes unpacking, it becomes **euchromatin,** a less compacted form of chromatin that contains active genes. You learned in Section 8.1 that, in eukaryotes, a *nucleosome* is a portion of DNA wrapped around a group of histone molecules. When DNA is transcribed, a chromatin remodeling complex pushes aside the histone portions of nucleosomes, so that transcription can begin (**Fig. 11.22**). In other words, even euchromatin needs further modification before transcription can begin. The presence of histones limits access to DNA, and euchromatin becomes genetically active when histones no longer bar access to DNA. Only then is it possible for a gene to be turned on and expressed in a eukaryotic cell.

Epigenetics

Scientists are beginning to recognize that some changes in gene regulation, especially those occurring at the cellular level, may be inherited from one generation to the next. For example, in X-inactivation (see Fig. 11.21), the pattern of inactivation is inherited as the cell divides. In an XX cell, the inactive chromosome in the first cell is the same in all of its daughter cells. This explains the patches of colored fur on the cat in Figure 11.21; the black patches are all inherited from a cell that expresses only the black allele because the orange is inactivated. This is an example of **epigenetic inheritance,** or the inheritance of changes in gene expression that are not the result of changes in the sequence of nucleotides on the chromosome. Epigenetic inheritance explains other physiological processes as well, such as genomic imprinting, in which the sex of the individual determines which alleles are expressed in a cell by suppressing genes on chromosomes inherited from the opposite sex.

DNA Transcription   Transcription in eukaryotes follows the same principles as in bacteria, except that many more regulatory proteins per gene are involved. The occurrence of so many regulatory proteins not only allows for greater control but also brings a greater chance of malfunction.

In eukaryotes, **transcription factors** are DNA-binding proteins that help RNA polymerase bind to a promoter. Several transcription factors are needed in each case; if one is missing, transcription cannot take place. All the transcription factors form a complex that also helps pull double-stranded

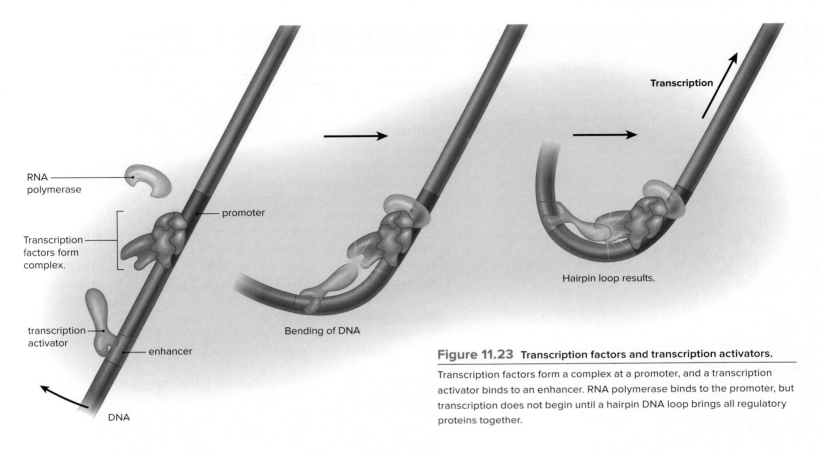

RNA polymerase

Transcription factors form complex.

promoter

transcription activator

enhancer

DNA

Bending of DNA

Hairpin loop results.

Transcription

Figure 11.23 **Transcription factors and transcription activators.**

Transcription factors form a complex at a promoter, and a transcription activator binds to an enhancer. RNA polymerase binds to the promoter, but transcription does not begin until a hairpin DNA loop brings all regulatory proteins together.

DNA apart and even acts to position RNA polymerase so that transcription can begin. The same transcription factors, in different combinations, are used over again at other promoters, so it is easy to imagine that, if one malfunctions, the result could be disastrous to the cell. The genetic disorder Huntington disease is a devastating psychomotor ailment caused by a defect in a transcription factor.

In eukaryotes, *transcription activators* are DNA-binding proteins that speed transcription dramatically. They bind to a DNA region, called an **enhancer,** that can be quite a distance from the promoter. A hairpin loop in the DNA can bring the transcription activators attached to enhancers into contact with the transcription factor complex (**Fig. 11.23**).

A single transcription activator can have a dramatic effect on gene expression. For example, investigators have found one DNA-binding protein, MyoD, that alone can activate the genes necessary for fibroblasts to become muscle cells in various vertebrates. Another DNA-binding protein, called Ey, can bring about the formation of not just a single cell type but a complete eye in flies (**Fig. 11.24**).

mRNA Processing   After transcription, the removal of introns and the splicing of exons occur before mature mRNA leaves the nucleus and passes into the cytoplasm. **Alternative mRNA processing** is a mechanism by which the same primary mRNA can produce different protein products according to which exons are spliced together to form mature mRNAs (**Fig. 11.25**).

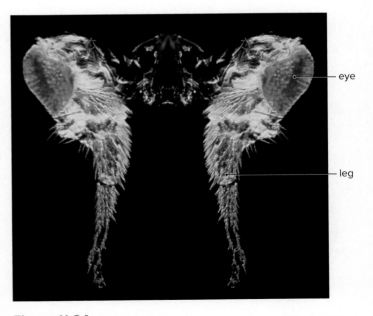

eye

leg

Figure 11.24 *Ey* gene.

In fruit flies, the expression of the *Ey* gene in the precursor cells of the leg triggers the development of an eye on the leg.

© Prof. Walter Gehring

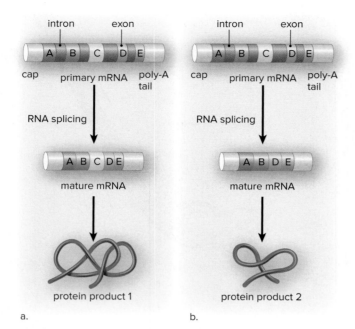

a.   b.

Figure 11.25 Processing of mRNA transcripts.

Because the primary mRNAs are processed differently in these two cells, distinct proteins (a and b) result.

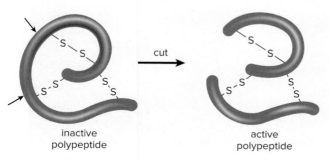

Figure 11.26 Protein activity.

The protein insulin is not active until an enzyme removes a portion of the initial polypeptide. The resulting two polypeptide chains are held together by two disulfide (S—S) bonds between like amino acids.

The fruit fly gene *DScam* offers a dramatic example of alternative mRNA processing. As many as 38,000 different proteins can be produced according to which of this gene's exons are combined in mature mRNAs. Evidence suggests that these proteins provide each nerve cell with a unique identity as it communicates with other nerve cells in a fruit fly's brain.

mRNA Translation   The cytoplasm contains proteins that can control whether translation of mRNA takes place. For example, an initiation factor, known as IF-2, inhibits the start of protein synthesis when it is phosphorylated by a specific protein kinase. Environmental conditions can also delay translation. Red blood cells do not produce hemoglobin unless heme, an iron-containing group, is available.

The longer an mRNA remains in the cytoplasm before it is broken down, the more of the gene product is produced. During maturation, mammalian red blood cells eject their nuclei, yet they continue to synthesize hemoglobin for several months. The necessary mRNAs must be able to persist the entire time. Differences in the length of the poly-A tail can determine how long a particular mRNA transcript remains active. Hormones may also affect the stability of certain mRNA transcripts. An mRNA called vitellin persists for 3 weeks, instead of 15 hours, if it is exposed to estrogen.

Protein Activity   Some proteins are not active immediately after synthesis. For example, insulin is a single, long polypeptide that folds into a three-dimensional structure. Only then is a sequence of about 30 amino acids enzymatically removed from the middle of the molecule. This leaves two polypeptide chains bonded together by disulfide (S—S) bonds, and an active insulin molecule results (**Fig. 11.26**). This mechanism allows a protein's activity to be delayed until it is needed.

Many proteins are short-lived in cells because they are degraded or destroyed. Cyclins, which are proteins involved in regulating the cell cycle, are destroyed by giant enzyme complexes called *proteasomes* when they are no longer needed. Proteasomes break down other proteins as well.

Signaling Between Cells in Eukaryotes

In multicellular organisms, cells are constantly sending out chemical signals that influence the behavior of other cells. During animal development, these signals determine the specialized role a cell will play in the organism. Later, the signals help coordinate growth and day-to-day functions. Plant cells also signal each other, so that their responses to environmental stimuli, such as direct sunlight, are coordinated.

Typically, **cell signaling** occurs because a chemical signal binds to a receptor protein in a target cell's plasma membrane. The signal causes the receptor protein to initiate a series of reactions within a **signal transduction pathway.** The end product of the pathway (not the signal) directly affects the metabolism of the cell. For example, growth is possible only if certain genes have been turned on by regulatory proteins.

In **Figure 11.27,** a signaling cell secretes a chemical signal that binds to a specific receptor located in the receiving cell's plasma membrane. The binding activates a series of reactions within a signal transduction pathway. The last reaction activates a transcription activator that enhances the transcription of a specific gene. Transcription leads to the translation of mRNA and a protein product that, in this case, stimulates the cell cycle, so that growth occurs.

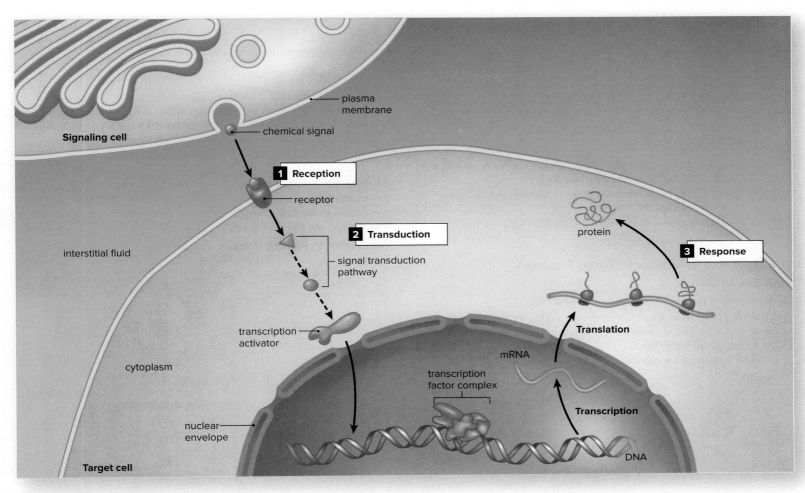

Figure 11.27 Cell signaling.

1. Reception: A chemical signal is received at a specific receptor protein. 2. Transduction: A signal transduction pathway terminates when a transcription activator is stimulated. It enhances transcription of a specific gene. 3. Response: Translation of the mRNA transcript follows, and a protein product results. The protein product is the response to the signal.

A protein called Ras functions in signal transduction pathways that lead to the transcription of many genes, several of which promote the cell cycle. Ras is normally inactive, but the reception of a growth factor leads to its activation. If Ras is continually activated, cancer will develop because cell division will occur continuously.

CONNECTING THE CONCEPTS

11.3

Gene expression may be controlled at a variety of locations in a cell.

Check Your Progress 11.3

1. Compare and contrast gene expression in prokaryotes and eukaryotes.

2. Describe the levels at which eukaryotes regulate gene expression.

3. Explain how transcription factors may fine-tune the level of gene expression.

SUMMARIZE

The DNA of a cell contains the genetic instructions that determine the characteristics of life. The processes of gene expression involve RNA and may be controlled a number of ways in the cell.

11.1 DNA is the genetic material, it contains the instructions for carrying out the processes of life. RNA is involved in the processing of that information.

11.2 Gene expression uses RNA molecules to process the information from the DNA to produce a gene product, such as protein.

11.3 Gene expression may be controlled at a variety of locations in a cell.

11.1 DNA and RNA Structure and Function

Structure of DNA

DNA, the genetic material, has the following structure:

- DNA is a polymer in which four nucleotides differ in their bases. These bases are **adenine (A)**, **guanine (G)**, **cytosine (C)**, and **thymine (T)**. A is always paired with T, and G is always paired with C, a pattern called **complementary base pairing:**

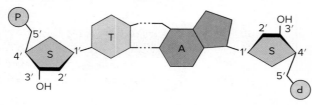

- DNA is a double helix. The deoxyribose sugar–phosphate units make up the sides of a twisted ladder and run counter to one another. The paired bases are the rungs of the ladder. Base pairs can be in any order in each type of species.

Replication of DNA

DNA replicates during the S phase of the cell cycle. During **DNA replication,** the two strands separate and each old strand (a **template**) gives rise to a new strand. Therefore, the process is called **semiconservative.** Because of complementary base pairing, all these double-helical molecules are identical:

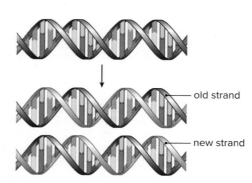

old strand

new strand

DNA replication uses a collection of enzymes, which include

- Helicase—unwinds the double helix
- DNA polymerase—synthesizes the new DNA strand
- DNA ligase—repairs breaks in the sugar-phosphate backbone of the DNA strand

RNA Structure and Function

Ribonucleic acid (RNA) has the following characteristics:

- It is found in the nucleus and cytoplasm of eukaryotic cells.
- It contains the sugar ribose and the bases adenine (A), **uracil (U)**, cytosine (C), and guanine (G).

- It is single-stranded.
- The three main forms are **messenger RNA (mRNA)**, which carries the DNA message to the ribosomes; **transfer RNA (tRNA)**, which transfers amino acids to the ribosomes, where protein synthesis occurs; and **ribosomal RNA (rRNA)**, which is found in the **ribosomes.**

11.2 Gene Expression

DNA specifies the sequence of amino acids in a protein. This information is represented by the genetic code. Gene expression requires two steps: transcription and translation.

From DNA to RNA to Protein

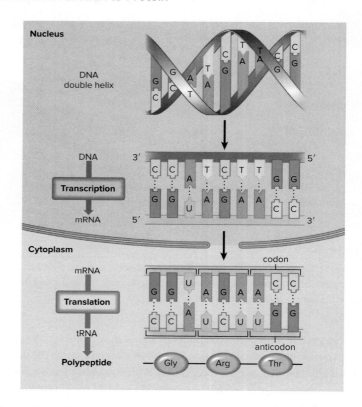

Transcription: During **transcription,** an RNA polymerase binds to the promoter, the regulatory region at the start of the gene. The RNA polymerase then makes a complementary copy of the nucleotides in the gene (substituting uracil for thymine). This copy is called an **mRNA transcript.**

Following transcription in eukaryotic cells, the mRNA is processed. Processing involves (1) the addition of a cap on one end, (2) the addition of a poly-A tail on the other end, and (3) the removal of introns, so only exons remain. The mature mRNA now contains instructions for assembling a protein.

Translation: Translation requires three types of RNA:

- mRNA contains codons in which every three bases code for a particular amino acid, except when the codon means "stop."
- tRNA brings amino acids to the ribosomes. One end of this molecule binds to the amino acid, and the other end is the **anticodon,** three bases that pair with a **codon.** Each tRNA binds with only 1 of the 20 types of amino acids that make up a protein.

- rRNA is located in the ribosomes. The P site of a ribosome contains a tRNA attached to a peptide, and the A site contains a tRNA attached to an amino acid. The E site represents the exit site for the tRNA once it has released its amino acid.

Translation Has Three Phases

The three phases of translation are initiation, the elongation cycle, and termination:

Initiation: During chain initiation at the start codon, the ribosomal subunits, the mRNA, and the tRNA–methionine complex come together.

Elongation cycle: The chain elongation cycle consists of the following events. A tRNA enters the A site. A peptide bond forms between the amino acids attached to the tRNAs in the P and A sites. As the ribosome translocates, the tRNA in the P site moves into the E site and exits, and the tRNA in the A site moves to the P site. The codon at the A site is ready for the next tRNA–amino acid.

Termination: During chain termination, a stop codon is reached on the mRNA. A release factor binds, the ribosome dissociates, the mRNA departs, and the polypeptide is released.

11.3 Gene Regulation

Levels of Gene Expression Control

The specialization of cells is not due to the presence or absence of genes; it is due to the activity or inactivity of genes.

In prokaryotes, gene expression is controlled at the level of transcription:

- In the *lac* operon, the **repressor,** a protein produced by a **regulatory gene,** usually binds to the operator. Then RNA polymerase cannot bind to the promoter, and the **operon** is turned off. When lactose is present, it binds to the repressor, and now RNA polymerase binds to the promoter and the genes for lactose metabolism are transcribed.
- In the *trp* operon, the genes are usually transcribed because the repressor is inactive. When tryptophan is present, the repressor becomes active, and RNA polymerase cannot bind to the promoter.

Eukaryotes regulate gene expression at various levels:

- Chromatin condensation—**heterochromatin** is converted to **euchromatin,** so that the genes can be expressed.
- **Epigenetic inheritance** represents an inheritable form of regulation of gene expression, as is the case with X-inactivation.
- DNA transcription—**transcription factors** present at the promoter and transcription activators (bound to **enhancers**) are brought into contact and promote transcription.
- mRNA processing—**alternative mRNA processing** leads to a sequence of different exons in mature RNA and, therefore, the production of different proteins.
- mRNA translation—various factors affect how long an mRNA stays active and how many times it is translated.
- Protein activity—activation of a protein may require further steps, and degradation of the protein can occur at once or after a while.

Eukaryotic cells constantly communicate with their neighbors through **cell signaling:**

- Signals are received at receptor proteins located in the plasma membrane.
- A **signal transduction pathway** (series of enzymatic reactions) leads to a change in the behavior of the cell. For example, when stimulated via a signal transduction pathway, a transcription activator stimulates transcription of a particular gene.

ASSESS

Testing Yourself

Choose the best answer for each question.

11.1 DNA and RNA Structure and Function

1. In the following diagram, label the indicated parts of the DNA molecule.

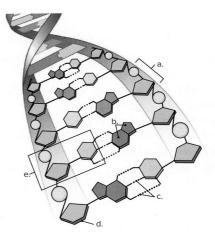

2. Chargaff's rules state that the amount of A, T, G, and C in DNA
   a. varies from species to species, and the amount of A = T and G = C.
   b. varies from species to species, and the amount of A = G and T = C.
   c. is the same from species to species, and the amount of A = T and G = C.
   d. is the same from species to species, and the amount of A = G and T = C.

3. The first picture of DNA, as a result of X-ray diffraction, was produced by
   a. Hershey and Chase.   c. Chargaff.
   b. Watson and Crick.   d. Franklin.

4. DNA replication is said to be semiconservative because
   a. each new DNA molecule contains one new strand and one old one.
   b. every other new DNA molecule contains two old strands.
   c. each new DNA strand contains two new strands.
   d. every other new DNA molecule contains one new strand and one old one.

5. Which of the following is not a feature of eukaryotic DNA replication?
   a. Replication bubbles move in opposite directions.
   b. A new strand is synthesized using an old one as a template.
   c. Complementary base pairing determines which nucleotides will be added to the new strand.
   d. Each chromosome has one origin of replication.

11.2 Gene Expression

For questions 6–10, match the items to those in the key. Answers can be used more than once.

Key:

    a. messenger RNA

    b. transfer RNA

    c. ribosomal RNA

  6. Contains an anticodon

  7. Carries amino acids to the ribosome

  8. Carries genetic information from DNA to ribosomes

  9. Along with proteins, forms the structure of the ribosome.

10. Is produced in the nucleolus

11. Which of the following statements about the genetic code is not true?

    a. The genetic code is almost universal.

    b. The genetic code is a doublet code.

    c. Multiple codons may encode the same amino acid.

    d. Some special codons mean "stop."

12. Transcription produces _____, while translation produces _____.

    a. DNA, RNA   **c.** polypeptides, RNA

    b. RNA, polypeptides   **d.** RNA, DNA

11.3 Gene Regulation

13. Which of the following is not a control mechanism for gene expression in eukaryotes?

    a. alternative mRNA processing   **c.** longevity of mRNA

    b. rate of ribosome synthesis   **d.** chromatin condensation

14. Which of the following may produce more than one functional protein from an mRNA transcript?

    a. chromatin condensation   **c.** epigenetics

    b. transcriptional regulation   **d.** alternative mRNA processing

15. The form of gene regulation that may be inherited from one generation to the next is

    a. chromatin condensation.   **c.** alternative mRNA splicing.

    b. epigenetic inheritance.   **d.** Mendelian inheritance.

ENGAGE

BioNOW

Want to know how this science is relevant to your life? Check out the BioNow videos below:

- Quail Hormones
- Metamorphosis

1. **Quail Hormones.** How might the environmental factors of light and temperature be acting as regulators of gene expression?

2. **Metamorphosis.** What factors may be regulating the expression of the developmental genes in the larvae?

Thinking Critically

1. Experiments with bacteria in the 1930s showed that exposing a nonvirulent (noninfectious) strain of bacteria to a heat-killed virulent strain can convert the nonvirulent strain into a virulent one. What might explain this phenomenon?

2. How might an understanding of epigenetic inheritance be used to develop new medicines?

3. The search for life in our solar system is an ongoing process. If life, or an early form of it, is discovered, one of the primary goals of researchers will be to better understand the genetic information that it contains.

    a. What basic requirements are needed for a molecule to be an effective genetic information system?

    b. Suppose that the genetic information in the newly discovered life-form turns out to be very similar to our own. What will this tell us about our DNA? What if the genetic information is very different?

(family): © JGI/Blend Images LLC; (DNA helix): © Zoonar GmbH/Alamy RF

12

Biotechnology and Genomics

CRISPR and Genome Editing

Biotechnology has come a long way since it first began in the early 1950s. Historically, biotech has focused on inserting modified genes into bacteria and plants to yield genetically modified organisms that can withstand changes in the climate, gain enhanced nutritional value, or produce new drugs to target human diseases.

Recently, a new form of genome editing called CRISPR is revolutionizing biotechnology. Based on a series of enzymes that protect bacteria from viruses, CRISPR acts as a "molecular scalpel," allowing researchers to target specific sequences of a genome for editing. Early experiments were able to correct defects in mice, and experiments have already been conducted on human embryos. Researchers are actively examining how to modify the pig genome to produce more hearts and organs for human transplants. Almost monthly, new applications of this technology are revealed in scientific journals.

These advances present tremendous opportunities to address human disease, but they also give rise to ethical questions. For example, to what extent is it ethical to improve the human genome, and should society allow couples to produce "designer babies"? In this chapter, we will explore some of these techniques and areas of study, so that you may be better informed to make decisions.

As you read through this chapter, think about the following questions:

1. What procedures are used to introduce a bacterial gene into a plant?
2. How does CRISPR allow for genome editing?
3. What is the difference between genetically modified and transgenic organisms?

OUTLINE

12.1 Biotechnology   208

12.2 Stem Cells and Cloning   212

12.3 Biotechnology Products   214

12.4 Genomics and Proteomics   216

BEFORE YOU BEGIN

Before beginning this chapter, take a few moments to review the following discussions.

Section 3.2 What are the roles of proteins and nucleic acids in a cell?

Section 11.1 What is the basic structure of a DNA molecule?

Section 11.1 How is the DNA of a cell replicated?

12.1 Biotechnology

Learning Outcomes

Upon completion of this section, you should be able to

1. Describe the steps involved in making a recombinant DNA molecule.
2. Explain the purpose of the polymerase chain reaction.
3. Explain how DNA fingerprinting may be used for identification.
4. Describe the principles of genome editing.

The term **biotechnology** refers to the use of natural biological systems to create a product or achieve some other end desired by humans. Today, **genetic engineering** allows scientists to modify the genomes of a variety of organisms, from bacteria to plants and animals, to either improve the characteristics of the organism or make biotechnology products. Such modification is possible because decades of research on how DNA and RNA function in cells has allowed for the development of new techniques. These techniques allow scientists not only to clone genes, but also to directly edit the genome of an organism.

A **genetically modified organism (GMO)** is one whose genome has been modified in some way, usually by using recombinant DNA technology. A **transgenic organism** is an example of a GMO that has had a gene from another species inserted into its genome. We will take a closer look at both GMOs and transgenics in Section 12.3, but first we need to explore some of the DNA techniques that are used in biotechnology.

Recombinant DNA Technology

Recombinant DNA (rDNA) contains genes from two or more different sources (**Fig. 12.1**). To make rDNA, a researcher needs a **vector,** a piece of DNA that acts as a carrier for the foreign DNA. One common vector is a *plasmid,* which is a small accessory ring of DNA found in bacterial cells.

Two enzymes are needed to introduce foreign DNA into plasmid DNA: (1) **restriction enzymes** that can cleave, or cut, DNA at specific places (for example, the restriction enzyme *Eco*RI always cuts DNA at the base sequence GAATTC) and (2) DNA ligase, which can seal the foreign DNA into an opening in a cut plasmid.

If a plasmid is cut with *Eco*RI, this creates a gap into which a piece of foreign DNA can be placed if that piece ends in bases complementary to those exposed by the restriction enzyme. To ensure that the bases are complementary, it is necessary to cleave the foreign DNA with the same restriction enzyme. The overhanging bases at the ends of the two DNA molecules are called "sticky ends," because they can bind a piece of foreign DNA by complementary base pairing. Sticky ends facilitate the insertion of foreign DNA into vector DNA, a process very similar to the way puzzle pieces fit together.

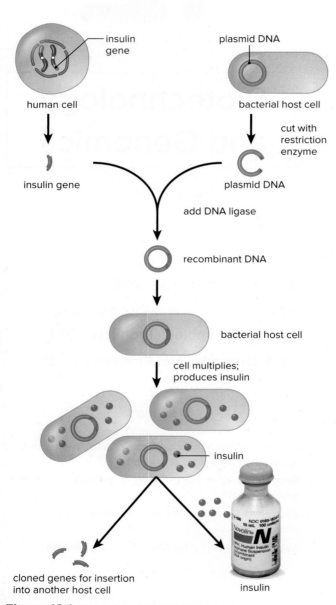

Figure 12.1 Recombinant DNA technology.

The production of insulin is one example of how recombinant DNA technology can benefit humans. In this process, human and plasmid DNA are spliced together and inserted into a bacterial host cell. As the cell undergoes cell division, the plasmid is replicated. This replication makes multiple copies, or clones, of the gene. The gene may be isolated for use in additional experiments, or left in the bacterial host cell to produce human insulin protein.

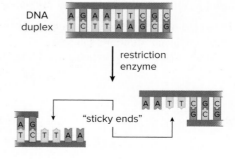

DNA ligase, the enzyme that functions in DNA replication to repair breaks in a double-stranded helix, seals the foreign piece of DNA into the plasmid. Molecular biologists often give the rDNA to bacterial cells, which readily take up recombinant plasmids if the cells have been treated to make them more permeable. Thereafter, as the bacteria replicate the plasmid, the gene is cloned. Cloned genes have many uses. A scientist may allow the genetically modified bacterial cells to express the cloned gene and retrieve the protein. Or copies of the cloned gene may be removed from the bacterial cells and then introduced into another organism, such as a corn plant, to produce a transgenic organism.

DNA Sequencing

DNA sequencing is a procedure used to determine the order of nucleotides in a segment of DNA, often within a specific gene. DNA sequencing allows researchers to identify specific alleles that are associated with a disease and thus facilitate the development of medicines or treatments. Information from DNA sequencing also serves as the foundation for the study of forensic biology and even contributes to our understanding of our evolutionary history.

When DNA technology was in its inception in the early 1970s, this technique was performed manually using dye-terminator substances or radioactive tracer elements attached to each of the four nucleotides during DNA replication, then deciphering the results from a pattern on a gel plate. Modern-day sequencing involves attaching dyes to the nucleotides and detecting the different dyes via a laser in an automated sequencing machine, which shows the order of nucleotides on a computer screen. To begin sequencing a segment of DNA, many copies of the segment are made, or replicated, using a procedure called the polymerase chain reaction.

Polymerase Chain Reaction

The **polymerase chain reaction (PCR)** can create billions of copies of a segment of DNA in a test tube in a matter of hours. PCR is very specific—it amplifies (makes copies of) a targeted DNA sequence, usually a few hundred bases in length. PCR requires the use of DNA polymerase, the enzyme that carries out DNA replication, and a supply of nucleotides for the new DNA strands. PCR involves three basic steps (**Fig. 12.2**), which occur repeatedly,

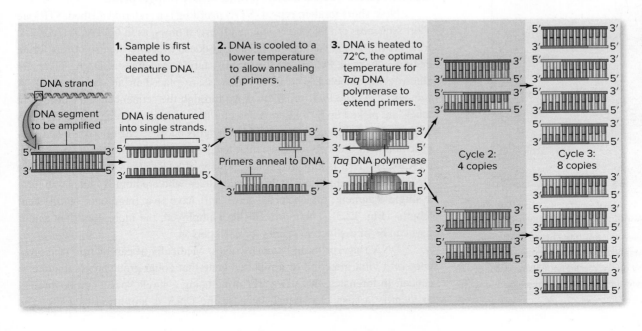

Figure 12.2 The polymerase chain reaction.

The polymerase chain reaction, or PCR, produces multiple copies of a segment of DNA. These segments may then be used by researchers in recombinant DNA studies.

usually for about 35 to 40 cycles: (1) a denaturation step at 95°C, where DNA is heated so that it becomes single-stranded; (2) an annealing step at a temperature usually between 50°C and 60°C, where a nucleotide primer hybridizes (or binds) to each of the single DNA strands; and (3) an extension step at 72°C, where a DNA polymerase adds complementary bases to each of the single DNA strands, creating double-stranded DNA.

PCR is a chain reaction because the targeted DNA is repeatedly replicated, much in the same way natural DNA replication occurs, as long as the process continues. Figure 12.2 uses color to distinguish the old strand from the new DNA strand. Notice that the amount of DNA doubles with each replication cycle. Thus, if you start with only one copy of DNA, after one cycle you will have two copies, after two cycles four copies, and so on. PCR has been in use since its development in 1985 by Kary Banks Mullis. The process relies on the discovery of a temperature-insensitive (thermostable) DNA polymerase that was extracted from the bacterium *Thermus aquaticus,* which lives in hot springs. The enzyme can withstand the high temperature used to denature double-stranded DNA. This enzyme can survive the high temperatures of a PCR reaction, which accelerates the production of copies of the selected DNA segment.

DNA amplified by PCR is often analyzed for various purposes. For example, mitochondrial DNA base sequences have been used to decipher the evolutionary history of human populations. Because so little DNA is required for PCR to be effective, it is commonly used as a forensic method for analyzing DNA found at crime scenes—only a drop of semen, a flake of skin, or the root of a single hair is necessary!

DNA Analysis

Analysis of DNA following PCR has improved over the years. At first, the entire genome was treated with restriction enzymes, and because each person has different restriction enzyme sites, this yielded a unique collection of DNA fragments of various sizes. During a process called *gel electrophoresis,* whereby an electrical current is used to force DNA through a porous gel material, these fragments were separated according to their size. Smaller fragments moved farther through the gel than larger fragments, resulting in a pattern of distinctive bands, called a DNA profile, or **DNA fingerprint.**

Now, **short tandem repeat (STR) profiling** is a preferred method. STRs are short sequences of DNA bases that recur several times, as in GATAGATAGATA. STR profiling is advantageous because it doesn't require the use of restriction enzymes. Instead, PCR is used to amplify target sequences of DNA, which are fluorescently labeled. The PCR products are placed in an automated DNA sequencer. As the PCR products move through the sequencer, the fluorescent labels are picked up by a laser. A detector then records the length of each DNA fragment. The fragments are different lengths because each person has a specific number of repeats at a particular location on the chromosome (i.e., at each STR locus). That is, the greater the number of STRs at a locus, the longer the DNA fragment amplified by PCR. Individuals who are homozygotes will have a single fragment, and heterozygotes will have two fragments of different lengths (**Fig. 12.3**). The more STR loci employed, the more confident scientists can be of distinctive results for each person.

DNA fingerprinting has many uses. Medically, it can identify the presence of a viral infection or a mutated gene that could predispose someone to cancer. In forensics, DNA fingerprinting using a single sperm can be enough to identify a suspected rapist, because the DNA is amplified by PCR. The

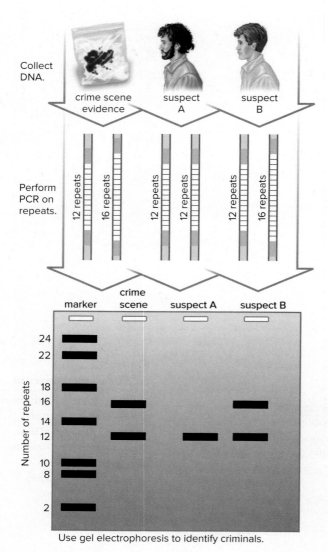

Collect DNA.

crime scene evidence   suspect A   suspect B

Perform PCR on repeats.

12 repeats   16 repeats   12 repeats   12 repeats   12 repeats   16 repeats

marker   crime scene   suspect A   suspect B

Number of repeats

24
22
18
16
14
12
10
8
2

Use gel electrophoresis to identify criminals.

Figure 12.3 **DNA fingerprinting.**

For DNA fingerprinting, PCR is used to generate copies of specific regions of DNA, which are then analyzed for length variations. These variations may be used to identify deceased individuals or suspects in a crime. In this example, the evidence suggests that suspect A is not the criminal.

fingerprinting technique can also be used to identify the parents of a child or identify the remains of someone who has died, such as a victim of a natural disaster. In the future, we will undoubtedly see more applications of recombinant DNA technology that will greatly enrich our lives and improve our health.

Genome Editing

A relatively new advance in DNA technology is **genome editing,** the targeting of specific sequences in the DNA for removal or replacement. There are several methods by which editing may be done; the most widely used is called CRISPR (*c*lustered *r*egularly *i*nterspaced *s*hort *p*alindromic *r*epeats).

CRISPR was first discovered in prokaryotes, where it acts as a form of immune defense against invading viruses. Viruses function by inserting their DNA into host cells, causing those cells to form new viruses (see Section 17.1). The CRISPR system is based on an endonuclease enzyme called Cas9, which is capable of identifying specific sequences of nucleotides in the genomic DNA of the invading virus and breaking both of the DNA strands, thus inactivating the virus.

Cas9 identifies the specific nucleotides to be cut using a guide RNA molecule that complementary base-pairs to the genomic DNA sequence (**Fig. 12.4**). To protect the bacteria from Cas9 activity against its own DNA, a sequence called PAM (which is not found in bacterial cells) must be adjacent to the target DNA sequence.

The CRISPR system can be used by researchers to target a specific sequence of nucleotides, in almost any organism, for editing. If the genomic sequence of the target is known, a complementary RNA strand can be used by Cas9 to produce a break in the DNA. This break can be used to inactivate the gene and thus study the role of the gene in the cell, or Cas9 can act as a form of molecular scissors to insert new nucleotides at specific DNA locations.

CRISPR and other genome editing technologies continue to develop. Scientists are investigating ways of making the processes more efficient, as well as new applications for genome editing in humans and other organisms.

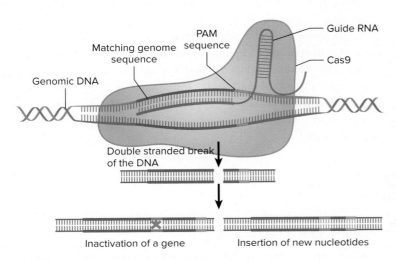

Figure 12.4 **CRISPR and genome editing.**

Genome editing involves using a nuclease, in this case, Cas9, to target specific sequences of DNA. Cas9 identifies specific sequences of genomic DNA using a guide RNA that complementary pairs with the target DNA sequence. The double-stranded break in the DNA created by Cas9 may be used to inactivate a gene to study its function, or to insert new sequences of nucleotides.

CONNECTING THE CONCEPTS

12.1   Biotechnology allows scientists to copy, edit, and analyze the genes of organisms.

Check Your Progress 12.1

1. Explain the role of the bacterial plasmid in recombinant DNA technology.

2. Summarize how the polymerase chain reaction (PCR) is used in biotechnology.

3. Describe the potential uses of DNA fingerprinting.

4. Explain why genome editing may be more efficient than recombinant DNA processes.

12.2 Stem Cells and Cloning

The events of the cell cycle and DNA replication ensure that every cell of the body receives a copy of all the genes. This means that every one of your cells has the potential to become a complete organism. While some cells, such as stem cells, retain their ability to form any other type of cell, most other cells differentiate, or specialize, to become specific types of cells, such as muscle cells. This specialization is based on the expression of certain groups of genes at specific times in development. One of the best ways to understand how specialization influences the fate of a cell is to take a look at the processes of reproductive and therapeutic cloning.

Reproductive and Therapeutic Cloning

In **reproductive cloning,** the desired end is an individual that is exactly like the original individual. The cloning of plants has been routine for some time and has been responsible for much of the success of modern agriculture. The cloning of some animals, such as amphibians, has been underway since the 1950s. However, at one time it was thought that the cloning of adult mammals would be impossible because investigators found it difficult to have the nucleus of an adult cell "start over," even when it was placed in another egg that had had its own nucleus removed (an enucleated egg cell).

In March 1997, Scottish investigators announced they had cloned a Dorset sheep, which they named Dolly. How was their procedure different from all the others that had been attempted? They began in the usual way, by placing an adult nucleus in an enucleated egg cell; however, the donor cells had been starved. Starving the donor cells caused them to stop dividing and go into a resting stage (the G_0 stage of the cell cycle). This was the change needed, because nuclei at the G_0 stage are open to cytoplasmic signals for the initiation of development (**Fig. 12.5**). Now it is common practice to clone farm animals that have desirable genetic traits, and even to clone rare animals that might otherwise become extinct.

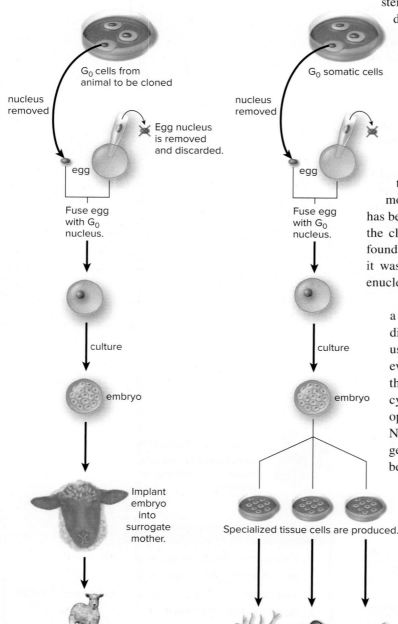

G_0 cells from animal to be cloned

nucleus removed

Egg nucleus is removed and discarded.

egg

Fuse egg with G_0 nucleus.

culture

embryo

Implant embryo into surrogate mother.

Clone is born.

a. Reproductive cloning

G_0 somatic cells

nucleus removed

egg

Fuse egg with G_0 nucleus.

culture

embryo

Specialized tissue cells are produced.

nervous   blood   muscle

b. Therapeutic cloning

Figure 12.5 Forms of cloning.

a. The purpose of reproductive cloning is to produce an individual that is genetically identical to the one that donated a diploid (2n) nucleus. The 2n nucleus is placed in an enucleated egg; after several divisions, the embryo comes to term in a surrogate mother. **b.** The purpose of therapeutic cloning is to produce specialized tissue cells. A 2n nucleus is placed in an enucleated egg; after several divisions, the embryonic cells (called embryonic stem cells) are separated and treated so that they become specialized cells.

Connections: Health

Have stem cells been used to cure human disease?

Stem cells have been used since the 1960s as a treatment for leukemia, a form of cancer in which the stem cells in bone marrow produce a large number of nonfunctional blood cells. In a bone marrow transplant, the patient's stem cells are destroyed using radiation or chemotherapy. Then, healthy stem cells from a donor are introduced into the patient's bone marrow. Stem cell therapy is available for a variety of human diseases, including type 2 diabetes, autoimmune diseases, heart conditions, and some forms of blindness.

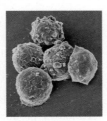
© Steve Gschmeissner/
SPL/Science Source

Currently in the United States, no federal funds can be used for experiments to clone humans. While there are advances in the science of cloning, many problems still exist. For example, in the case of Dolly, it took 247 tries before the process was successful. Also, there are concerns that cloned animals may not have the same life expectancy as noncloned animals. Some, but not all, cloned animals have demonstrated symptoms of abnormal aging. For example, Dolly was put down by lethal injection in 2003 because she was suffering from lung cancer and crippling arthritis. She had lived only half the normal life span for a Dorset sheep.

In **therapeutic cloning,** the desired end is not an individual organism but various types of mature cells. The purposes of therapeutic cloning are (1) to learn more about how cell specialization occurs and (2) to provide cells and tissues that could be used to treat human illnesses, such as diabetes, spinal cord injuries, and Parkinson disease.

Therapeutic cloning can be carried out in several ways. The most common method is to isolate **embryonic stem cells** and subject them to treatments that cause them to become particular cell types, such as red blood cells, muscle cells, or nerve cells (Fig. 12.5b). Because embryonic stem cells have the potential to become any other type of cell in an organism, they are said to be **totipotent.** Eventually, it may be possible to produce entire tissues and organs from totipotent stem cells. Ethical concerns exist about this type of therapeutic cloning because, if the embryo had been allowed to continue development, it would have become an individual.

Another way to carry out therapeutic cloning is to use **adult stem cells,** found in many organs of an adult's body. Adult stem cells are said to be **multipotent,** since they have already started to specialize and are not able to produce every type of cell in the organism. For example, the skin has stem cells that constantly divide and produce new skin cells, while the bone marrow has stem cells that produce new blood cells. Currently, adult stem cells are limited in the number of cell types that they may become. However, scientists have been able to coax adult stem cells from skin into becoming more like embryonic stem cells by adding only four genes. Researchers are investigating ways of controlling gene expression in adult stem cells, so that they can be used in place of the more controversial embryonic stem cells in fighting disease in humans.

Connections: Health

Are food products from cloned animals nutritious?

A comprehensive study by the Food and Drug Administration (FDA) in 2008 indicated that meat and milk from cloned animals, such as cows, goats, and pigs, are just as nutritious as products from noncloned animals. Many people confuse cloned with "transgenic" organisms—those that contain genes from multiple species. While scientific studies are still underway to assess the health aspects of transgenics, a cloned animal is no different from its parent. The meat and milk from cloned animals have the same nutritional content as those of noncloned animals.

CONNECTING THE CONCEPTS
12.2   Various types of stem cells are involved in the different types of cloning.

Check Your Progress 12.2

1. Describe the different goals of reproductive and therapeutic cloning.
2. Summarize the origins of both embryonic and adult stem cells.
3. Explain what is meant when a cell is described as totipotent or multipotent.

12.3 Biotechnology Products

Learning Outcomes

Upon completion of this section, you should be able to

1. Define the terms *genetically modified* and *transgenic*.
2. Summarize the uses of transgenic bacteria, plants, and animals.

Today, bacteria, plants, and animals are genetically engineered to produce biotechnology products. Recall that a *genetically modified organism* (GMO) is one whose genome has been modified in some way, usually by using recombinant DNA technology. Organisms that have had a foreign gene inserted into their genome are called *transgenic organisms.*

Genetically Modified Bacteria

Recombinant DNA technology is used to produce transgenic bacteria, grown in huge vats called bioreactors (**Fig. 12.6***a*). The gene product from the bacteria is collected from the medium. Such biotechnology products include insulin, clotting factor VIII, human growth hormone, tissue plasminogen activator (t-PA), and hepatitis B vaccine. Transgenic bacteria have many other uses as well. Some have been produced to promote the health of plants. For example, bacteria that normally live on plants and encourage the formation of ice crystals have been changed from frost-plus to frost-minus strains. As a result, new crops, such as frost-resistant strawberries and oranges, have been developed.

Bacteria can be selected for their ability to degrade a particular substance, and this ability can then be enhanced by genetic engineering. For instance, naturally occurring bacteria that eat oil have been genetically modified to do this even more efficiently and used in cleaning up beaches after oil spills (Fig. 12.6*b*). Further, these bacteria are given "suicide" genes, which cause them to self-destruct when the job has been accomplished.

Genetically Modified Plants

Corn, potato, soybean, and cotton plants have been engineered to be resistant to either insect predation or commonly used herbicides (**Fig. 12.7**). Some corn and cotton plants have been developed that are both insect- and herbicide-resistant. In 2015, 94% of the soybeans and 89% of the corn planted in the United States had been genetically engineered. If crops are resistant to a broad-spectrum herbicide and weeds are not, then the herbicide can be used to kill the weeds. When herbicide-resistant plants are planted, weeds are easily controlled, less tillage is needed, and soil erosion is minimized.

One of the main focuses of genetic engineering of plants has been the development of crops with improved qualities, especially improvements that reduce waste from food spoilage. For example, by knocking out a gene that causes browning in apples, a company called Okanagan Specialty Fruits produced the Arctic Apple, a genetically modified apple with an increased shelf life. Innate is a genetically modified potato in which a process called RNA interference turns off the expression of genes associated with bruising.

Figure 12.6 Uses of genetically modified bacteria.

a. Transgenic bacteria can be used to produce medicines.
b. Bacteria may be genetically modified to enhance their ability to clean up environmental spills.

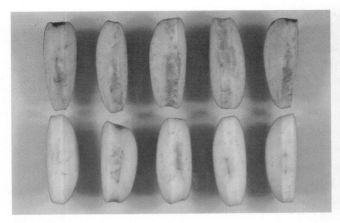

a.

b.

Figure 12.7 Uses of genetically modified plants.

Genetic modifications allow these (**a**) apples to resist browning and (**b**) corn to resist herbicides that control weeds.

(a): © Okanagan Specialty Fruits Inc.; (b): © AGStockUSA, Inc./Alamy

Progress has also been made to increase the food quality of crops. Soybeans have been developed that mainly produce the monounsaturated fatty acid oleic acid, a change that may improve human health. Other types of crop plants are genetically engineered to increase their productivity. Leaves can be engineered to lose less water and take in more carbon dioxide. This type of modification helps a range of crops grow successfully in various climates, including those that are more likely to experience drought or have a higher average temperature than the plant's normal growing climate. Other single-gene modifications allow plants to produce various products, including human hormones, clotting factors, antibodies, and vaccines.

Genetically Modified Animals

Biotechnology techniques have been developed to insert genes into the eggs of animals. For example, many types of animal eggs have taken up the gene for bovine growth hormone (BGH). The procedure has been used to produce larger fish, cows, pigs, rabbits, and sheep. Transgenic pigs supply many transplant organs for humans, a process called *xenotransplantation*.

Like plants, animals can be genetically modified to increase their value as food products. A new form of transgenic salmon (**Fig. 12.8**) contains genes from two other fish species; these genes produce a growth hormone that allows

a.

b.

Figure 12.8 Uses of genetically modified animals.

a. Salmon may be genetically modified to grow faster. **b.** Pest species, such as the mosquito *Aedes aegypti*, may be genetically modified to reduce population size.

(a): © AquaBounty Technologies; (b): Source: CDC/Frank Hadley Collins, Dir, Center for Global Health and Infectious Diseases, University of North Dakota

CONNECTING THE CONCEPTS

12.3 DNA technology can produce genetically modified and transgenic organisms for a variety of purposes.

Check Your Progress 12.3

1. Distinguish between transgenic and genetically modified organisms.

2. List some examples of the uses of genetically modified bacteria.

3. Describe how genetically modified plants and animals may help feed a growing population.

---

the salmon to grow quicker. Interestingly, these salmon are also engineered to be triploid females, which makes them sterile. The *Aedes aegypti* mosquito acts as a vector for several human diseases, including dengue fever and chikungunya. A transgenic form of the mosquito being released in Florida contains a genetic "kill switch," which produces proteins that kill the offspring, thus reducing the size of the population.

Gene pharming, the use of transgenic farm animals to produce pharmaceuticals, is being pursued by a number of firms. Genes that code for therapeutic and diagnostic proteins are incorporated into an animal's DNA, and the proteins appear in the animal's milk. It is possible to produce not only drugs but also vaccines by this method.

Transgenic mice are routinely used in medical research to study human diseases. For example, the allele that causes cystic fibrosis can be cloned and inserted into mouse embryonic stem cells. Occasionally, a mouse embryo homozygous for cystic fibrosis will result. This embryo develops into a mutant mouse that has a phenotype similar to that of a human with cystic fibrosis. New drugs for the treatment of cystic fibrosis can then be tested in such mice. A similar research animal is the OncoMouse, which carries genes for the development of cancer.

12.4 Genomics and Proteomics

Learning Outcomes

Upon completion of this section, you should be able to

1. Define the term *genomics*.
2. Describe how genome comparisons are used by researchers.
3. Explain the various uses of proteomics and bioinformatics.

In the twentieth century, researchers discovered the structure of DNA, how DNA replicates, and how protein synthesis occurs. Genetics in the twenty-first century is focusing on **genomics,** the study of all types of **genomes,** which consist of genes and **intergenic DNA.** Researchers now know the sequence of all the base pairs along the lengths of the human chromosomes. The enormity of this accomplishment can be appreciated by knowing that, at the very least, our DNA contains 3.2 billion base pairs and approximately 23,000 genes. Many organisms have even larger genomes. While genes that code for proteins make up only 3–5% of the human genome, we now know that the other regions are also important. Historically considered "junk" DNA, the regions between genes play an important role in generating the small RNA molecules that are involved in gene regulation. Many of these intergenic regions also have evolutionary significance and may provide us with useful hints on the evolution of our species.

Sequencing the Bases of the Human Genome

The Human Genome Project, completed in 2003, represented a 13-year effort to determine the sequence of the 3.2 billion base pairs in the human genome. The project involved both university and private laboratories around the world. It was made possible by the rapid development of innovative technologies to sequence segments of DNA. Over the project's 13-year span, DNA sequencers were constantly improving, and modern automated sequencers can analyze and sequence up to 120 million base pairs of DNA in a 24-hour period.

Variations in Base Sequence

Scientists studying the human genome found that many small regions of DNA vary among individuals. For instance, there may be a difference in a single base within a gene (**Fig. 12.9***a* and *b*) or within an intergenic sequence (Fig. 12.9*c*). One surprising finding is that some individuals even have additional copies of some genes (Fig. 12.9*d*)! Many of these differences have no ill effects, but some may increase or decrease an individual's susceptibility to disease.

Genome Comparisons

Researchers are comparing the human genome with the genomes of other species for clues to our evolutionary origins. One surprising discovery is that the genomes of all vertebrates are similar. Researchers were not surprised to find that the genes of chimpanzees and humans are over 90% alike, but they did not expect to find that the human genome is also an 85% match with that of a mouse. Scientists also discovered that we share a number of genes with much simpler organisms, including bacteria! As the genomes of more organisms are sequenced, genome comparisons will likely reveal never previously expected evolutionary relationships between organisms. Genome comparison studies are also providing insight into our evolutionary heritage. By comparing the genome of modern-day humans with those of some of our more recent ancestors, such as the Neandertals and Denisovans (see Section 19.6), we are developing a better understanding of how our species has evolved over time.

For example, several studies have compared the genes on chromosome 22 in humans and chimpanzees. Among the many genes that differed in sequence were several of particular interest: a gene for proper speech development, several for hearing, and several for smell. The gene necessary for proper speech development is thought to have played an important role in human evolution. Recent genomic studies have suggested that a change in intergenic sequences between 1 and 6 million years ago may have fundamentally changed the way the brain processes speech. Changes in genes affecting hearing may have

Figure 12.9 Variations in DNA sequence.

a. A section of DNA containing three genes (A, B, and C) and two intergenic sequences may vary between individuals. **b.** Mutation may introduce variation in the sequence of bases within a gene. This may alter the sequence of amino acids in the protein specified by the gene. **c.** Mutations in the intergenic regions may introduce variation that does not change the amino acid sequence of a protein. **d.** In addition to changes in the base sequence, DNA is also subject to changes in the number of copies of a gene. Note that in this example there are two copies of gene B.

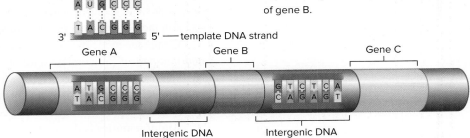

a. Normal chromosomal DNA

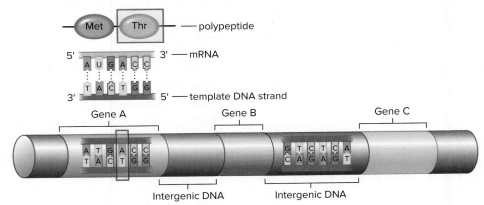

b. Variation in the order of the bases within a gene due to a mutation

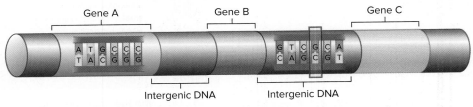

c. Variation in the order of the bases within an intergenic sequence due to a mutation

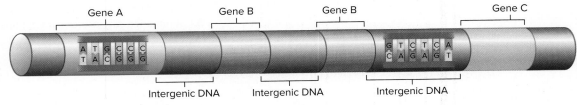

d. Variation in the gene copy number

facilitated the use of language for communication between people. Changes in smell genes are a little more problematic, but it is believed that the olfaction genes may have affected dietary changes or sexual selection. It was a surprise to find that many of the other genes that were located and studied are known to cause human diseases if abnormal. Perhaps comparing genomes is a way of finding genes associated with human diseases.

Proteomics and Bioinformatics

The known sequence of bases in the human genome predicts that the approximately 23,000 genes are translated into over 200,000 different proteins due to alternative mRNA splicing; collectively, all of the proteins produced by an organism's genome are referred to as a **proteome.** The field of **proteomics** explores the structure and function of these cellular proteins and examines how they interact to contribute to the production of traits. Because drugs tend to be proteins or small molecules that affect the behavior of proteins, proteomics is crucial in the development of new drugs for the treatment of disease.

Computer modeling of the three-dimensional shapes of proteins is an important part of proteomics, since structure relates to function. Researchers may then use the models to predict which molecules will be effective drugs. For example, a compound that fits the active site of an enzyme without being converted into a product may be an effective drug to inhibit that enzyme.

Bioinformatics is the application of computer technologies to the study of the genome and proteome. Genomics and proteomics produce raw data, and these fields depend on computer analysis to find significant patterns in the data. As a result of bioinformatics, scientists hope to find cause-and-effect relationships between an individual's overall genetic makeup and resulting genetic disorders.

More than half of the human genome consists of uncharacterized sequences that contain no genes and have no known function. Bioinformatics might be used to find the function of these regions by correlating any sequence changes with resulting phenotypes, or it might be used to discover that some sequences are an evolutionary relic that once coded for a protein that we no longer need. New computational tools will most likely be needed in order to accomplish these goals.

CONNECTING THE CONCEPTS

12.4

The application of genomics, proteomics, and bioinformatics enables scientists to understand evolutionary relationships.

Check Your Progress 12.4

1. Describe the goals of genomics research.
2. Explain why it is important to compare the genomes of different organisms.
3. Summarize why proteomics is important in medical research.

SMARTBOOK® Maximize your study time with McGraw-Hill SmartBook®, the first adaptive textbook.

SUMMARIZE

An understanding of DNA structure and function has allowed scientists to study biotechnology, or the use of technology to enhance or change the genetic composition of a species.

12.1 Biotechnology allows scientists to copy, edit and analyze the genes of organisms.

12.2 Various types of stem cells are involved in the different types of cloning.

12.3 DNA technology can produce genetically modified and transgenic organisms for a variety of purposes.

12.4 The application of genomics, proteomics and bioinformatics enables scientists to understand evolutionary relationships.

12.1 Biotechnology

Biotechnology is the use of natural biological systems to create a product or achieve some other end desired by humans. Biotechnology often involves genetic engineering, or changing the genetic structure of a species. The resulting organism is called a **genetically modified organism (GMO)**. If a gene from one species is inserted into the genome of another, the result is called a **transgenic organism.**

Recombinant DNA Technology

Recombinant DNA (rDNA) is created using **restriction enzymes** to cleave DNA, so that a foreign gene can be inserted. The foreign gene is carried by a **vector,** such as a plasmid. The recombinant DNA can then be inserted into a bacterium, which then expresses the foreign gene.

plasmid DNA

foreign gene

Polymerase Chain Reaction

The **polymerase chain reaction (PCR)** occurs in a test tube and is often used in conjunction with recombinant DNA technology. Heat separates double-stranded DNA. Primers flank the target DNA, and heat-insensitive DNA polymerase copies the target DNA. The cycle is repeated until millions of copies are produced.

DNA Analysis

DNA fingerprinting uses a process called **short tandem repeat (STR) profiling** and the polymerase chain reaction to analyze small variations in the copy number of repeat sequences in the human genome. DNA fingerprinting can help in forensic investigations and in the identification of remains from natural disasters.

Genome Editing

Genome editing is a form of DNA technology that targets specific sequences in a genome for inactivation or the insertion of nucleotides. CRISPR is one of the leading methods of genome editing.

12.2 Stem Cells and Cloning

The study of stem cells has provided insight into how gene expression results in the formation of an organism's tissues. **Embryonic stem cells** are **totipotent,** meaning that they have the potential to form any other type of cell in an organism. **Adult stem cells** are **multipotent,** meaning that they have undergone some level of specialization and are restricted in the types of cells that they can form.

Reproductive and Therapeutic Cloning

Cloning, or the production of genetically identical cells and organisms, is possible with an understanding of gene expression. In **reproductive cloning,** the desired end is an individual genetically identical to the original individual. To achieve this type of cloning, the early embryo is placed in a surrogate mother until it comes to term. In **therapeutic cloning,** the cells of the embryo are separated and treated, so that they develop into specialized tissues that can be used to treat human disorders. Alternately, adult stem cells can be treated to become specialized tissues.

12.3 Biotechnology Products

A genetically modified organism (GMO) is created when the genome of the organism has been modified in some way, usually by using recombinant DNA technology. An organism that has had a foreign gene inserted into its genome is called a transgenic organism.

12.4 Genomics and Proteomics

Genomics is the study of a **genome**—the sum of an organism's genes and **intergenic DNA.** The study of genomics often involves making genome comparisons. **Proteomics** is the study of the proteins produced by the genome of a species. Collectively, these proteins are called a **proteome. Bioinformatics** involves the application of computer technologies to analyzing the structure and function of the proteome and genome.

ASSESS

Testing Yourself

Choose the best answer for each question.

12.1 Biotechnology

1. Which of the following procedures makes multiple copies of a piece of DNA for use in recombinant DNA studies?
   a. use of restriction enzymes
   c. genome editing
   b. DNA sequencing
   d. polymerase chain reaction

2. Which enzyme has the ability to cut DNA at specific regions?
   a. DNA ligase
   c. DNA polymerase
   b. helicase
   d. restriction enzymes

3. The CRISPR method is used in
   a. DNA sequencing.
   c. genome editing.
   b. DNA fingerprinting.
   d. the polymerase chain reaction.

12.2 Stem Cells and Cloning

4. Reproductive cloning differs from therapeutic cloning in that the goal of reproductive cloning is
   a. the production of specific cell types for medical purposes.
   b. an individual that is genetically identical to the original individual.
   c. an individual that is genetically different from the original individual.
   d. the production of specific tissues and organs for medical transplantation.

5. The major challenge in therapeutic cloning using adult stem cells is
   a. finding appropriate cell types.
   c. controlling gene expression.
   b. obtaining enough tissue.
   d. keeping cells alive in a culture.

6. Adult stem cells are _____; embryonic stem cells are _____.
   a. totipotent, totipotent
   c. multipotent, totipotent
   b. totipotent, multipotent
   d. multipotent, multipotent

12.3 Biotechnology Products

7. A researcher inserts a gene from a drought-resistant species of plant into another species of plant. This procedure is an example of
   a. a vector.
   c. bioinformatics.
   b. polymerase chain reaction.
   d. transgenics.

8. The use of genetic engineering to silence or enhance a gene within an organism, without introducing genes from another species, produces a
   a. transgenic species.
   c. vector.
   b. genetically modified organism.
   d. clone.

12.4 Genomics and Proteomics

9. The application of computer techniques to analyze a genome is called
   a. biotechnology.
   c. bioinformatics.
   b. biogenomics.
   d. bioengineering.

10. The sum of all the proteins produced by an organism's genes is called the
    a. genome.
    c. proteome.
    b. intergenic DNA.
    d. chromosomes.

ENGAGE

BioNOW

Want to know how this science is relevant to your life? Check out the BioNow video below:

- Glowing Fish Genetics

Are the glowing fish examples of transgenic or genetically modified organisms? What DNA technologies might have been used to produce these fish?

Thinking Critically

1. Transgenic viruses that infect humans can be used to help treat genetic disorders. The viruses are genetically modified to contain "normal" human genes that will replace nonfunctional human genes. Using this type of gene therapy, a person is infected with a particular virus, which then delivers the "normal" human gene to cells by infecting them. What are some pros and cons of this viral gene therapy?

2. What are the advantages of using a recombinant DNA product from a human source instead of a product isolated from another organism? (For example, what advantage would there be to using human recombinant insulin versus insulin produced from cows or pigs?)

3. In a genomic comparison between humans and yeast, what genes would you expect to be similar?

© Belinda Images/Superstock

Mutations and Genetic Testing

Trisomy 21: Down Syndrome

Down syndrome is a genetic disorder that affects over 400,000 Americans. Down syndrome occurs when an individual inherits an extra copy of chromosome 21, causing what is called a trisomy. Trisomy 21 is the most common chromosomal anomaly in humans, and the incidence of the condition is slowly rising in Western cultures. From 1983 to 2003, the total number of cases of Down syndrome increased by 24.2%. People with Down syndrome have similar characteristics, including an upward slant to the eyes, low muscle tone, a deep crease across the center of the palm, and an increased risk of cardiac defects. There is some degree of intellectual disability, but most individuals with Down syndrome have IQs in the mild to moderate disability range.

Despite these health concerns, the life expectancy of Down syndrome individuals has increased tremendously in recent years, from 25 years in 1983 to over 60 years today. This increase occurred primarily because scientists identified not only how the extra copy of the chromosome ends up in the cell, but also the specific gene that, when present in extra copies, is responsible for the symptoms of Down syndrome.

In this chapter, we will not only explore the causes of mutations, but also investigate how these mutations are detected.

As you read through the chapter, think about the following questions:

1. What causes a change in the number of chromosomes?
2. What is the genetic basis of the symptoms of Down syndrome?
3. How can the chromosome number of an individual be tested before he or she is born?

OUTLINE

13.1 Gene Mutations   222

13.2 Chromosomal Mutations   224

13.3 Genetic Testing   226

13.4 Gene Therapy   232

BEFORE YOU BEGIN

Before beginning this chapter, take a few moments to review the following discussions.

Section 9.4 How does nondisjunction cause a change in chromosome number?

Section 10.1 How is probability used in analyzing patterns of inheritance?

Section 12.1 How are restriction enzymes and the polymerase chain reaction used in the study of DNA?

13.1 Gene Mutations

Learning Outcomes

Upon completion of this section, you should be able to

1. Explain how mutations cause variation in genes.
2. Describe how transposons may cause mutation.
3. Distinguish between point and frameshift mutations.

A **gene** is a sequence of DNA bases that codes for a cellular product, most often a protein (see Section 11.2). The information within a gene may vary slightly—these variations are called **alleles.** In our discussion of patterns in inheritance (see Section 10.1), we examined a few examples of alleles in traits (for instance, those associated with flower color or wing shape). However, many alleles code for variations in proteins that are not easily observed, such as susceptibility to a medication.

How are new alleles formed? A new allele is the result of a **mutation,** or a change in the nucleotide sequence of DNA. Mutations may have negative consequences, as is the case with the alleles for cystic fibrosis (see Section 10.2) However, mutations can increase the diversity of organisms by creating an entirely new gene product with a positive function for the organism. In fact, mutations play an important role in the process of evolution (see Section 15.1)

Causes of Gene Mutations

A gene mutation can be caused by a number of factors, including errors in DNA replication, a transposon, or an environmental mutagen. Mutations due to DNA replication errors are rare: They occur with a frequency of 1 in 100 million cell divisions on average in most eukaryotes. This average is low due to DNA polymerase, the enzyme that carries out replication (see Section 11.1) and proofreads the new strand against the old strand, detecting and correcting any mismatched pairs.

Mutagens are environmental influences that cause mutations. Different forms of radiation, such as radioactivity, X-rays, ultraviolet (UV) light, and chemical mutagens, such as pesticides and compounds in cigarette smoke, may cause breaks or chemical changes in DNA. If mutagens bring about a mutation in the DNA in an individual's gametes, the offspring of that individual may be affected. If the mutation occurs in the individual's body cells, cancer may result. The overall rate of mutation is low, however, because DNA repair enzymes constantly monitor for any irregularity and remedy the problem.

Transposons are specific DNA sequences that have the remarkable ability to move within and between chromosomes (**Fig. 13.1**). Their movement often disrupts genes, rendering them nonfunctional. These "jumping genes" have now been discovered in almost every group of organisms, including bacteria, plants, fruit flies, and humans.

Figure 13.1 Transposons may cause gene mutations.

a. A purple-coding gene ordinarily codes for a purple pigment. **b.** A transposon "jumps" into the purple-coding gene. This mutated gene is unable to code for purple pigment, and a white kernel results. **c.** Indian corn displays a variety of colors and patterns due to transposon activity.

(c): © Mondae Leigh Baker

Normal gene

codes for purple pigment

a.

Mutated gene

cannot code for purple pigment

transposon

b.

c.

Types and Effects of Mutations

The effects of mutations vary greatly. The severity of a mutation usually depends on whether it affects one or more codons on a gene. In general, we know a mutation has occurred when the organism has a malfunctioning protein that leads to a genetic disorder or to the development of cancer. But many mutations go undetected because they have no observable effect or have no detectable effect on a protein's function. These are called silent mutations (**Fig. 13.2**).

Point mutations involve a change in a single DNA nucleotide, and the severity of the results depends on the particular base change that occurs. A single base change can result in a change in the amino acid at that location of the gene. For example, in hemoglobin, the oxygen-transporting molecule in blood, a point mutation causes the amino acid glutamic acid to be replaced by a valine (Fig. 13.2). This changes the structure of the hemoglobin protein. The abnormal hemoglobin stacks up inside the red blood cells, causing them to become sickle-shaped, resulting in sickle-cell disease.

A **frameshift mutation** is caused by an extra or missing nucleotide in a DNA sequence. Frameshift mutations are usually much more severe than point mutations, because codons are read from a specific starting point. Therefore, all downstream codons are affected by the addition or deletion of a nucleotide. For instance, if the letter *C* is deleted from the sentence THE CAT ATE THE RAT, the "reading frame" is shifted. The sentence now reads THE ATA TET HER AT—which doesn't make sense. Likewise, a frameshift mutation in a gene often renders the protein nonfunctional, because its amino acid sequence no longer makes sense (Fig. 13.2). The movement of a transposon can cause a frameshift mutation.

Check Your Progress 13.1

1. Explain how mutations are related to alleles.
2. Summarize the role of transposons in causing mutations.
3. Explain why point mutations may be silent, but frameshift mutations rarely are.

CONNECTING THE CONCEPTS

13.1 Mutagens, replication errors, and transposons can produce mutations.

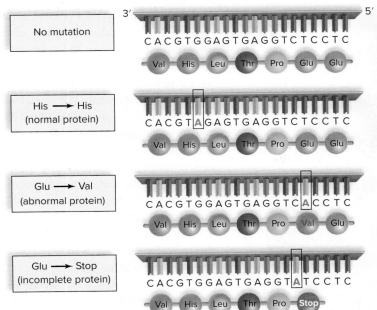

a.

b. Normal red blood cell   7,400×   c. Sickled red blood cell   7,400×

Figure 13.2 Varying effects of a point mutation.

a. Starting at the top: a normal sequence of bases in hemoglobin; next, the mutation is silent and has no effect; next, due to point mutation, the DNA now codes for valine instead of glutamic acid, and the result is that **(b)** normal red blood cells become **(c)** sickle-shaped; next, a frameshift mutation (notice the missing C) causes the DNA to code for termination, and the protein will be incomplete.

(b and c): © Eye of Science/Science Source

13.2 Chromosomal Mutations

Learning Outcomes

Upon completion of this section, you should be able to

1. Distinguish between a chromosomal deletion and duplication.
2. Explain the consequences of a translocation or an inversion.
3. List examples of the effects of chromosomal mutations in humans.

We have just reviewed the consequences of mutation at the level of the gene, but it is also possible for events, such as nondisjunction and mutation, to cause changes in the number of each chromosome in the cells or the structure of individual chromosomes.

In humans, only a few variations in chromosome number, such as Down syndrome, Turner syndrome, and Klinefelter syndrome, are typically seen. These are usually caused by nondisjunction events during meiosis (see Section 9.4). Changes in chromosome structure, or chromosomal mutations, are much more common in the population. Syndromes that result from changes in chromosome structure are due to the breakage of chromosomes and their failure to reunite properly. Various environmental agents—radiation, certain organic chemicals, and even viruses—can cause chromosomes to break apart. Ordinarily, when breaks occur in chromosomes, the segments reunite to give the same sequence of genes. But their failure to do so results in one of several types of mutations: deletion, duplication, translocation, or inversion. Chromosomal mutations can occur during meiosis, and if the offspring inherits the abnormal chromosome, a syndrome may result.

Deletions and Duplications

A **deletion** occurs when a single break causes a chromosome to lose an end piece or when two simultaneous breaks lead to the loss of an internal segment of a chromosome. An individual who inherits a normal chromosome from one parent and a chromosome with a deletion from the other parent no longer has a pair of alleles for each trait, and a syndrome can result.

Williams syndrome occurs when chromosome 7 loses a tiny end piece (**Fig. 13.3**). Children with this syndrome have a turned-up nose, a wide mouth, a small chin, and large ears. Although their academic skills are poor, they exhibit excellent verbal and musical abilities. The gene that governs the production of the protein elastin is missing, and this affects the health of the cardiovascular system and causes their skin to age prematurely. Such individuals are very friendly but need an ordered life, perhaps because of the loss of a gene for a protein that is normally active in the brain.

Cri du chat (cat's cry) syndrome occurs when chromosome 5 is missing an end piece. The affected individual

Figure 13.3 Deletion.

a. When chromosome 7 loses an end piece, the result is Williams syndrome. **b.** These children, although unrelated, all experience the appearance, health, and behavioral problems that are characteristic of Williams syndrome.

(b): © The Williams Syndrome Association

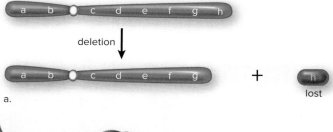

deletion

lost

a.

b.

has a small head, mental disabilities, and facial abnormalities. Abnormal development of the glottis and larynx results in the most characteristic symptom—the infant's cry resembles that of a cat.

In a **duplication,** a chromosome segment is repeated, so the individual has more than two alleles for certain traits. An inverted duplication is known to occur in chromosome 15. The term inverted indicates that a segment runs in the direction opposite from normal. Children with this syndrome, called *inv dup 15 syndrome,* have poor muscle tone, mental disabilities, seizures, a curved spine, and autistic characteristics that include poor speech, hand flapping, and lack of eye contact (**Fig. 13.4**).

Translocation

A **translocation** is the movement of a segment from one chromosome to another, nonhomologous chromosome or the exchange of segments between nonhomologous chromosomes (**Fig. 13.5***a*). A person who has both of the involved chromosomes has the normal amount of genetic material and a normal phenotype, unless the translocation breaks an allele into two pieces or fuses two genes together. The person who inherits only one of the translocated chromosomes will have only one copy of certain alleles and three copies of other alleles. A genetic counselor begins to suspect a translocation has occurred when spontaneous abortions are commonplace and family members suffer from various syndromes. A special microscopic technique allows a technician to determine that a translocation has occurred.

In 5% of Down syndrome cases, a translocation that occurred in a previous generation between chromosomes 21 and 14 is the cause. As long as the two chromosomes are inherited together, the individual is normal. But in future generations a person may inherit two normal copies of chromosome 21 and the abnormal chromosome 14 that contains a segment of chromosome 21. In these cases, Down syndrome is not related to parental age but instead tends to run in the family of either the father or the mother.

Some forms of cancer are also associated with translocations. One example is called chronic myeloid leukemia (CML). This translocation was first discovered in the 1970s when new staining techniques revealed a translocation of a portion of chromosome 22 to chromosome 9. This translocated chromosome is commonly called a Philadelphia chromosome. Individuals with

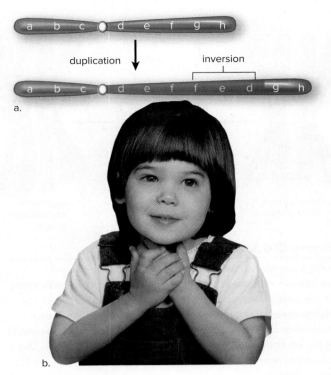

a.

b.

Figure 13.4 **Duplication.**

a. When a piece of chromosome 15 is duplicated and inverted, inv dup 15 syndrome results. **b.** Children with this syndrome have poor muscle tone and autistic characteristics.

(b): © Kathy Wise

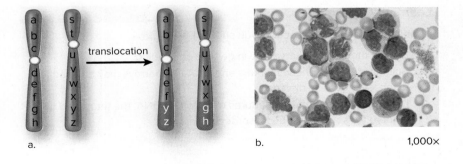

a.

b.   1,000×

Figure 13.5 **Translocation.**

a. Translocations represent the movement of genetic material between nonhomologous chromosomes. **b.** One example occurs between chromosomes 22 and 9, resulting in chronic myeloid leukemia (CML). The pink cells in this micrograph are rapidly dividing white blood cells.

(b): © Jean Secchi/Dominique Lecaque/Roussel-Uclaf/CNRI/Science Source

Connections: Health

Do we know what genes cause Down syndrome?

Because individuals who inherit Down syndrome by a translocation event receive only a portion of chromosome 21, it has been possible to determine which  genes on chromosome 21 are contributing to the symptoms of Down syndrome. One of the more significant genes appears to be *GART*, a gene involved in the processing of a type of nucleotide called a purine. Extra copies of *GART* are believed to be a major contributing factor in the symptoms of intellectual disability shown in Down syndrome patients. Another gene that has been identified (*COL6A1*) encodes the protein collagen, a component of connective tissue. Too many copies of this gene causes the heart defects common in Down syndrome.

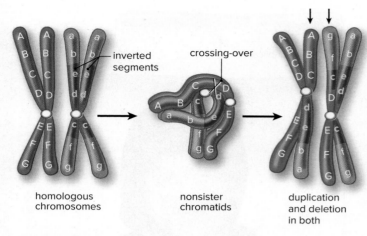

Figure 13.6 labels: inverted segments, crossing-over, homologous chromosomes, nonsister chromatids, duplication and deletion in both

Figure 13.6 Inversion.

(*Left*) A segment is inverted in the sister chromatids of one homologue. Notice that, in the red chromosome, *edc* occurs instead of *cde*. (*Middle*) The nonsister chromatids can align only when the inverted sequence forms an internal loop. After crossing-over, a duplication and a deletion can occur. (*Right*) The nonsister chromatid at the left arrow has *AB* and *ab* alleles but not *fg* or *FG* alleles. The nonsister chromatid at the right arrow has *gf* and *GF* alleles but not *AB* or *ab* alleles.

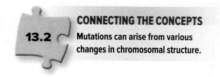

CONNECTING THE CONCEPTS

13.2 Mutations can arise from various changes in chromosomal structure.

CML have a rapid growth of white blood cells (Fig. 13.5*b*), which often prevents the ability of the body to form red blood cells and reduces the effectiveness of the immune system.

Another example is Alagille syndrome, caused by a translocation between chromosomes 2 and 20. This translocation also often produces a deletion on chromosome 20. The syndrome may also produce abnormalities of the eyes and internal organs. The symptoms of Alagille syndrome range from mild to severe, so some people may not be aware they have the syndrome. In Burkitt's lymphoma, translocations between chromosome 8 and either chromosomes 2, 14, or 22 disrupt a gene that is associated with regulating the cell cycle, resulting in the formation of very fast-growing tumors.

Inversion

An **inversion** occurs when a segment of a chromosome is turned 180° (**Fig. 13.6**). You might think this is not a problem because the same genes are present, but the reversed sequence of alleles can lead to altered gene activity if it disrupts the control of gene expression.

Inversions usually do not cause problems, but they can lead to an increased occurrence of abnormal chromosomes during sexual reproduction. Crossing-over between an inverted chromosome and the noninverted homologue can lead to recombinant chromosomes that have both duplicated and deleted segments. This happens because alignment between the two homologues is only possible when the inverted chromosome forms a loop (Fig. 13.6).

Check Your Progress 13.2

1. Explain why a deletion can potentially have a greater effect on an organism than a duplication.

2. Distinguish between a translocation and an inversion.

3. Describe the types of chromosomal mutations that produce cri du chat and Alagille syndromes.

13.3 Genetic Testing

Learning Outcomes

Upon completion of this section, you should be able to

1. Explain the role of a karyotype in genetic counseling.
2. Summarize how genetic markers and DNA microarrays may be used to diagnose a genetic disorder.
3. Distinguish among the procedures used to test DNA, the fetus, and the embryo for specific genetic disorders.

Potential parents are becoming aware that many illnesses are caused by abnormal chromosomal inheritance or by gene mutations. Therefore, more couples are seeking **genetic counseling,** which helps determine the risk of inherited disorders in a family. For example, a couple might be prompted to seek counseling after several miscarriages, when several relatives have a particular medical condition, or if they already have a child with a genetic defect. The counselor helps the couple understand the mode of inheritance, the medical consequences of a particular genetic disorder, and the decisions they might wish to make.

Various human disorders may result from abnormal chromosome number or structure. When a pregnant woman is concerned that her unborn child might have a chromosomal defect, the counselor may recommend karyotyping the fetus's chromosomes.

Analyzing the Chromosomes

A **karyotype** is a visual display of pairs of chromosomes arranged by size, shape, and banding pattern. Any cell in the body except red blood cells, which lack a nucleus, can be a source of chromosomes for karyotyping. In adults, it is easiest to use white blood cells separated from a blood sample for this purpose. In fetuses, whose chromosomes are often examined to detect a syndrome, cells can be obtained by either amniocentesis or chorionic villus sampling. For example, the karyotype of a person who has Down syndrome usually has three copies of chromosome number 21 instead of two (**Fig. 13.7**).

Amniocentesis is a procedure for obtaining a sample of amniotic fluid from the uterus of a pregnant woman. A long needle is passed through the abdominal and uterine walls to withdraw a small amount of fluid, which also contains fetal cells (**Fig. 13.8***a*). Tests are done on the amniotic fluid, and the cells are cultured for karyotyping. Karyotyping the chromosomes may be delayed as long as 4 weeks, so that the cells can be cultured to increase their number.

Blood tests and the age of the mother are considered when determining whether the procedure should be done. There is a slight risk of spontaneous abortion (about 0.6%) due to amniocentesis, with the greatest risk occurring in the first 15 weeks of pregnancy.

Chorionic villus sampling (CVS) is a procedure for obtaining chorionic villi cells in the region where the placenta will develop. This procedure can be done as early as the fifth week of pregnancy. A long, thin suction tube is inserted through the vagina into the uterus (Fig. 13.8*b*). Ultrasound, which gives a picture of the uterine contents, is used to place the tube between the uterine lining and the chorionic villi. Then, a sampling of the chorionic villi cells is obtained by suction. The cells do not have to be cultured, and karyotyping can be done immediately. But testing amniotic fluid is not possible, because no amniotic fluid is collected. Also, CVS carries a greater risk of spontaneous abortion than amniocentesis—0.7% compared with 0.6%. The advantage of CVS is getting the results of karyotyping at an earlier date.

After a cell sample has been obtained, the cells are stimulated to divide in a culture medium. A chemical is used to stop mitosis during metaphase when chromosomes are the most highly compacted and condensed. The cells are then killed, spread on a microscope slide, and dried. In a traditional karyotype, stains are applied to the slides, and the cells are photographed. Staining causes the chromosome to have dark and light cross-bands of varying widths, and these can be used, in addition to size and shape, to help pair up the chromosomes. Today, technicians use fluorescent dyes and computers to arrange the chromosomes in pairs.

Following genetic testing, a genetic counselor can explain to prospective parents the chances that a child of theirs will have a disorder that runs in the family. If a woman is already pregnant, the parents may want to know whether the unborn child has the disorder. If the woman is not pregnant, the parents may opt for testing of the embryo or egg before she does become pregnant, as described shortly.

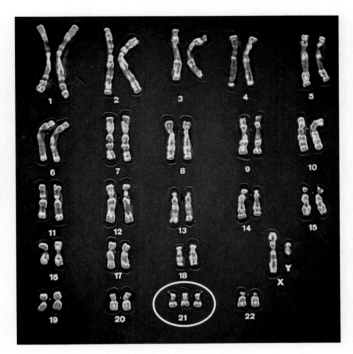

Figure 13.7 **Karyotype analysis.**

A karyotype can reveal chromosomal mutations. In this case, the karyotype shows that the newborn will have Down syndrome (trisomy 21).

© CNRI/SPL/Science Source

Connections: Scientific Inquiry

What do the colored bands on chromosomes in a karyotype represent?

The colored bands in a karyotype (Fig. 13.7) are not the natural color of chromosomes. Instead, a procedure called *fluorescent immunohistochemistry in situ hybridization* (*FISH*) is used to label short pieces of DNA with a fluorescent marker. These pieces of DNA bind to their complementary sequences on the chromosomes. When exposed to specific wavelengths of light, the markers emit different colors of light. This helps genetic counselors more easily identify the chromosomes. A tagged piece of DNA can also be developed for a specific gene, which helps researchers identify the chromosomal location of a gene of interest.

Figure 13.8 Testing for chromosomal mutations.

To test a fetus for an alteration in the chromosome number or structure, fetal cells can be acquired by (**a**) amniocentesis or (**b**) chorionic villus sampling.

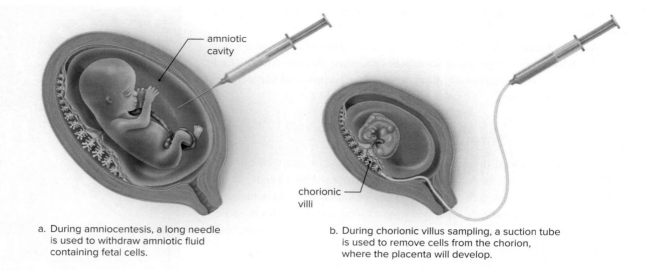

amniotic cavity

chorionic villi

a. During amniocentesis, a long needle is used to withdraw amniotic fluid containing fetal cells.

b. During chorionic villus sampling, a suction tube is used to remove cells from the chorion, where the placenta will develop.

Testing depends on the genetic disorder of interest. In some instances, it is appropriate to test for a particular protein, and in others, to test for the mutated gene.

Testing for a Protein

Some genetic mutations lead to disorders caused by a lack of enzyme activity. For example, in the case of methemoglobinemia, it is possible to test for the quantity of the enzyme diaphorase in a blood sample and, from that, determine whether the individual is likely homozygous normal, is a carrier, or has methemoglobinemia. If the parents are carriers, each child has a 25% chance of having methemoglobinemia. This knowledge may lead prospective parents to opt for testing of the embryo or egg, as described later in this section.

Testing the DNA

There are several methods of analyzing DNA for specific mutations, including testing for a genetic marker, using a DNA microarray, and direct sequencing of an individual's DNA.

Genetic Markers

Testing for a genetic marker relies on a difference in the DNA due to the presence of the abnormal allele. As an example, consider that individuals with sickle-cell disease or Huntington disease have an abnormality in a gene's base sequence. This abnormality in sequence is a **genetic marker.** The presence of specific genetic markers can be detected using DNA sequencing. Another option is to use restriction enzymes to cleave DNA at particular base sequences (see Section 12.1). The fragments that result from the use of a restriction enzyme may be different for people who are normal than for those who are heterozygous or homozygous for a mutation (**Fig. 13.9**).

DNA Microarrays

New technologies have made DNA testing easy and inexpensive. For example, it is now possible to place thousands of known disease-associated mutant alleles onto a DNA microarray, also called a gene chip—a small silicon chip containing many DNA samples, in this case, the mutant alleles (**Fig. 13.10**). Genomic DNA

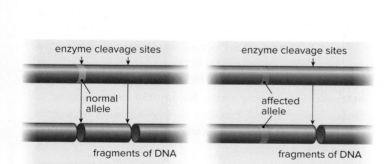

enzyme cleavage sites

normal allele

fragments of DNA

enzyme cleavage sites

affected allele

fragments of DNA

a. Normal fragmentation pattern

b. Genetic disorder fragmentation pattern

Figure 13.9 Use of a genetic marker to test for a genetic mutation.

a. In this example, DNA from a normal individual has certain restriction enzyme cleavage sites. **b.** DNA from another individual lacks one of the cleavage sites, and this loss indicates that the person has a mutated gene. In heterozygotes, half of their DNA would have the cleavage site and half would not have it. (In other instances, gaining a cleavage site could be an indication of a mutation.)

from the subject to be tested is labeled with a fluorescent dye, then added to the microarray. The spots on the microarray fluoresce if the DNA binds to the mutant alleles on the chip, indicating that the subject may have a particular disorder or is at risk of developing it later in life. An individual's complete genotype, including all the various mutations, is called a genetic profile.

With the help of a genetic counselor, individuals can be educated about their genetic profile. It's possible that a person has or will have a genetic disorder caused by a single pair of alleles. However, polygenic traits are more common, and in these instances, a genetic profile can indicate an increased or decreased risk for a disorder. Risk information can be used to design a program of medical surveillance and to foster a lifestyle aimed at reducing the risk. For example, suppose an individual has mutations common to people with colon cancer. It will be helpful for him or her to have an annual colonoscopy, so that any abnormal growths can be detected and removed before they became invasive.

DNA Sequencing

Recent advances in the processes of DNA sequencing (see Section 12.1) have made it much more feasible economically to sequence the genome of an individual to detect specific mutations associated with a disease. Whereas a few years ago the cost of sequencing an individual's genome could be as high as $100,000, that cost has decreased to almost $1,000. This decrease has ushered in an era of personal genomics, sometimes also referred to as personalized medicine.

There are several approaches to personal genomics. While it is possible to sequence the entire genome of an individual, it is often more practical to target specific genes and look for alleles that are known to increase the risk associated with a specific disease. This approach is called a *genome-wide association study,* and it has become more prevalent in the field of personalized medicine due to the increase in large genomic population studies.

One of the more interesting possibilities arising from personal genomics is pharmacogenomics, or the selection of a drug based on information coming directly from an individual's genome. In many cases, such as cancer, certain alleles associated with a gene will respond more effectively to a specific class of drugs. Thus, knowledge of the allelic combination of a patient can prove to be very beneficial to the physician.

Testing the Fetus

If a woman is already pregnant, ultrasound can detect serious fetal abnormalities, and it is possible to obtain and test the DNA of fetal cells for genetic defects.

Ultrasound

Ultrasound images help doctors evaluate fetal anatomy. An ultrasound probe scans the mother's abdomen, and a transducer transmits high-frequency sound waves, which are transformed into a picture on a video screen. This picture shows the fetus inside the uterus. Ultrasound can be used to determine a fetus's age and size, as well as the presence of more than one fetus. Also, some chromosomal abnormalities, such as Down syndrome, Edwards syndrome (three copies of chromosome 18), and Patau syndrome (three copies of chromosome 13), cause anatomical abnormalities during fetal development that may be detected by ultrasound by the twentieth week of pregnancy. For this reason, a routine ultrasound at this time is considered an essential part of prenatal care.

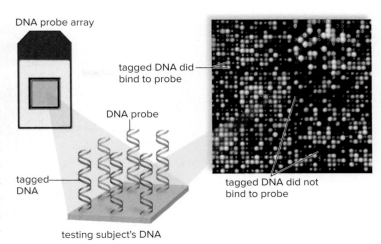

DNA probe array

tagged DNA did bind to probe

DNA probe

tagged DNA

tagged DNA did not bind to probe

testing subject's DNA

Figure 13.10 **Use of a DNA microarray to test for mutated genes.**

This DNA microarray contains many disease-associated mutant alleles. Fluorescently labeled genomic DNA from an individual has been added to the microarray. Any fluorescent spots indicate that binding has occurred and that the individual may have the genetic disorder or is at risk for developing it later in life.

(photo): © Deco/Alamy

Connections: Health

Are over-the-counter (OTC) genetic tests accurate?

© Images by Morgana/ Alamy RF

With advances in technology that allow DNA microarrays to be assembled inexpensively, a number of companies are now offering OTC genetic tests. The Food and Drug Administration (FDA) does not regulate OTC genetic tests, and there are a number of concerns regarding the validity of the information obtained from these tests. In addition, some people worry that the test results may be used to discriminate against individuals applying for insurance or jobs. It is recommended that genetic tests be performed by licensed labs, and only after all of the patient's rights have been discussed.

Many other conditions, such as spina bifida, can be diagnosed by an ultrasound. Spina bifida results when the backbone fails to close properly around the spinal cord during the first month of pregnancy. Surgery to close a newborn's spine in such a case is generally performed within 24 hours after birth.

Testing Fetal Cells

Fetal cells can be tested for various genetic disorders. If the fetus has an incurable disorder, the parents may wish to consider an abortion.

For testing purposes, fetal cells may be acquired through amniocentesis or chorionic villus sampling, as described earlier in this section. In addition, fetal cells may be collected from the mother's blood. As early as 9 weeks into the pregnancy, a small number of fetal cells can be isolated from the mother's blood using a cell sorter. Whereas mature red blood cells lack a nucleus, immature red blood cells do have a nucleus, and they have a shorter life span than mature red blood cells. Therefore, if nucleated fetal red blood cells are collected from the mother's blood, they are known to be from this pregnancy.

Only about one of every 70,000 blood cells in a mother's blood are fetal cells, and therefore the polymerase chain reaction (PCR) is used to amplify the DNA from the few cells collected. The procedure poses no risk to the fetus.

Testing the Embryo and Egg

As discussed in Section 29.2, in vitro fertilization (IVF) is carried out in laboratory glassware. A physician obtains eggs from the prospective mother and sperm from the prospective father and places them in the same receptacle, where fertilization occurs. Following IVF, now a routine procedure, it is possible to test the embryo. Prior to IVF, it is possible to test the egg for any genetic defect. In any case, only normal embryos are transferred to the uterus for further development.

Testing the Embryo

If prospective parents are carriers for one of the genetic disorders discussed earlier, they may want assurance that their offspring will be free of the disorder. Genetic diagnosis of the embryo will provide this assurance.

Following IVF, the zygote (fertilized egg) divides. When the embryo has six to eight cells, one of these cells can be removed for diagnosis, with no effect on normal development (**Fig. 13.11**). Only embryos that test negative for the genetic disorders of interest are placed in the uterus to continue developing.

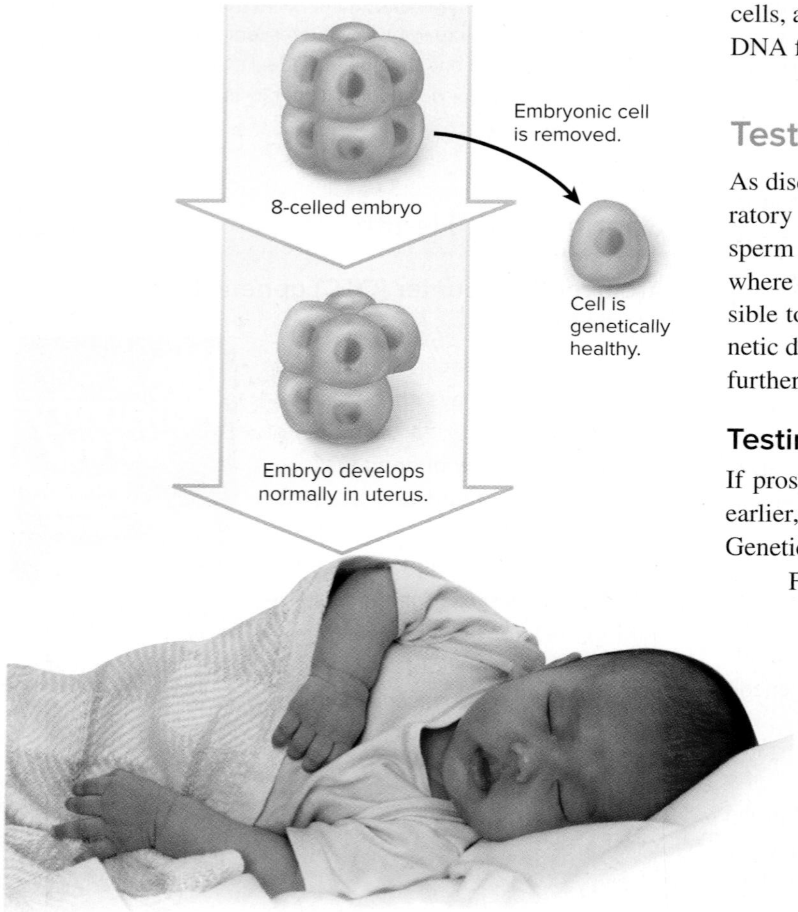

So far, thousands of children worldwide have been born free of alleles for genetic disorders that run in their families following embryo testing. In the future, embryos that test positive for a disorder could be treated by gene therapy, so that those embryos, too, would be allowed to continue to term.

Testing the Egg

Unlike males, who produce four sperm cells following meiosis, meiosis in females results in the formation of a single egg and at least two nonfunctional cells called polar bodies. Polar bodies, which later disintegrate, receive very little cytoplasm, but they do receive a haploid number of chromosomes and thus can be useful for genetic diagnosis. When a woman is heterozygous for a recessive genetic disorder, about half the polar bodies receive the mutated

8-celled embryo

Embryonic cell is removed.

Cell is genetically healthy.

Embryo develops normally in uterus.

Figure 13.11 **Testing the embryo.**

Genetic diagnosis is performed on one cell removed from an eight-celled embryo. If this cell is found to be free of the genetic defect of concern, and the seven-celled embryo is implanted in the uterus, it develops into a newborn with a normal phenotype.

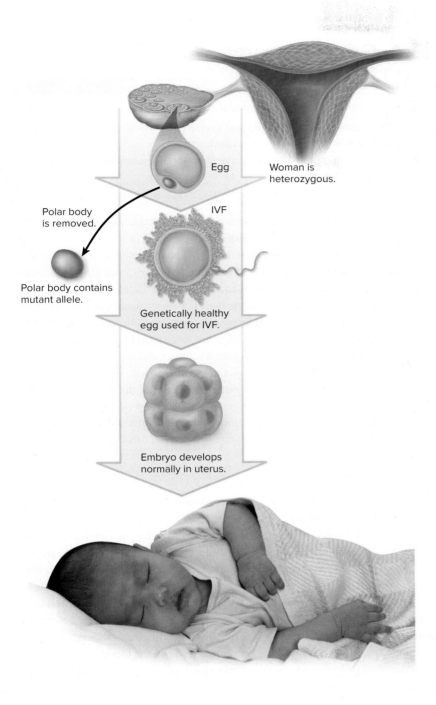

© Elyse Lewin/Exactostock-1555/Superstock RF

Figure 13.12 **Testing the egg.**

Genetic diagnosis is performed on a polar body removed from an egg. If the egg is free of a genetic defect, it is used for IVF, and the embryo is transferred to the uterus for further development.

Egg

Woman is heterozygous.

Polar body is removed.

IVF

Polar body contains mutant allele.

Genetically healthy egg used for IVF.

Embryo develops normally in uterus.

CONNECTING THE CONCEPTS

13.3   A variety of methods are available to analyze an individual for the presence of specific mutations.

Check Your Progress 13.3

1. Summarize how a karyotype can indicate potential problems in a fetus.

2. Explain the differences between testing for a protein and testing the DNA for genetic disorders.

3. Describe the differences between ultrasound, amniocentesis, and chorionic villus sampling (CVS). List the limitations of each.

4. Predict how changes in base sequence can be used to test for genetic disorders.

allele, and in these instances the egg receives the normal allele. Therefore, if a polar body tests positive for a mutated allele, the egg probably received the normal allele (**Fig. 13.12**). Only normal eggs are then used for IVF. Even if the sperm should happen to carry the mutation, the zygote will, at worst, be heterozygous. But the phenotype will appear normal.

If gene therapy becomes routine in the future, it's possible that an egg will be given genes that control traits desired by the parents, such as musical or athletic ability, prior to IVF. Such genetic manipulation, called *eugenics,* carries many ethical concerns.

13.4 Gene Therapy

Gene therapy is the insertion of genetic material into human cells for the treatment of a disorder. It includes procedures that give a patient healthy genes to make up for faulty genes, as well as the use of genes to treat various other human illnesses, such as cardiovascular disease and cancer. Gene therapy can be **ex vivo** (outside the body) or **in vivo** (inside the body). Viruses genetically modified to be safe can be used to ferry a normal gene into cells (**Fig. 13.13**), and so can liposomes, which are microscopic globules that form when lipoproteins are put into a solution and are specially prepared to enclose the normal gene. On the other hand, sometimes the gene is injected directly into a particular region of the body. This section discusses examples of ex vivo gene therapy (the gene is inserted into cells that have been removed and then returned to the body) and in vivo gene therapy (the gene is delivered directly into the body).

Ex Vivo Gene Therapy

Children who have severe combined immunodeficiency (SCID) lack the enzyme adenosine deaminase (ADA), which is involved in the maturation of cells that produce antibodies. In order to carry out gene therapy to treat this disorder, bone marrow stem cells are removed from the blood and infected with a virus that carries a normal gene for the enzyme. Then the cells are returned to the patient. Bone marrow stem cells are preferred for this procedure, because they divide to produce more cells with the same genes. Patients who have undergone this procedure show significantly improved immune function associated with a sustained rise in the level of ADA enzyme activity in the blood.

Ex vivo gene therapy is also used in the treatment of familial hypercholesterolemia, a genetic disorder in which high levels of plasma cholesterol make the patient subject to a fatal heart attack at a young age. A small portion of the liver is surgically excised and then infected with a retrovirus containing a normal gene for a cholesterol receptor before the tissue is returned to the patient. Several patients have experienced lowered plasma cholesterol levels following this procedure.

Some cancers are being treated with ex vivo gene therapy procedures. In one procedure, immune system cells are removed from a cancer patient and genetically engineered to display tumor antigens. After these cells are returned to the patient, they stimulate the immune system to kill tumor cells.

In Vivo Gene Therapy

Most cystic fibrosis patients lack a gene that codes for a regulator of a transmembrane carrier for the chloride ion. In gene therapy trials, the gene needed

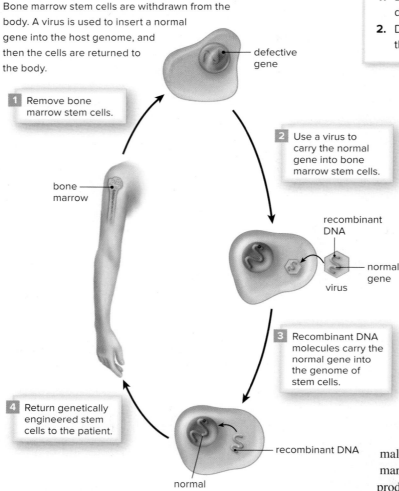

Figure 13.13 Ex vivo gene therapy in humans.
Bone marrow stem cells are withdrawn from the body. A virus is used to insert a normal gene into the host genome, and then the cells are returned to the body.

1. Remove bone marrow stem cells.

2. Use a virus to carry the normal gene into bone marrow stem cells.

3. Recombinant DNA molecules carry the normal gene into the genome of stem cells.

4. Return genetically engineered stem cells to the patient.

defective gene

bone marrow

recombinant DNA

normal gene

virus

recombinant DNA

normal gene

Connections: Scientific Inquiry

What is RNA interference?

RNA interference, or RNAi, is a procedure in which small pieces of RNA are used to "silence" the expression of specific alleles. These RNA sequences are designed to be complementary to the mRNA transcribed by a gene of interest. Once the complementary RNA sequences enter the cell, they bind with the target RNA, producing double-stranded RNA molecules. These double-stranded RNA molecules are then broken down by a series of enzymes within the cell. First discovered in worms, RNAi is believed to have evolved in eukaryotic organisms as a protection against certain types of viruses. Research into developing RNAi treatments for a number of human diseases, including cancer and hepatitis, is currently underway.

to cure cystic fibrosis is sprayed into the nose or delivered to the lower respiratory tract by adenoviruses or in liposomes. Investigators are trying to improve uptake of this gene and are hypothesizing that a combination of all three vectors might be more successful.

Genes are also being used to treat medical conditions such as poor coronary circulation. It has been known for some time that vascular endothelial growth factor (VEGF) can cause the growth of new blood vessels. The gene that codes for this growth factor can be injected alone, or within a virus, into the heart to stimulate branching of coronary blood vessels. Patients who have received this treatment report that they have less chest pain and can run longer on a treadmill.

Rheumatoid arthritis, a crippling disorder in which the immune system turns against a person's own body and destroys joint tissue, has recently been treated with in vivo gene therapy methods. Clinicians inject adenoviruses that contain an anti-inflammatory gene into the affected joint. The added gene reduces inflammation within the joint space and lessens the patient's pain and suffering. Clinical trials have been promising, and animal studies have even shown that gene therapy may stave off arthritis in at-risk individuals.

Check Your Progress 13.4

1. Explain the difference between ex vivo and in vivo gene therapy.

2. List the methods by which genes may be delivered to cells using in vivo gene therapy.

3. List some of the diseases that may be treated by ex vivo and in vivo gene therapy.

CONNECTING THE CONCEPTS

13.4 Gene therapy is being used to help treat specific genetic disorders.

STUDY TOOLS http://connect.mheducation.com

SMARTBOOK® Maximize your study time with McGraw-Hill SmartBook®, the first adaptive textbook.

SUMMARIZE

Various techniques are available to detect gene mutations and changes to chromosome structure and to test for and treat genetic disorders.

13.1 Mutagens, replication errors, and transposons can produce mutations.

13.2 Mutations can arise from various changes in chromosomal structure.

13.3 A variety of methods are available to analyze an individual for the presence of specific mutations.

13.4 Gene therapy is being used to help treat specific genetic disorders.

13.1 Gene Mutations

Mutations are changes in the sequence of nucleotides in the DNA. Although some mutations are silent, others produce changes in the sequence of amino acids in the encoded protein. Some causes of mutations are

- **Transposons:** small sequences of DNA that have the ability to move within the genome, sometimes disrupting gene function
- **Mutagens:** environmental influences (e.g., chemicals or radiation) that cause mutations

Two types of mutation occur at the level of the genes.

- **Point mutation:** a change in a single nucleotide in the DNA; it may be silent (no effect) or may change the structure of the encoded protein
- **Frameshift mutation:** removal or insertion of nucleotides that changes the way the information within the gene's codons is read

13.2 Chromosomal Mutations

Chromosomal mutations involve changes in chromosome number or structure. Abnormal chromosome number results from nondisjunction during meiosis. Changes in chromosome structure include **deletions, duplications, translocations,** and **inversions:**

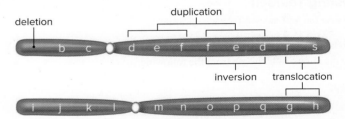

In Williams syndrome, one copy of chromosome 7 has a deletion; in cri du chat syndrome, one copy of chromosome 5 has a deletion; and in inv dup 15 syndrome, chromosome 15 has an inverted duplication. Down syndrome can be due to a translocation between chromosomes 14 and 21 in a previous generation.

An inversion can lead to chromosomes that have a deletion and a duplication. The homologue with the inverted sequence must loop back to align with the normal homologue. Following crossing-over, one nonsister chromatid has a deletion and the other has a duplication.

13.3 Genetic Testing

Analyzing the Chromosomes

A counselor can detect chromosomal mutations by studying a **karyotype** of the individual. A karyotype is a display of the chromosomes arranged by pairs; the autosomes are numbered from 1 to 22. The sex chromosomes are not numbered. **Genetic counseling** can use a variety of techniques to detect an inherited disorder.

Testing for a Protein

Blood or tissue samples can be tested for enzyme activity. Lack of enzyme activity can sometimes indicate that a genetic disorder exists.

Testing the DNA

DNA sequencing techniques may be used to look for **genetic markers.** Restriction enzymes can also be used to cut the DNA and then compare the fragment pattern to the normal pattern. DNA can be labeled with fluorescent tags to see if fragments bind to DNA probes on a DNA microarray that contains the mutation. Personal genomics allows information about a person's genome to be used to tailor preventative actions and treatments and for pharmacogenomics.

Testing the Fetus

An ultrasound can detect some disorders that are due to chromosomal abnormalities and other conditions, such as spina bifida, that are due to the inheritance of mutant alleles. Fetal cells can be obtained by **amniocentesis** or **chorionic villus sampling,** or by sorting out fetal cells from the mother's blood.

Testing the Embryo and Egg

Following in vitro fertilization (IVF), it is possible to test the embryo. A cell is removed from an eight-celled embryo, and if it is found to be genetically healthy, the embryo is implanted in the uterus, where it completes development. Before IVF, a polar body can be tested. If the woman is heterozygous, and the polar body has the genetic defect, the egg does not have it. Following fertilization, the normal embryo is implanted in the uterus.

13.4 Gene Therapy

During **gene therapy,** a genetic defect is treated by giving the patient a normal gene.

- **Ex vivo therapy:** Cells are removed from the patient, treated with the normal gene, and returned to the patient.
- **In vivo therapy:** A normal gene is given directly to the patient via injection, nasal spray, liposomes, or adenovirus.

ASSESS

Testing Yourself

Choose the best answer for each question.

13.1 Gene Mutations

For questions 1–4, choose from the following:

Key:

- **a.** transposon
- **b.** point mutation
- **c.** frameshift mutation

1. causes the information within a gene to be read incorrectly
2. may be silent
3. DNA sequence that can move within the genome
4. causes a change in a single nucleotide in the DNA

13.2 Chromosomal Mutations

5. Which of the following disorders is not caused by a change in chromosome structure?
   - **a.** cri du chat syndrome
   - **b.** inv dup 15 syndrome
   - **c.** Klinefelter syndrome
   - **d.** Alagille syndrome

6. In which of the following does the amount of genetic material on the chromosome remain the same?
   - **a.** inversion
   - **b.** duplication
   - **c.** deletion
   - **d.** All of these are correct.

13.3 Genetic Testing

7. Which of the following may be detected using a karyotype?
   - **a.** point mutations
   - **b.** a specific genetic marker
   - **c.** an abnormal number of chromosomes
   - **d.** the location of a transposon

8. Pharmacogenomics has been made possible by recent advances in
   - **a.** gene therapy.
   - **b.** DNA sequencing.
   - **c.** amniocentesis.
   - **d.** karyotype analysis.

13.4 Gene Therapy

9. Ex vivo gene therapy involves
   - **a.** the delivery of genes to cells in the body.
   - **b.** only the use of liposomes to deliver genes to cells.
   - **c.** the removal of cells from the body before treatment.
   - **d.** All of these are correct.

10. Which of the following is a vector that can be used in gene therapy to deliver normal genes directly into the cells of the body?
    - **a.** transposons
    - **b.** viruses
    - **c.** amniocentesis
    - **d.** mutagens

ENGAGE

Thinking Critically

1. As a genetic counselor, you request DNA sequencing for one of your patients to look for the presence of certain alleles that are associated with a pattern of disease in the person's family. During that procedure, you detect another allele that greatly increases the risk factor for another disease. How would you disclose this information to the patient? What ethical concerns do you have about knowing this information if the patient does not want to be informed?

2. The use of adenoviruses to deliver genes in gene therapy has sometimes proven problematic. Recently, adenoviruses were used in gene therapy trials performed on ten infants with X-linked severe combined immunodeficiency syndrome (XSCID), also known as "bubble boy disease." Although the trial was considered a success because the gene was expressed and the children's immune systems were restored, researchers were shocked and disappointed when two children developed leukemia. How might you explain the development of leukemia in these gene therapy patients?

3. Genetic counselors note that certain debilitating disorders are maintained in a population even when they are disadvantageous. Why wouldn't these disorders be removed by the evolutionary process? Can you think of any advantage to their remaining in the gene pool?

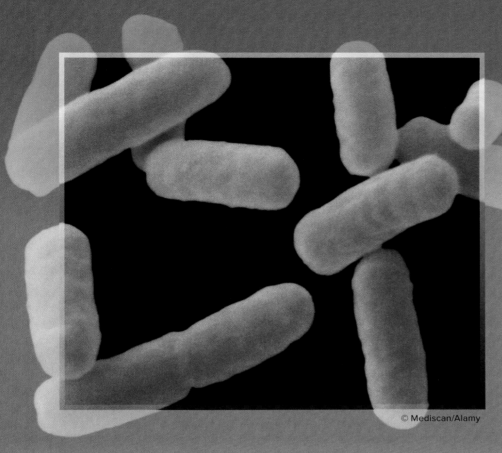

© Mediscan/Alamy

14

Darwin and Evolution

OUTLINE

14.1 Darwin's Theory of Evolution   236

14.2 Evidence of Evolutionary
Change   244

BEFORE YOU BEGIN

Before beginning this chapter, take a few moments to review the following discussions.

Section 1.2 Why is evolution a core concept of biology?

Section 9.1 What is an allele?

Section 9.2 How does meiosis increase variation?

Evolution of Antibiotic Resistance

You have probably heard of antibiotic-resistant bacteria such as MRSA (methicillin-resistant *Staphylococcus aureus*), a concern in hospitals and medical facilities. Unfortunately, examples of antibiotic-resistant bacteria are becoming more common. One of these is *Shigella sonnei* (shown above), a species of bacteria that is commonly found in human feces but can cause diarrhea and intestinal problems if ingested. Globally, around 100 million people per year are infected with *Shigella,* and over 600,000 die from such infections. Normally, regular hand washing with soap and warm water prevents most *Shigella* infections. Historically, antibiotics have been used to treat *Shigella,* but an antibiotic-resistant strain of *Shigella* has developed and is becoming a concern of the health industry.

Use and overuse of antibiotics have resulted in the evolution of resistant bacterial strains. Although we tend to think of evolution as happening over long time frames, human activities can accelerate the process of evolution quite rapidly. In fact, evolution of resistance to the antibiotic methicillin occurred in just a year! Antibiotic-resistant strains of bacteria are generally hard to treat, and treatment of infected patients can cost thousands of dollars.

Some scientists believe that "superbugs," or bacteria that have evolved antibiotic resistance, will be a far bigger threat to human health than viral diseases such as H1N1 flu and Ebola. The good news is that our understanding of evolutionary biology has helped change human behavior in dealing with superbugs. For example, doctors no longer prescribe antibiotics unless they are relatively certain a patient has a bacterial infection. Antibiotic resistance is an example of why evolution is important in people's everyday lives. In this chapter, you will learn about evidence that indicates evolution has occurred and about how the evolutionary process works.

As you read through this chapter, think about the following questions:

1. How does natural selection play a role in the evolution of antibiotic resistance?
2. What type of adaptation are the *Shigella* bacteria exhibiting?

235

14.1 Darwin's Theory of Evolution

Learning Outcomes

Upon completion of this section, you should be able to

1. Summarize how nineteenth-century scientists contributed to the study of evolutionary change.
2. Explain how Darwin's study of fossils and biogeography contributed to the development of the theory of natural selection.
3. Describe the steps in the theory of natural selection.
4. Distinguish between natural and artificial selection.

Charles Darwin was only 22 in 1831 when he accepted the position of naturalist aboard the HMS *Beagle,* a British naval ship about to sail around the world (**Fig. 14.1**). Darwin had a suitable background for this position. He was a dedicated student of nature and had long been a collector of insects. His sensitive nature had prevented him from studying medicine, and he went to divinity school at Cambridge instead. Even so, he had an intense interest in science and attended many lectures in both biology and geology. Darwin spent the summer of 1831 doing geology fieldwork at Cambridge, before being recommended to the captain of the *Beagle* as the ship's naturalist. The voyage was supposed to take 2 years, but it took 5. Along the way, Darwin had the chance to collect and observe a tremendous variety of lifeforms, many of which were very different from those in his native England.

Figure 14.1 **Voyage of the HMS *Beagle.***

The map shows the journey of the HMS *Beagle* around the world. As Darwin traveled along the east coast of South America, he noted that a bird called a rhea looked like the African ostrich. On the Galápagos Islands, marine iguanas, found no other place on Earth, use their large claws to cling to rocks and their blunt snouts for eating marine algae.

(Darwin & ship): © DEA Picture Library/Getty Images; (iguana): © FAN Travelstock/Alamy RF; (rhea): © Nicole Duplaix/National Geographic/Getty Images

marine iguana

Charles Darwin

HMS *Beagle*

Galápagos Islands

rhea

Initially, Darwin was a supporter of the long-held idea that species had remained unchanged since the time of creation. During the 5-year voyage of the *Beagle,* Darwin's observations challenged his belief that species do not change over time. His observations of geological formations and species variation led him to propose a new mechanism by which species arise and change. This process, called **evolution,** proposes that species arise, change, and become extinct due to natural, not supernatural, forces.

Before Darwin

Prior to Darwin, most people had an entirely different way of viewing the world. They believed that the Earth was only a few thousand years old and that, since the time of creation, species had remained exactly the same. Even so, studying the anatomy of organisms and then classifying them interested many investigators who wished to show that species were created to be suitable to their environment. Explorers and collectors traveled the world and brought back currently existing species and fossils to be classified. **Fossils** are the remains of once-living species often found in strata. *Strata* are layers of rock formed from sedimentary material (**Fig. 14.2**). The ages of the strata allow scientists to estimate when these species lived, and their characteristics indicate the environment that was found in the region during that time. For example, a fossil showing that snakes had hip bones and hindlimbs some 90 million years ago (MYA) indicates that these ancestors of today's snakes evolved in a land environment.

A noted zoologist of the early nineteenth century, Georges Cuvier, founded the science of *paleontology,* the study of fossils. Cuvier was faced with a problem. He believed in the fixity of species (the idea that species do not change over time), yet Earth's strata clearly showed a succession of different life-forms over time. To explain these observations, he hypothesized that a local catastrophe had caused a mass extinction whenever a new stratum of that region showed a new mix of fossils. After each catastrophe, the region was repopulated by species from surrounding areas, which accounted for the appearance of new fossils in the new stratum. The result of all these catastrophes was change appearing over time. Some of Cuvier's followers, who came to be called catastrophists, even suggested that worldwide catastrophes had occurred and that, after each of these events, new sets of species had been created.

In contrast to Cuvier, Jean-Baptiste de Lamarck, another biologist, hypothesized that evolution occurs and that adaptation to the environment is the cause of diversity. Therefore, after studying the succession of life-forms in strata, Lamarck concluded that more complex organisms are descended from less complex organisms. To explain the process of adaptation to the environment, Lamarck proposed the idea of *inheritance of acquired characteristics,* in which the use or disuse of a structure can bring about inherited change. One example Lamarck gave—and the one for which he is most famous—is that the long neck of giraffes developed over time because giraffes stretched their necks to reach food high in trees and then passed on a long neck to their offspring (see Fig. 14.9). However, the inheritance of acquired characteristics is not the primary mechanism that drives the change in species over time.

Visible strata

Figure 14.2 Strata in rock.

Due to erosion, it is often possible to see a number of strata, layers of rock or sedimentary material that contain fossils. The oldest fossils are in the lowest stratum.

© Doug Sherman/Geofile

If this were the case, then the knowledge you acquire over your lifetime would be passed on to your offspring. Still, in recent years scientists have unveiled some conditions under which acquired characteristics may be passed on to the next generation. For example, in *epigenetic inheritance,* changes in the expression of a gene that are not associated with mutations or other nucleotide changes may be passed from one generation to the next (see Section 11.3). These cases are relatively rare and usually relate only to the inheritance of specific genes.

Darwin's ideas were close to those of Lamarck. For example, Darwin said that living organisms share characteristics because they have a common ancestry. One of the most unfortunate misinterpretations of this statement was that humans are descendants of apes; we know that humans and apes shared a common ancestor, just as, say, you and your cousins can trace your ancestry back to the same grandparents. In contrast to Lamarck, Darwin's observations led him to conclude that species are suited to the environment through no will of their own but by natural selection. He saw the process of natural selection as the means by which different species come about (see Fig. 14.9).

Darwin's Conclusions

Darwin's conclusions that organisms are related through common descent and that adaptation to various environments results in diversity were based on several types of data, including his study of geology, fossils, and biogeography. **Biogeography** is the study of the distribution of life-forms on Earth.

Darwin's Study of Geology and Fossils

Darwin took Charles Lyell's book *Principles of Geology* with him on the *Beagle* voyage. In contrast to former beliefs, this book gave evidence that Earth is subject to slow but continuous cycles of erosion and uplift. Weathering causes erosion; thereafter, dirt and rock debris are washed into the rivers and transported to oceans. When these loose sediments are deposited, strata result (**Fig. 14.3***a*). Then the strata, which often contain fossils, are uplifted over long periods of time from below sea level to form land. Given enough time, slow natural processes can account for extreme geological changes. Lyell went on to propose the theory of *uniformitarianism,* which stated that these slow changes occurred at a uniform rate. Even though uniformitarianism has been rejected, modern geology certainly substantiates a hypothesis of slow and continual geological change. Darwin, too, was convinced that Earth's massive geological changes are the result of slow processes and therefore Earth must be very old.

On his trip, Darwin observed massive geological changes firsthand. When he explored what is now Argentina, he saw raised beaches for great distances along the coast. In Chile, he witnessed the effects of an earthquake that caused the land to rise several feet and left marine shells inland, well above sea level. When Darwin also found marine shells high in the cliffs of the impressive Andes Mountains, he became even more convinced that Earth is subject to slow geological changes. While Darwin was making geological observations, he also was collecting fossil specimens that differed somewhat from modern species (**Fig. 14.4**). Once Darwin accepted the supposition that Earth must be very old, he began to think that there would have been enough time for descent

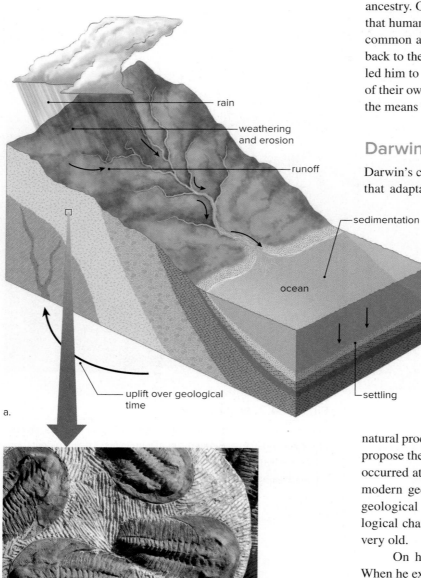

rain

weathering and erosion

runoff

sedimentation

ocean

uplift over geological time

settling

a.

b.

Figure 14.3   **Fossils in strata.**

a. This diagram shows how water takes sediments into the sea; the sediments then become compacted to form a stratum. Fossils are often trapped in strata, and as a result of later geological upheaval, the strata may be located on land. **b.** Fossil remains of extinct marine arthropods called trilobites.

(b): © Doug Sangster/Getty RF

with modification to occur. Living forms could be descended from extinct forms known only from the fossil record. It would seem that species are not fixed; instead, they change over time.

Darwin's Study of Biogeography

Darwin could not help but compare the animals of South America with those he observed in England. For example, instead of rabbits, he found Patagonian hares in the grasslands of South America. The Patagonian hare is a rodent that has long legs and ears and the face of a guinea pig (**Fig. 14.5***a*).

   While the Patagonian hare resembles a rabbit, or even a small deer, it is actually more closely related to the rodents. Its correct name is the Patagonian mara (*Dolichotis patagonum*). Darwin wondered whether the similarities between these two animals were not due to a common ancestor, but rather arose because the two types of animals were adapted to the same type of environment. Scientists call this process *convergent evolution,* and it explains how how distantly related species may converge on the same overall body form because they live in similar habitats and have similar behaviors (see Section 16.3).

   As Darwin sailed southward along the eastern coast of South America, he saw how similar species replaced one another. For example, the greater rhea (an ostrichlike bird) found in the north was replaced by the lesser rhea in the south. Therefore, Darwin reasoned that related species can be modified according to environmental differences caused by change in latitude. When he reached the Galápagos Islands, he found further evidence of this. The Galápagos Islands are a small group of volcanic islands located 965 km (600 miles) off the western coast of South America. These islands are too far from the mainland for most terrestrial animals and plants to colonize, yet life was present. The types of plants and animals found there were slightly different from species Darwin had observed on the mainland; even more important, they also varied from island to island according to each unique environment. Where did the species inhabiting these islands come from, and what caused the islands to have different species?

a. *Glyptodon*

b. *Mylodon*

Figure 14.4 Fossils discovered by Darwin.

Darwin discovered fossils of extinct mammals during his exploration of South America. **a.** A giant armadillo-like glyptodont, *Glyptodon,* is known only by the study of its fossil remains. Darwin found such fossils and came to the conclusion that this extinct animal must be related to living armadillos. The glyptodont weighed 2,000 kg (4,400 lb). **b.** Darwin also observed the fossil remains of an extinct giant ground sloth, *Mylodon.*

a. Patagonian hare

b. European rabbit

Figure 14.5 Patagonian hare and European rabbit.

a. The Patagonian hare, native to South America, has long legs and other adaptations similar to those of a rabbit but has a face similar to that of a guinea pig. **b.** The characteristics of the Patagonian hare resemble those of the European rabbit, which does not occur naturally in South America.

(a): © Juan & Carmecita Munoz/Science Source; (b): © Michael Maconachie/ Papilio /Corbis

Connections: Scientific Inquiry

What has happened to Darwin's finches since Darwin's time?

The finches of the Galápagos Islands have continued to provide a wealth of information on evolutionary processes. Starting in 1973, a team of researchers led by Drs. Peter and Rosemary Grant began a 30-year study of a species of ground finch on the island of Daphne Major in the Galápagos. Through detailed measurements and observations of over 19,000 birds, the Grants were able to document the evolutionary change of a species in response to changes in the environment. The details of their study were reported in the Pulitzer Prize–winning book *The Beak of the Finches: A Story of Evolution in Our Time* by Jonathan Weiner.

Figure 14.6 Galápagos finches.

Each of the present-day 13 species of finches has a beak adapted to a particular way of life. **a.** The heavy beak of the large ground-dwelling finch is suited to a diet of seeds. **b.** The beak of the woodpecker finch is suited for using tools to probe for insects in holes or crevices. **c.** The long, somewhat decurved beak and split tongue of the cactus finch are suited to probing cactus flowers for nectar.

(a): © Miguel Castro/Science Source; (b): © David Hosking/Alamy; (c): © Michael Stubblefield/Alamy RF

a. Ground-dwelling finch

b. Woodpecker finch

c. Cactus finch

Finches   Although some of the finches on the Galápagos Islands seemed like mainland finches, others were quite different (**Fig. 14.6**). Today, there are ground-dwelling finches with beak sizes dependent on the sizes of the seeds they eat. Tree-dwelling finches have beak sizes and shapes dependent on the sizes of their insect prey. A cactus-eating finch possesses a more pointed beak, enabling access to nectar within cactus flowers. The most unusual of the finches is a woodpecker-type finch. A woodpecker normally has a sharp beak to chisel through tree bark and a long tongue to probe for insects. The Galápagos woodpecker-type finch has the sharp beak but lacks the long tongue. To make up for this, the bird carries a twig or cactus spine in its beak and uses it to poke into crevices. Once an insect emerges, the finch drops this tool and seizes the insect with its beak.

Later, Darwin speculated as to whether all the different species of finches he had seen could have descended from a mainland finch. In other words, he wondered if a mainland finch was the common ancestor of all the types on the Galápagos Islands. In Darwin's time, the concept of **speciation,** the formation of a new species, was not well understood. To Darwin, it was possible that the islands allowed isolated populations of birds to evolve independently and that the present-day species had resulted from accumulated changes occurring within each of these isolated populations over time.

Tortoises   Each of the Galápagos Islands also seemed to have its own type of tortoise, and Darwin began to wonder if this could be correlated with a difference in vegetation among the islands. Long-necked tortoises seemed to inhabit only dry areas, where food was scarce, and most likely the longer neck was helpful in reaching tall-growing cactuses. In moist regions with relatively abundant ground foliage, short-necked tortoises were found. Had an ancestral tortoise from the mainland of South America given rise to these different types, each adapted to a different environment? An **adaptation** is any characteristic that makes an organism more suited to its environment. It often takes many generations for an adaptation to become established in a population.

Natural Selection and Adaptation

Once Darwin recognized that adaptations develop over time, he began to think about a mechanism by which adaptations might arise. Eventually, he proposed **natural selection** as the mechanism of evolutionary change. According to Darwin, natural selection requires the following steps:

1. The members of a population have heritable variations (**Fig. 14.7**).
2. The population produces more offspring than the resources of an environment can support.
3. The individuals that have favorable traits survive and reproduce to a greater extent than those that lack these traits.

4. Over time, the proportion of a favorable trait increases in the population, and the population becomes adapted to the environment.

Due to the fact that natural selection utilizes only variations that happen to be provided by genetic changes (such as mutations), it lacks any directedness or anticipation of future needs. Natural selection is an ongoing process, because the environment of living organisms is constantly changing. Extinction, or the complete loss of a species, can occur when previous adaptations are no longer suitable to a changed environment.

Organisms Vary in Their Traits

Darwin emphasized that the members of a population vary in their functional, physical, and behavioral characteristics. Prior to Darwin, variations were considered imperfections that should be ignored, since they were not important to the description of a species. Darwin, on the other hand, realized that variations are essential to the natural selection process. Darwin suspected that the occurrence of variations is completely random; they arise by accident and for no particular purpose.

We now know that genes, together with the environment, determine the phenotype of an organism. The genes that each organism inherits are based on the assortment of chromosomes during meiosis and fertilization. New alleles of a gene are the result of inheritable mutations in the DNA of the organism. These mutations are more likely to be harmful than beneficial to an organism. However, occasionally, a mutation occurs that produces a benefit to the organism, such as a change in camouflage (**Fig. 14.8**). This type of variation, which makes adaptation to an environment possible, is passed on from generation to generation.

Organisms Struggle to Exist

In Darwin's time, a socioeconomist named Thomas Malthus stressed the reproductive potential of humans. He proposed that death and famine are inevitable, because the human population tends to increase faster than the supply of food. Darwin applied this concept to all organisms and saw that the available resources were not sufficient to allow all members of a population to survive. He calculated the reproductive potential of elephants and concluded that, after only 750 years, the descendants of a single pair of elephants would number about 19 million! Obviously, no environment has the resources to support an elephant population of this magnitude. Because each generation has the same reproductive potential as the previous generation, there is a constant struggle for existence, and only certain members of a population survive and reproduce each generation.

Figure 14.7 Variations in shells of a tree snail, *Liguus fasciatus*.

For Darwin, variations, such as these in the coloration of snail shells, were highly significant and were required in order for natural selection to result in adaptation to the environment.

© James H. Robinson/Science Source

Organisms Differ in Fitness

Fitness is the reproductive success of an individual relative to other members of the population. Fitness is determined by comparing the numbers of surviving fertile offspring that are produced by each member of the population. The most fit individuals are the ones that capture a disproportionate amount of resources and convert these resources into a larger number of viable offspring. Because organisms vary anatomically and physiologically, and because the challenges of local environments vary, what determines fitness varies for different populations. For

eye — false head — false eyespot

Figure 14.8 Variation can lead to adaptation.

The alligator bug of the Brazilian rain forest has antipredator adaptations. The insect blends into its background, but if discovered, the false head, which resembles a miniature alligator, may frighten a predator. If the predator is not frightened, the insect suddenly reveals huge false eyespots on its hindwings.

© Danita Delimont/Alamy

example, among western diamondback rattlesnakes living on lava flows, the most fit are those that are black. Among those living on desert soil, the most fit are typically light-colored with brown blotching. Background matching helps an animal both capture prey and avoid being captured; therefore, it is expected to lead to survival and increased reproduction.

Darwin noted that, when humans carry out **artificial selection,** they breed selected animals that have particular traits they want to reproduce. For example, prehistoric humans probably noted desirable variations among wolves and selected particular individuals for breeding. Therefore, the desired traits increased in frequency in the next generation. The same process was repeated many times, resulting in today's numerous varieties of dogs, all descended from the wolf. In a similar way, several varieties of vegetables can be traced to a single ancestor. Chinese cabbage, brussels sprouts, and kohlrabi are all derived from a single species, *Brassica oleracea.*

In nature, interactions with the environment determine which members of a population reproduce to a greater degree than other members. In contrast to artificial selection, the result of natural selection is not predesired. Natural selection occurs because certain members of a population happen to have a variation that allows them to survive and reproduce to a greater extent than other members. For example, any variation that increases the speed of a hoofed animal helps it escape predators and live longer; a variation that reduces water loss is beneficial to a desert plant; and one that increases the sense of smell helps a wild dog find its prey. Therefore, we expect the organisms with these characteristics to have increased fitness. **Figure 14.9** contrasts Lamarck's ideas with those of Darwin.

Organisms Become Adapted

An adaptation may take many generations to evolve. We can especially recognize an adaptation when unrelated organisms living in a particular environment display similar characteristics. For example, manatees, penguins, and sea turtles all have flippers, which help them move through the water—also an example of convergent evolution. Adaptations also account for why organisms are able to escape their predators (see Fig. 14.8) and why they are suited to their way of life (**Fig. 14.10**). Natural selection causes adaptive traits to be increasingly represented in each succeeding generation. There are other processes at work in the evolution of populations (see Chapter 15), but it is the process of natural selection that allows a population to adapt to its environment.

Connections: Scientific Inquiry

Are there examples of artificial selection in animals?

Almost all animals that are currently used in modern agriculture are the result of thousands of years of artificial selection by humans. But perhaps the greatest example of artificial selection in animals is the modern dog. Analysis of canine DNA indicates that the dog (*Canis familiaris*) is a direct descendant of the gray wolf (*Canis lupus*). The wolf's domestication and

© Brand X Pictures/Getty RF

the subsequent selection for desirable traits appear to have begun over 130,000 years ago. Artificial selection of dogs continues to this day, with over 150 variations (breeds) currently known.

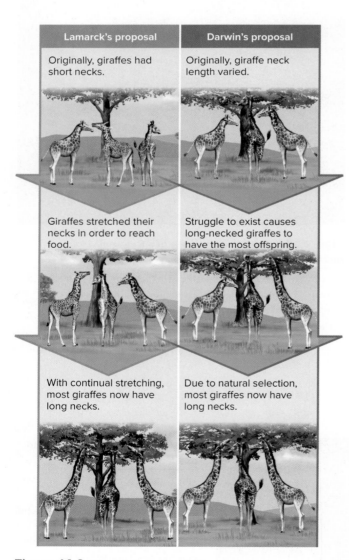

Lamarck's proposal	Darwin's proposal
Originally, giraffes had short necks.	Originally, giraffe neck length varied.
Giraffes stretched their necks in order to reach food.	Struggle to exist causes long-necked giraffes to have the most offspring.
With continual stretching, most giraffes now have long necks.	Due to natural selection, most giraffes now have long necks.

Figure 14.9 **Mechanism of evolution.**

This diagram contrasts Jean-Baptiste Lamarck's proposal of acquired characteristics with Charles Darwin's proposal of natural selection. Only natural selection is supported by data.

Darwin and Wallace

After the HMS *Beagle* returned to England in 1836, Darwin waited more than 20 years to publish his book *On the Origin of Species.* During the intervening years, he used the scientific process to support his hypothesis that today's diverse life-forms arose by descent from a common ancestor and that natural selection is a mechanism by which species can change and new species can arise. Darwin was prompted to publish his book after reading a similar hypothesis formulated by Alfred Russel Wallace.

Wallace was an English naturalist who, like Darwin, was a collector at home and abroad. He went on collecting trips, each of which lasted several years, to the Amazon and the Malay Archipelago. After studying the animals of every island within the Malay Archipelago, he divided the islands into a western group, with organisms like those found in Asia, and an eastern group, with organisms like those of Australia. The sharp line dividing these two island groups within the archipelago is now known as Wallace's Line (**Fig. 14.11**). A narrow but deep strait occurs along Wallace's Line. At times during the past 50 million years, this strait persisted even when sea levels were low and land bridges appeared between the other islands. Therefore, the strait served as a permanent barrier to the dispersal of organisms between the two groups of islands.

While traveling, Wallace wrote an essay called "On the Law Which Has Regulated the Introduction of New Species." In this essay, he said that "every species has come into existence coincident both in time and space with a preexisting closely allied species." A year later, after reading Malthus's treatise on human population increase, Wallace conceived the idea of "survival of the fittest." He quickly completed an essay proposing natural selection as an agent for evolutionary change and sent it to Darwin for comment. Darwin was stunned. Here was the hypothesis he had formulated but had never dared to publish. He told his friend and colleague Charles Lyell that Wallace's ideas were so similar to his own that even Wallace's "terms now stand as heads of my chapters."

forelimb (wing)

enlarged ears

pointed incisors

grasping hindlimbs

Figure 14.10 **Adaptations of the vampire bat.**

Vampire bats of the rain forests of Central and South America have the adaptations of a nocturnal, winged predator. The bat uses its enlarged ears and echolocation to locate prey in the dark. It bites its prey with two pointed incisors. Saliva containing an anticoagulant, called draculin, runs into the bite; the bat then licks the flowing blood. The vampire bat's forelimbs are modified to form wings, and it roosts by using its grasping hindlimbs.

(top): © Chewin/Getty RF; (bottom): © Haroldo Palo, Jr./NHPA/Photoshot

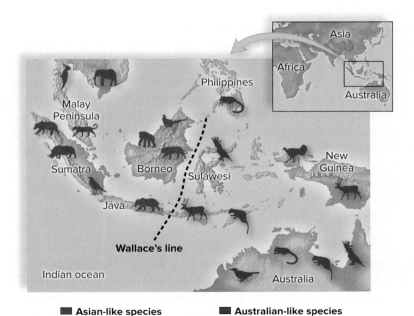

■ **Asian-like species**   ■ **Australian-like species**

Figure 14.11 **Wallace's line and biogeography.**

Wallace described how a deep ocean strait divides the islands of the Malay Archipelago in the southern Pacific into two regions. The animal populations on either side of Wallace's line evolved separately in response to different environmental influences.

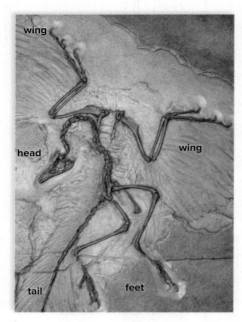

a.

b.

Figure 14.12 **Re-creation of *Archaeopteryx*.**

a. The fossil record suggests that *Archaeopteryx* had a feather-covered, reptilian-type tail. **b.** This tail shows up well in this artist's representation. (Red labels: reptilian characteristics; green labels: bird characteristics.)

(a): © Jason Edwards/Getty RF; (b): © Joe Tucciarone

Darwin suggested that Wallace's paper be published immediately, even though Darwin himself as yet had nothing in print. However, Lyell and others who knew of Darwin's detailed work substantiating the process of natural selection suggested that a joint paper be read to the renowned Linnean Society. On July 1, 1858, Darwin presented an abstract of *On the Origin of Species,* which was published in 1859. The title of Wallace's section was "On the Tendency of Varieties to Depart Indefinitely from the Original Type." This professional presentation served as the announcement to the world that species share a common descent and have diverged through natural selection.

Check Your Progress 14.1

1. Summarize Cuvier's and Lamarck's views on evolutionary change.
2. Distinguish between the process of natural selection and the inheritance of acquired characteristics.
3. Explain how natural selection can lead to adaptation.

14.2 Evidence of Evolutionary Change

Learning Outcome

Upon completion of this section, you should be able to

1. Explain how the fossil record, biogeographical evidence, comparative anatomy, and biochemistry support evolutionary theory.

Evolutionary theory, which includes the theories of natural selection and descent with modification, states that all living organisms have a common ancestor, but each has become adapted to its environment by the process of natural selection. A hypothesis becomes a scientific theory only when a variety of evidence from independent investigators supports the hypothesis. It has been over 150 years since Darwin first suggested that natural selection represents the mechanism by which species change over time. Since then, countless scientific studies have supported the ideas that species (1) change over time, (2) are related through common descent, and (3) are adapted to their environment by the process of natural selection.

Evolution is recognized as the unifying principle in biology because it can explain so many different observations in various fields of biology. From the study of microbes to the function of ecosystems, evolutionary theory helps biologists understand the complex nature of life. In this section, we will explore the evidence that evolutionary change has occurred. With the exception of biochemical data, Darwin was aware of much of this evidence when he formulated his ideas.

Fossil Evidence

The fossils trapped in rock strata are the **fossil record** that tell us about the history of life. One of the most striking patterns in the fossil record is a succession of life-forms from the simple to the more complex. Occasionally, this pattern is reversed, showing that evolution is not unidirectional. Particularly interesting are the fossils that serve as *transitional links* between groups. Even in Darwin's day, scientists knew of the *Archaeopteryx* fossils, which show that birds have reptilian features, including jaws with teeth and long, jointed tails. *Archaeopteryx* also had feathers and wings (**Fig. 14.12**).

In 2004, a team of paleontologists discovered fossilized remains of *Tiktaalik roseae,* nicknamed the "fishapod" because it is the transitional form between fish and four-legged animals, the tetrapods. *Tiktaalik* fossils are estimated to be 375 million years old and are from a time when the transition from fish to tetrapods is likely to have occurred. As expected of an intermediate fossil, *Tiktaalik* has a mix of fishlike and tetrapod-like features that illustrate the steps in the evolution of tetrapods from a fishlike ancestor (**Fig. 14.13**). For example, *Tiktaalik* has a very fishlike set of gills and fins, with the exception of the pectoral (front) fins, which have the beginnings of wrist bones (or legs) similar to those of a tetrapod. Unlike a fish, however, *Tiktaalik* has a flat head, a flexible neck, eyes on the top of its head like a crocodile, and interlocking ribs that suggest it had lungs. These transitional features suggest that it had the ability to push itself along the bottom of shallow rivers and see above the surface of the water—features that would come in handy in its river habitat.

In 2006, a snake fossil dated to 90 MYA was discovered that showed hip bones and hindlimbs—a trait absent in all living snakes. Some snakes, such as pythons, have vestigial hindlimbs, but these snakes lack the hip bones present in this fossil. Since all lizards have hip bones and most have limbs, this fossil is considered a transitional link between lizards and snakes.

The fossil record also provides important insights into the evolution of whales from land-living, hoofed ancestors (**Fig. 14.14**). The fossilized whale *Ambulocetus* may have been amphibious, walking on land and swimming in the sea. *Rodhocetus* swam with an up-and-down tail motion, as modern whales do; its reduced hindlimbs could not have helped in swimming.

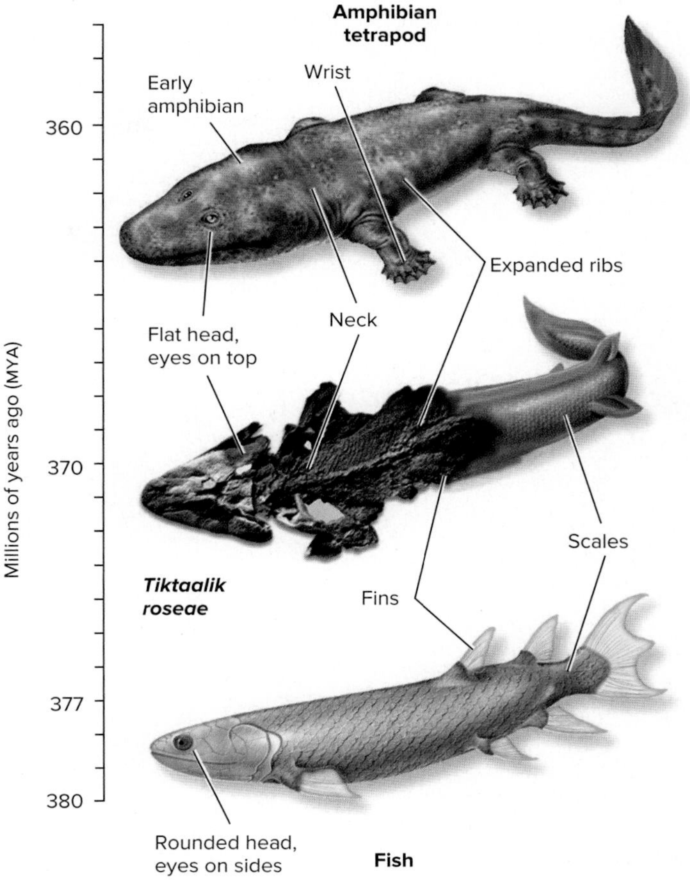

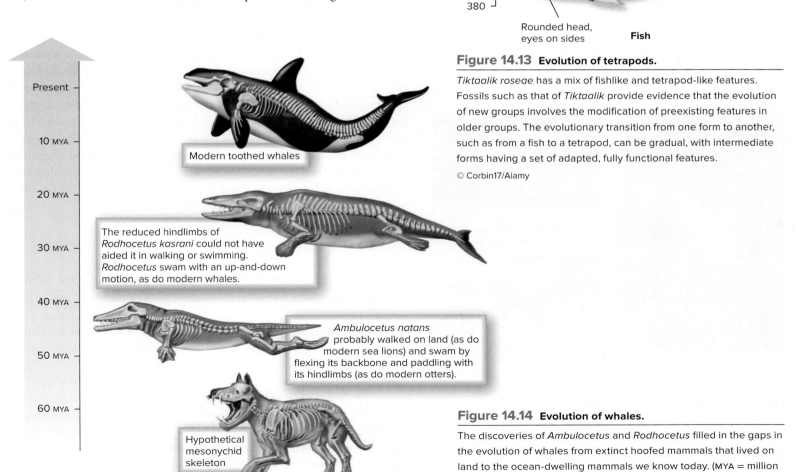

Figure 14.13 **Evolution of tetrapods.**

Tiktaalik roseae has a mix of fishlike and tetrapod-like features. Fossils such as that of *Tiktaalik* provide evidence that the evolution of new groups involves the modification of preexisting features in older groups. The evolutionary transition from one form to another, such as from a fish to a tetrapod, can be gradual, with intermediate forms having a set of adapted, fully functional features.

© Corbin17/Alamy

The reduced hindlimbs of *Rodhocetus kasrani* could not have aided it in walking or swimming. *Rodhocetus* swam with an up-and-down motion, as do modern whales.

Ambulocetus natans probably walked on land (as do modern sea lions) and swam by flexing its backbone and paddling with its hindlimbs (as do modern otters).

Modern toothed whales

Hypothetical mesonychid skeleton

Figure 14.14 **Evolution of whales.**

The discoveries of *Ambulocetus* and *Rodhocetus* filled in the gaps in the evolution of whales from extinct hoofed mammals that lived on land to the ocean-dwelling mammals we know today. (MYA = million years ago.)

Biogeographical Evidence

Biogeography is the study of the distributions of organisms throughout the world. Such distributions are consistent with the hypothesis that, when forms are related, they evolve in one locale and then spread to accessible regions. Therefore, you would expect a different mix of plants and animals wherever geography separated continents, islands, or seas. As previously mentioned, Darwin noted that South America lacked rabbits, even though the environment was quite suitable for them. He concluded that rabbits evolved elsewhere and had no means of reaching South America.

To take another example, both cactuses and euphorbia (a type of spurge) are plants adapted to a hot, dry environment—both are succulent, spiny, flowering plants. Why do cactuses grow in American deserts and most euphorbia grow in African deserts, when each would do well on the other continent? It seems obvious that they evolved similar adaptations on their respective continents because they lived in similar environments.

The islands of the world are home to many unique species of animals and plants found nowhere else, even when the soil and climate are the same. Why do so many species of finches live on the Galápagos Islands and so many species of honeycreepers, a type of finch, live in the Hawaiian Islands when these species are not found on the mainland? The reasonable explanation is that an ancestral finch migrated to all the different islands. Then geographic isolation allowed the ancestral finch to evolve into a different species on each island.

Also, long ago, South America, Antarctica, and Australia were connected (Fig. 16.14). Marsupials (pouched mammals) and placental mammals arose at this time, but today marsupials are plentiful only in Australia, and placental mammals are plentiful in South America. Why are marsupials plentiful only in Australia (**Fig. 14.15**)? After marsupials arose, Australia separated and drifted away, and marsupials were free to evolve into many different forms because they had no competition from placental mammals. In the Americas, the placental mammals competed successfully against the marsupials, and the opossum is the only marsupial in the Americas. In some cases, marsupial and placental mammals physically resemble one another—two such cases are the marsupial wombat and the marmot and the marsupial Tasmanian wolf and the wolf. This supports the hypothesis that evolution is influenced by the environment and by the mix of plants and animals on a particular continent—by biogeography, not by design.

Anatomical Evidence

Darwin was able to show that a hypothesis of common descent offers a plausible explanation for vestigial structures and anatomical similarities among organisms.

Vestigial structures are anatomical features that are fully developed in one group of organisms but reduced and nonfunctional in other, similar groups. Most birds, for example, have well-developed wings used for flight. However, some bird species (e.g., ostrich) have greatly reduced wings and do not fly. Similarly, whales (see Fig. 14.14) and snakes have no use for hindlimbs, yet extinct whales and snakes had remnants of hip bones and legs. Humans have a tailbone but no tail. The presence of vestigial structures can be explained by the common descent hypothesis. Vestigial structures occur because organisms inherit their anatomy from their ancestors; they are traces of an organism's evolutionary history.

The Australian wombat, *Vombatus*, is nocturnal and lives in burrows. It resembles the placental woodchuck.

The Tasmanian wolf (now extinct) was a carnivore that resembled the American wolf.

Sugar glider, *Petaurus breviceps*, is a tree-dweller and resembles the placental flying squirrel.

Figure 14.15   Marsupials of Australia.

Marsupials in Australia and placental mammals in the rest of the world often have similar characteristics, even though the marsupials all evolved from a common ancestor that entered Australia some 60 MYA.

(sugar glider): © A.N.T. Photo Library/Science Source; (wombat): © Photodisc Collection/Getty RF; (Tasmanian tiger): © World History Archive/Alamy

Connections: Scientific Inquiry

What are some other vestigial organs in humans?

The human body is littered with vestigial organs from our evolutionary past—for example, the tiny muscles (called piloerectors) that surround each hair follicle. During times of stress, these muscles cause the hair to stand straight up—a useful defense mechanism for small mammals trying to frighten predators but one that has little function in humans. Wisdom teeth are also considered to be vestigial organs, since most people now retain their teeth for the majority of their lives.

© fotographixx/Getty RF

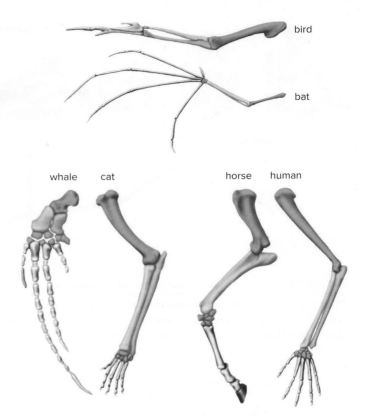

Figure 14.16 **Significance of structural similarities.**

Although the specific details of vertebrate forelimbs are different, the same basic bone stucture and position are present (color-coded here). This unity of anatomy is evidence of a common ancestor.

Vertebrate forelimbs are used for flight (birds and bats), orientation during swimming (whales and seals), running (horses), climbing (arboreal lizards), and swinging from tree branches (monkeys). However, all vertebrate forelimbs contain the same sets of bones organized in similar ways, despite their dissimilar functions (**Fig. 14.16**). The most plausible explanation for this unity of anatomy is that the basic forelimb plan belonged to a common ancestor, and then the plan became modified in the succeeding groups as each continued along its own evolutionary pathway. Anatomically similar structures explainable by inheritance from a common ancestor are called **homologous structures.** In contrast, **analogous structures** serve the same function but are not constructed similarly, and therefore could not have a common ancestry. The wings of birds and insects are analogous structures.

The homology shared by vertebrates extends to their embryological development (**Fig. 14.17**). At some point during development, all vertebrates have a postanal tail and exhibit paired pharyngeal pouches supported by cartilaginous bars. In fishes and amphibian larvae, these pouches develop into functioning gills. In humans, the first pair of pouches become the cavity of the middle ear and the auditory tube. The second pair becomes the tonsils; the third and fourth pairs become the thymus and parathyroid glands. Why should pharyngeal pouches, which have lost their original function, develop and then become modified in terrestrial vertebrates? The most likely explanation is that new structures (or structures with unique functions) originate by "modifying" the preexisting structures of an organism's ancestors.

Chick embryo

Pig embryo

eye

pharyngeal pouches

postanal tail

Figure 14.17 **Significance of developmental similarities.**

At this comparable developmental stage, a chick embryo and a pig embryo have many features in common, which suggests the two animals evolved from a common ancestor.

(pig & chick): © Carolina Biological Supply/Phototake

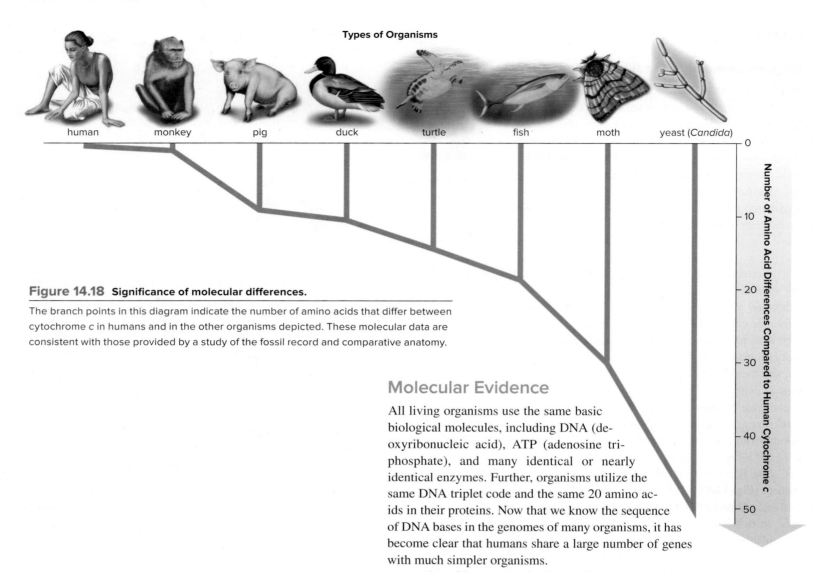

Types of Organisms

human   monkey   pig   duck   turtle   fish   moth   yeast (*Candida*)

Number of Amino Acid Differences Compared to Human Cytochrome *c*

Figure 14.18 Significance of molecular differences.

The branch points in this diagram indicate the number of amino acids that differ between cytochrome *c* in humans and in the other organisms depicted. These molecular data are consistent with those provided by a study of the fossil record and comparative anatomy.

Molecular Evidence

All living organisms use the same basic biological molecules, including DNA (deoxyribonucleic acid), ATP (adenosine triphosphate), and many identical or nearly identical enzymes. Further, organisms utilize the same DNA triplet code and the same 20 amino acids in their proteins. Now that we know the sequence of DNA bases in the genomes of many organisms, it has become clear that humans share a large number of genes with much simpler organisms.

Also of interest, evolutionary developmental biologists have found that many developmental genes, called *Hox* genes, are shared in animals ranging from worms to humans. It appears that life's vast diversity has come about through only slight differences in the same genes. The results have been widely divergent body plans. For example, a similar gene in arthropods and vertebrates determines the back-to-front axis. However, although the base sequences are similar, the genes have opposite effects. In arthropods, such as fruit flies and crayfish, the nerve cord is toward the front; in vertebrates, such as chickens and humans, the nerve cord is toward the back.

When the degree of similarity in DNA base sequences of genes or in amino acid sequences of proteins is examined, the data are as expected, assuming common descent. Cytochrome *c* is a molecule used in the electron transport chain of all the organisms shown in **Figure 14.18**. Data regarding differences in the amino acid sequence of cytochrome *c* show that the sequence in a human differs from that in a monkey by only 1 amino acid, from that in a duck by 11 amino acids, and from that in *Candida,* a yeast, by 51 amino acids. These data are consistent with other data regarding the anatomical similarities of these organisms, and therefore demonstrate their relatedness.

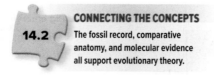

CONNECTING THE CONCEPTS

14.2 The fossil record, comparative anatomy, and molecular evidence all support evolutionary theory.

Check Your Progress 14.2

1. Explain how biogeographical information about Galápagos finches supports the theory of evolution.

2. Describe how vestigial structures support the theory of evolution.

3. Contrast homologous structures with analogous structures.

STUDY TOOLS http://connect.mheducation.com

SMARTBOOK® Maximize your study time with McGraw-Hill SmartBook®, the first adaptive textbook.

SUMMARIZE

The work of Charles Darwin helped shape modern evolutionary thought. All species on Earth, including humans, have a common ancestry due to the process of evolution. Scientific evidence strongly supports evolutionary theory.

Both Darwin and Wallace used multiple lines of scientific evidence to formulate the theory of natural selection.

The fossil record, comparative anatomy, and molecular evidence all support evolutionary theory.

14.1 Darwin's Theory of Evolution

Charles Darwin is recognized for developing the theory of natural selection, which explains the process of **evolution,** or how species change over time.

Before Darwin

Before Darwin, people believed that the Earth was young, species did not change, and variations were imperfections.

- Cuvier was an early paleontologist who believed that species do not change. He observed species come and go in the fossil record, and he said these changes were due to catastrophic events.
- Lamarck was a zoologist who hypothesized that evolution and adaptation to the environment do occur. He suggested that acquired characteristics could be inherited. For example, he said giraffes stretched their necks to reach food in trees, and then longer necks were inherited by the next generation.

Darwin's Conclusions

Darwin's conclusions based on geology and **fossils** are

- The Earth is very old, giving time for evolution to occur.
- Living organisms are descended from extinct life-forms known only from the fossil record.

Darwin's conclusions based on **biogeography** are

- Living organisms evolve where they are. This explains, for example, why South America has the Patagonian hare, whereas England has the European rabbit.
- Living organisms are adapted to their environments. This explains why there are many types of finches and tortoises in the Galápagos Islands.

Natural Selection and Adaptation

According to Darwin, the results of **natural selection** are **adaptations** that allow a species to be better suited to its local environment than it was in previous generations. These adaptations improve the **fitness,** or reproductive success, of the species:

Observation	Result	Conclusion
1 a. Organisms have variations.	b. New adaptations to the environment arise.	Organisms become more adapted with each generation.
2 a. Organisms struggle to exist.	b. More organisms are present than can survive.	
3 a. Organisms differ in fitness.	b. Organisms best suited to the environment survive and reproduce.	

Over time, these adaptations may result in **speciation,** or the formation of a new species. Darwin used examples of **artificial selection** to help explain the process of natural selection.

Darwin and Wallace

Alfred Russel Wallace was a naturalist who, like Darwin, traveled to other continents in the Southern Hemisphere. He also collected evidence of common descent, and his reading of Malthus caused him to propose the same mechanism for adaptation (natural selection) as Darwin. Darwin's work was more thorough, as evidenced by his book *On the Origin of Species.*

14.2 Evidence of Evolutionary Change

A theory in science is a concept supported by much evidence, and evolutionary theory is supported by several types of evidence:

- The **fossil record** indicates the history of life in general and allows us to trace the descent of particular groups. Transitional fossils play an important role in documenting the change in an organism over time.
- Biogeography shows that the distributions of organisms on Earth are explainable by assuming organisms evolved in one locale.
- Comparative anatomy reveals homologies among organisms that are explainable only by common ancestry.
  - **Vestigial structures** are the nonfunctional remnants of once functional structures.
  - **Homologous structures** are similar structures that may be explained by inheritance from a common ancestor.
  - **Analogous structures** are structures that have a similar function but different evolutionary origins.
- Molecular evidence compares biochemical molecules (DNA, proteins) to discover minor differences that may indicate the degree of relatedness between different types of organisms.

ASSESS

Testing Yourself

Choose the best answer for each question.

14.1 Darwin's Theory of Evolution

1. The idea that acquired characteristics can be inherited from one generation to the next was proposed by:
   - **a.** Darwin
   - **b.** Lamarck
   - **c.** Wallace
   - **d.** Sedgwick
   - **e.** Cuvier

2. Which of the following is not an example of natural selection?
   - **a.** Insect populations exposed to pesticides become resistant to the chemicals.
   - **b.** Plant species that produce fragrances to attract pollinators produce more offspring.
   - **c.** Rabbits that sprint quickly are more likely to escape predation.
   - **d.** On a tree, leaves that grow in the shade are larger than those that grow in the sun.

3. Natural selection is the only process that results in
   - **a.** genetic variation.
   - **b.** adaptation to the environment.
   - **c.** phenotypic change.
   - **d.** competition among individuals in a population.

4. Why was it helpful to Darwin to learn that Lyell had concluded the Earth is very old?
   a. An old Earth would have more fossils than a new Earth.
   b. It meant there was enough time for evolution to have occurred slowly.
   c. There was enough time for the same species to spread out into all continents.
   d. Darwin said artificial selection occurs slowly.
   e. All of these are correct.

5. New alleles for a trait arise by
   a. mutation.
   b. the needs of the species.
   c. sexual reproduction.
   d. mitosis.

14.2 Evidence of Evolutionary Change

6. Differences in DNA nucleotides between organisms
   a. indicate how closely related organisms are.
   b. indicate that evolution occurs.
   c. explain why there are phenotypic differences.
   d. are to be expected.
   e. All of these are correct.

7. The fossil record offers direct evidence for common descent because we can
   a. see that the types of fossils change over time.
   b. sometimes find common ancestors.
   c. trace the ancestry of a particular group.
   d. trace the biological history of living organisms.
   e. All of these are correct.

8. For there to be homologous structures,
   a. a common ancestor had to have existed.
   b. analogous structures also have to exist.
   c. the bones have to be used similarly.
   d. All of these are correct.

For questions 9–12, match the description with the type of evidence for evolution it supports, as listed in the key. Answers can be used more than once.

Key:
   a. biogeographical
   b. anatomical
   c. biochemical

9. The genetic code is the same for all organisms.

10. The human knee bone and spine were derived from ancestral structures that supported four-legged animals.

11. The South American continent lacks rabbits, even though the environment is quite suitable.

12. The amino acid sequence of hemoglobin in humans is more similar to that of rhesus monkeys than to that of mice.

13. Fossils that serve as transitional links allow scientists to
   a. determine how prehistoric animals interacted with each other.
   b. deduce the order in which various groups of animals arose.
   c. relate climate change to evolutionary trends.
   d. determine why evolutionary changes occur.

14. Among vertebrates, the flipper of a dolphin and the fin of a tuna are
   a. homologous structures.
   b. homogeneous structures.
   c. analogous structures.
   d. reciprocal structures.

ENGAGE

BioNOW

Want to know how this science is relevant to your life? Check out the BioNow video below:
   • Quail Evolution

Explain how the experiment in this video relates to Darwin's theory of natural selection.

Thinking Critically

1. The human appendix, a vestigial extension of the large intestine, is homologous to a structure called a caecum in other mammals. A caecum, generally larger than our appendix, houses bacteria that aid in digesting cellulose, the main component of plants. How might the presence of the appendix be used to show our common ancestry with other mammals, and what might it tell us about the dietary history of humans?

2. Geneticists compare DNA base sequences among organisms and from these data determine a gene's rate of evolution. Different genes have been found to evolve at different rates. Explain why some genes have faster rates of evolution than other genes as populations adapt to their environments.

3. Both Darwin and Wallace, while observing life on islands, concluded that natural selection is the mechanism for biological evolution. The Hawaiian and nearby islands once had at least 50 species of honeycreepers, and they lived nowhere else on Earth. Natural selection occurs everywhere and in all species. What characteristics of islands allow the outcome of natural selection to be so obvious?

© Michael Freeman/Getty RF

15

Evolution on a Small Scale

OUTLINE

15.1 Natural Selection   252

15.2 Microevolution   257

BEFORE YOU BEGIN

Before beginning this chapter, take a few moments to review the following discussions.

Section 1.2 Why is evolution considered to be the core concept of biology?

Section 9.3 How do sexual reproduction and meiosis increase variation in a population?

Section 14.1 How does natural selection act as the mechanism of evolutionary change?

Life at High Elevations

Normally, if a person moves to a higher altitude, where the level of oxygen in the air is lower, his or her body responds by making more hemoglobin, the component of blood that carries oxygen. For minor elevation changes, this increase in hemoglobin does not present much of a problem. But for people who move to extreme elevations (as in the Himalayas, where some people live at elevations of over 13,000 ft, or close to 4,000 m), this can present a number of health problems, including chronic mountain sickness, a disease that affects people who live at high altitudes for extended periods of time. When the amount of hemoglobin is increased substantially, the blood thickens and becomes more viscous. This can cause hypertension and an increase in the formation of blood clots, both of which have negative physiological effects.

Because high hemoglobin levels can be a detriment to people at high elevations, it makes sense that natural selection will favor individuals who produce less hemoglobin at these heights. This has been found to be the case with Tibetans. Researchers have identified an allele of a gene (*EPSA1*) that reduces hemoglobin production at high elevations. Comparisons between Tibetans living at high and low elevations strongly suggest that selection has played a role in the prevalence of the high-elevation allele.

Interestingly the *EPSA1* gene in Tibetans is identical to a similar gene found in an ancient group of humans called the Denisovans. Scientists now believe that the *EPSA1* gene entered the Tibetan population around 40,000 years ago, most likely through interbreeding between the early Tibetans and Denisovans.

This chapter explores how natural selection and microevolution influence a population's gene pool over time.

As you read through this chapter, think about the following questions:

1. What is the link between genes, populations, and evolution?
2. How do scientists determine whether a population is evolving?

15.1 Natural Selection

Learning Outcomes

Upon completion of this chapter, you should be able to

1. Describe the three types of natural selection—directional, stabilizing, and disruptive.
2. Explain how heterozygotes maintain variation in a population, and summarize the concept of a heterozygote advantage.

Natural selection is the process that results in adaptation of a population to the biotic (living) and abiotic (nonliving) components of the environment. In responding to the biotic components, organisms acquire resources through competition, predation, and parasitism. The abiotic environment includes weather conditions, dependent chiefly on temperatures and precipitation. Charles Darwin became convinced that species evolve with time and suggested natural selection as the mechanism for adaptation to the environment (see Section 14.1). In **Table 15.1,** Darwin's hypothesis of natural selection is stated in a way that is consistent with modern genetics.

As a result of natural selection, the most *fit* individuals become more prevalent in a population, and in this way, a population changes over time. The most fit individuals are those that reproduce more than others. In most cases, these individuals are those that are better adapted to the environment.

Table 15.1 Natural Selection

Evolution by natural selection involves:

1. Variation. The members of a population differ from one another.

2. Inheritance. Many of these differences are heritable genetic differences.

3. Increased fitness. Individuals that are better adapted to their environment are more likely to reproduce, and their fertile offspring will make up a greater proportion of the next generation.

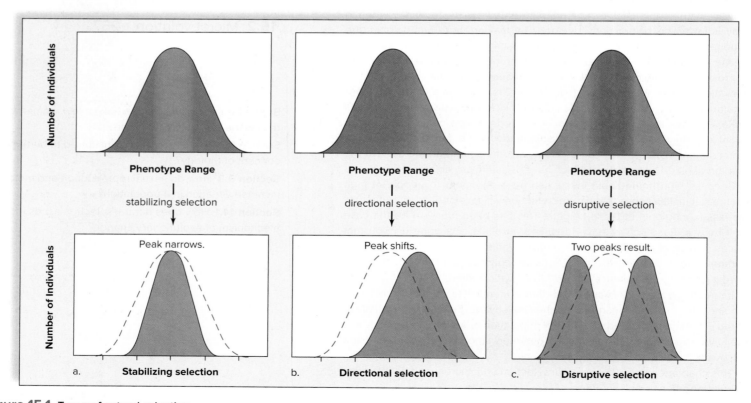

Figure 15.1 Types of natural selection.

Natural selection shifts the average value of a phenotype over time. **a.** During stabilizing selection, the intermediate phenotype increases in frequency. **b.** During directional selection, an extreme phenotype is favored, which changes the average phenotype value. **c.** During disruptive selection, two extreme phenotypes are favored, creating two new average phenotype values, one for each phenotype.

Types of Selection

For many traits, there are multiple alleles that may produce a range of phenotypes. The frequency distributions of these phenotypes in a population often resemble the bell-shaped curves shown in **Figure 15.1.** Natural selection works to decrease the prevalence of detrimental phenotypes and to favor those phenotypes that are better adapted to the environment. There are three basic types of natural selection: stabilizing selection, directional selection, and disruptive selection.

Stabilizing Selection

Stabilizing selection occurs when an intermediate phenotype is favored (see Fig. 15.1*a*). With stabilizing selection, extreme phenotypes are selected against, and individuals near the average are favored. This is the most common form of selection because the average individual is well adapted to its environment.

As an example, consider that when Swiss starlings (*Sturnus vulgaris*) lay four or five eggs, more young survive than when the female lays more or less than this number **(Fig. 15.2)**. Genes determining physiological characteristics, such as the production of yolk, and behavioral characteristics, such as how long the female will mate, are involved in determining clutch size.

Directional Selection

Directional selection occurs when an extreme phenotype is favored and the frequency distribution curve shifts in that direction (see Fig. 15.1*b*). Such a shift can occur when a population is adapting to a changing environment.

Resistance to antibiotics and insecticides provides a classic example of directional selection. As you may know, the widespread use of antibiotics and pesticides results in populations of bacteria and insects that are resistant to these chemicals. When an antibiotic is administered, some bacteria may

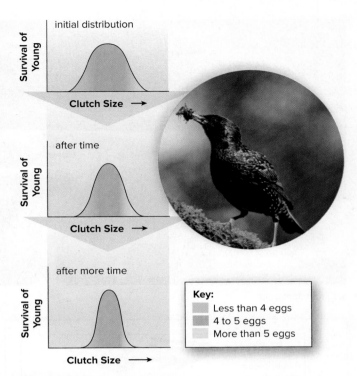

Figure 15.2   Stabilizing selection.

Stabilizing selection occurs when natural selection favors the intermediate phenotype over the extremes. For example, Swiss starlings that lay four or five eggs (usual clutch size) have more young survive than those that lay fewer than four eggs or more than five eggs.

© blickwinkel/Alamy

Connections: Scientific Inquiry

Are there examples of stabilizing selection in humans?

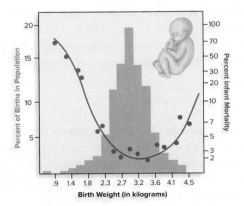

Perhaps the best example of stabilizing selection in humans is related to birth weight. Studies in England and the United States in the mid-twentieth century indicated that infants with birth weights between 6 and 8 pounds had a higher rate of survival. Interestingly, advances in medical care for premature babies with low birth weights and the increased use of cesarean sections to deliver high-birth-weight babies have lessened the effects of this stabilizing selection in some parts of the world.

a. Experimental site

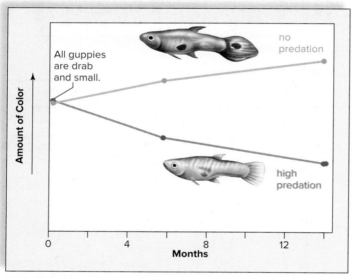

b. No predation results in colorful guppies.

Figure 15.3 Directional selection.

a. The experimental site used by the researchers. b. In the presence of selection (predation), the phenotype favored smaller, drab-colored male guppies. However, when the selective force was removed, the phenotype of the male guppies shifted to larger, more colorful individuals.

© Helen Rodd

survive because they are genetically resistant to the antibiotic. These are the bacteria that are likely to pass on their genes to the next generation. As a result, the number of resistant bacteria keeps increasing. Drug-resistant strains of bacteria that cause tuberculosis have become a serious threat to the health of people worldwide.

Another example of directional selection is the human struggle against malaria, a disease caused by an infection of the liver and the red blood cells. The *Anopheles* mosquito transmits the disease-causing protozoan *Plasmodium* from person to person. In the early 1960s, international health authorities thought malaria would soon be eradicated. A new drug, chloroquine, seemed effective against *Plasmodium,* and spraying of DDT (an insecticide) had reduced the mosquito population. But by the mid-1960s, *Plasmodium* was showing signs of chloroquine resistance, and worse yet, mosquitoes were becoming resistant to DDT. A few drug-resistant parasites and a few DDT-resistant mosquitoes had survived and multiplied, shifting the frequency distribution curve toward the resistant type of parasite.

Another example of directional selection was observed in an experiment performed with guppies. The environment included two areas, one below a waterfall and stocked with pike (a fish predator of guppies) and the other above the waterfall and lacking pike (**Fig. 15.3***a*). Over time, in the lower area, natural selection favored male guppies that were small and drab-colored so that they could avoid detection by the pike. However, when the researchers moved male guppies to the area above the waterfall, the absence of such selection caused a change in the phenotype toward larger, more colorful guppies (Fig. 15.3*b*).

Disruptive Selection

In **disruptive selection,** two or more extreme phenotypes are favored over any intermediate phenotype. Therefore, disruptive selection favors *polymorphism,* the occurrence of different forms in a population of the same species. For example, British land snails (*Cepaea nemoralis*) are found in low-vegetation areas (grass fields and hedgerows) and in forests. In low-vegetation areas, thrushes feed mainly on snails with dark shells that lack light bands; in forest areas, they feed mainly on snails with light-banded shells. Therefore, these two distinctly different phenotypes, each adapted to its own environment, are found in this population (**Fig. 15.4**).

Sexual Selection

The term **sexual selection** refers to adaptive changes in males and females that lead to an increased ability to secure a mate. Each sex has a different strategy with regard to sexual selection. Since females produce few eggs, the choice of a mate is a serious consideration. However, males can father many offspring because they continuously produce sperm in great quantity. Therefore, males often compete in order to inseminate as many females as possible. Because of this, sexual selection in males usually results in an increased ability to compete with other males for a mate. On the other hand, sexual selection in females favors the choice of a single male with the best **fitness,** or the ability to produce surviving offspring. Males often demonstrate their fitness by coloration or elaborate mating rituals (**Fig. 15.5**). By choosing a male with optimal fitness, the female increases the chances that her traits will be passed on to the next generation. Because of this, many consider sexual selection a form of natural selection.

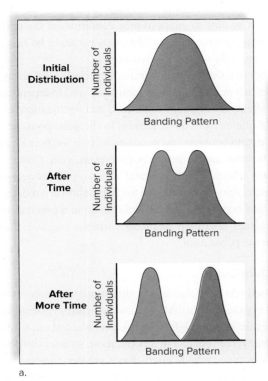

a.

b.

Figure 15.4 Disruptive selection.

a. Disruptive selection favors two extreme phenotypes among snails, no banding and banding. **b.** Today, British land snails mainly comprise these two different phenotypes, each adapted to a different habitat.

b: (left) © Graeme Teague; (right) © IT Stock Free/Alamy RF

Adaptations Are Not Perfect

Natural selection doesn't always produce organisms that are perfectly adapted to their environment. Why not? First, it is important to realize that evolution doesn't start from scratch. Just as you can only bake a cake with the ingredients available to you, evolution is constrained by the available variations. Each species must build upon its own evolutionary history, which limits the amount of variation that may be acted on by natural selection. Second, as adaptations are evolving in a species, the environment may also be changing. Most adaptations provide a benefit to the species for a specific environment for a specific time. As the environment changes, the benefit of a certain adaptation may be minimized. It is also important to recognize that imperfections are common because of necessary compromises. The success of humans is attributable to their dexterous hands, but the spine is subject to injury because the vertebrate spine did not originally evolve to stand erect. A feature that evolves has a benefit that is worth the cost. For example, the benefit of freeing the hands must have been worth the increased cost of spinal injuries from assuming an erect posture.

Maintenance of Variations

A population always shows some genotypic variation. The maintenance of variation is beneficial because populations

Figure 15.5 Sexual selection.

The elaborate coloration in the males of some species is a form of sexual selection that is intended to demonstrate an increased level of fitness.

© Ernest A. Janes/Bruce Coleman/Photoshot

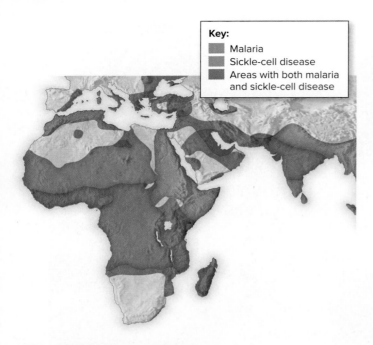

Key:
- Malaria
- Sickle-cell disease
- Areas with both malaria and sickle-cell disease

Figure 15.6 Sickle-cell disease.

Red shows the areas where malaria was prevalent in Africa, the Middle East, southern Europe, and southern Asia in 1920, before eradication programs began; shown in blue are the areas where sickle-cell disease most often occurred. The overlap of these two distributions (purple) suggested a connection.

Table 15.2 Example of Heterozygote Advantage

Genotype	Phenotype	Result
$Hb^A Hb^A$	Normal	Dies due to malarial infection
$Hb^A Hb^S$	Carrier of sickle-cell disease	Lives due to protection from both
$Hb^S Hb^S$	Sickle-cell disease	Dies due to sickle-cell disease

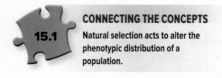

CONNECTING THE CONCEPTS

15.1 Natural selection acts to alter the phenotypic distribution of a population.

with limited variation may not be able to adapt to new conditions if the environment changes, and thus may become extinct. How can variation be maintained in spite of selection constantly working to reduce it?

First, we must remember that the forces that promote variation are always at work: Mutation still generates new alleles, recombination and independent assortment still shuffle the alleles during gametogenesis, and fertilization still creates new combinations of alleles from those present in the gene pool. Second, gene flow might be occurring between two populations (see Section 15.2). If the receiving population is small and is mostly homozygous, gene flow can be a significant source of new alleles. Finally, natural selection favors certain phenotypes, but the other phenotypes may remain in the population at a reduced frequency. Disruptive selection even promotes polymorphism in a population. In diploid species, heterozygotes also help maintain variation because they conserve recessive alleles in the population.

The Heterozygote Advantage

Only alleles that are expressed (cause a phenotypic difference) are subject to natural selection. In diploid organisms, this fact makes the heterozygote a potential protector of recessive alleles that might otherwise be weeded out of the gene pool. Because the heterozygote remains in a population, so does the possibility of the recessive phenotype, which might have greater fitness in a changed environment. When, over time, environmental conditions cause natural selection to maintain two different alleles of a gene at a certain ratio, the situation is called *balanced polymorphism*. Sickle-cell disease offers an example of a balanced polymorphism.

Sickle-Cell Disease Individuals with sickle-cell disease have the genotype $Hb^S Hb^S$ (Hb = hemoglobin, the oxygen-carrying protein in red blood cells; S = sickle cell) and tend to die at an early age due to hemorrhaging and organ destruction. Those who are heterozygous ($Hb^A Hb^S$; A = normal) have sickle-cell trait and are better off because their red blood cells usually become sickle-shaped only when the oxygen content of the environment is low. Ordinarily, those with a normal genotype ($Hb^A Hb^A$) are the most fit.

Geneticists studying the distribution of sickle-cell disease in Africa have found that the recessive allele (Hb^S) has a higher frequency (from 0.2, or 20%, to as high as 0.4, or 40%, in a few areas) in regions where malaria is also prevalent (**Fig. 15.6**).

What is the connection between higher frequency of the recessive allele and malaria? Malaria is caused by a parasite that lives in and destroys the red blood cells of the normal homozygote ($Hb^A Hb^A$). However, the parasite is unable to live in the red blood cells of the heterozygote ($Hb^A Hb^S$) because the infection causes the red blood cells to become sickle-shaped. Sickle-shaped red blood cells lose potassium, and this causes the parasite to die. In an environment where malaria is prevalent, the heterozygote is favored. Each of the homozygotes is selected against, but the recessive allele is maintained in the population. **Table 15.2** summarizes the effects of the three possible genotypes.

Check Your Progress 15.1

1. Distinguish between directional, stabilizing, and disruptive selection.

2. Explain how sexual selection represents a form of natural selection.

3. List the forces that help maintain genetic variability in a population.

15.2 Microevolution

Learning Outcomes

Upon completion of this section, you should be able to

1. Define the term *microevolution*.
2. Understand how the Hardy-Weinberg principle is used to explain the process of microevolution.
3. Describe how mutations, gene flow, nonrandom mating, genetic drift, and natural selection can cause changes in the frequency of an allele in a population.

Many traits can change temporarily in response to a varying environment. For example, the color change in the fur of an Arctic fox from brown to white in winter, the increased thickness of your dog's fur in cold weather, or the bronzing of your skin when exposed to the sun lasts only for a season.

These are not evolutionary changes. Changes to traits over an individual's lifetime are not evidence that an individual has evolved, because these traits are not heritable. In order for traits to evolve, they must have the ability to be passed on to subsequent generations. Evolution causes change in a heritable trait within a **population,** not within an individual, over many generations.

Darwin observed that populations, not individuals, evolve, but he could not explain how traits change over time. Now we know that genes interact with the environment to determine traits. Because genes and traits are linked, evolution is really about genetic change—or more specifically, evolution is the *change in allele frequencies in a population over time*. This type of evolution is called **microevolution.**

Evolution in a Genetic Context

It was not until the 1930s that biologists were able to apply the principles of genetics to populations and thereafter to develop a way to recognize when evolution has occurred and measure how much a population has changed.

In **population genetics,** the various alleles at all the gene loci in all individuals make up the **gene pool** of the population. It is customary to describe the gene pool of a population in terms of genotype and allele frequencies. The genotype frequency is the percentage of a specific genotype—for example, homozygous dominant individuals—in a population. The allele frequency represents how much a specific allele is represented in the gene pool of the population. Let's take an example based on peppered moths, which can be light-colored or dark-colored (**Fig. 15.7**).

Suppose you research the literature and find that the color of peppered moths is controlled by a single set of alleles and you decide to use the following key:

$$D = \text{dark color} \qquad d = \text{light color}$$

Furthermore, you find that, in one population of these moths in Great Britain before pollution fully darkened the trees (Fig. 15.7*a*), only 4% (0.04) of the moths were homozygous dominant (*DD*); 32% (0.32) were heterozygous (*Dd*), and 64% (0.64) were homozygous recessive (*dd*). From these genotype frequencies, you can calculate the allele frequencies in the population:

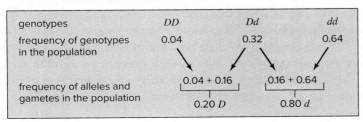

Connections: Scientific Inquiry

Why don't individuals evolve?

Evolution results in genetic change in a population over periods of time. While individual organisms, such as humans, may develop new skills and abilities (such as learning a new language or playing a guitar), their genetic material remains unchanged. These new abilities are not passed on to the next generation and do not change the genetic composition of the population.

a.

b.

Figure 15.7 Industrial melanism and microevolution.

Coloration in the peppered moth (*Biston betularia*) is due to two alleles in the gene pool. **a.** Before widespread air pollution due to industrial development in Great Britain, the light-colored phenotype was more frequent in the population, because birds were unable to see the light-colored moths on the light tree trunks. **b.** After pollution darkened the trunks of the trees, the dark-colored phenotype became more frequent in the population. Microevolution occurred, bringing changes in gene pool frequencies—in this case, due to natural selection.

(both): © Michael Tweedie/Science Source

In this population, the frequency of the *D* allele (dark) in the gene pool is 20% (0.20), and the frequency of the *d* allele (light) is 80% (0.80). Therefore, the gametes (sperm and egg) produced by this population will have a 20% chance of carrying the *D* allele and an 80% chance of carrying the *d* allele. Assuming random mating (all possible gametes have an equal chance to combine with any other), we can use these frequencies to calculate the ratio of genotypes in the next generation by using a Punnett square (**Fig. 15.8**). For example, to produce a homozygous dominant (*DD*) moth, both parents must contribute the *D* allele. Since this allele is present in only 20% of the gene pool, the chances that the male will contribute a sperm cell with the *D* allele is 20% (0.20) and that the female will contribute an egg with the *D* allele is 20% (0.20). The chances that both of these events will occur is 0.20 times 0.20, or 0.04 (4%). Therefore, if the moths are randomly mating, 4% of the next generation should be homozygous dominant (*DD*).

There is an important difference between a Punnett square that represents a cross between individuals (as was the case with one-trait and two-trait inheritance; see Section 10.1) and the one shown in Figure 15.8. In Figure 15.8, we are using the gamete frequencies in the *population* to determine the genotype frequencies in the next generation. As you can see, the results show that the genotype frequencies (and therefore the allele frequencies) in the next generation are the same as they were in the previous generation. In other words, the homozygous dominant moths (*DD*) are still 0.04 (4%), the heterozygous moths (*Dd*) are still 0.32 (32%), and the homozygous recessive moths (*dd*) are still 0.64 (64%) of the population. This remarkable finding tells us that *sexual reproduction alone cannot bring about a change in genotype and allele frequencies.* Also, the dominant allele need not increase from one generation to the next. Dominance does not cause an allele to become a common allele.

The fact that the allele frequencies of the gene pool appear to remain at equilibrium from one generation to the next, as demonstrated in Figure 15.8, was independently recognized in 1908 by G. H. Hardy, an English mathematician, and W. Weinberg, a German physician. They developed a binomial equation to calculate the genotype and allele frequencies of a population (Fig. 15.8). In this equation,

p = frequency of the dominant allele (in the case of the moths, the D allele)

q = frequency of the recessive allele (the d allele for the moths)

The *Hardy-Weinberg principle* states that an equilibrium (balance) of genotype frequencies exists in a gene pool and may be represented by the equation:

$$p^2 + 2pq + q^2 = 1$$

Let's take a look at this equation in relation to our example of the peppered moths (Fig. 15.8). In our example, the dark allele (D) was the dominant allele, and it was present in 20% (0.20) of the population. Therefore, $p = 0.20$, and the probability that both parents will contribute the allele is $p \times p$ (p^2), or 0.04 (4%). In order for an individual to be homozygous recessive (dd), he or she must inherit a recessive allele from both parents. Since the recessive allele (d) has a frequency of 0.80 in the gene pool (represented by q), the probability is 0.8×0.8 (q^2), or 0.64 (64%). Notice from the Punnett square in Figure 15.8 that there are two ways that an individual may be heterozygous, which in the equation is represented by $2pq$. Therefore, the probability of being heterozygous is $2 \times 0.2 \times 0.8$, or 0.32 (32%).

The mathematical relationships of the Hardy-Weinberg principle will remain in effect in each succeeding generation of a sexually reproducing population as long as five conditions are met:

1. No mutations: Allelic changes do not occur, or changes in one direction are balanced by changes in the opposite direction.
2. No gene flow: Migration of alleles into or out of the population does not occur.
3. Random mating: Individuals pair by chance, not according to their genotypes or phenotypes.
4. No genetic drift: The population is very large, and changes in allele frequencies due to chance alone are insignificant.
5. Selection: Natural selection is not occurring or does not favor any allele or combination of alleles over another.

These conditions are rarely, if ever, met, and genotype and allele frequencies in the gene pool of a population do change from one generation to the next. Therefore, microevolution does occur, and the extent of change can be measured. The significance of the Hardy-Weinberg principle is that it tells us what factors cause evolution—those that violate the conditions listed. Microevolution can be detected and measured by noting the amount of deviation from a Hardy-Weinberg equilibrium of genotype frequencies in the gene pool of a population.

For genotype frequencies to be subject to natural selection, they must result in a change of phenotype frequencies. *Industrial melanism,* an increase in the frequency of a dark phenotype due to pollution, provides us with an example. We supposed that only 36% of our moth population was dark-colored (homozygous dominant plus heterozygous). Why might that be? Before the rise of industry, dark-colored moths rested on light tree trunks, where they were seen and eaten by birds. However, with industrial development, the trunks of

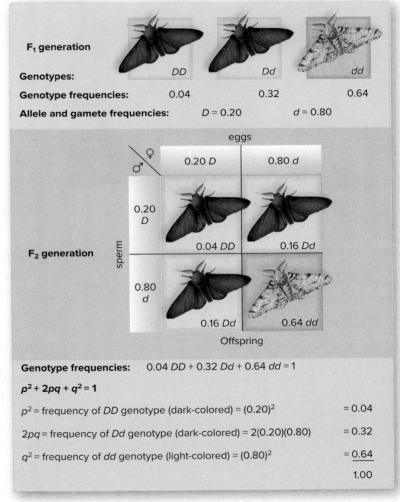

Figure 15.8 **The relationship between genotype and phenotype frequencies in a population.**

Using the gamete (allele) frequencies in a population, it is possible to employ a Punnett square to calculate the genotype frequencies of the next generation. This calculation indicates that sexual reproduction alone does not alter the genotype and allele frequencies.

trees darkened as a result of air pollution, and the light-colored moths became visible and were eaten more often (see Fig. 15.7*b*). Predatory birds acted as a selective agent, and microevolution occurred—in the mid-1950s, the number of dark-colored moths in some Great Britain populations exceeded 80%. Aside from showing that natural selection can occur within a short period of time, our example illustrates that a change in gene pool frequencies does take place as microevolution occurs.

Causes of Microevolution

Any conditions that cause a change in the equilibrium of alleles within a population can cause evolutionary change. Thus, the following five factors can cause a divergence from the Hardy-Weinberg equilibrium: genetic mutation, gene flow, nonrandom mating, genetic drift, and natural selection (see Section 15.1).

Genetic Mutation

Mutations, which are permanent genetic changes, are the raw material for evolutionary change. Without mutations, there can be no new variations among members of a population on which natural selection can act. However, the rate of mutations is generally very low—on the order of 1 mutation per 100,000 cell divisions. In addition, many mutations are neutral (**Fig. 15.9**), meaning that they are not selected for or against by natural selection. Prokaryotes do not reproduce sexually and therefore are more dependent than eukaryotes on mutations to introduce variations. All mutations that occur and result in phenotypic differences can be tested by the environment. However, in sexually reproducing organisms, mutations, if recessive, do not immediately affect the phenotype.

In a changing environment, even a seemingly harmful mutation that results in a phenotypic difference can be the source of an adaptive variation. For example, the water flea *Daphnia* ordinarily thrives at temperatures around 20°C, but there is a mutation that requires *Daphnia* to live at temperatures between 25°C and 30°C. The adaptive value of this mutation is entirely dependent on environmental conditions.

Gene Flow

Gene flow, also called *gene migration,* is the movement of alleles among populations by migration of breeding individuals. Gene flow can increase the variation within a population by introducing novel alleles that were produced by mutation in another population. Continued gene flow due to migration of individuals makes gene pools similar and reduces the possibility of allele frequency differences among populations now and in the future. Indeed, gene flow among populations can prevent speciation from occurring. Due to gene flow, the snake populations featured in **Figure 15.10** are *subspecies*—different populations within the same species. Despite somewhat distinctive characteristics, there is enough genetic similarity between the populations that these subspecies of *Pantherophis obsoleta* can readily interbreed when they come in contact with one another.

Nonrandom Mating

Random mating occurs when individuals select mates and pair by chance, not according to their genotypes or phenotypes. Inbreeding, or mating between relatives, is an example of **nonrandom mating.** Inbreeding does not change allele frequencies, but it does gradually increase the proportion of homozygotes, because the homozygotes that result must produce only homozygotes.

Assortative mating occurs when individuals tend to mate with those that have the same phenotype with respect to a certain characteristic. In humans,

Figure 15.9 Freckles.

A dominant allele causes freckles, so why doesn't everyone have freckles? The Hardy-Weinberg principle, which states that sexual reproduction in and of itself doesn't change allele frequencies, explains why dominant alleles don't become more prevalent with each generation.

© Corbis RF

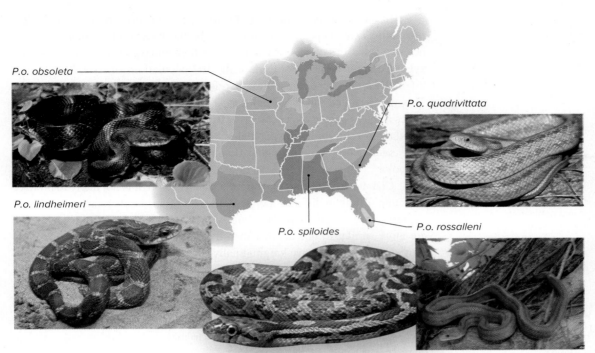

Figure 15.10 Gene flow.

Each rat snake shown here represents a separate population of snakes. Because the populations are adjacent to one another, interbreeding occurs, and so does gene flow between the populations. This keeps their gene pools somewhat similar, and each of these populations is considered a subspecies of the species *Pantherophis obsoleta* (as indicated by the three-part names).

(*P. o. obsoleta*): © Robert Hamilton/Alamy RF; (*P. o. quadrivittata*): © Millard H. Sharp/ Science Source; (*P. o. rossalleni*): © Graeme Teague; (*P. o. spiloides*): © F. Teigler/ blickwinkel/age fotostock; (*P. o. lindheimeri*): © Michelle Gilders/age fotostock/SuperStock

cultural differences often cause individuals to select members of their own group. Assortative mating causes the population to subdivide into two phenotypic classes, between which gene exchange is reduced. Homozygotes for the gene loci that control the trait in question increase in frequency, and heterozygotes for these loci decrease in frequency.

Sexual selection favors characteristics that increase the likelihood of obtaining mates, and in this way it promotes nonrandom mating. In most species, males that compete best for access to females and/or have a phenotype that attracts females are more apt to mate and have increased fitness (see Section 15.1).

Genetic Drift

Genetic drift refers to changes in the allele frequencies of a gene pool due to chance. This mechanism of evolution is called genetic drift because allele frequencies "drift" over time. They can increase or decrease depending on which members of a population die, survive, or reproduce with one another. Although genetic drift occurs in both large and small populations, a larger population is expected to suffer less of a sampling error than a smaller population. Suppose you had a large bag containing 1,000 green balls and 1,000 blue balls, and you randomly drew 10%, or 200, of the balls. Because of the large number of balls of each color in the bag, you can reasonably expect to draw 100 green balls and 100 blue balls, or at least a ratio close to this. It is extremely unlikely that you would draw 200 green or 200 blue balls. But suppose you had a bag containing only 10 green balls and 10 blue balls, and you drew 10%, or only 2 balls. You could easily draw 2 green balls or 2 blue balls, or 1 of each color.

When a population is small, random events may reduce the ability of one genotype with regard to the production of the next generation. Suppose that, in a small population of frogs, certain frogs by chance do not pass on their traits. Certainly, the next generation will have a change in allele frequencies (**Fig. 15.11**). When genetic drift leads to a loss of one or more alleles, other alleles over time become *fixed* in the population.

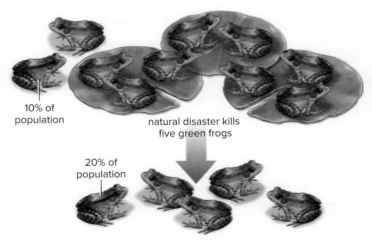

10% of population

natural disaster kills five green frogs

20% of population

Figure 15.11 Genetic drift.

Genetic drift occurs when a random event changes the frequency of alleles in a population. The allele frequencies of the next generation's gene pool may be markedly different from those of the previous generation.

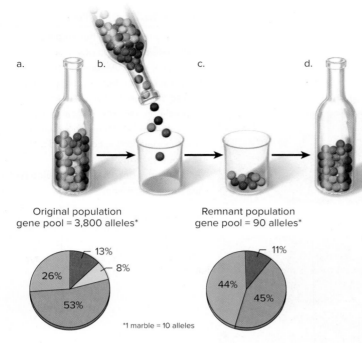

a.   b.   c.   d.

Original population
gene pool = 3,800 alleles*

Remnant population
gene pool = 90 alleles*

13%

8%

26%

53%

11%

44%

45%

*1 marble = 10 alleles

Figure 15.12 Bottleneck and founder effects.

a. In this example, the gene pool of a population contains four different alleles, each represented by a different color of marble in the bottle. Each allele has a different frequency in the population. **b.** A bottleneck event occurs, limiting the number of individuals in the resulting population. **c.** The gene pool has changed from the initial population. Note the absence of yellow marbles. **d.** The population may return to its original size, but the frequency of each allele has changed. A founder effect is similar to a bottleneck effect, except that the reduced population is simply isolated from the original population, which continues to exist.

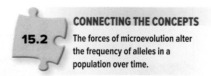

CONNECTING THE CONCEPTS

15.2 The forces of microevolution alter the frequency of alleles in a population over time.

Check Your Progress 15.2

1. Define the term *gene pool,* and explain how it relates to allele frequencies in a population.

2. Explain the Hardy-Weinberg principle. What happens to the equilibrium of allele frequencies when microevolution occurs?

3. List the five factors that prevent microevolution in a population.

4. Describe the significance of mutations in terms of evolution.

5. Explain how gene flow and nonrandom mating cause microevolution.

6. Describe the consequences of genetic drift, and explain why it is more likely to happen in a small population.

In an experiment involving brown eye color, each of 107 *Drosophila* populations was kept in its own culture bottle. Every bottle contained eight heterozygous flies of each sex. There were no homozygous recessive or homozygous dominant flies. For each of the 107 populations of flies, 8 males and 8 females were chosen from the offspring and placed in a new culture bottle. This was repeated for 19 generations. The random selection of males and females acted as a form of genetic drift. By the nineteenth generation, 25% of the populations (culture bottles) contained only homozygous recessive flies, and 25% contained only homozygous dominant flies having the allele for brown eye color.

Genetic drift is a random process, and therefore it is not likely to produce the same results in different populations. In California, there are a number of cypress groves, each a separate population. The phenotypes within each grove are more similar to one another than they are to the phenotypes in the other groves. Some groves have longitudinally shaped trees, and others have pyramidally shaped trees. The bark is rough in some colonies and smooth in others. The leaves are gray to bright green or bluish green, and the cones are small or large. Because the environmental conditions are similar for all the groves, and no correlation has been found between phenotype and environment across groves, scientists hypothesize that these variations among the populations are due to genetic drift.

Bottleneck Effect   Sometimes a species is subjected to near extinction because of a natural disaster (e.g., earthquake or fire) or because of overharvesting and habitat loss. It is as though most of the population has stayed behind and only a few survivors have passed through the neck of a bottle (**Fig. 15.12**). Called a **bottleneck effect,** such an event prevents the majority of genotypes from participating in the production of the next generation.

The extreme genetic similarity found in cheetahs is believed to be due to a bottleneck. In a study of 47 different enzymes, each of which can occur in several different forms in other types of cats, all the cheetahs studied had exactly the same form. This demonstrates that genetic drift can cause certain alleles to be lost from a population. Exactly what caused the cheetah bottleneck is not known. Several hypotheses have been proposed, including that cheetahs were slaughtered by nineteenth-century cattle farmers protecting their herds, were captured by Egyptians as pets 4,000 years ago, or were decimated by a mass extinction event tens of thousands of years ago. Today, cheetahs suffer from relative infertility because of the intense inbreeding that occurred after the bottleneck.

Founder Effect   The **founder effect** is a mechanism of genetic drift in which rare alleles, or combinations of alleles, occur at a higher frequency in a population isolated from the general population. After all, founding individuals contain only a fraction of the total genetic diversity of the original gene pool. The alleles carried by their founder or founders are dictated by chance alone. The Amish of Lancaster County, Pennsylvania, are an isolated group founded by German settlers. Today, as many as 1 in 14 individuals in this population carry a recessive allele that causes an unusual form of dwarfism (affecting only the lower arms and legs) and polydactylism (extra fingers). In most populations, only 1 in 1,000 individuals has this allele.

STUDY TOOLS   http://connect.mheducation.com

 SMARTBOOK® Maximize your study time with McGraw-Hill SmartBook®, the first adaptive textbook.

SUMMARIZE

Populations change over time due to the effects of natural selection; one such change resulted in antibiotic-resistant bacteria. Microevolution leads to changes in allele frequencies within a population.

> **15.1** Natural selection acts to alter the phenotypic distribution of a population.
>
> **15.2** The forces of microevolution alter the frequency of alleles in a population over time.

15.1 Natural Selection

Natural selection results in adaptation of a species to its environment. Adaptation occurs when the most fit individuals reproduce more than others. These individuals usually possess traits better suited for survival in the environment, and over generations the frequency of the adaptive traits increases within the population.

Types of Selection

Most of the traits of evolutionary significance are under the control of multiple genes, and the range of phenotypes in a population can be represented by a bell-shaped curve. Three types of selection occur:

- **Stabilizing selection:** The peak of the curve increases, as when most human babies have an intermediate birth weight. Babies that are very small or very large are less fit than those of intermediate weight.
- **Directional selection:** The curve shifts in one direction, as when dark-colored peppered moths become prevalent in polluted areas.
- **Disruptive selection:** The curve has two peaks, as when British land snails vary because a wide geographic range causes selection to vary.

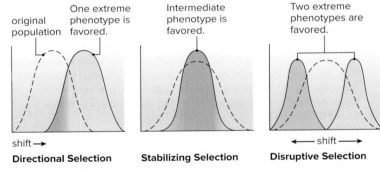

original population | One extreme phenotype is favored. | Intermediate phenotype is favored. | Two extreme phenotypes are favored.

shift → | ← shift →
Directional Selection | **Stabilizing Selection** | **Disruptive Selection**

Sexual Selection

Sexual selection is different between males and females. Sexual selection is associated with choosing a mate having the best **fitness,** or ability to produce surviving offspring.

Adaptations Are Not Perfect

Adaptations are not perfect because evolution builds on the variation that exists. Only certain types of variations are available, and developmental processes tend toward the same types of outcomes. The result is often a compromise between benefit and cost.

Maintenance of Variations

Despite constant natural selection, variation is maintained because

- Mutations and recombination still occur, gene flow among populations can introduce new alleles, and natural selection may not eliminate less favored phenotypes.
- In sexually reproducing diploid organisms, the heterozygote acts as a repository for recessive alleles whose frequency in the population is low. With respect to the sickle-cell alleles, the heterozygote is more fit in areas where malaria occurs; therefore, both homozygotes are maintained in the population.

15.2 Microevolution

Microevolution is the process by which small changes in genotype frequencies occur in a population over time. The study of microevolution is often referred to as **population genetics.**

Evolution in a Genetic Context

Microevolution involves several elements:

- All the various genes of a population make up its **gene pool.**
- Hardy-Weinberg equilibrium is present when allele frequencies in a gene pool remain the same from generation to generation. Certain conditions have to be met to achieve this equilibrium.
- The conditions are (1) no mutations, (2) no gene flow, (3) random mating, (4) no genetic drift, and (5) no selection. Since these conditions are rarely met, a change in gene pool frequencies is likely.
- When gene pool frequencies change, microevolution has occurred. Deviations from Hardy-Weinberg equilibrium allow us to determine when microevolution has taken place and to measure the extent of the change.

Causes of Microevolution

Microevolution occurs because of the following factors:

- **Mutations** are the ultimate source of variation. Certain genotypic variations may be of evolutionary significance only if the environment changes. Genetic diversity is promoted when there are several alleles for each gene locus.
- **Gene flow** is the movement of alleles that occurs when breeding individuals migrate to another population.
- **Nonrandom mating** occurs when relatives mate (inbreeding) or when assortative mating takes place. Sexual selection, which occurs when a characteristic that increases the chances of mating is favored, promotes random mating.
- **Genetic drift** occurs when allele frequencies are altered by chance. Genetic drift may occur through a **bottleneck effect** or **founder effect,** both of which change the frequency of alleles in the gene pool.
- Natural selection (see Section 15.1)

ASSESS

Testing Yourself

Choose the best answer for each question.

15.1 Natural Selection

For questions 1–5, choose the type of selection that best matches the statement. Each answer may be used more than once.

Key:

    a. sexual selection
    b. stabilizing selection
    c. directional selection
    d. disruptive selection

1. Selection acts to decrease the most common phenotype.
2. Choice is made based on the fitness of the mate.
3. Selection favors the extreme range of a phenotype.
4. Selection favors the intermediate phenotype.
5. Examples are antibiotic and insecticide resistance.

15.2 Microevolution

6. A population consists of 48 *AA*, 54 *Aa*, and 22 *aa* individuals. What is the frequency of the *A* allele?
    a. 0.60   **d.** 0.42
    b. 0.40   **e.** 0.58
    c. 0.62

7. Which of the following is the binomial equation expressing the Hardy-Weinberg principle?
    a. $2p^2 + 2pq + 2q^2 = 1$
    b. $p^2 + pq + q^2 = 1$
    c. $2p^2 + pq + 2q^2 = 1$
    d. $p^2 + 2pq + q^2 = 1$

8. The offspring of better-adapted individuals are expected to make up a larger proportion of the next generation. The most likely explanation for this is
    a. mutations and nonrandom mating.
    b. gene flow and genetic drift.
    c. mutations and natural selection.
    d. mutations and genetic drift.

9. A small, reproductively isolated religious sect called the Dunkers was established by 27 families that came to the United States from Germany 200 years ago. The frequencies for blood group alleles in this population differ significantly from those in the general U.S. population. This is an example of
    a. negative assortative mating.
    b. natural selection.
    c. the founder effect.
    d. the bottleneck effect.
    e. gene flow.

10. When a population is small, there is a greater chance of
    a. gene flow.
    b. genetic drift.
    c. natural selection.
    d. mutations.
    e. sexual selection.

11. The recessive sickle-cell allele is maintained in the populations in regions where malaria is prevalent because
    a. the allele confers resistance to the parasite.
    b. gene flow is high in those regions.
    c. disruptive selection is occurring.
    d. genetic drift randomly selects for the allele.

ENGAGE

BioNOW

Want to know how this science is relevant to your life? Check out the BioNow video below:

- Quail Evolution

What forms of microevolution are at work in this experiment?

Thinking Critically

1. A farmer uses a new pesticide. He applies the pesticide as directed by the manufacturer and loses about 15% of his crop to insects. A farmer in the next state learns of these results, applies three times as much pesticide, and loses only around 3% of the crop to insects. Each farmer follows this pattern for 5 years. At the end of 5 years, the first farmer is still losing about 15% of his crop to insects, but the second farmer is now losing around 40%.
    a. Explain how natural selection may be causing the effect observed at the second farm.
    b. Describe the form of selection that is occurring in the insect population at the second farm.
    c. Which of these insect populations is still in equilibrium? How do you know?

2. You are observing a grouse population in which two feather phenotypes are present in males. One is relatively dark and blends into shadows well, and the other is relatively bright and is more obvious to predators. All of the females are uniformly dark-feathered. Observing the frequency of mating between females and the two types of males, you record the following:

Matings with dark-feathered males: 13

Matings with bright-feathered males: 32
    a. Propose a hypothesis that explains why females may prefer bright-feathered males.
    b. Explain the selective advantage that might be associated with choosing a bright-feathered male.
    c. Outline an experiment to test your hypothesis.

(fossil): © Alan Morgan; (bird): © MIMOTITO/Digital Vision/Getty RF

16

Evolution on a Large Scale

Evolution of the Feathered Reptile

The discovery of a feathered reptile, named *Archeopteryx,* in 1860 forever changed the view of evolutionary change. *Archeopteryx* represented a transitional species; it possessed characteristics of both reptiles and birds. The fact that it was discovered shortly after the publication of Darwin's *On the Origin of Species* validated the idea that species change over time and this change powers the formation of new species.

However, the interesting part of this story is not *Archeopteryx,* it is the role of the feather in reptile and bird evolution. Additional fossil records have indicated that the feathers of these reptiles were probably not for flight. So what was their purpose? Competing hypotheses exist, but some maintain that the feathers acted as insulation to retain body heat, while others propose that they were designed to attract the attention of the opposite sex.

Over time, the adaptation of feathers began to have another advantage—allowing flight. The repurposing of feathers for flight led to other physiological adaptations that allowed for more efficient flight. Bird evolution represents an amazing story of adaptation and speciation. Today, there are more than 10,000 known species of birds, all of which are descended from a feathered reptile ancestor.

In this chapter, we go a step beyond microevolution and look at how a population, over time, accumulates differences large enough to become a new species. The origin of species is the key to understanding the origin of the diversity of all life on Earth.

As you read through this chapter, think about the following questions:

1. How do scientists determine whether an organism is a new species?
2. What processes drive the evolution of new species? Are they different from those that drive the evolution of populations?
3. What can the fossil record tell us about the origin and extinction of species over time?

OUTLINE

16.1 Speciation and Macroevolution   266

16.2 The Fossil Record   272

16.3 Systematics   277

BEFORE YOU BEGIN

Before beginning this chapter, take a few moments to review the following discussions.

Section 14.2 What roles do the fossil records and the study of comparative anatomy have in understanding evolutionary change?

Section 15.1 How does natural selection act as the mechanism of evolutionary change?

Section 15.2 What is microevolution?

16.1 Speciation and Macroevolution

In Chapter 15, we explored the process of **microevolution,** or the small changes in the allele frequencies of a population that occur over a relatively short period of time. In this chapter, we turn our attention to the process called **macroevolution,** which represents larger-scale changes in a population over very long periods of time. The history of life on Earth is a reflection of the process of macroevolution. Macroevolution often results in **speciation,** or the formation of new species. As we will see, speciation is due to changes in the gene pool and the divergence of two populations genetically, all of which is based on the principles of microevolution.

As populations change over time, they evolve adaptations to their environments. Over time, these changes may accumulate, allowing the population to undergo speciation and become different from other members of its species. The history of life on Earth, as recorded in the fossil record (**Fig 16.1***a*) and our genetic information (Fig 16.1*b*), is the documentation of the processes of microevolution, macroevolution, and speciation.

Defining Species

Before we consider the origin of species, we first need to define a species. Recall from Section 1.1 that the species is a level of biological organization between an organism and a population. In biology, appearance is not always a good way of distinguishing between two species. The members of different species can look quite similar, while the members of a single species can be diverse in appearance. For our purposes, we will state that a **species** represents a group of organisms that are capable of interbreeding and producing fertile offspring.

There are many variations on the concept of a species. The **biological species concept** states that the members of a species interbreed and have a shared gene pool, and each species is reproductively isolated from every other species. For example, the flycatchers in **Figure 16.2** are members of separate species because they do not interbreed in nature.

According to the biological species concept, gene flow occurs between the populations of a species, but not between populations of different species. The red maple and the sugar maple are found over a wide geographic range in

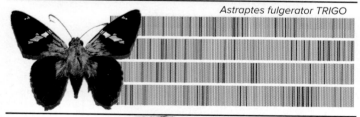

Astraptes fulgerator TRIGO

Bubo virginianus

Tyto alba

Figure 16.1 History of life.

The history of life is recorded in the (**a**) fossil record and in the (**b**) DNA of every organism.

pit-see

fitz-bew

che-bek or che-bek

Acadian flycatcher, *Empidonax virescens*   Willow flycatcher, *Empidonax trailli*   Least flycatcher, *Empidonax minimus*

Figure 16.2 **Three species of flycatchers.**

Although these flycatcher species are nearly identical in appearance, we know they are separate species because they are reproductively isolated—the members of each species reproduce only with one another. Each species has a characteristic song and its own habitat during the mating season as well.

(Acadian): © James Mundy/Alamy; (Willow): © All Canada Photos/Alamy; (Least): © Rick & Nora Bowers/Alamy

the eastern half of the United States, and each species is made up of many populations. However, the members of each species' populations rarely hybridize in nature. Therefore, these two types of plants are separate species. In contrast, the human species has many populations, which certainly differ in physical appearance (**Fig. 16.3**). We know, however, that all humans belong to one species because the members of these populations can produce fertile offspring.

The biological species concept is useful, as we will see, but even so, it has its limitations. For example, it applies only to sexually reproducing organisms and cannot apply to asexually reproducing organisms. Then, too, sexually reproducing organisms are not always as reproductively isolated as we would expect. Some North American orioles live in the western half of the continent, some in the eastern half, yet even the two most genetically distant oriole species, as recognized by analyzing their mitochondrial DNA, will hybridize where they meet in the middle of the continent.

There are other definitions of species aside from the biological definition. Several of these are based on studies of the evolutionary relationships between species. As we will see later in this chapter (Section 16.3), a species is a category of classification ranked below genus. Species in the same genus share a recent common ancestor. A **common ancestor** is a single ancestor shared by two or more different groups. For example, your father's mother is the common ancestor for you, your siblings, and your paternal cousins. By studying the relationships of species within a genus and those between closely related genera, scientists are able to gather a better understanding of how species evolve over time.

Figure 16.3 **Human populations.**

(**a**) The Maasai of East Africa and (**b**) the Kuna Indians from the San Blas Islands of Panama are both members of the species *Homo sapiens* because individuals from the two groups can produce fertile offspring.

(a): © Sylvia Mader; (b): © Adam Crowley/Getty Images

a.

b.

Connections: Scientific Inquiry

How can we determine if an organism that does not reproduce sexually is a distinct species?

Many organisms either do not reproduce sexually or do so very rarely. For example, there are species of moss that reproduce sexually only every 200 to 300 years! To determine if two populations of asexual organisms are distinct species, scientists rely on DNA analysis, morphological studies, and a close examination of the organisms' ecology to determine whether the two populations could reproduce naturally. Often, scientists have to revisit the classification of a species as research unveils new information.

© Nigel Cattlin/Science Source

Reproductive Barriers

As mentioned, for two species to be separate, they must be reproductively isolated. This means that gene flow must not occur between them. Reproductive barriers are isolating mechanisms that prevent successful reproduction (**Fig. 16.4**). In evolution, reproduction is successful only when it produces fertile offspring.

Prezygotic isolating mechanisms are those that occur before the formation of a zygote. In general, they prevent reproductive attempts and make it unlikely that fertilization will be successful if mating is attempted. Habitat isolation, temporal isolation, behavioral isolation, mechanical isolation, and gamete isolation make it highly unlikely that particular genotypes will contribute to the gene pool of a population.

Habitat isolation: When two species occupy different habitats, even within the same geographic range, they are less likely to meet and attempt to reproduce. This is one of the reasons that the flycatchers in Figure 16.2 do not mate and the red maple and sugar maple do not exchange pollen. In tropical rain forests, many animal species are restricted to a particular level of the forest canopy; in this way, they are isolated from similar species.

Temporal isolation: Two species can live in the same locale, but if they reproduce at different times of year, they do not attempt to mate. Five species of frogs of the genus *Rana* are all found near Ithaca, New York. The species remain separate because the period of peak mating activity differs, and so do the breeding sites. For example, wood frogs breed in woodland ponds or shallow water, leopard frogs in lowland swamps, and pickerel frogs in streams and ponds on high ground. Having different dispersal times often helps prevent fertilization of the gametes from different species.

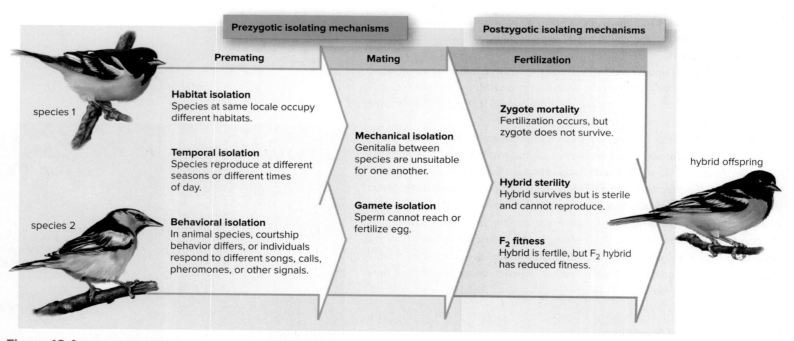

Figure 16.4 Reproductive barriers.

Prezygotic isolating mechanisms prevent mating attempts or a successful outcome if mating does take place—for example, between two species of orioles. No zygote is ever formed if these mechanisms are successful. Postzygotic isolating mechanisms prevent offspring from reproducing—that is, if a hybrid oriole should result, it would be unable to breed successfully.

Behavioral isolation: Many animal species have courtship patterns that allow males and females to recognize one another (**Fig. 16.5**). Female fireflies recognize males of their species by the pattern of the males' flashings; similarly, female crickets recognize males of their species by the males' chirping. Many males recognize females of their species by sensing chemical signals, called pheromones. For example, female gypsy moths secrete chemicals from abdominal glands. These chemicals are detected downwind by receptors on the antennae of males.

Mechanical isolation: When animal genitalia or plant floral structures are incompatible, reproduction cannot occur. Inaccessibility of pollen to certain pollinators can prevent cross-fertilization in plants, and the sexes of many insect species have genitalia that do not match, or other characteristics that make mating impossible. For example, male dragonflies have claspers that are suitable for holding only the females of their own species.

Gamete isolation: Even if the gametes of two different species meet, they may not fuse to become a zygote. In animals, the sperm of one species may not be able to survive in the reproductive tract of another species, or the egg may have receptors only for sperm of its species. Also, in each type of flower, only certain pollen grains can germinate, so that sperm successfully reach the egg.

Postzygotic isolating mechanisms are those that occur after the formation of a zygote. In general, they prevent hybrid offspring (reproductive product of two different species) from developing or breeding, even if reproduction attempts have been successful.

Zygote mortality: The hybrid zygote may not be viable, so it dies. A zygote with two different chromosome sets may fail to go through mitosis properly, or the developing embryo may receive incompatible instructions from the maternal and paternal genes, so that it cannot continue to exist.

Hybrid sterility: The hybrid zygote may develop into a sterile adult. As is well known, a cross between a horse and a donkey produces a mule, which is usually sterile—it cannot reproduce (**Fig. 16.6**). Sterility of hybrids generally results from complications in meiosis, which lead to an inability to produce viable gametes. A cross between a cabbage and a radish produces offspring that cannot form gametes, most likely because the cabbage chromosomes and the radish chromosomes cannot align during meiosis.

F_2 fitness: Even if hybrids can reproduce, their offspring may be unable to reproduce. In some cases, mules are fertile, but their offspring (the F_2 generation) are not fertile.

Models of Speciation

DNA comparisons suggest that iguanas of South America may be the common ancestor for both the marine iguana on the Galápagos Islands (to the west of South America) and the rhinoceros iguana on Hispaniola (the Caribbean island containing the countries of Haiti and the Dominican Republic). If so, how could it have happened? Green iguanas are strong swimmers, so by chance a few could have migrated to these islands, where they formed populations separate from each other and from the parent population in South America. Each population continued on its own evolutionary path as new mutations, genetic drift, and natural selection occurred. Eventually, reproductive isolation developed, and there were three species of iguanas. A speciation model based on geographic isolation of populations is called **allopatric speciation** (*allo,* different; *patria,* homeland).

Figure 16.5 Prezygotic isolating mechanism.

An elaborate courtship display allows the blue-footed boobies of the Galápagos Islands to select a mate. The male lifts his feet in a ritualized manner that shows off their bright blue color.

© Henri Leduc/Moment Open/Getty RF

Figure 16.6 Postzygotic isolating mechanism.

Mules are horse-donkey hybrids. Mules are infertile due to a difference in the chromosomes inherited from their parents.

(horse): © Creatas/PunchStock RF; (donkey): © Photodisc Collection/Getty RF; (mule): © Radius Images/Alamy RF

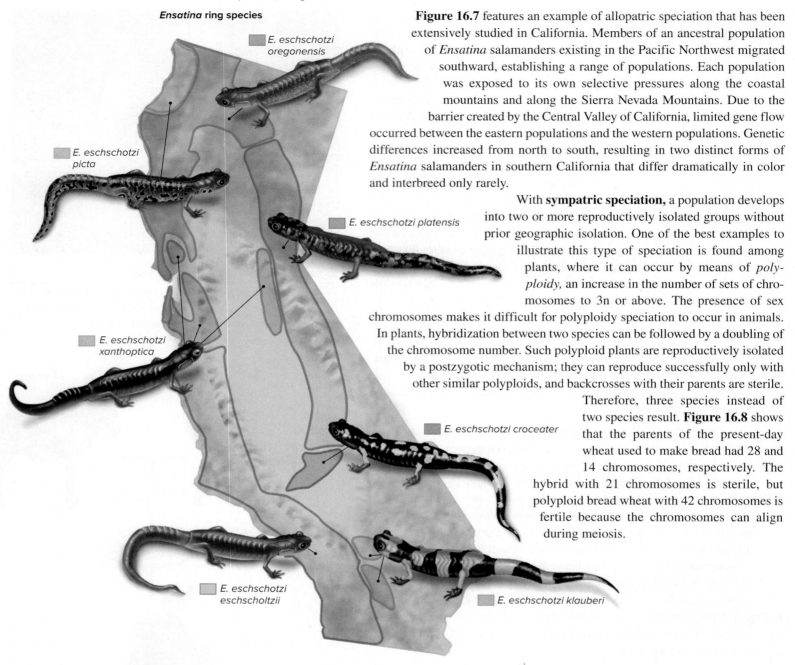

Ensatina ring species

E. eschschotzi oregonensis

E. eschschotzi picta

E. eschschotzi platensis

E. eschschotzi xanthoptica

E. eschschotzi croceater

E. eschschotzi eschscholtzii

E. eschschotzi klauberi

Figure 16.7 features an example of allopatric speciation that has been extensively studied in California. Members of an ancestral population of *Ensatina* salamanders existing in the Pacific Northwest migrated southward, establishing a range of populations. Each population was exposed to its own selective pressures along the coastal mountains and along the Sierra Nevada Mountains. Due to the barrier created by the Central Valley of California, limited gene flow occurred between the eastern populations and the western populations. Genetic differences increased from north to south, resulting in two distinct forms of *Ensatina* salamanders in southern California that differ dramatically in color and interbreed only rarely.

With **sympatric speciation,** a population develops into two or more reproductively isolated groups without prior geographic isolation. One of the best examples to illustrate this type of speciation is found among plants, where it can occur by means of *polyploidy,* an increase in the number of sets of chromosomes to 3n or above. The presence of sex chromosomes makes it difficult for polyploidy speciation to occur in animals. In plants, hybridization between two species can be followed by a doubling of the chromosome number. Such polyploid plants are reproductively isolated by a postzygotic mechanism; they can reproduce successfully only with other similar polyploids, and backcrosses with their parents are sterile. Therefore, three species instead of two species result. **Figure 16.8** shows that the parents of the present-day wheat used to make bread had 28 and 14 chromosomes, respectively. The hybrid with 21 chromosomes is sterile, but polyploid bread wheat with 42 chromosomes is fertile because the chromosomes can align during meiosis.

Figure 16.7 Allopatric speciation.

In this example of allopatric speciation, the Central Valley of California separates a range of populations descended from the same northern ancestral species. The limited contact between the populations on the west and those on the east allow genetic changes to build up to such an extent that members of the two southern populations rarely reproduce with each other and are designated as subspecies.

Figure 16.8 Sympatric speciation.

In this example of sympatric speciation, two populations of wild wheat hybridized many years ago. The hybrid is sterile, but chromosome doubling allowed some plants to reproduce. These plants became today's bread wheat.

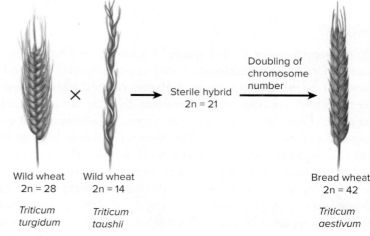

Wild wheat
2n = 28

Triticum turgidum

×

Wild wheat
2n = 14

Triticum taushii

→ Sterile hybrid
2n = 21

Doubling of chromosome number

→ Bread wheat
2n = 42

Triticum aestivum

Adaptive Radiation

A clear example of speciation through adaptive radiation is provided by the finches on the Galápagos Islands, which are often called Darwin's finches because Darwin first realized their significance as an example of how evolution works. During **adaptive radiation,** many new species evolve from a single ancestral species. The many species of finches that live on the Galápagos Islands are hypothesized to be descendants of a single type of ancestral finch from the mainland (**Fig. 16.9**). The populations on the various islands were subjected to the founder effect involving genetic drift, genetic mutations, and the process of natural selection. Because of natural selection, each population became adapted to a particular habitat on its island. In time, the various populations became so genotypically different that now, when by chance members of different groups reside on the same island, they do not interbreed and are therefore separate species. There is evidence that the finches use beak shape to recognize members of the same species during courtship. Rejection of suitors with the wrong type of beak is a behavioral prezygotic isolating mechanism.

Similarly, inhabiting the Hawaiian Islands is a wide variety of honeycreepers, all descended from a common goldfinchlike ancestor that arrived from Asia or North America about 5 MYA. Today, honeycreepers have a range of beak sizes and shapes (see Fig. 1.7) for feeding on various food sources, including seeds, fruits, flowers, and insects.

CONNECTING THE CONCEPTS

16.1 Speciation occurs due to an interruption of gene flow between two populations.

Connections: Scientific Inquiry

Are there examples of polyploid species in animals?

In general, polyploidy is rarer in animals than in plants. However, there are examples of polyploid insects and fish, and polyploidy appears to occur frequently in the amphibians, specifically in salamanders. In 1999, scientists reported a polyploid rat species (*Tympanoctomys barrerae*) in Argentina, but later genetic analysis refuted this claim.  © Carol Wolfe, photographer

Most geneticists believe that polyploidy in mammals is unlikely due to the well-defined role of mammalian sex chromosomes and the balance between the number of autosomes and sex chromosomes.

Check Your Progress 16.1

1. Explain how the biological species concept can be used to define a species.
2. Describe limitations of the biological species concept.
3. Explain the difference between a prezygotic and postzygotic isolation mechanism and give an example of each.
4. Compare and contrast allopatric speciation with sympatric speciation. Give an example of each.
5. Explain how adaptive radiation relates to variation.

Figure 16.9 Darwin's finches.

Each of Darwin's finches is adapted to gathering and eating a different type of food. Tree finches have beaks largely adapted to eating insects and, at times, plants. Ground finches have beaks adapted to eating the flesh of the prickly pear cactus or different-sized seeds.

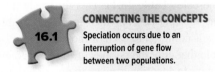

Tree Finches

Ground Finches

Warbler finch

Cactus ground finch

Woodpecker finch

Sharp-beaked ground finch

Small insectivorous tree finch

Small ground finch

Large insectivorous tree finch

Medium ground finch

Probing beaks

Grasping beaks

Vegetarian tree finch

Crushing beaks

Large ground finch

Parrot-like beaks

a.

b.

c.

Figure 16.10 Fossils.

a. A fern leaf from 245 MYA (million years ago) retains its form because it was buried in sediment that hardened to rock. **b.** This midge (40 MYA) became embedded in amber (hardened resin from a tree). **c.** Most fossils, such as this early insectivore mammal (47 MYA) are remains of hard parts because they do not decay as the soft parts do.

(a): © Carolina Biological Supply/Phototake; (b): © Alfred Pasieka/SPL/Science Source; (c): © Gary Retherford/Science Source

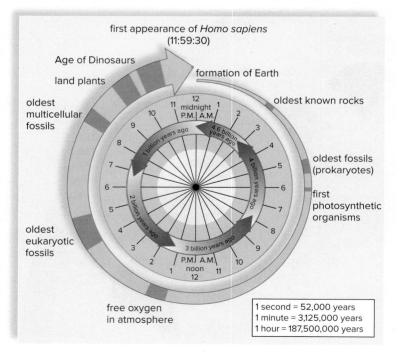

first appearance of *Homo sapiens*
(11:59:30)

Age of Dinosaurs

land plants

oldest
multicellular
fossils

formation of Earth

oldest known rocks

oldest fossils
(prokaryotes)

first
photosynthetic
organisms

oldest
eukaryotic
fossils

free oxygen
in atmosphere

| 1 second = 52,000 years |
| 1 minute = 3,125,000 years |
| 1 hour = 187,500,000 years |

16.2 The Fossil Record

Learning Outcomes

Upon completion of this section, you should be able to

1. Understand how the geologic time scale reflects the history of life on Earth.
2. Contrast the gradualistic model of evolution with the punctuated equilibrium model of evolution.
3. Summarize the causes of mass extinctions in the history of life on Earth.

The history of the origin and extinction of species on Earth is best discovered by studying the fossil record (**Fig 16.10**). **Fossils** are the traces and remains of past life or any other direct evidence of past life. **Paleontology** is the science dedicated to discovering and studying the fossil record and, from it, making decisions about the history of species.

The Geological Timescale

Because life-forms have evolved over time, the strata (layers of sedimentary rock, see Fig. 14.2) of the Earth's crust contain different fossils. By studying the strata and the fossils they contain, geologists have been able to construct a geological timescale (**Table 16.1**). This timescale divides the history of life on Earth into eras, then periods, and then epochs. The table includes descriptions of the types of fossils common to each of these divisions of time. Notice in the geological timescale that only the periods of the Cenozoic era are divided into epochs, meaning that more attention is given to the evolution of primates and flowering plants than to the earlier evolving organisms. Despite an epoch being assigned to modern civilization, humans have been around for only about 0.04% of the history of life.

It is often easier to visualize the vastness of the geological timescale by placing it in reference to a single day. **Figure 16.11** shows the history of the Earth as if it had occurred during a 24-hour time span that started at midnight. The actual time frames are shown on an inner ring of the diagram. If the Earth formed at midnight, prokaryotes did not appear until about 5 A.M., eukaryotes at approximately 4 P.M., and multicellular forms not until around 8 P.M. Invasion of the land didn't occur until about 10 P.M., and humans didn't appear until 30 seconds before the end of the day. This timescale has been worked out by studying the fossil record. In addition to sedimentary fossils, more recent fossils can be found in tar, ice, bogs, and amber. Shells, bones, leaves, and even footprints are commonly found in the fossil record.

In contrast to the brief amount of time that humans have been on the planet, prokaryotes existed for some 2 billion years before the eukaryotic cell and multicellularity arose during Precambrian time. Some prokaryotes became the first photosynthetic organisms to add oxygen to the atmosphere. The presence of oxygen may have spurred the evolution of the eukaryotic cell and of multicellularity during the Precambrian. All major groups of animals evolved during what is sometimes called the Cambrian explosion.

Figure 16.11 The history of life in 24 hours.

The blue ring of this diagram shows the history of life as it would be measured on a 24-hour timescale starting at midnight. The red ring shows the actual years going back in time to around 4.6 BYA.

Table 16.1 The Geological Timescale: Major Divisions of Geological Time and Some of the Major Evolutionary Events of Each Time Period

Era	Period	Epoch	Millions of Years Ago (MYA)	Plant Life	Animal Life
Cenozoic	Quaternary	Holocene	current	Humans influence plant life.	Age of *Homo sapiens*
			**Significant Extinction Event Underway**		
	Quaternary	Pleistocene	0.01	Herbaceous plants spread and diversify.	Presence of ice age mammals. Modern humans appear.
	Neogene	Pliocene	2.6	Herbaceous angiosperms flourish.	First hominids appear.
	Neogene	Miocene	5.3	Grasslands spread as forests contract.	Apelike mammals and grazing mammals flourish; insects flourish.
	Neogene	Oligocene	23.0	Many modern families of flowering plants evolve; appearance of grasses.	Browsing mammals and monkeylike primates appear.
	Paleogene	Eocene	33.9	Subtropical forests with heavy rainfall thrive.	All modern orders of mammals are represented.
	Paleogene	Paleocene	55.8	Flowering plants continue to diversify.	Ancestral primates, herbivores, carnivores, and insectivores appear.
			Mass Extinction: 50% of all species, dinosaurs and most reptiles		
Mesozoic	Cretaceous		65.5	Flowering plants spread; conifers persist.	Placental mammals appear; modern insect groups appear.
	Jurassic		145.5	Flowering plants appear.	Dinosaurs flourish; birds appear.
			Mass Extinction: 48% of all species, including corals and ferns		
	Triassic		199.6	Forests of conifers and cycads dominate.	First mammals appear; first dinosaurs appear; corals and molluscs dominate seas.
			Mass Extinction ("The Great Dying"): 83% of all species on land and sea		
Paleozoic	Permian		251.0	Gymnosperms diversify.	Reptiles diversify; amphibians decline.
	Carboniferous		299.0	Age of great coal-forming forests: ferns, club mosses, and horsetails flourish.	Amphibians diversify; first reptiles appear; first great radiation of insects.
			Mass Extinction: Over 50% of coastal marine species, corals		
	Devonian		359.2	First seed plants appear. Seedless vascular plants diversify.	First insects and first amphibians appear on land.
	Silurian		416.0	Seedless vascular plants appear.	Jawed fishes diversify and dominate the seas.
			Mass Extinction: Over 57% of marine species		
	Ordovician		443.7	Nonvascular land plants appear.	Invertebrates spread and diversify; first jawless and then jawed fishes appear.
	Cambrian		488.3	Marine algae flourish.	All invertebrate phyla present; first chordates appear.
			630		First soft-bodied invertebrates evolve.
			1,000	Protists diversify.	
			2,100	First eukaryotic cells evolve.	
			2,700	O_2 accumulates in atmosphere.	
			3,500	First prokaryotic cells evolve.	
			4,570	Earth forms.	

Connections: Scientific Inquiry

What is the Burgess Shale?

© Alan Morgan

The Burgess Shale is the name for a rock formation in the Canadian Rocky Mountains near the Burgess Pass. Around 525 MYA, this region was located along the coast. It is believed that an earthquake caused a landslide that almost instantly buried much of the marine life in the shallow coastal waters. Unlike many fossil beds, the Burgess Shale contains the remains of soft-shelled organisms, such as worms and sea cucumbers, as well as other organisms from the Cambrian explosion—a period of rapid diversification in marine life around 545 MYA. Over 60,000 unique types of fossils have been found in the Burgess Shale (including the trilobite fossils shown here), making this fossil bed one of our most valuable assets for studying the early evolution of life in the oceans.

The fossil record for Precambrian time is meager, but the fossil record for the Cambrian period is rich. The evolution of the invertebrate external skeleton accounts for this increase in the number of fossils. Perhaps this skeleton, which impedes the uptake of oxygen, couldn't evolve until oxygen was plentiful. Or perhaps the external skeleton was merely a defense against predation.

The origin of life on land is another interesting development. During the Paleozoic era, plants were present on land before animals. Nonvascular plants preceded vascular plants, and among these, cone-bearing plants (gymnosperms) preceded flowering plants (angiosperms). Among vertebrates, the fishes were aquatic, and the amphibians invaded land. The reptiles, including dinosaurs and birds, shared an amniote ancestor with the mammals. The number of species on Earth has continued to increase until the present time, despite the occurrence of five mass extinctions, including one significant mammalian extinction, during the history of life.

The Pace of Speciation

Darwin theorized that evolutionary changes occur gradually. In other words, he supported a *gradualistic model* to explain the pace of evolution. Speciation probably occurs after populations become isolated, with each group continuing slowly on its own evolutionary pathway. The gradualistic model often shows the evolutionary history of groups of organisms using a diagram called an **evolutionary tree,** as shown in **Figure 16.12***a*. In this diagram, note that an ancestral species has given rise to two separate species, represented by a slow change in plumage color. The gradualistic model suggests that it is difficult to indicate when speciation has occurred because there would be so many transitional links between species (see Section 14.2). In some cases, it has been possible to trace the evolution of a group of organisms by finding transitional links.

More often, however, species appear quite suddenly in the fossil record, and then they remain essentially unchanged phenotypically until they undergo

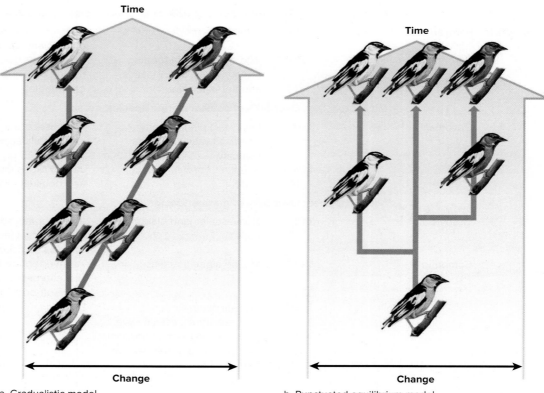

Figure 16.12 **Pace of evolution.**

a. According to the gradualistic model, new species evolve slowly from an ancestral species. **b.** According to the punctuated equilibrium model, new species appear suddenly and then remain largely unchanged until they become extinct.

a. Gradualistic model

b. Punctuated equilibrium model

extinction. Paleontologists have therefore developed a *punctuated equilibrium model* to explain the pace of evolution. The model says that a period of equilibrium (no change) is punctuated (interrupted) by speciation. Figure 16.12*b* shows the type of diagram paleontologists prefer to use when representing the history of evolution over time. This model suggests that transitional links are less likely to become fossils and less likely to be found. Speciation probably involves only an isolated subpopulation at one locale. Only when this new subpopulation expands and replaces existing species is it apt to show up in the fossil record.

The differences between these two models are subtle, especially when we consider that the "sudden" appearance of a new species in the fossil record could represent many thousands of years because geological time is measured in millions of years.

Causes of Mass Extinctions

As researchers have noted, most species exist for only a limited amount of time (measured in millions of years), and then they die out (become extinct). **Mass extinctions** are disappearances of a large number of species within a relatively short period of time. The geological timescale in Table 16.1 shows the occurrence of five mass extinctions: at the ends of the Ordovician, Devonian, Permian, Triassic, and Cretaceous periods. Also, there was a significant mammalian extinction at the end of the Pleistocene epoch. While many factors contribute to mass extinctions, scientists now recognize that continental drift, climate change, and meteorite impacts have all played a role.

Continental drift—the movement of continents—has contributed to several extinctions. You may have noticed that the coastlines of several continents are mirror images of each other. For example, the outline of the west coast of Africa matches that of the east coast of South America. Also, the same geological structures are found in many of the areas where the continents touched at one time. A single mountain range runs through South America, Antarctica, and Australia, for example. But the mountain range is no longer continuous because the continents have drifted apart. The reason the continents drift is explained by a principle of geology known as *plate tectonics,* which has established that the Earth's crust is fragmented into slablike plates that float on a lower, hot mantle layer (**Fig. 16.13**).

Connections: Scientific Inquiry

What is the "sixth mass extinction event"?

© Designpics.com/PunchStock RF

Many ecologists maintain that we are currently involved in the Earth's sixth mass extinction event. However, unlike the first five major events, this one is caused not by geological or astronomical events but by human actions. Pollution, land use, invasive species, and global climate change associated with the burning of fossil fuels are all recognized as contributing factors. The exact rate of species loss can be difficult to determine, but international agencies report that the current loss of species is between 100 and 1,000 times faster than the pre-human rates recorded by the fossil record.

Figure 16.13 Plate tectonics.

The Earth's surface is divided into several solid tectonic plates floating on the fluid magma beneath them. At rifts in the ocean floor, two plates gradually separate as fresh magma wells up and cools, enlarging the plates. Mountains, including volcanoes, are raised where one plate pushes beneath another at subduction zones. Where two plates slowly grind past each other at a fault line, tension builds up, which is released occasionally in the form of earthquakes.

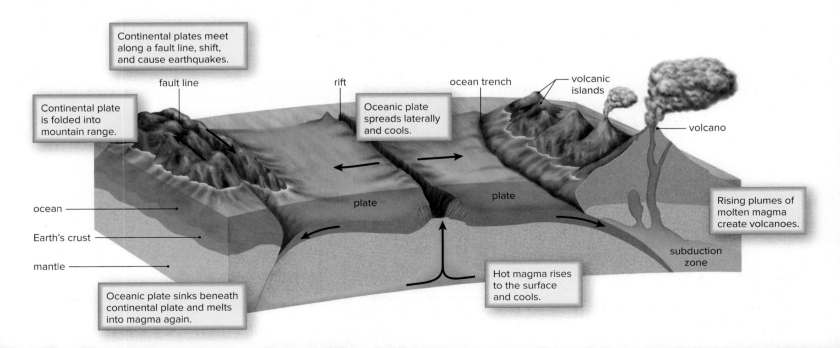

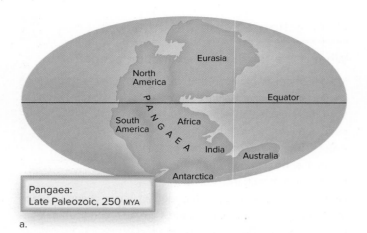

Pangaea:
Late Paleozoic, 250 MYA

a.

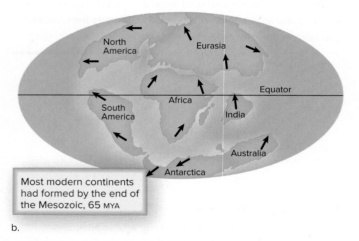

Most modern continents
had formed by the end of
the Mesozoic, 65 MYA

b.

Figure 16.14 Continental drift.

a. About 250 MYA, all the continents were joined into a supercontinent called Pangaea. **b.** By 65 MYA, all the continents had begun to separate. This process is continuing today. North America and Europe are separating at a rate of about 2 centimeters per year.

The continents and the ocean basins are a part of these rigid plates, which move like conveyor belts.

The loss of habitat is a significant cause of extinctions, and continental drift can lead to massive habitat changes. We know that 250 million years ago, at the time of the Permian mass extinction, all the Earth's continents came together to form one supercontinent called Pangaea (**Fig. 16.14**a). The result was dramatic environmental change through the shifting of wind patterns, ocean currents, and most importantly the amount of available shallow marine habitat. Marine organisms suffered as the oceans merged, and the amount of shoreline, where many marine organisms lived, was drastically reduced. Species diversity did not recover until some continents drifted away from the poles, shorelines increased, and warmth returned (Fig. 16.14b). Terrestrial organisms were affected as well because the amount of interior land, which tends to have a drier and more erratic climate, increased. Immense glaciers developing at the poles withdrew water from the oceans and chilled even once tropical regions.

Meteor impacts. The other event that is known to have contributed to mass extinctions is the impact of a meteorite as it crashed into the Earth. The result of a large meteorite striking Earth could have been similar to that of a worldwide atomic bomb explosion: A cloud of dust would have mushroomed into the atmosphere, blocking out the sun and causing plants to freeze and die. This type of event has been proposed as a primary cause of the Cretaceous extinction that saw the demise of the dinosaurs. Cretaceous clay contains an abnormally high level of iridium, an element that is rare in the Earth's crust but more common in meteorites. A layer of soot has been identified in the strata alongside the iridium, and a huge crater that could have been caused by a meteorite was found on and adjacent to the Yucatán peninsula of Mexico.

Climate change. Increasingly, scientists are finding evidence that mass extinction events are correlated to changes in the global climate. Some of these changes may be due to the long-term effects of continental drift, and some seem to be sudden changes due to meteor impacts, but evidence also suggests that global warming events, specifically those that changed the temperatures of the oceans (both warming and cooling), are linked to mass extinctions of species. The cause of these climate change events is still not well understood. We will explore the influence of climate change on ecosystems again in Section 31.2.

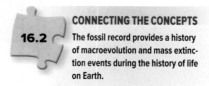

CONNECTING THE CONCEPTS

16.2 The fossil record provides a history of macroevolution and mass extinction events during the history of life on Earth.

Check Your Progress 16.2

1. Explain why the fossil record provides the best evidence for macroevolution.

2. Describe the punctuated equilibrium model of evolution.

3. Identify the causes of mass extinction events in Earth's history.

16.3 Systematics

Learning Outcomes

Upon completion of this section, you should be able to

1. List the hierarchical levels of Linnaean classification from the most inclusive to the least inclusive.
2. Describe the information that can be learned from a phylogenetic tree, and list some of the types of information that are used in constructing such trees.
3. Contrast homologous structures with analogous structures.
4. Define *cladistics*, and explain how this method may be used to study the evolutionary relationships between groups of organisms.
5. List the three domains of living organisms, and describe the general characteristics of organisms included within each domain.

All fields of biology, but especially **systematics,** are dedicated to understanding the evolutionary history of life on Earth. Systematics is very analytical and relies on a combination of data from the fossil record and comparative anatomy and development, with an emphasis today on molecular data, to determine **phylogeny,** the evolutionary history of a group of organisms. Classification is a part of systematics because ideally organisms are classified according to our present understanding of evolutionary relationships.

Linnaean Classification

Taxonomy is the branch of biology concerned with identifying, naming, and classifying organisms. A **taxon (pl., taxa)** is a group of organisms at a particular level in a classification system. The binomial system of nomenclature assigns a two-part name to each type of organism. For example, the plant in **Figure 16.15** has been named *Cypripedium acaule.* This name means that the plant is in the genus *Cypripedium* and that the specific epithet is *acaule.* Notice that the scientific name is in italics and only the genus is capitalized. The genus can be abbreviated to a single letter if the full name has been given previously and if it is used with a specific epithet. Thus, *C. acaule* is an acceptable way to designate this plant. The name of an organism usually tells you something about the organism. In this instance, the genus name, *Cypripedium,* refers to the slipper shape of the flower, and the specific epithet, *acaule,* says that the flower has no independent stem.

Why do organisms need scientific names? And why do scientists use Latin, rather than common names, to describe organisms? There are several reasons. First, a common name varies from country to country because different countries use different languages. Second, even people who speak the same language sometimes use different common names to describe the same organism—for example, bowfin, grindle, choupique, and cypress trout are all common names for a species of fish, *Amia calva.* Furthermore, the same common name is sometimes given to different organisms in two countries. A robin in England is very different from a robin in the United States, for example. Latin, on

Figure 16.15 **Taxonomy hierarchy.**

A domain is the most inclusive of the classification categories. The plant kingdom is in the domain Eukarya. In the plant kingdom are several phyla, each represented here by lavender circles. The phylum Anthophyta has only two classes (the monocots and eudicots). The class Monocotyledones encompasses many orders. In the order Orchidales are many families; in the family Orchidaceae are many genera; and in the genus *Cypripedium* are many species—for example, *Cypripedium acaule.* (This illustration is diagrammatic and doesn't necessarily show the correct number of subcategories.)

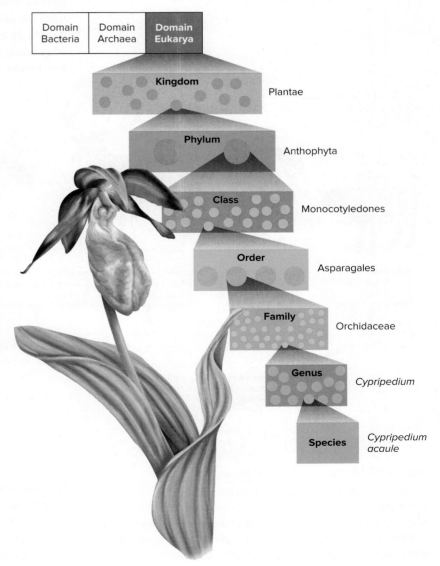

Domain Bacteria Domain Archaea **Domain Eukarya**

Kingdom — Plantae

Phylum — Anthophyta

Class — Monocotyledones

Order — Asparagales

Family — Orchidaceae

Genus — *Cypripedium*

Species — *Cypripedium acaule*

the other hand, is a universal language that not too long ago was well known by most scholars, many of whom were physicians or clerics. When scientists throughout the world use a scientific binomial name, they know they are speaking of the same organism.

Today, taxonomists use several categories of classification created by Swedish biologist Carl Linnaeus in the eighteenth century to show varying levels of similarity: **species, genus, family, order, class, phylum, kingdom, domain.** There can be several species within a genus, several genera within a family, and so forth. In this hierarchy, the higher the category, the more inclusive it is (Fig. 16.15). Therefore, species in the same domain have general traits in common, while those in the same genus have quite specific traits in common.

Taxonomists often subdivide each category of classification into additional categories, such as superorder, order, suborder, and infraorder. This allows for a further level of distinction within each major category.

Phylogenetic Trees

Figure 16.16 shows how the classification of groups of organisms allows us to construct a **phylogenetic tree,** a diagram that indicates common ancestors and lines of descent (lineages). The common ancestor at the base of the tree has traits that are shared by all the other groups in the tree. For example, the Artiodactyla are characterized by having hoofs with an even number of toes. On the other hand, notice that the Cervidae have antlers but the Bovidae have no antlers. Finally, among the Cervidae, the antlers are highly branched in red deer but palmate (having the shape of a hand) in reindeer. As the lineage moves from common ancestor to common ancestor, the traits become more specific to just particular groups of animals. It is this progression in specificity that allows classification categories to serve as a basis for constructing a phylogenetic tree.

Tracing Phylogeny

Figure 16.16 makes use of morphological data, but systematists today use several types of data to discover the evolutionary relationships between species. They rely heavily on a combination of fossil record, morphological data, and molecular data to determine the correct sequence of common ancestors in any group of organisms.

Morphological data include homologies, which are similarities among organisms that stem from having a common ancestor. Comparative anatomy, including embryological evidence and fossil data, provides information regarding homology. **Homologous structures** are related to each other through common descent. The forelimbs of vertebrates are homologous because they contain the same bones organized in the same general way as in a common ancestor (Fig. 14.16). This is the case even though a horse has only a single digit and toe (the hoof), while a bat has four lengthened digits that support its membranous wings.

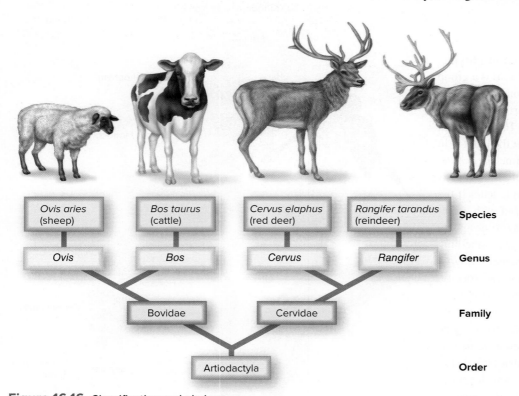

Figure 16.16 Classification and phylogeny.

The phylogenetic tree for a group of organisms is ideally constructed to reflect their classifications and phylogenetic history. A species is most closely related to other species in the same genus, more distantly related to species in other genera of the same family, and so forth on through order, class, phylum, and kingdom.

Deciphering homology is sometimes difficult because of convergent evolution. **Convergent evolution** is the acquisition of the same or similar traits in distantly related lines of descent. Similarity due to convergence is termed **analogy.** The wings of an insect and the wings of a bat are analogous. **Analogous structures** have the same function in different groups but organisms with these structures do not have a recent common ancestor (see Section 14.2). *Analogous structures arise because of adaptations to the same type of environment.* Both cactuses and spurges are adapted similarly to a hot, dry environment, and both are succulent, spiny, flowering plants. However, the details of their flower structure indicate that these plants are not closely related.

Speciation occurs when mutations bring about changes in the base-pair sequences of DNA. Systematists, therefore, assume that the more closely species are related, the fewer differences there will be in their DNA base-pair sequences. Because molecular data are straightforward and numerical, they can sometimes be used to clarify relationships obscured by inconsequential anatomical variations or convergence. Computer software breakthroughs have made it possible to analyze nucleotide sequences quickly and accurately. Also, these analyses are available to anyone doing comparative studies through the Internet, so each investigator doesn't have to start from scratch. The combination of accuracy and availability of vast amounts of data, even entire genomes, has made molecular systematics a standard way to study the relatedness of organisms.

All cells have ribosomes essential for protein synthesis, and the genes that code for ribosomal RNA (rRNA) have changed very slowly during evolution because drastic changes lead to malfunctioning cells. Therefore, comparative rRNA sequencing provides a reliable indicator of the similarity between organisms. Ribosomal RNA sequencing helped investigators conclude that all living organisms can be divided into the three domains.

One study involving DNA differences produced the data shown in **Figure 16.17**. Notice the close relationship between chimpanzees and humans. This relationship has recently been recognized by the designation of a new subfamily, Homininae, that includes not only chimpanzees and humans

Figure 16.17 Interpretation of molecular data.

The relationships of certain primate species are based on a study of their genomes. The length of the branches indicates the relative number of DNA base-pair differences between the groups. These data, along with knowledge of the fossil record for one divergence, make it possible to suggest a date for the other divergences in the tree.

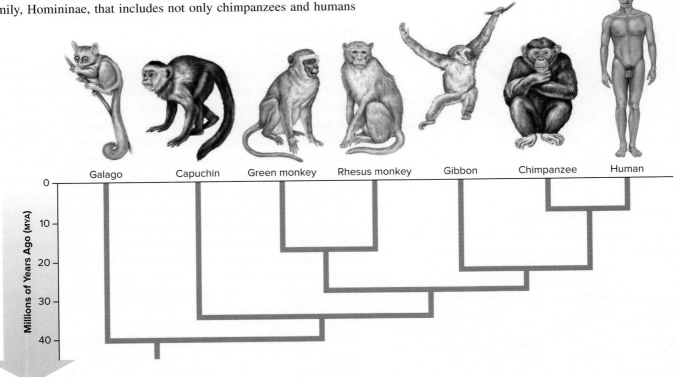

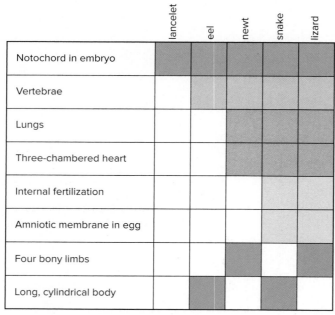

	lancelet	eel	newt	snake	lizard
Notochord in embryo					
Vertebrae					
Lungs					
Three-chambered heart					
Internal fertilization					
Amniotic membrane in egg					
Four bony limbs					
Long, cylindrical body					

a.

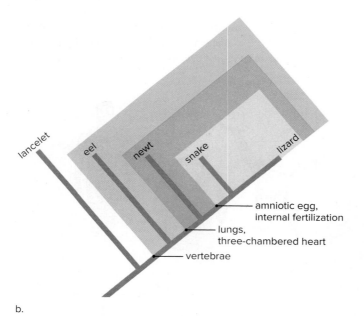

amniotic egg, internal fertilization

lungs, three-chambered heart

vertebrae

b.

Figure 16.18 Constructing a cladogram.

a. First, a table is drawn up, listing characters for all the taxa. An examination of the table shows which characters are ancestral (aqua) and which are derived (lavender, orange, and yellow). The shared derived characters distinguish the taxa. **b.** In a cladogram, the shared derived characters are sequenced in the order they evolved and are used to define clades. A clade includes a common ancestor and all the species that share the same derived characters (homologies). Four bony limbs and a long, cylindrical body (light red) were not used in constructing the cladogram because they are in scattered taxa.

but also gorillas. Molecular data indicate that gorillas and chimpanzees are more closely related to humans than they are to orangutans. Below the taxon subfamily, humans and chimpanzees are placed together in their own tribe, a rarely used classification category.

Cladistics and Cladograms

Cladistics is a way to trace the evolutionary history of a group by using traits derived from a common ancestor to determine relationships. These traits are then used to construct phylogenetic trees called cladograms. A **cladogram** depicts the evolutionary history of a group based on the available data.

The first step in constructing a cladogram is to draw up a table that summarizes the traits of the species being compared. At least one but preferably several species are considered an *outgroup*. The outgroup is not part of the study group, also called the *ingroup*. In **Figure 16.18**a, lancelets are the outgroup because, unlike the species in the ingroup, they are not vertebrates. Any trait, such as a notochord, found in both the outgroup and the ingroup is a shared *ancestral trait,* presumed to have been present in an ancestor common to both the outgroup and the ingroup. Ancestral traits are not shared derived traits and therefore are not used to construct a cladogram. They merely help determine which traits will be used to construct the cladogram.

A rule that many cladists follow is the principle of *parsimony,* which states that the least number of assumptions is the most probable. Thus, they construct a cladogram that minimizes the number of assumed evolutionary changes or that leaves the fewest number of derived traits unexplained. Therefore, any trait in the table found in scattered species (in this case, four bony limbs and a long, cylindrical body) is not used to construct the cladogram because we would have to assume that these traits evolved more than once among the species of the study group. The other differences are designated as *shared derived traits*—that is, they are homologies shared by only certain species of the study group. Combining the data regarding shared derived traits will tell us how the members of the ingroup are related to one another.

The Cladogram

A cladogram contains several clades; each **clade** includes a common ancestor and all of its descendant species. The cladogram in Figure 16.18b has three clades, which differ in size because the first includes the other two and so forth. All the species in the study group belong to a clade that has vertebrae; only newts, snakes, and lizards are in a clade that has lungs and a three-chambered heart; and only snakes and lizards are in a clade that has an amniotic egg and internal fertilization. (An amniotic egg has a sac that surrounds and protects the embryo—fish and amphibian eggs do not have this sac.) Following the principle of parsimony, this is the sequence in which these traits must have evolved during the evolutionary history of vertebrates. Any other arrangement of species would produce a less parsimonious evolutionary sequence—that is, a tree that would be more complicated.

A cladogram is typically constructed using as much morphological, fossil, and molecular data as are available at the time. Still, any cladogram should be viewed as a hypothesis. Whether the tree is consistent with the one, true evolutionary history of life can be tested, and modifications can be made on the basis of additional data.

Linnaean Classification Versus Cladistics

Figure 16.19 illustrates the types of problems that arise when trying to reconcile Linnaean classification with the principles of cladistics. Figure 16.19, which is based on cladistics, shows that birds are in a clade with crocodiles, with which they share a recent common ancestor. This ancestor had a gizzard. An examination of the skulls of crocodiles and birds shows other derived traits that they share. Birds have scaly skin and share this derived trait with other reptiles as well. However, Linnaean classification places birds in their own group, separate from crocodiles and from reptiles in general. In many other instances, Linnaean classification is not consistent with new understandings about phylogenetic relationships. Therefore, some cladists have proposed a different system of classification, called the International Code of Phylogenetic Nomenclature, or PhyloCode, which sets forth rules for the naming of clades. Other biologists are hoping to modify Linnaean classification to be consistent with the principles of cladistics.

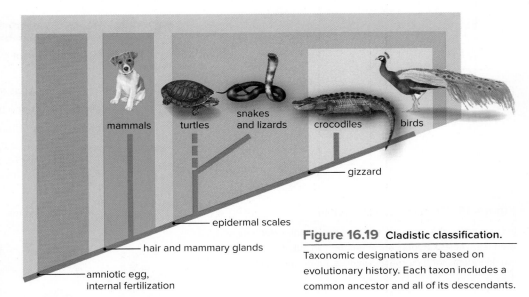

Figure 16.19 Cladistic classification.

Taxonomic designations are based on evolutionary history. Each taxon includes a common ancestor and all of its descendants.

Two major problems may be unsolvable: (1) Clades are hierarchical, as are Linnaean categories. However, there may be more clades than Linnaean taxonomic categories, and it is therefore difficult to equate clades with taxons. (2) The taxons are not necessarily equivalent in the Linnaean system. For example, the family taxon within kingdom Plantae may not be equivalent to the family taxon in kingdom Animalia. Because of such problems, some cladists recommend abandoning Linnaeus altogether.

The Three-Domain System

Classification systems change over time. Historically, most biologists utilized the **five-kingdom system** of classification, which contains kingdoms for the plants, animals, fungi, protists, and monerans (bacteria). Organisms were placed into these kingdoms based on type of cell (prokaryotic or eukaryotic), level of organization (single-celled or multicellular), and type of nutrition. In the five-kingdom system, the monerans were distinguished by their structure—they were prokaryotic (lack a membrane-bound nucleus)—whereas the organisms in the other kingdoms were eukaryotic (have a membrane-bound nucleus). The kingdom Monera contained all the prokaryotes, which evolved first, according to the fossil record.

Sequencing the genes associated with the production of ribosomal rRNA (rRNA) challenged the five-kingdom system of classification. As research progressed, it became apparent that there was a category of classification above the level of a kingdom, called the *domain,* and that all life may be grouped into one of three domains (**Fig. 16.20**).

In the **three-domain system,** the prokaryotes were recognized as belonging to two groups so fundamentally different from each other that they have been assigned to separate domains, called **domain Bacteria** and **domain Archaea. Domain Eukarya** contains kingdoms for protists, animals, fungi, and plants. Systematists have determined that domain Bacteria arose first, followed by domain Archaea and then domain Eukarya (Fig. 16.20). The Archaea and Eukarya are more closely related to each other than either is to the Bacteria. We will explore each of these domains, and their associated kingdoms, in later chapters.

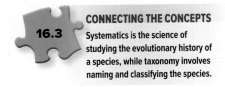

CONNECTING THE CONCEPTS

16.3 Systematics is the science of studying the evolutionary history of a species, while taxonomy involves naming and classifying the species.

Check Your Progress 16.3

1. List the categories of classification in order from least inclusive to most inclusive.

2. Contrast homologous structures with analogous structures.

3. Explain the difference between systematics and taxonomy.

4 Summarize the structure of the three-domain system.

Figure 16.20 **Three-domain system.**

In this system, the prokaryotes are in the domains Bacteria and Archaea. The eukaryotes are in the domain Eukarya, which contains four kingdoms for the protists, animals, fungi, and plants.

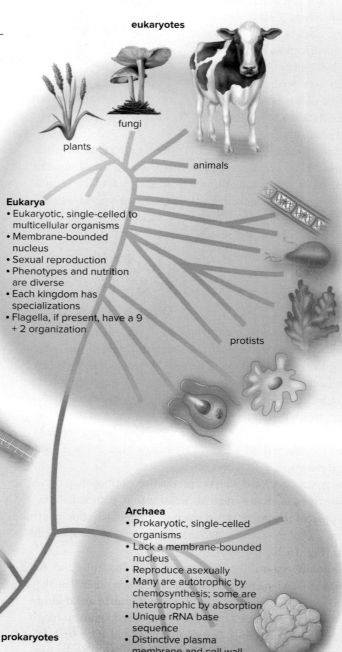

eukaryotes

fungi

plants

animals

Eukarya
• Eukaryotic, single-celled to multicellular organisms
• Membrane-bounded nucleus
• Sexual reproduction
• Phenotypes and nutrition are diverse
• Each kingdom has specializations
• Flagella, if present, have a 9 + 2 organization

protists

cyanobacteria

Bacteria
• Prokaryotic, single-celled organisms
• Lack a membrane-bounded nucleus
• Reproduce asexually
• Heterotrophic by absorption
• Autotrophic by chemosynthesis or by photosynthesis
• Move by flagella

heterotrophic bacteria

Archaea
• Prokaryotic, single-celled organisms
• Lack a membrane-bounded nucleus
• Reproduce asexually
• Many are autotrophic by chemosynthesis; some are heterotrophic by absorption
• Unique rRNA base sequence
• Distinctive plasma membrane and cell wall chemistry

prokaryotes

common ancestor

SUMMARIZE

When evolution occurs on a large scale, the result is the appearance of new species. The fossil record shows how species have evolved during the history of life on Earth.

16.1 Speciation occurs due to an interruption of gene flow between two populations.

16.2 The fossil record provides a history of macroevolution and mass extinction events during the history of life on Earth.

16.3 Systematics is the science of studying the evolutionary history of a species, while taxonomy involves naming and classifying the species.

16.1 Speciation and Macroevolution

Whereas **microevolution** is based on small changes in a population over short periods of time, **macroevolution** occurs over geological time frames, and the history of life on Earth is a reflection of this process. Macroevolution often results in **speciation,** or the origin of new species.

Defining Species

The **biological species concept**

- recognizes a species by its inability to produce viable fertile offspring with members of another group.
- is useful because species can look similar and members of the same species can have different appearances.
- has its limitations, including the facts that hybridization does occur between some species and that the concept applies only to sexually reproducing organisms.

Reproductive Barriers

Speciation occurs when prezygotic and postzygotic barriers keep species from reproducing with one another.

- **Prezygotic isolating mechanisms** include habitat isolation, temporal isolation, and behavioral isolation.
- **Postzygotic isolating mechanisms** prevent hybrid offspring from developing or breeding if reproduction has been successful.

Models of Speciation

Allopatric speciation and sympatric speciation are two models of speciation.

- In **allopatric speciation,** a geographic barrier keeps two populations apart. Meanwhile, prezygotic and postzygotic isolating mechanisms develop, and these prevent successful reproduction if these two groups come into contact in the future.
- In **sympatric speciation,** a geographic barrier is not required for speciation to occur.
- Following speciation, **adaptive radiation** allows the new species to adapt to changes in its unique environment.

16.2 The Fossil Record

The fossil record, as outlined by the geological timescale, traces the history of life in broad terms. **Paleontology** is the science dedicated to using the fossil record to establish the history of species. Scientists are able to date **fossils** using radioactive dating techniques.

Pace of Speciation

- Gradualistic model: Two groups of organisms arise from an ancestral species and gradually become two different species.
- Punctuated equilibrium model: A period of equilibrium (no change) is interrupted by speciation that occurs within a relatively short period of time.

Mass Extinctions

- The fossil record shows that at least five mass extinctions, including one significant mammalian extinction, have occurred during the history of life on Earth. Major contributors to mass extinctions are the loss of habitat due to continental drift, climate change, and the disastrous results of meteorite impacts.

16.3 Systematics

- **Systematics** is the study of the evolutionary relationships, or **phylogeny,** among all organisms, past and present. Systematics relies on the fossil record, comparative anatomy and development, and molecular data to determine relationships among organisms.
- **Taxonomy** involves the identification, naming, and classification of organisms. In the Linnaean system of classification, every organism belongs to a **taxon** and is assigned a scientific name, which indicates its **genus** and specific epithet. The combination of genus and specific epithet forms the name of the **species.** Species are also assigned to a **family,** an **order,** a **class,** a **phylum,** a **kingdom,** and a **domain** according to their molecular and structural similarities, as well as evolutionary relationships to other species.
- **Phylogenic trees** depict the evolutionary history of a group of organisms. Phylogenetic trees can help distinguish between **homologous structures** (those related by common descent) and **analogous structures** (those that are a result of **convergent evolution**). Cladists use shared derived characteristics to construct cladograms. In a **cladogram,** a **clade** consists of a common ancestor and all of its descendant species, which share derived traits.
- Linnaean classification has come under severe criticism because it does not always follow the principles of cladistics in the grouping of organisms.

Classification Systems

- The **five-kingdom system** was historically based on criteria such as cell type, level of organization, and type of nutrition.
- The **three-domain system** uses molecular data to designate three evolutionary domains: Bacteria, Archaea, and Eukarya.
- **Domain Bacteria** and **domain Archaea** contain the prokaryotes.
- **Domain Eukarya** contains kingdoms for the protists, animals, fungi, and plants.

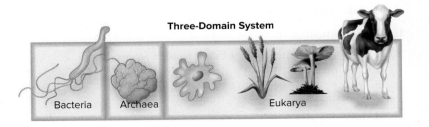

Three-Domain System

Bacteria   Archaea   Eukarya

ASSESS

Testing Yourself

Choose the best answer for each question.

16.1 Speciation and Macroevolution

1. A biological species
   a. always looks different from other species.
   b. always has a different chromosome number than other species.
   c. is reproductively isolated from other species.
   d. never occupies the same ecological niche as other species.

For questions 2–9, indicate the type of isolating mechanism described in each scenario.

Key:

   a. habitat isolation
   b. temporal isolation
   c. behavioral isolation
   d. mechanical isolation

   e. gamete isolation
   f. zygote mortality
   g. hybrid sterility
   h. low F_2 fitness

2. Females of one species do not recognize the courtship behaviors of males of another species.
3. One species reproduces at a different time of year than another species.
4. A cross between two species produces a zygote that always dies.
5. Two species do not interbreed because they occupy different areas.
6. A hybrid between two species produces gametes that are not viable.
7. Two species of plants do not hybridize because they are visited by different pollinators.
8. The sperm of one species cannot survive in the reproductive tract of another species.
9. The offspring of two hybrid individuals exhibit poor vigor.
10. Allopatric, but not sympatric, speciation requires
    a. reproductive isolation.
    b. geographic isolation.
    c. spontaneous differences between males and females.
    d. prior hybridization.
    e. a rapid rate of mutation.

16.2 The Fossil Record

11. One benefit of the fossil record is
    a. that hard parts of bodies are more likely to fossilize.
    b. that fossils can be dated.
    c. its completeness.
    d. that fossils congregate in one place.
    e. All of these are correct.
12. Which of the following contributed to mass extinctions?
    a. climate change
    b. continental drift
    c. meteor impacts
    d. All of these are correct.

16.3 Systematics

13. Which of the following is the scientific name of an organism?
    a. *Rosa rugosa*
    b. *Rosa*
    c. *rugosa*
    d. *rugosa rugosa*
    e. Both a and d are correct.

14. Which of these statements best pertains to taxonomy?
    a. Species always have three-part names, such as *Homo sapiens sapiens*.
    b. Species are always reproductively isolated from other species.
    c. Species share ancestral traits but may have their own unique derived traits.
    d. Species always look exactly alike.
    e. Both c and d are correct.
15. Answer these questions about the following cladogram.

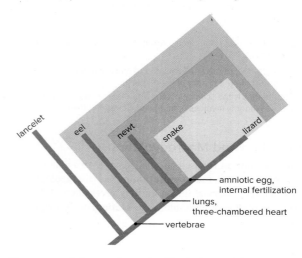

   a. How many clades are shown in this cladogram? How are they designated in the diagram?
   b. What trait is shared by all animals in the study group? What traits are shared by only snakes and lizards?
   c. Which animals share the most recent common ancestor? How do you know?

ENGAGE

Thinking Critically

1. Fill in the proposed phylogenetic tree for vascular plants with both the names of the groups (across the top of the table) and the shared traits (along the left side).

	Ferns	Conifers	Ginkgos	Monocots	Eudicots
vascular tissue	X	X	X	X	X
produce seeds		X	X	X	X
naked seeds		X	X		
needlelike leaves		X			
fan-shaped leaves			X		
enclosed seeds				X	X
one embryonic leaf				X	
two embryonic leaves					X

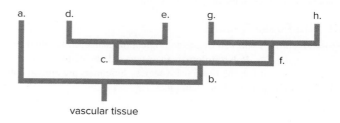

vascular tissue

2. The Hawaiian Islands are located thousands of kilometers from any mainland. Each island arose from the sea bottom and was colonized by plants and animals that drifted in on ocean currents or winds. Each island has a unique environment in which its inhabitants have evolved. Consequently, most of the plant and animal species on the islands do not exist anywhere else in the world.

   In contrast, on the islands of the Florida Keys, there are no unique or indigenous species. All of the species on those islands also exist on the mainland. Suggest an explanation for these two different patterns of speciation.

3. Reconstructing evolutionary relationships can have important benefits. For example, an emerging virus is apt to jump to a related species rather than to an unrelated species. We now know, for example, that the HIV virus jumped from chimpanzees—with which we share over 90% of our DNA sequence—to humans. Chimpanzees don't become ill from the virus; therefore, studying their immune system might help us develop strategies to combat AIDS in humans. But what about the study of evolutionary relationships between organisms that are not closely related to humans?

   a. Should the public be willing to fund all types of research on systematics or only research that has immediate medical benefits, like HIV research?

   b. Would you be willing to fund systematics research that helps us understand our evolutionary past, even if the medical benefit is not immediately known?

17

The Microorganisms: Viruses, Bacteria, and Protists

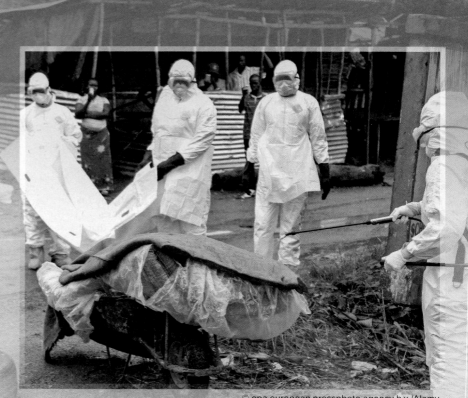

© epa european pressphoto agency b.v./Alamy

OUTLINE

17.1 The Viruses 287

17.2 Viroids and Prions 292

17.3 The Prokaryotes 293

17.4 The Protists 301

BEFORE YOU BEGIN

Before beginning this chapter, take a few moments to review the following discussions.

Figure 4.2 How much smaller is a virus than a bacterial cell? Than an animal or a plant cell?

Section 4.3 What are the differences between the structure of a prokaryotic and that of a eukaryotic cell?

Figure 16.20 What two domains of life contain prokaryotic organisms?

The West Africa Ebola Outbreak

In 2013, an outbreak of Ebola, one of the most feared viruses on the planet, began in the West African nation of Guinea. It is believed that a 1-year-old boy contracted the disease while playing near a tree that housed a species of bat that is known to carry the virus. By early 2014, the disease had become widespread in the neighboring countries of Sierra Leone and Liberia, and cases had been recorded in Nigeria, Mali, and Senegal. According to the CDC, there have been around 28,000 confirmed cases of Ebola in West Africa and over 11,000 confirmed deaths. But most agencies believe that these numbers are underestimates and that the complete toll from this outbreak may never be known.

What makes Ebola so feared is that it belongs to a family of viruses that cause hemorrhagic fever, a disease that targets several different cell types in the body, including macrophages of the immune system and the endothelial cells in the circulatory system and liver. While Ebola is frequently described as a disease the causes widespread bleeding, most deaths are actually due to fluid loss, organ failure (for example, of the liver), or an overall failure of the immune system. The virus is transmitted through direct contact with the body fluids of an infected person.

Like many viruses, the Ebola virus has been characterized by many misconceptions. These include that the virus is airborne, that you can get the disease from contact with cats and dogs, and that antibiotics are an effective treatment. In fact, in many ways Ebola is similar to any virus, it must invade specific cells of the body in order to hijack the cell's metabolic machinery to make more copies of itself.

In this chapter, we will examine the interaction of viruses with living organisms, including other members of the microbial world, such as bacteria and protists.

As you read through this chapter, think about the following questions:

1. Based on the characteristics shared by all living organisms, should viruses be considered living?
2. Although some cause disease, why are microorganisms essential to life?

17.1 The Viruses

Learning Outcomes

Upon completion of this section, you should be able to

1. Describe the structure of a virus.
2. Explain the basis of viral host specificity.
3. Describe the process of viral reproduction.

In Section 1.1, we described the general characteristics of life, and in Section 4.1, we introduced the concept that the cell is the fundamental unit of life. Viruses are *obligate intracellular parasites,* because they can reproduce only inside a living cell (*obligate* means "restricted to a specific form"). They lack the ability to acquire nutrients, or to use energy. Outside a living cell, viruses can be stored independently of living cells or even synthesized in the laboratory from chemicals. They are also incredibly small, most viruses are only about 0.2 micrometer (μm) long, or one-tenth the size of a bacterium.

Connections: Scientific Inquiry

How big can a virus get?

Generally, viruses are around one-tenth the size of a bacterium. However, several recent discoveries have challenged the idea that all viruses must be small.

In 2013, researchers in Siberia uncovered a giant virus (called *Pithovirus sibericum*) in the permafrost that was almost 1.5 μm long, or about the size of a small bacterium. However, this virus is not unique, other giant viruses that are similar in size have been discovered, and scientists have begun placing them into their own classification categories (e.g., *Megavirus, Pandoravirus*). Studies of these viruses may provide insight into viral evolution and the relationship of viruses to bacteria.

© ZJAN/Supplied by WENN.com/Newscom

While viruses have historically not been classified as living, that issue is still open to debate among biologists. Although small compared to bacteria, viruses do possess a genome, usually consisting of a few hundred genes that are used to manufacture new viruses in the host cell. The origins of viruses are also unclear, with some research suggesting that they may be remnants of bacteria. Viruses have also played an important role in the evolution of life on the planet. For these reasons, the debate over whether viruses are living or not continues in the scientific community.

Structure of a Virus

Each type of virus always has at least two parts: an outer **capsid,** composed of protein subunits, and an inner core containing its genetic material, which may be either DNA or RNA (**Fig. 17.1**).

The influenza virus shown in Figure 17.1 also has *spikes* (formed from a glycoprotein), which are involved in attaching the virus to the host cell. In the case of influenza viruses, there are two types of spikes, and the variations in their structure give each type of influenza virus its name, for example, H5N1

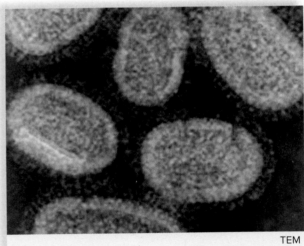

TEM

Influenza virus: RNA virus with a spherical capsid surrounded by an envelope with spikes.

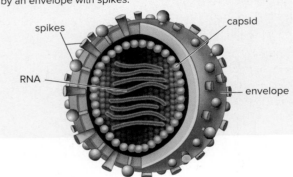

Figure 17.1 **Anatomy of an influenza virus.**

Typical of viruses, influenza has a nucleic acid core (in this case, RNA) and a coat of protein called the capsid. The projections, called spikes, help the influenza virus enter a cell and account for how some of these viruses are named (H1N1, H3N5, etc.)

(photo): *Source:* CDC/Cynthia Goldsmith

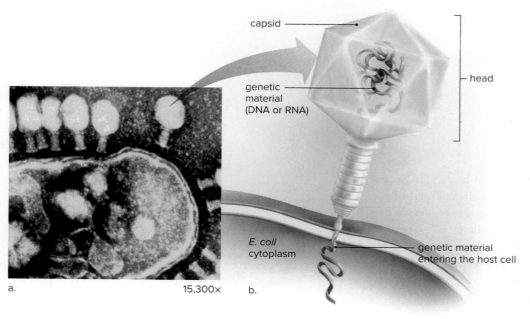

a. 15,300×   b.

Figure 17.2 Bacteriophage lambda (λ).

a. A micrograph shows the relative size of a phage compared to a bacterium. **b.** How DNA from a virus enters a bacterium, such as *E. coli*.

(a): © Lee D. Simon/Science Source

or H7N9. In some viruses that attack animals, the capsid is surrounded by an outer membranous *envelope* with glycoprotein spikes. The envelope is actually a piece of the host's plasma membrane, which also contains proteins produced by the virus. The interior of a virus can contain various enzymes that assist in the manufacture of new viruses. The viral genome may be either DNA or RNA, but it has at most several hundred genes; by contrast, a human cell contains around 23,000 genes.

Viral Reproduction

Viruses are specific to a particular host cell because a spike, or some portion of the capsid, adheres in a lock-and-key manner to a specific molecule (called a receptor) on the host cell's outer surface. A virus cannot infect a host cell to which it is unable to attach. For example, the tobacco mosaic virus cannot infect an exposed human because its capsid cannot attach to the receptors on the surfaces of human cells. Once inside a host cell, the viral genome takes over the metabolic machinery of the host cell. In large measure, the virus uses this machinery, including the host's enzymes, ribosomes, transfer RNA (tRNA), and ATP, to reproduce itself.

Reproduction of Bacteriophages

A **bacteriophage,** or simply *phage,* is a virus that reproduces in a bacterium. Bacteriophages are named in different ways; the one shown in **Figure 17.2** is called phage lambda (λ). Phages are often used as model systems for the study of viral reproduction.

When phage λ reproduces, it can do so via the lytic cycle or the lysogenic cycle. The **lytic cycle (Fig. 17.3***a*) may be divided into five stages: attachment, penetration, biosynthesis, maturation, and release. **1** During *attachment,* the capsid combines with a receptor in the bacterial cell wall. **2** During *penetration,* a viral enzyme digests away part of the cell wall, and viral DNA is injected into the bacterial cell. **3** *Biosynthesis* of viral components begins after the virus inactivates host genes not necessary to viral replication. The machinery of the host cell then carries out viral DNA replication and production of multiple copies of the capsid protein subunits. **4** During *maturation,* viral DNA and capsids assemble to produce several hundred viral particles. Lysozyme, an enzyme coded for by a viral gene, disrupts the cell wall, and **5** *release* of phage particles occurs. The bacterial cell dies as a result.

In the **lysogenic cycle** (Fig. 17.3*b*), the infected bacterium does not immediately produce phages, but it may do so in the future. In the meantime, the phage is *latent*—not actively reproducing. Following attachment and penetration, *integration* occurs: Viral DNA becomes incorporated into bacterial DNA with no destruction of host DNA. While latent, the viral DNA is called a *prophage*. The prophage is replicated along with the host DNA, and all subsequent cells, called lysogenic cells, carry a copy of the prophage. Certain environmental factors, such as ultraviolet radiation, can induce the prophage to enter the lytic stage of biosynthesis, followed by maturation and release.

Plant Viruses

Crops and garden plants are also subject to viral infections. Plant viruses tend to enter through damaged tissues and then move about the plant through *plasmodesmata,* cytoplasmic strands that extend between plant cell walls. The best studied plant virus is tobacco mosaic virus, a long, rod-shaped virus with only one type of protein subunit in its capsid (**Fig. 17.4**). Not all plant viruses are deadly, but over time, they often debilitate the host plant.

Viruses are often passed from one plant to another by insects and gardening tools, which move sap from one plant to another. Viral particles are also transmitted by way of seeds and pollen. Unfortunately, no chemical can control viral diseases. Until recently, the only way to deal with viral diseases was to destroy symptomatic plants and to control the insect vector, if there is one. Now that biotechnology and genetic engineering are routine (see Section 12.1), it is possible to transfer genes conferring disease resistance between plants. One of the most successful examples is the creation of papaya plants resistant to papaya ring spot virus (PRSV) in Hawaii. One transgenic line is now completely resistant to PRSV.

Animal Viruses

Viruses that cause disease in animals, including humans, reproduce in a manner similar to that of bacteriophages. However, there are modifications. In particular,

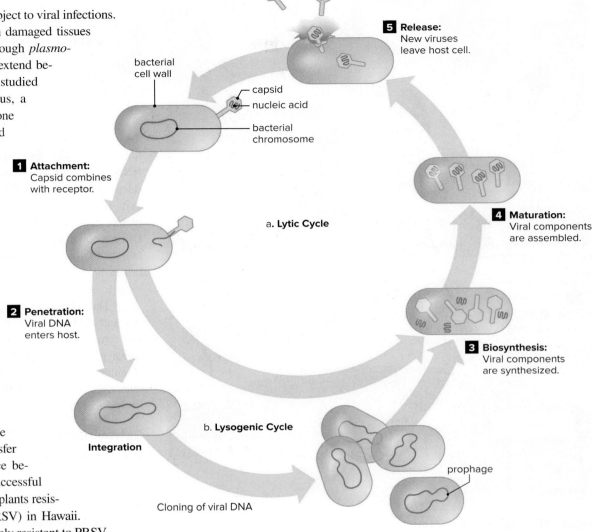

1 **Attachment:** Capsid combines with receptor.

2 **Penetration:** Viral DNA enters host.

Integration

bacterial cell wall

capsid
nucleic acid
bacterial chromosome

a. **Lytic Cycle**

5 **Release:** New viruses leave host cell.

4 **Maturation:** Viral components are assembled.

3 **Biosynthesis:** Viral components are synthesized.

b. **Lysogenic Cycle**

prophage

Cloning of viral DNA

Figure 17.3 Lytic and lysogenic cycles of phage lambda.

a. In the lytic cycle, viral particles escape when the cell is lysed (broken open). **b.** In the lysogenic cycle, viral DNA is integrated into host DNA. At a future time, the lysogenic cycle can be followed by the last three steps of the lytic cycle.

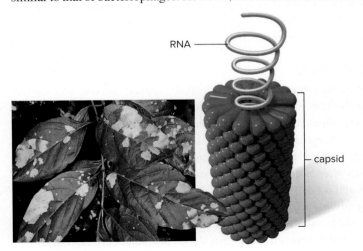

RNA

capsid

Figure 17.4 Infected tobacco plant.

Mottling and a distorted leaf shape are typical of a tobacco mosaic virus infection.

© Steven P. Lynch

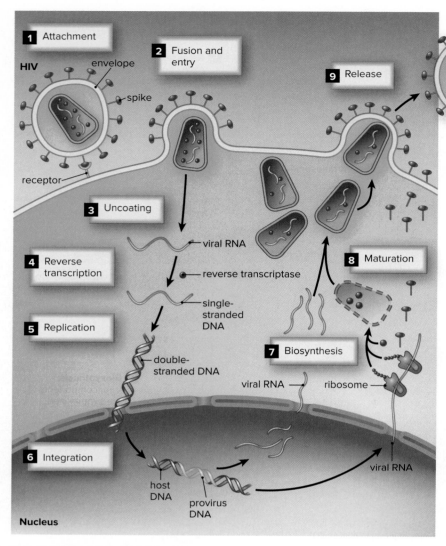

Figure 17.5 **Reproduction of HIV.**

HIV, the virus that causes AIDS, reproduces itself in several steps, as noted in the boxes. Because HIV is a retrovirus with an RNA genome, the enzyme reverse transcriptase is utilized to produce a single-stranded DNA copy of the genome. After synthesis of a complementary strand, the double-stranded viral DNA integrates into the host DNA.

some, but not all, animal viruses have an outer membranous envelope beyond their capsid. After attachment to a receptor in the plasma membrane, viruses with an envelope either fuse with the plasma membrane or enter the host cell by endocytosis.

The general life cycle of an animal virus is shown in **Figure 17.5.** After an enveloped virus enters, uncoating follows—that is, the capsid is removed by enzymes within the host cell (see **1** to **3**). Once the viral genome, either DNA or RNA, is free of its covering, biosynthesis, maturation, and release may occur (see **7** to **9**). While a naked animal virus exits the host cell in the same manner as does a bacteriophage, animal viruses with an envelope bud from the cell. During budding, the virus picks up its envelope, consisting mainly of lipids and proteins, from the host plasma membrane. Spikes, those portions of the envelope that allow the virus to enter a host cell, are coded for by viral genes.

The herpesviruses, which cause cold sores, genital herpes, and chickenpox in humans, are examples of animal viruses that remain latent much of the time. Herpesviruses linger in spinal ganglia until stress, excessive sunlight, or some other stimulus causes them to undergo the lytic cycle. The human immunodeficiency virus (HIV), the cause of AIDS, remains relatively latent in lymphocytes, only slowly releasing new viruses.

Retroviruses

Retroviruses are viruses that use RNA as their genetic material. Figure 17.5 illustrates the reproduction of the retrovirus HIV. A retrovirus contains an enzyme called reverse transcriptase, which carries out transcription of RNA to DNA. The enzyme synthesizes one strand of DNA using the viral RNA as a template, as well as a DNA strand that is complementary to the first one. Using host enzymes, the resulting double-stranded DNA is integrated into the host genome. The viral DNA (*provirus*) remains in the host genome and is replicated when the host DNA is replicated (see steps **4** to **6** in Fig. 17.5). When or if this DNA is transcribed, new viruses are produced by the steps we have already cited: biosynthesis, maturation, and release. Being an animal virus with an envelope, HIV buds from the cell. More information on HIV, and the disease AIDS, is provided in Section 26.5.

Emerging Viruses

HIV is an **emerging virus,** the causative agent of a disease that only recently has infected large numbers of people. Other examples of emerging viruses are West Nile virus, SARS virus, hantavirus, and avian influenza (H5N1) virus (**Fig. 17.6**).

Infectious diseases emerge in several different ways. In some cases, the virus is simply transported from one location to another. The West Nile virus is an emerging virus because it changed its range: It was transported into the United States and is taking hold in bird and mosquito populations. Severe acute respiratory syndrome (SARS) was clearly transported from Asia to Toronto, Canada. A world in which you can begin your day in Bangkok and end it in Los Angeles is a world in which diseases can spread at an unprecedented rate.

Other factors can also cause infectious viruses to emerge. Viruses are well known for their high mutation rates. Some of these mutations affect the structure of the spikes or capsids, so a virus that previously could infect only a particular animal species acquires the ability to infect humans. For example, the diseases AIDS and Ebola fever are caused by viruses that at one time infected only monkeys and apes.

The virus that causes Middle East respiratory syndrome, or MERS, is another example of an emerging virus that caused concern in the medical community. Like SARS, MERS is a coronavirus (MERS-CoV). These classes of viruses are known to cause respiratory problems, including shortness of breath, coughing, and fever. What makes MERS unique is the fact that it appears to be a novel class of coronavirus that had not previously been detected in humans

Bird flu remains a constant concern of world health experts. Wild ducks are resistant to avian influenza viruses that can spread from them to chickens, which increases the likelihood that the disease, another type of influenza, will spread to humans. Another way a virus could emerge is by a change in the mode of transmission—for example, from requiring contact for transmissoin to being transmitted through the air. Health officials are concerned that this may occur for viruses such as H5N1, thus greatly increasing the number of people who may be exposed to the disease bird flu.

Drug Control of Human Viral Diseases

Antibiotics do not work against viruses since those drugs are designed to interfere with the function of living bacterial cells. Historically, a viral infection was only treatable by allowing the immune system to target the virus using the adaptive immunity response (see Section 26.3). Immunization (see Section 26.4) is also effective in preventing an infection by a specific virus.

Understanding that viruses reproduce using the metabolic machinery of the cell has made it possible to develop antiviral drugs that target specific aspects of the viral life cycle. Most antiviral compounds, such as ribavirin and acyclovir, are structurally similar to nucleotides; therefore, they interfere with viral genome synthesis. Compounds related to acyclovir are commonly used to suppress herpes outbreaks. HIV is treated with antiviral compounds specific to a retrovirus. The well-publicized drug AZT and others block reverse transcriptase. And HIV protease inhibitors block the enzymes required for the maturation of viral proteins.

Check Your Progress 17.1

1. Describe the general structure of a virus.

2. Describe the two ways that a virus may infect a cell.

3. Explain why a virus is host-specific—that is, will infect only a certain type of cell.

Connections: Health

What is a norovirus?

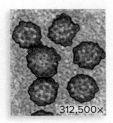

312,500×

© James Cavallini/ Science Source

The norovirus is one of the types of viruses that can cause gastroenteritis, or an inflammation of the stomach and intestines, sometimes called "stomach flu" or "food poisoning." It is a highly contagious virus that can be contracted by coming in contact with food, surfaces, or other people that have been exposed to it. While most people recover from a norovirus infection in a few days, the virus can cause several forms of serious illness and even death. The most effective way of protecting yourself from norovirus is by thorough hand-washing, since most hand sanitizers do not destroy the virus.

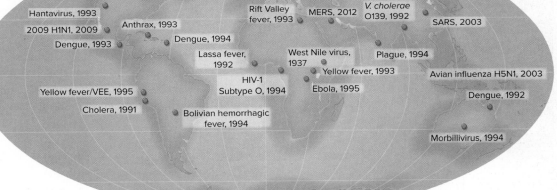

Figure 17.6 Emerging diseases.

Emerging diseases, such as those noted here according to their country of origin, are new or demonstrate increased prevalence. The agents causing such diseases may have acquired new virulence factors, or environmental factors may have encouraged their spread to an increased number of hosts.

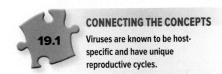

CONNECTING THE CONCEPTS

19.1 Viruses are known to be host-specific and have unique reproductive cycles.

a.

b.

Figure 17.7 Viroid diseases.

a. Viroids infect plants. For example, PSTVd causes potatoes to be become elongated and fibrous compared to normal potatoes.
b. Cannibalistic tribesmen in Papua, New Guinea.

(a, diseased potato): *Source:* Barry Fitzgerald/USDA; (a, normal potato): © McGraw-Hill Education/Mark Dierker, photographer; (b): © Bettmann/Corbis

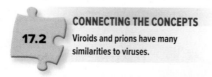

CONNECTING THE CONCEPTS
17.2 Viroids and prions have many similarities to viruses.

Check Your Progress 17.2

1. Contrast a viroid with a virus.
2. Describe the difference between a prion and a viroid.
3. List a few prion disorders.

17.2 Viroids and Prions

Learning Outcomes

Upon completion of this section, you should be able to

1. Distinguish between a viroid and a prion.

About a dozen crop diseases have been attributed not to viruses but to **viroids,** which are strands of RNA that are naked (not covered by a capsid). Like viruses, though, viroids direct the host cell to produce more viroids. Viroids are a concern in agriculture, since they can not only cause the host plant to be unhealthy but also alter the plant's physical characteristics, thus reducing its economic value.

Some diseases in humans have been attributed to **prions,** a name coined from the term *proteinaceous infectious* particles. The discovery of prions began with the observation that members of a primitive tribe, the Fore, in the highlands of Papua New Guinea died from a disease called kuru (meaning trembling with fear) after participating in the cannibalistic practice of eating a deceased person's brain (**Fig. 17.7**). The causative agent of kuru was smaller than a virus—it was a misshapen protein. It appears that a normal protein changes shape, so that its polypeptide chain is in a different configuration. The result is a prion, capable of causing a fatal infection and a neurodegenerative disorder.

It is believed that a prion can interact with normal proteins, causing them to change shape and, in some cases, to act as infectious agents themselves. The process has been best studied in a disease called scrapie, which attacks sheep. Other prion diseases include the widely publicized mad cow disease; human maladies, such as Creutzfeldt-Jakob disease (CJD); and a variety of chronic wasting syndromes in several other animal species.

Connections: Health

Do prions cause Alzheimer disease?

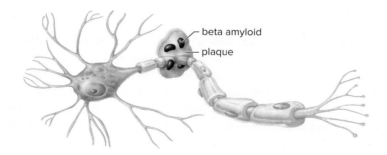

— beta amyloid
— plaque

One of the symptoms of Alzheimer disease is an accumulation of a protein called beta amyloid. This accumulation forms tangled structures in the brain called plaques. In normal brain cells, a protein prevents plaques of beta amyloid from forming. In some individuals, a mutation in a gene (*PrP*) causes this protein to malfunction and, in effect, become a prion. For people who have this mutation, beta amyloid accumulates and causes the death of nerve cells that is associated with Alzheimer disease.

17.3 The Prokaryotes

What was the first life like on Earth? What features in modern cells are the most primitive? The answers to these questions are coming from the study of **prokaryotes,** single-celled organisms that lack a nucleus. Prokaryotes also lack the internal organelle structures found within the cells of eukaryotes, although some prokaryotes contain internal membranes that possess similar functions. While structurally simple, these cells can be metabolically complex, and some species are able to live in the most inhospitable places on the planet. In fact, in terms of sheer numbers and biomass, prokaryotes are one of Earth's dominant life-forms.

In this chapter, we will examine the two types of prokaryotes—the bacteria and the archaea. We will discuss the origin of the first cells and then examine the bacteria, the best-known prokaryotes, followed by the archaea.

The Origin of the First Cells

Until the nineteenth century, many thought that prokaryotes could arise spontaneously. But in 1850, Louis Pasteur showed that previously sterilized broth cannot become cloudy with growth unless it is exposed directly to the air, where bacteria are abundant. The **cell theory** formulated about this time states that all organisms are composed of cells, cells are capable of self-reproduction, and cells come only from preexisting cells. How, then, did the first cells arise? Much work has gone into studying this question, which pertains to the origin of life. This research is not just confined to Earth. Many of our exploratory missions to planets and moons in our solar system are looking for signs of early life.

The first living cells were prokaryotes, possessing DNA but lacking a nucleus. Fossilized prokaryotes have been found in rocks that are 3.5 billion years old; scientists think that prokaryotes are likely to have existed for millions of years before those found in these ancient fossils. The first cells were preceded by biological macromolecules, such as proteins and nucleic acids. Today, amino acids, nucleotides, and other building blocks for biological macromolecules are routinely produced by living cells; this process is known as *biotic synthesis.* Prior to cellular life, macromolecules must have formed by *abiotic synthesis.*

Conditions on the early Earth were very different than they are today. Initially, temperatures were very high. Although there was little free oxygen (O_2) in the atmosphere, there would have been an abundance of other gases, such as water vapor (H_2O), carbon dioxide (CO_2), and nitrogen (N_2), along with smaller amounts of hydrogen (H_2), methane (CH_4), ammonia (NH_3), hydrogen sulfide (H_2S), and carbon monoxide (CO). As the early Earth cooled, water vapor condensed to liquid water, and rain began to fall, producing the oceans. The process of chemical evolution, or the abiotic synthesis of organic monomers under these special conditions, may have occurred with the input of energy from any of a variety of possible sources, including lightning, sunlight, meteorite impacts, volcanic activity, and radioactive decay (**Fig. 17.8**).

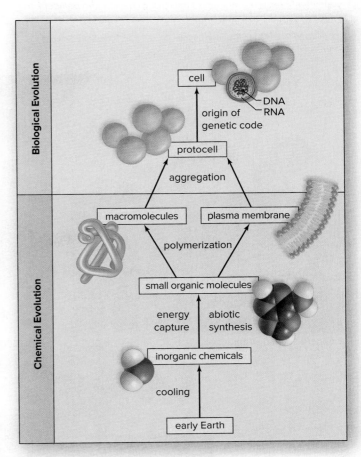

Figure 17.8 Origin of the first cell(s).

There was an increase in the complexity of macromolecules, leading to a self-replicating system (DNA → RNA → protein) enclosed by a plasma membrane. This protocell then underwent biological evolution, becoming a true cell.

In the 1950s, Stanley Miller conducted a now famous experiment in which he attempted to replicate these conditions. His results indicated that it was possible to form amino acids under the conditions that existed on the early Earth. Scientists are still debating whether the first reactions of chemical evolution occurred within the atmosphere, deep in the ocean, or elsewhere in the solar system. However, they widely agree that the first macromolecules arose via abiotic synthesis.

Protocells, cell-like structures complete with an outer membrane, may have resulted from the self-assembly of macromolecules and eventually given rise to cellular life. In fact, researchers have studied cell-like structures that arise from collections of biological macromolecules under laboratory conditions. Some are water-filled spheres with an outer layer similar to that of a cell. If enzymes are trapped inside such a sphere, they can catalyze chemical reactions, like those of cellular metabolism.

An important development in the biological evolution of life would have been a molecule capable of passing information about metabolism and structure from one generation to the next, the role fulfilled by DNA in today's cells. Some researchers think the first hereditary molecule may have been RNA. It is hypothesized that, over time, the more stable DNA replaced RNA as a long-term repository for genetic information, leading to the self-replicating system seen in all living cells today.

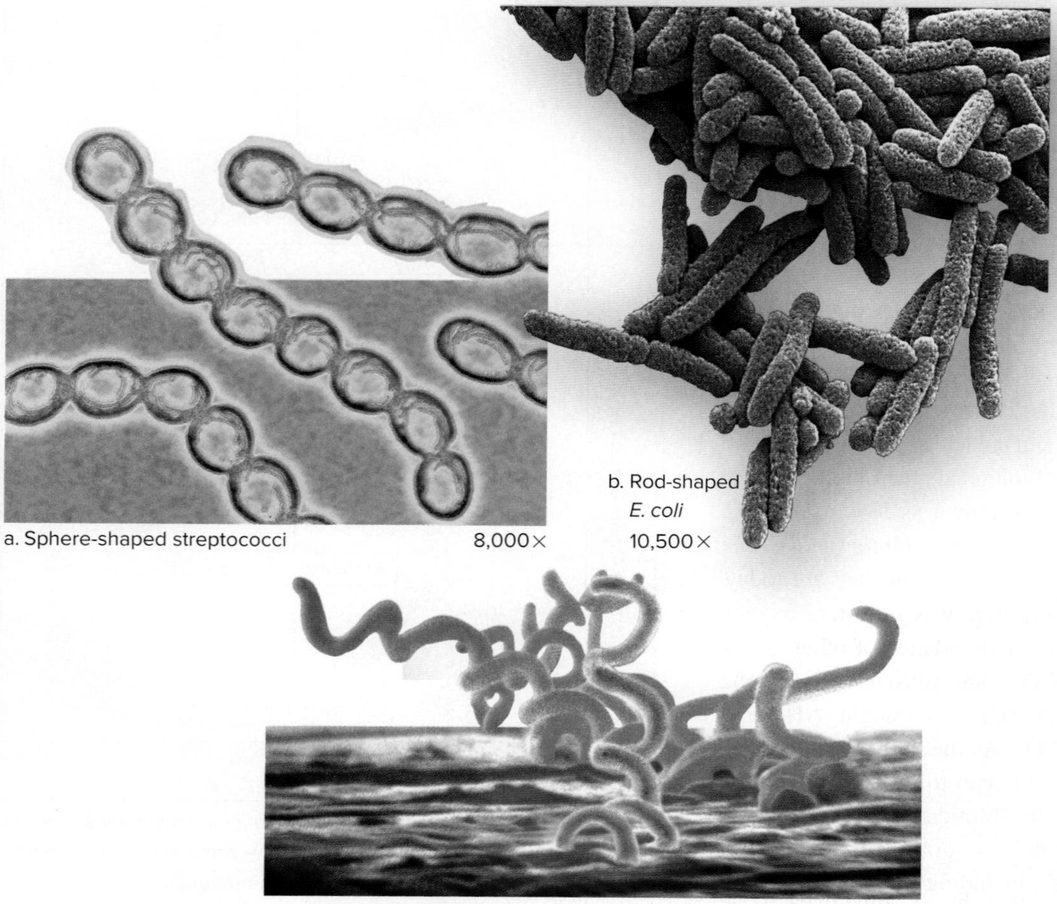

Figure 17.9 **Shapes of bacteria.**

a. Streptococci, which exist as chains of cocci, cause a number of illnesses, including strep throat. **b.** *Escherichia coli*, which lives in your intestines, is a rod-shaped bacillus. **c.** *Treponema pallidum*, the cause of syphilis, is a spirochete.

(a): © Alfred Pasieka/SPL/Science Source; (b): © David Scharf/SPL/Science Source; (c): © Science Source

a. Sphere-shaped streptococci   8,000×

b. Rod-shaped E. coli   10,500×

c. Spirochete, *T. pallidum*

Bacteria

Bacteria are the most diverse and prevalent organisms on Earth. Billions of bacteria exist in nearly every square meter of soil, water, and air. They also make a home on our skin and in our intestines. Although tens of thousands of different bacteria have been identified, they likely represent only a very small fraction of the total number of bacterial species on the planet. For example, less than 1% of bacteria in the soil can be grown in the laboratory. However, molecular genetic techniques are being used to discover the extent of bacterial diversity.

General Biology of Bacteria

Bacterial Structure   Bacteria have a variety of shapes. However, most bacteria are spheres (called cocci), rods (called bacilli), or spirals (called spirilla [sing., spirillum] if they are rigid or spirochetes if they are flexible) (**Fig. 17.9**).

There are a number of variations in bacterial shape. For example, a slightly curved rod is called a vibrio. While many bacteria grow as single cells, some form doublets. Others

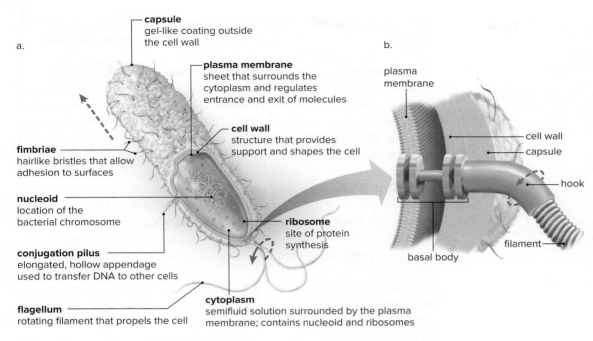

a.

capsule
gel-like coating outside
the cell wall

plasma membrane
sheet that surrounds the
cytoplasm and regulates
entrance and exit of molecules

cell wall
structure that provides
support and shapes the cell

fimbriae
hairlike bristles that allow
adhesion to surfaces

nucleoid
location of the
bacterial chromosome

conjugation pilus
elongated, hollow appendage
used to transfer DNA to other cells

flagellum
rotating filament that propels the cell

ribosome
site of protein
synthesis

cytoplasm
semifluid solution surrounded by the plasma
membrane; contains nucleoid and ribosomes

b.

plasma
membrane

cell wall

capsule

hook

filament

basal body

Figure 17.10 Flagella.

a. The structure of a prokaryotic cell.
b. Each flagellum of a bacterium
consists of a basal body, a hook, and a
filament. The red-dashed arrows
indicate that the hook and filament
rotate 360°.

form chains, such as the streptococci that are the cause of strep throat. A third growth habit produces a shape that resembles a bunch of grapes, seen in staphylococci, which cause food poisoning.

Figure 17.10*a* shows the structure of a bacterium. A bacterium, being a prokaryote, has no nucleus. A single, closed circle of double-stranded DNA constitutes the chromosome, which occurs in an area of the cell called the **nucleoid.** In some cases, extrachromosomal DNA molecules called **plasmids** are also found in bacterial cells.

Bacteria have ribosomes but not membrane-bound organelles, such as mitochondria and chloroplasts. However, some bacteria have internal folds of membranes, which contain enzymes for cellular functions. These do not enclose spaces in the same way as eukaryotic organelles do. An example in photosynthetic bacteria are the thylakoid membranes, which contain the pigments needed for the light reactions. Motile bacteria generally use **flagella** for locomotion. The bacterial flagellum is not structured like a eukaryotic flagellum (Fig. 17.10*b*). It has a filament composed of three strands of the protein flagellin wound in a helix. The filament is inserted into a hook that is anchored in the plasma membrane by a basal body. The fully reversible 360° rotation of the flagella causes the bacterium to spin as it moves forward and backward.

Bacteria have an outer cell wall strengthened not by cellulose but by **peptidoglycan,** a complex of polysaccharides linked by amino acids. The cell wall prevents bacteria from bursting or collapsing due to osmotic changes. Parasitic bacteria are further protected from host defenses by a polysaccharide capsule that surrounds the cell wall.

Bacterial Reproduction Bacteria (and archaea) reproduce asexually by means of **binary fission.** The single, circular chromosome replicates, and then the two copies separate as the cell enlarges. The newly formed plasma membrane and cell wall partition the two new cells, with a chromosome in each one **(Fig. 17.11).** Mitosis, which requires the formation of a spindle apparatus, does

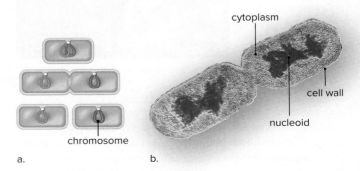

cytoplasm

cell wall

nucleoid

chromosome

a.

b.

Figure 17.11 Binary fission.

a. Binary fission produces two cells with identical genetic
information. **b.** A prokaryotic cell undergoing binary fission.

(b): © CNRI/SPL/Science Source

not occur in prokaryotes. Binary fission turns one cell into two cells, two cells into four cells, four cells into eight cells, and so on, potentially until billions of cells have been produced.

In eukaryotes, genetic recombination occurs as a result of sexual reproduction. Sexual reproduction does not occur among prokaryotes, but three means of genetic recombination have been observed in bacteria. **Conjugation** occurs when a donor cell passes DNA directly to a recipient cell. During conjugation, the donor and recipient are temporarily linked together, often by means of a *conjugation pilus*. While they are linked, the donor cell passes DNA to the recipient cell.

Transformation occurs when a bacterium picks up (from the surroundings) free pieces of DNA secreted by live prokaryotes or released by dead prokaryotes. During **transduction,** bacteriophages carry portions of bacterial DNA from one cell to another. Plasmids, which sometimes carry genes for resistance to antibiotics, can be transferred between infectious bacteria by any of these means.

When faced with unfavorable environmental conditions, some bacteria form **endospores.** A portion of the cytoplasm and a copy of the chromosome dehydrate and are then encased by three heavy, protective spore coats. The rest of the bacterial cell deteriorates, and the endospore is released. Spores survive in the harshest of environments—desert heat and desiccation, boiling temperatures, polar ice, and extreme ultraviolet radiation. They also survive for very long periods. When anthrax spores 1,300 years old germinate, they can still cause a severe infection (usually seen in cattle and sheep). Humans also fear a deadly but uncommon type of food poisoning called botulism, which is caused by the germination of endospores inside cans of food. Spore formation is not a means of reproduction, but it does allow the survival of bacteria and their dispersal to new places.

Bacterial Nutrition   Prokaryotes are more metabolically diverse than eukaryotes. For example, plants are **photoautotrophs** (often called autotrophs or photosynthesizers) that can perform oxygenic photosynthesis: They depend on solar energy to split water and energize electrons for the reduction of carbon dioxide. An example of photosynthetic bacteria are the **cyanobacteria.** The cyanobacteria may well represent the oldest lineage of oxygenic organisms (**Fig. 17.12**a). Some fossil cyanobacteria have been dated at 3.5 billion years old. The evolution of cyanobacteria drastically altered the atmosphere of the early Earth by adding vast amounts of oxygen. Many cyanobacteria are capable of fixing atmospheric nitrogen and reducing it to an organic form. Therefore, they need only minerals, air, sunlight, and water for growth.

Other bacterial photosynthesizers don't release oxygen because they take electrons from a source other than water; some split hydrogen sulfide (H_2S) and release sulfur (S) in marshes, where they live anaerobically.

The **chemoautotrophs** (also called chemosynthesizers) don't use solar energy at all. They reduce carbon dioxide using energetic electrons derived from inorganic molecules, such as ammonia or hydrogen gas. Electrons can also be extracted from certain minerals, such as iron. Some chemoautotrophs oxidize sulfur compounds spewing from deep-sea vents 2.5 kilometers below sea level. The organic compounds they produce support the growth of communities of organisms found at such vents, where darkness prevails (Fig. 17.12b).

a. Cyanobacterium

tube worm

b. Hosts for chemoautotrophic bacteria

Figure 17.12 Bacterial autotrophs.

a. Cyanobacteria are photoautotrophs that photosynthesize in the same manner as plants—they split water and release oxygen.
b. Certain chemoautotrophic bacteria live inside tube worms, where they produce organic compounds without the need of sunlight. In this way, they help support ecosystems at hydrothermal vents deep in the ocean.

(a): © Eric Grave/Science Source; (b): *Source:* NOAA Okeanos Explorer Program, Galapagos Rift Expedition 2011

Most bacteria are **chemoheterotrophs** (often referred to as simply heterotrophs); like animals, they take in organic nutrients, which they use as a source of energy and as building blocks to synthesize macromolecules. The first cells were most likely chemoheterotrophs that fed on the abundant organic molecules in their environment. Autotrophs would have appeared later as the nutrient supply was depleted and the ability to make one's own food became advantageous. Unlike animals, chemoheterotrophic bacteria are **saprotrophs** that send enzymes into the environment and decompose almost any large organic molecule to smaller ones that are absorbable. There is probably no natural organic molecule that cannot be digested by at least one bacterial species. Bacteria play a critical role in recycling matter and making inorganic molecules available to photosynthesizers.

Heterotrophic bacteria may be either free-living or **symbiotic,** meaning that they form relationships that are (1) mutualistic (both partners benefit), (2) commensalistic (one partner benefits and the other is not harmed), or (3) parasitic (one partner benefits but the other is harmed). Mutualistic bacteria that live in human intestines release vitamins K and B_{12}, which our bodies use to help produce blood components. In the stomachs of cows and goats, mutualistic prokaryotes digest cellulose, enabling these animals to feed on grass. Commensalism often occurs when one population modifies the environment in such a way that a second population benefits. Obligate anaerobes can live in our intestines because the bacterium *Escherichia coli* uses up the available oxygen, creating an anaerobic environment. The parasitic bacteria cause diseases, as discussed next.

Environmental and Medical Importance of Bacteria

Bacteria in the Environment   For an ecosystem to sustain its populations, the chemical elements available to living organisms must eventually be recycled. A fixed and limited number of elements are available to living organisms; the rest are either buried too deep in the Earth's crust or present in forms that are not usable. All living organisms, including producers, consumers, and decomposers, are involved in the important process of cycling elements to sustain life (see Fig. 1.4). Bacteria are *decomposers* that digest dead organic remains and return inorganic nutrients to producers. Without the work of decomposers, life would soon come to a halt.

In the process of decomposing organic remains, bacteria perform the reactions needed for biogeochemical cycling, such as for the carbon and nitrogen cycles. Let's examine how bacteria participate in the nitrogen cycle (see Section 31.2). Plants are unable to fix atmospheric nitrogen (N_2), but they need a source of ammonia or nitrate in order to produce proteins. Bacteria in the soil can fix atmospheric nitrogen and/or change nitrogen compounds into forms that plants can use. In addition, mutualistic bacteria live in the root nodules of soybean, clover, and alfalfa plants, where they reduce atmospheric nitrogen to ammonia, which is used by plants (**Fig. 17.13**). Without the work of bacteria, nitrogen would not be available for plants to produce proteins or available to animals that feed on plants or other animals.

root

nodule

Figure 17.13 Nodules of a legume.

Although some free-living bacteria carry on nitrogen fixation, those of the genus *Rhizobium* invade the roots of legumes, with the resultant formation of nodules. Here the bacteria convert atmospheric nitrogen to an organic nitrogen that the plant can use. These are nodules on the roots of a soybean plant.

a.

Figure 17.14
Bioremediation.

a. Bacteria have been used for many years in sewage treatment plants to break down human wastes. **b.** Increasingly, the ability of bacteria to break down pollutants, such as oil from spills, is being researched and enhanced.

(a): © Kent Knudson/Photolink/Getty RF; (b): © Accent Alaska.com/Alamy

b.

Bioremediation is the biological cleanup of an environment that contains harmful chemicals called *pollutants.* To deal with pollution, the vast ability of bacteria to break down almost any substance, including sewage, is being exploited (**Fig. 17.14***a*). People have added thousands of tons of slowly degradable pesticides and herbicides, nonbiodegradable detergents, and plastics to the environment. Strains of bacteria are being developed specifically for digesting these types of pollutants. Some strains have been used to remove agent orange, a potent herbicide, from soil samples, and dual cultures of two types of bacteria have been shown to degrade PCBs, chemicals formerly used as coolants and industrial lubricants. Oil spills, such as the Deep Water Horizon spill in the Gulf of Mexico in 2010, are very damaging to aquatic ecosystems. The damage can be lessened by the use of genetically modified bacteria that can break the carbon-carbon bonds within the oil. However, the effectiveness of these bacteria is based on weather conditions and the availability of certain nutrients, such as nitrogen and phosphates.

Bacteria in Food Science and Biotechnology

A wide variety of food products are created through the action of bacteria (**Fig. 17.15**). Under anaerobic conditions, bacteria carry out fermentation, which results in a variety of acids. One of these acids is lactate, a product that pickles cucumbers, curdles milk into cheese, and gives these foods their characteristic tangy flavor. Other bacterial fermentations can produce flavor compounds, such as the propionic acid in swiss cheese. Bacterial fermentation is also useful in the manufacture of such products as vitamins and antibiotics—in fact, most antibiotics known today were discovered in soil bacteria.

As you know, biotechnology can be used to alter the genome and the products generated by bacterial cultures. Bacteria can be genetically engineered to produce medically important products, such as insulin, human growth hormone, antibiotics, and vaccines against a number of human diseases. The natural ability of bacteria to perform all manner of reactions is also enhanced through biotechnology.

Bacterial Diseases in Humans

Microbes that can cause disease are called **pathogens.** Pathogens are able to (1) produce a toxin, (2) adhere to surfaces, and sometimes (3) invade organs or cells.

Toxins are small organic molecules, small pieces of protein, or parts of the bacterial cell wall that are released when bacteria die. Toxins are poisonous, and bacteria that produce a toxin usually cause serious diseases. In almost all cases, the growth of the microbes themselves does not cause disease; the toxins they release cause it. When someone steps on a rusty nail, bacteria may be introduced deep into damaged tissue. The damaged area does not have good blood flow and can become anaerobic. *Clostridium tetani,* the cause of tetanus (lockjaw), proliferates under these conditions. The bacteria never leave the site of the wound, but the tetanus toxin they produce does move throughout the body. This toxin prevents the relaxation of muscles. In time, the body contorts because all the muscles have contracted. Eventually, the person suffocates.

Adhesion factors allow a pathogen to bind to certain cells, and this determines which organs or cells of the body will be its host. Like many bacteria

Figure 17.15 **Bacteria in food processing.**

Bacteria are used to help produce food products. Bacterial fermentation results in acids that give some types of cheeses their characteristic taste.

© McGraw-Hill Education/John Thoeming, photographer

that cause dysentery (severe diarrhea), *Shigella dysenteriae* is able to stick to the intestinal wall. In addition, *S. dysenteriae* produces a toxin called Shiga toxin, which makes it even more life-threatening. Also, invasive mechanisms that give a pathogen the ability to move through tissues and into the bloodstream result in a more medically significant disease than if the pathogen were localized. Usually, a person can recover from food poisoning caused by *Salmonella*. But some strains of *Salmonella* have virulence factors—including a needle-shaped apparatus that injects toxin into body cells—that allow the bacteria to penetrate the lining of the colon and move beyond this organ. Typhoid fever, a life-threatening disease, can then result.

While some bacteria are pathogens that cause immediate diseases, such as dysentery, the presence of some bacteria in our bodies may cause long-term health problems. Scientists are beginning to recognize that the composition of the bacterial population in our intestines, sometimes called the *natural flora* or *microbiota,* may be associated with medical conditions such as obesity and cardiovascular disease. These bacteria cause health problems not by attacking our cells directly but by altering the metabolism of compounds in our food, so that the by-products cause health problems.

Antibiotics are often used to treat medical conditions known to be associated with bacteria. However, since bacteria are cells in their own right, antibiotics need to target the differences between the prokaryotic bacterial cells and the eukaryotic cells in our bodies. In general, antibiotics target the cell walls of bacteria (which are absent in eukaryotic cells) or the prokaryotic-specific enzymes that are involved in protein and DNA synthesis. Although a number of antibiotic compounds are active against bacteria and are widely prescribed, the overuse of antibiotics in both medicine and agriculture has led to an increasing level of bacterial resistance to antibiotics.

Archaea

In Section 16.3, we explained how the tree of life is organized into three domains: domain Archaea, domain Bacteria, and domain Eukarya. Because many **archaea** and some bacteria are found in extreme environments (hot springs, thermal vents, salt basins), it is believed that they may have diverged from a common ancestor relatively soon after life began. Later, the eukarya are believed to have split off from the archaeal line of descent. Archaea and eukarya share some of the same ribosomal proteins (not found in bacteria), initiate transcription in the same manner, and have similar types of tRNA.

Structure of the Archaea

The plasma membranes of archaea contain unusual lipids that allow many of them to function at high temperatures. The archaea have also evolved diverse types of cell walls, which facilitate their survival under extreme conditions. The cell walls of archaea do not contain peptidoglycan, as do the cell walls of bacteria. In some archaea, the cell wall is largely composed of polysaccharides; in others, the wall is pure protein. A few have no cell wall.

Types of Archaea

Archaea are often discussed in terms of their unique habitats. The **methanogens** (methane makers) are found in anaerobic environments, such as in swamps, marshes, and the intestinal tracts of animals. Those found in animal intestines exist as mutualists or commensals, not as parasites—that is, archaea are not known to cause infectious diseases. Methanogens are chemoautotrophs that

Figure 17.16 Methanogen habitat and structure.

a. A swamp where methanogens live. **b.** Micrograph of *Methanosarcina mazei*, a methanogen.

(a): © Susan Rosenthal/Corbis; (b): © Dr. M. Rohde, GBF/Science Source

Figure 17.17 Halophile habitat and structure.

a. Great Salt Lake, Utah, where halophiles live. **b.** Micrograph of *Halobacterium salinarium*, a halophile.

(a): © Marco Regalia Sell/Alamy RF; (b): © Eye of Science/Science Source

couple the production of methane (CH_4) from hydrogen gas (H_2) and carbon dioxide to the formation of ATP (**Fig. 17.16**). This methane, which is also called *biogas,* is released into the atmosphere, where it contributes to the greenhouse effect and global warming. About 65% of the methane in Earth's atmosphere is produced by methanogenic archaea.

The **halophiles** require high-salt concentrations for growth (usually 12–15%; by contrast, the ocean is about 3.5% salt). Halophiles have been isolated from highly saline environments, such as the Great Salt Lake in Utah, the Dead Sea, solar salt ponds, and hypersaline soils (**Fig. 17.17**). These archaea have evolved a number of mechanisms to survive in high-salt environments. They depend on a pigment related to the rhodopsin in our eyes to absorb light energy for pumping chloride and another, similar pigment for synthesizing ATP.

A third major type of archaea are the **thermoacidophiles** (**Fig. 17.18**). These archaea are isolated from extremely hot, acidic environments, such as hot springs, geysers, submarine thermal vents, and the areas around volcanoes. They reduce sulfur to sulfides and survive best at temperatures above 80°C; some can even grow at 105°C (remember that water boils at 100°C). The metabolism of sulfides results in acidic sulfates, and these bacteria grow best at pH 1 to 2.

Figure 17.18 Thermoacidophile habitat and structure.

a. Boiling springs and geysers in Yellowstone National Park, where thermoacidophiles live. **b.** Micrograph of *Sulfolobus acidocaldarius*, a thermoacidophile.

(a): © Alfredo Mancia/Getty RF; (b): © Eye of Science/Science Source

CONNECTING THE CONCEPTS

17.3

The prokaryotes are single-celled organisms that lack a nucleus. Both the bacteria and the archaea are prokaryotes.

Check Your Progress 17.3

1. Summarize the steps of chemical and biological evolution.

2. List and describe the three common shapes a bacterial cell can take.

3. Summarize the three means of introducing variation in bacterial cells.

4. How do archaea differ from bacteria?

17.4 The Protists

Learning Outcomes

Upon completion of this section, you should be able to

1. Explain the role of endosymbiosis in the origin of the eukaryotic cell.
2. Summarize the general characteristics of all protists.
3. Distinguish the main supergroups of protists, and provide an example of each.

Like ancient creatures from another planet, the **protists** inhabit the oceans and other watery environments of the world. Their morphological diversity is their most outstanding feature. Single-celled diatoms may be encrusted in silica "petri dishes," while dinoflagellates have plates of armor and ciliates are shaped like slippers.

Evolution of Protists

Many protists are single-celled, but all are eukaryotes with a nucleus and a wide range of organelles. There is now ample evidence that the organelles of eukaryotic cells arose from close symbiotic associations between bacteria and primitive eukaryotes (**Fig. 17.19**). The **endosymbiotic theory** is supported by the presence of double membranes around mitochondria and chloroplasts. Also, these organelles have their own genomes, although incomplete, and their ribosomal genes point to bacterial origins. The mitochondria appear closely related to certain bacteria, and the chloroplasts are most closely related to cyanobacteria.

To explain the diversity of protists, we can well imagine that, once the eukaryotic cell arose, it provided the opportunity for many different lineages to begin. Some single-celled protists have organelles not seen in other eukaryotes. For example, food is digested in food vacuoles, and excess water is expelled when contractile vacuoles discharge their contents.

Protists also possibly give us insight into the evolution of a multicellular organism with differentiated tissues. Some protists take the form of a colony of single cells, with certain cells specialized to produce eggs and sperm; others are multicellular, with tissues specialized for various purposes. Perhaps the first type of organization preceded the second in a progression toward multicellular organisms.

Classification of Protists

Many different classification schemes have been devised to define relationships between the protists. Protists can have a combination of characteristics not seen in other eukaryotic groups, which makes them difficult to classify. Traditionally, protists were classified by their source of energy and nutrients: The algae are photosynthetic (but use a variety of pigments); the protozoans are heterotrophic by ingestion; and the water molds and slime molds are heterotrophic by absorption. Newer data, especially gene-sequencing information, have resulted in some controversy, however. The result of these studies has been the formation of **supergroups** to classify the eukaryotes. The number of supergroups is under revision, but the current consensus has settled on five supergroups (**Table 17.1**), each representing a separate evolutionary lineage.

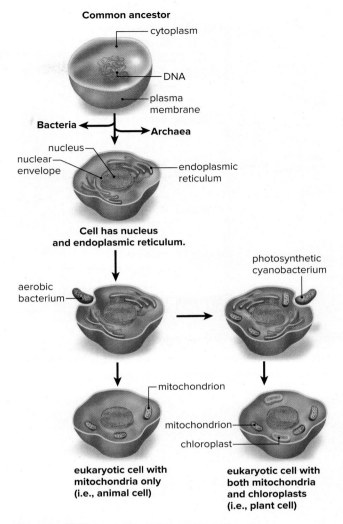

Figure 17.19 Evolution of the eukaryotic cell.
Invagination of the plasma membrane accounts for the formation of the nucleus and certain organelles. The endosymbiotic theory states that mitochondria and chloroplasts are derived from prokaryotes that were taken up by a much larger eukaryotic cell.

Table 17.1 Eukaryotic Supergroups

Supergroup	Types of Organisms
Archaeplastids	Plants as well as green and red algae
SAR supergroup	Stramenopiles (diatoms), Alveolata, Rhizaria
Excavates	Euglenids and certain other flagellates
Amoebozoans	Amoeboids, as well as plasmodial and cellular slime molds
Opisthokonts	Animals, fungi, and certain flagellates

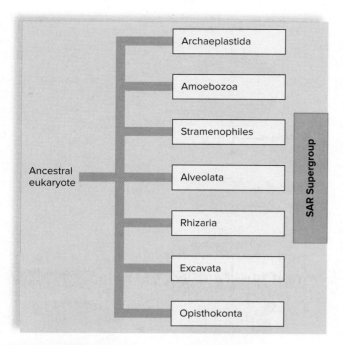

Figure 17.20 Eukaryotic supergroups.

A general representation of the major eukaryotic supergroups. See Table 17.1 for the types of organisms in each group.

Advances in DNA technology, especially the rapid advances in the ability to sequence the genomes of organisms (see Section 12.1), are constantly providing new information on the relationships between the supergroups. A simplified diagram of these relationships is provided in **Figure 17.20.** We will use the supergroup level of classification as our basis for understanding the diversity of protists.

The Archaeplastids

The archaeplastids include land plants and other photosynthetic organisms, such as green and red algae, that have chloroplasts (also called *plastids*) derived from endosymbiotic cyanobacteria (see Figure 17.19).

The term **algae** (sing., **alga**) traditionally indicated an aquatic organism that conducts photosynthesis. At one time, botanists classified algae as plants because they contain chlorophyll *a* and carry on photosynthesis. In aquatic environments, algae are a part of the phytoplankton, photosynthesizers that lie suspended in the water. They are producers, which serve as a source of food for other organisms and pour oxygen into the environment. In terrestrial systems, algae are found in soils, on rocks, and in trees. One type of algae is a symbiote of animals called corals (see Section 19.2), which depend on the algae for food as they build the coral reefs of the world. Other algae partner with fungi in lichens, which are capable of living in harsh terrestrial environments.

The traditional way to classify algae is based on the color of the pigments in their chloroplasts: green algae, red algae, and brown algae. However, not all algae belong to the same supergroup, we now know that the brown algae are actually Stramenopiles (discussed in the next subsection). The green algae are most closely related to plants, and they are commonly represented by three species: *Chlamydomonas;* colonial *Volvox,* a large, hollow sphere with dozens to hundreds of cells; and *Spirogyra,* a filamentous alga in which the chloroplasts form a green spiral ribbon (**Fig. 17.21**).

The green alga *Chlamydomonas* serves as our model for algal structure (**Fig. 17.22**). The most conspicuous organelle in the algal cell is the chloroplast. Algal chloroplasts share many features with those of plants, and

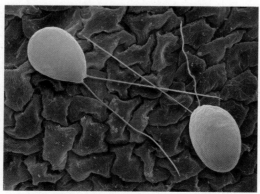

a. *Chlamydomonas*   SEM 1,770×

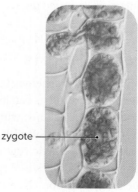

zygote

b. *Volvox*

c. *Spirogyra*

Figure 17.21 Examples of green algae.

a. Placeholder figure caption text. **b.** Placeholder caption. **c.** Figure captions in this pattern will extend full width.

the two groups likely have a common origin; for example, the photosynthetic pigments of both are housed in thylakoid membranes. Not surprisingly, then, algae perform photosynthesis in the same manner as plants. Pyrenoids are structures found in algae that are active in starch storage and metabolism. Vacuoles are seen in algae, along with mitochondria. Algae generally have a cell wall, and many produce a slime layer that can be harvested and used for food processing. Some algae are nonmotile, while others possess flagella.

Most algae can reproduce asexually or sexually. Asexual reproduction can occur by binary fission, as in bacteria. Some algae proliferate by forming flagellated spores called zoospores, while others simply fragment, with each fragment becoming a progeny alga. Sexual reproduction generally requires the formation of gametes that combine to form a zygote.

The SAR Supergroup

As previously mentioned, researchers are actively researching the evolutionary relationships between some of the protists. Genomic analyses have suggested that the groups of organisms called stramenopiles, alveolates, and rhizarians, many of which were previously assigned to other supergroups, are actually more closely related to one another. The name *SAR supergroup* uses the first letters of the names of these groups.

Stramenopiles   The stramenopiles represent a very large, diverse group of protists. Members of this group are typically photosynthetic algae and diatoms, but they have a different evolutionary lineage than the green and red algae.

Brown algae (**Fig. 17.23***a*) are the conspicuous multicellular seaweeds that dominate rocky shores along cold and temperate coasts. The color of brown algae is due to accessory pigments that actually range from pale beige to yellow-brown to almost black. These pigments allow the brown algae to extend their range down into deeper waters because the pigments are more

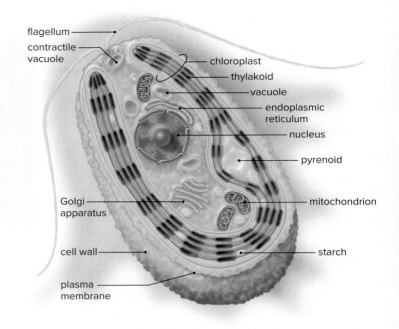

Figure 17.22 *Chlamydomonas.*

Chlamydomonas, a green alga, has the organelles and other structures typical of a motile algal cell.

a.

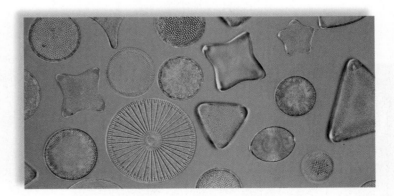

b.

Figure 17.23 Examples of stramenopiles.

a. Brown algae may form large, multicellular masses commonly called kelp. **b.** Diatoms are photosynthetic organisms that have a silica coating.

(a): © D.P. Wilson & Eric David Hosking/Science Source;
(b): © Darlyne A. Murawski/Getty Images

Sporozoan

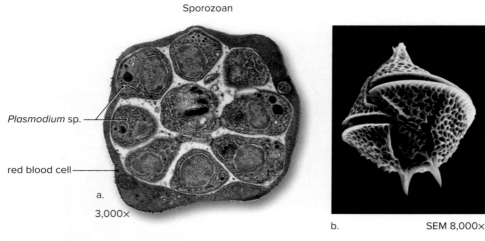

Plasmodium sp.

red blood cell

a.

3,000×

b.   SEM 8,000×

Figure 17.24 **Examples of alveolates.**

a. Plasmodium is the protist responsible for malaria.
b. Dinoflagellates have flagella, and they produce a toxin associated with red tides.

(a): © Omikron/Science Source; (b): © Biophoto Associates/Science Source

Figure 17.25 **A paramecium.**

A paramecium is a ciliate, a type of complex protozoan that moves by cilia.

© Dr. David Patterson/SPL/Science Source

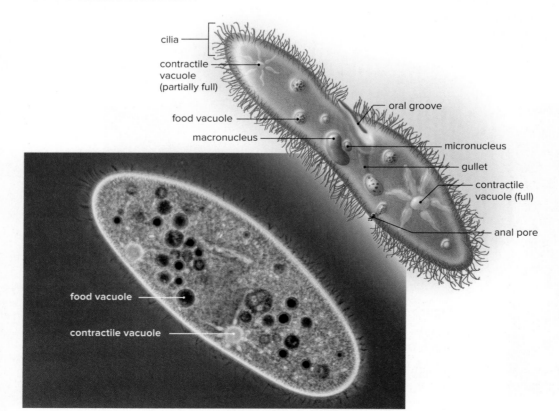

cilia

contractile vacuole (partially full)

food vacuole

macronucleus

oral groove

micronucleus

gullet

contractile vacuole (full)

anal pore

food vacuole

contractile vacuole

efficient than green chlorophyll at absorbing sunlight away from the ocean surface. Brown algae produce a slimy matrix that retains water when the tide is out and the seaweed is exposed. This gelatinous material, algin, is used in ice cream, cream cheese, and some cosmetics.

Diatoms are tiny, single-celled organisms that have an ornate silica shell (Fig 17.23*b*). The shell is made up of upper and lower shelves, called valves, that fit together. Diatoms have a photosynthetic accessory pigment that gives them an orange-yellow color. Diatoms make up a significant part of plankton, which serves as a source of oxygen and food for heterotrophs in both freshwater and marine ecosystems.

Alveolata The alveolates are another diverse group of single-celled protists, but they all share a common characteristic of having an internal series of cavities (*alveolate* means "cavity") that support the plasma membrane. Examples of alveolates include the apicomplexans, dinoflagellates, and ciliates.

Apicomplexans, commonly called sporozoans because they produce spores, are unlike other protozoan groups because they are not motile. One genus, *Plasmodium,* causes malaria, the most widespread and dangerous protozoan disease (**Fig. 17.24***a*). According to the Centers for Disease Control and Prevention (CDC), there are approximately 350–500 million cases of malaria worldwide every year; more than a million people die of the infection.

Dinoflagellates (Fig 17.24*b*) are best known for causing red tides when they greatly increase in number, an event called an algal bloom. *Gonyaulax,* the genus implicated in the red tides, produces a very potent toxin. This can be harmful by itself, but it also accumulates in shellfish. The shellfish are not damaged, but people can become quite ill when they eat them.

Ciliates are one of the largest groups of animal-like protists, also called protozoans. All of them have cilia, hairlike structures that rhythmically beat, moving the protist forward or in reverse. The cilia also help capture prey and particles of food and then move them toward the oral groove (mouth). After phagocytic vacuoles engulf the food, they combine with lysosomes, which supply the enzymes needed for digestion. A contractile vauole helps to maintain water balance with the surrounding environment.

Paramecium is the most widely known ciliate, and it is commonly used for research and teaching (**Fig. 17.25**). It is shaped like a slipper and has visible contractile vacuoles. Members of the genus *Paramecium* have a large *macronucleus* and a small *micronucleus.* The macronucleus produces mRNA and directs metabolic functions. The micronucleus is important during sexual reproduction. *Paramecium* has been important for studying ciliate sexual reproduction, which involves *conjugation,* with interactions between micronuclei and macronuclei.

Rhizaria   The rhizarians are all amoeba-like organisms, meaning that they do not have a defined body shape. The difference between the rhizarians and other amoeba-like protists is that the rhizarians generally produce external shells. For example, the foraminifera (**Fig.17.26***a*) are marine organisms that produce a hard external shell, called a test. When these organisms die, the tests accumulate on the ocean floor. Over geologic time, continental drift may cause these deposits to come to the surface, as in the case of the White Cliffs of Dover. The radiolarians are another group of rhizarians. These are also marine organisms (Fig. 7.26*b*) that produce intricate mineral-rich outer shells. Many radiolarians are zooplankton (animal-like protists), but some also form symbiotic relationships with algae.

The Excavata

The organisms in supergroup Excavata often lack mitochondria and possess distinctive flagella and/or deep (excavated) oral grooves. Historically, they were referred to as flagellates, since they frequently propel themselves using one or more flagella.

The Euglenozoa are freshwater single-celled organisms that typify the problem of classifying protists (**Fig. 17.27***a*). Many euglenozoa have chloroplasts, but some do not. Those that lack chloroplasts ingest or absorb their food. Those that have chloroplasts are believed to have originally acquired them by ingestion and subsequent endosymbiosis of a green algal cell. Three, rather than two, membranes surround these chloroplasts. The outermost membrane is believed to represent the plasma membrane of an original host cell that engulfed a green alga. Euglenids have two flagella, one of which is typically much longer than the other and projects out of the anterior, vase-shaped invagination. Near the base of this flagellum is an eyespot, which is a photoreceptive organelle for detecting light.

Another member of the Excavata is the genus *Giardia*. These interesting organisms have two nuclei and multiple flagella and do not possess the mitochondria and Golgi apparatus common to most eukaryotes. One species, called *Giardia lamblia* (or *G. intestinalis*) is a parasite of the intestinal tract of many animals, including humans (Fig. 17.27*b*). It is commonly spread via drinking water that has been contaminated by the feces of an infected animal, although it is possible to contract *Giardia* from the soil or food. Symptoms of giardiasis include diarrhea, stomach and/or intestinal cramping, and excess gas (flatulence). For most people, a *Giardia* infection lasts several weeks, but the parasite has been known to cause complications in the elderly, young children, and individuals with compromised immune systems. Since *Giardia* is a eukaryote, antibiotics are ineffective, although other medications are available.

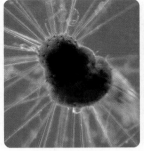

a. A foraminiferan   160×

b. Radiolarian tests   SEM 150×

Figure 17.26 Examples of rhizarians.

a. A foraminiferan, *Globigerina*. **b.** The mineralized tests of radiolarians. (a): © NHPA/Superstock; (b): © Eye of Science/Science Source

a.   960×

circular marking   *Giardia*   surface

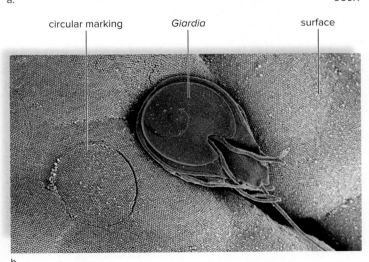

b.

Figure 17.27 Examples of Excavata.

a. *Euglena* has both animal- and plant-like characteristics. **b.** *Giardia* is an intestinal parasite of a number of mammals, including humans. (a): © Kage Mikrofotograffie/Phototake; (b): © Cultura RM/Alamy

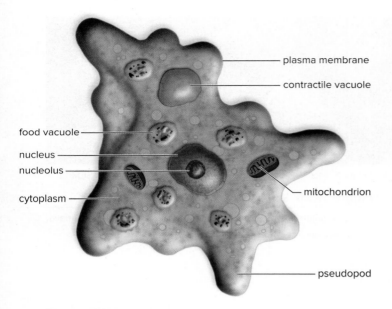

Figure 17.28 An amoeba.

Amoebas are common in freshwater ponds. Bacteria and other microorganisms are digested in food vacuoles, and contractile vacuoles rid the body of excess waste.

The Amoebozoa

As their name implies, the Amoebozoa are amoeba-like protists that move by *pseudopods,* processes that form when cytoplasm streams forward in a particular direction (**Fig. 17.28**). They usually live in aquatic environments, such as oceans and freshwater lakes and ponds, where they are part of the zooplankton. When these protists feed, their pseudopods surround and engulf their prey, which may be algae, bacteria, or other protists. Digestion then occurs within a food vacuole. An example of this group is *Amoeba proteus,* which due to its large size (up to 800 μm in length), is frequently used in science labs (μm).

Another type of amoebozoan is the slime mold. Like amoebas, slime molds are chemoheterotrophs. In forests and woodlands, slime molds feed on, and therefore help dispose of, dead plant material. They also feed on bacteria, keeping their population sizes under control. Usually, plasmodial slime molds exist as a plasmodium—a diploid, multinucleated, cytoplasmic mass enveloped by a slime sheath that creeps along, phagocytizing decaying plant material in a forest or an agricultural field. At times that are unfavorable for growth, such as during a drought, the plasmodium develops many sporangia. A sporangium is a reproductive structure that produces spores resistant to dry conditions. When favorable moist conditions return, the spores germinate, each one releasing a flagellated cell or an amoeboid cell. Eventually, two of these cells fuse to form a zygote that feeds and grows, developing into a multinucleated plasmodium once again. **Figure 17.29** shows the life cycle of a plasmodial slime mold.

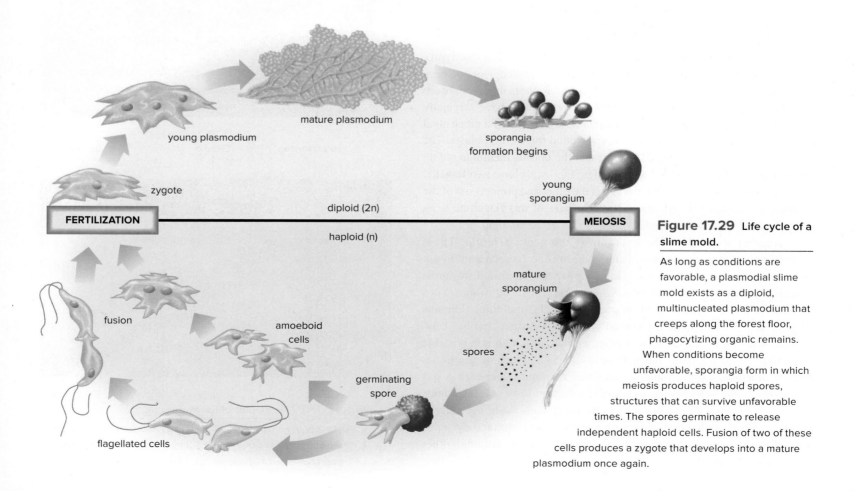

Figure 17.29 Life cycle of a slime mold.

As long as conditions are favorable, a plasmodial slime mold exists as a diploid, multinucleated plasmodium that creeps along the forest floor, phagocytizing organic remains. When conditions become unfavorable, sporangia form in which meiosis produces haploid spores, structures that can survive unfavorable times. The spores germinate to release independent haploid cells. Fusion of two of these cells produces a zygote that develops into a mature plasmodium once again.

The Opisthokonta

Supergroup Opisthokonta contains animals, fungi, and several closely related protists. The organisms in this group all have the common characteristic of being chemoheterotrophs with flagellated cells. This supergroup includes both single-celled and multicellular protozoans. Among the opisthokonts are the choanoflagellates, animal-like protozoans that are closely related to sponges. These protozoans, including single-celled as well as colonial forms, are filter-feeders with cells that bear a striking resemblance to the choanocytes that line the insides of sponges (see Section 19.2). Each choanoflagellate has a single posterior flagellum surrounded by a collar of slender microvilli (**Fig. 17.30**). Beating of the flagellum creates a water current that flows through the collar, where food particles are taken in by phagocytosis.

Figure 17.30 Choanoflagellate.

The choanoflagellates are the flagellated protists most closely related to the animals.

Check Your Progress 17.4

1. Describe the features of mitochondria and chloroplasts that support the endosymbiotic theory.
2. Identify the general characteristics of protists.
3. For each eukaryotic supergroup, provide a general characteristic and name a representative organism.

CONNECTING THE CONCEPTS

17.4 The protists evolved by a process called endosymbiosis. Protists and other eukaryotes are classified into supergroups.

STUDY TOOLS http://connect.mheducation.com

SMARTBOOK® Maximize your study time with McGraw-Hill SmartBook®, the first adaptive textbook.

SUMMARIZE

Viruses, bacteria, and protists are commonly called microorganisms. The classification of many microorganisms is under revision by the scientific community.

 17.1 Viruses are known to be host-specific and have unique reproductive cycles.

 17.2 Viroids and prions have many similarities to viruses.

17.3 The prokaryotes are single-celled organisms that lack a nucleus. Both the bacteria and archaea are prokaryotes.

 17.4 The protists evolved by a process called endosymbiosis. Protists and other eukaryotes are classified into supergroups.

17.1 The Viruses

Viruses are obligate intracellular parasites that can reproduce only inside living cells.

Structure of a Virus

- Viruses have at least two parts: an outer **capsid** composed of protein subunits and an inner core of genetic material, either DNA or RNA.
- Animal viruses either are naked (no outer envelope) or have an outer membranous envelope.

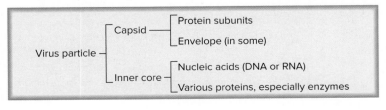

Viral Reproduction

Bacteriophages can have a lytic or lysogenic life cycle. The **lytic cycle** consists of five steps:

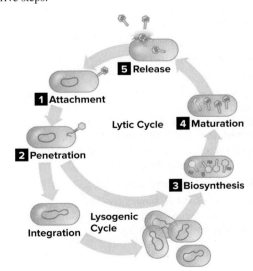

In the **lysogenic cycle,** viral DNA is integrated into bacterial DNA for an indefinite period of time, but it can undergo the last three steps of the lytic cycle at any time.

Plant Viruses

Crops and garden plants are subject to viral infections. Not all viruses are deadly, but over time they debilitate the host plant.

Animal Viruses

The reproductive cycle of animal viruses has the same steps as for bacteriophages, with the following modifications if the virus has an envelope:

- Fusion or endocytosis brings the virus into the cell.
- Uncoating is needed to free the genome from the capsid.
- Budding releases the viral particles from the cell.

HIV, the AIDS virus, is a **retrovirus.** These viruses use RNA as their genetic material and have an enzyme, reverse transcriptase, that carries out reverse transcription. This transcription produces single-stranded DNA, which replicates, forming a double helix that becomes integrated into the host DNA.

Drug Control of Human Viral Diseases

Emerging viruses are relatively new to humans. Antiviral drugs are structurally similar to nucleotides and interfere with viral genome synthesis.

17.2 Viroids and Prions

Viroids are naked (not covered by a capsid) strands of RNA that can cause disease. **Prions** are protein molecules that have changed shape and can cause other proteins to do so. Prions cause such diseases as Creutzfeld-Jakob disease (CJD) in humans and mad cow disease in cattle.

17.3 The Prokaryotes

The bacteria and the archaea are prokaryotes. **Prokaryotes** lack a nucleus and most of the other cytoplasmic organelles found in eukaryotic cells.

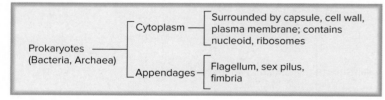

The Origin of the First Cells

The **cell theory** states that all life is derived from cells. The first cells were prokaryotes. Before the first cells appeared, biological macromolecules formed spontaneously in the unique conditions of early Earth. Scientists have shown that collections of biological macromolecules can assemble to form nonliving, cell-like structures, or **protocells,** under laboratory conditions.

General Biology of Bacteria

Bacterial Structure

- Structural variations include rods (bacilli), spheres (cocci), and spirals (spirilla or spirochetes).
- A single, closed circle of double-stranded DNA (chromosome) is in a **nucleoid. Plasmids** are small pieces of DNA in the cytoplasm.
- Each **flagellum** rotates, causing the organism to spin.
- The cell wall contains **peptidoglycan.**

Bacterial Reproduction and Survival

- Reproduction is asexual by **binary fission.**
- Genetic recombination occurs by means of **conjugation, transformation,** and **transduction.**

- **Endospore** formation allows bacteria to survive in unfavorable environments.

Bacterial Nutrition

- Bacteria can be autotrophic. **Cyanobacteria** are **photoautotrophs**—they photosynthesize, like plants. **Chemoautotrophs** oxidize inorganic compounds, such as hydrogen gas, hydrogen sulfide, and ammonia, to acquire energy to make their own food. Like animals, most bacteria are **chemoheterotrophs** (heterotrophs), but they are **saprotrophs** (decomposers). Many heterotrophic prokaryotes are **symbiotic,** such as the mutualistic nitrogen-fixing bacteria that live in nodules on the roots of legumes.

Environmental and Medical Importance of Bacteria

Bacteria in the Environment

- As decomposers, bacteria keep inorganic nutrients cycling in ecosystems.
- The reactions they perform keep the nitrogen cycle going.
- Bacteria play an important role in **bioremediation.**

Bacteria in Food Science and Biotechnology

- Bacterial fermentation is important in the production of foods.
- Genetic engineering allows bacteria to produce medically important products.

Bacterial Diseases in Humans

- Bacterial **pathogens** that can cause diseases are able to (1) produce a toxin, (2) adhere to surfaces, and (3) sometimes invade organs or cells.
- Indiscriminate use of antibiotics has led to bacterial resistance to some of these drugs.

Archaea

The **archaea** (domain Archaea) are the second type of prokaryote. The following are some characteristics of archaea:

- They appear to be more closely related to the eukarya than to the bacteria.
- They do not have peptidoglycan in their cell walls, as do the bacteria, and they share more biochemical characteristics with the eukarya than do bacteria.
- Some are well known for living under harsh conditions, such as anaerobic marshes (**methanogens**), salty lakes (**halophiles**), and hot sulfur springs (**thermoacidophiles**).

17.4 The Protists

General Biology of Protists

- **Protists** are eukaryotes. The **endosymbiotic theory** accounts for the presence of mitochondria and chloroplasts in eukaryotic cells. Some protists are multicellular with differentiated tissues.
- Protists have great ecological importance; in largely aquatic environments, they are the producers (algae) or sources (algae and protozoans) of food for other organisms.

- Protists, along with the other eukaryotes, are classified into **supergroups,** a classification level just under the domains.

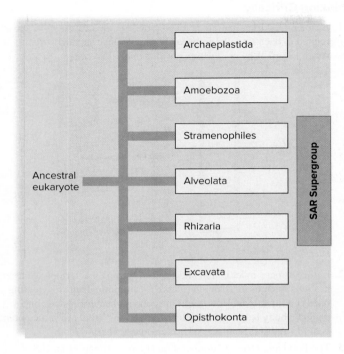

The Archaeplastids

- **Algae** possess chlorophylls, which they use for photosynthesis, just as plants do. They store reserves of food such as starch and have cell walls, as in plants.
- Examples are *Chlamydomonas, Volvox,* and *Spirogyra.*

The SAR Supergroup

The SAR supergroup is a diverse group of protists that includes:

- Stramenopiles—photosynthetic algae and diatoms that have a different evolutionary lineage from that of the Archaeplastids. Examples are brown algae and diatoms.
- Alveolata—protists with an internal series of cavities. Examples are apicomplexans, ciliates, and dinoflagellates.
- Rhizaria—amoeba-like organisms that produce hard external shells (tests). Examples are the foraminifera and radiolarians.

The Excavata

Members of this supergroup lack mitochondria and possess distinctive flagella and/or deep (excavated) oral grooves. Examples are *Euglena* and *Giardia.*

The Amoebozoa

Amoebozoa are ameoba-like heterotrophic protists that move by the use of pseudopods. Examples are amoebas and slime molds.

The Opisthokonta

The organisms in this group all have the common characteristic of being chemoheterotrophs with flagellated cells. Examples are animals, fungi, and choanoflagellates.

ASSESS

Testing Yourself

Choose the best answer for each question.

17.1 The Viruses

1. A virus contains which of the following?
   a. a cell wall
   b. a plasma membrane
   c. nucleic acid
   d. cytoplasm

2. The five stages of the bacteriophage lytic cycle occur in this order:
   a. penetration, attachment, release, maturation, biosynthesis.
   b. attachment, penetration, release, biosynthesis, maturation.
   c. biosynthesis, attachment, penetration, maturation, release.
   d. attachment, penetration, biosynthesis, maturation, release.
   e. penetration, biosynthesis, attachment, maturation, release.

3. The enzyme that is unique to retroviruses is
   a. reverse transcriptase.
   b. DNA polymerase.
   c. DNA gyrase.
   d. RNA polymerase.

17.2 Viroids and Prions

4. Which of the following statements about viroids is false?
   a. They are composed of naked RNA.
   b. They cause plant diseases.
   c. They die once they reproduce.
   d. They cause infected cells to produce more viroids.

5. Prions cause disease when they
   a. enlarge.
   b. break into small pieces.
   c. cause normal proteins to change shape.
   d. interact with DNA.

17.3 The Prokaryotes

6. A bacterium contains all of the following, except
   a. ribosomes.
   b. DNA.
   c. mitochondria.
   d. cytoplasm.
   e. a plasma membrane.

For questions 7–12, determine which organisms are described by each characteristic. Each answer in the key may be used more than once.

Key:

    a. bacteria
    b. archaea
    c. both bacteria and archaea
    d. neither bacteria nor archaea

7. have peptidoglycan in the cell wall
8. are methanogens
9. are sometimes parasitic
10. contain a nucleus
11. have lipids in the plasma membrane
12. reproduce by binary fission

17.4 The Protists

13. The endosymbiotic theory explains the:
    a. origin of the first prokaryotic cells.
    b. origins of mitochondria and chloroplasts in eukaryotes.
    c. evolutionary relationship between animals, plants, and fungi.
    d. method of reproduction in protists.
14. Ciliates and dinoflagellates belong to which of the following supergroups?
    a. SAR supergroup
    b. Excavata
    c. Opisthokonta
    d. Amoebozoa
    e. None of these are correct.
15. Which of the following protists was the ancestor of the animals?
    a. *Euglena*
    b. *Giardia*
    c. the choanoflagellates
    d. an amoeba
    e. None of these are correct.

ENGAGE

Thinking Critically

1. Based on the tree diagram shown below, are the protists more closely related to bacteria or to archaea? What evidence supports this relationship? Why are viruses not shown on this tree?

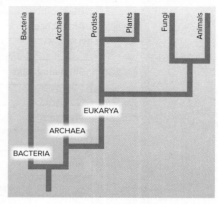

2. In the former Soviet Union and Eastern Europe, bacteriophage therapy has long been used as an alternative to antibiotic drugs for treating bacterial infections. How could introducing bacteriophages into a patient's body help fight a bacterial infection? What are some potential benefits and shortcomings of bacteriophage therapy?

3. The Food and Drug Administration (FDA) estimates that U.S. physicians annually write 50 million unnecessary prescriptions for antibiotics to treat viral infections. Of course, the antibiotics are ineffective against viruses. Furthermore, the frequent exposure of bacteria to these drugs has resulted in the development of numerous strains with antibiotic resistance. Doctors may prescribe antibiotics because patients demand them. In addition, if the cause of the illness is not known, it may be safer to prescribe an antibiotic that turns out to be ineffective than to withhold the antibiotic when it would have helped. The FDA has, therefore, initiated a policy requiring manufacturers to label antibiotics with precautions about their misuse.
    a. Do you think this type of labeling information will educate consumers enough to significantly reduce the inappropriate use of antibiotics?
    b. Physicians already know about the dangers of antibiotic resistance, but many prescribe them anyway. This is probably due, in large part, to the fact that they would be held liable for withholding an antibiotic if it could have helped, but not for inappropriately prescribing one. Should physicians be held more accountable for prescribing antibiotics? If so, how?

© Nakano Masahiro/amanaimages/Getty RF

Peppers Versus Fungi—an Evolutionary War

Have you ever thought about a plant being at war? In fact, plants are constantly at war with a variety of organisms, such as plant-eating animals (herbivores), bacteria, and fungi. In response, plants have evolved a number of adaptive strategies to protect themselves. In the case of chili peppers, the plant produces a pain-inducing chemical, called capsaicin, within the peppers. Capsaicin stimulates pain receptors in mammals, giving their brains the "mouth on fire" sensation. While capsaicins deter some animals, their real target are fungi, the ancient enemies of plants. Fungi are decomposers, which provide an important ecological service by digesting and breaking down dead organisms and recycling nutrients back into the soil. However, some fungi do not wait for organisms to die and instead prey upon living ones, including pepper plants.

The chemical warfare begins when a chili pepper is attacked by a fungus attempting to penetrate the pepper. The pepper responds by producing capsaicin, which wards off the fungus. The amount of capsaicinoids the pepper plants produce varies with the environment. Since fungi prefer moist environments, farmers grow milder chilis in drier environments, where there is less fungal attack, and hotter chilis in wetter environments, where fungi thrive and the plants are in full antifungal defense mode!

In this chapter, we will explore the evolution and diversity of both the plants and the fungi and will examine some of the evolutionary adaptations in these two kingdoms.

As you read through this chapter, think about the following questions:

1. How did the physical appearance of plants change as they evolved from living in water to living on land?
2. What aspects of the fungal body make fungi so successful as decomposers?

18

The Plants and Fungi

OUTLINE

18.1 Overview of the Plants   321

18.2 Diversity of Plants   315

18.3 The Fungi   325

BEFORE YOU BEGIN

Before beginning this chapter, take a few moments to review the following discussions.

Section 6.2 What process produces oxygen during photosynthesis?

Section 9.3 What are the differences between mitosis and meiosis with regard to chromosome number?

Section 16.3 What domain of life do the plants and fungi belong to?

18.1 Overview of the Plants

Learning Outcomes

Upon completion of this section, you should be able to

1. Describe the similarities and differences between the green algae and land plants.
2. List the significant events in the evolution of land plants.
3. Describe the alternation of generations in the life cycle of plants.

Plants (kingdom Plantae) are multicellular, photosynthetic eukaryotes. Although plants are well adapted to a land environment, the evolutionary history of plants begins in the water. Evidence indicates that plants evolved from a form of freshwater green algae some 500 MYA (million years ago). Green algae are members of the same eukaryotic supergroup as plants (the archaeplastids, see Section 17.4), and thus share some characteristics with plants. For example, green algae: (1) contain chlorophylls *a* and *b* and various accessory pigments, (2) store excess carbohydrates as starch, and (3) have cellulose in their cell walls.

A comparison of DNA and RNA base sequences suggests that land plants are most closely related to a group of freshwater green algae known as **charophytes.** Although *Spirogyra* (see Fig. 17.21) is a charophyte, molecular scientists tell us that the ancestors of land plants were more closely related to the charophytes shown in **Figure 18.1.** Although the common ancestor of modern charophytes and land plants no longer exists, if it did, it would have features resembling members of the genera *Chara* and *Coleochaete*.

Let's take a look at these filamentous green algae. Those in the genus *Chara* are commonly known as stoneworts because they are encrusted with calcium carbonate deposits. The body consists of a single file of very long cells anchored in mud by thin filaments. Whorls of branches occur at regions called nodes, located between the cells of the main axis. Male and female reproductive structures grow at the nodes. A *Coleochaete* looks flat, like a pancake, but the body is actually composed of long, branched filaments of cells. Most important to the evolution of plants, charophytes protect the zygote. Land plants not only protect the zygote but also protect and nourish the resulting embryo—an important feature that separates land plants from green algae.

Over their evolutionary history, plants have become well-adapted to a land existence. Although a land environment offers advantages, such as plentiful light, it also has challenges, such as the constant threat of drying out (desiccation). Most importantly, all stages of reproduction—gametes, zygote, and embryo—must be protected from the drying effects of air. To keep the internal environment of cells moist, a land plant must acquire water and transport it to all parts of the body, while keeping the body in an erect position. We will see how plants have adapted to these problems by evolving an internal vascular system. **Figure 18.2** traces the evolutionary history of plants. It is possible to associate each group with an evolutionary event that represents a major adaptation to existence on land.

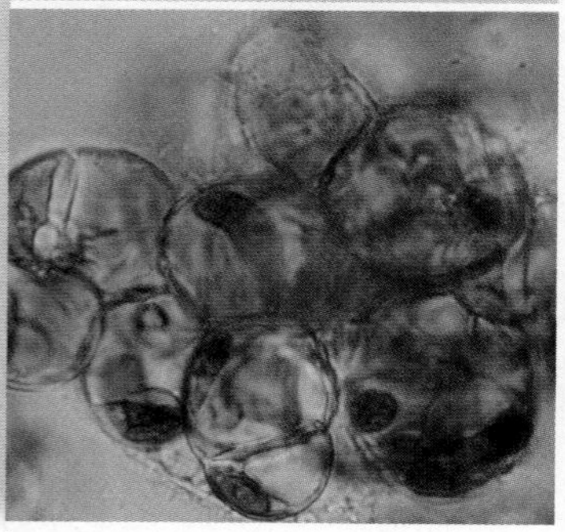

An Overview of Plant Evolution

Mosses represent the closest plant link between the green algae and the remainder of the plant kingdom (Fig. 18.2). Mosses are low-lying plants that generally lack vascular tissue and therefore have no means of transporting water, but they do have means to prevent the plant body from drying out and they protect the embryo within a special structure.

Figure 18.1 Close algal relatives of plants.

The closest living relatives of land plants are green algae known as charophytes. **a.** Members of the genus *Chara,* commonly called stoneworts, live in shallow freshwater lakes and ponds. **b.** The body of this *Coleochaete* is a flat disk found on wet stones or on other aquatic plants; it is only the size of a pinhead and one cell layer thick.

The lycophytes, which evolved around 420 MYA, were among the first plants to have a vascular system that transports water and solutes from the roots to the leaves of the plant body. Plants with vascular tissue have true roots, stems, and leaves. The leaves of lycophytes, called **microphylls,** are very narrow.

Ferns are well-known plants with large leaves called **megaphylls.** The evolution of branching and leaves allowed a plant to increase the amount of exposure to sunlight, thus increasing photosynthesis and the production of sugars. Without adequate food production, a plant can't increase in size. The next evolutionary event was the evolution of **seeds.** A seed contains an embryo and stored organic nutrients within a protective coat (look ahead to Fig. 18.10). Seeds are highly resistant structures well suited to protecting the plant embryos from drying out until conditions are favorable for germination. The gymnosperms were the first seed plants to appear, about 360 MYA. The final evolutionary event of interest to us is the evolution of the **flower,** a reproductive structure found in angiosperms. Flowers attract pollinators, such as insects, and they give rise to

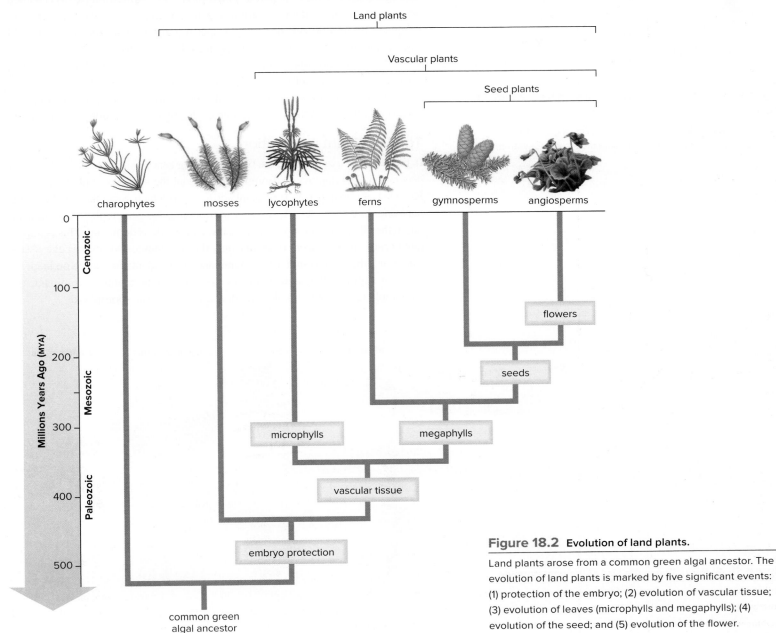

Figure 18.2 Evolution of land plants.

Land plants arose from a common green algal ancestor. The evolution of land plants is marked by five significant events: (1) protection of the embryo; (2) evolution of vascular tissue; (3) evolution of leaves (microphylls and megaphylls); (4) evolution of the seed; and (5) evolution of the flower.

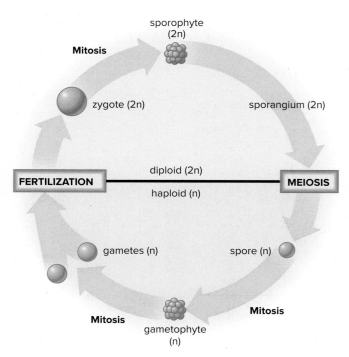

Figure 18.3 Alternation of generations in the life cycle of plants.

Plants alternate between a sporophyte (2n) stage and a gametophyte (n) stage. The dominant stage is photosynthetic.

fruits that cover seeds. Plants with flowers evolved between 120 and 140 MYA. In this chapter, we will use the angiosperms as our model organism for the discussion of many aspects of plant biology.

Alternation of Generations

The life cycle of land plants is quite different from that of animals. Unlike humans and other animals, which exhibit a diploid life cycle (see Fig. 9.2), all plants have a life cycle that features an **alternation of generations.** In a plant's life cycle, two multicellular individuals alternate, each producing the other (**Fig. 18.3**). The two individuals are (1) a sporophyte, which represents the diploid (2n) generation; and (2) a gametophyte, which represents the haploid (n) generation.

The **sporophyte** (2n) is the structure that produces spores by meiosis. A spore is a haploid reproductive cell that develops into a new organism without needing to fuse with another reproductive cell. In the plant life cycle, a spore undergoes mitosis and becomes a gametophyte. The **gametophyte** (n) is named because of its role in the production of gametes. In plants, eggs and sperm are produced by mitotic cell division. A sperm and egg fuse, forming a diploid zygote that undergoes mitosis and becomes the sporophyte.

There are two important aspects of the plant life cycle. First, in plants, meiosis produces haploid spores. This is consistent with the realization that the sporophyte is the diploid generation, and spores are haploid. Second, mitosis is involved in the production of the gametes during the gametophyte generation.

The Dominant Generation

In the plant life cycle, one of the two alternating generations acts as the dominant generation. The dominant generation carries out the majority of photosynthesis. However, the major groups of plants differ as to which generation is the dominant one. In the mosses (the nonvascular plants), the gametophyte is dominant, but in the other three groups of plants, the sporophyte is dominant. This is important because in the history of plants, only the sporophyte evolves vascular tissue. Therefore, the shift to sporophyte dominance is an adaptation to life on land.

In **Figure 18.4,** notice that as the sporophyte becomes dominant, the gametophyte becomes smaller and dependent upon the sporophyte.

Figure 18.4 Reduction in the size of the gametophyte.

Notice the reduction in the size of the gametophyte among these representatives of today's plants. This trend occurred as plants became adapted for life on land.

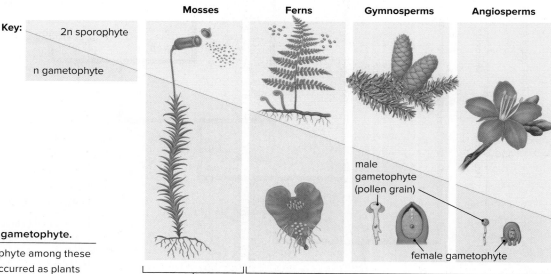

In mosses (bryophytes), the gametophyte is much larger than the sporophyte. In lycophytes and ferns, the gametophyte is a small, independent structure. In contrast, the female gametophyte of cone-bearing plants (gymnosperms) and flowering plants (angiosperms) is microscopic—it is retained within the body of the sporophyte plant. This protects the female gametophyte from drying out. Also, the male gametophyte of seed plants lies within a pollen grain. Pollen grains have strong, protective walls and are transported by wind, insects, and birds to reach the egg. In the life cycle of seed plants, the spores, the gametes, and the zygote are protected from drying out in the land environment.

CONNECTING THE CONCEPTS
18.1

Plants are multicellular, photosynthetic eukaryotes with an alternation-of-generations life cycle.

18.2 Diversity of Plants

Learning Outcomes

Upon completion of this section, you should be able to

1. Characterize and give examples of the various groups of plants.
2. Describe the life cycle and reproductive strategy of each group of plants.
3. Summarize the economic and ecological significance of plants.

Plants can be grouped into two general categories: nonvascular and vascular. **Nonvascular plants** receive water and nutrients through diffusion and osmosis directly into the plant body. The plant needs to stay wet and remains very short. **Vascular plants** have an internal transport system that facilitates the movement of water and nutrients throughout the body. This allows the plant to not only live in drier conditions but also increase in height to maximize photosynthesis.

Nonvascular Plants

The nonvascular plants include the liverworts, hornworts, and mosses (**Fig. 18.5**). Collectively, they are often called the **bryophytes.** Bryophytes, in general, do not have true roots, stems, and leaves—all of which, by definition, must contain well-developed vascular tissue. In bryophytes, the gametophyte is the dominant generation.

The most familiar bryophytes are the liverworts and mosses, which are both low-lying plants. In bryophytes, the gametophyte is the green, "leafy" part, which produces the gametes. The gametophyte stage of a bryophyte is completely dependent on water for reproduction. Flagellated sperm swim in a film of water to reach an egg. After a sperm fertilizes an egg, the resulting zygote becomes an embryo that develops into a sporophyte. The sporophyte is attached to, and derives its nourishment from, the photosynthetic gametophyte. The sporophyte produces spores in a structure called a *sporangium*. The spores are released into the air, where they can be dispersed by the wind, an adaptation to life on land. The spores will germinate if they land in moist surroundings. Upon germination, male and female gametophytes develop.

The common name of several organisms implies that they are mosses, when they are not. Irish moss is an edible red alga that grows in leathery tufts along northern seacoasts. Reindeer moss, a lichen, is the dietary mainstay of reindeer and caribou in northern lands. Club mosses, discussed later in this section, are vascular plants, and Spanish moss, which hangs in grayish clusters from trees in the southeastern United States, is a flowering plant of the pineapple family.

Hornwort

Liverwort gametophyte

Figure 18.5 Bryophytes.

In bryophytes, the gametophyte is the dominant generation. Bryophytes are short and need to stay wet. Although their habitats are varied, bryophytes can easily be found on tree trunks or wet rocks or in the cracks in a sidewalk.

(hornwort): © Steven P. Lynch;
(liverwort): © Hal Horwitz/Corbis;
(moss): © Steven P. Lynch

Moss

Adaptations and Uses of Bryophytes

The lack of well-developed vascular tissue and the presence of swimming sperm largely account for the short height of bryophytes, such as mosses. Still, mosses can be found from the Antarctic through the tropics to parts of the Arctic. Although most mosses prefer damp, shaded locations in the temperate zone, some survive in deserts and some inhabit bogs and streams. In forests, they frequently form a mat that covers the ground and rotting logs. In dry environments, they may become shriveled, turn brown, and look dead. As soon as it rains, the plant becomes green and resumes metabolic activity. Mosses are much better than flowering plants at living on stone walls, on fences, and in shady cracks of hot, exposed rocks. When bryophytes colonize bare rock, the rock is degraded to soil that they can use for their own growth and that other organisms can also use.

In areas such as bogs, where the ground is wet and acidic, dead mosses, especially those of the genus *Sphagnum,* do not decay. The accumulated moss, called peat or bog moss, can be used as fuel. Peat moss is also commercially important in another way. Because it has special, nonliving cells that can absorb moisture, peat moss is often used in gardens to improve the water-holding capacity of the soil. In addition, *Sphagnum* moss has antiseptic qualities and was used as bandages in World War I when medics ran out of traditional bandages.

Vascular Plants

All the other plants we will study are vascular plants. **Vascular tissue** consists of **xylem,** which conducts water and minerals up from the roots, and **phloem,** which conducts organic solutes such as sucrose from one part of a plant to another. The walls of conducting cells in xylem are strengthened by lignin, an organic compound that makes them stronger, more waterproof, and resistant to attack by parasites and predators. Only because of strong cell walls and vascular tissue can plants reach great heights.

The vascular plants usually have true roots, stems, and leaves. The roots absorb water from the soil, and the stem conducts water to the leaves. The leaves are fully covered by a waxy cuticle, except where it is interrupted by stomata, little pores for gas exchange, the opening and closing of which can be regulated to control water loss.

Seedless Vascular Plants

Certain vascular plants (e.g., lycophytes and ferns) are seedless; the other two groups of vascular plants (gymnosperms and angiosperms) are seed plants. In seedless vascular plants, the dominant sporophyte produces windblown spores, and the independent gametophyte produces flagellated sperm that require outside moisture to swim to an egg (look ahead to Fig. 18.8).

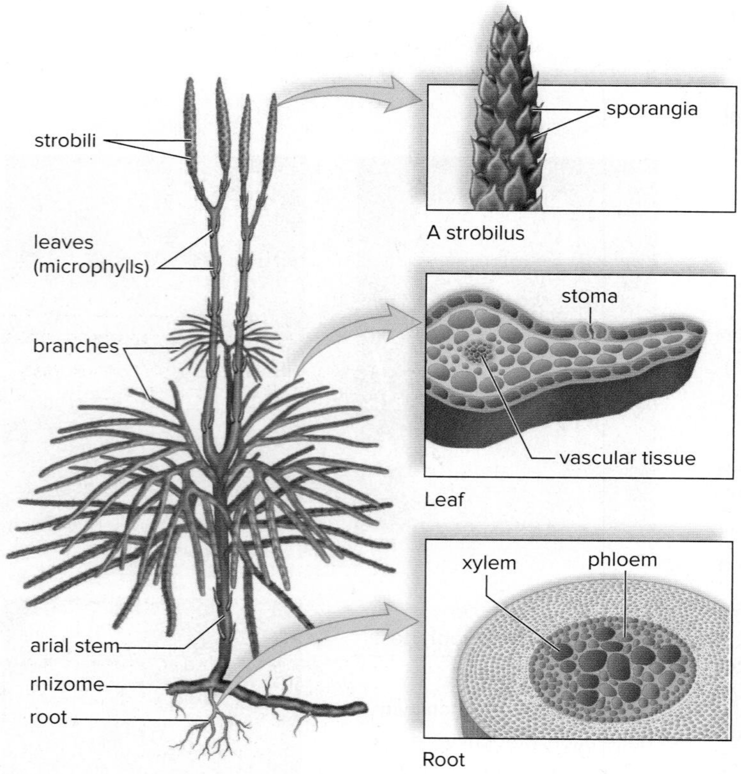

Lycophytes

Lycophytes, also called *club mosses,* were among the first land plants to have vascular tissue. Unlike true mosses (bryophytes), the lycophytes have well-developed vascular tissue in roots, stems, and leaves (**Fig. 18.6**). Typically, a fleshy underground and horizontal stem, called a rhizome, sends up upright aerial stems. Tightly packed, scalelike leaves cover the stems and branches, giving the plant a mossy look. The small leaves, termed *microphylls,* each have a single vein composed of xylem and phloem. The sporangia are borne on terminal clusters of leaves (individually called a strobilus), which are club-shaped. The spores are sometimes harvested and

Figure 18.6 *Lycopodium,* a type of club moss.

Lycophytes, such as this *Lycopodium,* have vascular tissue and thus true roots, stems, and leaves. *Top right,* the *Lycopodium* sporophyte develops a conelike strobilus, where sporangia are located. *Bottom left, Lycopodium* develops an underground rhizome system. A rhizome is an underground stem; this rhizome produces roots along its length.

sold as lycopodium powder, or vegetable sulfur, for use in pharmaceuticals and in fireworks because it is highly flammable. The *Lycopodium* featured in Figure 18.6 is common in moist woodlands in temperate climates, where they are called ground pines; they are also abundant in the tropics and subtropics.

Ferns

Ferns are a widespread group of plants that are well known for their attractiveness. Unlike lycophytes, ferns have *megaphylls,* or large leaves with branched veins. Megaphylls provide a large surface area for capturing the sunlight needed for photosynthesis, and the veins conduct water and minerals throughout the leaf tissue. Ferns and other plants with megaphylls are better able to produce food and thus can grow and reproduce more efficiently than plants with microphylls. Fern megaphylls are called *fronds*. The leatherleaf fern (found in flower arrangements) has fronds that are broad, with subdivided leaflets; the fronds of a tree fern can be about 1.4 m long; and those of the hart's tongue fern are straplike and leathery (**Fig. 18.7**). Sporangia are often located in clusters, called *sori* (sing., sorus), on the undersides of the fronds.

The life cycle of a typical temperate fern is shown in **Figure 18.8.** The dominant sporophyte produces windblown spores. When the spores germinate, a tiny green gametophyte develops, separate from the sporophyte. The gametophyte is water-dependent because it lacks vascular tissue. Also, flagellated sperm produced within antheridia (male gametophytes) require an outside source of moisture to swim to the eggs in the archegonia (female gametophyte).

Figure 18.7 Diversity of ferns.

All ferns are vascular plants that do not utilize seeds for reproduction.

(leatherleaf fern): © Gregory Preest/Alamy; (tree fern): © Danita Delimont/Getty Images; (hart's tongue fern): © Organics image library/Alamy RF

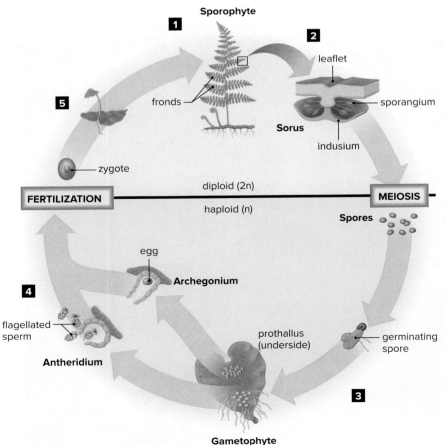

Figure 18.8 Fern life cycle.

1 The sporophyte is dominant in ferns. **2** In the fern shown here, sori are on the underside of the leaflets. Each sorus, protected by an indusium, contains sporangia, in which meiosis occurs and spores are produced and then released. **3** A spore germinates into a prothallus (the gametophyte), which has sperm-bearing (antheridium) and egg-bearing (archegonium) structures on its underside. **4** Fertilization takes place when moisture is present, because the flagellated sperm must swim in a film of water to the egg. **5** The resulting zygote begins its development inside an archegonium, and eventually a young sporophyte becomes visible. The young sporophyte develops, and fronds appear.

Connections: Health

What do horsetail dietary supplements do?

Horsetail belongs to a genus of plants called *Equisetum,* which are close relatives of the ferns. The use of horsetail supplements to treat a variety of illnesses, including tuberculosis and ulcers, goes back to the time of the ancient Romans. While some people have suggested that horsetail supplements may be used to prevent osteoporosis (since horsetails contain the mineral silicon), there have been very few studies on the long-term effects of the use of these supplements. For unknown reasons, horsetail causes a reduction in some B vitamins in the body, and it may interact with other medications and supplements. You should always consult a physician before you start taking any new dietary supplements.

© JoSon/The Image Bank/Getty Images

Figure 18.9 The Carboniferous period.

Growing in the swamp forests of the Carboniferous period were treelike club mosses (*left*), treelike horsetails (*right*), and lower, fernlike foliage (*left*). When the trees fell, they were covered by water and did not decompose completely. Sediment built up and turned to rock, which exerted pressure that caused the organic material to become coal, a fossil fuel that helps run our industrialized society.

© Field Museum Library/Contributor/Archive Photos/Getty Images

Upon fertilization, the zygote develops into a sporophyte. In nearly all ferns, the leaves of the sporophyte first appear in a curled-up form called a fiddlehead, which unrolls as it grows.

Adaptations and Uses of Ferns Ferns are most often found in moist environments because the small, water-dependent gametophyte, which lacks vascular tissue, is separate from the sporophyte. Also, flagellated sperm require an outside source of moisture in which to swim to the eggs. Once established, some ferns, such as the bracken fern, can spread into drier areas because their rhizomes, which grow horizontally in the soil, produce new plants.

At first, it may seem that ferns do not have much economic value, but they are frequently used by florists in decorative bouquets and as ornamental plants in the home and garden. Wood from tropical tree ferns is often used as a building material because it resists decay, particularly by termites. Ferns, especially the ostrich fern, are used as food—in the northeastern United States, many restaurants feature fiddleheads (the season's first growth) as a special treat. Ferns also have medicinal value; many Native Americans use ferns as an astringent during childbirth to stop bleeding, and the maidenhair fern is the source of a cold medicine.

Plants and Coal

Ferns and the other seedless vascular plants we have been discussing were as large as trees and more abundant during the Carboniferous period, when a great swamp forest encompassed what is now northern Europe, the Ukraine, and the Appalachian Mountains in the United States (**Fig. 18.9**). A large number of these plants died but did not decompose completely.

Instead, they were compressed to form the coal that we still mine and burn today; therefore, seedless vascular plants are sometimes called the *Coal Age* plants. (Oil had a similar origin, but it most likely formed in marine sedimentary rocks and includes animal remains.)

Seed Plants

Seed plants are the most plentiful land plants in the biosphere. Most trees, bushes, and garden plants are seed plants. The major parts of a seed are shown in **Figure 18.10.** The seed coat and stored food protect the sporophyte embryo and allow it to survive harsh conditions during a period of dormancy (arrested state), until environmental conditions become favorable for growth. Seeds can remain dormant for hundreds of years. When a seed germinates, the stored food is a source of nutrients for the growing seedling. This evolutionary adaptation has contributed to the success of the seed plants. In fact, most of the plant species on the planet are seed plants.

Seed plants have two types of spores that produce two types of microscopic gametophytes—male and female (**Fig. 18.11**). The male gametophyte is the **pollen grain** and produces sperm. The female gametophyte is the **ovule,** which contains the egg. **Pollination** occurs when the pollen lands on the female reproductive structure. The pollen will grow a pollen tube, and the sperm migrates toward the egg. This represents a major adaptation to the land environment, since the sperm does not need water to swim to the egg. **Fertilization** occurs when the sperm reaches the egg and they combine to form a diploid zygote. The zygote eventually becomes the embryo, and the ovule becomes the seed. When the seed germinates, the embryo will become the new sporophyte generation. In gymnosperms, the seeds remain "naked" and lie within the grooves of cones. In angiosperms, the seeds are "covered" and can be found inside a fruit.

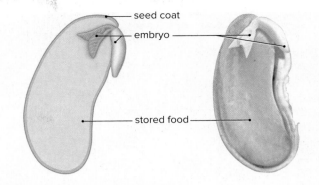

Figure 18.10 Seed anatomy.

A split bean seed showing the seed coat, sporophyte embryo, and stored food.

© David Moyer

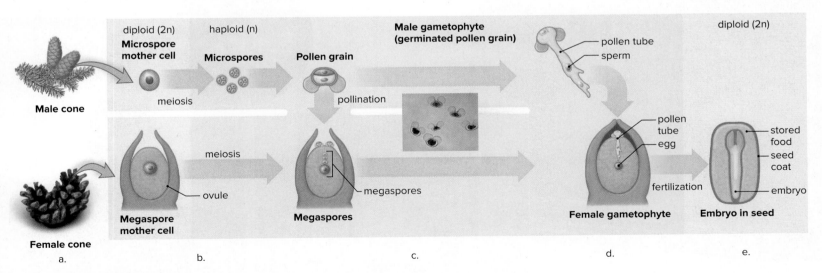

Figure 18.11 Production of a seed.

Development of the male gametophyte (*upper row*) begins when a microspore mother cell undergoes meiosis to produce microspores, each of which becomes a pollen grain. Development of the female gametophyte (*lower row*) begins in an ovule, where a megaspore mother cell undergoes meiosis to produce megaspores, only one of which will undergo mitosis to become the female gametophyte. During pollination, the pollen grain is carried to the vicinity of the ovule. The pollen grain germinates, and a nonflagellated sperm travels in a pollen tube to the egg produced by the female gametophyte. Following fertilization, the zygote becomes the sporophyte embryo, tissue within the ovule becomes the stored food, and the ovule wall becomes the seed coat.

(c): © Ed Reschke/Photolibrary/Getty Images

Figure 18.12 Cycads.

Cycads are an ancient group of gymnosperms that are threatened today because they grow slowly. They are typical landscape plants in tropical climates.

© DEA/RANDOM/De Agostini Picture Library/Getty Images

Gymnosperms

The term **gymnosperm** means "naked seed." In gymnosperms, ovules and seeds are exposed on the surface of a cone scale (modified leaf). Ancient gymnosperms, including cycads (**Fig. 18.12**), were present in the swamp forests of the Carboniferous period. The conifers are gymnosperms that have become a dominant plant group.

Conifers

Pines, spruces, firs, cedars, hemlocks, redwoods, and cypresses are all **conifers** (**Fig. 18.13**). The name *conifer* signifies plants that bear **cones** containing the reproductive structures of the plant, but other types of gymnosperms are also cone-bearing. The coastal redwood, a conifer native to northwestern California and southwestern Oregon, is the tallest living vascular plant; it can attain heights of nearly 100 m. Another conifer, the bristlecone pine of the White Mountains of California, is the oldest living tree. One living specimen is over 4,500 years old, and there is evidence that some bristlecone pines have lived as long as 4,900 years.

Adaptations and Uses of Pine Trees   Pine trees are well adapted for dry conditions. For instance, vast areas of northern temperate regions are covered

Figure 18.13 Conifers.

a. Pine trees are the most common of the conifers. The pollen cones (male) are smaller than the seed cones (female) and produce plentiful pollen. A cluster of pollen cones may produce more than a million pollen grains. Other conifers include (**b**) the spruces, which make beautiful Christmas trees, and (**c**) the junipers, which possess fleshy seed cones.

(a): (forest): © Steven P. Lynch; (pollen cones): © Maria Mosolova/Photolibrary/Getty RF; (seed cones): © Steven P. Lynch; (b): © Ed Reschke/Peter Arnold/Getty Images; (c): (tree): © T. Daniel/Bruce Coleman/Photoshot; (juniper berries): © Evelyn Jo Johnson

in evergreen coniferous forests. The tough, needlelike leaves of pines conserve water because they have a thick cuticle and recessed stomata. This type of leaf helps them live in areas where frozen topsoil makes it difficult for the roots to obtain plentiful water.

A substance called resin protects leaves and other parts of pine trees from insect and fungal attacks. The resin of certain pines is harvested; the liquid portion, called turpentine, is a paint thinner, while the solid portion is used on stringed instruments. The wood of pines is used extensively in construction, and vast forests of pines are planted for this purpose. The wood consists primarily of xylem tissue that lacks some of the more rigid cell types found in flowering trees. Therefore, it is considered a "soft" rather than a "hard" wood.

Angiosperms

The **angiosperms** (the name means "covered seeds") are an exceptionally large and successful group of land plants, with over 270,000 known species—six times the number of species of all other plant groups combined. Angiosperms, also called the flowering plants, live in all sorts of habitats, from fresh water to desert, and from the frigid north to the torrid tropics. They range in size from the tiny, almost microscopic duckweed to *Eucalyptus* trees over 35 m tall. Most garden plants produce flowers and therefore are angiosperms. In northern climates, the trees that lose their leaves are flowering plants. In subtropical and tropical climates, flowering trees as well as gymnosperms tend to keep their leaves all year.

Although the first fossils of angiosperms are no older than about 135 million years, the angiosperms probably arose much earlier. Indirect evidence suggests that the possible ancestors of angiosperms may have originated as long ago as 160 MYA. To help solve the mystery of their origin, botanists have turned to DNA comparisons to find a living plant that is most closely related to the first angiosperms. Their data point to *Amborella trichopoda* as having the oldest lineage among today's angiosperms (**Fig. 18.14**). This shrub, which has small, cream-colored flowers, lives only on the island of New Caledonia in the South Pacific.

The Flower

Most flowers have certain parts in common, despite their dissimilar appearances. The flower parts, called **sepals, petals, stamens,** and **carpels,** occur in whorls (circles) (**Fig. 18.15**). The sepals, collectively called the **calyx,** protect the flower bud before it opens. The sepals may drop off or may be colored like the petals. Usually, however, sepals are green and remain in place. The petals, collectively called the **corolla,** are quite diverse in size, shape, and color. The petals often attract a particular pollinator. Each stamen consists of a stalk, called a **filament,** and an **anther,** where pollen is produced in pollen sacs. In most flowers, the anther is positioned where the pollen can be carried away by wind or a pollinator. One or more carpels are at the center of a flower. A carpel has three major regions: the ovary, style, and stigma. The swollen base is the **ovary,** which contains from one to hundreds of ovules. The **style** elevates the **stigma,** which is sticky or otherwise adapted for the reception of pollen grains. Glands located in the region of the ovary produce nectar, a nutrient that is gathered by pollinators as they go from flower to flower.

Figure 18.14 *Amborella trichopoda.*

Molecular data suggest that this plant is most closely related to the first flowering plants.

© Stephen McCabe

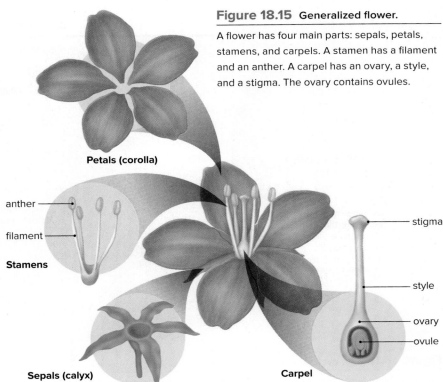

Figure 18.15 Generalized flower.

A flower has four main parts: sepals, petals, stamens, and carpels. A stamen has a filament and an anther. A carpel has an ovary, a style, and a stigma. The ovary contains ovules.

Petals (corolla)

anther
filament
Stamens

stigma
style
ovary
ovule

Sepals (calyx)

Carpel

Flowering Plant Life Cycle

In angiosperms, the flower produces seeds enclosed by fruit. The ovary of a carpel contains several ovules, and each of these eventually holds an egg-bearing female gametophyte called an embryo sac. During pollination, a pollen grain is transported by various means from the anther of a stamen to the stigma of a carpel, where it germinates. The **pollen tube** carries the two sperm into a small opening of an ovule. During **double fertilization,** one sperm unites with an egg nucleus, forming a diploid zygote, and the other sperm unites with two other nuclei, forming a triploid (3n) **endosperm** (**Fig. 18.16**). In angiosperms, the endosperm is the stored food.

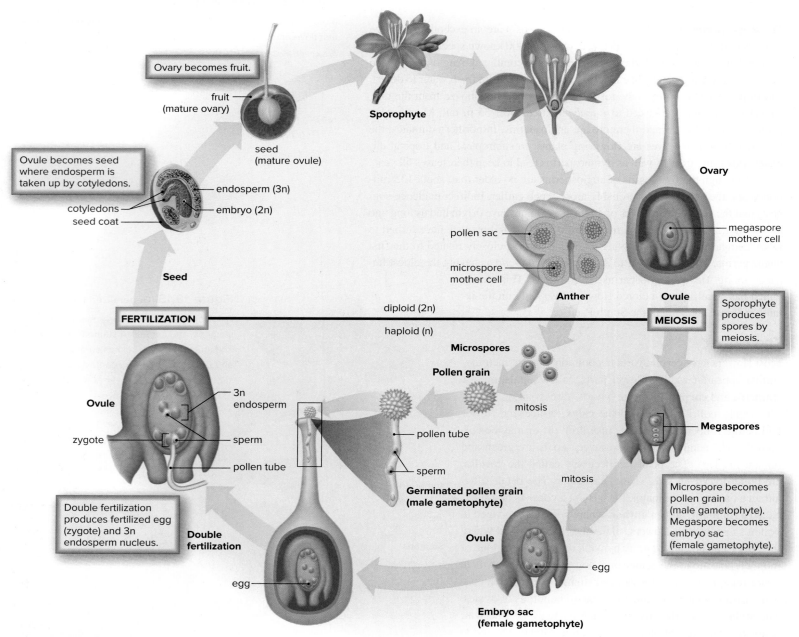

Figure 18.16 Life cycle of a flowering plant.

Flowering plants use double fertilization to produce a diploid zygote and a triploid endosperm.

Connections: Ecology

Do carnivorous plants capture insects for food?

All plants, including carnivorous plants, such as the sundew plant, are autotrophic organisms that get their energy from photosynthesis. However, some plants that live in mineral-poor soils have evolved adaptations to allow them to capture insects and sometimes small amphibians. The goal of these plants is not to extract energy from their prey. Instead the plant is after a mineral or nutrient (often nitrogen) that is lacking in the soil of their environment. There are over 600 known species of carnivorous plants.

Ultimately, the ovule becomes a seed that contains a sporophyte embryo. In some seeds, the endosperm is absorbed by the seed leaves, called **cotyledons;** whereas in other seeds, endosperm is digested as the seed germinates. When you open a peanut, the two halves are the cotyledons. If you look closely, you will see the embryo between the cotyledons. A **fruit** is derived from an ovary and possibly accessory parts of the flower. Some fruits (e.g., apple) provide a fleshy covering for their seeds, and other fruits provide a dry covering (e.g., pea pod, peanut shell).

Adaptations and Uses of Angiosperms

Successful completion of sexual reproduction in angiosperms requires the effective dispersal of pollen and then seeds. Adaptations have resulted in various means of dispersal of pollen and seeds. Wind-pollinated flowers are usually not showy, whereas many insect- and bird-pollinated flowers are colorful (**Fig. 18.17***a–c*). Night-blooming flowers attract nocturnal mammals and insects; these flowers are usually aromatic and white or cream-colored (Fig. 18.17*d*). Although some flowers disperse their pollen by wind, many are adapted to attract specific pollinators, such as bees, wasps, flies, butterflies, moths, and even bats, which carry only particular pollen from flower to flower. For example, bee-pollinated flowers are usually blue or yellow and have ultraviolet shadings that lead the pollinator to the location of nectar at the base of the flower. In turn, the mouthparts of bees are fused into a long tube that is able to obtain nectar from this location. Today, there are some 270,000 species of flowering plants and well over 1 million species of insects. Insects and the flowers they pollinate have *coevolved,* that is, have become dependent on each other for survival.

The fruits of flowers protect and aid in the dispersal of seeds. Dispersal occurs when seeds are transported by wind, gravity, water, and animals to another location. Fleshy fruits may be eaten by animals, which transport the seeds to a new location and then deposit them when they defecate. Because animals live in particular habitats and/or have particular migration patterns, they are apt to deliver the fruit-enclosed seeds to a suitable location for seed germination (initiation of growth) and development of the plant.

Figure 18.17 Pollinators.

a. A bee-pollinated flower is typically a color other than red (bees cannot see red). **b.** Butterfly-pollinated flowers are wide, allowing the butterfly to land. **c.** Hummingbird-pollinated flowers are curved back, allowing the bird's beak to reach the nectar. **d.** Bat-pollinated flowers are large and sturdy and able to withstand rough treatment.

Figure 18.18 Pea flower and the development of a pea pod.

Plants have flowers and develop fruit. **a.** Pea flower. **b.** Pea pod (the fruit) develops from the ovary.

(b): © blickwinkel/Kottmann/Alamy

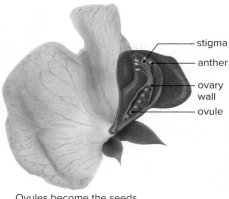

Ovules become the seeds.

a.

b.

Connections: Environment

Are all pollinators attracted to flowers for a food reward?

All flowers need a strategy to attract pollinators to their flowers. Instead of beautiful colors and sweet, tempting nectar, some species of *Ophyrys* orchids use "sexual mimicry" to attract their wasp pollinators. The orchid flowers look like female wasps and emit pheromones to attract male wasps. A male wasp will engage in "pseudocopulation" with a "female" and end up with pollen attached to his head. Frustrated, the male leaves the flower and attempts to mate with another flower. The male then deposits pollen in the second flower, thereby completing the cross-pollination that the orchid needs.

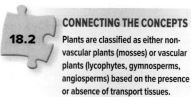

CONNECTING THE CONCEPTS

18.2 Plants are classified as either nonvascular plants (mosses) or vascular plants (lycophytes, gymnosperms, angiosperms) based on the presence or absence of transport tissues.

Check Your Progress 18.2

1. Identify the differences in structure between nonvascular and vascular plants.

2. Explain the difference between a moss and a fern with regard to dominant generation and physiology.

3. Summarize the major steps in the life cycle of a flowering plant.

4. Explain the importance of pollinators to the life cycle of an angiosperm.

Economic Benefits of Plants

One of the primary economic benefits of plants is the use of their fruits as food. Botanists use the term *fruit* in a much broader way than do laypeople. You would have no trouble recognizing an apple as a fruit, but a coconut is also a fruit, as are grains (corn, wheat, rice) and pods that contain beans or peas (**Fig. 18.18**). Cotton is derived from the cotton boll, a fruit containing seeds with seed hairs that become textile fibers used to make cloth.

Other economic benefits of plants include foods and commercial products made from roots, stems, and leaves. Cassava and sweet potatoes are edible roots; white potatoes are the tubers of underground stems. Most furniture, paper, and rope is made from the wood of a tree trunk or fibers from woody stems. Also, the many chemicals produced by plants make up 50% of all pharmaceuticals and various other types of products we can use. The cancer drug taxol originally came from the bark of the Pacific yew tree. Today, plants are even bioengineered to produce certain substances of interest (see Section 12.3).

Indirectly, the economic benefits of land plants are often dependent on pollinators (see Fig. 18.17). Only if pollination occurs can these plants produce a fruit and propagate themselves. In recent years, the populations of honeybees and other pollinators have been declining worldwide, principally due to a parasitic mite but partly because of the widespread use of pesticides. Consequently, some plants are endangered because they have lost their normal pollinators. Because of our dependence on flowering plants, we should protect pollinators!

Ecological Benefits of Plants

The ecological benefits of flowering plants are so important that we could not exist without them. Plants produce food for themselves and directly or indirectly for all other organisms in the biosphere. And all organisms that carry out cellular respiration use the oxygen that plants produce through photosynthesis.

Forests are an important part of the water cycle and the carbon cycle. In particular, the roots of trees hold soil in place and absorb water, which returns to the atmosphere. Without these functions of trees and other plants, rainwater runs off and contributes to flooding. Plants' absorption of carbon dioxide lessens the amount in the atmosphere. CO_2 in the atmosphere contributes to global warming because it and other gases trap heat near the surface of the Earth. The burning of tropical rain forests is a double threat with respect to global warming because it adds CO_2 to the atmosphere and removes trees that otherwise would absorb CO_2. Some plants can also be used to clean up toxic messes. For example, poplar, mustard, and mulberry species take up lead, uranium, and other pollutants from the soil.

18.3 The Fungi

Learning Outcomes

Upon completion of this section, you should be able to

1. Describe the general biology of a fungus.
2. Compare and contrast fungi with animals and plants.
3. Summarize the life cycle of a fungus.
4. Describe the economic and ecological significance of fungi.
5. Provide examples of fungal diseases.

Asked whether members of the kingdom Fungi are more closely related to animals or plants, most people would choose plants. But this would be wrong, because **fungi** do not have chloroplasts, and they can't photosynthesize. Then, too, fungi are not animals, even though they are **chemoheterotrophs,** like animals. Animals ingest their food, but fungi must grow into their food. Fungi release digestive enzymes into their immediate environment and then absorb the products of digestion. Also, animals are motile, but most fungi are nonmotile and do not have flagella at any stage in their life cycle. The fungal life cycle differs from that of both animals and plants because fungi produce windblown spores during both an asexual and a sexual life cycle.

Connections: Scientific Inquiry

Why are fungi and plants often studied together?

Before the recognition that plants and fungi have very different evolutionary histories, the study of fungi, called mycology, was conducted by scientists who also studied plants (botanists). Early classification systems placed both the plants and the fungi in the same groups, mainly because they were nonmotile, had cell walls, and shared other similar characteristics. Furthermore, as indicated in the chapter opener, the evolution of plants and fungi are very much intertwined. Evidence suggests that plants and fungi moved onto the land environment at about the same time, and the success of the plants on land is very much due to their relationship with the fungi.

Table 18.1 contrasts fungi, plants, and animals. The many unique features of fungi indicate that, although fungi are multicellular eukaryotes (except for the single-celled yeasts and chytrids), they are not closely related to any other group of organisms. DNA sequencing data suggest that fungi belong to the same eukaryotic supergroup as animals (Opisthokonts, see Section 17.4) and are more closely related to animals than to plants. Like animals, fungi are believed to be the descendants of a flagellated protist.

General Biology of a Fungus

The evolutionary relationships of the major groups of fungi are illustrated in **Figure 18.19.** Our description of general fungal structure will focus on the zygospore fungi, sac fungi, and club fungi.

All parts of a typical fungus are composed of **hyphae** (sing., hypha), which are thin filaments of cells. The hyphae are packed closely together to form a complex structure, such as a mushroom. However, the main body of

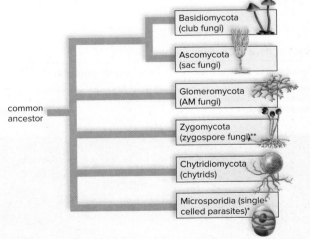

common ancestor

Basidiomycota (club fungi)

Ascomycota (sac fungi)

Glomeromycota (AM fungi)

Zygomycota (zygospore fungi)**

Chytridiomycota (chytrids)

Microsporidia (single-celled parasites)*

* Recently placed in the kingdom Fungi
** Molecular data suggests the Zygomycota may have multiple evolutionary origins.

Figure 18.19 Evolutionary relationships of the major groups of fungi.

The AM fungi consist of those species that form mycorrhizal relationships with plants.

Table 18.1 How Fungi Differ from Plants and Animals

Feature	Fungi	Plants	Animals
Nutrition	Chemoheterotrophic by absorption	Photosynthetic	Chemoheterotrophic by ingestion
Movement	Most nonmotile	Nonmotile	Motile
Body	Mycelium of hyphae	Specialized tissues/organs	Specialized tissues/organs
Adult chromosome number	Haploid	Haploid/diploid	Diploid
Cell wall	Composed of chitin	Composed of cellulose	No cell wall
Reproduction	Most have spores/mating hyphae	Spores/gametes	Gametes

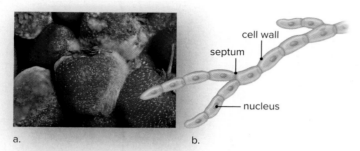

Figure 18.20 Body of a fungus.

a. This white mass of fungal hyphae is called a mycelium. **b.** A mycelium contains many individual strands, and each strand is called a hypha.

(a): © Matt Meadows/Peter Arnold/Getty Images

Figure 18.21 Chytrids.

Chytriomyces hyalinus, a chytrid, attacking an algal protist.

Reproduced with permission of the Freshwater Biological Association on behalf of The Estate of Dr Hilda Canter-Lund. © The Freshwater Biological Association

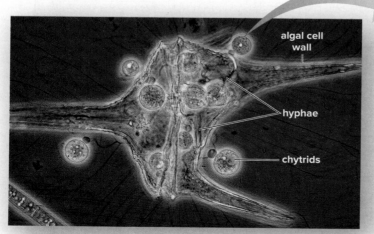
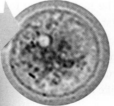

a fungus is not the mushroom, puffball, or morel; these are just temporary reproductive structures. The main body of a fungus is the mass of hyphae, called a **mycelium** (**Fig. 18.20**). Mycelia can penetrate the soil, wood, or our perishable foods. In Michigan, mycelia in the soil have been found covering 38 acres, making up the largest organism on Earth and earning the label "The Humongous Fungus."

Fungal cells have cell walls, but unlike the cell walls in plants, fungal cell walls do not contain cellulose. They are made of another polysaccharide in which the glucose monomers contain amino groups (amino sugars) and form a polymer called chitin. This polymer is also the major structural component of the exoskeletons of insects and arthropods, such as lobsters and crabs. Walls, or *septa* (sing., septum), divide the cells of a hypha in many types of fungi. Septa have pores that allow the cytoplasm to pass from one cell to another along the length of the hypha. The hyphae give the mycelium quite a large surface area, which facilitates the ability of the mycelium to absorb nutrients. Hyphae extend toward a food source by growing at their tips, and the hyphae of a mycelium absorb and then pass nutrients to the growing tips.

Fungal Diversity

Fungi are traditionally classified based on their mode of sexual reproduction. Major fungal groups include the microsporidians, chytrids, zygospore fungi, sac fungi, club fungi, and AM fungi.

Microsporidians

The single-celled microsporidians are parasites of animal cells, most often seen in insects but also found in vertebrates, such as fish, rabbits, and humans. Biologists once believed that these fungi were an ancient line of protist. However, genome sequencing of these organisms indicates that they are more closely related to the other fungi than to protists.

Chytrids

Chytrids, the most primitive of the fungi, are a unique group characterized by their motility (**Fig. 18.21**). Both spores and gametes of chytrids have flagella, a feature that was lost at some point in the evolution of the other fungi. Some of the chytrids are single-celled, while others form hyphae without septa. Another oddity distinguishes the chytrids: Some have an alternation-of-generations life cycle, much like that of green plants and certain algae, but very uncommon among fungi. Most chytrids inhabit water or soil, although some live as parasites of plants and animals.

Zygospore Fungi—Black Bread Mold

Multicellular organisms are characterized by specialized cells. Black bread mold, a type of zygospore fungi, demonstrates that the hyphae of a fungus may be specialized for various purposes (**Fig. 18.22**). In this fungus, horizontal hyphae exist on the surface of the bread; other hyphae grow into the bread, anchoring the mycelium and carrying out digestion; and some form stalks that bear sporangia.

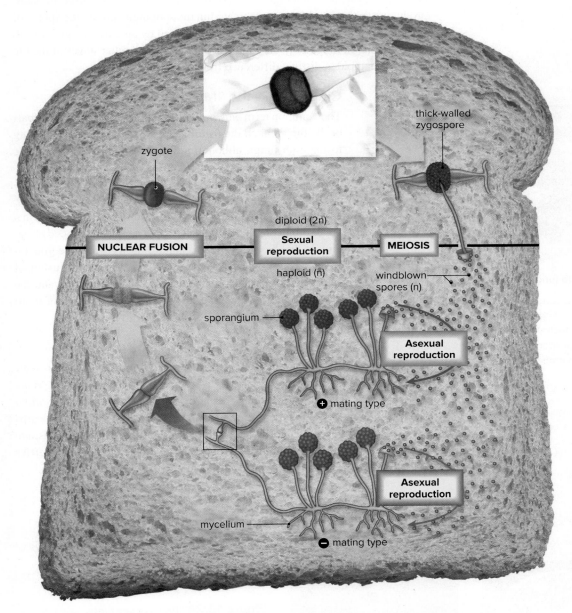

Figure 18.22 Life cycle of black bread mold.

During asexual reproduction, sporangia produce asexual spores. During sexual reproduction, two hypha tips fuse, and then two nuclei fuse, forming a zygote that develops a thick, resistant wall (zygospore). When conditions are favorable, the zygospore germinates, and meiosis within a sporangium produces windblown spores.

(bread): © Jules Frazier/Getty RF; (zygospore): © John Hardy, University of Washington, Stevens Point Department of Biology

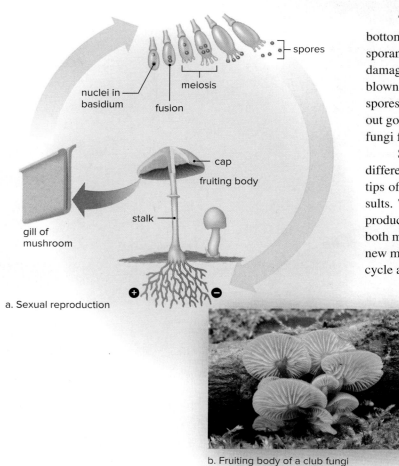

a. Sexual reproduction

Figure 18.23 **Sexual reproduction in club fungi produces mushrooms.**

a. For club fungi, fusion of tips of + and − hyphae results in hyphae that form the mushroom (a fruiting body). The nuclei fuse in clublike structures (basidia) attached to the gills of a mushroom, and meiosis produces spores. **b.** A club fungus, the white button mushroom.

(b): © Mantonature/Getty RF

b. Fruiting body of a club fungi

The mycelia of two different mating types are featured in the center and bottom of Figure 18.22. During asexual reproduction, each mycelium produces sporangia, where spore formation occurs. Spores are resistant to environmental damage, and they are often made in large numbers. Fungal spores are windblown, a distinct advantage for a nonmotile organism living on land. When spores encounter a moist environment, they germinate into new mycelia without going through any developmental stages, another feature that distinguishes fungi from animals.

Sexual reproduction in fungi involves the conjugation of hyphae from different mating types (usually designated + and −). In black bread mold, the tips of + and − hyphae join, the nuclei fuse, and a thick-walled zygospore results. The zygospore undergoes a period of dormancy before it germinates, producing sporangia. Meiosis occurs within the sporangia, producing spores of both mating types. The spores, which are dispersed by air currents, give rise to new mycelia. In fungi, only the zygote is diploid, and all other stages of the life cycle are haploid, a distinct difference from animals.

Club Fungi—the "Mushrooms"

The common term *mushroom* is often used to describe the part of the fungi called the *fruiting body*. Most mushrooms belong to the club fungi, although a few, such as the morel are sac fungi (see below). The function of the fruiting body is to produce spores (**Fig. 18.23***a*). In mushrooms, when the tips of + and − hyphae fuse, the haploid nuclei do not fuse immediately. Instead, what are called dikaryotic (two-nuclei) hyphae form a mushroom consisting of a stalk and a cap. Club-shaped structures called basidia (sing., basidium) project from the gills located on the underside of the cap (Fig. 18.23b). Fusion of nuclei inside these structures is followed by meiosis and the production of windblown spores. Each mushroom cap produces tens of thousands of spores.

Sac Fungi

Approximately 75% of all known fungi are sac fungi, named for their characteristic cuplike sexual reproductive structure, called an ascocarp. Many sac fungi reproduce by producing chains of asexual spores called *conidia* (sing., conidium). Cup fungi, morels, and truffles have conspicuous ascocarps (**Fig. 18.24**). Truffles, underground symbionts of hazelnut and oak trees, are highly prized as gourmet delights. Pigs and dogs are trained to sniff out truffles in the woods, but truffles are also cultivated on the roots of seedlings.

a. Morel

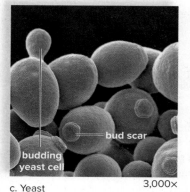

b. Cup fungi

c. Yeast   3,000×

Figure 18.24 **Types of sac fungi.**

a. A morel is a sac fungi. **b.** The ascocarp of many sac fungi resembles a cup. **c.** Yeast are sac fungi that are used in baking and brewing.

(a): © Robert Marien/Corbis RF; (b): © Carol Wolfe; (c): © Science Photo Library RF/Getty RF

Morels (Fig. 18.24a) are often collected by chefs and other devotees because of their unique and delicate flavor. Collectors must learn, however, to avoid "false morels," which can be poisonous. Interestingly, even true morels should not be eaten raw. As saprophytes, these organisms secrete digestive enzymes capable of digesting wood or leaves. If consumed prior to being inactivated by cooking, these enzymes can cause digestive problems.

Yeasts (Fig. 18.24c) are probably one of the better-known sac fungi. The term *yeast* is generally applied to single-celled fungi, and many of these organisms are sac fungi. *Saccharomyces cerevisiae,* brewer's yeast, is representative of budding yeasts. When unequal binary fission occurs in yeast, a small cell gets pinched off and then grows to full size. Asexual reproduction occurs when the food supply runs out, and it produces spores.

When some yeasts ferment, they produce ethanol and carbon dioxide. In the wild, yeasts grow on fruits, and historically, the yeasts already present on grapes were used to produce wine. Today, selected yeasts are added to relatively sterile grape juice in order to make wine. Also, yeasts are added to grains to make beer and liquor. Both the ethanol and carbon dioxide are retained in beers and sparkling wines, while carbon dioxide is released from wines. In breadmaking, the carbon dioxide produced by yeasts causes the dough to rise, and the ethanol quickly evaporates. The gas pockets are preserved as the bread bakes.

Ecological Benefits of Fungi

Most fungi are **saprotrophs** that decompose the remains of plants, animals, and microbes in the soil. Fungal enzymes can degrade cellulose and even lignin in the woody parts of plants. That is why fungi so often grow on dead trees. This degradational ability also means that fungi can be used to remove excess lignin from paper pulp. Ordinarily, lignin is difficult to extract from pulp and ends up being a pollutant once it is removed.

Along with bacteria, which are also decomposers, fungi play an indispensable role in the environment by returning inorganic nutrients to the soil. Many people take advantage of the activities of bacteria and fungi by composting their food scraps or yard waste. When a gardener makes a compost pile and provides good conditions for decomposition to occur, the result is a dark, crumbly material that serves as an excellent fertilizer. And while the material may smell bad as decomposition is occurring, the finished compost looks and smells like rich, moist earth.

Some fungi eat animals that they encounter as they feed on their usual meals of dead organic remains. For example, the oyster fungus (**Fig. 18.25**) secretes a substance that anesthetizes any nematodes (roundworms) feeding on it. After the worms become inactive, the fungal hyphae penetrate and digest their bodies, absorbing the nutrients. Other fungi snare, trap, or fire projectiles into nematodes and other small animals before digesting them. The animals serve as a source of nitrogen for the fungus.

Mutualistic Relationships

In a mutualistic relationship, two different species live together and help each other. **Lichens** are mutualistic associations between particular fungi and cyanobacteria or green algae (**Fig. 18.26**). In a lichen, the fungal partner provides water and minerals for the photosynthetic partner. The fungus uses organic acids to release these minerals from rocks or trees. The photosynthesizing partner in turn provides organic molecules, such as sucrose, to the fungus.

Figure 18.25 **Carnivorous fungus.**

Fungi like this oyster fungus, a type of bracket fungus, grow on trees because they can digest cellulose and lignin. If this fungus meets a roundworm in the process, it immobilizes the worm and digests it also. The worm acts as a source of nitrogen for the fungus.

© L. West/Science Source

Figure 18.26 **Lichens.**

a. Morphology of a lichen.
b. Examples of lichens.

(b): (crustose): © James Cade/123RF; (fruticose): © Steven P. Lynch; (foliose): © Yogesh More/Alamy RF

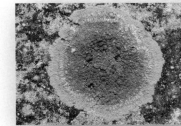

Crustose lichen

reproductive unit

fungal hyphae   algal cells

Fruticose lichen

fungal hyphae

Foliose lichen

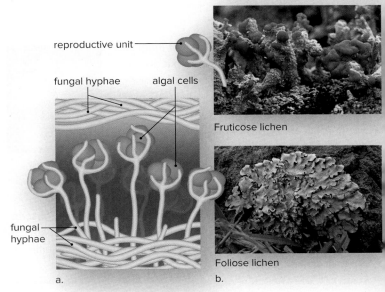

a.   b.

Figure 18.27 Plant roots showing mycorrhizae.

Mycorrhizal fungi increase the surface area for the absorption of water and nutrients.

© Dr. Jeremy Burgess/Science Source

Connections: Health

Are all fungi edible?

No! In fact, very few species of fungi are edible. Each year, many people are poisoned and some die from eating poisonous mushrooms. One of the most common poisonous mushrooms is the death cap, or *Amanita phalloides*. The poison in the death cap is a polypeptide that causes liver and kidney failure, and most people who ingest these fungi die within a few weeks. At the cellular level, the death cap's poison shuts down the process of transcription, causing cell death. The color and shape of the death cap are similar to those of many edible mushrooms, resulting in accidental poisonings.

© Holmes Garden Photos/Alamy

Overall, lichens are ecologically important because they produce organic matter and create new soil, allowing plants to invade the area.

Lichens occur in three varieties: compact crustose lichens, often seen on bare rocks or tree bark; shrublike fruticose lichens; and leaflike foliose lichens. Regardless, the body of a lichen has three layers. The fungal hyphae form a thin, tough upper layer and a loosely packed lower layer. These layers shield the photosynthetic cells in the middle layer. Specialized fungal hyphae that penetrate or envelop the photosynthetic cells transfer organic nutrients to the rest of the mycelium. The fungus not only provides minerals and water to the photosynthesizer but also offers protection from predation and desiccation. Lichens can reproduce asexually by releasing fragments that contain hyphae and an algal cell. At first, the relationship between the fungi and algae was likely a parasite-and-host interaction. Over evolutionary time, the relationship apparently became more mutually beneficial, although how to test this hypothesis is a matter of debate.

Mycorrhizal fungi form mutualistic relationships with the roots of most plants, helping the plants grow more successfully in dry or poor soils, particularly those deficient in inorganic nutrients (**Fig. 18.27**). The fungal hyphae greatly increase the surface area from which the plant can absorb water and nutrients. It has been found beneficial to encourage the growth of mycorrhizal fungi when restoring lands damaged by strip mining or chemical pollution.

Mycorrhizal fungi may live on the outside of roots, enter between root cells, or penetrate root cells. *Arbuscular mycorrhizal (AM) fungi* (see Fig. 18.20) penetrate root cells with clumps of bushy hyphae and are associated with 80% of all vascular plants. Ultimately, the fungus and plant cells exchange nutrients, with the fungus bringing water and minerals to the plant and the plant providing organic carbon to the fungus. Early plant fossils indicate that the relationship between fungi and plant roots is an ancient one, and therefore it may have helped plants adapt to life on dry land. The general public is not familiar with mycorrhizal fungi, but some people relish truffles, the fruiting bodies of a mycorrhizal fungus that grows in oak and beech forests.

Economic Benefits of Fungi

Fungi help us produce medicines and many types of food. The mold *Penicillium* was the original source of penicillin, a breakthrough antibiotic that led to the important class of cillin antibiotics. Cillin antibiotics have saved millions of lives.

In the United States, the average person consumes over 2.6 pounds of mushrooms annually. Today it is common to see up to a dozen varieties in specialty markets. In addition to adding taste and texture to soups, salads, and omelets and being used in stir-fries, mushrooms are an excellent low-calorie meat substitute with great nutritional value and lots of vitamins. Although there are thousands of mushroom varieties in the world, the white button mushroom, *Agaricus bisporus,* dominates the U.S. market. However, in recent years, brown-colored variants have surged in popularity and have been one of the fastest-growing segments of the mushroom industry. *Portabella* is a marketing name used by the mushroom industry for the more flavorful brown strains of *A. bisporus*. This mushroom is brown because it is allowed to open, exposing the mature gills with their brown spores; the *crimini* mushroom is the same brown strain, but it is not allowed to open before it is harvested. Non-*Agaricus* varieties, especially shiitake and oyster, have slowly gained in popularity over the past decade. Shiitake is touted for lowering cholesterol levels and having antitumor and antiviral properties.

Fungi as Disease-Causing Organisms

Fungi cause diseases in both plants and animals, including humans.

Fungi and Plant Diseases

Fungal pathogens, which usually gain access to plants by way of the stomata or a wound, are a major concern for farmers. Serious crop losses occur each year due to fungal diseases (**Fig. 18.28**). For example, the typical banana you see in the grocery store is threatened by a fungi that causes Panama disease. As much as a third of the world's rice crop is destroyed each year by rice blast disease. Corn smut is a major problem in the midwestern United States. Rusts, a type of fungi, infect a variety of plants, from fruit trees to grains.

The life cycle of rusts may be particularly complex, since it requires two different host species to complete the cycle. Black stem rust of wheat uses barberry bushes as an alternate host. Eradication of barberry bushes in areas where wheat is grown helps control this rust. Fungicides are regularly applied to crops to limit the negative effects of fungal pathogens. Wheat rust can also be controlled by producing new resistant strains of wheat.

Fungi and Animal Diseases

As is well known, certain mushrooms are poisonous. The ergot fungus that grows on grain can result in ergotism when a person eats contaminated bread. Ergotism is characterized by hysteria, convulsions, and sometimes death.

Mycoses are diseases caused by fungi. Mycoses have three possible levels of invasion: Cutaneous mycoses affect only the epidermis; subcutaneous mycoses affect deeper skin layers; and systemic mycoses spread their effects throughout the body by traveling in the bloodstream. Fungal diseases that can be contracted from the environment include ringworm from soil, rose gardener's disease from thorns, Chicago disease from old buildings, and basketweaver's disease from grass cuttings. Opportunistic fungal infections seen in AIDS patients stem from fungi that are always present in the body but take the opportunity to cause disease when the immune system becomes weakened.

a.   b.   c.

Figure 18.28 **Plant fungal diseases.**

a. Panama disease infects banana plants. **b.** Corn smut invades corn kernels. **c.** Cedar apple rust invades cedar trees (shown), then apple trees later.

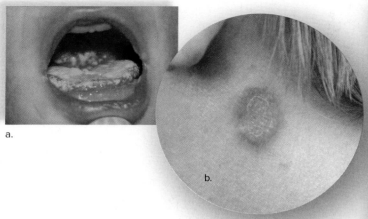

a.

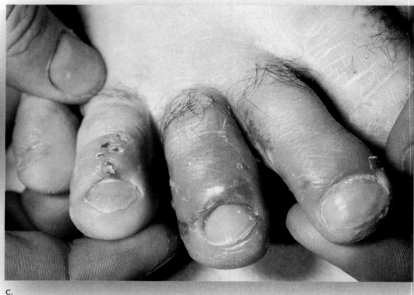

b.

c.

Figure 18.29 Human fungal diseases.

a. Thrush, or oral candidiasis, is characterized by the formation of white patches on the tongue. **b.** Ringworm and (**c**) athlete's foot are caused by species of *Tinea.*

(a): © Dr. M. A. Ansary/Science Source; (b): © John Hadfield/SPL/Science Source; (c): © P. Marazzi/SPL/Science Source

Candida albicans, a yeast, causes the widest variety of fungal infections. Disease occurs when antibacterial treatments kill off the microflora community, allowing *Candida* to proliferate. Vaginal *Candida* infections are commonly called "yeast infections" in women. Oral thrush is a *Candida* infection of the mouth common in newborns and AIDS patients (**Fig. 18.29***a*). In individuals with inadequate immune systems, *Candida* can move throughout the body, causing a systemic infection that can damage the heart, the brain, and other organs.

Ringworm is a group of related diseases caused, for the most part, by fungi in the genus *Tinea.* Ringworm is a cutaneous infection that does not penetrate the skin. The fungal colony grows outward, forming a ring of inflammation. The center of the lesion begins to heal, giving the lesion its characteristic appearance, a red ring surrounding an area of healed skin (Fig. 18.29*b*). Athlete's foot is a *Tinea* infection that affects the foot, mainly causing itching and peeling of the skin between the toes (Fig. 18.29*c*).

Batrachochytrium dendrobatidis is a parasitic chytrid that causes a cutaneous infection called chytridiomycosis in frogs around the world. The disease is thought to have originated in South Africa. It began to spread in the 1930s after African clawed frogs were captured and sold as pets and for use in medicine and research. Chytridiomycosis has recently decimated frog populations in Australia and Central and South America.

Bats found in the eastern United States and Canada are currently under attack by *Geomyces destructans*—a filamentous fungus causing white nose syndrome (**Fig. 18.30**). In winter, hibernating cave-dwelling bats have a slow metabolism and live off their fat reserves. The fungus grows on their muzzles and wings, irritating the bats and essentially "waking them up." The irritated bats fly out of the cave and starve to death. To date, 6.7 million bats have died from this disease, and efforts by the U.S. Fish and Wildlife Service are underway to control this fungus and save North American bats.

Because fungi are eukaryotes and more closely related to animals than to bacteria, it is hard to design an antibiotic against fungi that does not also harm animals. Thus, researchers exploit any biochemical differences they can discover between animals and fungi.

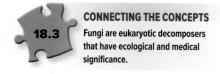

CONNECTING THE CONCEPTS

18.3   Fungi are eukaryotic decomposers that have ecological and medical significance.

Figure 18.30 White nose syndrome.

Brown bats in a hibernation cave exhibiting fungal growth on their muzzles.

© New York State Department of Environmental Conservation/AP Images

---

Check Your Progress 18.3

1. Describe the general body structure of a fungus.

2. List the major groups of fungi, and provide a characteristic of each group.

3. Identify the organisms that make up a lichen, and describe what each partner in the mutualistic relationship provides to the other.

4. List and describe two examples of fungal diseases affecting humans or other animals.

STUDY TOOLS http://connect.mheducation.com

SMARTBOOK® Maximize your study time with McGraw-Hill SmartBook®, the first adaptive textbook.

SUMMARIZE

While plants and fungi have separate evolutionary histories, they have coevolved over time and now participate in both beneficial and harmful relationships. Both plants and fungi are important to life on Earth.

18.1 Plants are multicellular, photosynthetic eukaryotes with an alternation-of-generations life cycle.

18.2 Plants are classified as either nonvascular plants (mosses) or vascular plants (lycophytes, gymnosperms, angiosperms) based on the presence or absence of transport tissues.

18.3 Fungi are eukaryotic decomposers that have ecological and medical significance.

18.1 Overview of the Plants

Plants (kingdom Plantae) evolved from a multicellular, freshwater green alga about 500 MYA. Whereas algae are adapted to life in the water, plants are adapted to living on land.

- The freshwater green algae known as **charophytes** appear to be the closest living relatives of land plants. The charophytes have some characteristics that became helpful to plants on land.
- During the evolution of plants, five significant events are associated with adaptation to a land existence: evolution of (1) embryo protection, (2) vascular tissue, (3) leaves, (4) seeds, and (5) flowers.

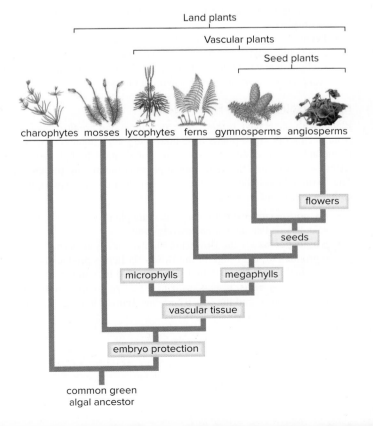

Plants Have a Unique Life Cycle

Land plants have a life cycle characterized by an **alternation of generations,** in which each type of plant exists in two forms: the **sporophyte** (2n) and the **gametophyte** (n).

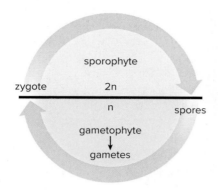

18.2 Diversity of Plants

Some Plants Are Nonvascular

Bryophytes, such as the mosses, are short plants with no **vascular tissue.** The plant must stay wet so that flagellated sperm can swim to the egg. The gametophyte is the dominant generation. Spores are windblown.

Most Plants Are Vascular Plants

In vascular plants, the sporophyte is the dominant generation. Vascular plants have two kinds of well-defined conducting tissues. **Xylem** is specialized to conduct water and dissolved minerals, and **phloem** is specialized to conduct organic solutes. Vascular plants thus have true roots, stems, and leaves, and they can grow taller due to the more efficient transport of water, minerals, and nutrients.

- **Lycophytes** are seedless and have narrow leaves called microphylls.
- **Ferns** are seedless and have large leaves with branching veins called megaphylls. Megaphylls increase the surface area for receiving sunlight to power photosynthesis. Ferns have windblown spores and still require water for sperm to swim to the egg.

Most Vascular Plants Have Seeds

Plants with seeds have reproductive structures that are protected from drying out. Seed plants have male and female gametophytes. Gametophytes are reduced in size. The female gametophyte is retained within an **ovule,** and the male gametophyte is the mature **pollen grain.** The fertilized ovule becomes the seed, which contains a sporophyte embryo, food, and a seed coat.

- **Gymnosperms,** including pine trees, are plants with cones that have naked seeds that are not enclosed by fruit.
- **Angiosperms** are the flowering plants. An angiosperm's reproductive organs are in the flower. Pollen is produced in pollen sacs inside an anther. Pollen is transported by wind or from flower to flower by birds, insects, or bats. Fertilized ovules in the **ovary** become seeds, and the ovary becomes **fruit.** Thus, angiosperms have covered seeds.

18.3 The Fungi

Kingdom Fungi includes the microsporidia, chytrids, zygospore fungi, club fungi, sac fungi and AM fungi.

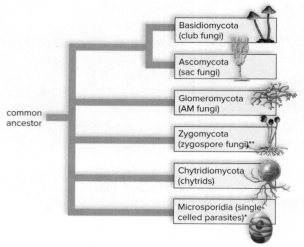

* Recently placed in the kingdom Fungi
** Molecular data suggests the Zygomycota may have multiple evolutionary origins.

General Biology of a Fungus

The body of a typical fungus is composed of thin filaments of cells, called **hyphae,** that form a mass called a **mycelium.** The cell wall contains chitin. Most fungi produce windblown spores during both asexual and sexual reproduction. Some fungi have a fruiting body (mushroom) for spore dispersal.

Fungi in the Environment

- Fungi are **saprotrophs** that carry on external digestion. As decomposers, fungi are key to ecological cycles in the biosphere.
- **Lichens** (fungi plus cyanobacteria or green algae) are primary colonizers in poor soils or on rocks.
- **Mycorrhizal fungi** grow on or in plant roots and help the plants absorb minerals and water.
- Fungi have economic benefits in helping to produce foods and medicines, as well as serving as food themselves.
- Fungi can cause diseases. Fungal pathogens of plants include blasts, smuts, and rusts that attack crops of great economic importance, such as rice and wheat. Animal diseases caused by fungi include thrush, ringworm, chytridiomycosis, and white nose syndrome.

ASSESS

Testing Yourself

Choose the best answer for each question.

18.1 Overview of the Plants

1. Which of the following is not a plant adaptation to land?
   a. ability to undergo photosynthesis
   b. protection of the embryo in maternal tissue
   c. development of flowers
   d. presence of vascular tissue
   e. seed production

2. Charophytes

    a. are freshwater green algae.

    b. lack vascular tissue.

    c. are the closest living relatives of land plants.

    d. enclose their zygotes within protective structures.

    e. All of these are correct.

3. The alternation-of-generations life cycle shows that

    a. plants are only haploid.

    b. plants are only diploid.

    c. plants spend part of their lives as haploid and diploid.

    d. None of these are correct.

18.2 Diversity of Plants

For items 4–8, identify the plant that fits the description. Use each answer only once.

    a. bryophytes   **d.** gymnosperms

    b. lycophytes   **e.** angiosperms

    c. ferns

4. short plants in which the gametophyte generation is the dominant generation

5. have true leaves in the form of microphylls

6. have seeds enclosed in a fruit

7. have cones with naked seeds

8. have windblown spores, swimming sperm, and megaphylls

9. Label the parts of the flower in the following illustration.

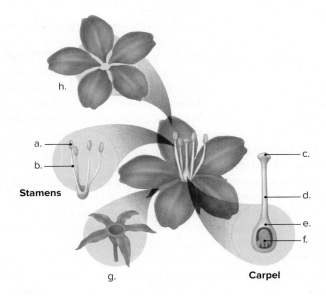

10. A flower

    a. captures spores.

    b. attracts pollinators.

    c. protects seeds.

    d. increases photosynthesis.

11. A fruit is derived from

    a. the corolla.   **c.** an ovule.

    b. an ovary.   **d.** the calyx.

12. A seed protects the

    a. embryo.   **c.** ovary.

    b. ovule.   **d.** pollen grain.

18.3 The Fungi

13. Fungi fulfill an ecological role by decaying mostly dead organic matter. Fungi are therefore called

    a. pathogens.   **c.** saprotrophs.

    b. carnivores.   **d.** photosynthesizers.

14. Which of the following statements about fungi is false?

    a. Most fungi are multicellular.

    b. Fungal cell walls are composed of cellulose.

    c. Most fungi are nonmotile.

    d. Fungi digest their food before ingesting it.

15. A mycelium is

    a. a mass of fungal filaments.

    b. a type of fungus with flagellated spores and gametes.

    c. the main body of a typical fungus.

    d. a mutualistic association between a fungus and a green alga or cyanobacterium.

    e. Both a and c are correct.

ENGAGE

BioNOW

Want to know how this science is relevant to your life? Check out the BioNow video below:

    • Enzymes and Fungi

How are the fungicides in this experiment interfering with the normal life cycle of the fungus?

Thinking Critically

1. Bare-root pine tree seedlings transplanted into open fields often grow very slowly. However, pine seedlings grow much more vigorously if they are dug from their native environment and then transplanted into a field, as long as some of the original soil is transported on the seedlings. Why is it so important to retain some native soil on the seedlings?

2. Evolutionary trees, such as the one that follows, indicate that members of kingdoms Plantae and Fungi had protist ancestors. To which domain do all three groups of organisms belong? What characteristics would distinguish the protist ancestors of plants from those of fungi?

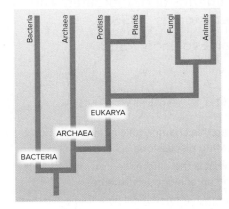

3. Many fungal infections of humans are considered to be opportunistic, meaning that fungi that are normally free-living (usually in soil) can sometimes survive, and even thrive, on or inside the human body. From the fungal "point of view," what unique challenges would be encountered when trying to survive on human skin? What about inside human lungs?

19

The Animals

© FLPA/SuperStock

OUTLINE

19.1 Evolution of Animals   337

19.2 Sponges and Cnidarians: The Early Animals   341

19.3 Flatworms, Molluscs, and Annelids: The Lophotrochozoans   343

19.4 Roundworms and Arthropods: The Ecdysozoans   347

19.5 Echinoderms and Chordates: The Deuterostomes   353

19.6 Human Evolution   363

BEFORE YOU BEGIN

Before beginning this chapter, take a few moments to review the following discussions.

Section 14.2 What is the difference between a homologous and an analogous structure?

Table 16.1 During what geological time frame did the first animals appear? The first land vertebrates?

Section 16.3 To what domain of life do the animals belong?

Canine Evolution

Dogs are probably the most familiar and beloved domesticated animals, "man's best friend." They are our companions as well as our working partners in war, law enforcement, herding, service, rescue, and therapy. But when and where did domestic dogs first evolve? Scientists have long suspected that *Canis familiaris* evolved from *Canis lupus,* the gray wolf, but genetic analyses now indicate that modern dogs evolved from a species of wolf that is now extinct. And while previous studies supported the idea that dogs were first domesticated in China or the Middle East, comparisons of mitochondrial DNA isolated from modern dogs, wolves, and ancient dog fossils now indicate that the first domesticated dogs most likely originated in Europe.

However, the time frame of this event still remains in question. While the domestication of the dog was believed to have begun around 16,000 years ago, newer molecular analytical techniques suggest that this may have happened as far back as 130,000 years ago. If this is the case, then humans and canines have been around each other for much longer than anyone thought.

The New Guinea singing dog is named for its habit of howling at different pitches. It is also called Stone Age dog, after the stone tool–using people who took it to New Guinea 6,000 years ago. Living on an island, it has remained geographically and reproductively isolated ever since. Therefore, it might be a "living fossil," an organism with the same genes and characteristics as its original ancestor. If so, the New Guinea singing dog offers an opportunity to study what early domesticated dogs were like.

However, like many other animals, which are the topic of this chapter, the New Guinea singing dog is threatened with extinction. Expeditions into the highlands of New Guinea have yielded only a few droppings, tracks, and haunting howls in the distance.

As you read through this chapter, think about the following questions:

1. What are the common characteristics of all animals?
2. What is the major difference between a vertebrate and an invertebrate?

19.1 Evolution of Animals

Learning Outcomes

Upon completion of this section, you should be able to

1. Explain how animals are distinguished from other groups of organisms.
2. Identify the key events in the evolution of animals.
3. List the characteristics that distinguish protostomes from deuterostomes.

The modern three-domain system places **animals** in the domain Eukarya and the kingdom Animalia. Within the Eukarya, they are placed in the supergroup Opisthokonta along with fungi and certain protists, notably the choanoflagellates (see Section 17.4). While animals are extremely diverse, they share some important differences from the other multicellular eukaryotes, the plants and fungi. Unlike plants, which make their food through photosynthesis, animals are heterotrophs and must acquire nutrients from an external source. Unlike fungi, which digest their food externally and absorb the breakdown products, animals ingest (eat) whole food and digest it internally.

In general, animals share the following characteristics (**Figure 19.1**):

- Multicellular with specialized cells that form tissues and organs.
- Possess nervous and muscular tissues that allow for mobility (locomotion).
- Have a life cycle in which the adult is typically diploid.
- Usually undergo sexual reproduction and produce an embryo that goes through developmental stages.
- Heterotrophs that acquire food by ingestion, followed by digestion.

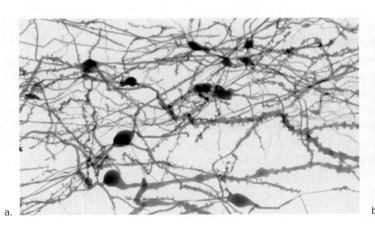
a.

b.

c.

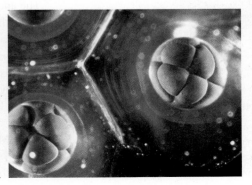
d.

Figure 19.1 General characteristics of animals.

a. Animals are multicellular, with specialized cells that form tissues and organs. **b.** Animals are heterotrophs, meaning they obtain nutrition from external sources. **c.** Animals are typically motile, due to their well-developed nervous and muscular systems. **d.** Most animals reproduce sexually, beginning life as a 2n zygote, which undergoes development to produce a multicellular organism that has specialized tissues.

(a): © Salvanegra/iStock/Getty RF; (b): © Mike Raabe/Getty RF; (c): © iStockphoto/Getty RF; (d): © Carolina Biological Supply/ Phototake

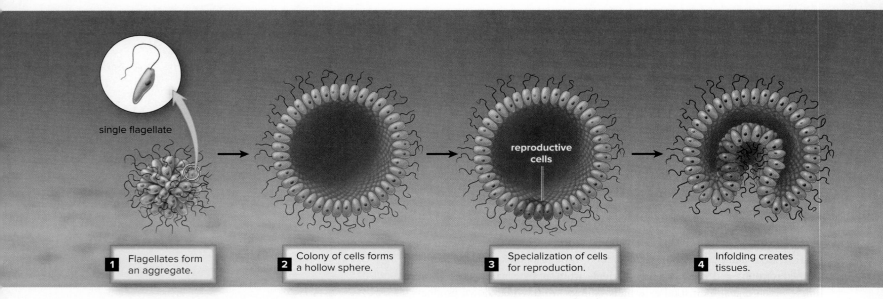

single flagellate

1 Flagellates form an aggregate.

2 Colony of cells forms a hollow sphere.

reproductive cells

3 Specialization of cells for reproduction.

4 Infolding creates tissues.

Figure 19.2 The colonial flagellate hypothesis.

The hypothesis explains how a colony of flagellated cells may have formed some of the specialized structures characteristic of the first animals.

Ancestry of Animals

In Section 18.1, we discussed evidence that plants most likely share a green algal ancestor with the charophytes. Most scientists agree that animals also evolved from a protist, most likely a protozoan. The *colonial flagellate hypothesis* states that animals are descended from an ancestor that resembled a hollow, spherical colony of flagellated cells. **Figure 19.2** shows how the process would have begun with an aggregate of a few flagellated cells. From there, a larger number of cells could have formed a hollow sphere. Individual cells within the colony would have become specialized for particular functions, such as reproduction. Two tissue layers could have arisen by an infolding of certain cells into a hollow sphere. Tissue layers do arise in this manner during the prenatal development of animals today.

Among the protists, choanoflagellates (see Section 17.4) most likely resemble the last single-celled ancestor of animals, and molecular data tell us that they are the closest living protist relative of animals. A choanoflagellate is a single cell, 3–10 μm in diameter, with a flagellum surrounded by a collar of 30–40 microvilli (**Fig. 19.3**).

The Evolutionary Tree of Animals

The animal kingdom is currently divided into 35 groups, or phyla. The majority of these animals are **invertebrates.** Invertebrates lack an internal skeleton, or endoskeleton, of bone or cartilage. The invertebrates evolved first, and they far outnumber the **vertebrates** (animals with an endoskeleton). Because early animals were simple invertebrates, the fossil record is sparse regarding the early evolution of animals. However, systematists have been able to establish a fairly clear record of the evolutionary history of animals (**Fig. 19.4**) based primarily on molecular, developmental, and anatomical data.

Molecular comparisons have played a major role in establishing evolutionary relationships, because the more closely related two organisms are, the more their DNA sequences will have in common. Advances in the study of

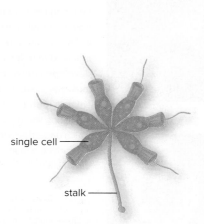

single cell

stalk

Figure 19.3 Choanoflagellates.

The choanoflagellates are the living protozoans most closely related to animals and may resemble animals' last single-celled ancestor. Some choanoflagellates live in colonies like these, in which a group of cells is attached to a surface by means of a stalk.

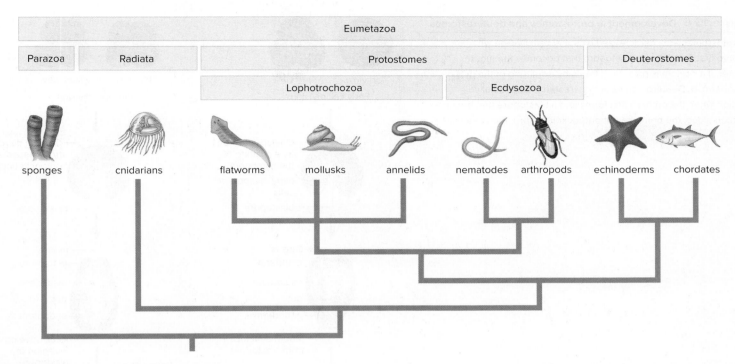

Figure 19.4 **The animal evolutionary tree.**

The relationships in this diagram are based on research into the developmental biology of each group, as well as molecular studies of DNA, RNA, and protein similarity.

genomics and proteomics (see Section 12.4) have greatly enhanced our ability to distinguish one group of animals from another.

Evolutionary Trends

Animals differ in biological organization. Sponges, like all animals, are *multicellular,* but they have no true tissues and therefore have the cellular level of organization. Because of these characteristics, sponges are classifed as parazoans, which means "beside the animals," referring to their position on the evolutionary tree alongside the "true animals," or eumetazoans. Cnidarians, such as *Hydra,* are eumetazoans. They have true tissues, which are formed from two germ layers when they are embryos. All other eumetazoans have three germ layers (ectoderm, mesoderm, endoderm) as embryos. The term **germ layers** refers to primary embryonic layers that give rise to all the other tissues and organs in an animal's body.

Animals differ in symmetry. Many sponges have no particular symmetry and are therefore asymmetrical. **Radial symmetry,** as seen in cnidarians, means that the animal is organized circularly, similar to a wheel. No matter where the animal is sliced longitudinally, two mirror images are obtained (**Fig. 19.5***a*). **Bilateral symmetry,** as seen in flatworms, means that the animal has definite right and left halves; only a longitudinal cut down the center of the animal will produce mirror images (Fig. 19.5*b*). During the evolution of animals, the trend toward bilateral symmetry has been accompanied by **cephalization,** localization of a brain and specialized sensory organs at the anterior end of an animal. Therefore, bilateral symmetry was an important precursor for the evolutionary trend of increasing complexity of the nervous system and sensory organs. The appearance of bilateral symmetry marked the evolution of animals that were able to exploit environmental resources in new ways through their more active lifestyles.

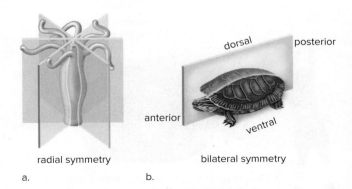

Figure 19.5 **Radial versus bilateral symmetry.**

a. With radial symmetry, two mirror images are obtained no matter how the animal is sliced longitudinally. Radially symmetrical animals tend to stay in one place and reach out in all directions to get their food. **b.** With bilateral symmetry, mirror images are obtained only if the animal is sliced down the middle. Bilaterally symmetrical animals tend to actively go after their food.

Figure 19.6 Development in protostomes and deuterostomes.

a. Protostomes are characterized by spiral cell division when the embryo first forms and a blastopore that becomes the mouth. Further, if a coelom is present, it forms by a splitting of the mesoderm. **b.** Deuterostomes are characterized by radial cell division when the embryo first forms and a blastopore that becomes the anus. Also, the primitive gut outpockets to form the coelom.

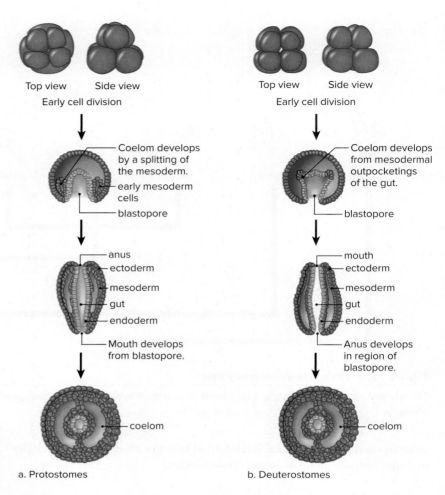

a. Protostomes

b. Deuterostomes

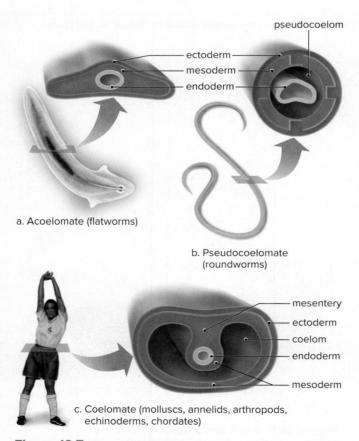

a. Acoelomate (flatworms)

b. Pseudocoelomate (roundworms)

c. Coelomate (molluscs, annelids, arthropods, echinoderms, chordates)

Figure 19.7 Type of body cavity.

a. Flatworms don't have a body cavity; they are acoelomates, and mesoderm fills the space between ectoderm and endoderm. **b.** Roundworms are pseudocoelomates; they have a body cavity, and mesoderm lies next to the ectoderm but not the endoderm. **c.** Other animals are coelomates, and mesoderm lines the entire body cavity.

Animals are also distinguished by the patterns of development as an early embryo. Notice in **Figure 19.6** that the cell division to form an embryo is different in **protostomes** (flatworms, roundworms, molluscs, annelids, and arthropods) than in **deuterostomes** (echinoderms and chordates). In both protostomes and deuterostomes, the first embryonic opening is called the blastopore. However, in protostomes the blastopore becomes the mouth, and in deuterostomes it becomes the anus. This difference explains their names: the mouth is the first opening in protostomes (*proto,* before) and the second opening in deuterostomes (*deutero,* second).

Molecular and developmental data suggest that the protostomes are more closely related to each other than they are to the deuterostomes. The protostomes are divided into several subgroups, including the lophotrochozoa and the ecdysozoa (see Fig. 19.4). The difference between these two subgroups is based on how they grow. Lophotrochozoans (flatworms, molluscs, and annelids) grow by adding mass to their existing body, while ecdysozoans (nematodes and arthropods) grow by molting. The protostomes also differ with regard to the presence of a **coelom,** or body cavity. Some protostomes have no body cavity and are **acoelomates;** an example is the flatworms (**Fig. 19.7***a*). Acoelomates are packed solid with mesoderm. In contrast, a body cavity provides a space for the various internal organs.

Roundworms are **pseudocoelomates,** and their body cavity is incompletely lined by mesoderm—that is, a layer of mesoderm exists beneath the body wall but not around the gut (Fig. 19.7*b*). The other protostomes and the deuterostomes are **coelomates** in which the body cavity is completely lined with mesoderm (Fig. 19.7*c*). In protostomes, the coelom develops by a splitting of the mesoderm, while in deuterostomes the coelom develops as outpocketings of the primitive gut (see Fig. 19.6). In animals with a coelom, mesentery, which is composed of strings of mesoderm, supports the internal organs. In coelomate animals, such as earthworms, lobsters, and humans, the mesoderm can interact not only with the ectoderm but also with the endoderm. Therefore, body movements are freer because the outer wall can move independently of the organs, and the organs have the space to become more complex. In animals without a skeleton, the coelom acts as a hydrostatic (fluid-filled) skeleton.

Animals can be nonsegmented or *segmented*. **Segmentation** is the repetition of body units along the length of the body. Annelids, such as earthworms; arthropods, such as lobsters; and chordates, such as ourselves, are segmented. To illustrate your segmentation, run your hand along your backbone, which is composed of a series of vertebrae. Segmentation leads to specialization of parts because the various segments can become differentiated for specific purposes. In the case of your backbone, the two small vertebrae just beneath your skull are specialized to permit you to move your head up and down and side to side. The vertebrae of your lower back are much larger and sturdier, as they support the weight of your upper body.

CONNECTING THE CONCEPTS

19.1 All animals have similar characteristics and can trace their ancestry back to a single-celled protist.

Check Your Progress 19.1

1. Summarize the key events in the evolution of animals.
2. Describe what distinguishes an animal from a plant or fungus.
3. Summarize the differences between protostomes and deuterostomes.

19.2 Sponges and Cnidarians: The Early Animals

Learning Outcomes

Upon completion of this section, you should be able to

1. Describe the importance of the sponges and cnidarians in the evolution of the animals.
2. Distinguish between a parazoan and a eumetazoan, and give an example of each.

This section describes the first two groups of invertebrate animals—the sponges and cnidarians. As we will see, although both of these groups are classified as simple animals, they differ significantly in their level of organization.

Sponges: Multicellularity

Sponges (phylum Porifera) have saclike bodies perforated by many pores (**Fig. 19.8**). Sponges are aquatic, largely marine animals that vary greatly in size, shape, and color. Sponges are multicellular but lack organized tissues. Therefore, sponges have a *cellular level of organization*. Molecular data classify sponges as parazoans, which are located at the base of the evolutionary tree of animals (see Fig. 19.4).

The body wall of a sponge is lined internally with flagellated cells called *collar cells,* or choanocytes (Fig. 19.8). The beating of the flagella produces

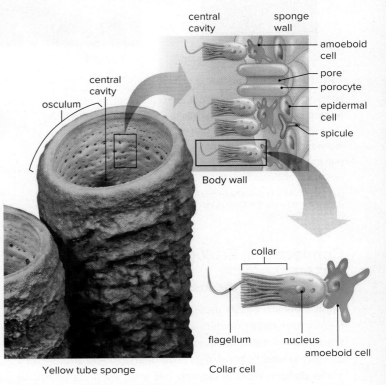

Yellow tube sponge   Collar cell

Figure 19.8 **Sponge anatomy.**

Water enters a sponge through pores and circulates past collar cells before exiting at the mouth, or osculum. Collar cells digest small particles that become trapped in the microvilli of their collar. Amoeboid cells transport nutrients from cell to cell.

© Andrew J. Martinez/Science Source

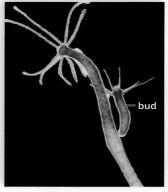

a. Hydra

tentacles

b. Sea anemone

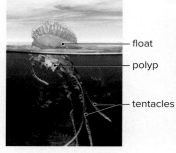

float

polyp

tentacles

c. Portuguese man-of-war

d. Cup coral

Figure 19.9 Cnidarians.

(**a**) Hydras and (**b**) sea anemones are solitary polyps that use tentacles laden with stinging cells to capture their food. **c.** The Portuguese man-of-war is a colony of polyp and medusa types of individuals. One polyp becomes a gas-filled float, and the other polyps are specialized for feeding. **d.** The calcium carbonate skeletons of corals form the coral reefs.

(a): © NHPA/M. I. Walker/Photoshot RF; (b): © Carolina Biological Supply/Phototake; (c): © NHPA/Charles Hood/Photoshot RF; (d): © Ron Taylor/Bruce Coleman/Photoshot

Connections: Ecology

What is causing the loss of the coral reefs?

Coral reefs are widely recognized as biodiversity hot spots—areas where large numbers of species can be found. However, scientists estimate that, over the past few decades, more than 25% of the world's coral reefs have been lost and another 33% are in danger. What is causing this loss? Destructive fishing practices, pollution and sediment from poor coastal land management, and the removal of coastal mangrove forests are all contributing factors. In addition, the elevation of ocean temperatures associated with global climate change is causing coral reefs to die, a phenomenon called "coral bleaching."

water currents that flow through the pores into the central cavity and out through the osculum, the upper opening of the body. Even a simple sponge only 10 centimeters tall is estimated to filter as much as 100 liters of water each day. It takes this much water to supply the needs of the sponge. A sponge is a sedentary filter feeder, an organism that filters its food from the water by means of a straining device—in this case, the pores of the walls and the microvilli making up the collar of collar cells. Microscopic food particles that pass between the microvilli are engulfed by the collar cells, which digest the food particles in food vacuoles.

Sponges can reproduce both asexually and sexually. They reproduce asexually by fragmentation or by *budding*. During budding, a small protuberance appears and gradually increases in size until a complete organism forms. Budding produces colonies of sponges that can become quite large. During sexual reproduction, eggs and sperm are released into the central cavity, and the zygote develops into a flagellated larva, which may swim to a new location. If the cells of a sponge are mechanically separated, they will reassemble into a complete and functioning organism! Like many less specialized organisms, sponges are also capable of regeneration, or growth of a whole from a small part.

Some sponges have an endoskeleton composed of *spicules,* small, needle-shaped structures with one to six rays. Most sponges have fibers of spongin, a modified form of collagen; a bath sponge is the dried spongin skeleton from which all living tissue has been removed. Today, however, commercial "sponges" are usually synthetic.

Cnidarians: True Tissues

Cnidarians (phylum Cnidaria) are an ancient group of invertebrates with a rich fossil record. Cnidarians are radially symmetrical and capture their prey with a ring of tentacles that bear specialized stinging cells, called cnidocytes (**Fig. 19.9**). Each cnidocyte has a capsule, called a **nematocyst,** containing a long, spirally coiled, hollow threadlike fiber. When the trigger of the cnidocyte is touched, the nematocyst is discharged. Some nematocysts merely trap a prey or predator; others have spines that penetrate and inject paralyzing toxins before the prey is captured and drawn into the gastrovascular cavity. Most cnidarians live in the sea, though there are a few freshwater species.

During development, cnidarians have two germ layers (ectoderm and endoderm); as adults, cnidarians have a *tissue level of organization.* Therefore, cnidarians are classified as the first of the eumetazoans. Two basic body forms are seen among cnidarians—the polyp and the medusa. The mouth of a polyp is directed upward from the substrate, while the mouth of the medusa is directed downward. A medusa has much jellylike packing material and is commonly called a "jellyfish." Polyps are tubular and generally attached to a rock (Fig. 19.9).

Cnidarians, as well as other marine animals, have been the source of medicines, particularly drugs that counter inflammation. Coral reefs are comprised of cnidarians whose calcium exoskeletons serve as the foundation for the reef. Corals form symbiotic relationships with photosynthetic protists called dinoflagellates.

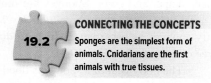

CONNECTING THE CONCEPTS

19.2

Sponges are the simplest form of animals. Cnidarians are the first animals with true tissues.

19.3 Flatworms, Molluscs, and Annelids: The Lophotrochozoans

Learning Outcomes

Upon completion of this section, you should be able to

1. List the distinguishing characteristics of a lophotrochozoan.

2. Distinguish among the flatworms, molluscs, and annelids based on body plan.

Flatworms, molluscs, and annelids all belong to the group of protostomes called lophotrochozoans. The *lopho* portion of this name is derived from a tentacle-like feeding structure called a lophophore. The *trocho* portion of the name refers to a larval stage, called a trochophore, that is characterized by a distinct band of cilia. While not all members of this group have both a lophophore and a trochophore larval stage, molecular analyses have supported the hypothesis that all lophotrochozoans share a common evolutionary ancestor. Lophotrochozoans are also distinct from other protostomes, such as the arthropods and nematodes, in that they increase their body mass gradually without molting.

Flatworms: Bilateral Symmetry

Flatworms (phylum Platyhelminthes) have *bilateral symmetry*. Like all the other animal phyla we will study, flatworms also have three germ layers. The presence of mesoderm in addition to ectoderm and endoderm gives bulk to the animal and leads to greater complexity. Flatworms are acoelomates, meaning that they lack a body cavity, or coelom.

Free-living flatworms, called **planarians,** have several body systems (**Fig. 19.10**), including a ladderlike nervous system. A small anterior brain and two lateral nerve cords are joined by cross-branches called transverse nerves. Planarians exhibit cephalization; aside from a brain, the "head" end has light-sensitive organs (the eyespots) and chemosensitive organs located on the auricles. The three muscle layers—an outer circular layer, an inner longitudinal layer, and a diagonal layer—allow for varied movement. Planarians' ciliated lower epidermis allows them to glide along on a film of mucus.

The animal captures food by wrapping itself around the prey, entangling it in slime, and pinning it down. Then the planarian extends a muscular pharynx and, by a sucking motion, tears up and swallows its food. The pharynx leads into a three-branched gastrovascular cavity, where digestion occurs. The digestive tract is incomplete because it has only one opening.

Planarians are **hermaphrodites,** meaning that they possess both male and female sex organs. The worms practice cross-fertilization: The penis of

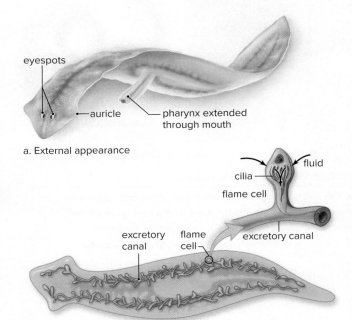

a. External appearance

b. Excretory system

c. Nervous system

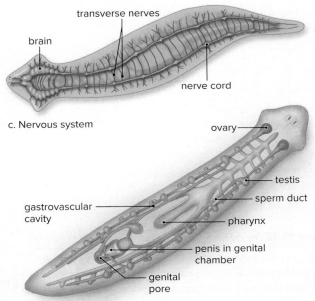

d. Reproductive and digestive systems

Figure 19.10 Anatomy of a planarian.

a. This drawing shows that flatworms are bilaterally symmetrical and have a head region with eyespots. **b.** The excretory system with flame cells is shown in detail. **c.** The nervous system has a ladderlike appearance. **d.** The reproductive system (shown in brown) has both male and female organs, and the digestive system (shown in pink) has a single opening. When the pharynx is extended, as shown in (a), the planarian sucks food up into a gastrovascular cavity, which branches throughout its body.

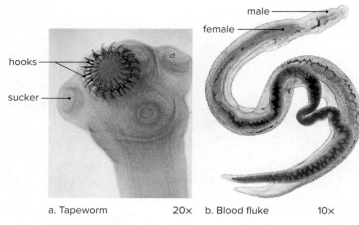

a. Tapeworm   20×   b. Blood fluke   10×

Figure 19.11 Parasitic flatworms.

a. The anterior end of a tapeworm has hooks and suckers for attaching itself to the intestinal wall of the host. **b.** Sexes are separate in blood flukes, which cause schistosomiasis.

(a): © C. James Webb/Phototake; (b): © NIBSC/SPL/Science Source

one is inserted into the genital pore of the other, and a reciprocal transfer of sperm takes place. The fertilized eggs hatch in 2–3 weeks as tiny worms.

The parasitic flatworms belong to two classes: the tapeworms and the flukes. As adults, tapeworms are endoparasites (internal parasites) of various vertebrates, including humans (**Fig. 19.11**a). They vary in size from a few millimeters to nearly 20 meters. Tapeworms have a well-developed anterior region called the scolex, which bears hooks and suckers for attaching to the intestinal wall of the host. Behind the scolex, a series of reproductive units called proglottids contain a full set of female and male sex organs. After fertilization, the organs within a proglottid disintegrate, and it becomes filled with mature eggs. The eggs, or the mature proglottids, are eliminated in the feces of the host.

Flukes are all endoparasites of various vertebrates. The anterior end of these animals has an oral sucker and at least one other sucker used for attachment to the host. Flukes are usually named for the organ they inhabit; for example, there are blood, liver, and lung flukes. Adults are small (approximately 2.5 cm long) and may live for years in their human hosts (Fig. 19.11b). Blood flukes (*Schistosoma* spp.) occur predominantly in the Middle East, Asia, and Africa. Several hundred million people each year require treatment for a blood fluke infection called schistosomiasis.

Molluscs

Molluscs (phylum Mollusca) are coelomate organisms with a complete digestive tract. Despite being a very large and diversified group, molluscs all have a body composed of at least three distinct parts: (1) The *visceral mass* is the soft-bodied portion that contains internal organs; (2) the *foot* is the strong, muscular portion used for locomotion; and (3) the *mantle* is a membranous or sometimes muscular covering that envelops, but does not completely enclose, the visceral mass. In addition, the *mantle cavity* is the space between the two folds of the mantle. The mantle may secrete an exoskeleton called a *shell*. Another feature often present is a rasping, tonguelike *radula*, an organ that bears many rows of teeth and is used to obtain food (**Fig. 19.12**).

In **gastropods** (the name means "stomach-footed"), which include nudibranchs, conchs, and snails, the foot is ventrally flattened, and the animal moves by muscle contractions that pass along the foot (**Fig. 19.13**). Many gastropods are herbivores that use their radula to scrape food from surfaces. Others are carnivores, using their radula to bore through surfaces, such as bivalve shells, to obtain food. In snails that are terrestrial, the mantle is richly supplied with blood vessels and functions as a lung.

Figure 19.12 Body plan of molluscs.

a. Molluscs have a three-part body: a muscular foot, a visceral mass, and a mantle. **b.** In the mouth, the radula is a tonguelike organ that bears rows of tiny, backward-pointing teeth.

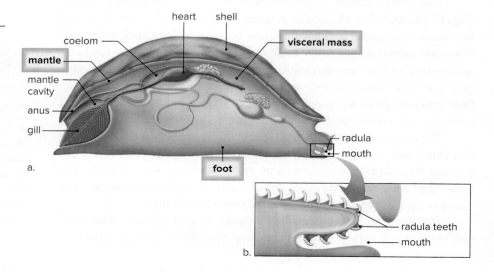

In **cephalopods** (the name means "head-footed"), which include octopuses, squids, and nautiluses, the foot has evolved into tentacles about the head (Fig. 19.13). The tentacles seize prey; then a powerful beak and radula tear it apart. Cephalopods possess well-developed nervous systems and complex sensory organs. The brain is formed from a fusion of ganglia, and nerves leaving the brain supply various parts of the body. An especially large pair of nerves control the rapid contraction of the mantle, allowing these animals to move quickly by jet propulsion of water. Rapid movement and the secretion of a brown or black pigment from an ink gland help cephalopods escape their enemies. In the squid and octopus, a well-developed eye resembles that of vertebrates, having a cornea, a lens, a retina, and an iris. Octopuses have no shell, and squids have only a remnant of one concealed beneath the skin. Scientific experiments have revealed that octopuses, in particular, are highly intelligent.

Clams, oysters, scallops, and mussels are called **bivalves** because of the two parts to their shells (Fig. 19.13). A muscular foot projects ventrally from the shell. In a clam, such as the freshwater clam, the calcium carbonate shell has an inner layer of mother-of-pearl. The clam is a filter feeder; food particles and water enter the mantle cavity by way of the incurrent siphon, a posterior opening between the two valves. Mucous secretions cause smaller particles to adhere to the gills, and ciliary action sweeps them toward the mouth.

Molluscs have some economic importance as a source of food and pearls. If a foreign body is placed between the mantle and the shell of a clam, concentric layers of shell are deposited about the particle to form a pearl.

Annelids: Segmented Worms

Annelids (phylum Annelida) are *segmented,* as can be seen externally by the rings that encircle the body of an earthworm. Partitions called septa divide the well-developed, fluid-filled coelom, which functions as a hydrostatic skeleton to facilitate movement. In annelids, the body plan includes a complete digestive tract, which has led to the specialization of parts (**Fig. 19.14**). For example, the digestive system may include pharynx, esophagus, crop, gizzard, intestine, and accessory glands. Annelids have an extensive closed circulatory system with blood vessels that run the length of the body and branch to every segment. The nervous system consists of a brain connected to a ventral nerve cord, with ganglia in each segment. The excretory system consists of nephridia in most segments. A *nephridium* is a tubule that collects waste material and excretes it through an opening in the body wall.

Most annelids are polychaetes (having many setae per segment) that live in marine environments. Setae are bristles that anchor the worm or help it move. A clam worm is a predator; it preys on crustaceans and other small animals, capturing them with a pair of strong, chitinous jaws that extend with a part of the pharynx (**Fig. 19.15***a*). In support of its predatory way of life, a clam worm has a well-defined head region, with eyes and other sense organs. Other polychaetes are sedentary (sessile) tube worms, with tentacles that form a funnel-shaped fan. Water currents created by the action of cilia carry food particles toward the mouth (Fig. 19.15*b*).

Figure 19.13 Molluscan diversity.

Top: Snails (*left*) and nudibranchs (*right*) are gastropods. *Middle:* Octopuses (*left*) and nautiluses (*right*) are cephalopods. *Bottom:* Scallops (*left*) and mussels (*right*) are bivalves.

(Land snail): © Ingram Publishing/Superstock RF; (Nudibranch): © Kenneth W. Fink/Bruce Coleman/Photoshot; (Octopus): © Alex Kerstitch/Bruce Coleman/Photoshot; (Nautilus): © Douglas Faulkner/Science Source; (Scallop): © ANT Photo Library/Science Source; (Mussels): © Michael Lustbader/Science Source

The oligochaetes (few setae per segment) include the earthworms (see Fig. 19.14). Earthworms do not have a well-developed head, and they reside in soil where there is adequate moisture to keep the body wall moist for gas exchange. They are scavengers, feeding on leaves and any other organic matter, living or dead, that can conveniently be taken into the mouth along with dirt.

Leeches have no setae but have the same body plan as other annelids. Most are found in fresh water, but some are marine or even terrestrial. The medicinal leech can be as long as 20 cm, but most leeches are much shorter. A leech has two suckers, a small one around the mouth and a large posterior one. While some leeches are free-living, most are fluid feeders that penetrate the surface of an animal by using a proboscis or their jaws and then suck in fluids with their powerful pharynx. Leeches are able to keep blood flowing and prevent clotting by means of a substance in their saliva known as hirudin, a powerful anticoagulant. This secretion adds to their potential usefulness in the field of medicine (Fig. 19.15c).

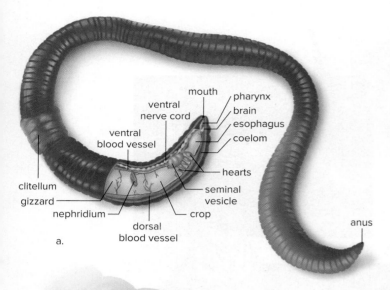
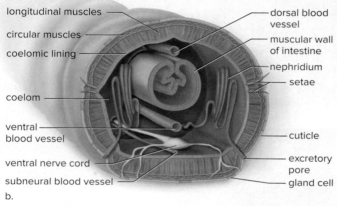

Check Your Progress 19.3

1. Explain why flatworms, molluscs, and annelids are classified as lophotrochozoans.
2. Contrast the body plans of a typical flatworm and an annelid.
3. Identify the types of body cavities found in the lophotrochozoans.

Figure 19.14 **Earthworm anatomy.**

a. Internal anatomy of the anterior part of an earthworm. **b.** Cross section of an earthworm.

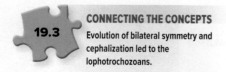

CONNECTING THE CONCEPTS
19.3 Evolution of bilateral symmetry and cephalization led to the lophotrochozoans.

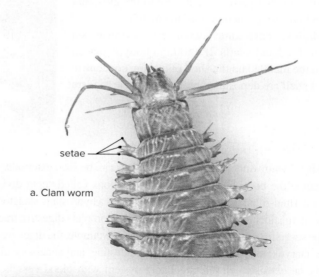

a. Clam worm

Figure 19.15 **Examples of annelids.**

A polychaete can be predatory, like (**a**) this marine clam worm, or live in a tube, like (**b**) this fan worm called the Christmas tree worm. **c.** The medicinal leech, also an annelid, is sometimes used to remove blood from tissues after surgery.

(a): © Heather Angel/Natural Visions; (b): © Diane R. Nelson; (c): © St. Bartholomews Hospital/Science Source

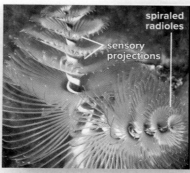

b. Christmas tree worm, *Spirobranchus giganteus*

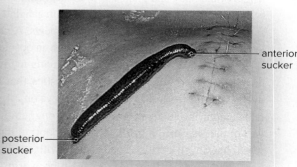

c. Medicinal leech

19.4 Roundworms and Arthropods: The Ecdysozoans

Learning Outcomes

Upon completion of this section, you should be able to

1. List the characteristics that classify an animal as an ecdysozoan.
2. Distinguish between roundworms and arthropods with regard to their body plans.
3. Explain the success of the arthropods.

Like the lophotrochozoans discussed in the previous section, the ecdysozoans are invertebrate protostomes. The name *ecdysozoan* is derived from the fact that these organisms secrete a nonliving exoskeleton (cuticle) that must be shed in order for growth to proceed. This process is called molting, or *ecdysis.* The ecdysozoans consist of the roundworms, or nematodes, and the highly successful arthropods.

Roundworms: Pseudocoelomates

The **roundworms** (phylum Nematoda) are nonsegmented, meaning that they have a smooth outside body wall. They possess a pseudocoelom that is incompletely lined with mesoderm (see Fig. 19.7). The fluid-filled pseudocoelom supports muscle contraction and enhances flexibility. The digestive tract is complete because it has both a mouth and an anus (**Fig. 19.16**). Roundworms are generally colorless and less than 5 cm long, and they occur almost everywhere—in the sea, in fresh water, and in the soil—in such numbers that thousands of them can be found in a small area. Many are free-living and feed on algae, fungi, microscopic animals, dead organisms, and plant juices, causing great agricultural damage. Parasitic roundworms live anaerobically in every type of animal and many plants. Several parasitic roundworms infect humans.

A female *Ascaris lumbricoides,* a human parasite, is very prolific, producing over 200,000 eggs daily. The eggs are eliminated with host feces, and under the right conditions, they can develop into a worm within 2 weeks. The eggs enter the body via uncooked vegetables, soiled fingers, or ingested fecal material and hatch in the intestines. The juveniles

Figure 19.16 Roundworm anatomy.

a. Roundworms, such as this male of the genus *Ascaris,* have a pseudocoelom and a complete digestive tract with a mouth and an anus. **b.** In *Ascaris* species, the sexes are separate. **c.** The larvae of the roundworm *Trichinella* penetrate striated muscle fibers, where they coil in a sheath formed from the muscle fiber.

(b): © Kim Scott/Ricochet Creative Productions LLC; (c): © John Burbidge/SPL/Science Source

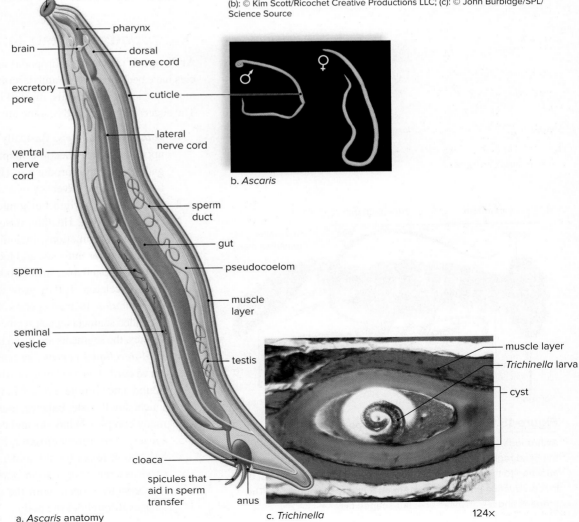

a. *Ascaris* anatomy

b. *Ascaris*

c. *Trichinella*

124×

make their way into the cardiovascular system and are carried to the heart and lungs. From the lungs, the larvae travel up the trachea, where they are swallowed and eventually reach the intestines. There the larvae mature and begin feeding on intestinal contents.

Trichinosis is a fairly serious human infection, rarely seen in the United States. The female trichina worm burrows into the wall of the host's small intestine; there she deposits live larvae, which are carried by the bloodstream to the skeletal muscles (Fig. 19.16c), where they encyst. Once the adults are in the small intestine, digestive disorders, fatigue, and fever occur. After the larvae encyst, the symptoms include aching joints, muscle pain, and itchy skin. Humans catch the disease when they eat infected meat.

Elephantiasis is caused by a roundworm called a filarial worm, which utilizes mosquitoes as a secondary host. Because the adult worms reside in lymphatic vessels, fluid return is impeded and the limbs of an infected human can swell to an enormous size, even resembling those of an elephant. When a mosquito bites an infected person, it can transport larvae to a new host.

Other roundworm infections are more common in the United States. Children are frequently infected by pinworm, and hookworm is seen in the southern states, as well as worldwide. A hookworm infection can be very debilitating because the worms attach to the intestinal wall and feed on blood. Good hygiene, proper disposal of sewage, thorough cooking of meat, and regular deworming of pets usually protect people from parasitic roundworms.

Arthropods: Jointed Appendages

Arthropods (phylum Arthropoda) are extremely diverse. Over 1 million species have been discovered and described, but some experts suggest that as many more species of arthropods, most of them insects, are waiting to be discovered. The success of arthropods can be attributed to the following six characteristics:

1. *Jointed appendages.* Basically hollow tubes moved by muscles, jointed appendages have become adapted to different means of locomotion, food gathering, and reproduction (**Fig. 19.17**). These modifications account for much of the diversity of arthropods.
2. *Exoskeleton.* A rigid but jointed exoskeleton is composed primarily of **chitin,** a strong, flexible, nitrogenous polysaccharide. The exoskeleton serves many functions, including protection, prevention of desiccation, attachment for muscles, and locomotion. Because an exoskeleton is hard and nonexpandable, arthropods must undergo molting, or shedding of the exoskeleton, as they grow larger.
3. *Segmentation.* In many species of arthropods, the repeating units of the body are called segments. Each segment has a pair of jointed appendages. In other species, the segments are fused into a head, a thorax, and an abdomen.
4. *Well-developed nervous system.* Arthropods have a brain and a ventral nerve cord. The head bears various types of sense organs, including compound and simple eyes. Many arthropods also have well-developed touch, smell, taste, balance, and hearing capabilities. Arthropods display many complex behaviors and communication skills.
5. *Variety of respiratory organs.* Marine forms utilize gills; terrestrial forms have book lungs (as do spiders) or air tubes called tracheae. Tracheae serve as a rapid way to transport oxygen directly to the cells. The circulatory system is open, with the dorsal heart pumping blood into various sinuses throughout the body.

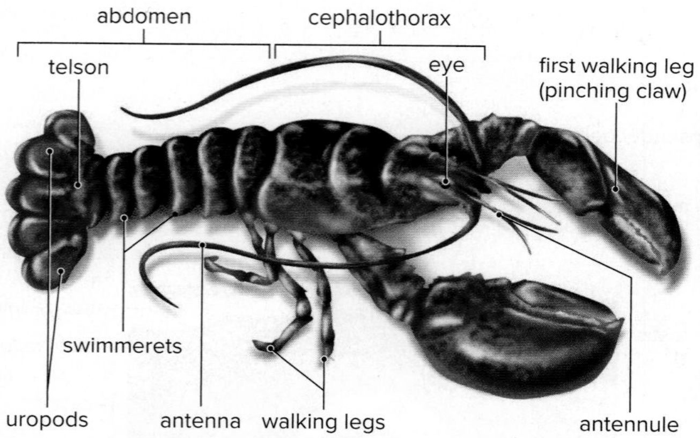

Figure 19.17   Body plan of an arthropod.

Arthropods, such as lobsters, have various appendages attached to the head region of a cephalothorax and five pairs of walking legs attached to the thorax region. Appendages called swimmerets, used in reproduction and swimming, are attached to the abdomen. The uropods and telson make up a fan-shaped tail.

6. *Metamorphosis.* Many arthropods undergo a change in form and physiology as the larva becomes an adult. Metamorphosis allows the larva to have a different lifestyle than the adult (**Fig. 19.18**). For example, larval crabs live among and feed on plankton, while adult crabs are bottom dwellers that catch live prey or scavenge dead organic matter. Among insects such as butterflies, the caterpillar feeds on leafy vegetation, while the adult feeds on nectar.

Crustaceans, whose name is derived from their hard, crusty exoskeleton, are a group of largely marine arthropods that include barnacles, shrimps, lobsters, and crabs (**Fig. 19.19**). There are also some freshwater crustaceans, including the crayfish, and some terrestrial ones, including the sowbug, or pillbug. Although crustacean anatomy is extremely diverse, the head usually bears a pair of compound eyes and five pairs of appendages. The first two pairs of appendages, called antennae and antennules, respectively, lie in front of the mouth and have sensory functions. The other three pairs are mouthparts used in feeding. In a lobster, the thorax bears five pairs of walking legs. The first walking leg is a pinching claw. The *gills* are situated above the walking legs. The head and thorax are fused into a cephalothorax, which is covered on the top and sides by a nonsegmented carapace. The abdominal segments, which are largely muscular, are equipped with swimmerets—small, paddlelike structures. The last two segments bear the uropods and the telson, which make up a fan-shaped tail to propel the lobster backward (see Fig. 19.17).

Crustaceans play a vital role in the food chain. Tiny crustaceans known as krill are a major source of food for baleen whales, sea birds, and seals. Countries such as Japan are harvesting krill for human use. Copepods and other small crustaceans are primary consumers in marine and aquatic ecosystems. Many species of lobster, crab, and shrimp are important in the seafood industry, and some barnacles are destructive to wharfs, piers, and boats.

The **arachnids** are a group of arthropods that includes spiders, scorpions, ticks, mites, and harvestmen (Fig. 19.20). Spiders have a narrow waist that separates the cephalothorax from the abdomen (**Fig. 19.20***a*). Most spiders inject venom into their prey and digest their food externally before sucking it into their stomach. Spiders use silk threads for all sorts of purposes, from lining their nests to catching prey. The internal organs of spiders also show that they are adapted to a terrestrial way of life. Malpighian tubules work in conjunction with rectal glands to reabsorb ions and water before a relatively dry nitrogenous waste (uric acid) is excreted. Invaginations of the inner body wall form

Figure 19.18 Monarch butterfly metamorphosis.

a. A caterpillar (larva) eats and grows. **b.** After the larva goes through several molts, it builds a chrysalis around itself and becomes a pupa. **c.** Inside the pupa, the larva undergoes changes in organ structure to become (**d**) an adult. **e.** The adult butterfly emerges from the chrysalis and reproduces, and the cycle begins again.

(a, c–e): © Creatas/Punchstock RF; (b): © PBNJ Productions/Getty RF

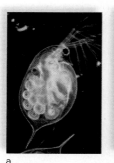
a.

b.

c.

d.

Figure 19.19 Crustacean diversity.

a. A copepod uses its long antennae for floating and its feathery maxillae for filter feeding. (**b**) Shrimp and (**c**) crabs are decapods—they have five pairs of walking legs. Shrimp resemble crayfish more closely than they do crabs, which have a reduced abdomen. Marine shrimp feed on copepods. **d.** The gooseneck barnacle is attached to an object by a long stalk. Barnacles have no abdomen and a reduced head; the thoracic legs project through a shell to filter feed. Barnacles often live on human-made objects, such as ships, buoys, and cables.

(a):© Melba Photo Agency/Punchstock RF; (b): © Ales Veluscek/Getty RF; (c): © Sandy Franz/Getty RF; (d): © L. Newman & A. Flowers/Science Source

a.
b.
c.
d.
e.

structures called book lungs, the respiratory surface of spiders and other arachnids.

Scorpions are among the oldest terrestrial arthropods (Fig. 19.20*b*) and may be the direct descendants of the first arthropods to leave aquatic environments. Today, they occur in the tropics, subtropics, and temperate regions worldwide. They are nocturnal and spend most of the day hidden under a log or rock. Ticks and mites are parasites. Ticks suck the blood of vertebrates and sometimes transmit diseases, such as Rocky Mountain spotted fever and Lyme disease. Chiggers, the larvae of certain mites, feed on the skin of vertebrates.

The horseshoe crab is grouped with the arachnids because its first pair of appendages are pincerlike structures used for feeding and defense. Horseshoe crabs have pedipalps, which they use as feeding and sensory structures, and four pairs of walking legs (Fig. 19.20*c*). Horseshoe

Figure 19.20 Arachnid diversity.

a. The black widow spider is venomous and spins a web. **b.** Scorpions have large pincers in front and a long abdomen, which ends with a stinger containing venom. **c.** Horseshoe crabs are common along the North American east coast. **d.** A millipede has two pairs of legs on most segments. **e.** A centipede has a pair of appendages on almost every segment.

(a): © Mark Kostich/Getty RF; (b): © John Bell/ Bruce Coleman/Photoshot; (c): (ventral) © Daniel Lyons/Bruce Coleman/Photoshot; (dorsal) © Ken Lucas/Ardea; (d): © Stuart Wilson/ Science Source; (e): © Larry Miller/Science Source

crabs are of great value in medical science. An extract from the blood cells of these crabs is used to ensure that vaccines are free of bacterial toxins. Other compounds from the horseshoe crab are being investigated for antibiotic, antiviral, and anticancer properties.

While millipedes (Fig. 19.20*d*), with two pairs of legs on most segments, are herbivorous, centipedes (Fig. 19.20*e*), with a pair of appendages on almost every segment, are carnivorous. The head appendages of these animals are similar to those of insects, which are the largest group of arthropods (and, indeed, of all animals).

Insects are so numerous (well over 1 million species) and so diverse that the study of this one group is a major specialty in biology called entomology (**Fig. 19.21**). Some insects show remarkable behavior adaptations, as exemplified by the social systems of bees, ants, termites, and other colonial insects.

Insects are adapted to an active life on land, although some have secondarily invaded aquatic habitats. The body is divided into a head, a thorax, and an abdomen. The head usually bears a pair of sensory antennae, a pair of compound eyes, and several simple eyes. The mouthparts are adapted to each species' particular way of life: A grasshopper has mouthparts that chew (**Fig. 19.22**), and a butterfly has a long tube for siphoning nectar from flowers.

Figure 19.21 Insect diversity.

Over 1 million insect species have been identified, but scientists estimate that many more species may exist on the planet.

(cottony cushion scale): © Nancy Nehring/Getty RF; (snout beetle): © Kjell Sandved/Bruce Coleman/Photoshot; (Housefly): © L. West/Bruce Coleman/Photoshot; (Green lacewing): © InsectWorld/Shutterstock RF; (Walking stick): © Creatas Images/PictureQuest RF; (Honeybee): © Daniel Prudek/Shutterstock RF; (Flea): © Kallista Images/Getty Images; (Aedes mosquito): Source: USDA; (Dragonfly): © Creatas Images/PictureQuest RF; (Grasshopper): © David Moyer; (Luna moth): © Charles Brutlag/Shutterstock RF

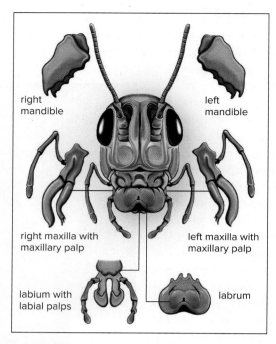

Figure 19.22 **Grasshopper mouthparts.**

The mouthparts of a grasshopper are specialized for chewing. The mouthparts of other insect species are adapted to their food source.

Connections: Health

Why are mosquitoes called vectors of disease?

A number of species of mosquitoes, such as *Aedes egypti*, are vectors of human diseases. A vector is an organism that transmits a disease between individuals. A mosquito does not directly cause disease, but rather moves the pathogen (a protist, virus, or bacterium) between individuals. Diseases such as Zika virus, chikungunya, dengue fever, and malaria all use mosquitoes as a vector.

The abdomen of an insect contains most of the internal organs; the thorax bears three pairs of legs and the wings—either one or two pairs, or none. Wings enhance an insect's ability to survive by providing a way of escaping enemies, finding food, facilitating mating, and dispersing the offspring. The exoskeleton of an insect is lighter and contains less chitin than that of many other arthropods. The male has a penis, which passes sperm to the female. The female, as in the grasshopper, may have an ovipositor for laying the fertilized eggs. Some insects, such as butterflies, undergo complete metamorphosis, involving a drastic change in form (see Fig. 19.18).

CONNECTING THE CONCEPTS

19.4 The ecdysozoans include roundworms and the most successful animals on the planet—the arthropods.

Check Your Progress 19.4

1. Describe the two anatomical features of roundworms that are not found in simpler organisms.

2. List the major characteristics of an arthropod.

3. Discuss why insects are the most numerous and diverse group of organisms on Earth.

19.5 Echinoderms and Chordates: The Deuterostomes

Learning Outcomes

Upon completion of this section, you should be able to

1. Recognize the common characteristics of echinoderms and chordates.
2. Describe the defining characteristics of echinoderms.
3. Describe the defining characteristics of the invertebrate and vertebrate chordates.
4. Describe the defining characteristics of each group of vertebrates: fishes, amphibians, reptiles, and mammals.

As shown in the evolutionary tree of animals (see Fig. 19.4), the deuterostomes (see Fig. 19.6) include the echinoderms and the chordates.

Echinoderms

Echinoderms (phylum Echinodermata) are the phylum of invertebrate animals represented by the sea stars and sea urchins (**Figure 19.23**). Although from a physical perspective they may not seem so, the echinoderms are the animals that are most closely related to the chordates, which includes vertebrates such as humans (see Fig. 19.4). This is because, like chordates, echinoderms are deuterostomes.

There are some definite differences, though. For example, the echinoderms are often radially, not bilaterally, symmetrical (Figure 19.23). The larva of an echinoderm is a free-swimming filter feeder with bilateral symmetry, but it metamorphoses into a radially symmetrical adult. Also, adult echinoderms do not have a head, a brain, or segmentation. The nervous system consists of nerves in a ring around the mouth and extending outward radially.

a. Sea lily (*above*), feather star (*right*)

b. Sea cucumber

c. Brittle star

d. Sea urchins (*left*), sand dollar (*right*)

Figure 19.23 Echinoderm diversity.

a. Sea lilies are immobile, but feather stars can move about. They usually cling to coral or sponges, where they feed on plankton. **b.** Sea cucumbers have a long, leathery body that resembles a cucumber, except for the feeding tentacles about the mouth. **c.** Brittle stars have a central disk from which long, flexible arms radiate. **d.** Sea urchins and sand dollars have spines for locomotion, defense, and burrowing.

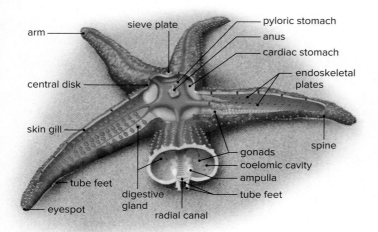

Figure 19.24 **Anatomy of a sea star.**

Many echinoderms, including this sea star, exhibit radial symmetry as adults.

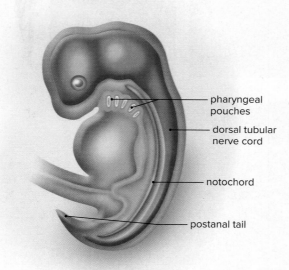

Figure 19.25 **The four chordate characteristics.**

All chordates, from tunicates to mammals, show these characteristics at some point in their lives.

Echinoderm locomotion depends on a water vascular system (**Fig. 19.24**). In the sea star, water enters this system through a sieve plate, or madreporite. Eventually, the water is pumped into many tube feet, expanding them. When the foot touches a surface, the center withdraws, producing suction that causes the foot to adhere to the surface. By alternating the expansion and contraction of its many tube feet, a sea star moves slowly along.

Echinoderms don't have complex respiratory, excretory, and circulatory systems. Fluids within the coelomic cavity and the water vascular system carry out many of these functions. For example, gas exchange occurs across the skin gills and the tube feet. Nitrogenous wastes diffuse through the coelomic fluid and the body wall.

Most echinoderms feed variously on organic matter in the sea or substratum, but sea stars prey upon crustaceans, molluscs, and other invertebrates. From the human perspective, sea stars cause extensive economic loss because they consume oysters and clams before they can be harvested. However, they are important in many ways. In many ecosystems, fishes and sea otters eat echinoderms, and scientists favor echinoderms for embryological research.

Chordates

At sometime in its life history, a **chordate** (phylum Chordata) has the four characteristics depicted in **Figure 19.25:**

1. A *dorsal supporting rod,* called a **notochord.** The notochord is located just below the nerve cord toward the back (i.e., dorsal side). Vertebrates have an endoskeleton of cartilage or bone, including a vertebral column that has replaced the notochord during development.

2. A *dorsal tubular nerve cord. Tubular* means that the cord contains a canal filled with fluid. In vertebrates, the nerve cord is protected by the vertebrae. Therefore, it is called the spinal cord because the vertebrae form the spine.

3. *Pharyngeal pouches.* These structures are seen only during embryonic development in most vertebrates. In the invertebrate chordates, the fishes, and some amphibian larvae, the pharyngeal pouches become functioning gills. Water passing into the mouth and the pharynx goes through the gill slits, which are supported by gill arches. In terrestrial vertebrates that breathe with lungs, the pouches are modified for various purposes. In humans, the first pair of pouches become the auditory tubes. The second pair become the tonsils, while the third and fourth pairs become the thymus gland and the parathyroids.

4. *A tail.* Because the tail extends beyond the anus, it is called a *postanal tail.*

The Invertebrate Chordates

In a few invertebrate chordates, the notochord is never replaced by the vertebral column. **Tunicates** live on the ocean floor and take their name from a tunic that makes the adults look like thick-walled, squat sacs. They are also called sea squirts because they squirt water from one of their siphons when disturbed (**Fig. 19.26***a*). The tunicate larva is bilaterally symmetrical and has the four chordate characteristics. Metamorphosis produces the sessile adult in which numerous cilia move water into the pharynx and out numerous gill slits, the only chordate characteristic that remains in the adult.

Lancelets are marine chordates only a few centimeters long. They have the appearance of a lancet—a small, two-edged surgical knife (Fig. 19.26*b*). Lancelets are found in the shallow water along most coasts, where they usually lie partly buried in sandy or muddy substrates with only their anterior mouth and gill apparatus exposed. They feed on microscopic particles filtered out of the constant stream of water that enters the mouth and exits through the gill slits. Lancelets retain the four chordate characteristics as adults. In addition, segmentation is present, as witnessed by the fact that the muscles are segmentally arranged and the dorsal tubular nerve cord has periodic branches.

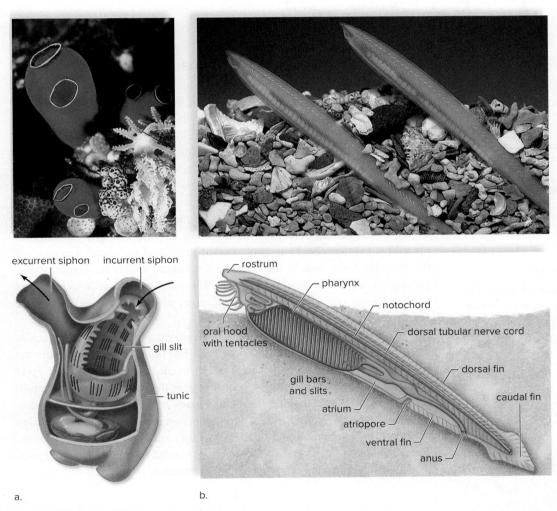

a.

b.

Figure 19.26 **Invertebrate chordates.**

a. Tunicates (sea squirts) have numerous gill slits, the only chordate characteristic that remains in the adult.

b. Lancelets have all four chordate characteristics as adults.

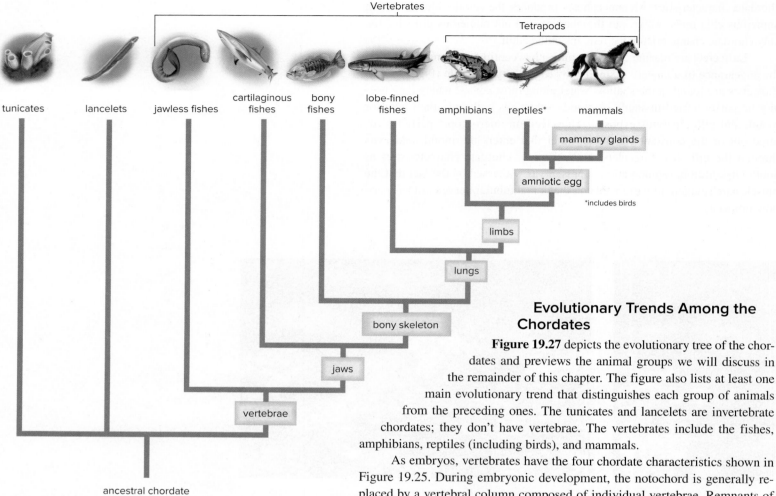

Figure 19.27   Evolutionary tree of the chordates.

Each of the innovations is shared by the classes beyond that point. For example, the bony fishes possess bony skeletons, jaws, and vertebrae.

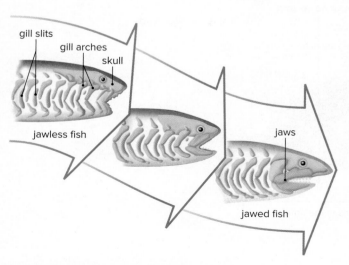

Figure 19.28   Evolution of jaws.

Jaws evolved from the anterior gill arches of ancient jawless fishes.

Evolutionary Trends Among the Chordates

Figure 19.27 depicts the evolutionary tree of the chordates and previews the animal groups we will discuss in the remainder of this chapter. The figure also lists at least one main evolutionary trend that distinguishes each group of animals from the preceding ones. The tunicates and lancelets are invertebrate chordates; they don't have vertebrae. The vertebrates include the fishes, amphibians, reptiles (including birds), and mammals.

As embryos, vertebrates have the four chordate characteristics shown in Figure 19.25. During embryonic development, the notochord is generally replaced by a vertebral column composed of individual vertebrae. Remnants of the notochord are seen in the intervertebral discs, which are compressible, cartilaginous pads between the vertebrae. The vertebral column, which is a part of the flexible but strong endoskeleton, gives evidence that vertebrates are segmented.

Fishes with an endoskeleton of cartilage and some bone in their scales were the first to have jaws. Early bony fishes had lungs. Amphibians were the first group to have jointed appendages and to invade land. The typical life cycle of amphibians includes a larval stage that lives in the water. However, amphibians exploit a wide range of habitats, and many reproduce on land. Reptiles, birds, and mammals have a means of reproduction more suited to land. During development, an amnion and other extraembryonic membranes are present. These membranes carry out all the functions needed to support the embryo as it develops into a young offspring, capable of feeding on its own.

Fishes: First Jaws and Lungs

The first vertebrates were jawless fishes, which wiggled through the water and sucked up food from the ocean floor. Today, there are three living classes of fishes: jawless fishes, cartilaginous fishes, and bony fishes. The last two groups have jaws, tooth-bearing bones of the head. Jaws are believed to have evolved from the first pair of gill arches, structures that ordinarily support gills (**Fig. 19.28**). The presence of jaws permits a predatory way of life.

Living representatives of the **jawless fishes** are cylindrical and up to a meter long. They have smooth, scaleless skin and no jaws or paired fins. The two groups of living jawless fishes are *hagfishes* and *lampreys* (**Fig. 19.29***a*). Although both are jawless, there are distinct differences. Hagfishes are an ancient group of fish. They possess a skull, but lack the vertebrae found in the other classes of vertebrates. However, molecular evidence suggests that these were once present in these fish, so they are traditionally classified with the vertebrates. Hagfishes are scavengers, feeding mainly on dead fishes. Lampreys possess a true vertebral column. Most are parasites which use their round mouth as a sucker to attach to another fish and tap into its circulatory system. Unlike other fishes, the lamprey cannot take in water through its mouth. Instead, water moves in and out through the gill openings.

The **cartilaginous fishes** are the sharks (Fig. 19.29*b*), the rays, and the skates, which have skeletons of cartilage instead of bone. The small dogfish shark is often dissected in biology laboratories. One of the most dangerous sharks inhabiting both tropical and temperate waters is the hammerhead shark. The largest sharks, the whale sharks, feed on small fishes and marine invertebrates and do not attack humans. Skates and rays are rather flat fishes that live partly buried in the sand and feed on mussels and clams.

Three well-developed senses enable sharks and rays to detect their prey: (1) They have the ability to sense electric currents in water—even those generated by the muscle movements of animals; (2) they, and all other types of fishes, have a lateral line system, a series of cells that lie within canals along both sides of the body and can sense pressure caused by a fish or another animal swimming nearby; and (3) they have a keen sense of smell—the part of the brain associated with this sense is twice as large as the other parts. Sharks can detect about one drop of blood in 115 liters (25 gallons) of water.

Bony fishes are by far the most numerous and diverse of all the vertebrates. Most of the bony fishes we eat, such as perch, trout, salmon, and haddock, are **ray-finned fishes** (**Fig. 19.30***a*). Their fins, which are used in balancing and propelling the body, are thin and supported by bony spikes. Ray-finned fishes have various ways of life. Some, such as herring, are filter feeders; others, such as trout, are opportunists; and still others, such as piranhas and barracudas, are predaceous carnivores.

Ray-finned fishes have a swim bladder (Fig 19.30*b*), which usually serves as a buoyancy organ. By secreting gases into the bladder or absorbing gases from it, these fishes can change their density, and thus go up or down in the water. The streamlined shape, fins, and muscle action of ray-finned fishes are all suited to locomotion in the water. Their skin is covered by bony scales that protect the body but do not prevent water loss. When ray-finned fishes respire, the gills are kept continuously moist by the passage of water through the mouth and out the gill slits. As the water passes over the gills, oxygen is absorbed by the blood, and carbon dioxide is given off. Ray-finned fishes have a single-circuit circulatory system. The heart is a simple pump, and the blood flows through the chambers, including a nondivided atrium and ventricle, to the gills. Oxygenated blood leaves the gills and goes to the body proper, eventually returning to the heart for recirculation.

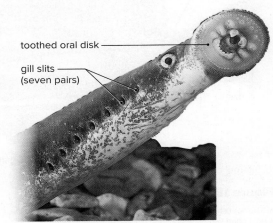

toothed oral disk

gill slits (seven pairs)

a. Lamprey, a jawless fish

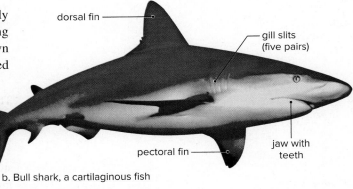

dorsal fin

gill slits (five pairs)

pectoral fin

jaw with teeth

b. Bull shark, a cartilaginous fish

Figure 19.29 **Jawless and cartilaginous fishes.**

a. The lamprey is a jawless fish. Note the toothed oral disk. **b.** The shark is a cartilaginous fish.

(a): © Heather Angel/Natural Visions; (b): © Ingram Publishing/Alamy RF

a. A lionfish

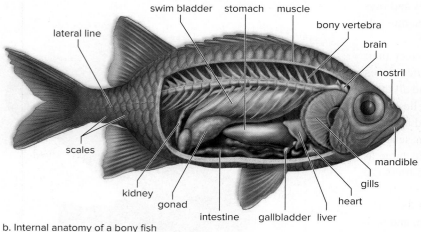

lateral line

swim bladder   stomach   muscle

bony vertebra

brain

nostril

scales

kidney

gonad

intestine   gallbladder   liver

mandible

gills

heart

b. Internal anatomy of a bony fish

Figure 19.30 Bony fish

a. The lionfish, an example of a ray-finned bony fish. **b.** The internal anatomy of a ray-finned fish.

(a): © powerofforever/Getty Rf

Another type of bony fish are the **lobe-finned fishes.** Ancient lobe-finned fishes gave rise to the amphibians (**Fig. 19.31a**). Not only did these fishes have fleshy append-ages that could be adapted to land locomotion, but most also had a **lung,** which was used for respiration.

Amphibians: Jointed Vertebrate Limbs

The first chordates to make their way to the land environment were the amphibians. **Amphibians,** whose name means that they live both on land and in the water, are represented today by frogs, toads, newts, and salamanders. Amphibians are variously col-ored. Some brightly colored ones have skin toxins that sicken or even kill predators. Usually, their color pattern is protective be-cause it allows them to remain unnoticed.

Aside from *jointed limbs* (Fig. 19.31a), amphibians have other features not seen in bony fishes: eyelids for keeping their eyes moist, a sound-producing larynx, and ears adapted to pick-ing up sound waves. Their brain is larger than that of a fish. Adult amphibians usually have small lungs: Air enters the mouth by way of nostrils, and when the floor of the mouth is raised, air is forced into the relatively small lungs. Respiration is supple-mented by gas exchange through the smooth, moist, glandular skin. The amphibian heart has only three chambers, compared with four in mammals. Mixed blood is sent to all parts of the body; some is sent to the skin, where it is further oxygenated.

Most members of this group lead an amphibious life—that is, the larval stage lives in the water and the adult stage lives on the land. While metamorphosis is a distinctive characteristic of amphib-ians, some do not demonstrate it. Some salamanders and even some frogs are direct developers; the adults do not return to the water to reproduce, and there is no tadpole stage. A great variety of reproduc-tive strategies are seen among amphibians. In the gastric-brooding frogs, the female ingests as many as 20 fertilized eggs, which un-dergo development in her stomach, and then she vomits them up as tadpoles or froglets, depending on the species. The great variety of life histories observed among amphibians made them successful colonizers of the land environment. However, water pollution and human-made chemicals, such as pesticides, have caused a drastic reduction in amphibian popula-tions worldwide. Many amphibian species are now in danger of extinction.

Reptiles: Amniotic Egg

The **reptiles** diversified and were most abundant between 245 and 66 MYA. These animals included the dinosaurs, which are remembered for their great size. *Brachiosaurus,* an herbivore, was about 23 m (75 feet) long and about 17 m (56 feet) tall. *Tyrannosaurus rex,* a carnivore, was 5 m (16 feet) tall when standing on its hind legs. The bipedal stance of some reptiles was preadaptive for the evolution of wings in birds.

The reptiles living today include turtles, crocodiles, snakes, lizards, and birds (discussed next). The typical reptilian body is covered with hard, keratin-ized scales, which protect the animal from desiccation and predators. Reptiles have well-developed lungs enclosed by a protective rib cage. Most reptiles have a three-chambered heart because a septum that divides the third chamber is incomplete. This allows some mixing of O$_2$-rich and O$_2$-poor blood in this chamber.

Perhaps the most outstanding adaptation of the reptiles is that they have a means of reproduction suitable to a land existence. The penis of the male passes sperm directly to the female. Fertilization is internal, and the female lays leathery, flexible, shelled eggs. The *amniotic egg* (**Fig. 19.32**) made development on land possible and eliminated the need for a water environment during development. The amniotic egg provides the developing embryo with atmospheric oxygen, food, and water; removes nitrogenous wastes; and protects the embryo from drying out and from mechanical injury. This is accomplished by the presence of extraembryonic membranes, such as the chorion.

Fishes, amphibians, and reptiles (other than birds) are **ectothermic,** meaning that their body temperature matches the temperature of the external environment. If it is cold externally, their internal body temperature drops; if it is hot externally, their internal body temperature rises. Most reptiles regulate their body temperature by exposing themselves to the sun if they need warmth and hiding in the shadows if they need cooling off.

Feathered Reptiles: Birds

Birds share a common ancestor with crocodiles and have traits such as a tail with vertebrae, clawed feet, and the presence of scales that show they are, in fact, reptiles. Perhaps you have noticed the scales on the legs of a chicken; in addition, a bird's *feathers* are actually modified reptilian scales. However, birds lay a hard-shelled amniotic egg rather than the leathery egg produced by other reptiles. The exact history of birds is still in dispute, but recent discoveries of fossils of feathered reptiles in China and other locations indicate that birds are closely related to bipedal dinosaurs and that they should be classified as such.

Nearly every anatomical feature of a bird can be related to its ability to fly (**Fig. 19.33**). The forelimbs are modified as wings. The hollow, very light bones are laced with air cavities. A horny beak has replaced jaws equipped with teeth, and a slender neck connects the head to a rounded, compact torso. Respiration is efficient, since the lobular lungs form anterior and posterior air sacs. The presence of these sacs means that the air moves one way through the lungs, and gases are continuously exchanged across respiratory tissues. Another benefit of air sacs is that they lighten the body and aid flying.

Birds have a four-chambered heart that completely separates O$_2$-rich blood from O$_2$-poor blood. Birds are **endothermic** and generate internal heat. Many endotherms can use metabolic heat to maintain a constant internal temperature. This may be associated with their efficient nervous, respiratory, and circulatory systems. Also, their feathers provide insulation. Birds have no bladder and excrete uric acid in a semidry state.

Birds have particularly acute vision and well-developed brains. Their muscle reflexes are excellent. These adaptations are suited to flight. An enlarged portion of the brain seems to be the area responsible for instinctive behavior. A ritualized courtship often precedes mating. Many newly hatched birds require parental care before they are able to fly away and seek food for

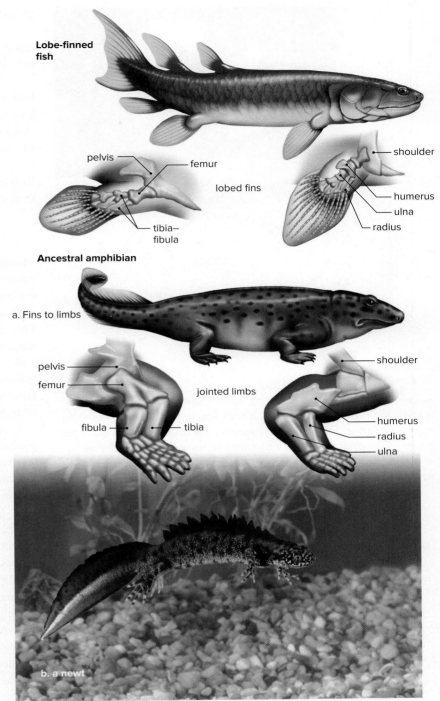

Figure 19.31 Evolution of amphibians.

a. A lobe-finned fish compared with an amphibian. A shift in the position of the bones in the forelimbs and hindlimbs lifted and supported the body. **b.** Newts (shown here) and frogs are types of living amphibians.

(b): © imageBroker/Alamy RF

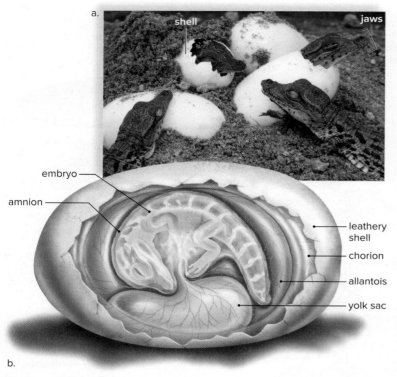

Figure 19.32 The amniotic egg.

a. A baby American crocodile hatching out of its shell. Note that the shell is leathery and flexible—not brittle, like a bird's egg. **b.** Inside the egg, the embryo is surrounded by extraembryonic membranes. The chorion aids gas exchange, the yolk sac provides nutrients, the allantois stores waste, and the amnion encloses a fluid that prevents drying out and provides protection.

(a): © Heinrich van den Berg/Gallo Images/Getty Images

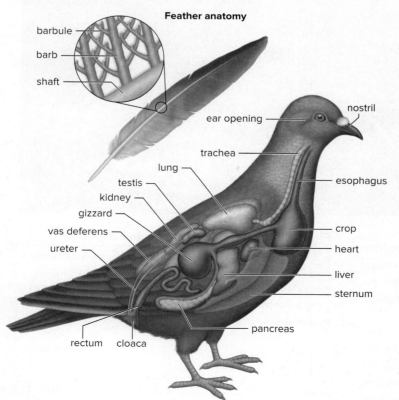

Figure 19.33 Birds are well adapted for flight.

Feathers and hollow bones are important adaptations for birds. This figure illustrates the internal anatomy of a typical bird.

(photo): © Tony Camacho/ SPL/Science Source

themselves. A remarkable aspect of bird behavior is the seasonal migration of many species over very long distances. Birds navigate by day and night, whether it's sunny or cloudy, by using the sun and stars and even the Earth's magnetic field to guide them.

Traditionally, the classification of birds was particularly based on beak and foot types (**Fig. 19.34**) and to some extent on habitat and behavior. The various orders include birds of prey with notched beaks and sharp talons; shorebirds with long, slender, probing bills and long, stilt-like legs; woodpeckers with sharp, chisel-like bills and grasping feet; waterfowl with webbed toes and broad bills; penguins with wings modified as flippers; and songbirds with perching feet. Genetic data are now used to determine relationships among birds.

Mammals: Hair and Mammary Glands

Mammals (class Mammalia) are amniotes, and as such, they share a common ancestor with reptiles. However, mammals represent a separate evolutionary lineage from the reptile lineage that led to the birds. The first mammals evolved during the Triassic period (about 199.6 MYA). They were small, about the size of mice. Due to many factors, including the dominance of the dinosaurs, mammals changed little during the Triassic and Jurassic (about 145.5 MYA) periods. Some of the earliest mammalian groups are still represented by the monotremes and marsupials. However, neither of these groups is as abundant as the placental mammals, which can be found in most environments on the planet.

The two chief characteristics of mammals are hair and milk-producing mammary glands. Almost all mammals are endotherms and generate internal heat. Many of the adaptations of mammals are related to temperature control. Hair, for example, provides insulation against heat loss and allows mammals to be active, even in cold weather.

Mammary glands enable females to feed (nurse) their young without leaving them to find food. Nursing also creates a bond between mother and offspring that helps ensure parental care while the young are helpless. In most mammals, the young are born alive after a period of development in the uterus, a part of the female reproductive system. Internal development shelters the young and allows the female to move actively about while the young are maturing.

Monotremes are mammals that, like birds, have a *cloaca,* a terminal region of the digestive tract serving as a common chamber for feces, excretory wastes, and sex cells. They also lay hard-shelled amniotic eggs. They are represented by the spiny anteater and the duckbill platypus, both of which live in Australia. The female duckbill platypus lays her eggs in a burrow in the ground. She incubates the eggs, and after hatching, the young lick up milk that seeps from mammary glands on her abdomen. The spiny anteater has a pouch on the belly side formed by swollen mammary glands and longitudinal muscle (**Fig. 19.35***a*). Hatching takes place in this pouch, and the young remain there for about 53 days. Then they stay in a burrow, where the mother periodically visits and nurses them.

a. Cardinal, *Cardinalis cardinalis*   b. Flamingo, *Phoenicopterus ruber*   c. Bald eagle, *Haliaetus leucocephalus*

Figure 19.34 Bird beaks.

a. A cardinal's beak allows it to crack tough seeds. **b.** A flamingo's beak strains food from the water with bristles that fringe the mandibles. **c.** A bald eagle's beak allows it to tear prey apart.

(a): © Graeme Teague; (b): © Medford Taylor/Getty Images; (c): © Dale DeGabriele/Getty Images

The young of **marsupials** begin their development inside the female's body, but they are born in a very immature condition. Newborns crawl up into a pouch on their mother's abdomen. Inside the pouch, they attach to the nipples of mammary glands and continue to develop. Frequently, more are born than can be accommodated by the number of nipples.

The Virginia opossum is the only marsupial north of Mexico (Fig. 19.35*b*). In Australia, however, marsupials underwent adaptive radiation for several

a.

Figure 19.35 Monotremes and marsupials.

a. The spiny anteater is a monotreme that lives in Australia. **b.** The opossum is the only marsupial in the United States. The Virginia opossum is found in a variety of habitats. **c.** The koala is an Australian marsupial that lives in trees.

(a): © B.G. Thompson/Science Source; (b): © John Macgregor/Photolibrary/Getty Images; (c): © John White Photos/Getty RF

b.   c.

Figure 19.36 Placental mammals.

Placental mammals have adapted to various ways of life. **a.** Deer are herbivores that live in forests. **b.** Lions are carnivores on the African plains. **c.** Monkeys typically inhabit tropical forests. **d.** Whales are sea-dwelling placental mammals.

(a): © Paul E. Tessier/Getty RF; (b): © Gallo Images–Gerald Hinde/Getty Images; (c): © David F. Cox, Howler Publications Ltd.; (d): © 2011 Tory Kallman/Getty RF

d.

a.

b.

c.

CONNECTING THE CONCEPTS

19.5

Echinoderms and chordates are both deuterostomes. Chordates have shared characteristics, including a dorsal notochord.

Check Your Progress 19.5

1. List the unique features of the echinoderms.
2. List the common characteristics of chordates.
3. Summarize the innovations of each class of chordates.
4. Explain the importance of the amniotic egg.
5. Identify the reason for the diversity of mammals.

million years without competition. Thus, marsupial mammals are now found mainly in Australia, with some in Central and South America as well. Among the herbivorous marsupials, koalas are tree-climbing browsers (Fig. 19.35c), and kangaroos are grazers. The Tasmanian wolf or tiger, now known to be extinct, was a carnivorous marsupial about the size of a collie dog.

The vast majority of living mammals are **placental mammals** (**Fig. 19.36**). In these mammals, the extraembryonic membranes of the amniotic egg (see Fig. 19.32) have been modified for internal development within the uterus of the female. The chorion contributes to the fetal portion of the placenta, while a part of the uterine wall contributes to the maternal portion. Here, nutrients, oxygen, and waste are exchanged between fetal and maternal blood.

Mammals are adapted to life on land and have limbs that allow them to move rapidly. The brain is well developed; the lungs are expanded not only by the action of the rib cage but also by the contraction of the diaphragm, a horizontal muscle that divides the thoracic cavity from the abdominal cavity; and the heart has four chambers. The internal temperature is constant, and hair, when abundant, helps insulate the body.

The mammalian brain is enlarged due to the expansion of the cerebral hemispheres that control the rest of the brain. The brain is not fully developed until after birth, and the young learn to take care of themselves during a period of dependency on their parents.

Mammals can be distinguished by their method of obtaining food and their mode of locomotion. For example, bats have membranous wings supported by digits; horses have long, hoofed legs; and whales have paddlelike forelimbs. The specific shape and size of the teeth may be associated with whether the mammal is an herbivore (eats vegetation), a carnivore (eats meat), or an omnivore (eats both meat and vegetation). For example, mice have continuously growing incisors; horses have large, grinding molars; and dogs have long canine teeth.

19.6 Human Evolution

The evolutionary tree in **Figure 19.37** shows that all primates share one common ancestor and that the other types of primates diverged from the human line of descent (called a **lineage**) over time. The **prosimians** include the lemurs, tarsiers, and lorises. The **anthropoids** include the monkeys, apes, and humans. The designation **hominid** includes the apes (African and Asian), chimpanzees, humans, and closest extinct relatives of humans. The term **hominin** refers to our species, *Homo sapiens,* and our close humanlike ancestors.

Figure 19.37 Evolutionary tree of primates.

Primates are descended from an ancestor that may have resembled a tree shrew. The descendants of this ancestor adapted to the new way of life and developed traits such as a shortened snout and nails instead of claws. The time when each type of primate diverged from the main line of descent is known from the fossil record. A common ancestor was living at each point of divergence. For example, a common ancestor for hominids (humans, apes, and chimpanzees) existed about 7 MYA and one for anthropoids about 45 MYA.

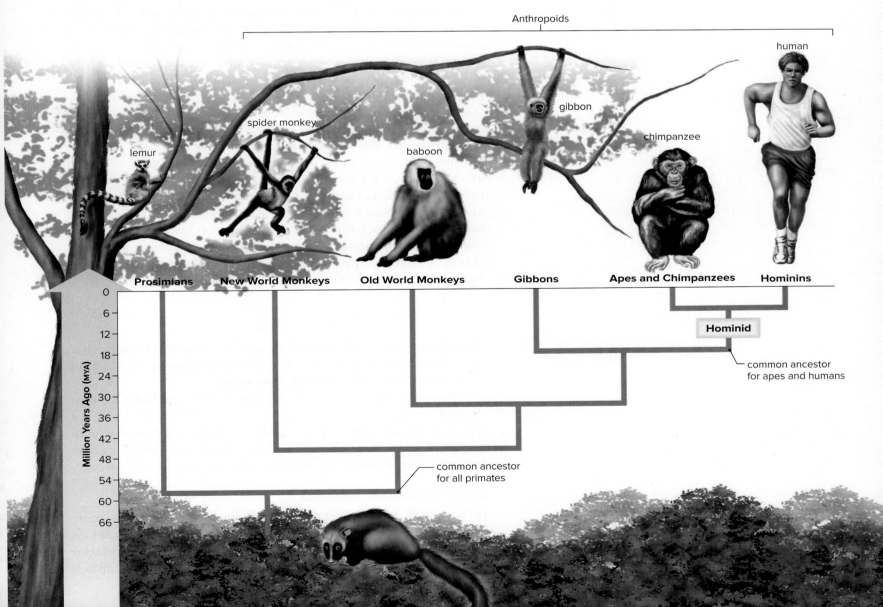

Primates are adapted to an arboreal life—life in trees. Primate limbs are mobile, and the hands and feet have five digits each. Many primates have both an opposable big toe and an opposable thumb—that is, the big toe and the thumb can touch each of the other toes or fingers. Humans don't have an opposable big toe, but the thumb is opposable, resulting in a grip that is both powerful and precise. The opposable thumb allows a primate to easily reach out and bring food, such as fruit, to the mouth. When mobile, primates grasp and release tree limbs freely because nails have replaced claws.

The evolutionary trend among primates is generally toward a larger and more complex brain—the brain is smallest in prosimians and largest in modern humans. In humans, the cerebral cortex has many association areas and has expanded so much that it has become extensively folded. The portion of the brain devoted to smell got smaller, and the portions devoted to sight increased in size and complexity during primate evolution. Also, more and more of the brain is involved in controlling and processing information received from the hands and the thumb. The result is good hand-eye coordination in humans.

Notice that prosimians were the first type of primate to diverge from the human line of descent, and apes and chimpanzees were the last group to diverge from our line of descent. The evolutionary tree also indicates that humans are most closely related to these primates. One of the most unfortunate misconceptions concerning human evolution is that Darwin and others suggested that humans evolved from apes. On the contrary, humans and apes share a common, apelike ancestor. Today's apes are our distant cousins, and we couldn't have evolved from our cousins, because we are contemporaries—living on Earth at the same time. Dating the last common ancestor for apes and humans is an active area of research, but most researchers estimate that this ancestor lived about 7 MYA.

There have been many recent advances in the study of the hominins, and recent discoveries of fossils in Africa are challenging our view of how early hominins evolved. Paleontologists use certain anatomical features when they try to determine if a fossil is a hominin. These features include **bipedalism** (walking on two feet), the shape of the face, and brain size. Today's humans have a flatter face and a more pronounced chin than do the apes, because the human jaw is shorter than that of the apes. Then, too, our teeth are generally smaller and less specialized. We don't have the sharp canines of an ape, for example. Chimpanzees have a brain size of about 400 cubic centimeters (cc), and modern humans have a brain size of about 1,360 cc.

Evolution of Humanlike Hominins

All the fossils shown in **Figure 19.38** are humanlike hominins. The bars in this figure extend from the approximate date of a species' appearance in the fossil record to the date it is believed to have gone extinct.

It is important to note that the evolution of our genus, *Homo,* has been studied extensively in the past several years. Increasingly, there is evidence that what were once considered separate species are, in fact, variations of a single species. Recent discoveries in Africa, specifically of *Homo neladi,* are challenging the dates for and the relationships between the species of our genus.

Early Humanlike Hominins

In Figure 19.38, early hominins are represented by orange-colored bars. Scientists have found several fossils dated around the time the ape lineage and the human lineage are believed to have split, and one of these is *Sahelanthropus tchadensis.* Only the braincase has been found and dated at 7 MYA. Although

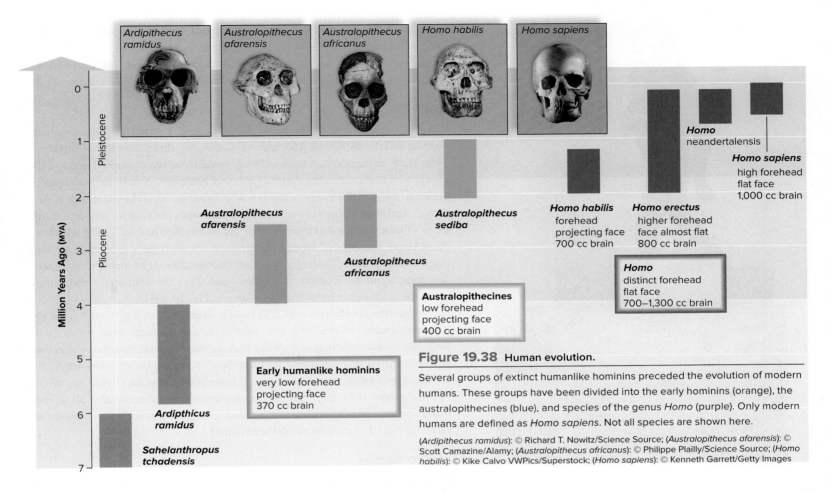

Figure 19.38 Human evolution.

Several groups of extinct humanlike hominins preceded the evolution of modern humans. These groups have been divided into the early hominins (orange), the australopithecines (blue), and species of the genus *Homo* (purple). Only modern humans are defined as *Homo sapiens*. Not all species are shown here.

(*Ardipithecus ramidus*): © Richard T. Nowitz/Science Source; (*Australopithecus afarensis*): © Scott Camazine/Alamy; (*Australopithecus africanus*): © Philippe Plailly/Science Source; (*Homo habilis*): © Kike Calvo VWPics/Superstock; (*Homo sapiens*): © Kenneth Garrett/Getty Images

the braincase is very apelike, the location of the opening for the spine at the back of the skull suggests bipedalism. Also, the canines are smaller and the tooth enamel is thicker than in an ape.

Another early hominin, *Ardipithecus ramidus,* is representative of the ardipithecines of 4.5 MYA. The fossil remains of this species are commonly called "Ardi." Reconstructions of the fossils (**Fig. 19.39**) suggest that the

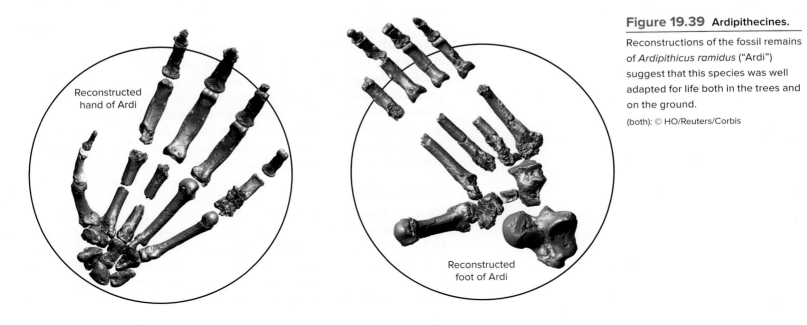

Figure 19.39 Ardipithecines.

Reconstructions of the fossil remains of *Ardipithicus ramidus* ("Ardi") suggest that this species was well adapted for life both in the trees and on the ground.

(both): © HO/Reuters/Corbis

a.

b.

Figure 19.40 *Australopithecus afarensis.*

a. A reconstruction of Lucy on display at the St. Louis Zoo. **b.** These fossilized footprints occur in ash from a volcanic eruption some 3.7 MYA. The larger footprints are double, and a third, smaller individual was walking to the side. (A female holding the hand of a youngster may have been walking in the footprints of a male.) The footprints suggest that *A. afarensis* walked bipedally.

(a): © Dan Dreyfus and Associates; (b): © John Reader/Science Source

species was bipedal and that some individuals may have been 122 cm tall. The teeth seem intermediate between those of earlier apes and those of later hominins. Recently, fossils dated at 4 MYA show a direct link between *A. ramidus* and the australopithecines, discussed next.

Australopithecines

It's possible that one of the **australopithecines,** a group of hominins that evolved and diversified in Africa about 4 MYA, is a direct ancestor of humans. More than 20 years ago, a team led by Donald Johanson unearthed nearly 250 fossils of a hominin called *Australopithecus afarensis.* A now famous female skeleton dated to 3.18 MYA is known worldwide by its field name, Lucy. Although her brain was quite small (400 cc), the shapes and relative proportions of Lucy's limbs indicate that she stood upright and walked bipedally (**Fig. 19.40***a*). Even better evidence of bipedal locomotion comes from a trail of footprints dated about 3.7 MYA. The larger prints are double, as though a smaller individual was stepping in the footfalls of another—and there are additional small prints off to the side, within hand-holding distance (Fig. 19.40*b*).

One species, *Australopithecus sediba,* demonstrates that the transition to bipedalism occurred gradually over time in the australopithecines. Analysis of limb size and the method of walking in *A. sediba* suggests that changes in the pelvis occurred at a different rate than those in the limbs. In many ways, *A. sediba* is apelike above the waist (small brain) and humanlike below the waist (walked erect), indicating that human characteristics did not evolve all at one time. The term *mosaic evolution* is applied when different body parts change at different rates and therefore at different times.

Homo habilis

Homo habilis, dated between 2.4 and 1.4 MYA, may be ancestral to modern humans. Some of these fossils have a brain size as large as 775 cc, which is about 45% larger than the brain of *A. afarensis.* The cheek teeth are smaller than those of any of the australopithecines. Therefore, it is likely that this early *Homo* was omnivorous and ate meat in addition to plant material. Bones at the campsites of *H. habilis* bear cut marks, indicating that these hominins, whose name means "handyman," used tools to strip the meat from bones. The rather crude tools, probably also used to scrape hide and cut tendons, consisted of sharp flakes of broken rocks.

The skulls of *H. habilis* suggest that the portions of the brain associated with speech areas were enlarged. We can speculate that the ability to speak may have led to hunting cooperatively. Some members of the group may have remained plant gatherers; if so, both hunters and gatherers most likely ate together and shared their food. In this way, society and culture could have begun. **Culture,** which encompasses human behavior and products (such as technology and the arts), is dependent on the capacity to speak and transmit knowledge. We can further speculate that the advantages that having a culture brought to *H. habilis* may have hastened the extinction of the australopithecines.

Homo erectus

Fossils of **Homo erectus** have been found in Africa, Asia, and Europe and dated between 1.9 and 0.15 MYA. Although all fossils assigned the name *H. erectus* are similar in appearance, there is enough discrepancy to suggest that several different species have been included in this group. Compared with *H. habilis, H. erectus* had a larger brain (about 1,000 cc) and a flatter face. The recovery of an almost complete skeleton of a 10-year-old boy indicates that

H. erectus was much taller than the earlier hominins. Males were 1.8 m tall (about 6 feet), and females were 1.55 m (approaching 5 feet). Indeed, these hominins were erect and most likely had a striding gait, like that of modern humans. The robust and most likely heavily muscled skeleton still retained some australopithecine features. Even so, the size of the birth canal indicates that infants were born in an immature state that required an extended period of care.

It is believed that *H. erectus* first appeared in Africa and then migrated into Asia and Europe (**Figure 19.41**). At one time, the migration was thought to have occurred about 1 MYA, but recently *H. erectus* fossil remains in Java and the Republic of Georgia have been dated at 1.9 and 1.6 MYA, respectively. These remains push the evolution of *H. erectus* in Africa to an earlier date that has yet to be determined. In any case, such an extensive population movement is a first in the history of humankind and a tribute to the intellectual and physical skills of the species.

H. erectus was most likely the first hominin to use fire, and it fashioned more advanced tools than earlier *Homo* species. These hominins used heavy, teardrop-shaped axes and cleavers, as well as stone flakes that were probably used for cutting and scraping. Some investigators believe *H. erectus* was a systematic hunter that took kills to the same site over and over again. In one location, researchers have found over 40,000 bones and 2,647 stones. These sites could have been "home bases," where social interaction occurred and a prolonged childhood allowed time for learning. Perhaps a language evolved and a culture more like our own developed.

Connections: Scientific Inquiry

What are the "hobbits"?

In 2004, a new piece of our human heritage, *Homo floresiensis,* was discovered on the Indonesian island of Flores. Sometimes called "hobbits" due to their very small size (less than a meter in height), *H. floresiensis* also had a very small brain, but it is believed that they used stone tools. Most scientists believe that *H. floresiensis* is more closely related to *H. erectus* than to *Homo sapiens.* Interestingly, some dating techniques suggest that *H. floresiensis* may have occupied Flores as little as 15,000 years ago. If so, then it most likely would have interacted with *H. sapiens* in the not too distant past.

Evolution of Modern Humans

The most widely accepted hypothesis for the evolution of modern humans from early humanlike hominins is referred to as the **replacement model,** or out-of-Africa hypothesis (**Fig. 19.42**), which proposes that modern humans evolved from archaic humans only in Africa and then migrated to Asia and Europe, where they replaced the archaic species about 100,000 years BP (before the present). However, even this hypothesis is being challenged as new genomic information on the Neandertals and Denisovans becomes available.

Neandertals

Neandertals (*Homo neandertalensis*) take their name from Germany's Neander Valley, where one of the first Neandertal skeletons, dated some 200,000 years BP, was discovered. The Neandertals had massive brow ridges, and their nose, jaws, and teeth protruded far forward. Their forehead was low and sloping, and their lower jaw lacked a chin. New fossils show that their pubic bone was longer than that of modern humans.

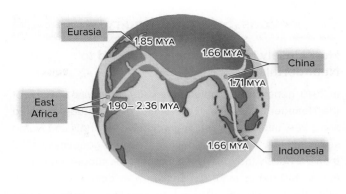

Figure 19.41 Migration of Homo erectus.

The dates indicate the migration of early *Homo erectus* from Africa.

Derived from "Evolution of Early Homo: An Integrated Biological Perspective," S. Antón et al., Science 4 July 2014: 345 (6192)

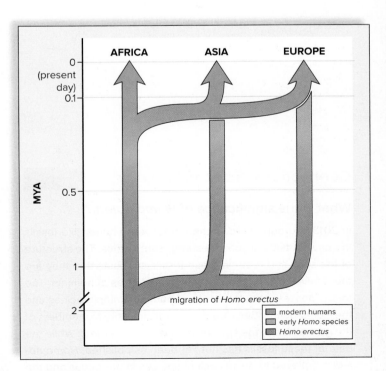

Figure 19.42 The replacement model.

According to the replacement model, modern humans evolved in Africa and then replaced early *Homo* species in Asia and Europe.

Connections: Scientific Inquiry

How closely is *Homo sapiens* related to Neandertals?

In 2010, geneticists completed their first sequence analysis of the Neandertal genome. The results of this study revealed some very interesting facts regarding Neandertals and *Homo sapiens*. First, these two species are more genetically alike than previously thought. The initial analysis suggested that as few as 100 genes in *H. sapiens* show evidence of evolution since the Neandertals died out, making the Neandertals one of our closest cousins. Second, some studies suggest that there is evidence that Neandertals and *H. sapiens* might have interbred. Although this is still being investigated, we know that these two species occupied overlapping territories in the Middle East and Europe for almost 14,000 years. Since Neandertals and *H. sapiens* were genetically similar, many scientists believe that interbreeding may have been possible, although research is continuing into other explanations for these similarities.

Connections: Scientific Inquiry

What is the significance of *Homo naledi?*

In 2013, a group of researchers discovered humanlike fossils in a cave outside of Johannesburg, South Africa. The structure of the skull and teeth in these fossils suggests that they are the remains of a previously unknown species of hominin. The fossils possess characteristics of both australopithecines and the *Homo* genus. Because of a closer similarity to members of the genus *Homo,* the species is called *Homo naledi.* While the date of these fossils has not yet been established, *Homo naledi* is believed to be an ancient species of our genus and the study of its characteristics could shed light on key events in the early evolution of our species.

According to the replacement model, Neandertals were eventually supplanted by modern humans. However, this traditional view is being challenged by studies of the Neandertal genome (completed in 2010), which suggests not only that Neandertals interbred with *Homo sapiens* but also that between 1% and 4% of the genomes of nonAfrican *H. sapiens* contain remnants of the Neandertal genome. Some scientists are suggesting that Neandertals were not a separate species but simply a race of *H. sapiens* that was eventually absorbed into the larger population. Research continues into these and other hypotheses that explain these similarities.

Physiologically, the Neandertal brain was, on the average, slightly larger than that of *H. sapiens* (1,400 cc, compared with 1,360 cc in most modern humans). The Neandertals were heavily muscled, especially in the shoulders and neck. The bones of the limbs were shorter and thicker than those of modern humans. It is hypothesized that a larger brain than that of modern humans was required to control the extra musculature. The Neandertals lived in Europe and Asia during the last Ice Age, and their sturdy build could have helped their bodies conserve heat.

The Neandertals give evidence of being culturally advanced. Most lived in caves, but those living in the open may have built houses. They manufactured a variety of stone tools, including spear points, which could have been used for hunting. Scrapers and knives could have helped in food preparation. They most likely successfully hunted bears, woolly mammoths, rhinoceroses, reindeer, and other contemporary animals. They used and could control fire, which probably helped them cook meat and keep themselves warm. They even buried their dead with flowers and tools and may have had a religion. Perhaps they believed in life after death. If so, they were capable of thinking symbolically.

Denisovans

In 2008, a fragment of a finger bone was discovered in Denisova cave in southern Siberia. Initially, scientists thought that it might be the remains of a species of early *Homo,* possibly related to *Homo erectus.* However, mitochondrial DNA studies indicate that the fossil belonged to a species that existed around 1 MYA, around the same time as Neandertals. The analyses suggest that the **Denisovans** and Neandertals had a common ancestor but did not interbreed with one another, possibly because of their geographic locations. However, what is interesting is the fact that *Homo sapiens* in the Oceania region (New Guinea and nearby islands) share around 5% of their genomes with the Denisovans. In 2014, researchers reported that an allele that allows for high-elevation living in Tibetans originated with the Denisovans. When coupled with the Neandertal data, this suggests that modern *H. sapiens* did not simply replace groups of archaic humans but rather may have assimilated them by inbreeding. Scientists are just beginning to unravel the implications of the Denisovan discovery.

Cro-Magnons

Cro-Magnons represent the oldest fossils to be designated *H. sapiens.* Cro-Magnons, who are named after a location in France where their fossils were first found, had a thoroughly modern appearance. They made advanced tools, including compound tools, such as stone flakes fitted to a wooden handle. They may have been the first to make knifelike blades and to throw spears, enabling them to kill animals from a distance. They were such accomplished hunters that some researchers believe they were responsible for the extinction of many larger mammals, such as the giant sloth, mammoth, saber-toothed tiger, and giant ox, during the late Pleistocene epoch.

Cro-Magnons hunted cooperatively and were perhaps the first to have a language. They are believed to have lived in small groups, with the men hunting by day while the women remained at home with the children, gathering and processing food items. Probably, the women also were engaged in maintenance tasks. The Cro-Magnon culture included art. They sculpted small figurines out of reindeer bones and antlers. They also painted beautiful drawings of animals, some of which have survived on cave walls in Spain and France.

CONNECTING THE CONCEPTS

19.6 Humans evolved from primate ancestors. Their success can be attributed to bipedalism and an increase in brain size.

Check Your Progress 19.6

1. Distinguish between hominins and anthropoids.

2. Explain the importance of bipedalism and brain size in the evolution of humans.

3. Summarize how the replacement model explains the evolution of modern humans.

STUDY TOOLS http://connect.mheducation.com

SMARTBOOK® Maximize your study time with McGraw-Hill SmartBook®, the first adaptive textbook.

SUMMARIZE

The evolution of animals follows trends that can be traced in the evolutionary histories of all animals, from sponges to humans.

19.1 All animals have similar characteristics and can trace their ancestors back to a single-celled protist.

19.2 Sponges are the simplest form of animals. Cnidarians are the first animals with true tissues.

19.3 Evolution of bilateral symmetry and cephalization led to the lophotrochozoans.

19.4 The ecdysozoans include roundworms and the most successful animals on the planet—the arthropods.

19.5 Echinoderms and chordates are both dueterostomes. Chordates have shared characteristics, including a dorsal notochord.

19.6 Humans evolved from primate ancestors. Their success can be attributed to bipedalism and an increase in brain size.

19.1 Evolution of Animals

Animals are motile, multicellular heterotrophs that ingest their food.

Ancestry of Animals

The choanoflagellates are protists that most likely resemble the last single-celled ancestor of animals. A colony of flagellated cells could have led to a multicellular animal that formed tissues by invagination.

The Evolutionary Tree of Animals

The evolutionary tree of animals is chiefly based on molecular and developmental data.

The majority of animals are **invertebrates** (lack an endoskeleton). Some, like humans are **vertebrates** (possess an endoskeleton).

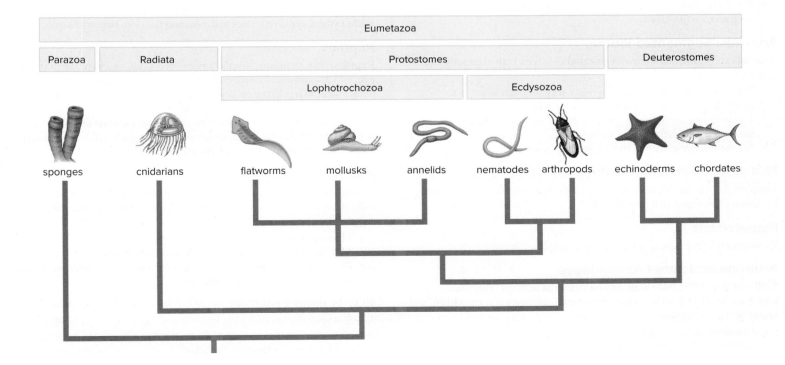

Evolutionary Trends

There are several important trends in the evolution of animals:

- In eumetazoans, multicellularity led to the formation of **germ layers** (ectoderm, mesoderm, endoderm).
- **Radial symmetry** preceded **bilateral symmetry.** Bilateral symmetry led to cephalization, or formation of a head containing a brain and sensory receptors.
- Bilaterally symmetrical animals undergo either **protostome** development or **deuterostome** development. In protostomes, the first embryonic opening becomes the mouth. In deuterostomes, the second embryonic opening becomes the mouth.
- Animals may be further identified as **coelomates, pseudocoelomates,** or **acoelomates,** depending on the structure of their body cavity.
- Some coelomate animals exhibit **segmentation,** or repetition of parts of the body.

19.2 Sponges and Cnidarians: The Early Animals

Sponges are multicellular (lack tissues) and have various symmetries. **Cnidarians** have two tissue layers, are radially symmetrical, have a saclike digestive cavity, and possess stinging cells and **nematocysts.**

19.3 Flatworms, Molluscs, and Annelids: The Lophotrochozoans

Lophotrochozoans are characterized by feeding structure and larval development. They increase their body mass gradually without molting.

Flatworms

Flatworms (such as planarians) have ectoderm, endoderm, and mesoderm, but no coelom. They are bilaterally symmetrical and have a saclike digestive cavity. Many are **hermaphrodites.**

Molluscs

The body of a **mollusc** typically contains a visceral mass, a mantle, and a foot.

- **Gastropods:** Snails, representatives of this group, have a flat foot, a one-part shell, and a mantle cavity that carries on gas exchange.
- **Cephalopods:** Octopuses and squids display marked cephalization, move rapidly by jet propulsion, and have a closed circulatory system.
- **Bivalves:** Bivalves, such as clams, have a muscular foot and a two-part shell and are filter feeders.

Annelids: Segmented Worms

Annelids are segmented worms; segmentation is seen both externally and internally.

- Polychaetes are worms that have many setae. A clam worm is a predatory marine worm with a well-defined head region.
- Earthworms are oligochaetes that scavenge for food in the soil and do not have a well-defined head region.
- Leeches are annelids that feed by sucking blood.

19.4 Roundworms and Arthropods: The Ecdysozoans

The ecdysozoans are invertebrate protostomes that increase their body mass by molting their exoskeleton, or cuticle.

Roundworms

Roundworms have a pseudocoelom and a complete digestive tract.

Arthropods: Jointed Appendages

Arthropods are the most varied and numerous of animals. Their success is largely attributable to a flexible exoskeleton (composed mostly of **chitin**) and specialized body regions.

- **Crustaceans** have a head that bears compound eyes, antennae, and mouthparts. Five pairs of walking legs are present.

- **Arachnids** include spiders, scorpions, ticks, mites, harvestmen, and horseshoe crabs. Spiders live on land and spin silk, which they use to capture prey as well as for other purposes.
- **Insects** have three pairs of legs attached to the thorax. Insects have adaptations to a terrestrial life.

19.5 Echinoderms and Chordates: The Deuterostomes

Echinoderms

Echinoderms have radial symmetry as adults (not as larvae) and endoskeletal spines. Typical echinoderms have tiny skin gills, a central nerve ring with branches, and a water vascular system for locomotion, as exemplified by the sea star.

Chordates

Chordates have a **notochord,** a dorsal tubular nerve cord, pharyngeal pouches, and a postanal tail at some time in their life history.

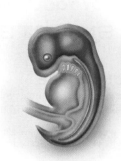

- Invertebrate chordates: Adult **tunicates** lack chordate characteristics except gill slits, but adult **lancelets** have the four chordate characteristics and show obvious segmentation.
- Vertebrate chordates: Vertebrate chordates include the fishes, amphibians, reptiles (and birds), and mammals.

Fishes

- The first vertebrates, the **jawless fishes** (represented by hagfishes and lampreys), lacked jaws and fins.
- **Cartilaginous fishes,** represented by sharks and rays, have jaws and a skeleton made of cartilage.
- **Bony fishes** have jaws and fins supported by bony spikes; the bony fishes include the **ray-finned fishes** and **lobe-finned fishes.** Some of the lobe-finned fishes have **lungs.**

Amphibians

Amphibians, such as frogs, toads, newts, and salamanders, evolved from ancient lobe-finned fishes and have two pairs of jointed vertebrate limbs. Frogs usually return to the water to reproduce and metamorphose into terrestrial adults.

Reptiles

Reptiles, such as snakes, lizards, turtles, crocodiles, and birds, lay a shelled egg, which contains extraembryonic membranes, including an amnion that allows them to reproduce on land. **Birds** are feathered reptiles, which helps them maintain a constant body temperature. They are adapted for flight; their bones are hollow, with air cavities; lungs form air sacs that allow one-way ventilation; and they have well-developed sense organs. All reptiles, except birds, are **ectothermic.** Birds are **endothermic.**

Mammals

Mammals are amniotes that have hair and mammary glands. The former helps them maintain a constant body temperature, and the latter allow them to nurse their young.

- **Monotremes** lay eggs.
- **Marsupials** have a pouch in which the newborn matures.
- **Placental mammals,** which are far more varied and numerous, retain offspring inside the uterus until birth.

19.6 Human Evolution

Arboreal primates are mammals adapted to living in trees. During the evolution of primates, various groups diverged in a particular sequence from the line of descent, or **lineage.** Important classifications of primates include

- Prosimians—lemurs and tarsiers
- Anthropoids—monkeys, apes, and humans
- Hominids—apes, chimpanzees, and humans
- Hominins—humans and their direct humanlike ancestors, all of which exhibit **bipedalism**

Evolution of Humanlike Hominins

Human evolution began in Africa, where it has been possible to find the remains of several early hominins that date back to 7 MYA. The evolution of hominins includes these innovations:

- Bipedal: The most famous **australopithecine,** Lucy (3.18 MYA), had a small brain but walked bipedally.
- Tool use: *Homo habilis,* present about 2 MYA, is certain to have made tools and may have been the first to exhibit **culture.**
- Increased brain size: *Homo erectus,* with a brain capacity of 1,000 cc and a striding gait, was the first to migrate out of Africa.

Evolution of Modern Humans

The **replacement model** proposes that modern humans originated in Africa and, after migrating into Europe and Asia, replaced the other *Homo* species (including the **Neandertals** and possibly the **Denisovans**) found there. **Cro-Magnons** were the first hominins to be considered *Homo sapiens.*

ASSESS

Testing Yourself

Choose the best answer for each question.

19.1 Evolution of Animals

1. Which of these protists is hypothesized to be ancestral to animals?
   a. a green algal protist
   b. a choanoflagellate
   c. an amoeboid protist
   d. a slime mold
2. Which of the following is not a feature of a coelomate?
   a. radial symmetry
   b. three germ layers
   c. a body plan including a complete digestive tract
   d. organ level of organization
3. The evolution of bilateral symmetry allowed for
   a. multicellularity.
   b. germ layers.
   c. the notochord.
   d. cephalization.

19.2 Sponges and Cnidarians: The Early Animals

4. Sponges are classified as
   a. eumetazoans.
   b. parazoans.
   c. protists.
   d. vertebrates.
5. Cnidarians are organized at the tissue level because they contain
   a. ectoderm and endoderm.
   b. ectoderm.
   c. ectoderm and mesoderm.
   d. endoderm and mesoderm.
   e. mesoderm.

19.3 Flatworms, Molluscs, and Annelids: The Lophotrochozoans

6. A feature of annelids is
   a. a segmented body.
   b. a coelom.
   c. a saclike body.
   d. radial symmetry.
7. What characteristic is common to the flatworms, molluscs, and annelids?
   a. All have bilateral symmetry.
   b. All are acoelomates.
   c. All are parazoans.
   d. All are vertebrates.

8. Gastropods and cephalopods are examples of
   a. molluscs.
   b. annelids.
   c. flatworms.
   d. None of these are correct.

19.4 Roundworms and Arthropods: The Ecdysozoans

9. Which of the following is not a feature of an insect?
   a. compound eyes
   b. eight legs
   c. antennae
   d. an exoskeleton
   e. jointed legs
10. Elephantiasis and trichinosis are diseases caused by
    a. mites.
    b. spiders.
    c. insects.
    d. roundworms.

19.5 Echinoderms and Chordates: The Deuterostomes

11. Which of the following is not a feature of mammals?
    a. hair
    b. milk-producing glands
    c. ectothermic capability
    d. four-chambered heart
    e. diaphragm to help expand lungs
12. Unlike bony fishes, amphibians have
    a. ears.
    b. jaws.
    c. a circulatory system.
    d. a heart.
13. Mammals are distinguished from other animals based on
    a. size and hair type.
    b. mode of reproduction.
    c. number of limbs and method of caring for young.
    d. number of mammary glands and number of offspring.

19.6 Human Evolution

14. The first humanlike feature to evolve in the hominins was
    a. a large brain.
    b. massive jaws.
    c. a slender body.
    d. bipedal locomotion.
15. Which of the following is an anthropoid but not a hominid?
    a. human
    b. gibbon
    c. orangutan
    d. chimpanzee
    e. gorilla
16. The replacement model supports which of the following in the evolution of modern humans?
    a. evolution of bipedalism
    b. evolution of brain size
    c. distribution of modern humans across the planet
    d. development of culture

ENGAGE

Thinking Critically

1. What do the discoveries of *Homo florensiensis, Homo naledi,* and the Denisovans tell you about our understanding of human evolution? Refer back to Figure 19.38 and propose periods of time on which researchers should focus their efforts in searching for rock deposits for new fossil remains in order to understand the process of human evolution.
2. Think of the animals in this chapter that are radially symmetrical (such as cnidarians and many adult echinoderms). How is their lifestyle different from that of bilaterally symmetrical animals? How does their body plan complement their lifestyle?
3. The development of more rapid DNA-sequencing techniques is changing our view of the organization of the animal kingdom. Why is genetic evidence a more powerful tool for systematics than the physical appearance or characteristics of organisms?

20

Plant Anatomy and Growth

© Graeme Teague

OUTLINE

20.1 Plant Cells and Tissues   373

20.2 Plant Organs   375

20.3 Organization of Leaves, Stems, and Roots   377

20.4 Plant Nutrition   385

20.5 Transport of Nutrients   387

BEFORE YOU BEGIN

Before beginning this chapter, take a few moments to review the following discussions.

Section 5.4 How do diffusion and osmosis affect the transport of water and solutes between cells?

Section 6.1 What are the inputs and outputs of photosynthesis?

Section 18.1 What are the major adaptations of plants to the land environment?

Carnivorous Plants Are Adapted to Harsh Conditions

Pitcher plants are unusual in that they like to grow in areas that lack nitrogen and phosphorus, which are required nutrients for plants. What's their secret? Pitcher plants feed on animals, particularly insects, to get those necessary nutrients. That's right—like a few other types of plants, pitcher plants are carnivorous.

A pitcher plant is named for its leaves, which are shaped like a container we call a pitcher. The pitcher attracts flying insects because it has a scent, is brightly colored, and provides nectar that insects can eat. When an insect lands on the lip of the pitcher, a slippery substance and downward-pointing hairs encourage it to slide into a pit. There, juices secreted by the leaf begin digesting the insect. The plant then absorbs these nutrients into its tissues. Although the pitcher is designed to attract insects, animals as large as rats have been found in pitcher plants! The unique adaptations of pitcher plants allow them to thrive in hostile environments where few plants can survive.

Some animals are able to turn the tables on the pitcher plant by taking advantage of the liquid it provides. Larvae of mosquitoes and flies have been known to develop safely inside a pitcher plant.

Carnivorous plants, such as the pitcher plant, are fairly unusual in the plant world, but so are many others, such as the coastal redwoods—the tallest trees on planet Earth. In this chapter, you will learn about the organs and structures of flowering plants, the nutrients they need, and how they transport those nutrients within their tissues.

As you read through this chapter, think about the following questions:

1. How does each plant organ contribute to the life of the plant?
2. What nutrients do plants need, and how are those nutrients acquired from the environment?
3. How are nutrients distributed throughout the plant body?

20.1 Plant Cells and Tissues

Learning Outcomes

Upon completion of this section, you should be able to

1. List the three types of specialized tissues in plants, and describe their functions.
2. Describe the role of epidermal tissue in a plant.
3. List the three types of ground tissue, and compare and contrast their structures and functions.
4. Explain how the vascular tissues of a plant are organized to move water and nutrients within a plant.

Plants have levels of biological organization similar to those of animals (see Fig. 1.2). As in animals, a *cell* is a basic unit of life, and a *tissue* is composed of specialized cells that perform a particular function. An *organ* is a structure made up of multiple tissues.

When a plant embryo first begins to develop, the first cells are called meristem cells (**Fig. 20.1**). Meristem cells organize into **meristem tissue,** allowing a plant to grow its entire life. Even a 5,000-year-old tree is still growing! Early on, meristem tissue is present at the very top and the very bottom of a plant. Because these areas of tissue are located at the ends, they are called the **apical meristems.** Apical meristems develop (differentiate) into the three types of specialized tissues of the plant body:

1. **Epidermal tissue** forms the outer protective covering of a plant.
2. **Ground tissue** fills the interior of a plant and helps carry out the functions of a particular organ.
3. **Vascular tissue** transports water and nutrients in a plant and provides support.

Plants not only grow up and down, but they can also grow wider. The *vascular cambium* is another type of meristem that gives rise to new vascular tissue called *secondary growth*. Secondary growth causes a plant to increase in girth.

Epidermal Tissue

The entire body of a plant is covered by an **epidermis,** a layer of closely packed cells that act as a barrier, similar to skin. The walls of epidermal cells that are exposed to air are covered with a waxy **cuticle** to minimize water loss. The cuticle also protects against bacteria and other organisms that might cause disease. Epidermal cells can be modified (changed) into other types of cells. **Root hairs** are long, slender projections of epidermal cells that increase the surface area of the root for absorption of water and minerals (**Fig. 20.2a**). In leaves, the epidermis often contains **stomata** (sing., stoma). A stoma is a small opening surrounded by two guard cells (Fig. 20.2b). When the stomata are open, gas exchange and water loss occur. *Trichomes* are another type of epidermal cell that make plant leaves and stems feel prickly or hairy and discourage insects from eating the plant (Fig. 20.2c).

In the trunk of a tree, the epidermis is replaced by cork, which is a part of bark (Fig. 20.2d). New cork cells are made by meristem tissue called *cork cambium.* As the new cork cells mature, they increase slightly in volume, and their walls become encrusted with *suberin,* a lipid material, so that they are

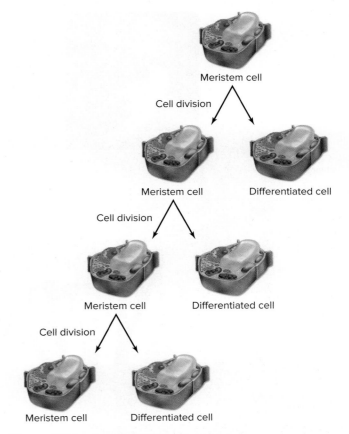

Figure 20.1 **Meristem cell division.**

Meristem cells are located in new and developing parts of a plant. Meristem cells divide and eventually differentiate, making up specialized plant tissues, such as the epidermal, ground, and vascular tissues.

Connections: Scientific Inquiry

How is paper made?

Most paper is made from the cellulose fibers of trees. Recall from Section 3.2 that cellulose is a component of the cell wall of plants. After a tree is harvested, it is debarked and cut into small chips to increase the surface area. However, before the cellulose can be extracted from the cells, the lignin and resins must be removed. This is done either mechanically or by using a combination of steam and sulfur-based chemicals to separate the lignin and resins from the cellulose fibers. In the next step, the lignins and resins are removed by bleaching with chlorine gas (or chlorine dioxide), which also makes the fibers white. In the final steps, fibers are dried and treated in a machine that forms rolls of paper for commercial use.

© Comstock Images/Jupiterimages

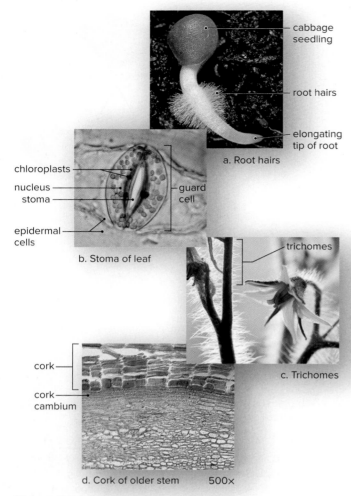

chloroplasts

nucleus

stoma

epidermal
cells

guard
cell

a. Root hairs

cabbage
seedling

root hairs

elongating
tip of root

b. Stoma of leaf

trichomes

c. Trichomes

cork

cork
cambium

d. Cork of older stem   500×

Figure 20.2   **Modifications of epidermal tissue.**

a. Root epidermis has root hairs, which help absorb water and
minerals. **b.** Leaf epidermis contains stomata (sing., stoma) for gas
exchange. **c.** Trichomes are hairy extensions of the leaf or stem and
protect the plant from herbivory. **d.** Cork, which is a part of bark,
replaces epidermis in older, woody stems.

(a): © Nigel Cattlin/Alamy; (b): © John Hardy, University of Washington, Stevens
Point Department of Biology; (c): © Anthony Pleva/Alamy; (d): © Biophoto
Associates/Science Source

waterproof and chemically inert. These nonliving cells protect the plant and
make it resistant to attack by fungi, bacteria, and animals.

Ground Tissue

Ground tissue forms the internal bulk of leaves, stems, and roots. Ground tissue
contains three types of cells (**Fig. 20.3**). *Parenchyma* cells are the least specialized of the cell types and are found in all the organs of a plant. They may contain chloroplasts and carry on photosynthesis, or they may contain colorless
organelles that store the products of photosynthesis. *Collenchyma* cells are like
parenchyma cells except that they have irregularly shaped corners and thicker
cell walls. Collenchyma cells often form bundles just beneath the epidermis
and give flexible support to immature regions of a plant body. The familiar
strands in celery stalks are composed mostly of collenchyma cells. *Sclerenchyma* cells have thick secondary cell walls containing *lignin,* which makes
plant cell walls tough and hard. If we compare a cell wall to reinforced concrete, cellulose fibrils would play the role of steel rods, and lignin would be
analogous to the cement. Most sclerenchyma cells are nonliving; their primary
function is to support the mature regions of a plant. The hard outer shells of
nuts are made of sclerenchyma cells. The long fibers in plants, composed of
strings of sclerenchyma, make them useful for a number of purposes. For example, cotton and flax fibers can be woven into cloth, and hemp fibers can
make strong rope.

Vascular Tissue

Vascular tissue extends from the root through the stem to the leaves, and vice
versa. In the root, the vascular tissue is located in a central cylinder; in the
stem, vascular tissue can be found in multiple vascular bundles; and in the
leaves, it is found in leaf veins. Although both types of vascular tissue are usually found together, they have different functions. **Xylem** transports water and
minerals from the roots to the leaves. **Phloem** transports sugar, in the form of
sucrose, and other organic compounds, such as hormones, often from the
leaves to the roots.

   Xylem contains two types of conducting cells: **vessel elements** and
tracheids (**Fig. 20.4***a*). Both types of conducting cells are hollow and nonliving,
but the vessel elements are larger, have perforated end walls, and are arranged
to form a continuous pipeline for water and mineral transport. The end walls
and side walls of tracheids have pits that allow water to move from one tracheid
to another.

Figure 20.3   **Ground tissue cells.**

a. Parenchyma cells are the least specialized of the
plant cells. **b.** Collenchyma cells have more irregular
corners and thicker cell walls than parenchyma cells.
c. Sclerenchyma cells have very thick walls and are
nonliving—their primary function is to give support
to mature regions of a plant.

(a-c): © Biophoto Associates/Science Source

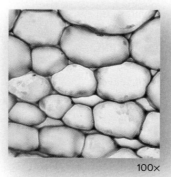

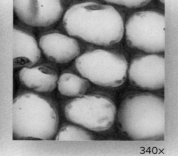

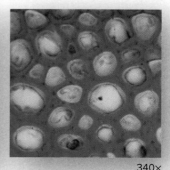

100×   340×   340×

a. Parenchyma cells   b. Collenchyma cells   c. Sclerenchyma cells

The conducting cells of phloem are **sieve-tube members,** which are named because they contain a cluster of pores in their end walls. The pores are collectively known as a *sieve plate*. The sieve-tube members are arranged to form a continuous sieve tube (Fig. 20.4*b*). Sieve-tube members contain cytoplasm but no nuclei. Each sieve-tube member has a **companion cell,** which does have a nucleus. The companion cells may very well be involved in the transport function of phloem.

CONNECTING THE CONCEPTS

20.1

Plants have meristem cells that form specialized epidermal, ground, and vascular tissues.

Check Your Progress 20.1

1. List the three types of tissue in a plant, and summarize the functions of each.
2. Contrast the function of xylem with that of phloem.
3. Describe the modifications that may occur in epidermal tissue.

20.2 Plant Organs

Learning Outcomes

Upon completion of this section, you should be able to

1. Define primary growth, and describe where this growth occurs.
2. Describe the root system and the shoot system.
3. Compare and contrast the seeds, roots, stems, leaves, and flowers of monocots and eudicots.

As you learned in Section 18.1, the earliest plants were simple and lacked true leaves, stems, and roots. As plants gained vascular tissue and began moving onto land away from water, organs developed to facilitate living in drier environments. Even though all vascular plants have vegetative organs, this chapter focuses on the organs commonly found in flowering plants, or angiosperms.

A flowering plant, whether a cactus, a daisy, or an apple tree, has a shoot system and a root system (**Fig. 20.5**). The **shoot system** consists of the stem, leaves, flowers, and fruit. A stem supports the leaves, transports materials between the roots and leaves, and produces new tissue.

Lateral (side) branches grow from a *lateral bud* located at the angle where a leaf joins the stem. A *node* is the location where leaves, or the buds for branches, are attached to the stem. An *internode* is the region between nodes. At the end of a stem, a *terminal bud* contains an apical meristem and produces new leaves and other tissues during *primary growth* (**Fig. 20.6**). Vascular tissue transports water and minerals from the roots through the stem to the leaves and transports the products of photosynthesis, usually in the opposite direction.

The **root system** simply consists of the roots. The root tip also contains an apical meristem, which produces primary growth downward. Primary growth at the terminal bud and root tip would be equivalent to your increasing in height by growing from your head and your feet! Ultimately, the three vegetative organs—the leaf, the stem, and the root—perform functions that allow a plant to live and grow. Flowers and fruit are reproductive organs and will be discussed in Section 21.3.

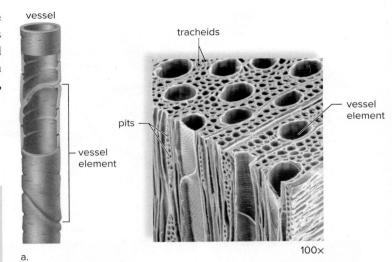

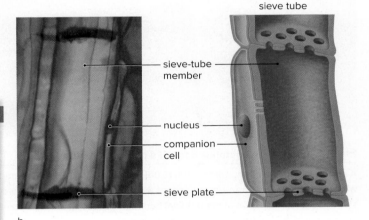

Figure 20.4 Vascular tissue.

a. Photomicrograph of xylem vascular tissue, with drawings of a vessel (composed of vessel elements) (*left*) and of tracheids (*right*).
b. Photomicrograph of phloem vascular tissue, with drawing of a sieve tube (*right*). Each sieve-tube member has a companion cell.

(a): © N.C. Brown Center for Ultrastructure Studies, SUNY, College of Environmental Science & Forestry, Syracuse, NY; (b): © Randy Moore/BioPhot

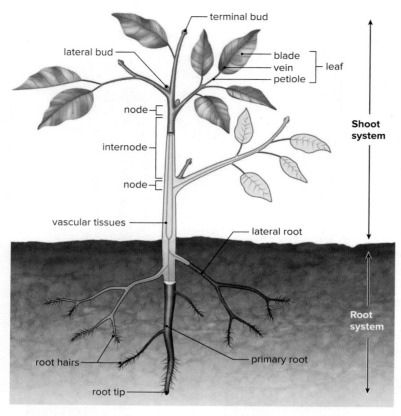

Figure 20.5 **The vegetative body of a plant.**

A plant has a root system, which extends below ground, and a shoot system, which is composed of the stem and leaves (and flowers and fruit, which are not shown here). Cell division at the terminal bud and the root tip allows a plant to increase in length, a process called primary growth.

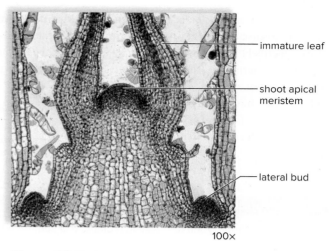

Figure 20.6 **A terminal bud.**

A terminal bud contains an apical meristem and produces new leaves and other tissues during primary growth. Lateral (side) branches grow from a lateral bud.

© Steven P. Lynch

Monocots Versus Eudicots

Plant organs are arranged in different patterns depending on the type of flowering plant. **Figure 20.7** shows how flowering plants can be divided into two major groups. One of the differences between the two groups concerns the **cotyledons,** which are embryonic leaves present in seeds. The cotyledons wither after the first true leaves appear. Plants whose embryos have one cotyledon are known as monocotyledons, or **monocots.** Other plant embryos have two

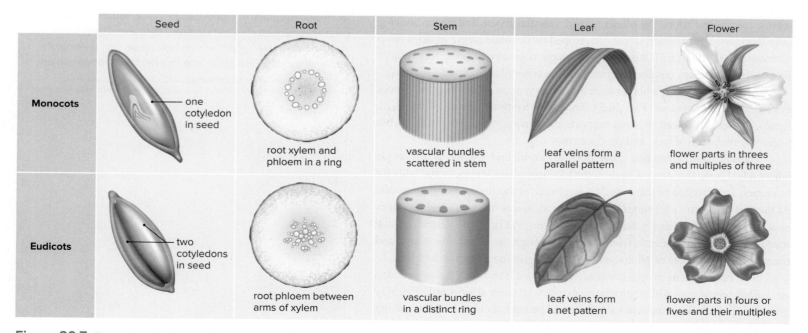

Figure 20.7 **Flowering plants are either monocots or eudicots.**

Five features are typically used to distinguish monocots from eudicots: number of cotyledons; arrangement of vascular tissue in roots, stems, and leaves; and number of flower parts.

cotyledons, and these plants are known as eudicotyledons, or **eudicots** (true dicots). The cotyledons of eudicots supply nutrients for seedlings, but the cotyledons of monocots store some nutrients and act as a transfer tissue for nutrients stored elsewhere.

Although the distinction between monocots and eudicots may seem minimal at first glance, there are many differences in their structures. For example, a microscopic view would show that the location and arrangement of vascular tissue differ between monocots and eudicots. Recall that vascular plants contain two main types of transport tissue: the *xylem* for water and minerals and the *phloem* for organic nutrients. In a sense, xylem and phloem are to plants what veins and arteries are to animals. In the monocot root, vascular tissue occurs in a ring around the center. But in the eudicot root, vascular tissue is located in the center. The xylem forms a star shape, and phloem is located between the points of the star. In a stem, vascular tissue occurs in **vascular bundles.** The vascular bundles are scattered in the monocot stem, and they occur in a ring in the eudicot stem.

To the naked eye, the vascular tissue forms *leaf veins*. In monocots, the veins are parallel, while in eudicots, the leaf veins form a netlike pattern. Monocots and eudicots also have different numbers of flower parts, and this difference will be discussed further in Section 21.3.

The eudicots are the larger group and include some of our most familiar flowering plants—from dandelions to oak trees. The monocots include grasses, lilies, orchids, and palm trees, among others. Some of our most significant food sources are monocots, including rice, wheat, and corn.

CONNECTING THE CONCEPTS

20.2 Plant tissues combine to form vegetative organs that make up the shoot system and the root system.

Connections: Scientific Inquiry

Do "magical" plants really exist?

© blickwinkel/Alamy

Strange and poisonous plants have been made famous by popular series such as *Harry Potter* and *The Hunger Games*. But do these types of plants really exist? Yes! Mandrake roots look like a human body, and giant hogweed leaves will cause painful blisters to sprout on the skin. Cuckoopint ("bloody man's finger") will cause the tongue to swell, and the black berries of the deadly nightshade are indeed lethal.

Check Your Progress 20.2

1. Describe the difference in function between the root and shoot systems of a plant.
2. Define *cotyledon,* and explain its function.
3. List the differences between monocots and eudicots.

20.3 Organization of Leaves, Stems, and Roots

Learning Outcomes

Upon completion of this section, you should be able to

1. List the structures found in a typical eudicot leaf.
2. Contrast nonwoody stems with woody stems.
3. Explain how secondary growth of a woody eudicot stem occurs, and contrast spring wood with summer wood.
4. List the three zones of a eudicot root tip.

Leaves

Leaves are the chief organs of photosynthesis, and as such they require a supply of solar energy, carbon dioxide, and water. Broad and thin foliage leaves maximize the surface area to collect sunlight and absorb carbon dioxide. Leaves receive water from the root system by way of vascular tissue that terminates in the leaves. *Deciduous* plants lose their leaves, often due to a yearly dry season or the onset of winter. Other trees, called *evergreens,* retain their leaves for the entire year.

Figure 20.8 shows the general structure of a leaf. The wide portion of a foliage leaf is called the *blade*. The *petiole* is a stalk that attaches the blade to the stem. The blade of a leaf is often undivided, or *simple,* such as a cottonwood

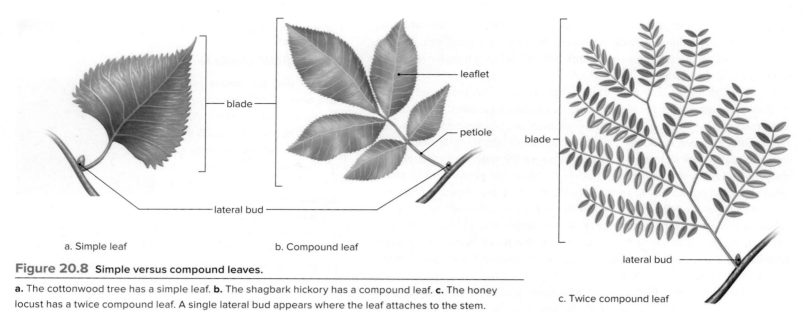

Figure 20.8 Simple versus compound leaves.

a. The cottonwood tree has a simple leaf. **b.** The shagbark hickory has a compound leaf. **c.** The honey locust has a twice compound leaf. A single lateral bud appears where the leaf attaches to the stem.

a. Simple leaf

b. Compound leaf

c. Twice compound leaf

leaf (Fig. 20.8*a*). But in some leaves, the blade is divided, or *compound,* as in a shagbark hickory or honey locust (Fig. 20.8*b,c*). Notice that it is possible to tell from the placement of the lateral bud whether you are looking at several individual leaves or one compound leaf.

Figure 20.9 shows a cross section of a typical eudicot leaf. The outermost structure, the waxy cuticle, prevents water loss. Next, the epidermal layer can be found above and below. The lower epidermis contains the stomata, allowing gas exchange. The interior area of a leaf is composed of **mesophyll,** the tissue that carries out photosynthesis. Vascular tissue terminating at the mesophyll transports water and minerals to a leaf and transports the product of photosynthesis, a sugar (sucrose), away from the leaf. The mesophyll has two distinct regions: *Palisade mesophyll* contains tightly packed cells, thereby increasing the surface area for the absorption of sunlight, and *spongy mesophyll* contains irregular cells surrounded by air spaces. The loosely packed arrangement of the cells in the spongy layer increases the amount of surface area for gas exchange and water loss. Water taken into a leaf by xylem tissue evaporates from spongy mesophyll and exits at the stomata.

Leaves may have several other functions aside from photosynthesis (**Fig. 20.10**). Leaves may be modified as tendrils that allow the plant to attach to objects (Fig. 20.10*a*) or as traps for catching insects (Fig. 20.10*b*). The leaves of a cactus are spines that reduce water loss and protect the plant from hungry animals (Fig. 20.10*c*).

Stems

If you carry a bouquet of daisies in one hand and lean against a tree with the other hand, you will be touching two different kinds of **stems.** The daisy has a nonwoody stem, and the tree trunk is a woody stem.

Nonwoody Stems

A stem that experiences only primary growth is nonwoody. Plants that have nonwoody stems, such as zinnias, mint, and daisies, are termed *herbaceous* plants. As in leaves, the outermost tissue of a herbaceous stem is the epidermis, which is covered by a waxy cuticle to prevent water loss. Beneath the epidermis

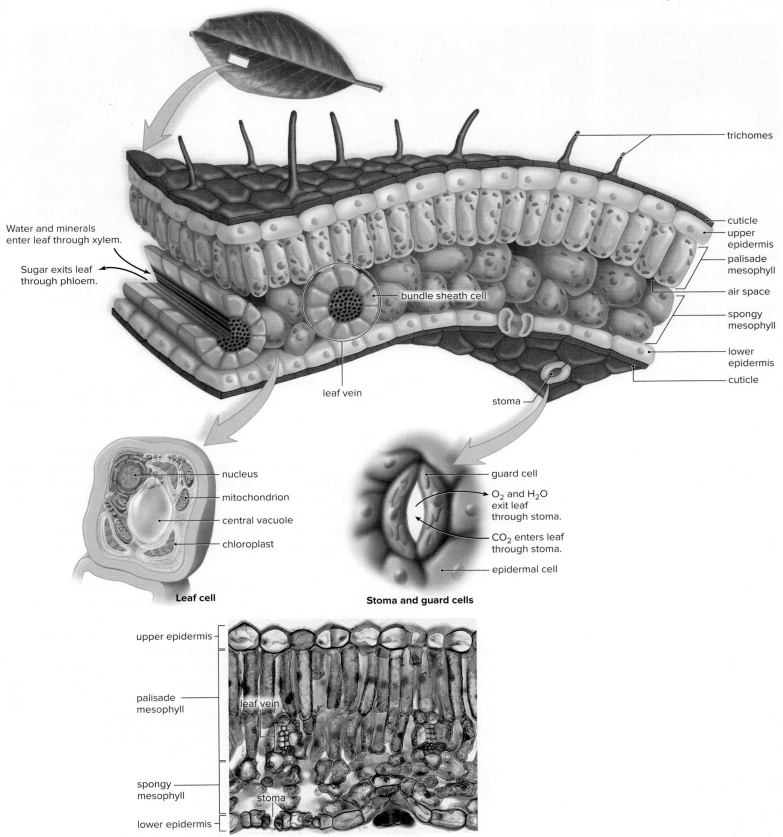

trichomes

cuticle

upper epidermis

palisade mesophyll

air space

spongy mesophyll

lower epidermis

cuticle

Water and minerals enter leaf through xylem.

Sugar exits leaf through phloem.

bundle sheath cell

leaf vein

stoma

nucleus

mitochondrion

central vacuole

chloroplast

Leaf cell

guard cell

O_2 and H_2O exit leaf through stoma.

CO_2 enters leaf through stoma.

epidermal cell

Stoma and guard cells

upper epidermis

palisade mesophyll

leaf vein

spongy mesophyll

stoma

lower epidermis

Figure 20.9 **Leaf structure.**

Photosynthesis takes place in mesophyll tissue of leaves. The leaf is enclosed by epidermal cells covered with a waxy layer, the cuticle. Water and solutes travel throughout the plant using the leaf veins, which contain xylem and phloem. A stoma is an opening in the epidermis that permits the exchange of gases.

© Ray F. Evert, University of Wisconsin

a. Cucumber   b. Venus's flytrap   c. Cactus   d. Potato

Figure 20.10 Leaf and stem diversity.

a. The tendrils of a cucumber are leaves modified to attach the plant to a support. **b.** The leaves of Venus's flytrap serve as the trap for unsuspecting insects. After digesting this fly, the plant will absorb the nutrients released. **c.** The spines of a cactus are leaves that protect the fleshy stem from herbivores. The cactus uses its fleshy stem for both food production and water storage. **d.** The familiar potato is actually an underground stem, or tuber, that is used for food storage.

(a): © Michael Gadomski/Science Source; (b): © Steven P. Lynch; (c): © Nature Picture Library; (d): © McGraw-Hill Education/Carlyn Iverson, photographer

of eudicot stems is the **cortex,** a narrow band of parenchyma cells. The cortex is sometimes green and carries on photosynthesis. The ground tissue in the center of a eudicot stem is called the **pith.** A monocot stem lacks an organized cortex or pith.

Herbaceous stems have distinctive vascular bundles, where xylem and phloem are found. In each bundle, xylem is typically found toward the inside of the stem, and phloem is found toward the outside. In the stems of herbaceous eudicots, the vascular bundles are arranged in a distinct ring that separates the cortex from the pith. In the stems of herbaceous monocots, the vascular bundles are scattered throughout the stem and have a characteristic "monkey face" appearance. **Figure 20.11** contrasts herbaceous eudicot and monocot stems.

Imagine placing a heavy object on top of a straw. The straw would bend and buckle under the weight. Stems resist breaking during growth because of the internal strength provided by vessel elements, tracheids, and sclerenchyma cells impregnated with lignin. The herbaceous stem of the mammoth sunflower holds up a flower head that can weigh up to 5 pounds!

Stems may have functions other than support and transport. In a cactus, the stem is the primary photosynthetic organ and serves as a water reservoir (see Fig. 20.10*c*), and the tuber of a potato plant is a food storage portion of an underground stem (Fig. 20.10*d*). *Perennial* plants are able to regrow each season from varied underground stems, such as tubers and rhizomes, all of which bear nodes that can produce a new shoot system.

Woody Stems

Plants with woody stems, such as trees and shrubs, experience both primary and secondary growth. *Secondary growth* increases the girth of stems, branches, and roots. It occurs because of a difference in the location and activity of vascular cambium, which, as you may recall from Section 20.1, is a type of meristem tissue. In herbaceous eudicots, *vascular cambium* is usually present between the xylem and phloem of each vascular bundle. In woody plants, the vascular cambium forms a ring of meristem that produces new xylem and

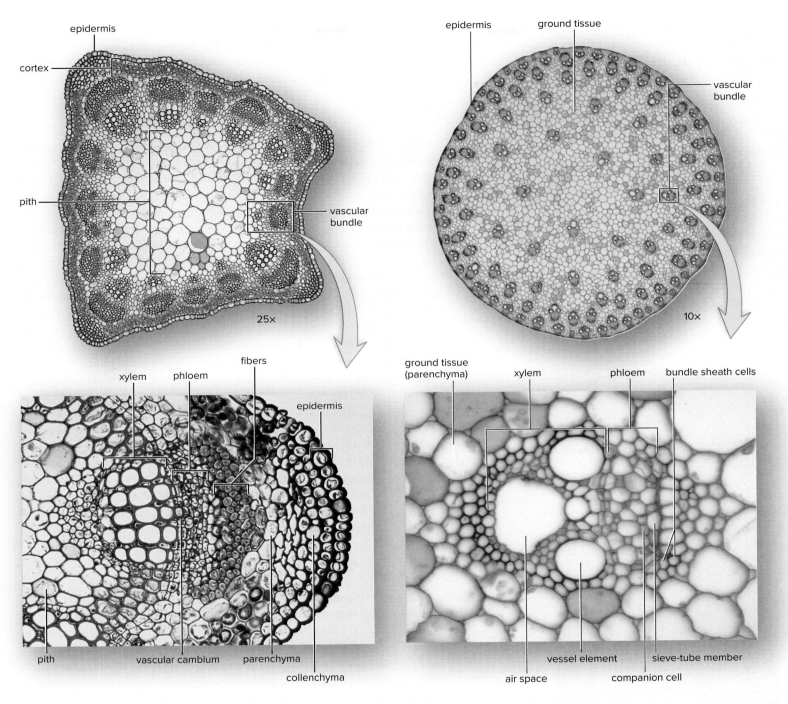

a. Eudicot stem: vascular
   bundles in a ring

b. Monocot stem: scattered
   vascular bundles

Figure 20.11 Nonwoody (herbaceous) stems.

a. In eudicot stems, the vascular bundles are arranged in a ring around well-defined ground tissue called pith. **b.** In monocot stems, the vascular bundles are scattered within the ground tissue.

(a): (top): © Ed Reschke; (bottom): © Ray F. Evert/University of Wisconsin Madison; (b): (top): © Carolina Biological Supply/Phototake; (bottom): © Kingsley Stern

phloem each year. In an older woody stem, vascular cambium occurs between the bark and the wood (**Fig. 20.12**).

Bark   The **bark** of a tree contains cork, cork cambium, cortex, and phloem. It is very harmful to remove the bark of a tree, because without phloem, organic nutrients cannot be transported. Although new phloem tissue is produced each year by vascular cambium, it does not build up in the same manner as xylem.

Connections: Ecology

What trees are used to produce cork bottle stoppers?

Cork stoppers, such as those tradition-ally found in wine bottles, are manufac-tured almost exclusively in Spain and Portugal from the cork oak tree (*Quercus suber*). Cork from these trees can be har-vested once the tree reaches an age of about 20 years. Then, every 9 years, the outer 1–2 inches of cork may be removed

© Peter Eastland/Alamy

from the tree without harming it. This yields about 16 kilograms (35 pounds) of cork per tree, and cork trees can continue to produce cork for around 150 years. Cork is considered to be an environmentally friendly forest product because the trees are not harmed during its production.

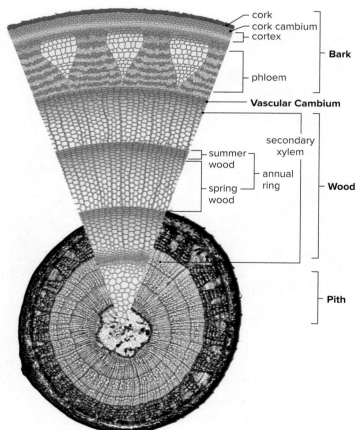

Figure 20.12 Organization of a woody stem.

In a woody stem, vascular cambium produces new secondary phloem and secondary xylem each year. Secondary xylem builds up to form wood consisting of annual rings. Counting the annual rings tells you that this woody stem is 3 years old.

© Ed Reschke

Cork cambium is located beneath the epidermis and is another region of active cell division. When cork cambium first begins to divide, it produces tissue that disrupts the epidermis and replaces it with cork cells. *Cork cells* are impreg-nated with suberin, a waxy layer that makes them waterproof but also causes them to die. In a woody stem, gas exchange is impeded, except at *lenticels,* which are pockets of loosely arranged cork cells not impregnated with suberin.

Wood When a plant first begins growing, the xylem is made by the apical meristem. Later, as the plant matures, xylem is made by the vascular cam-bium and is called *secondary xylem*. **Wood** is secondary xylem that builds up year after year, thereby increasing the girth of a tree. In trees that have a growing season, vascular cambium is dormant during the winter and be-comes active again in the spring when temperatures increase and water be-comes more available. In the spring, vascular cambium produces secondary xylem tissue that contains wide vessels with thin walls. This is called *spring wood,* and the wider vessels transport sufficient water to the growing leaves. In summer, there is less rain and the wood at this time, called *summer wood,* has a lower proportion of vessels. Strength is required because the tree is growing larger, and summer wood contains numerous thick-walled tra-cheids. At the end of the growing season, just before the vascular cambium becomes dormant again, only heavy fibers with especially thick secondary walls may develop. When the trunk of a tree has spring wood followed by summer wood, the two together make up one year's growth, or an **annual ring** (see Fig. 20.12). A dendrochronologist is a scientist who studies annual rings to determine the ages of trees. Annual rings can also give clues on water availability and forest fires in the past, and they support predictions about global climate change.

Wood is one of the most useful and versatile materials known. It is used for building structures and for making furniture. Much wood is also used for heating and cooking, as well as for producing paper, chemicals, and pharma-ceuticals. The resins derived from wood are used to make turpentine and rosin.

Roots

A plant's **roots** support the plant by anchoring it in the soil, as well as absorb-ing water and minerals from the soil for the entire plant. As a rule of thumb, the root system is at least equivalent in size and extent to the plant's shoot system. Therefore, an apple tree has a much larger root system than a corn plant. Also, the extent of a root system depends on the environment. A single corn plant may have roots that extend as deep as 2.5 meters (m), while a mesquite tree in the desert may have roots that penetrate to a depth of 20 m.

Roots have a cylindrical shape and a slimy surface. These features allow roots to penetrate the soil as they grow and permit water to be absorbed from all sides. In a special zone near the root tip, there are many root hairs that greatly increase the absorptive capacity of the root. *Root hairs* are so numerous that they increase the absorption of water and minerals tremendously. It has been estimated that a single rye plant has about 14 billion root-hair cells! Root-hair cells are constantly being replaced, so a rye plant forms about 100 million new root-hair cells every day. You probably know that a plant yanked out of the soil will not fare well when transplanted. This is because small lateral roots and root hairs are torn off. Transplantation is more apt to be successful if you take a part of the surrounding soil along with the plant, leaving as many of the lat-eral roots and root hairs intact as possible.

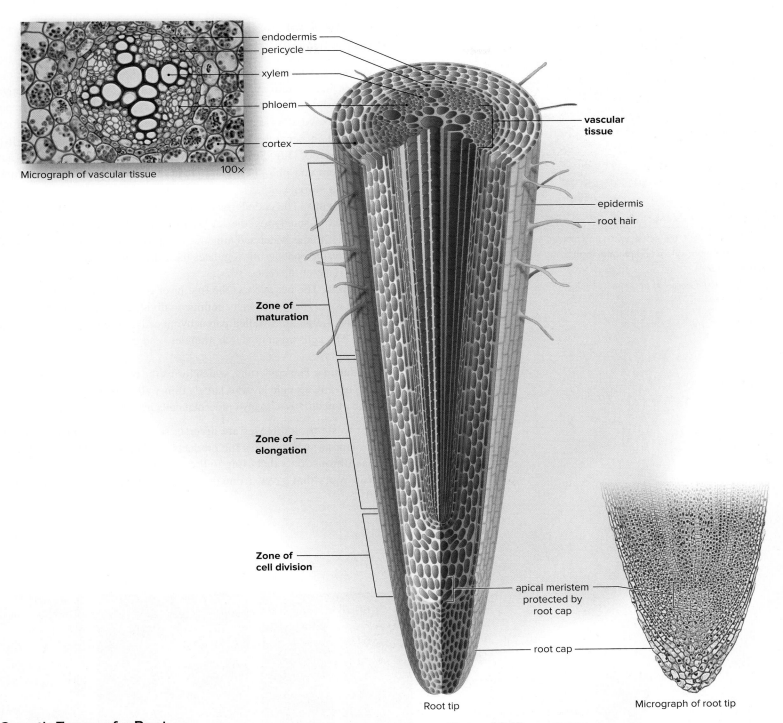

endodermis
pericycle
xylem
phloem
cortex

Micrograph of vascular tissue   100×

vascular tissue

epidermis
root hair

Zone of maturation

Zone of elongation

Zone of cell division

apical meristem protected by root cap

root cap

Root tip

Micrograph of root tip

Figure 20.13 Eudicot root tip.

The root tip is divided into three zones, best seen in a longitudinal section, such as this. In eudicots, xylem is typically star-shaped, and phloem lies between the points of the star. Water and minerals must eventually pass through the cytoplasm of endodermal cells in order to enter the xylem; these endodermal cells regulate the passage of minerals into the xylem.

(left): © Ed Reschke/Getty Images; (right): © Ray F. Evert/University of Wisconsin, Madison

Growth Zones of a Root

Both monocot and eudicot roots contain the same growth zones. In **Figure 20.13**, a longitudinal section of a eudicot root tip reveals these zones, where cells are in various stages of differentiation as primary growth occurs. The apical meristem of a root contains actively dividing cells and is surrounded by a *root cap*. The root cap is covered with a slimy substance to help it penetrate downward into the abrasive soil, and it is meant to be damaged to protect the apical meristem. The dividing cells of the apical meristem are found in the zone of cell division. As more cells are made, older cells are pushed into the zone of elongation, where cells lengthen as they become specialized. In the zone of maturation, which contains fully differentiated cells, many of the epidermal cells have root hairs.

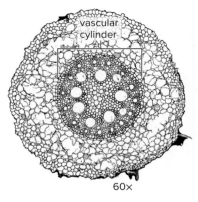

Figure 20.14 Monocot root.

In a monocot root, the same zones and tissues are present as in a eudicot root, but xylem and phloem form a ring surrounding the pith in the center.

© McGraw Hill Education, Al Telser, photographer

Tissues of a Root

The five tissues of a root are as follows:

1. **Vascular tissue.** Both monocot and eudicot roots have *vascular cylinders* that contain xylem and phloem, but the cylinders are arranged differently. In a eudicot root (Fig. 20.13), the xylem is star-shaped because several xylem arms radiate from a common center. Phloem is found in separate regions between the points of the star. In a monocot root (**Fig. 20.14**), the vascular cylinder consists of alternating xylem and phloem bundles that surround a pith. Pith can function as a storage site.

2. **Endodermis.** The endodermis of a root is a single layer of rectangular cells that fit snugly together. A layer of impermeable material on all but two sides forces water and minerals to pass through endodermal cells. In this way, the endodermis regulates the entrance of minerals into the vascular tissue of the root.

3. **Pericycle.** The pericycle is the first layer of cells inside the endodermis of the root. These cells can continue to divide and form lateral roots.

4. **Cortex.** Large, thin-walled parenchyma cells make up the cortex of the root. The cells contain starch granules, and the cortex may function in food storage.

5. **Epidermis.** The epidermis, which forms the outer layer of the root, consists of only a single layer of largely thin-walled, rectangular cells. In the zone of maturation, many epidermal cells have root hairs.

Just as stems and leaves are diverse, so are roots. A carrot plant has one main taproot, which stores the products of photosynthesis (**Fig. 20.15***a*). Grass has fibrous roots that cling to the soil (Fig. 20.15*b*), and corn plants have prop roots that grow from the stem to provide additional support (Fig. 20.15*c*).

CONNECTING THE CONCEPTS

20.3 Roots, stems, and leaves all work together to make the plant functional.

Check Your Progress 20.3

1. Explain how the structure of a leaf aids in photosynthesis.

2. List the components that make up wood and bark.

3. List the functions of the tissues of a eudicot root.

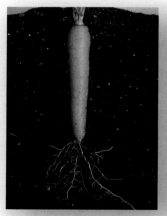

a. Taproot   b. Fibrous root system   c. Prop roots

Figure 20.15 Root diversity.

a. A taproot has one main root. **b.** A fibrous root has many slender roots with no main root. **c.** Prop roots are organized for support.

(a): © Jonathan Buckley/Getty Images; (b): © Evelyn Jo Johnson; (c): © NokHoOkNoi/iStock/360/Getty RF

20.4 Plant Nutrition

Learning Outcomes

Upon completion of this section, you should be able to

1. Differentiate between macronutrients and micronutrients, and list the macronutrients and micronutrients that plants require.
2. Describe some adaptations of roots for obtaining minerals from the soil.

Plant nutrition is remarkable to us because plants require only inorganic nutrients, and from these they make all the organic compounds that compose their bodies. Of course, they require carbon, hydrogen, and oxygen, which they can acquire from carbon dioxide and water, but the other elements, or minerals, that they need they obtain from the environment. An element is termed an essential nutrient if a plant cannot live without it. The essential nutrients are divided into *macronutrients* and *micronutrients,* according to their relative concentrations in plant tissue. The following diagram indicates which are the macronutrients and which are the micronutrients:

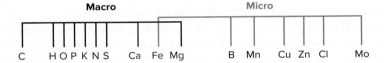

Notice that iron (Fe) is considered to be a micronutrient for some plants and a macronutrient for others. All of these elements play vital roles in plant cells, but deficiencies of the macronutrients tend to be the most devastating. Nitrogen, for example, is necessary for protein production. The carnivorous plants featured in the chapter opener have evolved to capture insects because of the lack of nitrogen in boggy soils. Traditional farming depletes the nitrogen in soil, and many farmers turn to crop rotation to keep soils healthy. **Figure 20.16** shows how an insufficient supply of the macronutrient nitrogen can affect the growth of a plant.

a.

b.

Figure 20.16 Nitrogen deficiency.

a. Experiment showing that plants respond poorly if their growth medium lacks nitrogen. **b.** A field showing crop rotation. One crop is the target crop, and the other is a legume plant with root nodules. The legume plant will leach nitrogen back into the soil, thus helping the target crop grow in the next season when it is planted in the strips that currently have the legume.

(a): © Photos Lamontagne/Photolibrary/Getty Images; (b): © David R. Frazier Photolibrary/Alamy

Connections: Scientific Inquiry

What do the numbers on a bag of fertilizer mean?

When you buy a bag of fertilizer, you will notice three numbers on the outside of the bag (for example, 20-20-20). These numbers are called the NPK ratio, and they refer to the amounts of nitrogen (N), phosphorus (P), and potassium (K) in the fertilizer. All three of these are macronutrients that are important for plant health and growth. Nitrogen promotes the vegetative growth of the plant, and the plant needs phosphorus to maintain a healthy root system. Potassium is also involved in the general health of a plant and is needed for the formation of the chloro-

© Ricochet Creative Productions LLC/Leah Moore, photographer

phyll molecules involved in photosynthesis. The optimal amount of each nutrient is dependent on the type of plant; thus, larger numbers do not always signify a better fertilizer.

CONNECTING THE CONCEPTS

20.4 Plants often have adaptations of their roots to enhance the uptake of macronutrients and micronutrients.

The micronutrients are often cofactors for enzymes in various metabolic pathways. *Cofactors* are elements that ensure that enzymes have the correct shape. It's interesting to observe that humans make use of a plant's ability to take minerals from the soil. For example, we depend on plants for supplies of iron to help carry oxygen to our cells. Eating plants provides minerals such as copper and zinc, which are also cofactors for our own enzymes.

Adaptations of Roots for Mineral Uptake

Minerals enter a plant at its root system, and two mutualistic relationships assist roots in fulfilling this function. Air contains about 78% nitrogen (N_2), but plants can't make use of it. Most plants depend on bacteria in the soil to *fix* nitrogen—that is, the bacteria change atmospheric nitrogen (N_2) to nitrate (NO_3^-) or ammonium (NH_4^+), both of which plants can take up and use. Legume plants, such as soybean and peas, have roots colonized by bacteria that are able to take up atmospheric nitrogen and reduce it to a form suitable for incorporation into organic compounds (**Fig. 20.17***a*). The bacteria live in **root nodules** (see Fig. 17.13), and the plant supplies the bacteria with carbohydrates, while the bacteria in turn furnish the plant with nitrogen compounds.

Another mutualistic relationship involves **mycorrhizal fungi** and almost all plant roots (Fig. 20.17*b*). The hyphae of the fungus increase the surface area available for water uptake and break down organic matter, releasing inorganic nutrients that the plant can use. In return, the root furnishes the fungus with sugars and amino acids. Plants are extremely dependent on mycorrhizal fungi. For example, orchid seeds, which are quite small and contain limited nutrients, do not germinate until a mycorrhizal fungus has invaded their cells.

Check Your Progress 20.4

1. Explain the difference between a macronutrient and a micronutrient.
2. List the macronutrients needed for plant growth.
3. Discuss the importance of root nodules and mycorrhizal fungi to plants.

Figure 20.17 Adaptations of roots for mineral uptake.

Plants with mycorrhizal fungi (*right*) grow much better than plants without mycorrhizal fungi (*left*).

(right): © Dr. Keith Wheeler/Science Source

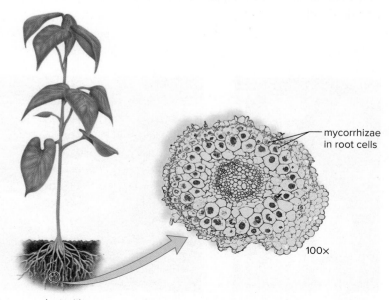

plant without mycorrhizae

plant with mycorrhizae

mycorrhizae in root cells

100×

20.5 Transport of Nutrients

Learning Outcomes

Upon completion of this section, you should be able to

1. Describe how the cohesion-tension model explains the movement of water in a plant.
2. Define *transpiration,* and explain the role of stomata in regulating this process.
3. Describe how the pressure-flow model explains the movement of nutrients in a plant.

Water Transport in Xylem

People have long wondered how plants, especially very tall trees, lift water from the roots to the leaves against gravity. The **cohesion-tension model** is an explanation of how water and minerals travel upward in xylem cells. To understand this proposed mechanism, one must start in the soil. Recall the concept of osmosis. In plants, water will move from an area of high concentration in the soil to an area of low concentration in the root hairs (**Fig. 20.18**). All of the water entering the roots creates root pressure. Root pressure is helpful for the upward movement of water but is not nearly enough to get it all the way up to the leaves.

Transpiration is the loss of water to the environment, mainly through evaporation from leaf stomata. This is the phenomenon that explains how water can completely resist gravity and travel upward. Focusing on the top of the tree (Fig. 20.18), notice the water molecules escaping from the spongy mesophyll and into the air through the stomata. The key is that it is not just one water molecule escaping but a chain of water molecules. Think about drinking water from a straw. Drinking exerts pressure on the straw and a chain of water molecules is drawn upward. Water molecules are polar and "stick" together with hydrogen bonds. Water's ability to stay linked in a chain is called *cohesion,* and its ability to stick to the inside of a straw is *adhesion* (see Fig. 2.10). In plants, evaporation of water at the leaves provides the *tension* that pulls on a chain of water molecules. Transpiration produces a constant tugging or pulling of the water column from the top due to evaporation. Cohesion of water molecules and their adhesion to the inside of xylem vessels facilitate this process. As transpiration occurs, the water column is pulled upward—first within the leaf, then from the stem, and finally from the roots.

The total amount of water a plant loses through transpiration over a long period of time is surprisingly large. At least 90% of the water taken up by roots is eventually lost at the leaves. A single corn plant loses between 135 and 200 liters of water through transpiration during a growing season.

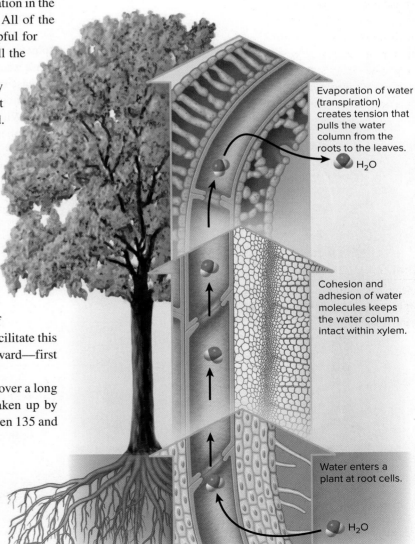

Evaporation of water (transpiration) creates tension that pulls the water column from the roots to the leaves.

H_2O

Cohesion and adhesion of water molecules keeps the water column intact within xylem.

Water enters a plant at root cells.

H_2O

Figure 20.18 Cohesion-tension model of xylem transport.

How does water rise to the top of a tall tree? Xylem vessels are water-filled pipelines from the roots to the leaves. When water evaporates from spongy mesophyll into the air spaces of leaves, the water columns in these vessels are pulled upward due to the cohesion of water molecules with one another and their adhesion to the sides of the vessels.

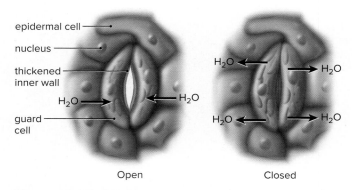

epidermal cell

nucleus

thickened inner wall

H_2O

guard cell

H_2O ← → H_2O

H_2O ← → H_2O

Open   Closed

Figure 20.19 **Opening and closing of stomata.**

Stomata open when water enters guard cells and turgor pressure increases. Stomata close when water exits guard cells and turgor pressure is decreased.

Opening and Closing of Stomata

A plant cell that is full of water will bulge (see Fig. 5.13). *Turgor pressure* is the force of the water creating this bulge. Stomata open and close due to changes in turgor pressure within guard cells (**Fig. 20.19**). When water enters the guard cells, turgor pressure increases, and the unique "banana" shape of the guard cells causes them to bow out and expose the pore (stoma); when water leaves the guard cells, turgor pressure decreases, and the pore is once again covered.

For transpiration to occur, the stomata must stay open. But when a plant is under stress and about to wilt from lack of water, the stomata close. Now the plant is unable to take up carbon dioxide from the air, and photosynthesis ceases.

Sugar Transport in Phloem

A plant must make and store sugar, then provide that sugar to its developing parts, such as the roots, flowers, and fruit. The location where the sugar is made or stored is called the *source,* and the location where the sugar will be used is the *sink*. How sugar is transported from the source to the sink can be explained using the **pressure-flow model** (**Fig. 20.20**).

Leaves undergoing photosynthesis make sugar and become the source. The sugar is actively transported from the cells in the leaf mesophyll into the sieve tubes of the phloem. Recall that, like xylem, phloem is a continuous "pipeline" throughout the plant. High concentrations of sugar in the sieve tubes cause water to follow by osmosis. Like turning the nozzle on a hose, there is an increase in positive pressure as water flows in. The sugar (sucrose) solution at the source is forced to move to areas of lower pressure at a sink, like a root.

When the sugar arrives at the root, it is actively transported out of the sieve tubes into the root cells. There, the sugar is used for cellular respiration or other metabolic processes. The high concentration of sugar in the root cells causes water to follow by osmosis, where it is later reclaimed by the xylem tissue.

Although leaves are generally the source, modified roots and stems such as carrots, beets, and potatoes are also examples of sources that provide much needed sugar during winter or periods of dormancy.

The high pressure of sucrose in phloem has resulted in a very interesting mutualistic relationship between aphids and some species of ants. Aphids are tiny insects with needlelike mouthparts. These mouthparts can poke a stem and tap into a sieve tube (**Fig. 20.21**). The high-pressure sucrose solution is forced through their digestive

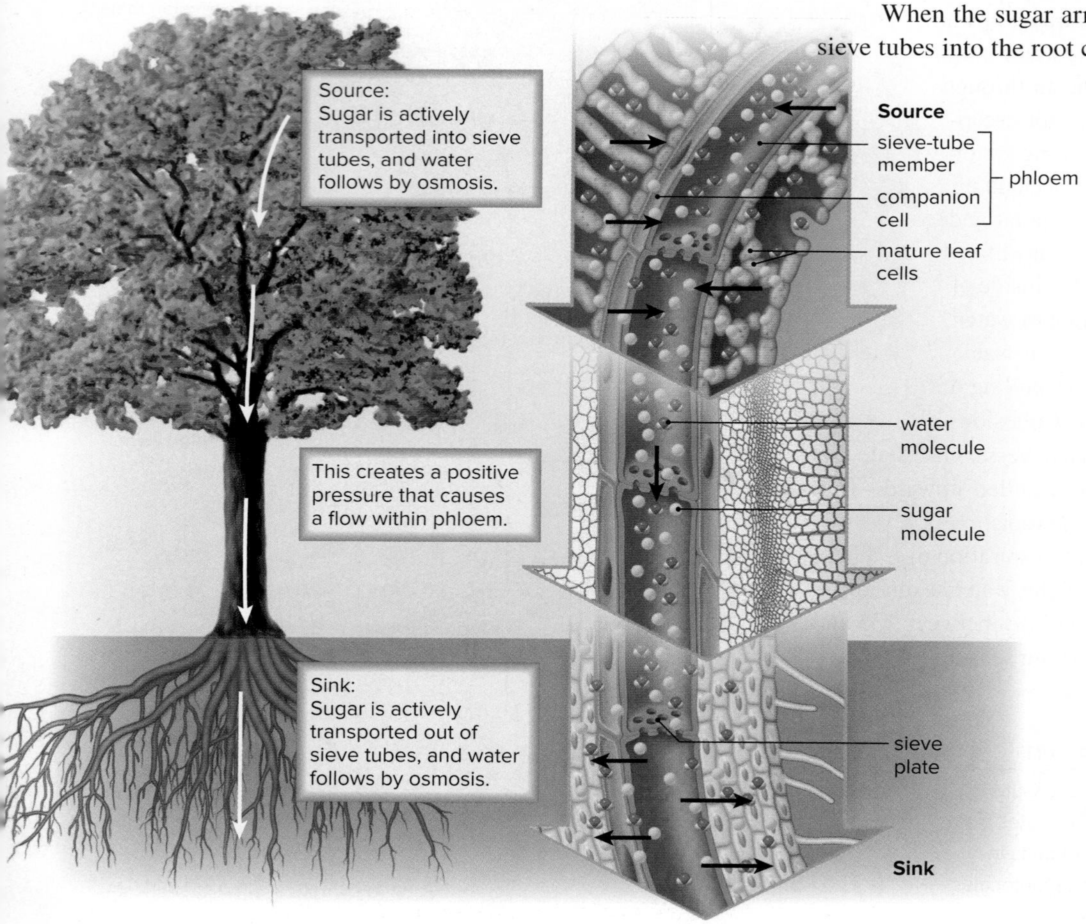

Source: Sugar is actively transported into sieve tubes, and water follows by osmosis.

This creates a positive pressure that causes a flow within phloem.

Sink: Sugar is actively transported out of sieve tubes, and water follows by osmosis.

Source

sieve-tube member

companion cell

mature leaf cells

phloem

water molecule

sugar molecule

sieve plate

Sink

Figure 20.20 **Pressure-flow model of phloem transport.**

Sugar and water enter sieve-tube members at a source. This creates a positive pressure, which causes the phloem contents to flow. Sieve-tube members form a continuous pipeline from a source to a sink, where sugar and water exit the sieve-tube members.

a. An aphid feeding on a plant stem

b. Some ants protect aphids and in turn consume honeydew.

Figure 20.21 Aphids acquire sucrose from phloem.

Aphids are small insects that remove sucrose from phloem by means of a needlelike mouthpart called a stylet. **a.** Excess sucrose appears as a droplet after passing through the aphid's body. **b.** A protective ant farming aphids and drinking honeydew.

(a): © M. H. Zimmermann/Harvard Forest, Harvard University; (b): © blickwinkel/Alamy

tracts very quickly, resulting in a droplet of sucrose at the rear called honeydew. Ants stroke the aphids with their antennae to induce honeydew production and then drink the beads of sucrose from the aphids' rear ends! Their "aphid farms" provide whole colonies of ants with all the sugar they need, and in turn the ants protect the aphids against predators.

CONNECTING THE CONCEPTS
Xylem and phloem conduct water and nutrients throughout the plant.

Check Your Progress 20.5

1. Explain the role of transpiration in water transport.
2. Explain how the cohesive and adhesive properties of water assist water transport in xylem.
3. Describe how the pressure-flow model explains the movement of sugar in the phloem.

STUDY TOOLS   http://connect.mheducation.com

 **SMARTBOOK®** Maximize your study time with McGraw-Hill SmartBook®, the first adaptive textbook.

SUMMARIZE

The organizational system of plants is as complex as our own. Plants use a variety of adaptations in order to survive in various environments.

20.1 Plants have meristem cells that form specialized epidermal, ground, and vasular tissues.

20.2 Plants tissues combine to form vegetative organs that make up the shoot system and the root system.

20.3 Roots, stems, and leaves all work together to make the plant functional.

20.4 Plants often have adaptations of their roots to enhance the uptake of macronutrients and micronutrients.

20.5 Xylem and phloem conduct water and nutrients throughout the plant.

20.1 Plant Cells and Tissues

The first cells to develop in a plant embryo are meristem cells that organize into **meristem tissue. Apical meristems** are located at growing ends and give rise to three types of specialized tissues:

- **Epidermal tissue** is composed of only epidermal cells and can be modified to form **stomata, root hairs,** and trichomes.

- **Ground tissue** makes up the bulk of the plant and contains parenchyma cells, which are thin-walled and capable of photosynthesis when they have chloroplasts; collenchyma cells, which have thicker walls for flexible support; and sclerenchyma cells, which are hollow, nonliving support cells with thick secondary cell walls.

- **Vascular tissue** transports water and nutrients and is made up of **xylem** and **phloem.** Xylem transports water and minerals and contains vessels composed of **vessel elements** and **tracheids.** Phloem transports sugar and other organic compounds and contains sieve tubes composed of **sieve-tube members,** each of which has a **companion cell.**

xylem   phloem

20.2 Plant Organs

A flowering plant has a **shoot system** and a **root system.** The three vegetative organs of a plant are the leaves, stems, and roots. A flowering plant has a **shoot system** and a **root system.**

Monocots Versus Eudicots

Flowering plants are divided into **monocots** and **eudicots** based on the number of **cotyledons** in the seed; the arrangement of vascular tissue in leaves, stems, and roots; and the number of flower parts.

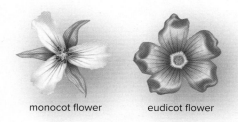

monocot flower   eudicot flower

20.3 Organization of Leaves, Stems, and Roots

Leaves

Leaves carry on photosynthesis but may be modified for other purposes. A leaf's surface is covered with a waxy cuticle, and the lower epidermis has stomata. Stomata allow water vapor and oxygen to escape and carbon dioxide to enter the leaf. **Mesophyll** (palisade and spongy) forms the body of a leaf and carries on photosynthesis.

Stems

Stems support leaves, conduct materials to and from roots and leaves, and produce new tissues. Stems can be nonwoody or woody.

- Nonwoody (herbaceous) eudicots have stems with an epidermis, **cortex** tissue, vascular bundles in a ring, and an inner **pith.** Monocot stems have scattered vascular bundles and lack a distinct cortex or pith.
- Woody stems have secondary growth due to vascular cambium, which produces new xylem and phloem every year. Cork replaces epidermis in woody plants. A woody stem has **bark** (containing cork, cork cambium, cortex, and phloem). **Wood** contains **annual rings** of xylem.

Roots

Roots anchor a plant, absorb water and minerals, and store the products of photosynthesis.

- In monocots and eudicots, a root tip has three zones: zone of cell division (contains root apical meristem) protected by the root cap; zone of elongation, where cells elongate and differentiate; and zone of maturation (has root hairs).

- Within the zone of maturation, monocot and eudicot roots contain the following tissues. The vascular tissue in eudicot roots shows xylem arranged in a star shape, with phloem between the arms of the xylem. In monocot roots, a central pith is surrounded by a ring of vascular tissue containing alternating bundles of xylem and phloem. The **endodermis** regulates the entrance of minerals into vascular tissue. The **pericycle** contains cells that can divide and form lateral roots. The **cortex** is made of parenchyma cells that function in food storage, and the **epidermis** forms the outer layer and may possess root hairs.

20.4 Plant Nutrition

Plants need only inorganic nutrients to make all the organic compounds that make up their bodies. Some nutrients are essential, and are classified as either macronutrients or micronutrients. Some roots have adaptations for mineral uptake, such as **root nodules,** where bacteria fix nitrogen into forms that plants can use, and most roots have **mycorrhizal fungi** that increase the surface area for water and mineral absorption.

20.5 Transport of Nutrients

- The **cohesion-tension model** of xylem transport states that **transpiration** (evaporation of water from spongy mesophyll through stomata) creates tension, which pulls water upward in xylem. This mechanism works only because water molecules are cohesive and adhesive and form a water column in xylem.
- The **pressure-flow model** of phloem transport states that sugar is actively transported into phloem at a source, and water follows by osmosis. The resulting increase in pressure creates a flow, which moves water and sugar through the phloem to a sink.

ASSESS

Testing Yourself

Choose the best answer for each question.

20.1 Plant Cells and Tissues

1. A region of active cell division where primary growth occurs is the
   - **a.** internode.
   - **b.** apical meristem.
   - **c.** root nodule.
   - **d.** cork cambium.
   - **e.** vascular cambium.

2. Hard materials in plants, such as the husks of nuts, are composed primarily of these dead cells.
   - **a.** vascular tissue
   - **b.** collenchyma cells
   - **c.** sclerenchyma cells
   - **d.** parenchyma cells
   - **e.** epidermal tissue

3. Which of the following cell types does not originate from epidermal tissue?
   - **a.** root hair
   - **b.** trichome
   - **c.** guard cells (stomata)
   - **d.** vessel elements

20.2 Plant Organs

4. Label the parts of a plant in the following diagram.

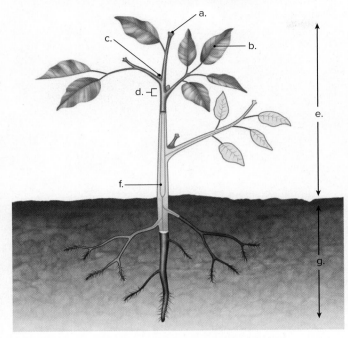

5. A monocot stem differs from a eudicot stem in that, in the monocot stem,
   a. xylem and phloem are in a ring surrounding the pith.
   b. xylem and phloem are in bundles scattered throughout the stem.
   c. there is an organized cortex and pith.
   d. xylem forms a star in the center, with phloem between the points of the star.
   e. the epidermis does not have a waxy cuticle covering it.

20.3 Organization of Leaves, Stems, and Roots

6. The main function of leaves is to
   a. transport water.
   b. support the weight of the plant.
   c. transport sugar.
   d. absorb sunlight.

7. During secondary growth, a tree adds more xylem and phloem through the activity of the
   a. apical meristems.
   b. vascular cambium.
   c. pericycle.
   d. cork cambium.

8. Which root zone contains the root cap and apical meristem?
   a. zone of maturation
   b. zone of elongation
   c. zone of cell division
   d. All of these are correct.

20.4 Plant Nutrition

9. Nitrogen is an example of a(n) _____ because plants need this element in large quantities.
   a. organic molecule
   b. micronutrient
   c. macronutrient
   d. deficiency

10. Root nodules are important because they
   a. encourage the growth of mycorrhizal fungi.
   b. represent areas of extensive branch root growth.
   c. contain nitrogen-fixing bacteria.
   d. provide extra oxygen to the plant root system.

20.5 Transport of Nutrients

11. Because of transpiration, water
   a. evaporates directly from the root surface.
   b. flows from leaf veins through the stem, toward the roots.
   c. exits the leaf through stomata, pulling water into the leaf from leaf veins.
   d. exits xylem in leaf veins and enters phloem by osmosis.

12. According to the pressure-flow model, sugar is actively transported into phloem and
   a. enters xylem, where it is moved toward the leaves due to transpiration.
   b. creates pressure to move water toward the roots.
   c. is transported out of the leaves through stomata.
   d. water follows by osmosis, providing pressure that moves the water and sugar through the phloem.

ENGAGE

BioNOW

Want to know how this science is relevant to your life? Check out the BioNow video below.

   • Saltwater Filter

Discuss how the vascular tissues xylem and phloem are involved in the experimental process in this video.

Thinking Critically

1. Scientists observe that the roots of legumes grow around nitrogen-fixing bacteria capable of forming nodules. Discuss how the roots might recognize these bacteria (see Fig. 20.17).

2. Plants of the genus *Welwitschia* live in the deserts in Africa. Annual rainfall averages only 2.5 centimeters per year, but every night a fog rolls in. Why might these plants have adapted so they open their stomata at night? What about the fog allows the plant to survive?

21

Plant Responses and Reproduction

OUTLINE

21.1 Plant Hormones   393

21.2 Plant Responses   396

21.3 Sexual Reproduction in Flowering Plants   399

21.4 Asexual Reproduction and Genetic Engineering in Plants   407

BEFORE YOU BEGIN

Before beginning this chapter, take a few moments to review the following discussions.

Figure 11.27 How does a signal transduction pathway relay information?

Figure 18.16 How does the sporophyte stage differ from the gametophyte stage in a flowering plant?

Section 20.2 What are an apical meristem and a terminal bud?

© David H. Wells/The Image Bank/Getty Images

Bananas—Mules of the Fruit World

The typical supermarket banana is sterile, meaning that it contains no viable seeds; there is nothing to plant if you want to grow more. This is because in the mid-nineteenth century two varieties of wild bananas were crossed to form the sweet, but sterile, banana that we all enjoy today. So how do farmers grow bananas if there are no seeds? The answer lies in three alternative techniques: asexual reproduction, tissue culture, and genetic engineering.

Sometimes, farmers use asexual reproduction—simply cutting a piece of the banana stem or root and planting it directly in the ground. This creates an identical plant, but it may pass on diseases from the first plant.

In an effort to obtain new disease-free banana plants, many plantations have turned to tissue culture. In this procedure, a technician scrapes a piece of the apical meristem and puts it in a petri dish containing nutrients and growth hormones. Eventually, new plants form and are transplanted in the field. However, fungi and insects often attack these disease-free plants. These pests can destroy entire banana plantations. In the 1950s, a fungus wiped out banana plantations throughout the Caribbean and Central America.

Another technique for growing disease-resistant bananas involves creating genetically engineered plants. Scientists are taking antifungal genes from rice plants and inserting them into banana plants. These modified banana plants are better able to fight off infections.

In this chapter, we will explore methods of sexual versus asexual reproduction in plants.

As you read through this chapter, think about the following questions:

1. How is genetic engineering used in asexual reproduction?
2. What are the advantages and disadvantages of sexual and asexual reproduction for plants?

21.1 Plant Hormones

Learning Outcomes

Upon completion of this section, you should be able to

1. List the five commonly recognized groups of plant hormones.
2. Describe the influence of auxin on apical dominance.
3. Understand how gibberellins are used in agriculture.
4. Explain the relationship between senescence and cytokinins.
5. Explain the role of abscisic acid in dormancy and in the closure of stomata.
6. Describe the effects of ethylene on both plants and fruits.

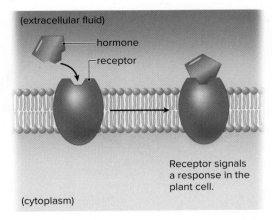

Figure 21.1 **Plant hormones.**

Hormones are chemical signals produced as a result of an environmental stimulus, such as sunlight, water loss, or temperature. The hormone binds receptor proteins, inducing a response in the specific cells or tissues of the plant.

Animals respond to environmental stimuli (danger, food, etc.) by *moving* toward or away from the stimulus. Plants respond to environmental stimuli (light, water, etc.) by *growing* toward or away from the stimulus. Most of these growth responses occur at the cellular level and are mediated by hormones. Much like our own hormones, plant **hormones** are small, organic molecules produced by a plant that serve as chemical signals between cells and tissues. **Figure 21.1** shows how most hormones are signals that bind to receptor proteins. Generally, the process involves the binding of a hormone to a receptor, which in turn signals the cell to respond to the stimulus.

The five commonly recognized groups of plant hormones are auxins, gibberellins, cytokinins, abscisic acid, and ethylene (**Table 21.1**).

Auxins

More than a century ago, an organic substance known as **auxin** was the first plant hormone to be discovered. All plant cells have a rigid cell wall. Auxin's role is to soften the cell wall so that plant growth can occur.

Auxin is involved in **phototropism,** a trait of plants that results in the bending of the stem in the direction of a light source (**Fig. 21.2***a*). When a plant is exposed to light on one side, auxin moves to the shady side. On that side, cells become longer, causing the stem to bend *toward* the light.

Elongation of cells in the shade occurs due to a series of events (Fig. 21.2*b*):

1. Auxin binds to a protein receptor.
2. Hydrogen ions (H^+) are actively pumped out of the cell (requiring ATP).
3. The increased concentration of H^+ ions creates an acidic environment.
4. The acid triggers other enzymes to soften the cell wall.
5. Growth and elongation of the cell takes place.

Auxin is also responsible for a phenomenon called **apical dominance** (**Fig. 21.3**). Experienced gardeners know that, to produce a bushier plant, they must remove the terminal bud. Normally, auxin is produced in the apical meristem of the terminal bud and is transported downward in the plant. The presence of auxin inhibits the growth of lateral buds. When the terminal bud is removed, auxin is not produced, allowing the lateral buds to grow and the plant to take on a fuller appearance. Interestingly, if auxin were to be applied to the broken terminal stem, apical dominance would be restored.

Synthetic auxins are used today in a number of applications. These auxins are sprayed on plants, such as tomatoes, to induce the development of fruit without pollination, creating seedless varieties. Synthetic auxins have been used as herbicides to control broadleaf weeds, such as dandelions and other plants. These substances have little effect on grasses. Agent Orange is a

Table 21.1 Functions of the Major Plant Hormones

Hormone	Functions
Auxins	Maintain apical dominance; are involved in phototropism and gravitropism; promote growth of roots in tissue culture; prevent leaf and fruit drop
Gibberellins	Promote stem elongation between nodes; break seed and bud dormancy and influence germination of seeds
Cytokinins	Promote cell division; prevent senescence; along with auxin, promote differentiation, leading to roots, shoots, leaves, or floral shoots
Abscisic acid	Initiates and maintains seed and bud dormancy; promotes formation of winter buds; promotes closure of stomata
Ethylene	Promotes abscission (leaf, fruit, or flower drop); promotes ripening of fruit

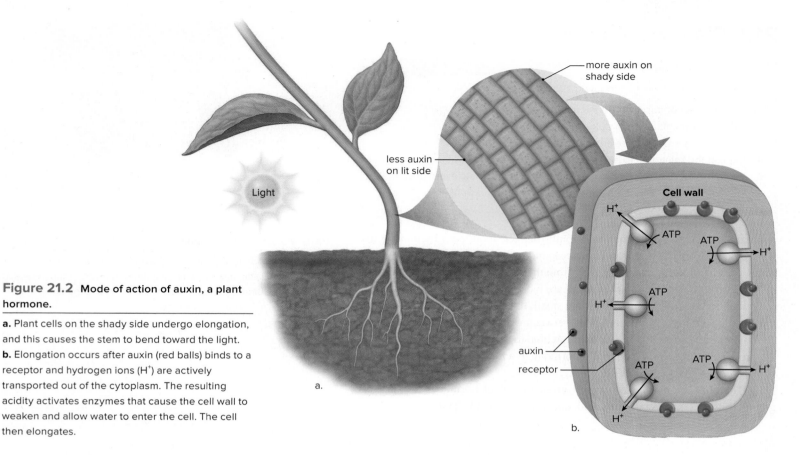

Figure 21.2 Mode of action of auxin, a plant hormone.

a. Plant cells on the shady side undergo elongation, and this causes the stem to bend toward the light.
b. Elongation occurs after auxin (red balls) binds to a receptor and hydrogen ions (H⁺) are actively transported out of the cytoplasm. The resulting acidity activates enzymes that cause the cell wall to weaken and allow water to enter the cell. The cell then elongates.

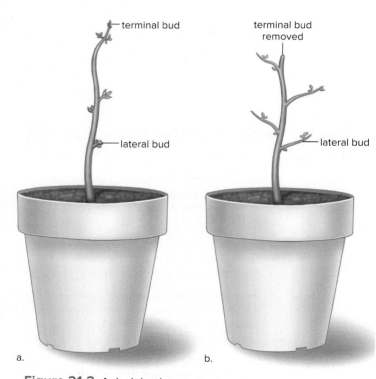

Figure 21.3 Apical dominance.

a. Lateral bud growth is inhibited when a plant retains its terminal bud. **b.** When the terminal bud is removed, lateral branches develop and the plant is bushier.

powerful synthetic auxin that was used in extremely high concentrations to defoliate the forests of Vietnam during the Vietnam War. This powerful auxin proved to be carcinogenic and harmed many of the local people.

Gibberellins

Gibberellins were discovered in 1926 when a Japanese scientist was investigating a fungal disease of rice plants called "foolish seedling disease" which caused rapid stem elongation that weakened the plants and caused them to collapse. The fungus infecting the plants produced an excess of a chemical called gibberellin, named after the fungus, *Gibberella fujikuroi*. It wasn't until 1956 that a form of gibberellin now known as gibberellic acid was isolated from a flowering plant rather than from a fungus. Over 130 different gibberellins have been identified. The most common of these is gibberellic acid, GA_3 (the subscript 3 distinguishes it from other gibberellins).

Young leaves, roots, embryos, seeds, and fruits are places where natural gibberellins can be found. Gibberellins are growth-promoting hormones that bring about elongation of cells. When gibberellins are applied externally to plants, the most obvious effect is stem elongation between the nodes (**Fig. 21.4***a*). Gibberellins can cause dwarf plants to grow, cabbage plants to become as much as 2 meters tall, and bush beans to become pole beans.

There are many commercial uses for gibberellins that promote growth in a variety of crops such as apples, cherries, and sugarcane. A notable example is their use on many of the table grapes grown in the United States. Commercial grapes are a genetically seedless variety that would naturally produce small fruit on very small bunches. Treatment with GA_3 substitutes for the presence of seeds, which would normally be the source of native gibberellins for fruit growth. Treatments increase both fruit stem length (producing looser clusters) and fruit size (Fig. 21.4*b*). In the brewing industry, the production of beer relies on the breakdown of starch in barley grains (seeds). Barley grains would naturally be dormant—a period in which a seed does not grow. Gibberellins are used to break the dormancy of barley grains, yielding fermentable sugars, mainly maltose, which are then fermented by yeast to produce ethanol.

Cytokinins

Cytokinins were discovered as a result of attempts to grow plant tissues and organs in culture vessels in the 1940s. It was found that cell division occurs when coconut milk (a liquid endosperm) and yeast extract are added to the culture medium. Although the effective agents could not be isolated at the time, they were collectively called cytokinins because, as you may recall, *cytokinesis* means "division of the cytoplasm." A naturally occurring cytokinin was not isolated until 1967. Because it came from the kernels of maize (*Zea*), it was called zeatin.

Cytokinins influence plant growth by promoting cell division. Cytokinins are found in plant meristems, young leaves, root tips, and seeds and fruits. Whenever a plant grows, cytokinins are involved.

Cytokinins also prevent **senescence,** or aging, of plant organs. As cytokinin levels drop within a plant organ, such as a leaf, growth slows or even stops. Then, the leaf loses its natural color as large molecules are broken down and transported to other parts of the plant. Senescence is a necessary part of a plant's growth. For example, as some plants grow taller, they naturally lose their lower leaves.

Researchers are aware that the ratio of auxin to cytokinin and the acidity of the culture medium determine whether a plant tissue forms an undifferentiated mass, called a callus (see Fig 21.23*a*), or differentiates to form roots, vegetative shoots, leaves, or floral shoots. Research indicates that the presence of auxin and cytokinins in certain ratios leads to the activation of an enzymatic pathway that releases from the cell walls chemicals that influence the specialization of plant cells.

Abscisic Acid

If environmental conditions are not favorable, a plant needs to protect itself. **Abscisic acid** is sometimes called the stress hormone because it initiates and maintains seed and bud dormancy and brings about the closure of stomata. Dormancy has begun when a plant stops growing and prepares for adverse conditions (even though conditions at the time may be favorable for growth). For example, it is believed that abscisic acid moves from leaves to vegetative buds in the fall, and thereafter these buds are converted to winter buds. A winter bud is covered by thick, hardened scales (**Fig. 21.5***a*). A reduction in the level of abscisic acid and an increase in the level of gibberellins are believed to

a.   b.

Figure 21.4 **Effect of gibberellins.**

a. The plant on the right was treated with gibberellins; the plant on the left was not treated. Gibberellins are often used to promote stem elongation in economically important plants. **b.** Grapes untreated (*left*) and treated (*right*) with gibberellins. The treated grapes have a longer stem length and larger fruit size.

(a): © Science Source; (b): © Amnon Lichter, The Volcani Center

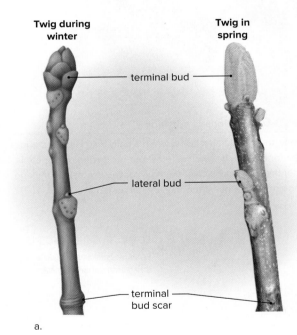

a.

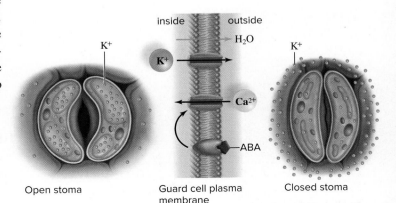

b.

Figure 21.5 Effects of abscisic acid.

a. Abscisic acid encourages the formation of winter buds (*left*), and a reduction in the amount of abscisic acid breaks bud dormancy (*right*). **b.** Abscisic acid also brings about the closing of a stoma by influencing the movement of potassium ions (K$^+$) out of the guard cells.

(a): © John Seiler/Virginia Tech Forestry Department

a. No abscission   b. Abscission

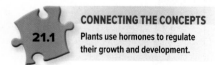

Green tomatoes harvested

Ethylene applied

c.

Figure 21.6 **Functions of ethylene.**

a. Normally, there is no abscission when a holly twig is placed under a glass jar for a week. **b.** When an ethylene-producing ripe apple is also under the jar, abscission of the holly leaves occurs. **c.** Ethylene is used to ripen tomatoes after they are harvested.

(a-b): © Kingsley Stern

CONNECTING THE CONCEPTS

21.1   Plants use hormones to regulate their growth and development.

break seed and bud dormancy. Seeds will then germinate, and buds will develop into leaves.

Abscisic acid brings about the closing of stomata when a plant is under water stress (Fig. 21.5b). Abscisic acid causes potassium ions (K^+) to leave guard cells. Thereafter, the guard cells lose water, and the stoma closes.

Ethylene

At one time, it was believed that abscisic acid functioned in the process of **abscission,** which is the dropping of leaves, fruits, or flowers from a plant. Although the external application of abscisic acid promotes abscission, this hormone is no longer believed to function naturally in this process. The hormone **ethylene** is now known to be responsible for fruit abscission and fruit ripening in plants.

Ethylene is a gas that can move freely in the air. The hormone stimulates certain enzymes, such as cellulase, that cause leaf, fruit, or flower drop (**Fig. 21.6**a,b). Cellulase hydrolyzes cellulose in plant cell walls and weakens that part of the plant.

In the early twentieth century, it was common practice to prepare citrus fruits for market by placing them in a room with a kerosene stove. Only later did researchers realize that ethylene, an incomplete combustion product of kerosene, was ripening the fruit. Because it is a gas, ethylene can act from a distance, and is often used commercially to ripen fruit just before it is delivered to the grocery store (Fig. 21.6c). A barrel of ripening apples can induce ripening in a bunch of bananas, even if they are in different containers. If a plant is wounded due to physical damage or infection, ethylene is released at the wound site. This is why one rotten apple spoils the whole barrel.

Check Your Progress 21.1

1. List the five commonly recognized groups of plant hormones.
2. Describe the main functions of each of the five groups of plant hormones.
3. Discuss why plant hormones that have some effect on plant growth can be found in different areas of a plant.

21.2 Plant Responses

Learning Outcomes

Upon completion of this section, you should be able to

1. Define *tropism,* and give examples of three common tropisms in plants.
2. Explain how and why seedlings are affected by positive and negative gravitropism.
3. Describe how phytochrome allows a flowering plant to detect the photoperiod.

Plant responses are strongly influenced by such environmental stimuli as light, day length, gravity, and touch. A plant's ability to respond to environmental signals enhances the survival of the plant in that environment.

Plant responses to environmental signals can be rapid, as when stomata open in the presence of light, or they can take some time, as when a plant flowers in season. Despite their variety, plant responses to environmental signals are most often exhibited in growth and sometimes changes in plant tissues, brought about at least in part by certain hormones.

Tropisms

A plant's growth response toward or away from a directional stimulus is called a **tropism.** Differential growth causes one side of an organ to elongate faster than the other, and the result is a curving toward or away from the stimulus. Growth toward a stimulus is called a positive tropism, and growth away from a stimulus is called a negative tropism. The following tropisms were each named for the stimulus that causes the response:

- Phototropism: growth in response to a light stimulus
- Gravitropism: growth in response to gravity
- Thigmotropism: growth in response to touch

Phototropism is the growth of plants toward a source of light. If the light is coming to the plant from a single direction, auxin migrates to the shady side of the plant and cell elongation causes the stem and leaves to bend toward the sunlight (**Fig. 21.7***a*). **Thigmotropism** is a response to touch from another plant, an animal, rocks, or the wind. Climbing vines such as English ivy use touch contact with rocks, tree trunks, or other supports for growth (Fig. 21.7*b*). This adaptation for thigmotropism supports leaf growth toward sunlight rather than investing energy in building supportive tissues of the stem.

Gravitropism is the growth response of plants to Earth's gravity. When a seed germinates, the embryonic shoot exhibits negative gravitropism by growing upward *against* gravity. Increased auxin concentration on the lower side of the young stem causes the cells in that area to grow more than the cells on the upper side, resulting in growth upward. The embryonic root exhibits positive gravitropism by growing *with* gravity downward into the soil (**Fig. 21.8***a*). Root cells know which way is down because of the presence of an organelle called an *amyloplast*. Imagine placing a few marbles in a tennis ball. No matter how you move the ball, the marbles will always settle to the bottom. The same holds true for amyloplasts, which settle at the bottom of endodermal root cells and signal downward growth (Fig. 21.8*b*).

a.　　　　　　　　　　　　　　b.

Figure 21.7 Examples of tropisms.

a. The stem of a plant curves toward the light, exhibiting phototropism. This response is due to the accumulation of auxin on the shady side of the stem. **b.** English ivy is thigmotropic and climbs a tree trunk.

(a): © Cathlyn Melloan/Stone/Getty Images; (b): © Alison Thompson/Alamy

Figure 21.8 Gravitropism.

a. The corn seed was germinated in a sideways orientation and in the dark. The shoot is growing upward (negative gravitropism) and the root downward (positive gravitropism). **b.** Amyloplasts settle toward the bottom of the cells and play a role in the perception of gravity by roots.

(a): © Martin Shields/Alamy; (b): © Randy Moore

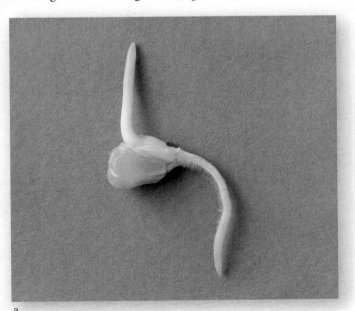

a.

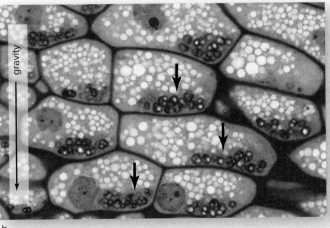

b.

Photoperiodism

Flowering in angiosperms is a striking response to environmental seasonal changes. In some plants, flowering occurs according to the **photoperiod,** which is the ratio of the length of day to the length of night over a 24-hour period.

You may know someone who suffers from seasonal allergies. In the spring, the days are longer and the nights are shorter, triggering long-day (short-night) plants to flower, produce pollen, and affect individuals with spring allergies. In the fall, the days are shorter and the nights are longer, triggering short-day (long-night) plants to flower, produce pollen, and affect individuals with fall allergies.

The spring-flowering and fall-flowering plants are responding to a *critical length*—a period of light specific in length for any given species, which appears to initiate flowering (**Fig. 21.9***a,b*). Experiments have shown that the length of continuous darkness, not light, is what actually controls flowering in many plants. Nurseries use these kinds of data to make all types of flowers available throughout the year (Fig. 21.9*b*).

Phytochrome

If flowering is dependent on night length, plants must have a way to detect these periods. In some plants, this appears to be the role of a leaf pigment called **phytochrome.** Phytochrome can detect the wavelengths of light from the sun (see Fig. 6.4) and can distinguish between red wavelengths and far-red wavelengths of light.

Figure 21.10 shows how the presence of red light or far-red light determines the particular form of phytochrome:

P_r (phytochrome red) absorbs red light and is converted into P_{fr} in the daytime.
P_{fr} (phytochrome far-red) absorbs far-red light and is converted into P_r in the evening

a. Short-day (long-night) plants

Long-day (short-night) plants

Fall flower critical length

Spring flower critical length

b.

Figure 21.9 The photoperiod controls flowering.

a. In the fall, the cocklebur plant flowers when the day is shorter (night is longer) than 8.5 hours; in the spring, the clover plant flowers when the day is longer (night is shorter) than 8.5 hours. **b.** Nurseries know how to regulate the photoperiod so that many types of flowers are available year-round.

During the day, sunlight contains more red light than far-red light, and P$_r$ is converted into P$_{fr}$. But at dusk, more far-red light is available, and P$_{fr}$ is converted to P$_r$. There is also a slow metabolic replacement of P$_{fr}$ by P$_r$ during the night.

Plant spacing is another interesting function of phytochrome. Next time you are at a garden center, read the instructions on a seed packet and notice the specific details on spacing the seeds placed in the ground. In nature, red and far-red light also signal spacing. Leaf shading increases the amount of far-red light relative to red light. Plants somehow measure the amount of far-red light bounced back to them from neighboring plants. The closer together plants are, the more far-red relative to red light they perceive and the more likely they are to grow tall, a strategy for outcompeting others for sunshine!

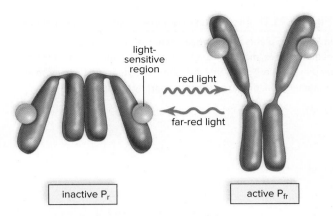

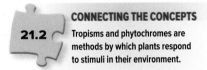

Figure 21.10 **Phytochrome control of growth.**

When phytochrome is inactive, the plant senses that it is evening. When phytochrome is in its active form, the plant senses that it is daytime.

Check Your Progress 21.2

1. Compare and contrast the three forms of tropisms.

2. Explain the difference between short-day and long-day plants.

3. Explain how phytochrome is affected by red wavelengths of light.

CONNECTING THE CONCEPTS

21.2 Tropisms and phytochromes are methods by which plants respond to stimuli in their environment.

21.3 Sexual Reproduction in Flowering Plants

Learning Outcomes

Upon completion of this section, you should be able to

1. Explain the alternation of generations life cycle of a flowering plant.

2. Identify the parts of a flower, and briefly define their functions.

3. Contrast a monoecious plant with a dioecious plant.

4. Describe the processes and result of double fertilization.

5. Understand the reason why plants have various methods of seed dispersal.

6. Compare and contrast seed germination in a eudicot versus in a monocot.

In Section 18.1, we discussed the two multicellular stages that alternate in the plant life cycle, which is called an **alternation of generations.** In this life cycle, a diploid (2n) sporophyte alternates with a haploid (n) gametophyte:

- The **sporophyte** (2n) produces haploid spores by meiosis. The spores develop into gametophytes.
- The **gametophytes** (n) produce gametes. Upon fertilization, the cycle returns to the 2n sporophyte.

Overview of the Plant Life Cycle

Flowering plants have an alternation-of-generations life cycle, but with the modifications shown in **Figure 21.11.** After this overview of the flowering plant life cycle, we will discuss the life cycle in more depth. In flowering plants, the sporophyte is dominant, and it is the generation that produces flowers. The flower is the reproductive organ of angiosperms. The flower of the sporophyte produces two types of spores: microspores and megaspores. A

Figure 21.11 **Alternation of generations in flowering plants.**

In flowering plants, there are two types of spores and two gametophytes, male and female. Flowering plants are adapted to a land existence: The spores, the gametophytes, and the zygote are protected from drying out, in large part by the sporophyte.

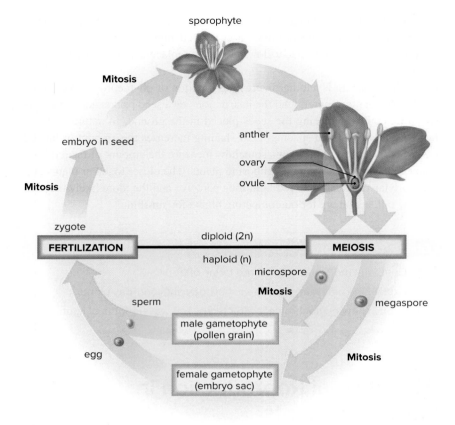

Figure 21.12 **Anatomy of a flower.**

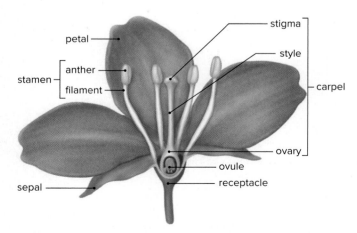

A complete flower has all these parts: sepals, petals, stamens, and at least one carpel.

microspore develops into a male gametophyte, which is a pollen grain. A *megaspore* develops into a female gametophyte, the embryo sac, which is microscopic and housed deep within the flower.

A pollen grain is either windblown or carried by an animal to the vicinity of the embryo sac. At maturity, a pollen grain contains two nonflagellated sperm. The embryo sac contains an egg.

A pollen grain develops a pollen tube, and the sperm move down the pollen tube to the embryo sac. After a sperm fertilizes an egg, the zygote becomes an embryo, still within the flower. The structure that houses the embryo develops into a seed. The seed also contains stored food and is surrounded by a seed coat. The seeds are often enclosed by a fruit, which aids in dispersing the seeds. When a seed germinates, a new sporophyte emerges and develops.

The life cycle of flowering plants is well adapted to a land existence (see Chapter 18). No external water is needed to transport the pollen grain to the embryo sac, or to enable the sperm to reach the egg. All stages of the life cycle are protected from drying out.

Flowers

The **flower** is a reproductive structure that is unique to the angiosperms (**Fig. 21.12**). Flowers produce the spores and protect the gametophytes. They often attract pollinators, which help transport pollen from plant to plant. Flowers also produce the fruits that enclose the seeds. The success of angiosperms, with over 270,000 species, is largely attributable to the evolution of the flower.

In monocots, flower parts occur in threes and multiples of three; in eudicots, flower parts are in fours or fives and multiples of four or five (**Fig. 21.13**).

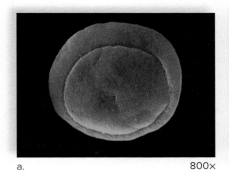
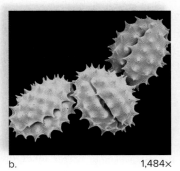
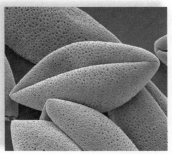

a.   800×   b.   1,484×   c.   2,000×

Figure 21.16 Pollen.

a. Pollen grains are so distinctive that a paleontologist can use fossilized pollen to date the appearance of a plant in a particular area. Pollen grains can become fossils because their strong walls are resistant to chemical and mechanical damage. **b.** Pollen grains of Canadian goldenrod. **c.** Pollen grains of a pussy willow.

(a): © Dee Breger/Science Source; (b): © Medical-on-line/Alamy; (c): © Steve Gschmeissner/Science Photo Library/Getty RF

Pollen grains are distinctive to the particular plant (**Fig. 21.16**), and **pollination** in flowering plants is simply the transfer of pollen from the anther to the stigma of a carpel. Plants often have adaptations that favor *cross-pollination,* which occurs when the pollen landing on the stigma is from a different plant of the same species. For example, the carpels may mature only after the anthers have released their pollen. Cross-pollination may also be brought about with the assistance of an animal pollinator. If a pollinator, such as a bee, goes from flower to flower of only one type of plant, cross-pollination is more likely to occur in an efficient manner. The secretion of nectar is one way that plants attract insects; over time, certain pollinators have become adapted to reach the nectar of only one type of flower. In the process, pollen is inadvertently picked up and taken to another plant of the same type. Plants attract particular pollinators in still other ways. For example, through the evolutionary process, some species have flowers that smell of rotting flesh; they are called carrion flowers, or stinking flowers. The putrid smell attracts flies, which in turn pick up the pollen and move it to another "rotting" plant!

Figure 21.15 also shows the development of the megaspore. In an ovule, within the ovary of a carpel, meiosis produces four megaspores. One of the cells develops into the female gametophyte, or so-called embryo sac, which is a seven-celled structure containing a single egg cell. Fertilization of the egg occurs after a pollen grain lands on the stigma of a carpel and develops a pollen tube. A pollen grain that has germinated and produced a pollen tube is the mature male gametophyte (see Fig. 21.15, *middle*). A pollen tube contains two sperm. Once the sperm reaches the ovule, **double fertilization** occurs. One sperm unites with the egg, forming a 2n (diploid) zygote. The other sperm unites with two nuclei centrally placed in the embryo sac, forming a 3n (triploid) endosperm cell.

Development of the Seed in a Eudicot

It is now possible to account for the three parts of a seed: seed coat, embryo, and endosperm. The *seed coat* is a protective covering that once was the ovule wall. Double fertilization results in an endosperm nucleus and a zygote. Cell division produces a multicellular *embryo* and a multicellular **endosperm,** which is the stored food of a seed.

Figure 21.17 shows the stages in the development of the seed. Tissues become specialized until eventually a shoot and root tip containing apical meristems develop.

Notice that the **cotyledons,** or embryonic leaves, absorb the developing endosperm and become large and fleshy (Fig. 21.17*f*). The food stored by the cotyledons will nourish the embryo when it resumes growth. Cotyledons wither when the first *true leaves* grow and become functional. The common garden bean is a good example of a eudicot seed with large cotyledons and no endosperm (see Fig. 21.20).

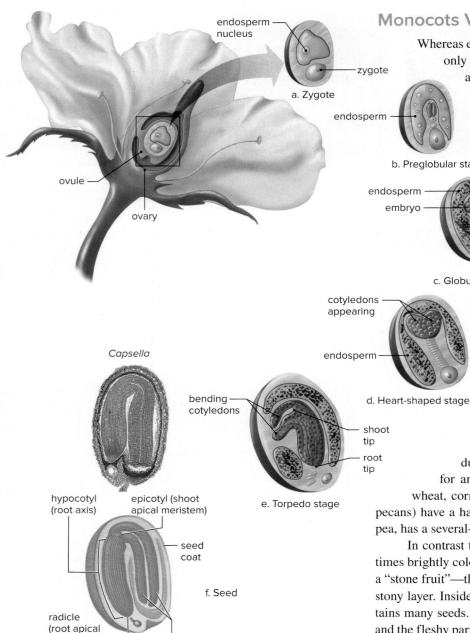

Capsella

Figure 21.17 **Development of the seed in a eudicot.**

Development begins with (**a**) the zygote and ends with (**f**) the seed. As the embryo develops, it progresses through several stages from (**b**) to (**e**).

(f): © Steven P. Lynch

Monocots Versus Eudicots

Whereas eudicot embryos have two cotyledons, monocot embryos have only one cotyledon. In monocots, the cotyledon stores food, and it absorbs food molecules from the endosperm and passes them to the embryo. In other words, the endosperm is retained in monocot seeds. In eudicots, the cotyledons usually store all the nutrient molecules the embryo uses. Therefore, the endosperm disappears, because it has been taken up by the two cotyledons.

A corn plant is a monocot; consequently, in a corn kernel, there is only one cotyledon and the endosperm is present (see Fig. 21.21).

Fruit Types and Seed Dispersal

Flowering plants have seeds enclosed by a **fruit** (**Fig. 21.18**). Seeds develop from ovules, and fruits develop from ovaries and sometimes other parts of a flower. Technically, market produce, such as cucumber, tomatoes, and sugar snap peas, are fruits—not vegetables—because they contain seeds. A *vegetable* is an edible plant part without seeds, such as celery (a stem), lettuce (leaves), and carrots (roots).

Fruits are quite diverse. *Dry fruits* are generally a dull color with a thin, dry ovary wall, so that the potential food for animals is largely confined to the seeds. In grains such as wheat, corn, and rice, the fruit looks like a seed. Nuts (e.g., walnuts, pecans) have a hard, outer shell covering a single seed. A legume, such as a pea, has a several-seeded fruit that splits open to release the seeds.

In contrast to dry *fruits, fleshy fruits* have a juicy portion that is sometimes brightly colored to attract animals. A drupe (e.g., peach, cherry, olive) is a "stone fruit"—the outer part of the ovary wall is fleshy, but there is an inner, stony layer. Inside the stony layer is the seed. A berry, such as a tomato, contains many seeds. An apple is a *pome,* in which a dry ovary covers the seeds, and the fleshy part is derived from the receptacle of the flower. A strawberry is an interesting fruit, because the flesh is derived from the receptacle, and what appear to be the seeds are actually dry fruits!

Dispersal of Seeds

In order for plants to be successful in their environments, their seeds have to be dispersed—that is, moved long distances from the parent plant. There are various methods of seed dispersal. For example, birds and mammals sometimes eat fruits, including the seeds, which then pass out of the digestive tract with the feces some distance from the parent plant (**Fig. 21.19***a*). Squirrels and other animals gather seeds and fruits, which they bury some distance away. Some plants have evolved unusual ways to ensure dispersal. The hooks and spines of clover, cocklebur, and burdock fruits attach to the fur of animals and the clothing of humans, which carry them far away from the parent plant (Fig. 21.19*b*). Other seeds are dispersed by wind. Woolly hairs, plumes, and wings are all adaptations for this type of dispersal. The dandelion fruit is attached to several

Figure 21.18 Examples of fruits.

Some fruits are dry, such as walnuts and peas. Some fruits are fleshy, such as peaches and apples. To a botanist, any plant product derived from an ovary plus perhaps other flower parts is a fruit.

(walnut half): © Jack Star/PhotoLink/Getty RF; (whole walnut): © Siede Preis/Getty RF; (peas and pods): © Martin Barraud/Getty Images; (peach half): © Peter Fakler/Alamy RF; (whole peach): © Jupiterimages/ Image Source RF; (both apples): © Photolink/Getty RF

hairs that function as a parachute and aid dispersal (Fig. 21.19*c*). The winged fruit of a maple tree, which contains two seeds, has been known to travel up to 10 kilometers from its parent (Fig. 21.19*d*). Different still, a touch-me-not plant has seed pods that swell as they mature. A passing animal may cause the swollen pods to burst, hurling the ripe seeds some distance away from the plant.

Germination of Seeds

Following dispersal, if all goes well, the seeds will **germinate.** As growth occurs, a seedling appears. Germination does not usually take place until there is sufficient water, warmth, and oxygen to sustain growth. In deserts, germination does not occur until there is adequate moisture. These requirements help ensure that seeds do not germinate until the most favorable growing season has arrived. Some seeds do not germinate until they have been dormant for a period of time. For seeds, dormancy is the time during which no growth occurs, even

a. Holly berry   b. Burdock

c. Dandelion   d. Maple

Figure 21.19 Methods of seed dispersal.

Many plants have adaptations to ensure that their seeds are spread some distance from the parent plant. **a.** The mockingbird will carry the seed of a holly some distance away. **b.** The spines of burdock fruit stick to a passerby. **c.** Dandelion and (**d**) maple fruits have adaptations that allow them to be carried long distances by wind.

(a): © Bill Draker/Getty RF; (b): © Scott Camazine/Science Source; (c): © Henrik Weis/Getty RF; (d): © Robert Llewellyn/Corbis RF

Connections: Health

Why is coconut part of so many processed foods?

© Foodcollection RF

A coconut is a large nut, and the white, milky substance on the inside is the endosperm. Humans have been using this versatile nut for food, oil, and shredded fiber from the husk. Health food stores sell electrolyte-packed coconut water as a hydrating energy drink, and coconut oil is favored by cooks who do not use dairy products to achieve a butterlike consistency. In the Philippines and Papua New Guinea, a coconut oil blend is being used to power ships, trucks, and cars!

CONNECTING THE CONCEPTS

21.3

Reproduction in angiosperms involves an alternation of generations and a reproductive structure called the flower.

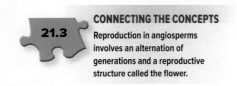

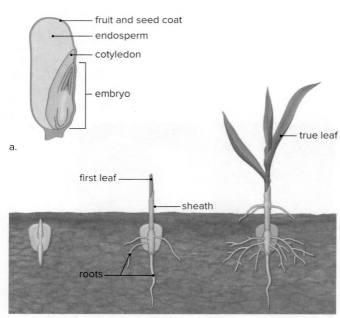

a.

b.

Figure 21.21 Corn, a monocot.

a. Corn kernel structure. **b.** Germination and development of the seedling. Notice that there is one cotyledon and that the leaves are parallel-veined.

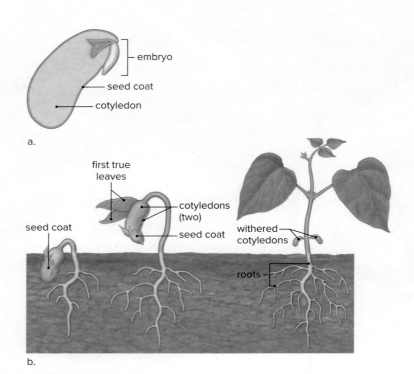

a.

b.

Figure 21.20 Common garden bean, a eudicot.

a. Seed structure. **b.** Germination and development of the seedling. Notice that there are two cotyledons and that the leaves are net-veined.

though conditions may be favorable for growth. In the temperate zone, seeds often have to be exposed to a period of cold weather before dormancy is broken. Fleshy fruits (e.g., apples, pears, oranges, and tomatoes) contain inhibitors, so that germination does not occur until the seeds are removed and washed. Aside from water, bacterial action and even fire can act on the seed coat, allowing it to become permeable to water. The uptake of water causes the seed coat to burst and germination to occur.

Eudicot Versus Monocot Germination

If the two cotyledons of a bean seed are parted, you can see the cotyledons and a rudimentary plant with immature leaves. As the eudicot seedling emerges from the soil, the shoot is hook-shaped to protect the immature leaves as they start to grow. The cotyledons shrivel up as the true leaves of the plant begin photosynthesizing (**Fig. 21.20**).

A corn kernel is actually the fruit of a monocot. The outer covering is the fruit and seed coat combined (**Fig. 21.21**). Inside is the single cotyledon. Also, both the immature leaves and the root are covered by sheaths. The sheaths are discarded when the seedling begins growing, and the immature leaves become the first true leaves of the corn plant.

Check Your Progress 21.3

1. Describe the anatomy of a flower.
2. Summarize the life cycle of a flowering plant.
3. Discuss the importance of seeds and seed dispersal.

21.4 Asexual Reproduction and Genetic Engineering in Plants

In asexual reproduction, there is only one parent and all the offspring are **clones**—genetically identical individuals. Clones are desirable for plant sellers, because the plants will look and behave exactly like the parent.

Propagation of Plants in a Garden

If you wanted to create a bed of tulips, irises, or gladiolas in your garden, you would not plant seeds but instead would rely on bulbs, rhizomes, or corms and reproduce the plants asexually (**Fig. 21.22**). Baking potatoes are modified stems called tubers, and each "eye" has a bud that can become a new plant. All of these structures are typically fleshy, underground food-storage tissues that contain buds that will sprout in the spring. Runners, or stolons, such as those found in strawberries, are horizontal stems that can also result in new clonal plants.

Propagation of Plants in Tissue Culture

One of the major disadvantages of most asexual propagation techniques is that they also propagate pathogens. Plant pathogens can be viruses, bacteria, or fungi, and clones created from an infected parent will also be infected. However, it is possible to maintain plants in a disease-free status if clones from an uninfected parent are made in sterile test tubes through tissue culture. Hence, **tissue culture** is simply plant propagation done in a laboratory under sterile conditions.

Figure 21.22 **Structures for asexual propagation.**

New plants can grow by the separation of parts from the original plant. **a.** A strawberry plant has aboveground horizontal stems called stolons. Every other node produces a new shoot system. **b.** An iris rhizome, (**c**) a potato tuber, and (**d**) a gladiola corm are all modified stems that can produce clonal offshoots.

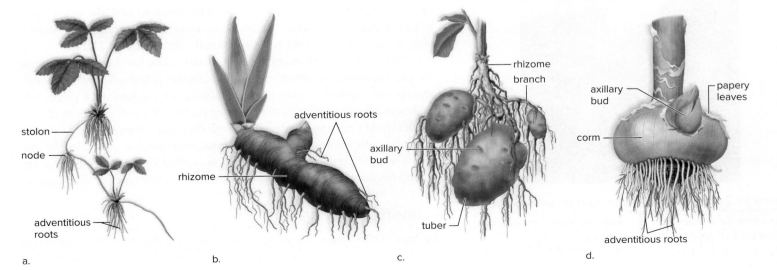

a.   b.   c.   d.

Figure 21.23 Tissue culture.

a. Meristem tissue is grown on a sterile medium, and an undifferentiated mass (*left*), called a callus, develops. From the callus, embryos (*right*) develop organs such as leaves and roots.
b. The embryos (*left*) develop into plantlets (*right*). The plantlets can be stored, then shipped in sterile containers and transferred to soil for growth into adult plants.

© Khwanchais/Getty RF

a.
b.

If you were to take one cell from an animal and try to grow it in a test tube, you would not be able to make another whole animal. On the other hand, if you take one cell from a plant you *can* grow an entirely new plant. This capacity of one plant cell to give rise to a mature plant is called *totipotency* and is the reason plant tissue culture is so successful.

Techniques for tissue culture vary, but most begin with cells from the meristem of the parent plant. Meristematic cells are grown in a sterile, jellylike substrate in flasks, tubes, or petri dishes. The substrate, also called a medium, contains growth hormones, vitamins, and macro- and micronutrients and provides dividing plant cells with all the support, nutrients and water they need. Initially, a mass of undifferentiated (unspecialized) cells, called a *callus,* forms (**Fig. 21.23***a*). The addition of more nutrients and hormones will initiate organ formation until a fully developed plantlet is formed (Fig. 21.23*b,c*). Plantlets can then be shipped off in their sterile containers to growers for transplantation into soil pots or the field (Fig. 21.23*d*).

Tissue culture is an important technique for propagating many fruits and vegetables found in local supermarkets. As described in the chapter opener, bananas are sterile fruit that do not produce seeds. The only way to provide this commercially important fruit for the whole world is through tissue culture. Asparagus is a dioecious plant, and all commercial stalks are male. The female stalks favor the production of flowers and are undesirable for eating. Tissue culture is therefore a more efficient means of producing disease-free male asparagus for growers.

Many botanical gardens and universities use tissue culture for plant conservation. The Atlanta Botanical Garden has a tissue culture lab that propagates native species of orchids and lilies. The propagated plants are shared with local nurseries, resulting in fewer plants being removed from the wild by collectors. Tissue culture of rare species is also used to help increase their populations in the wild, as they are replanted in native habitats (**Fig. 21.24**).

When the desired product is not an entire plant but merely a substance the plant produces, scientists can use a technique called *cell suspension culture.* In one method, rapidly growing calluses are cut into small pieces and shaken in a liquid nutrient medium, so that single cells or small clumps of cells break off and form a suspension. The target chemical is then extracted from the liquid the cells are growing in. Cell suspension cultures of cells from the quinine tree produce quinine, a drug used to combat malaria, and those of the woolly foxglove produce digitoxin, used to treat certain types of heart disease.

Figure 21.24 Tissue culture helps plant conservation.

This rare Kentucky ladyslipper orchid was grown in a tissue culture lab and will be replanted into a native habitat.

© Grace Clementine/Getty RF

Genetic Engineering of Plants

At least 10,000 years ago, humans began to leave the hunting and gathering lifestyle and settled down to farm. Plants were selected and hybridized to obtain the most desirable crops for food and textiles. Following simple Mendelian genetics, *hybridization* is the crossing of different varieties of plants to produce plants with desirable traits. Farming in this manner was suitable for feeding small communities of people.

Today, there are many more people who need to be fed. Providing enough food and textiles rests on scientists' ability to modify plants through genetic engineering. The invention of large-scale DNA sequencing and advances in modern molecular genetics have increased the understanding on how genes work and how they can be incorporated into plant DNA for very precise additions of desirable traits. A plant that has had its DNA altered in some way is said to be *transgenic,* or *genetically modified* (*GM*); more generally, any organism that has been altered in this way is a **genetically modified organism (GMO).**

Figure 21.25 shows two ways genetic engineering is accomplished. The parent plant lacks a desirable trait; a gene from another organism is studied, isolated, and used to transform the parent plant. A common *transformation* technique relies on a natural genetic engineer, *Agrobacterium tumefaciens.* This bacterium infects plant cells and then inserts DNA from its plasmids into the plant's chromosomes. Another technique uses a particle gun that shoots DNA into plant cells for transformation. The transformed plant is then isolated for its new trait, and plantlets are generated. Depending on the purpose, the GM plant can then be cloned through tissue culture or can be involved in sexual reproduction to produce seeds.

GM plants are often modified with genes that improve agricultural, food-quality, and medicinal traits (**Table 21.2**). A notable GM plant is Bt corn. In the past, corn plants were attacked by a butterfly larva called a corn borer. Growers sprayed their fields with insecticides that were costly and had a negative impact on the environment. The Bt gene, isolated from the soil bacterium *Bacillus thuringiensis,* produces proteins that kill corn borers. The Bt gene was inserted into corn chromosomes, resulting in corn plants resistant to the larvae.

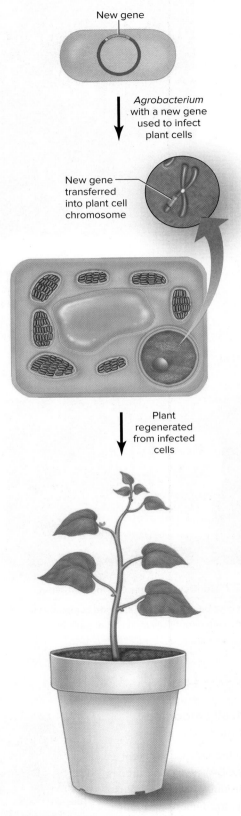

Figure 21.25   Transformation of plants.

Scientists can place a gene of interest into the bacterium *Agrobacterium tumefaciens,* which in turn can infect a plant cell, causing transformation of the plant.

Table 21.2   GM Plants

IMPROVED AGRICULTURAL TRAITS
Herbicide resistance: wheat, rice, sugar beets, canola
Salt tolerance: cereals, rice, sugarcane, canola
Drought tolerance: cereals, rice, sugarcane
Cold tolerance: cereals, rice, sugarcane
Improved yield: cereals, rice, corn, cotton
Disease protection: wheat, corn, potatoes, cotton
IMPROVED FOOD-QUALITY TRAITS
Fatty acid/oil content: corn, soybeans
Protein/starch content: cereals, potatoes, soybeans, rice, corn
Amino acid content: corn, soybeans
Vitamin A content: rice
MEDICINAL TRAITS
Vaccine production: corn, soybeans, tomatoes
Antibody production: tobacco

a.   b.

Figure 21.26 Genetically engineered Bt plants.

a. A field of Bt corn genetically modified to be resistant to corn borer larvae (inset). **b.** Cotton field with both unmodified (*left*) and modified (*right*) Bt cotton plants.

(a): (cornfield): © Bill Barksdale/Agstockusa/age fotostock; (corn ear worm): © Scott Camazine/Science Source; (b): Source: Colin Tann, Courtesy of Commonwealth Scientific and Industrial Research Organization (CSIRO)

Figure 21.27 Controversy over GM food labeling.

Activists want laws passed that require food companies to indicate if foods are made with genetically modified organisms.

© Norma Jean Gargasz/Alamy

Connections: Health

Are genetically modified crops organic?

According to the U.S. Department of Agriculture (USDA), an organic crop is one that is produced without the use of pesticides, irradiation, hormones, antibiotics, or bioengineering. There-  © Brand X Pictures/ Getty RF

fore, a genetically modified crop may not be marketed as an organic crop, since it is a product of artificial technology.

Currently, 65% of the total U.S. corn crop is Bt corn. Other GM crops, such as Bt potato and Bt cotton, also have this gene (**Fig. 21.26**).

One of the recent successes of GM crops is the development of golden rice. This rice has been genetically modified with daffodil and bacterial genes to produce beta-carotene (provitamin A). The World Health Organization (WHO) estimates that vitamin A deficiency affects between 140 and 250 million preschool children worldwide. The deficiency is especially severe in developing countries where the major staple food is rice. Provitamin A in the diet can be converted by enzymes in the body to vitamin A, alleviating the deficiency.

Environmental concerns about GM plants have focused on possible cross-hybridization with wild plants, the possibility of creating herbicide-resistant "super weeds," and the effects of GM pollen on pollinators. The biggest social concern is about the human and livestock consumption of GM plants, and laws have been proposed to place labels on all foods made with GM crops (**Fig. 21.27**). Currently, new GM crops under development must undergo rigorous analysis and testing before being approved in the United States. After citing a study that reviewed 25 years of GM crop research, the American Association for the Advancement of Science (AAAS) released a statement in 2012 stating that genetic engineering, resulting in crop improvement, is safe and that food labeling would only falsely alarm consumers.

Pharmaceutical Products

Genetic engineering has resulted in a new wave of research into the development of plant-made pharmaceuticals. Plants can produce antigens, antibodies, hormones, and therapeutic proteins. The advantages of using plants to produce pharmaceuticals include decreased cost, increased amount of protein produced, and decreased risk of contamination with animal and human pathogens. One type of antibody made by tobacco plants is being developed to combat tooth decay. Already approved for veterinary use, a plant-made antibody against certain forms of the cancer lymphoma is being developed for humans. In addition to producing antibodies, plants excel at creating "hard-to-make" proteins, such as anticoagulants, growth hormones, blood substitutes, collagen replacements, and antimicrobial agents, to name a few.

Check Your Progress 21.4

1. List the structures that some plants use for asexual reproduction.
2. Give three examples of applications for plant tissue culture.
3. Summarize the pros and cons of genetically modified plants.

CONNECTING THE CONCEPTS
21.4 Scientists use asexual reproduction to produce identical clones of a plant.

STUDY TOOLS http://connect.mheducation.com

SMARTBOOK® Maximize your study time with McGraw-Hill SmartBook®, the first adaptive textbook.

SUMMARIZE

By understanding plant hormones, the responses of plants to stimuli, and the forms of sexual and asexual reproduction found in plants, scientists have been able to increase food supplies and provide an abundance of plant-based materials.

21.1 Plants use hormones to regulate their growth and development.

21.2 Tropisms and phytochromes are methods by which plants respond to stimuli in their environment.

21.3 Reproduction in angiosperms involves an alternation of generations and a reproductive structure called the flower.

21.4 Scientists use asexual reproduction to produce identical clones of a plant.

21.1 Plant Hormones

Plant **hormones** are chemical signals that cause responses within plant cells and tissues. The five commonly recognized groups of plant hormones are

- **Auxins:** affect growth patterns and cause **apical dominance** and **phototropism**
- **Gibberellins:** promote stem elongation and break seed dormancy
- **Cytokinins:** promote cell division, prevent senescence of leaves, and influence differentiation of plant tissues
- **Abscisic acid:** initiates and maintains seed and bud dormancy and closing of stomata
- **Ethylene:** causes **abscission** of leaves, fruits, and flowers and ripens fruits

21.2 Plant Responses

Environmental signals play a significant role in plant growth and development.

Tropisms

Tropisms are growth responses toward (positive) or away from (negative) unidirectional stimuli.

phototropism:	response to a light
gravitropism:	response to gravity
thigmotropism:	response to touch

Auxins cause shoots to exhibit negative **gravitropism,** and amyloplasts cause roots to exhibit positive gravitropism. Other tropisms include **phototropism,** or growth toward a light source, and **thigmotropism,** or growth toward contact with an object (touch).

Photoperiodism

Flowering is a response to a seasonal change—namely, length of the **photoperiod.**
- Short-day plants flower when nights are longer than a critical length.
- Long-day plants flower when nights are shorter than a critical length.

Phytochrome is a plant pigment that responds to red and far-red light in sunlight. Phytochrome in plant cells brings about flowering and affects plant spacing.

Short-day (long-night) plants

Long-day (short-night) plants

21.3 Sexual Reproduction in Flowering Plants

The life cycle of flowering plants is adapted to a land existence.
- Flowering plants have a life cycle that involves an **alternation of generations,** with distinct **sporophyte** and **gametophytes.**

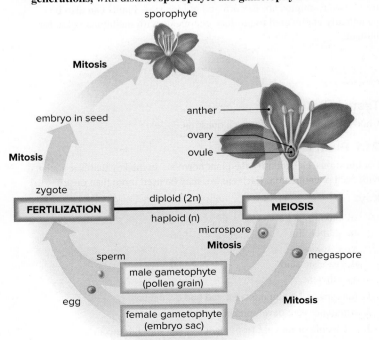
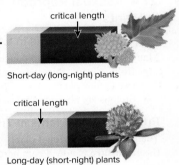

- **Flowers** are the reproductive structures of angiosperms. Most flowers contain sepals, which protect the bud, and petals, which help attract pollinators. The male portions of a flower are the stamen, and the female portions are the carpel.
- The **pollen grain** is the male gametophyte. Pollination brings together the male and female gametophytes.
- The embryo sac, located within the ovule of a flower, is the female gametophyte.
- **Double fertilization** occurs, and the zygote and **endosperm** result.

Development of the Seed in a Eudicot

The zygote undergoes a series of developmental stages to become an embryo. In eudicots, the embryo has two cotyledons, which absorb the endosperm. In monocots, the embryo has a single cotyledon and endosperm. In addition to the embryo and stored food, a seed has a seed coat.

Fruit Types and Seed Dispersal

A **fruit** is a mature, ripened ovary and may include other flower parts. Some fruits are dry (e.g., nuts, legumes), and some are fleshy (e.g., apples, peaches). In general, fruits aid the dispersal of seeds. Following dispersal, a seed **germinates.**

21.4 Asexual Reproduction and Genetic Engineering in Plants

Propagation of Plants in a Garden

Many flowering plants reproduce asexually.
- Bulbs, corms, rhizomes, and tubers are fleshy, modified stems that give rise to new plants.
- Stolons are aboveground stems that create new plants.

Propagation of Plants in Tissue Culture

The production of clonal plants of many fruits and vegetables utilizing **tissue culture** is now a commercial venture. Plant cells from tissue cultures can also produce chemicals of medical and commercial importance.

Genetic Engineering of Plants

Genetic engineering produces **genetically modified organisms (GMOs).** Genetically modified plants have improved agricultural or food-quality traits; two examples are Bt corn and golden rice. Plants can also be genetically engineered to produce chemicals with medicinal value for humans.

ASSESS

Testing Yourself

Choose the best answer for each question.

21.1 Plant Hormones

For questions 1–8, identify the plant hormone in the key that is associated with each phenomenon. Each answer may be used more than once.

Key:

    a. auxin
    b. gibberellin
    c. cytokinin
    d. abscisic acid
    e. ethylene

1. initiates and maintains seed and bud dormancy
2. stimulates root development
3. is capable of moving from plant to plant through the air

4. causes phototropism in stems
5. is responsible for apical dominance
6. stimulates leaf, fruit, and flower drop
7. is needed to break seed and bud dormancy
8. prevents senescence

21.2 Plant Responses

9. A tropism is defined as
  a. a growth response to a directional stimulus.
  b. a plant response to water stress.
  c. a hormonal response to overcrowding by other plants.
  d. abscission of leaves due to plant hormones.
10. Root cells exhibit positive gravitropism with the help of
  a. auxin in the root cells.
  b. amyloplasts in root cells.
  c. exposure to sunlight.
  d. a nearby substrate.
11. Phytochrome
  a. can be affected by red and far-red wavelengths of light.
  b. affects a plant's photoperiod.
  c. helps plants sense proximity to neighboring plants.
  d. All of these are correct.

21.3 Sexual Reproduction in Flowering Plants

12. Stigma is to carpel as anther is to
  a. sepal.   **c.** ovary.
  b. stamen.   **d.** style.
13. The megaspore is similar to the microspore in that both
  a. have the diploid number of chromosomes.
  b. become an embryo sac.
  c. become a gametophyte that produces a gamete.
  d. are necessary for seed production.
  e. Both c and d are correct.
14. Double fertilization is the formation of a _____ and a(n) _____.
  a. zygote, zygote   **c.** zygote, megaspore
  b. zygote, pollen grain   **d.** zygote, endosperm

21.4 Asexual Reproduction and Genetic Engineering in Plants

15. Which of the following is a natural method of asexual reproduction in plants?
  a. meristem culture
  b. propagation from stolons or rhizomes
  c. infection with *Agrobacterium*
  d. cell suspension culture
16. Plant tissue culture takes advantage of
  a. phototropism.
  b. gravitropism.
  c. asexual reproduction from tubers.
  d. totipotency.
17. Plant biotechnology can lead to
  a. increased crop production.
  b. disease-resistant plants.
  c. new or improved treatments for human diseases.
  d. more nutritious crops.
  e. All of these are correct.

ENGAGE

Thinking Critically

1. In late November every year, growers ship truckloads of poinsettia plants to stores around the country. Typically, the plants are individually wrapped in plastic sleeves. If the plants remain in the sleeves for too long during shipping and storage, their leaves begin to curl under and eventually fall off. What plant hormone do you think causes this response? How do you suppose the plastic sleeves affect the response?

2. Snow buttercups (*Ranunculus adoneus*) live in alpine regions and produce sun-tracking flowers. The flowers face east in the morning to absorb the sun's warmth in order to attract pollinators and speed the growth of fertilized ovules. The flowers track the sun all day, continually turning to face it. How might you determine whether the flower or the stem is responsible for sun tracking? Assuming you have determined that the stem is responsible, how would you determine which region of the stem follows the sun?

3. Witchweed is a parasitic weed that destroys 40% of Africa's cereal crop annually. Because it is intimately associated with its host (the cereal crop), witchweed is difficult to selectively destroy with herbicides. One control strategy is to create genetically modified cereals with herbicide resistance. Then, herbicides will kill the weed without harming the crop. Herbicide-resistant sorghum was created to help solve the witchweed problem. However, scientists discovered that the herbicide resistance gene could be carried via pollen into johnson grass, a relative of sorghum and a serious problem in the United States. Farmers would have a new challenge if johnson grass could no longer be effectively controlled with herbicides. Consequently, efforts to create herbicide-resistant sorghum were put on hold. Do you think farmers in Africa should be able to grow herbicide-resistant sorghum, allowing them to control witchweed? Or should the production of herbicide-resistant sorghum be banned worldwide in order to avoid the risk of introducing the herbicide resistance gene into johnson grass?

22

Being Organized and Steady

OUTLINE

22.1 The Body's Organization   415

22.2 Organs and Organ Systems   423

22.3 Homeostasis   426

BEFORE YOU BEGIN

Before beginning this chapter, take a few moments to review the following discussions.

Section 1.1 How do tissues and organs fit into the levels of biological organization?

Section 4.2 What are the functions of proteins in the plasma membrane?

Section 4.5 What types of junctions link cells together to form tissues?

© David M. Benett/Getty Images

The Biology of Performing

A performance of Taylor Swift is a complex production, and not only from the perspective of the people with her on the stage. Whether she is singing or dancing, she is demonstrating how the human body is able to perform very complicated feats. Her nervous system must coordinate a variety of tasks, including monitoring her breathing rate and remembering the words and tempo to the song. This is influenced by sensory input from her eyes and ears. At the same time, her brain is instructing her muscles how to move onstage, as well as working to maintain her balance while interacting with the other dancers onstage.

Of course, in biological terms, the same observations can be made about humans engaging in almost any other type of physical activity, such as playing sports, building a house, performing surgery, or driving a car. Even while you are sitting quietly, perhaps reading a textbook, your body systems are involved in a flurry of activity. Your muscles and bones work to keep your body upright. Your respiratory and circulatory systems provide oxygen to your tissues while transporting wastes for elimination. Your digestive system is providing the nutrients needed to power all of these activities while your immune system is keeping a watch out for harmful microbes. Each of these activities is being coordinated by the nervous and endocrine systems. Even when you are asleep, these body systems are at work.

In this chapter, we will explore the basic organization of the human body and see how the body maintains a stable internal environment in the face of changing external conditions.

As you read through this chapter, think about the following questions:

1. What types of tissue make up each of the human body's major organ systems?
2. How does the type of tissue found in each organ relate to the organ's function?
3. What is the role of feedback in the maintenance of the internal conditions of the body?

22.1 The Body's Organization

Learning Outcomes

Upon completion of this section, you should be able to

1. List in order of increasing complexity the levels of organization of an animal body.
2. Describe epithelial tissue, and explain its functions.
3. Describe the primary characteristics of connective tissue.
4. Compare and contrast the three types of muscle tissue.
5. Describe the function of nervous tissue.

In Section 4.4, you studied the general structure and function of a plant cell and an animal cell. Here we will take that knowledge of animal cell structure and see how millions of individual cells of different types come together to make an organism. **Cells** of the same structural and functional type occur within a **tissue.** An **organ** contains different types of tissues, each performing a function to aid in the overall action of the organ. In other words, the structure and function of an organ are dependent on the tissues it contains. That is why it is sometimes said that tissues, not organs, are the structural and functional units of the body. An organ system contains multiple organs that work together to perform a specific physiological function within the organism.

Let's look at an example (**Fig. 22.1**). In humans, the functions of the urinary system are to filter wastes out of the blood, produce urine from those waste products, and permanently remove the urine from the body. The urinary system is composed of several individual organs—the kidneys, ureters, bladder, and urethra—each playing a particular role in the overall process.

The kidneys are composed of several different tissue types; one type in particular, the epithelial tissue, contains cells that function in filtration. These cells can filter blood and remove the waste products (forming urine) into another organ, the ureters. The tissues in the ureters form tubelike structures that allow urine to pass from the kidneys into the bladder. The bladder, also composed of many different types of tissues, has one tissue made of cells that allow the entire organ to distend, or expand, when full of urine. And finally, the urethra, like the ureters, is composed of tissues that form

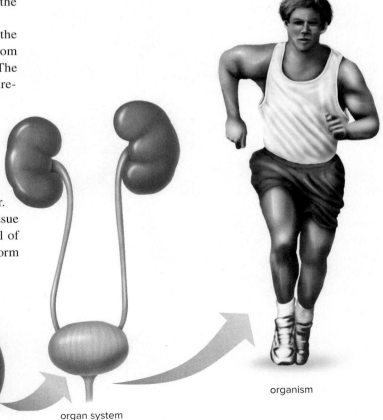

organism

organ system
(urinary system)

organ
(kidney)

cell

tissue
(cuboidal epithelium)

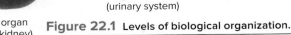

Figure 22.1 **Levels of biological organization.**

A tissue is composed of specialized cells, all having the same structure and performing the same functions. An organ is composed of the types of tissues that help it perform particular functions. An organ system contains several organs and has the functions necessary for the continued existence of an organism.

a cylindrical structure, allowing urine to flow from the bladder out of the body. The overall functions of the urinary system—to produce, store, and rid the body of metabolic wastes—are dependent on the cells that make up the tissues, the tissues that make up the organs, and the organs that make up the organ system.

The structure of the cells, the tissues, and the organs they compose also directly aid function. A common saying in biology is "structure equals function," meaning that the structure of an organ (and hence the tissues and cells that compose it) dictates its function. For example, the small intestine functions in the absorption of nutrients from the digestive tract. The larger the surface area of each cell, the more absorption can occur. Some intestinal cells have areas covered in microvilli (small, fingerlike projections that are extensions of the plasma membrane) that increase the surface area of the cell, thus increasing areas where absorption can occur, without increasing the overall size of the cell itself. A skeletal muscle cell has an internal arrangement of contractile fibers that can slide past each other when the cell contracts, instead of having an arrangement of fibers that would curl or twist in order for a contraction to occur. This cellular structure enables the entire muscle tissue to contract, without wear and tear on the fibers that might increase the possibility of damage.

From the many different types of animal cells, biologists have been able to categorize tissues into just four major types:

- *Epithelial tissue* (epithelium) covers body surfaces and lines body cavities.
- *Connective tissue* binds and supports body parts.
- *Muscular tissue* moves the body and its parts.
- *Nervous tissue* receives stimuli and conducts nerve impulses.

Except for nervous tissue, each type of tissue is subdivided into even more types (**Fig. 22.2**). This chapter looks at the structure and function of each of these tissue types, as well as the organs and organ systems where they are used.

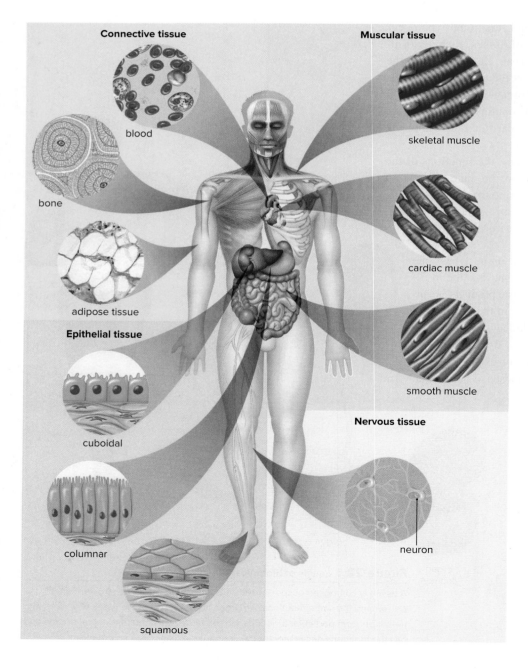

Figure 22.2 Classes of tissues.

The four classes of tissues are epithelial (pink), connective (blue), muscular (tan), and nervous (yellow).

(adipose tissue): © McGraw-Hill Education/Al Telser, photographer

Epithelial Tissue Protects

Epithelial tissue, also called *epithelium,* forms the external coverings and internal linings of many organs and covers the entire surface of the body (**Fig. 22.3**). Therefore, for a substance to enter or exit the body—for example, in the digestive tract, the lungs, or the genital tract—it must cross an epithelial tissue. Epithelial cells adhere to one another, but an epithelium is generally only one cell layer thick. This enables an epithelium to serve a protective function, as substances have to pass through epithelial cells in order to reach a tissue beneath them.

Notice in Figure 22.3 that the epithelial cells differ in shape. *Cuboidal epithelium,* which lines the kidney tubules and the *lumen* (cavity) of a portion of the kidney, contains cube-shaped cells that are roughly the same height as width. *Columnar epithelium* has cells resembling rectangular pillars or

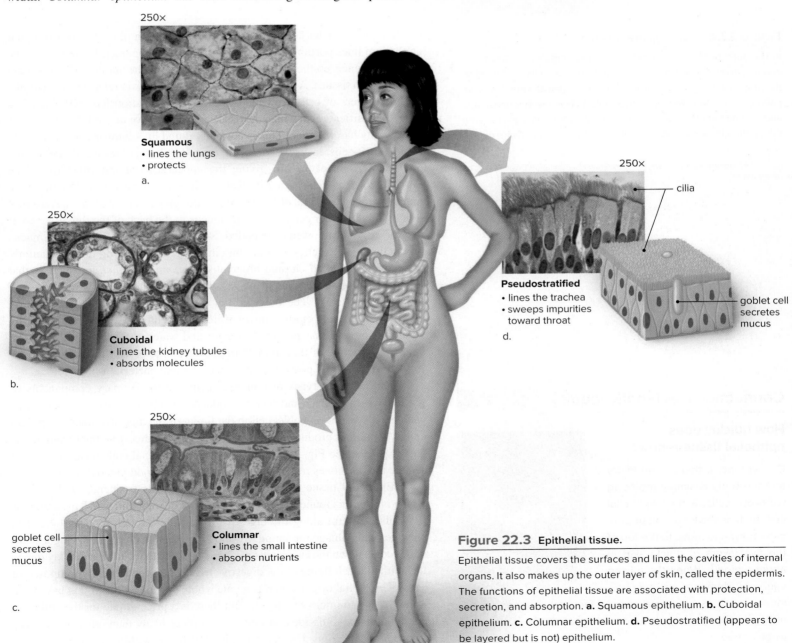

250×

Squamous
• lines the lungs
• protects

a.

250×

Cuboidal
• lines the kidney tubules
• absorbs molecules

b.

250×

goblet cell
secretes
mucus

Columnar
• lines the small intestine
• absorbs nutrients

c.

250×

cilia

Pseudostratified
• lines the trachea
• sweeps impurities
  toward throat

d.

goblet cell
secretes
mucus

Figure 22.3 Epithelial tissue.

Epithelial tissue covers the surfaces and lines the cavities of internal organs. It also makes up the outer layer of skin, called the epidermis. The functions of epithelial tissue are associated with protection, secretion, and absorption. **a.** Squamous epithelium. **b.** Cuboidal epithelium. **c.** Columnar epithelium. **d.** Pseudostratified (appears to be layered but is not) epithelium.

(a-d): © Ed Reschke

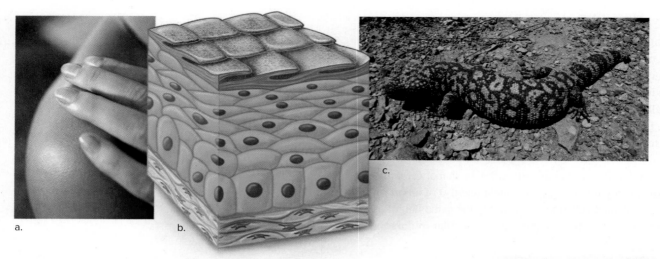

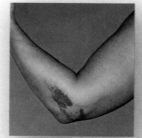

Figure 22.4 Skin as epithelial tissue.

a. The outer portion of skin, called the epidermis, is a stratified epithelium (**b**). The many layers of tightly packed cells reinforced by the protein keratin make the skin protective against water loss and pathogen invasion. New epidermal cells arise in the innermost layer and are shed at the outer layer. **c.** In reptiles, such as a gila monster, the epidermis forms scales, which are simply projections hardened with keratin.

(a): © Thinkstock/PunchStock RF; (c): © McGraw-Hill Education/David Moyer, photographer

Connections: Scientific Inquiry

How quickly does epithelial tissue renew?

On average, if there is no injury and the body is simply replacing worn-out cells, a new epithelial cell, such as the kind in your skin, can renew and move to the top of the five layers of (thick) skin in about a month. When there is an injury or damage, various hor-

© Eric Bean/Getty Images

mones, such as epidermal growth factor, will speed up the renewal process during wound healing, and it may take only a week or two.

columns, with nuclei usually located near the bottom of each cell. Columnar epithelium lines portions of the lumen of the digestive tract. In addition to the shape difference, epithelial cells can be classified by the number of layers these cells make in tissues. A layer that is only one cell thick is referred to as simple. Multiple layers of cells are called stratified. Pseudostratified epithelium is a special classification in which the tissue appears to have multiple layers of cells but actually has only one. This is normally found in columnar cells, where the nuclei of the cells, instead of all being near the bottom of each cell, are in various locations in each cell, giving the appearance of multiple layers (see Fig. 22.3d). *Pseudostratified epithelium* lines the trachea (windpipe), where mucus secreted by some of its cells traps foreign particles and the upward motion of cilia on other cells carries the mucus to the back of the throat, where it may be either swallowed or expelled. Smoking can cause a change in mucus production and secretion and inhibit ciliary action, resulting in an inflammatory condition called chronic bronchitis. *Squamous epithelium,* such as that lining blood vessels and areas of gas exchange in the lungs, is composed of thin, flattened cells. The outer region of the skin, called the epidermis, is stratified squamous epithelium in which the cells have been reinforced by keratin, a protein that provides strength and waterproofing (**Fig. 22.4**). The stratified structure of this epithelium allows skin to protect the body from injury, drying out, and possible pathogen (virus and bacterium) invasion.

One or more types of epithelial cells are the primary components of **glands,** which produce and secrete products (mainly hormones). For example, each mucus-secreting goblet cell in the lining of the digestive tract is a single-celled gland that produces mucus that protects the digestive tract from acidic gastric juices (see Fig. 22.3c). In the pancreas, special cells form glands that secrete the hormones responsible for maintaining blood glucose levels.

Epithelial tissue cells can go through mitosis frequently and quickly, which is why it is found in places that get a lot of wear and tear. This feature is particularly useful along the digestive tract, where rough food particles and enzymes can damage the lining. Swallowing a potato chip and having a sharp edge scrape down the esophagus is a typical injury that can heal quickly due to the epithelial cell lining of the digestive tract. The liver, which is composed of cells of epithelial origin, can regenerate whole portions of itself that have been removed due to injury or surgery. But there is a price to pay for the ability of epithelial tissue to divide constantly—it is more likely than other tissue types to become cancerous. Cancers of epithelial tissue within the digestive tract, lungs, and breast are called carcinomas.

Connective Tissue Connects and Supports

The many types of **connective tissue (Fig. 22.5)** are all involved in binding structures of the body together and providing support and protection. As a rule, connective tissue cells are widely separated by a **matrix,** a noncellular material that varies from solid to semifluid to fluid. The matrix usually has fibers—notably, collagen fibers. Collagen, used mainly for structural support, is the most common protein in the human body, which gives you some idea of the prevalence of connective tissue.

Loose Fibrous and Related Connective Tissues

Let's consider *loose fibrous connective tissue* first and then compare the other types with it. This tissue occurs beneath an epithelium and connects it to the other tissues within an organ. It also forms a protective covering for many internal organs, such as muscles, blood vessels, and nerves. Its cells are called *fibroblasts* because they produce a matrix that contains fibers, including collagen fibers and elastic fibers, that stretch under tension and return to their original shape when released. The presence of loose fibrous connective tissue in the walls of lungs and arteries gives these organs resilience, the ability to expand and then return to their original shape without damage.

Adipose tissue (see Fig. 22.2) is a type of loose connective tissue in which the fibroblasts enlarge and store fat, and there is limited matrix. Adipose tissue is located beneath the skin and around organs, such as the heart and kidneys, where it cushions and protects the organs and serves as long-term energy storage.

Compared with loose fibrous connective tissue, *dense fibrous connective tissue* contains more collagen fibers, which are packed closely together. This type of tissue has more specific functions than does loose fibrous connective tissue. For example, dense fibrous connective tissue is found in **tendons,** which connect skeletal muscles to bones, and in **ligaments,** which connect bones to other bones at joints.

Figure 22.5 Types of connective tissue.

The human knee provides examples of most types of connective tissue. (hyaline, adipose, bone): © Ed Reschke; (dense fibrous): © McGraw-Hill Education

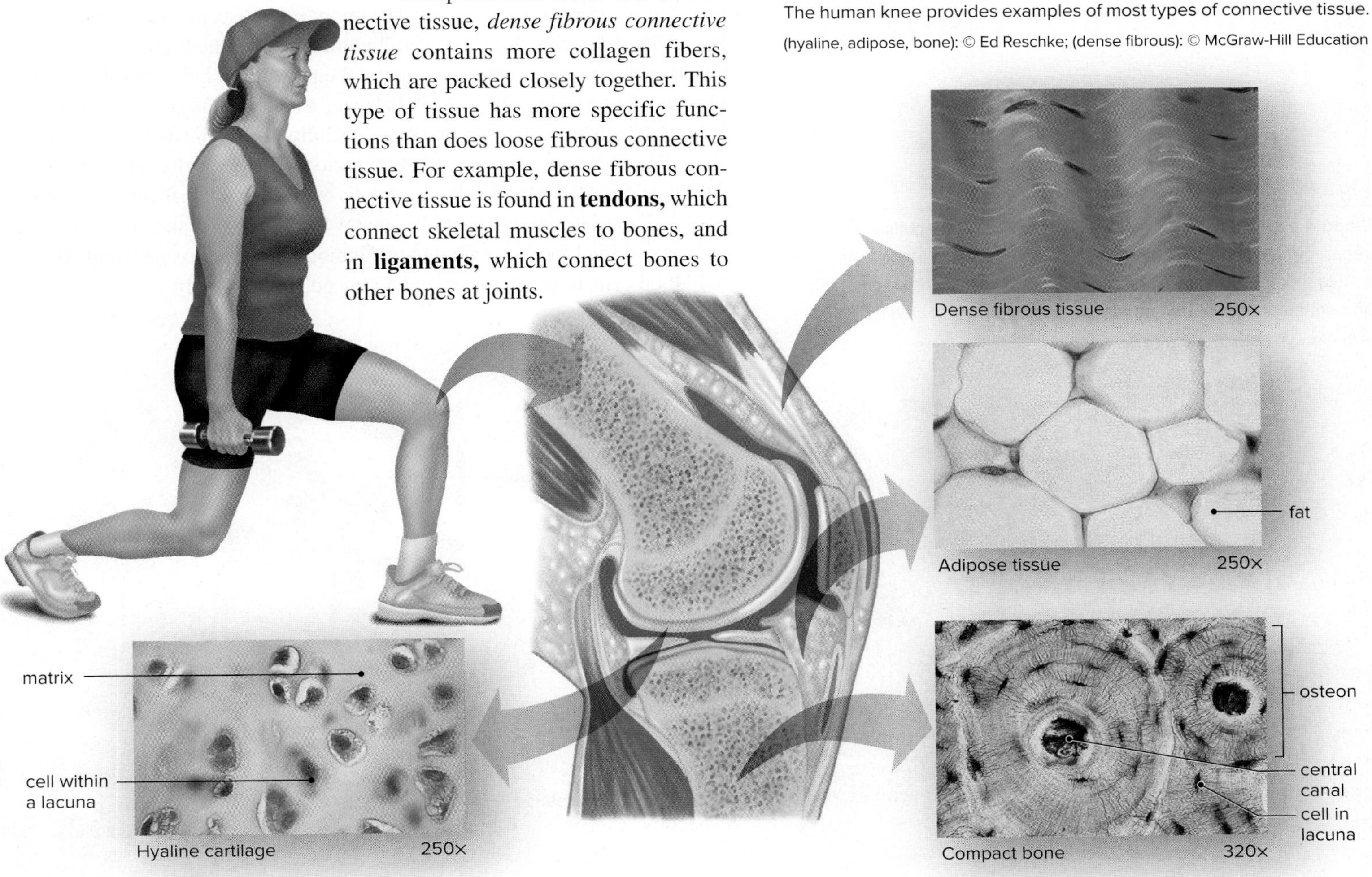

Dense fibrous tissue 250×

Adipose tissue 250×

fat

Compact bone 320×

osteon

central canal

cell in lacuna

matrix

cell within a lacuna

Hyaline cartilage 250×

In **cartilage,** the cells lie in small, open cavities called *lacunae,* separated by a matrix that is semisolid yet flexible. *Hyaline cartilage,* the most common type of cartilage, contains only very fine collagen fibers (Fig. 22.5). The matrix has a white, translucent appearance when unstained. Hyaline cartilage is found in the nose and at the ends of the long bones and the ribs, and it forms rings in the walls of respiratory passages. The human fetal skeleton is also made of this type of cartilage, which makes it easier for the baby to pass through the birth canal. Most of the cartilage is later replaced by bone. Cartilaginous fishes, such as sharks, have a cartilaginous skeleton throughout their lives.

Bone is the most rigid connective tissue (Fig. 22.5). It consists of an extremely hard matrix of inorganic salts, primarily calcium salts, which are deposited around collagen fibers. The inorganic salts give bone rigidity, and the collagen fibers provide elasticity and strength, much as steel rods do in reinforced concrete. The inorganic salts found in bone also act as storage for calcium and phosphate ions for the entire body. *Compact bone,* the most common type of bone in humans, consists of cylindrical structural units called osteons. The central canal of each osteon is surrounded by rings of hard matrix. Bone cells are located in lacunae between the rings of matrix. Blood vessels in the central canal carry nutrients that allow bone to renew itself.

Blood

Blood is composed of several types of cells suspended in a liquid matrix called plasma. Blood is unlike other types of connective tissue in that the matrix is not made by the cells of the connective tissue (**Fig. 22.6**). Even though the blood is liquid, it is a connective tissue by definition; it consists of cells within a matrix. Blood has many functions for overall homeostasis of the body. It transports nutrients and oxygen to cells and removes their wastes. It helps distribute heat and plays a role in fluid, ion, and pH balance. Also, various components of blood help protect us from disease, and blood's ability to clot prevents fluid loss.

Blood contains three formed elements: red blood cells, white blood cells, and platelets. **Red blood cells** are small, biconcave, disk-shaped cells without nuclei. The presence of the red pigment hemoglobin makes the cells red, as well as making blood as a whole red. Hemoglobin binds oxygen and allows the red blood cells to transport oxygen to the cells of the body.

Figure 22.6 Blood, a liquid tissue.

Blood is classified as connective tissue because the cells are separated by a matrix—plasma, the liquid portion of blood. The cellular components of blood are red blood cells, white blood cells, and platelets (which are actually fragments of larger cells).

© Biophoto Associates/Science Source

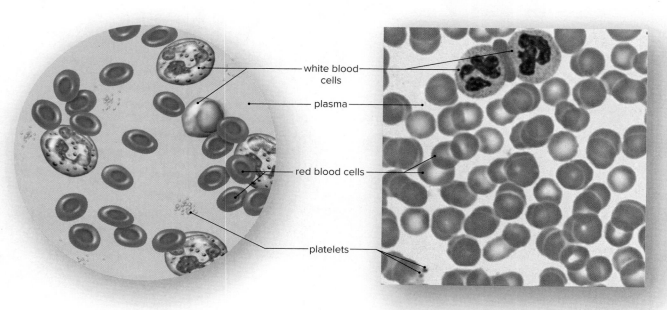

white blood cells

plasma

red blood cells

platelets

White blood cells can be distinguished from red blood cells by the fact that they are usually larger, have a nucleus, and would appear translucent without staining. White blood cells fight infection in two primary ways: (1) Some are phagocytic and engulf infectious pathogens; (2) others produce antibodies, molecules that combine with foreign substances to inactivate them.

Platelets are another component of blood, but they are not complete cells; rather, they are fragments of giant cells present only in bone marrow. When a blood vessel is damaged, platelets form a plug that seals the vessel; along with injured tissues, platelets release molecules that help the clotting process.

Muscular Tissue Moves the Body

Muscular tissue and nervous tissue work together to enable animals to move. **Muscular tissue** contains contractile protein filaments, called actin and myosin filaments, that interact to produce movement. The three types of vertebrate muscles are skeletal, cardiac, and smooth.

Skeletal muscle, which works under voluntary movement, is attached by tendons to the bones of the skeleton, and when it contracts, bones move. Contraction of skeletal muscle, being under voluntary control, occurs faster than in the other two muscle types. The cells of skeletal muscle, called fibers, are cylindrical and quite long—sometimes they run the entire length of the muscle (**Fig. 22.7***a*). They arise during development when several cells fuse, resulting in one fiber with multiple nuclei. The nuclei are located at the edge of the cell, just inside the plasma membrane. The fibers have alternating light and dark bands running across the cell, giving them a striated appearance. These bands are due to the arrangement of actin filaments and myosin filaments in the cell.

Cardiac muscle is found only in the walls of the heart, and its contraction pumps blood and accounts for the heartbeat. Like skeletal muscle, cardiac muscle has striations, but the contraction of the heart is autorhythmic (occurring at a set pace) and involuntary (Fig. 22.7*b*). Cardiac muscle cells also differ from skeletal muscle cells in that they have a single, centrally placed nucleus. The cells are branched and seemingly fused with one another. The heart appears to be composed of one large, interconnected mass of muscular cells. Actually, cardiac muscle cells are separate and individual, but they are bound end to end at intercalated disks, areas where folded plasma membranes allow the contraction impulse to spread from cell to cell.

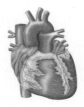
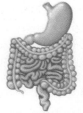

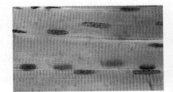
250×

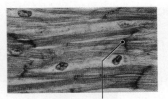

intercalated disk   100×

100×

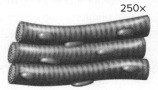

Figure 22.7 Types of muscular tissue.

The three types of muscular tissue are (**a**) skeletal, (**b**) cardiac, and (**c**) smooth.

(a-c): © Ed Reschke

a. **Skeletal muscle**
- has striated, tubular, multinucleated fibers
- is usually attached to skeleton
- is voluntary

b. **Cardiac muscle**
- has striated, branched, uninucleated fibers
- occurs in walls of heart
- is involuntary

c. **Smooth muscle**
- has spindle-shaped, nonstriated, uninucleated fibers
- occurs in walls of internal organs
- is involuntary

Figure 22.8 Nervous tissue.

Neurons are surrounded by neuroglia, such as Schwann cells, which envelop axons and form the myelin sheath.

(Nervous tissue): © Ed Reschke

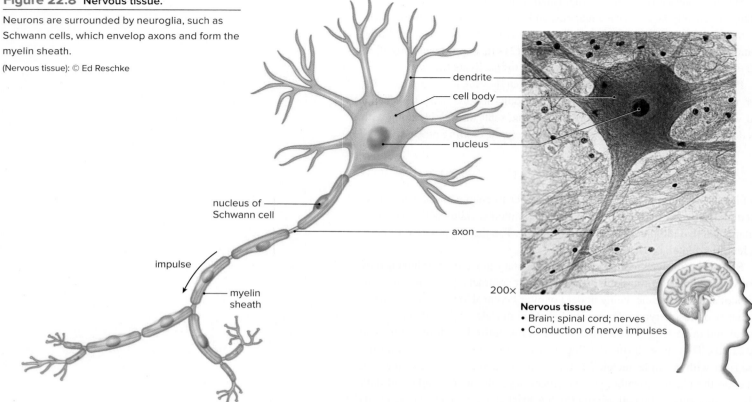

dendrite

cell body

nucleus

nucleus of Schwann cell

axon

impulse

myelin sheath

200×

Nervous tissue
• Brain; spinal cord; nerves
• Conduction of nerve impulses

Smooth muscle is named because the cells lack striations. The spindle-shaped cells form layers in which the thick middle portion of one cell is opposite the thin ends of adjacent cells. Consequently, the nuclei form an irregular pattern in the tissue (Fig. 22.7c). Like cardiac muscle, smooth muscle is involuntary. Smooth muscle is also sometimes called *visceral muscle* because it is found in the walls of the viscera (intestines, stomach, and other internal organs) and blood vessels. Smooth muscle contracts more slowly than skeletal muscle but can remain contracted for a longer time. When the smooth muscle of the intestines contracts, food moves along the lumen. When the smooth muscle of the blood vessels contracts, the vessels constrict, helping raise blood pressure.

Connections: Scientific Inquiry

How big are neurons and how quickly do they communicate?

The average adult human brain weighs about 3 pounds and contains over 100 billion neurons. Some of the neurons—specifically, the axons of the neurons—are less than a millimeter in length; others, such as the axons from the spinal cord to a muscle in the foot, can be 3 feet long or more. The speed at which transmission occurs can vary as well. It can be as slow as 0.5 meter/second or as fast as 120 meters/second—268 mph!

© Steve Gschmeissner/ Science Source

Nervous Tissue Communicates

Nervous tissue coordinates the functions of body parts and allows an animal to respond to external and internal environments. The nervous system depends on (1) sensory input, (2) integration of data, and (3) motor output to carry out its functions. Nerves conduct impulses from sensory receptors, such as pain receptors in the skin, to the spinal cord and brain, where integration occurs. The phenomenon called sensation occurs only in the brain, however. Nerves then conduct nerve impulses away from the spinal cord and brain to the muscles and glands, causing them to contract or secrete in response. In this way, a coordinated response to both internal and external stimuli is achieved.

A nerve cell is called a **neuron.** Every neuron has three parts: dendrites, a cell body, and an axon (**Fig. 22.8**). A *dendrite* is an extension of the neuron cell body that conducts signals toward the cell body. The *cell body* contains the major concentration of the cytoplasm and the nucleus of the neuron. An *axon* is an extension that conducts nerve impulses away from the neuron cell body to other cells. The brain and spinal cord contain many neurons, whereas **nerves** contain only bundles of axons of neurons. The dendrites and cell bodies of these neurons are located in the spinal cord or brain, depending on whether the nerve is a spinal nerve or a cranial nerve.

In addition to neurons, nervous tissue contains **neuroglia,** cells that support and nourish neurons. They outnumber neurons nine to one and take up more than half the volume of the brain. Although their primary function is support, research is currently being conducted to determine how much neuroglia directly contribute to brain function. Schwann cells are a type of neuroglia that encircle long nerve fibers within nerves, forming a protective coating on the axons called a myelin sheath. The presence of myelin sheaths insulates the axon and thus allows nerve impulses to travel much more quickly down its length.

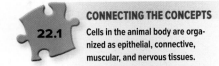

CONNECTING THE CONCEPTS
22.1 Cells in the animal body are organized as epithelial, connective, muscular, and nervous tissues.

Check Your Progress 22.1

1. Explain the difference between an organ and a tissue.
2. Describe the functions of epithelial tissue in the human body.
3. Summarize why connective tissue is needed in the body.
4. Describe the three types of muscular tissue, and state the function of each.
5. Explain how nervous tissue functions in the relay of information.

22.2 Organs and Organ Systems

Learning Outcomes

Upon completion of this section, you should be able to

1. Classify each organ system according to its involvement in transport and protection, body maintenance, control, sensory input and motor output, or reproduction.
2. List the general functions of each organ system.

Organs are composed of a number of tissues, and the structure and function of an organ are dependent on those tissues. Since an organ has many types of tissues, it can perform a function that none of the tissues can do alone. For example, the function of the bladder is to store urine and to expel it when convenient. But this function is dependent on the individual tissues making up the bladder. The epithelium lining the bladder helps the organ store urine by stretching, while preventing urine from leaking into the internal body cavity. The muscles of the bladder propel the urine forward into the urethra, so that it can be removed from the body. Similarly, the functions of an organ system are dependent on its organs. The function of the urinary system is to produce urine, store it, and then transport it. The kidneys produce urine, the bladder stores it, and various tubes transport it from the kidneys to the bladder (the ureters) and out of the body (through the urethra).

This text divides the systems of the body into those involved in (1) the transport of fluids throughout the body and the protection of the body, (2) maintenance of the body, (3) control of the body's systems, (4) sensory input and motor output, and (5) reproduction. All these systems have functions that contribute to homeostasis, the relative constancy of the internal environment (see Section 22.3).

Transport and Protection

The **cardiovascular system** (**Fig. 22.9***a*) consists of blood, the heart, and the blood vessels that carry blood throughout the body. The body's cells are surrounded by a liquid called *interstitial fluid.* Blood transports nutrients and oxygen to interstitial fluid for the cells and removes waste molecules, excreted by cells, from the interstitial fluid. The **lymphatic system** (Fig. 22.9*b*) consists of lymphatic vessels, lymph, lymph nodes, and other lymphatic organs. Lymphatic vessels absorb fat from the digestive system and collect excess interstitial fluid, which is returned to the blood in the cardiovascular system.

The cardiovascular and lymphatic systems are also involved in the protection of the body against disease. Along with the thymus and spleen, certain cells in the lymph and blood are part of the **immune system,** which specifically protects the body from disease.

Maintenance of the Body

Three systems (respiratory, urinary, and digestive) either add or remove substances from the blood. If the composition of the blood remains constant, so does that of the interstitial fluid.

The **respiratory system** (**Fig. 22.10***a*) consists of the lungs, the trachea, and other structures that take air to and from the lungs. The respiratory system brings oxygen into the body and takes carbon dioxide out through the lungs. It also exchanges gases with the blood.

The **urinary system** (Fig. 22.10*b*) consists of the kidneys and the urinary bladder, along with the structures that transport urine. This system rids blood of wastes and helps regulate the fluid level and chemical content of the blood.

The **digestive system** (Fig. 22.10*c*) consists of the organs along the digestive tract, together with associated organs, including the teeth, salivary glands, liver, and pancreas. This system receives food and digests it into nutrient molecules, which then enter the blood.

Control

The **nervous system** (**Fig. 22.11***a*) consists of the brain, the spinal cord, and associated nerves. The nerves conduct nerve impulses from receptors to the brain and spinal cord. They also conduct nerve impulses from the brain and

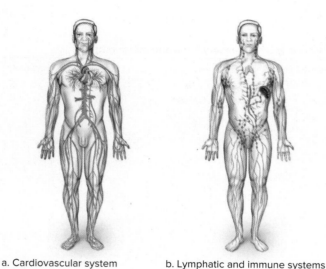

a. Cardiovascular system   b. Lymphatic and immune systems

Figure 22.9 Transport systems.

Both the cardiovascular and lymphatic systems are involved in transport. The lymphatic system is also involved in defense against disease.

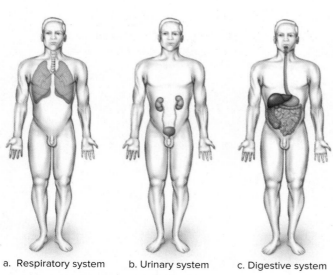

a. Respiratory system   b. Urinary system   c. Digestive system

Figure 22.10 Maintenance systems.

The respiratory, urinary, and digestive systems keep the body's internal environment constant.

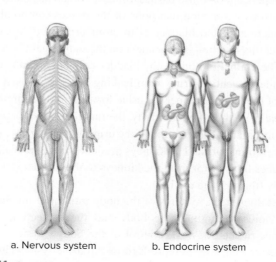

a. Nervous system   b. Endocrine system

Figure 22.11 Control systems.

The nervous and endocrine systems coordinate the other systems of the body.

spinal cord to the muscles and glands, allowing us to respond to both external and internal stimuli. Sensory receptors and sense organs are sometimes considered a part of the nervous system.

The **endocrine system** (Fig. 22.11*b*) consists of the hormonal glands, such as the thyroid and adrenal glands, which secrete hormones, chemicals that serve as messengers between body parts. Both the nervous and endocrine systems coordinate and regulate the functions of the body's other systems. The nervous system tends to cause quick responses in the body, while the body's responses to hormones released by the endocrine system tend to last much longer. The endocrine system also helps maintain the proper functioning of the male and female reproductive organs.

Sensory Input and Motor Output

The **integumentary system** (**Fig. 22.12***a*) consists of the skin and its accessory structures. The sensory receptors in the skin, and in organs such as the eyes and ears, respond to specific external stimuli and communicate with the brain and spinal cord by way of nerve fibers. These messages may cause the brain to respond to a stimulus.

The **skeletal system** (Fig. 22.12*b*) and the **muscular system** (Fig. 22.12*c*) enable the body and its parts to move as a result of motor output. The skeleton, as a whole, serves as a place of attachment for the skeletal muscles. Contraction of the muscles in the muscular system accounts for the movement of body parts.

These three systems protect and support the body. The skeletal system, consisting of the bones of the skeleton, protects body parts. For example, the skull forms a protective encasement for the brain, as does the rib cage for the heart and lungs. The skin serves as a barrier between the outside world and the body's tissues.

Reproduction

The **reproductive system** (**Fig. 22.13**) involves different organs in the male and female. The male reproductive system consists of the testes, other glands, and various ducts, such as the ductus deferens, that conduct semen to and through the penis. The testes produce sex cells called sperm. The female reproductive system consists of the ovaries, uterine tubes, uterus, vagina, and external genitals. The ovaries produce sex cells called eggs. When a sperm fertilizes an egg, an offspring begins development.

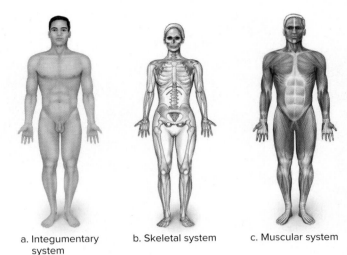

a. Integumentary system   b. Skeletal system   c. Muscular system

Figure 22.12 Sensory input and motor output.

Sensory receptors in the skin (integumentary system) and the sense organs send input to the skeletal and muscular systems, the control systems that cause a response to stimuli.

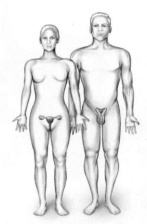

Reproductive system

Figure 22.13 The reproductive system.

The reproductive organs in the female and male differ. The reproductive system ensures the survival of the species.

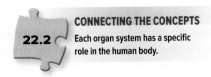

CONNECTING THE CONCEPTS

22.2 Each organ system has a specific role in the human body.

Check Your Progress 22.2

1. Distinguish between an organ and an organ system.
2. Describe the common function of the nervous and endocrine systems.
3. Summarize the function of each body system.

22.3 Homeostasis

To understand homeostasis, there are two types of environments to consider: the external environment, which includes everything outside the body, and the internal environment, which includes our cells, tissues, fluids, and organs. Recall that cells live in a liquid environment called the interstitial fluid. This fluid is constantly renewed with nutrients and gases via exchanges with the blood. Therefore, blood and interstitial fluid constitute part of the body's internal environment. The volume and composition of interstitial fluid remain relatively constant only as long as blood composition remains near normal levels. We say "relatively constant" because the composition of both interstitial fluid and blood varies within an acceptable range. Maintenance of the relatively constant condition of the internal environment within these ranges is called **homeostasis.**

Because of homeostasis, even though external conditions may change dramatically, internal conditions stay within a narrow range. For example, the temperature of the body is maintained near 37°C (97°F to 99°F), even if the surrounding temperature is lower or higher. If you eat acidic foods, the pH of your blood still stays about 7.4, and even if you eat a candy bar, the amount of sugar in your blood remains at just about 0.1%.

Organ Systems and Homeostasis

All the systems of the body contribute to maintaining homeostasis (**Fig. 22.14**). The cardiovascular system conducts blood to and away from capillaries, where exchange occurs. Red blood cells transport oxygen and participate in the transport of carbon dioxide. White blood cells fight infection, and platelets participate in the clotting process. Lymphatic capillaries collect excess interstitial fluid and return it via lymphatic vessels to the cardiovascular system. Lymph nodes help purify lymph and keep it free of pathogens.

The digestive system takes in and digests food, providing nutrient molecules that enter the blood to replace those that are constantly being used by the body's cells. The respiratory system removes carbon dioxide from and adds oxygen to the blood. The kidneys are extremely important in homeostasis, not only because they remove metabolic wastes but also because they regulate blood volume, salt balance, and pH. The liver, among other functions, regulates the glucose concentration of the blood. After a meal, the liver removes excess glucose from the blood for storage as glycogen. Later, the glycogen is broken down to replace the glucose that was used by body cells. The liver makes urea, a nitrogenous end product of protein metabolism.

The nervous and endocrine systems regulate the other systems of the body. In the nervous system, sensory receptors send nerve impulses to control centers in the brain, which then direct effectors (muscles or glands) to become active. Muscles bring about an immediate change. Endocrine glands secrete hormones that bring about slower, more lasting changes that keep the internal environment relatively stable.

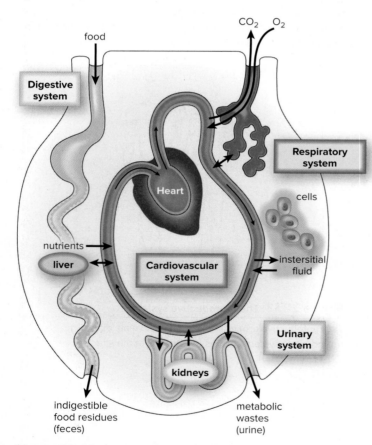

Figure 22.14 Major organ systems and homeostasis.

The organ systems of the body interact with the internal and external environments and with one another. These interactions can alter the composition of the interstitial fluid. For example, the respiratory system exchanges gases with the external environment and with the blood. The digestive system takes in food and adds nutrients to the blood, and the urinary system removes metabolic wastes from the blood and excretes them. In the respiratory system, the blood exchanges nutrients and oxygen for carbon dioxide and other wastes with the interstitial fluid, thus allowing the composition of the interstitial fluid to stay within normal limits.

Negative Feedback

Negative feedback is the primary homeostatic mechanism that allows the body to keep the internal environment relatively stable. A negative feedback mechanism has at least two components: a sensor and a control center (**Fig. 22.15**). The sensor detects a change in the internal environment (a stimulus); the control center initiates an effect that brings conditions back to normal. At that point, the sensor is no longer activated. In other words, a negative feedback mechanism is present when the output of the system dampens (reduces) the original stimulus.

Consider a simple example. When the pancreas detects that the blood glucose level is too high, it secretes insulin, a hormone that causes cells to take up glucose. Then the blood sugar level returns to normal, and the pancreas is no longer stimulated to secrete insulin.

When conditions exceed their limits and feedback mechanisms cannot compensate, illness results. For example, if the pancreas is unable to produce insulin, as in diabetes mellitus, the blood sugar level becomes dangerously high and the individual can become seriously ill. The study of homeostatic mechanisms is therefore medically important.

Mechanical Example

A home heating system is often used to illustrate how a more complicated negative feedback mechanism works (**Fig. 22.16**). Suppose you set the thermostat at 20°C (68°F). This is the *set point*. The thermostat contains a thermometer, a sensor that detects when the room temperature is above or below the set point. The thermostat also contains a control center; it turns the furnace off when the room is warm and turns it on when the room is cool. When the furnace is off, the room cools a bit, and when the furnace is on, the room warms a bit. In other words, a negative feedback system results in fluctuation above and below the set point.

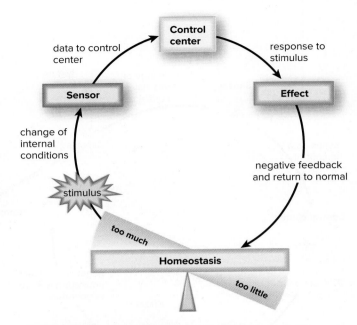

Figure 22.15   Negative feedback mechanism.

This diagram shows how the basic elements of a negative feedback mechanism work.

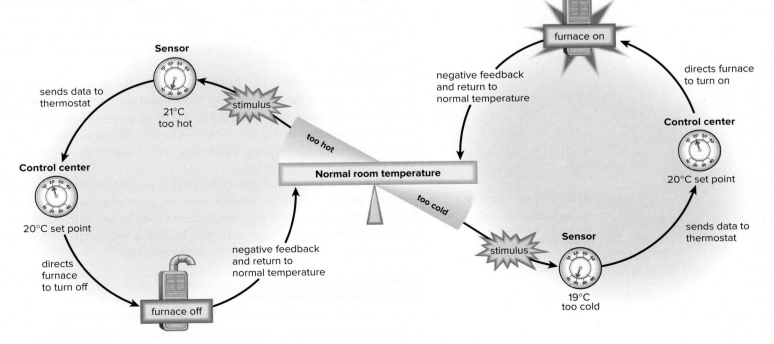

Figure 22.16   Regulation of room temperature.

This diagram shows how room temperature is returned to normal when the room becomes too hot. A contrary cycle in which the furnace turns on and gives off heat returns the room temperature to normal when the room becomes too cold.

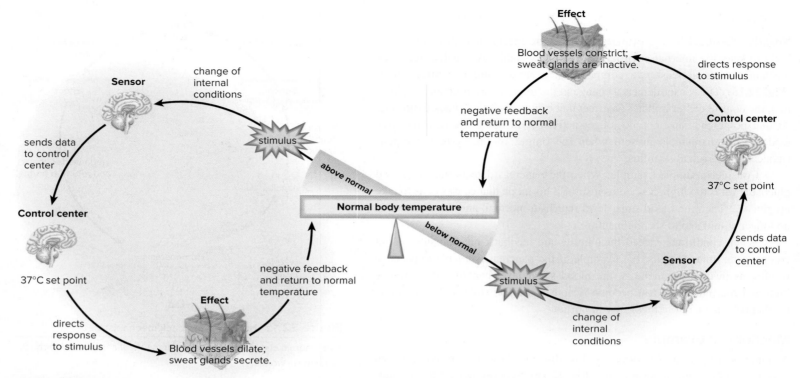

Figure 22.17 Regulation of body temperature.

This diagram shows how normal body temperature is maintained by a negative feedback system.

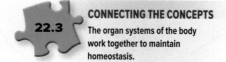

CONNECTING THE CONCEPTS
22.3
The organ systems of the body work together to maintain homeostasis.

Human Example: Regulation of Body Temperature

The thermostat for body temperature is located in the part of the brain called the hypothalamus. When the core body temperature falls below normal, the control center directs (via nerve impulses) the blood vessels of the skin to constrict (**Fig. 22.17**). This action conserves heat. If the core body temperature falls even lower, the control center sends nerve impulses to the skeletal muscles, and shivering occurs. Shivering generates heat, and gradually body temperature rises toward 37°C (98.6°F). When the temperature rises to normal, the control center is inactivated.

When body temperature is higher than normal, the control center directs the blood vessels of the skin to dilate. More blood is then able to flow near the surface of the body, where heat can be lost to the environment. In addition, the nervous system activates the sweat glands, and the evaporation of sweat helps lower body temperature. Gradually, body temperature decreases to 37°C (98.6°F).

Notice that a negative feedback mechanism prevents continued change in the same direction; body temperature does not get warmer and warmer, because warmth stimulates changes that decrease body temperature. Also, body temperature does not get colder and colder, because a body temperature below normal causes changes that bring body temperature up.

Check Your Progress 22.3

1. Summarize why homeostasis is necessary to living organisms.
2. Summarize the components of a negative feedback mechanism.
3. Describe the negative feedback system that maintains body temperature.

SUMMARIZE

Tissues work together in organs, which form organ systems, all of which must maintain homeostasis for the body's overall well-being.

22.1 Cells in the animal body are organized as epithelial, connective, muscular, and nervous tissues.

22.2 Each organ system has a specific role in the human body.

22.3 The organ systems of the body work together to maintain homeostasis.

22.1 The Body's Organization

Levels of organization of an animal body:

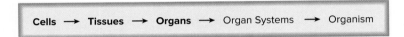

Cells → Tissues → Organs → Organ Systems → Organism

Four major types of animal tissues:

| Epithelial | Connective | Muscular | Nervous |

Epithelial Tissue Protects

Epithelial tissue covers the body and lines its cavities.

- The types of epithelial tissue are squamous, cuboidal, and columnar.
- Epithelial cells sometimes form **glands** that secrete either into ducts or into the blood.

Connective Tissue Connects and Supports

In **connective tissue,** cells are separated by a matrix that contains fibers (e.g., collagen fibers). The four types are

- Loose fibrous connective tissue, including adipose tissue
- Dense fibrous connective tissue (tendons and ligaments)
- **Cartilage** and **bone;** the matrix of cartilage is more flexible than that of bone
- **Blood;** the matrix is a liquid called plasma, and the cells are **red blood cells, white blood cells,** and **platelets** (cell fragments)

Muscular Tissue Moves the Body

Muscular tissue is of three types: skeletal, cardiac, and smooth.

- Both **skeletal muscle** (attached to bones) and **cardiac muscle** (forms the heart wall) are striated.
- Both cardiac muscle (wall of the heart) and **smooth muscle** (walls of internal organs) are involuntary.

Nervous Tissue Communicates

- Nervous tissue is composed of neurons and several types of neuroglia.
- Each neuron has dendrites, a cell body, and an axon. Axons form nerves that conduct impulses.

22.2 Organs and Organ Systems

Organs make up organ systems, which are summarized in **Table 22.1.**

Table 22.1 Organ Systems

Transport and Protection	Sensory
Cardiovascular system (heart and blood vessels) **Lymphatic and immune system** (thymus, spleen, lymphatic vessels and lymph nodes)	**Integumentary system** (skin)
Maintenance	*Motor*
Digestive system (e.g., stomach, intestines) **Respiratory system** (lungs and trachea) **Urinary system** (urinary bladder and kidneys)	**Skeletal system** (bones) **Muscular system** (muscles)
Control	*Reproduction*
Nervous system (brain, spinal cord, and nerves) **Endocrine system** (hormonal glands—e.g., thyroid and adrenal glands)	**Reproductive system** (testes in males; uterus and ovaries in females)

22.3 Homeostasis

Homeostasis is the maintenance of relative stability in the body's internal environment. The internal environment includes the blood and interstitial fluid. Due to the exchange of nutrients and wastes with the blood, interstitial fluid remains nearly constant in volume and composition:

Blood — nutrients → Interstitial fluid
← wastes

Organ Systems and Homeostasis

All organ systems contribute to the relative constancy of interstitial fluid and blood.

- The cardiovascular system transports nutrients to cells and wastes from cells.
- The lymphatic system absorbs excess interstitial fluid and functions in immunity.
- The digestive system takes in food and adds nutrients to the blood.
- The respiratory system carries out gas exchange with the external environment and the blood.
- The urinary system (i.e., the kidneys) removes metabolic wastes and regulates the pH and salt balance of the blood.
- The nervous system and endocrine system regulate the other systems.

Negative Feedback

Negative feedback keeps the internal environment relatively stable. In a negative feedback mechanism, when a sensor detects a change above or below a set point, a control center brings about an effect that reverses the change, brings conditions back to normal, and shuts off the response.

ASSESS

Testing Yourself

Choose the best answer for each question.

22.1 The Body's Organization

1. Label the levels of biological organization in the following illustration.

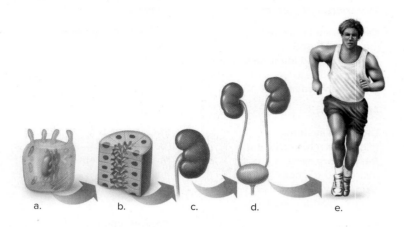

a.   b.   c.   d.   e.

For questions 2–6, identify the tissue in the key that matches the description. Each answer may be used more than once.

Key:

  a. epithelial tissue   **c.** muscular tissue
  b. connective tissue   **d.** nervous tissue

2. contains cells that are separated by a matrix
3. has a protective role
4. covers the body surface and forms the linings of organs
5. allows animals to respond to their environment
6. contains actin and myosin filaments
7. Label the parts of a neuron in the following illustration.

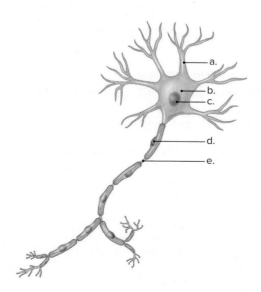

a.
b.
c.
d.
e.

8. Blood is a(n) _____ tissue because it has a _____.
  a. connective, gap junction   **c.** epithelial, gap junction
  b. muscular, matrix   **d.** connective, matrix

22.2 Organs and Organ Systems

9. The skeletal system functions in
  a. blood cell production.   **c.** movement.
  b. mineral storage.   **d.** All of these are correct.

10. The _____ system produces hormones.
  a. nervous   **c.** skeletal
  b. urinary   **d.** endocrine

11. The _____ system is responsible for processing information in the body.
  a. endocrine   **c.** nervous
  b. skeletal   **d.** lymphatic

22.3 Homeostasis

12. A major function of interstitial fluid is to
  a. provide nutrients and oxygen to cells and remove wastes.
  b. keep cells warm.
  c. prevent cells from touching each other.
  d. provide flexibility by allowing cells to slide over each other.

13. Which of these is a result of homeostasis?
  a. Muscular tissue is specialized to contract.
  b. Normal body temperature is always about 37°C.
  c. There are more red blood cells than white blood cells.
  d. Receptors in the eyes detect light.

14. Which of these body systems contribute to homeostasis?
  a. digestive and urinary systems
  b. respiratory and nervous systems
  c. nervous and endocrine systems
  d. immune and cardiovascular systems
  e. All of these are correct.

15. Homeostasis maintains the body's internal conditions within a narrow range.
  a. True
  b. False

ENGAGE

Thinking Critically

1. Patients with muscular dystrophy often develop heart disease. However, heart disease in these patients may not be detected until it has progressed past the point at which treatment options are effective. New research indicates that the link between muscular dystrophy and heart disease is so strong that muscular dystrophy patients should be screened for heart disease early, so that treatment can be started. What do you suppose is the link between muscular dystrophy and heart disease?

2. If homeostasis of the body's internal environment is so important to our survival, why do we interact with the outside environment at the risk of an imbalance in the internal environment? Would it be possible to survive with no interactions between the internal and external environments? Give a few examples of what could occur if no interactions were possible.

3. Not all animals have organ systems or even organs. What are the advantages to having tissues, organs, and organ systems that perform specific functions in the body? In other words, what might have been some of the selective pressures that led to the evolution of more complex organisms?

4. Explain why death can be described as a failure of the body to maintain homeostasis.

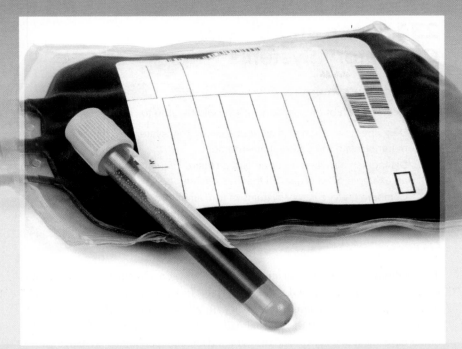

© Nyvlt-art/Shutterstock RF

23

The Transport Systems

Synthetic Blood

Every 2 seconds, someone in the United States needs a transfusion of blood. The need may have been caused by an injury, or a surgical procedure, or even a disease that requires regular transfusions. To answer these needs, blood donors provide over 36,000 units of blood every day, or 15 million units of blood every year.

Historically, the need for blood transfusions has been met by blood donors. Unfortunately, supply does not always meet demand. Natural disasters may place a strain on supplies, especially considering that blood can be stored for only about 42 days, which makes stockpiling difficult. Furthermore, some people are not able to receive blood from donors due to religious beliefs or medical conditions.

To meet the demand for transfusions, scientists have been developing synthetic, or artificial, blood. The use of an oxygen-carrying blood substitute, called oxygen therapeutics, has the goal of providing a patient's tissues with enough oxygen. In some cases, the blood substitute is completely synthetic, and contains chemicals that mimic the oxygen-carrying hemoglobin found in red blood cells. In other cases, scientists are applying biotechnology to manufacture replacement red blood cells. While several biotech companies are currently conducting clinical trials using synthetic blood, most medical professionals believe that we are still several years away from being able to produce a synthetic blood supply to meet the needs of society.

As you read through this chapter, think about the following questions:

1. What is the role of blood in the human body?
2. Besides red blood cells, what other elements need to be present in synthetic blood?
3. What is the role of hemoglobin in the blood?

OUTLINE

23.1 Open and Closed Circulatory Systems 432

23.2 Transport in Humans 435

23.3 Blood: A Transport Medium 442

BEFORE YOU BEGIN

Before beginning this chapter, take a few moments to review the following discussions.

Section 22.1 Why is blood considered to be a connective tissue?

Section 22.2 What is the overall function of the cardiovascular system in the body?

Section 22.3 How does the cardiovascular system play a central role in maintaining homeostasis?

23.1 Open and Closed Circulatory Systems

Learning Outcomes

Upon completion of this section, you should be able to

1. Describe how organisms without a circulatory system exchange materials with the external environment.
2. Distinguish between an open and a closed circulatory system, and give an example of an organism that has each type.
3. Compare the circulatory systems of fish, amphibians, birds, and mammals.
4. Summarize the differences between the pulmonary and systemic circulatory systems.

The **circulatory system** of an animal delivers oxygen and nutrients to cells and removes carbon dioxide and waste materials. In some animals, the body plan makes a circulatory system unnecessary. In a hydra (**Fig. 23.1***a*), cells either are part of a single layer of external cells or line the gastrovascular cavity. In either case, each cell is exposed to water and can independently exchange gases and rid itself of wastes. The cells that line the gastrovascular cavity are specialized to carry out digestion. They pass nutrient molecules to other cells by diffusion. In a planarian (Fig. 23.1*b*), the digestive cavity branches throughout the

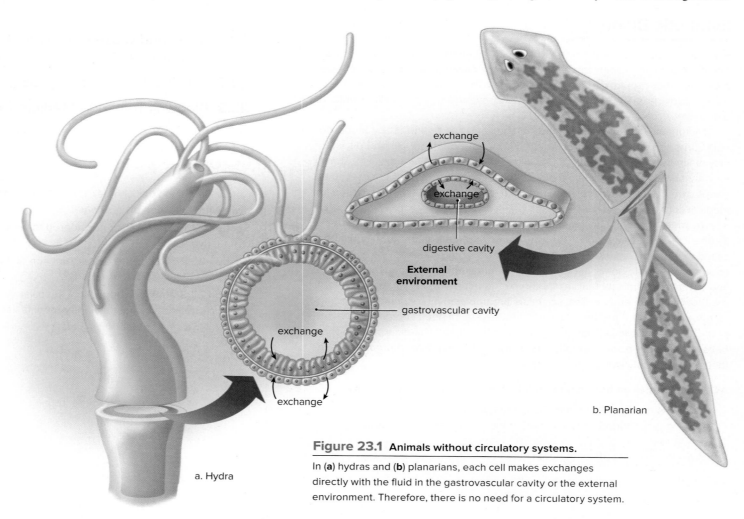

exchange

digestive cavity

External environment

gastrovascular cavity

exchange

exchange

b. Planarian

a. Hydra

Figure 23.1 **Animals without circulatory systems.**

In (**a**) hydras and (**b**) planarians, each cell makes exchanges directly with the fluid in the gastrovascular cavity or the external environment. Therefore, there is no need for a circulatory system.

small, flattened body. No cell is very far from one of the digestive branches, so nutrient molecules can diffuse from cell to cell. Similarly, diffusion meets the respiratory and excretory needs of the cells.

Some other animals, such as nematodes and echinoderms, rely on the movement of fluid within a body cavity (or coelom) to circulate gases, nutrients, and wastes.

Open Circulatory Systems

A circulatory system consists of a heart and associated vessels. The role of the **heart** is to keep the fluid (blood) moving within the vessels. Circulatory systems may be classified as being either open or closed.

The grasshopper is an example of an invertebrate animal that has an **open circulatory system** (**Fig. 23.2**), meaning that the fluid is not always confined to the vessels. A tubular heart pumps a fluid called hemolymph through a network of channels and cavities in the body. Collectively known as the **hemocoel,** these cavities, or **sinuses,** contain the animal's organs.

Eventually, hemolymph (a combination of blood and interstitial fluid) drains back to the heart. When the heart contracts, openings called ostia (sing., ostium) are closed; when the heart relaxes, the hemolymph is sucked back into the heart by way of the ostia. The hemolymph of a grasshopper is colorless, because it does not contain hemoglobin or any other respiratory pigment that combines with and carries oxygen. Oxygen is taken to cells, and carbon dioxide is removed from them by way of air tubes, called tracheae, which are found throughout the body. The tracheae provide efficient transport and delivery of respiratory gases while restricting water loss.

Closed Circulatory Systems

All vertebrates and some invertebrates have a **closed circulatory system** (**Fig. 23.3**), which is more commonly called a **cardiovascular system** because it consists of a strong, muscular heart and blood vessels. In a closed circulatory system, the blood remains within the blood vessels at all times. In humans, the heart has two receiving chambers, called atria (sing., atrium), and two pumping chambers, called ventricles. There are three kinds of vessels: arteries, which carry blood away from the heart; capillaries, which exchange materials with interstitial fluid; and veins, which return blood to the heart. Blood is always contained within these vessels and never runs freely into the body unless an injury occurs.

As blood passes through capillaries, the pressure of blood forces some water out of the blood and into the interstitial fluid. Some of this fluid returns directly to a capillary, and some is picked up by lymphatic capillaries in the vicinity. The fluid, now called lymph, is returned to the cardiovascular system by lymphatic vessels. The function of the lymphatic system is discussed in Section 23.2.

Comparison of Vertebrate Circulatory Pathways

Two types of circulatory pathways are seen among vertebrate animals. In fishes, blood follows a one-circuit (single-loop) pathway through the body. The heart has a single atrium and a single ventricle (**Fig. 23.4**a). The pumping action of the ventricle sends blood under pressure to the gills, where gas exchange occurs. After passing through the gills, blood is under reduced

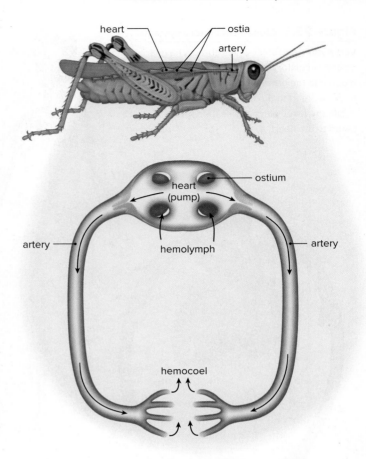

Figure 23.2 Open circulatory system.

The grasshopper has an open circulatory system. Hemolymph freely bathes the internal organs. The heart, a pump, keeps the hemolymph moving, but an open system cannot supply oxygen rapidly enough to wing muscles. These muscles receive oxygen directly from tracheae (air tubes).

Figure 23.3 Closed circulatory systems.

Vertebrates and some invertebrates have a closed circulatory system. The heart pumps blood into the arteries, which take blood away from the heart to the capillaries, where exchange takes place. Veins then return blood to the heart.

vein   heart   artery

capillaries

Key:
- O$_2$-rich blood
- O$_2$-poor blood
- mixed blood

pressure and flow. However, this single circulatory loop has advantages in that the gill capillaries receive oxygen-poor (O$_2$-poor) blood and the systemic capillaries receive O$_2$-rich blood.

As a result of evolutionary adaptations to life on land, the other vertebrates have a two-circuit (double-loop) circulatory pathway. The heart pumps blood to the tissues through the **systemic circuit,** and it pumps blood to the lungs through the **pulmonary circuit.** This double pumping action is seen in terrestrial animals that utilize lungs to breathe air.

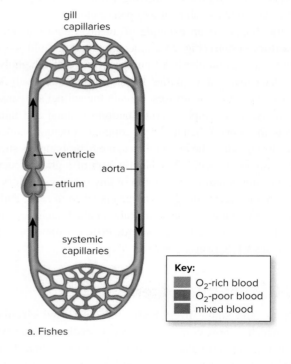

gill capillaries

ventricle

aorta

atrium

systemic capillaries

Key:
- O$_2$-rich blood
- O$_2$-poor blood
- mixed blood

a. Fishes

Figure 23.4 Comparison of circulatory circuits in vertebrates.

a. In a fish, the blood moves in a single circuit. The heart has a single atrium and ventricle, and it pumps the blood into the gill capillaries, where gas exchange takes place. Blood pressure created by the pumping of the heart is dissipated after the blood passes through the gill capillaries. **b.** Amphibians and most reptiles have a two-circuit system, in which the heart pumps blood to both the lungs and the body itself. There is a single ventricle, and some mixing of O$_2$-rich and O$_2$-poor blood takes place. **c.** The pulmonary and systemic circuits are completely separate in crocodilians, birds, and mammals. The right side pumps blood to the lungs, and the left side pumps blood to the rest of the body.

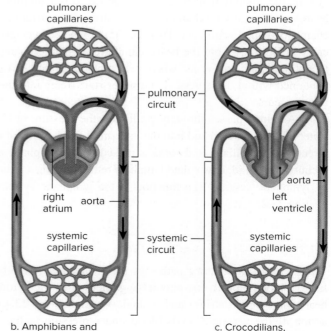

pulmonary capillaries

pulmonary capillaries

pulmonary circuit

right atrium   aorta

aorta

left ventricle

systemic capillaries

systemic circuit

systemic capillaries

b. Amphibians and most reptiles

c. Crocodilians, birds, and mammals

In amphibians, the heart has two atria, but only a single ventricle (Fig. 23.4*b*), and some mixing of O_2-rich and O_2-poor blood does occur. The same holds true for most reptiles, except that the ventricle has a partial septum, so this mixing is reduced. The hearts of some reptiles (for example, crocodilians and birds) and mammals are divided into right and left halves (Fig. 23.4*c*). The right ventricle pumps blood to the lungs, and the left ventricle, which is larger than the right ventricle, pumps blood to the rest of the body. This arrangement provides adequate blood pressure for both the pulmonary and systemic circuits.

CONNECTING THE CONCEPTS

23.1 While some animals lack a circulatory system, the majority possess either open or closed systems that exchange materials with the environment.

Check Your Progress 23.1

1. Explain why some animals do not require a circulatory system.
2. Compare and contrast an open circulatory system with a closed circulatory system.
3. Describe the function of the pulmonary and systemic systems.
4. Summarize the difference between the circulatory system of a fish and that of a mammal.

23.2 Transport in Humans

Learning Outcomes

Upon completion of this section, you should be able to

1. Describe the anatomy of the human heart.
2. Trace the flow of blood through the heart and the pulmonary and systemic circuits.
3. Explain the cardiac cycle, and describe the electrical activity associated with it.
4. Describe the types of blood vessels.
5. Explain the role of the lymphatic system in circulation within the human body.
6. Identify the forces that cause the movement of substances into and out of the capillaries.

In the human cardiovascular system, like that of other vertebrates, the heart pumps blood into blood vessels, which take it to capillaries, where exchanges take place. In the lungs, carbon dioxide is exchanged for oxygen; in the tissues, nutrients and oxygen are exchanged for carbon dioxide and other wastes. These exchanges in the lungs and tissues are so important that, if the heart stops pumping, death results.

The Human Heart

In humans, the heart is a double pump: The right side of the heart pumps O_2-poor blood to the lungs, and the left side of the heart pumps O_2-rich blood to the tissues (**Fig. 23.5**). The heart acts as a double pump because a **septum** separates the right side from the left. Further, the septum prevents O_2-poor blood from mixing with O_2-rich blood.

Each side of the heart has two chambers. The upper, thin-walled chambers are called atria (sing., **atrium**), and they receive blood. The lower chambers are the thick-walled **ventricles,** which pump the blood away from the heart.

Valves are located between the atria and the ventricles, and between the ventricles and attached vessels. Because these valves close after the blood moves through, they keep the blood moving in the correct direction. The valves between the atria and ventricles are

Figure 23.5 **Heart anatomy and the path of blood through the heart.**

The right side of the heart receives O_2-poor blood (blue arrows) and pumps it to the lungs; The left side of the heart receives O_2-rich blood (red arrows) and pumps it to the body tissues. The directions "right side" and "left side" of the heart refer to how the heart is positioned in your body, not to the left and right sides of the illustration.

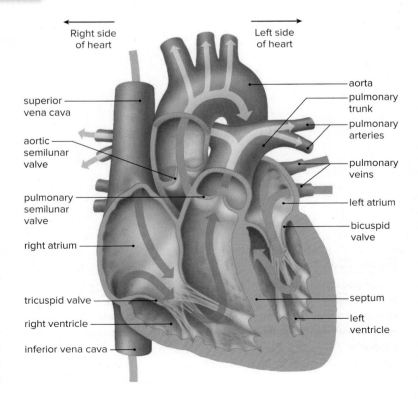

Right side of heart

Left side of heart

superior vena cava

aortic semilunar valve

pulmonary semilunar valve

right atrium

tricuspid valve

right ventricle

inferior vena cava

aorta
pulmonary trunk
pulmonary arteries
pulmonary veins
left atrium
bicuspid valve
septum
left ventricle

Connections: Health

What is a heart murmur?

Like mechanical valves, the heart valves are sometimes leaky; they may not close properly, and there is a backflow of blood. A *heart murmur* is often due to leaky atrioventricular valves, which allow blood to pass back into the atria after the valves have closed. These may be caused by a number of factors: high blood pressure (hypertension), heart disease, or rheumatic fever. Rheumatic fever is a bacterial infection that begins in the throat and spreads throughout the body. The bacteria attack various organs, including the heart valves. When damage is severe, the valve can be replaced with a synthetic valve or one taken from a pig's heart.

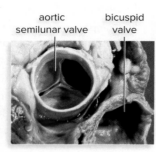

aortic semilunar valve   bicuspid valve

© Biophoto Associates/Science Source

called the **atrioventricular valves,** and the valves between the ventricles and their attached vessels are called **semilunar valves** because their cusps look like half-moons.

The following sequence traces the path of blood through the heart: The right atrium receives blood from the attached veins, called the **venae cavae,** that are returning O₂-poor blood to the heart from the tissues. After the blood passes through an atrioventricular valve (also called the *tricuspid valve* because of its three flaps), the right ventricle pumps it through the *pulmonary semilunar valve* into the pulmonary trunk and **pulmonary arteries,** which take it to the lungs. The **pulmonary veins** take O₂-rich blood to the left atrium. After this blood passes through an atrioventricular valve (also called the *bicuspid valve*), the left ventricle pumps it through the *aortic semilunar valve* into the **aorta,** which takes it to the tissues.

O₂-poor blood is often associated with all veins and O₂-rich blood with all arteries, but this idea is incorrect: Pulmonary arteries and pulmonary veins are just the reverse. That is why the pulmonary arteries in Figure 23.5 are colored blue and the pulmonary veins are colored red. The correct definitions are that an **artery** is a vessel that takes blood away from the heart, and a **vein** is a vessel that takes blood toward the heart.

The Cardiac Cycle

The heart's pumping action, known as the heartbeat or **cardiac cycle,** consists of a series of events: First the atria contract, then the ventricles contract, and then they both rest. **Figure 23.6** lists and depicts the events of

Figure 23.6 **The cardiac cycle.**

The cardiac cycle (or heartbeat) is a cycle of events. *Phase 1:* The atria contract and pass blood to the ventricles. *Phase 2:* The ventricles contract and blood moves into the attached arteries. *Phase 3:* Both the atria and the ventricles relax while the heart fills with blood.

Time	Atria	Ventricles
0.15 sec	Systole	Diastole
0.30 sec	Diastole	Systole
0.40 sec	Diastole	Diastole

superior vena cava
pulmonary veins
right atrium
inferior vena cava

Phase 3: atrial and ventricular diastole

semilunar valves (closed)
left atrium
left ventricle
right ventricle

Phase 1: atrial systole

aorta
pulmonary trunk
atrioventricular valves (closed)

Phase 2: ventricular systole

a heartbeat, using the term **systole** to mean contraction and **diastole** to mean relaxation. When the heart beats, the familiar *lub-dub* sound is caused by the closing of the heart valves. The longer and lower-pitched *lub* occurs when the atrioventricular valves close, and the shorter and sharper *dub* is heard when the semilunar valves close. The **pulse** is a wave effect that passes down the walls of the arterial blood vessels when the aorta expands and then recoils following the ventricular systole. Because there is one pulse per ventricular systole, the pulse rate can be used to determine the heart rate. The heart beats about 70 times per minute, although a normal adult heart rate can vary from 60 to 100 beats per minute.

The beat of the heart is regular because it has an intrinsic pacemaker, called the **SA (sinoatrial) node.** The nodal tissue of the heart, located in two regions of the atrial wall, is a unique type of cardiac muscle tissue. Every 0.85 second, the SA node automatically sends out an excitation impulse that causes the atria to contract (**Fig. 23.7***a*). When this impulse is picked up by the **AV (atrioventricular) node,** it passes to the *Purkinje fibers,* which cause the ventricles to contract. If the SA node fails to work properly, the ventricles still beat, due to impulses generated by the AV node, but the beat is slower (40–60 beats per minute). To correct this condition, it is possible to implant an artificial pacemaker, which automatically gives an electrical stimulus to the heart every 0.85 second. In self-adjusting pacemakers, sensors generate variable electrical signals depending on the person's level of activity; the pacemakers change their output based on these signals.

Although the beat of the heart is intrinsic, it is regulated by the nervous system and various hormones. Activities such as yoga and meditation lead to activation of the vagus nerve, which slows the heart rate. Exercise or anxiety leads to the release of the hormones norepinephrine and epinephrine by the adrenal glands, which causes the heart rate to speed up.

An **electrocardiogram (ECG)** is a recording of the electrical changes that occur in the wall of the heart during a cardiac cycle. Body fluids contain ions that conduct electrical currents, and therefore the electrical changes in heart muscle can be detected on the skin's surface. When an electrocardiogram is being taken, electrodes placed on the skin are connected by wires to an instrument that detects these electrical changes (Fig. 23.7*b*). Various types of abnormalities can be detected by an electrocardiogram. One of them, called ventricular fibrillation, is due to uncoordinated contraction of the ventricles (Fig. 23.7*c*). Ventricular fibrillation is found most often in individuals with heart disease, but it may also occur as the result of an injury or a drug overdose. It is the most common cause of sudden cardiac death in a seemingly healthy person. To stop fibrillation, a defibrillator can be used to apply a strong electrical current for a short period. Then, the SA node may be able to reestablish a coordinated beat. Easy-to-use defibrillators are becoming increasingly available in public places, such as airports and college campuses.

Blood Vessels

Arteries transport blood away from the heart. When the heart contracts, blood is sent under pressure into the arteries; thus, **blood pressure** accounts for the flow of blood in the arteries. Arteries have a much thicker wall than veins because of a well-developed middle layer composed of smooth muscle and elastic

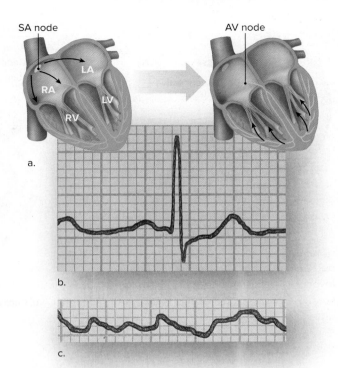

Figure 23.7   **Control of the heartbeat.**

a. The heart beats regularly because the SA node (called the pacemaker) automatically sends out an impulse that causes the atria (RA, LA) to contract and is picked up by the AV node. Thereafter, the ventricles (RV, LV) contract. **b.** An electrocardiogram records the electrical changes that occur as the heart beats. The large spike is associated with ventricular activation. **c.** Ventricular fibrillation produces an irregular ECG.

(b-c): © Ed Reschke

Figure 23.8 Blood vessels.

a. Arteries have well-developed walls with a thick middle layer of elastic fibers and smooth muscle. **b.** Capillary walls are composed of an epithelium only one cell thick. **c.** Veins have weaker walls, particularly because the middle layer is not as thick as in arteries.

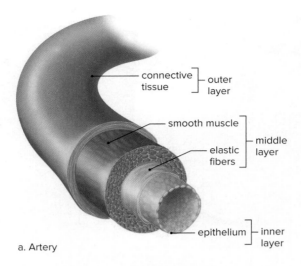

a. Artery

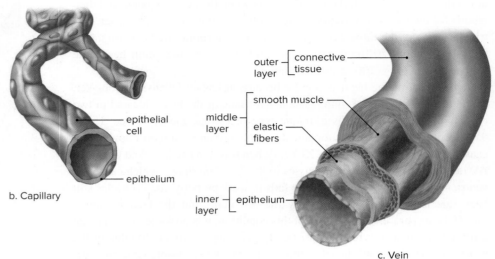

b. Capillary

c. Vein

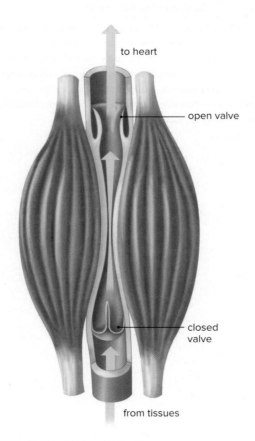

Figure 23.9 Movement of blood in a vein.

Veins have valves, which point toward the heart. Pressure on the walls of a vein, exerted by skeletal muscles, increases blood pressure within the vein and forces a valve open. When the muscles relax, gravity pulls the blood downward, closing the valves and preventing the blood from flowing in the opposite direction.

fibers. The elastic fibers allow arteries to expand and accommodate the sudden increase in blood volume that results after each heartbeat. The smooth muscle strengthens the wall and prevents overexpansion (**Fig. 23.8***a*).

Arteries branch into **arterioles,** small arteries just visible to the naked eye. Their diameter can be regulated by the nervous system, depending on the needs of the body. When arterioles are dilated, more blood flows through them; when they are constricted, less blood flows. The constriction of arterioles can also raise blood pressure. Arterioles branch into **capillaries,** which are extremely narrow, microscopic tubes with a wall composed of only epithelium, often called endothelium (Fig. 23.8*b*). Capillaries, which are usually so narrow that red blood cells pass through in single file, allow the exchange of nutrient and waste molecules across their thin walls. Capillary beds (many interconnected capillaries) are so widespread that, in humans, all cells are less than a millimeter from a capillary. Because the number of capillaries is so extensive, blood pressure drops and blood flows slowly along. The slow movement of blood through the capillaries also facilitates efficient exchange of substances between the blood and the interstitial fluid. The entrance to a capillary bed is controlled by bands of muscle called precapillary sphincters. During muscular exercise, these sphincters relax, and the capillary beds of the muscles are open. Also, after an animal has eaten, the capillary beds in the digestive tract are open. Otherwise, blood moves through a shunt that takes the blood from arteriole to venule (see Fig. 23.10*b*).

Venules collect blood from capillary beds and join as they deliver blood to veins. Veins carry blood back to the heart. Blood pressure is much reduced by the time blood reaches the veins. The walls of veins are much thinner and their diameters are wider than those of arteries (Fig. 23.8*c*). The thin walls allow skeletal muscle contraction to push on the veins, forcing the blood past a valve (**Fig. 23.9**). Valves within the veins point, or open, toward the heart, preventing a backflow of blood when they close. When inhalation occurs and the chest expands, the thoracic pressure falls and abdominal pressure rises. This action also aids the flow of venous blood back to the heart, because blood flows in the direction of reduced pressure.

Connections: Health

What are varicose veins?

Veins are thin-walled tubes, divided into many separate chambers by vein valves. Excessive stretching occurs if veins are overfilled with blood. For example, if a person stands in one place for a long time, leg veins can't drain properly and blood pools in them. As the vein expands, vein valves become distended and fail to function. These two mechanisms cause the veins to bulge and be visible on the skin's sur-

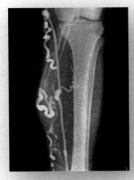

© BSIP/Science Source

face. Hemorrhoids are varicose veins in the rectum. Obesity, a sedentary lifestyle, female gender, genetic predisposition, and increasing age are risk factors for varicose veins.

The Pulmonary and Systemic Circuits

The human cardiovascular system includes two major circulatory pathways, the pulmonary circuit and the systemic circuit. The *pulmonary circuit* moves blood to and from the lungs, where O_2-poor blood becomes O_2-rich blood. The *systemic circuit* moves blood to and from the other tissues of the body. The function of the systemic circuit is to serve the metabolic needs of the body's cells. **Figure 23.10** traces the path of blood in both circuits.

Pulmonary Circuit O_2-poor blood from all regions of the body collects in the right atrium and then passes into the right ventricle. The pulmonary circuit begins when the right ventricle pumps blood to the lungs via the pulmonary trunk and the pulmonary arteries. As blood passes through pulmonary capillaries, carbon dioxide is given off and oxygen is picked up. O_2-rich blood returns to the heart via the pulmonary veins. The pulmonary veins enter the left atrium.

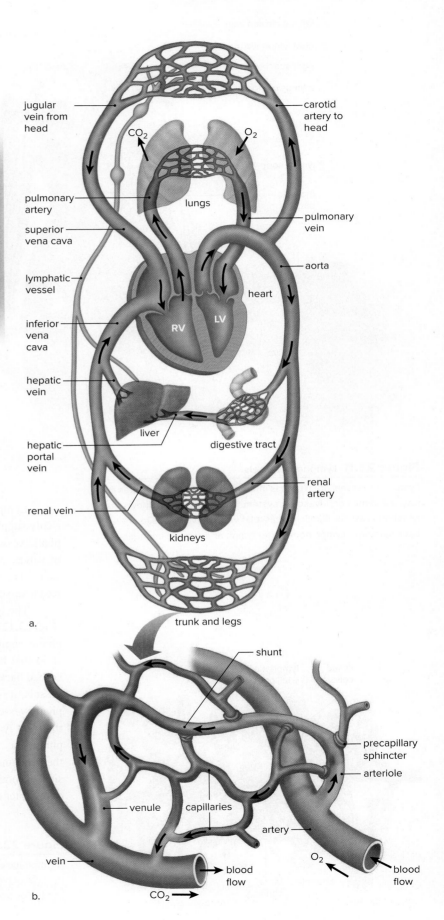

a.

b.

Figure 23.10 The path of blood.

a. Overview of the cardiovascular system. When tracing the flow of blood from the right to the left side of the heart in the pulmonary circuit, you must include the pulmonary vessels. When tracing the flow of blood from the digestive tract to the right atrium in the systemic circuit, you must include the hepatic portal vein, the hepatic vein, and the inferior vena cava. **b.** To move from an artery to a vein, blood must move through a capillary bed, where exchange occurs between blood and interstitial fluid. When precapillary sphincters shut down a capillary bed, blood moves through a shunt from arteriole to venule.

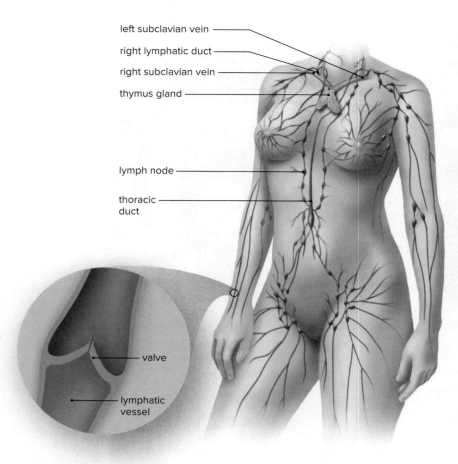

- left subclavian vein
- right lymphatic duct
- right subclavian vein
- thymus gland
- lymph node
- thoracic duct
- valve
- lymphatic vessel

Figure 23.11 Lymphatic vessels.

Lymphatic vessels drain excess interstitial fluid from the tissues and return it to the cardiovascular system. The enlargement shows that lymphatic vessels, like cardiovascular veins, have valves to prevent backward flow. Lymph nodes filter lymph and remove impurities.

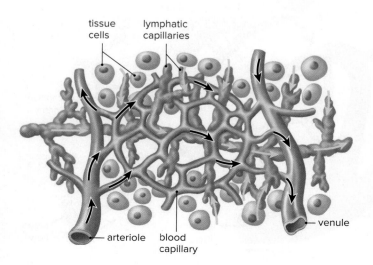

- tissue cells
- lymphatic capillaries
- venule
- arteriole
- blood capillary

Systemic Circuit O_2-rich blood enters the left atrium from the lungs and passes into the left ventricle. The systemic circuit begins when the left ventricle pumps the blood into the aorta. Arteries branching from the aorta carry blood to all areas and organs of the body, where it passes through capillaries and collects in veins. Veins converge on the venae cavae, which return the O_2-poor blood to the right atrium. In the systemic circuit, arteries contain O_2-rich blood and are bright red; veins contain O_2-poor blood and appear dull red or, when viewed through the skin, blue.

A **portal system** begins and ends in capillaries. For example, the hepatic portal vein takes blood from the intestines to the liver. The liver, an organ of homeostasis, modifies substances absorbed by the intestines and monitors the normal composition of the blood. The hepatic veins (see Fig. 25.12) carry blood out of the liver into the inferior vena cava.

Lymphatic System

The **lymphatic system** consists of lymphatic vessels and various lymphatic organs (**Figure 23.11**). The lymphatic system serves many functions in the body. The lymphatic vessels take up fat in the form of lipoproteins from the digestive tract and transport it to the circulatory system. As you will see in Section 26.1, the lymphatic system also works with the immune system to help defend the body against disease. In this chapter, we are interested in the lymphatic vessels that take up excess interstitial fluid and return it to cardiovascular veins in the shoulders, namely the subclavian veins.

Lymphatic vessels are quite extensive; most regions of the body are richly supplied with lymphatic capillaries. The construction of the larger lymphatic vessels is similar to that of cardiovascular veins, including the presence of valves. Also, the movement of lymph within these vessels is dependent on skeletal muscle contraction. When the muscles contract, the lymph is squeezed past a valve that closes, preventing the lymph from flowing backward.

The lymphatic vessels work very closely with the cardiovascular system (**Fig. 23.12**). The lymphatic system is a one-way system that begins at the lymphatic capillaries. These capillaries take up excess interstitial fluid. This is fluid that has diffused from the cells and capillaries, but has not been reabsorbed back into the capillaries. Once the interstitial fluid enters the lymphatic vessels, it is called **lymph.** The lymphatic capillaries join to form larger lymphatic vessels that merge before entering one of two ducts: the thoracic duct or the right lymphatic duct. The thoracic duct is much larger than the right lymphatic duct. It serves the lower limbs, abdomen, left arm, and left sides of both the head and the neck. The right lymphatic duct serves the right arm, the right sides of both the head and the neck, and the right thoracic area. The lymphatic ducts enter the subclavian veins.

Figure 23.12 Lymphatic capillary bed.

A lymphatic capillary bed lies near a blood capillary bed. The heavy black arrows show the flow of blood. The yellow arrows show that lymph is formed when lymphatic capillaries take up excess interstitial fluid.

Capillary Exchange in the Tissues

Figure 23.1 showed that in some animals, exchange is carried out by each cell individually because there is no cardiovascular system. When an animal has a cardiovascular system, the interstitial fluid is involved with exchanging materials between the capillaries and the cells. Notice in **Figure 23.13** that amino acids, oxygen, and glucose exit a capillary and enter interstitial fluid, to be used by cells. On the other hand, carbon dioxide and wastes leave the interstitial fluid and enter a capillary, to be taken away and excreted from the body. A chief purpose of the cardiovascular system is to take blood to the capillaries, where exchange occurs. Without this exchange, homeostasis is not maintained, and the cells of the body perish.

Figure 23.13 illustrates certain mechanics of capillary exchange. Blood pressure and osmotic pressure are two opposing forces at work along the length of a capillary. Blood pressure is caused by the beating of the heart, while osmotic pressure is due to the salt and protein content of the blood. Blood pressure holds sway at the arterial end of a capillary and water exits. Blood pressure is reduced by the time blood reaches the venous end of a capillary, and osmotic pressure now causes water to enter. Midway between the arterial and venous ends of a capillary, blood pressure pretty much equals osmotic pressure, and passive diffusion alone causes nutrients to exit and wastes to enter. Diffusion works because interstitial fluid always contains fewer nutrients and more wastes than blood does. After all, cells use nutrients and thereby create wastes.

The exchange of water at a capillary is not exact, and the result is always excess interstitial fluid. Excess interstitial fluid is collected by lymphatic capillaries, and in this way it becomes lymph (see Fig. 23.12). Lymph contains all the components of plasma, except it has much less protein. Lymph is returned to the cardiovascular system when the major lymphatic vessels enter the subclavian veins in the shoulder region (see Fig. 23.11).

In addition to nutrients and wastes, blood distributes heat to body parts. When you are warm, many capillaries that serve the skin are open, and your face is flushed. This helps rid the body of excess heat. When you are cold, skin capillaries close, conserving heat.

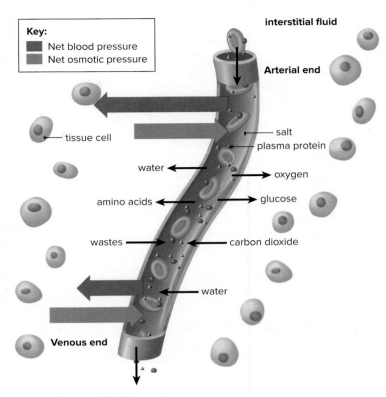

Figure 23.13 Capillary exchange.

At the arterial end of a capillary (*top*), the blood pressure is higher than the osmotic pressure; therefore, water tends to leave the bloodstream. In the midsection, molecules, including oxygen and carbon dioxide, follow their concentration gradients. At the venous end of a capillary (*bottom*), the osmotic pressure is higher than the blood pressure; therefore, water tends to enter the bloodstream. Notice that the red blood cells and the plasma proteins are too large to exit a capillary.

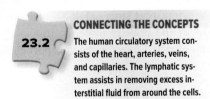

CONNECTING THE CONCEPTS

23.2 The human circulatory system consists of the heart, arteries, veins, and capillaries. The lymphatic system assists in removing excess interstitial fluid from around the cells.

Check Your Progress 23.2

1. Describe the events that occur in the generation of a heartbeat.

2. Describe the structure and function of each type of blood vessel, and explain why capillaries are where gas and nutrient exchange occurs.

3. Describe the flow of blood in the human circulatory system, starting with the right atrium of the heart.

4. Explain the role of the lymphatic system as a transport system in the body.

5. Summarize how fluids and nutrients are exchanged across the capillaries.

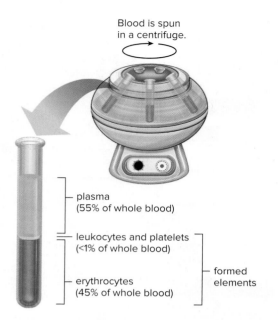

Blood is spun in a centrifuge.

plasma (55% of whole blood)

leukocytes and platelets (<1% of whole blood)

erythrocytes (45% of whole blood)

formed elements

Figure 23.14 **Components of blood.**

After blood is centrifuged and settling occurs, it is apparent that blood is composed of plasma and formed elements.

23.3 Blood: A Transport Medium

Learning Outcomes

Upon completion of this section, you should be able to

1. List the functions of blood.
2. Distinguish between plasma and formed elements.
3. Detail the types of blood cells and their functions.
4. Outline the steps in the formation of a blood clot.
5. Summarize the causes of common cardiovascular system disorders.

Blood is a form of fluid connective tissue that has a number of important roles in the body. These include:

1. Transport of substances to and from the capillaries, where exchange with the interstitial fluid takes place
2. Defense of the body against invasion by pathogens (e.g., disease-causing viruses and bacteria)
3. Assisting in homeostasis by regulating body temperature and pH

In humans, blood has two main portions: the liquid portion, called *plasma,* and the *formed elements,* consisting of various cells and platelets. These portions can be separated by spinning blood in a centrifuge (**Fig. 23.14**).

Plasma

Plasma is composed mostly of water (90–92%) and proteins (7–8%), but it also contains smaller quantities of many types of molecules, including nutrients, wastes, and salts. The salts and proteins are involved in buffering the blood, effectively keeping the pH near 7.4, slightly basic. They also maintain the blood's osmotic pressure, so that water has an automatic tendency to enter blood capillaries. Several plasma proteins are involved in blood clotting, and others transport large organic molecules in the blood. Albumin, the most plentiful of the plasma proteins, transports bilirubin, a breakdown product of hemoglobin. Lipoproteins transport cholesterol.

Formed Elements

The **formed elements** are red blood cells, white blood cells, and platelets. Among the formed elements, **red blood cells,** also called *erythrocytes,* transport oxygen. Red blood cells are small, biconcave disks that at maturity lack a nucleus and contain the respiratory pigment hemoglobin. There are 6 million red blood cells in a small drop of whole blood, and each one of these cells contains about 250 million hemoglobin molecules. **Hemoglobin** contains iron, which combines loosely with oxygen; in this way, red blood cells transport oxygen. If the number of red blood cells is insufficient, or if the cells do not have enough hemoglobin, the individual suffers from *anemia* and has a tired, run-down feeling. At high altitudes, where oxygen levels are lower, red blood cell formation is stimulated. If an individual lives and works in a high altitude long enough, he or she will develop a greater number of red blood cells.

Red blood cells are manufactured continuously within certain bones, namely the skull, the ribs, the vertebrae, and the ends of the long bones. The hormone *erythropoietin* (*EPO*) stimulates the production of red blood cells. The kidneys produce erythropoietin by acting on a precursor made by the liver. Now available as a drug, erythropoietin is helpful to persons with anemia, but it has also been abused by athletes to enhance their performance.

Connections: Health

What is blood doping?

Blood doping is any method of increasing the normal supply of red blood cells for the purpose of delivering oxygen more efficiently, reducing fatigue, and giving an athlete a competitive edge. To accomplish blood doping, athletes can inject themselves with EPO some months before the competition. These injections will increase the number of red blood cells in their blood. Several

© Prof P. M. Motta, and S. Correr/SPL/ Science Source

weeks later, units of their blood are removed and centrifuged to concentrate the red blood cells. The concentrated cells are reinfused shortly before the athletic event. Blood doping is a dangerous, illegal practice. Several cyclists died in the 1990s from heart failure, probably due to blood that was too thick with cells for the heart to pump.

Before they are released from the bone marrow into the blood, red blood cells synthesize hemoglobin and lose their nuclei. After living about 120 days, they are destroyed, chiefly in the liver and spleen, where they are engulfed by large, phagocytic cells. When red blood cells are destroyed, hemoglobin is released. The iron is recovered and returned to the red bone marrow for reuse. Another portion of the molecules (i.e., heme) undergoes chemical degradation and is excreted by the liver as bile pigments in the bile. The bile pigments are primarily responsible for the color of feces.

White blood cells, also called *leukocytes,* help fight infections. White blood cells differ from red blood cells in that they are usually larger and have a nucleus, they lack hemoglobin, and they appear translucent if unstained. **Figure 23.15** shows the appearance of the various types of white blood cells. When they are stained, white blood cells appear light blue unless they have granules that bind with certain stains. There are approximately 5,000–11,000 white blood cells in a small drop of whole blood. Growth factors are available to increase the production of all white blood cells, and these are helpful to people with low immunity, such as AIDS patients.

Red blood cells are confined to the blood, but white blood cells are able to squeeze between the cells of a capillary wall. Therefore, they are found in interstitial fluid, lymph, and lymphatic organs. When an infection is present, white blood cells greatly increase in number. Many white blood cells live only a few days—they probably die while engaging pathogens. Others live months or even years.

Formed Elements	Function
Red Blood Cells (erythrocytes) 4 million–6 million per mm^3 blood	Transport O$_2$ and help transport CO$_2$
White Blood Cells* (leukocytes) 5,000–11,000 per mm^3 blood	Fight infection
Granular leukocytes Neutrophils 40–70%	Phagocytize pathogens and cellular debris
Eosinophils 1–4%	Use granule contents to digest large pathogens, such as worms, and reduce inflammation
Basophils 0–1%	Promote blood flow to injured tissues and the inflammatory response
Agranular leukocytes Lymphocytes 20–45%	Are responsible for specific immunity; B cells produce antibodies; T cells destroy cancer and virus-infected cells
Monocytes 4–8%	Become macrophages that phagocytize pathogens and cellular debris
Platelets (thrombocytes) 150,000–300,000 per mm^3 blood	Aid clotting

*Appearance with Wright's stain.

Figure 23.15 Formed elements.

Red blood cells are involved in the transport of oxygen. White blood cells are quite varied and have different functions, which are all associated with defense of the body against infections. Platelets play an important role in blood clotting.

When microbes enter the body due to an injury, the body's response is called an *inflammatory response* because swelling, reddening, heat, and pain occur at the injured site (see Section 26.2). Damaged tissue releases kinins, which dilate capillaries, and histamines, which increase capillary permeability. White blood cells called **neutrophils,** which are amoeboid, squeeze through the capillary wall and enter the interstitial fluid, where they phagocytize foreign material. White blood cells called **monocytes** come on the scene next and are transformed into **macrophages**—large, phagocytizing cells that release white blood cell growth factors. Soon the number of white blood cells increases explosively. A thick, yellowish fluid called pus contains a large proportion of dead white blood cells that have fought the infection.

Lymphocytes, another type of white blood cell, also play an important role in fighting infection. Lymphocytes called T cells attack the body's cells that are infected with viruses. Lymphocytes called B cells produce **antibodies** to protect the body against certain types of **antigens,** which don't belong to the body. An antigen is most often a protein but sometimes a polysaccharide. Antigens are present in the outer covering of parasites or in their toxins. When antibodies combine with antigens, the complex is often phagocytized by a macrophage. An individual is actively immune when a large number of B cells are all producing the antibody needed to fight a particular infection. The role of lymphocytes in the immune response will be explored in greater detail in Section 26.3.

Platelets and Blood Clotting

Platelets (also called thrombocytes) result from the fragmentation of certain large cells, called megakaryocytes, in the red bone marrow. Platelets are produced at a rate of 200 billion a day, and a small drop of whole blood contains 150,000 to 300,000.

Platelets are involved in blood clotting, or coagulation. Blood contains at least 12 clotting factors that participate in clot formation. *Hemophilia* is an inherited clotting disorder in which the liver is unable to produce one of the clotting factors. The slightest bump can cause the affected person to bleed into the joints, and this leads to degeneration of the joints. Bleeding into muscles can lead to nerve damage and muscular atrophy. The most frequent cause of death due to hemophilia is bleeding into the brain.

Prothrombin and *fibrinogen,* two proteins involved in blood clotting, are manufactured and deposited in blood by the liver. Vitamin K, found in green vegetables and formed by intestinal bacteria, is necessary for the production of prothrombin; if this vitamin is missing from the diet, hemorrhagic disorders can develop.

A series of reactions leads to the formation of a blood clot (**Fig. 23.16**). When a blood vessel in the body is damaged, platelets clump at the site of the puncture and form a plug, which temporarily seals the leak. Platelets and the injured tissues release a clotting factor, called prothrombin activator, that converts prothrombin to *thrombin.* This reaction requires calcium ions (Ca^{2+}). Thrombin, in turn, acts as an enzyme that severs two short amino acid chains from each fibrinogen molecule. These activated fragments then join end to end, forming long threads of *fibrin.* Fibrin threads wind around the platelet plug in the damaged area of the blood vessel and provide the framework for the clot. Red blood cells also are trapped within the fibrin threads; these cells make a clot appear red. Clot retraction follows, during which the clot gets smaller as platelets contract. A fluid called serum is squeezed from the clot. A fibrin clot is present only temporarily. As soon as blood vessel repair is

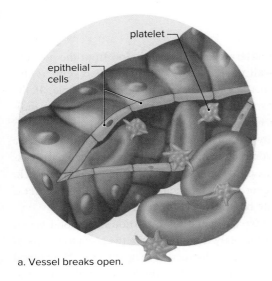

epithelial cells

platelet

a. Vessel breaks open.

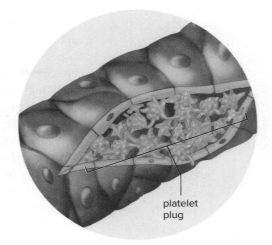

platelet plug

b. Platelet plug forms.

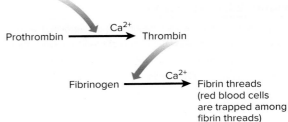

Platelets and damaged tissue cells release prothrombin activator.

Prothrombin $\xrightarrow{Ca^{2+}}$ Thrombin

Fibrinogen $\xrightarrow{Ca^{2+}}$ Fibrin threads (red blood cells are trapped among fibrin threads)

c. Clotting occurs.

initiated, an enzyme called plasmin destroys the fibrin network and restores the fluidity of the plasma.

Cardiovascular Disorders

Cardiovascular disease (CVD) is the leading cause of untimely death in Western countries. In the United States, it is estimated that about 31% of the population suffers from **hypertension,** which is high blood pressure. Normal blood pressure is 120/80 mm Hg. The top number is called *systolic* blood pressure because it is due to the contraction of the ventricles, and the bottom number is called *diastolic* blood pressure because it is measured during the relaxation of the ventricles. Hypertension occurs when blood pressure readings are higher than these numbers—say, 160/100. Hypertension is sometimes called a silent killer because it may not be detected until a stroke or heart attack occurs.

Heredity and lifestyle contribute to hypertension. For example, hypertension is often seen in individuals who have **atherosclerosis**, an accumulation of soft masses of fatty materials, particularly cholesterol, beneath the inner linings of arteries (**Fig. 23.17**). Such deposits, called plaque, tend to protrude into the lumen of the vessel, interfering with the flow of blood. Atherosclerosis begins in early adulthood and develops progressively through middle age, but symptoms may not appear until an individual is 50 or older. To prevent its onset and development, the American Heart Association and other organizations recommend a diet low in saturated fat and rich in fruits and vegetables. Smoking, alcohol or other drug abuse, obesity, and lack of exercise contribute to the risk of atherosclerosis.

Plaque can cause a clot to form on the irregular arterial wall. As long as the clot remains stationary, it is called a *thrombus,* but if it dislodges and moves along with the blood, it is called an *embolus.* If *thromboembolism* is not treated, serious health problems can result. A cerebrovascular accident, also called a **stroke,** often occurs when a small cranial arteriole bursts or is blocked by an embolus. Lack of oxygen causes a portion of the brain to die, and paralysis or death can result. A person is sometimes forewarned of a stroke by a feeling of numbness in the hands or face, difficulty speaking, or temporary blindness in one eye. If a coronary artery becomes completely blocked due to thromboembolism, a heart attack can result.

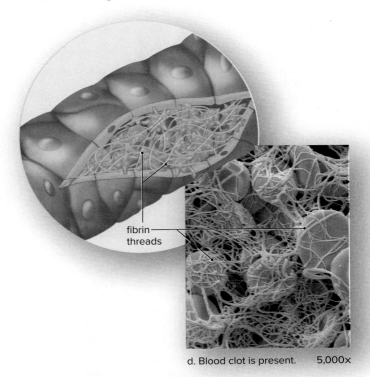

fibrin threads

d. Blood clot is present.   5,000×

Figure 23.16   Blood clotting.

a. When a capillary is injured, blood begins leaking out. **b.** Platelets congregate to form a platelet plug, which temporarily seals the leak. **c.** Platelets and damaged tissue cells release an activator, which sets in motion a series of reactions, ending with (**d**) a blood clot.

(d): © Science Photo Library RF/Getty RF

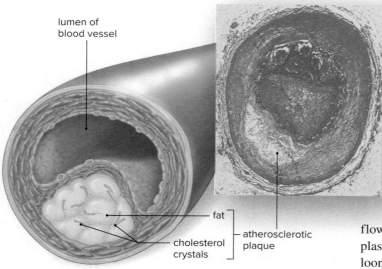

lumen of
blood vessel

fat

atherosclerotic
plaque

cholesterol
crystals

Figure 23.17 Atherosclerosis.

Plaque is an irregular accumulation of cholesterol and fat in the wall of
an artery that interferes with the flow of blood, causing atherosclerosis.
Of special concern is the blocking of coronary arteries.

© Biophoto Associates/Science Source

The coronary arteries take O_2-rich blood from the aorta to
capillaries in the wall of the heart, and the cardiac veins return
O_2-poor blood from the capillaries to the right ventricle. If the
coronary arteries narrow due to cardiovascular disease, the individual may first suffer from angina pectoris, chest pain that is often accompanied by a radiating pain in the left arm. When a
coronary artery is completely blocked, a portion of the heart muscle dies due to a lack of oxygen. This is known as a **heart attack.**
Two surgical procedures are frequently performed to correct a
blockage or facilitate blood flow (**Figure 23.18**).

In a coronary bypass operation, a portion of a blood vessel
from another part of the body is sutured from the aorta to the coronary artery, past the point of obstruction (Fig. 23.18*a*). Now blood
flows normally again from the aorta to the wall of the heart. In balloon angioplasty, a plastic tube is threaded through an artery to the blockage, and a balloon attached to the end of the tube is inflated to break through the blockage.
A stent is often used to keep the vessel open (Fig. 23.18*b*).

CONNECTING THE CONCEPTS

23.3 Blood consists of plasma and
formed elements and plays an
important role in maintaining
homeostasis.

Check Your Progress 23.3

1. Describe the role of each of the major components of blood.
2. Compare and contrast the types of white blood cells.
3. Describe the process involved in forming a blood clot.
4. Identify common cardiovascular disorders, and list some available
   treatments.

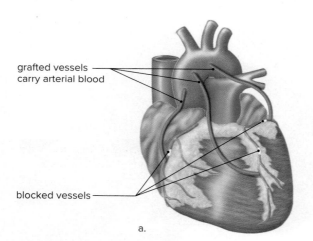

grafted vessels
carry arterial blood

blocked vessels

a.

Figure 23.18 Treatment for blocked
coronary arteries.

Many heart procedures, such as coronary
bypass and stent insertion, can be
performed using robotic surgery
techniques. **a.** During a coronary bypass
operation, blood vessels (usually veins from
the leg) are stitched to the heart, taking
blood past the region of obstruction.
b. During stenting, a cylinder of expandable
metal mesh is positioned inside the
coronary artery using a catheter. Then, a
balloon is inflated, so that the stent
expands and opens the artery.

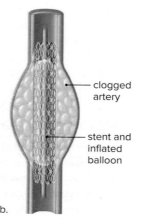

clogged
artery

stent and
inflated
balloon

b.

STUDY TOOLS   http://connect.mheducation.com

▣▣SMARTBOOK®   Maximize your study time with McGraw-Hill SmartBook®, the first adaptive textbook.

SUMMARIZE

The cardiovascular system has the essential function of moving oxygen and nutrients to the cells of the body. This system is regulated by the nervous system and hormones to ensure proper function.

While some animals lack a circulatory system, the majority possess either open or closed systems that exchange materials with the environment.

The human circulatory system consists of the heart, arteries, veins and capillaries. The lymphatic systems assists in removing excess interstitial fluid from around the cells.

Blood consists of plasma and formed elements and plays an important role in maintaining homeostasis.

23.1 Open and Closed Circulatory Systems

The **circulatory system** is responsible for supplying the cells of an animal with oxygen and nutrients and removing carbon dioxide and other waste materials.

- Some invertebrates do not have a circulatory system, because their body plan allows each cell to exchange molecules with the external environment.
- Other invertebrates do have a circulatory system that uses a **heart** to move the blood.
- In an **open circulatory system,** the fluid, called hemolymph, is not confined to the blood vessels but accumulates in cavities called **sinuses,** or collectively, the **hemocoel.**
- In a **closed circulatory system,** or **cardiovascular system,** the blood remains within the blood vessels.

Comparison of Vertebrate Circulatory Pathways

- Fishes have a one-circuit pathway of circulation because the heart, with a single atrium and ventricle, pumps blood only to the gills.
- The cardiovascular system of other vertebrates consists of a **pulmonary circuit** (moves blood to lungs) and a **systemic circuit** (moves blood to tissues). Amphibians have two atria but a single ventricle. Crocodilians, birds, and mammals, including humans, have a heart with two atria and two ventricles, in which O_2-rich blood is kept separate from O_2-poor blood.

23.2 Transport in Humans

The human cardiovascular system consists of the heart and the blood vessels.

The Human Heart

The heart has a right and a left side separated by a **septum.** Each side has an **atrium** (receives blood) and a **ventricle** (pumps blood). **Atrioventricular valves** and **semilunar valves** keep the blood moving in the correct direction. **Arteries** move blood away from the heart; **veins** move blood toward the heart.

CO_2 O_2

CO_2 O_2

- **Right side of the heart:** The right atrium receives O_2-poor blood from the tissues via the **venae cavae.** The ventricle pumps it to the lungs via the **pulmonary artery.**
- **Left side of the heart:** The left atrium receives O_2-rich blood from the lungs via the **pulmonary vein.** The ventricle pumps it to the tissues via the **aorta.**
- **Heartbeat:** The heartbeat, or **cardiac cycle,** consists of a series of contractions (**systole**) followed by relaxation (**diastole**). During a heartbeat first the atria contract and then the ventricles contract. The heart sounds, *lub-dub,* are caused by the closing of valves. A **pulse** is caused by the movement of the blood in the blood vessels.
- **SA node:** The **SA (sinoatrial) node** (pacemaker) causes the two atria to contract. The SA node also stimulates the AV node.
- **AV node:** The **AV (atrioventricular) node** causes the two ventricles to contract.
- An **electrocardiogram (ECG)** can be used to detect the activity of the SA and AV nodes.

Blood Vessels

Arteries, with thick walls, take blood away from the heart to **arterioles,** which take blood to **capillaries** that have thin walls composed only of epithelial cells. **Venules** take blood from capillaries and merge to form **veins,** which have thinner walls than arteries have.

- Blood pressure, created by the beat of the heart, accounts for the flow of blood in the arteries.
- Skeletal muscle contraction is largely responsible for the flow of blood in the veins, which have valves preventing backward flow.

Blood Has Two Circuits

- In the pulmonary circuit, blood travels to and from the lungs.
- In the systemic circuit, blood travels to and from the other tissues of the body. A **portal system** is unique in that it begins and ends with capillaries.

Lymphatic System

The **lymphatic system** is a one-way system that consists of lymphatic vessels and lymphatic organs. The lymphatic vessels receive fat at the digestive tract and excess interstitial fluid (**lymph**) at blood capillaries; the lymphatic vessels carry these to the subclavian veins (cardiovascular veins in the shoulders).

Capillary Exchange in the Tissues

Capillary exchange in the tissues helps keep the internal environment constant. When blood reaches a capillary, the following events occur:

- Water moves out at the arterial end due to blood pressure.
- Water moves in at the venous end due to osmotic pressure.
- Nutrients diffuse out of and wastes diffuse into the capillary in the midsection between the arterial end and the venous end.
- Lymphatic capillaries in the vicinity pick up excess interstitial fluid and return it to cardiovascular veins.

23.3 Blood: A Transport Medium

Blood has two main components: plasma and formed elements.

Plasma

Plasma contains mostly water (90–92%) and proteins (7–8%) but also nutrients, wastes, and salts.

- The proteins and salts buffer the blood and maintain its osmotic pressure.
- The proteins also have specific functions, such as participating in blood clotting and transporting large organic molecules.

Formed Elements

- **Formed elements** include red blood cells, white blood cells, and platelets.
- Red blood cells contain **hemoglobin** and function in oxygen transport.
- White blood cells defend against disease.
- **Neutrophils, monocytes,** and **macrophages** are phagocytic cells.
- Lymphocytes are involved in the development of specific immunity to disease. Some lymphocytes (B cells) respond to **antigens** by producing **antibodies;** others (T cells) attack infected cells of the body.
- **Platelets** and the plasma proteins (prothrombin, fibrinogen) are involved in the formation of a blood clot.

Cardiovascular Disorders

Hypertension and **atherosclerosis** are two conditions that often lead to a **heart attack** or **stroke.** Following a heart-healthy diet, getting regular exercise, maintaining a proper weight, and not smoking cigarettes are protective against these conditions.

ASSESS

Testing Yourself

Choose the best answer for each question.

23.1 Open and Closed Circulatory Systems

1. In insects with an open circulatory system, oxygen is taken to cells by
   a. blood.
   b. hemolymph.
   c. tracheae.
   d. capillaries.
2. The _____ circuit takes blood to and from the tissues.
   a. systemic
   b. pulmonary
   c. coronary
   d. atrial
3. Which of the following organisms do not possess a pulmonary circuit?
   a. amphibians
   b. reptiles
   c. mammals
   d. fish

23.2 Transport in Humans

4. Label the components of the cardiovascular system in the following diagram.

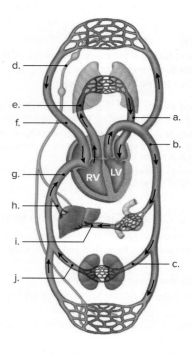

5. Which of the following statements is true?
   a. Arteries carry blood away from the heart, and veins carry blood toward the heart.
   b. Arteries carry blood toward the heart, and veins carry blood away from the heart.
   c. All arteries carry O_2-rich blood, and all veins carry O_2-poor blood.
   d. Arteries usually carry O_2-poor blood, and veins usually carry O_2-rich blood.

6. Label the following diagram of the heart.

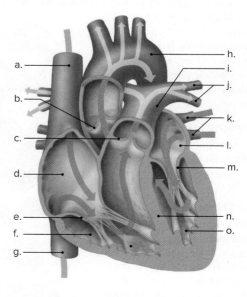

7. Which of the following lists the events of the cardiac cycle in the correct order?
   a. contraction of atria, rest, contraction of ventricles
   b. contraction of ventricles, rest, contraction of atria
   c. contraction of atria, contraction of ventricles, rest
   d. contraction of ventricles, contraction of atria, rest

8. An electrocardiogram measures
   a. chemical signals in the brain and heart.
   b. electrical activity in the brain and heart.
   c. chemical signals in the heart.
   d. electrical changes in the wall of the heart.

9. Water enters the venous end of capillaries because of
   a. osmotic pressure that is higher than blood pressure.
   b. an osmotic pressure gradient.
   c. higher blood pressure on the venous side.
   d. higher blood pressure on the arterial side.
   e. higher red blood cell concentration on the venous side.

10. Lymph is formed from
    a. damaged tissue.
    b. red blood cells.
    c. excess interstitial fluid.
    d. white blood cells.

23.3 Blood: A Transport Medium

11. Which association is not correct?
    a. white blood cells—infection fighting
    b. red blood cells—blood clotting
    c. plasma—water, nutrients, proteins, and wastes
    d. red blood cells—hemoglobin
    e. platelets—blood clotting

12. Plasma
    a. is composed mostly of proteins.
    b. has a pH of 12.
    c. is the liquid portion of blood.
    d. is approximately 10% water.

13. A decrease in lymphocytes would result in problems associated with
    a. clotting.
    b. immunity.
    c. oxygen transportation.
    d. All of these are correct.

14. Which of the following conditions is characterized by the accumulation of plaque in a blood vessel?
    a. hypertension
    b. atherosclerosis
    c. stroke
    d. heart attack

15. Which of the following are not involved in blood clotting?
    a. calcium ions
    b. lymphocytes
    c. platelets
    d. prothrombin and fibrinogen

ENGAGE

BioNOW

Want to know how this science is relevant to your life? Check out the BioNow video below:
- Deer Autopsy

Given the location of the wound in the video, what blood vessel was most likely involved?

Thinking Critically

1. Assume your heart rate is 70 beats per minute (bpm) and your heart pumps 5.25 liters of blood to your body each minute. Based on your age to the nearest day, about how many times has your heart beat so far, and what volume of blood has it pumped?

2. Explain why the evolution of the four-chambered heart was critical for the development of an endothermic lifestyle (the generation of internal heat) in birds and mammals.

3. Provide a physiological explanation for the benefit gained by athletes who train at high altitudes.

4. The cardiovascular system is an elegant example of the concept that structure supports function. Each type of blood vessel has a specific job. Each vessel's physical characteristics enable it to do that job. The muscle walls of the right and left ventricles vary in thickness depending on where they pump the blood. When organ structure is damaged or changed (as arteries are in atherosclerosis), the organ's ability to perform its function may be compromised. Homeostatic conditions, such as blood pressure, may be affected as well. Dietary and lifestyle choices can either prevent damage or harm the cardiovascular system.
   a. What do you think are the long-term effects of hypertension on the heart? What about other organ systems?
   b. Why do you think a combination of hypertension and atherosclerosis is particularly dangerous?

24

The Maintenance Systems

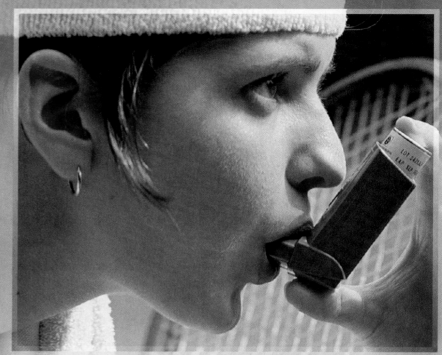

© Paul Windsor/Taxi/Getty Images

OUTLINE

24.1 Respiratory System   451

24.2 Urinary System   457

BEFORE YOU BEGIN

Before beginning this chapter, take a few moments to review the following discussions.

Section 5.4 What influences the diffusion of a gas across a membrane?

Section 22.1 What are the four types of tissue, and what is the general function of each?

Section 22.3 How does negative feedback assist in the maintenance of homeostasis?

When Breathing Becomes Difficult

Asthma is a disease in which the airways become constricted (narrowed) and inflamed (swollen), both of which can result in difficulty breathing. The symptoms often include wheezing and shortness of breath, a frequent cough (often at night), and a feeling of being very tired or weak, especially when exercising. Over 34 million children and adults in the United States have asthma, and the incidence seems to be increasing. Experts offer various explanations for this. One hypothesis is that we may be "too clean," in the sense that we are not exposed to enough common bacteria, viruses, and parasites as children. As a result, our immune systems may react to harmless material we inhale. There are many harmful forms of air pollution, but of increasing concern are tiny, "ultrafine" particles, those less than 0.1 micrometer (μm) across, which are produced at high levels by diesel engines. These particles can bypass the normal defenses of the upper respiratory tract and end up lodging deep in the lungs, with damaging effects.

Recently, there have been breakthroughs in asthma research. In addition to environmental factors, there appear to be a number of genes that increase susceptibility to asthma. For example, one set of genes on chromosome 5 has been linked to asthma and is associated with the body's inflammatory response. Inflammation can constrict the airways and increase fluid flow to the lungs. Variations in a gene on chromosome 17 have also been shown to increase susceptibility, especially in individuals who have been exposed to respiratory viruses. Although there is not currently a cure, the identification of these genes sheds hope for those with asthma.

As you read through this chapter, think about the following questions:

1. How would narrowing and swelling of the airways affect the respiratory volumes?
2. How would asthma indirectly affect the other systems of the body?

24.1 Respiratory System

Learning Outcomes

Upon completion of this section, you should be able to

1. List the three primary steps of respiration in animals.
2. List the components of the upper and lower respiratory tracts.
3. Compare inspiration and exhalation.
4. Compare and contrast the process of respiration in humans, insects, and fish.
5. Explain how oxygen and carbon dioxide are transported in the blood.

The cells of your body are bathed in a fluid, called interstitial fluid. The cells acquire oxygen and nutrients and get rid of carbon dioxide and other wastes through exchanges with the interstitial fluid. In turn, the interstitial fluid exchanges these compounds with the blood (see Section 23.2). Blood is refreshed because the respiratory, urinary, and digestive systems make exchanges with the external environment. Only in this way is blood cleansed of waste molecules and supplied with the oxygen and nutrients the cells require.

In this section we will focus on the role of the **respiratory system** in the exchange of gases. Notice in **Figure 24.1** that, when blood enters the lungs of the respiratory system, it gives up carbon dioxide (CO_2) and picks up oxygen (O_2). Carbon dioxide exits the body through exhalation, and oxygen, obtained through inhalation, is delivered to the body's cells. In this way, **respiration,** also referred to as ventilation, contributes to homeostasis. Respiration in terrestrial vertebrates (including humans) requires these steps:

1. Breathing: inspiration (entrance of air into the lungs) and expiration (exit of air from the lungs)
2. External exchange of gases between the air and the blood within the lungs
3. Internal exchange of gases between blood and interstitial fluid and the exchange of gases between the cells and interstitial fluid

Regardless of the particular gas-exchange surfaces of animals and the manner in which gases are delivered to the cells, in the end oxygen enters mitochondria, where aerobic cellular respiration takes place. Without the delivery of oxygen to the body's cells, ATP is not produced, and life ceases. Carbon dioxide, a waste molecule given off by cells, is a by-product of cellular respiration (see Section 7.1).

The Human Respiratory Tract

The human respiratory system includes all of the structures that conduct air in a continuous pathway to and from the **lungs** (**Fig. 24.2**), the major organ of gas exchange in the body.

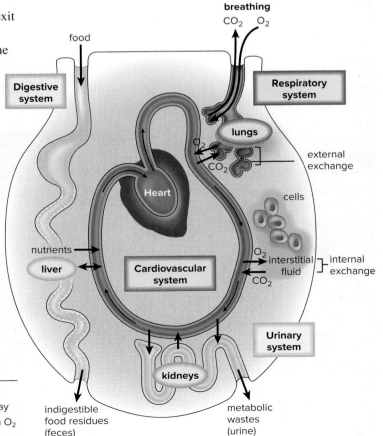

Figure 24.1 The respiratory system and homeostasis.

The respiratory system functions in gas exchange. CO_2-rich blood enters the pulmonary capillaries, and then CO_2 diffuses into the lungs and exits the body by way of respiratory passages. O_2-rich air enters the respiratory passages and lungs. Then O_2 diffuses into the blood at the pulmonary capillaries.

As air moves through the respiratory tract, it is filtered, so that it is free of debris, warmed, and humidified. By the time the air reaches the lungs, it is at body temperature and saturated with water. In the nose, hairs and cilia act as filtering devices. In the respiratory passages, cilia beat upward (**Fig. 24.3**), carrying mucus, dust, and other particles in the air into the throat, where the accumulation may be swallowed or spit out. Smoking cigarettes and cigars inactivates and eventually destroys these cilia, so that the lungs become laden with soot and debris. This is the first step toward various lung disorders.

As the air moves out of the respiratory tract, it cools and loses its moisture. As air cools, it deposits its moisture on the lining of the tract, and the nose may even drip as a result of this condensation. However, the air still retains so much moisture that, on a cold day, it forms a small cloud when we breathe out.

Connections: Health

Why does your nose run when you are cold?

One of the functions of the mucosal lining in the nose is that the mucus helps to warm the air entering the lungs. The tissue lining the nose is highly vascularized (lots of blood vessels), and combining the circulation of the blood with the mucus warms the cold air in the nostrils, so that it will be closer to body temperature when it gets inside the lungs. Warming the inhaled air means that the internal body temperature does not fluctuate too greatly and cause adverse effects on homeostasis.

© Poncho/Getty Images

Upper Respiratory Tract

The upper respiratory tract consists of the nasal cavities, pharynx, and larynx (see Fig. 24.2). The nose, a prominent feature of the face, is the only external portion of the respiratory system. The nose contains the **nasal cavities,** narrow canals separated from one another by a septum composed of bone and cartilage. Tears from the eyes drain into the nasal cavities by way of tear ducts. For this reason, crying produces a runny nose. The nasal cavities are connected to the **sinuses,** air-filled spaces that reduce the weight of the skull and act as resonating chambers for the voice. If the ducts leading from the sinuses become inflamed, fluid may accumulate, causing a sinus headache. The nasal cavities are separated from the mouth by a partition called the palate. The palate has two portions. Anteriorly, the hard palate is supported by bone; posteriorly, the soft palate is made solely of soft tissue and muscle.

The **pharynx** is a funnel-shaped passageway that connects the nasal cavity and mouth to the **larynx,** or voice box. The tonsils form a protective ring of lymphatic tissue at the junction of the mouth and the pharynx. *Tonsillitis* occurs when the tonsils become inflamed and enlarged. If tonsillitis occurs frequently and enlargement makes breathing difficult, the tonsils can be removed surgically, a procedure called a *tonsillectomy.* In the pharynx, the air passage and food passage cross because the larynx, which receives air, is anterior to the

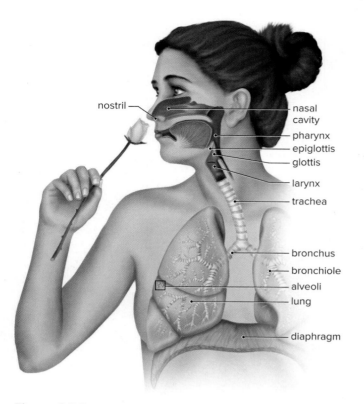

nostril —

nasal cavity

pharynx
epiglottis
glottis

larynx

trachea

bronchus

bronchiole
alveoli
lung

diaphragm

Figure 24.2 **The human respiratory tract.**

To reach the lungs, air moves from the nasal cavities through the pharynx, larynx, trachea, bronchi, and bronchioles, which end in the alveoli of the lungs.

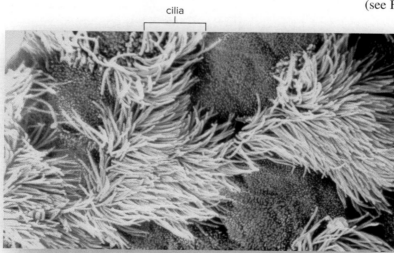

cilia

800×

Figure 24.3 **Ciliated cells of the respiratory passages.**

These cilia sweep impurities up, away from the lungs and toward the throat, where they may be swallowed. Smoking first inactivates and then destroys the cilia.

© Photo Insolite Realite/SPL/Science Source

esophagus, which receives food. This arrangement may seem inefficient, since there is danger of choking if food accidentally enters the trachea, but it does have the advantage of letting you breathe through your mouth if your nose is plugged. In addition, it permits greater intake of air during heavy exercise, when a higher rate of gas exchange is required. When swallowing occurs, the epiglottis, an elastic flap of cartilage, covers the glottis, the opening into the larynx, and this helps prevent choking.

Air passes from the pharynx through the **glottis.** The larynx is always open because it is formed by a complex of cartilages, among them the one that forms the "Adam's apple." At the edges of the glottis, embedded in mucous membrane, are the **vocal cords.** These flexible bands of connective tissue vibrate and produce sound when air is expelled past them through the glottis from the larynx. *Laryngitis* is an infection of the larynx with accompanying hoarseness, leading to the inability to speak audibly.

Lower Respiratory Tract

The lower respiratory tract contains the respiratory tree, consisting of the trachea, bronchi, and bronchioles (see Fig. 24.2). The **trachea,** commonly called the windpipe, is a tube connecting the larynx to the bronchi. The trachea is held open by a series of C-shaped, cartilaginous rings that do not completely meet in the rear. This arrangement allows food to pass down through the esophagus, which lies right behind the trachea in the neck, without the rings of cartilage damaging the outer tissue of the esophagus. The trachea divides into two primary **bronchi,** which enter the right and left lungs. *Bronchitis* is an infection of the bronchi. As bronchitis develops, a nonproductive cough becomes a deep cough that produces mucus and perhaps pus. The deep cough of smokers indicates that they have bronchitis and the respiratory tract is irritated. When a person stops smoking, this progression reverses and the airways become healthy again. Chronic bronchitis is the second step toward *emphysema* and lung cancer caused by smoking cigarettes. Lung cancer often begins in the bronchi, and from there it spreads to the lungs.

The bronchi continue to branch until there are a great number of smaller passages called **bronchioles.** The two bronchi resemble the trachea in structure, but as the passages divide and subdivide, their walls become thinner, the rings of cartilage disappear, and the amount of smooth muscle increases. During an attack of *asthma,* the smooth muscle of the bronchioles contracts, causing constriction of the bronchioles and characteristic wheezing (see the chapter opener). Each bronchiole terminates in an elongated space enclosed by a multitude of little air pockets, or sacs, called **alveoli** (sing., alveolus), which make up the lungs (see Fig. 24.8).

Respiration in Other Animals   Whereas humans have one trachea, insects have many tracheae, little air tubes supported by rings of chitin that branch into every part of the body (**Fig. 24.4**). The tracheal system begins at spiracles, openings that perforate the insect's body wall, and ends in very fine, fluid-filled tubules, which may actually indent the plasma membranes of cells to come close to mitochondria. Ventilation is assisted by the presence of air sacs that draw in the air. Since no cell is very far from the site of gas exchange, the bloodstream does not transport oxygen, and no oxygen-carrying pigment is required.

Breathing

In humans, the **diaphragm** is a muscular, membranous partition that divides the upper *thoracic cavity* from the lower *abdominal cavity* of the body.

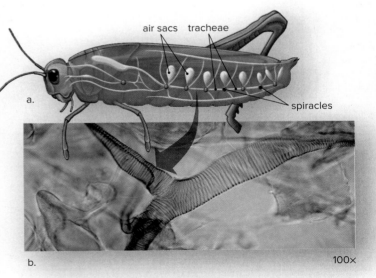

Figure 24.4   **Tracheae of insects.**

A system of air tubes extending throughout the body of an insect carries oxygen to the cells. Air enters the tracheae at openings called spiracles. From there, the air moves to smaller tubes, which take it to the cells where gas exchange takes place. **a.** Diagram of tracheae. **b.** The photomicrograph shows how the walls of the tracheae are stiffened with bands of chitin.

(b): © Ed Reschke

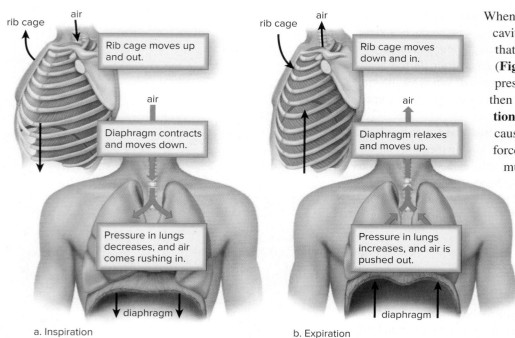

a. Inspiration
b. Expiration

Figure 24.5 Inspiration versus expiration.

a. During inspiration, the thoracic cavity and lungs expand, so that air is drawn in. **b.** During expiration, the thoracic cavity and lungs resume their original positions and pressures, forcing air out.

When humans breathe, the volume of the thoracic cavity and lungs is increased by muscle contractions that both lower the diaphragm and raise the ribs (**Fig. 24.5**). These movements create a negative pressure in the thoracic cavity and lungs, and air then flows into the lungs, a process called **inspiration.** It is important to realize that air comes in because the lungs have already opened up; air does not force the lungs open. When the rib and diaphragm muscles relax, the lungs recoil and air moves out as a result of increased pressure in the lungs, a process called **expiration.**

Breathing in Other Animals

In mammals and most reptiles, air moves in and out by the same route; therefore, some residual air, low in oxygen, is always left in the lungs of humans. Birds, on the other hand, use a one-way ventilation mechanism (**Fig. 24.6**). Incoming air is carried past the lungs by a trachea, which takes it to a set of abdominal air sacs. Then air passes forward through the lungs into a set of thoracic air sacs. Fresh air never mixes with used air in the lungs of birds, and thereby gas-exchange efficiency is greatly improved.

Control of Breathing

Increased concentrations of hydrogen ions (H^+) and carbon dioxide (CO_2) in the blood are the primary stimuli that increase the breathing rate in humans. The chemical content of the blood is monitored by chemoreceptors called **aortic bodies** and **carotid bodies,** specialized structures in the walls of the aorta, and common carotid arteries. These receptors are very sensitive to changes in H^+ and CO_2 concentrations, but they are only minimally sensitive to lower oxygen (O_2) concentrations. The need to breathe comes from a buildup of CO_2 in the bloodstream, not necessarily from a lack of oxygen. Information from the chemoreceptors goes to the *breathing center* in the brain, which increases the breathing rate when concentrations of hydrogen ions and carbon dioxide rise (**Fig. 24.7**). The breathing center is also directly sensitive to the chemical content of the blood, including its oxygen content.

Lungs and External Exchange of Gases

The lungs of humans and other mammals are more elaborately subdivided than those of amphibians and reptiles. Frogs and salamanders have a moist skin that allows them to use the surface of their body for gas exchange in addition to the lungs. It has been estimated that human lungs have a total surface area at least 50 times the skin's surface area because of the presence of alveoli.

An alveolus (**Fig. 24.8**), like the capillary that surrounds it, is bounded by squamous epithelium. Diffusion alone accounts for gas exchange between the alveolus and the capillary. Carbon dioxide, being more plentiful in the pulmonary vein, diffuses from a pulmonary capillary to enter an alveolus, while oxygen, being more plentiful in the lungs, diffuses from an alveolus into a pulmonary capillary. The process of diffusion requires a gas-exchange region that is not only large but also moist and thin. The alveoli are lined with surfactant, a film of lipoprotein

Figure 24.6 Breathing in birds.

The path of airflow in a bird is different than that in a human. When a bird inhales, most of the air enters the abdominal air sacs and then enters the lungs. When a bird exhales, air moves through the lungs to the thoracic air sacs before exiting the bird. This one-way flow of air through the lungs allows more fresh air to be present in the lungs with each breath, leading to greater uptake of oxygen from one breath of air.

that lowers the surface tension of water, thereby preventing the sides of the alveoli from sticking together during exhalation. Some newborns, especially if premature, lack this film. Surfactant replacement therapy is used to treat this condition.

The blood within pulmonary capillaries is indeed spread thin, and the red blood cells are pressed up against their narrow walls. The alveolar epithelium and the capillary epithelium are so close that together they are called the *respiratory membrane*. Hemoglobin in the red blood cells quickly picks up oxygen molecules as they diffuse into the blood.

Emphysema is a serious lung condition in which the walls of many alveoli have been destroyed. This greatly reduces the surface area for gas exchange to occur. Individuals with emphysema are unable to supply their cells with enough oxygen to conduct cellular respiration, and therefore usually the individual has very low energy levels. Emphysema is nonreversible and usually fatal, although individuals with early emphysema can use supplemental oxygen to increase oxygen delivery to their cells.

Gills of Fish

In contrast to the lungs of terrestrial vertebrates, fish and other aquatic animals use gills as their respiratory organ (**Fig. 24.9**). In fish, water is drawn into the mouth and out from the pharynx across the gills. The flow of blood in gill capillaries is opposite the flow of water across the gills; therefore, the blood is always exposed to water having a higher oxygen content. In the end, about 80–90% of the dissolved oxygen in the water is absorbed.

Transport and Internal Exchange of Gases

Recall from Section 23.3 that hemoglobin molecules within red blood cells carry oxygen to the body's tissues. If hemoglobin did not transport oxygen, it would take about 3 years for an oxygen molecule to move from your lungs to

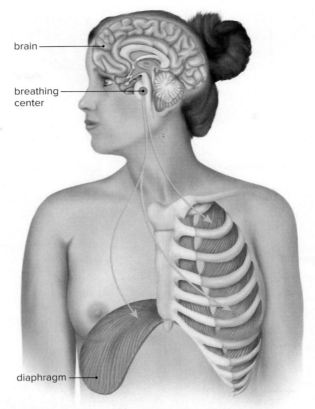

Figure 24.7 **Neural control of breathing rate.**

The brain regulates breathing rate by controlling contraction of the rib cage muscles and the diaphragm. When the breathing rate increases, H^+ lowers as CO_2 is removed from the blood, and the pH of the blood returns to normal.

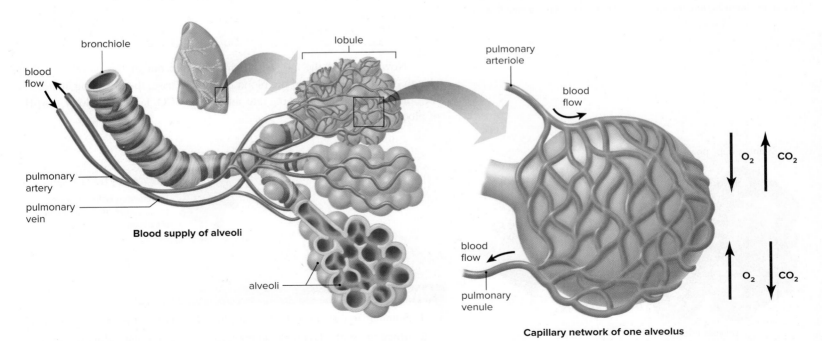

Figure 24.8 **Gas exchange in the lungs.**

Bronchioles lead to the alveoli, each of which is surrounded by an extensive capillary network. The pulmonary artery and arteriole carry O_2-poor blood (colored blue), and the pulmonary venule and vein carry O_2-rich blood (colored red).

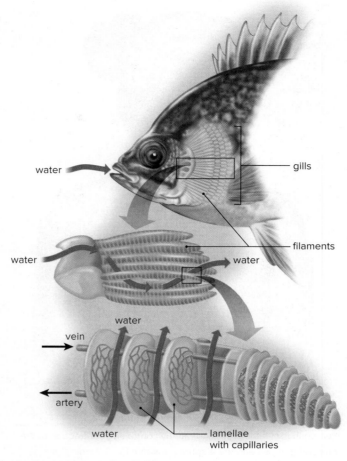

Figure 24.9 Gills of bony fishes.

The structure of a gill. Gases are exchanged between the capillaries inside the lamellae and the water that flows between the lamellae.

your toes by simple diffusion. Each hemoglobin molecule contains four polypeptide chains: two α and two β chains. Each polypeptide is folded around an iron-containing group called **heme.** It is actually the iron that bonds with oxygen and carries it to the tissues (**Fig. 24.10**). Since there are about 250 million hemoglobin molecules in each red blood cell, each cell is capable of carrying at least 1 billion molecules of oxygen. Hemoglobin gives up its oxygen in the tissues during internal exchange primarily because interstitial fluid always has a lower oxygen concentration than blood does. This difference occurs because cells take up and utilize oxygen when they carry on cellular respiration. Another reason hemoglobin gives up oxygen is the warmer temperature and lower pH in the tissues, environmental conditions that are also caused by cellular respiration. When cells respire, they give off heat and carbon dioxide as by-products.

Carbon dioxide enters the blood during internal exchange because the interstitial fluid always has a higher concentration of carbon dioxide. Most of the carbon dioxide is transported in the form of the **bicarbonate ion** (HCO_3^-). First, carbon dioxide combines with water, forming carbonic acid, and then this acid dissociates to a hydrogen ion (H^+) and HCO_3^-:

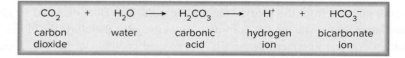

$$CO_2 + H_2O \longrightarrow H_2CO_3 \longrightarrow H^+ + HCO_3^-$$

| carbon dioxide | water | carbonic acid | hydrogen ion | bicarbonate ion |

The H^+ does cause the pH to lower, but only slightly because much of the H^+ is absorbed by the globin portions of hemoglobin. The HCO_3^- is carried in the plasma.

What happens to the HCO_3^- in the lungs? As blood enters the pulmonary capillaries, carbon dioxide diffuses out of the blood into the alveoli. Hemoglobin gives up the H^+ it has been carrying as this reaction occurs:

$$H^+ + HCO_3^- \longrightarrow H_2CO_3 \longrightarrow H_2O + CO_2$$

Now, much of the carbon dioxide diffuses out of the blood into the alveoli of the lungs. Should the blood level of H^+ rise, the breathing center in the brain increases the breathing rate, and as more CO_2 leaves the blood, the pH of blood is corrected.

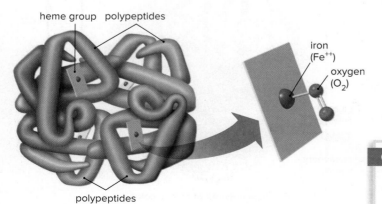

Figure 24.10 Hemoglobin.

Hemoglobin has two each of two different polypeptides, each with a heme group. Oxygen bonds with the central iron atom of a heme group.

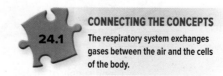

CONNECTING THE CONCEPTS

24.1 The respiratory system exchanges gases between the air and the cells of the body.

Check Your Progress 24.1

1. Summarize the three steps of respiration in a terrestrial vertebrate.

2. Trace the path of air through the upper and lower respiratory tracts.

3. Explain what causes oxygen to diffuse into the blood from the lungs and what happens when carbon dioxide enters the blood.

24.2 Urinary System

Learning Outcomes

Upon completion of this section, you should be able to

1. Explain how the urinary system participates in maintaining homeostasis.
2. Describe the anatomy of the kidney and nephron.
3. Explain the three processes involved in urine formation.
4. Compare and contrast excretion in insects with excretion in mammals.
5. Describe the causes and treatment of kidney disease.

In vertebrate animals, such as humans, the **urinary system** plays an important role in homeostasis. This role is primarily conducted by the kidneys (**Fig. 24.11**), the organs that are primarily responsible for the following functions of the urinary system:

1. Excretion of nitrogenous wastes, such as urea and uric acid
2. Maintenance of the water-salt balance of the blood
3. Maintenance of the acid-base balance of the blood

In humans, the **kidneys** are bean-shaped, reddish-brown organs, each about the size of a fist. They are located on each side of the vertebral column, just below the diaphragm, where they are partially protected by the lower rib cage. **Urine** made by the kidneys is conducted from the body by the other organs in the urinary system. Each kidney is connected to a **ureter,** a tube that takes urine from the kidney to the **urinary bladder,** where it is stored until it is voided from the body through the single **urethra** (**Fig. 24.12**). In males, the urethra passes through the penis; in females, it opens in front of the opening of the vagina. In females, there is no connection between the genital (reproductive) and urinary systems, but there is a connection in males, since the urethra also carries sperm during ejaculation.

In amphibians, birds, reptiles, and some fishes, the bladder empties into the cloaca, a common chamber and outlet for the digestive, urinary, and genital tracts.

Human Kidney

If a kidney is sectioned longitudinally, three major parts can be distinguished (**Fig. 24.13***a*). The *renal cortex,* the outer region of a kidney, has a somewhat granular appearance. The *renal medulla* consists of the cone-shaped renal pyramids, which lie inside the renal cortex. The innermost part of the kidney is a hollow chamber called the *renal pelvis.* Urine collects in the renal pelvis and then is carried to the bladder by a ureter. Microscopically, the cortex and medulla of each kidney are composed of about 1 million tiny tubules called **nephrons** (Fig. 24.13*b*). The nephrons of a kidney produce urine.

Nephrons

Each nephron is made up of several parts (**Fig. 24.14**). The blind, or closed, end of a nephron is pushed in on itself to form the **glomerular capsule.** Filtration of the blood occurs at this portion of the nephron. The inner layer of the capsule is composed of specialized cells that allow the easy passage of molecules. Leading from the capsule is a portion of the nephron called the **proximal convoluted tubule,** which is lined by cells with many mitochondria and tightly packed microvilli. Then comes the **nephron loop,** with a descending limb and

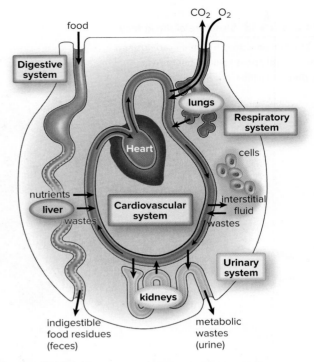

Figure 24.11 **The urinary system and homeostasis.**

Excretion carried out by the kidneys serves three functions: excretion of urea (produced by the liver) and other nitrogenous wastes, maintenance of water-salt balance, and maintenance of blood pH.

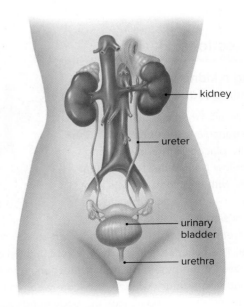

Figure 24.12 **The human urinary system.**

The organs of the human urinary system produce, store, and expel urine from the body.

Figure 24.13 Structure of the kidney

a. The human kidney has three regions and contains (**b**) about 1 million nephrons. Nephrons do the work of the kidney.

two nephrons

renal cortex

renal medulla

pyramid
renal artery
renal vein
renal pelvis
ureter

a. Kidney

glomeruli capsule
proximal convulated tubule
distal convoluted tubule
peritubular capillary
loop of the nephron
collecting duct

b. Two nephrons

Connections: Health

During a kidney transplant, is the failing kidney removed?

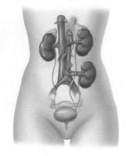

Normally, no. When a patient needs a kidney transplant, the donor kidney is normally placed in the lower abdomen, below one of the two original kidneys, with the blood vessels from the donor kidney connected to the recipient's arteries and veins and the ureter from the donor kidney connected to the recipient's bladder. The patient's failing kidney is kept in most cases because, even if the nephrons are not functioning in cleansing the blood and producing urine, the kidney can still produce other chemicals and substances the body needs to function. The failing kidney is removed in cases of severe infection, cancer, or polycystic kidney disease (PKD).

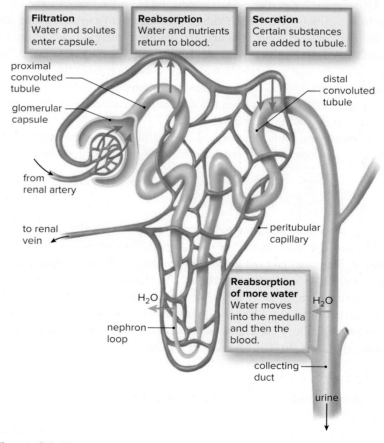

Filtration Water and solutes enter capsule.	**Reabsorption** Water and nutrients return to blood.	**Secretion** Certain substances are added to tubule.

proximal convoluted tubule

glomerular capsule

distal convoluted tubule

from renal artery

to renal vein

peritubular capillary

Reabsorption of more water Water moves into the medulla and then the blood.

H_2O

H_2O

nephron loop

collecting duct

urine

Figure 24.14 Urine formation.

Urine formation requires three steps: filtration, reabsorption, and secretion. Production of a hypertonic urine occurs when water is reabsorbed into the blood along the length of the nephron, and especially at the nephron loop and collecting duct.

an ascending limb. This is followed by the **distal convoluted tubule.** Several distal tubules enter one **collecting duct.** A collecting duct delivers urine to the renal pelvis. The nephron loop and the collecting duct give the pyramids of the renal medulla their striped appearance (see Fig. 24.13*a*).

Each nephron has its own blood supply, and various exchanges occur between parts of the nephron and a blood capillary as urine forms.

Urine Formation

Urine formation requires three steps: filtration, reabsorption, and secretion (Fig. 24.14).

Filtration   Filtration occurs whenever small substances pass through a filter and large substances are left behind. During urine formation, **filtration** is the movement of small molecules from a blood capillary to the inside of the glomerular capsule as a result of blood pressure. Small molecules, such as water, nutrients, salts, and urea, move to the inside of the capsule. Plasma proteins and blood cells are too large to be part of this filtrate, so they remain in the blood.

If the composition of the filtrate were not altered in other parts of the nephron, death from loss of nutrients (starvation) and loss of water (dehydration) would quickly follow. The next step, reabsorption, helps prevent this.

Reabsorption of Solutes   **Reabsorption** takes place when water and other substances move from the proximal convoluted tubule into the blood. Nutrients such as glucose and amino acids also return to the blood. This process is selective because some molecules, such as glucose, are both passively and actively reabsorbed. The cells of the proximal convoluted tubule have numerous microvilli, which increase the surface area, and numerous mitochondria, which supply the energy needed for active transport. However, if there is more glucose in the filtrate than there are carriers to handle it, glucose will appear in the urine. Glucose in the urine is a sign of *diabetes mellitus,* sometimes caused by a lack of the hormone insulin.

Sodium ions (Na^+) are also actively pumped into the peritubular capillary, and then chloride ions (Cl^-) follow passively. Water moves by osmosis from the tubule into the blood. Protein channels, called aquaporins, in the membrane of the epithelial cells lining the proximal convoluted tubule allow for the rapid reabsorption of water. Overall, about 60–70% of salt and water are reabsorbed at the proximal convoluted tubule.

Some of the urea, the primary nitrogenous waste product of human metabolism, and other types of nitrogenous wastes excreted by humans are passively reabsorbed, but most of these wastes remain in the filtrate.

Secretion   **Secretion** is the transport of substances into the nephron by means other than filtration. For our purposes, secretion may be particularly associated with the distal convoluted tubule. Substances such as uric acid, hydrogen ions, ammonia, and penicillin are eliminated by secretion. The process of secretion helps rid the body of potentially harmful compounds that were not filtered into the capsule.

Regulation of Water-Salt Balance and pH   All animals have some means of maintaining the water-salt balance and pH of the internal environment. In humans, the long nephron loop allows for the secretion of a hypertonic urine (Fig. 24.14). The ascending limb of the nephron loop pumps salt and urea into the renal medulla, and water follows by osmosis both at the descending limb of

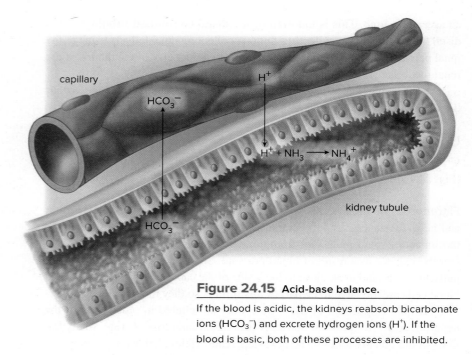

Figure 24.15 Acid-base balance.

If the blood is acidic, the kidneys reabsorb bicarbonate ions (HCO_3^-) and excrete hydrogen ions (H^+). If the blood is basic, both of these processes are inhibited.

the nephron loop and at the collecting duct. As you will see in Section 27.2, several hormones are involved in regulating water-salt reabsorption by the kidneys. Drinking coffee interferes with one of these hormones, and that is why coffee is a *diuretic,* a substance that causes the production of more urine.

The human kidney also assists in the regulation of the pH of the blood. Both the bicarbonate (HCO_3^-) buffer system of the blood and the regulation of breathing rate help rid the body of CO_2 and contribute to maintaining blood pH. As helpful as these mechanisms might be, only the kidneys can excrete a wide range of acidic and basic substances. The kidneys are slower-acting than the buffer/breathing mechanism, but they have a more powerful effect on pH. For the sake of simplicity, we can think of the kidneys as reabsorbing bicarbonate ions and excreting hydrogen ions as needed to maintain the normal pH of the blood (**Fig. 24.15**). If the blood is acidic, hydrogen ions are excreted and bicarbonate ions are reabsorbed. If the blood is basic, hydrogen ions are not excreted and bicarbonate ions are not reabsorbed. The fact that urine is most often acidic shows that usually an excess of hydrogen ions is produced by the body and excreted. Ammonia (NH_3) provides a means for buffering these hydrogen ions in urine: ($NH_3 + H^+ \rightarrow NH_4^+$). Ammonia (the presence of which is quite obvious in a diaper pail or kitty litter box) is produced in tubule cells by the breakdown of amino acids. Phosphate provides another means of buffering hydrogen ions in urine.

Regulation of Water-Salt Balance in Other Animals

Insects have a unique excretory system consisting of long, thin tubules called **Malpighian tubules** attached to the gut (**Fig. 24.16**). Uric acid, the primary nitrogenous waste product of insects, simply

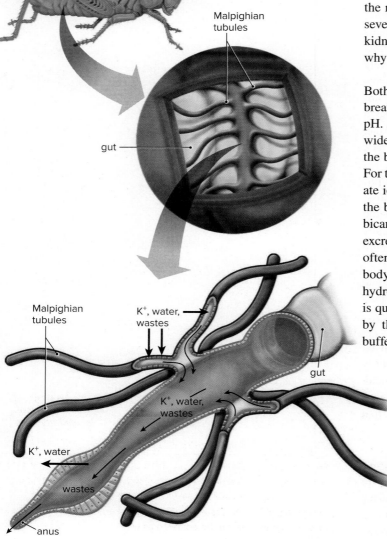

Figure 24.16 Malpighian tubules.

The Malpighian tubules of insects are attached to the gut and surrounded by the hemolymph of the open circulatory system. Wastes diffuse into tubules. K^+ is secreted into these tubules, drawing in water by osmosis. Much of the water and K^+ are reabsorbed across the wall of the rectum.

diffuses from the surrounding hemolymph into these tubules, and water follows a salt gradient established by active transport of potassium (K^+). Water and other useful substances are reabsorbed at the rectum, but the uric acid leaves the body at the anus. Insects that live in water or eat large quantities of moist food reabsorb little water. But insects in dry environments reabsorb most of the water and excrete a dry, semisolid mass of uric acid. Most animals can regulate the blood levels of both water and salt. For example, freshwater fishes take up salt in the digestive tract and gills and produce large amounts of dilute urine. Saltwater fishes take in salt water by drinking, but then they pump ions out at the gills and produce only small amounts of concentrated urine.

Problems with Kidney Function

Many types of illnesses, especially diabetes, hypertension, and inherited conditions, cause progressive renal disease and renal failure. Urinary tract infections, an enlarged prostate gland, pH imbalances, or simply an intake of too much calcium can lead to kidney stones. Kidney stones form in the renal pelvis and usually pass unnoticed in the urine flow. If they grow to several centimeters and block the renal pelvis or ureter, back pressure builds up and destroys nephrons. One of the first signs of nephron damage is the presence of albumin, white blood cells, or even red blood cells in the urine, as detected by a urinalysis. If damage is so extensive that more than two-thirds of the nephrons are inoperative, urea and other waste substances accumulate in the blood. Although nitrogenous waste in the blood is a threat to homeostasis, the retention of water and salts is of even greater concern. **Edema,** fluid accumulation in the body tissues, may occur. Imbalance in the ionic composition of body fluids can lead to loss of consciousness and even heart failure.

Hemodialysis and Kidney Transplants

Patients with renal failure can undergo *hemodialysis,* utilizing either an artificial kidney machine or *continuous ambulatory peritoneal dialysis* (*CAPD*). **Dialysis** is defined as the diffusion of dissolved molecules through a semipermeable membrane with pore sizes that allow only small molecules to pass through. In an artificial kidney machine, the patient's blood is passed through a membranous tube, which is in contact with a dialysis solution, or dialysate (**Fig. 24.17**). Substances more concentrated in the blood diffuse into the dialysate, and substances more concentrated in the dialysate diffuse into the blood. The dialysate is continuously replaced to maintain favorable concentration gradients. In this way, the artificial kidney can be utilized either to extract substances from the blood, including waste products or toxic chemicals and drugs, or to add substances to the blood—for example, bicarbonate ions (HCO_3^-) if the blood is acidic. In the course of a three- to six-hour hemodialysis procedure, from 50 to 250 grams of urea can be removed from a patient, which greatly exceeds the amount excreted by our kidneys within the same time frame. Therefore, a patient needs to undergo treatment only about twice a week.

CAPD is so named because the peritoneum, the epithelium that lines the abdominal cavity, is the dialysis membrane. A fresh amount of dialysate is introduced directly into the abdominal cavity from a bag that is temporarily attached to a permanently implanted plastic tube. The dialysate flows into the peritoneal cavity by gravity. Waste and salt molecules pass from the blood vessels in the abdominal wall into the dialysate before the fluid is collected

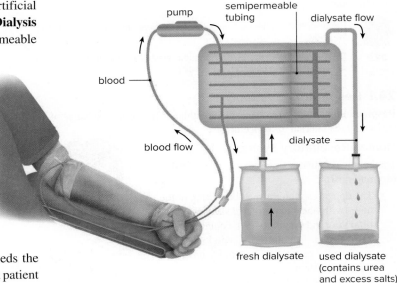

Figure 24.17 Artificial kidney machine.

As the patient's blood is pumped through dialysis tubing, the tubing is exposed to a dialysate (dialysis solution). Wastes exit from blood into the solution because of a preestablished concentration gradient. In this way, not only is blood cleansed but its water-salt and acid-base balances can be adjusted as well.

CONNECTING THE CONCEPTS

24.2 The urinary system excretes nitrogenous wastes and maintains the water-salt and acid-base balances of the body.

4 or 8 hours later. The solution is drained into a bag from the abdominal cavity by gravity, and then it is discarded. One advantage of CAPD over an artificial kidney machine is that the individual can go about his or her normal activities during CAPD.

Patients with renal failure may undergo a transplant operation, receiving a functioning kidney from a donor. A person needs only one functioning kidney; however, the possibility of organ rejection exists, as it does with all organ transplants. Receiving a kidney from a close relative has the highest chance of success. The current 1-year survival rate is 98%, and the 5-year survival rate is more than 91%. In the future, it may be possible to use either kidneys from pigs or kidneys created in the laboratory for transplant operations.

Check Your Progress 24.2

1. List the three functions of the urinary system

2. Describe the parts of a nephron and the function of each part.

3. Summarize the three stages of urine formation.

4. Summarize the importance of water-salt balance and pH balance to homeostasis.

STUDY TOOLS   http://connect.mheducation.com

SMARTBOOK® Maximize your study time with McGraw-Hill SmartBook®, the first adaptive textbook.

SUMMARIZE

The maintenance systems play vital roles in the optimal functioning of the human body. These systems supply nutrients and/or remove waste material from the body.

24.1 The respiratory system exchanges gases between the air and the cells of the body.

24.2 The urinary system excretes nitrogenous wastes and maintains the water-salt and acid-base balances of the body.

24.1 Respiratory System

Respiration involves three steps: (1) breathing, (2) external exchange of gases, and (3) internal exchange of gases.

Human Respiratory System

The human **respiratory system** can be divided into two major parts:

- The upper respiratory tract, consisting of the nose (with **nasal cavities** and **sinuses**), **pharynx,** and **larynx** (which contains the vocal cords). The **glottis** and epiglottis separate the respiratory system from the digestive system.
- The lower respiratory tract, made up of the **trachea, bronchi, bronchioles,** and lungs. The **lungs** contain many **alveoli,** which are air sacs surrounded by a capillary network. The base of each lung contacts the diaphragm, a muscle that separates the thoracic and abdominal cavities.

Breathing

- **Inspiration:** During **inspiration,** the **diaphragm** lowers and the rib cage moves up and out; the lungs expand and air rushes in.

- **Expiration:** During **expiration,** the diaphragm relaxes and moves up; the rib cage moves down and in; pressure in the lungs increases; air is pushed out of the lungs.
- **Control of breathing:** Levels of carbon dioxide and hydrogen ions (H^+) act as the stimuli to adjust breathing rates. **Aortic bodies** and **carotid bodies** are chemoreceptors that send information to the breathing center in the brain, which adjusts the rate of breathing.

Lungs and External Exchange of Gases

- Alveoli are surrounded by pulmonary capillaries.
- CO_2 diffuses from the blood into the alveoli, and O_2 diffuses into the blood from the alveoli, because of the respective concentration gradients of these gases.

Transport and Internal Exchange of Gases

Oxygen for cellular respiration follows this path:

- Transported by the iron portion (**heme**) of hemoglobin
- Exits blood at tissues by diffusion
- Given up by hemoglobin in capillaries because cellular respiration makes tissues warmer and more acidic

Carbon dioxide, from cellular respiration, follows this path:

- Enters blood by diffusion
- Taken up by red blood cells or joins with water to form carbonic acid
- Carbonic acid breaks down to H^+ and **bicarbonate ion** (HCO_3^-):

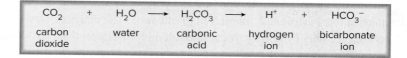

$$CO_2 + H_2O \longrightarrow H_2CO_3 \longrightarrow H^+ + HCO_3^-$$

| carbon dioxide | water | carbonic acid | hydrogen ion | bicarbonate ion |

Bicarbonate ions are carried in plasma. H$^+$ combines with globin of hemoglobin.

- In the lungs, H$^+$ joins with bicarbonate ion to form carbonic acid, which breaks down to water and carbon dioxide:

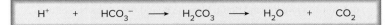

$$H^+ \quad + \quad HCO_3^- \quad \longrightarrow \quad H_2CO_3 \quad \longrightarrow \quad H_2O \quad + \quad CO_2$$

24.2 Urinary System

The **urinary system** performs the following functions:

- Excretes nitrogenous wastes, such as urea and uric acid
- Maintains the normal water-salt balance of blood
- Maintains the acid-base balance of blood

The urinary system consists of these parts:

- **Kidneys:** Produce **urine**
- **Ureters:** Take urine to the bladder
- **Urinary bladder:** Stores urine
- **Urethra:** Releases urine to the outside

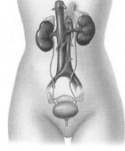

Kidneys

The kidneys have three regions: renal cortex, renal medulla, and renal pelvis. At the microscopic level, the cortex and medulla contain the **nephrons.** A nephron has a **glomerular capsule, proximal convoluted tubule, nephron loop, distal convoluted tubule,** and **collecting duct.**

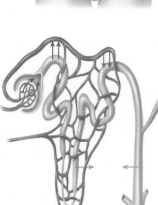

Urine formation has three steps:

- **Filtration:** Water, nutrients, and wastes move from the blood to the inside of the glomerular capsule.
- **Reabsorption:** Primarily salts, water, and nutrients are reabsorbed at the proximal convoluted tubule.
- **Secretion:** Certain substances (e.g., hydrogen ions) are transported into the distal convoluted tubule from blood.

Regulation of Water-Salt Balance and pH The kidneys play a role in homeostasis by regulating water-salt levels and blood pH. The reabsorption of water and the production of a hypertonic urine involve the establishment of a solute gradient that pulls water from the descending limb of the nephron loop and from the collecting duct. The kidneys keep blood pH at about 7.4 by reabsorbing HCO$_3^-$ and excreting H$^+$ as needed. Ammonia buffers H$^+$ in the urine. The Malpighian tubules of insects and the specialized organs of saltwater versus freshwater fishes are other examples of how animals regulate water-salt balance.

Problems with Kidney Function

Various medical conditions, including diabetes, kidney stones, and kidney infections, can lead to renal failure. Renal failure may cause **edema,** an accumulation of fluid in the tissues of the body. Renal failure can be treated by **dialysis** using a kidney machine or by a kidney transplant.

ASSESS

Testing Yourself

Choose the best answer for each question.

24.1 Respiratory System

1. Label the components of the human respiratory system in the following illustration.

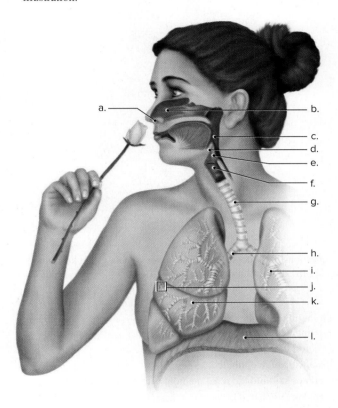

2. Gas exchange occurs in the _____, small air sacs at the ends of the bronchioles.
   a. alveoli
   b. tracheae
   c. tonsils
   d. nephrons
3. The larynx is
   a. the site of gas exchange in the human respiratory system.
   b. located between the trachea and the bronchi.
   c. a flap of cartilage that keeps food out of the respiratory tract during swallowing.
   d. the voice box.
4. How is carbon dioxide primarily transported in the cardiovascular system?
   a. as biocarbonate ions
   b. attached to the heme of hemoglobin
   c. as carbon dioxide gas
   d. as uric acid

24.2 Urinary System

5. Label the components of the human urinary system in the following illustration.

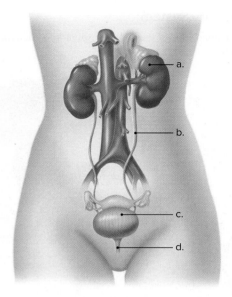

For questions 6–9, identify the kidney component in the key that matches the description.

Key:

   a. glomerular capsule
   b. proximal convoluted tubule
   c. nephron loop
   d. distal convoluted tubule

6. This pumps salt into the renal medulla, and water follows by osmosis.
7. Uric acid is eliminated by secretion here.
8. Microvilli reabsorb molecules, such as glucose, here.
9. Filtration occurs here.
10. The urinary system participates in the body's regulation of
   a. water levels.
   b. salt levels.
   c. blood pH.
   d. All of these are correct.

ENGAGE

Thinking Critically

1. Carbon monoxide (CO) binds to hemoglobin 230 to 270 times more strongly than oxygen does, which means less O_2 is delivered to tissues when CO is present in the blood. What would be the specific cause of death if too much hemoglobin became bound to CO instead of to O_2?

2. High blood pressure (hypertension) often is accompanied by kidney damage. In some people, the kidney damage is subsequent to the high blood pressure, but in others the kidney damage is what caused the high blood pressure. Explain how a low-salt diet would enable you to determine whether the high blood pressure or the kidney damage came first.

3. Grasshoppers, like most other insects, are highly adapted to life on land. What adaptations to life on land are present in their respiratory and urinary systems?

© Ricochet Creative Productions LLC/Synde Mass, photographer

25

Digestion and Human Nutrition

Gluten and Celiac Disease

On any trip to the grocery store, you may notice a wide abundance of foods that are labeled "gluten-free," and almost every restaurant has a gluten-free section of the menu. So, what exactly is gluten?

Contrary to what most people believe, gluten is not a carbohydrate. Instead it is a type of protein that is commonly found in wheat, rye, and barley. For most of us, gluten is processed by the digestive system in the same manner as any other protein, meaning that it is broken down into amino acids and absorbed into the circulatory system.

However, for about one out of every 100 individuals, the body misidentifies gluten as a foreign pathogen. The result is celiac disease, a disorder of the lining of the small intestine. In celiac disease, an example of an autoimmune response, the body's reaction to gluten causes inflammation and loss of the intestinal linings. The damage to the intestines causes malnutrition and other health problems. Individuals with celiac disease must eliminate gluten from their diet. Interestingly, a gluten-free diet does not seem to have health benefits for individuals who do not suffer from celiac disease.

In this chapter, you will learn about the function of the digestive system as well as the basic principles of nutrients and nutrition.

As you read through this chapter, think about the following questions:

1. How are carbohydrates and proteins processed by the digestive system?
2. What is the role of the small intestine in digestion?

OUTLINE

25.1 Digestive System   466

25.2 Nutrition   475

25.3 The Classes of Nutrients   476

25.4 Understanding Nutrition Guidelines   484

25.5 Nutrition and Health   487

BEFORE YOU BEGIN

Before beginning this chapter, take a few moments to review the following discussions.

Section 2.2 What are the properties of water that make it an important nutrient for all life-forms?

Section 3.2 What are the roles of carbohydrates, lipids, and proteins in a cell?

Section 7.1 How does aerobic respiration release the energy found in organic molecules, such as glucose?

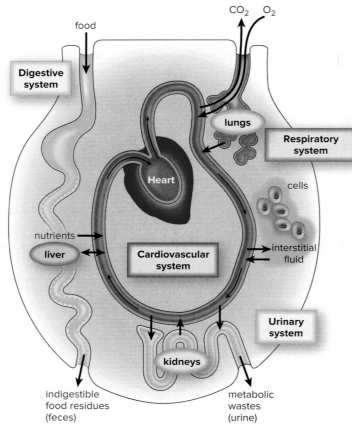

Figure 25.1 **The digestive system and homeostasis.**

The digestive system takes in food and digests it to nutrient molecules that enter the blood. The blood transports nutrients to the tissues, where exchange occurs with tissue fluid.

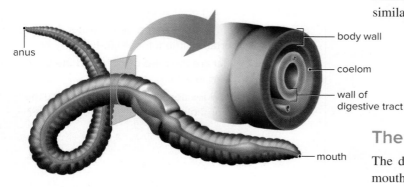

Figure 25.2 **Structure of a complete digestive system.**

A complete digestive system consists of a tube-within-a-tube configuration that allows for the specialization of organs along the digestive tract.

25.1 Digestive System

Learning Outcomes

Upon completion of this section, you should be able to

1. Distinguish between a complete and an incomplete digestive system.
2. Identify the organs of the human digestive system, and provide a function for each.
3. List the accessory organs of the digestive system, and explain their functions.
4. Describe the digestion and absorption of specific nutrients (e.g., carbohydrates) by the digestive tract.

One of the basic characteristics of life (see Section 1.1) is the ability to acquire nutrients for the energy to conduct the activities of living, such as responding to stimuli. In animals, the **digestive system** consists of the organs involved in the following processes:

- Ingesting food
- Breaking down food into smaller molecules that can be transported
- Absorbing nutrient molecules
- Eliminating indigestible materials

The digestive system interacts with the other organ systems of the body (**Fig. 25.1**) to maintain homeostasis by providing the body's cells with the nutrients they need to continue functioning.

Complete and Incomplete Digestive Systems

The hydras and planarians (see Chapter 19) are both examples of animals that have an incomplete digestive system, meaning that a single opening serves as both an entrance and an exit. Most other animals, such as the earthworm, have a complete digestive tract. A complete digestive tract has a tube-within-a-tube configuration (**Fig. 25.2**). The inner tube (the digestive tract, sometimes called the alimentary canal) has both an entrance (the mouth) and an exit (the anus). Notice that the inner tube is separated from the outer tube (the body wall) by the **coelom.** Specialized organs that assist with digestion are located within the coelom. While you might think that the digestive tract of humans is very different from that of the earthworms, there are actually a number of important similarities.

In all animals, including earthworms and humans, the digestion of food is an extracellular process. Digestive enzymes, produced by glands in the wall of the digestive tract or by accessory glands that lie nearby, are released into the tract. Food is never found within these accessory glands, only within the digestive tract itself.

The Digestive Tract

The digestive tract of humans, and most other vertebrates, consists of the mouth, esophagus, stomach, small intestine, and large intestine (**Fig. 25.3**).

Mouth

In humans, the digestive system begins with the **mouth,** which chews food into pieces, beginning the process of mechanical digestion. Many vertebrates have teeth, an exception being birds, which lack teeth and depend on the churning of small pebbles within a gizzard to break up their food. The teeth (dentition) of mammals reflect their diet (**Fig. 25.4**). **Carnivores** eat other animals, and meat is easily digestible because the cells do not have a cellulose wall. **Herbivores** eat plant material, which needs a lot of chewing and other processing to break up the cellulose walls. Humans are **omnivores;** they eat both meat and plant material. The four front teeth (top and bottom) of humans are sharp, chisel-shaped incisors used for biting. On each side of the incisors are the pointed canines used for tearing food. The premolars and molars grind and crush food. It is as though humans were carnivores in the front of their mouths and herbivores in the back.

Food contains cells composed of molecules of carbohydrates, proteins, nucleic acids, and lipids. Digestive enzymes break down these large molecules to smaller molecules. In the mouth, three pairs of **salivary glands** send saliva by way of ducts to the mouth, to begin the process of chemical digestion. One of these digestive enzymes is **salivary amylase,** which breaks down starch, a carbohydrate, to maltose, a disaccharide. While in the mouth, food is manipulated by the muscular tongue (mechanical digestion), mixing the chewed food with saliva (chemical digestion) and then forming this mixture into a mass called a **bolus,** which is swallowed.

Swallowing

The human digestive and respiratory passages come together in the pharynx and then separate (**Fig. 25.5a**). When food is swallowed, the soft palate, the rear portion of the mouth's roof, moves up to close off the nasal cavities (Fig. 25.5b). A flap of tissue called the **epiglottis** covers the glottis, an opening into the larynx. Ordinarily, the bolus must move through the pharynx and into the esophagus because the air passages are blocked. Unfortunately, we have all had the unpleasant experience of having food "go the wrong way." The wrong way may be either into the nasal cavities or into the trachea. If it is the latter, coughing will most likely force the food up out of the trachea and into the pharynx again.

The **esophagus** is a muscular tube that takes food to the stomach, which lies below the diaphragm. When food enters the esophagus, peristalsis begins.

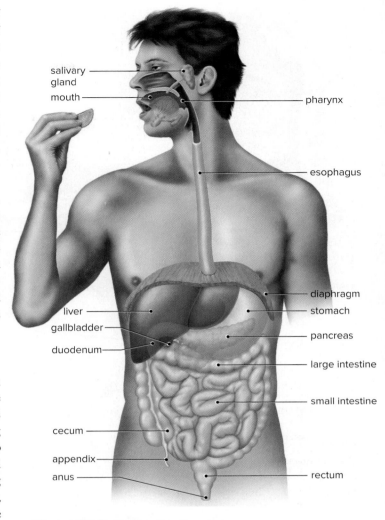

Figure 25.3 Human digestive system.

Humans have a closed digestive system that consists of digestive and accessory organs.

Figure 25.4 Dentition among mammals.

a. Carnivores have enlarged canines for killing prey, short incisors for scraping bones, and jagged molars for tearing flesh. **b.** Herbivores have sharp incisors for clipping vegetation and flat molars and premolars for grinding. **c.** Omnivores have front teeth like a carnivore's and back teeth like a herbivore's.

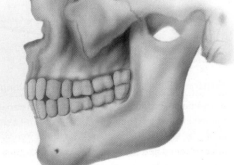

Key:

☐ incisors	☐ premolars
☐ canines	☐ molars

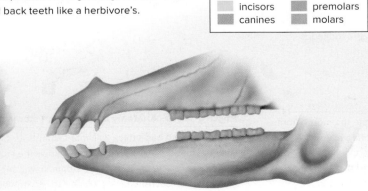

a. Carnivore   b. Herbivore   c. Omnivore

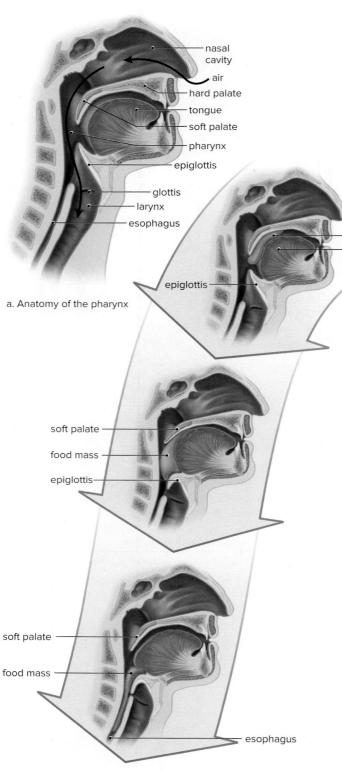

a. Anatomy of the pharynx

b. Swallowing

Figure 25.5 **The pharynx and swallowing.**

a. The palates (both the hard and the soft) separate the mouth from the nasal cavities. The pharynx leads to the esophagus and the larynx. **b.** When food is swallowed, the soft palate closes off the nasal cavities and the epiglottis closes off the larynx.

Peristalsis is a series of rhythmic contractions of smooth muscles that move the contents along in tubular organs—in this case, those of the digestive tract.

Stomach

The human **stomach** is a thick-walled, J-shaped organ (**Fig. 25.6**) on the left side of the abdominal cavity below the liver. The stomach is continuous with the esophagus above and the duodenum of the small intestine below. The cardiac sphincter separates the esophagus from the stomach. A sphincter is a muscle that surrounds a tube and closes or opens it by contracting and relaxing. The stomach is about 25 cm (10 in.) long, regardless of the amount of food it holds, but the diameter varies, depending on how full it is. The stomach receives food from the esophagus, stores food, starts the digestion of proteins, and moves food into the small intestine.

The wall of the stomach has deep folds, which disappear as the stomach fills to an approximate capacity of 1 liter. Therefore, humans can periodically eat relatively large meals, freeing the rest of their time for other activities. But the stomach is much more than a mere storage organ. The wall of the stomach contains three muscle layers: one is longitudinal, another is circular, and the third is obliquely arranged. The muscular walls mechanically digest food by contracting vigorously to mix it with digestive juices, which are secreted whenever food enters the stomach.

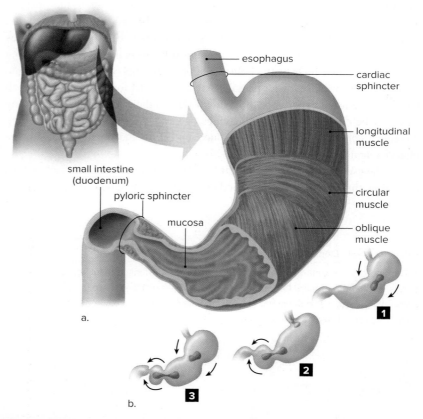

Figure 25.6 **Anatomy of the human stomach.**

a. The stomach has a thick wall that expands as it fills with food. The wall contains three layers of muscle, and their presence allows the stomach to churn and mix food with gastric juices. The mucosa of the stomach wall secretes mucus and contains gastric glands, which secrete gastric juices active in the digestion of proteins. **b.** Peristalsis, a series of rhythmic contractions, occurs along the length of the digestive tract.

The epithelial lining of the stomach, called a mucosa, has millions of gastric glands. These gastric glands produce gastric juice containing so much hydrochloric acid that the stomach routinely has a pH of about 2. Such strong acidity is usually sufficient to kill any microbes that might be in food. This low pH also promotes the activity of **pepsin,** a hydrolytic enzyme that acts on proteins to produce peptides. Sometimes, the stomach's strong acidity causes *heartburn* or even *gastric reflux disease* when gastric juice backs up into the esophagus.

As with the rest of the digestive tract, a thick layer of mucus protects the wall of the stomach from enzymatic action. *Ulcers* are open sores in the wall caused by the gradual destruction of tissues. Most ulcers are due to an infection by an acid-resistant bacterium, *Helicobacter pylori,* which is able to attach to the epithelial lining. Wherever the bacterium attaches, the lining stops producing mucus and the area becomes exposed to digestive action, allowing an ulcer to develop.

Alcohol and other liquids are absorbed in the stomach, but other nutrients are not. Peristalsis pushes food along in the stomach, as it does in other digestive organs (see Fig. 25.6b). At the base of the stomach is a narrow opening called the pyloric sphincter. When food leaves the stomach, it is a thick, soupy liquid called chyme. Whenever the pyloric sphincter relaxes, a small quantity of chyme squirts through the opening into the small intestine.

Ruminants Ruminants, a type of mammal that includes cattle, sheep, goats, deer, and buffalo, are named for a part of their stomach, the rumen (**Fig. 25.7**). The rumen contains symbiotic bacteria and protozoans that produce enzymes that can digest cellulose, an ability that other mammals lack. After herbivores feed on grass, it goes to the rumen, where it is broken down by the enzymes released by the microbes, and then it becomes small balls of cud. The cud returns to the mouth, where the animal "chews the cud." The cud may return to the rumen for a second go-round before passing through the other chambers of the stomach. The rumen is an adaptation to a diet rich in fiber that may have been promoted by competition among the many types of animals that feed on grass. The last chamber in ruminants is analogous to the human stomach, being the place where proteins are digested to peptides.

Small Intestine

Processing of food in the human digestive tract is more complicated than one might think. Food is chewed in the mouth and worked on by the enzyme salivary amylase, which digests starch to maltose. In addition, the digestion of proteins begins in the stomach as pepsin digests these molecules to peptides. At this point, the contents of the digestive tract are called **chyme.** Chyme passes from the stomach to the **small intestine,** a long, coiled tube that has two functions: (1) digestion of all the molecules in chyme, including polymers of carbohydrates, proteins, nucleic acids, and fats, and (2) absorption of the nutrient molecules into the body.

The first part of the small intestine is called the duodenum. Two important accessory glands—the **pancreas,** located behind the stomach, and the **liver**—send secretions to the duodenum by way of ducts (**Fig. 25.8**). The liver produces **bile,** which is stored in the **gallbladder.** Bile looks green because it contains pigments that are the products of hemoglobin breakdown. This green color is familiar to anyone who has observed how bruised tissue changes color. Hemoglobin within the bruised area breaks down into the same types of pigments found in bile. Bile also contains bile salts, which break up fat into fat

Connections: Scientific Inquiry

Why does your stomach "growl" when you are hungry?

Borborygmi is the medical term for the "growl" sound in your stomach. It is produced when the stomach walls squeeze together in an attempt to mix digestive juices and gases for digestion. If your stomach is empty, the result is the sound of these juices bouncing off the walls of the hollow stomach. The "hunger center" of the brain will send a message to your stomach to begin the process of digestion, sometimes initiating borborygmi and signaling the need to eat.

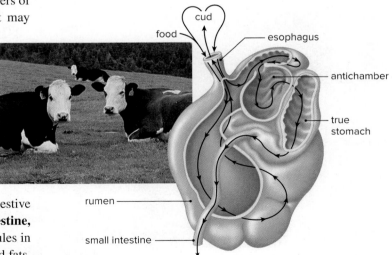

Figure 25.7 A ruminant's stomach.

Ruminants eat grass, which is made of cells with strong cellulose walls. The first chamber of a ruminant's stomach, called the rumen, contains symbiotic bacteria and protozoans, which release enzymes that digest cellulose. After a first pass through the rumen, the "cud" returns to the mouth, where it is leisurely chewed. Then, it may return to the rumen for a second round of digestion before passing through to the true stomach.

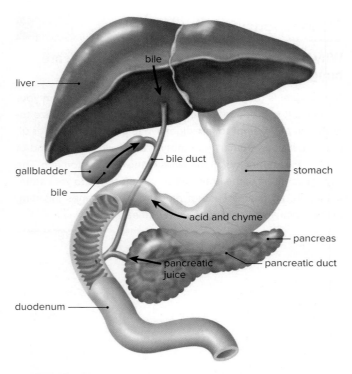

Figure 25.8   **The pancreas, liver, and gallbladder.**

Bile, made by the liver and stored in the gallbladder, and pancreatic juice, which contains enzymes, enter the duodenum by way of ducts.

droplets by a process called **emulsification.** Fat droplets mix with water and have more surface area where digestion by enzymes can occur.

The pancreas produces pancreatic juice, which contains sodium bicarbonate ($NaHCO_3$) and digestive enzymes. Sodium bicarbonate neutralizes chyme and makes the pH of the small intestine slightly basic. The higher pH helps prevent autodigestion of the intestinal lining by pepsin and is the optimal pH for the action of pancreatic enzymes. **Pancreatic amylase** digests starch to maltose; **trypsin** digests proteins to peptides; **lipase** digests fat droplets to glycerol and fatty acids; and **nuclease** digests nucleic acids to nucleotides.

Still more digestive enzymes are present in the small intestine. The wall of the small intestine contains fingerlike projections called **villi (Fig. 25.9).** The epithelial cells of the villi produce intestinal enzymes, which remain attached to them. These enzymes complete the digestion of peptides and sugars. Peptides, which result from the first step in protein digestion, are digested by peptidase to amino acids. Maltose, which results from the first step in starch digestion, is digested by maltase to glucose. Other disaccharides, each of which is acted upon by a specific enzyme, are digested to simple sugars as well.

Finally, these small nutrient molecules are absorbed into cells throughout the body from the bloodstream. Our cells use these molecules as a source of energy and as building blocks to make their own macromolecules.

Absorption by Villi   The wall of the small intestine is adapted to absorbing nutrient molecules because it has an extensive surface area—approximately

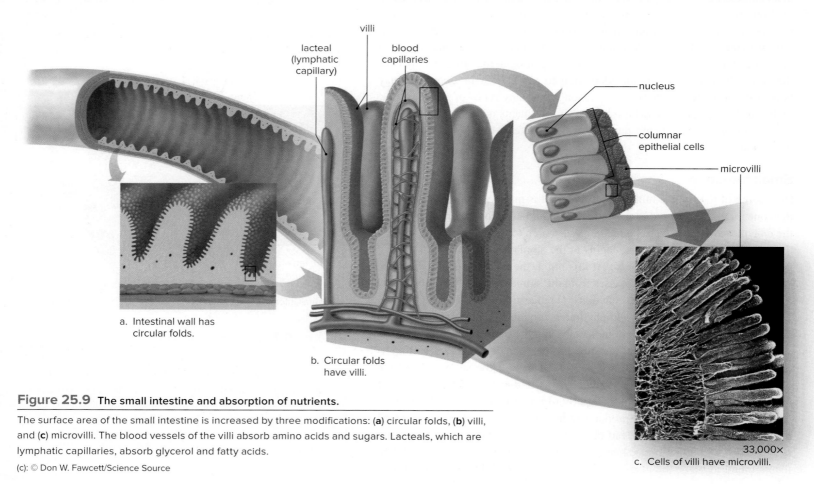

Figure 25.9   **The small intestine and absorption of nutrients.**

The surface area of the small intestine is increased by three modifications: (**a**) circular folds, (**b**) villi, and (**c**) microvilli. The blood vessels of the villi absorb amino acids and sugars. Lacteals, which are lymphatic capillaries, absorb glycerol and fatty acids.

(c): © Don W. Fawcett/Science Source

that of a tennis court! First, the mucous membrane layer of the small intestine has circular folds that give it an almost corrugated appearance (Fig. 25.9*a*). Second, on the surface of these circular folds are the villi (Fig. 25.9*b*). Finally, the cells on the surface of the villi have minute projections called **microvilli** (Fig. 25.9*c*). If the human small intestine were simply a smooth tube, it would have to be 500 to 600 meters long to have a comparable surface area for absorption. Carnivores have a much shorter digestive tract than herbivores because meat is easier to process than plant material (**Fig. 25.10**).

The villi of the small intestine absorb small nutrient molecules into the body. Each villus contains an extensive network of blood capillaries and a lymphatic capillary called a **lacteal.** As discussed in Section 23.2, the lymphatic system is an adjunct to the cardiovascular system—its vessels carry fluid, called lymph, to the cardiovascular veins. Sugars and amino acids enter the blood capillaries of a villus and are carried to the liver by way of the hepatic portal system. In contrast, glycerol and fatty acids (digested from fats) enter the epithelial cells of the villi and, within them, are joined and packaged as lipoprotein droplets, which enter a lacteal. Absorption continues until almost all nutrient molecules have been absorbed. Absorption occurs by diffusion, as well as by active transport, which requires an expenditure of cellular energy. Lymphatic vessels transport lymph to cardiovascular veins. Eventually, the bloodstream carries the nutrients absorbed by the digestive system to all the cells of the body.

Large Intestine

The word *bowel* technically means the part of the digestive tract between the stomach and the anus, but it is sometimes used to mean only the large intestine. The **large intestine** (also called the *colon*) absorbs water, salts, and some vitamins. It also stores indigestible material until it is eliminated at the anus. The large intestine takes its name from its diameter rather than its length, which is shorter than that of the small intestine. The large intestine has a blind pouch, the cecum, below the entry of the small intestine, with a small projection containing lymphatic tissue called the **appendix.** In humans, the appendix may play a role in fighting infections by acting as a reservoir of beneficial bacteria. In the condition called *appendicitis,* the appendix becomes infected and so filled with fluid that it may burst. If an infected appendix bursts before it can be removed, a serious, generalized infection of the abdominal lining, called *peritonitis,* may result.

The large intestine has a large population of bacteria, notably *Escherichia coli.* The bacteria break down indigestible material, and they produce some vitamins, including vitamin K. Vitamin K is necessary for blood clotting. Digestive wastes (feces) eventually leave the body through the **anus,** the opening of the anal canal. Feces are about 75% water and 25% solid matter. Almost one-third of this solid matter is made up of intestinal bacteria. The remainder is indigestible plant material (also called fiber), fats, waste products (such as bile pigments), inorganic material, mucus, and dead cells from the intestinal lining. A diet that includes fiber adds bulk to the feces, improves the regularity of elimination, and prevents *constipation.*

About 1.5 liters of water enter the digestive tract daily as a result of eating and drinking. An additional 8.5 liters also enter the digestive tract each day,

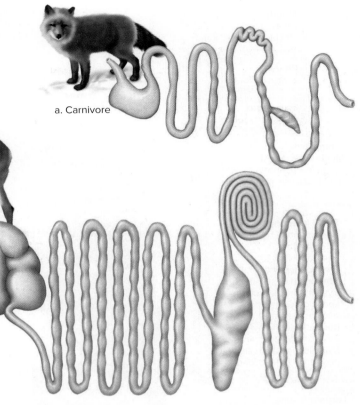

a. Carnivore

b. Herbivore

Figure 25.10 **Comparing the digestive tracts of a carnivore and a ruminant herbivore.**

The digestive tract of a carnivore (**a**) is much shorter than that of a ruminant herbivore (**b**) because proteins can be more easily digested than plant matter.

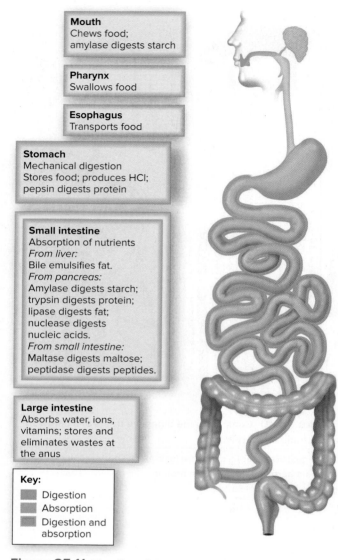

Mouth
Chews food;
amylase digests starch

Pharynx
Swallows food

Esophagus
Transports food

Stomach
Mechanical digestion
Stores food; produces HCl;
pepsin digests protein

Small intestine
Absorption of nutrients
From liver:
Bile emulsifies fat.
From pancreas:
Amylase digests starch;
trypsin digests protein;
lipase digests fat;
nuclease digests
nucleic acids.
From small intestine:
Maltase digests maltose;
peptidase digests peptides.

Large intestine
Absorbs water, ions,
vitamins; stores and
eliminates wastes at
the anus

Key:
  Digestion
  Absorption
  Digestion and
  absorption

Figure 25.11 The digestive organs and their functions.

A review of the processing of food to nutrient molecules and their absorption into the body.

carrying the various substances secreted by the digestive glands. About 95% of this water is absorbed by the small intestine, and much of the remaining portion is absorbed by the large intestine. If this water is not reabsorbed, **diarrhea** can occur, leading to serious dehydration and ion loss, especially in children.

Because the cells of the large intestine have a longer exposure to chemicals in our foods, this area is more subject to the development of **polyps,** which are small growths arising from the mucosa. Polyps, whether benign or cancerous, can be removed surgically.

Figure 25.11 reviews the process of digestion and the roles of the digestive organs.

Accessory Organs

The pancreas and the liver are the main accessory organs of digestion, along with the salivary glands and gallbladder.

Pancreas

The pancreas (see Fig. 25.8) functions as both an endocrine gland and an exocrine gland. **Endocrine glands** are ductless and secrete their products into the blood. The pancreas acts as an endocrine gland when it produces and secretes the hormones **insulin** and **glucagon** into the bloodstream. These hormones are involved in the regulation of blood glucose levels (see Section 27.2). **Exocrine glands** secrete their products into ducts. The pancreas is an exocrine gland when it produces and secretes pancreatic juice (see the content on "Digestive Enzymes" below) into the duodenum of the small intestine through the common bile duct.

Liver

The liver (see Fig. 25.8) has numerous functions, including the following:

1. Detoxifies the blood by removing and metabolizing poisonous substances
2. Produces plasma proteins, such as albumin and fibrinogen
3. Destroys old red blood cells and converts hemoglobin to the breakdown products in bile (bilirubin and biliverdin)
4. Produces bile, which is stored in the gallbladder before entering the small intestine, where it emulsifies fats
5. Stores glucose as glycogen and breaks down glycogen to glucose between meals to maintain a constant glucose concentration in the blood
6. Produces urea from amino acids and ammonia

Blood vessels from the large and small intestines, as well as others, merge to form the hepatic portal vein, which leads to the liver (**Fig. 25.12**). The liver helps maintain the glucose concentration in blood at about 0.1% by removing excess glucose from the hepatic portal vein and storing it as glycogen. Between meals, glycogen (see Section 3.2) is broken down to glucose, and glucose enters the hepatic veins. If the supply of glycogen and glucose runs short, the liver converts amino acids to glucose molecules.

Amino acids contain nitrogen in their amino groups, whereas glucose contains only carbon, oxygen, and hydrogen. Therefore, before amino acids can be converted to glucose molecules, deamination, the removal of amino groups from the amino acids, must occur. By a complex metabolic pathway, the liver converts the amino groups to urea, the most common nitrogenous (nitrogen-containing) waste product of humans. After urea is formed in the liver, it is transported in the bloodstream to the kidneys, where it is excreted.

Liver Disorders   When a person has *jaundice,* the skin has a yellowish tint due to an abnormally large quantity of bile pigments in the blood. In hemolytic jaundice, red blood cells are broken down in abnormally large amounts; in obstructive jaundice, the bile duct is obstructed or liver cells are damaged. Obstructive jaundice often occurs when crystals of cholesterol precipitate out of bile and form gallstones.

Jaundice can also result from a viral infection of the liver, called *hepatitis.* Hepatitis A is most often caused by eating contaminated food. Hepatitis B and C are commonly spread by blood transfusions, kidney dialysis, or injection with an unsterilized needle. These three types of hepatitis can also be spread by sexual contact.

Cirrhosis is a chronic liver disease in which the organ first becomes fatty, and then the liver tissue is replaced by inactive, fibrous scar tissue. Many alcoholics get cirrhosis, most likely due at least in part to the excessive amounts of alcohol the liver is forced to break down.

Digestive Enzymes

Does your mouth water when you smell food cooking? Even the thought of food can sometimes cause the nervous system to order the secretion of digestive juices. The secretion of these juices is also under the influence of several peptide hormones (whose structure consists of a small sequence of amino acids). When you eat a meal rich in protein, the stomach wall produces a peptide hormone that enters the bloodstream and doubles back to cause the stomach to produce more gastric juices. When protein and fat are present in the small intestine, another peptide hormone made in the intestinal wall stimulates the secretion of bile and pancreatic juices. In this way, the organs of digestion regulate their own needs.

The various digestive enzymes present in the digestive juices help break down carbohydrates, proteins, nucleic acids, and fats, the major components of food. Starch is a carbohydrate, and its digestion begins in the mouth. Saliva from the salivary glands has a neutral pH and contains salivary amylase, the first enzyme to act on starch.

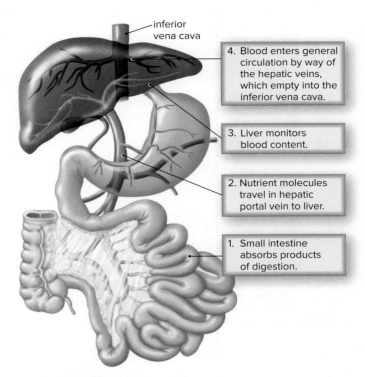

Figure 25.12   Hepatic portal system.

The hepatic portal vein takes the products of digestion from the digestive system to the liver, where they are processed before entering the hepatic veins.

---

	salivary amylase	
starch + H_2O	$\longrightarrow$	maltose

Maltose, a disaccharide, cannot be absorbed by the intestine; additional digestive action in the small intestine converts maltose to glucose, which can be absorbed.

Protein digestion begins in the stomach. Gastric juice secreted by gastric glands has a very low pH—about 2—because it contains hydrochloric acid (HCl). Pepsin, which is also present in gastric juice, acts on a protein molecule to produce peptides.

---

	pepsin	
protein + H_2O	$\longrightarrow$	peptides

Peptides are usually too large to be absorbed by the intestinal lining, but later they are broken down to amino acids in the small intestine.

Starch, proteins, fats, and nucleic acids are all enzymatically broken down in the small intestine (**Fig. 25.13**). Pancreatic juice, which enters the

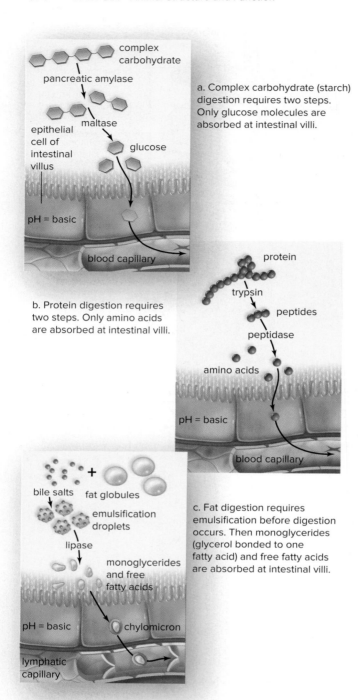

a. Complex carbohydrate (starch) digestion requires two steps. Only glucose molecules are absorbed at intestinal villi.

b. Protein digestion requires two steps. Only amino acids are absorbed at intestinal villi.

c. Fat digestion requires emulsification before digestion occurs. Then monoglycerides (glycerol bonded to one fatty acid) and free fatty acids are absorbed at intestinal villi.

Figure 25.13 **Digestion and absorption of nutrients in the small intestine.**

This figure provides a summary of (**a**) complex carbohydrate digestion, (**b**) protein digestion, and (**c**) fat digestion.

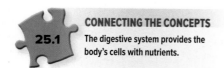

CONNECTING THE CONCEPTS
25.1
The digestive system provides the body's cells with nutrients.

duodenum, has a basic pH because it contains sodium bicarbonate ($NaHCO_3$). One pancreatic enzyme, pancreatic amylase, digests starch (Fig. 25.13a).

pancreatic amylase

starch + H_2O ⟶ maltose

Another pancreatic enzyme, trypsin, digests proteins (Fig. 25.13b).

trypsin

protein + H_2O ⟶ peptides

Maltase and peptidases, enzymes produced by the small intestine, complete the digestion of starch to glucose and proteins to amino acids, respectively. Glucose and amino acids are small molecules that are absorbed into the cells of the villi and enter the blood (Fig. 25.13a,b).

Maltose, a disaccharide that results from the first step in starch digestion, is digested to glucose by maltase.

maltase

maltose + H_2O ⟶ glucose + glucose

Other disaccharides have their own enzyme and are digested in the small intestine. The absence of any one of these enzymes can cause illness.

Peptides, which result from the first step in protein digestion, are digested to amino acids by peptidases.

peptidases

peptides + H_2O ⟶ amino acids

Lipase, a third pancreatic enzyme, digests fat molecules in fat droplets that have been emulsified by bile salts.

lipase

fat droplets + H_2O ⟶ glycerol + 3 fatty acids

Specifically, the end products of lipase digestion are monoglycerides (whose structure consists of glycerol and one fatty acid) and fatty acids. These products enter the cells of the villi, where they are rejoined and packaged as lipoprotein droplets, called chylomicrons. Chylomicrons enter the lacteals (Fig. 25.13c).

Check Your Progress 25.1

1. List the functions of a digestive system.

2. Contrast the structure and functions of the small and large intestines.

3. Describe the relationship among the duodenum, the liver, and the pancreas.

4. Explain the role of each accessory organ in digestion.

5. Specify the activity of each of the following digestive enzymes: salivary amylase, pepsin, trypsin, peptidase, and lipase.

25.2 Nutrition

Learning Outcomes

Upon completion of this section, you should be able to

1. Distinguish among macronutrients, micronutrients, and essential nutrients.
2. Explain why *empty-calorie food* is a better term than *junk food*.

The vigilance of your immune system, the strength of your muscles and bones, the ease with which your blood circulates—all aspects of your body's functioning—depend on proper nutrition. A **nutrient** is a substance in food that performs a physiological function in the body. Nutrients provide us with energy, promote growth and development by supplying the material for cellular structures, and regulate cellular metabolism. They are also involved in homeostasis. For example, nutrients help maintain the fluid balance and proper pH of blood. Your body can make up for a nutrient deficiency to a degree, but eventually signs and symptoms of a *deficiency disorder* will appear. As an example, vitamin C is needed to synthesize and maintain *collagen,* the protein that holds tissues together. When the body lacks vitamin C, collagen weakens and capillaries break easily. Gums may bleed, especially when the teeth are brushed, or tiny bruises may form under the skin when it is gently pressed. In other words, early signs of vitamin C deficiency are gums that bleed and skin that bruises easily (**Fig. 25.14**).

By learning about nutrition, you can improve your diet and increase the likelihood of enjoying a longer, more active, and more productive life. Conversely, poor diet and lack of physical activity are responsible for seven of the leading ten causes of death in the United States annually. Such lifestyle factors may soon overtake smoking as the major cause of preventable death. We all can benefit from learning what constitutes a poor diet versus a healthy diet, which will allow us to choose foods that supply all the nutrients in proper balance.

Introducing the Nutrients

A person's **diet** is his or her typical food choices. Several factors, including cultural and ethnic backgrounds, financial situations, environmental conditions, and psychological states, influence what we eat. A *balanced diet* supplies all the nutrients in the proper proportions necessary for a healthy, functioning body.

There are six classes of nutrients: carbohydrates, lipids, proteins, minerals, vitamins, and water. An **essential nutrient** must be supplied by the diet because the body is not able to produce it, or at least not in sufficient quantity to meet the body's needs. For example, amino acids are needed for protein synthesis, and the body is unable to produce several of them. Therefore, there are essential amino acids that are needed in the diet. Most vitamins, including vitamin C, and all of the minerals are considered to be essential nutrients.

We can also describe nutrients based on the quantities that are needed daily by our bodies. Carbohydrates, lipids, and proteins are called **macronutrients** because the body requires relatively large quantities of them. **Micronutrients,** such as vitamins and minerals, are needed in small quantities only. Macronutrients, not micronutrients, supply our energy needs. Although

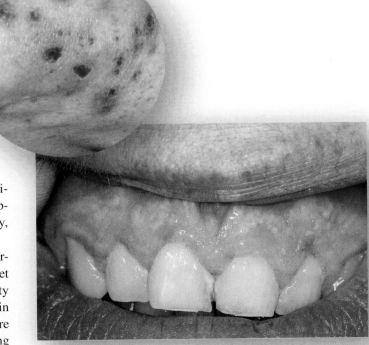

a.

b.

Figure 25.14 **Vitamin C deficiency.**

a. Pinpoint hemorrhages (tiny bruises that appear as red spots in the skin) are an early indication of vitamin C deficiency. **b.** Bleeding gums are another early sign of vitamin C deficiency.

(a): © Biophoto Associates/Science Source; (b): © ISM/Phototake

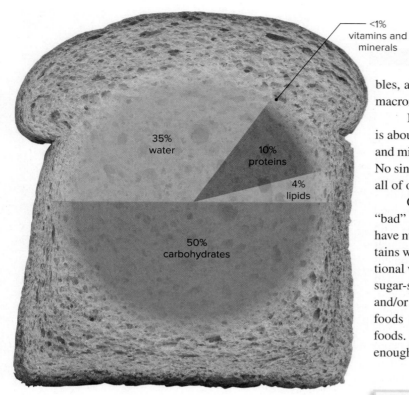

<1%
vitamins and
minerals

35%
water

10%
proteins

4%
lipids

50%
carbohydrates

Figure 25.15 **Nutrient composition of a slice of bread.**

Most foods are mixtures of nutrients.

© Jules Frazier/Getty RF

advertisements often imply that people can boost their energy levels by taking vitamin or mineral supplements, the body does not metabolize these nutrients for energy. Water does not provide energy, either. Therefore, foods with high water content, such as fruits and vegetables, are usually lower in energy content than foods with less water and more macronutrient content.

Nearly every food is a mixture of nutrients. A slice of bread, for example, is about 50% carbohydrates, 35% water, 10% proteins, and 4% lipids. Vitamins and minerals make up less than 1% of the bread's nutrient content (**Fig. 25.15**). No single, naturally occurring food contains enough essential nutrients to meet all of our nutrient needs.

Contrary to popular belief, "bad" foods, or "junk" foods, do have nutritional value. If a food contains water, sugar, or fat, it has nutritional value. However, such foods as sugar-sweetened soft drinks, cookies, and pastries have high amounts of fat and/or sugar in relation to their vitamin and mineral content. Therefore, these foods are more appropriately called *empty-calorie foods,* rather than junk foods. Diets that contain too many empty-calorie foods will assuredly lack enough vitamins and minerals.

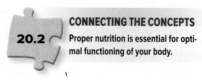

CONNECTING THE CONCEPTS

20.2 Proper nutrition is essential for optimal functioning of your body.

Check Your Progress 25.2

1. List examples of macro- and micronutrients.
2. Explain why the term *junk food* may be inappropriate.
3. Explain why essential nutrients must be obtained in the diet.

25.3 The Classes of Nutrients

Learning Outcomes

Upon completion of this section, you should be able to

1. List the dietary sources of carbohydrates.
2. Explain the importance of fiber in the diet.
3. List the types of lipids and their dietary sources.
4. List the dietary sources of protein.
5. Summarize the functions of major and trace minerals in the body.
6. Define the term *vitamin*, and list the essential vitamins.
7. Explain why the intake of water is important.

The human diet must contain the energy nutrients (macronutrients) in the correct proportions as well as adequate amounts of micronutrients. Generally, each type of nutrient has more than one function in the body and can be supplied by several different food sources (**Table 25.1**).

Carbohydrates

Carbohydrates are present in food as sugars, starch, and fiber. Fruits, vegetables, milk, and honey are natural sources of sugars. Glucose and fructose are

Table 25.1 Summarizing the Classes of Nutrients

Class of Nutrient	Major Food Sources	Primary Physiological Roles	Macronutrient or Micronutrient
Carbohydrates	Sugars and starches in fruits, vegetables, cereals and other grains	Glucose is metabolized for energy; fiber adds to fecal bulk, preventing constipation.	Macronutrient
Lipids	Oil, margarine, salad dressings, meat, fish, poultry, nuts, fried foods, dairy products made with whole milk	Triglycerides are metabolized for energy and stored for insulation and protection of organs. Cholesterol is used to make certain hormones, bile, and vitamin D.	Macronutrient
Proteins	Meat, fish, poultry, eggs, nuts, dried beans and peas, soybeans, cereal products, milk, cheese, yogurt	Proteins are needed for building, repairing, and maintaining tissues; synthesizing enzymes and certain hormones; and producing antibodies.	Macronutrient
Minerals	Widespread in vegetables and other foods	Minerals regulate energy metabolism, maintain fluid balance, and produce certain body structures, enzymes, and hormones.	Micronutrient
Vitamins	Widespread in foods; plants are good sources of antioxidants	Vitamins regulate metabolism and physiological development.	Micronutrient
Water	Widespread in foods, beverages	Water participates in many chemical reactions; it is needed for maintenance of body fluids, temperature regulation, joint lubrication, and transportation of material in cells and the body.	(Not applicable)

monosaccharide sugars, and lactose (milk sugar) and sucrose (table sugar) are disaccharides. After absorption into the body, all sugars are converted to glucose for transport in the blood. Glucose is the preferred direct energy source in cells.

Plants store glucose as starch, and animals store glucose as glycogen. High-starch foods are beans, peas, cereal grains, and potatoes. Starch is digested to glucose in the digestive tract, and any excess glucose is stored as glycogen. Although other animals likewise store glucose as glycogen in liver or muscle tissue (meat), it has broken down by the time an animal is eaten for food. Except for honey and milk, which contain sugars, animal sources of food do not contain carbohydrates.

Fiber

Fiber includes various nondigestible carbohydrates derived from plants. Foods rich in fiber include beans, peas, nuts, fruits, and vegetables. Whole-grain products are also a good source of fiber and are therefore more nutritious than food products made from refined grains. During the refinement of grains, fiber and vitamins and minerals are removed, so primarily starch remains. A slice of bread made from whole-wheat flour, for example, contains 3 grams (g) of fiber; a slice of bread made from refined wheat flour contains less than a gram of fiber.

Technically, fiber is not a nutrient for humans because it cannot be digested to small molecules that enter the bloodstream. Insoluble fiber, however, adds bulk to fecal material, which stimulates movements of the large intestine, preventing constipation. Soluble fiber combines with bile acids and cholesterol in the small intestine and prevents them from being absorbed. In this way, high-fiber diets may protect against heart disease. The typical American consumes only about 15 g of fiber each day; the recommended daily intake is 25 g for women and 38 g for men. To increase your fiber intake, eat whole-grain foods, snack on fresh fruits and raw vegetables, and include nuts and beans in your diet (**Fig. 25.16**).

Can Carbohydrates Be Harmful?

If you or someone you know has lost weight by following the Atkins or Paleo diet, you may think "carbs" are unhealthy and should be avoided. According to

Figure 25.16 Fiber-rich foods.

Plants are a good source of carbohydrates. They also provide fiber when they are not processed (refined).

© Cole Group/Getty RF

Connections: Health

What is the glycemic index?

The glycemic index is a reference used to indicate how a food item influences blood glucose levels. The index uses glucose as a reference value (100). In general, foods having a high amount of complex carbohydrates have a lower glycemic index than foods containing simple carbohydrates. For example, lentils and kidney beans have a glycemic index of 29, whereas most crackers have an index of 80. Originally designed to plan diets for diabetics, the glycemic index is useful for anyone who is interested in increasing the amount of complex carbohydrates in the diet.

Connections: Health

Does "0% trans fat" on a food label really mean that the item is free of trans fat?

When the Food and Drug Administration required that trans fat information be added to food labels in 2006, many food companies created labels touting their products as "trans fat free." A check of the label details would reveal 0 grams (0 g) of trans fat listed under "Nutrition Facts." But a more thorough check of the list of ingredients might reveal some trans fats lurking in the food. If you see "partially hydrogenated oil" listed with the ingredients, there are some trans fats in the food. Trans fats have to be listed with the breakdown of fat grams only when they constitute 0.5 g or more per serving. Limiting trans fats to 1% of daily calories is recommended by the American Heart Association. Unfortunately, eating more than one serving of a food with "hidden" trans fats might push some people over the recommended daily intake of trans fats.

Nutrition Facts
Serving Size: 1 cup (228g)
Servings Per Container about 2

Amount Per Serving

Calories 260 **Calories from Fat** 120

	% Daily Value*
Total Fat 13g	20%
Saturated Fat 5g	25%
Trans Fat 0g	0%
Cholesterol 30mg	10%
Sodium 660mg	28%
Total Carbohydrate 31g	10%
Dietary Fiber 0g	0%
Sugars 5g	
Protein 5g	

Table 25.2 How to Reduce Dietary Sugars

TO REDUCE DIETARY SUGARS:

1. Eat fewer sweets, such as candy, soft drinks, ice cream, and pastry.
2. Eat fresh fruits or fruits canned without heavy syrup.
3. Use less sugar—white, brown, or raw—and less honey and syrups.
4. Avoid sweetened breakfast cereals.
5. Eat less jelly, jam, and preserves.
6. Drink pure fruit juices, not imitations.
7. When cooking, use spices such as cinnamon, instead of sugar, to flavor foods.
8. Do not use refined sugar in tea or coffee.
9. Choose foods with a low glycemic index.

nutritionists, however, carbohydrates should supply a large portion of your energy needs. Evidence suggests that Americans are not eating the right kind of carbohydrates. In some countries, the traditional diet is 60–70% high-fiber carbohydrates, and these people have a low incidence of the diseases that plague Americans.

Obesity is associated with *type 2 diabetes* and *cardiovascular disease,* as discussed in Section 25.5. Some nutritionists hypothesize that the high intake of foods rich in refined carbohydrates and fructose sweeteners processed from cornstarch may be responsible for the prevalence of obesity in the United States. These are empty-calorie foods that provide sugars but no vitamins or minerals. **Table 25.2** tells you how to reduce dietary sugars in your diet. Other nutritionists point out that consuming too much energy from any source contributes to body fat, which increases a person's risk of obesity and associated illnesses. Because many foods, such as donuts, cakes, pies, and cookies, are high in both refined carbohydrates and fat, it is difficult to determine which dietary component is responsible for the current epidemic of obesity among Americans.

Many people mistakenly believe that children become hyperactive after eating sugar. There is no scientific basis for this belief, because sucrose is broken down into glucose and fructose in the small intestine, and these sugars are absorbed into the bloodstream. The spike caused by their absorption does not last long. Excess glucose and fructose enters the liver, and fructose is converted to glucose. As you know, the liver stores glucose as glycogen, and glycogen is broken down to maintain the proper glucose level.

Lipids

Like carbohydrates, **triglycerides** (the main components of fats and oils) supply energy for cells, but **fat** is stored for the long term in the body. Fat deposits under the skin, called *subcutaneous fat,* insulate the body from cold temperatures; deeper fat deposits in the trunk protect organs against injury.

Nutritionists generally recommend that unsaturated rather than saturated fats (see Fig. 3.13) be included in the diet. Two unsaturated fatty acids (alpha-linolenic and linoleic acids) are *essential dietary fatty acids.* Delayed growth and skin problems can develop when the diet lacks these essential fatty acids, which can be supplied by eating fatty fish and by including plant oils, such as canola and soybean oils, in the diet.

Animal foods such as butter, meat, whole milk, and cheeses contain saturated fats. Plant oils contain unsaturated fats. The differences between saturated and unsaturated fats are shown in Figure 3.13. Each type of oil has a particular percentage of monounsaturated and polyunsaturated fatty acids.

Cholesterol, a lipid, can be synthesized by the body in sufficient quantities to meet daily needs. Cells use cholesterol to make various compounds, including bile, steroid hormones, and vitamin D. Cholesterol is also an important component of the plasma membrane. Plant foods do not contain cholesterol; only animal foods, such as cheese, egg yolks, liver, and certain shellfish (shrimp and lobster), are rich in cholesterol.

Can Lipids Be Harmful?

Elevated blood cholesterol levels are associated with an increased risk of cardiovascular disease, the number one killer of Americans. A diet rich in cholesterol and saturated fats increases the risk of cardiovascular disease (see Section 25.5). Statistical studies suggest that trans fats are even more harmful than saturated fats. Trans fats are formed when unsaturated oils are hydrogenated to produce solid fats, found largely in processed foods. Trans fatty acids may reduce the function of the plasma membrane receptors that clear cholesterol from the bloodstream. Trans fatty acids are found in commercially packaged foods, such as cookies and crackers; in commercially fried foods, such as french fries; in packaged snacks, such as microwave popcorn; and in vegetable shortening and some margarines. **Table 25.3** tells you how to reduce harmful lipids in the diet.

Proteins

Dietary **proteins** are digested to amino acids, which cells use to synthesize hundreds of cellular proteins. Of the 20 different amino acids, 9 are *essential amino acids* that must be present in the diet. Children will not grow if their diets lack the essential amino acids. Eggs, milk products, meat, poultry, and most other foods derived from animals contain all 9 essential amino acids and are "complete," or "high-quality," protein sources.

Foods derived from plants generally do not have as much protein per serving as those derived from animals, and each type of plant food generally lacks one or more of the essential amino acids. Therefore, most plant foods are "incomplete," or "low-quality," protein sources. Vegetarians, however, do not have to consume animal sources of protein. To meet their protein needs, total vegetarians (vegans) can eat grains, beans, and nuts in various combinations. Also, tofu, soymilk, and other foods made from processed soybeans are complete protein sources. A balanced vegetarian diet is quite possible with a little planning.

Can Proteins Be Harmful?

According to nutritionists, proteins should not supply the bulk of dietary calories. The average American eats about twice as much protein as he or she needs, and some people may be on a diet that encourages the intake of proteins instead of carbohydrates as an energy source. Also, body builders, weight lifters, and other athletes may include amino acid or protein supplements in the diet because they think these supplements will increase muscle mass. However, excess amino acids are not always converted into muscle tissue. When they are used as an energy source, the liver removes the nitrogen portion (in the process called deamination) and uses it to form urea, which is

Table 25.3 Reducing Certain Lipids in the Diet

TO REDUCE DIETARY SATURATED FAT:

1. Choose poultry, fish, or dry beans and peas as a protein source.

2. Remove skin from poultry and trim fat from red meats before cooking; place on a rack, so that fat drains off.

3. Broil, boil, or bake rather than frying.

4. Limit your intake of butter, cream, trans fats, shortening, and tropical oils (coconut and palm oils).*

5. To season vegetables, use herbs and spices instead of butter, margarine, or sauces. Use lemon juice instead of salad dressing.

6. Drink skim milk instead of whole milk, and use skim milk in cooking and baking.

TO REDUCE DIETARY CHOLESTEROL:

1. Eat white fish and poultry in preference to cheese, egg yolks, liver, and certain shellfish (shrimp and lobster).

2. Substitute egg whites for egg yolks in both cooking and eating.

3. Include soluble fiber in the diet. Oat bran, oatmeal, beans, corn, and fruits, such as apples, citrus fruits, and cranberries, are high in soluble fiber.

*Although coconut and palm oils are from plant sources, they are mostly saturated fat.

excreted in urine. The water needed for the excretion of urea can cause dehydration when a person is exercising and losing water by sweating. High-protein diets can also increase calcium loss in the urine and encourage the formation of kidney stones. Furthermore, many high-protein foods contain a high amount of fat.

Minerals

The body needs about 20 elements called **minerals** for numerous physiological functions, including regulation of biochemical reactions, maintenance of fluid balance, and incorporation into certain structures and compounds. **Major minerals** are needed in the body at higher levels than **trace minerals.** More than 100 milligrams (mg) per day of each major mineral is required in the diet, and less than 100 mg per day of each trace mineral is required in the diet. **Table 25.4** lists the major minerals and a sample of the trace minerals. Their

Table 25.4 Minerals

Mineral	Function(s)	Food Source(s)	CONDITIONS CAUSED BY:	
			Too Little in the Diet	Too Much in the Diet
Major Minerals				
Calcium (Ca^{2+})	Strong bones and teeth, nerve conduction, muscle contraction	Dairy products, leafy green vegetables	Stunted growth in children, low bone density in adults	Kidney stones; interferes with iron and zinc absorption
Phosphorus (PO_4^{3-})	Bone and soft tissue growth; part of phospholipids, ATP, and nucleic acids	Meat, dairy products, sunflower seeds, food additives	Weakness, confusion, pain in bones and joints	Low blood and bone calcium levels
Potassium (K^+)	Nerve conduction, muscle contraction	Many fruits and vegetables, bran	Paralysis, irregular heartbeat, eventual death	Vomiting, heart attack, death
Sodium (Na^+)	Nerve conduction, pH and water balance	Table salt	Lethargy, muscle cramps, loss of appetite	High blood pressure, calcium loss
Chloride (Cl^-)	Water balance	Table salt	Not likely	Vomiting, dehydration
Magnesium (Mg^{2+})	Part of various enzymes for nerve and muscle contraction, protein synthesis	Whole grains, leafy green vegetables	Muscle spasms, irregular heartbeat, convulsions, confusion, personality changes	Diarrhea
Sulfur (S^{2-})	Part of some amino acids and some vitamins	Meats, legumes, whole grains	Not likely	Not likely
Trace Minerals				
Zinc (Zn^{2+})	Protein synthesis, wound healing, fetal development and growth, immune function	Meats, legumes, whole grains	Delayed wound healing, night blindness, diarrhea, mental lethargy	Anemia, diarrhea, vomiting, renal failure, abnormal cholesterol levels
Iron (Fe^{2+})	Hemoglobin synthesis	Whole grains, meats, prune juice	Anemia, physical and mental sluggishness	Iron toxicity disease, organ failure, eventual death
Copper (Cu^{2+})	Iron metabolism; part of various antioxidant enzymes	Meat, nuts, legumes	Anemia, stunted growth in children	Damage to internal organs if not excreted
Iodine (I^-)	Thyroid hormone synthesis	Iodized table salt, seafood	Thyroid deficiency	Depressed thyroid function, anxiety
Selenium (SeO_4^{2-})	Part of antioxidant enzyme	Seafood, meats, eggs	Vascular collapse, possible cancer development	Hair and fingernail loss, discolored skin

functions and food sources are given, as well as the health effects of too little or too much intake.

Some individuals (especially women) do not get enough iron, calcium, magnesium, or zinc in their diets. Adult females need more iron in the diet than males (18 mg compared to 10 mg) if they are menstruating each month. *Anemia,* characterized by a run-down feeling due to insufficient red blood cells, results when the diet lacks iron. Also, many people take calcium supplements as directed by a physician to counteract *osteoporosis,* a degenerative bone disease (**Fig. 25.17**) that affects an estimated one-quarter of older men and one-half of older women in the United States.

One mineral that people consume too much of is sodium. The recommended amount of sodium intake per day is 2,300 mg (1,500 mg if you have high blood pressure), although the average American takes in over 4,000 mg each day. The American Heart Association recommends that you aim to decrease your sodium intake to less than 1,500 mg per day. **Table 25.5** gives recommendations for reducing the amount of sodium in the diet.

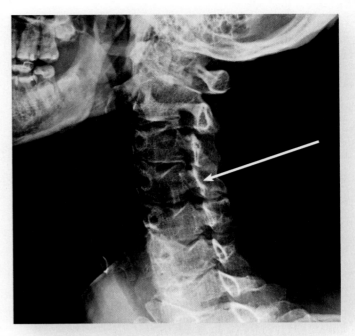

Figure 25.17 An X-ray of an individual with osteoporosis.

The fractures in this X-ray are caused by osteoporosis, which reduces the thickness and strength of bones.

© Lilli Day/Getty RF

Connections: Health

Where does most of the sodium in the diet come from?

Contrary to popular belief, the majority of the sodium in your diet does not come from the salt you put on your food when you are eating. Instead, most dietary sodium (over three-quarters!) comes from processed foods and condiments. Sodium is used in these items both to preserve the food or condiment and to make it taste better. But sometimes the amount of sodium is phenomenal. A single teaspoon of soy sauce contains almost 1,000 mg of sodium, and a half-cup of prepared tomato sauce typically has over 400 mg of sodium. Websites such as nutritiondata.com can help you track your daily sodium intake.

6% 5%
12%
77%

Key:
- From processed foods
- In foods naturally
- Added at eating
- Added at cooking

© Image Club RF

Table 25.5 Reducing Dietary Sodium

TO REDUCE DIETARY SODIUM:
1. Use spices instead of salt to flavor foods.
2. Add little or no salt to foods at the table, and add only small amounts of salt when you cook.
3. Eat unsalted crackers, pretzels, potato chips, nuts, and popcorn.
4. Avoid hot dogs, ham, bacon, luncheon meats, smoked salmon, sardines, and anchovies.
5. Avoid processed cheese and canned or dehydrated soups.
6. Avoid brine-soaked foods, such as pickles and olives.
7. Read labels to avoid high-salt products.

Vitamins

Vitamins are organic compounds (other than carbohydrates, fats, and proteins, including amino acids) that regulate various metabolic activities. Vitamins are classified as either water-soluble (**Table 25.6**) or fat-soluble (**Table 25.7**). Generally, water soluble vitamins are readily removed in the urine, and most must be replenished daily. Fat-soluble vitamins are stored in adipose tissue and persist longer in the body.

Although many people think vitamins can enhance health dramatically, prevent aging, and cure diseases such as arthritis and cancer, there is no scientific evidence that vitamins are "wonder drugs." However, vitamins C, E, and A have been shown to defend the body against free radicals, and therefore they are termed **antioxidants.** These vitamins are especially abundant in fruits and vegetables, so it is suggested that we eat about 4½ cups of fruits and vegetables per day. To achieve this goal, we should consume salad greens, raw or cooked vegetables, dried fruit, and fruit juice in addition to traditional apples and oranges and other fresh foods.

Vitamin deficiencies can lead to disorders, and even death, in humans. Although many foods in the United States are now enriched, or fortified, with vitamins, some individuals, especially the elderly, young children, alcoholics, and people with low incomes, are still at risk for vitamin deficiencies, generally as a

Table 25.6 Water-Soluble Vitamins

Vitamin	Functions	Food Sources	CONDITIONS CAUSED BY:	
			Too Little in the Diet	Too Much in the Diet
Vitamin C	Antioxidant; needed for forming collagen; helps maintain capillaries, bones, and teeth	Citrus fruits, leafy green vegetables, tomatoes, potatoes, cabbage	Scurvy, delayed wound healing, infections	Gout, kidney stones, diarrhea, decreased copper
Thiamine (vitamin B_1)	Part of coenzyme needed for cellular respiration; promotes activity of the nervous system	Whole-grain cereals, dried beans and peas, sunflower seeds, nuts	Beriberi, muscular weakness, enlarged heart	Can interfere with absorption of other vitamins
Riboflavin (vitamin B_2)	Part of coenzymes, such as FAD; aids cellular respiration, including oxidation of protein and fat	Nuts, dairy products, whole-grain cereals, poultry, leafy green vegetables	Dermatitis, blurred vision, growth failure	Unknown
Niacin (nicotinic acid)	Part of coenzymes NAD and NADP; needed for cellular respiration, including oxidation of protein and fat	Peanuts, poultry, whole-grain cereals, leafy green vegetables, beans	Pellagra, diarrhea, mental disorders	High blood sugar and uric acid, vasodilation
Folacin (folic acid)	Coenzyme needed for production of hemoglobin and formation of DNA	Leafy dark green vegetables, nuts, beans, whole-grain cereals	Megaloblastic anemia, spina bifida	May mask B_{12} deficiency
Vitamin B_6	Coenzyme needed for synthesis of hormones and hemoglobin; CNS control	Whole-grain cereals, bananas, beans, poultry, nuts, leafy green vegetables	Rarely, convulsions, vomiting, seborrhea, muscular weakness	Insomnia, neuropathy
Pantothenic acid	Part of coenzyme A needed for oxidation of carbohydrates and fats; aids in the formation of hormones and certain neurotransmitters	Nuts, beans, dark green vegetables, poultry, fruits, milk	Rarely, loss of appetite, mental depression, numbness	Unknown
Vitamin B_{12}	Complex, cobalt-containing compound; part of the coenzyme needed for synthesis of nucleic acids and myelin	Dairy products, fish, poultry, eggs, fortified cereals	Pernicious anemia	Unknown
Biotin	Coenzyme needed for metabolism of amino acids and fatty acids	Generally in foods, especially eggs	Skin rash, nausea, fatigue	Unknown

Table 25.7 Fat-Soluble Vitamins

Vitamin	Functions	Food Sources	CONDITIONS CAUSED BY: Too Little in the Diet	Too Much in the Diet
Vitamin A	Antioxidant needed for normal vision, immune system function, and cellular growth	Yellow-orange and dark green fruits and vegetables, milk, cereals	Night blindness, poor immune system functioning	Nausea, headache, bone fractures, hair loss
Vitamin D	A group of steroids needed for development and maintenance of bones and teeth	Milk fortified with vitamin D, fish liver oil	Rickets, bone decalcification and weakening	Calcification of soft tissues, diarrhea, possible renal damage
Vitamin E	Antioxidant that prevents oxidation of vitamin A and polyunsaturated fatty acids	Leafy green vegetables, fruits, vegetable oils, nuts, whole-grain breads and cereals	Unknown	Diarrhea, nausea, headache, fatigue, muscle weakness
Vitamin K	Synthesis of substances active in clotting of blood	Leafy green vegetables, cabbage, cauliflower	Easy bruising and bleeding	Can interfere with anticoagulant medication

result of poor food choices. For example, skin cells normally contain a precursor cholesterol molecule that is converted to vitamin D after UV exposure. But a vitamin D deficiency leads to a condition called rickets, in which defective mineralization of the skeleton causes bowing of the legs. Most milk today is fortified with vitamin D, which helps prevent the occurrence of rickets. Another example is vitamin C deficiency, the effects of which are illustrated in Figure 25.14. If a diet involves high alcohol consumption, then vitamin deficiencies may occur even if the intake of vitamins is adequate. Deficiencies occur because alcohol interferes with the absorption of certain vitamins, such as vitamin B_{12}, folacin, and vitamin A, and it increases the excretion of other vitamins, such as vitamin C.

Water

Water constitutes about 60% of an adult's body. Water participates in many chemical reactions; in addition, watery fluids lubricate joints, transport other nutrients, and help maintain body temperature. Beverages, soups, fruits, and vegetables are sources of water, and most solid foods contain some water. The amount of *total* water (water from beverages and foods) that you need to consume depends on your physical activity level, your diet, and environmental conditions. On average, men should consume about 125 ounces (oz) and women should consume about 90 oz of total water each day. Thirst is a healthy person's best guide for meeting water needs and avoiding dehydration. Too much water can also be a problem. In water toxication (*hyponatremia*), individuals who consume excessive amounts of water upset the balance of electrolytes, usually sodium and potassium, in their blood. This can lead to irregular heartbeat, and in some cases, death.

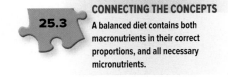

CONNECTING THE CONCEPTS

25.3 A balanced diet contains both macronutrients in their correct proportions, and all necessary micronutrients.

Check Your Progress 25.3

1. List the classes of nutrients, and give a source of each.
2. Describe the potential health consequences of overindulging in each of the following: carbohydrates, lipids, and proteins.
3. Summarize the potential health problems associated with deficiencies of vitamins, minerals, and water.

25.4 Understanding Nutrition Guidelines

Learning Outcomes

Upon completion of this section, you should be able to

1. Explain the purpose of dietary guidelines.
2. Evaluate the nutrition labels on food packaging.
3. Discuss the value of dietary supplements.
4. Summarize the components of a healthy diet.

Planning nutritious meals and snacks involves making daily food choices based on a wide variety of information about recommended amounts of nutrients. A day's food intake should provide the proper balance of nutrients—neither too much nor too little of each nutrient. Food guides can be helpful in planning your diet. Additionally, reading the "Nutrition Facts" panel on packaged foods can help you choose healthier sources of nutrients.

Updating Dietary Guidelines

Dietary guidelines are typically revised by the U.S. government every 5 years to reflect changes in nutrition science. The latest guidelines were released in 2015 by the Departments of Agriculture and Health and Human Services. The overall purposes of these guidelines were to:

- Promote health
- Prevent chronic long-term disease
- Assist people in reaching and maintaining a healthy weight

The new guidelines focus less on prescribing quantitative levels for nutrients and more on establishing healthy eating patterns. A healthy eating pattern includes the following foods:

1. A variety of vegetables, including leafy vegetables, beans, red and yellow vegetables, and starches
2. Fruits
3. Grains, at least half of which should be whole grains
4. Fat-free or low-fat dairy products (including soy)
5. Proteins in the form of seafood, lean meats, poultry, eggs, legumes, nuts, and soy products
6. Oils

To establish these healthy eating patterns, specific recommendations were made to limit certain nutrients that are recognized as raising health concerns. These are outlined in **Table 25.8.**

Visualizing Dietary Guidelines

ChooseMyPlate.gov

The U.S. Department of Agriculture (USDA) has developed a guideline called MyPlate (**Fig. 25.18**). This graphical representation replaced the older pyramids because most people found it easier to interpret. It can be used to help you decide how your daily calorie intake should be distributed among your food choices. MyPlate emphasizes the proportions of each food group that should be consumed daily.

Table 25.8 2015–2020 Dietary Guidelines

GENERAL GUIDELINES
Consume less than 10 percent of calories per day from sugar.
Consume less than 10 percent of calories per day from saturated fats.
Consume less than 2,300 milligrams (mg) per day of sodium.
Alcohol should be consumed in moderation: a maximum of one drink per day for women and two drinks per day for men (and only by adults of legal drinking age).

health.gov/dietaryguidelines/2015/guidelines/.

In addition, the USDA provides (on the ChooseMyPlate.gov website) recommendations concerning the minimum quantity of foods in each group that should be eaten daily. The site also contains an interactive component, Super-Tracker, that allows you to track your own diet and set personal weight and activity goals. To support these decisions, the USDA also provides examples of daily food plans and information on how to follow a healthy diet on a budget.

Making Sense of Nutrition Labels

A "Nutrition Facts" panel, shown in **Figure 25.19,** provides specific dietary information about the product and general information about the nutrients the product contains.

Serving Size and Calories   The serving size is based on the typical serving size for the product. If you are comparing Calories (kcal)[1] and other data about products of the same type, you want to be sure the serving size is the same for each product. The total number of Calories is based on the serving size. Obviously, if you eat twice the serving size, you have taken in twice the number of Calories. The new food labels (Fig. 25.19b) are designed to provide a better indication of a realistic serving size for most people and to increase the emphasis on the total Calories per serving.

% Daily Value   The % daily value (the percentage of the total amount needed in a 2,000-Calorie diet) is calculated by comparing the specific information about this product with the information given at the bottom of the panel. For example, the product in Figure 25.19a has a fat content of 13 g, and the total daily recommended amount is less than 65 g, so 13/65 = 20%. *The % daily values are not applicable for people who require more or less than 2,000 Calories (kcal) per day.*

A % daily value for protein is generally not given because determining such a value would require expensive testing of the protein quality of the product by the manufacturer. Also, notice there is a % daily value for carbohydrates but not sugars, because there is no recommended daily value for sugar.

How to Use the Panel   If the serving sizes are the same, you can use "Nutrition Facts" panels to compare two products of the same type. For example, if you wanted to reduce your Caloric intake and increase your fiber and vitamin C intakes, comparing the panels from two different food products would allow you to see which one is lowest in Calories and highest in fiber and vitamin C.

Dietary Supplements

Dietary supplements are nutrients and plant products (such as herbal teas) that are used to enhance health. The U.S. government does not require dietary supplements to undergo the same safety and effectiveness testing that new prescription drugs must complete before they are approved. Therefore, many herbal products have not been tested scientifically to determine their benefits. Although people often think herbal products are safe because they are "natural," many plants, including lobelia, comfrey, and kava kava, can be poisonous.

Dietary supplements that contain nutrients can also cause harm. Most fat-soluble vitamins are stored in the body and can accumulate to toxic levels, particularly vitamins A and D. Although excesses of water-soluble vitamins can be excreted, cases involving toxic amounts of vitamin B$_6$, thiamine, and

Figure 25.18 **MyPlate food guidelines.**

The U.S. Department of Agriculture (USDA) developed this visual representation of a food plate as a guide to better health. The size differences on the plate for each food group suggest what portion of your meal should consist of each category. The five different colors illustrate that foods in correct proportions are needed each day for good health.

Source: USDA, ChooseMyPlate.gov

Connections: Health

What health benefits are associated with drinking green tea?

All tea is derived from a plant native to China and India called *Camellia sinensis*. Green tea is made by steaming the leaves of this plant. Typically, green tea has less caffeine than black teas, which are made by fermenting the tea leaves. Research studies on the health benefits of green tea have shown that, overall, it  © D. Hurst/Alamy RF has a very positive effect on the body. Antioxidants in green tea (called flavonoids) inhibit the growth of certain forms of cancer, prevent plaque buildup in the blood vessels, and may help improve blood cholesterol levels. However, most nutritionists still warn against taking green tea supplements. Instead, substitute a cup of tea for soda in your daily diet.

[1] Nutritionists use the word *Calorie,* while scientists prefer *kcal.*

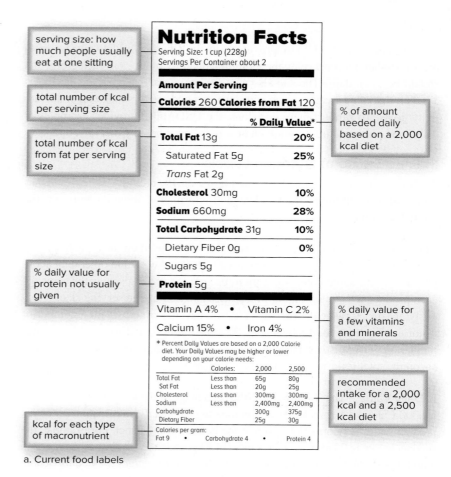

serving size: how much people usually eat at one sitting

total number of kcal per serving size

total number of kcal from fat per serving size

% daily value for protein not usually given

kcal for each type of macronutrient

% of amount needed daily based on a 2,000 kcal diet

% daily value for a few vitamins and minerals

recommended intake for a 2,000 kcal and a 2,500 kcal diet

a. Current food labels

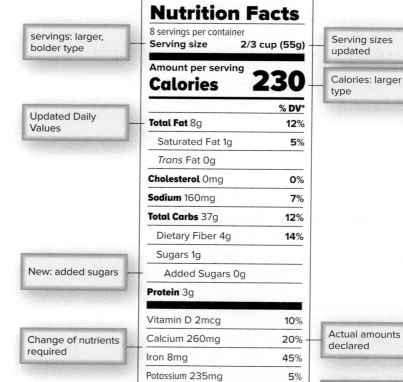

servings: larger, bolder type

Updated Daily Values

New: added sugars

Change of nutrients required

Serving sizes updated

Calories: larger type

Actual amounts declared

New footnote to come

b. Proposed changes to food labels

Figure 25.19 Nutrition labels on foods.

a. The current Nutrition Facts panel provides information about the amounts of certain nutrients and other substances in a serving of the food. **b.** The proposed changes to the Nutrition Facts panel based on the 2015–2020 guideline changes. www.fda.gov

vitamin C have been reported. Minerals can be harmful, even deadly, when ingested in amounts that exceed the body's needs.

Healthy people can take a daily supplement that contains recommended amounts of vitamins and minerals. Some people have metabolic diseases or physical conditions that interfere with their ability to absorb or metabolize certain nutrients. These individuals may need to add certain nutrient supplements to their diet. However, people should not take high doses of dietary supplements without checking with their physician.

The Bottom Line

Most nutritionists agree that a healthy diet:

- Has a moderate total fat intake and is low in saturated fats, trans fats, and cholesterol (see Table 25.3 for help in achieving this goal)
- Is rich in whole-grain products, vegetables, and legumes (e.g., beans and peas) as sources of complex carbohydrates and fiber
- Is low in refined carbohydrates, such as starches and sugars (see Table 25.2 for help in achieving this goal)
- Is low in salt and sodium content (see Table 25.5 for help in achieving this goal)
- Contains only adequate amounts of protein, largely from poultry, fish, and plants
- Includes only moderate amounts of alcohol
- Contains adequate amounts of minerals and vitamins but avoids questionable food additives and supplements

CONNECTING THE CONCEPTS

25.4   Planning nutritious meals is essential for a healthy diet.

Check Your Progress 25.4

1. Describe how the 2015–2020 dietary guidelines from the U.S. government differ from the previous guidelines.

2. Summarize the information provided by MyPlate.

3. Summarize the nutrition information provided by the current and proposed food labels.

25.5 Nutrition and Health

Learning Outcomes

Upon completion of this section, you should be able to

1. Calculate a body mass index, and determine if this value reflects a healthy weight.
2. Discuss the balance between energy intake and energy output.
3. Describe diseases associated with obesity.
4. List and describe eating disorders.

Many serious disorders in Americans are linked to a diet that results in excess body fat. In the United States, the number of people who are overweight or obese has reached epidemic proportions. Nearly two-thirds of adult Americans have too much body fat.

Excess body fat increases the risk of type 2 diabetes, cardiovascular disease, and certain cancers. These conditions are among the leading causes of disability and death in the United States. Therefore, it is important for us all to stay within the recommended weight for our height.

Body Mass Index

Medical researchers use the **body mass index (BMI)** to determine if a person is overweight or obese. On the whole, our height is determined genetically, while our weight is influenced by other factors as well, particularly diet and lifestyle. BMI is a number that reflects the relationship between a person's weight and height. Here's how to calculate your BMI:

$$BMI = \frac{weight\ (lb)}{height^2\ (in.^2)} \times 703$$

$$BMI = \frac{weight\ (kg)}{height^2\ (m^2)}$$

For example, a woman whose height is 63 inches and whose weight is 133 pounds has a BMI of 23.56.

$$\frac{133 \times 703.1}{63^2} = \frac{93512.3}{3969} = 23.56\ (BMI)$$

Underweight BMI	<	18.5
Healthy BMI	=	18.5 to 24.9
Overweight BMI	=	25.0 to 29.9
Obese BMI	=	30.0 to 39.9
Morbidly obese BMI	=	40.0 or more

Energy Intake Versus Energy Output

While genetics and physiological factors such as the types of bacteria in the large intestine are known to contribute to being overweight, a person cannot become fat without taking in more food energy (calories) than are expended.

Energy Intake   The energy value of food is often reported in kilocalories (kcal). A kilocalorie is the amount of heat that raises the temperature of a liter of water by 1°C.

For practical purposes, you can estimate a food's caloric value if you know how many grams of carbohydrate, fat, protein, and alcohol it contains. Each gram (g) of carbohydrate or protein supplies 4 kcal, and each gram of fat supplies 9 kcal. Although alcohol is not a nutrient, it is considered a food, and each gram supplies 7 kcal. Therefore, if a serving of food contains 30 g of carbohydrate, 9 g of fat, and 5 g of protein, it supplies 221 kcal:

carbohydrate	30 g × 4 kcal	=	120 kcal
fat	9 g × 9 kcal	=	81 kcal
protein	5 g × 4 kcal	=	20 kcal
Total			221 kcal

Energy Output   The body expends energy primarily for (1) metabolic functions; (2) physical activity; and (3) digestion, absorption, and processing of nutrients from food. Scientists can assess a person's energy expenditure for a particular physical activity by measuring oxygen intake and carbon dioxide output during performance of the activity (**Fig. 25.20**).

a.

b.

c.

Figure 25.20 Measuring energy needed for a physical activity.

a. A sedentary job requires much less energy than (**b**) a physical job. **c.** Scientists measure oxygen intake and carbon dioxide output to determine the energy expended in a particular physical activity.

(a): © Getty Images/BrandX RF; (b): © Aaron Roeth; (c): © Sean Bagshaw/ Science Source

For practical purposes, here's how to estimate your daily energy needs:

1. *Kcal needed daily for metabolic functions:* Multiply your weight in kilograms (weight in pounds divided by 2.2) times 1.0 if you are a man, and times 0.9 if you are a woman. Then multiply that number by 24.

   Example: Meghan, a woman who weighs 130 pounds (about 59 kg), would calculate her daily caloric need for metabolic functions as follows:

   0.9 kcal × 59 kg × 24 hours = approximately 1,274 kcal/day

2. *Kcal needed daily for physical activity:* Choose a multiplication factor from one of the follow categories:

   Sedentary (little or no physical activity) = 0.20 to 0.40
   Light (walk daily) = 0.55 to 0.65
   Moderate (daily vigorous exercise) = 0.70 to 0.75
   Heavy (physical labor/endurance training) = 0.80 to 1.20

   Multiply this factor times the kcal value you obtained in step 1.

   Example: Meghan performs light physical activity daily. She multiplies 0.55 × 1,274 kcal to determine her daily caloric need for physical activity, which is 701 kcal.

3. *Kcal needed for digestion, absorption, and processing of nutrients:*

   Multiply the total kcal from steps 1 and 2 by 0.1, and add that value to the total kcal from steps 1 and 2 to get your total daily energy needs.

   Example: Meghan adds 1,274 and 701 and then multiplies 1,975 kcal by 0.1 and adds that value (197.5) to 1,975 to obtain her total daily energy needs of 2,172 kcal.

   Therefore, Meghan will maintain her weight of 130 pounds if she continues to consume about 2,170 kcal a day and to perform light physical activities.

Maintaining a Healthy Weight

To maintain weight at an appropriate level, the daily kcal intake (from eating) should not exceed the daily kcal output (metabolism + physical activity + processing food). For many Americans, this ratio is out of sync; they take in more calories than they need. The extra energy is converted to fat stored in *adipose*

Connections: Health

What is the 10,000-step program?

© McGraw-Hill Education/Christopher Kerrigan

Research suggests that a minimum of 10,000 (10K) steps per day is necessary for weight maintenance and good health. That number of steps per day is roughly equivalent to the recommended 30 minutes of daily exercise. To get started, you'll need a pedometer. You'll probably find you need to increase the amount of walking you do to reach the goal of 10K steps a day. There are a number of easy ways to add steps to your routine. Park a little farther away from your office or the store. Take the stairs instead of the elevator, or go for a walk after a meal. If your goal is to lose weight, 12–15K steps a day have been shown to promote weight loss.

Intake	Output	Weight Change

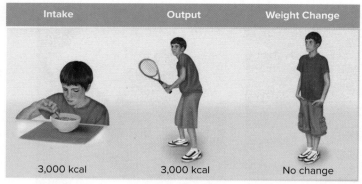

3,000 kcal | 3,000 kcal | No change

Energy balance (equilibrium)

4,000 kcal | 2,000 kcal | Increase

Positive energy balance

2,000 kcal | 3,000 kcal | Decrease

Negative energy balance

Figure 25.21 Changes in body weight.

These illustrations show the relationships among food intake, energy use, and weight change.

tissue, and they become overweight. To lose weight, an overweight person needs to lower the kcal intake and increase the kcal output in the form of physical activity. Only then does the body metabolize its stored fat for energy needs, allowing the person to lose weight. **Figure 25.21** illustrates how body weight changes in relation to kcal intake and kcal output.

Dieting   Fad weight-reduction diets—high-protein, low-carb, high-fiber, and even cabbage soup diets—come and go. During the first few weeks of a fad diet, overweight people often lose weight rapidly, because they consume fewer calories than usual, and excess body fat is metabolized for energy needs. In most cases, however, dieters become bored with eating the same foods and avoiding their favorite foods, which may be high in fat and sugars. When most dieters go off their diets, they regain the weight they lost, and they often feel frustrated and angry at themselves for failing to maintain the weight loss.

There are no quick and easy solutions for losing weight. The typical fad diet is nutritionally unbalanced and difficult to follow over the long term. Weight loss and weight maintenance require permanent lifestyle changes, such as increasing the level of physical activity and reducing portion sizes. Behavior modification allows an overweight person to lose weight safely, generally at a reasonable rate of about ½ to 2 pounds per week. Once body weight is under control, it needs to be maintained by continuing to eat sensibly.

Connections: Health

What is the Paleo diet?

The Paleo diet, also known as the caveman or warrior diet, is based on the idea that many modern nutritional problems, such as diabetes and cardiovascular disease, are due to the fact that humans are eating foods that their bodies have not had time to adjust to on an evolutionary scale. Supporters of the caveman diet suggest that a diet rich in nuts, meat, shellfish, vegetables, and berries—foods that were available to cavemen—reduces the risk for these diseases. While diets rich in these foods do reduce the risk for certain diseases, critics of the caveman diet state that the diet omits

© Phillippe Plailly & Atelier Daynes/Look Sciences/Science Source

important foods, such as low-fat dairy, grains, and beans—all of which are important in a healthy diet. In all cases, diets should remove low-nutrient foods and replace them with healthier choices. Individuals should consult with their physician before undertaking any new diet.

Disorders Associated with Obesity

Type 2 diabetes and cardiovascular disease are often seen in people who are obese.

Type 2 Diabetes

As discussed in Section 27.2, diabetes comes in two forms, type 1 and type 2. When a person has type 1 diabetes, the pancreas does not produce insulin, and the patient has to have daily insulin injections.

In contrast to type 1 diabetes, children and more often adults with type 2 diabetes are usually obese and display impaired insulin production and insulin resistance. Normally, the presence of insulin causes the cells of the body to take up and metabolize glucose. In a person with insulin resistance, the body's cells fail to take up glucose even when insulin is present. Therefore, the blood glucose value exceeds the normal level, and glucose appears in the urine.

Type 2 diabetes is increasing rapidly in most industrialized countries of the world. Because type 2 diabetes is most often seen in people who are obese, dietary factors are generally believed to contribute to its development. Further, a healthy diet, increased physical activity, and weight loss have been seen to improve insulin's ability to function properly in type 2 diabetics. How might diet contribute to the occurrence of type 2 diabetes? Simple sugars in foods, such as candy and ice cream, immediately enter the bloodstream, as do sugars from the digestion of starch in white bread and potatoes. When the blood glucose level rises rapidly, the pancreas produces an overload of insulin to bring the level under control. Chronically high insulin levels apparently lead to insulin resistance, increased fat deposition, and a high level of fatty acids in the blood. Over the years, the body's cells become insulin resistant, and thus type 2 diabetes can occur. In addition, high fatty acid levels can lead to increased risk for cardiovascular disease.

It is well worth the effort to control type 2 diabetes, because all diabetics, whether type 1 or type 2, are at risk for blindness, kidney disease, and cardiovascular disease.

Cardiovascular Disease

In the United States, cardiovascular disease, which includes hypertension, heart attack, and stroke, is among the leading causes of death. Cardiovascular disease is often due to blockage of arteries by plaque, which contains saturated fats and cholesterol. Cholesterol is carried in the blood by two types of lipoproteins: low-density lipoprotein (LDL) and high-density lipoprotein (HDL). LDL is thought of as "bad" because it carries cholesterol from the liver to the cells, while HDL is thought of as "good" because it carries cholesterol from the cells to the liver, which takes it up and converts it to bile salts.

Saturated fats, including trans fats, tend to raise LDL cholesterol levels, while unsaturated fats lower LDL cholesterol levels. Beef, dairy foods, and coconut oil are rich sources of saturated fat. Foods containing partially hydrogenated oils (e.g., vegetable shortening and stick margarine) are sources of trans fats. Unsaturated fatty acids in olive and canola oils, most nuts, and coldwater fish tend to lower LDL cholesterol levels. Furthermore, coldwater fish (e.g., herring, sardines, tuna, and salmon) contain polyunsaturated fatty acids, and especially *omega-3 unsaturated fatty acids,* which can reduce the risk for cardiovascular disease. Taking fish oil supplements to obtain omega-3 fatty acids is not recommended without a physician's approval, because too much of these fatty acids can interfere with normal blood clotting. Overall, dietary saturated fats and trans fats raise LDL cholesterol levels more than dietary cholesterol.

A physician can determine if blood lipid levels are normal. If a person's cholesterol and triglyceride levels are elevated, modifying the fat content of the diet, losing excess body fat, and exercising regularly can reduce them. If life-style changes do not lower blood lipid levels enough to reduce the risk for cardiovascular disease, a physician may prescribe medication.

Eating Disorders

People with eating disorders are dissatisfied with their body image. Social, cultural, emotional, and biological factors all contribute to the development of an eating disorder. These serious conditions can lead to malnutrition, disability, and death. Regardless of the eating disorder, early recognition and treatment are crucial. Treatment usually includes psychological counseling and antidepressant medications.

Anorexia nervosa is a severe psychological disorder characterized by an irrational fear of getting fat, causing a refusal to eat enough food to maintain a healthy body weight (**Fig. 25.22***a*). A self-imposed starvation diet is often accompanied by occasional binge eating, followed by purging and extreme physical activity to avoid weight gain. Binges usually include large amounts of high-calorie foods, and purging episodes involve self-induced vomiting and laxative abuse. About 90% of people suffering from anorexia nervosa are young women; an estimated 1 in 200 teenage girls is affected.

A person with **bulimia nervosa** binge eats, then purges to avoid gaining weight (Fig. 25.22*b*). The binge-purge cyclic behavior can occur several times a day. People with bulimia nervosa can be difficult to identify because their body weight is often normal, and they tend to conceal their bingeing and purging practices. Women are more likely than men to develop bulimia; an estimated 4% of young women suffer from this condition.

Other abnormal eating practices include binge-eating disorder and muscle dysmorphia. Many obese people suffer from **binge-eating disorder,** a condition characterized by episodes of overeating that are not followed by purging. Stress, anxiety, anger, and depression can trigger food binges. A person suffering from **muscle dysmorphia** thinks his or her body is underdeveloped. Body-building activities and a preoccupation with diet and body form accompany this condition. Each day, the person may spend hours in the gym, working out on muscle-strengthening equipment. Unlike anorexia nervosa and bulimia, muscle dysmorphia affects more men than women.

Figure 25.22 Eating disorders.

a. People with anorexia nervosa have a mistaken body image and think they are fat even though they are thin. **b.** Those with bulimia nervosa overeat and then purge their bodies of the food they have eaten.

(a): © Tony Freeman/PhotoEdit; (b): © Jack Star/PhotoLink/Getty RF

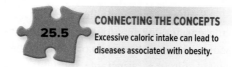

CONNECTING THE CONCEPTS

25.5   Excessive caloric intake can lead to diseases associated with obesity.

Check Your Progress 25.5

1. Calculate the BMI of an individual who is 5 foot 6 inches (1.67 meters) tall and weighs 162 pounds (73.4 kg).

2. Describe two disorders associated with obesity.

3. Explain why it is important to balance your daily caloric input and output.

SUMMARIZE

The digestive system is responsible for processing the nutrients needed by cells to conduct the activities of life. Understanding the principles of nutrition allows us to avoid the consequences of having an unhealthy diet and can help us to lead a more active and productive life.

25.1 The digestive system provides the body's cells with nutrients.

25.2 Proper nutrition is essential for optimal functioning of your body.

25.3 A balanced diet contains both macronutrients in their correct proportions, and all necessary micronutrients.

25.4 Planning nutritious meals is essential for a healthy diet.

25.5 Excessive caloric intake can lead to diseases associated with obesity.

25.1 Digestive System

The **digestive system,** along with the respiratory and urinary systems, makes exchanges between the external environment and the blood, thereby supplying the blood with nutrients and oxygen and cleansing it of waste molecules. Then the blood makes exchanges with the interstitial fluid, and in this way cells acquire nutrients and oxygen and rid themselves of wastes. The functions of the digestive system are to (1) ingest food, (2) break food down to smaller molecules for transport, (3) absorb these nutrient molecules, and (4) eliminate indigestible remains.

The digestive tract consists of several specialized parts:

- The **mouth** contains teeth for mechanical digestion. The teeth of **herbivores, carnivores,** and **omnivores** are adapted to their diets.
- **Salivary glands** produce saliva, which contains **salivary amylase** for digesting starch, and the tongue forms a **bolus** for swallowing.
- Pharynx: The air and food passages cross in the pharynx. During swallowing, the air passage is blocked off by the soft palate and epiglottis; **peristalsis** begins.
- **Esophagus:** The esophagus connects the pharynx with the stomach.
- **Stomach:** The **stomach** uses mechanical digestion to churn and mix food with the acidic gastric juices, producing **chyme.** The gastric juices contain **pepsin,** an enzyme that digests protein.
- **Small intestine:** The duodenum of the **small intestine** receives **bile** from the **gallbladder.** The bile, produced by the liver, is used in the **emulsification** of fats. The small intestine also receives pancreatic juice from the **pancreas,** which contains **trypsin** (digests protein), **lipase** (digests fat), **nuclease** (digests nucleic acids), and **pancreatic amylase** (digests starch). The small intestine produces enzymes that finish digestion, breaking food down to small molecules that are absorbed into the villi through the **microvilli.** Amino acids and glucose enter blood capillaries. Glycerol and fatty acids are joined and packaged as lipoproteins before entering lymphatic capillaries, called **lacteals.**
- **Large intestine:** The **large intestine** stores the waste materials from digestion until they can be eliminated through the **anus.** The **appendix** may protect against infections. The large intestine absorbs water, salts, and some vitamins. Reduced water absorption results in **diarrhea.** The intake of water and fiber helps prevent constipation. Small growths, called **polyps,** often occur in the large intestine.

Accessory Organs

The two main accessory organs are the pancreas and the liver.

- **Pancreas:** The pancreas is both an **exocrine gland** that produces pancreatic juice and an **endocrine gland** that produces the hormones **insulin** and **glucagon.**
- Liver: The liver produces bile, which is stored in the gallbladder.
- The gallbladder and salivary glands are also accessory organs.

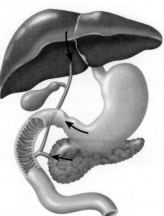

25.2 Nutrition

Nutrients perform various physiological functions in the body. A healthy diet can lead to a longer and more active life.

Introducing the Nutrients

Balanced **diets** supply **nutrients** in proportions necessary for health. **Essential nutrients** must be supplied by the diet, or else deficiency disorders result. **Macronutrients** (carbohydrates, lipids, and proteins) supply energy. **Micronutrients** (vitamins and minerals) and water do not supply energy.

25.3 The Classes of Nutrients

The six classes of nutrients are carbohydrates, lipids, proteins, minerals, vitamins, and water.

Carbohydrates (Macronutrient)

- **Carbohydrates** in the form of sugars and starch provide energy for cells. Glucose is a simple carbohydrate the body uses directly for energy.
- Sources of carbohydrates are fruits, vegetables, cereals, and other grains.
- **Fiber** consists of nondigestible carbohydrates from plants. Fiber in the diet can prevent constipation and may protect against cardiovascular disease, diabetes, and colon cancer. Large quantities of refined carbohydrates, from which fiber, vitamins, and minerals have been removed, in the diet may lead to **obesity.**

Lipids (Macronutrient)

- **Triglycerides** (from fats and oils) supply energy and are stored as **fat** for insulation and protection of organs. Alpha-linolenic and linoleic acids are essential fatty acids.
- **Cholesterol** is used in plasma membranes and to make bile, steroid hormones, and vitamin D. The only food sources of cholesterol are foods derived from animals. Elevated blood cholesterol levels are associated with cardiovascular disease.
- Sources of lipids include oils, fats, whole-milk dairy products, meat, fish, poultry, and nuts.
- High intake of saturated fats, trans fats, and cholesterol is harmful to health.

Proteins (Macronutrient)

- The body uses the 20 different amino acids to synthesize hundreds of **proteins.** Nine amino acids are essential to the diet. Most animal foods are complete sources of protein because they contain all of these essential amino acids.
- Sources of proteins include meat, fish, poultry, eggs, nuts, soybeans, and cheese.

- Healthy vegetarian diets rely on various sources of plant proteins.
- Consumption of excess protein can be harmful because the excretion of excess urea taxes the kidneys and can lead to kidney stones.

Minerals (Micronutrient)

- **Minerals** regulate metabolism and are incorporated into structures and compounds in the body.
- About 20 minerals, obtained from most foods, are needed by the body. Minerals are classified as either **major minerals** (more than 100 mg per day) or **trace minerals** (less than 100 mg per day).
- Deficiencies of calcium can lead to osteoporosis, and excesses of sodium can lead to health problems.

Vitamins (Micronutrient)

- **Vitamins,** obtained from most foods, regulate metabolism and physiological development. Vitamins C, E, and A also serve as **antioxidants.**
- Lack of any of the vitamins can lead to certain disorders.

Water

- Water participates in chemical reactions, lubricates joints, transports other nutrients, and helps maintain body temperature.
- Too little water leads to dehydration; too much water may lead to hyponatremia.

25.4 Understanding Nutrition Guidelines

Nutrition guidelines are designed to assist in the planning of nutritious meals and to allow you to make healthy and informed food choices. These nutrition guidelines are periodically updated to reflect new research and knowledge about nutrition.

ChooseMyPlate.gov

The USDA's ChooseMyPlate.gov program provides guidelines on the proportion of each food group in the diet. The program emphasizes the need for exercise to prevent weight gain.

Source: USDA, ChooseMyPlate.gov

Nutrition Labels

The information in the "Nutrition Facts" panel on packaged foods can be useful when comparing foods for nutrient content, especially serving size, total Calories, and % daily value.

Dietary Supplements

Dietary supplements are nutrients and plant products that are taken to enhance health. A multiple vitamin and mineral supplement that provides recommended amounts of nutrients can be taken daily, but herbal and nutritional supplements can be harmful if misused.

25.5 Nutrition and Health

Excess body fat increases the risk of type 2 diabetes, cardiovascular disease, and certain cancers.

Body Mass Index

Medical researchers use the **body mass index** (**BMI**) to determine if a person has a healthy weight, is overweight, or is obese.

Energy Intake Versus Energy Output

- Energy intake: The number of Calories (kcal) consumed daily is based on grams of carbohydrate, fat, and protein in the foods eaten.
- Energy output: The number of Calories (kcal) used daily is based on the amounts needed for metabolic functions, physical activity, and digestion, absorption, and processing of nutrients.

Maintaining a Healthy Weight

- To maintain a healthy weight, kcal intake should not exceed kcal output. To lose weight, decrease caloric intake and increase physical activity.
- Fad diets, in general, are nutritionally unbalanced and difficult to follow over the long term.

Disorders Associated with Obesity

- Type 2 diabetes: A healthy diet, increased physical activity, and weight loss improve insulin function in type 2 diabetics.
- Cardiovascular disease: Saturated fats and trans fats are associated with high blood levels of LDL cholesterol. To reduce risk of cardiovascular disease, cholesterol intake should be limited, and the diet should include sources of unsaturated, polyunsaturated, and omega-3 fatty acids to reduce cholesterol levels.

Eating Disorders

People suffering from eating disorders are dissatisfied with their body image. **Anorexia nervosa** and **bulimia nervosa** are serious psychological disturbances that can lead to malnutrition and death. Other disorders include **binge-eating disorder** and **muscle dysmorphia.**

ASSESS

Testing Yourself

Choose the best answer for each question.

25.1 Digestive System

1. Label the components of the human digestive system in the following illustration.

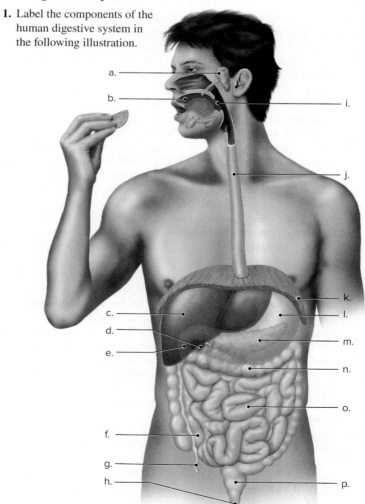

2. Pepsin
   a. breaks down protein in the small intestine.
   b. breaks down protein in the stomach.
   c. is found in saliva.
   d. breaks down fats in the stomach.

3. Which of the following is not a function of the liver?
   a. removal of poisonous substances from the blood
   b. secretion of digestive juices
   c. production of albumin
   d. storage of glucose
   e. production of bile

4. Mechanical digestion occurs in the
   a. mouth and stomach
   b. large and small intestines
   c. liver and small intestine
   d. esophagus and small intestine

5. The breakdown of nutrients for absorption occurs primarily in the
   a. large intestine.
   b. mouth.
   c. stomach.
   d. small intestine.

25.2 Nutrition

6. Vitamins are considered
   a. micronutrients because they are small in size.
   b. micronutrients because they are needed in small quantities.
   c. macronutrients because they are large in size.
   d. macronutrients because they are needed in large quantities.

7. A _____ is a component of food that performs a physiological function.
   a. macromolecule
   b. Calorie
   c. nutrient
   d. chemical

8. The amino acids that must be consumed in the diet are called essential. Nonessential amino acids
   a. can be produced by the body.
   b. are needed only occasionally.
   c. are stored in the body until needed.
   d. can be taken in via supplements.

25.3 The Classes of Nutrients

For questions 9–13, choose the class of nutrient from the key that matches the description. Each answer may be used more than once or not at all.

Key:
   a. carbohydrates
   b. lipids
   c. proteins
   d. minerals
   e. vitamins
   f. water

9. preferred source of direct energy for cells
10. constitutes the majority of human body mass
11. include antioxidants
12. an example is cholesterol
13. includes calcium, phosphorus, and potassium

14. Vitamins
    a. are inorganic compounds necessary in the diet.
    b. are needed in large quantities by the body and used as building blocks for body tissues.
    c. are organic compounds, needed in small quantities by the body, that regulate metabolic activities.
    d. can all be synthesized by the human body.

25.4 Understanding Nutrition Guidelines

15. A % daily value for sugar is not included in a nutrition label because
    a. sugar is not a nutrient.
    b. it is too difficult to determine the caloric value of sugar.
    c. there is a daily value given for carbohydrates, but not for sugars.
    d. sugar quality varies from product to product.

16. Which of the following is correct regarding the % daily values on a food label?
    a. They provide precise values for an individual's diet.
    b. They are based on a 2,000-Calorie-per-day diet.
    c. They are provided for every nutrient on the food label.
    d. They take serving size into consideration.

25.5 Nutrition and Health

17. The body mass index (BMI) is
    a. a measure of height relative to weight to determine whether a person is overweight.
    b. a measure of height relative to age to determine whether a person is of normal height.
    c. relatively low if a person is overweight.
    d. relatively high if a person is underweight.

18. The body's inability to regulate blood glucose levels is a characteristic of
    a. hypertension.
    b. cancer.
    c. cardiovascular disease.
    d. diabetes.

ENGAGE

Thinking Critically

1. Bariatric surgery is a medical procedure that reduces the size of the stomach and enables food to bypass a section of the small intestine. The surgery is generally done when obese individuals have unsuccessfully tried numerous ways to lose weight and their health is compromised by their weight. There are many risks associated with the surgery, but it helps a number of people lose a considerable amount of weight and ultimately improve their overall health. Based on your understanding of the digestive system and nutrition, list some nutritional deficiencies that may occur as a result of this surgery.

2. Some people believe that food and drink manufacturers are at least partly to blame for the obesity epidemic in the United States because of misleading advertising. For example, their advertisements show portion sizes that encourage excess consumption and make people believe that certain unhealthy substitutes are the equivalent of a well-balanced meal. Advertisements also lead children to desire sugar-coated cereals and fructose-loaded drinks. How do you think advertisements should be changed to address the actual nutritional content of a product? What obstacles might make it difficult to implement your plan?

26

Defenses Against Disease

OUTLINE

26.1 Overview of the Immune System 497

26.2 Nonspecific Defenses and Innate Immunity 499

26.3 Specific Defenses and Adaptive Immunity 502

26.4 Immunizations 506

26.5 Disorders of the Immune System 508

BEFORE YOU BEGIN

Before beginning this chapter, take a few moments to review the following discussions.

Section 17.1 What is the general structure of a virus?

Section 22.2 What is the role of the immune system in the body?

Section 23.2 What is the role of the lymphatic system with regard to circulation?

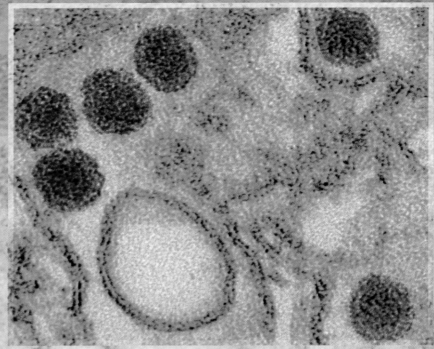

Source: Cynthia Goldsmith/CDC

The Search for a Vaccine Against Zika

Although Zika virus was first reported in Africa in 1952, the virus did not make an appearance in the Western hemisphere until 2015, when cases occurred in Brazil. The most common way the virus is transmitted between people is via an infected *Aedes* mosquito. However, it can be sexually transmitted from infected males to females. In a relatively short period of time, the virus has spread throughout South and Central America, and there have already been cases of infected travelers returning to the United States.

For most people, infection with the Zika virus produces mild symptoms, such as fevers, rashes, or joint pain. Some individuals do not experience any symptoms at all, and thus may not know that they have been infected. However, in a small number of cases, pregnant females who have been infected with the Zika virus have given birth to children with microcephaly. Microcephaly is a birth defect that causes the head and brain of an infant to be much smaller than normal. This can cause a number of developmental problems, including seizures, intellectual disabilities, and vision problems. Since there is no cure for microcephaly, researchers have been actively looking at ways to develop a vaccine against the virus.

In order to create a vaccine, researchers must identify the parts of a virus that will cause our immune system to react as if it has been infected and build up antibodies to the actual virus. Later, if an individual is exposed to the virus, these antibodies are used by the body to provide immunity. In this chapter, we will explore how our immune system protects us, not only from viruses such as Zika, but from a wide variety of pathogens.

As you read through this chapter, think about the following questions:

1. How does your body respond to its first exposure to a new pathogen?
2. How do vaccines provide immunity?
3. What is the difference between an antibody and an antibiotic?

26.1 Overview of the Immune System

Learning Outcomes

Upon completion of this section, you should be able to

1. Explain the function of the immune system in the body.
2. List the organs and tissues of the immune system, and provide a function for each.

The **immune system** plays an important role in keeping us healthy, because it fights infections and cancer. The immune system contains the lymphatic organs: the red bone marrow, thymus, lymph nodes, and spleen (**Fig. 26.1**). Lymphoid tissue may also be found in the tonsils and appendix.

Lymphatic Organs

Each of the lymphatic organs has a particular function in immunity, and each is rich in lymphocytes, one of the types of white blood cells.

Red Bone Marrow

In a child, most bones have red bone marrow, and in an adult, it is still present in the bones of the skull, the sternum (breastbone), the ribs, the clavicle, the pelvic bones, the vertebral column, and the ends of the humerus and femur nearest their attachment to the body. **Red bone marrow** produces all types of blood cells, but in this chapter we are interested in those cells that are directly associated with the immune system (see **Table 26.1**).

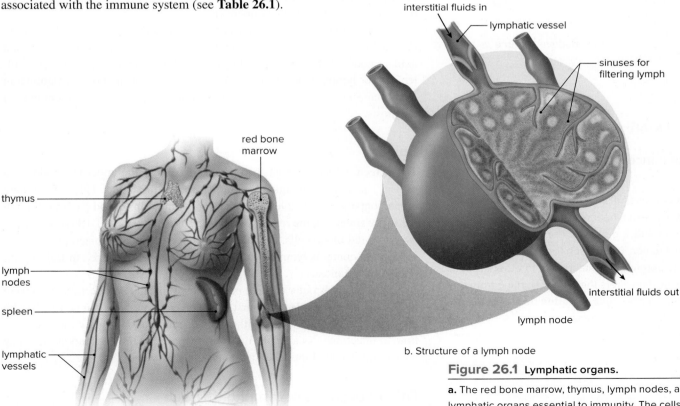

a. Location of lymph organs

b. Structure of a lymph node

Figure 26.1 **Lymphatic organs.**

a. The red bone marrow, thymus, lymph nodes, and spleen are lymphatic organs essential to immunity. The cells of the immune system, including lymphocytes, are found in these organs and in the lymph of the lymphatic vessels. **b.** The structure of a lymph node allows for the filtering of interstitial fluid.

Table 26.1 Cells of the Immune System

Cells	Function(s)	
Macrophages	Phagocytize pathogens; inflammatory response and specific immunity	
Mast cells	Release histamine, which promotes blood flow to injured tissues; inflammatory response	
Neutrophils	Phagocytize pathogens; inflammatory response	
Natural killer cells	Kill virus-infected and tumor cells by cell-to-cell contact	
B lymphocytes	Involved in the process of specific immunity by producing plasma cells and memory B cells	
Plasma cells	Produce specific antibodies	
Memory cells	Long-lived cells that may produce new B and T cells in the future	
T lymphocytes	Regulate immune response; produce cytotoxic T cells and helper T cells	
Cytotoxic T cells	Kill virus-infected and cancer cells	
Helper T cells	Coordinate the adaptive immune responses	

Connections: Health

Why do tests for cancer often take biopsies of the lymph nodes?

In a biopsy, a physician uses a small needle to take a sample of a tissue for additional examination. For individuals with cancer, the doctor often takes a biopsy of the lymph nodes surrounding the original tumor. The reason for this is to see if the cancer has begun to spread, or metastasize, to other tissues. Since the lymphatic system filters the fluids returning from the tissues, metastasizing cancer cells can often be first detected in the lymph nodes closest to the tumor.

Lymphocytes differentiate into either **B lymphocytes (B cells)** or **T lymphocytes (T cells),** which are discussed at length in Section 26.3. Bone marrow is not only the source of B lymphocytes but also the place where B lymphocytes mature. T lymphocytes mature in the thymus. As we will see, the B lymphocytes produce antibodies, while the T lymphocytes kill antigen-bearing cells outright.

Thymus

The soft, bilobed **thymus** varies in size, but it is larger in children and shrinks as we get older. The thymus plays a role in the maturing of T lymphocytes. Immature T lymphocytes migrate from the bone marrow through the bloodstream to the thymus, where they develop, or "mature," into functioning T lymphocytes. Only about 5% of T lymphocytes ever leave the thymus. These T lymphocytes have survived a critical test: If any show the ability to react with the cells of our own body, or "self," they die. If they have the potential to attack a foreign cell, they leave the thymus and enter lymphatic vessels and organs.

Lymph Nodes

Lymph nodes are small, ovoid structures along lymphatic vessels (Fig. 26.1b). Lymph nodes filter lymph and keep it free of pathogens and antigens. Lymph is filtered as it flows through a lymph node because the node's many sinuses (open spaces) are lined by **macrophages**—large, phagocytic cells that engulf and then devour as many as a hundred pathogens and still survive (see Fig. 26.4a). Lymph nodes are also instrumental in fighting infections and cancer, because they contain many lymphocytes.

Some lymph nodes are located near the surface of the body and are named for their location. For example, inguinal nodes are in the groin, and axillary nodes are in the armpits. Physicians often feel for the presence of swollen, tender lymph nodes in the neck as evidence that the body is fighting an infection. This method is a noninvasive, preliminary way to help them make a diagnosis.

Spleen

The **spleen,** which is about the size of a fist, is in the upper left abdominal cavity. The spleen's unique function is to filter the blood. This soft, spongy organ contains tissue called red pulp and white pulp. Blood passing through the many sinuses in the red pulp is filtered of pathogens and debris, including worn-out red blood cells, because the sinuses are lined by macrophages. The white pulp contains lymphocytes, which are actively engaged in fighting infections and cancer.

The spleen's outer capsule is relatively thin, and an infection or a severe blow can cause the spleen to burst. The spleen's functions can be fulfilled by other organs, but individuals without a spleen are often slightly more susceptible to infections. As a result, they will receive certain vaccinations and may have to receive antibiotic therapy indefinitely.

Other Locations of Lymphoid Tissue

The **tonsils,** which are located in the pharynx, and the **appendix,** which is attached to a portion of the large intestine, are lymphatic tissue structures that also belong to the immune system.

Cells of the Immune System

The immune system not only contains a network of lymphatic organs and lymphatic tissues but also includes a wide variety of cells (Table 26.1).

These cells play a role in the immune system's ability to distinguish between the cells of our body (self) and pathogens in the body (nonself). Pathogens are identified by the presence of antigens. An **antigen** is any molecule, usually a protein or carbohydrate from a pathogen, that stimulates the immune system. By being able to distinguish between self and nonself, the cells of the immune system provide us with immunity. **Immunity** is the body's ability to repel foreign substances, pathogens, and cancer cells. There are different levels of immunity: nonspecific immunity indiscriminately repels pathogens, while specific (adaptive) immunity requires that a certain antigen be present.

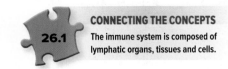

CONNECTING THE CONCEPTS

26.1   The immune system is composed of lymphatic organs, tissues and cells.

Check Your Progress 26.1

1. Explain the role of the immune system.
2. List the lymphatic organs, and provide a general function for each.
3. Explain the relationship between antigens and immunity.

26.2 Nonspecific Defenses and Innate Immunity

Learning Outcomes

Upon completion of this section, you should be able to

1. Describe the barriers to entry that keep pathogens out of the body.
2. Summarize the inflammatory response.
3. Describe the roles of the complement system and natural killer cells in nonspecific immunity.

The body has an **innate immunity** composed of the various types of nonspecific defenses—our first line of defense against most types of infections. The nonspecific defenses are the barriers to entry, the inflammatory response, the complement system, and natural killer cells.

Barriers to Entry

Skin and the mucous membranes lining the respiratory, digestive, reproductive, and urinary tracts serve as mechanical barriers to entry by pathogens. Oil gland secretions contain chemicals that weaken or kill certain bacteria on the skin (**Fig. 26.2**). The upper respiratory tract is lined by ciliated cells that sweep mucus and trapped particles up into the throat, where they can be swallowed or expectorated (spit out). The stomach has an acidic pH, which inhibits the growth of or kills many types of bacteria. The various bacteria that normally reside in the large intestine and other areas, such as the vagina, prevent pathogens from taking up residence.

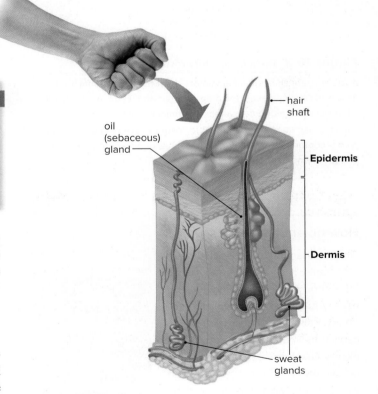

oil (sebaceous) gland

hair shaft

Epidermis

Dermis

sweat glands

Figure 26.2 Structure of the skin.

The cells of the epidermis harden and die as they progress to the outer layer of skin. These outer dead cells form a protective barrier against invasion by pathogens. The sweat and oil from glands in the dermis are acidic enough to inhibit invasion by bacteria.

© Ingram Publishing/Alamy RF

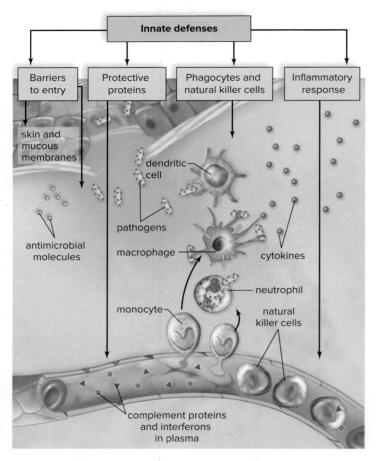

Figure 26.3 shows the innate defenses flowchart with the following labeled categories:

Innate defenses
- Barriers to entry
  - skin and mucous membranes
  - antimicrobial molecules
- Protective proteins
- Phagocytes and natural killer cells
  - dendritic cell
  - pathogens
  - macrophage
  - monocyte
  - neutrophil
  - natural killer cells
  - complement proteins and interferons in plasma
- Inflammatory response
  - cytokines

Figure 26.3 **The inflammatory response.**
If you cut your skin, the inflammatory response occurs immediately. As blood flow increases, the area gets red and warm. As mast cells, a type of white blood cell, release chemicals (such as histamine), the capillary becomes more permeable, localized swelling occurs, and pain receptors are stimulated. Neutrophils and macrophages begin phagocytizing the bacteria.

Connections: Health

How do antihistamines work?

Once histamine is released from mast cells, it binds to receptors on other body cells. This signals the nearby capillaries to become more "leaky," allowing fluid to leave the capillaries and enter the tissue. This excess fluid is the cause of the familiar symptoms of a runny nose and watery eyes. Antihistamines work by blocking the receptors on the cells, so that histamine can no longer bind. For allergy relief, antihistamines are most effective when taken before exposure to the allergen.

© McGraw-Hill Education/Jill Braaten, photographer

The Inflammatory Response

The **inflammatory response** plays an important role in the defense against invasion by pathogens. Inflammation employs mainly neutrophils and macrophages to surround and kill (engulf by phagocytosis) pathogens trying to get a foothold inside the body. Protective proteins are also involved. Inflammation is usually recognized by its four hallmark symptoms: redness, heat, swelling, and pain (**Fig. 26.3**).

Connections: Health

Is a fever always bad?

At the first sign of a fever, most people reach for over-the-counter (OTC) medicines to bring it under control. However, in many cases, using OTC medicines does more harm than good. Medical professionals now widely regard a low fever as a beneficial aspect of the immune system. When your body runs a fever, the elevated temperature increases your metabolic rate and slows down bacterial and viral reproduction. A low fever also promotes the release of chemicals, called interferons, that prevent the infection from spreading. Of course, a high fever or one that lasts for several days should immediately be brought to the attention of your physician.

© David Buffington/Getty RF

The four signs of the inflammatory response are due to capillary changes in the damaged area, and all serve to protect the body. Chemical mediators, such as **histamine,** released by damaged tissue cells and **mast cells,** cause the capillaries to dilate and become more permeable. Excess blood flow due to enlarged capillaries causes the skin to redden and become warm. Increased temperature in an inflamed area tends to inhibit the growth of some pathogens. Increased blood flow brings white blood cells to the area. Increased permeability of capillaries allows fluids and proteins, including blood-clotting factors, to escape into the tissues. Clot formation in the injured area prevents blood loss. The excess fluid in the area presses on nerve endings, causing the familiar pain associated with swelling. Together, these events summon white blood cells to the area. As soon as the white blood cells arrive, they move out of the bloodstream into the surrounding tissue. The **neutrophils** are first and actively phagocytize debris, dead cells, and bacteria they encounter. The many neutrophils attracted to the area can usually localize any infection and keep it from spreading. If neutrophils die off in great quantity, they become a yellow-white substance called pus.

When an injury is not serious, the inflammatory response is short-lived and the healing process quickly returns the affected area to a normal state. Nearby cells secrete chemical factors to ensure the growth (and repair) of blood vessels and new cells to fill in the damaged area. If, on the other hand, the neutrophils are overwhelmed, they call for reinforcements by secreting chemical mediators called cytokines. Cytokines attract more white blood cells to the area, including monocytes. Monocytes are longer-lived cells that become **macrophages,** even more powerful phagocytes than neutrophils. Macrophages can enlist the help of lymphocytes to carry out specific defense mechanisms.

Inflammation is the body's natural response to an irritation or injury and serves an important role. Once the healing process has begun, inflammation rapidly subsides. However, in some cases, chronic inflammation lasts for weeks, months, or even years if an irritation or infection cannot be overcome. Inflammatory chemicals may cause collateral damage to the body, in addition to killing the invaders. Should an inflammation persist, anti-inflammatory medications, such as aspirin, ibuprofen, or cortisone, can minimize the effects of various chemical mediators.

The Complement System

The **complement system,** often simply called complement, is composed of a number of blood plasma proteins designated by the letter C and a number. The complement proteins "complement" certain immune responses, which accounts for their name. For example, they are involved in and amplify the inflammatory response, because certain complement proteins can bind to mast cells and trigger histamine release. Others can attract phagocytes to the scene. Some complement proteins bind to the surface of pathogens already coated with antibodies, which ensures that the pathogens will be phagocytized by a neutrophil or macrophage (**Fig. 26.4**a). Certain other complement proteins join to form a membrane attack complex, which produces holes in the surfaces of microbes (Fig 26.4b). Fluids and salts then enter the pathogen to the point that it bursts.

Natural Killer Cells

Natural killer (NK) cells are large, granular lymphocytes that kill virus-infected cells and tumor (cancer) cells by cell-to-cell contact (see Table 26.1).

What makes an NK cell attack and kill a cell? The cells of your body ordinarily have self proteins on their surface that bind to receptors on NK cells. Sometimes virus-infected cells and cancer cells undergo alterations and lose their ability to produce self proteins. When NK cells can find no self proteins to bind to, they kill the cell, using the same method as T lymphocytes (see Fig. 26.8).

NK cells are not specific—their numbers do not increase when exposed to a particular antigen, and they have no means of "remembering" an antigen from previous contact with it.

CONNECTING THE CONCEPTS

26.2 Nonspecific defenses, such as physical barriers to entry, the inflammatory response, complement proteins, and NK cells, make up our first line of defense against pathogens.

Check Your Progress 26.2

1. List the physical barriers to entry that protect the body from pathogens.

2. Summarize the steps in the inflammatory response.

3. Explain why the complement system and natural killer cells are nonspecific.

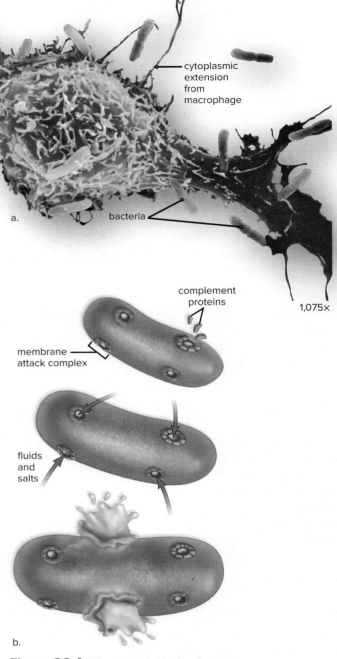

Figure 26.4 Ways to get rid of pathogens.

a. Macrophages, the body's scavengers, engulf pathogens and chop them up inside lysosomes. **b.** Complement proteins come together and form a membrane attack complex on the surface of a pathogen. Fluids and salts enter, and the pathogen bursts.

(a): © Dennis Kunkel Microscopy, Inc./Phototake

26.3 Specific Defenses and Adaptive Immunity

Learning Outcomes

Upon completion of this section, you should be able to

1. Summarize the role of lymphocytes in adaptive immunity.
2. Distinguish between the antibody response and the cellular response.
3. Explain what factors of the immune system contribute to transplant rejection.

Table 26.2 Characteristics of B Cells

• They provide an antibody response to a pathogen.
• They are produced and become mature in bone marrow.
• They reside in lymph nodes and the spleen; they circulate in blood and lymph.
• They directly recognize antigens and then undergo cell division.
• Cell division produces antibody-secreting plasma cells, as well as memory B cells.

When nonspecific defenses have failed to prevent an infection, specific defenses, or **adaptive immunity,** come into play. Specific defenses respond to antigens, which may be components of a pathogen or cancer cell. An antigen acts as a "marker" that a pathogen may be present in the body. If the marker is detected by the immune system, the adaptive immune responses begin to actively look for cells that possess the antigen, such as bacterial cells, viruses, or even the cells of our own body that may be infected by the pathogen. When the body learns to destroy a particular antigen, it develops immunity to that pathogen. Because our immune system does not ordinarily respond to the proteins on the surface of our own cells (as if they were antigens), the immune system is able to distinguish self from nonself. Only in this way can the immune system aid, rather than disrupt, homeostasis.

Lymphocytes are capable of recognizing antigens because their plasma membranes have receptor proteins whose shapes allow them to combine with specific antigens. Because we encounter millions of different antigens during our lifetime, we need a great diversity of lymphocytes to protect us against them. Remarkably, diversification occurs to such an extent during the maturation process that a lymphocyte type potentially exists for any possible antigen.

Immunity usually lasts for some time. For example, once we recover from the measles, we usually do not get the illness a second time. Immunity is primarily the result of the action of the B lymphocytes and T lymphocytes. B lymphocytes mature in the bone marrow, and T lymphocytes mature in the thymus. B lymphocytes (also called B cells) give rise to plasma cells, which produce antibodies. These antibodies are secreted into the blood, lymph, and other body fluids. In contrast, T lymphocytes (also called T cells) do not produce antibodies. Some T lymphocytes regulate the immune response, and other T lymphocytes directly attack cells that bear antigens (see Table 26.1).

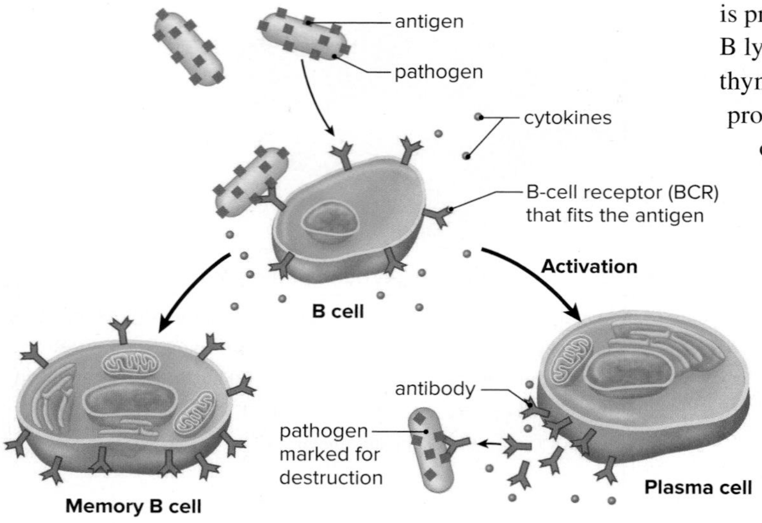

Figure 26.5 **B cells and the antibody response.**

When an antigen binds to a BCR, the B cell divides to produce plasma cells and memory B cells. Plasma cells produce antibodies but eventually undergo apoptosis. Memory B cells remain in the body, ready to produce the same antibody in the future.

B Cells and the Antibody Response

The general characteristics of a B cell are presented in **Table 26.2.** It is important to note that each B cell can bind only to a specific antigen—the antigen that fits the binding site of its receptor. The receptor is called a **B-cell receptor (BCR).** Some B cells never have anything to do, because an antigen that fits the binding site of their type of receptor is never encountered by the body. But if an antigen does bind to a BCR, that B cell is activated, and it divides, producing many plasma cells and memory B cells (**Fig. 26.5**). This mechanism is called the **clonal selection model,** since the antigen selects which B cell is activated and this cell then divides to produce many clones of itself. B cells are stimulated by cytokines to divide and produce **plasma cells.**

Cytokines are chemical signals that may be released by nonspecific defense mechanisms, such as the cells involved in the inflammatory response. Plasma cells are larger than regular B cells, because they have extensive rough endoplasmic reticulum for the mass production and secretion of antibodies specific to the antigen. The antibodies produced by plasma cells and secreted into the blood and lymph are identical to the BCR of the activated B cell.

Memory B cells are the means by which long-term immunity is established. If the same antigen enters the system again, memory B cells quickly divide and give rise to more plasma cells capable of producing the correct antibodies.

Once the threat of an infection has passed, the development of new plasma cells ceases, and those present undergo apoptosis. Apoptosis is a process of programmed cell death involving a cascade of cellular events, leading to the death and destruction of the cell (see Fig. 8.10).

The Function of Antibodies

Antibodies are immunoglobulin proteins that are capable of combining with a specific antigen (**Fig. 26.6**). The antigen-antibody reaction can take several forms, but quite often the reaction produces complexes of antigens combined with antibodies. Such antigen-antibody complexes, sometimes called immune complexes, mark the antigens for destruction (see Fig. 26.5). An antigen-antibody complex may be engulfed by neutrophils or macrophages, or it may activate the complement system (see Section 26.2), making the pathogens more susceptible to phagocytosis.

ABO Blood Type

One of the best ways to understand the role of antibodies is by examining human blood types. You are familiar with the concept of specific antibodies through blood types (A, B, AB, or O). These letters stand for antigens on your red blood cells. If you have type O blood, you do not have either antigen A or antigen B on your red cells. Some blood types have antibodies in the plasma (**Table 26.3**). As an example, type O blood has both anti-A and anti-B antibodies in the plasma. You cannot give a person with type O blood a transfusion from an individual with type A blood. If you do, the antibodies in the recipient's plasma will react to type A red blood cells, and agglutination will occur. **Agglutination,** the clumping of red blood cells, causes blood to stop circulating and red blood cells to burst. On the other hand, you can give type O blood to a person with any blood type, because type O red blood cells bear neither A nor B antigens.

It is possible to determine who can give blood to whom based on the ABO system. However, other red blood cell antigens, in addition to A and B, are used in typing blood. Therefore, it is best to put the donor's blood on a slide with the recipient's blood to observe whether the two types match (no agglutination occurs) to determine whether blood can be safely transfused from one person to another.

T Cells and the Cellular Response

When T cells leave the thymus, they have unique **T-cell receptors (TCRs),** just as B cells do. Unlike B cells, however, T cells are unable to recognize an antigen without help. The antigen must be presented to them by an **antigen-presenting cell (APC),** such as a macrophage. A macrophage becomes an APC by ingesting and destroying a pathogen. The APC then travels to a lymph node or the spleen, where T cells also congregate. When a macrophage phagocytizes a virus, it is digested in a lysosome. An antigen from the virus is combined with

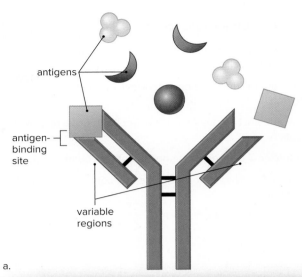

a.

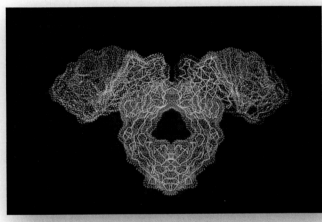

b.

Figure 26.6 Why an antibody is specific.

a. During a lifetime, a person will encounter a million different antigens, which vary in shape. An antibody has two variable regions that end in antigen-binding sites. The variable regions differ so much that the binding site of each antibody has a shape that will fit only one specific antigen. **b.** Computer model of an antibody molecule. The antigen combines with the two side branches.

(b): © Dr. Arthur J. Olson, Scripps Institute

Table 26.3 Blood Types

Blood Type	Antigens on Red Blood Cells	Antibodies in Plasma	Compatible Donor(s)
A	A	Anti-B	A, O
B	B	Anti-A	B, O
AB	A, B	Neither anti-A nor anti-B	A, B, AB, O
O	—	Both anti-A and anti-B	O

Table 26.4 Characteristics of T Cells

- They provide a cellular response to virus-infected cells and cancer cells.
- They are produced in bone marrow and mature in the thymus.
- Antigen must be presented to a T cell in groove of an MHC[1] protein.
- Cytotoxic T cells destroy nonself antigen-bearing cells.
- Helper T cells secrete cytokines that control the immune response.

[1]MHC = major histocompatibility complex.

a protein called a **major histocompatability complex (MHC).** MHC proteins play a major role in identifying self cells. After the antigen binds to the MHC protein, the complex appears on the cell surface. Then the combined MHC + antigen complex is presented to a T cell. The importance of self proteins in plasma membranes was first recognized when it was discovered that they contribute to the specificity of tissues and make it difficult to transplant tissue from one human to another.

The general characteristics of T cells are provided in **Table 26.4.** The two main types of T cells are **helper T cells** (T_H cells) and **cytotoxic T cells** (T_C cells). Each of these types has a TCR that can recognize an antigen fragment in combination with an MHC molecule. However, the major difference in antigen recognition by these two types of cells is that the T_H cells recognize only antigens presented by APCs with MHC class II molecules on their surface, while T_C cells recognize only antigens presented by APCs with MHC class I molecules on their surface. In **Figure 26.7,** the antigen is represented by a red triangle, and the helper T cell that binds to the antigen has the specific TCR that can combine with this particular MHC II + antigen. Now the helper T cell is activated and divides to produce more helper T cells.

As an illness disappears, the immune reaction wanes and the activated T cells become susceptible to apoptosis. Apoptosis contributes to homeostasis by regulating the number of cells present in an organ, or in this case, in the immune system. When apoptosis does not occur as it should, T-cell cancers (e.g., lymphomas and leukemias) can result. Also, in the thymus gland, any T cell that has the potential to destroy the body's own cells undergoes apoptosis.

Functions of Cytotoxic T Cells and Helper T Cells

Cytotoxic T cells specialize in cell-to-cell combat. They have storage vacuoles containing proteins called perforins or enzymes called granzymes. After a cytotoxic T cell binds to a virus-infected or cancer cell presenting the antigen it has learned to recognize, it releases perforin molecules, which perforate the target cell's plasma membrane, forming a pore. The cytotoxic T cell then delivers a supply of granzymes into the pore, and these cause the cell to undergo

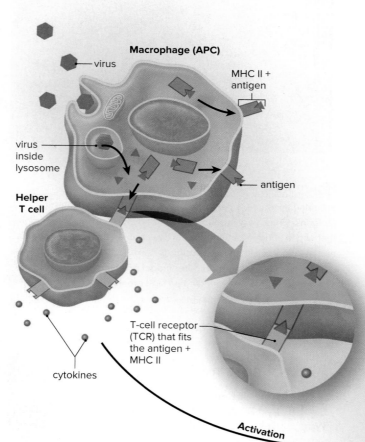

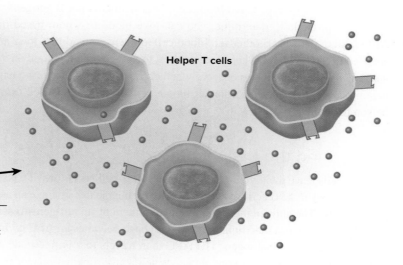

Figure 26.7 Activation of a T cell.

For a T cell to be activated, the antigen must be presented to it, along with an MHC protein, by an APC, often a macrophage. Each type of T cell bears a specific receptor, and if that TCR fits the MHC II + antigen complex, the helper T cell is activated to divide and produce more helper T cells, which also have this type of TCR. The helper T cells then produce and release cytokines.

Figure 26.8 Cytotoxic T cells and the cellular response.

a. Scanning electron micrograph of cytotoxic T cell attacking a target cell, which is either a virus-infected or cancer cell. **b.** A cytotoxic T cell attacks any cell that presents it with an MHC I + antigen complex that it has learned to recognize. First, vesicles release perforins, which form a pore in the target cell. Then vesicles release granzymes, which cause the cell to undergo apoptosis.

(a): © Steve Gschmeissner/Science Source

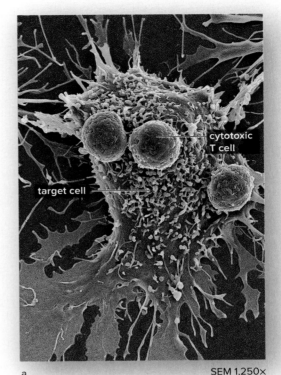

apoptosis. Once cytotoxic T cells have released their perforins and granzymes, they move on to the next target cell. Cytotoxic T cells are responsible for a cellular response to virus-infected and cancer cells (**Fig. 26.8**).

Helper T cells specialize in regulating immunity by secreting cytokines that, in particular, stimulate B cells and cytotoxic T cells. Similar to B cells, cloned T cells include memory T cells that live for many years and can jump-start an immune response to an antigen that was dealt with before. Because HIV, the virus that causes AIDS, infects helper T cells and other cells of the immune system, it inactivates the immune response and makes HIV-infected individuals susceptible to opportunistic infections. Infected macrophages serve as reservoirs for the HIV virus. AIDS is discussed in Section 26.5.

Tissue Rejection

Certain organs, such as the skin, the heart, and the kidneys, could be transplanted easily from one person to another if the body did not attempt to reject them. Rejection occurs because cytotoxic T cells and antibodies bring about the destruction of foreign tissues in the body. When rejection occurs, the immune system is correctly distinguishing between self and nonself.

Organ rejection can be controlled by carefully selecting the organ to be transplanted and administering immunosuppressive drugs. It is best if the transplanted organ has the same type of MHC proteins as those of the recipient; otherwise, the transplanted organ is antigenic to the recipient's T cells. Several immunosuppressive drugs act by inhibiting the response of T cells to cytokines. Without cytokines, all types of immune responses are weak.

CONNECTING THE CONCEPTS

26.3 Specific defenses and adaptive immunity that make up the body's second line of defense against pathogens.

Check Your Progress 26.3

1. Explain the role of antibodies in adaptive immunity.
2. Compare and contrast B cells with T cells.
3. Distinguish between the antibody and cellular responses, and identify the type(s) of cells involved in each.
4. Explain the role of MHC markers in tissue rejection.

26.4 Immunizations

After you have had an infection, you may be immune to it. Good examples of diseases against which you can acquire immunity are the childhood diseases measles and mumps. Unfortunately, few sexually transmitted diseases stimulate lasting immunity; for example, a person can get gonorrhea over and over again. If lasting immunity is possible, a vaccine for the disease likely exists or can be developed. **Vaccines** are substances that usually do not cause illness, even though the immune system responds to them. Traditionally, vaccines are the pathogens themselves, or their products, that have been treated so that they are no longer virulent (able to cause disease). Today, it is possible to genetically engineer bacteria to mass-produce a protein from pathogens, and this protein can be used as a vaccine. This method has produced a vaccine against hepatitis B, a virus-induced disease, and it is being used to prepare a vaccine against malaria, a protozoan-induced disease.

Immunization promotes **active immunity.** After a vaccine is given, it is possible to follow an active immune response by determining the amount of antibody present in a sample of plasma; this is called the antibody titer. After the first exposure to a vaccine, a primary response occurs. For a period of several days, no antibodies are present; then their concentration rises slowly, levels off, and gradually declines as the antibodies bind to the antigen or simply break down (**Fig. 26.9**). After a second exposure to the vaccine, a secondary response is expected. The concentration then rises rapidly to a level much greater than before; then it slowly declines. The second exposure is called a "booster" because it boosts the antibody concentration to a high level. The high antibody concentration is expected to help prevent disease symptoms if the individual is exposed to the disease-causing antigen.

Active immunity is dependent on the presence of memory B cells and possibly memory T cells that are capable of responding to lower doses of antigen. Active immunity is usually long-lasting, but a booster may be required after a certain number of years.

Figure 26.9 **Active immunity due to immunization.**

Immunization often requires more than one injection. A minimal primary response occurs after the first vaccine injection, but after a second injection, the secondary response usually shows a dramatic rise in the amount of antibody present in plasma.

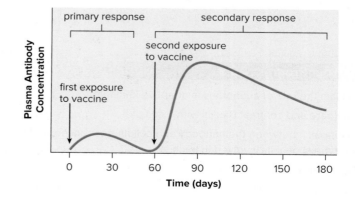

Connections: Health

Does the MMR vaccine cause autism?

Several large research studies have failed to find any connection among the measles, mumps, and rubella (MMR) vaccine; thimerosal (a preservative that was used for vaccines); and an increased risk of autism in children. While the cause of autism has not yet been identified, the evidence does not suggest that the vaccine or thimerosal is causing autism. The problem is that the first MMR vaccine is usually administered between 12 and 15 months of age, which is typically the age that an autistic child first presents symptoms. Most autism researchers believe that the factors causing autism are in place before this time frame and the timing with the MMR vaccine is coincidental. Several large-scale studies, including those sponsored by the Institute of Medicine and the National Alliance for Autism Research, have found no link between the MMR vaccine and an increased risk of autism. Most pediatricians agree that the threat of measles and rubella far outweighs the unsubstantiated risk of autism.

© Saturn Stills/
Science Source

Although the body usually makes its own antibodies, it is possible to give an individual prepared antibodies (immunoglobulins) to combat a disease. Because these antibodies are not produced by the individual's plasma cells, this is called **passive immunity,** and it is temporary. For example, newborns are passively immune to some diseases, because antibodies have crossed the placenta from the mother's blood. These antibodies soon disappear, however, so that within a few months infants become more susceptible to infections. Breast-feeding prolongs the natural passive immunity an infant receives from the mother, because antibodies are present in the mother's milk.

Even though passive immunity does not last, it is sometimes used to prevent illness in a patient who has been unexpectedly exposed to an infectious disease. Usually, the patient receives an injection of gamma globulin, a portion of blood that contains antibodies, preferably taken from an individual who has recovered from the illness. In the past, horses were immunized and gamma globulin was taken from them to provide the needed antibodies against such diseases as diphtheria, botulism, and tetanus. Unfortunately, patients who received these antibodies became ill about 50% of the time, because the serum contained proteins that the individual's immune system recognized as foreign. This condition was called serum sickness.

CONNECTING THE CONCEPTS
26.4
Vaccines are substances that help build our immunity against various pathogens.

Check Your Progress 26.4

1. Define the term *vaccine*.
2. Explain why two doses of a vaccine are sometimes needed for immunity.
3. Distinguish between active and passive immunity, and explain how each occurs.

26.5 Disorders of the Immune System

Learning Outcomes

Upon completion of this section, you should be able to

1. Distinguish between an immediate and a delayed allergic response.
2. List common autoimmune diseases, and provide the causes of each.
3. Outline the effects of HIV infection, and summarize the available treatments.

The immune system usually protects us from disease because it can distinguish self from nonself. Sometimes, however, it responds in a manner that harms the body, as when individuals develop allergies or have an autoimmune disease.

Allergies

SEM of pollen

Allergies are hypersensitivities to substances in the environment, such as pollen, food, or animal hair, that ordinarily would not cause an immune reaction. The response to these antigens, called **allergens,** usually includes some unpleasant symptoms (**Fig. 26.10**). An allergic response is regulated by cytokines secreted by both T cells and macrophages.

Immediate allergic responses are caused by receptors attached to the plasma membranes of mast cells in the tissues. When an allergen attaches to receptors on mast cells, they release histamine and other substances that bring about the symptoms.

An immediate allergic response can occur within seconds of contact with the antigen. The symptoms can vary, but a dramatic example, anaphylactic shock, is a severe reaction characterized by a sudden and life-threatening drop in blood pressure.

Allergy shots, injections of the allergen in question, sometimes prevent the onset of an allergic response. It has been suggested that injections of the allergen may cause the body to build up large quantities of antibodies released by plasma cells, and these combine with allergens received from the environment before they have a chance to reach the receptors located in the membranes of mast cells.

Delayed allergic responses are probably initiated by memory T cells at the site of allergen contact in the body. A classic example of a delayed allergic response is the skin test for tuberculosis (TB). When the test result is positive, the tissue where the antigen was injected becomes red and hardened. This shows that the person has been previously exposed to tubercle bacillus, the cause of TB. Contact dermatitis, which occurs when a person is allergic to poison ivy, jewelry, cosmetics, and so forth, is also an example of a delayed allergic response.

Figure 26.10 Allergies.

When people are allergic to pollen, they develop symptoms that include watery eyes, sinus headache, increased mucus production, labored breathing, and sneezing.

(pollen): © David Scharf/SPL/Science Source; (girl): © Damien Lovegrove/SPL/Science Source

Autoimmune Diseases

When cytotoxic T cells or antibodies mistakenly attack the body's own cells as if they bore antigens, the resulting condition is known as an **autoimmune disease.** Exactly what causes autoimmune diseases is not known. However, sometimes they occur after an individual has recovered from an infection.

In the autoimmune disease *myasthenia gravis,* neuromuscular junctions do not work properly, and muscular weakness results. In *multiple sclerosis* (*MS*), the myelin sheath of nerve fibers breaks down, causing various neuromuscular disorders. A person with *systemic lupus erythematosus* has various symptoms prior to death due to kidney damage. In *rheumatoid arthritis,* the joints are affected (**Fig. 26.11**). Researchers suggest that rheumatic fever and type 1 diabetes are autoimmune illnesses. As yet, there are no cures for autoimmune diseases, but they can be controlled with drugs.

AIDS

Understanding why patients with **acquired immunodeficiency syndrome (AIDS)** are so sick gives us a whole new level of appreciation for the workings of a healthy immune system. **Human immunodeficiency virus (HIV),** which causes AIDS, lives in and destroys helper T cells, which promote the activity of all the other cells in the immune system. Initially, the body of the individual infected with HIV is able to maintain an adequate number of T cells, but over time the helper T cells are destroyed faster than they can be produced and the virus gains the upper hand. Without helper T cells, the ravaged immune system can no longer fight off the onslaught of viruses, fungi, and bacteria that the body encounters every day. Without drug therapy, the number of T cells eventually drops from thousands to hundreds as the immune system becomes helpless (**Fig. 26.12**). More information on the life cycle of the HIV virus is provided in Section 17.1.

The symptoms of AIDS begin with weight loss, chronic fever, cough, diarrhea, swollen glands, and shortness of breath and progress to those of rare diseases. *Pneumocystis pneumonia,* a respiratory disease found in cats, and *Kaposi sarcoma,* a very rare type of cancer, are often observed in patients with advanced AIDS. Death approaches rapidly and certainly.

As of 2014, an estimated 36.9 million people were living with HIV infection. Among the 2.0 million new HIV infections, nearly 11% are in people under the age of 15. Although the number of deaths due to HIV/AIDS is declining, in 2014 the disease still claimed 1.2 million lives, bringing the total number of deaths attributed to HIV/AIDS to over 36 million. As of 2014, at least 0.8% of the adults in the world had an HIV infection.

HIV is transmitted by sexual contact with an infected person, including vaginal or rectal intercourse and oral/genital contact. Also, needle sharing among intravenous drug users is a high-risk behavior. Babies born to HIV-infected women may become infected before or during birth, or through breast-feeding after birth. Even though male-to-male sexual contact still accounts for the greatest number of new HIV cases in the United States, heterosexual contact and intravenous drug use account for the greatest percentage of increase in new cases.

Advances in treatment have reduced the serious complications of an HIV infection and have prolonged life. The sooner drug therapy begins after infection, the better the chances that HIV will not destroy the immune system. Also, medication must be continued indefinitely. Unfortunately, new strains of the virus have emerged that are resistant to the new drugs used for treatment. The likelihood of transmission from mother to child at birth can be lessened if the mother receives medication prior to birth and the child is delivered by cesarean section.

Figure 26.11 Rheumatoid arthritis.

Rheumatoid arthritis is due to recurring inflammation in skeletal joints. Complement proteins, T cells, and B cells all participate in deterioration of the joints, which eventually become immobile.

© Southern Illinois University/Science Source

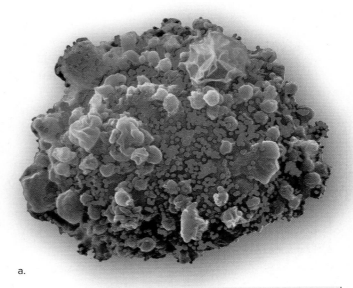

a.

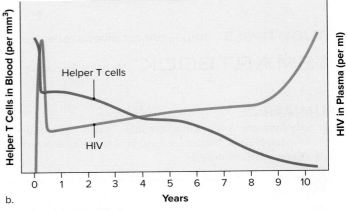

b.

Figure 26.12 HIV infection.

a. The HIV virus attacks the T cells and macrophages of the body. **b.** At first, the body produces enough helper T cells to keep the HIV infection under control, but then as the number of helper T cells declines, the HIV infection takes over.

(a): © NIBSC/Science Source

Although there are many difficulties in vaccine development, AIDS vaccine trials are underway. The process can take many years. After a vaccine has been tested in animals, it must pass through three phases of clinical trials before it is marketed or administered to the public.

In Phases I and II of a clinical trial, the vaccine is tested from one to two years in a small number of HIV-uninfected volunteers. The most effective vaccines move into Phase III. In Phase III, the vaccine is tested three to four years in thousands of HIV-uninfected people. A phase III trial of a preventative vaccine called RV144 concluded in 2009 in Thailand; though there is some evidence that the vaccine may help reduce HIV infection rates, researchers are still analyzing the data and working on follow-up studies.

The success of RV144 has encouraged researchers to believe that a preventative vaccine may be developed in the near future. A program called the HIV Vaccine Trials Network (HVTN) has been created to coordinate and analyze the data from all of the efforts currently under way to develop a vaccine. But the most compelling reason for optimism is the human body's ability to suppress the infection. The immune system is able to decrease the HIV viral load in the body, helping delay the onset of AIDS an average of 10 years in 60% of people who are HIV-infected in the United States. Studies have shown that a small number of people remain HIV-uninfected after repeated exposure to the virus, and a few HIV-infected individuals maintain a healthy immune system for over 15 years. It is these stories of the human body's ability to fight the infection that keep scientists hopeful that there is a way to help the body overcome HIV infection.

HIV infection is preventable. Suggestions for preventing this infection are to (1) abstain from sexual intercourse or develop a long-term, monogamous (always the same partner) sexual relationship with a person who is free of HIV; (2) be aware that having relations with an intravenous drug user is risky behavior; (3) avoid anal-rectal intercourse, because the lining of the rectum is thin and infected T cells easily enter the body there; (4) always use a latex condom during sexual intercourse if you do not know that your partner has been free of HIV for the past five years; (5) avoid oral sex, because this can be a means of transmission; and (6) be cautious about the use of alcohol or any other drug that may prevent you from being able to control your behavior.

CONNECTING THE CONCEPTS

26.5 Allergies and autoimmune diseases are due to an incorrect response by the immune system.

Check Your Progress 26.5

1. Describe the relationship between allergies and allergens.

2. Contrast an immediate allergic response with a delayed allergic response.

3. Explain why AIDS patients cannot fight pathogens.

STUDY TOOLS   http://connect.mheducation.com

SMARTBOOK®   Maximize your study time with McGraw-Hill SmartBook®, the first adaptive textbook.

SUMMARIZE

Our body's immune system has an incredible array of defenses that help keep us free from infections. Vaccinations and immunizations also assist in our fight against infectious diseases.

26.1 The immune system is composed of lymphoid organs, tissues, and cells.

26.2 Nonspecific defenses, such as physical barriers to entry, the inflammatory response, complement proteins, and NK cells, make up our first line of defense against pathogens.

26.3 Specific defenses and adaptive immunity that make up the body's second line of defense against pathogens.

26.4 Vaccines are substances that help build our immunity against various pathogens.

26.5 Allergies and autoimmune diseases are due to an incorrect response by the immune system.

26.1 Overview of the Immune System

The **immune system** consists of lymphatic organs and tissues as well as a variety of cells. The lymphatic organs are

- **Red bone marrow,** where all blood cells are made and the **B lymphocytes (B cells)** mature
- **Thymus,** where **T lymphocytes (T cells)** mature
- **Lymph nodes,** where lymph is cleansed by **macrophages**
- **Spleen,** where blood is cleansed of pathogens and debris
- Other organs, such as the **tonsils** and **appendix,** which are structures made of lymphatic tissue

The cells of the lymphatic system are responsible for responding to **antigens** in the body. We develop **immunity** to a pathogen as our immune system responds to its antigen.

26.2 Nonspecific Defenses and Innate Immunity

Immunity involves nonspecific and specific defenses. Nonspecific defenses, or **innate immunity,** include the following:

- Barriers to entry (e.g., skin)
- **Inflammatory response,** involving **mast cells,** which release **histamine** to increase capillary permeability, resulting in redness, warmth, swelling, and pain, and **neutrophils** and **macrophages,** which enter tissue fluid and engulf pathogens
- The **complement system,** which has many functions. One is to attack microbes outright by forming pores in their surface. Fluids and salts then enter, and the microbe bursts.
- **Natural killer (NK) cells** can distinguish between self and nonself proteins and cause virus-infected cells to undergo apoptosis.

26.3 Specific Defenses and Adaptive Immunity

Specific defenses, or **adaptive immunity,** require B lymphocytes and T lymphocytes, also called B cells and T cells.

B Cells and the Antibody Response

The **B-cell receptor (BCR)** of each type of B cell is specific to a particular antigen. When the antigen binds to a BCR, that B cell divides according to the **clonal selection model** to produce plasma cells and memory B cells.

- **Plasma cells** secrete **antibodies** and eventually undergo apoptosis. Plasma cells are responsible for antibody response to an antigen.

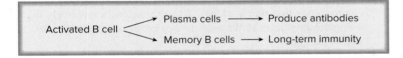

- **Memory B cells** remain in the body and produce antibodies if the same antigen enters the body at a later date.

T Cells and the Cellular Response

T cells are responsible for cellular response to an infection. Each **T-cell receptor (TCR)** is specific to a particular antigen. For a T cell to recognize an antigen, the antigen must be presented to it by an **antigen-presenting cell,** usually a macrophage, along with a protein called a **major histocompatability complex (MHC).**

Cytotoxic T cell (activated by MHC I + antigen) ⟶ Kills virus-infected cell

Helper T cells (activated by MHC II + antigen) ⟶ Regulate immune response

- **Cytotoxic T cells** kill virus-infected or cancer cells on contact, because these cells bear the MHC I antigen the T cells have learned to recognize. First, perforins are secreted, and these molecules form a pore in the cell's plasma membrane; then granzymes are delivered into the cell and cause it to undergo apoptosis.
- **Helper T cells** produce cytokines and stimulate other immune cells.

26.4 Immunizations

Immunity occurs after an infection or a vaccination.

- A **vaccine** brings about **active immunity** to a particular pathogen. Two injections may be required, because the number of antibodies is higher after the second injection, called a booster. A booster shot at some time in the future also increases the number of antibodies.

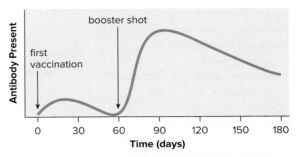

- **Passive immunity** (receiving prepared antibodies) is short-lived, because the antibodies are administered to, not made by, the vaccinated individual.

26.5 Disorders of the Immune System

Allergies

Hypersensitive responses, or **allergies,** to various substances can be immediate or delayed. Anaphylactic shock is a dangerous immediate allergic response.

Autoimmune Diseases

In an autoimmune disease (e.g., rheumatoid arthritis, multiple sclerosis, or perhaps type 1 diabetes), the immune system mistakenly attacks the body's own cells.

AIDS

The **human immunodeficiency virus (HIV)** lives in and destroys helper T cells. Without drug treatment, the number of T cells falls off, and the individual develops **acquired immunodeficiency syndrome (AIDS)** and eventually dies from infections that are rare in healthy individuals. Sexual contact and needle sharing transmit AIDS from person to person. Combined drug therapy has prolonged the lives of some people infected with HIV. So far, vaccine research has met with limited success. However, AIDS is a preventable disease if certain behaviors are avoided.

ASSESS

Testing Yourself

Choose the best answer for each question.

26.1 Overview of the Immune System

For questions 1–5, identify the lymphatic organ in the key that matches the description. Some answers may be used more than once or not at all.

Key:

    a. red bone marrow   **d.** spleen
    b. thymus   **e.** appendix
    c. lymph nodes

1. produces stem cells
2. located in every body cavity except the dorsal cavity
3. produces lymphocytes
4. site of maturation of T cells
5. structure of lymphatic tissue attached to the large intestine

26.2 Nonspecific Defenses and Innate Immunity

6. Which of the following is not a barrier to pathogen entry?
   a. oil gland secretions of the skin
   b. acidic pH in the stomach
   c. cilia in the upper respiratory tract
   d. nonpathogenic bacteria in the digestive tract
   e. saliva in the mouth

7. During the inflammatory response,
   a. T cells move to the site of injury.
   b. capillaries become constricted.
   c. histamine is produced.
   d. capillaries become less permeable.
   e. More than one of these occurs.

8. Complement
   a. is a general defense mechanism.
   b. is involved in the inflammatory response.
   c. is a series of proteins present in the plasma.
   d. plays a role in destroying bacteria.
   e. All of these are correct.

26.3 Specific Defenses and Adaptive Immunity

9. Which of the following characteristics apply to T lymphocytes?
   a. mature in bone marrow
   b. mature in the thymus
   c. Both a and b are correct.
   d. None of these are correct.

10. Antibodies combine with antigens
    a. at specific regions of the antibodies.
    b. at multiple regions of the antibodies.
    c. only if macrophages are present.
    d. Both a and c are correct.

11. An antigen-presenting cell (APC)
    a. presents antigens to T cells.
    b. secretes antibodies.
    c. marks each human cell as belonging to that person.
    d. secretes cytokines only.

12. Which of the following cells participate in the specific defense against an infection?
    a. mast cells   d. natural killer cells
    b. macrophages   e. B lymphocytes
    c. neutrophils

26.4 Immunizations

13. Vaccines are associated with
    a. active immunity.   c. passive immunity.
    b. long-lasting immunity. d. Both a and b are correct.

14. Passive immunity
    a. is permanent.
    b. may result from immunoglobulin injections.
    c. requires memory B cells.
    d. may be induced by vaccines.

26.5 Disorders of the Immune System

15. AIDS is caused by which of the following viruses?
    a. SIV   d. AIDS
    b. HSV   e. HIV
    c. HPV

16. _____ is a condition that results when cytotoxic T cells attack the body's own cells.
    a. An allergic response   c. Passive immunity
    b. Autoimmune disease   d. Active immunity

17. Allergens
    a. cause an immune response in most people.
    b. are antigens to those people who are allergic to them.
    c. reduce the production of histamine by mast cells.
    d. are molecules that are usually rare in the environment.

18. Which of the following is not an autoimmune disease?
    a. multiple sclerosis
    b. myasthenia gravis
    c. contact dermatitis
    d. systemic lupus erythematosus
    e. rheumatoid arthritis

ENGAGE

Thinking Critically

1. At one time, tonsillectomies (removal of the tonsils) were more commonplace than they are today. Why are the tonsils seemingly so susceptible to infection? Why should they not be removed at the first sign of infection?

2. The transplantation of organs from one person to another was impossible until the discovery of immunosuppressive drugs. Now, with the use of drugs such as cyclosporine, organs can be transplanted without rejection. Transplant patients must take immunosuppressive drugs for the remainder of their lives. What are the potential risks associated with long-term use of immunosuppressive drugs?

3. AIDS is deadly without proper medical care, but it can be simply chronic if treated. Drug companies typically charge a high price for AIDS medications because Americans and their insurance companies can afford them. However, these drugs are out of reach for many people in other countries, such as those in Africa, where AIDS is a widespread problem.
   a. Do drug companies have a moral obligation to provide low-cost AIDS drugs, even if they have to do so at a loss of revenue?
   b. Is it right for governments to ignore patent laws in order to provide their citizens with affordable drugs?

© Anderson Ross/Blend Images/Getty RF

27

The Control Systems

Multiple Sclerosis (MS)

Multiple sclerosis, or MS, is a disease that affects a major control system of the body, the nervous system. The first symptoms of MS tend to be weakness or tingling in the arms and legs, fatigue, a loss of coordination, and blurred vision. As the disease progresses, the individual may experience problems with speech and vision, tremors that make coordinated movement difficult, and numbness in the extremities. We now know that MS is an inflammatory disease that affects the myelin sheaths, which wrap parts of some nerve cells like insulation around an electrical cord. As these sheaths deteriorate, the nerves no longer conduct impulses normally. For unknown reasons, MS often attacks the optic nerves first, before spreading to other areas of the brain. Most researchers think MS results from a misdirected attack on myelin by the body's immune system, although other factors may be involved.

Almost 400,000 people in the United States have MS, and it is the most common disease of the nervous system in young adults. Typically, the first symptoms occur between the ages of 20 and 40. The disease is not contagious, is rarely fatal, and does not appear to be inherited, although some studies suggest a genetic component associated with susceptibility to MS. Like many diseases that affect the nervous system, there is no cure, so affected individuals must deal with their condition for the rest of their lives. The good news is that the disease is not severe in almost 45% of cases, and its symptoms can be controlled with medication.

As you read through this chapter, think about the following questions:

1. What is the function of myelin?
2. What specific type of neurological process is affected if myelin is damaged?
3. Why does MS produce such a variety of symptoms?

OUTLINE

27.1 Nervous System   514

27.2 Endocrine System   526

BEFORE YOU BEGIN

Before beginning this chapter, take a few moments to review the following discussions.

Section 5.4 How does active transport move molecules and ions against their concentration gradients?

Section 19.1 What are the key events in the evolution of the animals?

Section 22.3 What is the role of negative feedback in the maintenance of homeostasis?

27.1 Nervous System

Learning Outcomes

Upon completion of this section, you should be able to

1. Describe the structure and function of a neuron.
2. Explain how a nerve impulse is generated and propagated.
3. Describe the structure and function of a chemical synapse.
4. Give examples of drugs of abuse, and explain how they affect the nervous system.
5. Describe the structure and function of the central and peripheral nervous systems.
6. Compare and contrast the somatic and autonomic systems as well as the sympathetic and parasympathetic divisions of the autonomic system.

The nervous system and the endocrine system work together to regulate the activities of the other systems. Both control systems use chemical signals when they respond to changes that might threaten homeostasis, but they have different means of delivering these signals (**Fig. 27.1**). The **nervous system** quickly sends a message along a nerve fiber directly to a target organ or tissue, such as skeletal or smooth muscle. Once a chemical signal is released, the muscle brings about an appropriate response.

The **endocrine system** uses the blood vessels of the cardiovascular system to send hormones as chemical messengers to a target organ, such as the liver. The endocrine system is slower-acting because it takes time for the hormone molecules to move through the bloodstream to the target organ. Also, hormones change the metabolism of cells, and this takes time; however, the response is longer-lasting. Cellular metabolism tends to remain the same for at least a limited period of time.

As we examine the human nervous system, we will also compare it with the nervous systems of other animals.

Figure 27.1 Modes of action of the nervous and endocrine systems.

a. Nerve impulses passing along an axon cause the release of a neurotransmitter. The neurotransmitter, a chemical signal, causes the wall of an arteriole to constrict. **b.** The hormone insulin, a chemical signal, travels in the cardiovascular system from the pancreas to the liver, where it causes liver cells to store glucose as glycogen.

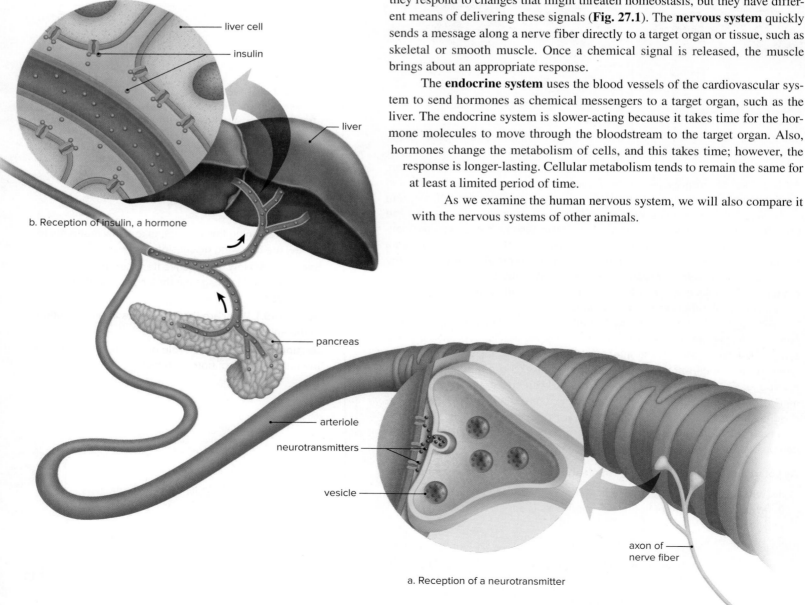

liver cell

insulin

liver

b. Reception of insulin, a hormone

pancreas

arteriole

neurotransmitters

vesicle

axon of nerve fiber

a. Reception of a neurotransmitter

Examples of Nervous Systems

While many organisms have mechanisms that enable them to respond to factors in the environment, the presence of a nervous system is a characteristic that is unique to animals. Most animals (except the parazoans—see Section 19.2) utilize a nervous system to detect stimuli in the environment and then perform coordinated reactions in response to the stimuli. For example, animals may use their nervous system to detect chemical signals that allow them to move toward a food source or away from a predator. While there are varying levels of complexity in animal nervous systems, they all receive sensory input, which is processed to direct a coordinated reaction.

An early example of a nervous system is found in the planarian, a bilateral organism with a simple body plan (see Section 19.3). Planarians possess two lateral nerve cords (bundles of nerves) joined together by transverse nerves. The arrangement is called a *ladderlike* nervous system. The simple brain receives sensory information from the eyespots and sensory cells in the auricles. The two lateral nerve cords allow a rapid transfer of information from the cerebral ganglia to the posterior end, and the transverse nerves between the nerve cords keep the movement of the two sides coordinated. The nervous organization in planarians is a foreshadowing of the central and peripheral nervous systems found in more complex invertebrates, such as earthworms, and in vertebrates, including humans (**Fig. 27.2**).

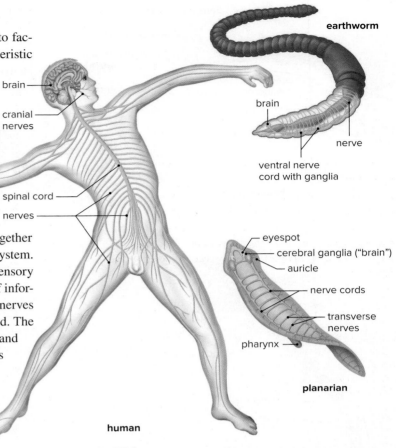

Figure 27.2 **Comparison of nervous systems.**

Invertebrates, such as a planarian and an earthworm, as well as vertebrates, such as a human, have a central nervous system (e.g., brain) and a peripheral nervous system (nerves).

The Human Nervous System

In humans, the nervous system controls the muscular system and works with the endocrine system to maintain homeostasis. The **central nervous system (CNS)** includes the brain and spinal cord, which have a central location along the midline of the body. The **peripheral nervous system (PNS)** consists of nerves that lie outside the central nervous system. The brain gives off paired cranial nerves (one on each side of the body), and the spinal cord gives off paired spinal nerves. The division between the central nervous system and the peripheral nervous system is arbitrary; the two systems work together and are connected to one another.

While based on similar principles, the human nervous system is much more complex than the planarian system. Over the course of animal evolution, a number of important events occurred in the development of the nervous system:

- A CNS developed that is able to summarize incoming messages before ordering outgoing messages.
- Nerve cells (neurons) became specialized to send messages to the CNS, between neurons in the CNS, or away from the CNS.
- A brain evolved that has special centers for receiving input from various regions of the body and for directing their activity.
- The CNS became connected to all parts of the body by peripheral nerves. Therefore, the central nervous system can respond to both external and internal stimuli.
- Complex sense organs, such as the human eye and ear, arose that can detect changes in the external environment.

This section primarily explores the evolution of the central nervous system (CNS). We will explore the structure and function of the sense organs in Section 28.1.

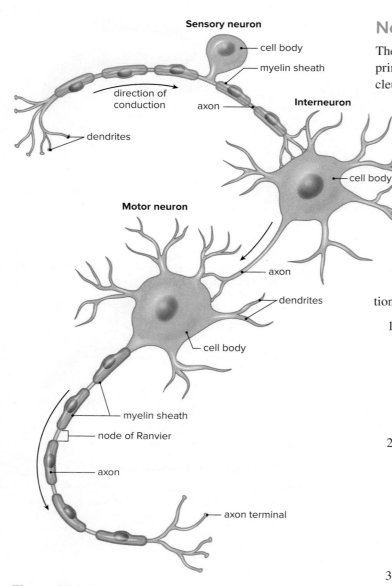

Figure 27.3 Types of neurons.

A sensory neuron, an interneuron, and a motor neuron are drawn here to show their arrangement in the body. Only axons conduct nerve impulses. In a sensory neuron, a process that extends from the cell body divides into an axon that takes nerve impulses all the way from the dendrites to the CNS. In a motor neuron and an interneuron, the axon extends directly from the cell body. The axon of sensory and motor neurons is covered by a myelin sheath. All long axons have a myelin sheath.

Neurons

The structure of a nerve cell, or **neuron,** is well suited to its function as the primary cell of the nervous system (**Fig. 27.3**). The **cell body** contains the nucleus and other organelles that allow a cell to function. The neuron's many short **dendrites** fan out to receive signals from sensory receptors or other neurons. These signals can result in nerve impulses carried by an axon. The **axon,** an extension of the neuron that is typically longer than a dendrite, is the location in the neuron that is responsible for conducting nerve impulses to their targets. Axons can deliver nerve impulses great distances, for example, there are axons that extend from your toes to your spinal cord. Long axons are covered by a white **myelin sheath** formed from the membranes of tightly spiraled cells that leave gaps called **nodes of Ranvier.** The axons of neurons are often organized as **nerves,** which frequently appear white due to the myelin sheaths.

The nervous system has three types of neurons specific to its three functions (Fig. 27.3):

1. The nervous system receives sensory input. **Sensory neurons** perform this function. They take nerve impulses from sensory receptors to the CNS. The sensory receptor, which is the distal end of the axon of a sensory neuron, may be as simple as a naked nerve ending (such as a pain receptor), or it may be built into a highly complex organ, such as the eye or ear. In any case, the axon of a sensory neuron can be quite long if the sensory receptor is far from the CNS.

2. The nervous system performs integration—in other words, the CNS sums up the input it receives from all over the body. **Interneurons** occur entirely within the CNS and take nerve impulses between its various parts. Some interneurons lie between sensory neurons and motor neurons, and some take messages from one side of the spinal cord to the other or from the brain to the spinal cord, and vice versa. Interneurons also form complex pathways in the brain, where processes that account for thinking, memory, and language occur.

3. The nervous system generates motor output. **Motor neurons** take nerve impulses from the CNS to muscles or glands. Motor neurons cause muscle fibers to contract or glands to secrete, and therefore they are said to *innervate* these structures.

The Nerve Impulse

Like some other cellular processes, a nerve impulse is dependent on concentration gradients. In neurons, these concentration gradients are maintained by the sodium-potassium pump. This pump actively transports sodium ions (Na^+) to the outside of the axon and actively transports potassium ions (K^+) inside. Aside from ion concentration differences across the axon's membrane, a charge difference also exists. The inside of an axon is negative compared with the outside. This charge difference is primarily due in part to an unequal distribution of Na^+ and K^+ ions across the membrane. The charge difference across the axon's membrane plays an important role in the generation of a **nerve impulse,** which is also called an **action potential.**

The nerve impulse is a rapid, short-lived, self-propagating reversal in the charge difference across the axon's membrane. **Figure 27.4** shows how it works. A nerve impulse involves two types of gated channel proteins in the axon's membrane. In contrast to ungated channel proteins, which constantly allow ions to move across the membrane, gated channel proteins open and close in response to a stimulus, such as a signal from another neuron. One type of gated channel protein allows sodium (Na^+) to pass through the membrane, and the other allows potassium (K^+) to pass through the membrane. As an axon is conducting a nerve impulse, the Na^+ gates open at a particular location, and the inside of the axon becomes positive as Na^+ moves from outside the axon to the inside. The Na^+ gates close, and then the K^+ gates open. Now K^+ moves from inside the axon to outside the axon, and the charge reverses.

In Figure 27.4, the axon is unmyelinated, and the action potential at one locale stimulates an adjacent part of the axon's membrane to produce an action potential. In myelinated axons, an action potential at one node of Ranvier causes an action potential at the next node (**Fig. 27.5**). This type of conduction, called *saltatory conduction,* is much faster than conduction by unmyelinated axons. Imagine running down a long hallway as quickly as you can; then picture yourself able to get to the end of the same hall in just a few leaping bounds. Leaping would enable you to travel the same distance in a much shorter time; likewise, saltatory conduction greatly speeds the conduction of nerve impulses. In thin, unmyelinated axons, the nerve impulse travels about 1.0 meter/second, but in thick, myelinated axons, the rate is more than 100 meters/second due to saltatory conduction. In any case, action potentials are self-propagating; each action potential generates another along the length of an axon.

The conduction of a nerve impulse (action potential) is an all-or-none event—that is, either an axon conducts a nerve impulse or it does not. The intensity of a message is determined by how many nerve impulses are generated within a given time span. An axon can conduct a volley of nerve impulses because only a small number of ions are exchanged with each impulse. As soon as an impulse has passed by each successive portion of an axon, it undergoes a short refractory period, during which it is unable to conduct an impulse. During a refractory period, the sodium gates cannot open. This period ensures that nerve impulses travel in only one direction and do not reverse.

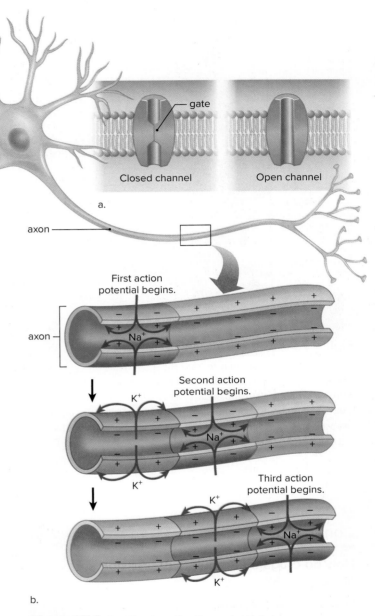

Figure 27.4 **Conduction of action potentials in an unmyelinated axon.**

a. Na^+ and K^+ each have their own gated channel protein through which they cross the axon's membrane. **b.** During an action potential, Na^+ enters the axon, and the charge difference between inside and outside reverses (blue); then K^+ exits, and the charge difference is restored (red). The action potential moves from section to section in an unmyelinated axon.

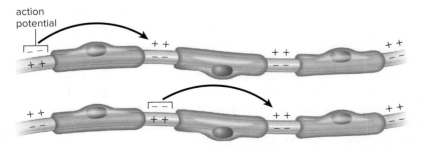

Figure 27.5 **Conduction of a nerve impulse in a myelinated axon.**

Action potentials can occur only at gaps in the myelin sheath, called nodes of Ranvier. This makes conduction much faster than in unmyelinated axons. In humans, all long axons are myelinated.

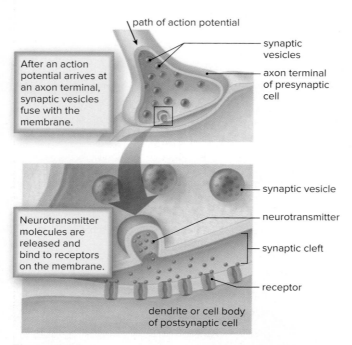

path of action potential

synaptic vesicles

axon terminal of presynaptic cell

After an action potential arrives at an axon terminal, synaptic vesicles fuse with the membrane.

synaptic vesicle

Neurotransmitter molecules are released and bind to receptors on the membrane.

neurotransmitter

synaptic cleft

receptor

dendrite or cell body of postsynaptic cell

Figure 27.6 Synapse structure and function.

Transmission across a synapse from one neuron (the presynaptic cell) to another occurs when a neurotransmitter is released, diffuses across a synaptic cleft, and binds to a receptor in the plasma membrane of the next neuron (the postsynaptic cell). Each axon releases only one type of neurotransmitter, symbolized here by a red ball.

Connections: Health

How do drugs that regulate depression and anxiety work?

In general, pharmaceutical drugs that regulate behavior work by regulating the amount of certain neurotransmitters in the synapses. For example, drugs such as Xanax and Valium increase the levels of gamma-aminobutyric acid. These medications are used for panic attacks and anxiety. Reduced levels of norepinephrine and serotonin are linked to depression. Drugs such as Prozac, Paxil, and Cymbalta allow norepinephrine and/or serotonin to accumulate at the synapses, usually by blocking their reabsorption. Increasing the levels of these neurotransmitters means that the postsynaptic cells receive a more constant chemical message, which explains the effectiveness of the antidepressant drugs.

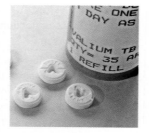

© Van D. Bucher/Science Source

The Synapse

Each axon has many axon terminals. In the CNS, a terminal of one neuron, known as the *presynaptic cell,* lies very close to the dendrite (or cell body) of another neuron, the *postsynaptic cell.* This region of close proximity is called a **synapse.** In the PNS, when the postsynaptic cell is a muscle cell, the region is called a neuromuscular junction. A small gap exists at a synapse, and this gap is called the **synaptic cleft.** While the synaptic cleft is very narrow, the nerve impulses are not able to cross it directly. Instead, transmission across a synaptic cleft is carried out by chemical signals called **neurotransmitters,** which are stored in synaptic vesicles. When nerve impulses traveling along an axon reach an axon terminal, synaptic vesicles release a neurotransmitter into the synaptic cleft. Neurotransmitter molecules diffuse across the cleft and bind to a specific receptor protein on the postsynaptic cell (**Fig. 27.6**).

Depending on the type of neurotransmitter and/or the type of receptor, the response of the postsynaptic cell can be toward excitation or toward inhibition. Several dozen different neurotransmitters have been identified, a few of the more widely used neurotransmitters are acetylcholine (ACh), norepinephrine, serotonin, and gamma-aminobutyric acid (GABA).

Once a neurotransmitter has been released into a synaptic cleft and has initiated a response, the neurotransmitter is removed from the cleft. In some synapses, the postsynaptic cell produces enzymes that rapidly inactivate the neurotransmitter. For example, the enzyme acetylcholinesterase (AChE) breaks down acetylcholine. In other synapses, the presynaptic cell rapidly reabsorbs the neurotransmitter. For example, norepinephrine is reabsorbed by the axon terminal. The short existence of neurotransmitters at a synapse prevents continuous stimulation (or inhibition) of the postsynaptic cell.

A single neuron has many dendrites plus the cell body, and both can have synapses with many other neurons; 1,000 to 10,000 synapses per single neuron are not uncommon. Therefore, a neuron is on the receiving end of many signals. An excitatory neurotransmitter produces a potential change that drives the neuron closer to an action potential, and an inhibitory neurotransmitter produces a potential change that drives the neuron farther from an action potential. Neurons integrate these incoming signals. **Integration** is the summing up of both excitatory and inhibitory signals. If a neuron receives many excitatory signals (either at different synapses or at a rapid rate from one synapse), chances are its axon will transmit a nerve impulse. On the other hand, if a neuron receives both inhibitory and excitatory signals, the integration of these signals may prohibit the axon from firing.

Drug Abuse

Many drugs that affect the nervous system act by interfering with or promoting the action of neurotransmitters. A drug can either enhance or block the release of a neurotransmitter, mimic the action of a neurotransmitter or block the receptor for it, or interfere with the removal of a neurotransmitter from a synaptic cleft. *Stimulants* are drugs that increase the activity of the CNS, and *depressants* decrease its activity. Increasingly, researchers are coming to believe that dopamine, a neurotransmitter in the brain, is responsible for mood. Many of the new medications developed to counter drug dependence and mental illness affect the release, reception, or breakdown of dopamine.

Drug abuse is apparent when a person takes a drug at a dose level and under circumstances that increase the potential for a harmful effect. A drug abuser often takes more of the drug than was intended. Drug abusers are apt to display a psychological and/or physical dependence on the drug. With physical dependence, formerly called "addiction," more of the drug is needed to get the same effect, and withdrawal symptoms occur when the user stops taking the drug.

Cocaine

Cocaine is an alkaloid derived from the shrub *Erythroxylon coca*. It is sold in powder form and as crack, a more potent extract. Because cocaine prevents the synaptic uptake of dopamine, the neurotransmitter remains in the synapse for a prolonged period of time and continues to stimulate the postsynaptic cell. As a result, the user experiences a "rush" sensation. The epinephrine-like effects of dopamine account for the state of arousal that lasts for several minutes after the rush experience.

A cocaine binge can go on for days, after which the individual suffers a crash. During the binge period, the user is hyperactive and has little desire for food or sleep but has an increased sex drive. During the crash period, the user is fatigued, depressed, and irritable; has memory and concentration problems; and displays no interest in sex.

Cocaine causes extreme physical dependence. With continued cocaine use, the postsynaptic cells become increasingly desensitized to dopamine. The user, therefore, experiences the withdrawal symptoms arising from physical dependence and an intense craving for cocaine. These are indications that the person is highly dependent on the drug.

Overdosing on cocaine can cause seizures and cardiac and respiratory arrest. It is possible that long-term cocaine abuse causes brain damage (**Fig. 27.7**). Babies born to cocaine addicts suffer withdrawal symptoms and may have neurological and developmental problems.

Methamphetamine

Methamphetamine is a synthetic drug made by adding a methyl group to amphetamine. Over 9 million people in the United States have used methamphetamine at least once in their lifetime; teenagers and young adults represent approximately one-fourth of these. The addition of the methyl group is fairly simple, so methamphetamine is often produced from amphetamine in makeshift home laboratories. It is available as a powder (speed) or as crystals ("crystal meth," "glass," or "ice"). The crystals are smoked, and the effects are almost instantaneous and nearly as quick as when methamphetamine is snorted. When the drug is smoked, the effects last 4–8 hours. Methamphetamine has a structure similar to that of dopamine, and its stimulatory effect mimics that of cocaine. It reverses fatigue, maintains wakefulness, and temporarily elevates the mood of the user. After the initial rush, there is typically a state of high agitation that, in some individuals, leads to violent behavior. Chronic use can lead to what is called an amphetamine psychosis, resulting in paranoia; auditory and visual hallucinations; self-absorption; irritability; and aggressive, erratic behavior. Drug tolerance, dependence, and addiction are common. Hyperthermia, convulsions, and death can occur.

Marijuana

The dried flowers, leaves, and stems of the Indian hemp plant, *Cannabis sativa*, contain and are covered by a resin that is rich in 9-tetrahydrocannabinol (THC).

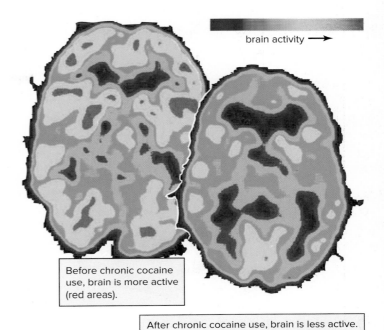

brain activity ⟶

Before chronic cocaine use, brain is more active (red areas).

After chronic cocaine use, brain is less active.

Figure 27.7 **Cocaine's effect on the brain.**

PET scans show that the usual activity of the brain is reduced after chronic use of cocaine. The red and yellow colors indicate areas of higher activity in the brain.

© Science Source

Connections: Health

What is medical marijuana used for?

Researchers continue to examine the potential medical benefits of THC, the active compound in marijuana. Initial studies have indicated that THC acts as an analgesic, or pain reducer. Medical marijuana is sometimes prescribed for patients in extreme pain, such as those in the later stages of AIDS or those who have spinal cord injuries. How-

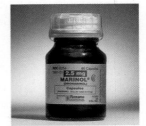

© James Keyser, The Life Images Collection/Getty Images

ever, it is important to note that it is the active compound in marijuana that has potential benefits. Smoking marijuana presents more health problems than the use of cigarettes. Marijuana smoke contains 50–70% more carcinogens than cigarette smoke. In addition, in the hour following marijuana use, there is almost a fivefold increase in the risk of a heart attack.

The names *cannabis* and *marijuana* apply to either the plant or THC. Usually, marijuana is smoked in a cigarette form called a "joint."

Recently, researchers have found that marijuana binds to a receptor for anandamide, a neurotransmitter that seems to create a feeling of peaceful contentment. The occasional marijuana user experiences a mild euphoria, along with alterations in vision and judgment, which result in distortions of space and time. Motor incoordination, including the inability to speak coherently, is experienced. Heavy use can result in hallucinations, anxiety, depression, rapid flow of ideas, body image distortions, paranoid reactions, and similar psychotic symptoms. Craving and difficulty in stopping usage can occur as a result of regular use.

The Central Nervous System

The organizations of vertebrate brains in reptiles (alligator), birds (goose) and mammals (horse and human) are compared in **Figure 27.8.** If we divide the brain into a hindbrain, midbrain, and forebrain, we can see that the forebrain is most prominent in humans. The forebrain of mammals also has an altered function in that it becomes the last depository for sensory information. This change accounts for why the forebrain carries on much of the integration for the entire nervous system before it sends out motor instructions to glands and muscles. In humans, the **spinal cord** provides a means of communication between the brain and the spinal nerves, which are a part of the PNS. Spinal nerves leave the spinal cord and take messages to and from the skin, glands, and muscles in all areas of the body, except the head and face. Long, myelinated fibers of interneurons in the spinal cord run together in bundles called tracts. These tracts connect the spinal cord to the brain. Because these tracts cross over, the left side of the brain controls the right side of the body, and vice versa. In our discussion of the peripheral nervous system, we will see that the spinal cord is involved in reflex actions, which are programmed, built-in circuits that allow for protection and survival. They are present at birth and require no conscious thought to take place.

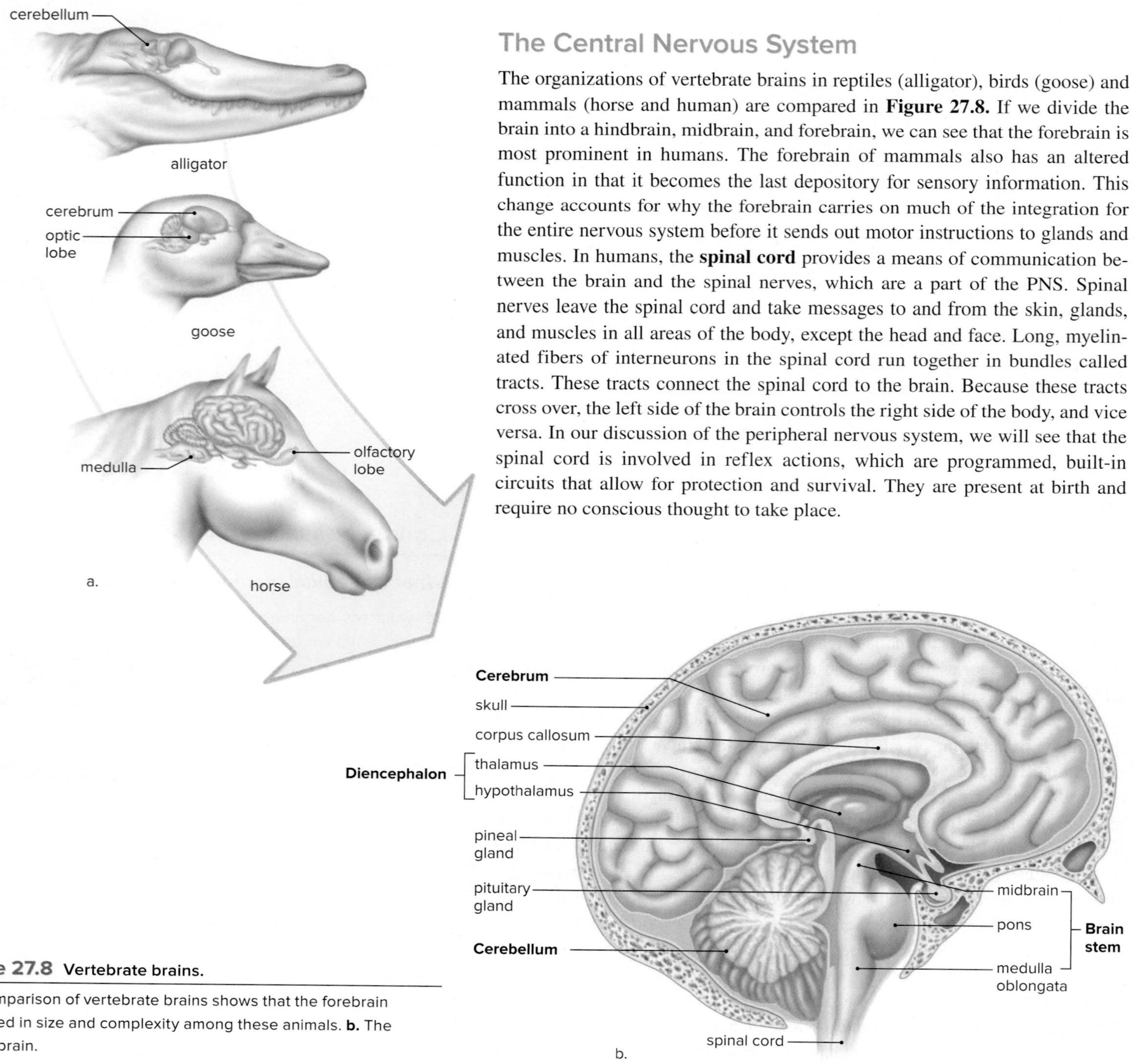

Figure 27.8 Vertebrate brains.

a. A comparison of vertebrate brains shows that the forebrain increased in size and complexity among these animals. **b.** The human brain.

The Brain

Our discussion will center on these parts of the **brain:** the cerebrum, the diencephalon, the cerebellum, and the brain stem (Fig. 27.8*b*).

Cerebrum The cerebrum communicates with and coordinates the activities of the other parts of the brain. The cerebrum has two halves, and each half has a number of lobes. Most of the cerebrum is white matter, where the long axons of interneurons are taking nerve impulses to and from the cerebrum. The highly convoluted outer layer of gray matter that covers the cerebrum is called the **cerebral cortex.** The cerebral cortex contains over a billion cell bodies, and it is the region of the cerebrum that interprets sensation, initiates voluntary movement, and carries out higher thought processes.

Investigators have found that each part of the cerebrum has specific functions (**Fig. 27.9**). To take an example, the **primary sensory area** located in the parietal lobe receives information from the skin, skeletal muscles, and joints. Each part of the body has its own receiving area in this region. The **primary motor area,** on the other hand, is in the frontal lobe just before the small cleft that divides the frontal lobe from the parietal lobe. Voluntary commands to skeletal muscles arise in the primary motor area, and the muscles in each part of the body are controlled by a certain section of the primary motor area.

The lobes of the cerebral cortex have a number of specialized centers to receive information from the sensory receptors for sight, hearing, and smell. The lobes also have *association areas,* where integration occurs. The **prefrontal area,** an association area in the frontal lobe, receives information from the other association areas and uses this information to reason and plan actions. Integration in this area accounts for our most cherished human abilities: to think critically and to formulate appropriate behaviors.

Diencephalon Beneath the cerebrum is the diencephalon, which contains the hypothalamus and the thalamus (see Fig. 27.8*b*). The **hypothalamus** is an integrating center that helps maintain homeostasis by regulating hunger, sleep, thirst, body temperature, and water balance. The hypothalamus controls the pituitary gland, thereby serving as a link between the nervous and endocrine systems. The **thalamus** is on the receiving end for all sensory input except smell. Information from the eyes, ears, and skin arrives at the thalamus via the cranial nerves and tracts from the spinal cord. The thalamus integrates this information and sends it on to the appropriate portions of the cerebrum. The thalamus is involved in arousal of the cerebrum; it participates in motor functions and higher mental processes, such as memory and emotions.

The pineal gland, which secretes the hormone melatonin, is located in the diencephalon. Presently, there is much popular interest in melatonin because it is released at night when we are sleeping. Supplements of melatonin have been effectively used to treat seasonal affective disorder, jet lag, and some sleep disorders.

Cerebellum The **cerebellum** has two portions joined by a narrow median strip. Each portion is primarily composed of white matter, which in

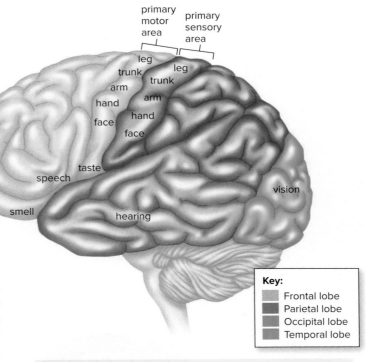

Function of lobes
- Frontal lobe—reasoning, planning, speech, movement, emotions, and problem solving
- Parietal lobe—integration of sensory input from skin and skeletal muscles, understanding speech
- Occipital lobe—seeing, perception of visual stimuli
- Temporal lobe—hearing, perception of auditory stimuli

Figure 27.9 **Functional regions of the cerebral cortex.**

Specific areas of the cerebral cortex receive sensory input from particular sensory receptors, integrate various types of information, or send out motor commands to particular areas of the body.

longitudinal section has a treelike pattern (see Fig. 27.8*b*). Overlying the white matter is a thin layer of gray matter, which forms a series of complex folds.

The cerebellum receives sensory input from the eyes, ears, joints, and skeletal muscles about the present position of body parts, and it receives motor output from the cerebral cortex that specifies where these parts should be located. After integrating this information, the cerebellum sends motor impulses by way of the brain stem to the skeletal muscles. In this way, the cerebellum maintains posture and balance. It also ensures that all of the muscles work together to produce smooth, coordinated voluntary movements. The cerebellum helps us learn new motor skills, such as playing the piano or hitting a baseball.

Brain Stem   The **brain stem,** which contains the midbrain, the pons, and the medulla oblongata, connects the rest of the brain to the spinal cord (see Fig. 27.8*b*). It contains tracts that ascend or descend between the spinal cord and higher brain centers. The **midbrain** contains important visual and auditory reflex centers, and it coordinates responses such as the startle reflex. This occurs when you automatically turn your head in response to a sudden, loud noise, trying to see its source. The **medulla oblongata** contains a number of reflex centers for regulating heartbeat, breathing, and vasoconstriction (blood pressure). It also contains the reflex centers for vomiting, coughing, sneezing, hiccuping, and swallowing. In addition, the medulla oblongata helps control various internal organs. The **pons** links the medulla oblongata with the midbrain, and it is vital for the control of breathing.

The Limbic System

The **limbic system** is a complex network that includes the diencephalon and areas of the cerebrum (**Fig. 27.10**). The limbic system blends higher mental functions and primitive emotions into a united whole. It accounts for why activities such as sexual behavior and eating seem pleasurable and why, for instance, mental stress can cause high blood pressure.

Two significant structures within the limbic system are the hippocampus and the amygdala, which are essential for learning and memory. The hippocampus, a seahorse-shaped structure that lies deep in the temporal lobe, is well situated in the brain to make the prefrontal area aware of past experiences stored in sensory association areas. The amygdala, in particular, can cause these experiences to have emotional overtones. A connection between the frontal lobe and the limbic system means that reason can keep us from acting out strong feelings.

Learning and Memory   **Memory** is the ability to hold a thought in mind or to recall events from the past, ranging from a word we learned only yesterday to an early emotional experience that has shaped our lives. Learning takes place when we retain and utilize past memories.

The prefrontal area in the frontal lobe is active during short-term memory, as when we temporarily recall a telephone number. Some telephone numbers go into long-term memory. Think of a telephone number you know by heart, and see if you can bring it to mind without also thinking about the place or person associated with that number. Most likely, you cannot, because typically long-term memory is a mixture of what is called *semantic memory* (numbers, words, and so on) and *episodic memory* (persons, events, and other

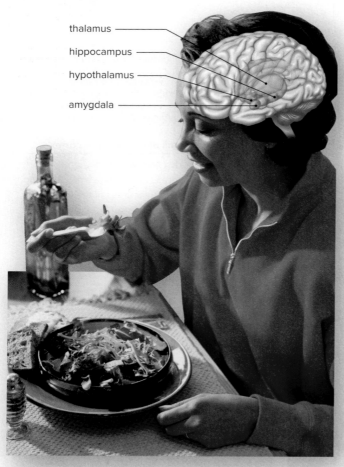

thalamus
hippocampus
hypothalamus
amygdala

Figure 27.10   The limbic system.

The limbic system includes the diencephalon and parts of the cerebrum. It joins higher mental functions, such as reasoning, with more instinctive feelings, such as fear and pleasure. Eating is a pleasurable activity for most people.

associations). *Skill memory* is a type of memory that can exist independently of episodic memory. Skill memory is what allows us to perform motor activities, such as riding a bike or playing ice hockey. A person who has **Alzheimer disease** experiences a progressive loss of memory, particularly for recent events. Gradually, the person loses the ability to perform any type of daily activity and becomes bedridden. In Alzheimer patients, clusters of abnormal tissue develop among degenerating neurons, especially in the hippocampus and amygdala. Major research efforts are devoted to seeking a cure for Alzheimer disease.

What parts of the brain are functioning when we remember something from long ago? Our long-term memories are stored in bits and pieces throughout the sensory association areas of the cerebral cortex. The hippocampus gathers this information for use by the prefrontal area of the frontal lobe when we remember, for example, Uncle George or a past summer's vacation. Why are some memories so emotionally charged? The amygdala is responsible for conditioning fear and for associating danger with sensory information received from the thalamus and the cortical sensory areas.

The Peripheral Nervous System

The purpose of the peripheral nervous system (PNS) is to relay sensory information to the CNS for processing and to relay motor responses to the tissues of the body. Some of this information comes from sensory neurons located within the sense organs (Section 28.1), while other information comes from sensory neurons within the tissues of the body.

The peripheral nervous system (PNS) lies outside the central nervous system and contains **nerves,** which are bundles of axons. The cell bodies of neurons are found in the CNS—that is, the brain and spinal cord—or in ganglia. Ganglia (sing., **ganglion**) are collections of cell bodies within the PNS. Humans have 12 pairs of **cranial nerves** that arise from the brain (**Fig. 27.11**). Cranial nerves are largely concerned with the head, neck, and facial regions of the body. However, the vagus nerve is a cranial nerve that has branches not only to the pharynx and larynx but also to most of the internal organs.

Humans also have 31 pairs of **spinal nerves,** and each of these contains many sensory and motor axons. The dorsal root of a spinal nerve contains the axons of sensory neurons, which conduct impulses to the spinal cord from sensory receptors. The cell body of a sensory neuron is in the *dorsal root ganglion.* The ventral root contains the axons of motor neurons, which conduct impulses away from the cord, largely to skeletal muscles. Each spinal nerve serves the region of the body in which it is located.

The Somatic System

The **somatic system** of the PNS includes the nerves that take information about external stimuli from sensory receptors to the CNS and motor commands away from the CNS to skeletal muscles. Voluntary control of skeletal muscles always originates in the brain. Involuntary responses to stimuli, called **reflexes,** can involve either the brain or just the spinal cord. Flying objects cause our eyes to blink, and sharp pins cause our hands to jerk away even without our having to think about it.

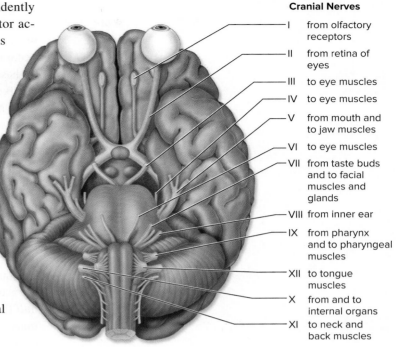

Cranial Nerves

I	from olfactory receptors
II	from retina of eyes
III	to eye muscles
IV	to eye muscles
V	from mouth and to jaw muscles
VI	to eye muscles
VII	from taste buds and to facial muscles and glands
VIII	from inner ear
IX	from pharynx and to pharyngeal muscles
XII	to tongue muscles
X	from and to internal organs
XI	to neck and back muscles

Figure 27.11 Cranial nerves.

Cranial nerves receive sensory inputs from, and send outputs to, the head region. Two important exceptions are the vagus nerve (X) and the spinal accessory nerve (XI). Spinal nerves (not shown) receive sensory input from and send outputs to the rest of the body.

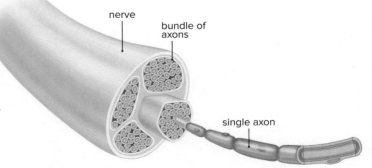

nerve

bundle of axons

single axon

Figure 27.12 illustrates the path of a reflex that involves only the spinal cord. If your hand touches a sharp pin, sensory receptors in the skin generate nerve impulses that move along sensory axons toward the spinal cord. Sensory neurons that enter the spinal cord pass signals on to many interneurons. Some of these interneurons synapse with motor neurons. The short dendrites and the cell bodies of motor neurons are in the spinal cord, but their axons leave the cord. Nerve impulses travel along motor axons to an effector, which brings about a response to the stimulus. In this case, a muscle contracts, so that you withdraw your hand from the pin. Various other reactions are possible—you will most likely look at the pin, wince, and cry out in pain. This whole series of responses is explained by the fact that some of the interneurons involved carry nerve impulses to the brain. Also, sense organs send messages to the brain that make us aware of our actions. Your brain makes you aware of the stimulus and directs these other reactions to it.

The Autonomic System

The **autonomic system** of the PNS automatically and involuntarily regulates the activity of glands and cardiac and smooth muscle. The system is divided into the parasympathetic and sympathetic divisions (**Fig. 27.13**). Reflex actions, such as those that regulate blood pressure and breathing rate, are especially important to the maintenance of homeostasis. These reflexes begin when the sensory neurons in contact with internal organs send information to the CNS. They are completed by motor neurons within the autonomic system.

The **parasympathetic division** includes a few cranial nerves (e.g., the vagus nerve) and axons that arise from the last portion of the spinal cord. The parasympathetic division, sometimes called the "housekeeper division," promotes all the internal responses we associate with a relaxed state. For example, it causes the pupil of the eye to constrict, promotes the digestion of food, and retards the heartbeat. The neurotransmitter used by the parasympathetic division is acetylcholine (ACh).

Figure 27.12 A reflex arc showing the path of a spinal reflex.

A stimulus (e.g., a pinprick) causes sensory receptors in the skin to generate nerve impulses, which travel in sensory axons to the spinal cord. Interneurons integrate data from sensory neurons and then relay signals to motor neurons. Motor axons convey nerve impulses from the spinal cord to a skeletal muscle, which contracts. Movement of the hand away from the pin is the response to the stimulus.

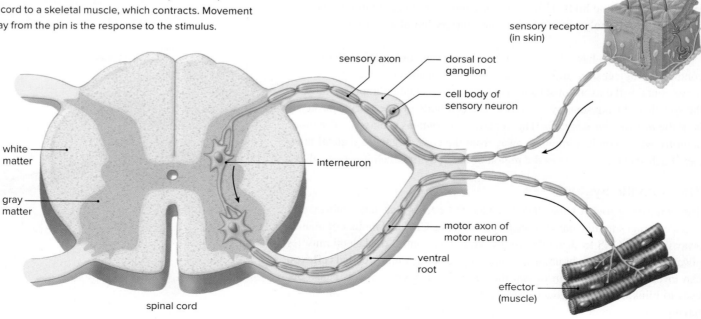

pin

sensory receptor (in skin)

sensory axon

dorsal root ganglion

cell body of sensory neuron

interneuron

white matter

gray matter

motor axon of motor neuron

ventral root

effector (muscle)

spinal cord

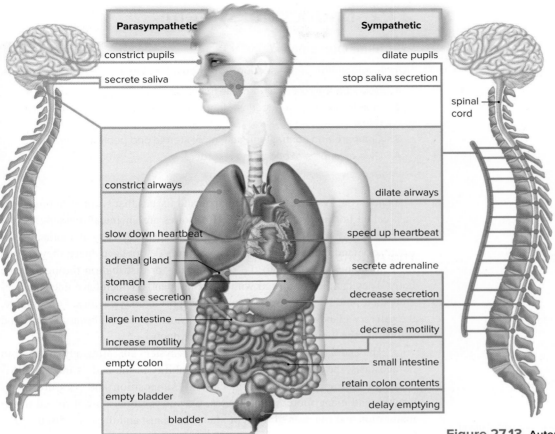

Parasympathetic	Sympathetic
constrict pupils	dilate pupils
secrete saliva	stop saliva secretion
	spinal cord
constrict airways	dilate airways
slow down heartbeat	speed up heartbeat
adrenal gland	secrete adrenaline
stomach	
increase secretion	decrease secretion
large intestine	
increase motility	decrease motility
empty colon	small intestine
	retain colon contents
empty bladder	delay emptying
bladder	

Figure 27.13 Autonomic system.

The parasympathetic and sympathetic motor axons go to the same organs, but they have opposite effects. The parasympathetic division is active when we, as mammals, feel comforted. The sympathetic division is active when we are stressed and feel threatened.

Axons of the **sympathetic division** arise from portions of the spinal cord. The sympathetic division is especially important during emergency situations and is associated with "fight or flight." If you need to fend off a foe or flee from danger, active muscles require a ready supply of glucose and oxygen. On one hand, the sympathetic division accelerates the heartbeat and dilates the airways. On the other hand, the sympathetic division inhibits the digestive tract, since digestion is not an immediate necessity if you are under attack. The sympathetic nervous system uses the neurotransmitter norepinephrine, which has a structure like that of epinephrine (adrenaline) released by the adrenal medulla.

Check Your Progress 27.1

1. List the three major types of neurons and their functions.

2. Describe the anatomy of a neuron.

3. List the steps in the generation of an action potential.

4. Explain how a signal is carried across the synaptic cleft.

5. Identify the structures of the central nervous system, and provide a function for each.

6. Summarize the events that occur during a reflex action.

7. Explain why the parasympathetic division of the autonomic system is called the "housekeeper division."

CONNECTING THE CONCEPTS

27.1 The nervous system uses nerve impulses and chemical signals to control the functions of the rest of the body.

27.2 Endocrine System

Learning Outcomes

Upon completion of this section, you should be able to

1. Describe the role of a hormone in the body.
2. Identify the major endocrine glands of the body, and summarize their roles.
3. Compare and contrast the actions of steroid and peptide hormones.
4. Give examples of endocrine disorders.

The endocrine system consists of glands and tissues that secrete chemical signals called hormones (**Fig. 27.14**). **Hormones** are chemical messengers that are released by one gland in the body and regulate the activity of another organ, tissue, or gland in the body. **Endocrine glands** do not have ducts; they secrete their hormones directly into the bloodstream for distribution throughout the body. They can be contrasted with exocrine glands, which have ducts and secrete their products into these ducts for transport to body cavities. For example, the salivary glands (see Section 25.1) are exocrine glands, because they send saliva into the mouth by way of the salivary ducts.

The endocrine system and the nervous system are intimately involved in homeostasis, the maintenance of relative stability in the body's internal environment. Hormones directly affect blood composition and pressure, body growth, and many more life processes. Certain hormones are involved in the maturation and function of the reproductive organs, and these are discussed in Section 29.2.

The Action of Hormones

The cells that can respond to a hormone have receptor proteins that bind to the hormone. Hormones cause these cells to undergo a metabolic change. The type of change is dependent on the chemical structure of the hormone. **Steroid hormones** are lipids, and they can pass through the plasma membrane. The hormone-receptor complex then binds to DNA, and gene expression follows—for example, a protein (such as an enzyme) is made by the cell. The enzyme goes on to produce a change in the target cell's function (**Fig. 27.15a**).

Since steroid hormones can pass through the membrane of every cell in the body, they are typically used to control changes that need to occur at an organismal level. For example, the steroid hormone testosterone is a sex hormone that is primarily involved in directing the development of male sexual characteristics.

Peptide hormones comprise peptides, proteins, glycoproteins, and modified amino acids. Peptide hormones can't pass through the plasma membrane, so they bind to a receptor protein in the plasma membrane (Fig. 27.15b). The peptide hormone is called the "first messenger," because a signal transduction pathway leads to a second molecule—that is, the "second messenger"—that changes the metabolism of the cell. The second messenger sets in motion an enzyme pathway, which is sometimes called an enzyme cascade because each enzyme in turn activates another. Because enzymes work over and over, every step in an enzyme cascade leads to more reactions—the binding of a single peptide hormone molecule can result in as much as a thousandfold response. Since peptide hormones interact with receptors, they may be specific with regard to the types of cells they target. For example, insulin, a

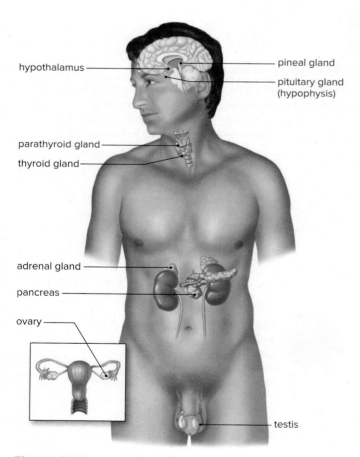

hypothalamus

pineal gland

pituitary gland (hypophysis)

parathyroid gland

thyroid gland

adrenal gland

pancreas

ovary

testis

Figure 27.14 The endocrine system.

Locations of major endocrine glands in the body. The endocrine functions of the pineal gland and sex organs (ovaries and testes) are covered in Sections 27.1 and 29.2, respectively.

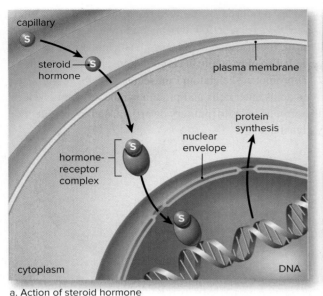

a. Action of steroid hormone

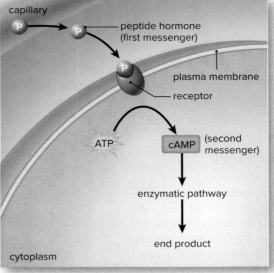

b. Action of peptide hormone

Figure 27.15 How hormones work.

a. A steroid hormone (S) is a chemical signal that is able to enter the cell through the plasma membrane. Reception of this messenger causes the cell to synthesize a product by way of the cellular machinery for protein synthesis. **b.** A peptide hormone (P) is a "first messenger," which is received by a cell at the plasma membrane. Reception of the first messenger starts a signal transduction pathway that leads to a "second messenger," which changes the metabolism of the cell.

peptide hormone, has a different effect on muscle cells than on adipose tissue cells due to the larger number of insulin receptors on the surfaces of muscle cells.

Hypothalamus and Pituitary Gland

The hypothalamus, a part of the brain (see Fig. 27.8*b*), helps regulate the internal environment. For example, it is on the receiving end of information about the heartbeat and body temperature. And to correct any abnormalities, the hypothalamus communicates with the medulla oblongata, where the brain centers that control the autonomic system are located. The hypothalamus is also a part of the endocrine system, containing specialized hormone-secreting neurons. It controls the glandular secretions of the **pituitary gland,** a small gland connected to the brain by a stalklike structure. The pituitary has two portions, the anterior pituitary and the posterior pituitary, which are distinct from each other.

Anterior Pituitary

The hypothalamus controls the **anterior pituitary** by producing *hypothalamic-releasing hormones,* most of which act by stimulating the action of other glands (**Fig. 27.16**).

1. *Thyroid-stimulating hormone* (*TSH*) stimulates the thyroid to produce triiodothyronine (T_3) and thyroxine (T_4).
2. *Adrenocorticotropic hormone* (*ACTH*) stimulates the adrenal cortex to produce the glucocorticoids.
3. *Gonadotropic hormones* (*FSH* and *LH*) stimulate the gonads—the testes in males and the ovaries in females—to produce gametes and sex hormones.

The hypothalamic-releasing hormones are kept in balance by a three-tiered control system that uses negative-feedback mechanisms (**Fig. 27.17**). For example, the secretion of thyroid-releasing hormone (TRH) by the hypothalamus stimulates the thyroid to produce the thyroid-stimulating hormone (TSH), and the thyroid produces its hormones (T_3 and T_4), which gives feedback to inhibit the release of the first two hormones mentioned.

Connections: Health

How is labor induced if a woman's pregnancy extends past her due date?

After the woman is given medication to prepare the birth canal for delivery, pitocin (a synthetic version of oxytocin) is used to induce labor. During labor, it may also be given to increase the strength of the contractions. Stronger contractions speed the labor process, if necessary (for example, if the woman's uterus is contracting poorly or if the health of the mother or child is at risk during delivery). Pitocin is routinely used following delivery to minimize postpartum bleeding by ensuring that strong uterine contractions continue. Administration of pitocin must be monitored carefully, because it may cause excessive uterine contractions. Should this occur, the uterus could tear itself. Further, a reduced blood supply to the fetus, caused by very strong contractions, may be fatal to the baby. Although it reduces the duration of labor, induction with pitocin can be very painful for the mother. Whenever possible, gentler and more natural methods should be used to induce labor and/or strengthen contractions.

Two other hormones produced by the anterior pituitary do not affect other endocrine glands. *Prolactin (PRL)* is produced in quantity during pregnancy and after childbirth. It causes the mammary glands in the breasts to develop and produce milk. It also plays a role in carbohydrate and fat metabolism. *Growth hormone (GH)* promotes skeletal and muscular growth. It stimulates the rate at which amino acids enter cells and protein synthesis occurs. Underproduction of growth hormone leads to pituitary dwarfism, and overproduction can lead to pituitary gigantism.

Figure 27.16   **The hypothalamus and pituitary.**

(*Left*) The hypothalamus controls the secretions of the anterior pituitary, and the anterior pituitary controls the secretions of the thyroid gland, adrenal cortex, and gonads, which are also endocrine glands. Growth hormone and prolactin are also produced by the anterior pituitary. (*Right*) The hypothalamus produces two hormones, ADH and oxytocin, which are stored and secreted by the posterior pituitary.

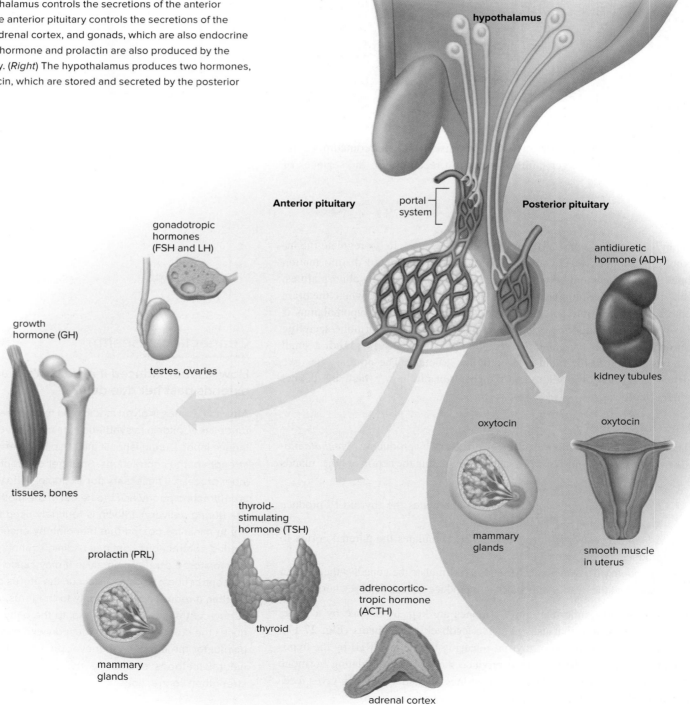

Posterior Pituitary

The hypothalamus produces two hormones, *antidiuretic hormone* (*ADH*) and *oxytocin* (see Fig. 27.16). The axons of hypothalamic secretory neurons extend into the **posterior pituitary,** where hormones are stored in the axon terminals. When the hypothalamus determines that the blood is too concentrated, ADH is released from the posterior pituitary. On reaching the kidneys, ADH causes water to be reabsorbed. As the blood becomes dilute, ADH is no longer released. This is also an example of control by negative feedback, because the effect of the hormone (to dilute blood) is to shut down the hormone's release. Negative feedback, as discussed in Section 22.3, maintains homeostasis.

Oxytocin, the other hormone made in the hypothalamus, causes uterine contraction during childbirth and milk letdown (ejection) when a baby is nursing.

Thyroid and Parathyroid Glands

The **thyroid gland** is a large gland located in the neck. It produces the hormones *calcitonin, triiodothyronine* (T_3), which contains three iodine atoms, and *thyroxine* (T_4), which contains four iodine atoms. As we will see below, calcitonin is involved in calcium homeostasis. Triiodothyronine and thyroxine are produced by the thyroid gland using iodine. The concentration of iodine in the thyroid gland can increase to as much as 25 times that in the blood. If iodine is lacking in the diet, the thyroid gland is unable to produce the thyroid hormones. In response to constant stimulation by the anterior pituitary, the thyroid enlarges, resulting in an **endemic goiter** (**Fig. 27.18***a*). Some years ago, it was discovered that the use of iodized salt (table salt to which iodine has been added) helps prevent endemic goiter.

Thyroid hormones increase the metabolic rate. They do not have a single, specific target organ; instead, they stimulate all the cells of the body to metabolize at a faster rate. More glucose is broken down, and more energy is used.

In the case of hyperthyroidism (oversecretion of thyroid hormone), or Graves disease, the thyroid gland is overactive, and bulging of the eyes, called *exophthalmos,* results (Fig. 27.18*b*). The eyes protrude because of swelling in the eye socket tissues and in the muscles that move the eyes. The patient usually becomes hyperactive, nervous, and irritable and suffers from insomnia. The removal or destruction of a portion of the thyroid by means of radioactive iodine is sometimes effective in curing the condition.

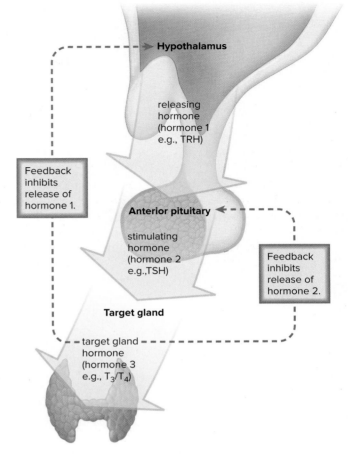

Figure 27.17 Negative feedback inhibition.

The hormones secreted by the thyroid (as well as the adrenal cortex and gonads) provide feedback to inhibit the anterior pituitary and hypothalamic-releasing hormones, so that their blood levels stay relatively constant.

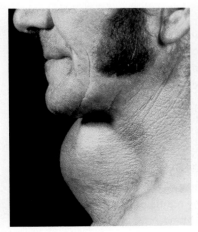
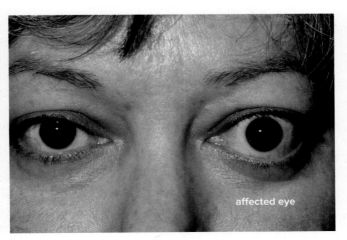

a.   b.

Figure 27.18 Conditions involving the thyroid gland.

a. Endemic goiter. An enlarged thyroid gland can result from too little iodine in the diet. **b.** Exophthalmos. Enlargement of the thyroid gland due to hyperthyroidism can cause the eyes to protrude, a condition called exophthalmos. In this individual, only the left eye is affected.

(a): © Biophoto Associates/Science Source; (b): © Dr. P. Marazzi/SPL/Science Source

Calcium Regulation

The thyroid gland also produces **calcitonin,** a hormone that helps regulate the blood calcium level. Calcium (Ca^{2+}) plays a significant role in both nervous conduction and muscle contraction. It is also necessary for blood clotting. Calcitonin temporarily reduces the activity and number of osteoclasts, cells that break down bone. Therefore, more calcium is deposited in bone. When the blood calcium level returns to normal, the thyroid's release of calcitonin is inhibited by negative feedback. However, a low level of blood calcium stimulates the parathyroid glands' release of **parathyroid hormone (PTH).** The parathyroid glands are embedded in the posterior surface of the thyroid gland. Many years ago, the four **parathyroid glands** were sometimes mistakenly removed during thyroid surgery because of their size and location.

Parathyroid hormone promotes the activity of osteoclasts and the release of calcium from the bones. PTH also promotes the reabsorption of calcium by the kidneys, where it activates vitamin D. Vitamin D, in turn, stimulates the absorption of calcium from the small intestine. These effects bring the blood calcium level back to the normal range, so that the parathyroid glands no longer secrete PTH.

When insufficient parathyroid hormone production leads to a dramatic drop in the blood calcium level, tetany results. In **tetany,** the body shakes from continuous muscle contraction. This effect is brought about by increased excitability of the nerves, which in turn initiates nerve impulses spontaneously and without rest.

Adrenal Glands

Two **adrenal glands** sit atop the kidneys. Each adrenal gland consists of an inner portion called the **adrenal medulla** and an outer portion called the **adrenal cortex.** These portions, like the anterior pituitary and the posterior pituitary, have no functional connection with one another.

The hypothalamus exerts control over the activity of both portions of the adrenal glands. It initiates nerve impulses that travel by way of the brain stem, spinal cord, and sympathetic nerve fibers to the adrenal medulla, which then secretes its hormones. The hypothalamus, by means of ACTH-releasing hormone, controls the anterior pituitary's secretion of ACTH, which in turn stimulates the adrenal cortex. Stress of all types—including emotional and physical trauma, and even vigorous exercise—prompts the hypothalamus to stimulate both the adrenal medulla and the adrenal cortex.

Adrenal Medulla

Epinephrine (adrenaline) and *norepinephrine* (noradrenaline) produced by the adrenal medulla rapidly bring about all the body changes that occur when an individual reacts to an emergency situation. In so doing, these two hormones complement the actions of the sympathetic autonomic system. The effects of these hormones are short-term. In contrast, the hormones produced by the adrenal cortex provide a long-term response to stress.

Adrenal Cortex

The two major types of hormones produced by the adrenal cortex are the mineralocorticoids, such as *aldosterone,* and the glucocorticoids, such as *cortisol.* Aldosterone acts on the kidneys and thereby regulates salt and water balance, leading to increases in blood volume and blood pressure. Cortisol regulates carbohydrate, protein, and fat metabolism, leading to an increase in the blood glucose level. It is also an anti-inflammatory agent. The adrenal cortex also secretes small amounts of both male and female sex hormones in both sexes.

When the level of adrenal cortex hormones is low due to hyposecretion, a person develops **Addison disease** (**Fig. 27.19**). ACTH may build up as more is secreted to attempt to stimulate the adrenal cortex. The excess can cause bronzing of the skin, because ACTH in excess stimulates melanin production. Without cortisol, glucose cannot be replenished when a stressful situation arises. Even a mild infection can lead to death. The lack of aldosterone results in the loss of sodium and water by the kidneys, low blood pressure, and possibly severe dehydration. Left untreated, Addison disease can be fatal.

When the level of adrenal cortex hormones is high due to hypersecretion, a person develops **Cushing syndrome.** The excess cortisol results in a tendency toward diabetes mellitus as muscle protein is metabolized and subcutaneous fat is deposited in the midsection. Excess production of adrenal male sex hormones in women may result in masculinization, including an increase in body hair, deepening of the voice, and beard growth. An excess of aldosterone and reabsorption of sodium and water by the kidneys lead to a basic blood pH and hypertension. The face swells and takes on a moon shape. Masculinization may occur in women because of excess adrenal male sex hormones.

Pancreas

The **pancreas** is composed of two types of tissue. Exocrine tissue produces and secretes digestive juices that pass through ducts to the small intestine. Endocrine tissue, called the **pancreatic islets** (islets of Langerhans), produces and secretes the hormones *insulin* and *glucagon* directly into the blood (**Fig. 27.20**).

Insulin is secreted when there is a high blood glucose level, which usually occurs just after eating. Insulin stimulates the uptake of glucose by cells, especially liver cells, muscle cells, and adipose tissue cells. In liver and muscle cells, glucose is then stored as glycogen. In muscle cells, glucose supplies energy for ATP production, leading to protein metabolism and muscle contraction. In fat cells, the breakdown of glucose supplies glycerol and acetyl groups for the formation of fat. In these ways, insulin lowers the blood glucose level.

Glucagon is usually secreted between meals, when the blood glucose level is low. The major target tissues of glucagon are the liver and adipose tissue. Glucagon stimulates the liver to break down glycogen to glucose and to use fat and protein in preference to glucose as energy sources. The use of fat and protein spares glucose and makes more available to enter the blood. In these ways, glucagon raises the blood glucose level.

Diabetes Mellitus

Diabetes mellitus is a fairly common hormonal disease in which the cells of the body do not take up and/or metabolize glucose. Therefore, the cells are in need of glucose, even though there is plenty in the blood. As the blood glucose level rises, water and glucose are excreted in the urine. The loss of water in this way causes the diabetic person to be extremely thirsty.

There are two types of diabetes mellitus. In *type 1 diabetes* (insulin-dependent diabetes), the pancreas is not producing insulin. The condition is believed to be brought on by exposure to an environmental agent, most likely a virus, whose presence causes cytotoxic T cells to destroy the pancreatic islets. The cells turn to the breakdown of protein and fat for energy. The metabolism of fat leads to acidosis (acid blood), which can eventually cause coma and death. As a result, the individual must have daily insulin injections. These injections control

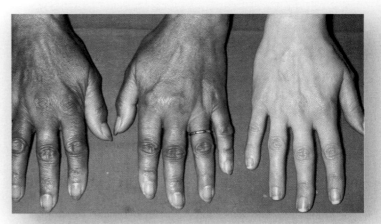

Figure 27.19 **Addison disease.**

Addison disease is characterized by a bronzing of the skin, particularly noticeable in light-skinned individuals. Note the color of the hands compared with the hand of an individual without the disease.

© BSIP/Science Source

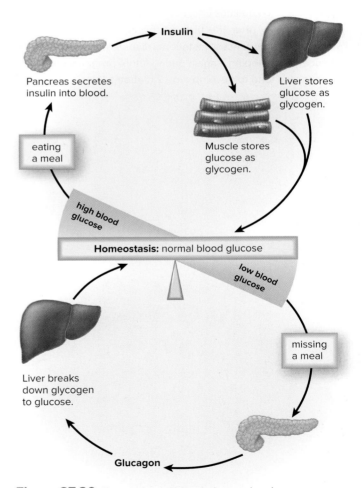

Insulin

Pancreas secretes insulin into blood.

Liver stores glucose as glycogen.

eating a meal

Muscle stores glucose as glycogen.

high blood glucose

Homeostasis: normal blood glucose

low blood glucose

missing a meal

Liver breaks down glycogen to glucose.

Glucagon

Figure 27.20 **Regulation of blood glucose level.**

Insulin is regulated by negative feedback. When blood glucose level is low, insulin is no longer secreted. The effect of insulin is countered by glucagon, which raises the blood glucose level. The two hormones work together to keep the blood glucose level relatively constant.

Connections: Health

What is gestational diabetes, and what causes it?

Women who were not diabetic prior to pregnancy but have high blood glucose levels during pregnancy have gestational diabetes. Gestational diabetes affects a small percentage of pregnant women. This form of diabetes is caused by insulin resistance—body insulin concentration is normal, but the cells fail to respond normally. Gestational diabetes and insulin resistance generally develop later in the pregnancy. Carefully planned meals and exercise often control this form of diabetes, but insulin injections may be necessary. If the woman is not treated, additional glucose crosses the placenta, causing high blood glucose in the fetus. The extra energy in the fetus is stored as fat, resulting in macrosomia, or a "fat" baby. Delivery of a very large baby can be dangerous for both the infant and the mother. Cesarean section is often necessary. Complications after birth are common for these babies. Further, there is a greater risk that the child will become obese and develop type 2 diabetes mellitus later in life. Usually, gestational diabetes goes away after the birth of the child. However, once a woman has experienced gestational diabetes, she has a greater chance of developing it again during future pregnancies. These women also tend to develop type 2 diabetes later in life.

the diabetic symptoms but can still cause inconveniences, since either taking too much insulin or failing to eat regularly can bring on the symptoms of hypoglycemia (low blood sugar). These symptoms include perspiration, pale skin, shallow breathing, and anxiety. Because the brain requires a constant supply of glucose, unconsciousness can result. The treatment is quite simple: Immediate ingestion of a sugar cube or fruit juice can very quickly counteract hypoglycemia.

Of the 29.1 million people who now have diabetes in the United States, most have *type 2 diabetes* (non-insulin-dependent diabetes). This type of diabetes mellitus usually occurs in people of any age who are obese and inactive (see Section 25.5). The pancreas produces insulin, but the liver and muscle cells do not respond to it in the usual manner. These cells are said to be insulin-resistant. If type 2 diabetes is untreated, the results can be as serious as those of type 1. People with diabetes of either type are prone to blindness, kidney disease, and circulatory disorders. It is usually possible to prevent or at least control type 2 diabetes by adhering to a low-fat and low-sugar diet and exercising regularly.

Table 27.1 summarizes the major organs and glands of the endocrine system, the hormones they produce, and their effects in the body.

Table 27.1 Summary of the Endocrine System

Hormones	Effects in the Body
HYPOTHALAMUS AND POSTERIOR PITUITARY	
Releasing and inhibiting hormones	Control anterior pituitary
Antidiuretic hormone (ADH)	Released by posterior pituitary; causes water uptake by kidneys
Oxytocin	Released by posterior pituitary; causes uterine contractions
ANTERIOR PITUITARY	
Gonadotropic hormones	Stimulate gonads
Thyroid-stimulating hormone (TSH)	Stimulates thyroid
Adrenocorticotropic hormone (ACTH)	Stimulates adrenal cortex
Prolactin	Causes milk production
Growth hormone (GH)	Causes cell division, protein synthesis, bone growth
THYROID GLAND	
Thyroxine (T_4) and triiodothyronine (T_3)	Increase metabolic rate
Calcitonin	Lowers blood calcium level
PARATHYROID GLANDS	
Parathyroid hormone (PTH)	Raises blood calcium level
ADRENAL MEDULLA	
Epinephrine and norepinephrine	Response to emergency situations
ADRENAL CORTEX	
Mineralocorticoids (aldosterone)	Cause kidneys to reabsorb Na^+
Glucocorticoids (cortisol)	Raises blood glucose level
PANCREAS	
Insulin	Causes uptake of glucose by cells and formation of glycogen in liver
Glucagon	Causes liver to break down glycogen

Check Your Progress 27.2

1. Explain how steroid and peptide hormones induce metabolic changes in cells.

2. Explain how the hypothalamus and pituitary gland work to control the endocrine system.

3. Summarize how negative feedback mechanisms are involved in the function of the endocrine system.

4. Summarize the role of each of the endocrine glands.

5. Explain how hormones aid in homeostasis.

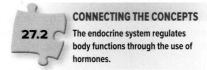

CONNECTING THE CONCEPTS

27.2 The endocrine system regulates body functions through the use of hormones.

STUDY TOOLS http://connect.mheducation.com

SMARTBOOK® Maximize your study time with McGraw-Hill SmartBook®, the first adaptive textbook.

SUMMARIZE

The nervous system controls the functions of other organ systems in the body. The endocrine system, controlled by the nervous system, regulates the body's internal environment.

The nervous system uses nerve impulses and chemical signals to control the functions of the rest of the body.

The endocrine system regulates body functions through the use of hormones.

27.1 Nervous System

The **nervous system** and **endocrine system** work to control the activities of the other organ systems of the body. The nervous system is organized into the **central nervous system (CNS)** and the **peripheral nervous system (PNS).**

Neurons

- A **neuron** is composed of a **cell body,** an **axon,** and **dendrites.** Only axons conduct nerve impulses. An axon may be covered by a **myelin sheath,** with small gaps called the **nodes of Ranvier. Nerves** are collections of axons from multiple neurons.

- Types of neurons are **sensory neurons,** which transmit nerve impulses to the CNS; **interneurons,** which carry nerve impulses between neurons of the CNS; and **motor neurons,** which carry nerve impulses away from the CNS.

The Nerve Impulse

- A **nerve impulse,** or **action potential,** moves a signal along an axon. The sodium-potassium pump transports Na^+ out of the axon and K^+ into the axon. The inside of the axon has a negative charge; the outside has a positive charge.

- During the nerve impulse, there is a reversal of charge as Na^+ flows in, then a return to the previous charge difference as K^+ flows out of an axon.

- The nerve impulse travels much faster in myelinated axons, because the impulse jumps from node to node.

The Synapse

- The **synapse** is a region of close proximity between an axon terminal and the next neuron (in the CNS) or between an axon terminal and a muscle cell (in the PNS).

- A nerve impulse causes the release of a **neurotransmitter** (excitatory or inhibitory) into the synaptic cleft. **Integration** of the excitatory and inhibitory signals determines whether the neuron will transmit the nerve impulse.

- Neurotransmitters are ordinarily removed quickly from the **synaptic cleft.**

- Many drugs affect the actions of neurotransmitters.

The Central Nervous System

The central nervous system consists of the brain and spinal cord. These are connected by tracts, long connections of interneurons.

The Brain

- The cerebrum functions in sensation, reasoning, learning and memory, language, and speech. The **cerebral cortex** has a **primary sensory area** in the parietal lobe, which receives sensory information from each part of the body, and a **primary motor area** in the frontal lobe, which sends out motor commands to skeletal muscles. Association areas, such as the **prefrontal area,** carry out integration.

- In the diencephalon, the **hypothalamus** helps control homeostasis; the **thalamus** specializes in sending sensory input to the cerebrum.

- The **cerebellum** primarily coordinates skeletal muscle contractions.

- In the **brain stem,** the **medulla oblongata** has centers for vital functions, such as breathing and the heartbeat, and helps control the internal organs. The midbrain contains visual and auditory areas, and the **pons** controls breathing.

The Limbic System

The **limbic system** blends higher functions into a united whole. The hippocampus and amygdala have roles in learning and **memory** and appear to be affected in **Alzheimer disease.**

The Peripheral Nervous System

The peripheral nervous system (PNS) connects the sensory neurons with the CNS. It contains nerves that are organized into ganglions. Examples of these nerves are cranial nerves and spinal nerves. The PNS is divided into two systems:

- **Somatic system: Reflexes** (automatic responses) involve a sensory receptor, a sensory neuron, interneurons in the spinal cord, and a motor neuron.
- **Autonomic system:** The **sympathetic division** is active during times of stress, and the **parasympathetic division** is active during times of relaxation. Both divisions control the same internal organs.

27.2 Endocrine System

Endocrine glands secrete **hormones** into the bloodstream for distribution to target organs or tissues. Hormones are classified as either **steroid hormones** or **peptide hormones.**

Major Organs and Glands

The major functions of the hormones secreted by the endocrine glands are listed in Table 27.1.

- Hypothalamus and pituitary gland: Together these control the internal environment of the body. The hypothalamus links the endocrine and nervous systems. The pituitary gland contains two parts, the **anterior pituitary** and the **posterior pituitary.** These release hormones that regulate the other endocrine glands, usually by negative feedback mechanisms.
- Thyroid and parathyroid glands: The **thyroid gland** regulates metabolism. Hyper- or hypothyroidism may result in **endemic goiter** or exophthalmos. The thyroid and **parathyroid glands** are involved in calcium homeostasis. Insufficient parathyroid gland activity may result in **tetany.**
- Adrenal glands: The adrenal glands consist of the **adrenal cortex** and the **adrenal medulla.** Problems with hormones secreted by the adrenal cortex may result in **Addison disease** or **Cushing syndrome.**
- Pancreas: The pancreas is both an exocrine and an endocrine gland. The **pancreatic islets** contain endocrine cells that secrete the hormones insulin and glucagon, both of which are involved in glucose homeostasis. **Diabetes mellitus** is a disease caused by a failure of glucose homeostasis.

ASSESS

Testing Yourself

Choose the best answer for each question.

27.1 Nervous Systems

1. The sympathetic division of the autonomic nervous system
   a. increases heart rate and digestive activity.
   b. decreases heart rate and digestive activity.
   c. causes pupils to constrict.
   d. None of these are correct.

2. When the action potential begins, Na^+ gates open, allowing Na^+ to cross the membrane. The charge difference across the axon membrane changes to
   a. negative outside and positive inside.
   b. positive outside and negative inside.
   c. neutral outside and positive inside.
   d. There is no difference in charge between outside and inside.

3. Transmission of a nerve impulse across a synapse is accomplished by the
   a. movement of Na^+ and K^+.
   b. release of a neurotransmitter by a dendrite.
   c. release of a neurotransmitter by an axon.
   d. release of a neurotransmitter by a cell body.
   e. All of these are correct.

4. The autonomic system has two divisions, called the
   a. CNS and PNS.
   b. somatic and skeletal systems.
   c. efferent and afferent systems.
   d. sympathetic and parasympathetic divisions.

5. This area of the brain is responsible for combining higher level brain functions with emotional responses.
   a. the cerebellum
   b. the brain stem
   c. the limbic system
   d. the prefrontal cortex
   e. the diencephalon

6. The _____ regulates heartbeat, breathing, and blood pressure.
   a. cerebrum   c. cerebellum
   b. diencephalon   d. brain stem

27.2 Endocrine System

7. Unlike the nervous system, the endocrine system
   a. uses chemical signals as a means of communication.
   b. helps maintain equilibrium.
   c. sends messages to target organs.
   d. changes the metabolism of cells.

8. Both the adrenal medulla and the adrenal cortex are
   a. endocrine glands.
   b. found in the same organ.
   c. involved in our response to stress.
   d. All of these are correct.

9. PTH causes the blood level of calcium to _____, and calcitonin causes it to _____.
   a. increase, not change
   b. increase, decrease
   c. decrease, also decrease
   d. decrease, increase
   e. not change, increase

10. Type 1 diabetes is thought to result from a virus that
    a. interferes with gene expression in pancreatic cells.
    b. causes T cells to destroy the pancreatic islets.
    c. breaks down insulin.
    d. prevents the secretion of insulin by the pancreatic islets.

11. The _____ acts as the link between the nervous system and the endocrine system.
    a. parathyroid gland
    b. hypothalamus
    c. cerebellum
    d. pancreas

12. Cushing syndrome and Addison disease are associated with the
    a. thyroid.   c. pancreas.
    b. adrenal glands.   d. anterior pituitary.

ENGAGE

BioNOW

Want to know how this science is relevant to your life? Check out the BioNow video below.

- Quail Hormones

From your understanding of the endocrine system in humans, what endocrine glands and hormones are most likely involved in this response of the quails to the changes in their environment?

Thinking Critically

1. Recent research indicates that Parkinson disease damages the sympathetic division of the peripheral nervous system. One test for sympathetic division function, called the Valsalva maneuver, requires the patient to blow against resistance. A functional nervous system will compensate for the decrease in blood output from the heart by constricting blood vessels. How do you suppose Parkinson patients respond to the Valsalva maneuver? How does this relate to a common condition in Parkinson patients called orthostatic hypotension, in which blood pressure falls suddenly when the person stands up, leading to dizziness and fainting?

2. Researchers have been trying to determine the reason for a dramatic rise in type 2 diabetes in recent decades. A sedentary lifestyle and poor eating habits certainly contribute to the risk of developing the disease. In addition, however, some researchers have observed a connection between childhood vaccinations and type 1 diabetes. Epidemiological data from countries that have recently initiated mass immunization programs indicate that the incidence of type 1 diabetes has increased there as well. What might be the connection between vaccination and diabetes?

3. Scientists studying orangutans on the Indonesian islands of Borneo and Sumatra recently found that the orangutans inhabiting the food-scarce island of Borneo have smaller brains than those on Sumatra, where food is abundant. From this observation, what can we infer about the relationship between brain size and diet in orangutan evolution?

4. The Food and Drug Administration (FDA) approved the use of human growth hormone (hGH) to treat "short stature." This decision implies that short stature is a medical condition presenting a legitimate need to be treated. In this case, medical care is not treating a disease but changing a feature of an otherwise healthy person. Proponents of the FDA decision say that administering hGH to short people will help them avoid discrimination and live a more normal life in a world designed for taller people. Opponents say that this decision may lead to slippery-slope scenarios in which medical treatments that make us smarter or faster are assumed to make us better. Do you think hGH should be used to "cure shortness"?

5. At the other end of the life span, hGH has been widely advertised as a cure for aging. Levels of hGH naturally decline with age, and there is evidence that treating older adults with hGH boosts muscle mass and reduces body fat. However, the only effective way to administer hGH is by injection; the hGH pills hyped on numerous websites do not work. Injections of hGH are by prescription only and very expensive, around $1,000 a month. Furthermore, hGH therapy can have undesirable side effects, such as elevated blood sugar, fluid retention, and joint pain. Nevertheless, there is keen interest in hGH therapy, not only from those who wish to delay or avoid the effects of aging but also from athletes who wish to improve their performance. Should the physical decline of aging be accepted, or should we try to maintain a youthful body using hGH?

28

Sensory Input and Motor Output

© Mike Hewitt/Getty Images Sport/Getty Images

100-Meter Dash World Record Holder

At the 2012 Summer Olympics, Usain Bolt set a new Olympic record of 9.63 seconds in the 100-meter dash. However, this was not his personal best. In Berlin in 2009, Bolt set a world record of 9.58 seconds for the 100-meter dash. When Bolt is running at these incredible speeds, his brain is coordinating sensory input from his eyes, ears, and internal sensors (the sensory input) to maximize the efforts of his skeletal and muscular systems (the motor output). Like Usain Bolt, we all perform these functions daily, when we are walking to class, exercising, or even working on the computer.

The senses, skeletal system, and muscular system all contribute to homeostasis. Our senses provide us with information about the external environment. Aside from giving our bodies shape and protecting our internal organs, the skeleton serves as a storage area for inorganic calcium and produces blood cells. The skeleton also protects internal organs while supporting the body against the pull of gravity. While contributing to body movement, the skeletal muscles give off heat, which warms the body. In this chapter, we will explore how the senses provide information to the brain, as well as how the skeletal and muscular systems are involved in movement and support.

As you read through this chapter, think about the following questions:

1. How do the senses provide information to the brain?
2. How does the nervous system specifically control the muscular system?
3. How do the muscular and skeletal systems allow for movement?

OUTLINE

28.1 The Senses 537

28.2 The Motor Systems 545

BEFORE YOU BEGIN

Before beginning this chapter, take a few moments to review the following discussions.

Figure 5.4 What are the steps in the ATP cycle?

Figure 6.4 How does color relate to the wavelength of light?

Section 27.1 What areas of the brain are responsible for integrating sensory input and generating motor output?

28.1 The Senses

Learning Outcomes

Upon completion of this section, you should be able to

1. Summarize how the human senses of taste and smell work.
2. Describe the structures of the human ear and their functions.
3. Differentiate between rotational and gravitational equilibrium, and identify the organs responsible for each.
4. Describe the structures of the human eye and their functions.
5. Explain the purpose of cutaneous receptors and proprioceptors.

All living organisms respond to stimuli. Stimuli are environmental signals that tell us about the external environment or the internal environment. In Chapter 21, you learned that plants often respond to external stimuli, such as light, by changing their growth pattern. An animal's response often results in motion. Complex animals rely on sensory receptors to provide information to the central nervous system (brain and spinal cord), which integrates sensory input before directing a motor response (**Fig. 28.1**). Section 28.2 will explore how the muscular and skeletal systems are involved in the motor response.

Sense organs, as a rule, are specialized to receive one kind of stimulus. The eyes ordinarily respond to light, ears to sound waves, pressure receptors to pressure, and chemoreceptors to chemical molecules. Sensory receptors transform the stimulus into nerve wimpulses that reach a particular section of the cerebral cortex. Those from the eye reach the visual areas, and those from the ears reach the auditory areas. Before sensory receptors initiate nerve signals, they also carry out **integration,** the summing up of signals. One type of integration is called **sensory adaptation,** a decrease in response to a stimulus. We have all had the experience of smelling an odor when we first enter a room and then later not being aware of it. When sensory adaptation occurs, sensory receptors send fewer impulses to the brain. Without these impulses, the sensation of the stimuli is decreased. However, the brain, not the sensory receptor, is ultimately responsible for sensation and perception, and each part of the brain interprets impulses in only one way. For example, if by accident the photoreceptors of the eye are stimulated by pressure and not light, the brain causes us to see "stars" or other visual patterns.

Peripheral nervous system

stimulus

Sensory receptor produces nerve impulses.

Nerve impulses move along sensory fiber.

Central nervous system

spinal cord

brain

Figure 28.1 **Sensory input.**

After detecting a stimulus, sensory receptors initiate nerve impulses within the peripheral nervous system (PNS). These impulses give the central nervous system (CNS) information about the external and internal environments. The CNS integrates all incoming information and then initiates a motor response to the stimulus.

Chemical Senses

The fundamental functions of sensory receptors include helping animals stay safe, find food, and find mates. **Chemoreceptors** give us the ability to detect chemicals in the environment, which is believed to be our most primitive sense. Chemoreceptors occur almost universally in animals. For example, they are present all over the bodies of planarians (flatworms) but found in higher concentrations on the auricles at the sides of the head. Male moths have receptors for a sex attractant on their antennae. The receptors on the antennae of the male silkworm moth are so sensitive that only 40 out of 40,000 receptor proteins need to be activated in order for the male to respond to a chemical released by the female.

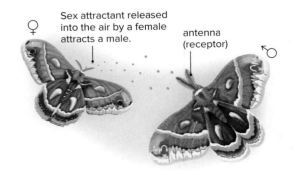

Sex attractant released into the air by a female attracts a male.

antenna (receptor)

Other insects, such as the housefly, have chemoreceptors largely on their feet—a fly tastes with its feet instead of its mouth. In mammals, the receptors for taste are located in the mouth, and the receptors for smell are in the nose.

Taste and Smell

In humans, *taste buds,* located primarily on the tongue, contain taste receptor cells, and the nose contains olfactory receptor cells (**Fig. 28.2***a*). Receptor proteins for chemicals are located on the microvilli of taste receptor cells and on the cilia of olfactory receptor cells (Fig. 28.2*b*). When molecules bind to these receptor proteins, nerve impulses are generated in sensory nerve fibers that go to the brain. When they reach the appropriate cortical areas, they are interpreted as taste and smell, respectively.

There are at least five primary types of tastes: bitter, sour, salty, sweet, and umami (Japanese for "savory" or "delicious"). Foods rich in certain amino acids, such as the common seasoning monosodium glutamate (MSG), as well as certain flavors of cheese, beef broth, and some seafood, produce the taste of umami. Taste buds for each primary taste are located throughout the tongue but may be concentrated in particular regions. A particular food can stimulate more than one of these types of taste buds. In this way, the response of taste buds can result in a range of sweet, sour, salty, umami, and bitter tastes. The brain appears to survey the overall pattern of incoming sensory impulses and take a "weighted average" of their taste messages as the perceived taste. Similarly, an odor contains many odor molecules, which activate a characteristic combination of receptor proteins. When this complex information is communicated to the cerebral cortex, we know we have smelled a rose—or an onion!

Our senses of taste and smell have evolved to meet our physiological needs. Foods that are rich in nutrients we require, such as fruits, have a favorable taste. The smell of food cooking triggers a reflex that starts the release of digestive juices. A revolting or repulsive substance in the mouth can initiate the gag reflex or even vomiting. Smell is even more important to our survival. Smells associated with danger, such as smoke, can trigger the fight-or-flight reflex. Unpleasant smells can cause us to sneeze or choke.

Have you ever noticed that a certain aroma vividly brings to mind a certain person or place? A person's perfume may remind you of someone else, or the smell of boxwood may remind you of your grandfather's farm. The olfactory bulbs have direct connections with the limbic system and its centers for emotions and memory. One researcher showed that, when subjects smelled an orange while viewing a painting, they not only remembered the painting when asked about it later but also had many deep feelings about it.

Hearing and Balance

The human ear has two sensory functions: hearing and balance (equilibrium). The sensory receptors for both of these consist of *hair cells* with long microvilli called stereocilia. These microvilli, unlike those of taste cells, are sensitive to mechanical stimulation. Therefore, these belong to a class of receptors called **mechanoreceptors.**

The similarity of the sensory receptors for balance and hearing and their presence in the same organ suggest an evolutionary relationship between them. In fact, the sense organs of the mammalian ear may have evolved from a type of sense organ in fishes.

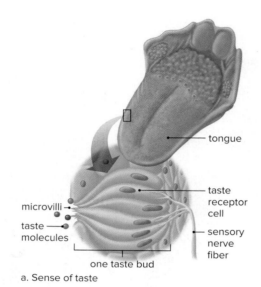

a. Sense of taste

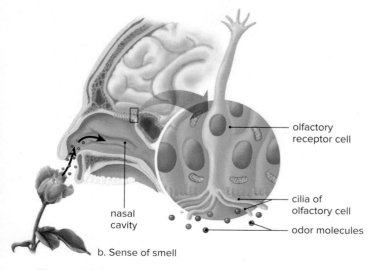

b. Sense of smell

Figure 28.2 **Senses of taste and smell.**

a. The sense of taste is due to receptor proteins on the microvilli of taste receptor cells on the tongue. **b.** The sense of smell is due to receptor proteins on the cilia of olfactory receptor cells.

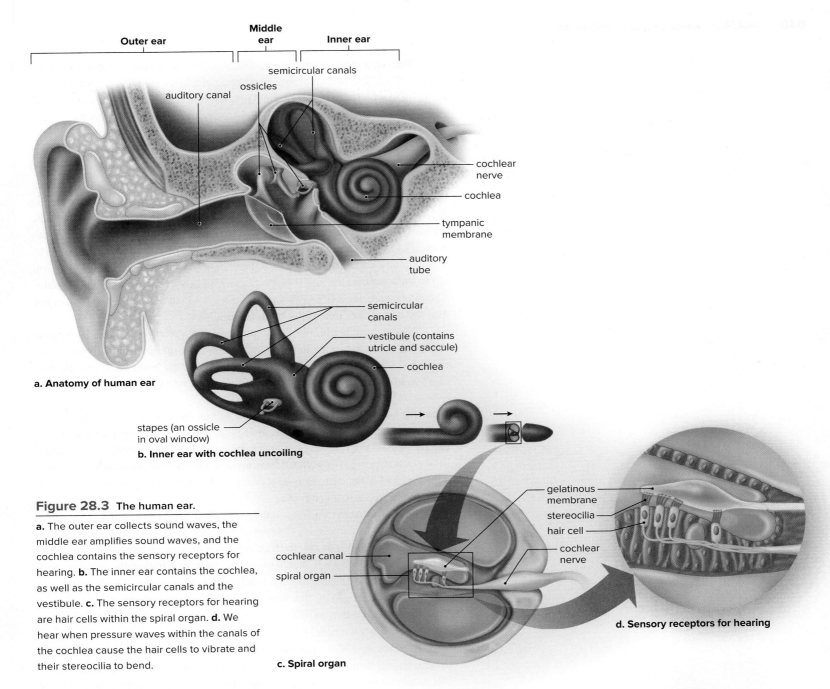

Outer ear

Middle ear

Inner ear

semicircular canals

ossicles

auditory canal

cochlear nerve

cochlea

tympanic membrane

auditory tube

a. Anatomy of human ear

semicircular canals

vestibule (contains utricle and saccule)

cochlea

stapes (an ossicle in oval window)

b. Inner ear with cochlea uncoiling

Figure 28.3 The human ear.

a. The outer ear collects sound waves, the middle ear amplifies sound waves, and the cochlea contains the sensory receptors for hearing. **b.** The inner ear contains the cochlea, as well as the semicircular canals and the vestibule. **c.** The sensory receptors for hearing are hair cells within the spiral organ. **d.** We hear when pressure waves within the canals of the cochlea cause the hair cells to vibrate and their stereocilia to bend.

cochlear canal

spiral organ

gelatinous membrane

stereocilia

hair cell

cochlear nerve

d. Sensory receptors for hearing

c. Spiral organ

Hearing

Most invertebrates cannot hear. Some arthropods, including insects, do have sound receptors but they are quite simple. In insects, the ear consists of a pair of air pockets, each enclosed by a membrane, called the tympanic membrane, that passes sound vibrations to sensory receptors. The human ear has a tympanic membrane also, but it is between the outer ear and middle ear (**Fig. 28.3***a*). The outer ear collects sound waves that cause the tympanic membrane to move back and forth (vibrate) ever so slightly. Three tiny bones in the middle ear (the ossicles) amplify the sound about 20 times as it moves from one to the other. The last of the ossicles (the stapes) strikes the membrane of the oval window, causing it to vibrate; in this way, the pressure is passed to a fluid within the hearing portion of the inner ear called the cochlea (Fig. 28.3*b*). The term *cochlea* means "snail shell." Specifically, the sensory receptors of hearing are located in the cochlear canal of the cochlea. The sensory receptors for hearing are hair cells whose stereocilia are embedded in a gelatinous membrane. Collectively, they are called the *spiral organ,* or *organ of Corti* (Fig. 28.3*c,d*).

The outer ear and middle ear, which collect and amplify sound waves, are filled with air. The auditory tube relieves pressure in the middle ear. But the inner ear is filled with fluid; therefore, fluid pressure waves actually stimulate the spiral organ. When the stapes strikes the membrane of the oval window, pressure waves cause the hair cells to move up and down, and the stereocilia of the hair cells embedded in the gelatinous membrane bend. The hair cells of the spiral organ synapse with the cochlear nerve, and when their stereocilia bend, nerve impulses begin in the cochlear nerve and travel to the brain stem. When these impulses reach the auditory areas of the cerebral cortex, they are interpreted as sound.

Each part of the spiral organ is sensitive to different wave frequencies, which correspond to the *pitch* of a sound. Near the tip, the spiral organ responds to low pitches, such as the sound of a tuba; near the base, it responds to higher pitches, such as a bell or whistle. The nerve fibers from each region along the length of the spiral organ lead to slightly different areas in the brain. The pitch sensation we experience depends on which region of the spiral organ vibrates and which area of the brain is stimulated.

Volume is a function of the amplitude of sound waves. Loud noises cause the spiral organ to vibrate to a greater extent. The brain interprets the resulting increased stimulation as volume. It is believed that the brain interprets the *tone* of a sound based on the distribution of the stimulated hair cells.

Hearing Loss   Especially when we are young, the middle ear is subject to infections that can lead to hearing impairment. It is quite common for youngsters to have "tubes" put into the tympanic membrane to allow the middle ear to drain, in an effort to prevent this type of hearing loss. The mobility of ossicles decreases with age, and if bone grows over the stapes, the only remedy is implantation of an artificial stapes that can move.

Deafness due to middle ear damage is called *conduction deafness*. Deafness due to spiral organ damage is called *nerve deafness*. In today's society, noise pollution is common, and even city traffic can be loud enough to damage the stereocilia of hair cells (**Fig. 28.4**). It stands to reason, then, that frequently attending rock concerts, constantly playing a stereo loudly, or using earphones at high volume can also damage hearing. The first hint of danger can be temporary hearing loss, a "full" feeling in the ears, muffled hearing, or tinnitus (ringing in the ears). If exposure to noise is unavoidable, noise-reduction earmuffs are available, as are earplugs made from compressible, spongelike material. These earplugs are not the same as those worn for swimming, and they should not be worn interchangeably. Finally, you should be aware that some medicines may damage the ability to hear. Anyone taking anticancer drugs, such as cisplatin, and certain antibiotics, such as streptomycin, should be especially careful to protect the ears from loud noises.

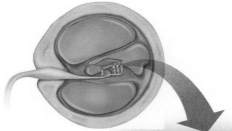

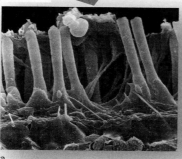

Figure 28.4 Effect of noise on hearing.

a. SEM of normal hair cells in the spiral organ. **b.** SEM of hair cells that have been damaged by excessive noise. Damaged cells cannot be replaced, so hearing is permanently impaired.

(a-b): © Dr. Goran Bredberg/Science Source

a.

b.

Balance

Humans have two senses of balance (equilibrium): rotational and gravitational. We are able to detect the rotational (angular) movement of the head as well as the straight-line movement of the head with respect to gravity.

Rotational equilibrium involves the semicircular canals (see Fig. 28.3*b*). In the base of each canal, hair cells have stereocilia embedded within a gelatinous membrane (**Fig. 28.5*a***). Because there are three semicircular canals, each responds to head movement in a different plane of space. As fluid within a semicircular canal flows over and displaces the gelatinous membrane, the stereocilia of the hair cells bend, and the pattern of impulses carried to the central nervous system (CNS) changes. These data, usually supplemented by vision, tell the brain how the head is moving. *Vertigo* is dizziness and a sensation of rotation. It is possible to bring on a feeling of vertigo by spinning rapidly and stopping suddenly. Now the person feels like the room is spinning because of sudden stimulation of stereocilia in the semicircular canals.

Gravitational equilibrium refers to the position of the head in relation to gravity. It depends on the *utricle* and *saccule,* two membranous sacs located in the inner ear (see Fig. 28.3*b*). Both of these sacs also contain hair cells with stereocilia in a gelatinous membrane (Fig. 28.5*b*). Calcium carbonate ($CaCO_3$) granules, called otoliths, rest on this membrane. When the head moves forward or back, up or down, the otoliths are displaced and the membrane moves, bending the stereocilia of the hair cells. This movement alters the frequency of nerve impulses to the CNS. These data, usually supplemented by vision, tell the brain the direction of the movement of the head.

Similar Receptors in Other Animals

Gravitational equilibrium organs, called *statocysts,* are found in several types of invertebrates, including cnidarians, molluscs, and crustaceans. These organs give information only about the position of the head; they are not involved in the sensation of movement (**Fig. 28.6*a***). When the head stops moving, a small particle called a statolith stimulates the cilia of the closest hair cells, and these cilia generate impulses, indicating the position of the head.

The *lateral line* system of fishes uses sense organs similar to those in the human inner ear (Fig. 28.6*b*). In bony fishes, the system consists of sense organs located within a canal that has openings to the outside. As you might expect, the sense organ is a collection of hair cells with cilia embedded in a gelatinous membrane. Water currents and pressure waves from nearby objects cause the membrane and the cilia of the hair cells to bend. Thereafter, the hair cells initiate nerve impulses that go to the brain. Fishes use these data not for hearing or balance but to locate other fish, including predators, prey, and mates.

Figure 28.6 Sensory receptors in other animals.

a. Invertebrates use statocysts to determine their position. When the statolith stops moving, cilia of the nearest hair cells are stimulated, telling the position of the head. **b.** The lateral line of fishes is not for hearing or balance, instead it is for knowing the location of other fishes.

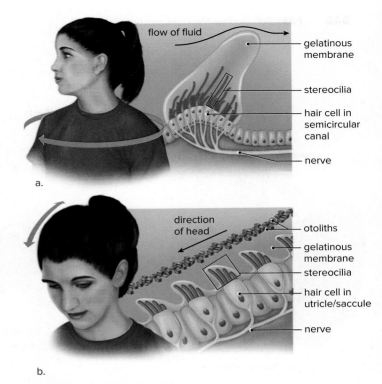

a.

b.

Figure 28.5 Sense of balance in humans.

a. Sensory receptors for rotational equilibrium. When the head rotates, the gelatinous membrane is displaced, bending the stereocilia. **b.** Sensory receptors for gravitational equilibrium. When the head bends, otoliths are displaced, causing the gelatinous membrane to sag and the stereocilia to bend.

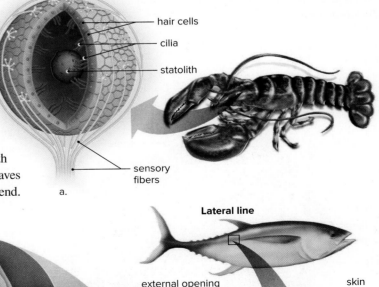

a.
b.

Vision

Sensory receptors that are sensitive to light are called **photoreceptors.** In planarians, eyespots allow these animals to determine only the direction of light. Other photoreceptors form actual images. Image-forming eyes provide information about an object, including how far away it is. Such detailed information is invaluable to an animal.

Among invertebrates, arthropods have *compound eyes* composed of many, independent visual units, each of which possesses a lens to focus light rays on photoreceptors (**Fig. 28.7**a). The image that results from all the stimulated visual units is crude because the small size of compound eyes limits the number of visual units, which still might number as many as 28,000.

Vertebrates (including humans) and certain molluscs, such as the squid and the octopus, have a *camera-type eye* (Fig. 28.7b). A single lens focuses light on photoreceptors, which number in the millions and are closely packed together within a retina. Since molluscs and vertebrates are not closely related, the camera-type eye evolved independently in each group.

Insects have color vision, but they make use of a slightly shorter range of the electromagnetic spectrum than humans do. However, they can see the longest of the ultraviolet rays, and this enables them to be especially sensitive to the reproductive parts of flowers, which have particular ultraviolet patterns. Some fishes and most reptiles are believed to have color vision, but among mammals, only humans and other primates have expansive color vision. It would seem, then, that this trait was adaptive for a diurnal lifestyle (being active during the day), which accounts for its retention in only a few mammals.

The Human Eye

When looking straight ahead, each of our eyes views the same object from a slightly different angle. This slight displacement of the images permits *binocular vision,* the ability to perceive three-dimensional images and to sense depth. Like the human ear, the human eye has numerous parts, many of which are involved in preparing the stimulus for the sensory receptors. In the case of the ear, sound wave energy is magnified before it reaches the sensory receptors. In the case of the eye, light rays are brought to a focus on the photoreceptors within the *retina.* As shown in **Figure 28.8,** the *cornea* and especially the *lens* are involved in focusing light rays on the photoreceptors. The *iris,* the colored part of the eye, regulates the amount of light that enters the eye by way of the *pupil.* The retina generates nerve impulses, which are sent to the visual part of the cerebral cortex; from this information, the brain forms an image of the object.

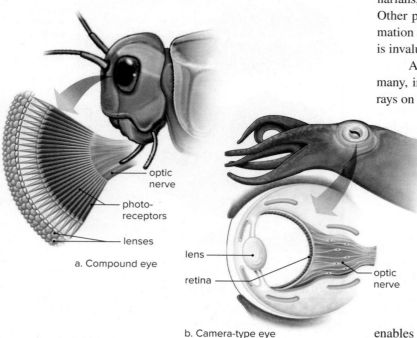

a. Compound eye

b. Camera-type eye

Figure 28.7 Eyes.

a. A compound eye has several visual units. Each visual unit has a lens that focuses light onto photoreceptor cells. **b.** A camera-type eye has one lens that focuses light onto a retina containing many photoreceptors.

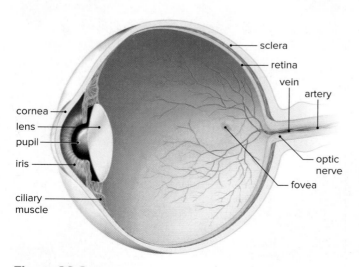

Figure 28.8 The human eye.

Light passes through the cornea and through the pupil (a hole in the iris) and is focused by the lens on the retina, which houses the photoreceptors.

The shape of the lens is controlled by the ciliary muscles. When we view a distant object, the lens remains relatively flat, but when we view a near object, the lens rounds up. With normal aging, the lens loses its ability to accommodate for near objects; therefore, many people need reading glasses when they reach middle age. Aging, and exposure to UV rays from the sun, also makes the lens subject to cataracts; the lens becomes opaque, incapable of transmitting light rays. Currently, surgery is the only viable treatment for cataracts.

Photoreceptors of the Eye

Figure 28.9 illustrates the structure of the photoreceptors in the human eye, which are called *rods* and *cones*. Both types of photoreceptors contain a visual pigment similar to that found in all types of eyes throughout the animal kingdom. The visual pigment in rods is a deep-purple pigment called rhodopsin. *Rhodopsin* is a complex molecule made up of the protein opsin and a light-absorbing molecule called *retinal,* which is a derivative of vitamin A. When a rod absorbs light, rhodopsin splits into opsin and retinal, leading to a cascade of reactions that ends in the generation of nerve impulses. Rods are very sensitive to light and therefore are suited to night vision. Because carrots are rich in vitamin A, it is true that eating carrots can improve your night vision. Rod cells are plentiful throughout the retina; therefore, they also provide us with peripheral vision and the perception of motion.

The cones, on the other hand, are located primarily in a part of the retina called the *fovea.* Cones are activated by bright light; they allow us to detect the fine detail and color of an object. Color vision depends on three different kinds of cones, which contain pigments called B (blue), G (green), and R (red) pigments. Each pigment is made up of retinal and opsin, but a slight difference in the opsin structure of each accounts for their specific absorption patterns. Various combinations of cones are believed to be stimulated by in-between shades of color.

Retina The retina has three layers of cells, and light has to penetrate through the first two layers to reach the photoreceptors (**Fig. 28.10**). The intermediate cells of the middle layer process and relay visual information from the photoreceptors to the ganglion cells that have axons forming the optic nerve. The relative sensitivity of cones versus rods is mirrored by how directly these two kinds of photoreceptors connect to ganglion cells. Information from several hundred rods may converge on a single ganglion cell, while cones show very little convergence. As signals pass through the layers of the retina, integration occurs. Integration improves the overall contrast and quality of the information sent to the brain, which uses the information to form an image of the object.

No rods and cones occur where the optic nerve exits the retina. Therefore, no vision is possible in this area. You can prove this to yourself by putting a very small dot to the right of center on a piece of paper. Close your left eye; then use your right hand to move the paper slowly toward your right eye while you look straight ahead. The dot will disappear at one point—this point is your *blind spot.*

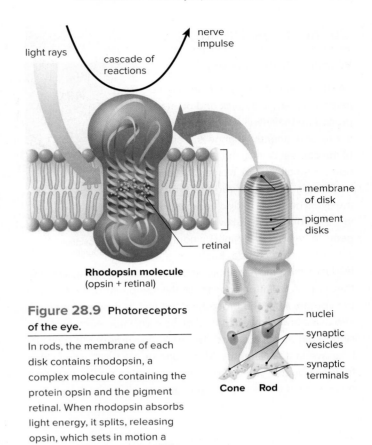

Figure 28.9 Photoreceptors of the eye.

In rods, the membrane of each disk contains rhodopsin, a complex molecule containing the protein opsin and the pigment retinal. When rhodopsin absorbs light energy, it splits, releasing opsin, which sets in motion a cascade of reactions that ends in nerve impulses.

(labels: nerve impulse, light rays, cascade of reactions, membrane of disk, pigment disks, retinal, Rhodopsin molecule (opsin + retinal), nuclei, synaptic vesicles, synaptic terminals, Cone, Rod)

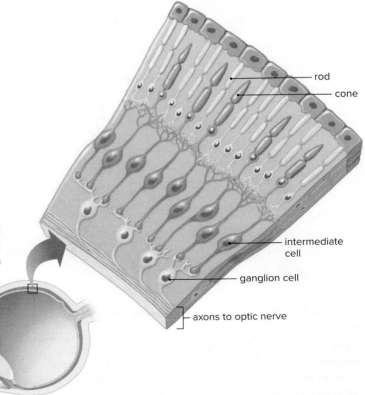

(labels: rod, cone, intermediate cell, ganglion cell, axons to optic nerve)

Figure 28.10 The retina.

The retina contains a layer of rods and cones, a layer of intermediate cells, and a layer of ganglion cells. Integration of signals occurs at the synapses between the layers, and much processing occurs before nerve impulses are sent to the brain.

Connections: Health

What is LASIK surgery?

LASIK, which stands for *laser in-situ keratomileusis*, is a quick and painless procedure that involves the use of a laser to permanently change the shape of the cornea. During the LASIK procedure, a small flap of tissue (the conjunctiva) is cut away from the front of the eye. The flap is folded back, exposing the cornea and allowing the surgeon to remove a defined amount of tissue from the cornea. Each pulse of

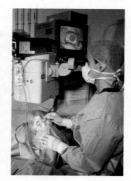

© Pascal Goetgheluck/ Science Source

the laser will remove a small amount of corneal tissue, allowing the surgeon to flatten or increase the steepness of the curve of the cornea. After the procedure, the flap of tissue is put back into place and allowed to heal on its own. Improvements to vision begin as soon as the day after the surgery but typically take two to three months. Most patients will have vision close to 20/20, but the chances for improved vision are based in part on how good the person's vision was before the surgery.

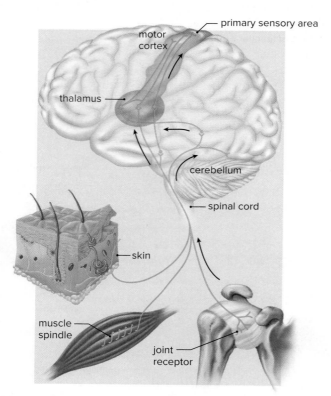

Figure 28.11 **Sensory input to the primary sensory area of the brain.**

Sensory receptors in the skin, muscles, and joints send nerve impulses to the CNS. Sensation and perception occur when these reach the primary sensory area of the cerebral cortex. A motor response can be initiated by the spinal cord and cerebellum without the involvement of the cerebrum.

Seeing in the Dark

Some nocturnal animals rely on vision, but others use sonar (sound waves) to find their way in the dark. Bats flying in a dark room easily avoid obstacles in their path because they echolocate, as submarines do. When searching for food, they emit ultrasonic sound (sound above the range humans can hear) in chirps that bounce off their prey. Bats are able to determine the distance to their dinner by timing the echo's return—a long delay means the prey is far away.

A bat-inspired sonar walking stick is being perfected to help visually impaired people sense their surroundings. It, too, emits ultrasonic chirps and picks up the echoes from nearby objects. Buttons on the cane's handle vibrate gently to warn a user to dodge a low ceiling or to sidestep objects blocking the path. A fast, strong signal means an obstacle is close by.

Cutaneous Receptors and Proprioceptors

Cutaneous receptors and proprioceptors provide sensory input from the skin, muscles, and joints to the primary sensory area of the cerebral cortex. Within the cerebral cortex are areas that represent each part of the body (**Fig. 28.11**).

Cutaneous Receptors

Skin, the outermost covering of our body, contains numerous sensory receptors, called **cutaneous receptors,** that help us respond to changes in our environment, be aware of dangers, and communicate with others. The sensory receptors in skin are for touch, pressure, pain, and temperature.

Skin has two layers, the epidermis and the dermis (**Fig. 28.12**). The epidermis is packed with cells that become keratinized as they rise to the surface. Among these cells are free nerve endings responsive to cold or to warmth. Cold receptors are far more numerous than warmth receptors, but there are no known structural differences between the two. Also in the epidermis are pain receptors (also called **nociceptors**) sensitive to extremes in temperature or pressure and to chemicals released by damaged tissue. Sometimes, the stimulation of internal receptors is felt as pain in the skin. This is called *referred pain*. For example, pain from the heart may be felt in the left shoulder and arm. This effect most likely is produced when nerve impulses from the pain receptors of internal organs travel to the spinal cord and synapse with neurons that are also receiving impulses from the skin.

The dermis of the skin contains sensory receptors for pressure and touch (Fig. 28.12). The Pacinian corpuscles are onion-shaped pressure receptors that lie deep inside the dermis. Several other cutaneous receptors detect touch. A free nerve ending, called a root hair plexus, winds around the base of a hair follicle and produces nerve impulses if the hair is touched. Touch receptors are concentrated in parts of the body essential for sexual stimulation: the fingertips, palms, lips, tongue, nipples, penis, and clitoris.

Proprioceptors

Proprioceptors help the body maintain equilibrium and posture, despite the force of gravity always acting on the skeleton and muscles. A *muscle spindle* consists of sensory nerve endings wrapped around a few muscle cells within a connective tissue sheath. **Golgi tendons** and other sensory receptors are located in the joints. The rapidity of nerve impulses from proprioceptors is proportional to the stretching of the organs they occupy. A motor response results in the contraction of muscle fibers adjoining the proprioceptor.

Connections: Health

How does aspirin work?

Aspirin is made of a chemical called acetylsalicylic acid (ASA). When tissue is damaged, it produces large amounts of a type of fatty acid called prostaglandin. Prostaglandin acts as a signal to the pain receptors that tissue damage has occurred, which the brain interprets as pain. Prostaglandins are manufactured in cells by an enzyme called COX. ASA reduces the capabilities of this enzyme, lowering the amount of prostaglandin produced and the perception of pain.

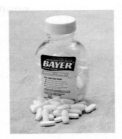

© McGraw-Hill Education/
Eric Misko/Elite Images

28.1 **CONNECTING THE CONCEPTS**
The sensory receptors of the PNS detect various stimuli in our environment and relay this information to the CNS for processing.

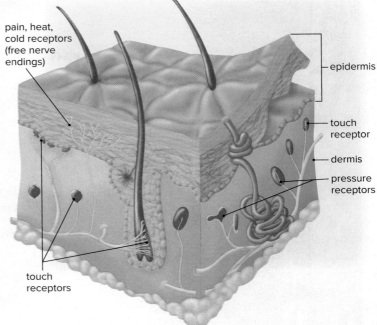

Figure 28.12 per **Cutaneous receptors.**

Numerous receptors are in the skin. Free nerve endings (yellow) in the epidermis detect pain, heat, and cold. Various touch and pressure receptors (red) are in the dermis.

Check Your Progress 28.1

1. Summarize the stimuli that are detected by chemoreceptors, mechanoreceptors, photoreceptors, cutaneous receptors, nociceptors, and proprioceptors.

2. Explain how the sensors in the skin provide sensory information to the body.

3. Summarize the types of sensory information provided by the eyes and ears.

28.2 The Motor Systems

Learning Outcomes

Upon completion of this section, you should be able to

1. Summarize the functions of the skeletal and muscular systems.
2. Distinguish among endoskeletons, exoskeletons, and hydrostatic skeletons, and give an example of each.
3. Describe the structure of a long bone.
4. Describe the structure and function of a muscle fiber, and summarize the process of muscle contraction.
5. List and provide examples of the types of joints in the human body.

Both the **muscular system** and the **skeletal system** of animals are involved in the nervous system's motor response to some form of stimuli. In many ways, the functions of the muscular and skeletal systems overlap, which is why some refer to these systems as the *musculoskeletal system*. In humans, the musculoskeletal system performs the following functions:

Both skeletal muscles and bones support the body and make the movement of body parts possible.

Figure 28.13 Types of skeletons.

a. Arthropods have an exoskeleton that must be shed as they grow.
b. Worms have a hydrostatic skeleton, in which muscle contraction
pushes against a fluid-filled internal cavity.

(a): © Pong Wira/Shutterstock RF

Both skeletal muscles and bones protect internal organs. Skeletal muscles
pad the bones that protect the heart and lungs, the brain, and the
spinal cord.

Both muscles and bones aid the functioning of other systems. Without the
movement of the rib cage, breathing would not occur. As an aid to digestion, the jaws have sockets for teeth; skeletal muscles move the jaws, so
that food can be chewed, and smooth muscle moves food along the digestive tract (see Section 25.1). Red bone marrow supplies the red blood
cells that carry oxygen, and the pumping of cardiac muscle in the wall of
the heart moves the blood to the tissues, where exchanges with tissue
fluid occur.

In addition to these shared functions, the muscles and bones have individual
functions:

*Skeletal muscle contraction assists the movement of blood in the veins and
lymphatic vessels* (see Section 23.2). Without the return of lymph to
the cardiovascular system and blood to the heart, circulation could not
continue.

Skeletal muscles help maintain a constant body temperature (see Section 22.3).
Skeletal muscle contraction causes ATP to break down, releasing heat
that is distributed about the body.

Bones store fat and calcium. Fat is stored in yellow bone marrow (see Fig. 28.15),
and the extracellular matrix of bone contains calcium. Calcium ions play
a major role in muscle contraction and nerve conduction.

Types of Skeletons

The **endoskeleton** of vertebrates can be contrasted with the **exoskeleton** of
arthropods. Both skeletons are jointed, which has helped members of both
groups of animals live successfully on land. The endoskeleton of humans is
composed of bone, which is living material and capable of growth. Like an
exoskeleton, an endoskeleton protects vital internal organs, but unlike an exoskeleton, it need not limit the space available for internal organs, because it
grows as the animal grows. The soft tissues that surround an endoskeleton
protect it, and injuries to soft tissues are apt to be easier to repair than is the
skeleton itself.

The exoskeleton of arthropods is composed of chitin—a strong, flexible,
nitrogenous polysaccharide. Besides providing protection against wear and
tear and against enemies, an exoskeleton also prevents the animal from drying
out. Although an arthropod exoskeleton provides support for muscle contractions, it does not grow with the animal, and arthropods molt to rid themselves
of an exoskeleton that has become too small (**Fig. 28.13***a*). This process makes
them vulnerable to predators.

One other type of skeleton is seen in the animal kingdom. In animals,
such as worms, that lack a hard skeleton, a fluid-filled internal cavity can act
as a **hydrostatic skeleton** (Fig. 28.13*b*). A hydrostatic skeleton offers support
and resistance to the contraction of muscles, so that the animal can move. As
an analogy, consider how a garden hose stiffens when filled with water and
how a water-filled balloon changes shape when squeezed at one end. Similarly,
an animal with a hydrostatic skeleton can change shape and perform a variety
of movements.

The Human Skeleton

The 206 bones of the human skeleton are arranged into an axial skeleton and an appendicular skeleton.

Axial Skeleton The bones of the **axial skeleton** are those that are in the midline of the body (blue shading in **Fig. 28.14**). The *skull* consists of the *cranium,* which protects the brain, and the facial bones. The most prominent of the facial bones are the lower and upper jaws, the cheekbones, and the nasal bones. The *vertebral column* (spine) extends from the skull to the *sacrum* (tailbone). It consists of a series of vertebrae separated by pads of fibrocartilage called the *intervertebral discs.* On occasion, discs can slip or even rupture. A damaged disc pressing against the spinal cord or spinal nerves causes pain, and removal of the disc may be required.

The *rib cage,* composed of the *ribs* and *sternum* (breastbone), demonstrates the dual function of the skeleton, providing protection and flexibility at the same time. In the United States, automobile accidents are the most common cause of sternum fractures; people with such injuries are typically examined for signs of heart damage. The rib cage protects the heart and lungs but moves when we breathe.

Appendicular Skeleton The **appendicular skeleton** contains the bones of two girdles and their attached limbs (unshaded bones in Fig. 28.14). The *pectoral girdle* (shoulder) and upper limbs are specialized for flexibility; the *pelvic girdle* (hip) and lower limbs are specialized for strength. The pelvic girdle also protects the internal organs.

A *clavicle* (collarbone) and a *scapula* (shoulder blade) make up the shoulder girdle. The shoulder pads worn by football players are designed to cushion direct blows that could cause clavicle fractures. The *humerus* of the arm articulates only with the scapula; the joint is stabilized by tendons and ligaments that form a *rotator cuff.* Vigorous circular movements of the arm can lead to rotator cuff injuries. Two bones (the *radius* and *ulna*) contribute to the easy twisting motion of the forearm. The flexible hand contains bones of the wrist, palm, and five fingers.

The pelvic girdle contains two massive *coxal bones,* which form a bowl, called the pelvis, and articulate with the longest and strongest bones of the body, the *femurs* (thighbones). The strength of the femurs allows these massive bones to support the weight of the upper half of the body. The kneecap protects the knee. The *tibia* of the leg is the shinbone. The *fibula* is the more slender bone in the leg. Each foot contains bones of the ankle, instep, and five toes.

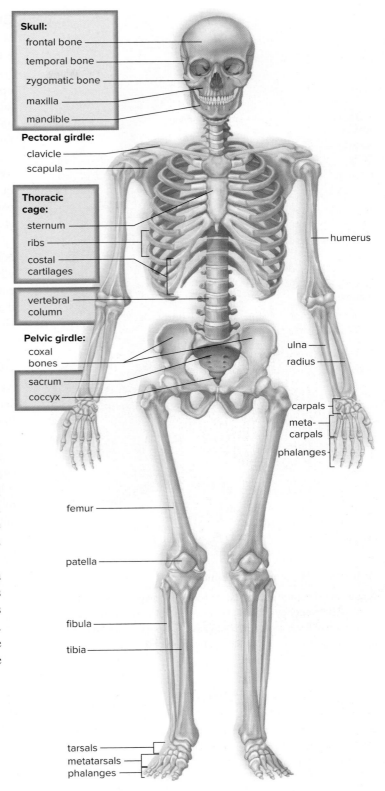

Figure 28.14 The human skeleton.

The bones of the axial skeleton (shaded in blue) are those along the midline of the body. The bones of the appendicular skeleton (unshaded) are those found in the girdles and appendages. Not all bones are labeled in this diagram.

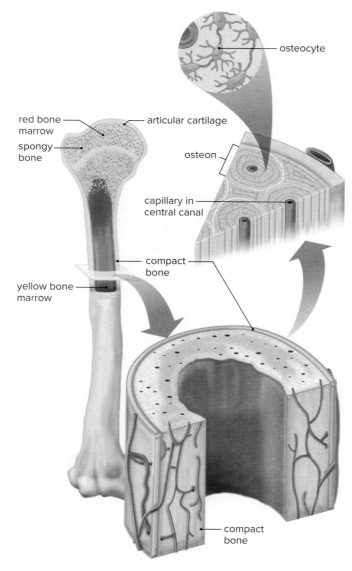

Figure 28.15 **Structure of bone.**

A bone has a central cavity that is usually filled with yellow bone marrow. Both spongy bone and compact bone are living tissues composed of bone cells within a matrix that contains calcium. The spaces of spongy bone contain red bone marrow.

Structure of a Bone

When a long bone is split open, as in **Figure 28.15,** a cavity is revealed that is bounded on the sides by compact bone. **Compact bone** contains many *osteons* (Haversian systems), where bone cells called **osteocytes** lie in tiny chambers arranged in concentric circles around central canals. The cells are separated by an extracellular matrix, which contains mineral deposits—primarily, calcium and phosphorus salts. Two other cell types are constantly at work in bones. *Osteoblasts* deposit bone, and *osteoclasts* secrete enzymes that digest the matrix of bone and release calcium into the bloodstream. When a person has **osteoporosis** (weak bones subject to fracture), osteoclasts are working harder than osteoblasts. Intake of high levels of dietary calcium, especially when a person is young and more active, encourages denser bones and lessens the chance of getting osteoporosis later in life.

A long bone has **spongy bone** at each end. Spongy bone has numerous bony bars and plates separated by irregular spaces. The spaces are often filled with red bone marrow, a specialized tissue that produces red blood cells. The cavity of a long bone is filled with yellow bone marrow and stores fat. Beyond the spongy bone are a thin shell of compact bone and a layer of hyaline cartilage, which is important to healthy joints.

Skeletal Muscle Structure and Physiology

As we explored in Section 22.1, the three types of muscle—smooth, cardiac, and skeletal—have different structures and functions in the body (see Fig. 22.7). Smooth muscles contain sheets of long, spindle-shaped cells, each with a single nucleus. Cardiac cells are striated (having a striped appearance) and typically possess a single nucleus. Cardiac muscle tissue contains branched chains of cells that interconnect, forming a lattice network. Skeletal muscle cells, called muscle fibers, are also striated. They are quite elongated, and they run the length of a skeletal muscle. Skeletal muscle fibers arise during development when several cells fuse, resulting in one long, multinucleated cell.

We will focus our discussion on skeletal muscles, since they represent the major type of muscle in the body and are the ones that are under voluntary control of the nervous system. **Figure 28.16** illustrates some of the major skeletal muscles in the body.

Connections: Scientific Inquiry

How many muscles are in the human body?

Most experts agree that there are over 600 muscles in the human body. The exact number varies, since some experts lump muscles together under one name, while others split them apart. The smallest muscle is the stapedius, a 1.27-millimeter-long muscle in the middle ear. The longest muscle is the sartorius, which starts at the hip and extends to the knee. The biggest muscle (in terms of mass) is the gluteus maximus, the muscle that makes up the majority of each buttock.

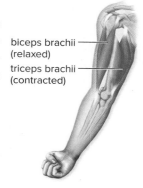

biceps brachii (relaxed)

triceps brachii (contracted)

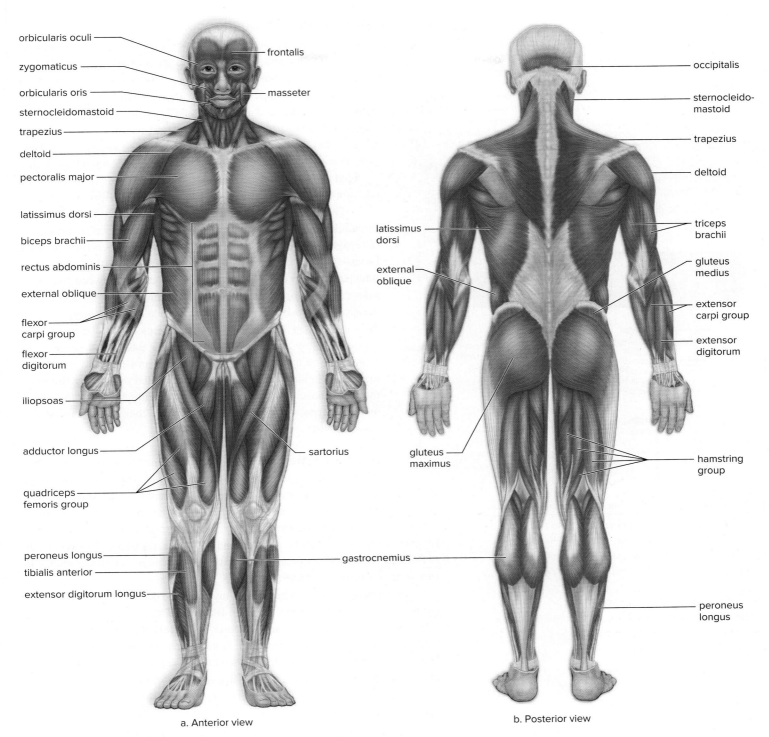

orbicularis oculi

frontalis

zygomaticus

orbicularis oris

masseter

sternocleidomastoid

trapezius

deltoid

pectoralis major

latissimus dorsi

biceps brachii

rectus abdominis

external oblique

flexor carpi group

flexor digitorum

iliopsoas

adductor longus

sartorius

quadriceps femoris group

peroneus longus

tibialis anterior

extensor digitorum longus

gastrocnemius

a. Anterior view

occipitalis

sternocleido-mastoid

trapezius

deltoid

triceps brachii

latissimus dorsi

gluteus medius

external oblique

extensor carpi group

extensor digitorum

gluteus maximus

hamstring group

peroneus longus

b. Posterior view

Figure 28.16 **Major skeletal muscles in the human body.**

The major muscles located near the surface in (**a**) front (anterior) and (**b**) rear (posterior) views.

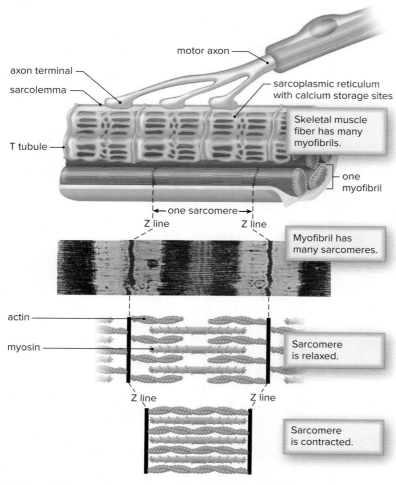

Figure 28.17 Skeletal muscle fiber structure and function.

A muscle fiber contains many myofibrils divided into sarcomeres, which are contractile. When innervated by a motor neuron, the myofibrils contract and the sarcomeres shorten because actin filaments slide past the myosin filaments.

© Hugh Huxley

Labels in figure:
motor axon
axon terminal
sarcolemma
sarcoplasmic reticulum with calcium storage sites
T tubule
Skeletal muscle fiber has many myofibrils.
one myofibril
one sarcomere
Z line
Z line
Myofibril has many sarcomeres.
actin
myosin
Sarcomere is relaxed.
Z line
Z line
Sarcomere is contracted.

Skeletal Muscle Contraction

When skeletal muscle fibers contract, they shorten. Let's look at details of the process, beginning with the motor axon (**Fig. 28.17**). When nerve impulses travel down a motor axon and arrive at an axon terminal, synaptic vesicles release the neurotransmitter acetylcholine (ACh) into a synaptic cleft. ACh quickly diffuses across the cleft and binds to receptors in the plasma membrane of a muscle fiber, called the sarcolemma. The sarcolemma generates impulses, which travel along its T tubules to the endoplasmic reticulum, which is called the *sarcoplasmic reticulum* in muscle fibers. The release of calcium from calcium storage sites causes muscle fibers to contract.

The contractile portions of a muscle fiber are many parallel, threadlike *myofibrils* (Fig. 28.17). An electron microscope shows that myofibrils (and therefore skeletal muscle fibers) are striated because of the placement of protein filaments within contractile units called **sarcomeres.** A sarcomere extends between two dark lines called Z lines. Sarcomeres contain thick filaments made up of the protein **myosin** and thin filaments made up of the protein **actin.** As a muscle fiber contracts, the sarcomeres within the myofibrils shorten because actin (thin) filaments slide past the myosin (thick) filaments and approach one another. The movement of actin filaments in relation to myosin filaments is called the **sliding filament model** of muscle contraction. During the sliding process, the sarcomere shortens, even though the filaments themselves remain the same length.

Why the Filaments Slide   The thick filament is a bundle of myosin molecules, each having a globular head with the capability of attaching to the actin filament when calcium (Ca^{2+}) is present. First, myosin binds to and hydrolyzes ATP. The energized myosin head attaches to an actin filament. The release of ADP and phosphorus Ⓟ causes myosin to shift its position and pull the actin filament to the center of the sarcomere. The action is similar to the movement of your hand when you flex your forearm (**Fig. 28.18**).

In the presence of another ATP, a myosin head detaches from actin. Now the heads attach farther along the actin filament. The cycle occurs again and again, and the actin filaments move nearer and nearer the center of the sarcomere each time the cycle is repeated. Contraction continues until nerve impulses cease and calcium ions are returned to their storage sites. The membranes of the sarcoplasmic reticulum contain active transport proteins that pump calcium ions back into the calcium storage sites, and the muscle relaxes. When a person or an animal dies, ATP production ceases. Without ATP, the myosin heads cannot detach from actin, nor can calcium be pumped back into the sarcoplasmic reticulum. As a result, the muscles remain contracted, a phenomenon called *rigor mortis*.

Connections: Scientific Inquiry

How do medical examiners use rigor mortis to estimate time of death?

Body temperature and the presence or absence of rigor mortis allow the time of death to be estimated. For example, the body of someone who has been dead for 3 hours or less will still be warm (close to body temperature, 98.6°F, or 37°C), and rigor mortis will be absent. After approximately 3 hours, the body will be significantly cooler than normal, and rigor mortis will begin to develop. The corpse of an individual dead at least 8 hours will be in full rigor mortis, and the temperature of the body will be the same as the surroundings. Forensic pathologists know that a person has been dead for more than 24 hours if the body temperature is the same as the environment and there is no longer a trace of rigor mortis.

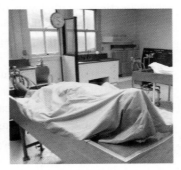
© Michael Matthews/Alamy

Skeletal Muscles Move Bones at Joints

Joints are classified as immovable, such as those of the cranium; slightly movable, such as those between the vertebrae; and freely movable (synovial joints), such as those in the knee and hip. In **synovial joints, ligaments** bind the two bones together, providing strength and support and forming a capsule containing lubricating synovial fluid. All of our movements, from those of graceful and agile ballet dancers to those of aggressive and skillful football players, occur because muscles are attached to bones by tendons that span movable joints. Because muscles shorten when they contract, they have to work in antagonistic pairs. If one muscle of an antagonistic pair flexes the joint and raises the limb, the other extends the joint and straightens the limb. **Figure 28.19** illustrates this principle with regard to the movement of the forearm at the elbow joint.

Figure 28.20*b* illustrates the anatomy of a freely movable synovial joint. Note that, in addition to a cavity filled with synovial fluid, a synovial joint may include additional structures—namely, menisci and bursae. Menisci (sing., meniscus), are C-shaped pieces of hyaline cartilage between the bones. These give added stability and act as shock absorbers. Fluid-filled sacs called bursae (sing., bursa) ease friction between bare areas of bone and overlapping muscles or between skin and tendons.

The **ball-and-socket joints** at the hips and shoulders are synovial joints that allow movement in all directions, even rotational movement (Fig. 28.20*c*). The elbow and knee joints are synovial joints called **hinge joints** because, like a hinged door, they largely permit movement in one direction only (Fig. 28.20*d*).

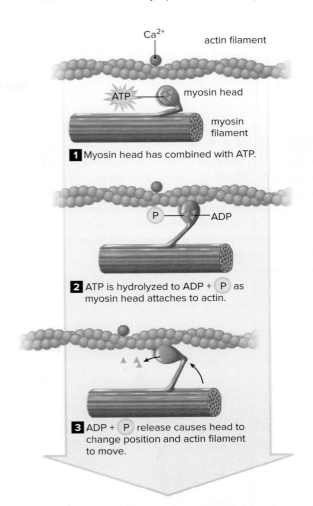

Figure 28.18 **Why a muscle shortens when it contracts.**

The presence of calcium (Ca^{2+}) sets in motion a chain of events (1–3) that causes myosin heads to attach to and pull an actin filament toward the center of a sarcomere. After binding to other ATP molecules, myosin heads return to their resting position. Then, the chain of events (1–3) occurs again, except that the myosin heads reattach farther along the actin filament.

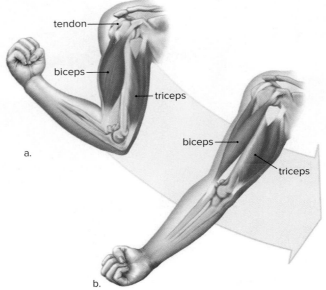

Figure 28.19 **Action of muscles.**

When muscles contract, they shorten. Therefore, muscles only pull—they cannot push. This limitation means that they need to work in antagonistic pairs; each member of the pair pulls on a bone in the opposite direction. For example, (**a**) when the biceps contracts, the forearm flexes (raises), and (**b**) when the triceps contracts, the forearm extends (lowers).

Figure 28.20 Synovial joints.

a. The synovial joints of the human skeleton allow the body to be flexible and move with precision even when bearing a weight. **b.** Generalized synovial joint. Problems arise when menisci or ligaments are torn, bursae become inflamed, and articular cartilage wears away. **c.** The shoulder is a ball-and-socket joint that permits movement in three planes. **d.** The elbow is a hinge joint that permits movement in a single plane.

(a): © Gerard Vandystadt/Science Source

Joint Disorders   *Sprains* occur when ligaments and tendons are over-stretched at a joint. For example, a sprained ankle can result if you turn your ankle too far. Overuse of a joint may cause inflammation of a bursa, called *bursitis*. Tennis elbow is a form of bursitis. A common knee injury is a torn meniscus. Because fragments of menisci can interfere with joint movements, most physicians believe they should be removed. Today, arthroscopic surgery is used to remove cartilage fragments or to repair ligaments or cartilage. A small instrument bearing a tiny lens and light source is inserted into a joint, as are the surgical instruments. Fluid is then added to distend the joint and allow visualization of its structure. Usually, the surgery is displayed on a monitor, so that the whole operating team can see the operation. Arthroscopy is much less traumatic than surgically opening the knee with long incisions. The benefits of arthroscopy are small incisions, faster healing, a more rapid recovery, and less scarring. Because arthroscopic surgical procedures are often performed on an outpatient basis, the patient is able to return home on the same day.

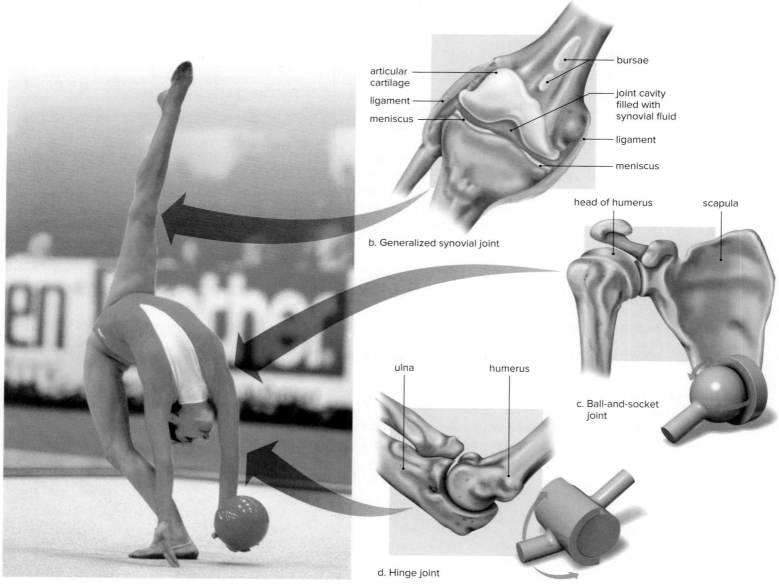

articular cartilage
ligament
meniscus
bursae
joint cavity filled with synovial fluid
ligament
meniscus

b. Generalized synovial joint

head of humerus
scapula

c. Ball-and-socket joint

ulna
humerus

d. Hinge joint

a. A gymnast depends on flexible joints.

Rheumatoid arthritis (see Section 26.5) is not as common as *osteoarthritis,* which is the deterioration of an overworked joint. Constant compression and abrasion continually damage articular cartilage, and eventually it softens, cracks, and in some areas wears away entirely. As the disease progresses, the exposed bone thickens and forms spurs that cause the bone ends to enlarge and restrict joint movement. Weight loss can ease arthritis. Taking off 3 pounds can reduce the load on a hip or knee joint by 9 to 15 pounds. A sensible exercise program helps build up muscles, which stabilize joints. Low-impact activities, such as biking and swimming, are best.

Today, the replacement of damaged joints with a prosthesis (artificial substitute) is often possible. Some people have found glucosamine-chondroitin supplements beneficial as an alternative to joint replacement. Glucosamine, an amino sugar, is thought to promote the formation and repair of cartilage. Chondroitin, a carbohydrate, is a cartilage component that is thought to promote water retention and elasticity and to inhibit enzymes that break down cartilage. Both compounds are naturally produced by the body.

Exercise   A sensible exercise program has many benefits. Exercise improves muscular strength, muscular endurance, and flexibility. It improves cardiorespiratory endurance and may lower blood cholesterol levels. People who exercise are less likely to develop various types of cancer. Exercise promotes the activity of osteoblasts; therefore, it helps prevent osteoporosis. It helps prevent weight gain, not only because of increased activity but also because, as muscle mass increases, the body is less likely to accumulate fat. Exercise even relieves depression and enhances mood.

CONNECTING THE CONCEPTS

28.2   In the musculoskeletal system of the body, the muscular system produces movement of body parts and the skeletal system provides support.

Check Your Progress 28.2

1. Summarize the roles of the muscular and skeletal systems.

2. Identify the components and relative positions of the axial and appendicular skeletons.

3. Explain how the structure of a long bone is suited for its function.

4. Explain how a muscle shortens when it contracts.

STUDY TOOLS   http://connect.mheducation.com

 SMARTBOOK®   Maximize your study time with McGraw-Hill SmartBook®, the first adaptive textbook.

SUMMARIZE

Our sensory structures detect various stimuli in our environment. The motor output of our nervous system enables us to respond appropriately to stimuli.

28.1   The sensory receptors of the PNS detect various stimuli in our environment and relay this information to the CNS for processing.

28.2   In the musculoskeletal system of the body, the muscular system produces movement of body parts and the skeletal system provides support.

28.1 The Senses

All living organisms respond to stimuli. In animals, stimuli generate nerve impulses at the sensory neurons. Often these signals undergo **integration** and **sensory adaptation** before being sent to the brain. The brain processes the information before initiating motor responses:

Stimulus ⟶ Nerve impulse ⟶ CNS ⟶ Motor response

Chemical Senses

Chemoreception is found universally in animals and is therefore believed to be the most primitive sense. Human taste buds and olfactory receptor cells are **chemoreceptors.**

- Taste buds have microvilli with receptor proteins that bind to chemicals in food.
- Olfactory receptor cells have cilia with receptor proteins that bind to odor molecules.

Hearing and Balance

The sensory receptors for hearing are **mechanoreceptors** (hair cells) with stereocilia that respond to pressure waves.

- Hair cells respond to stimuli that have been received by the outer ear and amplified by the ossicles in the middle ear.
- Hair cells are found in the spiral organ and are located in the cochlear canal of the cochlea. The spiral organ generates nerve impulses that travel to the brain.

The sensory receptors for balance (equilibrium) are also hair cells with stereocilia.

- Hair cells in the base of the semicircular canals provide **rotational equilibrium.**
- Hair cells in the utricle and saccule provide **gravitational equilibrium.**

Vision

Vision in animals uses **photoreceptors** that detect light. In humans, the photoreceptors

- Respond to light that has been focused by the cornea and lens.
- Consist of two types, rods and cones. In rods, rhodopsin splits into opsin and retinal.
- Communicate with the next layer of cells in the retina. Integration occurs in the three layers of the retina before nerve impulses go to the brain.

Cutaneous Receptors and Proprioceptors

These receptors communicate with the primary sensory area of the brain. They consist of receptors for hot, cold, touch, pressure (**cutaneous receptors**), and stretching (**proprioceptors** and **Golgi tendons**). **Nociceptors** are involved with detecting pain.

28.2 The Motor Systems

The **skeletal system** and **muscular system** are responsible for generating movement as directed by the central nervous system. Together, the muscles and bones

- Support the body and allow parts to move
- Help protect internal organs
- Assist the functioning of other systems

In addition,

- Skeletal muscle contraction assists movement of blood in cardiovascular veins and lymphatic vessels.
- Skeletal muscles provide heat that warms the body.
- Bones are storage areas for calcium and phosphorus salts, as well as sites for blood cell formation.
- There are several different types of skeletons in animals:
  - an endoskeleton is an internal skeleton; an example is the human skeleton
  - an exoskeleton is an external skeleton; an example is the insect skeleton
  - a hydrostatic skeleton is a fluid-filled internal cavity; an example is found in the earthworm

The Human Skeleton

The human skeleton is divided into two parts:

- The **axial skeleton** is made up of the skull, vertebral column, sternum, and ribs.
- The **appendicular skeleton** is composed of the shoulder and pelvic girdles and their attached appendages.

Bone contains

- **Compact bone** (site of calcium storage) and **spongy bone** (site of red bone marrow)
- A cavity (site of fat storage)
- Cells called **osteocytes**

Osteoporosis is a bone disease characterized by a lack of calcium in the bones.

Muscles

The three types of muscles are smooth muscle, cardiac muscle, and skeletal muscle:

- Smooth muscle is composed of spindle-shaped cells that form a sheet.
- Cardiac muscle has striated cells that form a lattice network.

- Skeletal muscle has striated, tubular cells (fibers) that run the length of the muscle.

In a skeletal muscle cell,

- Myofibrils, **myosin** filaments, and **actin** filaments are arranged in a **sarcomere.**
- Myosin filaments pull actin filaments, and sarcomeres shorten. This explanation of muscle function is called the sliding filament model.
- ATP is required to move the filaments.

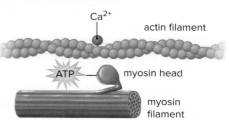

Skeletal muscles

- Move bones at joints
- Work in antagonistic pairs

At **synovial joints,** ligaments hold the bones together. Two types of synovial joints are

- **Ball-and-socket joints,** which allow movement in all directions
- **Hinge joints,** which allow movement in one direction only

ASSESS

Testing Yourself

Choose the best answer for each question.

28.1 The Senses

1. The human eye focuses by
   a. changing the thickness of the lens.
   b. changing the shape of the lens.
   c. opening and closing the pupil.
   d. rotating the lens.

2. The _____ is the region of the eye that controls the amount of light that enters.
   a. cornea   d. fovea
   b. pupil   e. retina
   c. lens

For questions 3–7, identify the type of sensory receptor that matches the description. Some answers may be used more than once.

Key:
   a. propioreceptor
   b. chemoreceptor
   c. mechanoreceptor
   d. nociceptor
   e. photoreceptor

3. detects pain
4. responds to wavelengths of light
5. involved in the sense of hearing
6. involved in the senses of taste and smell
7. helps the body maintain posture

8. Label the parts of the ear in the following illustration.

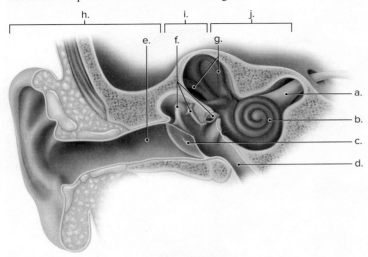

9. Label the parts of the eye in the following illustration.

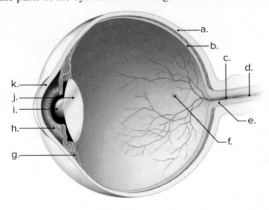

28.2 The Motor Systems

10. The thick filaments in a sarcomere are composed mainly of
   a. calcium.   **d.** rhodopsin.
   b. actin.   **e.** Both b and c are correct.
   c. myosin.

11. A freely movable, fluid-filled joint is a(n) _____ joint.
   a. synovial   **d.** otolithic
   b. proprioceptive   **e.** hydrostatic
   c. Golgi

12. The contractile portion of a skeletal muscle fiber is the
   a. myofibril.   **c.** thick filament.
   b. thin filament.   **d.** sarcoplasmic reticulum.

13. Which cell type deposits bone tissue?
   a. osteocyte   **c.** osteoblast
   b. osteoclast   **d.** None of these are correct.

14. The basic unit of a skeletal muscle contraction is called the
   a. osteoclast.   **c.** sarcomere.
   b. myosin filament. **d.** ligament.

15. Which type of skeleton is found in humans?
   a. endoskeleton
   b. exoskeleton
   c. hydrostatic skeleton

ENGAGE

Thinking Critically

1. The two leading causes of blindness are age-related macular degeneration and diabetic retinopathy. Both are characterized by the development of an abnormally high number of blood vessels (angiogenesis) in the retina. Why do you suppose angiogenesis impairs vision? Progress in cancer research has led to new strategies for the treatment of these eye diseases. Do you see a connection between these causes of blindness and cancer?

2. Human genome researchers have found a family of approximately 80 genes that encode receptors for bitter-tasting compounds. The genes encode proteins made in taste receptor cells of the tongue. Typically, many types of taste receptors are expressed per cell on the tongue. However, in the olfactory system, each cell expresses only 1 of the 1,000 olfactory receptor genes. Different cells express different genes. How do you suppose this difference affects our ability to taste versus smell different chemicals? Why might the two systems have evolved such different patterns of gene expression?

29

Reproduction and Embryonic Development

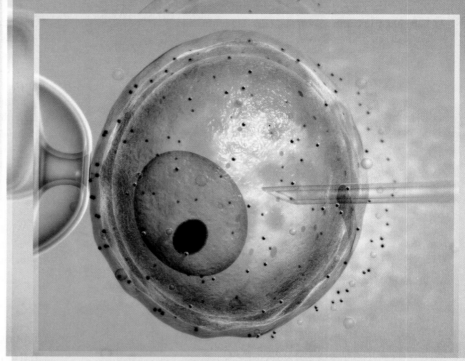

© MedicalRF.com

OUTLINE

29.1 How Animals Reproduce   557

29.2 Human Reproduction   559

29.3 Human Embryonic
Development   570

BEFORE YOU BEGIN

Before beginning this chapter, take a few moments to review the following discussions.

Figure 9.2 What are the roles of mitosis and meiosis in the human life cycle?

Section 9.1 What are the differences between spermatogenesis and oogenesis?

Section 27.2 Which of the hormones produced by the pituitary gland are associated with sexual reproduction?

Three Parents—One Baby

In vitro fertilization (IVF) technology has been used for many years to help couples conceive a child. The process involves taking an egg from the mother and fertilizing it with paternal sperm to create embryos that are then implanted into the mother (or a surrogate) to carry the child to term. A new technique has been proposed that allows for a donor egg to be used in cases where the mother's cells contain defective genes in the mitochondria.

Mitochondria are organelles inside a cell that generate the ATP that provides the energy the cell requires for its metabolism. These organelles contain genetic material (DNA) that encodes 37 genes, 14 of which are related to specific proteins involved in efficient ATP production. Any defect in these genes leads to serious conditions such as mitochondrial myopathies (muscle disorders). An innovative therapy to prevent genetic diseases directly tied to these mitochondrial genes is being developed.

The new technique for in vitro fertilization creates what is sometimes called a "three-parent baby" because a third party donates an egg to be fertilized with a maternal nucleus and paternal sperm. The donor's nuclear genes are removed but the genes inside the mitochondria remain behind. The resulting baby has all of the nuclear genes of the mother and father but the 37 genes found in the mitochondria belong to the donor of the original egg cell.

In this chapter, we will explore the structure and function of the reproductive system in humans, as well as some of the diseases that may cause individuals to need processes such as IVF.

As you read through this chapter, think about the following questions:

1. Which sex normally contributes the mitochondrial genes to the offspring?
2. How does IVF differ from other assisted reproductive technologies?

29.1 How Animals Reproduce

> **Learning Outcomes**
>
> Upon completion of this section, you should be able to
>
> 1. Distinguish between asexual and sexual reproduction in animals.
> 2. Explain the adaptations necessary for animals to reproduce on land versus in water.

Asexual Versus Sexual Reproduction

Animals usually reproduce sexually, but some can reproduce asexually. In **asexual reproduction,** there is only one parent and the offspring are usually genetically the same as the parents. In **sexual reproduction,** there are two parents, and the genetic material is shuffled, so the offspring usually have combinations of genes that are not like those of either parent.

Forms of Asexual Reproduction

An example of asexual reproduction occurs in the hydra, a type of cnidarian (see Section 19.2). Hydras can reproduce by budding. A new individual arises as an outgrowth (bud) of the parent (**Fig. 29.1**). Many flatworms can constrict into two halves; each half regenerates to become a new individual. Fragmentation, followed by regeneration, is also seen among sponges, echinoderms, and corals. Chopping up a sea star does not kill it; instead, each fragment can grow into another animal if a portion of the oral disk is retained with the cut fragment.

Parthenogenesis is a modification of sexual reproduction in which an unfertilized egg develops into a complete individual. In honeybees, the queen bee can fertilize eggs or allow eggs to pass unfertilized as she lays them. The fertilized eggs become diploid females called workers, and the unfertilized eggs become haploid males called drones. Parthenogenesis is also observed in some fish (including sharks) and in a number of amphibian and reptile species.

Sexual Reproduction

In sexual reproduction, animals usually produce gametes in specialized organs called **gonads.** The male gonads are called **testes,** which produce sperm. The female gonads, or **ovaries,** produce eggs. Eggs or sperm are derived from germ cells that become specialized for this purpose during early development. During sexual reproduction, the egg of one parent is usually fertilized by the sperm of another, and a **zygote** (fertilized egg) results. Even among earthworms, which are **hermaphrodites** (each worm has both male and female sex organs), cross-fertilization frequently occurs.

In some animals, the gonads may change function due to environmental conditions. For example, in coral reef fishes called wrasses, a male has a harem of several females. If the male dies, the largest female becomes a male. Reproduction in the slipper snail involves forming a stack of individuals. The individual that is currently at the top of the pile is the male, and all beneath are females.

Figure 29.1 Asexual reproduction.

A new hydra can bud from an adult hydra. This is a form of asexual reproduction.
© NHPA/M. I. Walker/Photoshot RF

Figure 29.2   Reproducing in water.

Animals that reproduce in water have no need to protect their eggs and embryos from drying out. Here, male and female frogs are mating. They deposit their gametes in the water, where fertilization takes place. The egg contains yolk, which nourishes the embryo until it is a free-swimming larva.

© Mode Images/Alamy

Figure 29.3   Reproducing on land.

Animals that reproduce on land need to protect their gametes and embryos from drying out. Here, the male passes sperm to the female by way of a penis, and the developing embryo/fetus will remain in the female's body until it is capable of living independently.

© Anup Shah/Getty Images

Many aquatic animals practice external fertilization, meaning the egg and sperm join in the water outside the body (**Fig. 29.2**). Following fertilization, many aquatic animals have a larval stage, an immature form capable of feeding. Over time, the larva develops into a new adult.

Internal fertilization involves **copulation,** or sexual union, to facilitate the reception of sperm by a female. Some aquatic animals have copulatory organs. Lobsters and crayfish have modified swimmerets. In terrestrial animals, males typically have a penis for depositing sperm into the vagina of females (**Fig. 29.3**). However, this is not the case for all terrestrial animals. For example, birds lack a penis or vagina. Instead they have a cloaca, a chamber that receives products from the digestive, urinary, and reproductive tracts. A male transfers sperm to a female after placing his cloacal opening against hers.

Following internal fertilization, animals either produce eggs or give birth to live offspring. Egg-laying animals are *oviparous,* producing eggs that will hatch after ejection from the body. Reptiles, particularly birds, are oviparous animals that provide their eggs with plentiful yolk, a rich nutrient material. Complete development takes place within a shelled egg. Special membranes, called *extraembryonic membranes,* serve the needs of the embryo and prevent it from drying out. The shelled egg frees these animals from the need to reproduce in the water—a significant adaptation to the terrestrial environment (see Section 19.5). Birds, in particular, tend their eggs, and newly hatched birds usually have to be fed before they are able to fly away and seek food for themselves. Complex hormones and neural regulation are involved in the reproductive behavior of parental birds.

Other animals follow a different course after internal fertilization. They do not deposit and tend their eggs; instead, these animals are *ovoviviparous,* meaning that the eggs are retained in the body until they hatch, releasing fully developed offspring that have a way of life like that of the parent. Oysters, which are molluscs, retain their eggs in the mantle cavity, and male sea horses, which are vertebrates, have a brood pouch in which the eggs develop. Garter snakes, water snakes, and pit vipers retain their eggs in their bodies until they hatch, thus giving birth to living young.

Connections: Scientific Inquiry

For viviparous mammals, what are typical gestation periods?

Gestation periods, the times of development in the uterus from conception to birth, vary greatly between species. The average gestation periods for some common mammals are as follows (listed in days of gestation): hamster, 16; mouse, 21; squirrel, 44; pig, 114; grizzly bear, 220; human, 270; giraffe, 425; killer whale, 500; Indian elephant, 624.

© Gemstone Images/Corbis

Finally, most mammals are *viviparous,* meaning that they produce living young. After offspring are born, the mother supplies the nutrients needed for further growth. Viviparity represents the ultimate in caring for the offspring. Some mammals, such as the duckbill platypus and the spiny anteater (monotremes), lay eggs. After hatching, the offspring are nourished by the mother. In contrast, marsupial offspring are born in a very immature state; they finish their development in a pouch, where they are supplied milk from the mother.

Placental mammals, including humans, provide nourishment during development via the placenta (see Section 29.3) and after birth by mammary glands. The evolution of the placenta allowed the developing offspring to exchange materials with the mother internally.

CONNECTING THE CONCEPTS
29.1
Animals have evolved asexual and sexual forms of reproduction.

Check Your Progress 29.1

1. Summarize the differences between sexual and asexual reproduction.
2. Describe the advantages and disadvantages of reproduction on land and in water.
3. Summarize the differences among oviparous, ovoviviparous, and viviparous animals.

29.2 Human Reproduction

Learning Outcomes

Upon completion of this section, you should be able to

1. Describe the structures of the human male and female reproductive systems and their functions.
2. Explain how hormones regulate the male and female reproductive systems.
3. Evaluate the effectiveness of various means of birth control, and explain how they work.
4. Identify the causes of male and female infertility.
5. Describe the assisted reproductive technologies.
6. Identify the causative agents of common sexually transmitted diseases.

In human males and females, the **reproductive system** consists of two components: (1) the gonads, either testes or ovaries, which produce gametes and sex hormones, and (2) accessory organs that conduct gametes and, in the female, house the embryo/fetus.

Male Reproductive System

The human male reproductive system includes the testes (sing., testis), the epididymis (pl., epididymides), the vas deferens (pl., vasa deferentia), and the urethra (**Fig. 29.4**a). The urethra in males is a part of both the urinary system and the reproductive system. The paired testes, which produce sperm, are suspended within the scrotum. The testes begin their development inside the abdominal cavity, but they descend into the scrotum as embryonic development proceeds. If the testes do not descend soon after birth, and the male does not receive hormone therapy or undergo surgery to place the testes in the scrotum, sterility results. **Sterility** is the inability to produce offspring. This type of sterility occurs because normal sperm production is inhibited at body temperature; a slightly cooler temperature is required.

Sperm produced by the testes mature within the *epididymis,* a coiled tubule lying just outside each testis. Maturation seems to be required for the sperm to swim to the egg. Once the sperm have matured, they are propelled into the *vas deferens* by muscular contractions. The vasa deferentia are severed

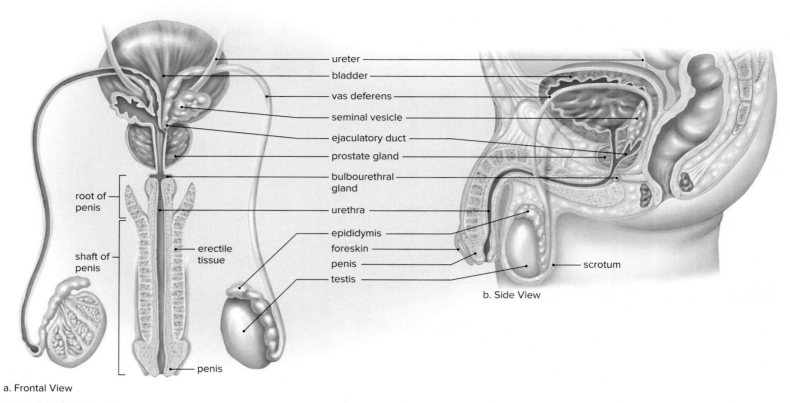

root of penis

shaft of penis

erectile tissue

penis

a. Frontal View

ureter
bladder
vas deferens
seminal vesicle
ejaculatory duct
prostate gland
bulbourethral gland
urethra
epididymis
foreskin
penis
testis

scrotum

b. Side View

Figure 29.4 Male reproductive system.

a. Frontal view. **b.** Side view. The testes produce sperm. The seminal vesicles, the prostate gland, and the bulbourethral glands provide a fluid medium for the sperm. Circumcision is the removal of the foreskin.

or blocked in a surgical form of birth control called a vasectomy (see Table 29.1). Sperm are stored in both the epididymis and the vas deferens. When a male becomes sexually aroused, sperm enter first the ejaculatory duct and then the urethra, part of which is located within the penis.

The **penis** is a cylindrical organ that hangs in front of the scrotum. Three cylindrical columns of spongy, erectile tissue containing distensible blood spaces extend through the shaft of the penis (Fig. 29.4b). During sexual arousal, nervous reflexes cause an increase in arterial blood flow to the penis. This increased blood flow fills the blood spaces in the erectile tissue, and the penis, which is normally limp (flaccid), stiffens and increases in size. These changes are called an erection. If the penis fails to become erect, the condition is called *erectile dysfunction* (*ED*). Drugs such as Viagra, Levitra, and Cialis work by increasing blood flow to the penis, so that when a man is sexually excited, he can achieve and keep an erection.

Semen (seminal fluid) is a thick, whitish fluid that contains sperm and secretions from three glands: the seminal vesicles, prostate gland, and bulbourethral glands. The *seminal vesicles* lie at the base of the bladder. Each joins a vas deferens to form an ejaculatory duct that enters the urethra. As sperm pass from the vas deferens into the ejaculatory duct, these vesicles secrete a thick, viscous fluid containing nutrients for use by the sperm. Just below the bladder is the *prostate gland,* which secretes a milky, alkaline fluid believed to activate or increase the motility of the sperm and neutralize the acidity of the urethra, which is due to urine. In older men, the prostate gland may become enlarged, thereby constricting the urethra and making urination difficult. Also, prostate cancer is the most common form of cancer in men. Slightly below the prostate gland, one on each side of the urethra, is a pair of small glands called *bulbourethral glands,* which have mucous secretions with a lubricating effect. Notice from Figure 29.4 that the urethra also carries urine from the bladder during urination.

If sexual arousal reaches its peak, ejaculation follows an erection. The first phase of ejaculation is called emission. During emission, the spinal cord sends nerve impulses via appropriate nerve fibers to the epididymides and vasa deferentia. Their muscular walls contract, causing sperm to enter the ejaculatory ducts, whereupon the seminal vesicles, prostate gland, and bulbourethral glands release their secretions. Secretions from the bulbourethral glands occur first and may or may not contain sperm.

During the second phase of ejaculation, called expulsion, rhythmic contractions of muscles at the base of the penis and within the urethral wall expel semen in spurts from the opening of the urethra. These contractions are an example of release from muscle tension. An erection lasts for only a limited amount of time. The penis then returns to its normal, flaccid state. Following ejaculation, a male may typically experience a time, called the refractory period, during which stimulation does not bring about an erection. The contractions that expel semen from the penis are a part of male orgasm, the physiological and psychological sensations that occur at the climax of sexual stimulation.

The Testes

A longitudinal section of a testis shows that it is composed of compartments, called lobules, each of which contains one to three tightly coiled *seminiferous tubules* (**Fig. 29.5**a,b). A microscopic cross section of a seminiferous tubule reveals that it is packed with cells undergoing spermatogenesis, a process that involves reducing the chromosome number from diploid (2n) to haploid (n). Also present are *Sertoli cells,* which support, nourish, and regulate the production of sperm. A sperm (Fig. 29.5c) has three distinct parts: a head, a middle piece, and a tail. The head contains a nucleus and is capped by a membrane-bound acrosome that contains digestive enzymes, so that the sperm can penetrate the outer membrane of an egg. The tail is a flagellum that allows sperm to swim toward the egg, and the middle piece contains energy-producing mitochondria.

The ejaculated semen of a normal human male contains 40 million sperm per milliliter, ensuring an adequate number for fertilization to take place. Fewer than 100 sperm ever reach the vicinity of the egg, however, and only one sperm normally enters an egg.

Hormonal Regulation in Males

The hypothalamus has ultimate control of the testes' sexual function, because it secretes a hormone called gonadotropin-releasing hormone, or GnRH, that stimulates the anterior pituitary to produce the gonadotropic hormones (see Section 27.2). Both males and females have two gonadotropic hormones— follicle-stimulating hormone (FSH) and luteinizing hormone (LH). In males, FSH promotes spermatogenesis in the seminiferous tubules. LH controls the production of testosterone by the interstitial cells, which are scattered in the spaces between the seminiferous tubules.

Testosterone, the main sex hormone in males, is essential for the normal development and functioning of the sexual organs. Testosterone is also necessary for the maturation of sperm. In addition, testosterone brings about and maintains the male secondary sex characteristics that develop at **puberty,** the

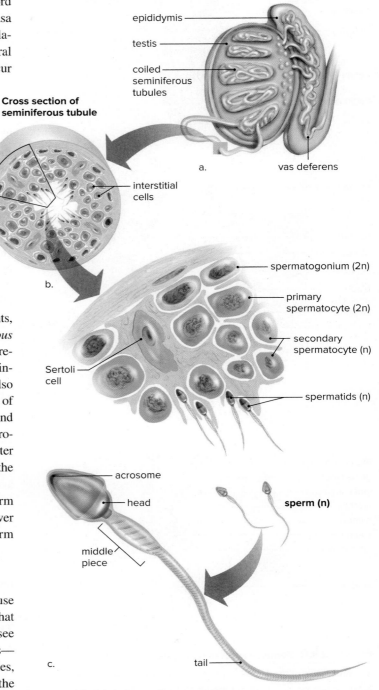

Cross section of seminiferous tubule

epididymis

testis

coiled seminiferous tubules

a.

vas deferens

interstitial cells

b.

spermatogonium (2n)

primary spermatocyte (2n)

secondary spermatocyte (n)

Sertoli cell

spermatids (n)

acrosome

head

sperm (n)

middle piece

c.

tail

Figure 29.5 Seminiferous tubules.

a. The testes contain seminiferous tubules, where sperm are produced. **b.** Cross section of a tubule. As spermatogenesis occurs, the chromosome number is reduced to the haploid number. **c.** A sperm has a head, a middle piece, and a tail. The nucleus is in the head, which is capped by an acrosome.

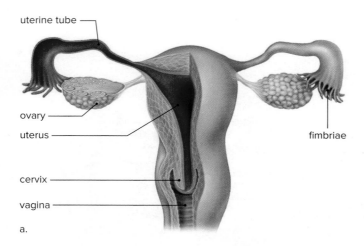

uterine tube

ovary

uterus

fimbriae

cervix

vagina

a.

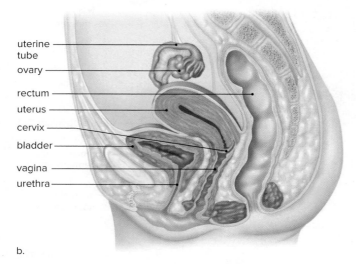

uterine tube

ovary

rectum

uterus

cervix

bladder

vagina

urethra

b.

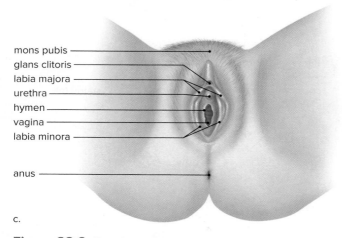

mons pubis
glans clitoris
labia majora
urethra
hymen
vagina
labia minora

anus

c.

Figure 29.6 **Female reproductive system.**

a. Frontal view of the female reproductive system. The ovaries produce one oocyte (egg) per month. Fertilization occurs in the uterine tube, and development occurs in the uterus. The vagina is the birth canal and organ of sexual intercourse. **b.** Side view of the female reproductive system plus nearby organs. **c.** Female external genitals, the vulva. At birth, the opening of the vagina is partially occluded by a membrane called the hymen. Physical activities or sexual intercourse can disrupt the hymen.

time of life when sexual maturity is attained. Males are generally taller than females and have broader shoulders and longer legs relative to trunk length. The deeper voice of males compared with females is due to males' having a larger larynx with longer vocal cords. Because the Adam's apple is a part of the larynx, it is usually more prominent in males than in females.

Testosterone causes males to develop noticeable hair on the face, chest, and occasionally other regions of the body, such as the back. Testosterone also leads to the receding hairline and contributes to the pattern baldness that frequently occurs in males. Testosterone is responsible for the greater muscular development in males.

Female Reproductive System

The human female reproductive system includes the ovaries, the uterine tubes, the uterus, and the vagina (**Fig. 29.6**a,b). The *uterine tubes,* also called the oviducts or fallopian tubes, extend from the ovaries to the uterus; however, the uterine tubes are not attached to the ovaries. Instead, the uterine tubes have fingerlike projections, called *fimbriae* (sing., fimbria), that sweep over the ovaries. When an oocyte, an immature egg cell, ruptures from an ovary during ovulation, it usually is swept into a uterine tube by the combined action of the fimbriae and the beating of cilia that line the uterine tube. Fertilization, if it occurs, normally takes place in the first one-third of the uterine tube, and the developing embryo is propelled slowly by ciliary movement and tubular muscle contraction to the uterus. The **uterus** is a thick-walled, muscular organ about the size and shape of an inverted pear. The narrow end of the uterus is called the **cervix.** An embryo completes its development after embedding itself in the uterine lining, called the *endometrium.* A small opening at the cervix leads to the vaginal canal. The **vagina** is a tube at a 45° angle with the body's vertical axis. The mucosal lining of the vagina lies in folds, and therefore the vagina can expand. This ability to expand is especially important when the vagina serves as the birth canal, and it can facilitate sexual intercourse, when the penis is inserted into the vagina.

The external genital organs of a female are known collectively as the *vulva* (Fig. 29.6c). The *mons pubis* and two folds of skin called *labia minora* and *labia majora* are on each side of the urethral and vaginal openings. At the juncture of the labia minora is the *clitoris,* which is homologous to the penis of males. The clitoris has a shaft of erectile tissue and is capped by a pea-shaped glans. The many sensory receptors of the clitoris allow it to function as a sexually sensitive organ. Most females are born with a thin membrane, called the hymen, which partially obstructs the vaginal opening and has no apparent biological function. The hymen is typically ruptured by physical activities, including sexual intercourse, tampon insertion, and even athletics.

The Ovaries

An oogonium, an undifferentiated germ cell, in the ovary gives rise to an oocyte surrounded by epithelium (**Fig. 29.7**). This is called a primary **follicle.** An ovary contains many primary follicles, each containing an oocyte. At birth, a female has as many as 2 million primary follicles, but the number has been reduced to 300,000–400,000 by the time of puberty. Only a small number of primary follicles (about 400) ever mature and produce a secondary oocyte. When mature, the follicle balloons out on the surface of the ovary and bursts, releasing the secondary oocyte surrounded by follicle cells. The release of a secondary oocyte from a mature follicle is termed **ovulation.** Oogenesis is

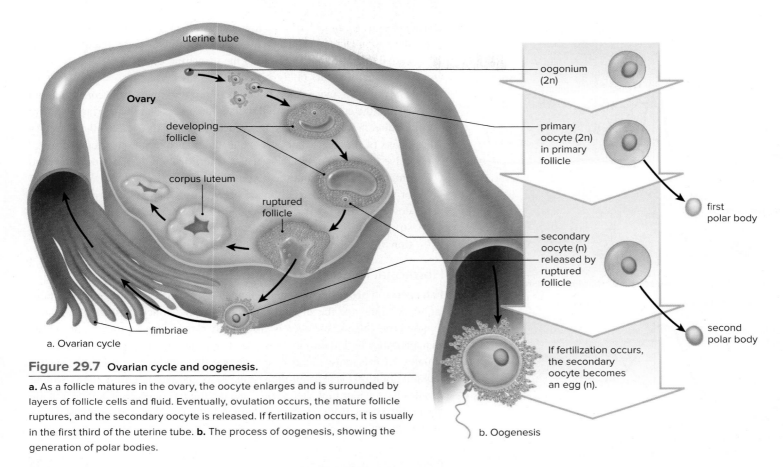

uterine tube

Ovary

developing follicle

corpus luteum

ruptured follicle

fimbriae

a. Ovarian cycle

oogonium (2n)

primary oocyte (2n) in primary follicle

first polar body

secondary oocyte (n) released by ruptured follicle

second polar body

If fertilization occurs, the secondary oocyte becomes an egg (n).

b. Oogenesis

Figure 29.7 Ovarian cycle and oogenesis.

a. As a follicle matures in the ovary, the oocyte enlarges and is surrounded by layers of follicle cells and fluid. Eventually, ovulation occurs, the mature follicle ruptures, and the secondary oocyte is released. If fertilization occurs, it is usually in the first third of the uterine tube. **b.** The process of oogenesis, showing the generation of polar bodies.

completed when and if the secondary oocyte is fertilized by a sperm. A follicle that has lost its oocyte develops into a *corpus luteum* and stays inside the ovary. If fertilization and pregnancy do not occur, the corpus luteum begins to degenerate after about 10 days.

The ovarian cycle is controlled by the gonadotropic hormones FSH and LH from the anterior pituitary gland (**Fig. 29.8**). During the first half, or **follicular phase,** of the cycle (pink in Fig. 29.8), FSH promotes the development of follicles that secrete estrogen. As the blood level of estrogen rises, it exerts feedback control over FSH secretion, and ovulation occurs. Ovulation marks the end of the follicular phase. During the second half, or **luteal phase,** of the ovarian cycle (yellow in Fig. 29.8), LH promotes the development of a corpus luteum, which secretes primarily progesterone. As the blood level of progesterone rises, it exerts feedback control over LH secretion, so that the corpus luteum begins to degenerate if fertilization does not occur. As the luteal phase comes to an end, menstruation occurs.

Notice that the female sex hormones estrogen and progesterone affect the endometrium of the uterus, causing the series of events known as the **menstrual cycle** (Fig. 29.8, *bottom*). The 28-day (on average) menstrual cycle in the nonpregnant female is divided as follows:

During *days 1–5,* female sex hormones are at a low level in the body, causing the endometrium to disintegrate and its blood vessels to rupture. A flow of blood, mucus, and degenerating endometrium, known as the **menses,** passes out of the vagina during **menstruation,** also known as the menstrual period.

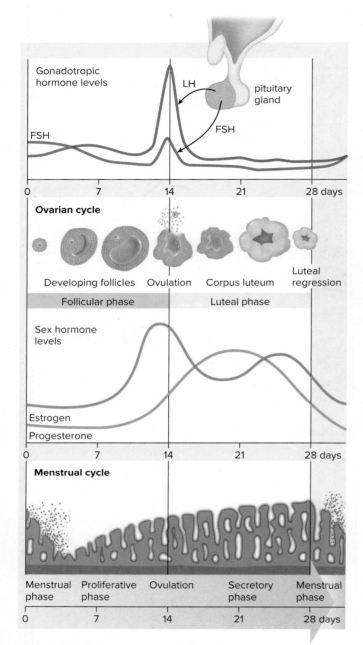

Figure 29.8 Ovarian and menstrual cycles.

During the follicular phase of the ovarian cycle (pink), FSH released by the anterior pituitary promotes the maturation of follicles in the ovary. Ovarian follicles produce increasing levels of estrogen, which cause the endometrium to thicken during the proliferative phase of the menstrual cycle (*bottom*). After ovulation and during the luteal phase of the ovarian cycle (yellow), LH promotes the development of a corpus luteum. This structure produces increasing levels of progesterone, which cause the endometrium to become secretory. Menstruation begins when progesterone production declines to a low level.

During *days 6–13,* increased production of estrogen by ovarian follicles causes the endometrium to thicken and become vascular and glandular. This is called the proliferative phase of the menstrual cycle.

Ovulation usually occurs on *day 14* of the 28-day cycle.

During *days 15–28,* increased production of progesterone by the corpus luteum causes the endometrium to double in thickness and the uterine glands to mature, producing a thick, mucoid secretion. This is called the secretory phase of the menstrual cycle. The endometrium now is prepared to receive the developing embryo. But if fertilization does not occur and no embryo embeds itself, the corpus luteum degenerates, and the low level of sex hormones in the female body causes the endometrium to break down. Menses begins, marking day 1 of the next cycle. Even while menstruation is occurring, the anterior pituitary begins to increase its production of FSH, and new follicles begin to mature.

Hormonal Regulation in Females

Estrogen and **progesterone** are the female sex hormones. Estrogen, in particular, is essential for the normal development and functioning of the female reproductive organs. Estrogen is also largely responsible for the secondary sex characteristics in females, including body hair and fat distribution. In general, females have a more rounded appearance than males because of a greater accumulation of fat beneath their skin. Also, the pelvic girdle aligns so that females have wider hips than males, and the thighs converge at a greater angle toward the knees. Both estrogen and progesterone are required for breast development as well.

Connections: Health

Do women make testosterone?

The adrenal glands and ovaries of women make small amounts of testosterone. Women's low testosterone levels may affect libido, or sex drive. The use of supplemental testosterone to restore a woman's libido has not been well researched.

By the way, men make estrogen, too. Some estrogen is produced by the adrenal glands. Androgens are also converted to estrogen by enzymes in the gonads and peripheral tissues. Estrogen may prevent osteoporosis in males.

Menopause, which usually occurs between ages 45 and 55, is the time in a woman's life when the ovarian and menstrual cycles cease. Menopause is not considered complete until menstruation has been absent for a year.

Control of Reproduction

Table 29.1 lists various means of birth control and gives their rates of effectiveness, assuming consistent and correct usage. **Figure 29.9** illustrates some birth control devices. **Abstinence**—that is, not engaging in sexual intercourse, is very reliable and has the added advantage of avoiding sexually transmitted diseases. An oral contraceptive, commonly called the **birth control pill,** often contains a combination of estrogen and progesterone and is taken on a daily basis. These hormones effectively shut down the pituitary production of both FSH and LH, so that no follicle in the ovary begins to develop; because ovulation does not occur, pregnancy cannot take place. Because of possible side effects, including headaches, blurred vision, chest or abdominal pain, and swollen legs, women taking birth control pills should see a physician regularly.

Contraceptive implants use a synthetic progesterone to prevent ovulation by disrupting the ovarian cycle. The older version of the implant consists of six match-sized, time-release capsules that are surgically implanted under the skin of a woman's upper arm. The newest version consists of a single capsule that remains effective for about 3 years.

Contraceptive injections are available as progesterone only or as a combination of estrogen and progesterone. The length of time between injections can vary from a few weeks to several months.

Interest in barrier methods of birth control, such as **condoms,** has increased, because they offer some protection against sexually transmitted diseases. A *female condom* consists of a large, polyurethane tube with a flexible ring that fits onto the cervix. The open end of the tube has a ring that covers the external genitals. A *male condom* is most often a latex sheath that fits over the erect penis. The ejaculate is trapped inside the sheath and thus does not enter the vagina. When a condum is used in conjunction with a spermicide, the protection is better than with the condom alone. The **diaphragm** is a soft, latex cup with a flexible rim that lodges behind the pubic bone and fits over the cervix. Each woman must be properly fitted by a physician, and the diaphragm can be inserted into the vagina no more than 2 hours before sexual relations. Also, it must be used with spermicidal jelly or cream and should be left in place for at least 6 hours after sexual relations. The **cervical cap** is a mini-diaphragm.

Connections: Health

What is hormone replacement therapy?

Hormone replacement therapy (HRT) is often begun after menopause. The fluctuation of hormone levels during menopause
(HRT): © Philippe Garo/Science Source

can cause symptoms such as hot flashes, mood swings, trouble sleeping, increased abdominal fat, and thinning hair, which HRT can alleviate. The use of HRT has its pros and cons; there is evidence that using HRT after menopause can help prevent bone loss, decrease the risk of colorectal cancer, and decrease certain types of heart disease. Studies also show that some types of HRT in certain patients can increase incidences of stroke and blood clots. Women on HRT should be evaluated every six months by their physician.

Table 29.1 Common Birth Control Methods

Name	Procedure	Effectiveness*	Name	Procedure	Effectiveness*
Abstinence	Refrain from sexual intercourse	100%	Diaphragm	Latex cap is inserted into the vagina to cover cervix before intercourse.	With jelly, about 90%
Sterilization Vasectomy Tubal ligation	Vas deferens cut and tied. Uterine tubes cut and tied.	Almost 100% Almost 100%	Cervical cap	Latex cap is held by suction over cervix.	Almost 85%
Combined estrogen/progesterone available as a pill, an injectable, or a vaginal ring and patch	Pill is taken daily; injectable and ring last a month; patch is replaced weekly.	About 100%	Male condom	Latex sheath is fitted over erect penis.	About 85%
			Female condom	Polyurethane liner is fitted inside vagina.	About 85%
Progesterone only available as a tube implant and an injectable	Implant lasts 3 years; injectable lasts 3 weeks.	About 95%	Coitus interruptus	Penis is withdrawn before ejaculation.	About 75%
			Jellies, creams, foams	Spermicidal product is inserted before intercourse.	About 75%
Intrauterine device (IUD)	Newest hormone-free device contains progesterone and lasts up to 10 years.	More than 90%	Natural family planning	Day of ovulation is determined by record keeping; various methods of testing are used.	About 70%
Vaginal sponge	Sponge permeated with spermicide is inserted into vagina.	About 90%	Douche	Vagina is cleansed after intercourse.	Less than 70%

*The percentage of sexually active women per year who will not get pregnant using this method.

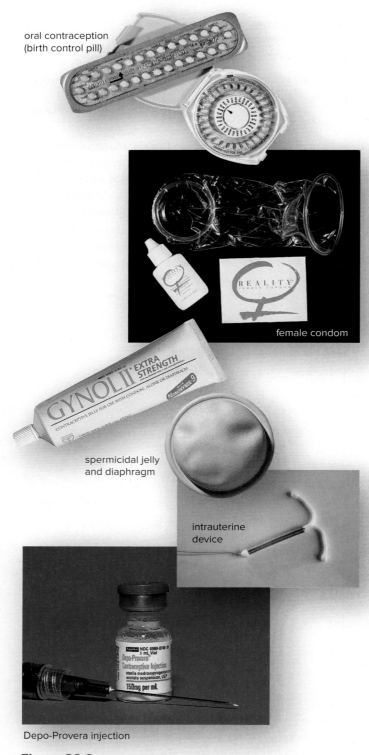

oral contraception
(birth control pill)

female condom

spermicidal jelly
and diaphragm

intrauterine
device

Depo-Provera injection

Figure 29.9 **Contraceptive devices.**

A sample of some of the more common forms of contraception.

(pill, female condom, jelly/diaphragm, Depo-Provera): © McGraw-Hill Education/
Bob Coyle, photographer; (intrauterine device): © imagebroker/Alamy

An **intrauterine device (IUD)** is a small piece of molded plastic that is inserted into the uterus by a physician or another qualified health-care practitioner. IUDs are believed to alter the environment of the uterus and uterine tubes, so that fertilization probably will not occur—but if it should occur, implantation cannot take place.

Contraceptive vaccines are currently being researched. For example, a vaccine intended to immunize women against human chorionic gonadotropin (hCG), a hormone necessary to maintain the implantation of the embryo, has been successful in a limited clinical trial. Since hCG is not normally present in the body, no autoimmune reaction is expected, but the immunization does wear off with time. Other researchers believe that it would also be possible to develop a safe antisperm vaccine for women.

Emergency contraception includes approaches to birth control that can prevent pregnancy after unprotected intercourse. The expression "morning-after pill" is a misnomer in that a woman can use these medications up to several days after unprotected intercourse.

The first FDA-approved medication produced for emergency contraception was a kit called Preven. Preven includes four synthetic progesterone pills; two are taken up to 72 hours after unprotected intercourse, and two more are taken 12 hours later. The hormone upsets the normal uterine cycle, making it difficult for an embryo to implant in the endometrium. One study estimated that Preven was 85% effective in preventing unintended pregnancies. The Preven kit also includes a pregnancy test; women are instructed to take the test first before using the hormone, because the medication is not effective on an established pregnancy.

In 2006 the FDA approved another drug, called Plan B One-Step, which is up to 89% effective in preventing pregnancy if taken within 72 hours after unprotected sex. It is available without a prescription to women age 17 and older. In August 2010, ulipristal acetate (also known as "ella") was also approved for emergency contraception. It can be taken up to 5 days after unprotected sex, and studies indicate it is somewhat more effective than Plan B One-Step. Unlike Plan B One-Step, however, a prescription is required.

Mifepristone, also known as RU-486 or the "abortion pill," can cause the loss of an implanted embryo by blocking the progesterone receptors of endometrial cells. This causes the endometrium to slough off, carrying the embryo with it. When taken in conjunction with a prostaglandin to induce uterine contractions, RU-486 is 95% effective at inducing an abortion up to the 49th day of gestation. Because of its mechanism of action, the use of RU-486 is more controversial compared to other medications, and while it is currently available in the United States for early medical abortion, it is not approved for emergency contraception.

Infertility

Control of reproduction does not only mean preventing pregnancy. It also involves methods of treating infertility. **Infertility** is generally defined as the inability to produce offspring; a medical definition is the failure of a couple to achieve pregnancy after 1 year of regular, unprotected intercourse. It is estimated that one in six couples in America face problems with infertility. The cause of infertility can be attributed to the male (40%), the female (40%), or both (20%).

The most frequent cause of infertility in males is low sperm count and/or a large proportion of abnormal sperm, which can be due to environmental influences. Physicians advise that a sedentary lifestyle coupled with smoking and alcohol consumption can lead to male infertility. When males spend most of the day driving or sitting in front of a computer or television, the testes' temperature remains too high for adequate sperm production.

In females, body weight is an important fertility factor. Only if a woman is of normal weight do fat cells produce a hormone, called leptin, that stimulates the hypothalamus to release GnRH, so that follicle formation begins in the ovaries. If a woman is overweight, the ovaries may contain many small, ineffective follicles, and ovulation does not occur. About 10% of women of childbearing age have an endocrine condition that may be due to an inability to utilize insulin properly. The condition is called *polycystic ovary syndrome* (*PCOS*), because the ovaries contain many cysts (small, fluid-filled sacs) but no functioning follicles, and ovulation does not occur. An impaired ability to respond to insulin is implicated because many women with PCOS eventually develop type 2 diabetes. Other causes of infertility in females are blocked uterine tubes due to *pelvic inflammatory disease* (see "Sexually Transmitted Diseases," later in this section) and *endometriosis,* the presence of uterine tissue outside the uterus, particularly in the uterine tubes and on the abdominal organs.

Sometimes the causes of infertility can be corrected by medical intervention, so that couples can have children. It is also possible to give females fertility drugs, gonadotropic hormones that stimulate the ovaries and bring about ovulation. Such hormone treatments have been known to cause multiple ovulations and multiple births.

When reproduction does not occur in the usual manner, many couples adopt a child. Others try one of the assisted reproductive technologies discussed next.

Assisted Reproductive Technologies

Assisted reproductive technologies (ART) are techniques used to increase the chances of pregnancy. Often, sperm and/or eggs are retrieved from the testes and ovaries, and fertilization takes place in a clinical or laboratory setting.

Artificial Insemination by Donor (AID)   During artificial insemination, harvested sperm are placed in the vagina by a physician. Sometimes a woman is artificially inseminated by her partner's sperm. This technique is especially helpful if the partner has a low sperm count, because the sperm can be collected over a period of time and concentrated, so that the sperm count is sufficient to result in fertilization. Often, however, a woman is inseminated by sperm acquired from a donor who is a complete stranger to her.

In Vitro Fertilization (IVF) and Intracytoplasmic Sperm Injection (ICSI)
During IVF and ICSI, conception occurs outside the body in a laboratory. Ultrasound machines can spot follicles in the ovaries that hold immature eggs; therefore, the latest method is to forgo the administration of fertility drugs and retrieve immature eggs by using a needle. In IVF, the immature eggs are then brought to maturity in glassware before concentrated sperm are added. In ICSI, a single sperm is injected into an egg, usually because a male has severe infertility problems. After about 2–4 days, embryos are ready to be transferred to the uterus of the woman, who is now in the secretory phase of her menstrual cycle. If desired, embryos can be tested for a genetic disease, and only those found to be free of disease will be used. If implantation is successful and development is normal, pregnancy continues to term.

If several embryos are produced and transferred at a time, multiple births are common. When excess embryos are left over, these may be frozen for transfer later, or they may be donated to other couples or used in research.

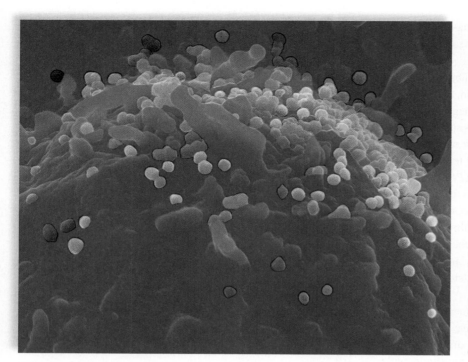

Figure 29.10 Cells infected by the HIV virus.

HIV viruses (yellow) can infect helper T cells (blue) as well as macrophages, which work with helper T cells to stem infection.

© Dr. Olivier Schwartz/Institut Pasteur/Science Source

Gamete Intrafallopian Transfer (GIFT)   The term **gamete** refers to a sex cell, either a sperm or an egg. Gamete intrafallopian transfer was devised to overcome the low success rate (15–20%) of in vitro fertilization. The method is similar to IVF, except the eggs and sperm are placed in the uterine tubes immediately after they have been brought together. GIFT has the advantage of being a one-step procedure for the woman— the eggs are removed and reintroduced at the same time. A variation on this procedure is to fertilize the eggs in the laboratory and then place the zygotes in the uterine tubes.

Sexually Transmitted Diseases

Abstinence is the best protection against the spread of sexually transmitted diseases (STDs). For those who are sexually active, a latex condom offers some protection. Among STDs caused by viruses, treatment is available for AIDS and genital herpes, but these conditions are not curable. Only STDs caused by bacteria (e.g., chlamydia, gonorrhea, and syphilis) are curable with antibiotics.

STDs Caused by Viruses

Acquired immunodeficiency syndrome (AIDS) is caused by a retrovirus called **human immunodeficiency virus (HIV).** HIV attacks the type of lymphocyte known as helper T cells (**Figure 29.10**). Helper T cells (see Section 26.3) stimulate the activities of B lymphocytes, which produce antibodies.

After an HIV infection sets in, helper T cells begin to decline in number, and the person becomes debilitated and more susceptible to other types of infection. AIDS has three stages of infection, called categories A, B, and C.

During the category A stage, which in untreated individuals typically lasts about a year, the individual is an asymptomatic carrier. He or she may exhibit no symptoms but can pass on the infection. Immediately after infection and before the blood test is positive, a large number of infectious viruses are present in the blood and can be passed on to another person. Even after the blood test is positive, the person remains well as long as the body produces sufficient helper T cells to keep the count higher than 500 per mm³. With a combination therapy of several drugs, AIDS patients can remain in this stage indefinitely.

During the category B stage, which may last 6–8 years, the lymph nodes swell and the person may experience weight loss, night sweats, fatigue, fever, and diarrhea. Infections such as thrush (white sores on the tongue and in the mouth) and herpes recur.

Finally, the person may progress to category C, which is full-blown AIDS characterized by nervous disorders and the development of an opportunistic disease, such as an unusual type of pneumonia or skin cancer. Opportunistic diseases are those that occur only in individuals who have little or no capability of fighting an infection. Without intensive medical treatment, the AIDS patient dies about 7–9 years after infection.

An estimated 20 million Americans are currently infected with the **human papillomavirus (HPV).** There are over 100 types of HPV. Most cause warts, and about 30 types cause genital warts, which are sexually transmitted. Genital warts may appear as flat or raised warts on the penis and/or foreskin of males, as well as on the vulva, vagina, and/or cervix of females. Note that if warts are only on the cervix, there may be no outward signs or symptoms of the disease. Newborns can also be infected with HPV during passage through the birth canal.

About 10 types of HPV can cause cervical cancer, the second leading cause of cancer death in women in the United States (approximately 500,000 deaths per year). These HPV types produce a viral protein that inactivates a host protein called p53, which normally acts as a "brake" on cell division (see Section 8.3). Once p53 has been inactivated in a particular cell, that cell is more prone to the uncontrolled cell division characteristic of cancer. Early detection of cervical cancer is possible by means of a Pap test, in which a few cells are removed from the region of the cervix for microscopic examination. If the cells are cancerous, a hysterectomy (removal of the uterus) may be recommended. In males, HPV can cause cancers of the penis, anus, and other areas. Several studies have indicated that HPV now causes as many cancers of the mouth and throat in U.S. males as does tobacco, perhaps due to an increase in oral sex as well as a decline in tobacco use.

Currently, there is no cure for an HPV infection, but the warts can be treated effectively by surgery, freezing, application of an acid, or laser burning. However, even after treatment, the virus can sometimes be transmitted. Therefore, once someone has been diagnosed with genital warts, abstinence or the use of a condom is recommended to help prevent transmission of the virus.

In June 2006, the U.S. Food and Drug Administration licensed Gardasil, an HPV vaccine that is effective against the four most common types of HPV found in the United States, including the two types that cause about 70% of cervical cancers. Because the vaccine doesn't protect those who are already infected, ideally children should be vaccinated before they become sexually active. In 2009, the U.S. Food and Drug Administration approved Gardasil for use in males. The Centers for Disease Control and Prevention recommends that 11- to 12-year-old girls and boys receive three doses of the vaccine. Nonpregnant females between ages 13 and 26, and males from age 13 to 21, can also be vaccinated if they did not receive any or all of the three recommended doses when they were younger. Older individuals should speak with their doctor to find out if getting vaccinated is right for them.

Genital herpes is characterized by painful blisters on the genitals. Once the blisters rupture, they leave painful ulcers, which may take as long as 3 weeks or as little as 5 days to heal. The blisters may be accompanied by fever, pain on urination, swollen lymph nodes in the groin, and in women, a copious discharge. After the ulcers heal, the disease is only latent, and blisters can recur, although usually at less frequent intervals and with milder symptoms. Fever, stress, sunlight, and menstruation are associated with the recurrence of symptoms.

Hepatitis is an infection of the liver and can lead to liver failure, liver cancer, and death. Several types of hepatitis exist, some of which can be transmitted sexually. Each type of hepatitis and the virus that causes it are designated by a letter. Hepatitis A is usually acquired from sewage-contaminated drinking water, but this infection can also be sexually transmitted through oral/anal contact. Hepatitis B, which is spread in the same manner as AIDS, is even more infectious. Fortunately, a vaccine is available for hepatitis B. Hepatitis C is spread when a person comes in contact with the blood of an infected person.

STDs Caused by Bacteria

Chlamydia is a bacterial infection of the lower reproductive tract that is usually mild or asymptomatic, especially in women. About 8–21 days after infection,

Connections: Health

What can you do to reduce your chances of contracting an STD?

1. Abstain from sexual intercourse or develop a long-term, monogamous (having intercourse with only one person) relationship with a person who is free of STDs.

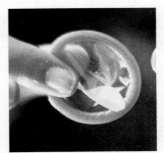
© PhotoAlto/PunchStock RF

2. Refrain from having multiple sex partners or a relationship with a person who does have multiple sex partners.

3. Be aware that having relations with an intravenous drug user is risky, because the behavior of this group puts them at risk for AIDS and hepatitis B.

4. Avoid anal intercourse, because HIV has easy access through the lining of the rectum.

5. Always use a latex condom if your partner has not been free of STDs for the past 5 years.

6. Avoid oral sex, because this may be a means of transmitting AIDS and other STDs.

7. Stop, if possible, the habit of injecting drugs; if you cannot stop, at least always use a sterile needle.

men may experience a mild burning sensation upon urination and a mucoid discharge. Women may have a vaginal discharge, along with the symptoms of a urinary tract infection. Chlamydia also causes cervical ulcerations, which increase the risk of acquiring AIDS. If the infection is misdiagnosed or if a woman does not seek medical help, there is a risk of the infection spreading from the cervix to the uterine tubes, and **pelvic inflammatory disease (PID)** results. This very painful condition can result in blockage of the uterine tubes, with the possibility of sterility or infertility.

Gonorrhea is easier to diagnose in males than in females because males are more likely to experience painful urination and a thick, greenish-yellow urethral discharge. In males and females, a latent infection leads to PID, which affects the vasa deferentia or uterine tubes. As the inflamed tubes heal, they may become partially or completely blocked by scar tissue, resulting in sterility or infertility.

Syphilis has three stages, which are typically separated by latent periods. During the final stage, syphilis may affect the cardiovascular system and/or nervous system. An infected person may become mentally retarded or blind, walk with a shuffle, or show signs of insanity. Gummas, which are large, destructive ulcers, may develop on the skin or in the internal organs. Syphilitic bacteria can cross the placenta, causing birth defects or stillbirth. Syphilis is easily diagnosed with a blood test.

STDs Caused by Other Organisms

Females very often have *vaginitis,* or infection of the vagina, caused by either the flagellated protozoan *Trichomonas vaginalis* or the yeast *Candida albicans.* **Trichomoniasis** is most often acquired through sexual intercourse, and the asymptomatic male is usually the reservoir of infection. *Candida albicans,* however, is an organism normally found in the vagina; its growth simply increases beyond normal under certain circumstances. For example, women taking birth control pills are sometimes prone to yeast infections. Also, the legitimate and indiscriminate use of antibiotics for infections elsewhere in the body can alter the normal balance of organisms in the vagina, so that a yeast infection flares up.

CONNECTING THE CONCEPTS

29.2 The human male and female reproductive systems produce gametes for reproduction. Methods exist to both prevent pregnancy and protect against STDs.

Check Your Progress 29.2

1. Identify the structures involved in sperm development and ejaculation.
2. Summarize the functions of testosterone in the male reproductive system.
3. Describe the process of ovulation in the ovaries.
4. Detail the events of the menstrual cycle in females.
5. Summarize the technologies that may be used to treat infertility.
6. Categorize STDs according to how they are transmitted (viruses, bacteria, and other causes).

29.3 Human Embryonic Development

Learning Outcomes

Upon completion of this section, you should be able to

1. Describe the processes of fertilization, embryonic development, fetal development, and birth.
2. Identify the stages of embryonic development, and explain the significance of gastrulation and neurulation.
3. Describe the structure and function of the placenta.

Embryonic development encompasses all the events that occur from the time of fertilization until an animal—in this case, a human—is fully formed. Humans and other mammals have developmental stages similar to those of all animals but with some marked differences, chiefly due to the presence of extraembryonic membranes (membranes outside the embryo). Developing mammalian embryos (and fetuses in placental mammals, such as humans), such as those of reptiles and birds, depend on these membranes to protect and nourish them. **Figure 29.11** shows what these membranes are and what they do in a mammal.

Fertilization

Fertilization", which results in a zygote, requires that the sperm and the secondary oocyte interact (**Fig. 29.12**). The plasma membrane of the secondary oocyte is surrounded by an extracellular material termed the *zona pellucida*. In turn, the zona pellucida is surrounded by a few layers of adhering follicle cells.

During fertilization, a sperm moves past the leftover follicle cells, and acrosomal enzymes released by exocytosis digest a route through the zona pellucida. A sperm binds and fuses to the secondary oocyte's plasma membrane, and the sperm's nucleus enters. Only then does the secondary oocyte complete meiosis II and become an egg. Finally, the haploid egg and sperm nuclei fuse. Only one sperm should fertilize an egg, or the

chorion
becomes part of the placenta where the embryo/fetus receives oxygen and nutrient molecules and rids itself of waste molecules

yolk sac
first site of blood cell formation

allantois
its blood vessels become the blood vessels of the umbilical cord

amnion
contains the amniotic fluid, which cushions and protects the embryo

Figure 29.11 The extraembryonic membranes.

Humans, like other animals that reproduce on land, are dependent on these membranes to protect and nourish the embryo (and later the fetus).

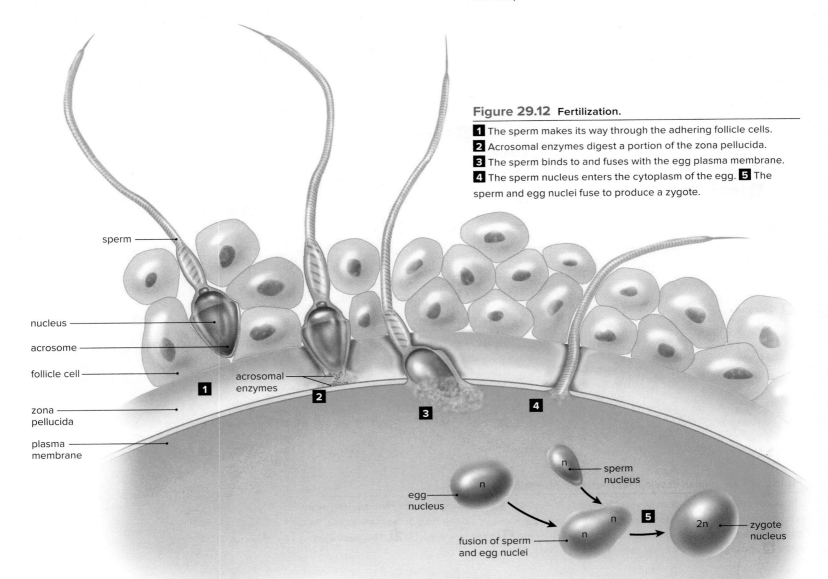

Figure 29.12 Fertilization.

1 The sperm makes its way through the adhering follicle cells.
2 Acrosomal enzymes digest a portion of the zona pellucida.
3 The sperm binds to and fuses with the egg plasma membrane.
4 The sperm nucleus enters the cytoplasm of the egg. **5** The sperm and egg nuclei fuse to produce a zygote.

sperm

nucleus

acrosome

follicle cell

acrosomal enzymes

zona pellucida

plasma membrane

egg nucleus

sperm nucleus

fusion of sperm and egg nuclei

zygote nucleus

zygote will have too many chromosomes and the resulting zygote will not be viable. Changes in the zona pellucida usually prevent the binding and penetration of additional sperm (called polyspermy).

Early Embryonic Development

The first two months of development are considered the embryonic period. Approximately 6 days of development occur in the uterine tube before the embryo implants itself in the uterine lining, or endometrium (**Fig. 29.13**). During the first stage of development, the embryo becomes multicellular. Following fertilization, the zygote undergoes **cleavage,** which is cell division without growth. DNA replication and mitotic cell division occur repeatedly, and the cells get smaller with each division. Notice that cleavage only increases the number of cells; it does not change the original volume of the egg cytoplasm. The resulting tightly packed ball of cells is called a **morula.**

The cells of the morula continue to divide, but they also secrete a fluid into the center of a ball of cells. A hollow ball of cells, called the **blastocyst,** is formed, surrounding a fluid-filled cavity called the blastocoel. Within the ball is an inner cell mass that will go on to become the embryo. The outer layer of cells is the first sign of the chorion, the extraembryonic membrane that will contribute to the development of the placenta. As the embryo implants itself in the uterine lining (endometrium), the placenta begins to form and to secrete the hormone human chorionic gonadotropin (hCG). This hormone is the basis for the pregnancy test, and it maintains the corpus luteum

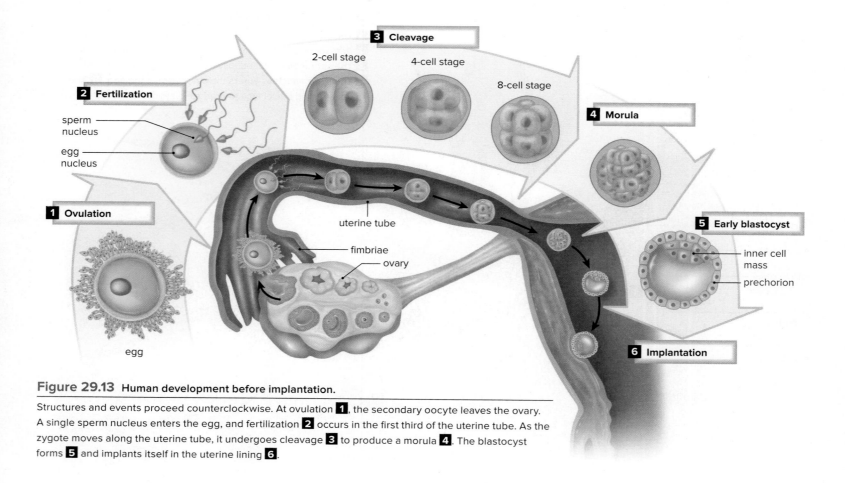

Figure 29.13 Human development before implantation.

Structures and events proceed counterclockwise. At ovulation **1**, the secondary oocyte leaves the ovary. A single sperm nucleus enters the egg, and fertilization **2** occurs in the first third of the uterine tube. As the zygote moves along the uterine tube, it undergoes cleavage **3** to produce a morula **4**. The blastocyst forms **5** and implants itself in the uterine lining **6**.

past the time it normally disintegrates inside the ovary. Because of hCG, the endometrium is maintained until this function is taken over by estrogen and progesterone, produced by the placenta. Ovulation and menstruation do not normally occur during pregnancy.

Later Embryonic Development

Gastrulation

A major event, called **gastrulation,** turns the inner cell mass into the embryonic disk. Gastrulation is an example of morphogenesis, during which cells move or migrate. In this case, cells migrate to become tissue layers called germ layers. By the time gastrulation is complete, the embryonic disk has become an embryo with three primary germ layers: ectoderm, mesoderm, and endoderm.

Gastrulation is complete when the three layers of cells that will develop into adult organs have been produced. The outer layer is the **ectoderm;** the inner layer is the **endoderm;** and the third, or middle, layer of cells is called the **mesoderm.** Ectoderm, mesoderm, and endoderm are called the embryonic **germ layers.** As shown in **Figure 29.14,** the organs of an animal's body develop from these three germ layers.

Neurulation

The first organs to form are those of the central nervous system. The newly formed mesoderm cells coalesce to form a dorsal supporting rod called the **notochord.** The central nervous system develops from midline ectoderm located just above the notochord. During **neurulation,** a thickening of cells, called the *neural plate,* is seen along the dorsal surface of the embryo. Then, neural folds develop on both sides of a neural groove, which becomes the *neural tube* when these folds fuse. The anterior portion of the neural tube becomes the brain, and the posterior portion becomes the spinal cord. Development of the neural tube is an example of **induction,** the process by which one tissue or organ influences the development of another. Induction occurs because the tissue initiating the induction releases a chemical that turns on genes in the tissue being induced.

Midline mesoderm cells that did not contribute to the formation of the notochord now become two masses of tissue. These two masses are then blocked off into *somites,* which are serially arranged on both sides along the length of the notochord. Somites give rise to the vertebrae and to muscles associated with the axial skeleton. The sequential order of the vertebrae and the muscles of the trunk testify that chordates are segmented animals. Lateral

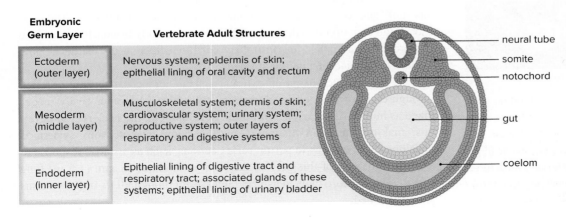

Embryonic Germ Layer	Vertebrate Adult Structures
Ectoderm (outer layer)	Nervous system; epidermis of skin; epithelial lining of oral cavity and rectum
Mesoderm (middle layer)	Musculoskeletal system; dermis of skin; cardiovascular system; urinary system; reproductive system; outer layers of respiratory and digestive systems
Endoderm (inner layer)	Epithelial lining of digestive tract and respiratory tract; associated glands of these systems; epithelial lining of urinary bladder

Figure 29.14 The germ layers.

Each germ layer is responsible for the development of specific structures in the body.

Figure 29.15 Human embryo.

This figure shows the development of a human embryo at about the fifth week.

© Anatomical Travelogue/Science Source

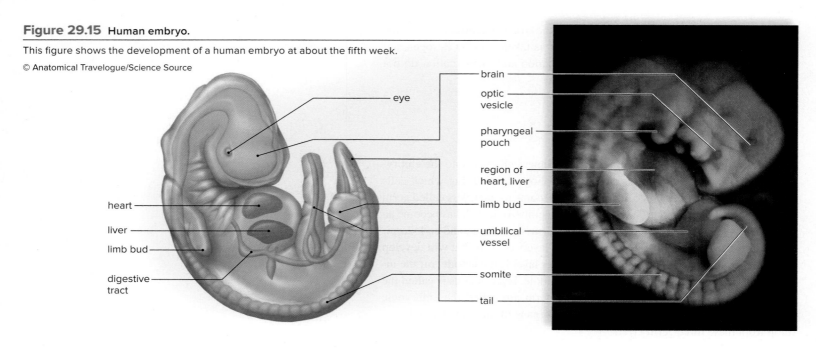

to the somites, the mesoderm splits and forms the mesodermal lining of the coelom. In addition, the neural crest consists of a band of cells that develops where the neural tube pinches off from the ectoderm. These cells migrate to various locations, where they contribute to the formation of skin and muscles, in addition to the adrenal medulla and the ganglia of the peripheral nervous system.

At the end of the third week, over a dozen somites are evident, and the blood vessels and gut have begun to develop. At this point, the embryo is about 2 millimeters (mm) long.

Organ Formation Continues

A human embryo at 5 weeks has little flippers called limb buds (**Fig. 29.15**). Later, the arms and legs develop from the limb buds, and even the hands and feet become apparent. During the fifth week, the head enlarges and the sense organs become more prominent.

The umbilical cord has developed from a bridge of mesoderm called the body stalk, which connects the caudal (tail) end of the embryo with the chorion. A fourth extraembryonic membrane, the allantois, is contained within this stalk, and its blood vessels become the umbilical blood vessels. The head and the tail then lift up as the body stalk moves anteriorly by constriction. Once this process is complete, the **umbilical cord,** which connects the developing embryo to the placenta, is fully formed.

A remarkable change in external appearance occurs during the sixth through eighth weeks of development—the embryo becomes easily recognized as human. Concurrent with brain development, the neck region develops, making the head distinct from the body. The nervous system is developed well enough to permit reflex actions, such as a startle response to touch. At the end of this period, the embryo is about 38 mm (1.5 in) long and weighs no more than an aspirin tablet, even though all its organ systems are established.

Connections: Scientific Inquiry

Why is the female gender sometimes referred to as the *default sex?*

The term *default sex* has to do with the presence or absence of the Y chromosome in the fetus. On the Y chromosome is a gene called *sex determining region Y* (SRY), which produces a protein that causes

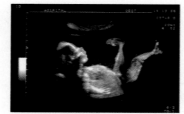

© Nic Cleave Photography/Alamy

Sertoli cells in the testes to produce Müllerian inhibiting substance (MIS). This causes Leydig cells in the testes to produce testosterone, which signals the development of the male sex organs (vas deferens, epididymis, penis, and so on). Without the SRY gene and this hormone cascade, female structures (uterus, fallopian tubes, ovaries, and so on) will begin to form.

Placenta

The **placenta** has a fetal side contributed by the chorion, the outermost extra-embryonic membrane, and a maternal side consisting of uterine tissues. Notice in **Figure 29.16** that the chorion has treelike projections, called *chorionic villi*. The chorionic villi are surrounded by maternal blood, yet maternal and fetal blood never mix, because exchange always takes place across the walls of the villi. Carbon dioxide and other wastes move from the fetal side to the maternal side of the placenta, and nutrients and oxygen move from the maternal side to the fetal side. The umbilical cord stretches between the placenta and the fetus. The umbilical blood vessels are an extension of the fetal circulatory system and simply take fetal blood to and from the placenta.

Harmful chemicals can also cross the placenta. This is of particular concern during the embryonic period, when various structures are first forming. Each organ or part seems to have a sensitive period, during which a substance can alter its normal development. For example, if a woman takes the drug thalidomide, a tranquilizer, between days 27 and 40 of her pregnancy, the infant is likely to be born with deformed limbs. After day 40, however, the limbs will develop normally.

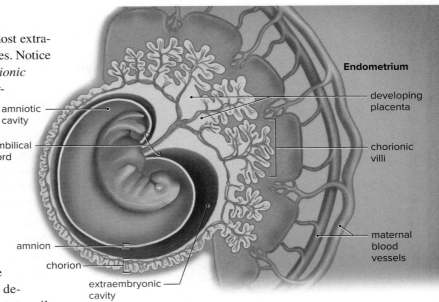

Figure 29.16 Placenta.

Blood vessels within the umbilical cord lead to the placenta, where exchange takes place between fetal blood and maternal blood.

Fetal Development and Birth

Fetal development encompasses the third to the ninth months (**Fig. 29.17**). Fetal development is marked by an extreme increase in size. The weight changes significantly, increasing from less than 28 grams to approximately 3 kilograms. During this time, too, the fetus grows to about 50 centimeters in length. The genitalia appear in the third month, so it is possible to tell if the fetus is male or female.

Soon after the third month, hair, eyebrows, and eyelashes add finishing touches to the face and head. In the same way, fingernails and toenails complete the hands and feet. Later, during the fifth through seventh months, a fine, downy hair (lanugo) covers the limbs and trunk, only to disappear later. The fetus looks very old because the skin is growing so fast that it wrinkles. A waxy, almost cheeselike substance (called vernix caseosa) protects the wrinkly skin from the watery amniotic fluid.

The fetus at first only flexes its limbs and nods its head, but later it can move its limbs vigorously to avoid discomfort. The mother feels these movements from about the fourth month on. The other systems of the body also begin to function. As early as 10 weeks, the fetal heartbeat can be heard through a stethoscope. A fetus born at 22 weeks has a chance of surviving, although the lungs are still immature and often cannot capture oxygen adequately. Weight gain during the last couple of months increases the likelihood of survival.

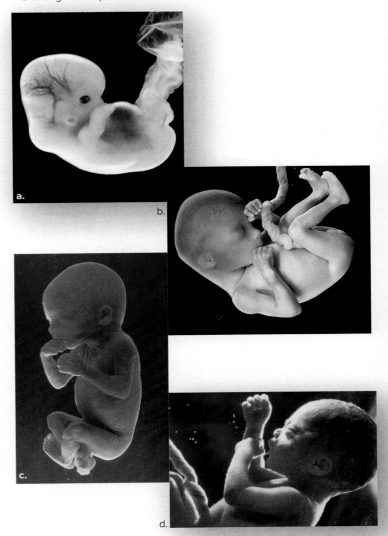

Figure 29.17 Development of the fetus.

This series shows the development of a human fetus at (**a**) 6 weeks, (**b**) 12–16 weeks, (**c**) 20–28 weeks, and (**d**) 28 weeks.

(a): © Neil Harding/Stone/Getty Images; (b): © John Watney/Science Source; (c): © James Stevenson/SPL/Science Source; (d): © Petit Format/Science Source

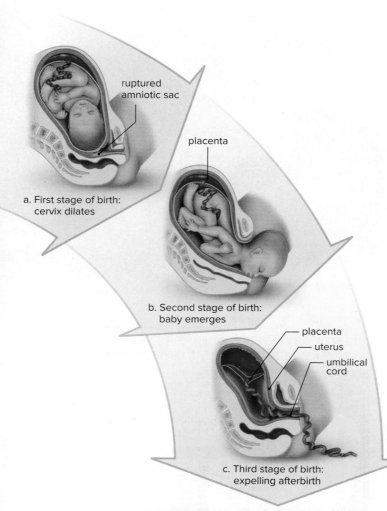

a. First stage of birth: cervix dilates

ruptured amniotic sac

placenta

b. Second stage of birth: baby emerges

placenta
uterus
umbilical cord

c. Third stage of birth: expelling afterbirth

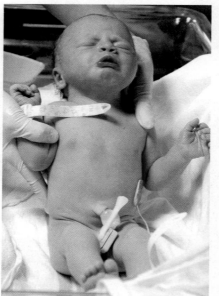

d.

Figure 29.18 Three stages of birth (parturition).

a. Dilation of the cervix. **b.** Birth of the baby. **c.** Expulsion of the afterbirth (placenta). **d.** The newborn at 40 weeks.

(d): © Dennis MacDonald/PhotoEdit

Connections: Health

How is Zika virus related to birth defects?

The first cases of Zika virus were reported in Africa in 1952, but the virus had been largely unknown in the Western Hemisphere until it was reported in Brazil in 2015.

The virus is transmitted by a bite from an infected *Aedes* mosquito, and it can also be sexually transmitted in humans from infected males to females. For most people, infection with the Zika virus produces mild symptoms, such as fevers, rashes, or joint pain. Some individuals do not experience any symptoms.

However, in a small number of cases, pregnant females who have been infected with the Zika virus have given birth to children with microcephaly. Microcephaly is a form of birth defect where the infant's head and brain are abnormally small. This can cause a number of developmental problems, including seizures, intellectual disabilities, and vision problems. There is no cure for microcephaly.

The exact method by which Zika virus may cause microcephaly is still being investigated. Research is focusing on a group of neural stems cells that are associated with brain development. For more information, visit www.cdc.gov/zika/.

The Stages of Birth

The latest findings suggest that, when the fetal brain is sufficiently mature, the hypothalamus causes the pituitary to stimulate the adrenal cortex, so that androgens are released into the bloodstream. The placenta uses androgens as a precursor for estrogen, a hormone that stimulates the production of oxytocin and prostaglandin (a molecule produced by many cells that acts as a local hormone). All three of these molecules—estrogen, oxytocin, and prostaglandin—cause the uterus to contract and expel the fetus.

The process of birth (parturition) has three stages (**Fig. 29.18**). During the first stage, the cervix dilates to allow passage of the baby's head and body. The amnion usually bursts about this time. During the second stage, the baby is born and the umbilical cord is cut. During the third stage, the placenta is delivered.

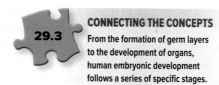

CONNECTING THE CONCEPTS

29.3

From the formation of germ layers to the development of organs, human embryonic development follows a series of specific stages.

Check Your Progress 29.3

1. Explain what is meant by the term *extraembryonic membrane,* and give an example.
2. List the three primary germ layers.
3. Explain the difference between gastrulation and neurulation.
4. Contrast embryonic development with fetal development.
5. Explain how the placenta is used during development.

STUDY TOOLS   http://connect.mheducation.com

⬛⬛|SMARTBOOK®   Maximize your study time with McGraw-Hill SmartBook®, the first adaptive textbook.

SUMMARIZE

Among animals, there are examples of both asexual and sexual reproduction. In humans, sexual reproduction is followed by a period of embryonic development.

29.1 Animals have evolved asexual and sexual forms of reproduction.

29.2 The human male and female reproductive systems produce gametes for reproduction. Methods exist to both prevent pregnancy and protect against STDs.

29.3 From the formation of germ layers to the development of organs, human embryonic development follows a series of specific stages.

29.1 How Animals Reproduce

Asexual reproduction involves a single parent and produces offspring with the same genetic information as the parent. **Sexual reproduction** involves two parents and produces offspring with different genetic combinations than the parents. In **parthenogenesis,** an unfertilized egg becomes a new individual.

Sexual Reproduction

- Animals have **gonads** (**testes** and **ovaries**) for the purpose of producing gametes. Most animals have separate male and females sexes, but some are **hermaphrodites,** or individuals with both male and female sex organs. **Copulation** is the sexual union of a male and female to produce offspring.
- Animals adapted to reproducing in water shed their gametes into the water; fertilization and zygote development occur there. Animals that reproduce on land protect their gametes and embryos from drying out.
- Oviparous animals reproduce using eggs. Ovoviviparous animals begin as eggs, but hatch within the body of the parent and then are born live. Viviparous animals are born live.

29.2 Human Reproduction

The **reproductive system** of humans contains the male and female gonads and is responsible for producing gametes for reproduction.

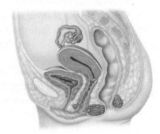

Male Reproductive System

- Sperm are produced in the testis, mature in the epididymis, and are stored in the epididymis and vas deferens.
- Sperm enter the urethra (in the penis) prior to ejaculation, along with seminal fluid (produced by seminal vesicles, the prostate gland, and bulbourethral glands).
- **Sterility** is the inability to produce offspring.
- Spermatogenesis occurs in the seminiferous tubules of the testes, which also produce testosterone in interstitial cells.
- **Testosterone** brings about the maturation of the sex organs during **puberty** and promotes the secondary sex characteristics of males.

Female Reproductive System

- Oocytes are produced in the ovary and move through the uterine tubes to the uterus. The **uterus** opens into the vagina at the **cervix.** The external genital area of women includes the clitoris, the labia minora, the labia majora, and the vaginal opening.
- Ovulation is the release of a secondary oocyte from the follicle. The follicle and later the corpus luteum produce estrogen and progesterone.

The ovarian cycle (including the **follicular phase** and **luteal phase**) and **menstrual cycle** last 28 days. The events of these cycles are as follows:

- Days 1–13: **Menstruation** of the **menses** occurs for 5 days; the anterior pituitary produces FSH, and follicles produce primarily estrogen. Estrogen causes the endometrium to increase in thickness.
- Day 14: Ovulation occurs.
- Days 15–28: LH from the anterior pituitary causes the corpus luteum to produce progesterone. Progesterone causes the endometrium to become secretory.

Estrogen and **progesterone** bring about the maturation of the sex organs during puberty and promote the secondary sex characteristics of females. **Menopause** results in the stopping of the ovarian and menstrual cycles.

Control of Reproduction

Numerous birth control methods and devices are available for those who wish to prevent pregnancy. The most effective method is abstinence. Others are birth control pills, contraceptive implants, contraceptive injections, condoms, diaphragms, intrauterine devices (IUDs), contraceptive vaccines, and emergency contraception.

Infertility

Infertility is the inability to produce offspring after 1 year of unprotected intercourse. Infertility may be treated using assisted methods of reproduction.

Sexually Transmitted Diseases

Sexually transmitted diseases include the following:

- **Acquired immunodeficiency syndrome (AIDS),** an epidemic disease caused by the **human immunodeficiency virus (HIV)** that destroys the immune system
- **Genital herpes,** which repeatedly flares up
- **Human papillomavirus (HPV),** which may cause genital warts and cancer of the cervix
- **Hepatitis,** especially types A and B
- **Chlamydia** and **gonorrhea,** which cause **pelvic inflammatory disease (PID)**
- **Syphilis,** which leads to cardiovascular and neurological complications if untreated
- **Trichomoniasis,** which is caused by a parasitic protozoan

29.3 Human Embryonic Development

Embryonic development encompasses all the events that occur from **fertilization** to a fully formed animal—in this case, a human.

Fertilization

The acrosome of a sperm releases enzymes that digest a pathway for the sperm through the zona pellucida. The sperm nucleus enters the egg and fuses with the egg nucleus.

Early Embryonic Development (Months 1 and 2)

- **Cleavage,** which occurs in the uterine tube, is cell division and formation of a **morula** and then a **blastocyst.** The blastocyst implants itself in the endometrium.
- **Gastrulation** is invagination of cells into the blastocoel to form the **gastrula,** which in turn results in formation of the **germ layers.** The germ layers are the **ectoderm, mesoderm,** and **endoderm.**
- Organ formation can be related to the germ layers. Organ development begins with the formation of the **notochord** and neural tube (**neurulation**). **Induction** helps account for the steady progression of organ formation during embryonic development.
- The **placenta** has a fetal side and a maternal side. Gases and nutrients are exchanged at the placenta. The blood vessels in the **umbilical cord** are an extension of the embryo/fetal cardiovascular system and carry blood to and from the placenta.

Fetal Development and Birth (Months 3–9)

- During fetal development, the refinement of organ systems occurs and the fetus adds weight.
- Birth has three stages: The cervix dilates, the baby is born, and the placenta is delivered.

ASSESS

Testing Yourself

Choose the best answer for each question.

29.1 How Animals Reproduce

1. Parthenogenesis is
   a. the process by which the germ layers form.
   b. involved in the formation of the notochord.
   c. the formation of sperm.
   d. the development of an unfertilized egg.
2. Which of the following is incorrect regarding sexual reproduction?
   a. It involves two parents.
   b. The offspring have the same genetic combinations as the parents.
   c. Gonads produce egg and sperm cells.
   d. Copulation brings the male and female gametes together.

29.2 Human Reproduction

3. Label the parts of the male reproductive system in the following illustration.

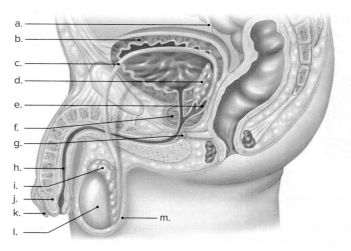

4. Label the parts of the female reproductive system in the following illustration.

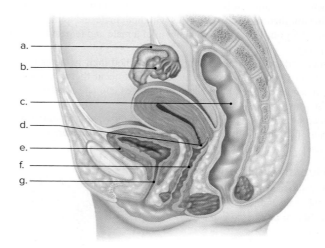

5. Spermatogenesis produces cells that are
   a. diploid and genetically identical to each other.
   b. diploid and genetically different from each other.
   c. haploid and genetically identical to each other.
   d. haploid and genetically different from each other.

6. Hepatitis is caused by a
   a. bacterium.
   b. fungus.
   c. protozoan.
   d. virus.

7. The assisted reproductive technology (ART) in which harvested sperm are placed in the vagina by a physician is
   a. in vitro fertilization (IVF).
   b. artificial insemination by donor (AID).
   c. gamete intrafallopian transfer (GIFT).
   d. intracytoplasmic sperm injection (ICSI).

8. The HPV virus can cause
   a. genital warts.
   b. AIDS.
   c. herpes.
   d. hepatitis.

29.3 Human Embryonic Development

9. The process by which one tissue or organ influences the development of another is
   a. induction.
   b. insemination.
   c. neurulation.
   d. cleavage.
   e. gastrulation.

For questions 10–13, identify the stage of embryonic development in the key that matches the description.

Key:

a. cleavage
b. gastrulation
c. neurulation
d. organ formation

10. The notochord is formed.
11. The zygote divides without increasing in size.
12. Ectoderm and endoderm are formed.
13. Reflex actions, such as response to touch, develop.

ENGAGE

Thinking Critically

1. Female athletes who train intensively often stop menstruating. The important factor appears to be the reduction of body fat below a certain level. Give a possible evolutionary explanation for a relationship between body fat in females and reproductive cycles.

2. Human females undergo menopause typically between the ages of 45 and 55. In most other animal species, both males and females maintain their reproductive capacity throughout their lives. Why do you suppose human females lose that ability in midlife?

3. In one form of prenatal testing for fetal genetic abnormalities, chorionic villi samples are taken and analyzed. Why is this possible?

4. At-home pregnancy tests check for the presence of hCG in a female's urine. Where does hCG come from? Why is hCG found in a pregnant woman's urine?

5. The average sperm count in males is now lower than it was several decades ago. The reasons for the lower sperm counts usually seen today are not known. What data might be helpful in order to formulate a testable hypothesis?

30

Ecology and Populations

© Jason Lindsey/Alamy

OUTLINE

30.1 The Scope of Ecology   581

30.2 The Human Population   583

30.3 Characteristics of Populations   588

30.4 Life History Patterns and
Extinction   595

BEFORE YOU BEGIN

Before beginning this chapter, take a few moments to review the following discussions.

Section 1.1 What is the relationship between populations, ecosystems, and the biosphere?

Section 16.2 What have been the causes of mass extinctions in the past?

The Uncontrollable Asian Carp

In the early 1970s, several species of Asian carp were imported into Arkansas for the biocontrol of algal blooms in aquaculture facilities. Shortly thereafter, they escaped into the middle and lower Mississippi drainage. Over time, in some areas they have become the most abundant species and have now spread throughout the majority of the Mississippi River system.

Asian carp mainly eat the microscopic algae and zooplankton in freshwater ecosystems. They can achieve weights up to a hundred pounds, grow to a length of more than 4 ft, and live up to 30 years. Because of their voracious appetite, they have the potential to reduce the populations of native fish species such as gizzard shad and bigmouth buffalo and of native mussels, via competition for the same food sources. The main fear is that the Asian carp will put extreme pressure on the zooplankton populations, which can lead to a dense planktonic algal bloom. Ultimately, they can produce a significant disruption of the freshwater ecosystems that they invade.

These fish also pose an economic threat by fouling the nets of commercial fishermen. Another major concern is the impact these fish may have on the Great Lakes' annual $7 billion fishing industry. Our knowledge of population growth and regulation has led to controls on fishing, including size limits and catch limits, to try and sustain the world's commercially valuable fish populations. This knowledge can also be applied to management of species that can be considered harmful to a specific ecosystem.

In this chapter, we will explore not only the science of ecology, but also the dynamics of population structure and growth.

As you read through this chapter, consider the following questions:

1. Why are the biotic potentials of exotic species often higher than those of native species?
2. Do population growth models apply to humans the same way they apply to other species?

30.1 The Scope of Ecology

Learning Outcomes

Upon completion of this section, you should be able to

1. Describe the levels of biological organization from the organism to the biosphere.
2. Define *ecology,* and state its relationship to environmental science.

Ecology is the scientific study of the interactions of organisms with each other and their physical environment. Ecology is one of the two biological sciences of most interest to the public today, the other being genetics. Ecological studies offer information key to the survival of all species, present and future. Understanding ecology will help us make informed decisions, ranging from what kind of car to drive to how to support the preservation of a forested area in our town. Current ecological decisions will affect not only our lives but also the lives of generations to come.

Ecology involves the study of several levels of biological organization (**Fig. 30.1**). Some ecologists study individual **organisms,** focusing on their adaptations to a particular environment. For example, organismal ecologists might investigate which features enable a clownfish to live in a coral reef.

While biologists have several different ways of defining a **species,** for the purposes of our discussions of ecology, we will focus on organisms that are reproductively compatible and are capable of interbreeding and producing viable offspring.

A **population** is a group of individuals of the same species occupying a given location at the same time. At the population level of study, ecologists describe changes in population size over time. For example, a population ecologist might compare the number of clownfishes living in a given location today with data obtained from the same location 20 years ago.

A **community** consists of all the various populations in a particular locale. A coral reef contains numerous populations of fishes, crustaceans, corals, algae, and so forth. At the community level, ecologists study how the interactions between populations affect the populations' well-being. For example, a community ecologist might study how a decrease in algal populations affects the population sizes of crustaceans and fishes living on the coral reef.

An **ecosystem** consists of a community of living organisms as well as their physical environment. As an example of how the physical environment affects a community, consider that the presence of suspended particles in the water decreases the amount of sunlight reaching algae living on the coral reef. Without solar energy, algae cannot produce the organic nutrients they and the other populations require.

Finally, the **biosphere** is the zone at the Earth's surface—air, water, and land—where life exists. Having information about the many levels of organization within a coral reef allows ecologists to understand how a coral reef contributes to the biodiversity and dynamics of the biosphere.

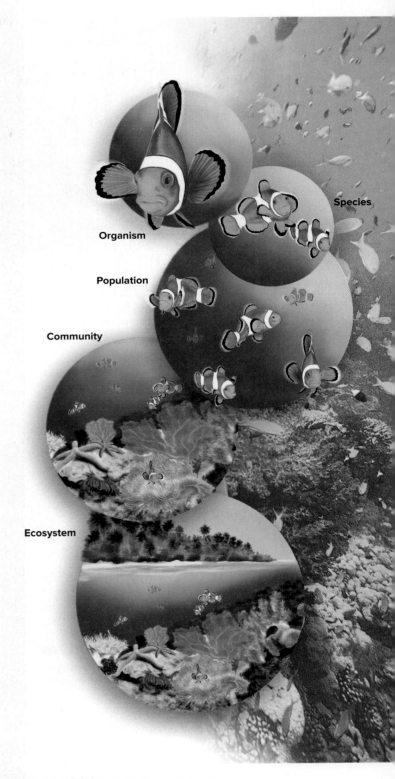

Figure 30.1 Levels of organization.

The study of ecology encompasses the organism, species, population, community, and ecosystem levels of biological organization.

© Frank & Joyce Burek/Getty RF

a.

b.

Figure 30.2 **Studying the effects of fire.**

Fire is a natural occurrence in lodgepole pine forests, such as these in Yellowstone National Park. **a.** Recently burned forest. **b.** The same forest, several years after the fire.

(a): © Dennis MacDonald/Age fotostock/Superstock; (b): © TMI/Alamy

Ecology: A Biological Science

Ecology began as a part of **natural history,** the discipline dedicated to observing and describing organisms in their environment, but today ecology is also an experimental science. A central goal of ecology is to develop models that explain and predict the distribution and abundance of populations based on their interactions within an ecosystem. Achieving such a goal involves testing hypotheses. For example, ecologists might formulate and test hypotheses about the role fire plays in maintaining a lodgepole pine forest (**Fig. 30.2**). To test these hypotheses, ecologists can compare the characteristics of a community before and after a burn. Ultimately, ecologists wish to develop models about the distribution and abundance of ecosystems within the biosphere.

Ecology and Environmental Science

The field of *environmental science* applies ecological principles to practical human concerns. Ecological principles help us understand why a functioning biosphere is critical to our survival and why our population of over 7 billion people and our overconsumption of resources pose threats to the biosphere.

Conservation biology is a discipline of ecology that studies all aspects of biodiversity, with the goal of conserving natural resources, including wildlife, for the benefit of this generation and future generations. Conservation biology recognizes that wildlife species are an integral part of a well-functioning biosphere, on which human life depends.

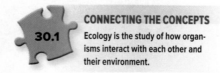

CONNECTING THE CONCEPTS

30.1

Ecology is the study of how organisms interact with each other and their environment.

Check Your Progress 30.1

1. List the levels of biological organization, starting with the organism.
2. Distinguish an ecosystem from a community.
3. Define *ecology.*

30.2 The Human Population

Learning Outcomes

Upon completion of this section, you should be able to

1. Identify the relationship among the birthrate, death rate, and annual growth rate of a population.

2. Distinguish between more-developed countries (MDCs) and less-developed countries (LDCs) with regard to population growth.

3. Compare the environmental impacts of MDCs and LDCs.

The human population covers the majority of the Earth. However, on both a large scale and a small scale, the human population has a clumped distribution; 59% of the world's people live in Asia, and most live in China and India. Mongolia has a population density of only 0.25 persons per square kilometer, while Bangladesh has a density of over 1,000 persons per square kilometer. On every continent, human population densities are highest along the coasts.

Present Population Growth

For most of the twentieth century, the growth of the global human population was relatively slow, but as more individuals achieved reproductive age, the growth rate began to increase. The Industrial Revolution (which began sometime before 1800) brought an increase in the production of food and medicines, along with a decrease in the death rate. At this time, the growth curve for the human population began to increase steeply, so that currently the population is undergoing rapid growth (**Fig. 30.3**).

Figure 30.3 Human population growth.

It is predicted that the world's population size may grow to between 9 and 10 billion by 2050, depending on the future growth rate. Close examination of the curve shows changes in growth that occurred in the fourteenth century during the Black Plague, in the nineteenth century with the Industrial Revolution, and in the twentieth century with advances in science and medicine.

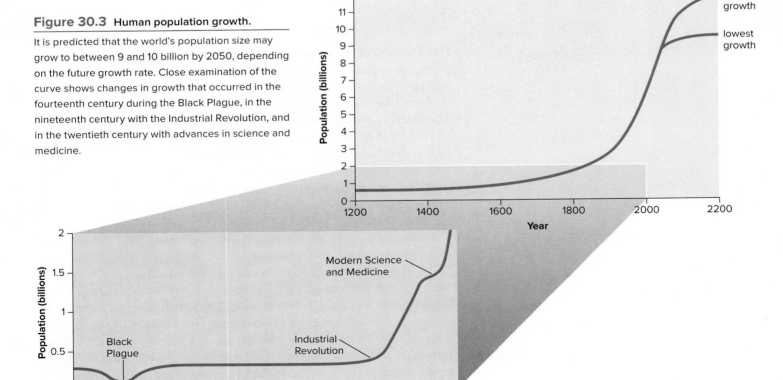

Just as with populations in nature, the growth rate of the human population is determined by the difference between the number of people born and the number of people who die each year. For example, the current global birthrate is an estimated 18.7 per 1,000 per year, while the global death rate is approximately 7.9 per 1,000 per year. From these numbers, we can calculate that the current annual growth rate of the human population is

$$(18.7 - 7.9)/1,000 = (10.8)/1,000 = 0.0108 \times 100 = 1.08\%$$

Connections: Health

How do birthrates and death rates vary among countries?

© ER Productions/Corbis

In 2015, the country with the highest projected birthrate was Niger (6.76 children born per woman), followed by Burundi (6.09 children per woman) and Mali (6.06 children per woman), whereas Singapore was the country with lowest birthrate (0.81 children born per woman). For death rates, Lesotho had the highest rate (14.89 deaths per 1,000 people) and Qatar had one of the lowest (1.53 deaths per 1,000). With respect to overall annual growth rates, the fastest-growing countries are projected to be South Sudan (4.02%) and Malawi (3.31%). Several countries are expected to experience declines, including Latvia (–1.06%) and Lithuania (–1.04%). In comparison, in the United States, the projected birthrate is 1.87 births per woman, the death rate is 8.15 per 1,000 people, and the annual growth rate 0.78%.

Future Population Growth

Future growth of the human population is of great concern. It's possible that population increases will exceed the rate at which resources can be supplied. The potential for dire consequences can be appreciated by considering the *doubling time*—the length of time it takes for a population to double in size. The doubling time of the human population is now estimated to be around 67 years. Will we be able to meet the extreme demands for resources by such a large population increase in such a short period of time? Already, there are areas across the globe where people have inadequate access to fresh water, food, and shelter, yet in just 67 years the world would need to double the amount of food, jobs, water, energy, and other resources to maintain the present standard of living.

Rapid growth usually begins to decline when resources, such as food and space, become scarce. Then population growth declines to zero, and the population levels off at a size appropriate to the carrying capacity of its environment. The **carrying capacity** is the number of individuals the environment can sustain for an indefinite period of time. The Earth's carrying capacity for humans has not been determined; some authorities think the Earth is potentially capable of supporting far more people than currently inhabit the planet, perhaps as many as 50 billion. Others believe that we may have already exceeded the number of humans the Earth can support and that the human population may undergo a catastrophic crash.

Projected Annual Growth Rate of Country Populations, 2010–2050

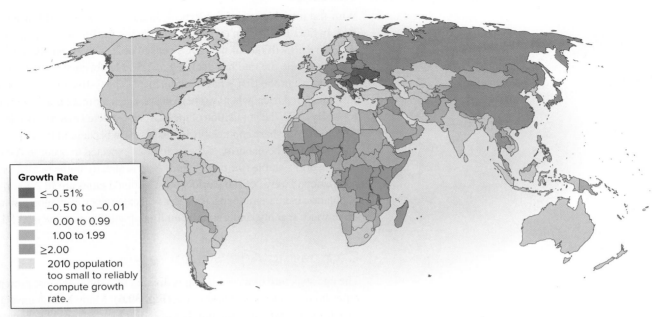

Growth Rate
- ≤−0.51%
- −0.50 to −0.01
- 0.00 to 0.99
- 1.00 to 1.99
- ≥2.00
- 2010 population too small to reliably compute growth rate.

Figure 30.4 Projected population growth rates 2010–2050.

The highest population growth rates are found in Africa and the Middle East, while the growth rates throughout Europe are much lower.

Source: GlobalHealthFacts.org. Based on 2012 data.

More-Developed Versus Less-Developed Countries

Complicating the issue of future growth for the human population is the fact that not all countries have the same growth rate (**Fig. 30.4**). The countries of the world today can be divided into two groups (**Fig. 30.5**). In the **more-developed countries (MDCs),** such as those in North America and Europe, population growth is modest and the people enjoy a fairly good standard of living. In the **less-developed countries (LDCs),** such as those in Latin America, Asia, and Africa, population growth is dramatic and the majority of the people live in poverty.

The MDCs

The MDCs did not always have such a low growth rate. With the development of modern medicine, improved socioeconomic conditions, and the decrease in the death rate, their populations doubled between 1850 and 1950. The decline in the death rate was followed shortly thereafter by a decline in birthrate, so populations in the MDCs have experienced only modest growth since 1950. This sequence of events (i.e., decreased death rate followed by decreased birthrate) is termed a *demographic transition.*

The growth rate for the MDCs as a whole is now about 0.1%, but populations in several countries are not growing at all or are actually decreasing. The MDCs are expected to increase by 52 million between 2002 and 2050, but this modest amount will still keep their total population at just about 1.2 billion. In contrast to the other MDCs, population growth in the United States is not leveling off and continues to increase. The United States currently has a growth rate of 0.78%. This higher rate is due to the fact that many people immigrate to the United States each year, and a large number of the women are still of reproductive age. Therefore, it is unlikely that the United States will experience a decline in its growth rate in the near future.

a.

b.

Figure 30.5 More-developed versus less-developed countries.

People in (**a**) more-developed countries have a relatively high standard of living and will contribute the least to world population growth, while people in (**b**) less-developed countries have a low standard of living and will contribute the most to world population growth.

(a): © Ariel Skelley/Blend Images LLC; (b): © Henk Badenhorst/Getty RF

The LDCs

With the introduction of modern medicine following World War II, the death rates in the LDCs began to decline sharply but the birthrates remained high. The *collective* growth rate of the LDCs peaked at 2.5% between 1960 and 1965, and since that time the rate has declined to 1.5%. Unfortunately, the growth rates in the majority of the LDCs have not declined. Thirty-five of these countries are in sub-Saharan Africa, where women on average bear more than five children each.

By 2050, the population of the LDCs is expected to jump to at least 8 billion. Africa will not share appreciably in this increase because of the many deaths there due to AIDS. The majority of the increase is expected to occur in Asia. With 31% of the world's arable (farmable) land, Asia is already home to 59% of the human population. Twelve of the world's most polluted cities are in Asia. If the human population increases as expected, Asia will experience even more urban pollution, acute water scarcity, and a significant loss of wildlife over the next 50 years.

Comparing Age Structures

The age structure of a population is divided into three groups: prereproductive, reproductive, and postreproductive (**Fig. 30.6**). Many MDCs have a stable age structure, meaning that the number of individuals is about the same in all three groups. Therefore, MDC populations will grow slowly if couples have two children, and they will slowly decline if each couple has fewer than two children. A birthrate of two children per couple is often called **replacement reproduction,** since if each couple produces only two children, they are replacing themselves. In reality, due to mortality, the replacement reproduction rate is often slightly greater than two children per couple. In contrast, the age-structure diagram of most LDCs has a pyramid shape, with the prereproductive group the largest. Therefore, LDC populations will continue to expand, even if replacement reproduction is attained. It might seem at first that replacement reproduction would cause an LDC population to undergo zero population growth and therefore no increase in population size. However, if there are more young women entering the reproductive years than there are older women leaving them, replacement reproduction still results in population growth. We will take a closer look at age-structure diagrams in Section 30.3.

Figure 30.6 **Age-structure diagrams.**

The diagrams predict that (**a**) the LDCs will expand rapidly due to their age distributions, while (**b**) the MDCs are approaching stabilization.

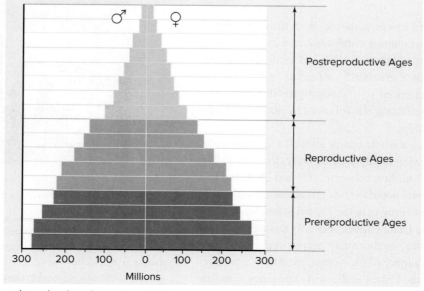

a. Less-developed countries (LDCs)

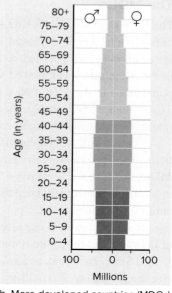

b. More-developed countries (MDCs)

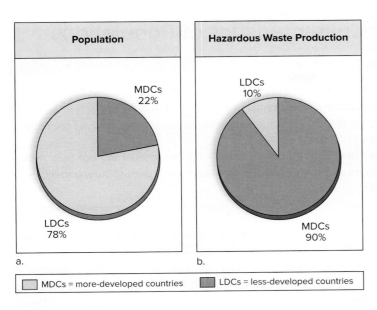

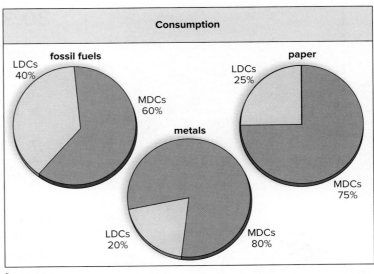

a.

b.

c.

MDCs = more-developed countries   LDCs = less-developed countries

Population Growth and Environmental Impact

Population growth is putting extreme pressure on each country's social organization, the Earth's resources, and the biosphere. Since the population of the LDCs is still growing at a significant rate, it might seem that their population increase is the main cause of environmental degradation. But this is not necessarily the case, because the MDCs consume a much larger proportion of the Earth's resources than do the LDCs. This excessive consumption leads to environmental degradation, which is of great concern.

Environmental Impact

The environmental impact of a population is measured not only in terms of population size but also in terms of resource consumption and pollution per capita. Therefore, there are two possible causes of environmental impact: population size and resource consumption. Overpopulation is more obvious in LDCs; resource consumption is more obvious in MDCs, because per capita consumption is so much higher in those countries. For example, an average American family, in terms of per capita energy consumption, is the equivalent of 20 people in India. We need to realize that only a limited number of people can be sustained anywhere near the standard of living enjoyed in the MDCs.

The comparative environmental impacts of MDCs and LDCs are shown in **Figure 30.7**. The MDCs account for only about one-fourth of the world population. However, compared with the LDCs, the MDCs account for 90% of the hazardous waste production, due to their high rate of consumption of such resources as fossil fuels, metals, and paper.

Figure 30.7 Environmental impacts.

a. The combined population of the MDCs is much smaller than that of the LDCs. The MDCs account for 22% of the world's population, and the LDCs account for 78%. **b.** MDCs produce most of the world's hazardous wastes—90% for MDCs compared with 10% for LDCs. The production of hazardous waste is tied to **(c)** the consumption of fossil fuels, metals, and paper, among other resources. The MDCs consume 60% of fossil fuels, compared with 40% for the LDCs, 80% of metals compared with 20% for the LDCs, and 75% of paper compared with 25% for the LDCs.

CONNECTING THE CONCEPTS

30.2 The human population is still growing, with the majority of that growth in the less-developed countries.

Check Your Progress 30.2

1. Calculate the annual growth rate of a population that is experiencing a birthrate of 18.5% and a death rate of 9.8%.

2. Compare the characteristics of an MDC with those of an LDC, and give an example of a country in each group.

3. Explain what is meant by the term *replacement reproduction*.

4. Contrast the environmental impact of an MDC with that of an LDC.

a. Mature desert shrubs

Young, small shrubs

b. Clumped

Medium shrubs

c. Random

Large shrubs

d. Uniform

Figure 30.8 Patterns of distribution within a population.

a. A population of mature desert shrubs. **b.** Young, small desert shrubs are clumped. **c.** Medium shrubs are randomly distributed. **d.** Large shrubs are uniformly distributed.

(a): © Evelyn Jo Johnson

Connections: Environment

Where are the highest and lowest population densities of humans?

Based on 2015 demographic data, the country with the highest population density is the Principality of Monaco (in Europe), with a density of 25,718 people per square kilometer. Greenland tech-

© Photodisk/PunchStock RF

nically has the lowest population density (0.03), followed by Mongolia (1.87) and the Western Sahara (2.27). Overall, the land masses of Earth have a population density of approximately 47 people per square kilometer.

30.3 Characteristics of Populations

Learning Outcomes

Upon completion of this section, you should be able to

1. Distinguish among the types of population distribution.
2. Describe how age-structure diagrams and survivorship curves are used to predict future growth.
3. Explain how biotic potential contributes to exponential and logistic growth.
4. Compare and contrast density-dependent and density-independent factors.

Various characteristics of populations change over time. Populations are periodically subject to environmental instability. During times of environmental instability, individuals are under the pressures of natural selection. Individuals better adapted to the environment will leave behind more offspring than those that are not as well adapted. Once the population becomes adapted to the new environment, an increase in size can occur. When an ecologist lists the characteristics of a population, it is a snapshot of the characteristics at a particular time and place.

Distribution and Density

Resources are the components of an environment that support its organisms. Food, water, shelter, and space are some of the important resources for populations. The availability of resources influences the spatial distribution of individuals in a given area. Three terms—**clumped distribution, random distribution,** and **uniform distribution**—are used to describe the observed patterns of distribution. In a study of desert shrubs, it was found that the distribution changes from clumped to random to uniform as the plants mature (**Fig. 30.8**). After sufficient study, competition for belowground resources was found to be the main cause for these distribution patterns.

Suppose we were to step back and consider the distribution of individuals not within a single desert, forest, or pond but within a species' range. A **range** is that portion of the globe where the species can be found. On this scale, all the members of a species within the range make up the population. Members of a population are clumped within the range because they are located in areas that contain the resources they require.

Population density is the number of individuals per unit area (or unit volume) of a particular habitat. Population density tends to be higher in areas with plentiful resources than in areas with scarce resources. Other factors can be involved, however. In general, population density declines with increasing body size. Consider that tree seedlings show a much higher population density than do mature trees. Therefore, more than one factor must be taken into account when explaining population densities.

Population Growth

As mentioned in our discussion on the human population (Section 30.2), the growth rate is based on the number of individuals born and the number of individuals who die each year within the population. Growth occurs when the number of births exceeds the number of deaths. To take another example, if the number of births is 30 per year per 1,000 individuals and the number of deaths

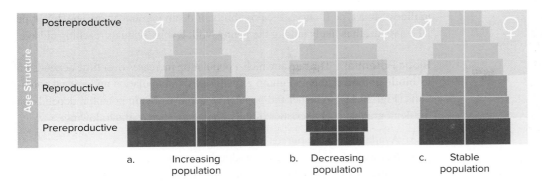

a. Increasing population b. Decreasing population c. Stable population

Figure 30.9 Age-structure diagrams.

Typical age-structure diagrams for hypothetical populations that are (**a**) increasing, (**b**) decreasing, and (**c**) stable. Different numbers of individuals in each age class create these distinctive shapes. In each diagram, the *left half* represents males, while the *right half* represents females.

is 10 per year per 1,000 individuals, the growth rate is 2%. In cases where the number of deaths exceeds the number of births, the value of the growth rate is negative—the population is shrinking.

Demographics and Population Growth

Availability of resources and certain characteristics of a population—called **demographics**—influence the population growth rate. One demographic characteristic of interest for all populations is the **age structure** of a population. Bar diagrams (like that shown in Figure 30.6) are typically used to depict the age structure of a population. The population is divided into members that are *prereproductive, reproductive,* or *postreproductive,* based on their age. If the prereproductive and reproductive individuals outnumber those that are postreproductive, the birthrate exceeds the death rate (**Fig. 30.9***a*). In contrast, if the postreproductive individuals exceed those that are prereproductive and/or reproductive, the number of individuals in the population will decrease over time because the death rate exceeds the birthrate (Fig. 30.9*b*). If the numbers of prereproductive, reproductive, and postreproductive individuals are approximately equal and the birthrate equals the death rate, then the number of individuals in the population will remain relatively stable over time (Fig. 30.9*c*).

Ecologists also study patterns of **survivorship**—how age at death influences population size. For example, if members of a population die young, the number of reproductive individuals will decrease. The first step is to construct a life table, which lists the number of individuals of each age or age range that are alive at a given time. The best way to arrive at a life table is to identify a large number of individuals that are born at about the same time and keep records on them from birth until death.

A life table for Dall sheep is given in **Figure 30.10***a*. This table was constructed by gathering a large number of skulls and counting the growth rings on the horns to estimate each sheep's age at death, given that one growth ring is produced annually. Plotting the number of survivors per 1,000 births against age produces a *survivorship curve* (Fig. 30.10*b*). Each species has its own pattern, but three types of survivorship curves are common. In a type I curve, typical of Dall sheep and humans, survival is high until old age, when deaths increase due

Figure 30.10 Life table and survivorship curves.

a. A life table for Dall sheep. **b.** Three typical survivorship curves. Among Dall sheep, with a type I curve, most individuals survive until old age, when they gradually die off. Among hydras, with a type II curve, there is an equal chance of death at all ages. Among oysters, with a type III curve, most die when they are young, and few become adults that are able to survive until old age.

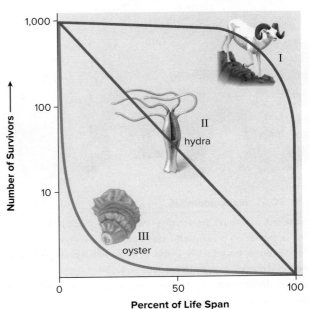

Age (years)	Number of survivors at beginning of year	Number of deaths during year
0–1	1,000	199
1–2	801	12
2–3	789	13
3–4	776	12
4–5	764	30
5–6	734	46
6–7	688	48
7–8	640	69
8–9	571	132
9–10	439	187
10–11	252	156
11–12	96	90
12–13	6	3
13–14	3	3
14–15	0	

1,000–199
801–12
etc.

a.

b.

to illness. In a type II curve, the possibility of death is equal at any age. In a type III curve, death is likely among the young, with few individuals reaching old age.

Biotic Potential   The rate at which a population increases over time depends on its **biotic potential,** the maximum growth rate of a population under ideal conditions (**Fig. 30.11**). Whether the population achieves biotic potential depends on the demographic characteristics of the population, such as the following:

- Availability of resources
- Number of offspring per reproductive event
- Chances of survival until age of reproduction
- How often each individual reproduces
- Age at which reproduction begins

Patterns of Population Growth

The patterns of population growth are dependent on (1) the biotic potential of the species and (2) the availability of resources. The two fundamental patterns of population growth are exponential growth and logistic growth.

Exponential Growth

Suppose ecologists are studying the growth of a population of insects that are capable of infesting and taking over an area. Under these circumstances, **exponential growth** is expected. An exponential pattern of population growth results in a J-shaped curve (**Fig. 30.12**). This growth pattern can be likened to compound interest at a bank: The more your money increases, the more interest you will get. If the insect population has 2,000 individuals and the per capita rate of increase is 20% per month, there will be 2,400 insects after one month, 2,880 after two months, 3,456 after three months, and so forth.

Notice that a J-shaped curve has two phases:

Lag phase: Growth is slow because the number of individuals in the population is small.
Exponential growth phase: Growth is accelerating.

Usually, exponential growth continues as long as there are sufficient resources available. When the number of individuals in the population approaches the maximum number that can be supported by available resources, competition for these resources will increase.

a.

b.

Figure 30.11 **Biotic potential.**

A population's maximum growth rate under ideal conditions—that is, its biotic potential—is greatly influenced by the number of offspring produced in each reproductive event. **a.** Mice, which produce many offspring that quickly mature to produce more offspring, have a much higher biotic potential than (**b**) the rhinoceros, which produces only one or two offspring per infrequent reproductive event.

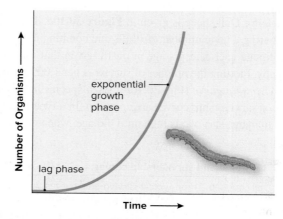

Figure 30.12 **Exponential growth.**

Exponential growth results in a J-shaped curve because the growth rate is positive.

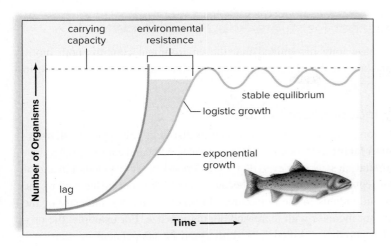

Figure 30.13 Logistic growth.

Logistic growth results in an S-shaped growth curve because environmental resistance causes the population size to level off and be in a steady state at the carrying capacity of the environment.

Logistic Growth

Usually, exponential growth cannot continue for long because of environmental resistance. Environmental resistance encompasses all those environmental conditions—such as limited food supply, accumulation of waste products, increased competition, and predation—that prevent populations from achieving their biotic potential. The amount of environmental resistance will increase as the population grows larger. As resources decrease, population growth levels off and a pattern of population growth called **logistic growth** occurs. Logistic growth results in an S-shaped growth curve (**Fig. 30.13**).

The characteristics of logistic growth include:

Lag phase: Growth is slow because the number of individuals in the population is small.

Exponential growth phase: Growth is accelerating due to biotic potential (see Fig. 30.11).

Logistic growth phase: The rate of population growth slows down due to environmental resistance.

Stable equilibrium phase: Little, if any, growth takes place because births and deaths are about equal.

The stable equilibrium phase is said to occur at the carrying capacity of the environment. This number is not a constant; it tends to fluctuate around a value that is determined by the current environmental resources. For example, the carrying capacity of an island may stay constant for years, but then decrease due to a period of drought.

Applications Our knowledge of logistic growth has practical applications. The model predicts that exponential growth will occur only when population size is much lower than the carrying capacity. Thus, if humans are using a fish population as a continuous food source, it is best to maintain the fish population size in the exponential phase of growth when biotic potential is having its full effect and the birthrate is the highest. If people overfish, the fish population will sink into the lag phase, and it may be years before exponential growth recurs (**Fig. 30.14**). On the other hand, if people are trying to limit the growth of a pest, it is best to reduce the carrying capacity, rather than the population size. Reducing the population size only encourages exponential growth to begin once again. Farmers can reduce the carrying capacity for a pest by alternating rows of different crops instead of growing one type of crop throughout the entire field.

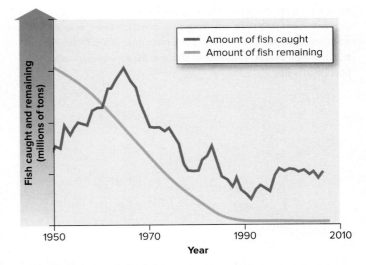

Figure 30.14 Sustainable population density.

As the amount of fish caught increased over the 1950s and 1960s, the remaining amount decreased. Because fishing still continues, fish population growth is maintained in its lag phase. Fishing must be reduced to allow the fish population to enter an exponential growth phase. Then a balance between fish population growth and fish caught could be maintained regularly at a higher amount.

a. Low density of mice

b. High density of mice

Figure 30.15 Density-independent effects.

The impact of a density-independent factor, such as weather or a natural disaster, is not influenced by population density. Two populations of field mice are in the path of a flash flood. **a.** In the low-density population, 3 out of the 5 mice drown, a 60% death rate. **b.** In the high-density population, 12 out of 20 mice drown—also a 60% death rate.

Factors That Regulate Population Growth

Ecologists have long recognized that the environment contains both biotic (living) and abiotic (nonliving) components that play an important role in regulating population size.

Density-Independent Factors

Abiotic factors, such as weather and natural disasters, are typically **density-independent factors.** Abiotic factors can cause sudden and catastrophic reductions in population size. However, density-independent factors cannot in and of themselves regulate population size because the effect is not influenced by the number of individuals in the population. In other words, the intensity of the effect does not increase with increased population size. For example, the proportion of a population killed in a flash flood is independent of density—floods don't necessarily kill a larger percentage of a dense population than of a less dense population. If a mouse population has only 5 members and 3 drown in a flash flood, the death rate is 60% (**Fig. 30.15a**). If a mouse population has 20 mice and 12 drown, the death rate is still 60% (Fig. 30.15b).

Density-Dependent Factors

Biotic factors are considered **density-dependent factors,** because the percentage of the population affected increases as the density of the population increases. Competition, predation, and parasitism are all biotic factors that increase in intensity as the density increases. We will discuss these interactions between populations again in Section 31.1 because they influence community composition and diversity.

Competition **Competition** results when organisms attempt to use resources (such as light, food, or space) that are in limited supply. Competition may occur between different species, or between members of the same species. Competition limits access to the resources necessary to ensure survival, reproduction, or some other aspect of an organism's life cycle.

Let's consider a woodpecker population in which members have to compete for nesting sites. Each pair of birds requires a tree hole in order to raise offspring. If the number of tree holes is the same as or greater than the number of breeding pairs, each pair can have a hole in which to lay eggs and rear young birds (**Fig. 30.16a**). But if there are fewer holes than breeding pairs, each pair must compete to acquire a nesting site (Fig. 30.16b). Pairs that fail to gain access to holes will be unable to contribute new members to the population.

A well-known example of competition for food affected the reindeer on St. Paul Island, Alaska. Overpopulation led to the overconsumption of resources, causing the population to crash—that is, it became drastically reduced.

Resource partitioning among different age groups is a way to reduce competition for food. During the life cycle of butterflies, the caterpillar stage requires different food than the adult stage. Caterpillars graze on leaves, while adult butterflies feed on nectar produced by flowers. Therefore, the parents are less apt to compete with their offspring for food.

Predation **Predation** occurs when one organism, the predator, eats another, the prey. In the broadest sense, predation includes not only animals such as lions that kill zebras but also filter-feeding blue whales, which strain krill from the ocean waters, and herbivorous deer, which browse on trees and bushes.

The effect of predation generally increases as the prey population increases in density, because prey are easier to find when their population is

a. Low density of birds   b. High density of birds

Figure 30.16 Density-dependent effects—competition.

The impact of competition is directly proportional to the density of a population. **a.** When density is low, every member of the population has access to the resource. **b.** When density is high, members of the population must compete to gain access to the available resources, and some fail to do so.

larger. Consider a field inhabited by a population of mice (**Fig. 30.17**). Each mouse must have a hole in which to hide to avoid being eaten by a hawk. If there are 102 mice but only 100 holes, 2 mice will be left out in the open. It might be hard for the hawk to find only 2 mice in the field. If neither mouse is caught, the predation rate is 0/2 = 0%. However, if there are 200 mice and only 100 holes, there is a greater chance that the hawk will be able to find some of the 100 mice without holes. If half of the exposed mice are caught, the predation rate is 50/100 = 50%. Therefore, increasing the density of the available prey has increased the proportion of the population preyed upon.

Predator-Prey Population Cycles Rather than remaining steady, some predator and prey populations experience an increase, then a decrease, in population size. Such a cycle occurs between the snowshoe hare and the Canada lynx, a species of wild cat (**Fig. 30.18**). The snowshoe hare is a common herbivore in the coniferous forests of North America, where it feeds on terminal twigs of various shrubs and small trees. Investigators first assumed that the predatory lynx brings about a decrease in the hare population, and this decrease in prey brings about a subsequent decrease in the lynx population. Once the hare population recovers, so does the lynx population, and the result is a boom-bust cycle. But some biologists noted that the decline in snowshoe hare abundance is accompanied by low growth and reproductive rates, which could be signs of food shortage. A field experiment showed that, if the lynx predator

CONNECTING THE CONCEPTS
30.3 Population growth influences the community and ecosystem structure.

a. Field with a low-density of mice.   b. Field with a high-density of mice.

Figure 30.17 Density-dependent effects—predation.

The impact of predation on a population is directly proportional to the density of the population. **a.** In a low-density population, the chances of a predator finding the prey are low, resulting in little predation (a mortality rate of 0/2, or 0%). **b.** In a higher-density population, there is a greater likelihood of the predator locating potential prey, resulting in a greater predation rate (a mortality rate of 50/100, or 50%).

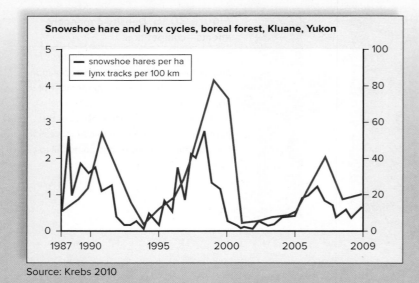

Snowshoe hare and lynx cycles, boreal forest, Kluane, Yukon

— snowshoe hares per ha
— lynx tracks per 100 km

Source: Krebs 2010

Figure 30.18 Predator-prey cycling of a lynx and a snowshoe hare.

Research indicates that the snowshoe hare population reaches a peak abundance before that of the lynx by a year or more. A study conducted from 1987 to 2009 shows a cycling of the lynx and snowshoe hare populations.

© Alan Carey/Science Source

lynx

snowshoe hare

was denied access to the hare population, the cycling of the hare population still occurred based on food availability. The results suggested that a hare-food cycle and a predator-hare cycle have combined to produce the pattern observed in Figure 30.18.

Check Your Progress 30.3

1. Explain how an age-structure diagram may be used to predict the future growth of a population.

2. Describe what survivorship curves can tell you about a species' reproductive strategies.

3. Compare and contrast an exponential growth curve and a logistic growth curve.

4. Provide examples of density-independent factors that regulate population growth.

5. Explain how competition and predation act as density-dependent factors for regulating population growth.

30.4 Life History Patterns and Extinction

Learning Outcomes

Upon completion of this section, you should be able to

1. Distinguish between opportunistic and equilibrium species.
2. Describe factors that can lead to the extinction of a species.

A **life history** consists of a particular mix of the characteristics we have been discussing. After studying many types of populations, from mayflies to humans, ecologists have discovered two fundamental and contrasting patterns: the opportunistic life history pattern and the equilibrium life history pattern.

Opportunistic species tend to exhibit exponential growth. The members of the population are small in size, mature early, have a short life span, and provide limited parental care for a great number of offspring (**Fig. 30.19***a*). Density-independent factors tend to regulate the population size, which is large enough to survive an event that threatens to annihilate it. These populations typically have a high dispersal capacity. Various types of insects and weeds are good examples of opportunistic species.

Equilibrium species exhibit logistic population growth, with the population size remaining close to or at the carrying capacity (Fig. 30.19*b*). Resources are relatively scarce, and the individuals best able to compete—those with phenotypes best suited to the environment—have the largest number of offspring. They allocate energy to their own growth and survival and to the growth and survival of a small number of offspring. Therefore, they are fairly large, are slow to mature, and have a fairly long life span. The population growth of equilibrium species tends to be regulated by density-dependent factors. Various birds and mammals are good examples of equilibrium species.

Extinction

Extinction is the total disappearance of a species or higher group. Which species shown in Figure 30.19, the dandelion or the bears, is apt to become extinct? Because the dandelion matures quickly, produces many offspring at one time, and has seeds dispersed widely by wind, it can withstand a local decimation more easily than the bears.

A study of equilibrium species shows that three other factors—the size of the geographic range, degree of habitat tolerance, and size of local populations—can help determine whether an equilibrium species is in danger of extinction. **Figure 30.20** compares 14 equilibrium species on the basis of these three factors. The mountain gorilla has a restricted geographic range, narrow habitat tolerance (few preferred places to live), and small local populations. This combination of characteristics makes the mountain gorilla very vulnerable to extinction. The possibility of extinction increases depending on whether a species is similar to the gorilla in one, two, or three ways. Such population studies can assist conservationists and others trying to preserve the biosphere's biodiversity.

Opportunistic Pattern

Small individuals
Short life span
Fast to mature
Many offspring
Little or no care of offspring

a.

Equilibrium Pattern

Large individuals
Long life span
Slow to mature
Few offspring
Much care of offspring

b.

Figure 30.19 Life history patterns.

a. Dandelions are an opportunistic species, whereas (**b**) bears are an equilibrium species.

(a): © Elena Elisseeva/Alamy RF; (b): © Barrett Hedges/National Geographic/Getty RF

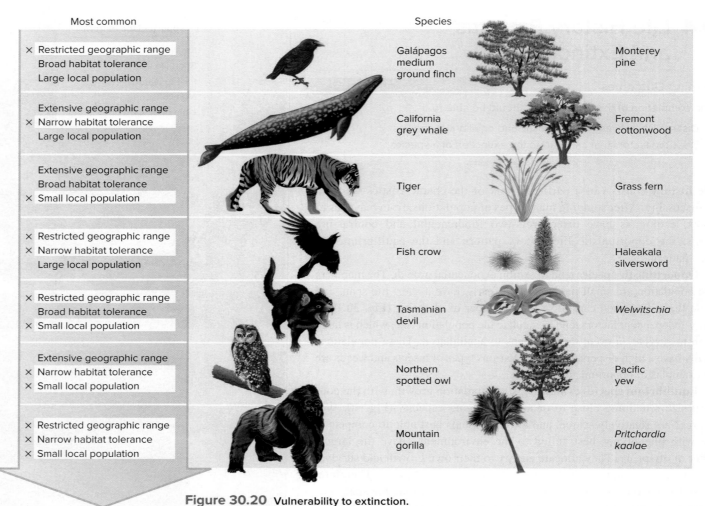

Most common

	Species	
× Restricted geographic range Broad habitat tolerance Large local population	Galápagos medium ground finch	Monterey pine
Extensive geographic range × Narrow habitat tolerance Large local population	California grey whale	Fremont cottonwood
Extensive geographic range Broad habitat tolerance × Small local population	Tiger	Grass fern
× Restricted geographic range × Narrow habitat tolerance Large local population	Fish crow	Haleakala silversword
× Restricted geographic range Broad habitat tolerance × Small local population	Tasmanian devil	*Welwitschia*
Extensive geographic range × Narrow habitat tolerance × Small local population	Northern spotted owl	Pacific yew
× Restricted geographic range × Narrow habitat tolerance × Small local population	Mountain gorilla	*Pritchardia kaalae*

Rarest

Figure 30.20 **Vulnerability to extinction.**

Vulnerability is particularly tied to range, habitat, and size of population. The number of strikes (X) in the boxes indicates the chances and causes of extinction.

Connections: Environment

How are humans influencing the rates of extinction?

Many conservation biologists believe that we are in the midst of one of the largest periods of mass extinction since the end of the Cretaceous, 65 mya (see Table 16.1). The major causes of this "sixth mass extinction" are human activity (population growth and resource use) and global climate change. While extinction is a normal aspect of life, the current extinction rates are between 100 and 1,000 times the normal levels of one to five species per year. Some estimates suggest that as much as 38% of all species on the planet, including most primates (such as the lemur shown in photo), birds, and amphibians, may be in danger of extinction before the end of this century.

© Goddard Photography/iStock/Getty Images Plus RF

Check Your Progress 30.4

1. Contrast opportunistic species with equilibrium species.
2. Define *extinction*.
3. List five factors that can determine whether an equilibrium population is in danger of extinction.

CONNECTING THE CONCEPTS

30.4

Knowledge of life history patterns helps us predict the threat of extinction for various species.

SUMMARIZE

Ecology is the study of the interactions between organisms, as well as between organisms and their environment. Ecologists study biological organization at a variety of levels: organism, species, population, community, ecosystem, and biosphere. Knowledge gained from ecology helps us understand all of the species on Earth.

30.1 Ecology is the study of how organisms interact with each other and their environment.

30.2 The human population is still growing, with the majority of that growth in the less-developed countries.

30.3 Population growth influences the community and ecosystem structure.

30.4 Knowledge of life history patterns helps us predict the threat of extinction for various species.

30.1 The Scope of Ecology

Ecology is an experimental science that studies the interactions of organisms with each other and with the physical environment. The general levels of biological organization studied by ecologists are

- Organism, species, population, community, ecosystem, biosphere

Environmental scientists and conservation biologists both utilize the principles of ecology in their work.

30.2 The Human Population

The human population exhibits clumped distribution (on both a large scale and a small scale) and is undergoing rapid growth.

Future Population Growth

At the present growth rate, the doubling time for the human population is estimated to be 67 years. This rate of growth will put extreme demands on resources, and growth will decline due to resource scarcity. Eventually, the population will most likely level off at its carrying capacity.

- **More-developed countries** (**MDCs**) have a lower growth rate but greater demand for resources per person.
- **Less-developed countries** (**LDCs**) have a higher growth rate and generally a lower demand for resources per person.

Comparing Age Structures

The age structure of a population is divided into three age groups: prereproductive, reproductive, and postreproductive. The MDCs are approaching stabilization, with just about equal numbers in each group. The LDCs will continue to expand, because their prereproductive group is the largest. Ideally, populations would exhibit **replacement reproduction** and each couple would produce only two children.

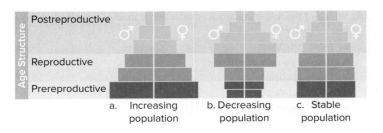

Population Growth and Environmental Impact

There are two causes of environmental impact: The LDCs put stress on the biosphere due to population growth, and the MDCs put stress on the biosphere due to resource consumption and waste production.

30.3 Characteristics of Populations

Patterns of populations in the environment, such as **clumped distribution, random distribution,** and **uniform distribution,** along with population density are dependent on resource availability. These factors determine the **range** that a species will use. The number of individuals per unit area is higher in areas with abundant resources and lower where resources are limited. Population growth can be calculated based on the annual birthrate and death rate. Population growth is determined by

- Resource availability
- **Demographics** (**age structure, survivorship,** and **biotic potential**— the highest rate of increase possible)

Patterns of Population Growth

The two patterns of population growth are exponential growth and logistic growth.

- **Exponential growth** results in a J-shaped curve. The two phases of exponential growth are the lag phase (slow growth) and exponential growth phase (accelerating growth).
- **Logistic growth** results in an S-shaped curve. The four phases of logistic growth are the lag phase (slow growth), exponential growth phase (accelerating growth), logistic growth phase (slowing growth), and stable equilibrium phase (relatively no growth).

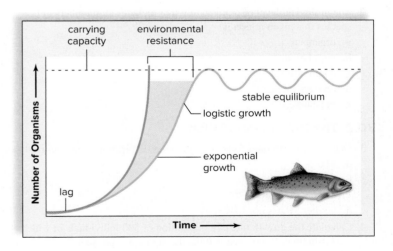

Factors That Regulate Population Growth

The two types of factors that regulate population growth are density-independent and density-dependent factors.

- **Density-independent factors** include abiotic factors, such as weather and natural disasters. The effect of the factor is not dependent on population density.
- **Density-dependent factors** include biotic factors, such as **competition** and **predation.** The effect of the factor is dependent on population density.

30.4 Life History Patterns and Extinction

The two fundamental life history patterns are exhibited by opportunistic species and equilibrium species.

- **Opportunistic species** are characterized by individuals that have short life spans, mature quickly, produce many offspring, have strong dispersal ability, provide little or no care to their offspring, and exhibit exponential popuation growth.
- **Equilibrium species** are characterized by individuals that have long life spans, mature slowly, produce few offspring, provide much care to the offspring, and exhibit logistic population growth.

Extinction

Extinction is the total disappearance of a species or higher group. Opportunistic species are less likely than equilibrium species to become extinct. Three factors, in particular, influence the vulnerability of equilibrium species to extinction: size of geographic range, degree of habitat tolerance, and size of local populations.

ASSESS

Testing Yourself

Choose the best answer for each question.

30.1 The Scope of Ecology

1. Which of these levels of ecological study involves both abiotic and biotic components?
   a. organisms
   b. populations
   c. communities
   d. ecosystem
   e. All of these are correct.

2. The biological level of organization that includes all members of a species in a given region is
   a. organism.
   b. population.
   c. community.
   d. ecosystem.
   e. biosphere.

30.2 The Human Population

3. Decreased death rates followed by decreased birthrates has occurred in
   a. MDCs.
   b. LDCs.
   c. MDCs and LDCs.
   d. neither MDCs nor LDCs.

4. Calculate the growth rate of a population of 500 individuals in which the birthrate is 10 per year and the death rate is 5 per year.
   a. 5%   c. 1%
   b. –5%   d. –1%

5. If the human birthrate were reduced to 15 per 1,000 per year and the death rate remained the same (9 per 1,000), what would be the growth rate?
   a. 9%   d. 0.6%
   b. 6%   e. 15%
   c. 10%

30.3 Characteristics of Populations

6. A population's maximum growth rate is also called its
   a. carrying capacity.   c. growth curve.
   b. biotic potential.   d. replacement rate.

7. If members of a population live only along a lake's shoreline, this population exhibits which type of spatial distribution?
   a. variable   c. random
   b. clumped   d. uniform

8. Label each of the following age-structure diagrams to indicate whether the population is stable, increasing, or decreasing.

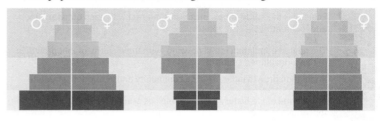

a. _____   b. _____   c. _____

9. An S-shaped growth curve indicates
   a. logistic growth.
   b. logarithmic growth.
   c. exponential growth.
   d. additive growth.

For questions 10–13, choose the term from the key that matches each scenario.

Key:
   a. density-independent factor
   b. competition
   c. predation
   d. predator-prey cycle

10. A severe drought destroys the entire food supply of a herd of gazelle.

11. A population of feral cats increases in size as the mouse population increases and then crashes regularly after the mouse population enters periods of decline.

12. Swift coyotes are able to catch rabbits and live to reproduce. Other coyotes lack a good food source and are not strong enough to reproduce.

13. Deer prefer to feed on a dense thicket of oak saplings rather than more widely spaced young oak trees.

30.4 Life History Patterns and Extinction

14. Which of the following features is present in organisms that exhibit an equilibrium life history pattern?
    a. long life span
    b. small body size
    c. many offspring
    d. fast to mature
    e. no parental care

15. Which of the following is not an adaptive feature of an opportunistic life history pattern?
    a. many offspring
    b. little or no care of offspring
    c. long life span
    d. small individuals
    e. individuals that mature quickly

ENGAGE

Thinking Critically

1. In Sri Lanka, the death rate is 6 per 1,000, while the birthrate is 17 per 1,000. Calculate the current population growth rate. The formula for the doubling time of a population is

    $t = 0.69/r$, where t is the doubling time and r is the growth rate

    Determine the doubling time (the number of years it will take to double the population size) in Sri Lanka. Be sure to convert your growth rate from a percentage (e.g., 2.1%) to a decimal value (0.021). If the birthrate drops to 10 per 1,000, what is the doubling time?

2. Upland game hunters in Illinois have been noticing a decrease in the rabbit and pheasant populations over the past 5 years. They have lobbied the Department of Natural Resources and the state legislature to allow them to shoot red-tailed hawks, one of the main predators of rabbits and pheasants. Explain why shooting hawks will not ultimately solve the problem of decreased rabbit and pheasant populations.

3. What type of data would you collect to show that a population is undergoing stabilizing selection, as in Figure 15.2?

4. Species that are more vulnerable to certain risk factors are more likely than others to become extinct. For example, species with a unique lineage, such as the giant panda, are likely to be at severe risk of extinction.
    a. Should our limited resources for species protection be focused on species that are at the highest risk of extinction?
    b. Do you support the idea that high-risk species may be less successful products of evolution and should not receive extraordinary protection?

31

Communities and Ecosystems

OUTLINE

31.1 Ecology of Communities   601

31.2 Ecology of Ecosystems   610

31.3 Ecology of Major Ecosystems   620

BEFORE YOU BEGIN

Before beginning this chapter, take a few moments to review the following discussions.

Section 1.1 What are the roles of producers, consumers, and decomposers in an ecosystem?

Section 6.1 How do photosynthetic organisms convert CO_2 and water to sugar?

Section 30.2 What is the environmental impact of human population growth?

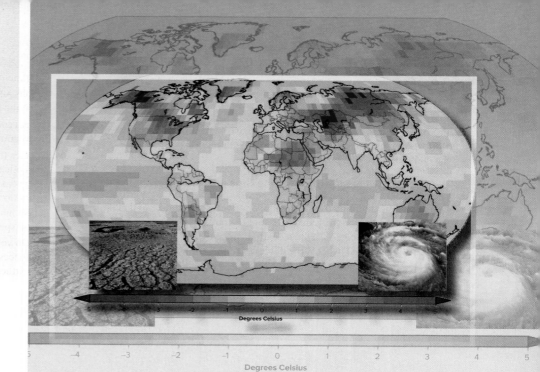

(map): Source: NOAA National Centers for Environmental Information, State of the Climate: Global Analysis for March 2016, published online April 2016, retrieved on June 7, 2016 from http://www.ncdc.noaa.gov/sotc/global/201603; (dry lake): © Vladislav T. Jirousek/Shutterstock RF; (hurricane): © Purestock/SuperStock RF

The Consequences of Climate Change

Almost every month, announcements from climate scientists present data supporting the observation that our planet is warming. Not only was 2015 the hottest year on record, but almost every month in 2016 broke previous heat records, and usually by significant amounts. After decades of studies and analyses, the scientific community has concluded that global warming, and the resulting climate changes, are not a result of natural cycles, but rather due to the emission of greenhouse gases.

For most of us, it is difficult to see how global changes impact us directly. However, in many cases, the evidence is already around us. Droughts are more severe, there are reductions in mountain snow packs, and precipitation events are more unpredictable. These are all indications that the climate is changing.

On a more personal level, climate change has the potential to increase our exposure to diseases that are not normally a part of our geographical area. For example, the spread of malaria, dengue fever, and even the Zika virus, is due to expansion of the range of mosquitoes that act as vectors for each disease. In the near future, heat warnings will limit outdoor activities, and water and air quality will be degraded. All of these events are tied to an imbalance in how our planet functions on a global scale.

In this chapter, we will explore how ecosystems function and how variations in these natural cycles, and human influences, affect the basic structure of an ecosystem.

As you read through this chapter, think about the following questions:

1. How does climate influence a community and an ecosystem?
2. What is the relationship between greenhouse gases and the carbon cycle?

31.1 Ecology of Communities

Learning Outcomes

Upon completion of this section, you should be able to

1. Define *community* and *coevolution,* and provide an example of each.
2. Distinguish between species richness and species diversity.
3. Compare and contrast primary and secondary succession.
4. Describe and provide examples of how a species' niche influences the types of interactions within a community.

A **community** is an assemblage of populations of multiple species, interacting with one another within a single environment. For example, the various species living in and on a fallen log, such as plants, fungi, worms, and insects, interact with one another and form a community. The fungi break down the log and provide food for the earthworms and insects living in and on the log. Those insects may feed on one another, too. If birds flying throughout the forest feed on the insects and worms living in and on the log, then the insects and worms are also part of the larger forest community.

Communities come in various sizes, and it is sometimes difficult to decide where one community ends and another one begins. The relationships and interactions between species in a community form as the products of **coevolution,** by which an evolutionary change in one species results in an evolutionary change in another. **Figure 31.1** gives examples of how flowering plants and their pollinators have coevolved. The Australian orchid, *Chiloglottis trapeziformis,* resembles the body of a wasp, and its odor mimics the pheromones of a female wasp. Therefore, male wasps are attracted to the flower, and when they attempt to mate with it, they become covered with pollen, which they transfer to the next orchid

a.

b.

c.

Figure 31.1 Coevolution.

Flowers and pollinators have evolved to be suited to one another. **a.** Hummingbird-pollinated flowers are usually red, a color these birds can see, and the petals are recurved to allow the stamens to dust the birds' heads. **b.** The reward offered by the flower is not always food. This orchid looks and smells like the female of this wasp's species. The male tries to copulate with flower after flower and in the process transfers pollen. **c.** Bats are nocturnal, and the flowers they pollinate are white or light-colored, making them visible in moonlight. The flowers smell like bats and are large and sturdy, enabling them to withstand insertion of the bat's head as it uses its long, bristly tongue to lap up nectar and pollen.

(a): © Ondrej Prosicky/Shutterstock RF; (b): © Perennou Nuridsany/Science Source; (c): © Dr. Merlin D. Tuttle/Bat Conservation International/Science Source

flower (Fig. 31.1*b*). The orchid is dependent on wasps for pollination, because neither wind nor other insects pollinate these flowers.

The species in each community have adaptations that are suited to the environmental conditions. An **ecosystem** consists of these species interacting with each other and with the physical environment. If the physical environment changes, the species that make up the community and the relationships among these species will also change. Extinction occurs when environmental change is too rapid for suitable adaptations to evolve. We will explore the structure of ecosystems in more detail in Section 31.2.

Rapid environmental changes are detrimental to humans, despite our use of technology to adapt. Sometimes the economy of an area is dependent on the species composition of an aquatic or terrestrial ecosystem. Therefore, human activities that negatively impact the biodiversity of the area can also negatively affect the economy of the area. Knowledge of community and ecosystem ecology will help you better understand how our human activities are detrimental to our own well-being.

Community Composition and Diversity

Ecologists have a variety of methods of analyzing a community. Frequently, these focus on determining what species are present in the community, their distribution, and their interactions. In this chapter, we will focus on comparing communities based on their species composition and diversity.

Species Composition and Richness

A comparison of the species composition of a coniferous forest and a tropical rain forest (**Fig. 31.2**) reveals some very distinctive differences. Narrow-leaved evergreen tree species are prominent in the coniferous forest, whereas broad-leaved evergreen tree species are numerous in the tropical rain forest. There are also differences in the types and number of plant and animal species in these two communities. Ecologists often use the term **species richness,** which is a listing of the various species found in a community. Ecologists have concluded that not only are the species compositions of these two communities different but the tropical rain forest has a higher species richness, meaning that it has a greater number of species.

Diversity

The **diversity** of a community includes both species richness and species distribution. The diversity of a community goes beyond species richness to include distribution and relative abundance, the number of individuals of each species in a given area. For example, a deciduous forest in West Virginia may contain 76 poplar trees but only 1 American elm. If you were simply walking through this forest, you could miss the lone American elm. If, instead, the forest had 36 poplar trees and 41 American elm trees, the forest would seem more diverse to you and indeed would have a higher diversity value, although both forests would have the same species richness. The greater the species richness and the more even the distribution of the species in the community, the greater the diversity.

Figure 31.2 Community species composition.

Communities differ in their species composition, as exemplified by the predominant plants and animals in (**a**) a coniferous forest and (**b**) a tropical rain forest.

(a): © Yi Jiang Photography/Flickr Open/Getty RF; (b): © Nejron Photo/Shutterstock RF

b.

a.

Ecological Succession

The species composition and diversity of communities change over time. It often takes decades for noticeable changes to occur. Natural forces, such as glaciers, volcanic eruptions, lightning-ignited forest fires, hurricanes, tornadoes, and floods are considered disturbances that can bring about community changes. Communities also change because of human activities, such as logging, road building, sedimentation, and farming. A more or less orderly process of community change is known as **ecological succession.**

Ecologists have developed models to explain why succession occurs and to predict patterns of succession following a disturbance. The climax-pattern model says that the climate of an area will always lead to the same assemblage of bacterial, fungal, plant, and animal species (**Fig. 31.3**) known as a **climax community.** For example, a coniferous forest community is expected in northern latitudes, a deciduous forest in temperate zones, and a tropical forest in areas with a tropical climate. Scientists know that disturbances influence community composition and diversity and that, despite the climate, the composition of a climax community in a given climate is not always the same.

Two Types of Succession

Ecologists define two types of ecological succession: primary and secondary (**Fig. 31.4**). **Primary succession** starts where soil has not yet formed. For example, hardened lava flows and the scraped bedrock that remains following a glacial retreat are subject to primary succession. **Secondary succession** begins, for example, in a cultivated field that is no longer farmed, where soil is already present. With both primary and secondary succession, a progression of species occurs over time. The spores of fungi and vascular seedless plants are usually the first to grow and then seeds of nonvascular plants, followed by seeds of gymnosperms and/or angiosperms, are carried into the area by wind, water, or animals from the surrounding regions (**Fig. 31.5**; see also Fig. 31.4).

a.

b.

Figure 31.3 Climax communities.

Does succession in a particular area always lead to the same climax community? For example, (**a**) temperate rain forests occur only where there is adequate rainfall and (**b**) deserts occur where rainfall is minimal. Even so, exactly the same mix of plants and animals may not always arise, because the assemblage of organisms depends on which organisms, by chance, migrate to the area.

(a): © Corbis RF; (b): © Anton Foltin/Shutterstock RF

Figure 31.4 Primary and secondary succession.

Primary succession begins on areas of bare rock. Secondary succession begins in areas where soil remains following natural or human-caused disturbance.

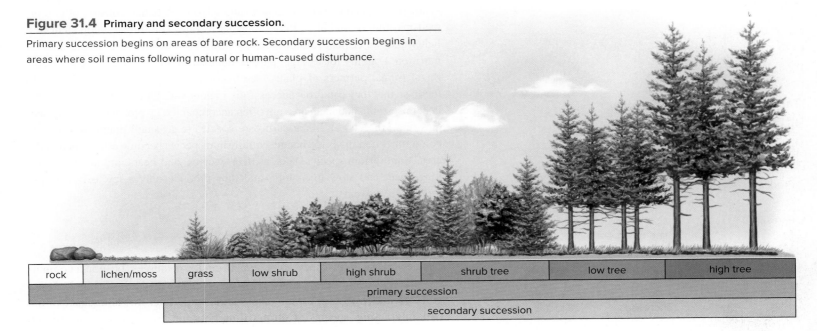

rock	lichen/moss	grass	low shrub	high shrub	shrub tree	low tree	high tree

primary succession

secondary succession

a. Glacier/lichen stage

Dryas plant

b. Shrub stage

c. Low tree stage

d. High tree stage

Figure 31.5 **Secondary succession.**

Secondary successional changes in a western Pennsylvania field: (**a**) first year, (**b**) second year, (**c**) fifth year, and (**d**) after 20 years.

(a): (glacier): © Kevin Smith/Design Pics; (a): (lichen): © Background Abstracts/ Getty RF; (b): (shrub stage): © Don Paulson/Alamy RF; (b): (Dryas plant): © Martin Fowler/Shutterstock RF; (c): © Bruce Heinemann/Getty RF; (d): © Don Paulson/ Alamy RF

Table 31.1 **Species Interactions**

Interaction	Expected Outcome
Competition (− −)	Abundance of both species decreases.
Predation (+ −)	Abundance of predator increases, and abundance of prey decreases.
Parasitism (+ −)	Abundance of parasite increases, and abundance of host decreases.
Commensalism (+ 0)	Abundance of one species increases, and the other is not affected.
Mutualism (+ +)	Abundance of both species increases.

The first species to appear in an area undergoing either primary or secondary succession are called opportunistic pioneer species. These tend to be small in stature, short-lived, and quick to mature, and they produce numerous offspring per reproductive event. The first pioneer species to arrive are photosynthetic organisms, such as lichens and mosses. A lichen consists of two organisms, fungus and alga, living together as one. The fungal component of a lichen plays a critical role in breaking down rock or lava into usable mineral nutrition, not only for its algal partner but also for pioneer plants. The mycorrhizal fungal partners of plants (see Section 18.3) pass minerals directly to them, so that they can grow successfully in poor soil. Pioneer plant species that become established in an area are often accompanied by pioneer herbivore species (e.g., insects) and then carnivore species (e.g., small mammals). As the community continues to change, equilibrium species become established in the area. Equilibrium species, such as deer, wolves, and bears, are larger in size, long-lived, and slow to mature, and they produce only a few offspring per reproductive event.

Interactions in Communities

Species interactions, especially competition for resources, fashion a community into a dynamic system of interspecies relationships. In **Table 31.1,** the plus and minus signs show how the relationship affects the abundance of the two interacting species. **Competition** between two species for limited resources has a negative effect on the abundance of both species. In **predation,** one animal, the predator, feeds on another, the prey (see Fig. 30.18); in **parasitism,** one species obtains nutrients from another species, called the host, but usually does not kill the host. **Commensalism** is a relationship in which one species benefits, while the second species is neither harmed nor benefited. Commensalism often occurs when one species provides a home or transportation for another. In **mutualism,** two species interact in such a way that both benefit.

Ecological Niche

Each species occupies a particular position in the community, in both a spatial and a functional sense. Spatially, species live in a particular area of the community, or **habitat,** such as underground, in the trees, or in shallow water. Functionally, each species plays a role, such as photosynthesizer, predator, prey, parasite, or decomposer. The **ecological niche** of a species incorporates the role the species plays in its community, its habitat, and its interactions with other species. The niche includes the living and nonliving resources that individuals in the population need to meet their energy, nutrient, and survival demands. For example, the habitat of an insect called a backswimmer is a pond or lake, where it eats other insects (**Fig. 31.6**). The pond or lake must contain vegetation where the backswimmer can hide from predatory fish and birds. The pond water must be clear enough for the backswimmer to see its prey and warm enough for it to maintain a good metabolic rate.

It is often difficult to describe and measure the whole ecological niche of a species in a community, so ecologists often focus on a certain aspect of a species' niche, as with the birds featured in **Figure 31.7.**

Competition

Competition for resources, such as light, space, or nutrients, contributes to the niche of each species and helps form the community structure. Laboratory

Figure 31.6 Niche of a backswimmer.

Backswimmers require warm, clear pond water containing insects that they can eat and vegetation where they can hide from predators.

© Steve Austin/Papilio/Corbis

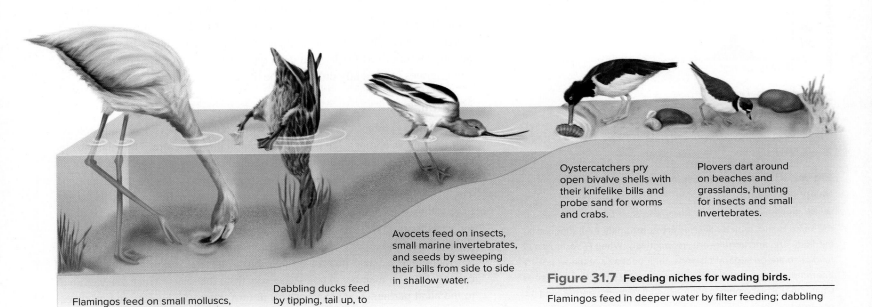

Oystercatchers pry open bivalve shells with their knifelike bills and probe sand for worms and crabs.

Plovers dart around on beaches and grasslands, hunting for insects and small invertebrates.

Avocets feed on insects, small marine invertebrates, and seeds by sweeping their bills from side to side in shallow water.

Flamingos feed on small molluscs, crustaceans, and vegetable matter strained from mud pumped through their bills by their powerful tongues.

Dabbling ducks feed by tipping, tail up, to reach aquatic plants, seeds, snails, and insects.

Figure 31.7 Feeding niches for wading birds.

Flamingos feed in deeper water by filter feeding; dabbling ducks feed in shallower areas by upending; avocets feed by sifting. Oystercatchers and plovers have adaptations, such as shorter legs, for feeding in shallows and on land.

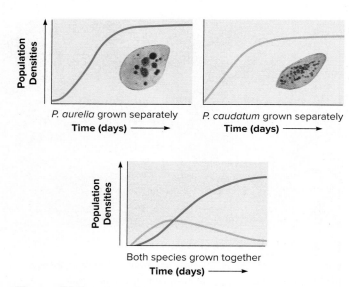

Figure 31.8 **Competitive exclusion principle demonstrated by *Paramecium*.**

The competitive exclusion principle states that no two species occupy exactly the same niche. When grown separately, *Paramecium caudatum* and *Paramecium aurelia* exhibit logistic growth. When grown together, *P. aurelia* excludes *P. caudatum*.

Data from G. F. Gause, The Struggle for Existence, 1934, Williams & Wilkins Company, Baltimore, MD, p. 557.

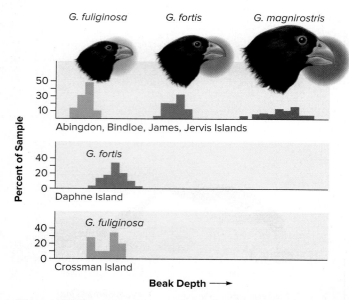

Figure 31.9 **Character displacement in finches on the Galápagos Islands.**

When *Geospiza fuliginosa, G. fortis,* and *G. magnirostris* coexist on the same island, their beak sizes are appropriate for eating small, medium-size, and large seeds, respectively. When *G. fortis* and *G. fuliginosa* are on separate islands, their beaks have the same intermediate size, which allows them to eat seeds of various sizes. Character displacement is evidence that resource partitioning has occurred.

experiments helped ecologists formulate the **competitive exclusion principle,** which states that no two species can occupy the same niche at the same time. In the 1930s, G. F. Gause grew two species of *Paramecium* in one test tube containing a fixed amount of bacterial food. Although populations of each species survived when grown in separate test tubes, only one species, *Paramecium aurelia,* survived when the two species were grown together (**Fig. 31.8**). *P. aurelia* acquired more of the food and had a higher population growth rate than did *P. caudatum.* Eventually, as the *P. aurelia* population grew and obtained an increasingly greater proportion of the food resource, the number of *P. caudatum* individuals decreased and the population died out.

Niche Specialization   Competition for resources does not always lead to localized extinction of a species. Multiple species coexist in communities by partitioning, or sharing, resources. In another laboratory experiment using other species of *Paramecium,* Gause found that the two species could survive in the same test tube if one species consumed bacteria at the bottom of the tube and the other ate bacteria suspended in solution in the middle of the tube. This **resource partitioning** decreased competition between the two species, leading to increased niche specialization. One niche was split into two due to differences in feeding behavior.

When three species of ground finches of the Galápagos Islands live on the same island, their beak sizes differ, and each feeds on different-sized seeds (**Fig. 31.9**). When the finches live on separate islands, their beaks tend to be the same intermediate size, enabling each to feed on a wider range of seeds. Such **character displacement** often is viewed as evidence that competition and resource partitioning have taken place.

The niche specialization that permits the coexistence of multiple species can be very subtle. Species of warblers that live in North American forests are all about the same size, and all feed on budworms, a type of caterpillar found on spruce trees. Robert MacArthur recorded the length of time each warbler species spent in different regions of spruce canopies to determine where each species did most of its feeding. He discovered that each species primarily used different parts of the tree canopy and, in that way, had a more specialized niche. As another example, consider that three types of birds—swallows, swifts, and martins—all eat flying insects and parachuting spiders. These birds even frequently fly in mixed flocks. But each type of bird has a different nesting site and migrates at a slightly different time of year.

Mutualism

Mutualism, a symbiotic relationship in which both members benefit, is at least as important as competition in shaping community structure. The relationship between plants and their pollinators mentioned previously is a good example of mutualism. Perhaps the relationship began when herbivores feasted on pollen. The plant's provision of nectar may have spared the pollen and, at the same time, allowed the animal to become an agent of pollination. Over time, pollinator mouthparts have become adapted to gathering the nectar of a particular plant species. This species has become dependent on the pollinator for dispersing its pollen. As also mentioned previously, lichens can grow on rocks because the fungal member extracts minerals from the rocks, which are provided to the algal partner. The algal partner, in turn, photosynthesizes and provides organic food for both members of the relationship.

In tropical America, ants form mutualistic relationships with certain plants. The bullhorn acacia tree is adapted to provide a home for ants of the species *Pseudomyrmex ferruginea* (**Fig. 31.10**). Unlike other acacias, this species has swollen thorns with a hollow interior, where ant larvae can grow and develop. In addition to housing the ants, acacias provide them with food. The ants feed from nectaries at the base of the leaves and eat fat- and protein-containing nodules, called Beltian bodies, at the tips of the leaves. The ants constantly protect the tree from herbivores that would like to feed on it. The ants are so critical to the trees' survival that, when the ants on experimental trees were poisoned, the trees died.

The outcome of mutualism is an intricate web of species interdependencies critical to the community. For example, in areas of the western United States, the branches and cones of whitebark pine are turned upward, meaning that the seeds do not fall to the ground when the cones open. Birds called Clark's nutcrackers eat the seeds of whitebark pine trees and store them in the ground (**Fig. 31.11**). Grizzly bears find the stored seeds and consume them. Thus, Clark's nutcrackers and grizzly bears are critical seed dispersers for the trees. Whitebark pine seeds do not germinate unless their seed coats are exposed to fire. When natural forest fires in the area are suppressed, whitebark pine trees decline in number, and so do Clark's nutcrackers and grizzly bears. When lightning-ignited fires are allowed to burn, or prescribed burning is used in the area, the whitebark pine populations increase, as do the populations of Clark's nutcrackers and grizzly bears.

a.

b.

Beltian bodies

c.

Figure 31.10 Mutualism.

The bullhorn acacia tree is adapted to provide nourishment for a mutualistic ant species. **a.** The thorns are hollow, and the ants live inside. **b.** The bases of the leaves have nectaries (openings) where ants can feed. **c.** The tips of the leaves of the bullhorn acacia have Beltian bodies, which ants harvest for larval food.

(a): © WILDLIFE GmBh/Alamy; (b): © Carol Farneti–Foster/Oxford Scientific/Getty Images; (c): © Bazzano Photography/Alamy

a.

b.

Figure 31.11 Interdependence of species.

a. The Clark's nutcracker feeds on the seeds of the whitebark pine. The nutcracker also stores seeds in the ground, and these seeds may be found and eaten by (**b**) the grizzly bear. Clark's nutcrackers and grizzly bears disperse the seeds of this species of pine.

(a): © Peter Chadwick/Alamy; (b): © ArCaLu/Shutterstock RF

Connections: Scientific Inquiry

Do humans form mutualistic relationships with other organisms?

One of the best examples of mutualism between humans and another species involves the bacteria *Escherichia coli* that live in our large intestine. While some strains of *E. coli* have a bad reputation for causing intestinal problems, the majority of the *E. coli* in our intestine are beneficial. In exchange for a warm environment, and plenty of food, the

© Science Photo Library RF/Getty RF

intestinal *E. coli* produce vitamin K and assist in the breakdown of fiber into glucose.

Community Stability

As demonstrated by the phenomenon of succession, community stability is fragile. However, some communities have one species that stabilizes the community, helps maintain its characteristics, and plays a critical role in holding the web of interactions together. Such a species is known as a **keystone species,** a species on which the majority of the community depends. The term *keystone* comes from the name for the center stone of an arch that holds the other stones in place, so that the arch can keep its shape.

Keystone species are not necessarily the most numerous in the community. However, the loss of a keystone species can lead to the extinction of other species and a loss of diversity. For example, bats are designated as keystone species in tropical forests. Bats are pollinators, and they disperse the seeds of certain tropical trees. When bats are killed off or their roosts are destroyed, many species of trees will fail to reproduce. The grizzly bear is a keystone species in the northwestern United States and Canada; they disperse as many as 7,000 berry seeds in one dung pile. Grizzly bears also kill the young of many herbivorous mammals, such as deer, thereby keeping their populations under control.

The sea otter is a keystone species of a kelp forest ecosystem. Kelp forests, created by large, brown seaweeds, provide a home for a vast assortment of organisms. The kelp forests occur just off the coast and protect coastline ecosystems from damaging wave action. Among other species, sea otters eat sea urchins, keeping their population size in check. Otherwise, sea urchins feed on the kelp, causing the kelp forest and its associated species to severely decline. Fishermen don't like sea otters because they also prey on abalone, a mollusc prized for its commercial value. Many do not realize that, without the otters, abalone and many other species would not be around, because the kelp forest would no longer exist.

Native Versus Exotic Species

Native species are species indigenous to an area because they have evolved to fit into the particular community. For example, you naturally find maple trees in Vermont and many other states of the eastern United States. The introduction of **exotic species** (also sometimes called alien or nonnative species), into a community greatly disrupts normal interactions, and therefore changes a community's web of species. Populations of exotic species may grow exponentially, because they outcompete the native species or because their population

Figure 31.12 Exotic species.

Human introduction of exotic species, such as **(a)** the Asian carp and **(b)** the emerald ash borer, have disrupted communities across the entire midwestern United States.

(a): © Jason Lindsey/Alamy; (b-1): © J. D. Pooley/AP Images; (b-2): Stephen Ausmus/USDA

a.

b.

size is not controlled by a natural predator or disease. The unique assemblage of native species on an island often cannot compete well against an exotic species. For example, myrtle trees, introduced to the Hawaiian Islands from the Canary Islands, are mutualistic with a type of bacterium capable of fixing atmospheric nitrogen. The bacterium provides the tree with a form of nitrogen it can use. This feature allows the trees to become established on nitrogen-poor volcanic soil, a distinct advantage in Hawaii. Once established, myrtle trees prevent the normal succession of native plants on volcanic soil.

Exotic species disrupt communities in continental areas as well. Asian carp were accidentally introduced into the Mississippi River system and have spread throughout the majority of the Midwest. They have significantly decreased the populations of many native species of fish (**Fig. 31.12a**). The emerald ash borer was accidentally introduced to the United States from Asia back in 2002 (Fig. 31.12b). Since then, it has spread throughout 19 states, wiping out large numbers of ash trees in the process. On the Galápagos Islands, black rats accidentally carried to the islands by ships have reduced populations of the giant tortoise. Goats and feral pigs have changed the vegetation on the islands from highland forest to pampaslike grasslands and destroyed stands of cactuses. In the United States, gypsy moths, zebra mussels, the Chestnut blight fungus, fire ants, and African bees are well-known exotic species that have killed native species. At least two species, fire ants and African bees, have attacked humans, with serious consequences.

CONNECTING THE CONCEPTS

31.1 Community interactions are complex and dynamic.

Check Your Progress 31.1

1. Describe an example of coevolution.
2. Contrast species richness with species diversity.
3. Contrast primary succession with secondary succession.
4. List the five major types of species interactions in a community that determine an organism's ecological niche.

Figure 31.13 Producers.

Green plants and algae are photoautotrophs.

(tree): © Authors Image/PunchStock RF; (diatoms): © Ed Reschke

a. Herbivores

b. Carnivores

Figure 31.14 Consumers.

a. Caterpillars and giraffes are herbivores. **b.** A praying mantis and a lion are carnivores.

(a): (caterpillar): © Corbis RF; (giraffes): © George W. Cox; (b): (mantis): © Kristina Postnikova/Shutterstock RF; (lion): © Sue Green/Shutterstock RF

31.2 Ecology of Ecosystems

Learning Outcomes

Upon completion of this section, you should be able to

1. Describe the role of autotrophs and heterotrophs in an ecosystem.
2. Compare how energy and chemicals interact with an ecosystem.
3. Describe the different trophic levels and the formation of ecological pyramids.
4. Summarize the biogeochemical cycles, and state how human activity is influencing each cycle.
5. Understand the consequences of global climate change and its relationship to the carbon cycle.

An ecosystem is broader than a community, because community ecology considers only how species interact with one another. When studying ecosystem ecology, interactions with the physical environment are also considered. For example, one important aspect of an ecological niche is how an organism acquires food. It is obvious that autotrophs interact with the physical environment, but so do heterotrophs.

Autotrophs

Autotrophs take in only inorganic nutrients (e.g., CO_2 and minerals) and rely on an outside energy source to produce organic nutrients for their own use and for all the other members of a community. They are called **producers** because they produce food. *Photoautotrophs* are photosynthetic organisms that produce most of the organic nutrients for the biosphere (**Fig. 31.13**). Algae of all types possess chlorophyll and carry on photosynthesis in fresh water and marine habitats. Algae make up the phytoplankton, which are photosynthesizing organisms suspended in water. Green plants are the dominant photosynthesizers on land. All photosynthesizing organisms release O_2 into the atmosphere.

Some bacteria are *chemoautotrophs*. They obtain energy by oxidizing inorganic compounds, such as ammonia, nitrites, and sulfides, and they use this energy to synthesize organic compounds. Chemoautotrophs have been found to support communities in some caves and at hydrothermal vents along deep-sea oceanic ridges.

Heterotrophs

Heterotrophs consume preformed organic nutrients and release CO_2 into the atmosphere. They are called **consumers** because they consume food. **Herbivores** are animals that graze directly on algae or plants (**Fig. 31.14a**). In aquatic habitats, zooplankton act as herbivores; in terrestrial habitats, many species of insects play that role. **Carnivores** eat other animals; for example, a praying mantis catches and eats caterpillars, and a lion hunts for game (Fig. 31.14b). These examples illustrate that there are primary consumers (e.g., caterpillars), secondary consumers (e.g., praying mantis), and tertiary consumers (e.g., lion). Sometimes tertiary consumers are called top predators. **Omnivores** are animals that eat both plants and animals. As you likely know, humans are omnivores.

Connections: Environment

How do chemoautotrophs on the ocean floor produce food?

The chemoautotrophs on the ocean floor use chemical energy instead of sunlight to make food. Volcanoes on the ocean floor release hydrogen sulfide gas through cracks called hydrothermal vents. (Hydrogen sulfide is the nasty-smelling gas we associate with the smell of rotten eggs.) Some chemoautotrophs split hydrogen sulfide to obtain the energy needed to link carbon atoms together to form glucose. The glucose contained in these chemoautotrophs sustains a variety of bizarre organisms, such as giant tube worms, anglerfish, and giant clams.

© B. Murton/Southampton Oceanography Centre/Science Source

The **decomposers** are heterotrophic bacteria and fungi, such as molds and mushrooms, that break down dead organic matter (**Fig. 31.15**). Decomposers perform a very valuable service, because they release inorganic nutrients (CO_2 and minerals), which are then taken up by plants once more. Otherwise, plants would rely on minerals to be slowly released from rocks. **Detritus** is composed of the remains of dead organisms plus the bacteria and fungi that aid in decay. Fanworms feed on detritus floating in marine waters, while clams take detritus from the sea bottom. Earthworms and some beetles, termites, and maggots are soil detritus feeders.

Figure 31.15 Decomposers.

Fungi and bacteria are decomposers.

(mushrooms): © BadZTuA/Getty RF; (bacteria): © Science Photo Library/Alamy RF

Energy Flow and Chemical Cycling

The living components of ecosystems process energy and chemicals. Energy flow through an ecosystem begins when producers absorb solar energy. Chemical cycling then begins when producers take in inorganic nutrients from the physical environment (**Fig. 31.16**). Thereafter, via photosynthesis, producers convert the solar energy and inorganic nutrients into chemical energy in the form of organic nutrients, such as carbohydrates. Producers synthesize organic nutrients directly for themselves and indirectly for the heterotrophs within the ecosystem. Energy flows through an ecosystem because, as organic nutrients pass from one component of the ecosystem to another, as when an herbivore eats a plant or a carnivore eats an herbivore, a portion is used as an energy source. Eventually, the energy dissipates into the environment as heat. Therefore, the vast majority of ecosystems cannot exist without a continual supply of solar energy.

Only a portion of the organic nutrients made by producers are passed on to consumers. Plants use some of the organic molecules to fuel their own cellular respiration. Similarly, only a small percentage of nutrients consumed by lower-level consumers, such as herbivores, are available to higher-level consumers, or carnivores. As **Figure 31.17** demonstrates, a certain amount of the nutrients eaten by an herbivore is eliminated as feces. Metabolic wastes are excreted as urine. Of the assimilated energy, a large portion is used during cellular respiration, which produces ATP, as well as heat. Only the remaining energy, which is converted into increased body weight or additional offspring, becomes available to carnivores.

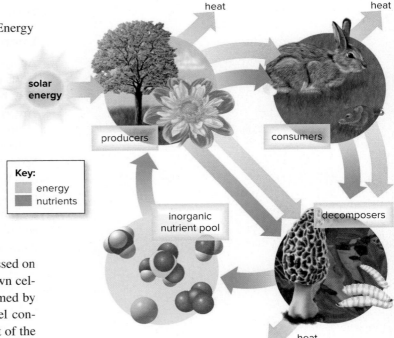

Figure 31.16 Chemical cycling and energy flow.

Chemicals cycle within, but energy flows through, an ecosystem. As energy is repeatedly passed from one component to another, all the chemical energy derived from solar energy dissipates as heat.

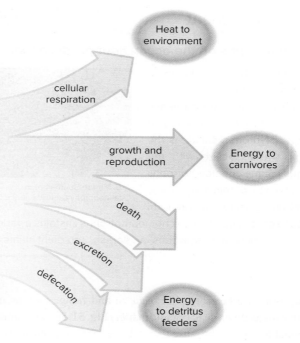

Figure 31.17 Energy balances.

Only about 10% of the nutrients and energy taken in by an herbivore is passed on to carnivores. A large portion goes to detritus feeders. Another large portion is used for cellular respiration.

© Brand X Pictures/PunchStock RF

The elimination of feces and urine by a heterotroph, and indeed the death of any organism, does not mean that organic nutrients are lost to an ecosystem. Instead, the organic nutrients are made available to decomposers. Decomposers convert the organic nutrients back into inorganic chemicals and release them to the soil or atmosphere. Chemicals complete their cycle in an ecosystem when producers absorb inorganic chemicals from the atmosphere or soil.

Energy Flow

Applying the principles discussed so far to a temperate deciduous forest, ecologists can draw a **food web** to represent the interconnecting paths of energy flow between the components of the ecosystem. In **Figure 31.18,** the green arrows are part of a **grazing food web,** because the energy flow begins with plants, such as the oak trees depicted. A **detrital food web** (orange arrows) begins with bacteria and fungi. In the grazing food web, caterpillars and other herbivorous insects feed on the leaves of the trees, while other herbivores, including mice, rabbits, and deer, feed on leaves at or near the ground. Various birds, chipmunks, and mice feed on fruits and nuts of the trees, but, in fact, they are omnivores because they also feed on caterpillars and other insects. These herbivores and omnivores all provide food for a number of different carnivores. In the detrital food web, detritus, which includes smaller decomposers (such as bacteria and fungi), is food for larger organisms. Because some of these organisms, such as shrews and salamanders, become food for aboveground animals, the detrital and grazing food webs are joined.

We tend to think that the aboveground parts of trees are the largest storage form of organic matter and energy, but this is not necessarily the case. In temperate deciduous forests, the organic matter lying on the forest floor and mixed into the soil, along with the underground roots of the trees, contains over twice the energy of the leaves, branches, and trunks of living trees combined. Therefore, more energy and matter in a forest may be stored in or funneled through the detrital food web than the grazing food web.

Key:
→ grazing food web
→ detrital food web

birds

hawks

chipmunks

fruits and nuts

owls

mice

leaf-eating
insects

snakes

leaves

rabbits

fishers

old leaves,
dead twigs

skunks

deer

shrews

foxes

salamanders

bacteria and fungi

invertebrates

carnivorous
invertebrates

Figure 31.18 **Food webs.**

The grazing and detrital food webs of an ecosystem are linked.

Trophic Levels and Ecological Pyramids   The arrangement of component species in **Figure 31.19** suggests that organisms are linked to one another in a straight line according to feeding relationships, or who eats whom. Diagrams that show a single path of energy flow in an ecosystem are called **food chains** (Fig. 31.19). A **trophic level** is a level of nourishment within a food chain or web. In the grazing food web (see Fig. 31.18), from left to right, the trees are

Figure 31.19 Food chain.

A food chain diagrams a single path of energy flow in an ecosystem. Most food chains have three or four links.

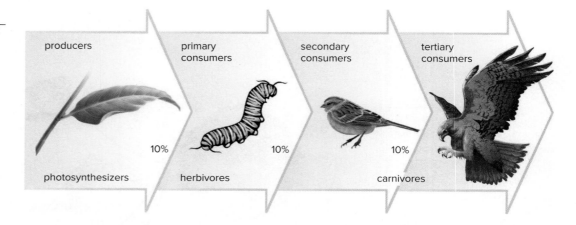

producers / photosynthesizers — 10% — primary consumers / herbivores — 10% — secondary consumers / carnivores — 10% — tertiary consumers

producers (the first trophic level), the first series of animals are herbivores (the second trophic level), and many of the animals in the next series are carnivores (the third and possibly fourth trophic levels).

Food chains are short, because energy is lost between trophic levels. In general, only about 10% of the energy of one trophic level is available to the next trophic level. Therefore, if an herbivore population consumes 1,000 kg of plant material, only about 100 kg is converted to herbivore tissue, 10 kg to first-level carnivores, and 1 kg to second-level carnivores. This 10% rule explains why few carnivores can be supported in a food web. The transfer of energy between successive trophic levels is sometimes depicted as an **ecological pyramid** (**Fig. 31.20**).

A pyramid based on the number of organisms can run into problems because, for example, one tree can support many herbivores. Pyramids based on **biomass,** which is the number of organisms multiplied by their weight, eliminate size as a factor. Even then, apparent inconsistencies can arise. In aquatic ecosystems, such as lakes and open seas where algae are the only producers, the herbivores at some point in time may have a greater biomass than the producers. Why? Even though the algae reproduce rapidly, they are also consumed at a high rate.

Chemical Cycling

The pathways by which chemicals cycle within ecosystems involve both living components (producers, consumers, decomposers) and nonliving components (rock, inorganic nutrients, atmosphere) and therefore are known as **biogeochemical cycles.** Biogeochemical cycles can be sedimentary or gaseous. In a *sedimentary cycle,* such as the phosphorus cycle, the element (chemical) is absorbed from the sediment by plant roots, passed through the food chain, and eventually returned to the soil by decomposers, usually in the same general area. In a *gaseous cycle,* such as the nitrogen and carbon cycles, the element returns to and is withdrawn from the atmosphere as a gas.

Chemical cycling of an element may involve **reservoirs** and exchange pools as well as the biotic community (**Fig. 31.21**). A *reservoir* is a source normally unavailable to organisms. For example, much carbon is found in calcium carbonate shells in sediments on the ocean floors. An *exchange pool* is a source from which organisms can obtain elements. For example, photosynthesizers can use carbon dioxide in the atmosphere for their carbon needs. The *biotic community* consists of the autotrophic and heterotrophic species of an ecosystem that feed on one another. Human activities, such as mining or burning fossil fuels, increase the amounts of chemical elements removed from reservoirs and cycling within ecosystems. As a result, the physical environment of the

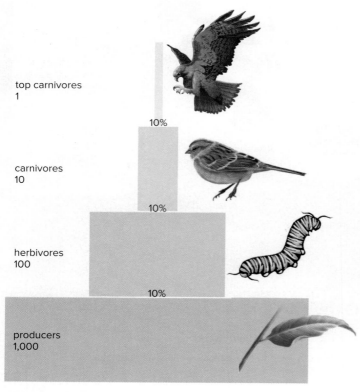

top carnivores
1

10%

carnivores
10

10%

herbivores
100

10%

producers
1,000

Figure 31.20 Ecological pyramid.

An ecological pyramid depicts the transfer of nutrients and energy from one trophic level to the next.

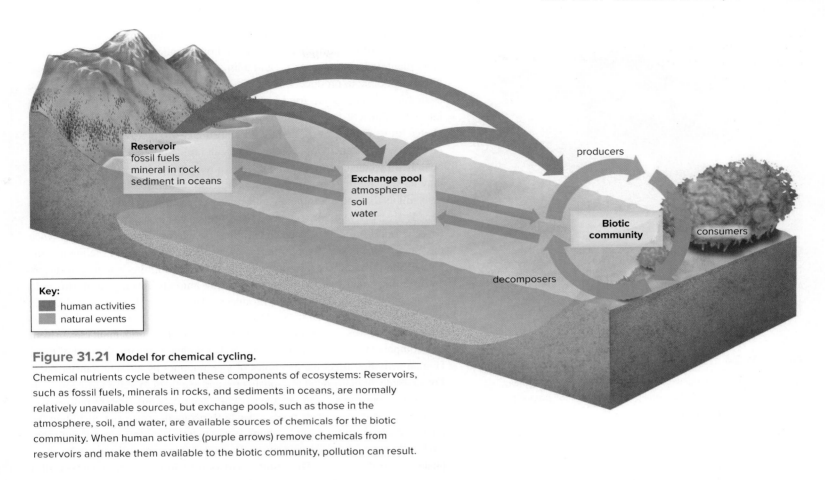

Figure 31.21 Model for chemical cycling.

Chemical nutrients cycle between these components of ecosystems: Reservoirs, such as fossil fuels, minerals in rocks, and sediments in oceans, are normally relatively unavailable sources, but exchange pools, such as those in the atmosphere, soil, and water, are available sources of chemicals for the biotic community. When human activities (purple arrows) remove chemicals from reservoirs and make them available to the biotic community, pollution can result.

ecosystem contains excess chemicals, which in turn may alter the species composition and diversity of the biotic community.

Phosphorus Cycle

On land, the very slow weathering of rocks fostered by fungi adds phosphates (PO_4^{2-} and HPO_4^{2-}) to the soil, some of which become available for uptake by terrestrial plants (**Fig. 31.22**). Phosphates made available by weathering also run off into aquatic ecosystems, where algae absorb the phosphates from the water before they become trapped in sediments. Phosphates in sediments become available again only when a geological upheaval exposes sedimentary rocks to weathering once more.

Producers use phosphates in a variety of molecules, including phospholipids, ATP, and the nucleotides

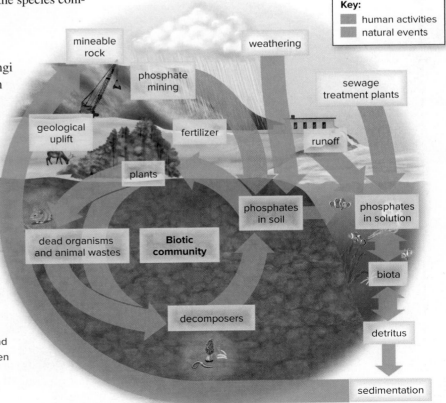

Figure 31.22 The phosphorus cycle.

The phosphorus cycle is a sedimentary biogeochemical cycle. Globally, phosphates flow into large bodies of water and become a part of sedimentary rocks. Thousands or millions of years later, the seafloor can rise; the phosphates are then exposed to weathering and become available. Locally, phosphates cycle within a community when plants on land and algae in the water take them up. Animals gain phosphates when they feed on plants or algae. Decomposers return phosphates to plants or algae, and the cycle begins again.

Figure 31.23 Lemmings.

The population size of lemmings influences the mineral cycles of the Arctic ecosystem.

© Tom McHugh/Science Source

Figure 31.24 The nitrogen cycle.

The nitrogen cycle is a gaseous biogeochemical cycle normally maintained by the work of several populations of soil bacteria.

(bacteria inset): © Perennou Nuridsany/Science Source

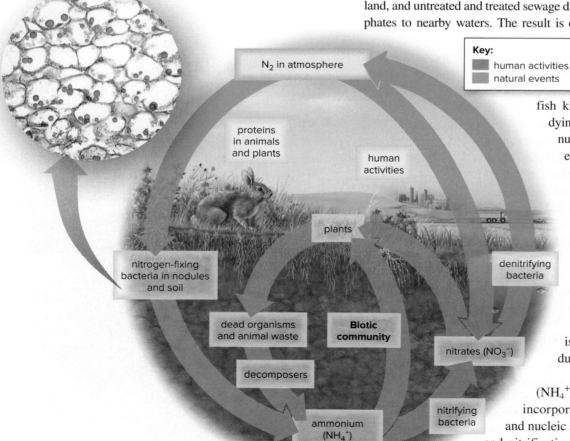

that become a part of DNA and RNA. Animals consume producers and incorporate some of the phosphates into teeth, bones, and shells. Decomposition of dead plant and animal material and animal wastes does, however, make phosphates available to producers faster than does weathering. Because most of the available phosphates are used in food chains, phosphate is usually a limiting inorganic nutrient for ecosystems. In other words, the finite supply of phosphates limits plant growth, and therefore primary productivity.

The importance of phosphates and calcium to population growth is demonstrated by considering the fate of lemmings every 4 years (**Fig. 31.23**). You may have heard that lemmings dash mindlessly over cliffs into the sea; ecologists tell us that these lemmings are actually migrating to find food. What happened? Every 4 years or so, grasses and sedges of the tundra (see Fig. 31.30) become rich in minerals, and the lemming population starts to explode. Once the lemmings number in the millions, the grasses and sedges of the tundra suffer a decline caused by a lack of minerals. Then the lemming population suffers a crash, but it takes about 4 years before the animals decompose in this cold region and minerals return to the producers. Then the cycle begins again.

Human Activities   A **transfer rate** is the amount of a nutrient that moves from one component of the environment to another within a specified period of time. Human activities affect the dynamics of a community by altering transfer rates. For example, humans mine phosphate ores and use them to make fertilizers, animal feed supplements, and detergents. Phosphate ores are slightly radioactive; therefore, mining phosphate poses a health threat to all organisms, including the miners. Animal wastes from livestock feedlots, fertilizers from lawns and cropland, and untreated and treated sewage discharged from cities all add excess phosphates to nearby waters. The result is **eutrophication,** or overenrichment of a body of water, which causes a rapid algal population growth called an algal bloom. When the algae die and decay, oxygen is consumed, causing fish kills. In the mid-1970s, Lake Erie was dying because of eutrophication. Control of nutrient phosphates, particularly in sewage effluent and household detergents, reversed the situation.

Nitrogen Cycle

Nitrogen, in the form of nitrogen gas (N_2), constitutes about 78% of the atmosphere by volume. But plants cannot make use of nitrogen gas. Instead, plants rely on various types of bacteria to make nitrogen available to them. Therefore, nitrogen, like phosphorus, is a limiting inorganic nutrient of producers in ecosystems.

Plants can take up both ammonium (NH_4^+) and nitrate (NO_3^-) from the soil and incorporate the nitrogen into amino acids and nucleic acids. Two processes, nitrogen fixation and nitrification, convert nitrogen gas, N_2, into NH_4^+ and NO_3^-, respectively (**Fig. 31.24**). *Nitrogen fixation* occurs when nitrogen gas is converted to ammonium. Some cyanobacteria in aquatic ecosystems and some free-living, nitrogen-fixing bacteria in soil are able to fix nitrogen in this way. Other nitrogen-fixing bacteria live in nodules on the roots of plants such as peas, beans, and alfalfa. They make organic compounds containing nitrogen available to the host plants.

Nitrification is the production of nitrates. Ammonium in the soil is converted to nitrate by certain nitrifying soil bacteria in a two-step process. First, nitrite-producing bacteria convert ammonium to nitrite (NO_2^-), and then nitrate-producing bacteria convert nitrite to nitrate. In Figure 31.24, notice that the biotic community subcycle in the nitrogen cycle does not depend on the presence of nitrogen gas.

Denitrification is the conversion of nitrate to nitrogen gas, which enters the atmosphere. Denitrifying bacteria are chemoautotrophs that live in the anaerobic mud of lakes, bogs, and estuaries; they carry out this process as part of their own metabolism. In the nitrogen cycle, denitrification counterbalanced nitrogen fixation until humans started making fertilizer.

Human Activities   When humans produce fertilizers from N_2, they are altering the transfer rates in the nitrogen cycle. In fact, humans nearly double the fixation rate. The nitrate in fertilizers, just like phosphate, can leach out of agricultural soils into waterways, leading to eutrophication. Deforestation by humans also causes a loss of nitrogen to groundwater and makes regrowth of the forest difficult. The underground water supplies in farming areas today are more apt to contain excess nitrate. A high concentration of nitrates interferes with blood oxygen levels, and infants below the age of six months who drink water containing excessive amounts of nitrate can become seriously ill and, if untreated, may die.

To cut back on fertilizer use, it might be possible to genetically engineer soil bacteria with increased nitrogen fixation rates. Also, farmers could grow various plants that increase the nitrogen content of the soil (**Fig. 31.25**). In one study, the rotation of legumes and winter wheat produced a better yield than fertilizers after several years.

Carbon Cycle

In the carbon cycle, organisms in both terrestrial and aquatic ecosystems exchange carbon dioxide with the atmosphere (**Fig. 31.26**). On land, plants take up carbon dioxide from the air. Then through photosynthesis, they incorporate carbon into organic nutrients, which are used by both autotrophs and heterotrophs. When aerobic organisms respire, a portion of this carbon is returned to the atmosphere as carbon dioxide, a waste product of cellular respiration.

In aquatic ecosystems, the exchange of carbon dioxide with the atmosphere is indirect. Carbon dioxide from the air combines with water to produce bicarbonate ion (HCO_3^-), a source of carbon for algae that produce food

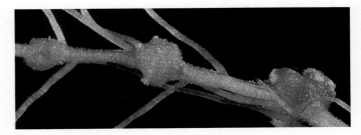

Figure 31.25   **Root nodules.**

Bacteria that live in nodules on the roots of plants in the legume family, such as pea plants, convert nitrogen in the air to a form that plants can use to make proteins.

© Biophoto Associates/Science Source

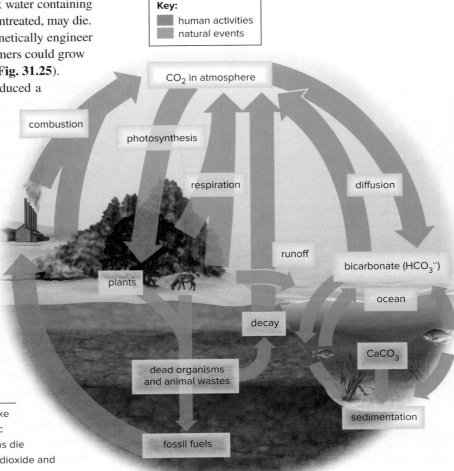

Figure 31.26   **The carbon cycle.**

The carbon cycle is a gaseous biogeochemical cycle. Producers take in carbon dioxide from the atmosphere and convert it to the organic molecules that feed all organisms. Fossil fuels arise when organisms die but do not decompose. The burning of fossil fuels releases carbon dioxide and causes environmental pollution.

Connections: Environment

What causes ocean acidification?

The release of excessive CO_2 into the atmosphere has led to a change in the basic chemistry of the ocean water. The ocean absorbs about 25% of the CO_2 in the atmosphere. As the CO_2 levels in the atmosphere increase, so do the levels in the ocean, leading to ocean acidification. As this occurs, many marine organisms do not have the minerals they need to produce and maintain their shells. The pH level of the ocean will shift to a more acidic environment, leading to stress and the possible extinction of many species of shellfish and corals. Ocean acidification is a global issue due to the potential impact this will have on the entire marine food chain.

for themselves and for heterotrophs. Similarly, when aquatic organisms respire, the carbon dioxide they give off becomes bicarbonate ion. The amount of bicarbonate in the water is in equilibrium with the amount of carbon dioxide in the air.

Living and dead organisms contain organic carbon and serve as one of the reservoirs for the carbon cycle. The world's biotic components, particularly trees, contain billions of tons of organic carbon, and additional tons are estimated to be held in the remains of plants and animals in the soil. If dead plant and animal remains fail to decompose, they are subjected to extremely slow physical processes that transform them into coal, oil, and natural gas, the **fossil fuels.** Most of the fossil fuels were formed during the Carboniferous period, 286–360 MYA, when an exceptionally large amount of organic matter was buried before decomposing. Another reservoir is the calcium carbonate ($CaCO_3$) that accumulates in limestone and shells. Many marine organisms have calcium carbonate shells that remain in bottom sediments long after the organisms have died. Over time, geological forces change these sediments into limestone.

Human Activities   The transfer rates of carbon dioxide due to photosynthesis and cellular respiration are just about even. However, more carbon dioxide is being deposited in the atmosphere than is being removed. In 1850, atmospheric CO_2 was at about 280 parts per million (ppm); today, it is over 400 ppm. This increase is largely due to the burning of fossil fuels and the destruction of forests to make way for farmland and pasture. Today, the amount of carbon dioxide released into the atmosphere is about twice the amount that remains in the atmosphere. CO_2 concentrations in the atmosphere would rise higher, except for the fact that the oceans take up CO_2, and so far they have been taking up more CO_2 than they vent. Despite a number of international agreements to address carbon dioxide emissions, it is believed that atmospheric concentrations will continue to rise until around 2100.

The increased amount of carbon dioxide (and other gases) in the atmosphere is causing **climate change** to occur. These gases allow the sun's rays to pass through, but they absorb and reradiate heat back to Earth, a phenomenon called the **greenhouse effect.**

We are already beginning to see the consequences of climate change. If the Earth's temperature continues to rise, more water will evaporate, forming more clouds. This sets up a positive feedback effect that could increase global warming still more. The global climate has already warmed about 0.6°C (1.1°F) since the Industrial Revolution. Enhancements in computer science are allowing scientists to explore the majority of the variables that influence the global climate. Most climate scientists agree that the Earth's temperature may rise 1.5–4.5°C (2.0–8.1°F) by 2100 if greenhouse emissions continue at the current rates (**Fig. 31.27**).

Figure 31.28 shows how the average temperature in the United States has steadily increased over the past two centuries. As climate change continues, other effects will possibly occur. It is predicted that, as the oceans warm, temperatures in the polar regions will rise to a greater degree than in other regions. As a result, sea levels will rise, because glaciers will melt, and water expands as it warms. Water evaporation will increase, and most likely there will be increased rainfall along the coasts and dryer conditions inland. The occurrence of droughts will reduce agricultural yields and cause trees to die off. Furthermore, weather pattern changes might cause the American Midwest to become a dust bowl. Expansion of forests into Arctic areas might not offset the loss of forests in the temperate zones. Coastal agricultural lands, such as the deltas of Bangladesh and China, will be inundated with water. Billions of

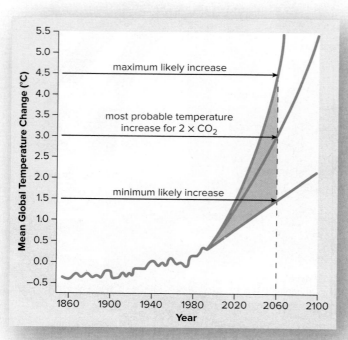

Figure 31.27 Global warming.

Mean global temperature is expected to rise due to the addition of greenhouse gases to the atmosphere.

dollars will have to be spent to keep coastal cities such as New Orleans, New York, Boston, Miami, and Galveston from disappearing into the sea.

As we explored in the chapter opener, the changing climate will also have an impact on our personal lives. Over the past few decades, the mosquito species *Anopheles* and *Aedes* have increased their range into the United States. These are tropical species of mosquitoes, but increase in global temperatures have reduced the severity of winters. These mosquitoes are vectors for diseases such as malaria, dengue fever, and Zika virus. As the climate warms, we will see more instances of these diseases.

There are efforts to reduce carbon dioxide emissions. Increasingly, you hear about alternative sources of energy (see Section 32.4), and increased energy efficiency of cars and appliances. Globally, international meetings, such as the recent Paris Climate Change Conference, are working to reduce carbon dioxide emissions to keep global temperatures from rising more than 2°C over pre-industrial levels.

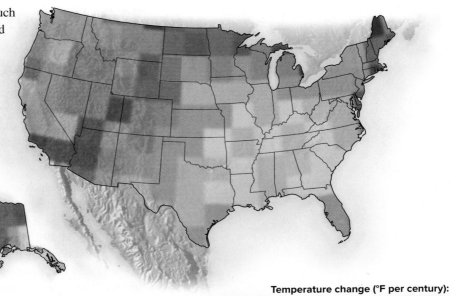

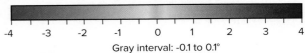

Temperature change (°F per century):

-4   -3   -2   -1   0   1   2   3   4

Gray interval: -0.1 to 0.1°

Figure 31.28 **Climate change in the United States.**

The average temperature in the United States has steadily increased over the past two centuries, leading to more severe droughts and more erratic periods of precipitation that are altering the composition of many communities.

Connections: Ecology

What are some other greenhouse gases?

In addition to carbon dioxide, the following gases also play a role in the greenhouse effect:

- *Methane (CH$_4$).* A single molecule of methane has 21 times the warming potential of a molecule of carbon dioxide, making it a powerful greenhouse gas. Methane is a natural by-product of the decay of organic material, but it is also released by landfills and the production of coal, natural gas, and oil.
- *Nitrous oxide (N$_2$O).* Nitrous oxide is released from the combustion of fossil fuels and as gaseous waste from many industrial activities.
- *Hydrofluorocarbons.* These were initially produced to reduce the levels of ozone-depleting compounds in the upper atmosphere. Unfortunately, although present in very small quantities, they are potent greenhouse gases.

CONNECTING THE CONCEPTS

31.2

Ecosystems are defined by the cycling of energy and nutrients. Global disruptions in nutrient cycles are causing problems with ecosystem function.

Check Your Progress 31.2

1. Compare how autotrophs and heterotrophs acquire energy.
2. Describe the differences between energy flow and chemical cycling in an ecosystem.
3. Compare an ecological pyramid with a food chain.
4. Explain how human activity is influencing each of the biogeochemical cycles.

31.3 Ecology of Major Ecosystems

Learning Outcomes

Upon completion of this section, you should be able to

1. List the two major types of ecosystems that make up the biosphere.
2. List the major types of terrestrial ecosystems.
3. Explain primary productivity, and relate this concept to the different types of aquatic and terrestrial ecosystems.

The **biosphere,** which encompasses all the ecosystems on Earth, is the final level of biological organization. **Aquatic ecosystems** are divided into those associated with fresh water and those associated with salt water (marine ecosystems; **Fig. 31.29**). The ocean, a marine ecosystem, covers 70% of the Earth's surface. Two types of freshwater ecosystems are those with standing water, such as lakes and ponds, and those with running water, such as rivers and streams. The richest marine ecosystems lie near the coasts. Coral reefs are located offshore, while estuaries occur where rivers meet the sea.

Figure 31.29 The major aquatic ecosystems.

Aquatic ecosystems are divided into those associated with salt water, such as (**a**) the ocean, and (**b**) those with fresh water, such as a river. Saltwater, or marine, ecosystems also include (**c**) coral reefs and (**d**) estuaries.

Scientists recognize several distinctive major types of **terrestrial ecosystems,** also called *biomes* (**Fig. 31.30**). Temperature and rainfall define the biomes, which contain communities adapted to the regional climate. The northernmost biome is the tundra. A permafrost persists year-round and prevents large plants from becoming established. The taiga is a very cold northern coniferous forest, and the tundra, which borders the North Pole, is also very cold, with long winters and a short growing season. Temperate grasslands receive less rainfall than temperate deciduous forests (in which trees lose their leaves during the winter) and more water than deserts, which lack trees. The savanna is a tropical grassland with a high temperature and alternating wet and dry seasons. The tropical rain forests, which occur near the equator, have a high average temperature and the greatest amount of rainfall of all the biomes. They are dominated by large, evergreen, broad-leaved trees.

Primary Productivity

One way to compare ecosystems is to look at **primary productivity,** the rate at which producers capture and store energy as organic nutrients over a certain length of time. Temperature and moisture, and secondarily the nature of the soil, influence the primary productivity, which determines the assemblage of species in an ecosystem. In terrestrial ecosystems, primary productivity is generally lowest in high-latitude tundras and deserts, and it is highest near the

Connections: Environment

What is an estuary?

An estuary is a region where fresh water, usually from a river, meets the ocean. The water in an estuary zone is often called "brackish," because it is a mixture of salt and fresh water. Estuaries are considered to be some of the most productive ecosystems in the world. Not only do they provide food sources, such as shrimp, clams, and other seafood, but they also play an important role in the

© Roger de la Harpe/Gallo Images/Getty Images

chemical cycling of nitrogen and phosphorus. Estuaries are fragile ecosystems; pollution, oil spills (such as the one that threatened Louisiana's Bartaria Bay in 2010), and human activity can easily reduce the productivity of an estuary.

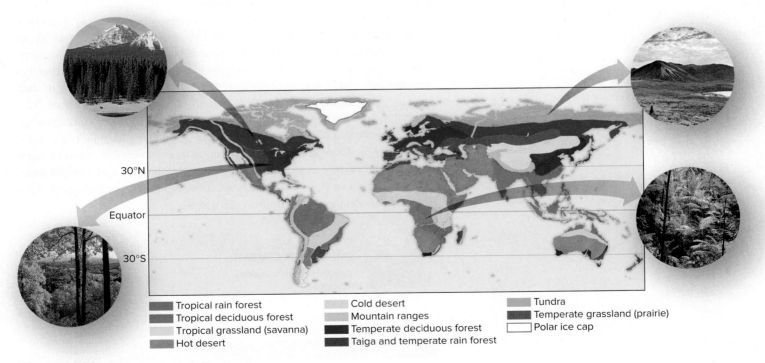

Tropical rain forest
Tropical deciduous forest
Tropical grassland (savanna)
Hot desert

Cold desert
Mountain ranges
Temperate deciduous forest
Taiga and temperate rain forest

Tundra
Temperate grassland (prairie)
Polar ice cap

Figure 31.30 **The major terrestrial ecosystems.**

The tundra is the northernmost terrestrial ecosystem and has the lowest average temperature of all the terrestrial ecosystems, with minimal to moderate rainfall. The taiga, a coniferous forest that encircles the globe, also has a low average temperature, but moderate rainfall. Temperate forests have moderate temperatures and occur where rainfall is moderate yet sufficient to support trees. Tropical rain forests, which generally occur near the equator, have a high average temperature and the greatest amount of rainfall of all the terrestrial ecosystems.

(taiga): © John E. Marriott/All Canada Photos/Getty Images; (temperate deciduous forest): © David Sucsy/E+/Getty RF; (tundra): © Stockbyte/Getty RF; (tropical rain forest): © Digital Vision/Getty RF

Figure 31.31 Primary productivity.

Ecologists can compare ecosystems based on primary productivity, the rate at which producers convert and store solar energy as chemical energy.

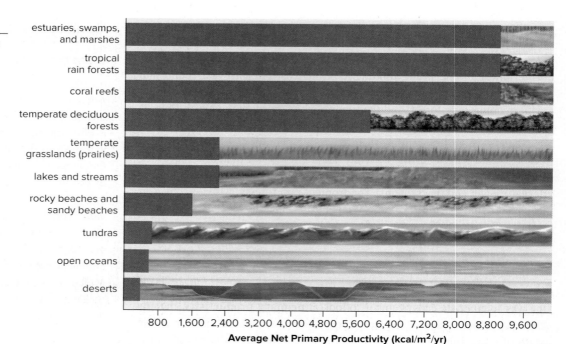

Average Net Primary Productivity (kcal/m²/yr)

equator, where tropical forests occur (**Fig. 31.31**). The high productivity of tropical rain forests provides varied niches and lots of food for consumers. The number and diversity of species in tropical rain forests are the highest of all the terrestrial ecosystems. Therefore, conservation biologists are interested in preserving as much of this biome as possible.

The primary productivity of aquatic communities is largely dependent on the availability of inorganic nutrients. Estuaries, swamps, and marshes are rich in organic nutrients and in decomposers that convert those organic nutrients into their inorganic chemical components. Estuaries, swamps, and marshes also contain a large variety of species, particularly in the early stages of their development before they venture forth into the sea. Therefore, all of these coastal regions are in great need of preservation. The open ocean has a productivity between that of a desert and that of the tundra, because it lacks a concentrated supply of inorganic nutrients. Coral reefs exist near the coasts in warm tropical waters, where currents and waves bring nutrients and sunlight penetrates to the ocean floor. Coral reefs are areas of remarkable biological abundance, equivalent to that of tropical rain forests.

CONNECTING THE CONCEPTS

31.3

Aquatic and terrestrial ecosystems are connected.

Check Your Progress 31.3

1. Describe the two major types of ecosystems that make up the biosphere.

2. List the major terrestrial ecosystems.

3. Explain why swamps have higher levels of primary productivity than do open oceans.

SUMMARIZE

All of the species in a given area make up a community. Energy flow and chemical cycling are essential for the stability of both aquatic and terrestrial ecosystems.

31.1 Community interactions are complex and dynamic.

31.2 Ecosystems are defined by the cycling of energy and nutrients. Global disruptions in nutrient cycles are causing problems with ecosystem function.

31.3 Aquatic and terrestrial ecosystems are connected.

31.1 Ecology of Communities

Knowledge of community and ecosystem ecology is important for understanding the impacts of human activities on the environment.

- A **community** is an assemblage of the populations of different species interacting with each other in a given area. **Coevolution** has shaped the interactions of many species. Communities are often characterized by their **diversity** and **species richness.**
- An **ecosystem** consists of species interacting with one another and with the physical environment.

Ecological Succession

bare rock → lichens/mosses → grasses → shrubs → trees

The two types of **ecological succession** are **primary succession** (begins on bare rock) and **secondary succession** (following a disturbance; begins where soil is present). Ecological succession leads to a stable **climax community.**

Interactions in Communities

Species in communities interact with one another in the following ways:

- **Competition.** Species vie with one another for resources, such as light, space, and nutrients. Aspects of competition are the **competitive exclusion principle, resource partitioning,** and **character displacement.**
- **Predation.** One species (predator) eats another species (prey).
- **Parasitism.** One species (parasite) obtains nutrients from another species (host) but does not kill the host species.
- **Commensalism.** One species benefits from the relationship, while the other species is neither harmed nor benefited.
- **Mutualism.** Two species interact in a way that benefits both.

Ecological Niche

The **ecological niche** of a species is defined by the role it plays in its community, its **habitat,** and its interactions with other species.

Keystone Species

The interactions of a **keystone species** in the community hold the community and its species together. Removal of a keystone species can lead to extinction of other species and loss of diversity. An example of a keystone species is the grizzly bear.

Native Versus Exotic Species

Native species are indigenous to a given area and thrive without assistance. **Exotic species** are introduced into an area and may disrupt the balance and interactions among native species in that area's community.

31.2 Ecology of Ecosystems

In a food chain in an ecosystem, some populations are autotrophs and some are heterotrophs.

Autotrophs

The **autotrophs** are the **producers.** They require only inorganic nutrients (e.g., CO_2 and minerals) and an outside energy source to produce organic nutrients for their own use and for the use of other members of the community. Examples of autotrophs are algae, cyanobacteria, and plants.

Heterotrophs

The **heterotrophs** are **consumers.** They require a preformed source of organic nutrients and give off CO_2. Examples of heterotrophs are **herbivores** (feed on plants), **carnivores** (feed on animals), and **omnivores** (feed on both plants and animals). Other heterotrophs are the **decomposers** (the bacteria and fungi that aid in decay by breaking down **detritus**).

Energy Flow and Chemical Cycling

Energy flows through an ecosystem, while chemicals cycle within an ecosystem.

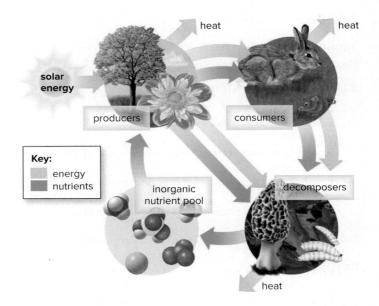

- Energy flows within an ecosystem through **food chains** and **detrital food webs** and **grazing food webs.**
- A **trophic level** is a level of nourishment in a food chain or web.
- An **ecological pyramid** diagrams the energy losses that occur between trophic levels. Only about 10% of the energy of one trophic level is available to the next trophic level. Top carnivores occupy the topmost and smallest trophic level. The **biomass** of a trophic level is the total weight of the living organisms at that level.

Biogeochemical Cycle

Chemicals cycle within ecosystems through various **biogeochemical cycles,** such as the phosphorus cycle, the nitrogen cycle, and the carbon cycle. The **transfer rate** is the amount of a nutrient that moves from one component of the environment to another within a specified time.

Human activities significantly alter the transfer rates in the biogeochemical cycles.

- In the phosphorus and nitrogen cycles, excess nutrients being released into the environment cause **eutrophication.**

• In the carbon cycle, the burning of **fossil fuels** has increased the level of greenhouse gases in the atmosphere, resulting in global warming and **climate change.**

31.3 Ecology of Major Ecosystems

The **biosphere** encompasses all the ecosystems on Earth.

Aquatic Ecosystems

The **aquatic ecosystems** are classified as freshwater ecosystems (rivers, streams, lakes, ponds) and marine ecosystems (oceans, coral reefs, saltwater marshes).

Terrestrial Ecosystems

The **terrestrial ecosystems** are called biomes. The major biomes are the tundra, taiga, temperate deciduous forest, tropical grassland (savanna), temperate grassland (prairie), desert, and tropical rain forest.

Primary Productivity

Primary productivity is the rate at which producers capture and store energy and convert it to organic nutrients over a certain length of time. The number of species in an ecosystem is positively related to its primary productivity.

ASSESS

Testing Yourself

Choose the best answer for each question.

31.1 Ecology of Communities

1. As diversity increases,
   a. species richness increases and the distribution of species becomes more even.
   b. species richness decreases and the distribution of species becomes more even.
   c. species richness increases and the distribution of species becomes less even.
   d. species richness decreases and the distribution of species becomes less even.

For statements 2–6, indicate the type of interaction described in each scenario.

Key:

   a. competition   d. commensalism
   b. predation   e. mutualism
   c. parasitism

2. An alfalfa plant gains fixed nitrogen from the bacterial species *Rhizobium* living on its root system, while *Rhizobium* gains carbohydrates from the plant.

3. Both foxes and coyotes in an area feed primarily on a limited supply of rabbits.

4. Roundworms live and reproduce within a cat's digestive tract.

5. A fungus captures nematodes as a food source.

6. An orchid plant lives in the treetops, gaining access to sunlight and pollinators but not harming the trees.

7. According to the competitive exclusion principle,
   a. one species is always more competitive than another for a particular food source.
   b. competition excludes multiple species from using the same food source.
   c. no two species can occupy the same niche at the same time.
   d. competition limits the reproductive capacity of species.

31.2 Ecology of Ecosystems

8. In the following diagram, fill in the components of chemical cycling and nutrient flow.

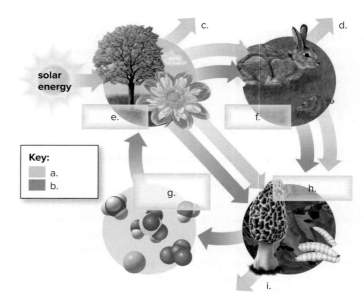

9. An ecological pyramid depicts the amount of _____ in various trophic levels.
   a. food
   b. organisms
   c. energy
   d. waste

10. Which of the following represents a grazing food chain?
    a. leaves → detritus feeders → deer → owls
    b. birds → mice → snakes
    c. nuts → leaf-eating insects → chipmunks → hawks
    d. leaves → leaf-eating insects → mice → snakes

11. Identify the components of the ecological pyramid in the following diagram.

12. Ecosystems include which of the following components that communities do not?
    a. energy transfer between members
    b. interaction with the physical environment
    c. intricate food webs
    d. herbivores and carnivores

31.3 Ecology of Major Ecosystems

13. Which biome is characterized by a coniferous forest with low average temperature and moderate rainfall?

a. taiga

b. savanna

c. tundra

d. temperate deciduous forest

e. tropical rain forest

14. Which biome has the lowest primary productivity?

a. tundra

b. lake

c. sandy beach

d. prairie

e. temperate deciduous forest

ENGAGE

BioNOW

Want to know how this science is relevant to your life? Check out the BioNow video below.

- Biodiversity

What was the effect of the exotic (alien) species on the biodiversity? Why do you think this was the case?

Thinking Critically

1. One of the most striking examples of coevolution is between insects and flowers. The earliest angiosperms produced large amounts of pollen on flowers that were wind-pollinated. The ovules were partially exposed and exuded tiny droplets of sugary sap to catch passing pollen. Outline a course of events that could have resulted in the coevolution we observe today between a flower and its pollinator.

2. Sea otters play an important role in maintaining the kelp forests of the Pacific coast. Otters maintain the sea urchin population, preventing the sea urchins from overgrazing the kelp beds. If the kelp beds are overgrazed, a multitude of other species will decline as a result. Identify the chain of events that could occur if the sea otter population were reduced along the Pacific coast.

3. Many exotic species, such as zebra mussels and sea lampreys, are so obviously troublesome that most people do not object to programs aimed at controlling their populations. However, some eradication programs directed toward exotic species meet with more resistance. For example, the mute swan, one of the world's largest flying birds, is beautiful and graceful, and it has an impressive presence. However, it is very aggressive and territorial. The mute swan was introduced to the United States from Asia and Europe in the nineteenth century as an ornamental bird but has since established a large wild population.

The birds consume large amounts of aquatic vegetation and displace native birds from feeding and nesting areas. The U.S. Fish and Wildlife Service has programs for reducing mute swans in Maryland in order to protect native bird populations. Attempts to limit the size of the mute swan populations in Maryland and other states have been met with opposition by citizens who find the birds beautiful. Do you feel that native populations need not be protected as long as the exotic species serves a suitable human purpose? Or do you feel that native species should be protected regardless?

32

Human Impact on the Biosphere

© Brett Carlsen/Stringer/Getty Images News

OUTLINE

32.1 Conservation Biology   627

32.2 Biodiversity   628

32.3 Resources and Environmental Impact   632

32.4 Sustainable Societies   643

BEFORE YOU BEGIN

Before beginning this chapter, take a few moments to review the following discussions.

Section 12.3 What is a transgenic organism?

Section 30.2 How do the environmental impacts of a more-developed country (MDC) compare with those of a less-developed country (LDC)?

Section 31.2 How much energy is transferred from one level of an ecological pyramid to the next level?

Flint Water Crisis

In 2014, the city of Flint, Michigan, in an attempt to save money due to a financial crisis, switched the source of its drinking water from Lake Huron to the Flint River. Almost immediately, the residents of Flint began to notice a change in their water quality. Foul odors, sediment, and bacteria were now present in the water, and residents were complaining of feeling sick after using the water.

But the real problem was lead. Portions of Flint's water system date back to the early decades of the twentieth century, when lead was used in the connections between pipes. Unfortunately, the Flint River water contains high levels of chloride, which is a corrosive material. The chloride caused the lead (and other heavy metals) in the pipes to leach out, exposing residents to high levels of lead in their drinking water.

The presence of lead in drinking water is a major health hazard. The amount of lead in water is limited by government regulations to less than 15 parts per billion (ppb), but even low levels of lead exposure can be dangerous. Lead exposure in adults can lead to kidney and liver problems, and an increased rate of illness. The effects of lead on children are more dangerous. Lead affects the developing nervous system and can cause permanent developmental problems. An estimated 7,000–12,000 children in Flint may have been exposed to unsafe levels of lead.

In this chapter, we will explore how humans are interacting with their ecosystems, and how those interactions may sometimes have negative consequences.

As you read through this chapter, think about the following questions:

1. In addition to chemical pollution, what other threats to the water cycle are associated with human activity?
2. Does the lead in the Flint water crisis represent a point or nonpoint source of pollution?

32.1 Conservation Biology

In order to understand the diversity of life on Earth, we need to know more about species other than their total numbers. **Conservation biology** is a field of biology that focuses on conserving natural resources for this and future generations. Conservation biology is concerned with developing new scientific concepts and applying them to our lives, along with sustainably managing the Earth's biodiversity for human use. Multiple subfields of biology blend together to form the concepts of conservation biology.

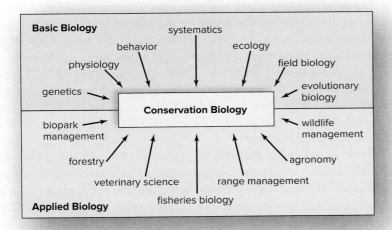

Conservation biology is unique among the life sciences, because it supports a variety of ethical principles: (1) Biodiversity is good for the Earth and therefore good for humans; (2) extinctions are undesirable; (3) the interactions within ecosystems support biodiversity and are beneficial to humans; and (4) biodiversity is the result of evolutionary change and has great value to humans.

Conservation biology is often referred to as a crisis discipline. In the next 20–50 years, approximately 10–20% of the biodiversity on Earth will disappear due to the extinction of species. It is extremely important that everyone realize the importance and value of biodiversity, as well as how human actions contribute to the extinction crisis before us.

CONNECTING THE CONCEPTS

32.1   Conservation biology seeks to balance the needs of humans with the needs of the rest of the species on Earth.

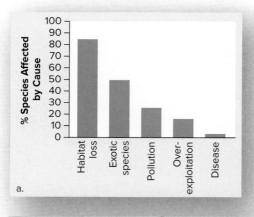

Figure 32.1 Habitat loss.

a. In a study examining records of imperiled U.S. plants and animals, habitat loss emerged as the greatest threat to wildlife. **b.** The Canada lynx is threatened by the loss of habitat and the overexploitation of forests for wood.

© Don Johnston/All Canada Photos/Getty Images

32.2 Biodiversity

Learning Outcomes

Upon completion of this section, you should be able to

1. Define *biodiversity,* and briefly list some of the threats to it.
2. List the direct values of biodiversity, and explain how they are beneficial to the human population.
3. List the indirect values of biodiversity, and briefly describe their economic benefits.

Biodiversity can be defined as the diversity of life on Earth, described in terms of the number of different species. We are presently in a biodiversity crisis—the number of extinctions (loss of species) expected to occur in the near future will, for the first time, be due to human activities. According to the U.S. Fish and Wildlife Service (FWS), as of 2016, there are over 694 animal species and 898 plant species in the United States that are threatened or in danger of extinction. The majority of these species (85%) are threatened by habitat loss (**Fig. 32.1***a*), usually associated with the sprawl of urban areas. Other factors contributing to the biodiversity crisis are the introduction of exotic species (50%), water and air pollution (24%), and the overexploitation of natural resources (17%). In many cases, endangered species are threatened by multiple factors. For example, the Canada lynx (*Lynx canadensis*) (Fig. 32.1*b*) lives in the northern forests of the United States. It is a relatively rare species that prefers dense forests. It is currently listed as a threatened species due to habitat loss caused by the creation of roads for snowmobiling and skiing, as well as increased harvesting of timber.

Biodiversity is not evenly distributed over the Earth. It is highest at the tropics and declines toward the poles, whether we are considering terrestrial, freshwater, or marine species. The regions of the world that contain the greatest concentration of species are known as **biodiversity hotspots** (**Fig. 32.2**). The hotspots contain over 50% of all known plant species and 42% of all terrestrial vertebrate species but cover only about 2.4% of the Earth's land area. Therefore, when trying to save the greatest number of species, hotspots should be prioritized.

Figure 32.2 Biodiversity hotspots.

Hotspots cover only 2.4% of the Earth, yet they contain over 50% of the world's plant species and 42% of terrestrial vertebrates.

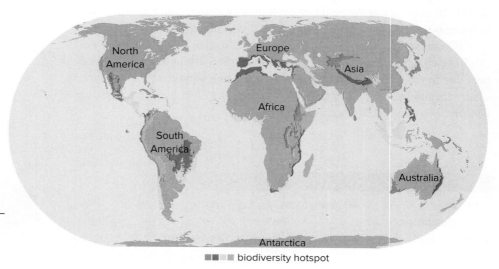

Conservation biology strives to reverse the trend toward the extinction of thousands of plants and animals. To bring this about, it is necessary to make all people aware that biodiversity is a resource with immense value—both direct and indirect.

Direct Values of Biodiversity

The direct values of biodiversity include medicines, foods, and other products that benefit humans.

Medicinal Value

Most of the prescription drugs used in the United States were originally derived from organisms. The rosy periwinkle from Madagascar is an excellent example of a tropical plant that has provided useful medicines (**Fig. 32.3a**). Potent chemicals from this plant are used to treat two forms of cancer: leukemia and Hodgkin's disease. Due to these drugs, the survival rate for childhood leukemia has gone from 10% to 90%, and Hodgkin's disease is now usually curable. Although the value of saving a life cannot be calculated, it is still sometimes easier for us to appreciate the worth of a resource if it is explained in monetary terms. Thus, researchers tell us that, judging from the success rate in the past, hundreds of additional types of drugs are yet to be found in tropical rain forests, and the value of this resource to society is probably in excess of several hundred billion dollars.

The popular antibiotic penicillin is derived from a fungus, and certain species of bacteria produce the antibiotics tetracycline and streptomycin. These drugs have proven to be indispensable in the treatment of diseases, including certain sexually transmitted diseases.

Leprosy is among the diseases for which there is, as yet, no cure. The bacterium that causes leprosy will not grow in the laboratory, but scientists discovered that it grows naturally in the nine-banded armadillo (Fig. 32.3b). Having a source for the bacterium may make it possible to find a potential cure for leprosy. The blood of horseshoe crabs contains a substance called limulus amoebocyte lysate, which is used to ensure that medical devices, such as pacemakers, surgical implants, and prostheses, are free of bacteria. Blood is taken from 250,000 horseshoe crabs a year, and then they are returned to the sea unharmed.

Agricultural Value

Crops such as wheat, corn, and rice were originally derived from wild plants that were modified to be high producers. The same high-yield, genetically similar strains tend to be grown worldwide. When cultivated rice crops in Africa were being devastated by a virus, researchers grew wild rice plants from thousands of seed samples until they found one that contained a gene for resistance to the virus. These wild plants were then used in a breeding program to transfer the gene into high-yield rice plants. If this variety of wild rice had become extinct before its resistance could be discovered, rice cultivation in Africa might have collapsed.

Biological pest controls—specifically, natural predators and parasites—are often preferable to chemical pesticides (**Fig. 32.4**). When a rice pest called the brown planthopper became resistant to pesticides, farmers began to use natural enemies of the brown planthopper instead. The economic savings were calculated at well over $1 billion. Similarly, cotton growers in Cañete Valley, Peru, found that pesticides were no longer working against the cotton aphid

a. Wild species, like the rosy periwinkle, are sources of many medicines.

b. Wild species, like the nine-banded armadillo, play a role in medical research.

Figure 32.3 **Medicinal value of biodiversity.**

Many wildlife species, such as the (**a**) rosy periwinkle and (**b**) nine-banded armadillo, are sources of medical benefits to humans.

(a): © Steven P. Lynch; (b): © Steve Bower/Shutterstock RF

Wild species, like ladybugs, play a role in biological control of agricultural pests.

Wild species, like the long-nosed bat, are pollinators of agricultural and other plants.

Wild species, like many marine species, provide us with food.

Wild species, like rubber trees, can provide a product indefinitely if the forest is not destroyed.

due to resistance. Research identified natural predators, which cotton farmers are now using to an even greater degree.

Most flowering plants are pollinated by animals, such as bees, wasps, butterflies, beetles, birds, and bats. The honeybee has been domesticated, and it pollinates almost $10 billion worth of food crops annually in the United States. The danger of this dependency on a single species is exemplified by mites that have wiped out more than 20% of the commercial honeybee population in the United States. Where can we get resistant bees? From the wild, of course; the value of wild pollinators to the U.S. agricultural economy has been calculated at $15 billion a year.

Consumptive Use Value

We have had much success in cultivating crops, keeping domesticated animals, growing trees on plantations, and so on. However, the environment provides all sorts of other products that are sold in marketplaces worldwide, including wild fruits and vegetables, skins, fibers, beeswax, and seaweed. Also, some people obtain their meat directly from the environment. In one study, researchers calculated that the economic value of wild pig in the diet of native hunters in Sarawak, East Malaysia, was about $40 million per year.

Similarly, many trees are still felled in the natural environment for their wood. Researchers have calculated that a species-rich forest in the Peruvian Amazon is worth far more if the forest is used for fruit and rubber production than for timber production. Fruit and the latex needed to produce rubber can be brought to market for an unlimited number of years, whereas once the trees are gone, no more forest products can be harvested.

Indirect Values of Biodiversity

All wild species play important roles in the ecosystems to which they belong. If we want to preserve them, it is more beneficial to save the entire ecosystem than the individual species. Ecosystems perform many indirect services that cannot always be measured economically. These services are said to be indirect values because they are wide-ranging and not easily perceptible (**Fig. 32.5**).

Maintaining Biogeochemical Cycles

We observed in Section 31.2 that ecosystems are characterized by energy flow and chemical cycling. The biodiversity within ecosystems contributes to the functioning of the water, carbon, nitrogen, phosphorus, and other biogeochemical cycles. We are dependent on these cycles for fresh water, the removal of carbon dioxide from the atmosphere,

Figure 32.4 Agriculture and the consumptive value of wildlife.

Many wildlife species help account for our bountiful harvests and are sources of food for the world.

(ladybug): © Anthony Mercieca/Science Source; (bat): © Dr. Merlin D. Tuttle/Bat Conservation International/Science Source; (fishing boat): © Herve Donnezan/Science Source; (rubber tree): © Bryn Campbell/Stone/Getty

the uptake of excess soil nitrogen, and the provision of phosphate. When human activities upset one aspect of a biogeochemical cycle, other parts within the cycle are also affected. Technology has been unable to artificially contribute to or replicate any of the biogeochemical cycles.

Waste Disposal

Decomposers break down dead organic matter and other types of wastes to inorganic nutrients, which are used by the producers within ecosystems. This function aids humans immensely, because we dump millions of tons of waste material into natural ecosystems each year. If not for decomposition, waste would soon cover the entire surface of our planet. We can build sewage treatment plants, but they are expensive, and few of them break down solid wastes completely to inorganic nutrients. It is less expensive and more efficient to water plants and trees with partially treated wastewater and let soil bacteria cleanse it completely.

Biological communities are also capable of breaking down and immobilizing pollutants, such as the heavy metals and pesticides that humans release into the environment. A review of wetland functions in Canada assigned a value of $50,000 per hectare (2.471 acres, or 10,000 square meters) per year to the ability of natural areas to purify water and take up pollutants.

Provision of Fresh Water

Few terrestrial organisms are adapted to living in a salty environment—they need fresh water. The water cycle continually supplies fresh water to terrestrial ecosystems. Humans use fresh water in innumerable ways, including drinking it and irrigating their crops with it. Freshwater ecosystems, such as rivers and lakes, also supply fish and other types of organisms for our consumption.

Unlike other commodities, there is no substitute for fresh water. We can remove salt from seawater to obtain fresh water, but the cost of desalination is about four to eight times the average cost of fresh water acquired via the water cycle.

Forests and other natural ecosystems exert a "sponge effect." They soak up water and then release it at a controlled rate. When rain falls in a natural area, plant foliage and dead leaves lessen its impact, and the soil slowly absorbs it, especially if the soil has been aerated by organisms. The water-holding capacity of forests reduces the possibility and degree of flooding. The value of a marshland outside Boston, Massachusetts, has been estimated at $72,000 per hectare per year solely on its ability to reduce floods. Forests release water slowly for days or weeks after rains have ceased. Comparing rivers in West African coffee plantations, those flowing through forests release twice as much water halfway through the dry season and three to five times as much at the end of the dry season. These data show the water-retaining ability of forests.

Prevention of Soil Erosion

Intact terrestrial ecosystems naturally retain soil and prevent soil erosion. The importance of this ecosystem attribute is especially noticeable following deforestation. In Pakistan, the world's largest dam, the Tarbela Dam, is losing 12 billion cubic meters of storage capacity sooner than expected because of the silt that is building up behind the dam due to deforestation of areas upriver. At one time, the Philippines were exporting $100 million worth of oysters, mussels, clams, and cockles each year. Now, silt carried down rivers following deforestation is smothering the mangrove ecosystem that serves as a nursery

Figure 32.5 Indirect values of ecosystems.

Forests and coral reefs perform many of the functions listed as the indirect values of ecosystems.

(deciduous forest): © BananaStock/PunchStock RF; (coral reef): © Vlad61/ Shutterstock RF

Figure 32.6 Ecotourism.

Whale watchers experience the thrill of seeing a humpback whale surfacing.

© Chase Dekker Wildlife Images/Getty RF

for many marine species. Most coastal ecosystems are not as bountiful as they once were because of deforestation upriver.

Regulation of Climate

At the local level, trees provide shade, block drying winds, and reduce the need for fans and air conditioners during the summer. Globally, forests regulate the climate, because they take up carbon dioxide, a greenhouse gas. When the leaves of trees photosynthesize, they use carbon dioxide, which is stored as the wood of the tree. When trees are cut and burned, carbon is no longer released into the soil through natural decomposition; instead, carbon dioxide is released into the atmosphere. Subsequently, the reduction of forests worldwide reduces the amount of carbon dioxide removed from the atmosphere, amplifying the buildup of that greenhouse gas. Deforestation often removes soil nutrients needed by future tree generations, which in turn limits reforestation.

Ecotourism

Almost everyone prefers to vacation in the natural beauty of an ecosystem (**Fig. 32.6**). In the United States, nearly 100 million people enjoy vacationing in a natural setting. To do so, they spend nearly $5 billion each year on fees, transportation, lodging, and food. Many tourists want to go sport fishing, whale watching, boat riding, hiking, bird watching, and the like. Some merely want to immerse themselves in the beauty of a natural environment. Some less-developed countries are realizing that there is significant economic potential in developing their ecotourism industry.

CONNECTING THE CONCEPTS

32.2 The Earth's biodiversity has tremendous value to humans, but it is under stress.

Check Your Progress 32.2

1. Summarize the direct and indirect values of biodiversity.
2. Explain how modern medicine and agriculture are dependent on biodiversity.
3. Explain how the preservation of biodiversity can help reduce the problems associated with overexploitation of water, land, food, or energy resources.

32.3 Resources and Environmental Impact

Learning Outcomes

Upon completion of this section, you should be able to

1. Describe the effects of human populations on land and water resources.
2. Describe how modern farming methods impact the environment, and explain how the "green revolution" may help alleviate these problems.
3. Explain how the use of fossil fuels and the production of pollution impact the environment.

Humans use resources to meet their basic needs. Land, water, food, energy, and minerals are the maximally used resources (**Fig. 32.7**).

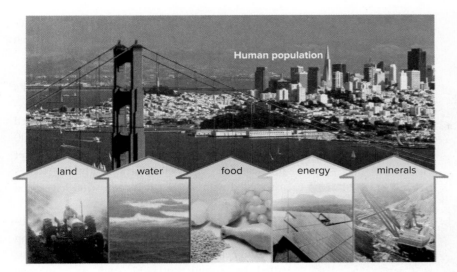

Figure 32.7 Resources.

Humans use land, water, food, energy, and minerals to meet their needs, including a place to live, food to eat, and products that make their lives easier.

(population): © Jan Hanus/Alamy; (land): © Doug Menuez/Getty RF; (water): © Evelyn Jo Johnson; (food): © McGraw-Hill Education, John Thoeming, photographer; (energy): © PhotoLink/Getty RF; (minerals): © T. O'Keefe/PhotoLink/Getty RF

Some resources are nonrenewable, and some are renewable. **Nonrenewable resources** are limited in supply. For example, the amount of land, fossil fuels, and minerals is finite and can be exhausted. Better extraction methods and efficient use can make the supply last longer, but eventually these resources will run out. **Renewable resources** are unlimited in supply. We can use water and certain forms of energy (e.g., solar energy) or harvest plants and animals for food, and more will always be forthcoming. However, even though these resources are renewable, we must be careful not to overconsume them. Consider that most species have a population threshold below which the population cannot recover or sustain its numbers, as when the huge herds of buffalo that once roamed the western United States disappeared after being overexploited.

Unfortunately, one side effect of resource consumption is pollution. **Pollution** is any alteration of the environment in an undesirable way. Pollution is often caused by human activities. The human impact on the environment is proportional to the size of the population and consumption of the resources. As the population grows, so does the need for resources and the amount of pollution generated by using these resources. Consider that seven people adding waste to the ocean might not be alarming, but 7 billion people doing so would certainly affect its cleanliness. In modern times, the consumption of mineral and energy resources has grown faster than population size. People in the more-developed countries (MDCs) use a disproportionate amount of these resources.

Land

People require a physical place to live. Worldwide, there are more than 47 people for each square kilometer of available land, including Antarctica, mountain ranges, jungles, and deserts. Naturally, land is also needed for a variety of uses aside from homes, such as agriculture, electric power plants, manufacturing plants, highways, hospitals, and schools.

Beaches and Human Habitation

At least 40% of the world population lives within 100 km (60 mi) of a coastline, with this number increasing each year. In the United States, over half of the population lives within 80 km (50 mi) of the coasts (including the Great

Figure 32.8 Beach erosion.

a. Most of the U.S. coastline is subject to beach erosion. **b.** Therefore, people who choose to live near the coast may eventually lose their homes.

(b): © Tyrone Turner/Getty Images

Lakes). Living right on the coast has a negative impact, because it accelerates natural beach erosion and loss of habitat for marine organisms.

Beach Erosion An estimated 70% of the world's beaches are eroding; **Figure 32.8***a* shows how extensive this problem is along coasts of the United States. The seas have been rising for the past 12,000 years, ever since the climate turned warmer after the last Ice Age, and consequently the forces associated with waves, tides, and wind have been amplified.

Humans often participate in activities that divert more water to the oceans, contributing to rising seas and beach erosion. For example, humans have filled in coastal wetlands, such as mangrove swamps in the southern United States and saltwater marshes in the northern United States formerly considered "wastelands." With growing recognition of the services provided by wetlands, wetland conservation and restoration has been growing in the United States during the past 40 years. There are multiple reasons to protect coastal wetlands. They serve as spawning areas for fish and other forms of marine life as well as buffers against hurricane storm surges, and they reduce shoreline erosion. They are also habitats for certain terrestrial species, including many types of birds.

Humans often try to stabilize beaches by building groynes (structures that extend from the beach into the water) and seawalls. Groynes trap sand and build beaches on one side as well as slowing the longshore currents, but erosion is increased on the other side. Seawalls, in the end, also increase erosion because ocean waves remove sand from in front of and to the sides of these walls. Importing sand is a better solution, but it is very costly and can disturb plant and animal populations. It's estimated that the U.S. shoreline loses 40% more sediment than it receives, because the building of dams prevents sediment from reaching the coast.

Coastal Pollution The coasts are particularly subject to pollution because toxic substances placed in freshwater lakes, rivers, and streams may eventually find their way to a coast. Oil spills at sea also cause localized harmful effects.

Semiarid Lands and Human Habitation

Forty percent of the Earth's lands are already deserts, and any land adjacent to a desert is in danger of becoming desert if humans manage it improperly. **Desertification** is the conversion of semiarid land to desertlike conditions (**Fig. 32.9**).

Quite often, desertification begins when humans allow animals to overgraze the land. The soil can no longer hold rainwater, which runs off instead of keeping the remaining plants alive or recharging wells. Humans then remove whatever vegetation they can find to use as fuel or fodder for their animals. The end result is a desert unable to support agriculture, which is then abandoned as people move on to continue the process someplace else. Some estimate that nearly three-quarters of all rangelands world-wide are in danger of desertification. The famine in Ethiopia during 2002–2003 was due, at least in part, to degradation of the land to the point that it could no longer support humans and their livestock.

Tropical Rain Forest and Human Habitation

Deforestation, the removal of trees that reduces the extent of a forest (**Fig. 32.10**), has long allowed humans to live in areas where forests once covered the land. The concern recently has been that people are settling in tropical rain forests, such as the Amazon, following the building of roads. This land, too, is subject to desertification. Soil in the tropics is often thin and nutrient-poor because all the nutrients are tied up in the trees and other vegetation. When the trees are felled and removed and the land is used for agriculture or grazing, it quickly loses its fertility and becomes subject to desertification.

Other consequences of deforestation are a loss of biodiversity and an increase in atmospheric carbon dioxide. The trees that once took in and stored carbon dioxide have been removed. The destruction of tropical rain forests causes the extinction of a large number of species that have value to humans.

Water

Access to clean drinking water is considered a human right, but actually most fresh water is used by agriculture and industry (**Fig. 32.11***a,b*). In the United States, approximately 39% of all fresh water is used to irrigate crops! Much of a recent surge in demand for water stems from increased industrial activity and irrigation-intensive agriculture, the type of agriculture that now supplies about 40% of the world's food crops. Domestically, in the more-developed countries (MDCs), more water is usually used for bathing, flushing toilets, and watering lawns than for drinking and cooking (Fig. 32.11*c*). In the water-poor areas of the world, people may not have ready access to drinking water, and if they do, the water may be unclean.

Increasing Water Supplies

Although the needs of the human population overall do not exceed the renewable supply, this is not the case in certain regions of the world. As illustrated in Figure 32.9, about 40% of the world's land is desert, and deserts are bordered by semiarid land. When necessary to support human population growth, the supply of fresh water is increased by damming rivers and withdrawing water from aquifers.

Dams The world's 45,000 large dams catch 14% of all precipitation runoff, provide water for up to 40% of irrigated land, and give some 65 countries more than half their electricity. Damming of certain rivers has been so extensive that they no longer flow as they once did. The Yellow River in China fails to reach the sea most years; the Colorado River barely makes it to the Gulf of California; and even the Rio Grande dries up before it can merge with the Gulf of Mexico. The first and third longest rivers on Earth, the Nile in Egypt and the Ganges in India, respectively, are

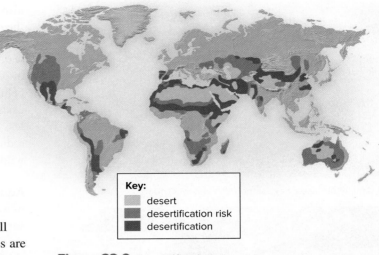

Key:
- desert
- desertification risk
- desertification

Figure 32.9 **Desertification.**

Desertification is a worldwide occurrence that reduces the amount of land suitable for human habitation.

Data from A. Goudie and I. Wilkinson, *The Warm Desert Environment.* Copyright 1977 by Cambridge University Press, New York

Key:

	10,000 yrs ago	today
temperate forests		
tropical forests		

a.

b.

Figure 32.10 **Deforestation.**

a. Nearly half of the world's forest lands have been cleared for farming, logging, and urbanization. **b.** The soil of tropical rain forests is not suitable for long-term farming.

(b): © Eco Images/Universal Images Group/Getty Images

a. Agriculture uses most of the fresh water consumed.

b. Industrial use of water is about half that of agricultural use.

c. Domestic use of water is about half that of industrial use.

Figure 32.11 Global water use.

a. Agriculture uses water primarily for irrigation. **b.** Industry uses water in various ways. **c.** People use water for drinking, bathing, flushing toilets, and watering lawns.

(a): © Corbis RF; (b): © Jeremy Samuelson/Photolibrary/Getty Images; (c): © Stockbyte/PunchStock RF

also so overexploited that, at some times of the year, they hardly make it to the ocean.

Dams have additional drawbacks: (1) Reservoirs behind the dam lose water due to evaporation and seepage into underlying rock beds. The amount of water lost sometimes equals the amount made available! (2) The salt left behind by evaporation and agricultural runoff can make a river's water unusable farther downstream. (3) Sediment buildup causes dams to hold back less water; with time, dams may become useless for storing water. (4) The reduced amount of water below the dam has a negative impact on native wildlife.

Aquifers   To meet their freshwater needs, people are pumping vast amounts of water from **aquifers,** which are natural reservoirs of water found below the surface of the Earth. Aquifers hold about 1,000 times the amount of water that falls on land as precipitation each year. This water accumulates from rain that fell over long periods of time. In the past 50 years, groundwater depletion has become a problem in many areas of the world. In substantial portions of the High Plains Aquifer, which stretches from South Dakota to the Texas Panhandle in the central United States, much water is pumped out for irrigation or for household use. In the 1950s, India had 100,000 motorized pumps in operation; today, India has 20 million pumps, a huge increase in groundwater pumping.

Environmental Consequences   Removal of water causes **subsidence,** settling of the soil as it dries out, which causes the surface to lower. In California's San Joaquin Valley, an area of more than 13,000 km^2 has subsided, and in the worst spot the surface has dropped more than 9 m! In some parts of Gujarat, India, the water table has dropped as much as 7 m. Subsidence damages canals, buildings, and underground pipes. Withdrawal of groundwater can cause sinkholes, as when an underground cavern collapses because water no longer holds up its roof.

Saltwater intrusion is another consequence of groundwater depletion. The flow of water from streams and aquifers usually keeps them fairly free of seawater. But as water is withdrawn, the water table can lower to the point that seawater backs up into streams and aquifers. Saltwater intrusion reduces the supply of fresh water along the coast.

Connections: Environment

How much water is required to produce your food?

Growing a single serving of lettuce takes about 6 gallons of water. Producing an 8-oz glass of milk requires 49 gallons. That includes the amount of water the cow drinks, the water used to grow the cow's food, and the water needed to process the milk. Producing a single serving of steak consumes more than 2,600 gallons of water.

© PhotoSpin, Inc./ Alamy RF

Conservation of Water

By 2025, two-thirds of the world's population may be living in countries that are facing serious water shortages. Many solutions for expanding water supplies are available (**Fig. 32.12**). Using drip irrigation delivers more water to crops and saves about 50% over traditional methods while increasing crop yields. Although the first drip systems were developed in 1960, they are used on less than 1% of irrigated land. Most governments subsidize irrigation so heavily that farmers have little incentive to invest in drip systems or other water-saving methods. Planting drought- and salt-tolerant crops would decrease the water required for agriculture. Recycling water and adopting conservation measures could help the world's industries cut their water demands by more than half.

Food

In 1950, the human population numbered 2.5 billion, and there was only enough food to provide less than 2,000 calories per person per day; now, with over 7.4 billion people on Earth, the world food supply provides more than 2,500 calories per person per day. Generally speaking, food comes from three activities: growing crops, raising animals, and fishing the seas. Unfortunately, modern farming methods, which have increased the food supply, include some harmful practices:

1. *Planting of a few genetic varieties.* The majority of farmers practice monoculture, meaning that they plant a single type (strain) of crop throughout their fields. Unfortunately, the resulting lack of genetic diversity means that a single type of parasite or disease can infect the entire crop. This scenario does not take into account any genetic variants that may exhibit disease resistance.

2. *Heavy use of fertilizers, pesticides, and herbicides.* Fertilizer production is energy-intensive, and fertilizer runoff contributes to water pollution. Pesticides reduce soil fertility, because they kill beneficial soil organisms as well as pests, and some pesticides and herbicides are linked to the development of cancer.

3. *Generous irrigation.* Water is sometimes taken from aquifers whose water content may in the future become so reduced that it could become too expensive to pump out any more.

4. *Excessive fuel consumption.* Irrigation pumps remove water from aquifers, and large farming machines are used to spread fertilizers, pesticides, and herbicides, as well as to sow and harvest the crops. In effect, modern farming methods transform fossil fuel energy into food energy.

Figure 32.13 shows ways to minimize the harmful effects of modern farming practices.

Soil Loss and Degradation

Land suitable for farming and for grazing animals is being degraded worldwide. Topsoil, the topmost portion of the soil, is the richest in organic matter and the most capable of supporting grass and crops. When bare soil is acted on by water and wind, soil erosion occurs and topsoil is lost. As a result, marginal rangeland becomes desert, and farmland loses its productivity.

The custom of planting the same crop in straight rows that facilitate the use of large farming machines has caused the United States and Canada to have one of the highest rates of soil erosion in the world. Conserving the nutrients

a. Drought-resistant plants

tubing

b. Drip irrigation

Figure 32.12 Conservation measures to save water.

a. Planting drought-resistant plants in parks and gardens and drought-resistant crops in fields cuts down on the need to irrigate. **b.** Drip irrigation delivers water directly to the roots.

(a): © Bruno Barbier/Getty Images; (b): © Inga Spence/Alamy

a. Polyculture

b. Contour with no-till farming

Figure 32.13 Conservation methods.

a. Polyculture reduces the ability of one parasite to wipe out an entire crop and reduces the need to use an herbicide to kill weeds. This farmer has planted alfalfa between strips of corn, which also replenishes the nitrogen content of the soil (instead of adding fertilizers). **b.** Contour with no-till farming conserves topsoil, because water has less tendency to run off. **c.** Instead of pesticides, it is sometimes possible to use a natural predator. Here a ladybug is eating aphids off a cultivated crop plant.

c. Biological pest control

now being lost could save farmers $20 billion annually in fertilizer costs. Much of the eroded sediment ends up in lakes and streams, where it reduces the ability of aquatic species to survive.

Almost all water contains dissolved salts, and these salts are left behind when either plants take up the water or the water evaporates. Between 25% and 35% of the irrigated western croplands are thought to have undergone **salinization,** an accumulation of mineral salts generated through irrigation. Salinization makes the land unsuitable for growing crops.

Green Revolutions

About 50 years ago, research scientists began to breed tropical wheat and rice varieties specifically for farmers in the LDCs. The dramatic increase in yield due to the introduction of these new varieties around the world was called "the green revolution." These plants helped the world food supply keep pace with the rapid increase in world population. However, most of these plants are called "high responders" because they need high levels of fertilizer, water, and pesticides in order to produce a high yield. In other words, they require the same subsidies and create the same ecological problems as do modern farming methods.

Genetic Engineering Genetic engineering can produce transgenic plants with new and different traits, among them resistance to both insects and herbicides (see Section 12.3). When herbicide-resistant crops are planted, weeds are easily controlled, less tillage is needed, and soil erosion is minimized. Researchers also want to produce crops that tolerate salt, drought, and cold. In addition, some progress has been made in increasing the food quality of crops, so that they will supply more of the proteins, vitamins, and minerals people need. Genetically engineered crops are resulting in another green revolution. Nevertheless, some people are opposed to the use of genetically engineered crops, fearing that they will damage the environment and lead to health problems in humans.

Domestic Livestock

A low-protein, high-carbohydrate diet consisting only of grains, such as wheat, rice, or corn, can lead to malnutrition. In the LDCs, kwashiorkor, a condition caused by a severe protein deficiency, is seen in infants and children ages 1–3; this deficiency tends to develop after a new baby arrives in the family and the older children are no longer breast-fed. Such children are lethargic and irritable, and they have bloated abdomens. The condition often results in intellectual disabilities.

In the MDCs, many people tend to have more than enough protein in their diet. Almost two-thirds of U.S. cropland is devoted to producing livestock feed. This means that a large percentage of the fossil fuel, fertilizer, water, herbicides, and pesticides we use are actually for the purpose of raising livestock. Typically, cattle are range-fed for about four months, and then they are taken to crowded feedlots, where they receive growth hormone and antibiotics while feeding on grain. Most pigs and chickens spend their entire lives in crowded pens and cages (**Fig. 32.14**).

Figure 32.14 Crowding of livestock.

Hogs milling in feedlot pens.

Just as livestock eat a large proportion of the crops in the United States, raising livestock accounts for much of the pollution associated with farming. Consider also that fossil fuel energy is needed not just to produce herbicides and pesticides and to grow food but also to make the food available to the livestock. Raising livestock is extremely energy-intensive in the MDCs. In addition, water is used to wash livestock wastes into nearby bodies of water, where they add significantly to water pollution. Raw animal wastes are not always regulated in the same fashion as human wastes.

For these reasons, it is prudent to recall the ecological pyramid (see Fig. 31.20), which shows that, as you move up a food chain, energy is lost. As a rule of thumb, for every 10 calories of energy from a plant, only 1 calorie is available for the production of animal tissue in an herbivore. In other words, it is possible to feed 10 times as many people on grain as on meat.

Fishing

Since 2000, the world fish catch has been on the decline. Worldwide, between 1970 and 1990, the number of large boats devoted to fishing doubled to 1.2 million. The U.S. fishing fleet participated in this growth due to the availability of federal loans for building fishing boats. The new boats have sonar and depth recorders, and their computers remember the sites of previous catches, so that the boats can go there again. Helicopters, planes, and even satellite data are used to help find fish. The number of North Atlantic swordfish caught in the United States declined 70% from 1980 to 1990, and the average weight fell from 115 to 60 pounds. The Atlantic bluefin tuna is so overfished that it may never recover and instead become extinct.

Modern fishing practices negatively impact biodiversity, because a large number of marine animals are caught by chance in the huge nets some fishing boats use (**Fig. 32.15**a). These animals are discarded. The world's shrimp fishery has an annual catch of 1.8 million tons, but the other animals caught and discarded in the process amount to 9.5 million tons. Raising tuna, shrimp, and other aquatic organisms in controlled settings, called aqua farming, has reduced the fishing pressures on wild populations (Fig. 32.15b). In fact, over 90% of the shrimp available in the United States are produced via aquaculture.

Energy

About 6% of the world's energy supply comes from nuclear power, and 81% comes from fossil fuels; both of these are finite, nonrenewable sources. Although it was once predicted that the nuclear power industry would fulfill a significant portion of the world's energy needs, this has not happened for two reasons: (1) People are very concerned about the potential for disasters at nuclear power plants, such as those that occurred at Chernobyl (1986) and Fukushima (2011). (2) Radioactive wastes from nuclear power plants remain a threat to the environment for thousands of years, and we still have not determined the best way to store them safely.

As you learned in Section 31.2, oil, natural gas, and coal are **fossil fuels**—the compressed remains of organisms that died millions of years ago. The MDCs presently consume more than twice as much fossil fuel as the LDCs, yet there are many more people in the LDCs than in the MDCs. It has been estimated that each person in the MDCs uses approximately as much energy in a day as a person in an LDC does in a year.

Among the fossil fuels, oil burns more cleanly than coal, which may contain a considerable amount of sulfur. Thus, despite the fact that the United

a. Fishing by use of a drag net

b. Aqua farming

Figure 32.15   **Fishing.**

The world fish catch has declined in recent years (**a**) because modern fishing methods are overexploiting fisheries. **b.** Aqua farming, the raising of aquatic organisms in controlled settings, is decreasing the demand placed on the oceans.

(a): © StrahilDimitrov/Getty RF; (b): © Michael Pole/Corbis

Figure 32.16 Effects of fossil fuel use.

Mean global temperature is rising due to the introduction of greenhouse gases into the atmosphere. A temperature rise of only a few degrees causes coral reefs to "bleach" and become lifeless. The majority of the Great Barrier Reef has already died off due to coral bleaching.

© Len Zell/Getty Images

States has a good supply of coal, imported oil is our preferred fossil fuel. Even so, the burning of any fossil fuel contributes to environmental problems, because, as it burns, pollutants are emitted into the air.

As discussed in Section 31.2, the burning of fossil fuels is elevating the concentration of greenhouse gases in the atmosphere. Countless scientific studies have determined that the elevation of carbon dioxide in the atmosphere is contributing to global warming and global climate change. Since the oceans absorb a large amount of the excess carbon dioxide in the atmosphere, they are susceptible to warming and acidification (**Figure 32.16**).

Renewable Energy Sources

Renewable energy sources include wind power, hydropower, geothermal energy, and solar energy (**Fig. 32.17**).

Wind Power Wind power is expected to account for a significant percentage of our energy needs in the future. A community that generates its own electricity using wind power can solve the problem of uneven energy production by selling electricity to a local public utility when an excess is available and buying electricity from the same facility when wind power is in short supply.

Hydropower Hydroelectric plants convert the energy of falling water into electricity. Hydropower accounts for about 6% of the electric power generated in the United States and almost 67% of the total renewable energy used. Much of the hydropower development in recent years has been due to the construction of enormous dams, which have detrimental environmental effects. An alternative choice may be small-scale dams that generate less power per dam but do not have the same environmental impact.

Geothermal Energy Elements such as uranium, thorium, radium, and plutonium undergo radioactive decay below the Earth's surface and then heat the surrounding rocks to hundreds of degrees Celsius. When the rocks are in contact with underground streams or lakes, huge amounts of steam and hot water are produced. This steam can be piped up to the surface to supply hot water for home heating or to run steam-driven turbogenerators. California's Geysers project is the world's largest geothermal electricity-generating complex.

Energy and the Solar-Hydrogen Revolution Solar energy is diffuse energy that must be (1) collected, (2) converted to another form, and

Figure 32.17 Renewable energy sources.

a. Wind power requires land on which to place enough windmills to generate energy. **b.** Hydropower dams provide a clean form of energy but can be ecologically disastrous in other ways. **c.** Photovoltaic cells on rooftops can collect diffuse solar energy more affordably than in the past.

(a): © Glen Allison/Getty RF (b): © Corbis RF; (c): © Danita Delimont/Getty Images

(3) stored if it is to compete with other available forms of energy. Passive solar heating of a house is successful when its windows face the sun, the building is well insulated, and heat can be stored in water tanks, rocks, bricks, or some other suitable material.

In a **photovoltaic (solar) cell,** a wafer of an electron-emitting metal is in contact with another metal that collects the electrons and passes them along into wires in a steady stream. The photovoltaic cells placed on roofs, for example, generate electricity that can be used inside a building or sold to a power company.

Scientists are working on improving technologies that allow solar energy to be used to extract hydrogen from water via electrolysis. The hydrogen can then be used as a clean-burning fuel; when it burns, water is produced. Presently, most cars have internal combustion engines that run on gasoline. Automobile manufacturers are working on building vehicles that will be powered by fuel cells, which use hydrogen to produce electricity (**Fig. 32.18**), and many gasoline–fuel cell hybrid vehicles now exist. The electricity runs a motor that propels the vehicle. Fuel cells are powering buses in Vancouver and Chicago, with more in the development phase.

The advantages of a solar-hydrogen revolution would be at least two-fold: (1) The world would no longer be dependent on oil, and (2) environmental problems, such as acid rain and smog, would begin to lessen.

Minerals

Minerals are nonrenewable raw materials in the Earth's crust that can be mined (extracted) and used by humans. Nonrenewable minerals include fossil fuels; nonmetallic raw materials, such as sand, gravel, and phosphate; and metals, such as aluminum, copper, iron, lead, and gold.

Nonrenewable resources are subject to depletion—that is, the supply will eventually run out. We can extend our supply of fossil fuels if we conserve our use, recycle metals, and find new reserves.

In the United States, huge machines can go as far as removing mountaintops in order to reach a mineral (**Fig. 32.19**). The land becomes devoid of vegetation, allowing runoff from rain to wash toxic waste deposits into nearby streams and rivers. Legislation now requires that strip miners restore the land to its original condition, a process that can take years to complete.

Other Sources of Pollution

Synthetic organic compounds and wastes are also pollutants of concern.

Synthetic Organic Compounds

Synthetic organic compounds are of considerable ecological concern due to their detrimental effects on the health of living organisms, including humans. Synthetic organic compounds play a role in the production of plastics, pesticides, herbicides, cosmetics, coatings, solvents, wood preservatives, and hundreds of other products.

Synthetic organic compounds include halogenated hydrocarbons, compounds in which halogens (chlorine, bromine, fluorine) have replaced some hydrogen atoms. **Chlorofluorocarbons (CFCs)** are a type of halogenated hydrocarbon in which both chlorine and fluorine atoms replace some of the hydrogen atoms. CFCs have brought about a thinning of the Earth's ozone shield, which protects terrestrial life from the dangerous effects of

a.

b.

Figure 32.18 **Solar-hydrogen revolution.**

a. This bus is powered by hydrogen fuel. **b.** The use of gasoline–fuel cell hybrid vehicles, such as this prototype, will reduce air pollution and dependence on fossil fuels.

(a): © LusoEnvironment/Alamy; (b): © Piroschka van de Wouw/EPA/Newscom

Figure 32.19 **Modern mining capabilities.**

Giant mining machines—some as tall as a 20-story building—can remove an enormous amount of the Earth's crust in one scoop in order to mine for coal or a metal ore.

© James P. Blair, National Geographic Creative

ultraviolet radiation. In most MDCs, legislation has been passed to prevent the production of any more CFCs. Hydrofluorocarbons, which contain no chlorine, are expected to take their place in coolants and other products. The ozone shield is predicted to recover by 2050; in the meantime, many more cases of skin cancer are expected to occur.

Other synthetic organic chemicals pose a direct and serious threat to the health of all living organisms. Rachel Carson's book *Silent Spring,* published in 1962, made the public aware of the deleterious effects of pesticides.

Wastes

Every year, the countries of the world discard billions of tons of solid wastes, some on land and some in fresh and marine waters. Solid wastes are visible wastes, some of which are hazardous to our health.

Industrial Wastes   Industrial wastes are generated during the mining and production of a product. Clean-water and clean-air legislation in the United States in the early 1970s prohibited venting of industrial wastes into the atmosphere and flushing them into waterways. Industry turned to land disposal, which was unregulated at the time. The use of deep-well injection, pits with plastic liners, and landfills led to much water pollution and human illness, including cancer. An estimated 5 billion metric tons of highly toxic chemicals were improperly discarded in the United States between 1950 and 1975. The public's concern was so great that the Environmental Protection Agency (EPA) came into existence. Using an allocation of monies called the Superfund, the EPA oversees the cleanup of hazardous waste disposal sites in the United States.

Connections: Health

What is methylmercury, and why is it dangerous?

Methylmercury is a molecule in which the element mercury has been bound to a methyl (CH_3) group. Because of this methyl group, methylmercury easily accumulates in the food chain by biological magnification, or the concentration of chemicals in a food chain. Methylmercury is released into the environment by the burning of coal, the mining of certain metals, and the incineration of medical waste. Methylmercury is a powerful neurotoxin that also inhibits the activity of the immune system. Because of this, the FDA and the EPA recommend that pregnant women and small children not eat shark, swordfish, tilefish, or king mackerel and that they limit their consumption of albacore tuna to less than 6 ounces per week. Most states have also posted warnings about eating local fish that have been caught from mercury-contaminated waters. For more information, visit the EPA website, www.epa.gov/waterscience/fish.

© Franco Banfi/WaterFrame/Getty Images

Among the most commonly found contaminants in industrial wastes are heavy metals (lead, arsenic, cadmium, chromium) and organic compounds (trichloroethylene, toluene, benzene, polychlorinated biphenyls [PCBs], and chloroform). Some of the chemicals used in pesticides, herbicides, plastics, food additives, and personal hygiene products are classified as endocrine disrupters.

These products can affect the endocrine system and interfere with reproduction. In the environment, they occur at a level 1,000 times greater than the hormone levels in human blood.

Decomposers are unable to break down these wastes. They enter and remain in the bodies of organisms because they are not excreted. Therefore, they become more concentrated as they pass along a food chain, a process termed **biological magnification** (**Fig. 32.20**). This effect is most apt to occur in aquatic food chains, which have more links than do terrestrial food chains. Humans are one of the final consumers in both types of food chains, and in some areas, human breast milk contains detectable amounts of polychlorinated hydrocarbons, DDT, and PCBs.

Sometimes industrial wastes accumulate in the mud in deltas and estuaries of highly polluted rivers and cause environmental problems if disturbed. Industrial pollution is being addressed in many MDCs, but it usually has low priority in LDCs.

Sewage   Raw sewage can contribute to oxygen depletion in lakes and rivers. Sewage serves as a fertilizer for plants, which can lead to eutrophication; first there is an algal bloom; then, when the algae use up all the nutrients, they die. Decomposition of dead algae robs the water of oxygen, which can result in a massive fish kill. Also, human feces can contain pathogenic microorganisms that cause cholera, typhoid fever, and dysentery. In regions of LDCs where sewage treatment is practically nonexistent, many children die each year from these diseases. Sewage treatment plants use bacteria to break down organic matter to inorganic nutrients, such as nitrates and phosphates, which then enter surface waters.

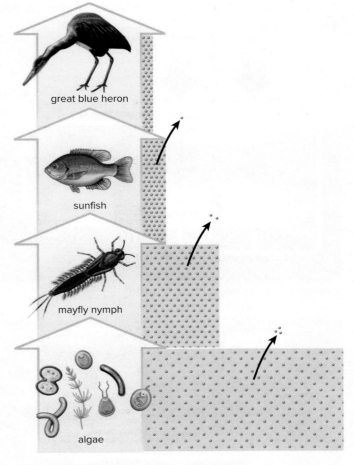

Figure 32.20 **Biological magnification.**

A poison, such as DDT, that is excreted in relatively small amounts (arrows) becomes maximally concentrated as it passes along a food chain due to the reduced size of the trophic levels.

Check Your Progress 32.3

1. Distinguish between renewable and nonrenewable resources.

2. Explain the causes of deforestation and desertification, and list environmental problems associated with each.

3. List the harmful effects of modern agricultural practices.

4. Summarize the influences of the "green revolution."

5. List four renewable energy sources, and identify a drawback of each.

CONNECTING THE CONCEPTS

32.3 The human population's need for resources is creating problems for the world's ecosystems.

32.4 Sustainable Societies

Learning Outcomes

Upon completion of this section, you should be able to

1. Explain why human society is unsustainable in its current form.

2. List some activities that may help make rural societies more sustainable.

3. List some activities that may help make urban societies more sustainable.

A **sustainable society** is able to provide the same goods and services for future generations of humans as it does now while preserving biodiversity.

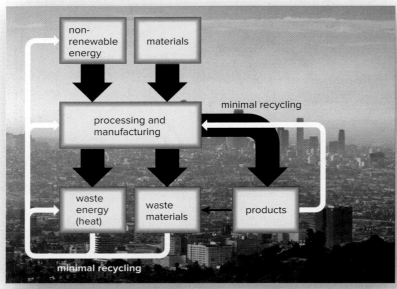

a. Human society at present

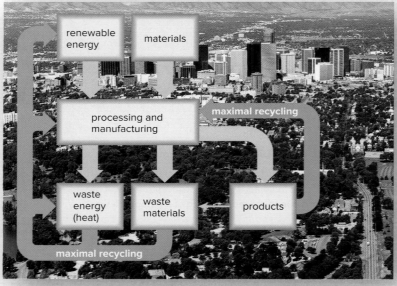

b. Sustainable society

Figure 32.21 Current human society versus a sustainable society.

a. Our "throw-away" society is characterized by high input of energy and raw material, large output of waste materials and energy in the form of heat, and minimal recycling (white arrows). b. A sustainable society is characterized by the use of only renewable energy sources, the reuse of heat and waste materials, and maximal recycling of products (blue arrows).

(a): © Kent Knudson/PhotoLink/Getty RF; (b): © Scenics of America/PhotoLink/Getty RF

Today's Society

The following evidence indicates that, at present, human society is not sustainable (**Fig. 32.21***a*):

- A considerable proportion of land, and therefore of natural ecosystems, is being used for human purposes (homes, agriculture, factories, etc.).
- Agriculture requires large inputs of nonrenewable fossil fuel energy, fertilizer, and pesticides, which creates pollution. More fresh water is used in agriculture than in homes.
- At least half the agricultural yield in the United States goes toward feeding animals. According to the 10-to-1 rule of thumb, it takes 10 pounds of grain to grow 1 pound of meat. Therefore, it is inefficient for citizens in MDCs to eat as much meat as they do. Also, animal sewage pollutes water.
- Even though fresh water is a renewable resource, we are running out of the available supply.
- Our society uses primarily nonrenewable fossil fuel energy, which leads to acid precipitation, smog, and various other pollutants entering the ecosystems.
- Minerals are nonrenewable resources, and the mining, processing, and use of minerals are responsible for much environmental pollution.

Characteristics of a Sustainable Society

A natural ecosystem can offer clues as to what a sustainable human society would be like. A natural ecosystem is characterized by the use of renewable solar energy, the cycling of materials through the various populations, and the return of nutrients and energy to the producers. If we want to develop a sustainable society, we should also use renewable energy sources and recycle materials (Fig. 32.21*b*). These principles can be applied to both rural and urban areas.

Rural Sustainability

In rural areas, we should put the emphasis on preservation. We should preserve ecosystems, including terrestrial ecosystems—such as forests and prairies—and aquatic ecosystems—both freshwater ones and brackish ones along the coast. We should also preserve agricultural land, groves of fruit trees, and other areas that provide us with renewable resources.

It is imperative that we take all possible steps to preserve what remains of our topsoil and replant areas with native grasses, as necessary. Native grasses stabilize the soil and rebuild soil nutrients while serving as a source of renewable biofuel. Trees can be planted to break the wind and protect the soil from erosion while providing a product. Creative solutions to today's ecological problems are very much needed.

Here are some other methods we can use to make rural areas more sustainable:

- Plant *cover crops,* which often are a mixture of legumes and grasses, to stabilize the soil between rows of cash crops or between seasonal plantings of cash crops.
- Practice *multiuse,* or *polyculture, farming* (see Fig. 32.13*a*) by planting a variety of crops and use a variety of farming techniques to increase the amount of organic matter in the soil.

- Replenish soil nutrients through composting, organic gardening, or other self-renewable methods.
- Use low-flow or trickle irrigation, retention ponds, and contour farming (see Fig. 32.13*b*) to conserve water.
- Increase the planting of *cultivars* (plants propagated vegetatively) that are resistant to blight, rust, insect damage, salt, drought, and encroachment by noxious weeds.
- Use *precision farming* (*PF*) techniques that rely on accumulated knowledge to reduce habitat destruction while improving crop yields.
- Use *integrated pest management* (*IPM*), which encourages the growth of competitive beneficial insects and uses biological controls to reduce the abundance of a pest (see Fig. 32.13*c*).
- Plant a variety of species, including native plants, to reduce our dependence on traditional crops.
- Plant *multipurpose trees*—trees with the ability to provide numerous products and perform a variety of functions, in addition to serving as windbreaks (**Fig. 32.22**). Mature trees can provide many types of products: Mature rubber trees provide rubber, and tagua nuts are an excellent substitute for ivory, for example.
- Maintain and restore wetlands, especially in hurricane- or tsunami-prone areas. Protect deltas from storm damage. By protecting wetlands, we protect the spawning grounds for many valuable fish species.
- Use renewable forms of energy, such as wind, hydropower, and solar energy.
- Support local farmers, fishermen, and feed stores by buying food products produced locally.

Urban Sustainability

More and more people are moving to cities. Much thought needs to be given to serving the needs of new arrivals without overexpanding the city. Here are some methods by which we can make cities more sustainable:

- Create energy-efficient transportation systems to efficiently move people about.
- Use solar or geothermal energy to heat buildings; cool them with an air-conditioning system that uses seawater; in general, use conservation methods to regulate the temperature of buildings.
- Utilize *green roofs*—a wild garden of grasses, herbs, and vegetables on the tops of buildings—to assist in temperature control, supply food, reduce the amount of rainwater runoff, and create visual appeal (**Fig. 32.23**).
- Improve storm-water management by using sediment traps for storm drains, artificial wetlands, and holding ponds. Increase the use of porous surfaces for walking paths, parking lots, and roads. These surfaces reflect less heat while soaking up rainwater runoff.
- Instead of traditional grasses, plant native species that attract bees and butterflies and require less water and fertilizers.
- Create *greenbelts* that suit the particular urban setting. Include plentiful walking and bicycle paths.
- Revitalize old sections of a city before developing new sections.
- Use lighting fixtures that hug the walls or ground and send light down; control noise levels by designing quiet motors.
- Promote sustainability by encouraging the recycling of business equipment; use low-maintenance building materials, rather than wood.

Figure 32.22 Rural sustainability.

Trees planted by a farmer to break the wind and prevent soil erosion can also have other purposes, such as supplying fruits and nuts.

Source: USDA Natural Resources Conservation Service/Erwin Cole, photographer

Figure 32.23 Urban sustainability.

Green roofs are ecologically sound and can be visually beautiful.

© Diane Cook and Len Jenshel/Getty Images

CONNECTING THE CONCEPTS

32.4

Current human society is not sustainable, however, progress is being made to increase sustainability.

Check Your Progress 32.4

1. List the characteristics of a sustainable society.
2. Summarize the steps that could increase rural sustainability.
3. Summarize the steps that could increase urban sustainability.

Connections: Environment

What are some simple things you can do to conserve energy and/or water and help solve environmental problems, such as climate change?

The following are a few things you can do:

- Change the lightbulbs in your home to compact fluorescent bulbs. They use 75–80% less electricity than incandescent bulbs.
- Walk, ride your bike, carpool, or use mass transit. It will save you a lot of money, too!
- Get cloth or mesh bags for groceries and other purchases. Plastic bags may take 10–20 years to degrade. They're also dangerous to wildlife if mistaken for food and consumed.
- Turn off the water while you brush your teeth. © Mike Kemp/Getty RF
  If you don't finish a bottle of water, use it to water your plants. In dry climates, plant native plants that won't require frequent watering.

STUDY TOOLS   http://connect.mheducation.com

SMARTBOOK® Maximize your study time with McGraw-Hill SmartBook®, the first adaptive textbook.

SUMMARIZE

Human actions are placing tremendous stress on the Earth's natural resources. Human-induced climate change is going to decrease Earth's biodiversity. Sustainable options are available if people are willing to make the necessary changes.

32.1 Conservation biology seeks to balance the needs of humans with the needs of the rest of the species on Earth.

32.2 The Earth's biodiversity has tremendous value to humans, but it is under stress.

32.3 The human population's need for resources is creating problems for the world's ecosystems.

32.4 Current human society is not sustainable, however, progress is being made to increase sustainability.

32.1 Conservation Biology

Conservation biology is a field of biology that is concerned with conserving the Earth's natural resources for this and future generations.

32.2 Biodiversity

Biodiversity is the variety of life on Earth. The majority of species are located in a relatively small number of **biodiversity hotspots.** Biodiversity has both direct values and indirect values.

Direct Values of Biodiversity

The direct values of biodiversity are:

- Medicinal value (medicines derived from living organisms)
- Agricultural value (crops derived from wild plants)
- Biological pest controls and animal pollinators
- Consumptive use values (food production)

Indirect Values of Biodiversity

Biodiversity in ecosystems contributes to:

- The functioning of biogeochemical cycles (water, carbon, nitrogen, phosphorus, and others)

- Waste disposal (through the action of decomposers and the ability of natural communities to purify water and take up pollutants)
- Fresh water provision through the water biogeochemical cycle
- Prevention of soil erosion, which occurs naturally in intact ecosystems
- Climate regulation (plants take up carbon dioxide)
- Ecotourism (human enjoyment of a beautiful ecosystem)

32.3 Resources and Environmental Impact

Five resources are maximally used by humans:

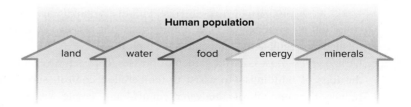

Resources are either nonrenewable or renewable. **Nonrenewable resources,** such as fossil fuels and minerals, are limited in supply. **Renewable resources,** such as solar energy and hydroelectric power, are not limited in supply.

Land

Human activities, such as habitation, farming, and mining, contribute to erosion, **pollution, desertification, deforestation,** and loss of biodiversity.

Water

Industry and agriculture use most of the freshwater supply. Water supplies are increased by damming rivers and drawing from **aquifers.** As aquifers are depleted, **subsidence,** sinkhole formation, and **saltwater intrusion** can occur. Irrigation of farmlands consumes large amounts of water and can lead to **salinization.** If used by industries, water conservation methods could cut world water consumption by half.

Food

Food comes from growing crops, raising animals, and fishing.

- Modern farming methods increase the food supply, but some methods harm the land, pollute the water, and consume fossil fuels excessively.
- Transgenic plants can increase the food supply and reduce the need for chemicals.
- Raising livestock contributes to water pollution and uses fossil fuel energy.
- The increased number and high efficiency of fishing boats have caused the world fish catch to decline.

Energy

Most of the world's energy is supplied by the burning of **fossil fuels,** a nonrenewable resource, which causes pollutants and greenhouse gases to enter the air. Renewable energy sources include wind power, hydropower, and geothermal energy. **Photovoltaic (solar) cells** are being used to access solar energy.

Minerals

Minerals are a nonrenewable resource that can be mined. These raw materials include sand, gravel, phosphate, and metals. The act of mining causes destruction of the land by erosion, loss of vegetation, and toxic runoff into bodies of water. Land ruined by mining can take many years to recover.

Synthetic Organic Compounds

Compounds such as **chlorofluorocarbons (CFCs)** are detrimental to the ozone shield and to the health of living organisms, including humans.

Wastes

Raw sewage and industrial wastes can pollute land and bodies of water. Some industrial wastes cause **biological magnification,** a process by which toxins become more concentrated as they are passed upward in a food chain.

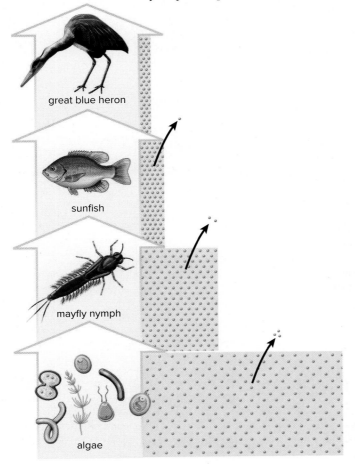

great blue heron

sunfish

mayfly nymph

algae

32.4 Sustainable Societies

A **sustainable society** would use only renewable energy sources, reuse heat and waste materials, and recycle almost everything. It would also provide the same goods and services presently provided and would preserve biodiversity.

ASSESS

Testing Yourself

Choose the best answer for each question.

32.1 Conservation Biology

1. Which subfield of biology helps support conservation biology?
   a. ecology
   b. forestry
   c. genetics
   d. evolutionary biology
   e. All of these are correct.

2. What is the main goal of conservation biology?
   a. conserving natural resources for current and future generations
   b. increasing agricultural yields
   c. decreasing human mortality
   d. identifying species that provide a monetary value to humans
   e. All of these are correct.

32.2 Biodiversity

3. Which of the following is an indirect value of biodiversity?
   a. participation in biogeochemical cycles
   b. participation in waste disposal
   c. provision of fresh water
   d. prevention of soil erosion
   e. All of these are correct.

4. The preservation of ecosystems indirectly provides fresh water because
   a. trees produce water as a result of photosynthesis.
   b. animals excrete water-based products.
   c. forests soak up water and release it slowly.
   d. ecosystems promote the growth of bacteria that release water into the environment.
   e. plants trap moisture with their leaves.

32.3 Resources and Environmental Impact

5. Desertification typically happens because
   a. deserts naturally expand in size.
   b. humans allow overgrazing.
   c. desert animals wander into adjacent areas for food.
   d. humans tap into limited water supplies for their water needs in the nearby desert.

6. Soils in tropical rain forests are typically nutrient-poor because
   a. they receive more water than other ecosystems.
   b. they are sandy.
   c. nitrogen-fixing bacteria are absent.
   d. nutrients are tied up in vegetation.

7. Most of the fresh water in the world is used for
   a. drinking.
   b. supporting industry.
   c. irrigating crops.
   d. cooking.
   e. bathing.

8. A major negative effect of the dumping of raw sewage into lakes and rivers is
   a. oxygen depletion.
   b. the buildup of carbon.
   c. a reduction of light penetration into the water.
   d. an increase in populations of small fish.

9. Which of the following will help conserve fresh water?
   a. saltwater intrusion
   b. drought-tolerant crops
   c. increased irrigation
   d. use of water contained in aquifers

10. The "green revolution" resulted from the development of
    a. high-responder wheat and rice varieties.
    b. effective irrigation systems.
    c. transgenic crop plants.
    d. crops designed for animal feed.

32.4 Sustainable Societies

11. Show how the following diagram must change in order to develop a sustainable society.

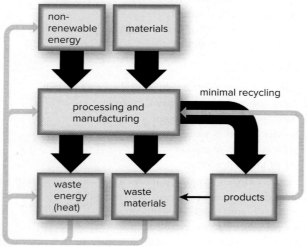

minimal recycling

12. Which of the following would help make both urban and rural areas more sustainable?
    a. plant multipurpose trees
    b. use low-flow irrigation to conserve water
    c. use more solar energy
    d. plant cover crops
    e. use green roofs

ENGAGE

Thinking Critically

1. In 1993, the U.S. Fish and Wildlife Service established Emiquon National Wildlife Refuge to restore and protect wetland habitats. The Illinois and Spoon River confluence was restored to its original floodplain. In order to achieve this, hundreds of acres of farmland were flooded. What might be the positive and negative environmental impacts of flooding the confluence of these two rivers? What direct and indirect benefits could a project like this provide for the residents of central Illinois?

2. Every time a species becomes extinct, we lose a potential source of medicine for human illness. Since we do not know which species are likely to provide valuable medicines, it is difficult to know where to focus conservation efforts. If you were in charge of determining high priorities for species conservation in a country, with the goal of identifying plants with medicinal value, what factors would you consider in making your decisions?

3. Knowing that organisms are closely adapted to their environment, why might global climate change cause the extinction even of mammals living in the tropics?

4. Visit footprint.wwf.org.uk to determine the size of your ecological footprint, and then discuss ways you can reduce that value.

Appendix A Periodic Table of the Elements & The Metric System

Periodic Table of the Elements

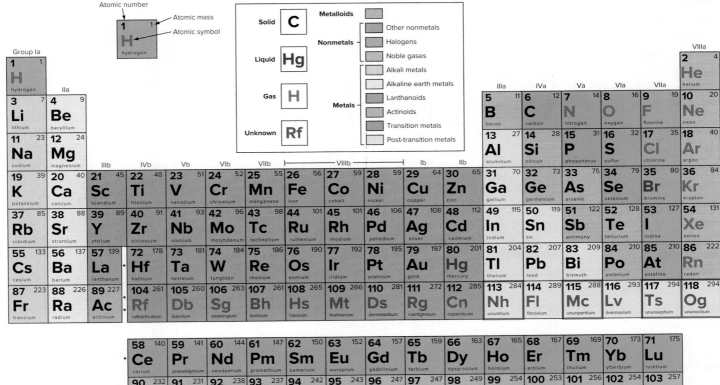

The names of elements 113, 115, 117 and 118 are pending approval by the IUPAC.

The Metric System

Unit and Abbreviation	Metric Equivalent	Approximate English-to-Metric Equivalents	Units of Temperature
Length			
nanometer (nm)	$= 10^{-9}$ m $(10^{-3} \mu m)$		
micrometer (μm)	$= 10^{-6}$ m $(10^{-3}$ mm$)$		
millimeter (mm)	$= 0.001 (10^{-3})$ m		
centimeter (cm)	$= 0.01 (10^{-2})$ m	1 inch = 2.54 cm 1 foot = 30.5 cm	
meter (m)	$= 100 (10^2)$ cm = 1,000 mm	1 foot = 0.30 m 1 yard = 0.91 m	
kilometer (km)	$= 1,000 (10^3)$ m	1 mi = 1.6 km	
Weight (mass)			
nanogram (ng)	$= 10^{-9}$ g		
microgram (μg)	$= 10^{-6}$ g		
milligram (mg)	$= 10^{-3}$ g		
gram (g)	= 1,000 mg	1 ounce = 28.3 g 1 pound = 454 g	
kilogram (kg)	$= 1,000 (10^3)$ g	= 0.45 kg	
metric ton (t)	= 1,000 kg	1 ton = 0.91 t	
Volume			
microliter (μl)	$= 10^{-6}$ l $(10^{-3}$ ml$)$		
milliliter (ml)	$= 10^{-3}$ l $= 1$ cm^3 (cc) = 1,000 mm^3	1 tsp = 5 ml 1 fl oz = 30 ml	
liter (l)	= 1,000 ml	1 pint = 0.47 l 1 quart = 0.95 l 1 gallon = 3.79 l	
kiloliter (kl)	= 1,000 l		

°C	°F	
100	212	Water boils at standard temperature and pressure.
71	160	Flash pasteurization of milk
57	134	Highest recorded temperature in the United States, Death Valley, July 10, 1913
41	105.8	Average body temperature of a marathon runner in hot weather
37	98.6	Human body temperature
13.7	56.66	Human survival is still possible at this temperature.
0	32.0	Water freezes at standard temperature and pressure.

To convert temperature scales:

$$°C = \frac{(°F - 32)}{1.8}$$

$$°F = 1.8(°C) + 32$$

Appendix B   Answer Key

CHAPTER 1

Check Your Progress

(1.1) 1. The basic characteristics of life are: living organisms are organized, they acquire materials and energy, they maintain an internal environment, they respond to stimuli, they can reproduce and develop, and they can adapt to their environment. **2.** From smallest to largest the levels of biological organization are: small molecule, large molecule, cells, tissues, organs, organ systems, complete organism. **3.** Energy flow and chemical cycling occur within organisms and in ecosystems. Each cell or organism must take in energy and materials including nutrients. Producers capture energy and produce food by photosynthesis. The chemicals in this food are cycled through food chains and eventually deposited or released upon death when they become available again to producers. Energy flows from the sun through plants and animals. It eventually dissipates as heat into the atmosphere. **(1.2) 1.** Species, genus, family, order, class, phylum, kingdom, domain. **2.** Natural selection is the process in which populations adapt to pressures imposed by their environments. Over time those individuals having adaptations that make them well suited to their environment will live and produce more offspring than others. Natural selection is an important biological concept that explains population changes, like those seen in antibiotic resistant bacteria. **3.** Descent with modification means that descent occurs from common ancestors to different species, each of which is adapted to a particular environment. Both common ancestry and diversity are important to the study of evolution. **(1.3) 1.** Making observations of our environment is the first step in the scientific method. Constructing a hypothesis based on existing knowledge to explain the observation is next. Making a prediction based on the hypothesis that can be tested by an experiment follows. The results from the experiment either support the hypothesis or not. Further testing and data analysis result in a conclusion being made. Ultimately, with repeated confirmation through experimentation, a scientific theory can be constructed. **2.** Including controls in an experimental design is essential to determine if the experimental variable can explain the hypothesis. **3.** Publishing the results of experiments in journals allows for critical review by others of the hypothesis, predictions, experimental design, and data. **(1.4) 1.** Science is a systematic way of acquiring knowledge. Technology is the application of scientific knowledge for human benefit. **2.** Ecosystem degradation, emerging diseases, and climate change are occurring at accelerated rates due to human influences on the world's ecosystems. Biodiversity is being reduced and this threatens all species.

Testing Yourself

1. c; **2.** b; **3.** c; **4.** d; **5.** c; **6.** c; **7.** b; **8.** b; **9.** c; **10.** b; **11.** d; **12.** d

BioNow

Our bodies continuously maintain a stable internal environment that is organized through the processes of homeostasis. We respond to our environment through our senses and motor skills. We consume nutrients and gather materials so that we can adapt to our changing environment and be fit to develop and reproduce.

Thinking Critically

1. Domain Archaea gave rise to Eukarya. Kingdom Protista gave rise to plants, animals, and fungi. **2.** Model organisms, perhaps mice, which exhibits the type of cancer being studied are identified. A hypothesis that the drug is effective against cancer is formed. The prediction is made that the drug administered to the mice will remove the cancer from them.The mice are divided into groups; a control group receiving a placebo, and test group(s) receiving levels of the new drug. The drug or placebo is administered during the experiment. The mice are monitored for the disappearance of cancer. The data collected are analyzed and conclusions are made about the drug's effectiveness against cancer. The experiment is repeated as needed to test any revised hypotheses and the results are published for review by others. **3.** The hypothesis would be that the potential life-form is alive. The prediction is that it exhibits the characteristics of behavior and adaptation found in living organisms. An experiment is constructed to observe or test if the life-form can acquire materials and nutrients, respond to its environment, or reproduce and develop and in so doing, adapt to its environment over generations.

CHAPTER 2

Check Your Progress

(2.1) 1. Nucleus with protons (+) and neutrons (neutral) surrounded by shells of electrons (−). **2.** Isotopes are atoms of the same element that differ in the number of neutrons. They have the same number of protons but have different mass numbers. Radioactive isotopes are unstable and emit radiation. They can be used in PET scans and to kill cancer cells. **3.** Both are types of chemical bonds. Covalent bonds share pairs of electrons; ionic bonds have one atom giving an electron to another atom. **4.** If an atom has two or more shells, the outer shell (the valence shell) is most stable when it has eight electrons. Elements with incomplete valence shells are more reactive. **5.** Reactants are molecules that participate in a reaction. Products are molecules formed by reactions. **(2.2) 1.** Water molecules are polar with positive hydrogen atoms in one molecule being attracted to the negative oxygen atoms in the other. This is a hydrogen bond. The properties of water are related to its polarity and its ability to form hydrogen bonds. **2.** Water has unique properties. It is a solvent, its molecules are adhesive and cohesive, it has a high surface tension, it has a high heat capacity, and it is less dense than ice. **3.** Water is a polar molecule and its special properties come from its ability to form hydrogen bonds. **(2.3) 1.** Acids in solution release hydrogen ions creating solutions with higher hydrogen ion concentrations. Bases in solution either take up hydrogen ions or release hydroxide ions. Basic solutions have lower hydrogen ion concentrations. **2.** The pH scale indicates the number of hydrogen ions in a solution. A value below 7 indicates an acid and a value above 7 indicates a base. **3.** Buffers in fluids of the body resist pH changes by taking up excess hydrogen or hydroxide ions.

Testing Yourself

1. a; **2.** c; **3.** c; **4.** a; **5.** b; **6.** a; **7.** b; **8.** c; **9.** b; **10.** c; **11.** c; **12.** a

BioNow

All the properties of water are important to living organisms. It is a great solvent for biomolecules. Its high heat capacity helps to maintain a constant internal temperature. Its properties of cohesion and adhesion help in moving water through a vascular system. Its high heat of vaporization helps in cooling by sweating and in maintaining hydration. Lakes do not freeze solid killing organisms because ice is less dense than water.

Thinking Critically

1. Surface tension produced by hydrogen bonding keeps the surface of the water smooth and continuous. Drops of water cling to your skin due to the adhesiveness of water. When this water evaporates it cools your skin because of water's high heat of vaporization. A lot of body heat is removed as water evaporates. **2.** Silicon, with one extra shell, is too big to form the same varied molecules as carbon. **3.** The chemicals in antacids act as buffers absorbing excess hydrogen ions in the stomach. **4.** The acids in acid precipitation can change the pH of the water environments of an ecosystem if they exceed the buffering capacity present there. Altered pH levels affect the homeostatic processes of the environment, and the organisms within it, causing damage.

CHAPTER 3

Check Your Progress

(3.1) 1. Organic molecules which make up cells and organisms contain carbon and hydrogen atoms. Inorganic molecules, such as water and salt, do not contain a combination of carbon and hydrogen. **2.** Carbon is small and can bond with up to four other molecules. The C—C bond is stable. **3.** The functional group of an organic molecule determines the reactivity and chemical properties of the molecule. **(3.2) 1.** Monosaccharides—glucose; disaccharides—maltose; polysaccharides—glycogen. Starch is a polysaccharide used for energy storage. Cellulose is a polysaccharide that provides structural components to cells. **2.** Fats and oils—olive oil as an energy source; phospholipids—primary component of plasma membrane; steroids—cholesterol as a component of plasma membrane and precursor to sex steroids. **3.** Structural proteins provide support. Enzymes are proteins that function as catalysts in metabolism. There are channel and carrier proteins in the plasma membrane. Antibodies are proteins which provide defense. Some hormones which function in regulation are proteins. **4.** Primary—sequence of amino acids; secondary—alpha helix and pleated sheets; tertiary—globular shape; quaternary—more than one polypeptide **5.** Nucleic acids are polymers of nucleotides each of which has a phosphate group, a 5-carbon sugar, and a nitrogen-containing base. **6.** DNA, a nucleic acid, encodes the information that determines the sequence of amino acids in a polypeptide. RNA, another nucleic acid, functions in the process of polypeptide synthesis.

Testing Yourself

1. c; **2.** d; **3.** b; **4.** b; **5.** a; **6.** a; **7.** a; **8.** a; **9.** c; **10.** b; **11.** d; **12.** b; **13.** d; **14.** c; **15.** a; **16.** a; **17.** c; **18.** d

Thinking Critically

1. The labeling of HDL as good and LDL as bad is an easy way to understand the different functions of the lipoproteins. HDL helps to move cholesterol from tissues to the liver which is good and LDL moves cholesterol to the tissues where it can possibly form plaques which is bad. However some cholesterol needs to be in the tissues and in the liver for proper metabolism. So it is the opposition of the functions of HDL and LDL that keeps cholesterol levels in the tissues and liver in balance. **2.** A protein's ability to function correctly is dependent upon its shape. In this case, the amino acid with the polar *R* group may interact with other polar *R* groups within the protein and change the enzyme's shape and its ability to function. **3.** Mutations in DNA that change the sequence of amino acids occur over time. The longer the time since two species diverged, the greater the chance for mutations to occur. Measuring the differences in the amino acid sequences of the protein in two species can be used as a clock to determine how long ago they diverged.

CHAPTER 4

Check Your Progress

(4.1) 1. Most cells are too small to be seen with the naked eye. Microscopes allow us to see these cells. **2.** frog egg—light; animal cell—light or electron; amino acid—electron; chloroplast—light or electron; plant cell—light or electron. **3.** Enough surface area of the cell is needed for the adequate exchange of gases, nutrients, and wastes. **(4.2) 1.** Phospholipid bilayer including cholesterol molecules and embedded proteins. **2.** The proteins embedded in the bilayer have a mosaic pattern and the entire bilayer is semi-fluid with the molecules being able to move within the structure. **3.** Channel—allow passage of certain molecules; Transport—passage of ions and molecules requiring energy; Cell recognition—glycoproteins which help to distinguish foreign from own cells; Receptor—bind signal molecules; Enzymatic—participate in metabolic reactions; Junction—form cell junctions between cells. **(4.3) 1.** Plasma membrane, semifluid interior called the cytoplasm, genetic material (DNA). **2.** Prokaryotic cells store their DNA in a region of the cytoplasm called the nucleoid.

Eukaryotic cells have a membrane-bound nucleus that houses the DNA. **3.** Nucleoid—region of the cytoplasm that stores DNA in prokaryotes; Cell wall—maintains the shape of the cell providing strength; Ribosomes—site of protein synthesis; Flagella—rotating filament that propels the cell. **(4.4) 1.** Nucleus: nuclear envelope—double membrane phospholipid that separates nucleus from cytoplasm; Nuclear pore—openings in the nuclear envelope that allow the cell nucleus to communicate with the cytoplasm; Chromatin—network of DNA and proteins that contains genetic information; Ribosomes: composed of a small and large subunit that together use the information of the RNA from the nucleus to synthesize proteins. **2.** Nuclear envelope—houses DNA; Rough endoplasmic membrane—contains ribosomes and forms transport vesicles that take proteins to other parts of the cell; Smooth endoplasmic reticulum—synthesizes tissue specific lipids; Golgi apparatus—functions as a transfer station which receives transport vesicles from the endoplasmic reticulum, modifies the molecules within the vesicles, packages modified molecules, and sends them out in outgoing transport vesicles; vesicles like lysozymes which digest molecules. **3.** Contractile vacuoles rid the cell of excess water; Digestive vacuoles break down nutrients; Storage vacuoles in plants can hold water, sugars, salts, pigments, or toxic chemicals; Adipocytes in animals store lipids. Without vacuoles cells would not be able to regulate water and ion content, break down certain nutrients, or store compounds. **4.** Chloroplasts contain pigments that capture solar energy which is used to synthesize carbohydrates. Mitochondria break down carbohydrates to produce ATP. Chloroplasts are larger and are found in plants and algae. Mitochondria are smaller and are in all eukaryotic cells. Both are bound by a double membrane. **5.** Cytoskeleton proteins maintain the shape of the cell and work with motor proteins to create movement. The cytoskeleton contains microtubules, intermediate filaments, and actin filaments. The motor protein myosin interacts with actin filaments to create amoeboid movement and muscle contraction. The motor proteins kinesin and dynein transport vesicles along the microtubules. **(4.5) 1.** Cell walls provide strength to the cell structure as well as defining the cell's shape. **2.** The extracellular matrix is a network outside the cell of fibrous proteins and polysaccharides produced by a cell. Collagen and elastin are proteins that are common to the matrix which can vary widely in its composition and rigidity. **3.** Adhesion junctions are formed when internal cytoplasmic plaques are joined by intercellular filaments. They are sturdy and flexible structures that hold cells together. Tight junctions are created when plasma membrane proteins attach to those of another cell producing a barrier between the cells. Gap junctions form when identical membrane channels join between two cells. The channels allow small molecules to pass between the cells and they add strength to the cell.

Testing Yourself

1. a; **2.** b; **3.** b; **4.** d; **5.** d; **6.** c; **7.** a; **8. a.** vesicle; **b.** centrioles; **c.** mitochondrion; **d.** rough endoplasmic reticulum; **e.** smooth endoplasmic reticulum; **f.** lysosome; **g.** Golgi apparatus; **h.** nucleus; **9.** a; **10.** c; **11.** c; **12.** a; **13.** c; **14.** d; **15.** e

BioNow

A larger surface to volume ratio means that there is relatively more surface area through which the molecules involved in metabolism can pass. Oxygen, water, and nutrients can move into the cell more rapidly and wastes and carbon dioxide can move out more quickly. The increased amounts and of exchange could lead to increased rates of metabolism.

Thinking Critically

1. The smaller hummingbird egg with its greater surface to volume ratio would be expected to have the higher metabolic rate. **2.** Without mitochondria a cell could not produce ATP from nutrients aerobically but would instead have to rely only on glycolysis which is less efficient, for ATP production. Giardia may have increased its ability to metabolize nutrients using glycolysis or have adapted to a lower metabolic rate. **3.** A cell line that already is involved in the secretion of proteins should be selected. This would be a cell line that efficiently produces proteins that are enclosed in vesicles which are transported to the plasma membrane where the vesicles fuse and release their contents to the outside. The genes for the production of the protein of interest would need to be inserted into the genome of the cell and these genes would need to be activated.

CHAPTER 5

Check Your Progress

(5.1) 1. Potential energy is stored energy; kinetic energy is energy of motion. Pulling back on an arrow in a bow is potential energy. When the arrow is shot from the bow, it turns into kinetic energy. **2.** First Law—Energy cannot be created or destroyed but it can be changed from one form to another. Second Law—Energy cannot be changed from one form to another without loss of usable energy. **3.** Cells carry out many energy requiring processes and reactions to maintain their organization while the entropy of the universe increases. **(5.2) 1.** ATP is produced through cellular respiration. ATP is unstable so it cannot be stored. **2.** Myosin combines with ATP prior to its breakdown. Release of ADP + ℗ causes myosin to change position and pull on the actin filament. **3.** During cellular respiration in the mitochondria, carbohydrates are broken down releasing energy that is used in the formation of ATP. Oxygen is used and carbon dioxide and water are released. During photosynthesis in the chloroplasts solar energy is captured and used to synthesize carbohydrates from carbon dioxide and water. Oxygen is released. **(5.3) 1.** The reactions in a pathway each produce a product that is usable in one or more other ways. The reactions of a metabolic pathway allow energy to be used or generated in increments in an organized way. **2.** The binding of the substrate to an enzyme results in a small change in shape of the molecule-substrate complex. This facilitates the reaction to produce product(s). **3.** It ensures that the enzyme is active only when product levels are low and more product is needed. **(5.4) 1.** With diffusion, facilitated diffusion, and osmosis substances move from an area of higher concentration to an area of lower concentration and no energy is required. Channel proteins function in facilitated diffusion. During osmosis water diffuses across a semipermeable membrane. **2.** In hypotonic environments cells gain water; in isotonic environments there is no net gain or loss of water; in hypertonic environments cells lose water. **3.** Active transport moves molecules or ions across the plasma membrane collecting or absorbing molecules against their concentration gradients. This is a mechanism for collecting molecules on one side of a membrane. **4.** Exocytosis transports materials out of a cell; endocytosis transports materials into the cell; phagocytosis transports large materials such as food into the cell; pinocytosis transports liquid and small particles into the cell; receptor-mediated endocytosis forms a vesicle around molecules and their receptors.

Testing Yourself

1. d; **2.** b; **3.** a; **4.** d; **5.** d; **6.** c; **7.** c; **8.** b; **9.** a,b,c; **10.** d; **11.** c; **12.** b,d; **13.** e; **14.** d; **15.** c

BioNow

1. The cells in the potato will gain water or lose water by diffusion depending on whether the environment is hypotonic or hypertonic. The cells will gain water faster the more hypotonic the solution is or lose water faster if the water is more hypertonic. **2.** Energy was transferred from the apple to the person, to his cells to produce ATP which was used by the muscles to power the bike which converted the energy to electrical energy which was stored in the battery. At each step energy was transferred and some lost as heat, but no energy was created or destroyed.

Thinking Critically

1. Knowing the shape of the active site of an enzyme would allow one to design a substrate (a drug in this case) which fits less well or better at the active site. The better the fit the more efficient the metabolic pathway. The worse the fit the pathway would be slowed down. If the drug could compete with the usual substrate in binding to the enzyme then you would have a mechanism to regulate the speed of the reaction in the pathway. **2.** A normal channel protein allows chloride to leave the cells. When chloride leaves, water follows by osmosis, keeping the outer surfaces hydrated. Since the defective cystic fibrosis chloride channel does not allow chloride to leave the cell, water remains inside, and the cell surface becomes covered with a thick mucus. **3.** The composition of the archaeal plasma membrane is different from that of bacteria and eukaryotes. It is more resistant to disruption from heat. In addition, archaeal enzymes have evolved to function at higher temperatures without being denatured.

CHAPTER 6

Check Your Progress

(6.1) 1. The stroma of a chloroplast is a fluid-filled area surrounded by a double membrane. Within the stroma are membrane-bound flattened sacs called thylakoids. Stacks of thylakoids are called grana. **2.** solar energy $+ H_2O + CO_2 \rightarrow (CH_2O) + O_2$ **3.** Reduction—when a molecule gains electrons (e^-) and hydrogen ions (H^+); Oxidation—when a molecule gives up electrons (e^-) and hydrogen ions (H^+). **(6.2) 1.** Because plants have accessory pigments as well as chlorophyll a and b grouped in the pigment complex, they are able to gather solar energy of a broader range of wavelengths. **2.** The pigment complexes of PS II and PS I receive solar energy and pass this energy from one pigment to another to

their reaction centers. In PS II the absorption of solar energy results in electrons being activated and passed on to an electron-acceptor molecule in an electron transport chain. Replacement electrons are supplied by the splitting of water. In PS I electrons are activated by the absorption of solar energy and passed down an electron transport chain to an $NADP^+$ molecule, producing NADPH. **3.** Electrons are removed from water as it is split by PS II. They are passed from PS II to PS I through an electron transport chain. Electrons from PSI are transferred to $NADP^+$ as it becomes NADPH. **4.** The splitting of water supplies replacement electrons to the reaction center in PS II. **(6.3) 1.** The Calvin cycle uses the energy from the light reactions to fix CO_2, reduce CO_2, and regenerate RuBP. It generates molecules of glyceraldehyde 3-phosphate which plants use to make glucose and other biomolecules. **2.** The ATP and NADPH used in the Calvin cycle are generated by the light reactions. **3.** Plants can use G3P to make all the molecules they need including lipids, glucose, sucrose, starch, and cellulose. **(6.4) 1.** In C_3 plants carbon dioxide is taken up by the Calvin cycle directly into the mesophyll cells and the first detectable molecule is a C_3 molecule. In C_4 plants such as corn, carbon dioxide fixation results in C_4 molecules in the mesophyll cells. The C_4 molecule releases carbon dioxide to the Calvin cycle in the bundle sheath cells. **2.** In C_4 plants the mesophyll cells are exposed to carbon dioxide. The bundle sheath cells, where the Calvin cycle occurs, are not. **3.** CAM plants are better able to conserve water and live in hot, arid regions.

Testing Yourself

1. c; **2.** a; **3.** d; **4.** a; **5.** a; **6.** d; **7.** b; **8.** d; **9.** c; **10.** b; **11.** c

BioNow

Wavelengths of light in the yellow and green range are not used in photosynthesis. Instead they are reflected giving plants their color. Blue and violet wavelengths as one end of the visible light spectrum and orange and red wavelengths at the other end of the spectrum are absorbed and used in photosynthesis.

Thinking Critically

1. The alga was producing O_2 wherever, along its length, its chlorophyll was able to absorb the wavelength of light energy it could use in photosynthesis. The bacteria were congregated wherever this O_2 was produced and were using it for cellular respiration. **2.** Artificial leaves would need to have a large surface area for absorbing solar energy and they would need to be thin for efficient gas exchange. They would need a vascular system for the transport of water to the leaf and nutrients away. Within the leaf there would need to be a surface for housing the pigments and across which the gradients of hydrogen ions used in ATP production are created. A location and mechanism for the synthesis of sugars would need to be created. **3.** Broad leaves have a large surface area for absorbing light energy and thin leaves promote efficient gas exchange, both of which promote photosynthesis. **4.** As temperatures rise in North America C_3 plants might not be able to survive while C_4 plants will have an advantage. **5.** As the world warms, C_4 weeds are proliferating. The cellulose in these weeds, as well as cellulose in grasses, wood waste, and crop residues, is a potential source of carbohydrate for ethanol production if

the technology for breaking down cellulose can be improved.

CHAPTER 7

Check Your Progress

(7.1) 1. Cellular respiration provides the energy in the form of ATP that is needed for cellular activities. **2.** During cellular respiration glucose is oxidized to form CO_2, O_2 is reduced to form H_2O, and energy is produced. This is the reverse of the reaction for photosynthesis. **3.** Glycolysis—breakdown of glucose to 2 molecules of pyruvate in the cytoplasm; prep reaction—pyruvate is broken down to a 2-carbon acetyl group carried by CoA in the matrix of the mitochondria with the production of CO_2 and NADH; citric acid cycle—in the matrix of the mitochondria oxidation of molecules occurs, NADH and $FADH_2$ are produced, and CO_2 is released; electron transport chain—electrons are passed down a series of carrier molecules in the cristae of the mitochondria, NADH and $FADH_2$ give up electrons, energy is captured as ATP, O_2 accepts electrons to become H_2O. **(7.2) 1.** Glycolysis is the first step in cellular respiration. It occurs in the cytoplasm of the cell and requires no oxygen. Alternatively, the citric acid cycle and the electron transport chain occur within the mitochondria. The electron transport chain requires oxygen. Pyruvate, produced by glycolysis, feeds into the citric acid cycle. **2.** During the energy-investment step, the breakdown of ATP provides the phosphate groups that activate substrates. During the energy-harvesting steps, NADH and ATP are produced. **3.** 2 ATP molecules per glucose molecule. **(7.3) 1.** Fermentation which does not require oxygen consists of glycolysis followed by a reduction of pyruvate by NADH to produce lactate or ethanol and CO_2. It occurs outside of the mitochondria. Cellular respiration uses oxygen, occurs in mitochondria, and results in a higher yield of ATP. **2.** During lactate fermentation pyruvate is reduced by NADH, producing NAD^+ which is recycled back to glycolysis to accept more H^+. This keeps glycolysis going, supplying some ATP. **3.** Humans produce lactate; yeast produces ethyl alcohol and carbon dioxide. **(7.4) 1.** The preparatory reaction produces a substrate that can enter the citric acid cycle. The citric acid cycle generates CO_2, NADH, $FADH_2$, and ATP at the substrate level. The electron transport chain passes electrons along a series of carriers and captures energy that is used in the production of ATP. **2.** Oxygen is the final electron acceptor of the electron transport chain. It becomes reduced to form water. **3.** Preparatory reactions—pyruvate is oxidized and CO_2 is given off, NAD^+ accepts a H^+ to make NADH, a C_2 acetyl group is made and attached to CoA; Citric acid cycle—the acetyl group is oxidized, CO_2 is released, NAD^+ and FAD accept H^+ producing NADH and $FADH_2$, ATP is produced at the substrate level; Electron transport chain—electrons are passed along a series of carriers, NADH and $FADH_2$ provide electrons, energy is captured for ATP production, O_2 is reduced making water. **(7.5) 1.** For each glucose molecule 2 ATPs are produced during glycolysis, 2 ATPs during the citric acid cycle, and a maximum of 34 ATPs as a result of the electron transport chain for a total of 38 ATPs. Various cells produce less ATP depending on metabolic differences. **2.** Each fat or oil molecule contains three long-chain fatty acids and one glycerol.

Many more acetyl-CoAs result from the breakdown of the three fatty acids than from the breakdown of one glucose molecule. Therefore many more ATPs result from the breakdown of a fat than from glucose. **3.** The amino groups are removed from the amino acids and excreted as urea. The hydrocarbon backbones can enter the cellular respiration pathway at pyruvate, at acetyl-CoA, or during the citric acid cycle.

Testing Yourself

1. a; **2.** c; **3.** a; **4.** c; **5.** b; **6.** b; **7.** a; **8.** b; **9.** c; **10.** b; **11.** c; **12.** c; **13.** c; **14.** c; **15.** d

BioNow

The CHO comes from the food Jason Carlson ate and the oxygen is present in the air he breathed. Cellular respiration produces the CO_2 and the energy used to charge the battery powering the lights needed by the plants.

Thinking Critically

1. Malonate is a product of one enzymatic reaction in the citric acid cycle and a substrate of the next enzyme in the cycle. Due to a feedback mechanism, excessive amounts of malonate will inhibit the enzyme that produced it, thereby stopping the citric acid cycle. **2.** As we age, our mitochondria lose efficiency and do not break down fats as quickly. This causes fats to accumulate in cells. Since fats and oils enter the cellular respiration pathway as acetyl-CoA, there are plenty of citric acid cycle substrates to create energy, and more glucose is not needed. Since the cells already have plenty of energy, they ignore the insulin signal and do not readily take up glucose. Thus, fatty acid accumulation can lead to insulin resistance. **3.** Mitochondria and some prokaryotes are similar in form and function. Mitochondria resemble free-living prokaryotic cells in that their inner membrane also has invaginations. The inner membrane of mitochondria like that of the prokaryote is the site of a metabolic pathway. **4.** Humans cannot live without the electron transport chains working in cells. Our metabolic need for ATP cannot be met by just glycolysis and the citric acid cycle.

CHAPTER 8

Check Your Progress

(8.1) 1. Single-celled organisms divide in order to reproduce. Multicellular organisms must undergo cellular reproduction in order to develop from a single cell. Mature organisms need to replenish worn-out cells and repair injuries. **2.** Before cell division the chromosomes are duplicated to produce two sister chromatids, each of which are held together at a centromere. Each chromatid is identical to the other. **3.** Chromatin is wound around histones to produce nucleosomes. The histones and the nucleosomes package the chromatin so that it fits inside the nucleus. **(8.2) 1.** The cell cycle is an orderly set of steps with checkpoints that is followed when cells duplicate. It ensures accuracy of the process. **2.** During interphase the cell is performing its normal functions. In the G_1 phase organelles are duplicated. During S phase DNA is duplicated, and in G_2 growth occurs and proteins are synthesized that are needed for cell division. **3.** Prophase—chromosomes condense and consist of two sister chromatids; Metaphase—chromosomes

align along the spindle equator; Anaphase—sister chromatids separate and move to each pole becoming daughter chromosomes. Telophase—spindle disappears and nuclear envelope forms around the daughter chromosomes. **4.** The cleavage furrow formed in animal cells between the two daughter nuclei contracts until the two daughter cells separate. In plants the rigid cell wall does not permit cytokinesis by furrowing. Instead the cell builds a new plasma membrane and cell wall between the two daughter cells. **(8.3) 1.** Checkpoints regulate the cell cycle by ensuring that each step is completed correctly before the next step begins. **2.** Internal and external signals are molecules that control the checkpoints of the cell cycle. Internal signals (cyclins) control the G_1 to S and the G_2 to M transitions. External signals such as growth factors and hormones stimulate the cell to go through the cell cycle. **3.** If apoptosis were not regulated the balance between controlled cell death and the cell cycle would be disrupted. Normal cell levels would be altered. **(8.4) 1.** Proto-oncogenes code for proteins that promote the cell cycle and inhibit apoptosis. Tumor suppressor genes code for proteins that inhibit the cell cycle and promote apoptosis. Mutations in these genes can lead to unregulated cell growth. **2.** Teleomere shortening regulates cell division by limiting the number of times a cell can divide. Without it cells continue to divide indefinitely. Rearrangement of segments of chromosomes results in translocations which can disrupt genes that regulate the cell cycle. This can lead to cancer. **3.** The *RB* gene is a tumor suppressor gene that when mutated leads to retinoblastoma. Mutations in both alleles are required for cancer to occur. The *RET* gene is a proto-oncogene that when mutated can lead to thyroid cancer. It is autosomal dominant so only 1 mutated allele is needed to increase the likelihood of cancer. **(8.5) 1.** Cancer cells lack differentiation, have abnormal nuclei, do not undergo apoptosis, form tumors, undergo metastasis, and promote angiogenesis. **2.** Treatments include: removing the tumor surgically, radiation therapy, chemotherapy, and immunotherapy. **3.** Evidence exists that behaviors including not smoking, avoiding sun exposure, moderating alcohol intake, and following a protective diet to maintain a healthy body weight and obtain proper nutrients, can reduce a person's risk of cancer.

Testing Yourself

1. b; **2.** a; **3.** d; **4.** a; **5.** c; **6.** b; **7. a.** G_1; **b.** S; **c.** G_2; **d.** prophase; **e.** metaphase; **f.** anaphase; **g.** telophase; **h.** cytokinesis; **8.** a; **9.** d; **10.** c; **11.** b; **12.** a; **13.** b; **14.** c; **15.** a; **16.** a; **17.** e

BioNow

The rooting hormone acts as an external signal stimulating cells to divide. Likely the hormone interacts with DNA in the S phase of interphase promoting DNA replication.

Thinking Critically

1.a Cancer results from a series of mutations that accumulate in cells over a lifetime. It could be that the appearance of different cancers at different times is due to some cancers requiring a larger number of mutations to occur before they develop. Cancers that require more mutations to occur would more likely occur after a longer period of time.

1.b. Documentation of the types of cancers that appeared over time in the population exposed to the radiation from Fukushima compared to a similar population not exposed to the radiation would test the idea. Care would be needed to restrict both populations from other cancer risks during the duration of the study. **2.** It could be that having multiple copies of tumor suppressor genes is an adaptation that elephants have developed against cancer. Those elephants with multiple copies survive longer and reproduce passing their genomes on to their offspring. **3.a.** BPA-like hormones could act as an external signal stimulating cells to divide. **3.b.** Because BPA acts like a hormone very few BPA molecules are needed to initiate the pathway. Also, the signaling pathway is amplified when executed.

CHAPTER 9

Check Your Progress

(9.1) 1. Meiosis is a type of nuclear division that reduces the number of chromosomes by half and reorganizes the chromosomes to produce genetically different gametes. This creates genetic diversity with each offspring being unique. **2.** Adult humans produce gametes by meiosis. A haploid sperm fertilizes a haploid egg and a diploid zygote is formed which divides by mitosis to develop into an adult. **3.** Chromosomes occur in pairs with each member called a homologous chromosome. One member comes from each parent. They carry the same types of genes but can vary as to the traits they code for. During the cell cycle each homologous chromosome duplicates to form two identical sister chromatids. **4.** Crossing-over shuffles genetic information between nonsister chromatids of a homologous pair, altering the types of alleles on the chromosomes. The chromosomes from each parent cell are shuffled during meiosis I to produce new combinations unique to the daughter cells. These processes create genetic variation that is the basis of evolution. **(9.2) 1.** During prophase I crossing-over between nonsister chromatids occurs. During metaphase I the tetrads align at the equator with each homologue facing a spindle pole. Whether the homologue came from the father or from the mother does not matter in how they line up. Genetic variability in the daughter cells is increased by these processes. **2.** During metaphase I tetrads (2 homologous chromosomes each consisting of 2 chromatids) line up at the spindle equator. During metaphase II the dyads (1 chromosome consisting of 2 sister chromatids) align at the spindle equator. **3.** During meiosis I the chromosome number is halved and the genetic information is reshuffled. During meiosis II the cells reproduce like in mitosis, except they are haploid. **(9.3) 1.** Mitosis produces 2 daughter cells while meiosis produces 4. **2.** Mitosis produces daughter cells which are diploid while meiosis produces haploid daughter cells. **3.** In meiosis I crossing-over during synapsis occurs and homologues are separated into the daughter nuclei irrespective of their being from the father or mother. Genetically different nuclei are created. The chromosome number is halved. During meiosis II the cells replicate like in mitosis except that they are haploid. Mitosis results in daughter cells which are identical to the parents. **(9.4) 1.** If nondisjunction occurs during meiosis I or meiosis II, a gamete and therefore a zygote, could have an abnormal chromosome number. **2.** Monosomy results when an individual is missing one chromosome. Trisomy results when an individual has one extra chromosome. **3.** Abnormalities in the sex chromosomes occur only in the X or Y chromosomes. Autosomal abnormalities occur in any of the other 22 pairs of chromosomes. Newborns with abnormal number of sex chromosomes are more likely to survive than those with abnormal numbers of autosomal chromosomes.

Testing Yourself

1. d; **2.** a; **3.** d; **4.** a; **5.** f; **6.** h; **7.** c; **8.** g; **9.** b; **10.** e; **11.** c; **12.** b; **13.** b; **14.** c; **15.** d; **16.** c

Thinking Critically

1. A mosaic such as (46, XY/47, XXY) may result from nondisjunction occurring during mitosis, especially at a very early stage of development. A fertile Klinefelter mosaic may arise from an individual in which normal (46, XY) embryonic cells gave rise to the testes. The symptoms usually associated with Klinefelter individuals would be less pronounced in mosaics and it is possible that they could produce normal sperm and reproduce. **2.** Hypotheses to explain the relationship between the increase in Down syndrome with advancing maternal age include: (1) The older a woman is, the longer her oocytes have been arrested in meiosis, and the higher the risk that these oocytes have been exposed to mutagens which might cause nondisjunction; (2) A woman may have a pool of oocytes resulting from nondisjunction which take longer to mature, and thus are more commonly released as she ages; (3) Estrogen levels (which control the rate of meiosis in developing oocytes) drop with advancing maternal age and may slow down the rate of meiosis. This might allow nondisjunction to occur more frequently in older women. **3.** Large amounts of genetic variability in a population allow the population to more quickly adapt to changing conditions. For example, if a population is suddenly exposed to a new pathogen, it is likely that at least a few individuals will be able to survive the pathogen and the species would continue.

CHAPTER 10

Check Your Progress

(10.1) 1. Mendel's experiments explained the patterns of inheritance of traits seen in families and the variation between offspring seen from one generation to the next. **2.** The offspring resulting from a cross between two individuals who are heterozygous for one trait (*Tt*) will produce three different genotypes, *TT, Tt, tt* in a 1:2:1 ratio. Because the homozygous dominant (*TT*) and the heterozygous (*Tt*) individuals look alike, the phenotypic ratio will be 3:1. **3.** The probability of producing a short plant with green pods is 3/16. **4.** Traits are carried by a homologous pair of chromosomes. One homologue carries the trait from the mother and the other carries the trait from the father. When following two traits, two homologous pairs are considered. During meiosis the homologues separate from each other independently and go into different gametes no matter whether they came from the father or the mother. All combinations of chromosomes are possible. This is what Mendel saw with his experiments with peas. **(10.2) 1.** An autosomal recessive pedigree (Fig. 10.9) shows that

affected children can have unaffected parents, that heterozygotes have a normal phenotype, and that both males and females are affected equally. The individual III-1 is heterozygous because she had affected children. An autosomal dominant pedigree (Fig. 10.10) shows that affected children have at least one affected parent, that heterozygotes are affected, and that males and females are equally affected. The individual II-2 is heterozygous because he had one unaffected child. The genotype of individual II-3 cannot be determined because there are no unaffected children present. **2.** With an autosomal recessive disorder like cystic fibrosis, individuals with normal phenotypes can be heterozygous (they can carry a dominant and recessive allele). Each parent is capable of producing gametes with the recessive allele which can combine in the offspring to produce the disorder. **3.** Sickle-cell disease is autosomal recessive. Homozygotes have sickle-shaped red blood cells as children but the heterozygotes appear normal unless stressed by low oxygen levels. Huntington disease is autosomal dominant but the disease does not appear until middle age when individuals may have already passed it on to their children. **(10.3) 1.** Incomplete dominance is exhibited when the heterozygote has an intermediate phenotype between that of either homozygote. An example is pink four-o'clock flowers. Codominance is exhibited when both alleles in a heterozygote are fully expressed. **2.** The type O child would need to have inherited a recessive *i* allele from both parents. The type AB parent does not have an *i* allele. **3.** Polygenic traits are governed by several sets of alleles often on different chromosomes. The dominant alleles have an additive effect creating a bell-shaped distribution curve of phenotypes. Multifactorial traits are controlled by polygenes that are subject, in varying degrees, to environmental influences. **4.** If genes are linked on one chromosome then they are inherited together. In a two-trait cross the assumption is that the alleles are on nonhomologous chromosomes and will be subject to independent assortment. It is possible that one gene can override the expression of another gene. This type of gene interaction appears as a variation of Mendel's laws of inheritance involving dominant and recessive traits. **(10.4) 1.** Since males are XY, some sperm will contain an X chromosome and others will contain a Y. **2.** With X-linked recessive inheritance (Fig. 10.25) more males than females are affected, affected sons can have parents with normal phenotypes, affected female offspring have an affected father, and the mother must be affected or be a carrier. With autosomal recessive inheritance, males and females are affected equally, heterozygotes have a normal phenotype, and affected children can have unaffected parents. **3.** Color blindness and Duchenne muscular dystrophy are sex-linked disorders.

Testing Yourself

1. a; **2.** d; **3.** d; **4. a.** *TG*; **b.** *tg*; **c.** *TtGg*; **d.** *TTGG*; **e.** *TtGg*; **f.** *TTgg*; **g.** *Ttgg*; **h.** *TtGG*; **i.** *ttGG*; **j.** *TtGg*; **k.** *ttGg*; **l.** *ttgg*; **5.** a; **6.** b; **7.** a; **8.** b; **9.** a; **10.** a; **11.** a; **12.** e; **13.** e; **14.** d; **15.** c.

BioNow

Multiple crosses of the glowfish with other glowfish would provide the evidence. If the color was due to a dominant trait then crosses between glowfish could produce two possible phenotypes: the colored fish

(homozygous dominant or heterozygous dominant) or the regular fish (homozygous recessive). If the color was a codominant trait then three different phenotypes would be produced from the genotypes of homozygous dominant, heterozygous, and homozygous recessive.

Thinking Critically

1. a. *CcPp* × *CcPp* would produce 9/16 purple (*C_P_*) and 7/16 white offspring (*C_pp*, *ccP_*, or *ccpp*). **2. a.** Set up a cross between a true-breeding strain of flies that have the trait with a true-breeding strain that does not. If no offspring of the cross have the trait it is recessive. If all the offspring have the trait it is dominant. **b.** *Drosophila* have lots of offspring, a short generation time, can be housed and raised inexpensively, and the ethics of treatment of the flies is broader than with humans. **3. a.** The technology could be used to produce only female offspring which reduces the probability of an X-linked genetic disorder showing as a phenotype. **b.** Parents could select the sex of their children based on cultural preferences thus distorting the balance of males to females in the population.

CHAPTER 11

Check Your Progress

(11.1) 1. DNA contains a series of nucleotides, each consisting of a deoxyribose sugar, a phosphate, and a nitrogen-containing base: cytosine, thymine, adenine, or guanine. Two strands of nucleotides are arranged antiparallel to each other and are twisted into a helix with cytosine hydrogen bonded to guanine and adenine to thymine. **2.** First, helicase unwinds the double-helix structure by breaking the hydrogen bonds between the nucleotides; DNA polymerase adds nucleotides to the complementary new strand in a 5′–3′ direction; DNA ligase then seals any breaks in the deoxyribose-phosphate backbone of the daughter strand. **3.** DNA contains the sugar deoxyribose; RNA has the sugar ribose. DNA has the base thymine; RNA has the base uracil. DNA is double-stranded; RNA is single-stranded. **4.** mRNA carries the genetic information from DNA in the nucleus to the ribosomes in the cytoplasm; rRNA combines with proteins in the cytoplasm to form ribosomes, where proteins are synthesized; tRNA transfers amino acids to a growing polypeptide at a ribosome. **(11.2) 1.** Genetic information flows from DNA to RNA to proteins. DNA is transcribed into mRNA, which moves from the nucleus to the ribosomes in the cytoplasm and undergoes processing before it is translated into a protein. **2.** The genetic code in DNA defines the sequence of amino acids in the proteins a cell makes. This stored information is translated from DNA to RNA and eventually results in a specific polypeptide synthesized at a ribosome. This process of protein synthesis in which genetic information flows from DNA to RNA to protein is known as the central dogma of molecular biology. **3.** The mRNA molecule must be processed by capping, the addition of a poly-A tail, and splicing in order for it to function properly in protein synthesis. **4.** Ribosomes are made up of ribosomal RNA and proteins arranged as large and small subunits. At the beginning of translation, the large ribosomal subunit joins the small unit and a tRNA-amino acid complex containing an anticodon binds to the codon on the mRNA. The amino acid is then added to the growing polypeptide. **(11.3) 1.** In prokaryotes, a single promoter controls transcription in a unit of several genes. In eukaryotes, there is a promoter for each gene. Prokaryotes regulate gene expression mostly by controlling transcription. Eukaryotes have many mechanisms for controlling gene expression, such as turning a gene on or off, regulating the speed at which a gene is expressed, and controlling how long it is expressed. **2.** Eukaryotes can regulate gene expression in the nucleus by chromatin condensation, control of DNA transcription, and mRNA processing. In the cytoplasm, translation of mRNA can be allowed or delayed, mRNA lifetimes can vary, and how long proteins exist can be controlled. **3.** There are many transcription factors in eukaryotes that form a complex that binds to DNA to help separate the DNA strands and promote the positioning of RNA polymerase. Because there are many transcription factors working together, a higher level of control over transcription is possible.

Testing Yourself

1. a. sugar-phosphate backbone; **b.** purine base; **c.** hydrogen bonds; **d.** deoxyribose sugar; **e.** nucleotide; **2.** a; **3.** d; **4.** a; **5.** d; **6.** b; **7.** b; **8.** a; **9.** c; **10.** c; **11.** b; **12.** b; **13.** b; **14.** d; **15.** b

BioNow

1. Light and increased temperature could influence brain function, causing hormone production and release. The hormones could influence gene expression, leading to ovulation and sperm production as well as behavior changes involved with mating. **2.** Regulators of the expression of the developmental genes could include increased temperature, exposure to light, and the passage of time. Also, perhaps only those larvae exposed to adequate food while in the wood could develop into adults.

Thinking Critically

1. DNA from the heat-killed virulent strain contains genes that produce the virulence. Somehow, the living nonvirulent bacteria obtained some of this DNA, converting them into a virulent strain. **2.** A medicine could be developed whose effect on gene regulation was inheritable. Such a medicine could be given just once, and its effect would be passed on to following generations. **3. a.** The basic requirements are being able to store genetic information that is inheritable, having the stored information be expressible, containing variations that allow for evolution, and being small enough to be packaged inside a cell. **b.** If the genetic information turns out to be very similar to our own, that implies that DNA has unique qualities that are universal. If the genetic information is very different, then different storage molecules are possible.

CHAPTER 12

Check Your Progress

(12.1) 1. A bacterial plasmid is a ring of accessory DNA that serves as a vector to carry foreign DNA in the production of recombinant DNA. **2.** PCR is used to produce many copies of a specific segment of DNA, which is usually a few hundred base pairs long. **3.** DNA fingerprinting can be used to identify sources of DNA, such as bacteria or viruses that are causing an infection, people involved in a crime scene,

parents of a child, or unidentified bodies. In addition, it can be used to detect mutations that can lead to an increased risk of disease. **4.** Genomic editing may be more efficient because the guide RNA molecule can locate the specific nucleotides of interest in the target DNA. The process can be applied to almost all organisms, and it can be used to either inactivate a gene or insert a new DNA sequence. Recombinant DNA processes may be less efficient because both an organism with the gene of interest and an organism housing the vector (plasmid) are required. Also, two enzymes, a specific restriction endonuclease and a DNA ligase, must be used. **(12.2) 1.** The goal of reproductive cloning is to create an identical copy of an individual organism. The goal of therapeutic cloning is to produce cells or tissues that can be used in the study of cell specialization or in treating human illnesses. **2.** Embryonic stem cells are obtained from an embryo. Adult stem cells are harvested from many organs in the body. **3.** Embryonic stem cells are totipotent because they can differentiate into any type of cell in the body. Adult stem cells are multipotent, meaning that they are limited as to the types of cells they can produce. **(12.3) 1.** Transgenic organisms have had a foreign gene inserted into their genome. Genetically modified organisms have had their genetic material modified using recombinant DNA technology. **2.** Transgenic bacteria have been used to produce insulin, clotting factor VIII, human growth hormone, tissue plasminogen activator, and hepatitis B vaccine. Frost-minus strains of bacteria have been engineered that can reduce the formation of ice crystals on strawberries and other fruits. Naturally-occurring bacteria that degrade oil have been engineered to work more efficiently. **3.** Transgenic plants resistant to herbicides have increased crop yields while reducing herbicide use. Transgenic food crops have been developed that have higher nutrient quality and contain more healthful fats. Transgenic animals that have the gene for bovine growth hormone grow bigger and provide more meat. These advances increase food production. Plants have been developed that have leaves that take in more carbon dioxide and lose less water, which helps the crops grow in a wider range of climate conditions. **(12.4) 1.** The goals include an understanding of what genes are present in a genome, the gene products and their functions, and the function of the intergenic DNA. **2.** Comparing different genomes identifies their differences and similarities, which provides information about evolutionary history. Insight into how the human species evolved can be gained by comparing the human genome with those of our recent ancestors. **3.** Proteomics is important to drug development because knowing the structure and function of cell proteins and examining how they interact gives insight into how drug molecules will react with them.

Testing Yourself

1. d; **2.** d; **3.** c; **4.** b; **5.** c; **6.** c; **7.** d; **8.** b; **9.** c; **10.** c

BioNow

The glowing fish are genetically modified, as indicated by their ability to pass the glowing trait on to their offspring. The gene for glowing may have been inserted into the fishes' genome, or CRISPR may have been used to edit the fishes' color genes to produce the glow.

Thinking Critically

1. Advantages include efficiency of the treatment, since many viruses could deliver the genes rapidly. Also, the process would be relatively noninvasive. A problem might be that the genes could be inserted in the wrong place in the genome, disrupting other genes and either making them nonfunctional or altering them to produce harmful gene products. Another possible problem is that the viruses could promote cell division, resulting in unregulated cell growth leading to cancer. **2.** Human recombinant insulin matches human insulin completely. Cow or pig insulin differs slightly from human insulin. Using human recombinant insulin would avoid any potential antibody response to the insulin. **3.** Genes whose products are involved in glycolysis, the citric acid cycle, and the electron transport chain would likely be similar.

CHAPTER 13

Check Your Progress

(13.1) 1. Mutations alter the sequence of nucleotides in DNA. Alleles are variations of a gene, and a mutation can create a new allele. **2.** When a transposon moves, it inserts its DNA sequence into another chromosome, altering the sequence of nucleotides. This can result in a mutation that disrupts a gene. **3.** The single base change that occurs in a point mutation may or may not change the one amino acid that is coded for, so the function of the protein may not be altered. A frameshift mutation alters all the downstream codons and often results in a dysfunctional protein. **(13.2) 1.** A deletion can result in the loss of one allele of a gene, possibly leading to a syndrome. With a duplication, a segment of the chromosome is repeated and thus there may be more than two alleles for a gene. This disrupts normal gene function but does not eliminate the allele. **2.** Inheritance of a chromosome with a translocation can leave an individual with one copy of certain alleles and three copies of other alleles. When a segment of DNA is inverted, duplication and deletion can occur in later generations, resulting in disrupted gene expression, **3.** Cri du chat syndrome results from a deletion; the end piece of chromosome 5 is missing. Alagille syndrome results from a translocation between chromosomes 2 and 20. **(13.3) 1.** Karyotyping can detect abnormal numbers and shapes of chromosomes as well as their banding patterns, which may indicate chromosomal mutations. **2.** Genetic disorders can be detected if a specific protein is present or absent in the blood. If a protein is absent, then the gene is altered or it is not being expressed. Alternatively, a protein might be present that is normally absent. Testing DNA involves determining if the sequence of bases in a person's DNA has been changed. Such changes can lead to disorders. **3.** Ultrasound images help doctors evaluate fetal anatomy, but not the genome of the fetus. Amniocentesis uses a long needle to withdraw amniotic fluid containing fetal cells. The process does have a low risk (0.6%) of causing a spontaneous abortion. In chorionic villus sampling (CVS), a suction tube is used to remove chorionic villi cells from the region where the placenta develops. This procedure can be performed as early as the 5th week of pregnancy, and it carries a 0.7% risk of spontaneous abortion. **4.** Changes in base sequences can be identified and used as genetic markers of diseases. The markers can be detected by conducting a complete genetic profile, using microarrays, or analyzing the products of restriction enzyme digestion of a subject's DNA using gel electrophoresis. **(13.4) 1.** Ex vivo gene therapy procedures are carried out on tissues removed from the body, treated with viruses carrying the gene of interest, and then reintroduced into the body. With in vivo gene therapy, the gene of interest is delivered directly into the body. **2.** Genes may be delivered by spraying them into the nose or by injecting them into particular tissues or organs. Liposomes or adenoviruses may also be used to carry the genes into cells and tissues. **3.** Ex vivo: SCID treated by inserting ADA gene into stem cells; familial hypercholesterolemia where liver cells are infected with virus carrying gene for the normal cholesterol receptor; cancer when patient cells are treated to express cancer antigens to stimulate the immune system. In vivo: cystic fibrosis treated when gene for normal chloride ion carrier is sprayed into the nose or carried into the lungs by liposomes; VEGF is injected into the heart to stimulate growth of coronary blood vessels; RA when anti-inflammatory gene is injected directly into the affected joint.

Testing Yourself

1. c; **2.** b; **3.** a; **4.** b; **5.** c; **6.** a; **7.** c; **8.** b; **9.** c; **10.** b

Thinking Critically

1. A genetic counselor works for the patient but may be uncomfortable with the ethics of not disclosing information. In light of this, the counselor needs to educate the patient about DNA sequencing and all the information it can divulge before the testing is performed. Information about other diseases, their risks, and the emotional and financial impact of having a child with the disease should be communicated in advance. Formal agreements about how to handle added information resulting from the testing should be established between the counselor and the patient. **2.** The adenovirus may have inserted itself into or near a gene that is involved in controlling the cell cycle, such as a proto-oncogene or a tumor suppressor gene. If so, disrupting the gene may have inactivated it. In the case of the infants with XSCID, the virus was found to have inserted itself near a proto-oncogene called *LMO-2*. **3.** When disorders are homozygous recessive, the recessive allele can be present in a portion of the population at all times. Only when two heterozygous carriers mate is there a chance that the offspring will have the disorder. Sometimes the recessive allele confers an evolutionary advantage. For example, carriers of sickle-cell disease are more resistant to malaria.

CHAPTER 14

Check Your Progress

(14.1) 1. Cuvier founded the science of paleontology. He thought that species were fixed and did not change over time. Instead, catastrophes occurred that led to mass extinctions. Repopulation of an area after a catastrophe led to new fossils appearing in a new stratum, which explains the changes seen in the fossil record. Lamarck proposed that animals change to adapt to their environments and these changes are passed on to their offspring. This inheritance of

acquired characteristics explains the changes seen over time. **2.** According to the idea of inheritance of acquired characteristics, an individual organism changes in response to its environment. It acquires new characteristics that it passes on to its offspring. In the process of natural selection, individuals' survival depends on how well they are adapted to their environment. Those who are well adapted survive to reproduce, and those who are not die out. **3.** The individuals best adapted to an environment are the fittest because they survive and reproduce. The adaptations that allow survival in an environment are selected for over generations. **(14.2) 1.** The finches show that a common ancestor can give rise to different species through adaptation to unique environments. In this case, the environments on different islands varied and put different pressures on the finches that arrived there. **2.** Vestigial structures show that present-day organisms have ancestors from which they have evolved. These structures are not functional now but they were in the ancestors. **3.** Homologous structures share an anatomical similarity that reflects a common ancestry. Analogous structures have a common function but are not constructed similarly and do not indicate a common ancestry.

Testing Yourself

1. b; **2.** d; **3.** b; **4.** b; **5.** a; **6.** e; **7.** e; **8.** a; **9.** c; **10.** b; **11.** a; **12.** c; **13.** b; **14.** c

BioNow

The quail eggs had variations in their coloration that influenced the degree to which they were camouflaged in their environment. In this experiment, it was supposed that the environment shifted to be either white or dark, which created a pressure on the quail population to evolve to adapt to this change. The whitest eggs would survive predation in the white environment, and the darkest eggs would survive in the dark environment. Those eggs would hatch and produce offspring, which in turn would produce eggs that shared the phenotypes of the parents. Under the artificial selection imposed by the experiment, the frequencies of the lightest and darkest eggs increased.

Thinking Critically

1. Humans and other mammals have these homologous structures because we share a common ancestor that had such a structure. The presence of this structure informs us about the dietary history of humans. It suggests that humans previously ate more plants than they currently do. **2.** Different genes are under varying levels of evolutionary pressure based on their function and importance for an organism's survival. **3.** Islands are isolated and provide unique environments for the organisms that live there. The particular conditions in these environments influence the evolution of these organisms in a specific direction.

CHAPTER 15

Check Your Progress

(15.1) 1. Directional selection favors one extreme phenotype. Stabilizing selection favors the intermediate phenotype. Disruptive selection favors two or more extreme phenotypes over any intermediate phenotype. **2.** In females, sexual selection favors the fittest males, meaning those that

have the traits that improve the ability to reproduce. By selecting the fittest male, the female is improving the chances that her traits will be passed on to her offspring. **3.** Mutation, recombination, independent assortment, gene flow, disruptive selection, and diploidy. **(15.2) 1.** The various alleles at all the gene loci in all individuals make up the gene pool of the population. The gene pool of a population is described in terms of allele frequencies. **2.** The Hardy-Weinberg principle states that the equilibrium of genotype frequencies in a gene pool will remain stable in each succeeding generation of a sexually reproducing population as long as five conditions are met. Microevolution can be detected and measured by noting the amount of deviation from Hardy-Weinberg equilibrium that is exhibited by the genotype frequencies in a population. **3.** To prevent microevolution in a population no mutations must occur, there is no gene flow, only random mating happens, there is no genetic drift, and no natural selection functions. **4.** Mutations are ultimately responsible for all the genetic variability on which evolution depends. **5.** Gene flow is the migration of alleles into or out of populations, changing their frequency and leading to microevolution. Nonrandom mating pairs individuals according to their phenotypes and ultimately their genotypes. Inbreeding increases the proportion of homozygotes in the population, and assortative mating causes the population to subdivide into phenotypes. Nonrandom mating can therefore result in microevolution. **6.** With genetic drift, allele frequencies in a gene pool change gradually over time due to chance. In a small population, the probability that an individual with a rare genotype might pass alleles to the next generation is greater. It will be more likely that a decrease or increase in the frequency of this genotype will result.

Testing Yourself

1. d; **2.** a; **3.** c; **4.** b; **5.** c; **6.** a; **7.** d; **8.** c; **9.** c; **10.** b; **11.** a

BioNow

The experimenter selected the eggs that were incubated to produce the next generation. This imposed a pressure on the population analogous to nonrandom mating. The distribution of egg coloration was altered over generations by this disruptive selection, increasing the frequencies of the lightest and darkest eggs.

Thinking Critically

1. a. By applying three times as much pesticide, the second farmer generates pesticide-resistant insects that are selected for over the five years. **b.** The resistant insects have an extreme phenotype that is being selected for, an example of directional selection. **c.** The insect population on the first farm is in equilibrium because the percentage of crop loss remains constant, indicating the same number of resistant insects in each generation. **2. a.** The females are selecting the bright-feathered males as mates because they perceive them to be more fit. This is an example of sexual selection. **b.** By selecting a bright-feathered male, a female is more likely to have bright-feathered offspring that will have greater success in generating more offspring. This promotes nonrandom mating. **c.** In their natural environment, mate dark-feathered females with light-feathered males and measure the reproductive success of the offspring. Compare that to the reproductive success of offspring from mating dark-feathered females and dark-feathered males.

CHAPTER 16

Check Your Progress

(16.1) 1. A biological species is a group of organisms that can interbreed, have a shared gene pool, and are reproductively isolated from members of other species. This is a testable concept that does not depend on physical appearance. **2.** The biological species concept cannot be applied to asexually reproducing organisms. In addition, reproductive isolation may not be complete, so some breeding between species may occur. **3.** Prezygotic isolating mechanisms do not allow zygotes to be formed. Examples are habitat, temporal, behavioral, mechanical, and gametic isolation. Postzygotic mechanisms allow zygotes to be formed, but the zygotes either die or develop into offspring that are infertile or that produce infertile offspring. Examples are zygote mortality, hybrid sterility, and low F_2 fitness. **4.** Allopatric speciation occurs when a geographic barrier stops gene flow. *Ensatina* salamanders in California are separated by the Central Valley and limited gene flow occurs between the eastern and western populations. Two distinct forms exist in southern California that interbreed only rarely. Sympatric speciation produces reproductively isolated groups without geographic isolation. Polyploidy in wheat plants resulted in the reproductive isolation of bread wheat. **5.** One mainland finch species was probably an ancestor to all existing species on the Galápagos Islands through adaptive radiation. Each new species acquired adaptations in response to the unique environment on the island it inhabits. **(16.2) 1.** Fossils provide durable evidence of life-forms that is relatively complete and can be dated. They display evolutionary changes that have occurred over time. **2.** The punctuated equilibrium model says that periods of no change in evolution are interrupted by periods of relatively rapid speciation. **3.** Continental drift, loss of habitat, the crash of a meteorite into the Earth, and climate change have contributed to mass extinctions. **(16.3) 1.** Species, genus, family, order, class, phylum, kingdom, domain. **2.** Homologous structures are similar anatomically because they are derived from a common ancestor. Analogous structures have the same function in different groups and do not share a common ancestry. **3.** Taxonomy is concerned with identifying, naming, and classifying organisms. It uses scientific naming rules to establish a hierarchy of categories. Systematics attempts to define the evolutionary history of organisms using fossils, comparative anatomy and development, and molecular data. **4.** The three-domain system adopts a category of classification above the level of kingdom. It is based on the sequencing of rRNA genes. It consists of the domains Bacteria and Archaea, which include the prokaryotes, and the domain Eukarya, which includes protists, animals, fungi, and plants.

Testing Yourself

1. c; **2.** c; **3.** b; **4.** f; **5.** a; **6.** g; **7.** d; **8.** e; **9.** h; **10.** b; **11.** e; **12.** d; **13.** a; **14.** c; **15. a.** 3, with purple, orange, and yellow shading; **b.** vertebrae, amniotic egg and internal fertilization; **c.** snake and lizard, because they share the most derived traits

Thinking Critically

1. a. ferns; **b.** produce seed; **c.** naked seeds; **d.** conifers, needle-like leaves; **e.** ginkgos, fan-shaped leaves; **f.** enclosed seeds; **g.** monocots, one embryonic leaf; **h.** eudicots, two embryonic leaves. **2.** On the Hawaiian Islands, species become isolated on islands that are far away from each other and the mainland. Over generations, they adapt to the island environment, which is often different from the mainland environment from which they came. The Florida Keys are close enough to each other and close enough to the Florida mainland to allow species to move among the islands and the mainland. Thus, they are not reproductively isolated and further speciation does not occur. **3. a.** Because humans share much of their genome with other organisms, basic research in systematics at all levels has the potential to provide beneficial information to humans and the organisms that make up their world. **b.** If a country has the financial capacity to fund basic systematics research while attending to the needs of its population, such research should be a priority because of the potential benefits that could result.

CHAPTER 17

Check Your Progress

(17.1) 1. Viruses are composed of an outer capsid of protein subunits and an inner core containing DNA or RNA. **2.** In the lytic cycle, a virus enters a host cell and replicates, producing many copies of itself. Alternatively, in the lysogenic cycle, the virus, after infecting the cell, may enter a latent stage in which it is dormant but ready to replicate in the future. **3.** In order to enter a cell, a virus must have a specific capsid protein that binds to a specific receptor protein on the cell's surface. **(17.2) 1.** A viroid is composed of naked RNA, not covered by a capsid. A virus has an inner core of DNA or RNA and a capsid composed of protein subunits. **2.** Prions are proteins that have changed shape and can interact with normal proteins to change their shapes. This causes fatal infections and neurological disorders. Viroids are strands of naked RNA. **3.** Prion disorders include kuru, mad cow disease, scrapie (which attacks sheep), and Creutzfeldt-Jacob disease in humans. **(17.3) 1.** Small organic molecules formed from inorganic chemicals as the Earth cooled. Macromolecules self-assembled from the small organic molecules and became surrounded by a barrier that defined the inside and the outside of a cell-like structure. Within the cell-like structure (protocell), the reactions of cellular metabolism evolved. **2.** Cocci (spheres), bacilli (rods), and spirilla or spirochetes (spirals). **3.** Conjugation—transfer of DNA from cell to cell; transformation—transfer of DNA from the environment to a cell; transduction—transfer of DNA from cell to cell by a bacteriophage. **4.** The plasma membrane of archaea have unusual lipids so they can survive in extreme environments such as conditions of high heat. They have diverse types of cell walls, sometimes containing proteins or polysaccharides but not peptidoglycan. They also have unusual metabolic characteristics such as the ability to generate methane or live in high-salt environments. Because of these characteristics, they do not compete with bacteria. **(17.4) 1.** Both mitochondria and chloroplasts have double membranes and distinct genomes. Their ribosomal genes originated in bacteria, with mitochondria being related to certain bacteria and chloroplasts related to cyanobacteria. **2.** Protists are eukaryotes with a nucleus and a wide range of organelles. Many are single-celled. They are a diverse group that represent the common ancestors of different multicellular organisms. **3.** Archaeplastids have chloroplasts derived from endosymbiotic cyanobacteria; green algae are an example. The SAR supergroup consists of stramenopiles, which are photosynthetic but evolved differently than green and red algae; alveolates, which have an internal series of cavities; and rhizarians which are amoeba-like organisms with external shells. Examples of these three groups are diatoms, ciliates, and foraminifera. Excavata lack mitochondria and have distinct flagella and/or oral grooves; euglenids are an example. Amoebozoa move by means of pseudopods; slime molds are an example. Opisthokonts are chemoheterotrophs with flagellated cells; choanoflagellates are an example.

Testing Yourself

1. c; **2.** d; **3.** a; **4.** c; **5.** c; **6.** c; **7.** a; **8.** b; **9.** a; **10.** d; **11.** c; **12.** c; **13.** d; **14.** a; **15.** d

Thinking Critically

1. Protists are members of the domain Eukarya and are more closely related to archaea than to bacteria. Archaea and eukarya have certain ribosomal proteins in common that are not shared with bacteria. They initiate transcription in the same way and have similar tRNAs. Viruses are not shown on the tree because they are acellular and are not considered to belong to any of the three domains of life. **2.** Bacteriophages can infect bacteria in a patient's body and initiate a lytic cycle that will kill the bacteria. The most important benefit is the ability to treat infections caused by antibiotic-resistant bacteria. The main shortcoming is that bacteria will almost certainly evolve resistance to bacteriophage infection. **3. a.** Labeling antibiotics as being ineffective against viruses would help educate consumers. Not everyone would read the label or respond to the information it contains, but some people would respond to the message and change their behaviors. **b.** The laws covering liability for withholding antibiotic use should be changed. It is likely that physicians would then become more accountable for prescribing antibiotics, and if they were protected from lawsuits for not prescribing antibiotics, they could afford to be more conservative about the prescriptions they write.

CHAPTER 18

Check Your Progress

(18.1) 1. The green algae and land plants contain chlorophylls *a* and *b* and accessory pigments, store carbohydrates as starch, have cellulose in their cell walls, and form structures that protect the zygote. **2.** (1) Protection of the embryo; (2) evolution of vascular tissue; (3) evolution of leaves (microphylls and megaphylls); (4) evolution of seeds; (5) evolution of flowers. **3.** Meiosis forms a haploid spore from a diploid sporangium, which undergoes mitosis and forms gametes. Following fertilization, a diploid zygote is formed that develops into a sporophyte by mitosis. **(18.2) 1.** Nonvascular plants do not have true roots, stems, or leaves like vascular plants do. **2.** The dominant generation of a moss is the gametophyte; for a fern, the gametophyte stage is tiny and the sporophyte generation is dominant. Mosses are nonvascular and lack true roots, stems, and leaves. They are short and require a moist environment to live in and reproduce. Ferns have a vascular system with true roots, stems, and leaves and can live in drier environments and grow taller. **3.** Following pollination, the pollen grain germinates and a nonflagellated sperm travels in a pollen tube to the egg produced by the female gametophyte. Following fertilization, the zygote becomes the sporophyte embryo, tissue within the ovule becomes the stored food, and the ovule becomes the seed coat. **4.** To reproduce, angiosperms require the dispersal of pollen from flower to flower. Wind and pollinators such as insects, birds, bats, and flies help disperse the pollen. **(18.3) 1.** A fungus is made up of hyphae packed together to form a mycelium. The cell walls are composed of chitin. The cells are divided by septa, which contain pores that allow the movement of cytoplasm from one cell to another. **2.** Basidiomycota—dikaryotic hyphae form a mushroom consisting of a stalk and a cap; Ascomycota—ascocarp shaped like a cup; Glomeromycota—form arbuscular mycorrhizae (AMs) with the roots of land plants; Zygomycota—have specialized hyphae for different functions; Chytridiomycota—have flagella and are mobile; Microsporidia—obligate parasitic fungi. **3.** A lichen is composed of a fungus and a cyanobacterium or a green alga. The fungus provides protection, water, and minerals, and the photosynthesizer provides organic nutrients. **4.** Oral candidiasis—*Candida albicans* infects the mouth and tongue producing painful blisters. It can occur after antibiotic use or in states of immunosuppression. White nose syndrome—an infection in hibernating bats caused by *Geomyces destructans*. The disease disturbs their hibernation, making them wake up and use up their fat stores, resulting in starvation.

Testing Yourself

1. a; **2.** e; **3.** c; **4.** a; **5.** b; **6.** e; **7.** d; **8.** c; **9. a.** anther; **b.** filament; **c.** stigma; **d.** style; **e.** ovary; **f.** ovule; **g.** sepals (calyx); **h.** petals (corolla); **10.** b; **11.** b; **12.** a; **13.** c; **14.** b; **15.** e

BioNow

The garlic, commercial fungicide, hydrogen peroxide, and rubbing alcohol all disrupt normal protein structure by denaturing proteins. These proteins may be the enzymes used by the fungi to break down the nuts into nutrients that the fungi can absorb. The fungi cannot grow and reproduce without the functional enzymes.

Thinking Critically

1. The native soil contains mycorrhizal fungi, which have mutualistic relationships with plant roots and enhance the growth of plants. **2.** Protists, plants, and fungi all belong to domain Eukarya. The protist ancestors of plants resembled certain green algae, possessing chlorophylls *a* and *b* and accessory pigments for photosynthesis, storing carbohydrates as starch, having cellulose in their cell walls, and utilizing an alternation-of-generations life cycle. The protist ancestors of fungi probably resembled modern chytrids in having flagellated gametes and spores, a feature that has since been lost in the

evolution of the other fungal groups. Cell walls of fungal ancestors may have contained chitin, a feature shared by all modern fungi, including chytrids. Fungal ancestors may have had an alternation-of-generations life cycle, as do some chytrids. **3.** Human skin is dry and the cells that make up the skin are constantly being shed. Humans keep their skin relatively clean which discourages fungal growth. Lungs, however, provide a moist and protected environment that is well vascularized and supportive of fungal growth.

CHAPTER 19

Check Your Progress

(19.1) 1. Multicellularity; development of tissues and cellular organization into germ layers; differences in symmetry—bilateral and radial; cephalization—localization of brain and specialized sensory receptors; development of a coelom; segmentation—repetition of body parts along the length of the body. **2.** Animals are chemoheterotrophs that digest their food with a digestive system. They reproduce sexually to produce a fertilized diploid egg that undergoes a complex order of developmental steps. Specialized tissues develop, such as muscles and nerves, which provide the ability to move. Plants can photosynthesize but cannot move, and fungi digest their food externally and absorb the nutrients. **3.** Protostomes are characterized by spiral cell division when the embryo first forms and by a blastopore that becomes the mouth. If a coelom is present, it is formed by the splitting of the mesoderm. Deuterostomes are characterized by radial cell division when the embryo first forms and a blastopore that becomes the anus. Outpocketing of the primitive gut forms the coelom. **(19.2) 1.** Both sponges and cnidarians are multicellular. Sponges have a cellular level of organization with no true tissues, and they exhibit no symmetry in body form. Cnidarians have a tissue level of organization with two germ layers (ectoderm and an endoderm), and they are radially symmetrical. **2.** Both sponges and cnidarians are multicellular invertebrates. Sponges have saclike bodies perforated by many pores through which water enters and circulates past collar cells before exiting through the osculum. They lack organized tissues. Cnidarians are radially symmetrical and have tentacles with cnidocytes. They have either a polyp or medusa body form, two germ layers, and tissues as adults. **3.** Sponges lack organized tissues and exhibit a cellular level of organization. They are regarded as parazoans at the base of the evolutionary tree of animals. Cnidarians are radially symmetrical, have two germ layers, and exhibit a tissue level organization. They are the first of the eumatazoans. **(19.3) 1.** The *lopho* portion of the name is derived from a tentacle-like feeding structure called a lophophore. The *trocho* portion of the name refers to a larval stage, called a trochophore that is characterized by a distinct band of cilia. Flatworms, molluscs, and annelids all share a common evolutionary ancestor that has these structures. **2.** Flatworms and annelids are bilaterally symmetrical and have three germ layers. Flatworms are acoelomates. Annelids are segmented and have a coelom and a closed circulatory system. **3.** Flatworms are acoelomates; molluscs and annelids are coelomates. **(19.4) 1.** Roundworms secrete a

nonliving cuticle and undergo molting. They have a pseudocoelom. **2.** (1) Jointed appendages; (2) exoskeleton; (3) segmentation; (4) well-developed nervous system; (5) variety of respiratory organs; (6) metamorphosis. **3.** Insects are so numerous and diverse because they show remarkable structural and behavioral adaptations that allow them to live in many different habitats. **(19.5) 1.** Echinoderms have radial symmetry as adults and a water vascular system with tube feet for locomotion. They lack a head and a brain and are not segmented. **2.** The chordates have a dorsal supporting rod, dorsal tubular nerve cord, pharyngeal pouches, and a postanal tail. Mammals are amniotes and share an ancestor with reptiles. **3.** Tunicates—four chordate characteristics as larvae; lancelets—four chordate characteristics as adults and segmentation; jawless fishes—vertebrae; cartilaginous fishes—jaws; bony fishes—skeleton; lobe-finned fishes—lungs; amphibians—limbs; reptiles and birds—amniotic egg; mammals—mammary glands. **4.** The amniotic egg eliminated the need for a water environment during development, which made development on land possible. It provides gas exchange, nutrients, removal of nitrogenous wastes, and protection from desiccation and mechanical harm. **5.** Mammals have exhibited multiple adaptations that have allowed them to live in many different environments. **(19.6) 1.** The anthropoids include the monkeys, apes, and humans; hominins just the species *Homo sapiens* and our close humanlike ancestors. **2.** Among primates, the evolutionary trend was toward larger and more complex brains. The development of the ability to stand erect and bipedalism is the defining event of the origins of the hominins. **3.** The replacement model proposes that modern humans evolved from archaic humans in Africa. After that, the modern humans migrated to Asia and Europe, where they replaced the archaic species already present there.

Testing Yourself

1. b; **2.** a; **3.** d; **4.** b; **5.** a; **6.** a; **7.** a; **8.** a; **9.** b; **10.** d; **11.** c; **12.** a; **13.** b; **14.** d; **15.** b; **16.** c

Thinking Critically

1. The recent discoveries support the idea that a lineage in Africa from primitive to modern humans existed with branches of this lineage continuing outside Africa. The discovery of new *Homo* fossils in Asia and Europe would provide more information on the process of human evolution. Efforts at expanding the fossil record from Asia, Europe, and Oceania, as well as Africa, especially from 1 million years ago to relatively recently, should be continued. **2.** Animals that are sessile tend to be radially symmetrical because their food comes to them from all directions. There is no need to have anterior and posterior body regions. Animals that move through their environment are bilaterally symmetrical, with the anterior portion containing sensory receptors. This allows the animal to sense and respond to the environment as it travels through it. **3.** Genetic information from DNA sequencing can more specifically identify relationships between animal species than can physical appearance or characteristics, which are subject to variations in genetic expression.

CHAPTER 20

Check Your Progress

(20.1) 1. Epidermal tissue, or epidermis, is a layer of closely packed cells that provides the outer protective covering of a plant. Root hairs, guard cells, and cork cells are specialized epidermal cells. Ground tissue contains three types of cells. Parenchyma cells are found in all organs of a plant. They sometimes contain chloroplasts and carry out photosynthesis, or they have vacuoles used to store the products of photosynthesis. They help to carry out the functions of a particular organ. Collenchyma cells have irregular corners and thicker cell walls. They form the internal bulk of leaves, stems, and roots and provide flexible support for immature regions of the plant. Sclerenchyma cells have a thick secondary wall, are nonliving, and contain lignin. They provide support for mature regions of the plant. Vascular tissue is xylem or phloem. Xylem contains vessel elements and tracheids, and phloem is made up of sieve-tube members and companion cells. Both types of vascular tissue transport water and nutrients as well as provide support for the plant. **2.** Xylem transports water and minerals from roots to leaves. Phloem transports organic compounds throughout the plant. **3.** In roots, epidermal cells may form root hairs, which increase the surface area of roots for increased absorption of water and minerals. In leaves, epidermal cells can be modified to form guard cells, which surround stomata. These openings regulate water loss and gas exchange. Trichomes are prickly projections formed by epidermal cells that deter insect herbivores. The epidermis in the trunk of a tree can be replaced by cork, which is a part of the bark. **(20.2) 1.** The root system is composed of roots, which stabilize the plant and are the site of water and mineral absorption. The root tips contain an apical meristem that produces primary growth downward. The shoot system is composed of the stem, leaves, flowers, and fruit. The stem supports the leaves, transports materials between roots and leaves, and produces new tissue via the apical meristem at the terminal bud. The leaves are the main site of photosynthesis, and the flowers and fruit are organs involved in reproduction. **2.** A cotyledon is an embryonic leaf present in seeds of flowering plants. In monocots, which have one cotyledon, the cotyledon stores some nutrients and acts as a transfer tissue for nutrients stored elsewhere. In eudicots, which have two cotyledons, the cotyledons supply nutrients for developing seedlings. **3.** In monocot roots, the vascular tissue is arranged in a ring around the center. In eudicot roots, the vascular tissue is in the center. The xylem forms a star shape and the phloem is located between the points of the star. In the monocot stem, bundles of vascular tissue are scattered throughout, but in eudicots the bundles are arranged in a ring close to the edge. Leaf veins are parallel in monocots and form a netlike pattern in eudicots. **(20.3) 1.** The mesophyll, which is the site of photosynthesis, has a large surface area for gas exchange and water loss. The leaf contains stomata for gas exchange and veins that carry nutrients to and from the leaf. Some leaves are broad and thin, providing large surface areas for light absorption and gas exchange. The waxy cuticle of the leaf prevents water loss. **2.** Wood is secondary xylem made by the vascular cambium. Bark is made up of cork, cork cambium, cortex, and phloem. **3.** The vascular tissue

of eudicot roots contains xylem that is arranged in a star in the center and transports water and minerals and phloem that is arranged between the arms of the star and transports nutrients. The endodermis forms a layer that forces water and minerals to pass through the endodermal cells. The pericycle is a layer of cells inside the endodermis that continue to divide and form lateral roots. The cortex of the root contains starch granules and may function in food storage. The epidermis forms the outer layer of the root, and in the zone of maturation, its cells may have root hairs. **(20.4) 1.** Essential nutrients for plants are divided into macronutrients and micronutrients according to their relative concentration in plant tissue. Plants need a larger amount of macronutrients than micronutrients. **2.** Macronutrients are C, H, O, P, K, N, S, Ca, Fe (in some plants), and Mg. **3.** The bacteria within root nodules fix atmospheric nitrogen, changing it to nitrate or ammonium, which the plant can use. The plant provides carbohydrates to the bacteria. The hyphae of the mycorrhizal fungi have a large surface area for water uptake, and they break down organic matter. The roots of the plant provide sugars and amino acids to the fungi. These mutualistic relationships are beneficial to both the plants and the bacteria or fungi. **(20.5) 1.** Transpiration, the evaporation of water at the leaves, causes water to move in the xylem from the roots upward throughout the plant. **2.** A continuous water column exists in xylem because of water's properties of cohesion and adhesion. Without a continuous water column, transpiration would be unable to pull water from roots to the leaves. **3.** Sugar is actively transported into phloem at a source. When water follows by osmosis, pressure builds and causes the phloem contents to flow toward a sink, where the sugar is transported out of the phloem.

Testing Yourself

1. b; **2.** c; **3.** d; **4. a.** terminal bud; **b.** blade (leaf); **c.** lateral bud; **d.** node; **e.** shoot system; **f.** vascular tissues; **g.** root system; **5.** b; **6.** d; **7.** b; **8.** c; **9.** c; **10.** c; **11.** c; **12.** d

BioNow

In this video, xylem tissue from oak and pine was used, which is involved in the transport of water and minerals. The question was whether the pits in the tracheids, which along with vessel elements make up xylem tissue, were small enough to filter salt out of water as it passed through the xylem. The conclusion was that the pits are not small enough to filter salt out of water.

Thinking Critically

1. There are unique proteins on the surfaces of the nitrogen-fixing bacteria that the plant recognizes. **2.** When the stomata open at night, both CO_2 and the water present in the fog move into the leaves, allowing the plant to acquire enough moisture to survive. The opening of the stomata at night is an adaption to the dry environment, in which moist fog is present only at night.

CHAPTER 21

Check Your Progress

(21.1) 1. Auxins, gibberellins, cytokinins, abscisic acid, and ethylene. **2.** Auxins—maintain apical dominance, involved in phototropism and gravitropism, promote root formation and fruit growth; gibberellins—elongation of cells, break

seed and bud dormancy; cytokinins—promote cell division, prevent aging, and with auxins promote cell differentiation; abscisic acid—maintains seed and bud dormancy and closes stomata; ethylene—promotes fruit ripening and abscission. **3.** The hormones are found in those areas of the plant where they are most effective at promoting organized growth that benefits the plant. **(21.2) 1.** Phototropism, which is a positive tropism, is the unidirectional growth of plants toward a source of light. Thigmotropism is a growth response by a plant to touch, whether from an animal, wind, rocks, or another plant. It can be positive or negative. Gravitropism is a growth response to gravity. Shoots exhibit negative gravitropism by growing upward against gravity, and roots exhibit positive gravitropism by growing downward with gravity. **2.** Short-day plants flower when the night becomes longer than a critical length. Long-day plants flower when the night becomes shorter than a critical length. **3.** In the daytime, more red light is present than far-red light. Inactive phytochrome exposed to red light is converted into its active form. In the evening, when far-red light is more available, the active phytochrome is converted into its inactive form. This is the mechanism by which plants can detect night length. **(21.3) 1.** Sepals protect the bud as the flower develops; petals attract pollinators; a stamen consists of anther and filament and is the "male" portion; a carpel has stigma, style, ovary, and ovules and is the "female" portion. **2.** After fertilization of haploid egg and sperm, the diploid embryo in the seed goes through mitosis and develops into a sporophyte. The sporophyte produces haploid microspores and megaspores by meiosis. Through mitosis, the megaspore develops into the female gametophyte, the embryo sac, which produces the egg, and the microspore develops into the male gametophyte, the pollen grain, which produces the sperm. **3.** The seed provides nourishment for the embryo, and the coat protects it from desiccation. It is often enclosed by a fruit. These characteristics help the seed stay viable and assist in its dispersal. Dispersal of seeds helps plants succeed in their environments. **(21.4) 1.** Bulbs, rhizomes, corms, buds on tubers, and runners. **2.** Propagation of commercial crops in sterile culture; conservation of plants normally harvested in the wild; propagation of rare species of plants to be put back into the wild. **3.** Pros—GM plants may be more tolerant of salt, drought, and cold; they may be more disease- and pest-resistant; they may have improved yields and nutritional quality; and they may also make products of medical importance to humans. Cons—people might be allergic to GM plants and have unforeseen reactions to them. GM plants could upset the interaction of species and harm the environment.

Testing Yourself

1. d; **2.** a; **3.** e; **4.** a; **5.** a; **6.** e; **7.** b; **8.** c; **9.** a; **10.** b; **11.** d; **12.** b; **13.** e; **14.** d; **15.** b; **16.** d; **17.** e

Thinking Critically

1. When plants are shipped long distances, they produce ethylene as a result of stress. Ethylene builds up within the shipping container and the plastic sleeves near the plant, amplifying the abscission response. **2.** You could determine whether flowers or stems are responsible by cutting the flowers off the stems and looking to see whether the stems still exhibit sun tracking. If so, then the stems rather than the flowers are responsible. You could determine which portion of the stem is responsible for sun tracking by shading different portions of the stem and determining which treatment blocks the response. **3.** Until the problem of acquisition of herbicide resistance by johnson grass is resolved, the production of herbicide-resistant sorghum should be banned. However, in the interim, African farmers should be supported in new ways to manage witchweed in their farming practices so as to ensure their well-being.

CHAPTER 22

Check Your Progress

(22.1) 1. A tissue is composed of cells that are of the same structural and functional type. An organ is made up of different tissues, each of which contributes to the functioning of the whole organ. **2.** Epithelial tissue covers and lines organs and covers the body's surface; it forms a barrier that can withstand a high amount of wear and tear, and its cells can divide frequently to renew it. **3.** Connective tissue is needed to hold organs and parts of the body together, giving the body structure, support, and protection against physical damage. **4.** Skeletal muscle is striated, is attached to the skeleton to produce movement, has fibers that are cylindrical and multinucleated, and is under voluntary control. Cardiac muscle is found in the walls of the heart. Its fibers are striated, branched, and uninucleated and are under involuntary control. Smooth muscle is found in the walls of the digestive system and other internal organs. Its fibers are spindle-shaped, nonstriated, and uninucleated, and it is under involuntary control. **5.** Nerves communicate information from sensory receptors to the spinal cord and brain. There the information is integrated and interpreted. Nerves then conduct impulses to the muscles and glands, causing a response of contraction or secretion. Information about the external and internal environments is coordinated by nervous tissue. **(22.2) 1.** Organ systems are composed of organs, which are composed of tissues. The functions of an organ system are dependent on its organs functioning together. **2.** Both the nervous and endocrine systems coordinate and regulate the functions of other systems in the body. **3.** Cardiovascular—transports nutrients and gases and removes wastes; lymphatic—absorbs fat from the digestive system and collects excess interstitial fluid; immune—protects the body from disease; respiratory—exchanges gases with blood; urinary—rids blood of wastes; digestive—receives food and digests into nutrient molecules for entry into the blood; nervous—receives stimuli and conducts impulses; endocrine—secretes hormones and regulates functions of other systems; integumentary—provides protection and houses sensory receptors; skeletal—protection, mineral storage, blood cell production, movement; muscular—movement; reproductive—production of sex cells and offspring. **(22.3) 1.** Homeostasis is required to maintain constant conditions in the internal environment as an organism is exposed to varying conditions in the external environment. **2.** A sensor detects a change in the internal environment and communicates that information to a control center, which initiates an effect that brings conditions back to normal. At that point, the sensor is no longer activated. **3.** With a drop in temperature, the control center (hypothalamus)

brings about constriction of the blood vessels in the skin. If the temperature continues to drop, shivering begins in order to generate heat. With a rise in temperature, blood vessels in the skin dilate and sweat glands are activated.

Testing Yourself

1. a. cell; **b.** tissue; **c.** organ; **d.** organ system; **e.** organism; **2.** b; **3.** a, b; **4.** a; **5.** c, d; **6.** c; **7. a.** dendrite; **b.** cell body; **c.** nucleus; **d.** nucleus of Schwann cell; **e.** axon **8.** d; **9.** d; **10.** d; **11.** c; **12.** a; **13.** b; **14.** e; **15.** a

Thinking Critically

1. Even though the heart is made of cardiac muscle instead of skeletal muscle, the link between muscular dystrophy and heart disease is that both types of muscle tissue deteriorate in a similar manner. **2.** Organisms have to interact with the external environment to obtain energy and nutrients for growth and reproduction, to find shelter, and, sometimes, to find mates. These interactions bring the risk of upsetting the internal environment. However, without interacting with the external environment, many of the functions that define a living organism would be lost. Without interaction with the environment, plants could not receive energy from the sun for photosynthesis; animals could not get nutrients from eating food or sense danger from predators. **3.** The advantage of having tissues, organs, and organ systems is that their functions can be separated, made more specific, and made more efficient. Organisms with organ systems are selected for because they have greater fitness. Just as a large company uses specialized divisions to carry out its functions, so animals use specialized organs that work together within a system complementing other systems. **4.** Upon death, cells, tissues, organs, and organ systems no longer function. Without functioning organ systems, the homeostasis that is essential to life cannot be maintained.

CHAPTER 23

Check Your Progress

(23.1) 1. Animals that have all their cells exposed to water do not need a circulatory system because they can adequately exchange gases, obtain nutrients, and get rid of wastes by diffusion across cell membranes. **2.** Both types of circulatory systems use a heart to pump fluid. An open system pumps hemolymph through channels and cavities that open up to the body cavity. The hemolymph eventually drains back to the heart. A closed system pumps blood through closed vessels that carry blood both away from and back to the heart. **3.** The pulmonary circuit carries blood to and from the lungs, and the systemic circuit carries blood to and from the tissues. **4.** In a fish, the blood moves in a single circuit through a heart that has a single atrium and ventricle. In this circuit, blood moves through the capillaries of the gills, where gas exchange occurs. In mammals, the pulmonary and systemic circuits are separate. The heart functions as two coordinated pumps. **(23.2) 1.** Phase 1 (atrial systole): the atria contract and pass blood to the ventricles. Phase 2 (ventricular systole): the ventricles contract and blood moves into the attached arteries. Phase 3 (atrial and ventricular diastole): both the atria and ventricles relax while the atria fill with blood. **2.** Arteries have thick walls with a middle layer of elastic fibers and smooth muscle

that allows for expansion and contraction in response to pulses of pressure created by the contraction of the heart. Capillary walls are composed of an epithelium only one cell thick, which allows for the diffusion of gases, nutrients, and wastes in and out of the blood. Veins have thinner, less elastic walls than arteries. They have valves and carry blood back to the heart under relatively low pressure. **3.** Blood flows from the right atrium into the right ventricle and then through the pulmonary artery to the lungs. The blood is oxygenated in the capillaries of the lungs and next passes through the pulmonary vein to the left atrium of the heart. With the contraction of the left atrium, the blood flows into the left ventricle and then out through the aorta to the tissues of the body, where it passes through capillary beds, delivering oxygen to the tissues. Veins receive the blood from the capillaries and return the blood back to the right atrium via the inferior and superior venae cavae. **4.** The lymphatic system takes up fat from the intestines, it works with the immune system, and it takes up excess interstitial fluid and returns it to cardiovascular veins via the subclavian veins. **5.** At the arterial end of a capillary, higher blood pressure forces water out of the blood. In the midsection of the capillary, oxygen and carbon dioxide diffuse across the capillary wall according to their concentration gradients. At the venous end of the capillary, the osmotic pressure inside the capillary is greater than the blood pressure and water moves back into the blood. **(23.3.) 1.** Plasma, the liquid portion of blood, contains salts and proteins that buffer the blood and maintain its osmotic pressure. Some of the plasma proteins are involved in blood clotting and the transport of large organic molecules. Nutrients are also carried in the plasma. Red blood cells transport oxygen to the tissues. White blood cells function to defend the body against infection. Platelets play a role in blood clotting. **2.** All white blood cells are involved in the immune response. Granular leukocytes include neutrophils, which phagocytize pathogens and cellular debris; eosinophils, which digest large pathogens, such as worms, and reduce inflammation; and basophils, which promote blood flow to injured tissues during the inflammatory response. Agranular leukocytes are lymphocytes, which are responsible for specific immunity and come in two types (B cells produce antibodies and T cells destroy cancer and virus-infected cells), or monocytes, which become macrophages that phagocytize pathogens and cellular debris. **3.** When a blood vessel is damaged, blood containing platelets leaks out. The platelets clump to form a plug at the injury site. Platelets and the injured tissue release prothrombin activator, which converts prothrombin to thrombin. Thrombin acts to convert fibrinogen to fibrin, which forms threads that trap red blood cells to form a clot. Clot retraction follows. **4.** Hypertension can be treated with medication and changes in diet and exercise. Atherosclerosis is an accumulation of soft masses of fatty materials beneath the lining of the arteries. Treatments include medication, surgery, and changes in diet and exercise.

Testing Yourself

1. c; **2.** a; **3.** d; **4. a.** pulmonary vein; **b.** aorta; **c.** renal artery; **d.** lymphatic vessel; **e.** pulmonary artery; **f.** superior vena cava; **g.** inferior vena cava; **h.** hepatic vein; **i.** hepatic portal vein; **j.** renal vein; **5.** a; **6. a.** superior vena cava; **b.** aortic semilunar valve; **c.** pulmonary semilunar valve; **d.** right atrium;

e. tricuspid valve; **f.** right ventricle; **g.** inferior vena cava; **h.** aorta; **i.** pulmonary trunk; **j.** pulmonary arteries; **k.** pulmonary veins; **l.** left atrium; **m.** bicuspid valve; **n.** septum; **o.** left ventricle; **7.** c; **8.** d; **9.** a; **10.** c; **11.** b; **12.** c; **13.** b; **14.** b; **15.** b

BioNow

The wound was on the neck and involved damage to the trachea and the carotid artery. Blood flow to the head and neck would be severely disrupted.

Thinking Critically

1. A 20-year-old on his or her birthday has been alive for (20 years × 365 days/year × 24 hours/day × 60 min/day) 1.05×10^7 minutes and experienced 7.4×10^8 heartbeats. In that lifetime, the individual's heart has pumped 5.5×10^7 liters of blood. **2.** The four-chambered heart keeps O_2-poor and O_2-rich blood separate. This allows the maximum amount of oxygen to be delivered to the cells. This in turn allows for the greatest amount of energy to be produced. A four-chambered heart works with greater efficiency to support the high metabolism of an endothermic lifestyle. **3.** Lower oxygen levels found at higher altitudes stimulate an increased production of red blood cells and hemoglobin. When athletes train at higher altitudes long enough for this to take effect, they will have the ability to carry a higher amount of oxygen in their bloodstream. This effect will continue even at lower altitudes for a few months after the high altitude training has stopped. More oxygen means more energy through aerobic cellular respiration for muscle contraction. **4. a.** Long-term hypertension damages arteries and increases the likelihood of plaque formation. The heart has to work harder to move the blood. The increased pressure is particularly damaging to capillary beds present in the kidneys and in the retina of the eye. **b.** The plaques that result from atherosclerosis interfere with blood flow and, combined with hypertension, further strain the heart. The plaques cause arteries to narrow and harden, and the tissues supplied by the arteries will get less blood, oxygen, and nutrients. If coronary arteries are narrowed or blocked, the heart muscle cannot function and some of its tissue can die during a heart attack.

CHAPTER 24

Check Your Progress

(24.1) 1. Breathing is the movement of air into the lungs (inspiration) and out of the lungs (expiration). External exchange of gases occurs between the air and the blood within the lungs. Internal exchange of gases occurs between blood and interstitial fluid and also between cells and interstitial fluid. **2.** In the upper respiratory tract, air enters the nose and moves through the nasal cavities to the pharynx and into the larynx. From there, the air moves to the lower respiratory tract, starting with the trachea, then to the bronchi, bronchioles, and finally the alveoli. **3.** Oxygen follows its concentration gradient as it moves from a higher concentration in the lungs to a lower concentration in the blood. When carbon dioxide from the tissues enters the blood, it combines with water to form carbonic acid, which breaks down to bicarbonate ion and H^+. The H^+ binds to the globin portion of hemoglobin, and the bicarbonate ion is carried in the plasma. **(24.2) 1.** The kidney functions are excretion of nitrogenous wastes, maintenance of

the water-salt balance of blood, and maintenance of the acid-base balance of blood. **2.** A nephron consists of the glomerular capsule, which functions in filtration; the proximal convoluted tubule and nephron loop, which function in reabsorption; and the distal tubule, which functions in secretion. **3.** During filtration, small molecules such as water, nutrients, salts, and urea move from the blood capillaries into the glomerular capsule as a result of blood pressure. During reabsorption, glucose, amino acids, and other nutrients move from the proximal convoluted tubule back into the blood capillaries by active transport. Salts and water are reabsorbed by a combination of active transport, osmosis, and protein channels. During secretion, substances such as uric acid, hydrogen ions, ammonia, and penicillin are moved from the blood capillaries into the distal convoluted tubule by means other than filtration. **4.** Water-salt balance and pH balance are important to homeostasis because blood volume and composition must be stable in order to maintain circulation and proper gas, nutrient, and waste exchange.

Testing Yourself

1. a. nostril; **b.** nasal cavity; **c.** pharynx; **d.** epiglottis; **e.** glottis; **f.** larynx; **g.** trachea; **h.** bronchus; **i.** bronchiole; **j.** alveoli; **k.** lung; **l.** diaphragm; **2.** a; **3.** d; **4.** a; **5. a.** kidney; **b.** ureter; **c.** bladder; **d.** urethra; **6.** c; **7.** d; **8.** b; **9.** a; **10.** d

Thinking Critically

1. If the tissues were not receiving much oxygen due to CO binding to hemoglobin, then the cause of death would be asphyxiation. The tissues would not be receiving enough oxygen to carry out cellular respiration. **2.** If a low-salt diet lowered the high blood pressure, that would indicate that the hypertension caused the kidney damage. If a low-salt diet failed to lower blood pressure, then it would appear that kidney damage caused the hypertension. **3.** Grasshoppers have a tracheal system consisting of air tubes to carry out gas exchange. Their exoskeleton is both airtight and waterproof. They excrete uric acid, which requires little water, and their Malphighian tubules absorb most of the water back into their bodies, which helps them avoid desiccation in dry conditions on land.

CHAPTER 25

Check Your Progress

(25.1) 1. Ingesting food; breaking down food into molecules that can be transported; absorbing nutrient molecules; eliminating indigestible materials. **2.** The small intestine is longer and narrower than the large intestine. Its folded lining is covered by villi containing microvilli. The pancreas, liver, and gallbladder are connected to the small intestine by ducts. In the small intestine, proteins and carbohydrates are broken down to smaller molecules, and fats are emulsified. The smaller molecules are then absorbed into the microvilli. The large intestine begins with a cecum to which the appendix is attached. The large intestine absorbs water, salts, and some vitamins. It houses bacteria that help break down food and produce some vitamins. It stores indigestible material that is then eliminated at the anus. **3.** The duodenum is the first part of the small intestine and is connected to the pancreas and liver by ducts. The liver produces bile, which is stored in

the gallbladder and eventually secreted into the small intestine. The pancreas produces pancreatic juice containing sodium bicarbonate and digestive enzymes, which is secreted into the small intestine. **4.** The salivary glands secrete saliva containing digestive enzymes into the mouth by way of ducts. This starts the chemical digestion of food. The bile produced by the liver and stored in the gallbladder enters the small intestine and emulsifies fats. The pancreatic juices enter the small intestine at the duodenum. The bicarbonate neutralizes the chyme, and the enzymes break down food molecules. **5.** Amylase digests starch to maltose; pepsin hydrolyses proteins to peptides in the stomach; trypsin hydrolyzes proteins to peptides in the small intestine; peptidases digest peptides to amino acids; lipase digests fat droplets to glycerol and fatty acids. **(25.2) 1.** Macronutrients are needed in large quantities to supply energy; they include carbohydrates, lipids, and proteins. Micronutrients are needed in small quantities and do not supply energy; they include vitamins and minerals such as vitamin D and calcium. **2.** Junk foods do contain nutrients; however, their sugar and/or fat content may be out of proportion to their vitamin and mineral content. They may provide just empty calories. **3.** Essential nutrients must be obtained from the diet because the body cannot synthesize them or cannot produce them in sufficient quantities to supply its needs. **(25.3) 1.** Carbohydrates, sugars and starches; lipids, oils; proteins, meat or fish; minerals, vegetables; vitamins, fruits or vegetables; water, beverages. **2.** Consuming too many carbohydrates can lead to obesity, which is associated with type 2 diabetes and cardiovascular disease; too many lipids can lead to elevated lipid levels and increased risk of cardiovascular disease as well as obesity; too much protein can lead to increased calcium loss in urine, formation of kidney stones, and obesity. **3.** Without vitamins and minerals, proper organ development, as well as proper protein synthesis and function, will not occur. Without enough water, the chemical reactions in the body, joint lubrication, nutrient transport, and temperature regulation, will be compromised. **(25.4) 1.** The guidelines released in 2015 focus on establishing healthy eating patterns for people rather than recommending specific amounts of nutrients. The goals are to promote health, prevent chronic long-term disease, and help people achieve and maintain a healthy weight. **2.** The MyPlate graphic depicts the desired proportions of the five food groups (fruits, vegetables, grains, protein, and dairy) for daily consumption. A related website, ChooseMyPlate.gov, provides detailed information about each of the food groups and guidance for following the dietary recommendations. **3.** The "Nutrition Facts" label states serving size, servings per container, calories per serving, calories from fat (current label), the % daily value in a serving and amounts for fats, cholesterol, sodium, and carbohydrates, the amount of protein, and the % daily value (and amounts in proposed label) for some vitamins and minerals. The current label recommends intake amounts for fat, cholesterol, sodium, and carbohydrate as well as the Calories/gram for each type of macronutrient. The proposed label will have a footnote on daily values and caloric values. **(25.5) 1.** BMI = 26.1 **2.** Type 2 diabetes often occurs in people who are obese and have impaired insulin production and insulin resistance. With insulin resistance, cells cannot effectively take up and metabolize glucose. The glucose builds up in the blood and appears in the urine. Obesity can lead to cardiovascular disease, which because of the buildup of plaque in arteries, leads to hypertension, heart attack, and stroke. **3.** Each person varies in metabolism, activity level, and energy needs, so it is important to balance your daily energy expenditure with your intake of food.

Testing Yourself

1. a. salivary gland; **b.** mouth; **c.** liver; **d.** gallbladder; **e.** duodenum; **f.** cecum; **g.** appendix; **h.** anus; **i.** pharynx; **j.** esophagus; **k.** diaphragm; **l.** stomach; **m.** pancreas; **n.** large intestine; **o.** small intestine; **p.** rectum; **2.** b; **3.** b; **4.** a; **5.** d; **6.** b; **7.** c; **8.** a; **9.** a; **10.** f; **11.** e; **12.** b; **13.** b; **14.** c; **15.** c; **16.** b; **17.** a; **18.** d

Thinking Critically

1. Bariatric surgery alters the physical structure and therefore the function of the stomach and small intestine. It can lead to incomplete digestion and malabsorption of nutrients. Protein digestion, which starts in the stomach and ends in the small intestine, can be affected as well as the absorption of amino acids in the small intestine. The availability of iron, vitamin B_{12}, folate, and calcium, and therefore their absorption, are often altered by bariatric surgery. **2.** Marketing strategies that focus on selling unhealthful foods to children should be regulated, similar to the way advertising of tobacco products and alcohol must follow certain regulations. However, because food must be consumed and not just eliminated, because people have enormous physiological and mental investments in food, and because marketing tactics are subtle and pervasive, innumerable obstacles to regulation of food advertising exist. The advertisements should be required to contain an educational component that informs the consumer of what the product contains and how much should be consumed. Due to children being particularly susceptible to marketing strategies, access to children's markets should be restricted.

CHAPTER 26

Check Your Progress

(26.1) 1. The immune system functions to maintain health by fighting infections and destroying cancer cells. **2.** Red bone marrow—produces all types of blood cells; thymus—aids in maturation of T lymphocytes and tests their ability to recognize self- versus nonself cells; spleen—filters pathogens and debris from the blood and contains lymphocytes; lymph nodes—house macrophages that remove pathogens and antigens from lymph and contain lymphocytes. **3.** Antigens are protein or carbohydrate molecules that stimulate the immune system to respond to foreign substances, pathogens, and cancer cells. This type of response constitutes immunity. **(26.2) 1.** Physical barriers include the skin with its sweat and oil glands and the mucous membranes of the respiratory, digestive, reproductive, and urinary tracts. The acidity of the stomach inactivates or kills harmful bacteria. Bacteria that reside in the large intestine and vagina prevent pathogenic organisms from becoming established there. **2.** Damaged cells and mast cells release chemicals that cause capillaries to dilate and

become more permeable. More blood moves to the area, and more fluid moves into the tissues. Neutrophils and macrophages migrate to the site of injury and phagocytize pathogens. **3.** Complement proteins bind to mast cells to trigger histamine release, attract phagocytes to the site of injury, bind to pathogens already coated with antibodies to ensure that they are phagocytized, and form a membrane attack complex, which produces a hole in the surface of a microbe. These functions "complement" certain immune responses and are not specific to one pathogen. NK cells kill virus-infected cells and cancer cells, and they are not specific. They bind to cells that have lost their ability to make self proteins. Their numbers do not increase, and they do not recognize antigens from previous exposure to them. **(26.3) 1.** The antibodies produced by plasma cells bind to specific antigens. Such antigen-antibody complexes mark the antigen for destruction. A person's immune system becomes adapted to recognize a particular antigen so that the person becomes immune to it. **2.** B cells are produced and mature in the bone marrow. They give rise to plasma cells that produce antibodies. The antigen-antibody complexes are engulfed by neutrophils or macrophages or may activate complement. T cells are produced in the bone marrow and mature in the thymus as cytotoxic T cells, which destroy virus-infected or cancer cells, or helper T cells, which regulate immune function. **3.** The antibody response uses proteins called antibodies produced by B cells to fight pathogens. The cellular response uses entire cells, T cells, to fight pathogens. **4.** MHC markers are unique to each individual, except identical twins, and allow the immune system to identify self from nonself cells. If the MHC markers are nonself, tissue rejection can occur. **(26.4) 1.** A vaccine is a substance prepared from a pathogen or its products that, when introduced into an individual, stimulates the immune system to create lasting immunity against the pathogen. **2.** A booster, or second dose of a vaccine, is sometimes needed to stimulate the immune response so as to raise the plasma antibody concentration to a level that is effective at fighting the pathogen. **3.** Active immunity involves the immune system being exposed to a pathogen or vaccine and responding to it by producing plasma cells and antibodies. Memory cells are also produced, ensuring long-term immunity. Passive immunity involves introducing into the body prepared antibodies against a pathogen, ones that the individual did not produce. Passive immunity is temporary. **(26.5) 1.** Allergies are hypersensitive responses of individuals to antigens called allergens, which produce the response. **2.** An immediate allergic response occurs within seconds of exposure to an allergen and is caused by chemicals, including histamine, that are released by mast cells. A delayed allergic response takes longer to develop and is probably initiated by memory T cells. **3.** An AIDS patient has a compromised immune system due to the HIV virus living in and destroying helper T cells, which are necessary for the activity of all other immune system cells.

Testing Yourself

1. a; **2.** c; **3.** a; **4.** b; **5.** e; **6.** e; **7.** c; **8.** e; **9.** b; **10.** a; **11.** a; **12.** e; **13.** d; **14.** b; **15.** e; **16.** b; **17.** b; **18.** c

Thinking Critically

1. The tonsils are composed of lymphatic tissue and help to capture pathogens as they enter the digestive and respiratory tracts. Because of their location and function, they often contain bacteria or viruses that at times can overwhelm the lymphatic tissue and cause an infection. Either through the body's usual defenses or through help from medication, individuals can usually recover from these infections. The tonsils, once healthy, will again aid the body in the fight against infection. Only if they become chronically infected so as to obstruct airways and no longer perform their functions should they be removed. **2.** Immunosuppressive drugs dampen the body's response not only to the transplanted tissue but to other foreign and infected cells as well. Therefore, the body's ability to fight infections is weakened and the transplant patient becomes more susceptible to all types of infections. **3. a.** Drug companies have an obligation to price their products fairly to avoid exploitation of consumers. However, they need to support their operations as well as conduct research and development of new pharmaceuticals. Nonetheless, they should not seek excessive profits for their shareholders or excessive compensation for their executives. **b.** It is best for everyone if patent laws are respected since they provide incentives for businesses to operate and develop new products. However, when human lives are being lost due to egregious corporate policies, then governments need to step in and regulate the drug companies.

CHAPTER 27

Check Your Progress

(27.1) 1. Sensory neurons transmit nerve impulses from sensory receptors to the central nervous system. Interneurons carry nerve impulses between parts of the central nervous system. Motor neurons carry nerve impulses from the central nervous system to muscles or glands. **2.** The cell body contains the nucleus and other organelles. Dendrites receive signals from sensory receptors or other neurons. An axon is a long extension that conducts nerve impulses and is often covered by a myelin sheath. **3.** First, Na^+ gates open, and Na^+ moves to the inside of the axon, which becomes positive, reversing the charge difference across the axon membrane. Second, K^+ moves to the outside, and the inside of the axon becomes negative again. **4.** Nerve impulses arriving at the axon terminal of the presynaptic cell cause synaptic vesicles to fuse with the axon membrane and release a neurotransmitter into the synaptic cleft. The neurotransmitter molecules then diffuse across the synaptic cleft and bind to receptors in the membrane of the postsynaptic cell. **5.** The central nervous system consists of the brain (cerebrum, diencephalon, cerebellum, brain stem) and spinal cord. The brain receives sensory information, integrates it, and sends out motor instructions to glands and muscles. The spinal cord conducts nerve impulses between the brain and spinal nerves. **6.** A sensory input generates a nerve impulse that moves along sensory axons to the spinal cord. The signal is passed on to many interneurons, some of which synapse with motor neurons. The nerve impulses travel along the axons of the motor neurons to an effector, which brings about a response to the stimulus. Other interneurons carry nerve impulses to the brain, which results in the sensation of the stimulus. **7.** The parasympathetic division "keeps house" by automatically and involuntarily maintaining the internal responses associated with a relaxed state, such as pupil contraction, food digestion, and slow heartbeat. **(27.2) 1.** Steroid hormones enter the cell through the plasma membrane, bind to a receptor within the cell, and then move to the nucleus to stimulate genes to produce proteins that alter cell activity. Peptide hormones bind to a receptor in the plasma membrane, which then induces a signal transduction pathway leading to the creation of a second messenger, which affects the metabolism of the cell. **2.** The hypothalamus is a part of the brain. It receives sensory information about the internal environment of the body and communicates with the autonomic nervous system to maintain homeostasis. It also contains hormone-secreting neurons that control the glandular secretions of the pituitary gland. **3.** In a negative-feedback mechanism, a hormone controls the level of a product. When that level varies from normal, the amount of hormone, whether too much or too little, is altered until the product's level returns to normal. **4.** The hypothalamus controls the release and inhibition of hormones in the pituitary gland. The anterior pituitary gland releases gonadotropic hormones, TSH, ACTH, prolactin, and GH, and the posterior pituitary gland releases ADH and oxytocin. The thyroid gland releases thyroxine, triiodothyronine, and calcitonin. The parathyroid gland releases PTH. The adrenal medulla releases epinephrine and norepinephrine, and the adrenal cortex releases aldosterone and cortisol. The pancreas releases insulin and glucagon. **5.** Hormones are chemical messengers that are released by a gland and regulate the activity of another organ, tissue, or gland. Often two hormones act together to maintain a constant internal environment. One hormone raises the level of the response and the other lowers it, creating a stable balance.

Testing Yourself

1. d; **2.** a; **3.** c; **4.** d; **5.** c; **6.** d; **7.** a; **8.** d; **9.** b; **10.** b; **11.** b; **12.** b

BioNow

Increasing the temperature to 60°F and the length of daylight to 16 hours triggered the hypothalamus in the quails to produce releasing hormones that stimulate the anterior pituitary to produce FSH and LH, which stimulate the testes to make sperm and the ovaries to make eggs. Release of androgens by the testes and estrogen and progesterone by the ovaries likely triggered mating behavior in the birds.

Thinking Critically

1. Parkinson's patients do not exhibit normal constriction of blood vessels, so their blood pressure slowly decreases. The loss of function of the sympathetic division prevents the body from regulating blood pressure, so rapid changes in blood pressure can occur upon standing, leading to orthostatic hypotension. **2.** Type 1 diabetes is thought to result when an environmental agent, such as a virus, causes T cells to destroy the pancreatic islets. Vaccines may introduce that agent and stimulate the autoimmune response. **3.** The brain is an energetically expensive organ, accounting for a substantial proportion of our daily calorie intake. The selective

pressure in an environment with scarce food, like that on Borneo, would tend to sacrifice brain size to conserve energy. In an environment with plentiful food like Sumatra, the orangutans can afford to spend extra energy on nervous tissue, and thus reap the advantage of larger brain size. **4.** Extreme shortness has been found to limit a person's ability to succeed in many ways, so giving hGH to extremely short people can be supported. However, defining "extreme shortness" is necessary. The FDA should set the limits of the normal range for height and decide at what height it is acceptable to administer hGH. Without such regulations, it is likely that hGH will be misused. **5.** It is natural to want to avoid death and improve performance; however, the use of hGH has too many negative side effects, the injections are very expensive, and the hormone should be considered a drug that falls under the anti-doping rules applied to competitive athletes. Though one is tempted to control one's fate, it is better to accept aging as a natural part of life than to use hGH to postpone the inevitable.

CHAPTER 28

Check Your Progress

(28.1) 1. Chemoreceptors detect chemicals in the environment that are interpreted as tastes and smells. Mechanoreceptors detect mechanical signals involved in hearing and balance. Photoreceptors detect light involved in sight. Cutaneous receptors detect touch, pain, temperature, and pressure. Nociceptors are pain receptors that detect extremes in temperature and pressure and chemicals released by damaged tissue. Proprioceptors detect changes in positions of joints, muscles, and bones. **2.** Sensors in the skin provide sensory input that is communicated to the primary sensory area of the cerebral cortex. In the epidermis, free nerve endings respond to cold and warmth as well as pain. The dermis contains sensory receptors for pressure and touch, including the Pacinian corpuscle and root hair plexus. **3.** The eyes have sensory receptors sensitive to light. These photoreceptors detect the amount of light, the direction from which it is coming, and the wavelength which is interpreted as color. Human ears detect sound pressure waves, the head's position relative to gravity, and the rotational movement of the head. **(28.2) 1.** Bones and muscles together support the body, allow body parts to move, protect internal organs, and help other systems, such as the respiratory system, to function. Contraction of skeletal muscle helps to move blood and lymph through their vessels. Skeletal muscle contraction generates heat, which helps to maintain body temperature. The skeleton stores fat and calcium. **2.** The axial skeleton is composed of the skull, vertebral column, ribs, and sternum and lies along the midline of the body. The appendicular skeleton is located more laterally and is composed of the shoulder girdle, the pelvic girdle, and their attached appendages. **3.** The middle part of long bones is a cavity surrounded by compact bone. This structure provides strength as well as some flexibility to support the body and tolerate the stresses put on bone by muscle contraction. The red bone marrow in the spongy bone located at the ends of long bones is the site of blood cell formation. Articular cartilage located on the end surfaces of long bones allows these bones to slide over other bones at the joints. **4.** The globular heads of myosin molecules attach to actin filaments

and pull them toward the center of the sarcomere. Then ATP allows the heads to be released, and they reattach at a new location along the actin filament, resulting in contraction or shortening of the muscle.

Testing Yourself

1. b; **2.** b; **3.** d; **4.** e; **5.** c; **6.** b; **7.** a; **8. a.** cochlear nerve; **b.** cochlea; **c.** tympanic membrane; **d.** auditory tube; **e.** auditory canal; **f.** ossicles; **g.** semicircular canals; **h.** outer ear; **i.** middle ear; **j.** inner ear; **9. a.** sclera; **b.** retina; **c.** vein; **d.** artery; **e.** optic nerve; **f.** fovea; **g.** ciliary muscle; **h.** iris; **i.** pupil; **j.** lens; **k.** cornea; **10.** c; **11.** a; **12.** a; **13.** c; **14.** c; **15.** a

Thinking Critically

1. The blood vessels interfere with the transmission of light to the back of the retina, consequently impairing the signals sent to the brain. Both cancer and these eye diseases involve the uncontrolled growth of new blood vessels. Anti-angiogenesis drugs that have been developed for cancer treatment are currently being studied for the treatment of these eye diseases. **2.** Our olfactory system allows us to recognize and discriminate among many different odors. Therefore, we can detect the difference between a dangerous odor, such as smoke, and a harmless one, such as the fragrance of a flower. On the other hand, every natural toxin tastes bitter, so it is more important to be able to sense bitterness than to be able to discriminate between bitter tastes. Therefore, humans have evolved to identify bitter compounds by the taste while not being able to distinguish between the different compounds.

CHAPTER 29

Check Your Progress

(29.1) 1. Asexual reproduction occurs when one parent produces offspring that are genetically identical to each other and to the parent. Sexual reproduction involves two parents, and the offspring have different combinations of the parents' genes. **2.** Animals that reproduce in the water have no need to protect their eggs and embryos from drying out, but they do have to worry more about predators destroying the zygotes. Animals that reproduce on land need to protect their gametes and embryos from desiccation while providing a stable environment for development. **3.** Egg-laying animals are oviparous. Animals that retain their eggs in the body until hatching has occurred are ovoviviparous. Mammals that produce living young are viviparous. **(29.2) 1.** Sperm are produced in the seminiferous tubules in the testis, mature in the epididymis, are propelled into the vas deferens, and travel out through the penis. **2.** Testosterone is the main sex hormone in males and is essential for the normal development and functioning of male sex organs. It is necessary for the maturation of sperm and maintaining male secondary sex characteristics. **3.** During the follicular phase of the ovarian cycle, FSH released by the anterior pituitary promotes the maturation of follicles in the ovary. Ovarian follicles produce increasing levels of estrogen, which exert feedback control over FSH secretion, resulting in ovulation. **4.** After ovulation and during the luteal phase of the ovarian cycle, LH promotes the development of a corpus luteum. This structure produces increasing levels of progesterone, which causes the endometrium to become secretory.

Menstruation begins when progesterone production declines to a low level. During the follicular phase, FSH promotes the development of ovarian follicles which produce estrogens that cause the endometrium to thicken. **5.** Artificial insemination by donor (AID)—sperm from a donor is placed in the vagina by a physician. In vitro fertilization (IVF)—immature eggs are removed from the ovaries and cultured outside the body. When mature, concentrated sperm is added to the eggs. After a few days, the embryos that have developed are transferred to the uterus. Intracytoplasmic sperm injection (ICSI)—a single sperm is injected into an egg outside the body. The embryo is cultured and eventually transferred to the uterus. Gamete intrafallopian transfer (GIFT)—the eggs and sperm are brought together and then immediately placed in the uterine tubes. **6.** STDs caused by bacteria include chlamydia, gonorrhea, and syphilis. STDs caused by a virus include HIV, genital warts, genital herpes, and hepatitis. Trichomonas is caused by a parasitic protozoan. **(29.3) 1.** Mammals develop extraembryonic membranes, or membranes outside the embryo. The developing embryo depends on these membranes for protection and nourishment. The chorion, which becomes part of the placenta, is an example. **2.** During gastrulation, three primary germ layers develop: ectoderm, endoderm, and mesoderm. **3.** During gastrulation, cells migrate to become germ layers, forming ectoderm, endoderm, and mesoderm. Neurulation occurs after the formation of the notochord. The neural plate thickens, then folds to form a neural tube, the anterior portion of which becomes the brain and the posterior portion the spinal cord. **4.** Embryonic development occurs during the first two months after fertilization and is characterized by the establishment of organ systems. Fetal development occurs during the third through ninth month and is characterized by an increase in size. **5.** The placenta has a fetal side and a maternal side, and exchange of molecules occurs between the sides during development. Carbon dioxide and wastes move from the fetal side to the maternal side, and nutrients and oxygen move from the maternal side to the fetal side. Maternal blood and fetal blood do not mix.

Testing Yourself

1. d; **2.** b; **3. a.** ureter; **b.** bladder; **c.** vas deferens; **d.** seminal vesicle; **e.** ejaculatory duct; **f.** prostate gland; **g.** bulbourethral gland; **h.** urethra; **i.** epididymis; **j.** penis; **k.** foreskin; **l.** testis; **m.** scrotum; **4. a.** uterine tube; **b.** ovary; **c.** rectum; **d.** cervix; **e.** bladder; **f.** vagina; **g.** urethra; **5.** d; **6.** d; **7.** b; **8.** a; **9.** a; **10.** c; **11.** a; **12.** b; **13.** d

Thinking Critically

1. Fetuses and infants place a high caloric toll on their mothers. Mothers with low percentages of body fat might starve during pregnancy or during the prolonged period of breast-feeding. They would be less successful at surviving and producing healthy children. If the reproductive cycle halts when the body has insufficient fat reserves, a woman will be better able to survive to a time when conditions are more favorable for reproduction. **2.** One hypothesis is that menopause in women developed early in human evolution. It allowed women to channel their efforts into caring for their existing children and grandchildren, increasing the survival rate of the next generation. This would give a woman the best chance

to pass her genes on to future generations, thus increasing her fitness. Since men were not the main caregivers for their children, their fitness would not be increased if spermatogenesis stopped later in life. **3.** The villi of the chorion of the embryo are numerous and spread through the developing placenta. Removing a small portion of them for analysis of the fetal cells is not known to harm the fetus. **4.** Human chorionic gonadotropin (hCG) is secreted by the placenta as it begins to form. It maintains the endometrium until enough estrogen and progesterone are produced by the placenta. Because hCG is a hormone, it is transported in the blood and is removed from the blood into the urine by the kidneys. **5.** In addition to sperm counts, information about individuals' ages, where they live, and their eating, drinking, and smoking habits, sleep patterns, clothing, and environmental exposures would be useful in forming a testable hypothesis.

CHAPTER 30

Check Your Progress

(30.1) 1. Organism, species, population, community, ecosystem, biosphere. **2.** An ecosystem consists of a community of living organisms as well as their physical environment. A community consists of all the various populations in a particular locale. **3.** Ecology is the scientific study of the interactions of organisms with each other and their physical environment. **(30.2) 1.** $(18.5 - 9.8)/1,000 \times 100 = 0.87\%$; **2.** MDCs such as those in Europe and North America have modest population growth and a fairly high standard of living. LDCs such as those in Latin America, Asia, and Africa have rapid population growth, and most of the people live in poverty. **3.** Replacement reproduction occurs when reproduction in a population results in no population growth. Due to mortality, this occurs when the replacement rate is slightly greater than two children per couple. **4.** MDCs strain available resources due to high resource consumption, which leads to the production of more hazardous waste. LDCs strain available resources due to large population growth rates. **(30.3) 1.** An age-structure diagram shows different numbers of individuals in each age class as a stacked set of bars. Large numbers of prereproductive and reproductive individuals compared to postreproductive individuals indicate that the population will increase. The smaller the relative number of individuals who will or can reproduce, the more likely that the population will be stable or decrease. **2.** Populations with a type I survivorship curve have fewer offspring and most of those offspring live to reproduce. Populations with a type III curve have large numbers of offspring and most of them will die before they reproduce. Members of populations with a type II curve have the same chance of death at all ages. **3.** Both curves plot number of organisms versus time. Exponential growth produces a J-shaped curve and occurs when resources are unlimited. Logistic growth produces an S-shaped curve and occurs after exponential growth, when resources become restricted. **4.** Density-independent factors include weather and natural disasters such as flooding, tornadoes, and wildfires that can cause injury and death. **5.** As population density increases, there are increased levels of competition for limited resources and higher levels of predation. Each individual competes to obtain a share of the resources.

(30.4) 1. Opportunistic species are small, have a short life span, are fast to mature, have many offspring, and take little or no care of their offspring. They exhibit exponential population growth, and populations are regulated by density-independent effects. Equilibrium species are large, have a long life span, are slow to mature, have few offspring, and care for the offspring. They exhibit logistic population growth, regulated by density-dependent factors, and populations remain near carrying capacity. **2.** Extinction is the total disappearance of a species or higher group of organisms. **3.** Long time to maturity; few offspring produced; restricted geographic range; narrow habitat tolerance; small size of local populations.

Testing Yourself

1. d; **2.** b; **3.** a; **4.** c; **5.** d; **6.** b; **7.** b; **8. a.** increasing; **b.** decreasing; **c.** stable; **9.** a; **10.** a; **11.** d; **12.** b; **13.** c; **14.** a; **15.** c

Thinking Critically

1. Population growth rate = 1.1%; doubling time = 62.7 years; 172.5 years if birthrate drops. **2.** As the numbers of rabbits and pheasants decrease, the impact of predation by red-tailed hawks also decreases. This is a density-dependent effect. Killing off the hawks will not bring back the rabbit and pheasant populations. **3.** The number of surviving individuals and a measurement of the variable being studied. In Figure 15.2, that variable is clutch size. **4. a.** It can be argued that the species with the highest risk of extinction are less successful products of evolution and should not receive extraordinary protection, which uses up limited resources. However, this position assumes humans can place value on a particular species' existence in an impartial manner. Often big animals with cute faces are given preference. **b.** Determining which species are successful products of evolution is subjective and therefore inexact. A more cautious approach would be to treat species in risk of extinction equally, distributing limited resources between them all.

CHAPTER 31

Check Your Progress

(31.1) 1. Flowers and pollinators have coevolved to be suited to one another. Hummingbirds pollinate red flowers that they can see, and the flower petals are specifically curved so that the stamens dust the bird's head, resulting in the spread of pollen. **2.** Species richness is the list of species in a community (species composition), while species diversity encompasses species richness, distribution, and relative abundance. **3.** Primary succession occurs where soil has not yet been formed, while secondary succession begins once soil is present. **4.** Competition, predation, parasitism, commensalism, mutualism. **(31.2) 1.** Autotrophs take in only inorganic nutrients and acquire energy from an outside source, most often the sun. Heterotrophs acquire energy from consuming preformed organic nutrients. **2.** Energy flows from the sun to the Earth and then through living organisms until it is lost as heat. Chemicals cycle throughout the environment, through both living and nonliving components. **3.** A food chain depicts a single path of energy flow in an ecosystem by displaying feeding relationships. An ecological pyramid depicts the flow of energy between successive trophic levels. **4.** Phosphorus—mining ore, fertilizing, animal feed supplements, detergents, animal waste, and sewage discharge leading to runoff, eutrophication, and algal blooms. Nitrogen—making fertilizers, leading to runoff; deforestation leading to loss of nitrogen to groundwater; groundwater contamination by nitrates, leading to health problems. Carbon—excess CO_2 release into the atmosphere by burning fossil fuels, and deforestation, leading to less carbon fixation. **(31.3) 1.** The biosphere is divided into aquatic and terrestrial ecosystems. Aquatic ecosystems include freshwater and saltwater systems. Terrestrial ecosystems can be divided into several major types depending on temperature and rainfall. **2.** Tropical rain forest, savanna, desert, temperate deciduous forest, temperate grasslands (prairie), taiga, tundra. **3.** Swamps have a more concentrated supply of organic nutrients and more decomposers than do open oceans and therefore have a higher primary productivity.

Testing Yourself

1. a; **2.** e; **3.** a; **4.** c; **5.** b; **6.** d; **7.** c; **8. a.** energy; **b.** nutrients; **c.** heat; **d.** heat; **e.** producers; **f.** consumers; **g.** inorganic nutrients; **h.** decomposers; **i.** heat; **9.** c; **10.** d; **11. a.** top carnivores; **b.** carnivores; **c.** herbivores; **d.** producers; **12.** b; **13.** a; **14.** a

BioNow

The exotic species, reed canary grass, outcompeted the native species for resources and took over the field. Contributing to this outcome is the fact that alien species often do not have natural predators such as insects, that limit their growth and reproduction. Because the reed canary grass comes to dominate the field and be the only plant species present, it lowers the biodiversity of consumers (such as insects) present there. With just one grass species present, consumers have limited choices for food and habitat and the variety of species able to live in the field is reduced.

Thinking Critically

1. Insects that are best adapted for obtaining the most food from flowers that they can easily access reproduce more. Those flowers best adapted for attracting insect pollinators produce more offspring in the next generation, and likewise those insects that obtain the most food produce more offspring than others. This process continues over generations and leads to the coevolution of specific adaptations. **2.** The otter population is reduced; sea urchin population rises; kelp plants are reduced; species living in the kelp forests are reduced or eliminated due to habitat destruction; productivity of the kelp forest is reduced because of fewer producers and consumers; less food is being produced and the food chain is weakened; chemical cycling is reduced. **3.** Native populations need to be protected in order to preserve the species richness and biodiversity of ecosystems. The argument that the mute swan should be allowed to displace native birds is fallacious if that argument is based on a human perception of beauty rather than on the health of the community or ecosystem.

CHAPTER 32

Check Your Progress

(32.1) 1. Because conservation biology has the broad focus of conserving natural resources while maintaining biodiversity, it draws on many subfields of basic biology, including genetics, physiology, behavior, systematics, ecology, field biology, and evolutionary biology. Applied biology subfields such as biopark management, forestry, veterinary science, fisheries biology, range management, agronomy, and wildlife management are also used in conservation biology. **2.** Biodiversity is good for the Earth and therefore good for humans. Extinctions are undesirable. Interactions within ecosystems support biodiversity and are beneficial to humans. Biodiversity is the result of evolutionary change and has great value to humans. **(32.2) 1.** Direct values include development of medicines, agriculture management, and products that humans consume. Indirect values include contributions to biogeochemical cycles, waste disposal mechanisms, provision of fresh water, prevention of soil erosion, regulation of climate, and ecotourism. **2.** Most modern medicines are derived from organisms, so more biodiversity increases the potential of developing more medicines. In agriculture, biodiversity can lead to the development of crop varieties able to live in varied climates, requiring less fertilizer and water, and able to withstand herbivores more successfully. **3.** Biodiversity gives us flexibility by providing options if any resource, whether it is water, land, food, or energy, becomes limited. If there are high levels of biodiversity, it may be possible to substitute another species adapted to the altered environment. **(32.3) 1.** Renewable resources like solar energy, wind, hydropower, and geothermal are unlimited in supply. More will be available as they are used. Nonrenewable resources such as land, fossil fuels, and minerals are limited in supply and can be exhausted. **2.** Deforestation occurs when humans harvest timber, clear the forest to create cropland, or settle in the forest. The thin poor soil is not well suited for farming and is quickly abandoned. Increased erosion and runoff result. Desertification occurs when humans convert semiarid land to desert, usually by overgrazing. The loss of vegetation increases runoff, depletes groundwater, and eventually leads to desert conditions that force the population to move on. **3.** Irrigation depletes water tables. Monoculture reduces biodiversity. Application of fertilizer, herbicides, and pesticides results in runoff into water sources. Excessive amounts of fossil fuels are used. Converting land to fields results in habitat destruction. **4.** The "green revolution" increased crop yields by introducing new varieties of crops, which led to an increased world food supply. Increasing the food supply is necessary to keep pace with the exponentially growing human population. However, the new varieties require increased amounts of fertilizers, pesticides, and water, impacting the environment negatively. **5.** Wind power: Land must be committed to the construction of windmills.

Hydropower: Dams which disrupt river ecosystems must be constructed. Geothermal energy: There is a large capital expenditure associated with the collection of geothermal energy, its storage, and its distribution. Solar energy: Solar collection systems that can convert and effectively store solar energy before distribution are costly. **(32.4) 1.** Characteristics of a sustainable society include use of renewable energy sources, reuse of heat and waste materials, and maximal recycling of materials. **2.** Plant cover crops, practice multiuse farming, replenish soil nutrients, conserve water, plant cultivars, use precision farming, use integrated pest management, plant a variety of species, plant multipurpose trees, maintain and restore wetlands, use renewable forms of energy, and buy locally. **3.** Create energy-efficient transportation systems, utilize green roofs, create greenbelts, use solar or geothermal energy to heat buildings, improve storm-water management, plant native species, revitalize old sections of the city before developing new sections, change lighting, reduce noise levels, and promote sustainable building practices.

Testing Yourself

1. e; **2.** a; **3.** e; **4.** c; **5.** b; **6.** d; **7.** c; **8.** a; **9.** b; **10.** a; **11.** use more renewable resources than nonrenewable resources, increase recycling, refine processing to use fewer materials, and reduce waste materials and heat production; **12.** c

Thinking Critically

1. Positive environmental impacts include increasing the diversity of animals and plants by increasing seasonal flooding and nutrient deposition. The negative environmental impacts include miscalculation of the economic impacts and the potential of introducing invasive species. The direct benefits for residents include removing them from the economic and personal threat of floodplain residency. The indirect benefits for residents include those provided by the diversity the Emiquon Refuge can offer. **2.** How rare the species is, how difficult it will be to preserve it, the cost of preserving it, the potential medicinal benefit to humans. **3.** The internal temperature of mammals must stay within a narrow normal range. As global change leads to climate warming in the tropics, the external temperature could become too high for the homeostatic mechanisms in mammals to control their internal temperature. If the core temperature of the mammals rises above normal, they will die off. **4.** Steps that can reduce one's ecological footprint include adopting a plant-based diet composed of local foods, reducing food waste, driving a smaller and more fuel-efficient car, using trains and buses, flying less, living in a smaller and more energy-efficient dwelling, reducing the number of material items purchased, recycling of all types of materials, and composting.

Glossary

A

abscisic acid Plant hormone that causes stomata to close and initiates and maintains dormancy.

abscission Dropping of leaves, fruits, or flowers from a plant.

abstinence Method of birth control; the practice of not engaging in sexual intercourse.

acetyl-CoA Molecule made up of a 2-carbon acetyl group attached to coenzyme A. During cellular respiration, the acetyl group enters the citric acid cycle for further breakdown.

acid Molecule tending to raise the hydrogen ion concentration in a solution and to lower its pH numerically.

acoelomate Animal without a coelom, as in flatworms.

acquired immunodeficiency syndrome (AIDS) Disease caused by a retrovirus and transmitted via body fluids; characterized by failure of the immune system.

actin Muscle protein making up the thin filaments in a sarcomere. Its movement shortens the sarcomere, yielding muscle contraction. Actin filaments play a role in the movement of the cell and its organelles.

actin filament Cytoskeletal filament in eukaryotic cells composed of the protein actin; also refers to the thin filaments of muscle cells.

action potential Electrochemical changes that take place across the axon's membrane; the nerve impulse.

active immunity Resistance to disease due to the immune system's response to a microorganism or a vaccine.

active site Region on the surface of an enzyme where the substrate binds and where the reaction occurs.

active transport Use of a plasma membrane carrier protein to move a molecule or an ion from a region of lower concentration to one of higher concentration. Active transport opposes equilibrium and requires energy.

adaptation A modification in an organism's structure, function, or behavior suitable to the environment.

adaptive immunity Form of immunity that involves B and T lymphocytes in targeting specific pathogens in the body.

adaptive radiation Rapid evolution of several species from a common ancestor as populations move into new ecological or geographical zones.

Addison disease Condition resulting from a deficiency of adrenal cortex hormones; characterized by low blood glucose, weight loss, and weakness.

adenine (A) One of four nitrogen-containing bases in nucleotides composing the structure of DNA and RNA. Pairs with thymine in DNA and uracil in RNA.

adenosine triphosphate (ATP) Nucleotide with three phosphate groups. The breakdown of ATP into ADP + P$_i$ makes energy available for energy-requiring processes in cells.

adhesion Ability of water molecules to be attracted to, or cling to, a polar surface, such as the inside of a transport vessel (capillary).

adhesion junction Junction between cells in which the adjacent plasma membranes do not touch but are held together by intercellular filaments attached to button-like thickenings.

adrenal cortex Outer portion of the adrenal gland; produces the glucocorticoid and mineralocorticoid hormones.

adrenal gland Gland that lies atop a kidney. The *adrenal medulla* produces the hormones epinephrine and norepinephrine, and the *adrenal cortex* produces the glucocorticoid and mineralocorticoid hormones.

adrenal medulla Inner portion of the adrenal gland; secretes the hormones epinephrine and norepinephrine.

adult stem cell Cell that can be found in many organs in an adult body and that may differentiate to form other cell types.

age structure In demographics, the distribution of various age groups in a given population, typically displayed on a bar diagram called an age-structure diagram. A growing population has a pyramid-shaped diagram.

agglutination Clumping of red blood cells due to a reaction between antigens on red blood cell plasma membranes and antibodies in the plasma.

algae (sing., alga) Type of protist that carries on photosynthesis. Single-celled forms are a part of phytoplankton, and multicellular forms are called seaweed.

allele Alternative form of a gene. Alleles occur at the same locus on homologous chromosomes.

allergen Foreign substance (antigen) capable of stimulating an allergic response.

allergy Hypersensitivity to a substance in the environment that ordinarily would not cause an immune reaction.

allopatric speciation Origin of new species due to geographic separation of populations.

alternation of generations Pattern that is characteristic of the life cycle of plants, in which a diploid sporophyte alternates with a haploid gametophyte.

alternative mRNA processing Mechanism that allows cells to produce multiple proteins from the same gene by changing the way exons are spliced together.

alveolus (pl., alveoli) In humans, terminal, microscopic air sac in the lungs.

Alzheimer disease Brain disorder characterized by a progressive loss of memory, particularly for recent events.

amino acid Organic molecule composed of an amino group and an acid group; covalently bonds to produce peptide molecules.

amniocentesis Procedure for removing amniotic fluid surrounding the developing fetus for testing of the fluid or fetal cells in it.

amphibian Member of a class of vertebrates that include frogs, toads, and salamanders and that are tied to a watery environment for reproduction.

analogous structure Structure that has a similar function in separate lineages but differs in anatomy and ancestry.

analogy Similarity of characteristics due to convergent evolution.

anaphase Mitotic phase during which daughter chromosomes move toward the poles of the spindle.

angiogenesis Formation of new blood vessels.

angiosperm Flowering plant. The seeds are borne within a fruit.

animal Multicellular, heterotrophic organism belonging to the kingdom Animalia.

annelid A segmented worm, such as an earthworm or a clam worm.

annual ring Layer of wood (secondary xylem) usually produced during one growing season.

anorexia nervosa Eating disorder characterized by an irrational fear of getting fat.

anterior pituitary Portion of the pituitary gland that is controlled by the hypothalamus. It produces six types of hormones, some of which control other endocrine glands.

anther In flowering plants, pollen-bearing portion of the stamen.

anthropoids Group of primates that includes monkeys, apes, and humans.

antibody Protein produced in response to the presence of an antigen. Each antibody combines with a specific antigen.

anticodon Three-base sequence in a transfer RNA molecule base that pairs with a complementary codon in mRNA.

antigen Foreign substance, usually a protein or carbohydrate molecule, that stimulates the immune system to react, often by producing antibodies.

antigen-presenting cell (APC) Cell that displays an antigen to certain cells of the immune system, so that they can defend the body against that antigen.

antioxidant Substance, such as vitamin A, C, or E, that defends the body against free radicals.

anus Outlet of the digestive tract at the end of the anal canal.

aorta In humans, the major systemic artery that takes blood from the heart to the tissues.

aortic body Structure located in the walls of the aorta; contains chemoreceptors sensitive to hydrogen ion and carbon dioxide concentrations in the blood.

apical dominance Influence of a terminal bud in suppressing the growth of lateral buds.

apical meristem In vascular plants, masses of cells in the root and shoot that develops into the epidermal, ground, and vascular tissues of a plant.

apoptosis Programmed cell death involving a cascade of specific cellular events, leading to death and destruction of the cell.

appendicular skeleton Part of the vertebrate skeleton containing the bones of the pectoral girdle and pelvic girdle and their attached limbs.

appendix In humans, small, tubular appendage containing lymphatic tissue, that extends outward from the cecum of the large intestine; may play a role in fighting infections.

aquatic ecosystem Freshwater ecosystem (river, stream, lake, pond) or saltwater (marine) ecosystem (ocean, coral reef, saltwater marsh).

aquifer Rock layers that contain water and release it in appreciable quantities to wells or springs.

arachnids Group of arthropods that includes spiders, scorpions, mites, harvestmen, and ticks.

archaea Prokaryotic members of the domain Archaea. Many live in unique habitats, such as swamps, highly saline environments, and submarine thermal vents.

arteriole Vessel that takes blood from an artery to capillaries.

artery Blood vessel that transports blood away from the heart.

arthropods Invertebrates with an exoskeleton and jointed appendages, such as crustaceans and insects.

artificial selection Selection by breeders of individual organisms with particular traits, or combinations of traits, over others lacking the traits.

asexual reproduction Form of reproduction that does not require two individuals; example is binary fission in bacteria.

atherosclerosis Form of cardiovascular disease characterized by the buildup of cholesterol and fatty acid plaques in the blood vessels.

atom Smallest particle of an element that displays the properties of the element.

atomic mass Average value of the mass numbers for all isotopes of an element.

atomic number Number of protons within the nucleus of an atom.

atomic symbol One or two letters that represent the name of an element—e.g., H stands for a hydrogen atom, and Na stands for a sodium atom.

ATP synthase Enzyme that is part of an *ATP synthase complex* and functions in the production of ATP in chloroplasts and mitochondria.

atrioventricular valve Heart valve located between an atrium and a ventricle.

atrium (pl., atria) Chamber; particularly an upper chamber of the heart lying above a ventricle.

australopithecine One of several species of *Australopithecus*, a genus that contains the first generally recognized hominins.

autoimmune disease Disease that results when the immune system mistakenly attacks the body's own cells.

autonomic system Portion of the peripheral nervous system that automatically and involuntarily regulates the activity of glands and cardiac and smooth muscle; divided into parasympathetic and sympathetic divisions.

autosome Any chromosome other than a sex (X or Y) chromosome.

autotroph Organism that can capture energy and synthesize organic molecules from inorganic nutrients.

auxin Plant hormone regulating growth, particularly cell elongation; also called indoleacetic acid (IAA).

AV (atrioventricular) node Small region of cardiac muscle tissue that transmits impulses received from the SA node to the ventricular walls.

axial skeleton Part of the vertebrate skeleton forming the vertical support or axis, including the skull, rib cage, and vertebral column.

axon Elongated portion of a neuron that conducts nerve impulses, typically from the cell body to the synapse.

B

B lymphocyte (B cell) Lymphocyte that matures in the bone marrow and, when stimulated by the presence of a specific antigen, gives rise to antibody-producing plasma cells.

B-cell receptor (BCR) Complex on the surface of a B cell that binds an antigen and stimulates the B cell.

bacteria Prokaryotic members of the domain Bacteria; most diverse and prevalent organisms on Earth.

bacteriophage Virus that reproduces in a bacterium; often referred to as a *phage*.

ball-and-socket joint Synovial joint (at hip or shoulder) that allows movement in all directions.

bark External covering of a tree stem, containing cork, cork cambium, cortex, and phloem.

base Molecule tending to lower the hydrogen ion concentration in a solution and raise the pH numerically.

benign tumor Tumor that does not invade adjacent tissue and stays at the site of origin.

bicarbonate ion Ion that participates in buffering the blood, and the form in which carbon dioxide is transported in the blood.

bilateral symmetry Body plan having two corresponding or complementary halves.

bile Secretion of the liver that is temporarily stored and concentrated in the gallbladder before being released into the small intestine, where it emulsifies fat.

binary fission Splitting of a parent cell into two daughter cells; serves as an asexual form of reproduction in bacteria.

binge-eating disorder Condition characterized by overeating episodes that are not followed by purging.

binomial name Scientific name of an organism, the first part of which designates the genus and the second part of which designates the specific epithet.

biodiversity Total number of different species living on Earth.

biodiversity hotspots Regions on Earth that contain an above-average number of species; examples are coral reefs and rain forests.

biogeochemical cycle Circulating pathway of an element, such as carbon or nitrogen, within an ecosystem, involving exchange pools, storage areas, and the biotic community.

biogeography Study of the geographical distribution of life-forms on Earth.

bioinformatics The application of computer technologies to study the genome and other molecular data.

biological magnification Process by which substances become more concentrated in organisms in the higher trophic levels of a food web.

biological molecule Organic molecule found in cells: carbohydrate, lipid, protein, or nucleic acid.

biological species concept Definition of a species that states that members of a species interbreed and have a shared gene pool and that each species is reproductively isolated from other species.

biology Scientific study of life.

biomass Number of organisms of a particular type multiplied by their weight.

bioremediation Cleanup of the environment using bacteria to break down pollutants, such as oil spills.

biosphere Zone of air, land, and water at the surface of the Earth in which organisms are found; encompasses all the ecosystems on Earth.

biotechnology Genetic engineering and other techniques that make use of natural biological systems to create a product or achieve a particular result desired by humans.

biotic potential Maximum growth rate of a population under ideal conditions.

bipedalism Walking erect on two feet.

bird Endothermic reptile that has feathers and wings, is often adapted for flight, and lays hard-shelled eggs.

birth control pill Oral contraceptive containing estrogen and progesterone.

bivalve Type of mollusc with a shell composed of two valves; includes clams, oysters, mussels, and scallops.

blastocyst Early stage of animal embryonic development that consists of a hollow, fluid-filled ball of cells.

blood Fluid connective tissue that is composed of several types of cells suspended in plasma and is circulated by the heart through a closed system of vessels.

blood pressure Force of blood pushing against the inside wall of blood vessels.

body mass index (BMI) A calculated number used to determine whether or not a person is overweight or obese.

bolus Mass of chewed food mixed with saliva.

bone Connective tissue having collagen fibers and an extremely hard matrix of inorganic salts, notably calcium salts.

bony fishes Fishes with a bony rather than cartilaginous skeleton.

bottleneck effect Evolutionary event in which a significant percentage of a population is prevented from reproducing, therefore reducing variation. Population bottlenecks increase the influence of *genetic drift*.

brain Primary organ of the central nervous system; site of the majority of signal integration.

brain stem In mammals, portion of the brain that consists of the medulla oblongata, pons, and midbrain and connects the rest of the brain to the spinal cord.

bronchiole In terrestrial vertebrates, small tube that conducts air from a bronchus to the alveoli.

bronchus (pl., bronchi) In terrestrial vertebrates, branch of the trachea that leads to one of the lungs.

bryophytes Nonvascular plants—the mosses, liverworts, and hornworts. These plants have no vascular tissue and occur in moist locations.

buffer Substance or group of substances that tends to resist pH changes of a solution, thus stabilizing its relative acidity and basicity.

bulimia nervosa Eating disorder characterized by binge eating followed by purging via self-induced vomiting or use of a laxative.

C

C₃ photosynthesis Photosynthetic process where the first stable product of photosynthesis is a 3-carbon molecule.

C₄ photosynthesis Photosynthetic process where the plant fixes carbon dioxide to produce a C_4 molecule that releases carbon dioxide to the Calvin cycle.

calcitonin Hormone secreted by the thyroid gland that helps regulate the blood calcium level.

calorie Amount of heat energy required to raise the temperature of 1 gram of water 1°C.

Calvin cycle Portion of photosynthesis that takes place in the stroma of chloroplasts and can occur in the dark. It uses the products of the light reactions to reduce CO_2 to a carbohydrate.

calyx The sepals, collectively; the outermost flower whorl.

CAM photosynthesis Form of photosynthesis that is based on crassulacean-acid metabolism. A plant fixes carbon dioxide at night to produce a C_4 molecule that releases carbon dioxide to the Calvin cycle during the day.

cancer Malignant tumor whose nondifferentiated cells exhibit loss of contact inhibition, uncontrolled growth, and the ability to invade tissue and metastasize.

capillary Microscopic blood vessel. Gases and other substances are exchanged across the walls of a capillary between blood and interstitial fluid.

capsid Protective protein container that surrounds the genetic material of a virus.

capsule Gelatinous layer surrounding the cells of green algae and certain bacteria.

carbohydrate Class of organic compounds that includes monosaccharides, disaccharides, and polysaccharides; present in food as sugars, starch, and fiber.

carcinogenesis Development of cancer.

cardiac cycle One complete cycle of systole and diastole for all heart chambers.

cardiac muscle Striated, involuntary muscle tissue found only in the heart.

cardiovascular system Organ system in which blood vessels distribute blood under the pumping action of the heart.

carnivore Consumer in a food chain that eats other animals.

carotenoid Yellow or orange pigment that serves as an accessory to chlorophyll in photosynthesis.

carotid body Structure located at the branching of the carotid arteries; contains chemoreceptors sensitive to hydrogen ion and carbon dioxide concentrations in blood.

carpel Ovule-bearing unit that is a part of a pistil.

carrying capacity Number of organisms of a particular species that can be maintained indefinitely by a given environment.

cartilage Connective tissue in which the cells lie within lacunae separated by a flexible matrix.

cartilaginous fishes Fishes with a cartilaginous rather than bony skeleton; include sharks, rays, and skates.

cell Fundamental unit of life. Lowest level of biological organization that has all of the characteristics of life.

cell body Portion of a neuron that contains a nucleus and from which dendrites and an axon extend.

cell cycle Repeating sequence of events in eukaryotes that involves cell growth and nuclear division; consists of the stages G_1, S, G_2, and M.

cell plate Structure across a dividing plant cell that signals the location of new plasma membranes and cell walls.

cell signaling The process by which cells communicate with other cells and detect changes in their environment.

cell theory One of the major theories of biology which states that all organisms are made up of cells, cells are capable of self-reproduction, and cells come only from preexisting cells.

cell wall Structure that surrounds a plant, protistan, fungal, or bacterial cell and maintains the cell's shape and rigidity.

cellular respiration Metabolic reaction that uses the energy from the breakdown of carbohydrates (primarily glucose), fatty acids, or amino acids to produce ATP molecules.

cellulose Polysaccharide that is the major complex carbohydrate in plant cell walls.

central nervous system (CNS) Portion of the nervous system consisting of the brain and spinal cord.

centriole Cell organelle, existing in pairs, that is located in the centrosome and may help organize a mitotic spindle for chromosome movement during animal cell division.

centromere Constricted region on a chromosome where two sister chromatids are joined.

centrosome Microtubule organizing center of cells. In animal cells, it contains two centrioles.

cephalization Having a well-recognized anterior head with a brain and sensory receptors.

cephalopod Type of mollusc in which a modified foot develops into the head region; includes squids, octopuses, and nautiluses.

cerebellum In terrestrial vertebrates, portion of the brain that coordinates skeletal muscles to maintain posture and balance and produce smooth, graceful motions.

cerebral cortex Outer layer of gray matter that covers the cerebrum; interprets sensation, initiates voluntary movement, and carries out higher thought processes.

cervical cap Birth control device made of latex in the shape of a cup, which covers the cervix; considered a mini-diaphragm.

cervix Narrow end of the uterus leading into the vagina.

character displacement Tendency for characteristics to be more divergent when similar species belong to the same community than when they are isolated from one another.

charophytes A group of freshwater green algae thought to be the closest relatives of land plants.

checkpoint In the cell cycle, one of several points where the cell cycle can stop or continue on, depending on the influence of internal and external signals; ensures that each step of the cell cycle is completed before the next one begins.

chemoautotroph Organism able to synthesize organic molecules by using carbon dioxide as the carbon source and the oxidation of an inorganic substance (such as hydrogen sulfide) as the energy source.

chemoheterotroph Organism that is unable to reproduce its own organic molecules and therefore requires organic nutrients in its diet.

chemoheterotrophs Organisms that get their energy and nutrients from digesting organic materials from other organisms.

chemoreceptors Receptors in the nervous system that detect chemicals in the environment.

chitin Strong but flexible nitrogenous polysaccharide found in the exoskeleton of arthropods.

chlamydia Bacterial STD of the lower reproductive tract that can result in pelvic inflammatory disease.

chlorofluorocarbons (CFCs) Organic compounds containing carbon, chlorine, and fluorine atoms. CFCs, such as freon, can deplete the ozone shield by releasing chlorine atoms in the upper atmosphere.

chlorophyll Green pigment that absorbs solar energy and is important in algal and plant photosynthesis; occurs as chlorophyll *a* and chlorophyll *b*.

chloroplast Membrane-bound organelle in algae and plants with chlorophyll-containing membranous thylakoids; where photosynthesis takes place.

cholesterol One of the major lipids found in animal plasma membranes; makes the membrane impermeable to many molecules.

chordate Animal that has a dorsal tubular nerve cord, a notochord, pharyngeal pouches, and a postanal tail at some point in its life cycle.

chorionic villus sampling (CVS) Prenatal test in which a sample of chorionic villi cells is removed for diagnostic purposes.

chromatid Single DNA strand of a chromosome. Chromosomes may consist of a pair of sister chromatids.

chromatin Network of fibrils consisting of DNA and associated proteins observed within a nucleus that is not dividing.

chromosome Structure, consisting of DNA complexed with proteins, that transmits genetic information from the previous generation of cells and organisms to the next generation.

chyme Thick, semiliquid food material that passes from the stomach to the small intestine.

cilium (pl., cilia) Short, hair-like projection from the plasma membrane, occurring usually in large numbers.

circulatory system Organ system of animals that is responsible for supplying oxygen and nutrients to the cells of the organism and removing carbon dioxide and waste materials.

citric acid cycle Cycle of reactions in mitochondria that begins with citric acid. It breaks down an acetyl group and produces CO_2, ATP, NADH, and $FADH_2$; also called the Krebs cycle.

clade A group of organisms that includes a common ancestor and all of its descendants and that has its own particular derived traits.

cladistics School of systematics that uses traits derived from a common ancestor to trace the evolutionary history of groups and construct cladograms.

cladogram Branching diagrammatic tree used to depict the evolutionary history of a group of organisms.

class One of the categories, or taxa, used by taxonomists to group species; the taxon above the order level.

cleavage Cell division without cytoplasmic addition or enlargement; occurs during the first stage of animal development.

climate change Overall change in global weather patterns caused by any number of human-related events.

climax community In ecology, the community that results when succession has come to an end.

clonal selection model Mechanism of antibody response whereby an antigen selects which B lymphocyte will undergo clonal expansion and produce more lymphocytes bearing the same type of receptor.

clone An identical genetic copy of a cell or organism.

closed circulatory system In all vertebrates and some invertebrates, cardiovascular system composed of a muscular heart and blood vessels.

clumped distribution Spatial distribution of individuals in a population in which individuals are more dense in one area than in another.

cnidarian Invertebrate existing as either a polyp or a medusa with two tissue layers and radial symmetry.

codominance Inheritance pattern in which both alleles of a gene are equally expressed in a heterozygote.

codon Three-base sequence in mRNA that causes the addition of a particular amino acid onto a protein or the termination of translation.

coelom Body cavity, lying between the digestive tract and body wall, that is completely lined by mesoderm.

coelomates Animals that possess a true coelom as their body cavity.

coenzyme Nonprotein organic molecule that aids the action of an enzyme.

coevolution Joint evolution in which one species exerts selective pressure on another species.

Cohesion Ability of water molecules to cling to each other due to hydrogen bonding.

cohesion-tension model Explanation that says that the upward transport of water in xylem is due to transpiration-created tension and the cohesive properties of water molecules.

collecting duct Duct in the kidneys that receives fluid from several nephrons. The reabsorption of water occurs here.

commensalism Interaction in which one species is benefited and the other is neither harmed nor benefited.

common ancestor Ancestor held in common by at least two lines of descent.

community All the populations in a particular locale.

compact bone The more dense form of bone; contains osteons containing concentric circles surrounding bone cells called osteocytes.

companion cell Type of cell found in the phloem of plants that aids in the transport of nutrients.

competition Interaction between two species in which both need the same limited resource, which results in harm to both.

competitive exclusion principle Theory that no two species can occupy the same niche at the same time.

complement system Series of proteins in plasma that provide a nonspecific defense mechanism against invasion by pathogens; this mechanism complements certain immune responses.

complementary base pairing Hydrogen bonding between particular purines and pyrimidines in DNA.

compound Substance having two or more different elements united chemically in a fixed ratio.

conclusion Statement made following an experiment as to whether or not the results support the hypothesis.

condom Plastic contraceptive that prevents the fertilization of an egg by a sperm. May be worn by male or female.

cone In vertebrate eyes, photoreceptor cell that responds to bright light and makes color vision possible. In conifers, structure that bears either pollen (male

gametophyte) or seeds (female gametophyte). The sporangia-bearing structures of certain seedless vascular plants are also termed cones.

conifer Member of a group of cone-bearing gymnosperm plants, including pine, cedar, and spruce trees.

conjugation Transfer of genetic material from one cell to another.

connective tissue Type of animal tissue that binds structures together, provides support and protection, fills spaces, stores fat, and forms blood cells.

conservation biology Field of biology that focuses on conserving natural resources and preserving biodiversity.

consumer Organism that feeds on another organism in a food chain; primary consumers eat plants, and secondary consumers eat animals.

contraceptive implant Birth control method utilizing synthetic progesterone; prevents ovulation by disrupting the ovarian cycle.

contraceptive injection Birth control method utilizing progesterone or estrogen and progesterone; prevents ovulation by disrupting the ovarian cycle.

contraceptive vaccine Under development, this birth control method immunizes against the hormone hCG, crucial to maintaining implantation of the embryo.

control group Sample that goes through all the steps of an experiment but is not exposed to the experimental variable; a standard against which the results of an experiment are checked.

convergent evolution Acquisition of the same or similar traits in distantly related lines of descent.

copulation Sexual union between a male and a female.

corolla The petals, collectively; usually the conspicuously colored whorl of a flower.

cortex In plant stems and roots, a band of ground tissue made up of parenchyma cells and bounded by the epidermis and vascular tissue; in animals, outer layer of an organ, such as the cortex of the kidneys or adrenal gland.

cotyledon Seed leaf of an embryo of a flowering plant; provides nutrient molecules for the developing plant before photosynthesis begins.

covalent bond Chemical bond in which atoms share one or more pairs of electrons.

cranial nerve Nerve that arises from the brain.

Cro-Magnons Common name for the oldest fossils to be designated *Homo sapiens*.

crossing-over Exchange of genetic material between nonsister chromatids of a tetrad during meiosis.

crustacean Member of a group of marine arthropods that contains, among others, shrimps, crabs, crayfish, and lobsters.

culture Total pattern of human behavior and products; includes technology and the arts and is dependent on the capacity to speak and transmit knowledge.

Cushing syndrome Condition resulting from hypersecretion of adrenal cortex hormones; characterized by masculinization in women, a "moon face" in both men and women, and an increase in blood glucose and sodium levels.

cutaneous receptor Sensory receptor for pressure, pain, touch, and temperature; found in the dermis of the skin.

cuticle Waxy layer covering the epidermis of plants; protects the plant against water loss and disease-causing organisms.

cyanobacteria Photosynthetic bacteria that contain chlorophyll and release O_2; formerly called blue-green algae.

cyclin Protein that cycles in quantity as the cell cycle progresses; combines with and activates the kinases that promote the events of the cycle.

cytokinesis Division of the cytoplasm following mitosis and meiosis.

cytokinin Plant hormone that promotes cell division; often works in combination with auxin during organ development in plant embryos.

cytosine (C) One of four nitrogen-containing bases in the nucleotides composing the structure of DNA and RNA; pairs with guanine.

cytoskeleton Internal framework of the cell, consisting of microtubules, actin filaments, and intermediate filaments.

cytotoxic T cell T cell that attacks and kills antigen-bearing cells.

D

data (sing., datum) Facts, or information, collected through observation and/or experimentation.

decomposer Organism, usually a bacterium or fungus, that breaks down dead organic matter into inorganic nutrients that can be recycled in the environment.

deforestation Removal of trees from a forest in a way that reduces the extent of the forest.

dehydration synthesis reaction Chemical reaction resulting in a covalent bond with the accompanying loss of a water molecule.

deletion Change in chromosome structure in which the end of a chromosome breaks off or two simultaneous breaks lead to the loss of an internal segment; often causes abnormalities—e.g., cri du chat syndrome.

demographics Study of the characteristics of a population and the changes in those characteristics over time.

denatured Loss of an enzyme's normal shape, so that it no longer functions; caused by a less than optimal pH and temperature.

dendrite Part of a neuron that sends signals toward the cell body.

Denisovans Recently discovered group of the genus *Homo* that existed in the region that is now Siberia around 1 MYA.

density-dependent factor Biotic factor, such as disease, competition, or predation, that affects population size according to the population's density.

density-independent factor Abiotic factor, such as fire or flood, that affects population size independent of the population's density.

deoxyribose Pentose monosaccharide sugar that is found in DNA and has one less hydroxyl group than ribose.

desertification Conversion of semiarid land to desertlike conditions, often beginning with overgrazing.

detrital food web Complex pattern of interlocking and crisscrossing food chains that begins with a population of decomposers.

detritus Organic matter produced by decomposition of substances, such as tissues and animal wastes.

deuterostome Coelomate animals in which the second embryonic opening is associated with the mouth. The first embryonic opening, the blastopore, is associated with the anus.

development Events that occur from fertilization until the formation of an adult organism.

diabetes mellitus Condition characterized by a high blood glucose level and the appearance of glucose in the urine, due to a deficiency of insulin production and failure of cells to take up glucose.

dialysis Movement of dissolved molecules through a semipermeable membrane by diffusion. Used to remove compounds from blood in kidney patients.

diaphragm In mammals, a muscular, membranous partition separating the thoracic cavity from the abdominal cavity; important in inhalation. Also, a birth control device consisting of a soft, latex cup that fits over the cervix.

diarrhea Excessively frequent and watery bowel movements.

diastole Relaxation period of a heart chamber during the cardiac cycle.

diet A person's typical food choices. A balanced diet contains all the nutrients in the right proportions to maintain a healthy body.

dietary supplement Nutrient or plant product (e.g., herbal tea, protein supplement) used to enhance health. Supplements do not undergo the same safety and effectiveness testing required for prescription drugs.

diffusion Passive movement of molecules or ions from an area of higher concentration to an area of lower concentration.

digestive system Organ system that includes the mouth, esophagus, stomach, small intestine, and large intestine (colon). It receives and digests food into nutrient molecules and has associated organs: teeth, tongue, salivary glands, liver, gallbladder, and pancreas.

dihybrid cross Genetic cross involving two traits, where the individuals involved in the cross are heterozygous for both traits; example is $AaBb \times AaBb$

diploid (2n) number Cell condition in which two of each type of chromosome are present.

directional selection Outcome of natural selection in which an extreme phenotype at one end of a population distribution is favored over all other phenotypes, leading to one distinct form.

disaccharide Sugar that contains two units of a monosaccharide. One example is maltose.

disruptive selection Outcome of natural selection in which both extreme phenotypes at the ends of a population distribution are favored over the average phenotype, leading to more than one distinct form.

distal convoluted tubule Final portion of a nephron that joins with a collecting duct; associated with tubular secretion.

diversity Measure of both species richness and species distribution in a community.

DNA (deoxyribonucleic acid) Nucleic acid polymer produced from covalent bonding of nucleotide monomers that contain the sugar deoxyribose; the genetic material of nearly all organisms.

DNA fingerprinting Use of DNA fragment lengths resulting from restriction enzyme cleavage to identify individuals.

DNA ligase Enzyme that links DNA fragments; used during production of recombinant DNA to join foreign DNA to vector DNA.

DNA polymerase During replication, the enzyme that joins the nucleotides so that their sequence is complementary to that in the parental strand.

DNA replication Synthesis of a new DNA double helix prior to mitosis or meiosis in eukaryotic cells and during prokaryotic fission in prokaryotic cells.

domain Largest of the categories, or taxa, used by taxonomists to group species. The three domains are Archaea, Bacteria, and Eukarya.

domain Archaea One of the three domains of life; contains prokaryotic cells that often live in extreme habitats and have unique genetic, biochemical, and physiological characteristics; its members are sometimes referred to as *archaea.*

domain Bacteria One of the three domains of life; contains prokaryotic cells that differ from archaea because they have unique genetic, biochemical, and physiological characteristics.

domain Eukarya One of the three domains of life, consisting of organisms with eukaryotic cells and further classified into the kingdoms Protista, Fungi, Plantae, and Animalia.

dominant allele Allele that exerts its phenotypic effect in the heterozygote. It masks the expression of the recessive allele.

double fertilization In flowering plants, event in which one sperm nucleus unites with the egg nucleus and the other sperm nucleus unites with the polar nuclei of an embryo sac.

Down syndrome Trisomy 21. The individual has three copies of chromosome 21 and the following characteristics: mental disabilities of varying degree, short stature, an eyelid fold, stubby fingers, and a palm crease.

duplication Change in chromosome structure in which a particular segment is present more than once in the same chromosome.

dyad Chromosome composed of two sister chromatids.

E

echinoderm Marine invertebrate, such as a sea star, sea urchin, or sand dollar; characterized by radial symmetry and a water vascular system.

ecological niche Role an organism plays in its community, including its habitat and its interactions with other organisms.

ecological pyramid A diagram that depicts the transfer of nutrients and energy among various trophic levels in a food chain or web.

ecological succession Gradual replacement of communities in an area following a disturbance (secondary succession) or the creation of new soil (primary succession).

ecology Scientific study of the interactions of organisms with each other and with their physical environment.

ecosystem A community of living organisms along with their physical environment; characterized by a flow of energy and a cycling of inorganic nutrients.

ectoderm Outermost primary germ layer of an animal embryo; gives rise to the nervous system and the outer layer of the integument.

ectothermic Having a body temperature that varies according to the environmental temperature.

edema Swelling due to fluid accumulation in the intercellular spaces in tissues.

electrocardiogram (ECG) Recording of the electrical activity associated with the heartbeat.

electron Negative subatomic particle moving about in an energy level around the nucleus of an atom.

electron shell Average location, or energy level, of an electron in an atom. Often shown as concentric circles around the nucleus of an atom. Also called *orbitals.*

electron transport chain Passage of electrons along a series of electron carriers from a higher to lower energy levels; the energy released is used to synthesize ATP.

electronegativity Ability of an atom to attract electrons toward itself in a chemical bond.

element Substance that cannot be broken down into substances with different properties; composed of only one type of atom.

embryo sac Female gametophyte of flowering plants.

embryonic development Period of animal development that encompasses all of the events from fertilization until the full formation of a new individual.

embryonic stem cell Undifferentiated cell, obtained from an embryo, that can be manipulated to become a specialized cell type, such as a muscle, nerve, or red blood cell.

emergency contraception Forms of contraception that are used after intercourse to stop a pregnancy.

emerging disease Infectious disease that has not previously been detected in humans.

emerging virus Causative agent of a disease that is new or is demonstrating increased prevalence, such as the viruses that cause AIDS, SARS, and avian influenza.

emulsification Breaking up of fat globules into smaller droplets by the action of bile salts or any other emulsifier.

endemic goiter Condition in which an enlarged thyroid produces low levels of thyroxine.

endocrine gland Ductless organ that secretes one or more hormones into the blood.

endocrine system Organ system that consists of hormonal glands and is involved in the coordination of body activities; secretes hormones as chemical messengers.

endocytosis Process by which substances are moved into the cell from the environment by phagocytosis (cellular eating) or pinocytosis (cellular drinking); includes receptor-mediated endocytosis.

endoderm Innermost primary germ layer of an animal embryo that gives rise to the linings of the digestive and respiratory tracts and their associated structures.

endodermis Internal tissue of a plant root forming a boundary between the cortex and the vascular cylinder.

endomembrane system Collection of membranous structures involved in transport within the cell.

endoplasmic reticulum (ER) System of membranous channels and saccules in the cytoplasm, often with attached ribosomes.

endoskeleton Protective internal skeleton, as in vertebrates.

endosperm In flowering plants, nutritive storage tissue derived from the union of a sperm nucleus and polar nuclei in the embryo sac.

endospore Spore formed by certain types of bacteria in response to unfavorable environmental conditions.

endosymbiotic theory Explanation for the evolution of eukaryotic organelles that involves the establishment of symbiotic relationships between bacteria to form the internal organelles of eukaryotic cells.

endothermic Maintaining a constant body temperature by generating internal heat.

energy Capacity to do work and bring about change; occurs in a variety of forms.

energy of activation (E$_a$) Energy that must be added in order for molecules to react with one another.

enhancer Region of DNA that stimulates transcription of nearby genes; functions by acting as a binding site for transcription factors.

entropy Measure of disorder, randomness or disorganization of a system.

enzyme Organic catalyst, usually a protein, that speeds a reaction in cells due to its shape.

epidermal tissue (epidermis) Tissue that forms the outer protective layer on a part of a plant.

epigenetic inheritance Transmission of genetic information by means that are not based on the coding sequences of a gene.

epiglottis Structure that covers the glottis and closes off the air tract during the process of swallowing.

epistatic interaction Pattern of inheritance where one gene overrides or cancels the instructions of another gene.

epithelial tissue Tissue that forms external coverings and internal linings of many organs and covers the entire surface of the body; also called epithelium.

equilibrium species Species demonstrating a life history pattern in which members exhibit logistic population growth and the population size remains at or near the carrying capacity. Its members are large and slow to mature, have a long life span and few offspring, and provide much care to offspring (e.g., bears, lions).

esophagus Muscular tube for moving swallowed food from the pharynx to the stomach.

essential nutrient Substance that contributes to good health and must be supplied by the diet because the body either cannot synthesize it or makes an insufficient amount.

estrogen Female sex hormone that helps maintain sexual organs and secondary sex characteristics.

ethylene Plant hormone that causes ripening of fruit and is involved in abscission.

euchromatin Chromatin that is extended and accessible for transcription.

eudicot Eudicotyledon; member of a flowering plant group that has two embryonic leaves (cotyledons), net-veined leaves, vascular bundles in a ring, and flower parts in fours or fives and their multiples.

eukaryotic cell Type of cell that has a membrane-bound nucleus and membranous organelles; found in organisms within the domain Eukarya.

eutrophication Overenrichment of a body of water with inorganic nutrients used by phytoplankton. Often, overenrichment caused by human activities leads to excessive bacterial growth and oxygen depletion.

evolution Inheritable changes in a population over time that make the population better adapted to their environment.

evolutionary tree Diagram that shows the evolutionary history of groups of organisms.

ex vivo gene therapy Gene therapy in which infected tissue is removed from an organism, injected with normal genes, and then returned to the organism. The normal genes divide, producing normal genes and tissues instead of infected ones.

exocrine gland Gland that discharges its secretion into one or more ducts. The pancreas is an exocrine gland when it secretes pancreatic juice into the duodenum.

exocytosis Process in which an intracellular vesicle fuses with the plasma membrane, so that the vesicle's contents are released outside the cell.

exoskeleton Protective external skeleton, as in arthropods.

exotic species Species that is new to a community (nonnative).

experiment Series of actions undertaken to collect data with which to test a hypothesis.

experimental design Artificial situation devised to test a hypothesis.

experimental variable In a scientific experiment, a condition of the experiment that is deliberately changed.

expiration Movement of air out of the lungs by positive pressure.

exponential growth Growth, particularly of a population, in which the increase occurs in the same manner as compound interest.

extinction Total disappearance of a species or higher group.

extracellular matrix (ECM) Meshwork of polysaccharides and proteins that provides support for an animal cell and affects its behavior.

F

family One of the categories, or taxa, used by taxonomists to group species; the taxon above the genus level.

fat Organic molecule that contains glycerol and fatty acids and is found in adipose tissue of vertebrates.

fatty acid Molecule that contains a hydrocarbon chain that ends with a carboxyl group.

feedback inhibition Mechanism for regulating metabolic pathways in which the concentration of the product is kept within a certain range until binding of the substrate to the enzyme's active site reduces (or stops) the activity of the pathway.

fermentation Anaerobic breakdown of glucose that results in a gain of 2 ATP and end products, such as alcohol and lactate.

fern Member of a group of plants that have large fronds. In the sexual life cycle, the independent gametophyte produces flagellated sperm, and the vascular sporophyte produces windblown spores.

fertilization Fusion of sperm and egg nuclei, producing a zygote that develops into a new individual.

fiber Structure resembling a thread; also carbohydrate plant material that is nondigestible.

filament End-to-end chain of cells that forms as cell division occurs in only one plane; in plants, the elongated stalk of a stamen.

filtration Movement of small molecules from a blood capillary into the glomerular capsule of a nephron due to the action of blood pressure.

fitness The reproductive success of an individual relative to other members of a population.

five-kingdom system System of classification that contains the kingdoms Monera, Protista, Plantae, Animalia, and Fungi.

flagellum (pl., flagella) Long, slender extension used for locomotion by some bacteria, protozoans, and sperm.

flatworm Invertebrate, such as a planarian or tapeworm, that has a thin body, three-branched gastrovascular cavity, bilateral symmetry, and ladderlike nervous system; member of phylum Platyhelminthes.

flower Reproductive organ of a flowering plant (an angiosperm), consisting of several kinds of modified leaves arranged in concentric rings and attached to a modified stem called a receptacle.

follicle In the ovary of an animal, structure that contains an oocyte; site of oocyte production.

follicular phase First half of the ovarian cycle, during which the follicle matures and much estrogen (and some progesterone) is produced.

food chain Order in which one population feeds on another in an ecosystem, thereby showing the flow of energy from a decomposer (detrital food chain) or a producer (grazing food chain) to the final consumer.

food web In an ecosystem, a complex pattern of linked and crisscrossing food chains, showing the energy flow between the components of the ecosystem.

formed elements Constituents of blood that are either cellular (red blood cells and white blood cells) or at least cellular in origin (platelets).

fossil Remains of a once-living organism that have been preserved in the Earth's crust.

fossil fuel Fuel such as oil, coal, and natural gas that is the result of partial decomposition of plants and animals coupled with exposure to heat and pressure for millions of years.

fossil record History of life recorded in remains from the past that are found in rock strata.

founder effect Mechanism of genetic drift that occurs when a new population is established by a small number of individuals, carrying only a small fraction of the original population's genetic variation. The alleles carried by these individuals often, by chance, do not occur in the same frequency as in the original population.

frameshift mutation Alteration in a DNA sequence due to addition or deletion of a nucleotide, so the reading "frame" is shifted; this type of mutation can result in a nonfunctional protein.

fruit In flowering plants, the structure that forms from an ovary and associated tissues and encloses seeds.

functional group Specific cluster of atoms attached to the carbon skeleton of organic molecules that enters into reactions and behaves in a predictable way.

fungus (pl., fungi) Saprotrophic decomposer. The body is made up of filaments, called hyphae, that form a mass called a mycelium. Member of the kingdom Fungi.

G

gallbladder Organ connected to the liver that stores and concentrates bile.

gamete Haploid sex cell: egg or sperm.

gametophyte Haploid generation of the life cycle of a plant; produces gametes that unite to form a diploid zygote.

ganglion Collection of neuron cell bodies found within the peripheral nervous system.

gap junction Junction between cells formed by the joining of two adjacent plasma membranes; lends strength and allows ions, sugars, and small molecules to pass between cells.

gastropod Mollusc with a broad, flat foot for crawling (e.g., snails and slugs).

gastrulation Formation of a gastrula from a blastula; characterized by an invagination of the cell layers to form a caplike structure.

gene Unit of heredity existing as alleles on the chromosomes. In diploid organisms, typically two alleles are inherited—one from each parent.

gene flow Movement of alleles among populations by migration of breeding individuals.

gene pool The various alleles at all the gene loci in all individuals in a population.

gene therapy Insertion of genetic material into human cells for the treatment of a disorder.

genetic counseling Consultation between prospective parents and a counselor who determines the genotype of each and whether an unborn child will have an inherited disorder.

genetic drift Mechanism of evolution due to random changes in the allele frequencies of a gene pool; more likely to occur in small populations or when only a few individuals of a large population reproduce.

genetic engineering Alteration of genomes for medical or industrial purposes.

genetic marker Abnormality in the sequence of bases at a particular location on a chromosome, signifying a disorder.

genetically modified organism (GMO) An organism whose genetic material has been changed, usually by using genetic engineering techniques such as recombinant DNA technology.

genital herpes Viral STD characterized by painful blisters on the genitals.

genome Sum of all of the genetic material in a cell or organism.

genome editing Form of DNA technology that uses nucleases that can target specific sequences of nucleotides in the genome of an organism for inactivation or insertion of new nucleotides.

genomics Study of genomes.

genotype Combination of alleles that determines a particular trait in an organism; often designated by letters—for example, *BB* or *Aa*.

genus One of the categories, or taxa, used by taxonomists to group species; contains those species that are most closely related through evolution.

germ layer One of the primary tissue layer of a vertebrate embryo—namely, the ectoderm, mesoderm, or endoderm.

germinate Beginning of growth of a seed, spore, or zygote, especially after a period of dormancy.

gibberellin Plant hormone promoting increased stem growth; also involved in flowering and seed germination.

gland Group of epithelial cells that are specialized to produce a substance.

global warming Predicted increase in the Earth's temperature due to human activities that promote the greenhouse effect.

glomerular capsule Cuplike structure that is the initial portion of the nephron.

glottis Opening for airflow in the larynx.

glucagon Hormone, secreted by the pancreas, that causes the liver to break down glycogen and raises the blood glucose level.

glucose Six-carbon sugar that organisms break down as a source of energy during cellular respiration.

glycerol Three-carbon carbohydrate with three hydroxyl groups attached; a component of fats and oils.

glycogen Storage polysaccharide found in animals; composed of glucose molecules joined in a linear fashion but having numerous branches.

glycolysis Anaerobic breakdown of glucose that results in a gain of 2 ATP and the end product pyruvate.

Golgi apparatus Organelle, consisting of saccules and vesicles, that processes, packages, and distributes molecules about or from the cell.

Golgi tendon Proprioceptive sensory receptor located where skeletal muscle inserts into tendons; sensitive to changes in tendon tension.

gonads Organs that produce gametes. The ovary produces eggs, and the testis produces sperm.

gonorrhea Bacterial STD that can lead to sterility or infertility.

gravitational equilibrium Maintenance of balance that depends on sensing the position of the head in relation to gravity; involves the utricle and saccule of the inner ear.

gravitropism Growth response of plant roots and stems to Earth's gravity. Roots demonstrate positive gravitropism, and stems demonstrate negative gravitropism.

grazing food web Complex pattern of linked and crisscrossing food chains that begins with a population of photosynthesizers serving as producers.

greenhouse effect Reradiation of solar heat toward the Earth, caused by gases such as carbon dioxide, methane, nitrous oxide, water vapor, and nitrous oxide in the atmosphere.

ground tissue Tissue that constitutes most of the body of a plant; consists of parenchyma, collenchyma, and sclerenchyma cells that function in storage, basic metabolism, and support.

growth factor Chemical signal that regulates mitosis and differentiation of cells that have receptors for it; important in such processes as fetal development, tissue maintenance and repair, and hematopoiesis; sometimes a contributing factor in cancer.

guanine (G) One of four nitrogen-containing bases in nucleotides composing the structure of DNA and RNA; pairs with cytosine.

gymnosperm Type of woody seed plant in which the seeds are not enclosed by fruit and are usually borne in cones, such as those of the conifers.

H

habitat Place where an organism lives and is able to survive and reproduce.

halophile Type of archaea that lives in extremely salty habitats.

haploid (n) number Cell condition in which only one of each type of chromosome is present.

heart Muscular organ whose contraction causes blood or other fluid to circulate in the body of an animal.

heart attack Myocardial infarction; damage to the myocardium due to blocked circulation in the coronary arteries.

heat Type of kinetic energy. Captured solar energy eventually dissipates as heat in the environment.

helicase The enzyme in DNA replication that separates DNA strands by breaking the hydrogen bonds between the strands.

helper T cell Cell that secretes cytokines, which stimulate all kinds of immune cells.

heme Iron-containing group found in hemoglobin.

hemocoel Body cavity in arthropods where exchange between hemolymph and tissues occurs.

hemoglobin Iron-containing respiratory pigment occurring in vertebrate red blood cells and in the blood plasma of some invertebrates.

hepatitis Viral infection of the liver; can be transmitted sexually and can lead to liver failure and liver cancer.

herbivore Primary consumer in a grazing food chain; plant eater.

hermaphrodite An animal that has both male and female sex organs.

heterochromatin Highly compacted chromatin that is not accessible for transcription.

heterotroph Organism that cannot synthesize organic compounds from inorganic substances and therefore must take in organic nutrients.

heterozygous Possessing unlike alleles for a particular trait.

hinge joint Synovial joint (at elbow or knee) that permits movement in one direction only.

histamine Substance, produced by basophils in blood and mast cells in connective tissue, that causes capillaries to dilate.

histone Protein molecule responsible for packing chromatin.

homeostasis Maintenance of the relatively constant condition of the internal environment of the body within an acceptable range.

hominid Family designation for humans, apes (African and Asian), and chimpanzees.

hominin Designation that includes humans and extinct species very closely related to humans.

Homo erectus Early species of the genus *Homo* that used fire and migrated out of Africa to Europe and Asia.

Homo habilis Early species of the genus *Homo,* dated between 2.4 and 1.4 MYA and believed to have been the first tool users.

homologous chromosome Homologue; member of a pair of chromosomes that are alike and come together in synapsis during prophase of the first meiotic division.

homologous structure In evolution, a structure that is similar in different types of organisms because these organisms are descended from a common ancestor.

homologue Homologous chromosome; member of a pair of chromosomes that are alike and come together in synapsis during prophase of the first meiotic division.

homozygous Possessing two identical alleles for a particular trait.

hormone Chemical messenger produced by a gland that controls the activity of other tissues, organs, or cells of the body.

human immunodeficiency virus (HIV) Virus responsible for AIDS.

human papillomavirus (HPV) Virus that is responsible for genital warts and cervical cancer.

hydrogen bond Weak chemical bond between a slightly positive hydrogen atom on one molecule and a slightly negative atom on another molecule.

hydrolysis reaction Splitting of a compound by the addition of water, with the H^+ being incorporated into one molecule and the OH^- into the other.

hydrophilic Type of molecule that interacts with water by dissolving in it and/or by forming hydrogen bonds with water molecules.

hydrophobic Type of molecule that does not interact with water because it is nonpolar.

hydrostatic skeleton Fluid-filled body compartment that provides support for muscle contraction resulting in movement; seen in cnidarians, flatworms, roundworms, and segmented worms.

hypertension Elevated blood pressure, particularly the diastolic pressure.

hypertonic solution Solution with a higher solute concentration (less water) than the cytoplasm of a cell; causes cells to lose water by osmosis.

hyphae (sing., hypha) Filaments of the vegetative body of a fungus.

hypothalamus In vertebrates, part of the brain that helps regulate the internal environment of the body—for example, heart rate, body temperature, and water balance.

hypothesis Supposition established by reasoning after consideration of available evidence. It can be tested by obtaining more data, often by experimentation.

hypotonic solution Solution having a lower solute (more water) concentration than the cytoplasm of a cell; causes cells to gain water by osmosis.

I

immune system Cells and organs that protect the body against foreign organisms and substances as well as cancerous cells.

immunity Ability of the body to protect itself from foreign substances and cells, including disease-causing agents.

in vivo gene therapy Gene therapy in which normal genes are delivered directly into an organism.

incomplete dominance Inheritance pattern in which the offspring has an intermediate phenotype, as when a red-flowered plant and a white-flowered plant produce pink-flowered offspring.

induced fit model Change in the shape of an enzyme's active site that enhances the fit between the active site and its substrate(s).

induction Ability of a tissue or organ to influence the development of another tissue or organ.

inductive reasoning Using specific observations and the process of logic and reasoning to arrive at a hypothesis.

infertility The failure of a couple to achieve pregnancy after 1 year of regular, unprotected intercourse.

inflammatory response Tissue response to injury or pathogen that is characterized by redness, heat, swelling, and pain.

innate immunity A function of the immune system that protects the body against pathogens in a nonspecific manner.

insect Type of arthropod. The head has antennae, compound eyes, and simple eyes; the thorax has three pairs of legs and often wings; and the abdomen has internal organs.

inspiration Movement of air into the lungs by negative pressure.

insulin Hormone, secreted by the pancreas, that lowers blood glucose level by promoting the uptake of glucose by cells and the conversion of glucose to glycogen by the liver and skeletal muscles.

integration Summing up of excitatory and inhibitory signals by a neuron or a part of the brain.

integumentary system Organ system consisting of the skin and its accessory structures.

intergenic DNA The sequences of nucleotides that lie between genes on a chromosome.

interkinesis Period of time between meiosis I and meiosis II during which no DNA replication takes place.

interneuron In the central nervous system, a neuron that conveys messages between parts of the central nervous system.

interphase Stages of the cell cycle (G_1, S, G_2) during which growth and DNA synthesis occur when the nucleus is not actively dividing.

intrauterine device (IUD) Birth control device consisting of a small piece of molded plastic inserted into the uterus; believed to alter the uterine environment, so that fertilization does not occur.

inversion Change in chromosome structure in which a segment of a chromosome is turned 180°. The reversed sequence of genes can lead to altered gene activity and abnormalities.

invertebrates Animals lacking an internal skeleton made of cartilage or bone.

ion Charged particle that carries a negative or positive charge.

ionic bond Chemical bond in which ions are attracted to one another by opposite charges.

isomer Molecule with the same molecular formula as another but having a different structure and therefore a different shape.

isotonic solution Solution that is equal in solute concentration to that of the cytoplasm of a cell; causes a cell to neither lose nor gain water by osmosis.

isotope Atom of the same element having the same atomic number but a different mass number due to the number of neutrons.

J

jawless fishes Group of fishes that have no jaws; includes hagfishes and lampreys.

K

karyotype Image of chromosomes arranged by pairs according to their size, shape, and banding pattern.

keystone species Species whose activities significantly affect community structure.

kidneys Paired organs of the vertebrate urinary system that regulate the chemical composition of the blood and produce the waste product urine.

kilocalorie Measure of the caloric value of food; equal to 1,000 calories.

kinase Enzyme that activates another enzyme by adding a phosphate group.

kinetic energy Energy associated with motion.

kingdom One of the categories, or taxa, used by taxonomists to group species; the taxon above phylum.

Klinefelter syndrome Condition caused by the inheritance of XXY chromosomes.

L

lacteal Lymphatic capillary in an intestinal villus; aids in the absorption of fats.

lancelet Invertebrate chordate with a body that resembles a lancet and has the four chordate characteristics as an adult.

large intestine In vertebrates, portion of the digestive tract that follows the small intestine; in humans, consists of the cecum, colon, rectum, and anal canal.

larynx Voice box; cartilaginous structure located between the pharynx and the trachea; in humans, contains the vocal cords.

law Universal principle that describes the basic functions of the natural world.

law of independent assortment Alleles of unlinked genes assort independently of each other during meiosis, so that the gametes contain all possible combinations of alleles.

law of segregation Separation of alleles from each other during meiosis, so that the gametes contain one from each pair. Each resulting gamete has an equal chance of receiving either allele.

leaf Lateral appendage of a stem, highly variable in structure, often containing cells that carry out photosynthesis.

less-developed country (LDC) Country that is becoming industrialized, with rapid population growth and the majority of people living in poverty.

lichen Symbiotic (mutualistic) relationship between certain fungi and cyanobacteria or green algae, in which the fungi provide minerals and water and the cyanobacteria or algae provide organic molecules, such as sucrose.

life cycle Recurring pattern of genetically programmed events by which individuals grow, develop, maintain themselves, and reproduce.

life history A particular mix of characteristics of a population or species, such as how many offspring it produces, its survivorship, and factors (such as age and size) that determine its reproductive maturity.

ligament Tough cord or band of dense fibrous tissue that binds bone to bone at a joint.

light reactions Portion of photosynthesis that captures solar energy and takes place in thylakoid membranes of chloroplasts; produces ATP and NADPH.

limbic system In humans, complex network that includes the diencephalon and areas of the cerebrum, including the amygdala and hippocampus; governs learning and memory and various emotions, such as pleasure, fear, and happiness.

lineage Evolutionary line of descent.

linkage group Genes on the same chromosome that tend to be inherited together.

lipase Enzyme secreted by the pancreas that breaks down fats to glycerol and fatty acids.

lipid Class of organic compounds that tends to be soluble in nonpolar solvents; includes fats and oils.

liver Large, dark red internal organ that produces urea and bile, detoxifies the blood, stores glycogen, and produces the plasma proteins, among other functions.

lobe-finned fishes Bony fishes with limblike fins.

locus Specific location of a particular gene on homologous chromosomes.

logistic growth Population increase that results in an S-shaped curve. Growth is slow at first, steepens, and then levels off due to environmental resistance.

lung Internal respiratory organ containing moist surfaces for gas exchange.

luteal phase Second half of the ovarian cycle, during which the corpus luteum develops and much progesterone (and some estrogen) is produced.

lycophytes Club mosses; seedless plants with microphylls and well-developed vascular tissue in roots, stems, and leaves.

lymph Fluid, derived from interstitial fluid, that is carried in lymphatic vessels.

lymph node Small mass of lymphatic tissue located along a lymphatic vessel; function is to filter excess interstitial fluid (lymph) before it is returned to the circulatory system.

lymphatic system Organ system consisting of lymphatic vessels, lymph nodes, and other lymphatic organs; absorbs fat from the digestive system, collects excess interstitial fluid (lymph), and aids the immune system.

lymphocyte Specialized white blood cell that functions in specific defense and occurs in two forms: T cells and B cells.

lysogenic cycle Bacteriophage life cycle in which the virus incorporates its DNA into that of a bacterium; occurs preliminary to the lytic cycle.

lysosome Membrane-bounded vesicle that contains hydrolytic enzymes for digesting macromolecules.

lytic cycle Bacteriophage life cycle in which a virus takes over the operation of a bacterium immediately upon entering it and subsequently destroys the bacterium.

M

macroevolution Large-scale evolutionary change, such as the formation of new species.

macronutrient Essential element or other substance needed in large amounts by plants or humans. In plants, nitrogen, calcium, and sulfur are needed for plant growth; in humans, carbohydrates, lipids, and proteins supply the body's energy needs.

macrophage In vertebrates, large, phagocytic cell derived from a monocyte that ingests microbes and debris.

major histocompatability complex (MHC) Cell surface proteins that recognize self cells and bind antigens for presentation to a T cell.

major mineral Essential inorganic nutrient (such as calcium, potassium, phosphorus, sodium, chloride, magnesium, or sulfur) required daily by humans to regulate metabolic activities and maintain health.

malignant tumor Form of invasive tumor that may spread to other tissues of the body.

Malpighian tubule Blind, threadlike excretory tubule near the anterior end of an insect's hindgut.

mammal Endothermic vertebrate characterized especially by the presence of hair and mammary glands.

marsupial Member of a group of mammals bearing immature young nursed in a marsupium, or pouch—for example, kangaroo or opossum.

mass extinction Episode of large-scale extinction in which large numbers of species disappear within a relatively short period of time.

mass number Mass of an atom equal to the number of protons plus the number of neutrons within the nucleus.

mast cell Type of white blood cell that releases histamine in the inflammatory response.

matrix Noncellular material that fills the spaces between cells in connective tissues and inside organelles.

matter Anything that takes up space and has mass.

mechanoreceptor Type of receptor that detects mechanical changes (such as movement).

medulla oblongata In vertebrates, part of the brain stem that is continuous with the spinal cord; controls heartbeat, blood pressure, breathing, and other vital functions.

megaphylls Large leaves with complex networks of veins.

meiosis, meiosis I, meiosis II Type of nuclear division that occurs as part of sexual reproduction in which the daughter cells receive the haploid number of chromosomes in varied combinations.

memory Capacity of the brain to store and retrieve information about past sensations, perceptions, and events; essential to learning.

Memory B cells B cells that are not used in the initial adaptive response but are held in reserve to provide long-term immunity to an antigen.

menopause Termination of the ovarian and menstrual cycles in older women.

menses Flow of blood during menstruation.

menstrual cycle Cycle that runs concurrently with the ovarian cycle. It prepares the uterus to receive a developing zygote.

menstruation Periodic shedding of tissue and blood from the inner lining of the uterus in primates.

meristem tissue Undifferentiated embryonic tissue in the active growth regions of plants.

mesoderm Middle primary germ layer of an animal embryo that gives rise to several internal organ systems and the dermis.

mesophyll Plant tissue where the majority of photosynthesis occurs.

messenger RNA (mRNA) Type of RNA formed from a DNA template and bearing coded information for the amino acid sequence of a polypeptide.

metabolism All of the chemical reactions that occur in a cell during growth and repair.

metaphase Mitotic phase during which chromosomes are aligned at the spindle equator.

metastasis Spread of cancer from the place of origin throughout the body; caused by the ability of cancer cells to migrate and invade tissues.

methanogen Type of archaea that lives in oxygen-free habitats, such as swamps, and releases methane gas.

microevolution Small, measurable evolutionary changes in a population from generation to generation; change in allele frequencies within a population over time.

micronutrient Essential element or other substance needed in small amounts by plants or humans. In plants, boron, copper, and zinc are needed for plant growth; in humans, vitamins and minerals help regulate metabolism and physiological development.

microphylls Small, narrow leaves with single, unbranched veins.

microtubule Small, cylindrical organelle composed of tubulin protein around an empty central core; present in the cytoplasm, centrioles, cilia, and flagella.

microvillus (pl., microvilli) Cylindrical process that extends from an epithelial cell of a villus; increases the surface area of the cell.

midbrain Part of the brain stem located between the diencephalon and the pons; contains visual and auditory reflex centers.

mineral An element that is needed in the diet for numerous physiological functions; a nonrenewable raw material in the Earth's crust that can be mined and used by humans.

mitochondrion (pl., mitochondria) Membrane-bounded organelle in which ATP molecules are produced during the process of cellular respiration.

mitosis Process in which a parent nucleus produces two daughter nuclei, each having the same number and kinds of chromosomes as the parent nucleus.

model system Simulation of a process; aids conceptual understanding until the process can be studied firsthand; a hypothesis that describes how a particular process might be carried out.

molecule Union of two or more atoms of the same element; also, the smallest part of a compound that retains the properties of the compound.

molluscs Animals characterized by a coelom, a complete digestive tract, a muscular foot, and a mantle; member of phylum Mollusca.

monocot Monocotyledon; member of a flowering plant group that has one embryonic leaf (cotyledon), parallel-veined leaves, scattered vascular bundles, and flower parts in threes or multiples of three.

monocyte Type of agranular leukocyte that functions as a phagocyte, particularly after it becomes a macrophage.

monohybrid cross Cross that involves individuals possessing two different alleles for a single trait; for example, *Aa*.

monomer Small molecule that is a subunit of a polymer—e.g., glucose is a monomer of starch.

monosaccharide Simple sugar; a carbohydrate that cannot be broken down by hydrolysis—e.g., glucose.

monotreme Egg-laying mammal—for example, duckbill platypus or spiny anteater.

more-developed country (MDC) Country that is industrialized, with modest population growth and a fairly good standard of living.

morula Spherical mass of cells resulting from cleavage during animal development prior to the blastula stage.

motor neuron Nerve cell that conducts nerve impulses away from the central nervous system and innervates effectors (muscle and glands).

mouth In humans, organ of the digestive tract where food is chewed and mixed with saliva.

mRNA transcript Complementary copy of the sequence of bases in the template strand of DNA.

multicellular Organism composed of many cells; usually has organized tissues, organs, and organ systems.

multifactorial trait Trait controlled by multiple genes subject to environmental influences. Each dominant allele contributes to the phenotype in an additive and like manner.

multipotent A term used to describe a stem cell that has the potential to form a limited number of cell types.

muscle dysmorphia Mental state in which a person thinks his or her body is underdeveloped and becomes preoccupied with body-building activities and diet; affects more men than women.

muscular system Organ system consisting of muscles that produce movement, both movement in the body and movement of its limbs. The principal components are skeletal muscle, smooth muscle, and cardiac muscle.

muscular tissue Type of animal tissue composed of fibers that shorten and lengthen to produce movements.

mutagen Agent, such as radiation or a chemical, that brings about a mutation in DNA.

mutation Change made in the nucleotide sequence of DNA. Such changes may be due to a replication error or the influence of external sources called mutagens. Mutations generate variation in the gene pool of a population.

mutualism Interaction in which both species benefit in terms of growth and reproduction.

mycelium Tangled mass of hyphal filaments composing the vegetative body of a fungus.

mycorrhizal fungi Fungi that form mutualistic relationships with the roots of vascular plants.

myelin sheath White, fatty material—derived from the membranes of tightly spiraled cells—that forms a covering for nerve fibers.

myosin Muscle protein making up the thick filaments in a sarcomere. It pulls actin to shorten the sarcomere, yielding muscle contraction.

N

nasal cavity One of two canals in the nose, separated by a septum.

native species Indigenous species that have colonized an area without human assistance.

natural history Study of how organisms are influenced by factors such as climate, predation, competition, and evolution; uses field observations instead of experimentation.

natural killer (NK) cell Large lymphocyte that causes a virus-infected or cancerous cell to burst.

natural selection Mechanism of evolution caused by environmental selection of organisms most fit to reproduce; results in adaptation to the environment.

Neandertal Later *Homo* species (*Homo neandertalensis*) with a sturdy build that lived during the last Ice Age in Europe and the Middle East; hunted large game and left evidence of being culturally advanced.

negative feedback Homeostatic mechanism that allows the body to keep the internal environment relatively stable.

nematocyst In cnidarians, a capsule that contains a threadlike fiber, the release of which aids in the capture of prey.

nephron Microscopic tubule in a kidney that regulates blood composition by filtration, reabsorption, and secretion; site of urine production.

nephron loop Portion of a nephron between the proximal and distal convoluted tubules; functions in water reabsorption.

nerve Bundle of long axons outside the central nervous system.

nerve impulse Electrochemical changes that take place across the axon's membrane; the action potential.

nervous system Organ system, consisting of the brain, spinal cord, and associated nerves; coordinates the other organ systems of the body.

nervous tissue Tissue that contains nerve cells (neurons), which conduct impulses, and neuroglia, which support, protect, and provide nutrients to neurons. Nervous tissue coordinates the functions of body parts and the body's responses to the external and internal environments.

neuroglia Nonconducting cells in nervous tissue that support and nourish neurons.

neuron Nerve cell, which characteristically has three parts: dendrites, a cell body, and an axon.

neurotransmitter Chemical stored at the ends of axons; responsible for transmission of a nerve impulse across a synapse.

neurulation Development of the central nervous system organs in an embryo.

neutron Neutral subatomic particle, located in the nucleus and assigned one atomic mass unit.

neutrophil Phagocytic, granular leukocyte that is the most abundant of the white blood cells; first to respond to infection.

nociceptor Type of pain receptor that is sensitive to extremes in temperature or pressure and detects chemical signals from damaged tisue.

nodes of Ranvier In the peripheral nervous system, gaps in the myelin sheath encasing each long axon.

nondisjunction Failure of homologous chromosomes or daughter chromosomes to separate during meiosis I or meiosis II, respectively.

nonrandom mating Mating that does not occur on a purely random basis; examples are inbreeding and sexual selection.

nonrenewable resources Minerals, fossil fuels, and other materials present in essentially fixed amounts (within the human timescale) in our environment.

nonsister chromatid A tetrad consists of four chromatids. Only the chromatids belonging to a homologue are sister chromatids; the others, belonging to the other homologue, are nonsister chromatids.

nonvascular plants Bryophytes, or plants that have no vascular tissue and either occur in moist locations or have adaptations for living in dry locations; mosses are an example.

notochord Cartilaginous-like, dorsal supporting rod in all chordates sometime in their life cycle; replaced by vertebrae in vertebrates.

nuclear envelope Double membrane that surrounds the nucleus in eukaryotic cells and is connected to the endoplasmic reticulum; has pores that allow substances to pass between the nucleus and the cytoplasm.

nuclease Enzyme that catalyzes decomposition of nucleic acids into nucleotides.

nucleic acid Polymer of nucleotides. Both DNA and RNA are nucleic acids.

nucleoid Region of a prokaryotic cell where the DNA is located. It is not bound by a nuclear envelope.

nucleolus In the nucleus, a dark-staining, spherical body that produces ribosomal subunits.

nucleosome In the nucleus of a eukaryotic cell, a unit composed of DNA wound around a core of eight histone proteins, giving the appearance of a bead on a string of beads.

nucleotide Monomer of DNA and RNA consisting of a 5-carbon sugar bonded to a nitrogenous base and a phosphate group.

nucleus Center of an atom, in which protons and neutrons are found; membrane-bound organelle within a eukaryotic cell that contains chromosomes and controls the structure and function of the cell.

nutrient Substance in food that performs a physiological function in the body and thus contributes to good health.

O

obesity Excess adipose tissue; exceeding desirable weight by more than 20%.

observation Step in the scientific method by which data are collected before a conclusion is drawn.

octet rule States that an atom other than hydrogen tends to form bonds until it has eight electrons in its outer shell. An atom that already has eight electrons in its outer shell does not react and is inert.

oil Triglyceride, usually of plant origin, that is composed of glycerol and three fatty acids and is liquid in consistency due to many unsaturated bonds in the hydrocarbon chains of the fatty acids.

omnivore Organism in a food chain that feeds on both plants and animals.

oncogene Cancer-causing gene.

oogenesis Production of eggs in females by the process of meiosis and maturation.

open circulatory system Circulatory system, such as that found in a grasshopper, in which a tubular heart pumps hemolymph through channels and body cavities.

operon Group of structural and regulating genes that function as a single unit.

opportunistic species Species demonstrating a life history pattern characterized by exponential population growth. Its members are small in size, mature early, have a short life span, produce many offspring, and provide little or no care to offspring (e.g., dandelions).

order One of the categories, or taxa, used by taxonomists to group species; the taxon above the family level.

organ Combination of two or more different tissues performing a common function.

organ system Group of related organs working together.

organelle Small, often membranous compartment in the cytoplasm having a specific structure and function.

organic Molecule that always contains carbon and hydrogen and often contains oxygen as well. Organic molecules are associated with living organisms.

organism An individual plant, fungi, animal, or single-celled organism.

osmosis Diffusion of water across a semipermeable membrane.

osteocyte Branched bone cell embedded in a calcium-containing extracellular matrix.

osteoporosis Condition in which bones break easily because calcium is removed from them faster than it is replaced.

ovary In animals, the female gonad that produces an egg and female sex hormones; in flowering plants, the enlarged, ovule-bearing portion of the carpel that develops into a fruit.

ovulation Release of a secondary oocyte from the ovary via the bursting of a mature follicle. If fertilization occurs, the secondary oocyte becomes an egg.

ovule In seed plants, a structure that contains the female gametophyte and has the potential to develop into a seed.

oxidation Loss of one or more electrons from an atom or a molecule; in biological systems, generally the loss of hydrogen atoms.

P

paleontology Study of the fossil record, which yields knowledge about the history of species and therefore of life on Earth.

pancreas Internal organ that produces digestive enzymes and the hormones insulin and glucagon.

pancreatic amylase Enzyme that digests starch to maltose.

pancreatic islets Masses of cells that constitute the endocrine portion of the pancreas; also called the islets of Langerhans.

parasitism Interaction in which one species (the *parasite*) obtains nutrients from another species (the *host*) but does not usually kill the host.

parasympathetic division Division of the autonomic system that is active under normal conditions; uses acetylcholine as a neurotransmitter.

parathyroid gland Gland embedded in the posterior surface of the thyroid gland; produces parathyroid hormone.

parathyroid hormone (PTH) Hormone, secreted by the four parathyroid glands, that increases the blood calcium level and decreases the phosphate level.

parthenogenesis Development of an egg cell into a whole organism without fertilization.

passive immunity Protection against infection acquired by transfer of antibodies to a susceptible individual.

pathogen Disease-causing agent, such as viruses, parasitic bacteria, fungi, and animals.

pedigree Chart showing a family's history with regard to a particular genetic trait.

pelvic inflammatory disease (PID) Latent infection of chlamydia or gonorrhea in the uterine tubes or the vasa deferentia.

penis Male copulatory organ; in humans, the male organ of sexual intercourse.

pepsin Enzyme, secreted by gastric glands, that digests proteins to peptides.

peptide Two or more amino acids joined together by covalent bonding.

peptide bond Type of covalent bond that joins two amino acids.

peptide hormone Type of hormone that is a protein or a peptide or is derived from an amino acid.

peptidoglycan Unique molecule found in bacterial cell walls.

pericycle Layer of cells surrounding the vascular tissue of roots; produces lateral roots.

peripheral nervous system (PNS) Nerves that lie outside the central nervous system.

peristalsis Rhythmic contractions that propel substances along a tubular structure, such as the esophagus.

petal Flower part just inside the sepals; often conspicuously colored to attract pollinators.

pH A logarithmic measure of the hydrogen ion concentration.

pH scale Numerical scale ranging from 0 to 14 that indicates the negative log of the hydrogen ion concentration in a solution.

pharynx In vertebrates, common passageway for both food intake and air movement; located between the mouth and the larynx.

phenotype Visible expression of a genotype—e.g., brown eyes or attached earlobes.

phloem Vascular tissue that conducts organic solutes in plants; contains sieve-tube members and companion cells.

phospholipid Molecule that forms the phospholipid bilayer of plasma membranes. It has a polar, hydrophilic head bonded to two nonpolar, hydrophobic tails.

photoautotroph Organism able to synthesize organic molecules by using carbon dioxide as a carbon source and sunlight as an energy source.

photoperiod Ratio of the lengths of day and night over a 24-hour period, which affects the physiology and behavior of organisms.

photoreceptor Sensory receptor that responds to light stimuli.

photosynthesis Process occurring usually within chloroplasts whereby chlorophyll-containing organelles trap solar energy to reduce carbon dioxide to carbohydrates.

photosystem Photosynthetic unit in which solar energy is absorbed and high-energy electrons are generated; contains a pigment complex and an electron acceptor; occurs as PS (photosystem) I and PS II.

phototropism Growth response of plant stems to light; stems demonstrate positive phototropism.

photovoltaic (solar) cell Energy-conversion device that captures solar energy and converts it to electrical current.

phylogenetic tree Diagram that indicates common ancestors and lines of descent among groups of organisms.

phylogeny Evolutionary history of a group of organisms.

phylum One of the categories, or taxa, used by taxonomists to group species; the taxon above the class level.

phytochrome Photoreversible leaf pigment involved in photoperiodism and other responses of plants, such as etiolation.

pith Parenchyma tissue in the center of some eudicot stems and roots.

pituitary gland Small gland that lies just inferior to the hypothalamus; consists of the anterior and posterior pituitary, both of which produce hormones.

placebo Treatment that contains no medication but appears to be the same treatment as that administered to the test groups in a controlled study.

placenta Organ formed during the development of placental mammals from the chorion and the uterine wall; allows the embryo, and then the fetus, to acquire nutrients and rid itself of wastes; produces hormones that regulate pregnancy.

placental mammal Mammal characterized by the presence of a placenta during the development of the offspring.

planarian Free-living flatworm with a ladderlike nervous system.

plant Multicellular, usually photosynthetic, organism belonging to the kingdom Plantae.

plants Multicellular, photosynthetic eukaryotes whose life cycle is characterized by having distinct diploid and haploid stages.

plasma In vertebrates, the liquid portion of blood, composed of water and proteins with smaller quantities of nutrients, wastes, and salts.

plasma cells B lymphocytes that have been activated to produce antibodies against a specific antigen.

plasma membrane Membrane surrounding the cytoplasm; consists of a phospholipid bilayer with embedded proteins. It regulates the entrance and exit of molecules from a cell.

plasmid Self-duplicating ring of accessory DNA in the cytoplasm of a bacterial cell.

plasmodesmata (sing., plasmodesma) In plants, cytoplasmic strands that extend through pores in the cell wall and connect the cytoplasm of two adjacent cells.

platelet In blood, a formed element that is necessary to blood clotting; also called thrombocytes.

pleiotropy Inheritance pattern in which one gene affects many phenotypic characteristics of the individual.

point mutation Alteration in a gene due to a change in a single nucleotide. The results of this mutation vary.

polar In chemistry, bond in which the sharing of electrons between atoms is unequal.

pollen grain In seed plants, structure that is derived from a microspore and develops into a male gametophyte.

pollen tube In seed plants, tube that forms when a pollen grain lands on the stigma and germinates. The tube grows, passing between the cells of the stigma and the style to reach the egg inside an ovule, where fertilization occurs.

pollination In gymnosperms, the transfer of pollen from pollen cone to seed cone; in angiosperms, the transfer of pollen from anther to stigma.

pollution Any environmental change that adversely affects the lives and health of living organisms.

polygenic The contribution of two or more genes to a phenotype.

polymer Macromolecule consisting of covalently bonded monomers. For example, a polypeptide is a polymer of monomers called amino acids.

polymerase chain reaction (PCR) Technique that uses the enzyme DNA polymerase to produce billions of copies of a particular piece of DNA.

polyp Small growth that arises from the epithelial lining of the large intestine.

polypeptide Polymer of many amino acids linked by peptide bonds.

polyribosome String of ribosomes simultaneously translating regions of the same mRNA strand during protein synthesis.

polysaccharide Polymer made from sugar monomers. The polysaccharides starch and glycogen are polymers of glucose monomers.

pons Part of the brain stem located between the midbrain and the medulla oblongata; controls breathing.

population Group of individuals of the same species occupying a given location at the same time.

population density Number of individuals per unit area (or unit volume) of a particular habitat.

population genetics Study of gene frequencies and their changes within a population.

portal system Pathway of blood flow that begins and ends in capillaries, such as the portal system between the intestines and the liver.

posterior pituitary Portion of the pituitary gland that stores and secretes oxytocin and antidiuretic hormone produced by the hypothalamus.

postzygotic isolating mechanism Anatomical or physiological difference between two species that prevents successful reproduction after mating has taken place.

potential energy Stored energy as a result of location or spatial arrangement.

predation Interaction in which one organism (the *predator*) uses another (the *prey*) as a food source.

prediction Step of the scientific process that follows the formation of a hypothesis and assists in creating the experimental design.

prefrontal area In the frontal lobe, association area that receives information from other association areas and uses it to reason and plan actions.

preparatory (prep) reaction Reaction that oxidizes pyruvate with the release of carbon dioxide; results in acetyl-CoA and connects glycolysis to the citric acid cycle.

pressure-flow model Explanation for the transport of sugar by phloem that says that osmotic pressure following active transport of sugar into phloem brings a flow of sap from a source to a sink.

prezygotic isolating mechanism Anatomical or behavioral difference between two species that prevents the possibility of mating.

primary motor area In the frontal lobe of the cerebrum, area where voluntary commands begin. Each section controls a part of the body.

primary productivity The rate at which producers in an ecosystem capture and store energy and organic nutrients over a certain length of time.

primary sensory area In the parietal lobe of the cerebrum, area where sensory information arrives from the skin, skeletal muscles, and joints.

primary succession Stage in ecological succession, which involves the creation of new soil.

principle Theory that is generally accepted by an overwhelming number of scientists, also called a law.

prion Infectious particle consisting of protein only and no nucleic acid.

producer An organism that makes its own food (an autotroph) and is thus at the base of a food chain.

product Anything that forms as a result of a chemical reaction.

progesterone Female sex hormone that helps maintain sexual organs and secondary sex characteristics.

prokaryote Single-celled organism that lacks the membrane-bound nucleus and membranous organelles typical of eukaryotes.

prokaryotic cell Cell lacking a membrane-bound nucleus and organelles; the cell type within the domains Bacteria and Archaea.

promoter A sequence of DNA that is involved in the regulation of gene expression. Located where the RNA polymerase binds prior to transcription.

prophase Mitotic phase during which chromatin condenses, so that chromosomes appear. Chromosomes are scattered.

proprioceptor Sensory receptor that helps the body maintain equilibrium and posture by detecting changes in the positions of the joints, muscles, and bones.

prosimians Group of primates that includes lemurs, lorises, and tarsiers and may resemble the first primates to have evolved.

protein Molecule consisting of one or more polypeptides; a macronutrient in the diet that is digested to the amino acids used by cells to synthesize cellular proteins.

proteome Collection of proteins resulting from the translation of all the genes in an organism's genome.

proteomics Study of the structure, function, and interaction of proteins.

protist Eukaryotic organisms that are usually single-celled; members of the kingdom Protista.

proto-oncogene Normal gene that can become an oncogene through mutation.

protocell Cell-like structure with an outer membrane, thought to have preceded the true cell.

proton Positive subatomic particle located in the nucleus and assigned one atomic mass unit.

protostome Group of coelomate animals in which the first embryonic opening (the blastopore) is associated with the mouth.

proximal convoluted tubule Portion of a nephron leading from the nephron capsule where reabsorption of filtrate occurs.

pseudocoelomate Animal, such as a roundworm, with a body cavity incompletely lined by mesoderm.

puberty Period of life when secondary sex changes occur in humans; marked by the onset of menses in females and sperm production in males.

pulmonary artery Blood vessel that takes blood away from the heart to the lungs.

pulmonary circuit Circulatory pathway between the lungs and the heart.

pulmonary vein Blood vessel that takes blood to the heart from the lungs.

pulse Vibration felt in arterial walls due to expansion of the aorta following ventricle contraction.

Punnett square Grid used to calculate the expected results of simple genetic crosses.

R

radial symmetry Body plan in which similar parts are arranged around a central axis, like spokes of a wheel.

random distribution Spatial distribution of individuals in a population in which individuals have an equal chance of living anywhere within an area.

range Portion of the globe where a certain species can be found.

ray-finned fishes Group of bony fishes with fins supported by parallel bony rays connected by webs of thin tissue.

reabsorption Movement of primarily nutrient molecules and water from the contents of the nephron into blood at the proximal convoluted tubule.

reactant Substance that participates in a chemical reaction.

receptor-mediated endocytosis Selective uptake of molecules into a cell by vesicle formation after they bind to specific receptor proteins in the plasma membrane.

recessive allele Allele that exerts its phenotypic effect only in the homozygote. Its expression is masked by a dominant allele.

recombinant DNA (rDNA) DNA that contains genes from more than one source.

red blood cell Erythrocyte; contains hemoglobin and carries oxygen from the lungs or gills to the tissues in vertebrates.

redox reaction Oxidation-reduction reaction; one molecule loses electrons (oxidation) while another molecule gains electrons (reduction).

reduction Gain of electrons by an atom or a molecule with a concurrent storage of energy. In biological systems, the electrons are accompanied by hydrogen ions.

reflex Automatic, involuntary response of an organism to a stimulus.

regulatory gene A gene outside an operon that codes for a protein that regulates the expression of other genes.

release factor Protein complex that binds to a stop codon on an mRNA at the A site of the ribosome, causing termination of the transcription.

renewable resources Resources normally replaced or replenished by natural processes and not depleted by moderate use. Examples include solar energy, biological resources (such as forests and fisheries), biological organisms, and some biogeochemical cycles.

replacement model Proposal that modern humans originated only in Africa, then migrated and supplanted populations of other *Homo* species in Asia and Europe about 100,000 years ago; also called the out-of-Africa hypothesis.

replacement reproduction A measure of population dynamics in which each person is replaced by only one child in the next generation.

repressor In an operon, a protein molecule that binds to an operator, preventing transcription of structural genes.

reproduce To produce a new individual of the same kind.

reproductive cloning The process that creates a new individual that is genetically identical to the original individual.

reproductive system Organ system that consists of different organs in males and females and specializes in the production of offspring.

reptiles Terrestrial vertebrates with internal fertilization, scaly skin, and a shelled egg; include snakes, lizards, turtles, crocodiles, and birds.

reservoir In chemical cycling, a source (e.g., fossil fuels, minerals in rocks, and ocean sediments) that is normally unavailable to organisms.

resource Any component of an environment that supports its organisms; includes food, water, shelter, and space.

resource partitioning Mechanism that increases the number of niches by apportioning the supply of a resource, such as food or living space, between species.

respiration Sequence of events that results in gas exchange between the cells of the body and the environment.

respiratory system Organ system consisting of the lungs, trachea, and other structures; brings oxygen into the body and takes carbon dioxide out.

responding variable In an experiment, the value that is obtained as a result of changing the experimental variable.

restriction enzyme Bacterial enzyme that stops viral reproduction by cleaving viral DNA; used to cut DNA at specific points during production of recombinant DNA.

retrovirus A virus that uses RNA as its genetic material and contains the enzyme reverse transcriptase, which carries out transcription of the RNA to DNA.

ribose Pentose monosaccharide sugar found in RNA.

ribosomal RNA (rRNA) Type of RNA found in ribosomes that translates messenger RNAs to produce proteins.

ribosome Structure consisting of rRNA and proteins in two subunits; site of protein synthesis in the cytoplasm.

RNA (ribonucleic acid) Nucleic acid produced from covalent bonding of nucleotide monomers that contain the sugar ribose. Three major forms are messenger RNA, ribosomal RNA, and transfer RNA.

RNA polymerase During transcription, an enzyme that joins nucleotides in a sequence complementary to that in a DNA template.

root hair Long, slender projection of a root epidermal cell that increases the surface area of the root for the absorption of water and minerals.

root nodule Plant root structure that contains nitrogen-fixing bacteria.

root system Main plant root and all of its lateral (side) branches.

roots System of a plant responsible for absorbing nutrients and water from the soil, as well as providing anchoring to the plant.

rotational equilibrium Maintenance of balance when the head and body are suddenly moved or rotated; involves the semicircular canals of the inner ear.

rough ER Membranous system of tubules, vesicles, and sacs in cells; has attached ribosomes.

roundworm Invertebrate with a cylindrical body covered by a cuticle that molts; also has a complete digestive tract and a pseudocoelom. Some forms are free-living in water and soil, and many are parasitic.

RuBP carboxylase (rubisco) Enzyme required for carbon dioxide fixation (atmospheric CO_2 attaches to RuBP) in the Calvin cycle.

S

SA (sinoatrial) node Pacemaker; small region of cardiac muscle tissue that initiates the heartbeat.

salinization Process in which mineral salts accumulate in the soil, killing crop plants; occurs when soils in dry climates are irrigated profusely.

salivary amylase In humans, enzyme in saliva that breaks down starch to maltose.

salivary gland In humans, gland associated with the mouth that secretes saliva.

salt Solid substance formed by ionic bonds that usually dissociates into individual ions in water.

saltwater intrusion Movement of salt water into freshwater aquifers in coastal areas where groundwater is withdrawn faster than it is replenished.

saprotroph Organism that secretes digestive enzymes and absorbs the resulting nutrients back across the plasma membrane; fungus or bacterium that decomposes the remains of plants, animals, and microbes in the soil.

sarcomere One of many identical units, arranged linearly in a myofibril, whose contraction produces muscle contraction.

saturated fatty acid Fatty acid molecule that lacks double bonds between the carbons of its hydrocarbon chain. The chain bears the maximum number of hydrogens possible.

scientific theory Concept supported by a broad range of observations, experiments, and data.

secondary succession In ecological succession, the stage at which there is gradual replacement of communities in an area following a disturbance.

secretion In the cell, release of a substance by exocytosis from a cell that may be a gland or part of a gland; in the urinary system, movement of certain molecules from blood into the distal convoluted tubule of a nephron, so that they are added to urine.

seed Mature ovule that contains an embryo, with stored food enclosed in a protective coat.

segmentation Repetition of body units, as in the earthworm.

semen (seminal fluid) Thick, whitish fluid consisting of sperm and secretions from several glands of the male reproductive tract.

semiconservative Duplication of DNA resulting in two double-helix molecules, each having one parental and one new strand.

semilunar valve Valve resembling a half-moon located between a ventricle of the heart and its attached vessels.

senescence Sum of the processes involving aging, decline, and eventual death of a plant or plant part.

sensory adaptation Decrease in the response to a sensory stimulus, usually from overexposure to the stimulus.

sensory neuron Nerve cell that transmits nerve impulses to the central nervous system after a sensory receptor has been stimulated.

sepal Outermost, leaf-like covering of a flower; usually green.

septum Partition that divides two areas. The septum in the heart separates the right side from the left side.

sex chromosomes Pair of chromosomes that determine the sex of an individual. In humans, females have two X chromosomes, and males have an X and a Y chromosome.

sexual reproduction Form of reproduction that involves an input of genetic material from two individuals; increases the genetic variation of the offspring.

sexual selection Adaptive changes in males and females of a species, often due to male competition and female selectivity, leading to increased fitness.

shoot system Aboveground portion of a plant consisting of the stem, leaves, flowers, and fruit.

short tandem repeat (STR) profiling Procedure for analyzing DNA in which PCR and gel electrophoresis are used to create a banding pattern, which is usually unique for each individual; process used in DNA barcoding.

sieve-tube member Cell that joins with others in the phloem tissue of plants to provide a means of transport for organic solutes.

signal transduction pathway A series of reactions that brings about activation and inhibition of intracellular targets after binding of growth factors.

sinus Cavity into which hemolymph flows and bathes the organs in an open circulatory system; also, an air-filled space connected to a nasal cavity.

sister chromatid One of two genetically identical chromosomal units that are the result of DNA replication and are attached to each other at the centromere.

skeletal muscle Striated, voluntary muscle tissue that comprises skeletal muscles; also called striated muscle.

skeletal system Organ system that consists of bones and works with the muscular system to protect the body and provide support.

sliding filament model Explanation of muscle contraction based on the movement of actin filaments in relation to myosin filaments.

small intestine In vertebrates, portion of the digestive tract that precedes the large intestine; in humans, the duodenum, jejunum, and ileum, which are responsible for most digestion and absorption of the resulting nutrients.

smooth ER Membranous system of tubules, vesicles, and sacs in eukaryotic cells; lacks attached ribosomes.

sodium-potassium pump Carrier protein in the plasma membrane that moves sodium ions out of and potassium ions into animal cells; important in nerve and muscle cells.

solute Substance dissolved in a solvent, forming a solution.

solution Fluid (the solvent) that contains a dissolved solid (the solute).

solvent Liquid portion of a solution that dissolves a solute.

somatic cell Body cell; excludes cells that undergo meiosis and become sperm or eggs.

somatic system Portion of the peripheral nervous system that includes the nerves that take information about external stimuli from sensory receptors to the central nervous system and motor commands away from the central nervous system to skeletal muscles.

speciation Formation of new species due to the evolutionary process of descent with modification.

species Group of similarly constructed organisms capable of interbreeding and producing viable offspring; organisms that share a gene pool; the taxon at the lowest level of classification.

species richness List of the different species found in a community.

spermatogenesis Production of sperm in males by the process of meiosis and maturation.

spinal cord In vertebrates, the nerve cord that is continuous with the base of the brain and housed within the vertebral column.

spinal nerve Nerve that arises from the spinal cord.

spindle Microtubule structure that brings about chromosomal movement during nuclear division.

spleen Large, glandular, lymphatic organ in the upper left abdominal cavity; stores and purifies blood.

sponge Invertebrate animal of the phylum Porifera; pore-bearing filter feeder whose inner body wall is lined by collar cells.

spongy bone Type of bone that has an irregular, meshlike arrangement of thin plates of bone.

sporophyte Diploid generation of the life cycle of a plant; produces haploid spores that develop into the haploid generation.

stabilizing selection Outcome of natural selection in which extreme phenotypes are eliminated and the intermediate (average) phenotype is conserved.

stamen In flowering plants, portion of the flower that consists of a filament and an anther containing pollen sacs where pollen is produced.

starch In plants, storage polysaccharide composed of glucose molecules joined in a linear fashion with few side chains.

stem Usually the upright, vertical portion of a plant that transports substances to and from the leaves.

sterility Inability to produce offspring.

steroid Type of lipid molecule having a complex of four carbon rings—e.g., cholesterol, estrogen, progesterone, and testosterone.

steroid hormone Type of hormone that is a lipid and can pass through a cell's plasma membrane.

stigma In flowering plants, portion of the carpel where pollen grains adhere and germinate before fertilization can occur.

stoma (pl., stomata) Small opening between two guard cells in the epidermis of a leaf through which gases pass; serves to minimize water loss.

stomach In vertebrates, thick-walled organ of the digestive tract that mixes food with gastric juices to form chyme, which enters the small intestine.

stroke Condition resulting when an arteriole in the brain bursts or becomes blocked by an embolus; cerebrovascular accident.

stroma Fluid-filled space within a chloroplast that contains enzymes involved in the synthesis of carbohydrates during photosynthesis.

style Elongated, central portion of the carpel between the ovary and stigma.

subsidence Gradual settlement of a portion of the Earth's surface.

substrate Reactant in a reaction controlled by an enzyme.

supergroup High-level taxonomic groups just below the domain level; used for classification of eukaryotes.

surface-area-to-volume ratio Ratio of a cell's outside area to its internal volume.

survivorship Term used to describe the probability of newborn individuals of a cohort surviving to particular ages and its influence on population size.

sustainable society A society that is able to continue to provide the same goods and services for future generations while preserving biodiversity.

symbiotic Type of relationship that occurs when two different species live together in a unique way. It may be beneficial, neutral, or detrimental to one or both species.

sympathetic division Division of the autonomic system that is active when an organism is under stress; uses norepinephrine as a neurotransmitter.

sympatric speciation Origin of new species without prior geographic isolation of a population.

synapse Region of close proximity between the axon terminal of one neuron and a dendrite or cell body of another neuron.

synapsis Pairing of homologous chromosomes during meiosis I.

synaptic cleft Small gap between parts of neurons at a synapse.

synovial joint Freely moving joint in which two bones, joined by ligaments, are separated by a fluid-filled cavity.

syphilis Bacterial STD with three stages separated by latent periods; may result in blindness, cause birth defects or stillbirth, and affect the cardiovascular and/or nervous system.

systematics Study of the diversity of organisms to classify them and determine their evolutionary relationships.

systemic circuit Circulatory pathway of blood flow between the tissues of the body and the heart.

systole Contraction period of a chamber of the heart during the cardiac cycle.

T

T lymphocyte (T cell) Lymphocyte that matures in the thymus and exists in four varieties, one of which kills antigen-bearing cells outright.

T-cell receptor (TCR) On the T-cell surface, receptor consisting of two antigen-binding peptide chains; associated with a large number of other glycoproteins. Binding of antigen to the TCR, usually in association with MHC, activates the T cell.

taxon (pl., taxa) Group of organisms that fills a particular classification category.

taxonomy Branch of biology concerned with identifying, naming, and classifying organisms.

technology Application of scientific knowledge for a practical purpose.

telomere Long, repeating DNA base sequence at the ends of chromosomes; functions as a cap and keeps chromosomes from fusing with each other.

telophase Mitotic phase during which daughter cells are located at each pole.

template Parental strand of DNA that serves as a guide for the complementary daughter strand produced during DNA replication.

tendon Strap of fibrous connective tissue that connects skeletal muscle to bone.

terrestrial ecosystem A biome, such as a tundra, taiga, temperate forest, tropical grassland (savanna), temperate grassland (prairie), desert, or tropical rain forest.

testcross Cross between an individual with the dominant phenotype and an individual with the recessive phenotype. The resulting phenotypic ratio indicates whether individual with the dominant phenotype is homozygous or heterozygous.

testis (pl., testes) Male gonad that produces sperm and the male sex hormones.

testosterone Male sex hormone that helps maintain sexual organs and secondary sex characteristics.

tetany Severe twitching caused by involuntary contraction of the skeletal muscles due to a low blood calcium level.

tetrad Association of two homologous chromosomes, each having two sister chromatids that are joined during meiosis; also called bivalent.

thalamus In vertebrates, portion of the diencephalon that receives and passes on to the cerebrum all sensory input except smell.

therapeutic cloning Cloning done to create mature cells of various types; also used to learn about specialization of cells and provide cells and tissue to treat human illnesses.

thermoacidophile Type of archaea that lives in hot, acidic, aquatic habitats (such as hot springs) or near hydrothermal vents.

thigmotropism Growth response of plants to touch.

three-domain system System of classification that recognizes three domains: Bacteria, Archaea, and Eukarya.

thylakoid Flattened sac within a granum whose membrane contains chlorophyll and where the light reactions of photosynthesis occur.

thymine (T) One of four nitrogen-containing bases in nucleotides composing the structure of DNA; pairs with adenine.

thymus Lymphatic organ involved in the development and functioning of the immune system. T cells mature in the thymus.

thyroid gland Large gland in the neck; produces several important hormones, including thyroxine, triiodothyronine, and calcitonin.

tight junction Junction between cells in which adjacent plasma membrane proteins join to form an impermeable barrier.

tissue Group of similar cells combined to perform a common function.

tissue culture Process of growing tissue artificially, usually in a liquid medium in laboratory glassware.

tonsils Partially encapsulated lymphatic nodules in the pharynx.

totipotent Cell that has the full genetic potential of the organism, including the potential to develop into a complete organism.

trace mineral Essential inorganic nutrient (such as zinc, iron, copper, iodine, or selenium) needed in a relatively small quantity in the daily diet of humans to regulate metabolic activities and maintain good health.

trachea (pl., tracheae) In tetrapod vertebrates, air tube (windpipe) that runs between the larynx and the bronchi.

tracheid In vascular plants, type of cell in xylem that has tapered ends and pits through which water and minerals flow.

trans fat A form of unsaturated fatty acid in which the hydrogen atoms on the carbon chain are on opposite sides of the double bond.

transcription Process whereby a DNA strand serves as a template for the formation of mRNA.

transcription factor In eukaryotes, protein required for the initiation of transcription by RNA polymerase.

transduction Transport of DNA between bacterial cells by a bacteriophage.

transfer rate Amount of a nutrient that moves from one component of the environment to another within a specified period of time.

transfer RNA (tRNA) Type of RNA that transfers a particular amino acid to a ribosome during protein synthesis. At one end, tRNA binds to the amino acid, and at the other end, it has an anticodon that binds to an mRNA codon.

transformation Taking up of extraneous genetic material from the environment by a bacterial cell.

transgenic organism An organism that has had a gene from another species inserted into its genome.

translation Process whereby ribosomes use the sequence of codons in mRNA to produce a polypeptide with a particular sequence of amino acids.

translocation The movement of a chromosomal segment from one chromosome to another, nonhomologous chromosome or the exchange of segments between nonhomologous chromosomes, leading to abnormalities—e.g., Down syndrome.

transpiration Plant's loss of water to the atmosphere, mainly through evaporation at leaf stomata.

transposon DNA sequence capable of randomly moving from one site to another in the genome.

trichomoniasis Sexually transmitted disease caused by the parasitic protozoan *Trichomonas vaginalis*.

triglyceride Molecule composed of glycerol and three fatty acids that is the main component of fats and oils.

trophic level A level of nourishment within a food chain or web.

tropism In plants, a growth response toward or away from a directional stimulus.

trypsin Enzyme secreted by the pancreas that breaks down proteins to peptides.

tumor Group of cells derived from a single, mutated cell that has repeatedly undergone cell division. Benign tumors remain at their site of origin, while malignant tumors metastasize.

tumor suppressor gene Gene that codes for a protein that ordinarily suppresses cell division. Inactivity can lead to a tumor.

tunicate Type of primitive invertebrate chordate.

Turner syndrome Condition caused by the inheritance of a single X chromosome.

U

umbilical cord Cord connecting the developing embryo and then the fetus to the placenta, through which blood vessels pass.

uniform distribution Spatial distribution of individuals in a population in which individuals are dispersed uniformly through the area.

unsaturated fatty acid Fatty acid molecule that has one or more double bonds between the carbons of its hydrocarbon chain. The chain bears fewer hydrogens than the maximum number possible.

uracil (U) Pyrimidine base that replaces thymine in RNA; pairs with adenine.

ureter Tubular structure conducting urine from the kidney to the urinary bladder.

urethra Tubular structure that receives urine from the bladder and carries it to the outside of the body.

urinary bladder Organ where urine is stored.

urinary system Organ system consisting of the kidneys and urinary bladder, along with other structures that transport urine; rids the body of wastes and helps regulate the fluid level and chemical content of the blood.

urine Liquid waste product made by the nephrons of the vertebrate kidney through the processes of filtration, reabsorption, and secretion.

uterus In mammals, expanded portion of the female reproductive tract through which eggs pass to the environment or in which an embryo develops and is nourished before birth.

V

vaccine Substance prepared from a pathogen or its products in such a way that it can promote active immunity without causing disease.

vagina Female sex organ that serves as the birth canal during pregnancy.

valence shell Outer energy shell of an atom.

vascular bundle In plants, primary phloem and primary xylem enclosed by a bundle sheath.

vascular plants Plants that have vascular tissue (xylem and phloem); include seedless vascular plants (e.g., ferns) and seed plants (gymnosperms and angiosperms).

vascular tissue Tissue in plants that provides transport and support; consists of xylem and phloem.

vector Piece of DNA that can have a foreign DNA attached to it. A common vector is a plasmid.

vein Blood vessel that arises from venules and transports blood toward the heart.

vena cava (pl., venae cavae) Large, systemic vein that returns blood to the right atrium of the heart; either the superior or inferior vena cava.

ventricle Cavity in an organ, such as a lower chamber of the heart or a ventricle in the brain.

venule Vessel that takes blood from capillaries to a vein.

vertebrate Chordate in which the notochord is replaced by a vertebral column made of cartilage or bone.

vesicle Small, membrane-bound sac that stores substances in a cell.

vessel element Cell that joins with others to form a major conducting tube in xylem.

vestigial structure Anatomical feature that is fully developed in one group of organisms but reduced and nonfunctional in other, similar groups.

villus (pl., villi) Small, fingerlike projection on the wall of the small intestine.

viroid Infectious strand of RNA devoid of a capsid and much smaller than a virus.

vitamin Organic compound required in the human diet. Vitamins are often part of coenzymes in the body.

vocal cord In the human larynx, flexible bands of connective tissue that create sounds when they vibrate.

W

white blood cell Leukocyte, of which there are several types, each having a specific function in protecting the body from invasion by foreign substances and organisms.

wood Secondary xylem that builds up year after year in woody plants, becoming annual rings.

X

X inactivation Process in females that causes one of the two X chromosomes to condense into an inactive form called a Barr body. An example of epigenetic inheritance.

X-linked Allele that is located on an X chromosome but related to a trait that has nothing to do with sex characteristics.

xylem Vascular tissue that transports water and mineral solutes upward through the plant body. It contains vessel elements and tracheids.

Z

zygote Diploid cell formed by the union of two gametes; the product of fertilization.

Index

Note: Page numbers followed by *f* and *t* indicate figures and tables, respectively.

A

Abdominal cavity, 453–454, 454*f*
Abiotic synthesis, 293
ABO blood type, 503, 503*t*
Abortion pill, 566
Abscisic acid, 393*t*, 395–396, 395*f*
Abscission, 396
Abstinence, 564, 565*t*
Acetyl-CoA, 117
Acetylcholine (ACh), 518
Acetylcholinesterase (AChE), 518
Acid-base balance, 459–460
Acidosis, 35*f*
Acids
  buffers and, 35
  defined, 33
  pH scale, 34, 34*f*
  stomach acid, 33
  strength of, 33, 34*f*
Acoelomates, 340, 340*f*
Acquired immunodeficiency
    syndrome (AIDS). *See* AIDS
    (acquired immunodeficiency
    syndrome)
Actin, 48*f*, 49, 550, 550*f*
Actin filaments, 72, 72*f*
Action potential, 516–517, 517*f*
Active immunity, 506, 506*f*
Active site, 86–87
Active transport, 91, 91*f*
Acyclovir, 291
Adaptation
  common descent with
    modification, 8, 8*f*
  defined, 6, 240, 241*f*
  natural selection and,
    240–242, 242*f*
  sensory, 537
Adaptive immunity, 502–505
Adaptive radiation, 271
Addison disease, 531, 531*f*
Adenine (A), 52, 52*f*, 186, 186*f*, 190
  structure of DNA and, 186–188
Adenosine triphosphate (ATP)
  ATP cycle, 82–83, 83*f*
  cellular respiration and, 111–112,
    111*f*, 112*f*, 113–114, 120
  coupled reaction, 83–84, 84*f*
  electron transport chain and, 120
  as energy carrier, 52

fermentation and, 115–116, 116*f*
glycolysis, 113–114
light reaction and, 102, 102*f*
production of, 69, 71, 102
structure of, 82, 82*f*
substrate-level ATP synthesis,
    114, 114*f*
use and production of, 82–84
Adhesion junction, 74*f*, 75
Adhesion of water, 30–31, 30*f*
Adipose tissue, 419, 419*f*, 489–490
Adrenal cortex, 530–531
Adrenal gland, 526*f*, 530–531
Adrenal medulla, 530
Adrenocorticotropic hormone
    (ACTH), 527
Adult stem cells, 213
Aerobic metabolism, 110
Age structure, 589, 589*f*
  comparing, 586, 586*f*
Agglutination, 503
Agricultural value of biodiversity,
    629–630, 630*f*
AIDS (acquired immunodeficiency
    syndrome), 568–569, 568*f*
  emerging virus and, 290–291
  transmission and treatment of,
    509–510, 509*f*
Alagille syndrome, 226
Aldosterone, 530
Algae, 302, 302*f*
Algal bloom, 616
Alimentary canal, 466
Alkaptonuria, 171, 171*f*
Allantois, 571*f*
Alleles, 147
  dominant, 165, 169
  linkage group, 177
  multiple-allele traits, 174
  population genetics and, 257
  recessive, 165
  X-linked, 178–179
Allergens, 508
Allergies, 508, 508*f*
Allopatric speciation, 269–270, 270*f*
Alternation of generations,
    314–315, 314*f*, 399, 400*f*
Alternative mRNA processing,
    201–202, 201*f*
Altitude, 251
Alveolates, 304–305, 304*f*

Alveoli, 453, 454–455, 455*f*
Alzheimer disease, 292, 523
Amanita, 10*t*
Amino, 40*f*
Amino acids, 479–480
  metabolic pathway,
    121–122, 121*f*
  structure and function of, 49–50
Ammonia, as base, 34
Amniocentesis, 227, 228*f*
Amnion, 571*f*
Amniotic egg, 359, 360*f*
Amoebozoa, 306, 306*f*
Amphibians, 358, 358*f*
  circulatory system, 435, 435*f*
Amygdala, 522, 522*f*, 523
Anabolic steroids, 48
Anaerobic metabolism, 110
Analogous structures, 247,
    279, 279*f*
Analogy, 279
Anaphase, 131*f*, 132*f*, 150*f*, 151,
    151*f*, 152, 153*f*, 154
Anatomy
  as evidence for evolution,
    246–248
  vestigial structures,
    246–248, 247*f*
Anemia, 442
Angiogenesis, 136, 140
Angiosperms, 312*f*
  adaptations and uses of, 323
  alternation of generations,
    314*f*, 315
  flower of, 321
  life cycle, 322–323
Anglerfish, 1
Animals
  amphibians, 358
  annelids, 345–346
  arthropods, 348–352
  birds, 359–360
  cells, 64, 64*f*, 74–75
  characteristics of, 10, 10*t*,
    337, 337*f*
  chordates, 354–356, 355*f*
  cloning, 212–213, 212*f*
  cnidarians, 342
  compared to fungi and
    plants, 326*t*
  cytokinesis, 132, 133*f*

disease of, and fungi, 331–333
echinoderms, 353–354
evolution of, 337–341
fishes, 356–358
flatworms, 343–344
human evolution, 363–369
invertebrate, 338
mammals, 360–362
molluscs, 344–345
reproduction, 557–559
reptiles, 358–360
roundworms, 347–348
sponges, 341–343
vertebrate, 338
viruses of, 289–291
Annelids, 339*f*, 345–346, 346*f*
Annual ring, 382
Anorexia nervosa, 492
Anterior pituitary, 527–528, 528*f*
Anther, 321, 321*f*, 400*f*, 401
Anthrax, 295
Anthropods, 339*f*
Anthropoids, 363
Antibiotics
  bacteria and, 299
  as enzyme inhibitor, 87
  evolution and antibiotic resistant
    bacteria, 235
  resistance to and directional
    selection, 253–254
Antibodies, 48, 444
  B cells and antibody response,
    502–503, 502*f*
  function of, 503, 503*f*
Anticodon, 194
Antidiuretic hormone (ADH), 529
Antigen-presenting cell (APC), 503
Antigens, 444, 499
  ABO blood type, 503
  adaptive immunity and, 502–505
Antihistamine, 500
Antioxidants, 95, 482
Antiviral drugs, 291
Ants, 388–389, 389*f*
Anus, 471
Aorta, 435*f*, 436
Aortic bodies, 454
Aortic semilunar valve, 436
Aphids, 388–389, 389*f*
Apical dominance, 393, 394*f*
Apical meristem, 373

Apoptosis, 134, 135, 135*f*
Appendicitis, 471
Appendicular skeleton, 547, 547*f*
Appendix, 471, 498
Aquafarming, 639
Aquaporins, 61, 89, 90
Aquatic ecosystems, 620, 620*f*
Aquifers, 636
Arachnids, 349–350, 350*f*
Archaea
　characteristics of, 9, 10*t*
　structure, 299
　types of, 299–300
Archaeopteryx, 244, 244*f*, 265
Archaeplastids, 302–303
Architeuthis dux, 59
Ardipithecines, 365–366, 365*f*
Ardipithecus ramidus,
　365–366, 365*f*
Arteries, 438, 438*f*
　plaque in, and cholesterol, 38
　plaque in, and fats, 46
Arterioles, 438
Arthritis
　osteoarthritis, 553
　rheumatoid, 553
Arthropods, 348–352, 348*f*–352*f*
Artificial insemination, 567
Artificial selection, 242
Ascaris lumbricoides, 347–348
Asexual reproduction, 126, 557,
　557*f*
　in flowering plants, 407–411,
　407*f*
Asian carp, 580
Aspartame, 160
Aspirin, 545
Assisted reproductive technologies,
　567–568
Association areas, 521
Assortative mating, 260–261
Aster, 129, 131*f*
Asthma, 450, 453
Atherosclerosis, 46, 445
Atomic mass, 23
Atomic number, 23
Atomic symbol, 23
Atomic theory, 23
Atoms, 2, 3*f*
　arrangement of electrons in,
　25–26, 25*f*, 26*f*
　defined, 23
　electronegativity of, 29
　period table, 23–24, 23*f*
　structure of, 23, 23*f*
　valence shell, 26
ATP. *See* Adenosine *t*riphospha*t*e
　(ATP)
ATP cycle, 82–83, 83*f*
Atrioventricular valves, 435–436
Atrium, 435, 435*f*

Attachment, of virus, 288, 289*f*,
　290, 290*f*
Australopithecines, 365*f*, 366
Australopithecus afarensis, 365,
　365*f*, 366*f*
Australopithecus africanus, 365*f*
Australopithecus sediba, 365*f*, 366
Autism, MMR vaccine and, 507
Autoimmune diseases, 509
Autonomic system of PNS, 524–525
Autosomal dominant pedigree,
　170, 170*f*
Autosomal recessive disorder,
　169–170, 170*f*
Autosomes, 146, 169
Autotrophs, 610
Auxins, 393–394, 393*t*, 394*f*
AV node, 437
Avian influenza (H5N1), 17,
　287–288
　as emerging virus, 290–291
Axial skeleton, 547, 547*f*
Axon, 422, 422*f*, 516, 516*f*
AZT, 291

B

B cells. *See* B lymphocy*t*es (B cells)
B lymphocytes (B cells),
　498, 498*t*
　antibody response and, 502–503,
　502*f*
　characteristics of, 502*t*
B-cell receptor (BCR), 502
Bacilli, 294
Bacteria
　antibiotic resistant and natural
　　selection, 7, 8, 235
　bacterial diseases in humans,
　　298–299
　characteristics of, 9, 10*t*
　diversity of, 294–299
　in environment, 297–298, 297*f*
　food irradiation, 25
　in food science and
　　biotechnology, 298, 298*f*
　nutrition, 296–297
　as prokaryotic cell, 62–63
　reproduction of, 295–296
　role of, 62
　in root nodules, 386
　sexually transmitted diseases,
　　569–570
　structure of, 62–63, 63*f*,
　　294–295, 294*f*
　transgenic, 214*f*
Bacteriophages, reproduction of,
　288
Balance, sense of, 541, 541*f*
Ball-and-socket joints, 551, 552*f*
Balloon angioplasty, 446, 446*f*

Bananas, 392
Bark, 381–382
Barr body, 156, 199
Barriers to entry, 499
Basal metabolic rate, 80
Bases
　buffers and, 35
　defined, 34
　pH scale, 34, 34*f*
　strength of, 34, 34*f*
Basophils, 443*f*
Batrachochytrium dendrobatidis, 332
BCR-ABL oncogene, 138, 138*f*
Beaches
　erosion, 634, 634*f*
　pollution, 634
Beadle, George, 191
Beer, 43, 43*f*
Behavior, 5
Behavioral isolation, 269
Benign tumor, 140
Beta-carotene, 95, 141
Bicarbonate ion, 456
Bicuspid valve, 436
Bilateral symmetry, 339, 339*f*
Bile, 469
Binary fission, 126, 295, 295*f*
Binge-eating disorder, 492
Binocular vision, 542
Binomial name, 10
Biodiversity, 632
　agricultural value of, 629–630,
　　630*f*
　as bioethical issue, 16–17, 17*f*
　biogeochemical cycles and,
　　630–631
　climate regulation, 632
　consumptive use value, 630, 630*f*
　defined, 628
　extinction and, 628
　fresh water provided by, 631
　medicinal value of, 629, 629*f*
　soil erosion prevention and,
　　631–632
　value of, 629–630
　waste disposal and, 631
Biodiversity hotspots, 628, 628*f*
Bioethics, 16–17
Biogas, 300
Biogeochemical cycles, 614,
　630–631
Biogeography, as evidence of
　　evolution, 239–240, 246
Bioinformatics, 218
Biological magnification, 643, 643*f*
Biological molecules, 39, 41–53
Biological organization
　categories of classification, 9, 9*t*
　in human body, 415–416, 415*f*
　levels of, 2, 3*f*
Biological species concept, 266–269

Biology, 2, 11
Biomass, 614
Bioremediation, 298, 298*f*
Biosphere, 2, 3*f*, 581, 620,
　620*f*, 621*f*
Biosynthesis, of virus, 288, 289*f*,
　290, 290*f*
Biotechnology
　genetic engineering, 208
　reproductive cloning,
　　212–213, 212*f*
　therapeutic cloning, 212*f*
　transgenic organisms, 208,
　　214–216, 214*f*
Biotic community, 614
Biotic potential, 590, 590*f*
Biotic synthesis, 293
Biotin, 482*t*
Bipedalism, 364
Bird flu, 287–288, 291
Birds, 359–360, 360*f*–361*f*
　anatomy of, 359, 360*f*
　beaks, 361*f*
　breathing in, 454, 454*f*
Birth, 576, 576*f*
Birth control methods, 564–566,
　565*t*, 566*f*
Birth control pill, 564, 565*t*
Birth weight, as stabilizing
　　selection, 253
Birthrate, 584
Bivalves, 345, 345*f*
Black bread mold, 327–328, 327*f*
Black widow spider, 350*f*
Blade (leaf), 377–378, 378*f*
Blastocyst, 572, 572*f*
Blind spot, 543
Blood
　ABO blood type, 503
　capillary exchange, 441, 441*f*
　components of, 420–421, 420*f*,
　　442–443, 442*f*, 443*f*
　as connective tissue, 420–421
　exchange of gas and, 455, 455*f*
　function of, 420, 442
　inheritance of blood type,
　　174, 174*f*
　movement in vein, 438–439, 438*f*
　path of, 439, 439*f*
　plasma, 442
　platelets and blood clotting,
　　444–445
　pulmonary and systemic circuits,
　　439–440
　sickle-cell disease, 52–53, 53*f*
　transportation and exchange of
　　gas, 455–456
　water as component of, 30*f*, 31
Blood doping, 443
Blood pressure, 437–438, 441
　high, 445

Blood vessels, 437–440, 438*f*
Body. *See* Human body
Body mass index, 488–490
Bolus, 467
Bonds
 carbon-carbon, 40
 covalent, 27, 27*f*
 hydrogen, 29, 29*f*
 ionic, 26–27, 26*f*
 peptide, 49, 49*f*
Bone marrow transplant, 213
Bone marrow, red bone marrow,
  497–498, 548
Bones. *See also* Musculoskeletal
  system
 compact, 420
 as connective tissue,
  419*f*, 420
 function of, 545–546
 skeletal muscle moving bones at
  joints, 551–553, 551*f*
 structure of, 548, 548*f*
Bony fish, 357, 358*f*
Borborygmi, 469
Bottleneck effect, 262, 262*f*
Botulism, 296
Bovine growth hormone (BGH), 215
Bowel, 471
Brain
 brain stem, 522
 cerebellum, 522
 cerebrum, 521
 comparison of, 520, 520*f*
 diencephalon, 521
Brain stem, 520*f*, 522
BRCA1/BRCA2 gene mutation, 138
Breast cancer, 140*f*
Breathing, 453–454, 454*f*
Bronchi, 453
Bronchioles, 453
Bronchitis, 453
Brown algae, 303–304, 303*f*
Bryophytes, 315–316, 315*f*
Budding, 342
Buffer, 35
Bulbourethral glands, 561
Bulimia nervosa, 492
Bullhorn acacia, 606–607, 607*f*
Burgess Shale, 27
Bursa, 551, 552*f*
Bursitis, 553
Butter, 46, 46*f*

C

C₃ plant, 105, 105*f*
C₄ plant, 105–107, 105*f*, 106*f*
Calcitonin, 529
Calcium
 as element of living organisms, 22
 muscle contraction and, 550
 regulation of, 530

Callus, 408
Calorie, 80
Calvin cycle reactions
 fixation of carbon dioxide,
  103–104
 overview of, 98*f*, 99, 103, 103*f*
 reduction of carbon dioxide, 104
 regeneration of RuBP, 104
Calvin, Melvin, 103
Calyx, 321, 321*f*
CAM photosynthesis, 106, 106*f*
Cambrian period, 273*t*
Camera-type eyes, 542, 542*f*
Cancer, 136–142
 absence of telomere shortening,
  138
 breast, 140*f*
 cell cycle and, 136–139
 characteristics of cancer cells,
  139–140, 139*f*
 chromosomal rearrangements
  and, 138, 138*f*
 classification of, 139
 defined, 136
 development of, 136, 136*f*
 of epithelial tissue, 418
 genes associated with, 125,
  138–139
 growth factor and, 137
 lymph nodes and, 498
 normal cells compared with, 139*f*
 prevention of, 141–142
 proto-oncogenes become
  oncogenes, 137, 137*f*
 transposons, 139
 treatment for, 140–141
 tumor suppressor genes become
  inactive, 137, 137*f*
Candida albicans, 332, 570
Canola oil, 46*f*
Capillaries, 438, 438*f*
Capillary exchange, 441, 441*f*
Capsaicin, 79, 311
Capsid, 287, 287*f*
Capsule, 63, 63*f*
Carbohydrates, 41*f*
 defined, 42
 digestion of, 473–474, 474*f*
 disaccharides, 43, 43*f*
 fiber, 44, 477, 477*f*
 health and, 478
 monosaccharides, 42, 42*f*
 polysaccharides, 43–45, 44*f*
 produced in photosynthesis,
  96, 97
 role of, 476–477
Carbon
 carbon skeleton and functional
  groups, 40, 40*f*
 characteristics of, 39–40
 as element of living organisms, 22*f*
 model of, 25*f*

Carbon cycle, 617–618, 617*f*
Carbon dioxide
 breathing rate and, 454
 carbon cycle and, 617–618, 617*f*
 cellular respiration, 111–112,
  111*f*, 112*f*
 exchange of gas and, 454–455,
  455*f*
 fixation of, in Calvin cycle,
  103–104
 forests absorption of, 104
 global warming, 640
 in photosynthesis equation, 28,
  97, 97*f*
 reduction of, in Calvin
  cycle, 104
 transportation and exchange of
  gas, 455–456
Carbon monoxide poisoning, 120
Carboniferous period, 273*t*,
  318–319, 318*f*
Carboxyl, 40, 40*f*
Carcinogenesis, 136
Carcinogens, 138
Carcinomas, 139
Cardiac cycle, 436–437, 436*f*
Cardiac muscle, 421, 421*f*, 548
Cardiovascular disease (CVD), 478,
  491–492
Cardiovascular system
 capillary exchange, 441, 441*f*
 as closed circulatory system, 433
 defined, 424, 424*f*
 disorders of, 445–446
 homeostasis and, 426
 pulmonary circuit, 439, 439*f*
 systemic circuit, 439*f*, 440
Carnivores, 467, 610, 610*f*
Carnivorous plants, 372
Carotenoids
 color of fall leaves, 95
 in photosynthesis, 100
Carotid bodies, 454
Carpels, 321, 321*f*, 400*f*, 401, 401*f*
Carriers, 170, 179
Carrying capacity, 584, 591
Cartilage, 420
Cartilaginous fish, 357, 357*f*
Caveman diet, 490
Celiac disease, 465
Cell body, 422, 422*f*, 516, 516*f*
Cell cycle
 apoptosis, 134, 135*f*
 cancer and, 136–139
 checkpoints, 134, 134*f*
 cytokinesis, 132–133
 internal and external signals,
  134–135
 interphase, 128–129
 mitosis, 129–133
Cell membrane, phospholipids and,
  47, 47*f*

Cell plate, 132, 133*f*
Cell recognition proteins, 61, 61*f*
Cell reproduction
 basics of, 126, 126*f*
 cancer and, 136–141
 cell cycle control systems,
  134–135
 cell cycle overview, 128, 128*f*
 chromatin to chromosomes, 127
 cytokinesis, 132–133
 mitosis, 129–133
Cell suspension culture, 408
Cell theory, 62, 126, 293–294
Cell transport, 88–91
 active transport, 91
 passive transport, 88–91
Cell wall
 animal, 74–75
 plant, 74
 prokaryotic cells, 63, 63*f*
Cell-signaling, 202–203, 203*f*
Cells, 2, 3*f*, 415
 animal cell, 64–65, 64*f*, 74–75
 cell respiration, 110–122
 cell transport, 88–91
 energy for, 82–85
 entropy and, 81, 81*f*
 eukaryotic, 64–73
 of immune system, 498*t*, 499
 junctions between, 75
 metabolic pathways and
  enzymes, 85–87
 origin of, 293–294, 293*f*
 plant cell, 65, 65*f*, 74, 373–375
 plasma membrane, 59–61,
  60*f*, 61*f*
 prokaryotic, 62–63
 structural components of, 44, 44*f*
 surface-area-to-volume ratio and,
  58–59, 58*f*, 59*f*
Cellular respiration, 71
 alternative metabolic pathways,
  121–122
 citric acid cycle, 112, 118
 electron transport chain, 112,
  119–120
 energy yield from, 121, 121*f*
 glycolysis, 112, 112*f*, 113–114
 overview of, 111–112, 111*f*, 112*f*
 preparatory reaction, 112, 117
Cellulose, 44
Cenozoic era, 273*t*
Central nervous system (CNS)
 brain, 520–522
 defined, 515
 limbic system, 522–523
 overview of, 520
Centrioles, 64*f*, 72–73, 73*f*, 129
Centromere, 127, 127*f*
Centrosome, 72, 72*f*
 spindles and, 129–130
Cephalization, 339

Cephalopods, 345
Cerebellum, 520f, 522
Cerebral cortex, 521, 521f
Cerebrum, 520f, 521
Cervical cap, 565
Cervix, 562, 562f
Channel proteins, 61, 61f
Chara, 311, 311f
Character displacement,
    606, 606f
Chargaff, Erwin, 186
Chargaff's rules, 186–187
Charophytes, 311, 313f
Chase, Martha, 185
Checkpoints, cell cycle, 134, 134f
Chemical bonds, 26–27
Chemical cycling, 614–619
    carbon cycle, 617–618
    in ecosystem, 4, 4f
    model for, 614–619, 615f
    nitrogen cycle, 616–617
    overview, 611–612
    phosphorus cycle, 615–616
Chemical reactions, 28
Chemoautotrophs, 296, 610, 611
Chemoheterotrophs, 297, 325
Chemoreceptors, 537–538, 538f
Chemotherapy, 140
Chickenpox, 290
Chiloglottis trapeziformis,
    601–602, 601f
Chitin, 44, 348
Chlamydia, 569–570
Chlamydomonas, 302, 303f
Chlorine, in sodium chloride
    formation, 26–27, 26f
Chlorofluorocarbons (CFCs),
    641–642
Chlorophyll
    color of leaves and, 95, 100f
    in photosynthesis, 97, 100
Chloroplasts, 65, 65f
    energy flow and, 84–85, 85f
    photosynthesis and, 97, 97f
    structure and function of,
        69–70, 70f
Choanocytes, 341
Choanoflagellates, 307, 307f
    338, 338f
Cholesterol, 479, 491–492
    fiber and, 44
    heart disease and, 38
    high-density lipoproteins
        (HDLs), 38
    hypercholesterolemia, familial,
        173, 173f
    low-density lipoproteins (LDLs),
        38
    as precursor of steroids, 48
    reducing, 479t
    role of, 38
    structure and function of, 47f, 48

Chordates, 339f
    characteristics of, 354
    evolutionary trends,
        356, 356f
    invertebrate, 354–355, 355f
Chorion, 571f, 572
Chorionic villi, 575, 575f
Chorionic villi sampling (CVS),
    227, 228f
Chromatid, 128
Chromatin, 66
    cell division and, 127, 127f
    condensation, 199
Chromium, 122
Chromosomal disorders
    chromosomal mutations,
        224–226
    testing for, 227–228
Chromosomal theory of
    inheritance, 165
Chromosomes
    abnormal chromosome
        inheritance, 156–157
    compaction of, 127, 127f
    composition of, 66
    defined, 126
    deletion, 224–225, 224f
    diploid number of, 146
    duplication, 225, 225f
    gender determination of humans
        and, 178
    haploid number of, 146
    homologous, 146–147
    inversion, 226, 226f
    karyotyping, 227–228
    linkage groups, 177, 177f
    meiosis, 146–157
    number of, and complexity of
        organism, 127
    rearrangements of, and cancer,
        138, 138f
    sex, 146, 146f, 178
    translocation, 138, 138f,
        225–226, 225f
    X chromosome, 146, 178–180
    Y chromosome, 146, 178–180
Chronic myelogenous leukemia
    (CML), 138, 138f
Chyme, 469
Chytrids, 326–327, 326f
Cialis, 560
Cigarette smoking, cancer and, 138,
    141
Cilia, 73–74, 73f, 452, 452f
Ciliates, 304
Circulatory systems
    closed, 433
    human cardiovascular system,
        435–446
    open, 433
    in vertebrates, 433–435
Cirrhosis, 473

Citric acid cycle, 112, 112f,
    118, 118f
Clade, 280
Cladistics, 280–281, 281f
Cladograms, 280–281, 280f
Clark's nutcracker, 607, 607f
Class, as category of classification,
    9, 9t, 277f, 278
Classification categories, 9, 9t,
    277–278, 277f
Clavicle, 547
Cleavage, 572
Climate
    biodiversity and regulation
        of, 632
    fossil fuels and global climate
        change, 639–640
Climate change, 18, 276, 618, 619f
    global, 600
Climax community, 603, 603f
Clitoris, 562
Cloaca, 360
Clonal selection model, 502
Cloning
    first cloned sheep, 212–213
    plants, 407
    reproductive, 212–213, 212f
    therapeutic, 212f
Closed circulatory system,
    433, 434f
Clostridium tetani, 298–299
Club fungi, 328, 328f
Club moss, 316–317
Clumped distribution, 588
Cnidarians, 339, 339f, 342, 342f
Coal age plants, 318–319, 318f
Cocaine, nervous system and,
    519, 519f
Cocci, 294, 294f
Cochlea, 539–540, 539f
Coconut, 406
Codominance, 174
Codon, 192
Coelom, 340, 466
Coelomate, 340–341, 340f
Coenzyme A (CoA), 112, 112f
Coenzymes, in cellular respiration,
    112
Coevolution, 601, 601f
Cofactors, 386
Cohesion-tension model, 387, 387f
Cohesion, of water, 30–31, 30f
Coleochaete, 311, 313f
Collagen, 74, 74f, 419, 475
    as fibrous protein, 51
    osteogenesis imperfecta and, 172
Collar cells, 341
Collecting duct, 458f, 459
Collenchyma cells, 374, 374f
Colon, 471
Colonial flagellate hypothesis,
    338, 338f

Color blindness, as X-linked
    recessive disorder, 179, 180
Colossal squid, 59
Columnar epithelium,
    417–418, 417f
Commensalism, 604t, 605
Common ancestor, 267
Community, 2, 3f, 581
    climax community, 603
    coevolution, 601–602
    competition, 605–606
    defined, 601
    diversity of, 602–603
    ecological niche, 605
    ecological succession, 603–604
    mutualism, 606–608
    native vs. exotic species,
        608–609
    species composition and, 602,
        602f
    species richness, 602
    stability of, 608–609
    succession, 603–604
Compact bone, 420, 548, 548f
Companion cell, 375
Competition, 592
    character displacement,
        606, 606f
    in community ecology,
        604t, 605–606
    niche specialization, 606
    resource partitioning, 606
Competitive exclusion principle,
    605–606, 606f
Complement system, 501, 501f
Complementary base pairing, 52,
    188, 188f
Compound eyes, 542, 542f
Compounds, 26
Conclusion, 11f, 13–14
Condoms, 565, 569
Conduction deafness, 540
Cones, 320, 543, 543f
Conifers, 320–321, 320f
Conjugation, 296
Conjugation pili, 296
Connective tissue, 416, 416f
    blood, 420–421
    defined, 419
    loose fibrous and related,
        419–420
Conservation biology, 582, 627
Conservation of energy, 80
Constipation, 471
Consumers, 4, 4f, 610, 610f
Contact inhibition, 135
Continental drift, 275–276, 276f
Continuous ambulatory peritoneal
    dialysis (CAPD), 461–462
Contraceptive implants, 565, 565t
Contraceptive injections, 565, 565t
Contraceptive vaccines, 566

Contractile proteins, 48
Control group, 13
Control systems of human body, 424–425, 424f
Controlled study, 14–15, 14f
Convergent evolution, 279–280
Copulation, 558
Coral reefs, 342
Cork cambium, 373, 382
Cork cells, 382
Cornea, 542, 542f
Corolla, 321, 321f
Coronary bypass operation, 446, 446f
Corpus luteum, 563, 563f
Cortex, plant, 379–380, 383f, 384
Cortisol, 530–531
Cotyledons, 323, 376, 404
Coupled reaction, 83–84, 84f
Covalent bond, 27, 27f
    electronegativity and, 29
    polar, 29, 29f
Coxal bones, 547
Cranial nerves, 523, 523f
Cranium, 547
Creatine, 84
Cretaceous period, 273t, 276
Cri du chat (cat's cry) syndrome, 224–225
Crick, Francis, 187–188, 187f
CRISPR, 207, 211, 211f
Cristae, 71, 117f
    organization of, 120, 120f
Cro-Magnons, 368–369
Cross-pollination, 403
Crossing-over, 149–150, 149f
Cruetzfeldt-Jakob syndrome, 292
Crustaceans, 349, 350f
Cuboidal epithelium, 417, 417f
Culture, 366
Cushing syndrome, 531
Cutaneous receptors, 544–545, 544f, 545f
Cuticle, 373
Cuvier, Georges, 237
Cyanide, as enzyme inhibitor, 87
Cyanobacteria, 96, 96f, 97, 296, 296f
Cyclins, 134, 202
Cymbalta, 518
Cystic fibrosis
    as autosomal recessive disorder, 170–171, 171f
    causes of, 88
    gene therapy, 232–233
Cytochrome c, 248, 248f
Cytokines, 502–503
Cytokinesis, 131, 132–133
    in animal cells, 132, 133f
    defined, 128
    in plant cells, 132–133
Cytokinins, 393t, 395

Cytoplasm, 63f
    glycolysis and, 113–114
Cytosine (C), 52, 52f, 186, 186f, 190
    structure of DNA and, 186–188
Cytoskeleton, 64f, 65, 72
Cytotoxic T cells (T_C cells), 498t, 504–505, 505f

D

Dams, 635–636
Darwin, Charles, 7–8
    background of, 236–237, 236f
    natural selection and adaptation, 240–242, 252, 252f
    study of geology, fossils, biogeography, 238–240
    Wallace and, 243–244
Darwin's finches, 271, 271f
Data, 13
Daughter cell, 126
Daughter chromosomes, 129, 129f, 130, 131f
Death rate, 584
Deciduous plants, 377
Decomposers, 4, 4f, 611, 611f
Default sex, 574
Deficiency disorder, 475
Deforestation, 635, 635f
Dehydration synthesis reaction
    defined, 42, 42f
    of fat, 45, 45f
    peptides, 49f
Delayed allergic response, 508
Deletion, chromosomes, 224–225, 224f
Demographic transition, 585
Demographics, 589–590
Denatured protein, 51
Dendrite, 422, 422f, 516, 516f
Dendrogamma, 1
Denisovans, 367–368
Denitrification, 617
Dense fibrous tissue, 419–420, 419f
Density-dependent factors, 592–594, 593f
Density-independent factors, 592, 592f
Deoxyribonucleic acid. See DNA (deoxyribonucleic acid)
Deoxyribose, 42
Depressants, 518
Desertification, 634–635, 635f
Detrital food web, 612, 613f
Detritus, 611
Deuterostomes, 340, 340f, 353–362
Development, 5
Devonian period, 273t
Diabetes
    gestational, 532
    type 1 diabetes, 491
    type 2 diabetes, 491

Diabetes mellitus, 459
    overview of, 531
    type 1 diabetes, 531–532
    type 2 diabetes, 532
Dialysate, 461–462
Dialysis, 461–462
Diaphragm (anatomical), 453, 454f
Diaphragm (birth control), 565, 566f
Diarrhea, 472
Diastole, 436–437
Diastolic blood pressure, 445
Diencephalon, 520f, 521
Diet, 475. See also Nutrition
    to prevent cancer, 141–142
Dietary supplements, 485–487
Dieting, 490
Diffusion
    facilitated, 89, 89f
    simple, 88, 89f
Digestive enzymes, 473–474, 474f
Digestive system, 424, 424f, 466–474, 466f–467f
    accessory organs, 472–473
    complete and incomplete, 466
    digestive juices, 473–474, 474f
    functions of, 466
    homeostasis and, 426
    organs of, 467–472
    tube-within-a-tube plan, 466
Dihybrid, 167
Dioecious, 401
Diploid number, 146
Directional selection, 252f, 253–254
Disaccharides, 43, 43f
Diseases
    bacterial diseases in humans, 298–299
    emerging diseases, 17–18
    emerging viruses and, 290–291
    fungi causing, 331–333
Disruptive selection, 252f, 254, 255f
Distal convoluted tubule, 458f, 459
Diuretic, 460
Diversity, of communities, 602–603
DNA (deoxyribonucleic acid), 5
    Chargaff's rules, 186–187
    compared to RNA, 191t
    as evidence for evolution, 248
    flow of genetic information, 192–197, 192f
    Franklin's X-ray diffraction data, 187
    gene mutation, 222–223
    genetic code, 192–193
    genetic marker, 228, 228f
    as genetic material, 185, 185f
    Human Genome Project, 216–218
    intergenic DNA, 216
    replication of, 189–190

    structure and function of, 51–52, 52f, 186–188
    testing for genetic disorders, 228–229, 228f
    transcription, 193–194
    viral reproduction and, 288, 289f
    Watson and Crick model, 187–188, 187f
DNA analysis, 210–211
DNA fingerprinting, 210–211, 210f
DNA ligase, 189
DNA microarray, 228–229, 229f
DNA polymerase, 189
DNA replication, 126, 128, 129, 129f, 189–190, 189f
DNA sequencing, 209, 217f, 229
DNA technology
    DNA fingerprinting, 210–211, 210f
    polymerase chain reaction (PCR), 209–210, 209f
    recombinant DNA (rDNA), 208–209, 208f
Dogs, 336
Domain
    as category of classification, 9, 9t, 10t, 277f, 278
    three-domain system, 281–282, 282f
Domain Archaea
    characteristics of, 7f, 9, 10t
    three-domain system and, 281–282, 282f
Domain Bacteria
    characteristics of, 7f, 9, 10t
    three-domain system and, 281–282, 282f
Domain Eukarya
    characteristics of, 7f, 9, 10t
    kingdoms of, 9, 10t
    three-domain system and, 281–282, 282f
Domestic livestock, 638–639, 638f
Dominant alleles, 165, 169
Dopamine, drug abuse and, 519
Dormancy, 395
Dorsal root ganglion, 523
Double fertilization, 322, 402f, 403
Double-blind study, 15
Doubling time, 584
Down syndrome, 155–156, 155f
    characteristics of, 221
    gene causing, 225
    karyotype of, 227
    maternal age and, 156
    translocation, 225
Drug abuse, nervous system and, 518–520
Dry fruit, 404, 405f
DScam gene, 202
Duchenne muscular dystrophy, as X-linked disorder, 180, 180f

Duodenum, 469, 472
Duplication, chromosomes, 225, 225f
Dyads, 148f, 149
Dynein, 72, 72f
Dysentery, 299

E

E. coli. See Escherichia coli
Earthworms, 346, 346f
Eating disorders, 492, 492f
Ebola virus, 286, 291
Ecdysis, 347
Ecdysozoans, 340, 347–352
Echinoderms, 339f, 353–354, 353f
Ecological niche, 605, 605f
Ecological pyramid, 613–614, 614f
Ecological succession, 603–604
Ecology
  basic concepts of, 581, 581f
  biodiversity, 628–632
  as biological science, 582
  of communities, 601–609
  defined, 581
  of ecosystems, 610–619
  extinction, 595–596
  life history patterns, 595–596, 595f
  population growth, 588–591
  resources and pollution, 632–643
  succession, 603–604
  sustainable society, 643–646
Ecosystems, 2, 3f
  aquatic ecosystems, 620
  autotrophs, 610
  biosphere, 620
  chemical cycling and energy flow in, 4, 4f
  coevolution and, 601–602
  defined, 581
  energy flow and chemical cycling, 611–614, 611f
  heterotrophs, 610–611
  human influence on, 17
  primary productivity, 621–622
  terrestrial, 621
Ecotourism, 632, 632f
Ectoderm, 573, 573f
Ectothermic, 359
Edema, 461
Egg, 557
  testing for genetic disorders, 230–231, 231f
Ejaculation, 561
Elastin, 74
Electrocardiogram (ECG), 437
Electron microscope, 57–58, 57f
Electron shell, 25–26, 25f, 26f
Electron transport chain
  in cellular respiration, 112, 112f, 119–120, 119f, 120f
  light reactions, 101–103, 102f

Electronegativity, 29
Electrons
  arrangement of, in atoms, 25–26, 25f, 26f
  covalent bonding, 27, 27f
  defined, 23, 23f
  electronegativity, 29
  ionic bonding, 26–27, 26f
  light reactions and, 100–101, 101f
  octet rule, 26
Elements
  atomic structure, 23, 23f
  defined, 22
  isotopes, 24
  of living organisms, 22, 22f
  periodic table, 23–24, 23f
  source of, 22
Elephant, 125
Elephantiasis, 348
Elongation cycle, 195–196, 195f
Embolus, 445
Embryo
  development of, 570–576, 572f
  gastrulation and neurulation, 573–574
  organ formation, 574, 574f
  plant, 403, 404f
  testing for genetic disorders, 230–231, 230f
Embryo sac, 402, 402f
Embryonic stem cells, 213
Emergency contraception, 566
Emerging virus, 290–291, 291f
Emphysema, 453, 455
Empty-calorie foods, 476
Emulsification, 469–470
Endemic goiter, 529, 529f
Endocrine glands, 472, 526–532
Endocrine system, 424f, 425, 526–532
  action of hormones, 514, 526–527
  adrenal gland, 530–531
  as control system, 514, 514f
  homeostasis and, 426
  hypothalamus and pituitary gland, 527–529
  overview of, 526
  pancreas, 531–532
  summary of, 532t
  thyroid and parathyroid gland, 529–530
Endoderm, 573, 573f
Endodermis, plants, 383f, 384
Endomembrane system, 64–65, 68, 69f
Endometriosis, 567
Endometrium, 562
Endoplasmic reticulum (ER), 64f, 65, 65f, 67, 67f, 68, 68f, 69f
Endoskeleton, 546

Endosperm, 322
Endospores, 296
Endosymbiosis, 70
Endosymbiotic theory, 301
Endotherms, 359
Energy
  of activation, 87
  ATP cycle, 82–83, 83f
  for cells, 82–85
  conservation law of, 80
  coupled reaction, 83–84, 84f
  defined, 2, 80
  energy flow in ecosystem, 4, 4f
  entropy and, 81
  flow of, 84–85, 85f
  fossil fuels and global climate change, 639–640
  intake and output of body, 488–489
  kinetic, 80, 80f
  laws of, 80–81
  measuring, 80
  polysaccharides as energy storage molecules, 43–44, 44f
  potential, 80, 80f
  radiant, 99f
  renewable sources of, 640–641
Energy flow, 612–614
Energy laws, 80–81
Energy of activation, 87
Enhancer, 201
Entropy, 81
Environmental factors
  gene mutations and, 222
  phenotype and, 175–176
Environmental impact, population growth and, 587, 587f
Environmental science, 582
Enzymatic proteins, 61, 61f
Enzyme inhibition, 87
Enzymes
  active site of, 86–87
  defined, 86
  energy of activation, 87, 87f
  enzymatic action, 86–87, 86f
  in eukaryotic cells, 64
  function of, 48
  as globular proteins, 51
  induced fit model, 86
  inhibition of, 87
  restriction enzymes, 208, 208f
Eocene epoch, 273t
Eosinophils, 443f
Epidermal growth factor, 135
Epidermal tissue, plants, 373–374, 374f, 383f, 384
Epidermis, 373, 417–418
Epididymis, 559, 560f
Epigenetic inheritance, 200, 238
Epiglottis, 452f, 453, 467, 467f
Epinephrine, 530
Episodic memory, 522–523

Epistatic interaction, 177
Epithelial tissue, 416, 416f, 417–418, 417f, 418f
Epithelium, 417
EPSA1 gene, 251
Equilibrium species, 595
Equisetum, 318
Erectile dysfunction (ED), 560
Erection, 560
Erosion
  beach, 634, 634f
  biodiversity and prevention of, 631–632
  food production and, 637–638
Erythrocytes, 442–443, 443f
Erythropoietin, 443
Escherichia coli, 9, 10t, 25, 62, 185, 185f, 294f, 297, 471, 608
  enzymes for lactose metabolism and, 197, 198f
Esophagus, 467, 467f, 472f
Essential amino acids, 479–480
Essential dietary fatty acids, 478
Essential nutrient, 475
Estrogen, 563–564, 564f
  structure and function of, 47–48, 47f
Estuary, 621
Ethyl alcohol, 43, 43f, 116
Ethylene, 393t, 396, 396f
Euchromatin, 200
Eudicots, 376–377, 376f
  development of seeds in, 403, 404f
  germination, 405–406, 406f
  root, 383f, 384
Eugenics, 231
Euglena, 10t
Eukarya, 10, 10t
Eukaryotes
  cell-signaling, 202–203, 203f
  evolution of, 301, 301f
  gene expression in, 199–203, 199f, 200f
Eukaryotic cells, 64–73
  centrioles, 72–73, 73f
  cilia and flagella, 73–74, 73f
  cytoskeleton, 65, 72
  defined, 62
  DNA replication, 189–190, 190f
  endomembrane system, 64–65, 68, 69f
  energy-related organelles, 9–71, 65, 70f
  gene expression in, 196f
  motor proteins and, 72
  nucleus, 64–65, 66–67, 66f, 67f
  organelles, 64
  plasma membrane, 59–61, 60f, 61f
  ribosomes, 67, 67f
  vacuoles, 69, 70f
  vesicles, 65

Eumetazoans, 339
Eutrophication, 616
Evergreens, 377
Evolution, 6–10
    adaptation, 240–242, 241*f*, 242*f*
    anatomical evidence, 246–248
    of animals, 337–341
    antibiotic resistant bacteria
        and, 235
    biogeographical evidence of, 246
    bottleneck effect and, 262
    chordates, 356, 356*f*
    coevolution, 601–602
    common descent with
        modification, 8
    convergent, 279–280
    before Darwin, 237–238
    Darwin's and, 236–237, 238–244
    defined, 7
    evolutionary tree, 7, 7*f*
    fitness and, 241–242
    fossil evidence, 244–245
    fungi, 325*f*
    gene flow, 260
    in genetic context, 257–260
    genetic drift, 261–262, 261*f*
    genetic mutation, 260
    gradualistic model of, 274–275
    Hardy-Weinberg principle, 259
    of humans, 363–369
    of jaws, 356–358, 356*f*
    of land plants, 312–314, 313*f*
    macroevolution, 266–272
    mass extinction, 275–276
    microevolution, 257–262
    molecular evidence of, 248
    mosaic evolution, 366
    natural selection and, 7–8,
        240–242, 252–256
    nonrandom mating and, 260–261
    pace of, 274–275, 274*f*
    photosynthesis and, 106–107
    as principle, 14
    speciation and, 240, 266–272
    systematics, 277–282
    taxonomy and systematics, 9–10
    Wallace and, 243–244
Evolutionary tree, 7, 7*f*, 10,
    274–275, 274*f*
Ex vivo gene therapy, 232, 232*f*
Excavata, 305, 305*f*
Exchange pool, 614
Exercise, benefits for joint
    disorder, 553
Exocrine gland, 472
Exophthalmos, 529, 529*f*
Exoskeleton, 348, 546, 546*f*
Exotic species, 608–609, 609*f*
Experiment, 11*f*, 12
Experimental design, 12–13
Experimental variable, 12–13
Expiration, 454, 454*f*

Exponential growth, 590, 590*f*
External fertilization, 558
Extinction
    biodiversity and, 628
    as bioethical issue, 16–17
    defined, 595
    humans influence on, 596
    mass, 275–276
    natural selection and, 240
    of vertebrates and invertebrates,
        336
    vulnerability to, 595, 596*f*
Extracellular matrix (ECM),
    74–75, 74*f*
Extraembryonic membrane, 558
Ey gene, 201, 201*f*
Eyes
    camera-type, 542
    color, 176–177, 176*f*
    compound, 542
    human, 542–544
    LASIK surgery, 544

F

F₂ fitness, 269
Facial bones, 547
Facilitated diffusion, 89, 89*f*
FAD, 112
FAD⁺, 119
FADH₂, 112, 119
Family, as category of classification,
    9, 9*t*, 277*f*, 278
Fat-soluble vitamins, 483*t*
Fat, body, 478
    glucose metabolized to, 43
    maintaining healthy weight,
        489–490, 490*f*
    synthesis and breakdown of, 45*f*
Fat, dietary, 45–46. *See also* Lipids
    cardiovascular disease and,
        491–492
    digestion of, 474, 474*f*
    reducing, 49*t*
    trans, 478
    types of, 478
Fatty acids
    metabolic pathway,
        121–122, 121*f*
    omega-3 fatty acid, 46
    saturated, 46, 46*f*
    structure and function, 45–46
    unsaturated, 46, 46*f*
Feathers, 359, 360*f*
Feedback inhibition, 87, 87*f*
Female condom, 565, 566*f*
Female gametophytes, 402*f*, 403
Female reproductive system,
    562–564, 562*f*–563*f*
Femur, 547
Fermentation, 43, 43*f*,
    115–116, 116*f*

Ferns, 312, 313*f*, 314*f*, 315,
    317–318, 317*f*
Fertilization, 146, 571–572, 571*f*
    double, 322
    external, 558
    seed plants, 319
Fertilizer, 386
Fetus
    development of, 575–576
    testing for genetic disorders,
        229–231
Fever, 500
Fiber, 477, 477*f*
    cellulose as, 44
    cholesterol and, 44
    insoluble, 45
    soluble, 45
Fibrinogen, 444
Fibroblasts, 419
Fibrous proteins, 51
Fibrous tissue, 419–420, 419*f*
Fibula, 547
Fight-or-flight response, 525
Filament, 321, 321*f*, 400*f*, 401
Filter feeder, 342
Filtration, in kidneys, 458*f*, 459
Fimbriae, 562
Finches, Darwin's study of, 240, 240*f*
Fire, 582, 582*f*
Fish, 356–358, 356*f*–358*f*
    circulatory system in, 433–434,
        435*f*
    gills, 455, 456*f*
Fishing, 639, 639*f*
Fitness, 241–242, 252
    F₂ fitness, 269
    sexual selection and, 254
Five-kingdom system, 281
Flagella, 73–74, 73*f*, 295, 295*f*
Flagellum, 63, 63*f*
Flatworms, 339*f*, 343–344, 344*f*
Fleming, Alexander, 12
Fleshy fruit, 404, 405*f*
Flint water crisis, 626
Flowering plants, 321–324,
    321*f*–324*f*
    asexual reproduction in,
        407–411, 407*f*
    life cycle overview, 402*f*
    sexual reproduction in, 399–406
Flowers
    anatomy of, 321, 321*f*,
        400–401, 400*f*
    in evolution of land plants,
        312–313, 313*f*
Fluid-mosaic model, 60, 60*f*
Flukes, 344
Fluorescent immunohistochemistry
    in situ hybridization
    (FISH), 227*f*
Folic acid, 482*t*
Follicle, 562

Follicle-stimulating hormone
    (FSH), 561, 563, 564*f*
Follicular phase, 563, 564*f*
Food, 4. *See also* Nutrition
    agricultural value of biodiversity,
        629–630
    biological magnification
        and, 643
    from cloned animals, 213
    domestic livestock, 638–639, 638*f*
    fishing, 639, 639*f*
    fruit, 324
    fungi as, 330
    genetic engineering, 638
    green revolutions, 638
    harmful farming methods, 637
    rural sustainability and, 644–645
    soil loss and degradation,
        637–638
    water used to produce, 636
Food chains, 614, 614*f*
Food irradiation, 25
Food web, 612–614, 613*f*
Formed elements, 442–445, 443*f*
Fossil fuels
    acid deposition and, 34
    carbon cycle and, 617*f*, 618
Fossil record, 244–245
Fossils, 238–239, 238*f*, 239*f*
    Darwin's study of, 238–239
    defined, 272
    as evidence of evolution,
        244–245
    geological timescale and,
        272–274, 272*f*, 273*t*
    transitional links, 244
Founder effect, 262, 262*f*
Fovea, 543
Frameshift mutation, 223
Franklin, Rosalind, 186, 187, 187*f*
Freckles, 260, 260*f*
Fronds, 317
Fructose, 42, 43
Fruit, 323, 324
    seed dispersal and, 404–405,
        404*f*, 405*f*
    types of, 404–405
Fruiting body, 328
Fukushima Nuclear Facility, 24, 24*f*
Functional groups, 40, 40*f*
Fungi
    characteristics of, 10, 10*t*,
        325–326, 326–329, 326*f*
    chytrids, 326–327
    club or sac (mushrooms),
        328–329
    compared to plants and
        animals, 326*t*
    as disease-causing organisms,
        331–333
    ecological benefits of, 329–330
    economic benefits of, 330

evolution of, 325f
mutualistic relationships, 329–330
mycorrhizal, 330
peppers and, 311
plant roots and, 386
zygospore (black bread mold), 327–328
Fusion and entry, of virus, 290f

G

G3P
    fate of, 104, 104f
    production of, in Calvin cycle, 103, 104
Galactose, 42
Galápagos Islands, 239–240, 606
    adaptive radiation and finches of, 271, 271f
Gallbladder, 469, 470f
Gamete, 568
Gamete intrafallopian transfer (GIFT), 568
Gamete isolation, 269
Gametes, 145
Gametophyte, 314, 314f, 399
Gamma globulin, 507
Gamma-aminobutyric acid (GABA), 518
Gap junction, 74f, 75
Garrod, Archibald, 191
Gaseous cycle, 614
Gastric reflux disease, 469
Gastropods, 344
Gastrula, 573
Gastrulation, 573, 573f
Gause, G. F., 606
Gender
    determination of, in humans, 178, 574
    sexual selection and, 254, 255f
Gene expression
    from DNA to RNA to protein, 192–197
    in eukaryotes, 199–203, 199f, 200f
    levels of control, 197–203
    in prokaryotes, 197–198
    in specialized cells, 197–198, 197f
    summary of, 196–197, 196f
    transcription, 193–194
    translation, 194–196
Gene flow, 260, 261f
Gene locus, 165
Gene migration, 260
Gene pool, 257
Gene therapy, 232–233
Genes, 5
    composed of DNA, 185, 185f
    expression of, 191–197

flow of genetic information, 192–197, 192f
Hox genes, 248
    mutations of, 222–223
    regulatory gene, 198, 198f
Genetic code, 192–193
Genetic counseling, 226–227
Genetic disorders
    chromosomal mutations, 224–226
    disorders of interest, 170–173
    Down Syndrome, 155–156, 155f, 221
    family pedigree and, 169–170
    gene mutation, 222–223
    gene therapy, 232–233
    testing for, 226–231
Genetic drift, 261–262, 261f
Genetic engineering, 208, 638
Genetic marker, 228, 228f
Genetic mutation, 260
Genetic profile, 228–229
Genetic testing
    for cancer associated genes, 139
    of DNA, 228–229
    DNA microarray, 228–229
    of egg, 230–231, 231f
    of embryo, 230–231, 230f
    of fetus, 229–230
    genetic marker, 228, 228f
    karyotyping, 227–228
    over-the-counter tests, 229
    for a protein, 228
Genetically engineered plants, 392, 408–411, 409f–410f, 409t
Genetically modified organisms (GMOs), 208, 214–216
Genetically modified plants (GMPs), 408–411, 409f–410f, 409t
Genetics
    abnormal chromosome inheritance, 156–157
    asthma susceptibility and, 450
    cellular reproduction, 126–139
    chromosomal mutations, 224–226
    cloning and stem cells, 212–213
    DNA and RNA structure and function, 191–196
    environment and phenotype, 175–176
    epigenetic inheritance, 200
    evolution in genetic context, 257–260
    gene expression, 185–197
    gene mutations, 222–223
    gene therapy, 232–233
    genotype versus phenotype, 165–166, 166t
    Hardy-Weinberg principle, 259
    homologous chromosomes, 146–147

Human Genome Project, 216–218
    incomplete dominance, 173–174
    law of independent assortment, 166
    law of segregation, 164
    linkage groups, 177
    meiosis, 146–157
    Mendel's laws, 161–173
    multiple-allele traits, 174
    one-trait inheritance, 162–166
    pleiotropy, 177
    polygenic inheritance, 174–176
    rule of multiplication, 167–168
    sex-linked inheritance, 178–180
    sexual reproduction, 146–157
    testing for genetic disorders, 226–231
    transgenic organisms, 214–216, 214f
    two-trait inheritance, 166–167
Genital herpes, 569
Genital warts, 569
Genome, 216
    Human Genome Project, 216–218
Genome editing, 207, 211, 211f
Genomics, 216–218
Genotype, 258, 259f
Genotype, 165
    defined, 165
    versus phenotype, 165–166, 166t
Genus, as category of classification, 9, 9t, 277f, 278
Geological timescale, 272–274, 272f, 273t
Geomyces destructans, 332
Geothermal energy, 640
Germ layers, 339, 573, 573f
Germinate, 405
Gestation period, 558
Gestational diabetes, 532
Giant squid, 59
Giardia, 305
Gibberellins, 393t, 394–395, 395f
Gills, 455, 456f
Glands, 418
Global warming, 18, 618, 618f
    fossil fuels and, 639–640
Globular proteins, 51
Glomerular capsule, 457, 458f
Glottis, 453
Glucagon, 472
    function of, 531
    regulation of blood glucose level, 531f
Glucosamine-chondroitin supplements, 553
Glucose, 43
    breakdown of, in cellular respiration, 111–120
    cellulose, 44, 44f

chitin, 44
disaccharides and, 43, 43f
energy yield from metabolism of, 121–122, 121f
fermentation, 115–116, 116f
metabolic pathway, 121–122, 121f
in photosynthesis equation, 28
polysaccharides and, 44, 44f
as product of Calvin cycle, 104
structure and characteristics of, 42, 42f
Glucose phosphate, 104
Gluten, 465
Glycemic index, 478
Glycerol
    metabolic pathway, 121–122, 121f
    structure and function, 45–46, 45f
Glycogen, 44, 44f
Glycolysis
    defined, 112, 112f
    energy investment step, 113, 113f
    energy-harvesting step, 114, 114f
    substrate-level ATP synthesis, 114, 114f
Glycoproteins, 60
Goiter, endemic, 529, 529f
Golgi apparatus, 65, 68, 69f
Golgi tendon, 544
Golgi, Camillo, 68
Gonadotropic hormone, 527, 561
Gonadotropin-releasing hormone, 561
Gonads, 557
Gonorrhea, 570
Gradualistic model of evolution, 274
Grana, 97, 97f
Grant, Peter, 240
Grant, Rosemary, 240
Granum, 70, 70f
Granzymes, 504–505
Grasshopper, 351–352, 352f
Gravitational equilibrium, 541
Gravitropism, 397, 398f
Grazing food web, 612, 613f
Green revolution, 638
Green tea, 485
Greenhouse, 99
Greenhouse effect, 18, 618
Greenhouse gases, 619, 640
Ground tissue, 373, 374, 374f
Grow-lights, 99
Growth factors
    cancer and, 137
    proto-oncogenes and cancer, 137
    as signal in cell cycle, 135
Growth hormone (GH), 528
Groynes, 634
Guanine (G), 52, 52f, 186, 186f, 190
    structure of DNA and, 186–188

Gymnosperms, 312, 313*f*, 320–321, 320*f*
    alternation of generations, 314, 314*f*

H

H1N1 virus, 17
H5N1 virus, 17, 287–288
    as emerging virus, 291
Habitat, 605
Habitat isolation, 268
Habitat loss, 16–17, 628*f*
Hagfish, 357
Hair, 360
    perms and relaxers, 51
Hair cells, 538–539
Halophiles, 300, 300*f*
Haploid number, 146
Hardy-Weinberg principle, 259
Hardy, G. H., 259
Hawaiian honeycreeper, 8, 8*f*
Hawks, 4, 4*f*, 6
Hearing, 538–542, 539*f*
Hearing loss, 540–541, 540*f*
Heart
    in closed circulatory system, 433, 434*f*
    human, 435–437, 436*f*–438*f*
    in open circulatory system, 433, 433*f*
    pulmonary circuit, 434, 434*f*
    systemic circuit, 434, 434*f*
Heart attack, 446
Heart disease, cholesterol and, 38
Heart murmur, 436
Heartbeat, 437–438, 437*f*
Heartburn, 469
Heat, as lost energy, 81
Height, as polygenic trait, 174, 174*f*
Helicase, 189
Helicobacter pylori, 14–15, 469
Helium, model of, 23, 23*f*
Helper T cells (TH cells), 498*t*, 504–505
Heme, 456
Hemocoel, 433
Hemodialysis, 461–462, 461*f*
Hemoglobin, 6, 420, 442
    carbon monoxide binding with, 120
    as globular proteins, 51
    mRNA translation, 202
    quaternary structure of, 51
    sickle-cell disease, 52–53, 53*f*
    as transport protein, 48*f*, 49
    transportation and exchange of gas, 455–456
Hemolysis, 90, 90*f*
Hemophilia, 444
    as X-linked trait, 179
Hepatic portal vein, 472, 473*f*

Hepatitis, 473, 569
Herbaceous plants, 378
Herbivores, 467, 610, 610*f*
Heredity. *See* Genetics
Hermaphroditic, 343–344, 557
Herpes viruses, 290
Herpes, genital, 290, 569
Hershey, Alfred, 185
Heterochromatin, 199
Heterotrophs, 610–611
Heterozygous, 165
    advantage of, 256
High-density lipoproteins (HDLs), 38, 491
High-fructose corn syrup, 43
Hinge joints, 551, 552*f*
Hippocampus, 522, 522*f*, 523
Histamine, 500
Histones, 127
HIV (human immunodeficiency virus), 505, 568
    as emerging virus, 290–291
    reproduction of, 290*f*
    transmission and treatment of, 509–510, 509*f*
Hobbits, 367
Holocene epoch, 273*t*
Homeostasis, 4
    body temperature and, 428, 428*f*
    defined, 426
    negative feedback mechanism, 427
    systems of body and, 426
Hominids, 363
Hominins, 363, 364–367
Homo erectus, 365*f*, 366–367, 367*f*
Homo floresiensis, 367
Homo habilis, 365*f*, 366
Homo naledi, 368
Homo sapiens, 10*t*, 365*f*, 368
Homogentisate oxygenase (HGD) gene, 171
Homologous chromosomes, 146–147, 146*f*, 165, 165*f*
Homologous structures, 247, 278
Homologues, 146–147
Homozygous, 165
Hookworm, 348
Hormone replacement therapy (HRT), 565
Hormones
    action of, 526–527, 527*f*
    in males, 561–562
    as messenger of endocrine system, 514
    peptide, 526–527
    plant, 393–396, 393*f*, 393*t*
    as regulatory protein, 48
    as signal in cell cycle, 135
    steroid, 526

Hornwort, 315, 315*f*
Horseshoe crab, 350–351, 350*f*
Horsetail dietary supplements, 318
Hox genes, 248
Human activities
    beaches and, 634
    carbon cycle and, 618–619
    nitrogen cycle and, 617
    phosphorous cycle and, 616
    semiarid lands, 634–635
    tropical rain forests, 635
Human body
    blood, 442–446
    body mass index, 488–490
    body temperature regulation, 428, 428*f*
    control systems of, 424–425, 424*f*
    development, 570–576, 572*f*
    digestive system, 466–474, 466*f*–467*f*
    healthy weight maintenance, 489–490, 490*f*
    heart and blood vessels, 435–441
    homeostasis, 426–428, 426*f*
    levels of biological organization, 415–416, 415*f*
    lymphatic system, 440
    maintenance system of, 424, 424*f*
    motor system, 545–553
    nervous system, 514–525
    organs and organ systems of, 423–425
    reproductive system of, 425, 425*f*, 559–570
    respiratory system, 451–456, 451*f*
    senses, 537–545, 537*f*
    sensory input and motor output system, 425, 425*f*
    tissue of, 415–423
    transport system, 424, 424*f*, 435–442
    urinary system, 457–462, 457*f*
Human chorionic gonadotropin (hCG), 572–573
Human evolution, 363–369
Human Genome Project, 52, 216–218
Human growth hormone (hGH), as regulatory protein, 48
Human immunodeficiency virus (HIV). *See* HIV (human immunodeficiency virus)
Human papillomavirus (HPV), 568–569
Human population, 583–587
    comparing age structures, 586, 586*f*
    growth and environmental impact, 587, 587*f*

more-developed *vs.* less-developed countries, 585–586, 585*f*
    present and future growth of, 583–584, 583*f*, 585*f*
Humerus, 547
Huntington disease, 172–173, 172*f*, 201
Hyaline cartilage, 420
Hybrid sterility, 269
Hybridization, 409
Hydra, 432, 432*f*
Hydrocarbons, 40, 40*f*
Hydrochloric acid
    dissociation of, 33
    as stomach acid, 33
    as strong acid, 33
Hydrofluorocarbons, as greenhouse gas, 619
Hydrogen
    breathing rate and, 454
    as element of living organisms, 22*f*
    model of, 25*f*
    molecular formula, 27, 27*f*
    structural formula, 27, 27*f*
Hydrogen bond, 29, 29*f*
Hydrogen gas, 27, 27*f*
Hydrogen ions (H$^+$), 27, 33
    in acids, 34
    bases and, 34
    pH and, 34, 34*f*
Hydrolysis reaction, 42, 42*f*
Hydrophilic, 29–30, 40
Hydrophobic, 29–30, 40
Hydropower, 640, 640*f*
Hydrostatic skeleton, 546, 546*f*
Hydroxide ions (OH-), 34
    in bases, 33
    pH and, 34, 34*f*
Hydroxyl, 40*f*
Hymen, 563
Hypercholesterolemia, familial, 173, 173*f*
Hypertension, 445
Hypertonic solution, 90*f*, 91
Hyphae, 325
Hyponatremia, 483
Hypothalamic-releasing hormone, 527
Hypothalamus, 521, 522*f*, 526*f*
    control of pituitary gland, 527–529
Hypothesis, in scientific method, 11*f*, 12
Hypotonic solution, 90–91, 90*f*

I

Ice, density of, 32, 32*f*
Iandumoema smeagol, 1
Immediate allergic response, 508

Immune system/immunity, 424, 424f, 496–510
  ABO blood type, 503
  active immunity, 506, 506f
  adaptive immunity, 502–505
  AIDS, 509–510, 509f
  allergies, 508, 508f
  antibody function, 503
  autoimmune disease, 509
  B cells and antibody response, 502–503, 502f
  barriers to entry, 499
  complement system, 501, 501f
  cytotoxic T cells, 504–505
  fever and, 500
  helper T cells, 504–505
  immunity defined, 499
  immunizations, 506–507
  inflammatory response, 500–501
  innate immunity, 499
  natural killer (NK) cells, 501
  nonspecific defenses, 499–501
  organs, tissues and cells of, 497–499
  passive immunity, 507
  problems with, 508–510
  specific defenses, 502–505
  T cells and cellular response, 503–505, 505f
  tissue rejection, 505
In vitro fertilization (IVF), 230–231, 556, 567
In vivo gene therapy, 232–233
Inbreeding, 260–261
Incomplete dominance, 173–174, 173f
Independent assortment, law of, 166
Induced fit model, 86
Induction, 57
Inductive reasoning, 12
Industrial melanism, 258, 258f, 259
Industrial waste, 642–643
Infertility, 566–568
Inflammatory response, 444, 500–501, 500f
Influenza
  anatomy of virus, 287–288, 287f
  H5N1 virus, 287–288, 291
  H7N9 virus, 287–288
Inheritance. See also Genetics
  abnormal chromosome inheritance, 156–157
  of acquired characteristics, 237–238
  chromosomal theory of, 165
  codominance, 174
  environment and phenotype, 175–176
  epigenetic inheritance, 200, 238
  genotype versus phenotype, 165–166, 166t

incomplete dominance, 173–174
  law of independent assortment, 166
  law of segregation, 164
  linkage groups, 177
  Mendel's laws, 161–173
  multiple-allele traits, 174
  one-trait inheritance, 162–166
  particulate theory of, 162
  pleiotropy, 177
  polygenic, 174–176
  rule of multiplication, 167–168
  sex-linked inheritance, 178–180
  two-trait inheritance, 166–167
Initiation, 195, 195f
Innate immunity, 499
Inner ear, 539–540, 539f
Inorganic chemistry, 39
Insecticides, resistance to and directional selection, 253–254
Insects, 351–352, 352f
  excretory system, 460–461, 460f
  respiration of, 453, 453f
  walk on water, 31
Insoluble fiber, 45
Inspiration, 454, 454f
Insulin, 459, 472, 491
  function of, 531
  quaternary structure of, 51
  regulation of blood glucose level, 531f
  as regulatory protein, 48
Integration
  neurons, 518
  senses and, 537
  of virus, 288, 289f, 290f
Integumentary system, 425, 425f
Intercalated disk, 421
Interferon, fever and, 500
Intergenic DNA, 216
Interkinesis, 152
Intermediate filament, 72, 72f
International Code of Phylogenetic Nomenclature, 281
Interneurons, 516, 516f
Internode, 375
Interphase, 128–129, 128f, 130f
Interstitial fluid, 424
Intervertebral discs, 547
Intestinal enzymes, 470
Intracytoplasmic sperm injection (ICSI), 567
Intrauterine device (IUD), 565t, 566
Inv dup 15 syndrome, 225, 225f
Inversion, chromosomes, 226, 226f
Invertebrate animals, 339
Ionic bond, 26–27, 26f
Ions, 26–27
Iris, 542, 542f
Iron, plant nutrition and, 385, 386
Isomers, 40

Isotonic solutions, 90, 90f
Isotopes
  defined, 24
  uses of radioactive isotopes, 24, 24f

J

Jacob, Francois, 197–198
Jacobs syndrome (XYY), 157
Jaundice, 473
Jawless fish, 357, 357f
Jaws, evolution of, 356–358, 356f
Joint appendages, 348–352, 348f–352f
Jointed limbs, 358, 358f
Joints
  ball-and-socket joints, 551, 552f
  disorders of, 552–553
  hinge joints, 551, 552f
  skeletal muscles moving bones at, 551
  synovial, 551, 552f
Junction proteins, 61, 61f
Jurassic period, 273t

K

Kaposi's sarcoma, 509
Karyotype, 146, 227f
Karyotyping, 227–228
Keratin, 51
Keystone species, 608
Kidney stones, 461
Kidneys
  function of, 457, 457f, 458f
  hemodialysis and transplants, 461–462
  nephrons, 457, 459
  urine formation, 458f, 459–461
Kilocalories, 80
Kinases, 134
Kinesin, 72, 72f
Kinetic energy, 80, 80f
Kingdom
  as category of classification, 9–10, 9t, 10t, 277f, 278
  of domain Eukarya, 7f, 9, 10t
  five-kingdom system, 281
Klinefelter syndrome, 156, 156f
Krebs cycle, 118, 118f
Kuru, 292
Kwashiorkor, 638

L

Labia majora, 562, 562f
Labia minora, 562, 562f
Lac operon, 197–198, 198f
Lactate, 114, 116
Lacteal, 471
Lactic acid, 115

Lactose, 43
Lactose metabolism, 197–198, 198f
Lacunae, 420
Ladder-like nervous system, 515
Lamarck, Jean-Baptiste de, 237–238
Lambda virus, 288, 288f, 289f
Lampreys, 357, 357f
Lancelets, 355
Land plants. See also Plants
  diversity of, 315–324
  overview of, 31–315
Land resources, 633–635
Land subsidence, 636
Land, soil loss and degradation, 637–638
Large intestine, 471–472, 472f
Laryngitis, 453
Larynx, 452, 452f, 467f
LASIK surgery, 544
Lasiognathus dinema, 1
Lateral bud, 375
Lateral line, 542
Law, 14
Learning, memory and, 522–523
Leaves
  color of, and chlorophyll, 95, 100f
  diversity of, 379f, 380f
  function of, 377–378
  organization of, 377–378, 378f
  structure of, 377–378, 379f
Leeches, 346, 346f
Legionnaires' disease, 17
Lemmings, 616, 616f
Lens, 542, 542f
Lenticels, 382
Leprosy, 629
Less-developed countries (LDCs), 585–586, 585f
  fossil fuel use, 639
Leukemias, 139
  myelogenous, 225–226
  stem cell therapy, 213
Leukocytes, 443, 443f
Levitra, 560
Lichen, 329–330, 329f, 604
Life cycle
  of humans, 147–148, 147f
  of plants, 399–400
Life history, 595, 595f
Life table, 589–590, 589f
Ligaments, 419, 551, 552f
Light microscope, 57, 57f
Light reactions, 99–103
  ATP production, 102, 102f
  electron pathway of, 100–101, 101f
  NADPH production, 102–103
  organization of thylakoid membrane, 102, 102f
  overview of, 98, 98f
  photosynthetic pigments, 100
Light, visible, 99, 99f
Lignin, 374

Limbic system, 522–523, 522*f*
Lineage, 363
Linkage groups, 177, 177*f*
Linnaean classification, 277–278, 280–281
Linnaeus, Carl, 278
Lipase, 470, 474
Lipids, 41*f*, 45–48, 478–479
  characteristics of, 45
  fats and oils, 45–46
  fatty acids, 45–46, 45*f*, 46*f*
  function of, 45
  omega-3 fatty acid, 46
  phospholipids, 47
  steroids, 47–48
  trans fat, 46, 46*f*
Liver, 469, 470*f*
  disorders of, 473
  functions of, 472
Liverwort, 315, 315*f*
Living things
  characteristics of, 2–6
  organization of, 2–3, 3*f*
Lobe-finned fish, 358, 358*f*
Logistic growth, 591, 591*f*
Loose fibrous connective tissue, 419
Lophotrochozoans, 340, 343–346
Loss of function mutation, 137
Low-density lipoproteins (LDLs), 38, 491
Lower respiratory tract, 452*f*, 453
Lumen, 417
Lungs, 358, 452, 452*f*
  exchange of gas and, 454–455, 455*f*
Luteal phase, 563, 564*f*
Luteinizing hormone (LH), 561, 563, 564*f*
Lycophytes, 312, 313*f*, 316–317, 316*f*
Lyell, Charles, 238, 243
Lymph, 440, 441
Lymph nodes
  cancer and, 498
  function of, 498
Lymphatic organs, 497–498, 497*f*
Lymphatic system, 424, 424*f*, 440, 440*f*
Lymphatic vessels, 440, 440*f*
Lymphocytes, 443*f*, 444, 498, 498*t*
Lysogenic cycle, 288, 289*f*
Lysosomes, 68, 69*f*
Lytic cycle, 288, 289*f*

M

MacArthur, Robert, 606
Macroevolution, 266–272
Macronutrients, 385, 475
Macrophages, 444, 498, 498*t*
  inflammatory response and, 500, 500*f*

Mad cow disease, 292
Maintenance system of human body, 424, 424*f*
Major histocompatability complex (MHC), 503–504
Major minerals, 480
Malaria
  directional selection and, 254
  sickle-cell disease and, 256, 256*f*
Male condom, 565
Male gametophytes, 402*f*, 403
Male reproduction system, 559–562, 560*f*–561*f*
Malignant tumors, 140
Malpighian tubules, 460–461, 460*f*
Maltase, 474
Malthus, Thomas, 241
Maltose, 43, 43*f*, 473–474
Mammals, 360–362, 361*f*–362*f*
Mammary glands, 360
Marfan syndrome, 177
Marijuana
  medical, 519
  nervous system and, 519–520
Mars, exploration of, 21
Marsupials, 246, 246*f*, 361, 361*f*
Mascroscelides micus, 1
Mass extinction, 273*t*, 275–276
Mass number, 23
Mast cells, 498*t*, 500
Mating
  assortative, 260–261
  nonrandom, 260–261
Matrix, 71, 117*f*, 419
Matter, 22
Maturation, of virus, 288, 289*f*, 290*f*
Mechanical isolation, 269
Mechanoreceptors, 539
Medicinal value of biodiversity, 629, 629*f*
Medulla oblongata, 522
Medusa, 342
Megaphylls, 312, 313*f*, 317
Megaspore, 400, 401–402
Meiosis, 146–157
  abnormal chromosome inheritance, 156–157
  basics of, 146–150
  compared to mitosis, 152–154, 153*f*
  crossing-over, 149–1501, 149*f*
  defined, 146
  function of, 146
  in human life cycle, 147–148, 147*f*
  importance of, 150
  Mendel's laws and, 168–169, 168*f*
  occurrence of, 152–154
  overview of, 148–150, 148*f*
  phases of, 150*f*–151*f*, 151–152
  in plants, 148

Meiosis I, 148–150, 148*f*, 150*f*, 151–152, 153–154, 153*f*
Meiosis II, 148–150, 148*f*, 151*f*, 152, 153*f*, 154
Melanin, 176
Melatonin, 521
Memory B cells, 503, 506
Memory cells, 498*t*
Memory, learning and, 522–523
Mendel, Gregor, 161–162, 162*f*–163*f*
  one-trait inheritance, 162–166
  two-trait inheritance, 166–167, 166*f*, 167*f*
Mendel's laws, 161–173
  family pedigrees, 169–170
  genetic disorders of interest, 170–173
  law of independent assortment, 166
  law of segregation, 164
  meiosis and, 168–169, 168*f*
  probability and, 167–168
  rule of multiplication, 167–168
Meniscus, 551, 552*f*
Menopause, 145, 564
Menses, 563–564
Menstruation, 563–564, 564*f*
Meristem tissue, 373, 373*f*
Mesoderm, 573, 573*f*
Mesonychoteuthis hamiltoni, 59
Mesophyll cells, 105, 105*f*, 378
Mesozoic era, 273*t*
Messenger RNA (mRNA), 66, 190
  alternative mRNA processing, 201–202, 201*f*, 202*f*
  formation of, 193, 193*f*
  processed, 193–194, 194*f*
  processing, 201–202, 201*f*, 202*f*
  translation, 202
Metabolic pathway, 85–86
Metabolism, 2
Metamorphosis, 349, 349*f*
Metaphase, 131*f*, 132*f*, 150*f*, 151–152, 151*f*, 152, 153, 153*f*
Metastasis, 136, 136*f*, 140, 140*f*
Meteorite, mass extinction and, 276
Methamphetamine, nervous system and, 519
Methanasarcina mazei, 10*t*
Methane
  as greenhouse gas, 619
  models of, 27, 27*f*
Methanogens, 299–300, 300*f*
Methemoglobinemia, 170, 171*f*, 228
Methicillin-resistant Staphylococcus aureus (MRSA), 235
Methylmercury, 642
Microevolution
  causes of, 260–262
  defined, 257

in genetic context, 257
  Hardy-Weinberg principle, 259
Micronutrients, 385, 475
Microphylls, 312, 313*f*, 316
Microscopes, 57–59, 57*f*
Microspore, 399–400, 402
Microsporidians, 326
Microtubules, 72, 72*f*
Microvilli, 471
Midbrain, 522
Middle ear, 539–540, 539*f*
Mifepristone, 566
Milk of magnesia, as base, 34
Miller, Stanley, 294
Millipedes, 350*f*, 351
Minerals
  dietary, 481
  as nonrenewable resource, 641
  plant nutrition and, 385
  transport in plants, 386, 386*f*
Miocene epoch, 273*t*
Mitochondria, 65, 65*f*
  cellular respiration, 111–112, 111*f*, 112*f*, 117–120, 117*f*
  energy flow and, 84–85, 85*f*
  intermembrane space, 120
  structure and function of, 69–71, 71*f*
Mitosis, 129–133
  compared to meiosis, 152–154, 153*f*
  defined, 128
  in human life cycle, 147*f*
  occurrence of, 152–154
  overview of, 129, 129*f*
  phases of, 129–133, 130*f*–131*f*, 132*f*
MMR vaccine, 507
Models, 13, 13*f*
Molecular formula, 27, 27*f*
Molecules, 2, 3*f*
  defined, 26
  as evidence for evolution, 248
  polar, 29, 29*f*
Molluscs, 339*f*, 344–345, 344*f*, 345*f*
Molting, 347, 348
Monocots, 376–377, 376*f*
  germination, 405–406, 406*f*
  root, 384, 384*f*
  seed development, 403
Monocytes, 443*f*, 444
Monod, Jacques, 197–198
Monoecious, 401, 401*f*
Monohybrid cross, 163
Monomers, 42
Monosaccharides, 42, 42*f*
Monosomy, 154
Monotremes, 360, 361*f*
Monounsaturated oils, 46–47
Mons pubis, 562
More-developed countries (MDCs), 585–586, 585*f*
  fossil fuel use, 639

Morning-after pill, 566
Morula, 572, 572f
Mosaic evolution, 366
Mosquitoes, 352
Moss, 312, 313f, 314f, 315, 315f
Motor neurons, 516, 516f
Motor output system of human body, 425, 425f
Motor proteins, 72, 72f
Motor system, 545–553
Mouth, 466–467, 467f, 472f
MRNA transcript, 193
MRSA infection, 7, 235
Mullis, Kary Banks, 210
Multicellular, 2
Multifactorial traits, 175
Multiple sclerosis, 509, 513
Multiple-allele traits, 174, 174f
Multipotent, 213
Muscle dysmorphia, 492
Muscle spindle, 544
Muscles. See also Musculoskeletal system
    contraction and calcium, 550
    function of, 545–546
    major muscles in body, 549f
    number of, in body, 548
    types of, 548
Muscular dystrophy, 180, 180f
Muscular system, 425, 425f
Muscular tissue, 416, 416f, 421–422, 421f
Musculoskeletal system, 545–553
    appendicular skeleton, 547
    axial skeleton, 547
    bone structure, 548, 548f
    function of, 545–546
    skeletal muscle contraction, 550
    skeletal muscle moving bones at joints, 551–553, 551f
Mushrooms, 328–329, 330
Mutagens, 222
Mutations, 5, 260
    chromosomal, 224–226
    gene, 222–223, 222f, 223f
Mutualism, 604t, 605, 606–608
Myasthenia gravis, 509
Mycelium, 326, 326f
Mycorrhizal association, 386
Mycorrhizal fungi, 330, 330f
Mycoses, 331
Myelin sheath, 422f, 423, 516, 516f
MyoD gene, 201
Myofibrils, 550, 550f
Myosin, 48f, 49, 72, 72f, 550, 550f
MyPlate food guidelines, 484–485, 485f

N

NAD+, 112, 115, 117, 118, 119
NADH, 112, 113, 114, 115, 117, 118, 119

NADPH, 101, 101f, 102–103
Nasal cavities, 452, 452f
National Geographic, 15f, 16
Native species, 608–609
Natural flora, 299
Natural history, 582
Natural killer (NK) cells, 498t, 501
Natural selection, 252–256
    adaptations not perfect, 255
    antibiotic resistant bacteria and, 7, 8, 235
    Darwin's hypothesis, 252, 252f
    defined, 240
    directional selection, 252f, 253–254, 254f
    discovery of, 7
    disruptive selection, 252f, 254, 255f
    fitness, 252
    heterozygote advantage, 256
    maintenance of variations, 255–256
    process of, 7–8
    sexual selection, 254, 255f
    stabilizing selection, 253, 253f
    steps of, 240–242
Neandertals, 367–368
Negative feedback mechanism, 427, 427f, 529, 529f
Nematocyst, 342
Nematodes, 339f
Nephridium, 345
Nephron loop, 457, 458f, 459
Nephrons, 457, 458f, 459
Nerve cells, 422, 422f
Nerve deafness, 540
Nerve impulse, 516–517, 517f
Nerves, 422
    cranial, 525
    spinal, 525
Nervous system, 424–425, 424f, 514–525, 514f
    central nervous system, 520–523
    comparison of, 515, 515f
    as control system, 514
    drug abuse, 518–520
    examples of, 515
    homeostasis and, 426
    nerve impulse, 516–517
    neurons, 516
    overview of, 514
    peripheral nervous system, 523–525
    synapse, 518
Nervous tissue, 416, 416f, 422–423, 422f
Neural plate, 573, 573f
Neural tube, 573
Neuroglia, 423
Neurons, 422, 422f
    structure and function of, 516
    types of, 516, 516f

Neurotransmitters
    drug abuse and, 518
    drugs that regulate depression and anxiety, 518
    nerve impulses and, 518
Neurulation, 573–574, 573f
Neutrons, 22–23, 23f
Neutrophils, 443f, 444, 498t
    inflammatory response and, 500, 500f
New Scientist, 15f, 16
Niacin, 482t
Niche specialization, 606
Nitrates, 142
Nitrification, 617
Nitrites, 142
Nitrogen
    as element of living organisms, 22, 22f
    model of, 25f
    plant nutrition and, 385f, 386, 387
Nitrogen cycle, 616–617, 616f
    bacteria and, 297–298
Nitrogen fixation, 616–617
Nitrogen gas, 27
Nitrous oxide, as greenhouse gas, 619
Noble gases, 26
Nociceptors, 544
Node, stems, 375
Nodes of Ranvier, 516, 516f
Nondisjunction, 154, 155f
Nonrandom mating, 260–261
Nonrenewable resources, 633
Nonsister chromatids, 149–150, 149f
Nonspecific defenses, 499–501
Nonvascular plants, 315–316, 315f
Nonwoody stems, 378–380, 381f
Noradrenaline, 530
Norepinephrine (NE), 518, 530
    drugs that regulate depression and anxiety, 518
Norovirus, 291
Notochord, 354, 354f, 573, 573f
Nuclear envelope, 67
Nuclear pores, 67
Nuclease, 470
Nucleic acids
    relationship with proteins, 52–53
    sickle-cell disease, 52–53, 53f
    structure and function of, 51–52, 52f
Nucleoid, 63, 63f, 295
Nucleolus, 66–67
Nucleosome, 127
Nucleotide, 52
Nucleus, 23f, 64–65, 64f, 65f
    structure and function, 66–67, 66f, 67f
Nutrients
    classes of, 476–483, 477t
    defined, 475

    essential, 475
    transport in phloem, 388–389
Nutrition
    body mass index, 488–490
    carbohydrates, 476–478
    dietary supplements, 485–487
    dieting, 490
    eating disorders, 492, 492f
    energy intake vs. energy output, 488–489
    fiber and, 477, 477f
    food labels, 485, 486f
    lipids, 478–479
    maintaining healthy weight, 489–490, 490f
    minerals, 480–481
    MyPlate food guidelines, 484–485, 485f
    nutrients, 475–476, 476f
    overview of, 475
    plant, 385–386
    proteins, 479–480
    Type 2 diabetes and cardiovascular disease, 491–492
    vitamins, 482–483
    water, 483
Nutrition labels, 485, 486f

O

Obesity
    cancer risk and, 141
    cardiovascular disease and, 491–492
    diet and, 478
    type 2 diabetes and, 491
Observation, in scientific method, 11–12, 11f
Ocean acidifcation, 618
Octet rule, 26
Oils, structure and function, 45–46, 45f
Oligocene epoch, 273t
Omega-3 fatty acid, 46, 491
Omnivores, 467, 610
On the Origin of Species (Darwin), 7, 243, 244
Oncogenes, 137
OncoMouse, 216
One-trait inheritance, 162–166
Oocyte, 563, 563f, 571
Oogenesis, 148
    compared to spermatogenesis, 145
Open circulatory system, 433f
Operon, 197–198, 198f
Ophyrys, 324
Opisthokonta, 307, 307f
Opportunistic pioneer species, 604
Opportunistic species, 595
Order, as category of classification, 9, 9t, 277f, 278

Ordovician period, 273t
Organ systems, 2, 3f
    homeostasis and, 426, 426f
Organ transplant, tissue
    rejection, 505
Organelles, 64, 65f
Organic chemistry, 39
Organic crops, 411
Organic farming, 39
Organic molecules, 39–53, 39f
    carbohydrates, 41f, 42–45
    carbon, 39–40
    carbon skeleton and functional
        groups, 40, 40f
    defined, 39
    lipids, 41f, 45–48
    nucleic acids, 41, 51–53
    proteins, 41f, 48–51
Organisms
    evolutionary tree of, 7, 7f
    organization of, 2–3, 3f
    study of, in ecology, 581
Organs, 2, 3f
    defined, 415
    formation of, 574, 574f
    of immune system, 497–499
    of plants, 375–377
    tissue and function of, 423
Osmosis
    defined, 90
    effect on cells, 90–91, 90f
Osmotic pressure, 441
Osteoarthritis, 553
Osteoblasts, 548
Osteoclasts, 548
Osteocytes, 548, 548f
Osteogenesis imperfecta, 172
Osteons, 420, 548, 548f
Osteoporosis, 481, 481f, 548
Out-of-Africa hypothesis, 367
Outer ear, 539–540, 539f
Ovarian cycle, 563–564, 563f, 564f
Ovaries
    of flower, 321, 321f, 400f, 401
    human, 526f, 557, 562–563,
        562f, 563–564, 563f
Oviducts. See uterine tubes
Oviparous, 558
Ovoviviparous, 558
Ovulation, 562, 563–564
Ovule, 319
    of flower, 400f, 401
Oxidation, 98
Oxygen
    cellular respiration, 111–112,
        111f, 112f, 119
    as element of living organisms,
        22, 22f
    exchange of gas and,
        454–455, 455f
    model of, 25f
    in photosynthesis equation, 28

production of, atmosphere of
    early Earth, 97
    transportation and exchange of
        gas, 455–456
Oxygen deficit, 116
Oxygen gas, 27, 27f
Oxytocin, 529

P

P53 gene, 125
Pacemaker, 437
Pain, referred, 544
Paleo diet, 490
Paleocene epoch, 273t
Paleontology, 237, 272
Paleozoic era, 273t, 274
Palisade mesophyll, 378
Pancreas, 469, 470f, 526f
    function of, 531
    type 1 diabetes, 531–532
Pancreatic amylase, 470, 474
Pancreatic islets, 531
Pangaea, 276, 276f
Pantothenic acid, 482t
Papaya ring spot virus (PRSV), 289
Paper, 373
Paramecium, 305–306, 305f
Parasitism, 604t, 605
Parasympathetic division, 524, 525f
Parathyroid gland, 526f, 530
Parathyroid hormone (PTH), 530
Parazoans, 339, 341
Parenchyma cells, 374, 374f
Parent cell, 126
Parsimony, 280
Parthenogenesis, 557
Particulate theory of inheritance, 162
Passive immunity, 507
Passive transport, 88–91
    facilitated diffusion, 89, 89f
    osmosis, 90–91
    simple diffusion, 88, 89f
Pasteur, Louis, 293
Pathogens, ways to get rid of, 501f
Paxil, 518
Peat moss, 316
Pectoral (shoulder) girdle, 547
Pedigree, 169–170, 170f
    autosomal dominant, 170, 170f
    autosomal recessive,
        169–170, 170f
    for sex-linked disorders,
        179–180
Pelvic (hip) girdle, 547
Pelvic inflammatory disease (PID),
    567, 570
Penetration, of virus, 288, 289f
Penicillin, 629
    antibiotic resistant bacteria
        and, 8
    discovery of, 12

Penicillium, 330
Penis, 560, 560f
Peppers, fungi and, 311
Pepsin, 469
Peptidase, 474
Peptide bond, 49, 49f
Peptide hormones, 526–527, 527f
Peptides
    structure and function of, 49
    synthesis and degradation of, 49f
Peptidoglycan, 295
Pericycle, 383f, 384
Periodic table, 23–24, 23f
Peripheral nervous system
    autonomic system, 524–525
    defined, 515
    parasympathetic division,
        524, 525f
    reflexes, 523–524
    somatic system, 523–524
    sympathetic division,
        524–525, 525f
Peristalsis, 468
Peritonitis, 471
Permian period, 273t
PET (positron-emission
    tomography) scan, 24, 24f
Petals, 321, 321f, 400f, 401, 401f
Petiole, 377, 378f
pH buffers and, 35
    defined, 34
    regulation of, in urine formation,
        459–460
pH scale, 34, 34f
Phage, 288
Pharmaceutical products, 410–411
Pharyngeal pouch, 354, 354f
Pharynx, 452, 452f, 468f, 469, 472f
Phenotype, 258, 259f
    defined, 166
    environmental factors and,
        175–176
    versus genotype, 165–166, 166t
Phenylketonurics, 160
Philadelphia chromosome, 138, 138f
Phloem, 316
    structure of, 374, 377, 380–381
    sugar nutrient transport in,
        388–389
Phosphate, 40f
Phospholipids, 47, 47f
    plasma membrane as, 59–60, 60f
Phosphorous cycle, 615–616, 615f
Phosphorus
    as element of living organisms,
        22, 22f
    model of, 25f
    plant nutrition and, 386
Photoautotrophs, 296, 296f, 610
Photoperiod, 398–399
Photoperiodism, 398–399, 398f
Photoreceptors, 542, 543, 543f

Photorespiration, 105
Photosynthesis, 96–106
    C₃ plant, 105, 105f
    C₄ plant, 105–107, 105f, 106f
    Calvin cycle reactions, 103–104
    CAM photosynthesis, 106, 106f
    chloroplasts role in, 70
    defined, 4, 96
    equation for, 28
    grow-lights and, 99
    light reactions, 98–103
    organisms conducting,
        96–97, 96f
    overview of, 96, 96f
    photosynthetic pigments,
        100, 100f
    plants as photosynthesizers,
        97, 97f
Photosynthetic pigments, 100, 100f
Photosystems, 100–101, 101f
Phototropism, 393, 394f, 397, 397f
Photovoltaic (solar) cells, 640f, 641
PhyloCode, 281
Phylogenetic tree, 278–280, 278f
Phylogeny, 277, 278f
    tracing, 278–280
Phylum, as category of
    classification, 9, 9t, 277f, 278
Phytochrome, 398, 399f
Pine trees, 320–321, 320f
Pineal gland, 521, 526f
Pinus longaeva, 10t
Pinworm, 348
Pitcher plant, 372
Pith, 380, 381f
Pitocin, 527
Pituitary gland, 526f
    anterior, 527–528
    control of, by hypothalamus,
        527–529
    posterior, 529
Placebo, 15
Placenta, 558, 575, 575f
Placental mammals, 362, 362f
Plan B One-Step, 566
Planarians, 343, 343f, 432–433, 432f
Plants
    alternation of generations,
        314–315, 314f, 399, 400f
    angiosperms, 321–324
    C₃ plant, 105, 105f
    C₄ plant, 105–107, 105f, 106f
    carnivorous, 372
    cell wall, 74
    cells, 64, 65f
    characteristics of, 10, 10t
    compared to fungi and
        animals, 326t
    cytokinesis, 132–133
    disease of, 331, 331f
    diversity of land plants, 315–324
    ecological benefit of, 324–325

economic benefits of, 324
eudicots, 376–377
evolution of, 312–314, 313f
flowering plants, 321–324, 399–406
genetic engineering, 638
genetically engineered plants, 392, 408–411, 409f–410f, 409t
gymnosperms, 320–321
hormones, 393–396, 393f, 393t
leaves, 375, 377–378
life cycle of, 399–400
monocots, 376–377
nonvascular, 315–316
nutrition, 385–386
organs of, 375–377
overview of land plants, 312–315
photoperiodism, 398–399
as photosynthesizers, 97, 97f
phototropism, 393, 397, 397f
propagating, 407–408
roots, 375, 382–384
seed plants, 319
seeds, 404–406
stems, 375, 378–382
sugar transport in phloem, 388–389
tissue of, 373–375
transgenic, 214–215, 214f, 215f
transpiration, 387
transport of nutrients, 387–389
transport of water in, 30f, 31
tropism, 397
vascular, 316–319
viruses of, 289, 290f
water and mineral transport in xylem, 387–388
Plaque, 445
Plasma, 420, 442
Plasma cells, 498t, 502
Plasma membrane, 63, 63f
    active transport, 91
    function of membrane proteins, 61, 61f, 79
    osmosis, 90–91
    structure and function of, 59–60, 60f
Plasmid, 208, 295
Plasmodesmata, 75, 75f, 289
Plasmolysis, 91
Plate tectonics, 275–276, 275f
Platelets, 420f, 421, 443f
    blood clotting and, 444–445, 445f
Pleiotropy, 177
Pleistocene epoch, 273t
Pliocene epoch, 273t
Pneumocystis pneumonia, 509
Point mutation, 223, 223f
Polar, 29, 29f
Pollen grains, 319, 400, 402, 402f, 403f

Pollen tube, 322
Pollination, 319, 403
Pollinators, 323, 323f
Pollution
    biological magnification and, 643
    chlorofluorocarbons (CFCs), 641–642
    coastal, 634
    defined, 633
    industrial waste, 642–643
    sewage, 643
Polygenic inheritance, 174–176
Polymerase chain reaction (PCR), 209–210, 209f
Polymers, 42, 42f
Polymorphism, 254
    balanced, 256
Polyp, 342, 472
Polypeptide, 49
Polyploidy, 270, 271
Polyribosome, 196, 196f
Polysaccharides, 43–45, 44f
    as energy storage molecules, 43–44, 44f
    as structural components of cells, 44–45, 44f
Polyunsaturated oils, 46–47
Pome, 404
Pons, 522
Population, 2, 3f
Population density, 588, 591f
Population genetics, 257
Population growth
    age structure and, 589, 589f
    biotic potential, 590
    demographics and, 589–590
    doubling time, 584
    exponential growth, 590
    factors that regulate, 592–594
    human, 583–587
    logistic growth, 591, 591f
Populations
    defined, 257, 581
    distribution and density, 588, 588f, 591f
    extinction, 595–596
    human, 267f, 583–587
    life history patterns, 595–596
    study of, in ecology, 581
Porphyria, as pleiotropic trait, 177
Portal system, 440
Postanal tail, 354, 354f
Posterior pituitary, 528f, 529
Postsynaptic cell, 518
Postzygotic isolating mechanisms, 269, 269f
Potassium, plant nutrition and, 386
Potential energy, 80, 80f
Precambrian time, 272–275, 273t
Predation, 592–593, 593t, 604t, 605

Predator-prey population cycle, 593–594, 594f
Prediction, scientific experiment, 11f, 12
Prefrontal area, 521
Preparatory (prep) reaction, 112, 112f, 117, 118f
Pressure-flow model, 388–389, 388f
Presynaptic cell, 518
Preven, 566
Prezygotic isolating mechanisms, 268–269, 269f
Primary growth, 375, 376f
Primary motor area, 521, 521f
Primary productivity, 621–622, 622f
Primary sensory area, 521, 521f
Primary structure, of proteins, 50, 50f
Primary succession, 603, 603f
Primates, evolutionary tree, 363–364, 363f
Principle, 14
Prions, 292
Probability, Mendel's laws and, 167–168
Producers, 4, 4f, 610, 610f
Product, 85
Products, 28
Progesterone, 564, 564f
Prokaryotes, 293–300
    archaea, 299–300
    bacteria, 294–299
    characteristics of, 10, 10t
    defined, 293
    gene expression in, 197–198
    importance of, 293
    origin of cells, 293–294, 293f
Prokaryotic cells, 62–63
    defined, 62
    plasma membrane, 59–61, 60f, 61f
    structure and function of, 62–63, 63f
Prolactin, 528
Prometaphase, 132f
Promoter, 193
Prop root, 384f
Prophage, of virus, 288, 289f
Prophase, 130f, 132f, 150f, 151, 151f, 152, 153, 153f
Proprioceptors, 544
Prosimians, 363
Prostaglandin, 545
Prostate gland, 560
Proteasomes, 202
Protein, 41f
    amino acids, 49–50, 49f
    denatured, 51
    dietary, 479–480
    digestion of, 473–474, 474f
    fibrous, 51
    functions of, 48, 48f
    globular, 51

peptide synthesis and degradation, 49, 49f
    in plasma membrane, 61, 61f
    relationship with nucleic acid, 52–53
    shape of, 50–51, 50f
    sickle-cell disease, 52–53, 53f
    structure of, 50–51, 50f
    testing for genetic disorder, 228
    types, 48
Protein activity, 202, 202f
Proteinaceous infectious particles, 292
Proteome, 218
Proteomics, 218
Prothrombin, 444
Protista, 10, 10t
Protists
    characteristics of, 301
    classification of, 301–307, 301t
    evolution of, 301
Proto-oncogenes, 137, 137f
Protocells, 294
Protons, 23, 23f
Protostomes, 340, 340f
Proximal convoluted tubule, 457, 458f
Prozac, 518
Pseudocoelomates, 340, 340f, 347–348
Pseudopods, 306
Pseudostratified epithelium, 417f, 418
Puberty, 561–562
Pulmonary arteries, 435–440, 435f
Pulmonary circuit, 434, 434f, 439, 439f
Pulmonary semilunar valve, 436
Pulmonary trunk, 436
Pulmonary veins, 435–440, 435f
Pulse, 437
Punctuated equilibrium model of, 275
Punnett square, 163, 163f, 166f, 167
Punnett, Reginald, 163
Pupil, 542, 542f
Purkinje fibers, 437
Pyruvate, 114, 116, 116f, 117

Q

Quaternary period, 273t
Quaternary structure, of proteins, 50f, 51

R

Radial symmetry, 339, 339f
Radiant energy, 99, 99f
Radiation, to treat cancer, 140
Radioactivity
    food irradiation, 25
    use of radioactive isotopes, 24, 24f

Radius, 547
Random distribution, 588
Range, 588
Ras functions, 203
RAS oncogenes, 137
Ray-finned fish, 357, 358*f*
RB gene, 138
Reabsorption, in kidneys, 458*f,* 459
Reactants, 28, 85
Receptacle, flower, 400*f,* 401
Receptor proteins, 61, 61*f*
Recessive allele, 165
Recombinant DNA (rDNA),
    208–209, 208*f*
Red blood cells, 420, 420*f*
    function and role of,
        442–443, 443*f*
    hypertonic solution, 90*f,* 91
    hypotonic solution and,
        90–91, 90*f*
    isotonic solution and, 90, 90*f*
Red bone marrow, 548
    function of, 497–498
Redox reaction, 98, 111
Reduction, 98
Referred pain, 544
Reflexes, 523–524, 524*f*
Regeneration, 557
Regulatory gene, 198, 198*f*
Regulatory proteins, 48
Release factor, 196
Release, of virus, 288, 289*f,* 290*f*
Renewable resources, 633,
    640–641, 640*f*
Replacement model, 367, 367*f*
Replacement reproduction, 586
Replication, of virus, 2902*f*
Repressor, 198
Reproduction, 5, 5*f. See also*
    Asexual reproduc*t*ion
    animals, 557–559
    asexual, 407–411, 407*f,* 557
    assisted reproductive
        technologies, 567–568
    bacteria, 295–296
    birth control methods, 564–566
    female reproductive system,
        562–564
    human, 559–570
    infertility, 566–568
    on land *vs.* water, 558, 558*f*
    male reproduction system,
        559–562
    replacement reproduction, 586
    sexual, 146–157, 399–406, 557
    sexually transmitted diseases
        (STDs), 568–570
    viruses, 288
Reproductive barriers, 268–269, 268*f*
Reproductive cloning, 212–213, 212*f*
Reproductive system of human
    body, 425, 425*f*

Reptiles, 358–360
    circulatory system,
        433–435, 434*f*
Reservoirs, in chemical cycling, 614
Resource partitioning, 606
Resources, 588
    energy, 639–641
    food, 637–639
    land, 633–635
    minerals, 641
    nonrenewable, 633
    renewable, 633, 640–641, 640*f*
    water, 635–637
Respiration, 451
    birds, 454, 454*f*
    fish, 455, 456*f*
    insects, 453, 453*f*
Respiratory membrane, 455
Respiratory system, 424, 424*f,*
    451–456, 451*f*
    breathing, 453–454
    lower respiratory tract, 452*f,* 453
    lungs and exchange of gas,
        454–455
    steps in respiration, 451
    upper respiratory tract,
        452–453, 452*f*
Restriction enzymes, 208, 208*f*
RET gene, 139
Retina, 542, 543, 543*f*
Retinal, 543
Retroviruses, 290
Reverse transcription, 290*f*
Rheumatoid arthritis, 233, 509,
    509*f,* 553
Rhizaria, 305, 305*f*
Rhodopsin, 543, 543*f*
Rib cage, 547
Ribose, 42
Ribosomal RNA (rRNA), 66, 191,
    194–195
Ribosomes, 63, 63*f,* 65*f,* 191
    structure and function of, 67, 67*f*
    in translation, 194–195, 195*f*
Rigor mortis, 550, 551
Ringworm, 332, 332*f*
RNA (ribonucleic acid)
    compared to DNA, 191*t*
    flow of genetic information,
        192–197, 192*f*
    genetic code, 192–193
    messenger, 66, 190, 193–194
    origin of cells and, 294
    ribosomal, 66, 191, 194–195
    structure and function of,
        51–52, 52*f*
    transcription, 193–194
    transfer, 190, 194, 194*f*
    translation, 194–196
RNA interference, 232
RNA polymerase, 193
Rods, 543, 543*f*

Root cap, 383, 383*f*
    diversity of, 384, 384*f*
Root hair, 373, 382, 383*f*
Root nodules, 386
Root pressure, 387
Root system, 375, 376*f*
Roots
    adaptation of, for mineral uptake,
        386, 386*f*
    growth of, 383
    tissues of, 384
    vascular tissue of, 384
Rotational equilibrium, 541
Rotator cuff, 547
Rough ER, 68, 68*f,* 69*f*
Roundworms, 347–348, 347*f*
RuBP carboxylase (rubisco),
    103–104, 105
Rule of multiplication, 167–168
Ruminants, 469, 469*f,* 471*f*
Rural sustainability, 644–645

S

SA node, 437
Sac fungi, 328–329, 328*f*
Saccule, 541
Sacromeres, 550, 550*f*
Sacrum, 547
Sagan, Carl, 22
Sahelanthropus tchadensis,
    364–365, 365*f*
Salinization, 638
Salivary amylase, 467
Salivary glands, 467
Salmonella, 25, 299
Salt
    ionic compounds as, 26*f,* 27
    regulation of, in urine formation,
        459–460
Saltatory conduction, 517, 517*f*
Saltwater intrusion, 636
Saprotrophs, 297, 329
Sarcomas, 139
Sarcoplasmic reticulum, 550, 550*f*
SARS virus, 17–18, 290–291
Saturated fatty acids, 46, 46*f,* 491
Scanning probe microscope, 58
Scapula, 547
Schistosomiasis, 344
Schwann cells, 423
Science, 15*f*
Scientific American, 15*f,* 16
Scientific method, 11–14, 11*f*
    conclusion, 12*f,* 13–14
    data, 13
    experiments, 11*f,* 12–13
    hypothesis, 11*f,* 12
    observation, 11*f,* 12
Scientific name, 10
Scientific studies, publication of,
    15–16, 15*f*

Scientific theories, 14
Sclerenchyma cells,
    374, 374*f*
Scorpions, 350, 350*f*
Sea star, 354, 354*f*
Seawater, 91
Secondary growth, 373, 380
Secondary structure, of proteins,
    50, 50*f*
Secondary succession, 603–604,
    603*f,* 604*f*
Secondary xylem, 382
Secretion, 68, 458*f,* 459
Sedimentary cycle, 614
Seed coat, 403
Seed plants, 319, 319*f*
Seeds
    anatomy of, 319, 319*f*
    development of, in eudicot,
        403, 404*f*
    dispersal of, 323, 404–405,
        404*f,* 405*f*
    in evolution of land plants,
        312, 313*f*
    germination, 405–406
    production of, 319, 319*f*
Segmentation, 341, 348
Segregation, law of, 164
Semantic memory, 522
Semen (seminal fluid), 560–561
Semiarid land, 634–635
Semiconservative replication,
    189, 189*f*
Semilunar valves, 436
Seminal vesicles, 560
Seminiferous tubules, 561, 561*f*
Senescence, 135, 395
Senses, 537–545, 537*f*
    balance, 541
    chemical, 537–538
    cutaneous receptors,
        544–545, 544*f*
    hearing, 538–542
    proprioceptors, 542
    sensory receptors in other
        animals, 541–542, 541*f*
    taste and smell, 538, 538*f*
    vision, 542–544
Sensory adaptation, 537
Sensory neurons, 516, 516*f*
Sensory output system of human
    body, 425, 425*f*
Sepals, 321, 321*f,* 400*f,* 401
Septum, 435
Serotonin, 518
    drugs that regulate depression
        and anxiety, 518
Sertoli cells, 561
Serum sickness, 507
Set point, 427
Severe acute respiratory syndrome
    (SARS), 17–18

Severe combined immunodeficiency (SCID), 232
Sewage, 643
Sex chromosomes, 146, 146f, 178
Sex-linked inheritance, 178–180, 178f–180f
Sexual mimicry, 324
Sexual reproduction, 5, 5f, 146–157
    abnormal chromosome inheritance, 156–157
    compared to asexual reproduction, 557
    in flowering plants, 399–406
    meiosis, 146–157
Sexual selection, 254, 255f
    nonrandom mating and, 260–261
Sexually transmitted diseases (STDs), 568–570
    causes of, 568–570
    preventing, 569
Shigella dysenteriae, 299
Shoot system, 375, 376f
Short tandem repeat (STR) profiling, 210
Sickle-cell disease, 52–53, 53f
    as autosomal disorder, 171–172, 171f, 172f
    as balanced polymorphism, 256
    gene mutation, 223, 223f
    incidence of, 256f
    Malaria and, 256, 256f
Sieve plate, 375
Sieve-tube members, 375
Signal molecule, 61
Signal transduction pathway, 202–203
Signals, in cell cycle, 134–135
Silurian period, 273t
Simple diffusion, 88, 89f
Sinuses, 433, 452, 452f
Sister chromatids, 127, 127f, 128, 129, 129f, 130, 148f, 149
Skeletal muscle, 421, 421f, 548
    contraction of, 550
    moving bones at joints, 551–553, 551f
Skeletal system, 425, 425f. See also Musculoskeletal system
Skill memory, 523
Skin, 417–418, 418f
    as barrier to entry, 499
    cutaneous receptors, 544–545, 544f, 545f
    genetic basis of skin color, 175, 175f
Skull, 547
Sliding filament model, 550
Slime mold, 306, 306f
Small intestine, 469–470, 470f, 472f
Smell, 538, 538f
Smooth ER, 68, 68f, 69f

Smooth muscle, 421f, 422, 548
Sodium
    dietary, 481, 481t
    in sodium chloride formation, 26–27, 26f
Sodium chloride
    formation of, 26–27, 26f
    water and, 30
Sodium hydroxide, dissociation of, 34
Sodium-potassium pump, 91
Solar energy, 640–641, 640f
Soluble fiber, 45
Solute, 88
Solution, 88
Solvent, 88
Somatic cells, 135
Somatic system of PNS, 523–524
Somites, 573
Sori, 317
Speciation, 240, 266–272
    adaptive radiation, 271
    allopatric, 269–270
    biological species concept, 266–269, 267f
    common ancestor, 267
    defined, 266
    defining species, 266–269
    geological timescale, 272–274, 273t
    mass extinction of species, 275–276
    models of, 269–271
    pace of, 274–275, 274f
    reproductive barriers, 268–269
    sympatric, 270
Species, 2, 3f, 581
    biological species concept, 266–269, 269f
    as category of classification, 9, 9t, 277f, 278
    communities and species composition, 602, 602f
    defining, 266–269
    exotic, 608–609, 609f
    interaction in communities, 604–608, 604t
    keystone, 608
    mass extinction of, 275–276
    native, 608–609
    systematics, 277–282
Species richness, 602
Specific defenses, 502–505
Sperm, 557, 559–560, 561, 561f, 571, 571f
Spermatogenesis, 148, 561
    compared to oogenesis, 145
Sphagnum, 316
Spicules, 342
Spinal cord, 520
Spinal nerves, 525
Spindle, 129–130, 131f

Spindle equator, 129, 131f
Spiral organ, 540
Spirilla, 294
Spirochetes, 294, 294f
Spirogyra, 302, 302f
Spleen, function of, 498
Sponge effect, 631
Sponges, 339, 339f, 341–342, 341f
Spongy bone, 548
Spongy mesophyll, 378
Sporangium, 315
Spores, 296, 314, 314f, 401–403
Sporophyte, 314, 314f, 315, 315f, 399
Sprains, 553
Spring wood, 382
Squamous epithelium, 417f, 418
SRY gene, 156
Stabilizing selection, 253, 253f
Stamen, 321, 321f, 400f, 401, 401f
Starches, 104
Starches function, 43–44, 44f
Statocysts, 541
Stem cells, 56, 2113
Stems
    diversity of, 380f
    function of, 380–381
    nonwoody, 378–380
    organization of, 378–382
    woody, 380–382, 382f
Stent, 446, 446f
Sterility, 559
Sternum, 547
Steroid hormones, 526, 527f
Steroids, 47–48, 47f
Stigma, 321, 321f, 400f, 401
Stimulants, 518
Stomach, 46f, 468–469, 472f
Stomach flu, 291
Stomata, 97, 373, 374f
    opening and closing, 388, 388f
Stramenophiles, 303–304, 303f
Strata, 237, 237f, 238f
    Darwin's study of, 238–239
Streptococci, 294f
Striated, 421
Stroke, 445
Stroma, 70, 70f, 97, 97f
Structural formula, 27, 27f
Structural proteins, 48
Style, 321, 321f
Style, flower, 400f, 401
Subcutaneous fat, 478
Suberin, 373–374
Subsidence, 636
Subspecies, 260, 261f
Substrate-level ATP synthesis, 114, 114f
Substrates, 86
Sucrose, 43

Sugar, 42–43
    high-fructose corn syrup, 43
    as nutrient transport in plants, 388–389, 388f
    reducing dietary, 478t
Sulfhydryl, 40f
Sulfur
    as element of living organisms, 22, 22f
    model of, 25, 25f
Summer wood, 382
Superbugs, 235
Supergroups, 301, 302f, 303–305
Supernova, 22
Surface-area-to-volume ratio, 58–59, 58f, 59f
Survivorship, 589–590, 589f
Sustainable society, 643–646, 644f
Swallowing, 467–468, 468f
Swine flu (H1N1), 17
Symbiotic, 297
Sympathetic division, 524–525, 525f
Sympatric speciation, 270, 270f
Synapse, 518, 518f
Synapsis, 148–150, 149f
Synaptic cleft, 518
Synovial joints, 551, 552f
Synthetic blood, 431
Synthetic organic compounds, 641–642
Syphilis, 570
Systematics
    cladistics and cladograms, 280–281, 281f
    defined, 9
    Linnaean classification, 277–278
    phylogenetic tree, 278–280, 278f
    three-domain system, 281–282, 282f
Systemic circuit, 434, 434f, 439f, 440
Systemic lupus erythematosus, 509
Systole, 436–437
Systolic blood pressure, 445

T

T cells. See T lymphocytes (T cells)
T lymphocytes (T cells), 498, 498t
    activation of, 503–504, 504f
    cellular response and, 503–505
    characteristics of, 504t
T-cell receptors (TCRs), 503
Tapeworm, 344, 344f
Taproot, 384f
Taste, 538, 538f
Tatum, Edward, 191
Taxon, 277
Taxonomy, 9, 277–278
Tay-Sachs disease, 68
Technology, 16

Teeth, 467, 467f
Telomerase, 138
Telomere, 135
    absence of shortening, and
        cancer, 138
Telophase, 131f, 132f, 150f, 151,
    151f, 152, 153f
Template, 189
Temporal isolation, 268
Tendons, 419
Terminal bud, 375, 376f
Termination, 196, 196f
Terrestrial ecosystems, 621, 621f
Tertiary structure, of proteins,
    50–51, 50f
Test group, 13
Testcross
    one-trait, 163f, 164
    two-trait, 166–167, 166f, 167f
Testes, 526f, 557, 559–560,
    560f, 561
Testosterone, 68, 561–562
    structure and function of,
        47–48, 47f
    women and, 564
Tetanus, 298–299
Tetany, 530
Tetrad, 148–150, 148f
Thalamus, 521, 522f
Therapeutic cloning, 212f
Thermoacidophiles, 300, 300f
Thermodynamics, laws of, 80–81
Thigmotropism, 397, 397f
Thoracic cavity, 453–454, 454f
Three-domain system,
    281–282, 282f
Thrombin, 444
Thrombus, 445
Thylakoid membrane, 295
    organization of, 102, 102f
Thylakoids, 70, 70f, 97, 97f
Thymine (T), 52, 52f, 186, 186f
    structure of DNA and,
        186–188
Thymus gland, 498, 526f
Thyroid gland, 526f, 529–530
Thyroid-stimulating hormone
    (TSH), 527
Thyroxine, 529
Tibia, 547
Tight junctions, 74f, 75
Tiktaalik fossils, 245, 245f
Tinea, 332, 332f
Tissue, 2, 3f
    capillary exchange in, 441, 441f
    connective, 416, 416f, 419–421
    defined, 415
    epithelial, 416, 416f, 417–418
    of immune system, 497–499
    muscular, 416, 416f, 421–422
    nervous, 416, 416f, 422–423
    plant, 373–375

    root, 384
    structure and function, 416, 423
Tissue culture, 407–408, 408f
Tissue fluid, 424, 451
Tissue rejection, 505
Tobacco mosaic virus, 289, 289f
Tonsillitis, 452
Tonsils, 498
Tortoises, Darwin's study of, 240
Totipotent, 408
Toxins, 298–299
Trace minerals, 480
Tracer, 24
Trachea, 452f, 453, 453f
Tracheids, 374
Trans fat, 46, 46f, 478
Transcription, 192, 193–194, 193f
    gene expression in eukaryotes,
        199–202, 200f
    gene expression in prokaryotes,
        197–198
Transcription activators, 201, 201f
Transcription factors, 201, 201f
Transduction, 296
Transfer rate, 616
Transfer RNA, 190, 194, 194f
Transformation, 296
Transgenic bacteria, 214f
Transgenic organisms, 208,
    214–216, 214f
Transgenic plants, 214–215,
    214f, 409
Translation, 192, 194–196,
    194f, 195f
Translocation, 138f, 225–226, 225f
Transpiration, 387
Transplants, kidney, 461–462
Transport proteins, 48, 61, 61f
Transport system of body, 424, 424f
Transposon, 139, 222, 222f
Trees
    annual ring, 382
    bark of, 381–382
    wood of, 382
Treponema pallidum, 294f
Triassic period, 273t
Trichinosis, 348
Trichomes, 373
Trichomoniasis, 570
Tricuspid valve, 436
Triglycerides, 45, 478
Triiodothyronine, 529
Trisomy, 154, 221
Trisomy 21, 155–156, 155f, 221
Trophic levels, 613–614
Tropical level, 614
Tropism, 397, 397f–398f
True-breeding plants, 161
Trypsin, 470, 474
Tumor, 140
Tumor suppressor genes,
    137, 137f

Tunicates, 354–355
Turgid, 91
Turgor pressure, 388
Turner syndrome, 156, 156f
Two-trait inheritance, 166–167
Type 1 diabetes, 491, 531–532
Type 2 diabetes, 478, 491, 532
Typhoid fever, 299

U

Ulcer, 469
Ulna, 547
Ultrasound, 229–230
Umbilical cord, 574
Uncoating, of virus, 290f
Uniform distribution, 588
Uniformitarianism, 238–239
Unsaturated fatty acids, 46, 46f
Upper respiratory tract,
    452–453, 452f
Uracil (U), 52, 52f, 190
Urban sustainability, 645
Ureter, 457, 457f
Urethra, 457, 457f
Urinary bladder, 457, 457f
Urinary system, 415, 415f, 424,
    424f, 457–462, 457f
    functions of kidneys, 457
    kidney problems, 461–462
    urine formation, 458f, 459–461
Urine, 457
    formation of, 458f, 459–461
Uterine tubes, 562
Uterus, 562, 562f
Utricle, 541

V

Vaccines
    contraceptive, 566
    immunity and, 506
    MMR vaccine, 507
Vacuoles, 69, 70f
Vagina, 562, 562f
Vaginitis, 570
Valence shell, 25–26, 25f, 26, 26f
Valine (Val), 53
Valium, 518
Vampire bat, 243f
Varicose veins, 439
Vas deferens, 559–560, 560f
Vascular bundles, 377
Vascular cambium, 373
Vascular cylinder, 384
Vascular endothelial growth factor
    (VEGF), 233
Vascular plants, 315, 316–319
Vascular tissue
    of plants, 316, 373,
        374–375, 375f
    of roots, 384

Vector, 208
Veins, 439
Venae cavae, 435f, 436
Ventricles, 435, 435f
Ventricular fibrillation, 437
Venules, 438–439
Vertebral column, 547
Vertebrate animals, circulatory
    pathways in, 433–435
Vertigo, 541
Vesicle formation, 65f
Vesicles, 65
Vessel elements, 374
Vestigial structures, 246–248, 247f
Viagra, 560
Villi, 470–471, 470f
Viroids, 292, 292f
Viruses, 287–291
    animal viruses, 289–291
    bacteriophage reproduction,
        288, 289f
    defined, 287
    drug control of, 291
    Ebola, 286
    emerging, 290–291
    norovirus, 291
    parts of, 287–288, 287f
    plant viruses, 289
    reproduction of, 288
    retroviruses, 290
    sexually transmitted diseases,
        568–570
Visceral muscle, 422
Visible light, 99, 99f
Vision, 542–544
    in the dark, 544
    LASIK surgery, 544
    photoreceptors in eye, 543
    types of eyes, 542
Vitamins, 482–483, 482t
    A, 141–142, 482–483, 483t
    B₁, 482t
    B₁₂, 482t
    B₂, 482t
    B₆, 482t
    C, 141–142, 475, 475f,
        482–483, 482t
    D, 482–483, 483t, 530
    E, 483t
    fat-soluble, 483t
    K, 471, 483t
    water-soluble, 482t
Viviparous, 558
Vocal cords, 453
Volvox, 302, 302f
Vulva, 562, 562f

W

Wallace, Alfred Russel, 243–244
Wallace, Alfred Russell, 7
Wallace's Line, 243, 243f

Waste
   biodiversity and disposal of, 631
   industrial waste, 642–643
   sewage, 643
Water
   acids and, 33
   biodiversity and freshwater
      provisions, 631
   cellular respiration, 111–112,
      111f, 112f
   cohesion and adhesion of,
      30–31, 30f
   conservation of, 637, 637f–638f
   dehydration synthesis reaction,
      42, 42f
   density of ice, 32, 32f
   in digestive process, 471–472
   dissociation of, 33
   heat capacity of, 31–32
   heat of vaporization,
      31–32, 31f
   hydrogen bond, 29, 29f

   hydrolysis reaction, 42, 42f
   increasing water supplies,
      635–636
   as nutrient, 483
   oxidation of, in photosynthesis, 98
   in photosynthesis equation, 28
   polarity of, 29, 29f
   to produce food, 636
   properties of, 29–32
   regulation of, in urine formation,
      459–460
   seawater, 91
   as solvent, 29–30
   structure of, 29, 29f
   surface tension of, 31
   transport in plants, 387–388
   use of, 635–636, 636f
Water-soluble vitamins, 482t
Watson, James, 187–188, 187f
Weight loss
   cancer prevention and, 141
   chromium and, 122

Weight, maintaining healthy,
      489–490, 490f
Weinberg, W., 259
West Nile virus, as emerging virus,
      290–291
Whales, evolution of, 245, 245f
White blood cells, 420f, 421
   structure and function of,
      443, 443f
Williams syndrome, 224, 224f
Wind power, 640, 640f
Wood, 382
Woody stems, 380–382, 382f

X

X chromosome, 146, 146f, 178–180,
      178f–179f
X-inactivation, 199, 200f
X-linked alleles, 178–179
X-linked dominant, 179
X-linked recessive disorder, 179, 180f

Xanax, 518
Xenotransplantation, 215
Xylem, 316
   structure of, 374, 375f, 377, 382
   water and mineral transport in,
      387–388

Y

Y chromosome, 146, 146f, 178–180,
      178f–179f
Y-linked pattern of inheritance, 180
Yeast, fermentation and, 116
Yolk sac, 571f

Z

Zika virus, 496, 576
Zona pellucida, 571, 571f
Zygospore fungi, 327–328, 327f
Zygote, 148, 557
Zygote mortality, 269